NATURAL RESPONSE OF A PARALLEL *RLC* CIRCUIT

1. **Determine the initial capacitor voltage (V_0) and inductor current (I_0)** from the circuit.
2. **Determine the values of α and ω_0** using the equations in Table 8.2.
3. **If $\alpha^2 > \omega_0^2$, the response is overdamped** and $v(t) = A_1e^{s_1t} + A_2e^{s_2t}$, $t \geq 0$;
If $\alpha^2 < \omega_0^2$, the response is underdamped and $v(t) = B_1e^{-\alpha t}\cos\omega_d t + B_2e^{-\alpha t}\sin\omega_d t$, $t \geq 0$;
If $\alpha^2 = \omega_0^2$ the response is critically damped and $v(t) = D_1te^{-\alpha t} + D_2e^{-\alpha t}$, $t \geq 0$.
4. **If the response is overdamped, calculate s_1 and s_2** using the equations in Table 8.2;
If the response is underdamped, calculate ω_d using the equation in Table 8.2.
5. **If the response is overdamped, calculate A_1 and A_2** by simultaneously solving the equations in Table 8.2;
If the response is underdamped, calculate B_1 and B_2 by simultaneously solving the equations in Table 8.2;
If the response is critically damped, calculate D_1 and D_2 by simultaneously solving the equations in Table 8.2.
6. **Write the equation for $v(t)$ from Step 3** using the results from Steps 4 and 5; find any desired branch currents.

TABLE 8.2 — Equations for analyzing the natural response of parallel *RLC* circuits

Characteristic equation	$s^2 + \dfrac{1}{RC}s + \dfrac{1}{LC} = 0$
Neper, resonant, and damped frequencies	$\alpha = \dfrac{1}{2RC} \quad \omega_0 = \sqrt{\dfrac{1}{LC}} \quad \omega_d = \sqrt{\omega_0^2 - \alpha^2}$
Roots of the characteristic equation	$s_1 = -\alpha + \sqrt{\alpha^2 - \omega_0^2}, \quad s_2 = -\alpha - \sqrt{\alpha^2 - \omega_0^2}$
$\alpha^2 > \omega_0^2$: overdamped	$v(t) = A_1e^{s_1t} + A_2e^{s_2t},\ t \geq 0$ $v(0^+) = A_1 + A_2 = V_0$ $\dfrac{dv(0^+)}{dt} = s_1A_1 + s_2A_2 = \dfrac{1}{C}\left(\dfrac{-V_0}{R} - I_0\right)$
$\alpha^2 < \omega_0^2$: underdamped	$v(t) = B_1e^{-\alpha t}\cos\omega_d t + B_2e^{-\alpha t}\sin\omega_d t,\ t \geq 0$ $v(0^+) = B_1 = V_0$ $\dfrac{dv(0^+)}{dt} = -\alpha B_1 + \omega_d B_2 = \dfrac{1}{C}\left(\dfrac{-V_0}{R} - I_0\right)$
$\alpha^2 = \omega_0^2$: critically damped	$v(t) = D_1te^{-\alpha t} + D_2e^{-\alpha t},\ t \geq 0$ $v(0^+) = D_2 = V_0$ $\dfrac{dv(0^+)}{dt} = D_1 - \alpha D_2 = \dfrac{1}{C}\left(\dfrac{-V_0}{R} - I_0\right)$

(Note that the equations in the last three rows assume that the reference direction for the current in every component is in the direction of the reference voltage drop across that component.)

STEP RESPONSE OF A PARALLEL *RLC* CIRCUIT

1. **Determine the initial capacitor voltage (V_0), the initial inductor current (I_0), and the final inductor current (I_f)** from the circuit.
2. **Determine the values of α and ω_0** using the equations in Table 8.3.
3. **If $\alpha^2 > \omega_0^2$, the response is overdamped** and $i_L(t) = I_f + A_1'e^{s_1t} + A_2'e^{s_2t}$, $t \geq 0^+$;
If $\alpha^2 > \omega_0^2$ the response is underdamped and $i_L(t) = I_f + B_1'e^{-\alpha t}\cos\omega_d t + B_2'e^{-\alpha t}\sin\omega_d t$, $t \geq 0^+$;
If $\alpha^2 = \omega_0^2$, the response is critically damped and $i_L(t) = I_f + D_1'te^{-\alpha t} + D_2'e^{-\alpha t}$, $t \geq 0^+$.
4. **If the response is overdamped, calculate s_1 and s_2** using the equations in Table 8.3;
If the response is underdamped, calculate ω_d using the equation in Table 8.3.
5. **If the response is overdamped, calculate A_1' and A_2'** by simultaneously solving the equations in Table 8.3;
If the response is underdamped, calculate B_1' and B_2' by simultaneously solving the equations in Table 8.3;
If the response is critically damped, calculate D_1' and D_2' by simultaneously solving the equations in Table 8.3.
6. **Write the equation for $i_L(t)$ from Step 3** using the results from Steps 4 and 5; find the inductor voltage and any desired branch currents.

TABLE 8.3 — Equations for analyzing the step response of parallel *RLC* circuits

Characteristic equation	$s^2 + \dfrac{1}{RC}s + \dfrac{1}{LC} = \dfrac{I}{LC}$
Neper, resonant, and damped frequencies	$\alpha = \dfrac{1}{2RC} \quad \omega_0 = \sqrt{\dfrac{1}{LC}} \quad \omega_d = \sqrt{\omega_0^2 - \alpha^2}$
Roots of the characteristic equation	$s_1 = -\alpha + \sqrt{\alpha^2 - \omega_0^2}, \quad s_2 = -\alpha - \sqrt{\alpha^2 - \omega_0^2}$
$\alpha^2 > \omega_0^2$: overdamped	$i_L(t) = I_f + A_1'e^{s_1t} + A_2'e^{s_2t},\ t \geq 0$ $i_L(0^+) = I_f + A_1' + A_2' = I_0$ $\dfrac{di_L(0^+)}{dt} = s_1A_1' + s_2A_2' = \dfrac{V_0}{L}$
$\alpha^2 < \omega_0^2$: underdamped	$i_L(t) = I_f + B_1'e^{-\alpha t}\cos\omega_d t + B_2'e^{-\alpha t}\sin\omega_d t,\ t \geq 0$ $i_L(0^+) = I_f + B_1' = I_0$ $\dfrac{di_L(0^+)}{dt} = -\alpha B_1' + \omega_d B_2' = \dfrac{V_0}{L}$
$\alpha^2 = \omega_0^2$: critically damped	$i_L(t) = I_f + D_1'te^{-\alpha t} + D_2'e^{-\alpha t},\ t \geq 0$ $i_L(0^+) = I_f + D_2' = I_0$ $\dfrac{di_L(0^+)}{dt} = D_1' - \alpha D_2' = \dfrac{V_0}{L}$

(Note that the equations in the last three rows assume that the reference direction for the current in every component is in the direction of the reference voltage drop across that component.)

ELECTRIC CIRCUITS

ELEVENTH EDITION

ELECTRIC CIRCUITS

ELEVENTH EDITION

James W. Nilsson

Professor Emeritus
Iowa State University

Susan A. Riedel

Marquette University

 Pearson

330 Hudson Street, NY NY 10013

Senior Vice President Courseware Portfolio Management, Engineering, Computer Science, Mathematics, Statistics, and Global Editions: Marcia J. Horton
Director, Portfolio Management, Engineering, Computer Science, and Global Editions: Julian Partridge
Specialist, Higher Ed Portfolio Management: Norrin Dias
Portfolio Management Assistant: Emily Egan
Managing Producer, ECS and Mathematics: Scott Disanno
Senior Content Producer: Erin Ault
Manager, Rights and Permissions: Ben Ferrini
Operations Specialist: Maura Zaldivar-Garcia

Inventory Manager: Ann Lam
Product Marketing Manager: Yvonne Vannatta
Field Marketing Manager: Demetrius Hall
Marketing Assistant: Jon Bryant
Project Manager: Rose Kernan
Cover Design: Black Horse Designs
Cover Art: © Leonardo Ulian, Matrix board series 06 - Resistance by abstraction, 2017.
Composition: Integra Publishing Services
Cover Printer: Phoenix Color/Hagerstown
Printer/Binder: LSC Communications, Inc.

Credits and acknowledgments borrowed from other sources and reproduced, with permission, in this textbook appear on appropriate page within text.

MATLAB is a registered trademark of The MathWorks, Inc., 3 Apple Hill Road, Natick, MA.

Library of Congress Cataloging-in-Publication Data
Names: Nilsson, James William, author. | Riedel, Susan A., author.
Title: Electric circuits / James W. Nilsson, professor emeritus Iowa State University, Susan A. Riedel, Marquette University.
Description: Eleventh edition. | Pearson, [2019] | Includes index.
Identifiers: LCCN 2017025128 | ISBN 9780134746968 | ISBN 0134746961
Subjects: LCSH: Electric circuits.
Classification: LCC TK454 .N54 2019 | DDC 621.319/2—dc23
LC record available at https://lccn.loc.gov/2017025128

1 18

ISBN-10:   0-13-474696-1
ISBN-13: 978-0-13-474696-8

Courtesy of Anna Nilsson

In Memoriam

We remember our beloved author, James W. Nilsson, for his lasting legacy to the electrical and computer engineering field.

The first edition of *Electric Circuits* was published in 1983. As this book evolved over the years to better meet the needs of both students and their instructors, the underlying teaching methodologies Jim established remain relevant, even in the Eleventh Edition.

Jim earned his bachelor's degree at the University of Iowa (1948), and his master's degree (1952) and Ph.D. (1958) at Iowa State University. He joined the ISU faculty in 1948 and taught electrical engineering there for 39 years.

He became an IEEE fellow in 1990 and earned the prestigious IEEE Undergraduate Teaching Award in 1992.

For Anna

Brief Contents

List of Examples   **xii**

List of Tables   **xvi**

List of Analysis Methods   **xvii**

Preface   **xx**

Chapter 1   Circuit Variables   2

Chapter 2   Circuit Elements   26

Chapter 3   Simple Resistive Circuits   58

Chapter 4   Techniques of Circuit Analysis   92

Chapter 5   The Operational Amplifier   150

Chapter 6   Inductance, Capacitance, and Mutual Inductance   182

Chapter 7   Response of First-Order *RL* and *RC* Circuits   220

Chapter 8   Natural and Step Responses of *RLC* Circuits   272

Chapter 9   Sinusoidal Steady-State Analysis   318

Chapter 10   Sinusoidal Steady-State Power Calculations   374

Chapter 11   Balanced Three-Phase Circuits   412

Chapter 12   Introduction to the Laplace Transform   444

Chapter 13   The Laplace Transform in Circuit Analysis   482

Chapter 14   Introduction to Frequency Selective Circuits   536

Chapter 15   Active Filter Circuits   572

Chapter 16   Fourier Series   618

Chapter 17   The Fourier Transform   660

Chapter 18   Two-Port Circuits   692

Appendix A   The Solution of Linear Simultaneous Equations   718

Appendix B   Complex Numbers   727

Appendix C   More on Magnetically Coupled Coils and Ideal Transformers   733

Appendix D   The Decibel   741

Appendix E   Bode Diagrams   743

Appendix F   An Abbreviated Table of Trigonometric Identities   757

Appendix G   An Abbreviated Table of Integrals   758

Appendix H   Common Standard Component Values   760

Answers to Selected Problems   761

Index   771

Contents

List of Examples   xii

List of Tables   xvi

List of Analysis Methods   xvii

Preface   xx

Chapter 1   Circuit Variables   2

Practical Perspective: Balancing Power   3
1.1   Electrical Engineering: An Overview   4
1.2   The International System of Units   9
1.3   Circuit Analysis: An Overview   11
1.4   Voltage and Current   12
1.5   The Ideal Basic Circuit Element   14
1.6   Power and Energy   15
   Practical Perspective: Balancing Power   18
   Summary   19
   Problems   20

Chapter 2   Circuit Elements   26

*Practical Perspective: Heating with Electric
   Radiators   27*
2.1   Voltage and Current Sources   28
2.2   Electrical Resistance (Ohm's Law)   32
2.3   Constructing a Circuit Model   36
2.4   Kirchhoff's Laws   39
2.5   Analyzing a Circuit Containing Dependent
   Sources   45
   *Practical Perspective: Heating with Electric
   Radiators   48*
   Summary   50
   Problems   50

Chapter 3   Simple Resistive
   Circuits   58

*Practical Perspective: Resistive Touch
   Screens   59*
3.1   Resistors in Series   60
3.2   Resistors in Parallel   61
3.3   The Voltage-Divider
   and Current-Divider Circuits   64
3.4   Voltage Division and Current Division   68
3.5   Measuring Voltage and Current   70
3.6   Measuring Resistance—
   The Wheatstone Bridge   73
3.7   Delta-to-Wye (Pi-to-Tee) Equivalent Circuits   75

*Practical Perspective: Resistive Touch
   Screens   78*
Summary   79
Problems   80

Chapter 4   Techniques of Circuit
   Analysis   92

*Practical Perspective: Circuits with Realistic
   Resistors   93*
4.1   Terminology   94
4.2   Introduction to the Node-Voltage Method   96
4.3   The Node-Voltage Method
   and Dependent Sources   98
4.4   The Node-Voltage Method:
   Some Special Cases   100
4.5   Introduction to the Mesh-Current Method   104
4.6   The Mesh-Current Method
   and Dependent Sources   107
4.7   The Mesh-Current Method: Some
   Special Cases   108
4.8   The Node-Voltage Method Versus the
   Mesh-Current Method   112
4.9   Source Transformations   115
4.10   Thévenin and Norton Equivalents   118
4.11   More on Deriving the Thévenin Equivalent   123
4.12   Maximum Power Transfer   126
4.13   Superposition   129
   *Practical Perspective: Circuits with Realistic
   Resistors   131*
   Summary   134
   Problems   136

Chapter 5   The Operational
   Amplifier   150

Practical Perspective: Strain Gages   151
5.1   Operational Amplifier Terminals   152
5.2   Terminal Voltages and Currents   152
5.3   The Inverting-Amplifier Circuit   156
5.4   The Summing-Amplifier Circuit   158
5.5   The Noninverting-Amplifier Circuit   160
5.6   The Difference-Amplifier Circuit   162
5.7   A More Realistic Model for the Operational
   Amplifier   167
Practical Perspective: Strain Gages   171
Summary   172
Problems   173

Chapter 6   Inductance, Capacitance, and Mutual Inductance   182

Practical Perspective: Capacitive Touch Screens   183

6.1   The Inductor   *184*
6.2   The Capacitor   *189*
6.3   Series-Parallel Combinations of Inductance and Capacitance   *194*
6.4   Mutual Inductance   *199*
6.5   A Closer Look at Mutual Inductance   *203*
Practical Perspective: Capacitive Touch Screens   209
Summary   211
Problems   212

Chapter 7   Response of First-Order *RL* and *RC* Circuits   220

Practical Perspective: Artificial Pacemaker   221

7.1   The Natural Response of an *RL* Circuit   *222*
7.2   The Natural Response of an *RC* Circuit   *228*
7.3   The Step Response of *RL* and *RC* Circuits   *233*
7.4   A General Solution for Step and Natural Responses   *241*
7.5   Sequential Switching   *246*
7.6   Unbounded Response   *250*
7.7   The Integrating Amplifier   *252*
Practical Perspective: Artificial Pacemaker   255
Summary   256
Problems   256

Chapter 8   Natural and Step Responses of *RLC* Circuits   272

Practical Perspective: Clock for Computer Timing   273

8.1   Introduction to the Natural Response of a Parallel *RLC* Circuit   *274*
8.2   The Forms of the Natural Response of a Parallel *RLC* Circuit   *278*
8.3   The Step Response of a Parallel *RLC* Circuit   *289*
8.4   The Natural and Step Response of a Series *RLC* Circuit   *296*
8.5   A Circuit with Two Integrating Amplifiers   *303*
Practical Perspective: Clock for Computer Timing   308
Summary   309
Problems   310

Chapter 9   Sinusoidal Steady-State Analysis   318

Practical Perspective: A Household Distribution Circuit   319

9.1   The Sinusoidal Source   *320*
9.2   The Sinusoidal Response   *323*
9.3   The Phasor   *324*
9.4   The Passive Circuit Elements in the Frequency Domain   *327*
9.5   Kirchhoff's Laws in the Frequency Domain   *332*
9.6   Series, Parallel, and Delta-to-Wye Simplifications   *333*
9.7   Source Transformations and Thévenin–Norton Equivalent Circuits   *340*
9.8   The Node-Voltage Method   *344*
9.9   The Mesh-Current Method   *345*
9.10   The Transformer   *347*
9.11   The Ideal Transformer   *351*
9.12   Phasor Diagrams   *357*
Practical Perspective: A Household Distribution Circuit   359
Summary   361
Problems   362

Chapter 10   Sinusoidal Steady-State Power Calculations   374

Practical Perspective: Vampire Power   375

10.1   Instantaneous Power   *376*
10.2   Average and Reactive Power   *377*
10.3   The rms Value and Power Calculations   *382*
10.4   Complex Power   *384*
10.5   Power Calculations   *386*
10.6   Maximum Power Transfer   *393*
Practical Perspective: Vampire Power   399
Summary   401
Problems   401

Chapter 11   Balanced Three-Phase Circuits   412

Practical Perspective: Transmission and Distribution of Electric Power   413

11.1   Balanced Three-Phase Voltages   *414*
11.2   Three-Phase Voltage Sources   *415*
11.3   Analysis of the Wye-Wye Circuit   *416*
11.4   Analysis of the Wye-Delta Circuit   *422*
11.5   Power Calculations in Balanced Three-Phase Circuits   *425*
11.6   Measuring Average Power in Three-Phase Circuits   *430*
Practical Perspective: Transmission and Distribution of Electric Power   433
Summary   435
Problems   436

Chapter 12 Introduction to the Laplace Transform 444

Practical Perspective: Transient Effects 445
12.1 Definition of the Laplace Transform 446
12.2 The Step Function 447
12.3 The Impulse Function 449
12.4 Functional Transforms 452
12.5 Operational Transforms 453
12.6 Applying the Laplace Transform 458
12.7 Inverse Transforms 460
12.8 Poles and Zeros of $F(s)$ 470
12.9 Initial- and Final-Value Theorems 472
Practical Perspective: Transient Effects 474
Summary 476
Problems 477

Chapter 13 The Laplace Transform in Circuit Analysis 482

Practical Perspective: Surge Suppressors 483
13.1 Circuit Elements in the s Domain 484
13.2 Circuit Analysis in the s Domain 486
13.3 Applications 488
13.4 The Transfer Function 500
13.5 The Transfer Function in Partial Fraction Expansions 502
13.6 The Transfer Function and the Convolution Integral 505
13.7 The Transfer Function and the Steady-State Sinusoidal Response 511
13.8 The Impulse Function in Circuit Analysis 514
Practical Perspective: Surge Suppressors 520
Summary 521
Problems 522

Chapter 14 Introduction to Frequency Selective Circuits 536

Practical Perspective: Pushbutton Telephone Circuits 537
14.1 Some Preliminaries 538
14.2 Low-Pass Filters 539
14.3 High-Pass Filters 545
14.4 Bandpass Filters 550
14.5 Bandreject Filters 560
Practical Perspective: Pushbutton Telephone Circuits 564
Summary 564
Problems 565

Chapter 15 Active Filter Circuits 572

Practical Perspective: Bass Volume Control 573

15.1 First-Order Low-Pass and High-Pass Filters 574
15.2 Scaling 577
15.3 Op Amp Bandpass and Bandreject Filters 580
15.4 Higher-Order Op Amp Filters 587
15.5 Narrowband Bandpass and Bandreject Filters 600
Practical Perspective: Bass Volume Control 605
Summary 608
Problems 609

Chapter 16 Fourier Series 618

Practical Perspective: Active High-Q Filters 619
16.1 Fourier Series Analysis: An Overview 621
16.2 The Fourier Coefficients 622
16.3 The Effect of Symmetry on the Fourier Coefficients 625
16.4 An Alternative Trigonometric Form of the Fourier Series 631
16.5 An Application 633
16.6 Average-Power Calculations with Periodic Functions 639
16.7 The rms Value of a Periodic Function 641
16.8 The Exponential Form of the Fourier Series 642
16.9 Amplitude and Phase Spectra 645
Practical Perspective: Active High-Q Filters 647
Summary 649
Problems 650

Chapter 17 The Fourier Transform 660

Practical Perspective: Filtering Digital Signals 661
17.1 The Derivation of the Fourier Transform 662
17.2 The Convergence of the Fourier Integral 664
17.3 Using Laplace Transforms to Find Fourier Transforms 666
17.4 Fourier Transforms in the Limit 668
17.5 Some Mathematical Properties 671
17.6 Operational Transforms 672
17.7 Circuit Applications 677
17.8 Parseval's Theorem 679
Practical Perspective: Filtering Digital Signals 685
Summary 686
Problems 686

Chapter 18 Two-Port Circuits 692

Practical Perspective: Characterizing an Unknown Circuit 693
18.1 The Terminal Equations 694
18.2 The Two-Port Parameters 695
18.3 Analysis of the Terminated Two-Port Circuit 703
18.4 Interconnected Two-Port Circuits 708
Practical Perspective: Characterizing an Unknown Circuit 711
Summary 712
Problems 713

Appendix A The Solution of Linear Simultaneous Equations 718

A.1 Preliminary Steps 718
A.2 Calculator and Computer Methods 719
A.3 Paper-and-Pencil Methods 721
A.4 Applications 723

Appendix B Complex Numbers 727

B.1 Notation 727
B.2 The Graphical Representation of a Complex Number 728
B.3 Arithmetic Operations 729
B.4 Useful Identities 730
B.5 The Integer Power of a Complex Number 731
B.6 The Roots of a Complex Number 731

Appendix C More on Magnetically Coupled Coils and Ideal Transformers 733

C.1 Equivalent Circuits for Magnetically Coupled Coils 733

C.2 The Need for Ideal Transformers in the Equivalent Circuits 737

Appendix D The Decibel 741

Appendix E Bode Diagrams 743

E.1 Real, First-Order Poles and Zeros 743
E.2 Straight-Line Amplitude Plots 744
E.3 More Accurate Amplitude Plots 747
E.4 Straight-Line Phase Angle Plots 748
E.5 Bode Diagrams: Complex Poles and Zeros 750
E.6 Straight-Line Amplitude Plots for Complex Poles 751
E.7 Correcting Straight-Line Amplitude Plots for Complex Poles 752
E.8 Phase Angle Plots for Complex Poles 754

Appendix F An Abbreviated Table of Trigonometric Identities 757

Appendix G An Abbreviated Table of Integrals 758

Appendix H Common Standard Component Values 760

Answers to Selected Problems 761

Index 771

List of Examples

Chapter 1

1.1 Using SI Units and Prefixes for Powers of 10 *11*
1.2 Relating Current and Charge *15*
1.3 Using the Passive Sign Convention *17*
1.4 Relating Voltage, Current, Power, and Energy *17*

Chapter 2

2.1 Testing Interconnections of Ideal Sources *30*
2.2 Testing Interconnections of Ideal Independent and Dependent Sources *31*
2.3 Calculating Voltage, Current, and Power for a Simple Resistive Circuit *34*
2.4 Constructing a Circuit Model of a Flashlight *36*
2.5 Constructing a Circuit Model Based on Terminal Measurements *38*
2.6 Using Kirchhoff's Current Law *41*
2.7 Using Kirchhoff's Voltage Law *42*
2.8 Applying Ohm's Law and Kirchhoff's Laws to Find an Unknown Current *42*
2.9 Constructing a Circuit Model Based on Terminal Measurements *43*
2.10 Analyzing a Circuit with a Dependent Source *45*
2.11 Applying Ohm's Law and Kirchhoff's Laws to Find an Unknown Voltage *46*
2.12 Applying Ohm's Law and Kirchhoff's Law in an Amplifier Circuit *47*

Chapter 3

3.1 Applying Series-Parallel Simplification *62*
3.2 Solving a Circuit Using Series-Parallel Simplification *63*
3.3 Designing a Simple Voltage Divider *65*
3.4 Adding a Resistive Load to a Voltage Divider *65*
3.5 The Effect of Resistor Tolerance on the Voltage-Divider Circuit *66*
3.6 Designing a Current-Divider Circuit *67*
3.7 Using Voltage Division and Current Division to Solve a Circuit *69*
3.8 Using a d'Arsonval Ammeter *71*
3.9 Using a d'Arsonval Voltmeter *72*
3.10 Using a Wheatstone Bridge to Measure Resistance *75*
3.11 Applying a Delta-to-Wye Transform *77*

Chapter 4

4.1 Identifying Node, Branch, Mesh, and Loop in a Circuit *94*
4.2 Using Essential Nodes and Essential Branches to Write Simultaneous Equations *95*
4.3 Using the Node-Voltage Method *97*
4.4 Using the Node-Voltage Method with Dependent Sources *99*
4.5 Node-Voltage Analysis of the Amplifier Circuit *102*
4.6 Using the Mesh-Current Method *106*
4.7 Using the Mesh-Current Method with Dependent Sources *107*
4.8 A Special Case in the Mesh-Current Method *108*
4.9 Mesh-Current Analysis of the Amplifier Circuit *111*
4.10 Understanding the Node-Voltage Method Versus Mesh-Current Method *113*
4.11 Comparing the Node-Voltage and Mesh-Current Methods *114*
4.12 Using Source Transformations to Solve a Circuit *116*
4.13 Using Special Source Transformation Techniques *117*
4.14 Finding a Thévenin Equivalent *120*
4.15 Finding a Norton Equivalent *121*
4.16 Finding the Thévenin Equivalent of a Circuit with a Dependent Source *122*
4.17 Finding the Thévenin Equivalent Resistance Directly from the Circuit *123*
4.18 Finding the Thévenin Equivalent Resistance Using a Test Source *124*
4.19 Finding the Thévenin Equivalent of a Circuit with Dependent Sources and Resistors *124*
4.20 Using a Thévenin Equivalent to Analyze the Amplifier Circuit *125*
4.21 Calculating the Condition for Maximum Power Transfer *127*
4.22 Using Superposition to Solve a Circuit *129*
4.23 Using Superposition to Solve a Circuit with Dependent Sources *130*

Chapter 5

5.1 Analyzing an Op Amp Circuit *155*
5.2 Designing an Inverting Amplifier *157*
5.3 Designing a Summing Amplifier *159*
5.4 Designing a Noninverting Amplifier *161*
5.5 Designing a Difference Amplifier *163*
5.6 Calculating the CMRR *167*
5.7 Analyzing a Noninverting-Amplifier Circuit using a Realistic Op Amp Model *169*

Chapter 6

6.1   Determining the Voltage, Given the Current, at the Terminals of an Inductor   *184*

6.2   Determining the Current, Given the Voltage, at the Terminals of an Inductor   *186*

6.3   Determining the Current, Voltage, Power, and Energy for an Inductor   *187*

6.4   Determining Current, Voltage, Power, and Energy for a Capacitor   *191*

6.5   Finding v, p, and w Induced by a Triangular Current Pulse for a Capacitor   *192*

6.6   Finding the Equivalent Inductance   *196*

6.7   Finding the Equivalent Capacitance   *197*

6.8   Finding Mesh-Current Equations for a Circuit with Magnetically Coupled Coils   *201*

6.9   Calculating the Coupling Coefficient and Stored Energy for Magnetically Coupled Coils   *209*

Chapter 7

7.1   Determining the Natural Response of an *RL* Circuit   *224*

7.2   Determining the Natural Response of an *RL* Circuit with Parallel Inductors   *227*

7.3   Determining the Natural Response of an *RC* Circuit   *230*

7.4   Determining the Natural Response of an *RC* Circuit with Series Capacitors   *231*

7.5   Determining the Step Response of an *RL* Circuit   *234*

7.6   Determining the Step Response of an *RC* Circuit   *239*

7.7   Using the General Solution Method to Find an *RL* Circuit's Natural Response   *242*

7.8   Using the General Solution Method to Find an *RC* Circuit's Step Response   *243*

7.9   Using the General Solution Method to Find an *RL* Circuit's Step Response   *244*

7.10   Determining the Step Response of a Circuit with Magnetically Coupled Coils   *245*

7.11   Analyzing an *RL* Circuit that has Sequential Switching   *247*

7.12   Analyzing an *RC* Circuit that has Sequential Switching   *249*

7.13   Finding the Unbounded Response in an *RC* Circuit   *251*

7.14   Analyzing an Integrating Amplifier   *253*

7.15   Analyzing an Integrating Amplifier that has Sequential Switching   *253*

Chapter 8

8.1   Finding the Roots of the Characteristic Equation of a Parallel *RLC* Circuit   *277*

8.2   Finding the Overdamped Natural Response of a Parallel *RLC* Circuit   *280*

8.3   Calculating Branch Currents in the Natural Response of a Parallel *RLC* Circuit   *281*

8.4   Finding the Underdamped Natural Response of a Parallel *RLC* Circuit   *284*

8.5   Finding the Critically Damped Natural Response of a Parallel *RLC* Circuit   *288*

8.6   Finding the Overdamped Step Response of a Parallel *RLC* Circuit   *293*

8.7   Finding the Underdamped Step Response of a Parallel *RLC* Circuit   *294*

8.8   Finding the Critically Damped Step Response of a Parallel *RLC* Circuit   *294*

8.9   Comparing the Three-Step Response Forms   *295*

8.10   Finding Step Response of a Parallel *RLC* Circuit with Initial Stored Energy   *295*

8.11   Finding the Natural Response of a Series *RLC* Circuit   *302*

8.12   Finding the Step Response of a Series *RLC* Circuit   *302*

8.13   Analyzing Two Cascaded Integrating Amplifiers   *305*

8.14   Analyzing Two Cascaded Integrating Amplifiers with Feedback Resistors   *307*

Chapter 9

9.1   Finding the Characteristics of a Sinusoidal Current   *321*

9.2   Finding the Characteristics of a Sinusoidal Voltage   *322*

9.3   Translating a Sine Expression to a Cosine Expression   *322*

9.4   Calculating the rms Value of a Triangular Waveform   *322*

9.5   Adding Cosines Using Phasors   *326*

9.6   Calculating Component Voltages Using Phasor Techniques   *331*

9.7   Using KVL in the Frequency Domain   *333*

9.8   Combining Impedances in Series   *334*

9.9   Combining Impedances in Series and in Parallel   *337*

9.10   Using a Delta-to-Wye Transform in the Frequency Domain   *339*

9.11   Performing Source Transformations in the Frequency Domain   *341*

9.12   Finding a Thévenin Equivalent in the Frequency Domain   *342*

9.13   Using the Node-Voltage Method in the Frequency Domain   *344*

9.14   Using the Mesh-Current Method in the Frequency Domain   *346*

9.15 Analyzing a Linear Transformer in the Frequency Domain *349*

9.16 Analyzing an Ideal Transformer Circuit in the Frequency Domain *355*

9.17 Using Phasor Diagrams to Analyze a Circuit *357*

9.18 Using Phasor Diagrams to Analyze Capacitive Loading Effects *358*

Chapter 10

10.1 Calculating Average and Reactive Power *380*

10.2 Making Power Calculations Involving Household Appliances *382*

10.3 Determining Average Power Delivered to a Resistor by a Sinusoidal Voltage *384*

10.4 Calculating Complex Power *385*

10.5 Calculating Power Using Phasor Voltage and Current *387*

10.6 Calculating Average and Reactive Power *389*

10.7 Calculating Power in Parallel Loads *390*

10.8 Balancing Power Delivered with Power Absorbed in an AC Circuit *391*

10.9 Determining Maximum Power Transfer without Load Restrictions *395*

10.10 Determining Maximum Power Transfer with Load Impedance Restriction *396*

10.11 Finding Maximum Power Transfer with Impedance Angle Restrictions *396*

10.12 Finding Maximum Power Transfer in a Circuit with an Ideal Transformer *397*

Chapter 11

11.1 Analyzing a Wye-Wye Circuit *420*

11.2 Analyzing a Wye-Delta Circuit *423*

11.3 Calculating Power in a Three-Phase Wye-Wye Circuit *428*

11.4 Calculating Power in a Three-Phase Wye-Delta Circuit *428*

11.5 Calculating Three-Phase Power with an Unspecified Load *429*

11.6 Computing Wattmeter Readings in Three-Phase Circuits *432*

Chapter 12

12.1 Using Step Functions to Represent a Function of Finite Duration *448*

12.2 Using Laplace Transforms to Predict a Circuit's Response *460*

12.3 Finding the Inverse Laplace Transform when $F(s)$ has Distinct Real Roots *462*

12.4 Finding the Inverse Laplace Transform when $F(s)$ has Distinct Complex Roots *465*

12.5 Finding the Inverse Laplace Transform when $F(s)$ has Repeated Real Roots *467*

12.6 Finding the Inverse Laplace Transform when $F(s)$ has Repeated Complex Roots *468*

12.7 Finding the Inverse Laplace Transform of an Improper Rational Function *470*

12.8 Finding and Plotting the Poles and Zeros of an s-Domain Function *471*

12.9 Applying the Initial- and Final-Value Theorems *474*

Chapter 13

13.1 Transforming a Circuit into the s Domain *488*

13.2 The Natural Response of an *RC* Circuit *489*

13.3 The Step Response of an *RLC* Circuit *489*

13.4 Analyzing a Circuit with a Sinusoidal Source *491*

13.5 Analyzing a Circuit with Multiple Meshes *493*

13.6 Creating a Thévenin Equivalent in the s Domain *495*

13.7 Analyzing a Circuit with Mutual Inductance *497*

13.8 Applying Superposition in the s Domain *499*

13.9 Deriving the Transfer Function of a Circuit *501*

13.10 Analyzing the Transfer Function of a Circuit *503*

13.11 Using the Convolution Integral to Find an Output Signal *509*

13.12 Using the Transfer Function to Find the Steady-State Sinusoidal Response *513*

13.13 A Series Inductor Circuit with an Impulsive Response *515*

13.14 A Circuit with Both Internally Generated and Externally Applied Impulses *518*

Chapter 14

14.1 Designing a Low-Pass Filter *543*

14.2 Designing a Series *RC* Low-Pass Filter *544*

14.3 Designing a Series *RL* High-Pass Filter *547*

14.4 Loading the Series *RL* High-Pass Filter *548*

14.5 Designing a Bandpass Filter *555*

14.6 Designing a Parallel *RLC* Bandpass Filter *555*

14.7 Determining Effect of a Nonideal Voltage Source on a *RLC* Bandpass Filter *557*

14.8 Designing a Series *RLC* Bandreject Filter *562*

Chapter 15

15.1 Designing a Low-Pass Op Amp Filter *575*

15.2 Designing a High-Pass Op Amp Filter *576*

15.3 Scaling a Series *RLC* Filter *578*

15.4 Scaling a Prototype Low-Pass Op Amp Filter *579*

15.5 Designing a Broadband Bandpass Op Amp Filter *583*

15.6 Designing a Broadband Bandreject Op Amp Filter *586*

15.7 Designing a Fourth-Order Low-Pass Active Filter *589*
15.8 Calculating Butterworth Transfer Functions *592*
15.9 Designing a Fourth-Order Low-Pass Butterworth Filter *594*
15.10 Determining the Order of a Butterworth Filter *597*
15.11 An Alternate Approach to Determining the Order of a Butterworth Filter *597*
15.12 Designing a Butterworth Bandpass Filter *599*
15.13 Designing a High-Q Bandpass Filter *602*
15.14 Designing a High-Q Bandreject Filter *604*

Chapter 16

16.1 Finding the Fourier Series of a Triangular Waveform *623*
16.2 Finding the Fourier Series of a Periodic Function with Symmetry *630*
16.3 Calculating Forms of the Trigonometric Fourier Series for Periodic Voltage *632*
16.4 Finding the Response of an RLC Circuit to a Square-Wave Voltage *637*
16.5 Calculating Average Power for a Circuit with a Periodic Voltage Source *640*
16.6 Estimating the rms Value of a Periodic Function *642*
16.7 Finding the Exponential Form of the Fourier Series *644*
16.8 Plotting the Amplitude and Phase Spectra for a Periodic Voltage *646*

Chapter 17

17.1 Finding the Fourier Transform of a Constant *665*
17.2 Finding the Fourier Transform from the Laplace Transform *667*
17.3 Deriving an Operational Fourier Transform *675*
17.4 Using the Fourier Transform to Find the Transient Response *677*
17.5 Using the Fourier Transform to Find the Sinusoidal Steady-State Response *678*
17.6 Applying Parseval's Theorem *681*
17.7 Applying Parseval's Theorem to an Ideal Bandpass Filter *682*
17.8 Applying Parseval's Theorem to a Low-Pass Filter *683*
17.9 Calculating Energy Contained in a Rectangular Voltage Pulse *684*

Chapter 18

18.1 Finding the z Parameters of a Two-Port Circuit *696*
18.2 Finding the a Parameters from Measurements *697*
18.3 Finding h Parameters from Measurements and Table 18.1 *700*
18.4 Determining Whether a Circuit Is Reciprocal and Symmetric *701*
18.5 Analyzing a Terminated Two-Port Circuit *707*
18.6 Analyzing Cascaded Two-Port Circuits *710*

List of Tables

1.1 The International System of Units (SI)   *10*

1.2 Derived Units in SI   *10*

1.3 Standardized Prefixes to Signify Powers of 10   *10*

1.4 Interpretation of Reference Directions in Fig. 1.5   *14*

1.5 Voltage and Current Values for the Circuit in Fig. 1.7   *19*

4.1 Terms for Describing Circuits   *95*

4.2 PSpice Sensitivity Analysis Results   *133*

4.3 Steps in the Node-Voltage Method and the Mesh-Current Method   *135*

6.1 Inductor and Capacitor Duality *198*

7.1 Value of $e^{-t/\tau}$ For t Equal to Integral Multiples of τ *226*

8.1 Natural-Response Parameters of the Parallel *RLC* Circuit   *276*

8.2 Equations for Analyzing the Natural Response of Parallel *RLC* Circuits   *288*

8.3 Equations for Analyzing the Step Response of Parallel *RLC* Circuits   *293*

8.4 Equations for Analyzing the Natural Response of Series *RLC* Circuits   *299*

8.5 Equations for Analyzing the Step Response of Series *RLC* Circuits   *301*

9.1 Impedance and Reactance Values   *331*

9.2 Admittance and Susceptance Values   *336*

9.3 Impedance and Related Values   *361*

10.1 Annual Energy Requirements of Electric Household Appliances   *381*

10.2 Three Power Quantities and Their Units   *385*

10.3 Average Power Consumption of Common Electrical Devices   *399*

12.1 An Abbreviated List of Laplace Transform Pairs   *453*

12.2 An Abbreviated List of Operational Transforms   *458*

12.3 Four Useful Transform Pairs   *469*

13.1 Summary of the s-Domain Equivalent Circuits   *486*

14.1 Input and Output Voltage Magnitudes for Several Frequencies   *543*

15.1 Normalized (so that $\omega_c = 1\,\text{rad/s}$) Butterworth Polynomials up to the Eighth Order   *593*

17.1 Fourier Transforms of Elementary Functions   *670*

17.2 Operational Transforms   *675*

18.1 Parameter Conversion Table   *699*

18.2 Two-Port Parameter Relationships for Reciprocal Circuits   *701*

18.3 Terminated Two-Port Equations   *704*

List of Analysis Methods

Analysis Method 4.1: The Basic Version of the Node-Voltage Method *97*

Analysis Method 4.2: Modified Step 3 for the Node-Voltage Method *99*

Analysis Method 4.3: Complete Form of the Node-Voltage Method *102*

Analysis Method 4.4: The Basic Version of the Mesh-Current Method *105*

Analysis Method 4.5: Modified Step 3 for the Mesh-Current Method *107*

Analysis Method 4.6: Complete Form of the Mesh-Current Method *110*

Analysis Method 5.1: Analyzing an Ideal Op Amp Circuit with a Negative Feedback Path *154*

Analysis Method 7.1: Finding the *RL* Natural Response *224*

Analysis Method 7.2: Finding the *RC* Natural Response *230*

Analysis Method 7.3: Finding the *RL* Step Response *234*

Analysis Method 7.4: Finding the *RC* Step Response *238*

Analysis Method 7.5: Finding the *RL* and *RC* Natural and Step Response *242*

Analysis Method 8.1: The Natural Response of an Overdamped Parallel *RLC* Circuit *280*

Analysis Method 8.2: The Natural Response of an Overdamped or Underdamped Parallel *RLC* Circuit *283*

Analysis Method 8.3: The Natural Response of Parallel *RLC* Circuits *287*

Analysis Method 8.4: The Step Response of Parallel *RLC* Circuits *292*

Analysis Method 8.5: The Natural Response of Series *RLC* Circuits *299*

Analysis Method 8.6: The Step Response of Series *RLC* Circuits *301*

Analysis Method 13.1: Laplace-Transform Circuit Analysis Method *487*

Combine this...

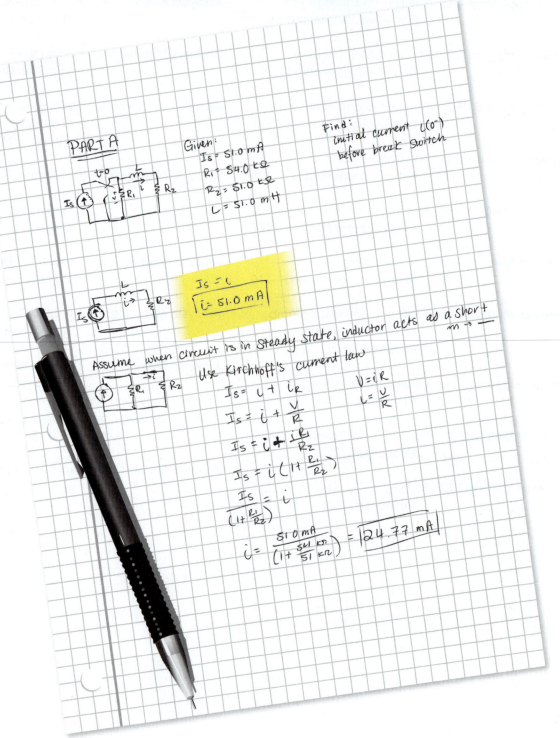

PART A

Given:
$I_S = 51.0\ mA$
$R_1 = 54.0\ k\Omega$
$R_2 = 51.0\ k\Omega$
$L = 51.0\ mH$

Find:
initial current $i(0^-)$
before break switch

$I_S = i$
$\boxed{i = 51.0\ mA}$

Assume when circuit is in steady state, inductor acts as a short
$m \to \underline{\quad}$

Use Kirchhoff's current law

$I_S = i + i_R$
$I_S = i + \dfrac{V}{R}$
$I_S = i + \dfrac{iR_1}{R_2}$
$I_S = i\left(1 + \dfrac{R_1}{R_2}\right)$
$\dfrac{I_S}{\left(1 + \dfrac{R_1}{R_2}\right)} = i$

$V = iR$
$i = \dfrac{V}{R}$

$i = \dfrac{51.0\ mA}{\left(1 + \dfrac{54\ k\Omega}{51\ k\Omega}\right)} = \boxed{24.77\ mA}$

With the Power of
Mastering Engineering for Electric Circuits 11/e

Mastering is the teaching and learning platform that empowers every student. By combining trusted authors' content with digital tools developed to engage students and emulate the office hours experience, Mastering personalizes learning and improves results for each student.

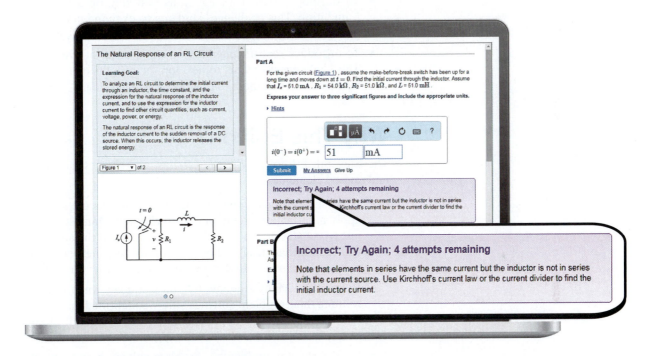

Empower each learner

Each student learns at a different pace. Personalized learning, including adaptive tools and wrong-answer feedback, pinpoints the precise areas where each student needs practice, giving all students the support they need — when and where they need it — to be successful.

Learn more at www.pearson.com/mastering/engineering

Preface

The Eleventh Edition of *Electric Circuits* represents the most extensive revision to the text since the Fifth Edition, published in 1996. Every sentence, paragraph, subsection, and chapter has been examined to improve clarity, readability, and pedagogy. Yet the fundamental goals of the text are unchanged. These goals are:

- To build new concepts and ideas on concepts previously presented. This challenges students to see the explicit connections among the many circuit analysis tools and methods.
- To develop problem-solving skills that rely on a solid conceptual foundation. This challenges students to examine many different approaches to solving a problem before writing a single equation.
- To introduce realistic engineering experiences at every opportunity. This challenges students to develop the insights of a practicing engineer and exposes them to practice of engineering.

Why This Edition?

The Eleventh Edition of *Electric Circuits* incorporates the following new and revised elements:

- Analysis Methods – This new feature identifies the steps needed to apply a particular circuit analysis technique. Many students struggle just to get started when analyzing a circuit, and the analysis methods will reduce that struggle. Some of the analysis methods that are used most often can be found inside the book's covers for easy reference.
- Examples – Many students rely on examples when developing and refining their problem-solving skills. We identified many places in the text that needed additional examples, and as a result the number of examples has increased by nearly 35% to 200.
- End-of-chapter problems – Problem solving is fundamental to the study of circuit analysis. Having a wide variety of problems to assign and work is a key to success in any circuits course. Therefore, some existing end-of-chapter problems were revised, and some new end-of-chapter problems were added. Approximately 30% of the problems in the Eleventh Edition were rewritten.
- Fundamental equations and concepts – These important elements in the text were previously identified with margin notes. In this edition, the margin notes have been replaced by a second-color background, enlarged fonts, and a descriptive title for each fundamental equation and concept. In additional, many equation numbers have been eliminated to make it easier to distinguish fundamental equations from the many other equations in the text.
- Circuit simulation software – The PSpice® and Multisim® manuals have been revised to include screenshots from the most recent versions of these software simulation applications. Each manual presents the simulation material in the same order as the material is encountered in the text. These manuals include example simulations of circuits from the text. Icons identify end-of-chapter problems that are good candidates for simulation using either PSpice or Multisim.

- Solving simultaneous equations – Most circuit analysis techniques in this text eventually require you to solve two or more simultaneous linear algebraic equations. Appendix A has been extensively revised and includes examples of paper-and-pencil techniques, calculator techniques, and computer software techniques.

- Student workbook – Students who could benefit from additional examples and practice problems can use the Student Workbook, which has been revised for the Eleventh Edition of the text. This workbook has examples and problems covering the following material: balancing power, simple resistive circuits, node voltage method, mesh current method, Thévenin and Norton equivalents, op amp circuits, first-order circuits, second-order circuits, AC steady-state analysis, and Laplace transform circuit analysis.

- The Student Workbook now includes access to Video Solutions, complete, step-by-step solution walkthroughs to representative homework problems.

- Learning Catalytics, a "bring your own device" student engagement, assessment, and classroom intelligence system is available with the Eleventh Edition. With Learning Catalytics you can:

  - Use open-ended questions to get into the minds of students to understand what they do or don't know and adjust lectures accordingly.

  - Use a wide variety of question types to sketch a graph, annotate a circuit diagram, compose numeric or algebraic answers, and more.

  - Access rich analytics to understand student performance.

  - Use pre-built questions or add your own to make Learning Catalytics fit your course exactly.

- Pearson Mastering Engineering is an online tutorial and assessment program that provides students with personalized feedback and hints and instructors with diagnostics to track students' progress. With the Eleventh Edition, Mastering Engineering will offer new enhanced end-of-chapter problems with hints and feedback, Coaching Activities, and Adaptive Follow-Up assignments. Visit www.masteringengineering.com for more information.

Hallmark Features

Analysis Methods

Students encountering circuit analysis for the first time can benefit from step-by-step directions that lead them to a problem's solution. We have compiled these directions in a collection of analysis methods, and revised many of the examples in the text to employ these analysis methods.

Chapter Problems

Users of *Electric Circuits* have consistently rated the Chapter Problems as one of the book's most attractive features. In the Eleventh Edition, there are 1185 end-of-chapter problems with approximately 30% that have been revised from the previous edition. Problems are organized at the end of each chapter by section.

Practical Perspectives

The Eleventh Edition continues using Practical Perspectives to introduce the chapter. They provide real-world circuit examples, taken from real-world devices. Every chapter begins by describing a practical application of the material that follows. After presenting that material, the chapter revisits the Practical Perspective, performing a quantitative circuit analysis using the newly introduced chapter material. A special icon identifies end-of-chapter problems directly related to the Practical Perspective application. These problems provide additional opportunities for solving real-world problems using the chapter material.

Assessment Problems

Each chapter begins with a set of chapter objectives. At key points in the chapter, you are asked to stop and assess your mastery of a particular objective by solving one or more assessment problems. The answers to all of the assessment problems are given at the conclusion of each problem, so you can check your work. If you are able to solve the assessment problems for a given objective, you have mastered that objective. If you need more practice, several end-of-chapter problems that relate to the objective are suggested at the conclusion of the assessment problems.

Examples

Every chapter includes many examples that illustrate the concepts presented in the text in the form of a numeric example. There are now nearly 200 examples in this text, an increase of about 35% when compared to the previous edition. The examples illustrate the application of a particular concept, often employ an Analysis Method, and exemplify good problem-solving skills.

Fundamental Equations and Concepts

Throughout the text, you will see fundamental equations and concepts set apart from the main text. This is done to help you focus on some of the key principles in electric circuits and to help you navigate through the important topics.

Integration of Computer Tools

Computer tools can assist students in the learning process by providing a visual representation of a circuit's behavior, validating a calculated solution, reducing the computational burden of more complex circuits, and iterating toward a desired solution using parameter variation. This computational support is often invaluable in the design process. The Eleventh Edition supports PSpice and Multisim, both popular computer tools for circuit simulation and analysis. Chapter problems suited for exploration with PSpice and Multisim are marked accordingly.

Design Emphasis

The Eleventh Edition continues to support the emphasis on the design of circuits in many ways. First, many of the Practical Perspective discussions focus on the design aspects of the circuits. The accompanying Chapter Problems continue the discussion of the design issues in these practical examples. Second, design-oriented Chapter Problems have been labeled explicitly, enabling students and instructors to identify those problems with a design focus. Third, the identification of problems suited to exploration with PSpice or Multisim suggests design opportunities using these software tools. Fourth, some problems in nearly every chapter focus on the use of realistic component values in achieving a desired circuit design. Once such a problem has been analyzed, the student can proceed to a laboratory to build and test the circuit, comparing the analysis with the measured performance of the actual circuit.

Accuracy

All text and problems in the Eleventh Edition have undergone our strict hall-mark accuracy checking process, to ensure the most error-free book possible.

Resources For Students

Mastering Engineering. Mastering Engineering provides tutorial homework problems designed to emulate the instructor's office hour environment, guiding students through engineering concepts with self-paced individualized coaching. These in-depth tutorial homework problems provide students with feedback specific to their errors and optional hints that break problems down into simpler steps. Visit www.masteringengineering.com for more information.

Learning Catalytics. Learning Catalytics is an interactive student response tool that encourages team-based learning by using student's smartphones, tablets, or laptops to engage them in interactive tasks and thinking. Visit www.learningcatalytics.com for more information.

Student Workbook. This resource teaches students techniques for solving problems presented in the text. Organized by concepts, this is a valuable problem-solving resource for all levels of students. The Student Workbook now includes access to Video Solutions, complete, step-by-step solution walkthroughs to representative homework problems.

Introduction to Multisim and Introduction to PSpice Manuals—Updated for the Eleventh Edition, these manuals are excellent resources for those wishing to integrate PSpice or Multisim into their classes.

Resources for Instructors

All instructor resources are available for download at www.pearsonhigh-ered.com. If you are in need of a login and password for this site, please contact your local Pearson representative.

Instructor Solutions Manual—Fully worked-out solutions to Assessment Problems and end-of-chapter problems.

PowerPoint lecture images—All figures from the text are available in Pow-erPoint for your lecture needs. An additional set of full lecture slides with embedded assessment questions are available upon request.

MasteringEngineering. This online tutorial and assessment program allows you to integrate dynamic homework with automated grading and person-alized feedback. MasteringEngineering allows you to easily track the per-formance of your entire class on an assignment-by-assignment basis, or the detailed work of an individual student. For more information visit www.masteringengineering.com.

Learning Catalytics—This "bring your own device" student engagement, assessment and classroom intelligence system enables you to measure student learning during class, and adjust your lectures accordingly. A wide variety of question and answer types allows you to author your own questions, or you can use questions already authored into the system. For more information visit www.learningcatalytics.com or click on the Learning Cat-alytics link inside Mastering Engineering.

Prerequisites

In writing the first 12 chapters of the text, we have assumed that the reader has taken a course in elementary differential and integral calculus. We have also assumed that the reader has had an introductory physics course, at either the high school or university level, that introduces the concepts of energy, power, electric charge, electric current, electric potential, and electromagnetic fields. In writing the final six chapters, we have assumed the student has had, or is enrolled in, an introductory course in differential equations.

Course Options

The text has been designed for use in a one-semester, two-semester, or a three-quarter sequence.

- *Single-semester course:* After covering Chapters 1–4 and Chapters 6–10 (omitting Sections 7.7 and 8.5) the instructor can develop the desired emphasis by covering Chapter 5 (operational amplifiers), Chapter 11 (three-phase circuits), Chapters 13 and 14 (Laplace methods), or Chapter 18 (Two-Port Circuits).

- *Two-semester sequence:* Assuming three lectures per week, cover the first nine chapters during the first semester, leaving Chapters 10–18 for the second semester.

- *Academic quarter schedule:* Cover Chapters 1–6 in the first quarter, Chapters 7–12 in the second quarter, and Chapters 13–18 in the third quarter.

Note that the introduction to operational amplifier circuits in Chapter 5 can be omitted with minimal effect on the remaining material. If Chapter 5 is omitted, you should also omit Section 7.7, Section 8.5, Chapter 15, and those assessment problems and end-of-chapter problems that pertain to operational amplifiers.

There are several appendixes at the end of the book to help readers make effective use of their mathematical background. Appendix A presents several different methods for solving simultaneous linear equations; complex numbers are reviewed in Appendix B; Appendix C contains additional material on magnetically coupled coils and ideal transformers; Appendix D contains a brief discussion of the decibel; Appendix E is dedicated to Bode diagrams; Appendix F is devoted to an abbreviated table of trigonometric identities that are useful in circuit analysis; and an abbreviated table of useful integrals is given in Appendix G. Appendix H provides tables of common standard component values for resistors, inductors, and capacitors, to be used in solving many end-of-chapter problems. Selected Answers provides answers to selected end-of-chapter problems.

Acknowledgments

I will be forever grateful to Jim Nilsson for giving me the opportunity to collaborate with him on this textbook. I started by revising the PSpice supplement for the Third Edition, and became a co-author of the Fifth Edition. Jim was a patient and gracious mentor, and I learned so much from him about teaching and writing and hard work. It is a great honor to be associated with him through this textbook, and to impact the education of the thousands of students who use this text.

There were many hard-working people behind the scenes at our publisher who deserve my thanks and gratitude for their efforts on behalf of the Eleventh Edition. At Pearson, I would like to thank Norrin Dias, Erin Ault, Rose Kernan, and Scott Disanno for their continued support and encouragement, their professional demeanor, their willingness to lend an ear, and their months of long hours and no weekends. The author would also like to acknowledge the staff at Integra Software Solutions for their dedication and hard work in typesetting this text.

I am very grateful for the many instructors and students who have done formal reviews of the text or offered positive feedback and suggestions for improvement more informally. I am pleased to receive email from instructors and students who use the book, even when they are pointing out an error I failed to catch in the review process. I have been contacted by people who use our text from all over the world, and I thank all of you for taking the time to do so. I use as many of your suggestions as possible to continue to improve the content, the pedagogy, and the presentation in this text. I am privileged to have the opportunity to impact the educational experience of the many thousands of future engineers who will use this text.

SUSAN A. RIEDEL

ELECTRIC CIRCUITS

ELEVENTH EDITION

CHAPTER

1

CHAPTER CONTENTS

1.1 **Electrical Engineering: An Overview** *p. 4*
1.2 **The International System of Units** *p. 9*
1.3 **Circuit Analysis: An Overview** *p. 11*
1.4 **Voltage and Current** *p. 12*
1.5 **The Ideal Basic Circuit Element** *p. 14*
1.6 **Power and Energy** *p. 15*

CHAPTER OBJECTIVES

1 Understand and be able to use SI units and the standard prefixes for powers of 10.
2 Know and be able to use the definitions of *voltage* and *current*.
3 Know and be able to use the definitions of *power* and *energy*.
4 Be able to use the passive sign convention to calculate the power for an ideal basic circuit element given its voltage and current.

Circuit Variables

Electrical engineering is an exciting and challenging profession for anyone who has a genuine interest in, and aptitude for, applied science and mathematics. Electrical engineers play a dominant role in developing systems that change the way people live and work. Satellite communication links, cell phones, computers, televisions, diagnostic and surgical medical equipment, robots, and aircraft represent systems that define a modern technological society. As an electrical engineer, you can participate in this ongoing technological revolution by improving and refining existing systems and by discovering and developing new systems to meet the needs of our ever-changing society.

This text introduces you to electrical engineering using the analysis and design of linear circuits. We begin by presenting an overview of electrical engineering, some ideas about an engineering point of view as it relates to circuit analysis, and a review of the International System of Units. We then describe generally what circuit analysis entails. Next, we introduce the concepts of voltage and current. We continue by discussing the ideal basic element and the need for a polarity reference system. We conclude the chapter by describing how current and voltage relate to power and energy.

■ Practical Perspective

Balancing Power

One of the most important skills you will develop is the ability to check your answers for the circuits you design and analyze using the tools developed in this text. A common method used to check for valid answers is to calculate the power in the circuit. The linear circuits we study have no net power, so the sum of the power associated with all circuit components must be zero. If the total power for the circuit is zero, we say that the power balances, but if the total power is not zero, we need to find the errors in our calculation.

As an example, we will consider a simple model for distributing electricity to a typical home. (Note that a more realistic model will be investigated in the Practical Perspective for Chapter 9.) The components labeled a and b represent the source of electrical power for the home. The components labeled c, d, and e represent the wires that carry the electrical current from the source to the devices in the home requiring electrical power. The components labeled f, g, and h represent lamps, televisions, hair dryers, refrigerators, and other devices that require power.

Once we have introduced the concepts of voltage, current, power, and energy, we will examine this circuit model in detail, and use a power balance to determine whether the results of analyzing this circuit are correct.

romakoma/Shutterstock

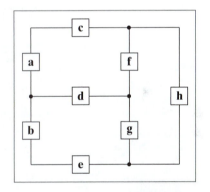

PhotoSerg/Shutterstock

AfricaStudio/Shutterstock

1.1 Electrical Engineering: An Overview

The electrical engineering profession focuses on systems that produce, transmit, and measure electric signals. Electrical engineering combines the physicist's models of natural phenomena with the mathematician's tools for manipulating those models to produce systems that meet practical needs. Electrical systems pervade our lives; they are found in homes, schools, workplaces, and transportation vehicles everywhere. We begin by presenting a few examples from each of the five major classifications of electrical systems:

- communication systems
- computer systems
- control systems
- power systems
- signal-processing systems

Then we describe how electrical engineers analyze and design such systems.

Communication systems are electrical systems that generate, transmit, and distribute information. Well-known examples include television equipment, such as cameras, transmitters, receivers, and monitors; radio telescopes, used to explore the universe; satellite systems, which return images of other planets and our own; radar systems, used to coordinate plane flights; and telephone systems.

Figure 1.1 depicts the major components of a modern telephone system that supports mobile phones, landlines, and international calling. Inside a telephone, a microphone turns sound waves into electric signals. These signals are carried to local or mobile exchanges, where they are combined with the

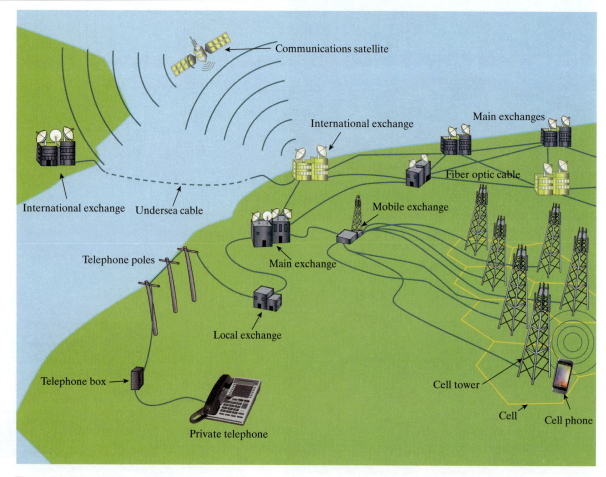

Figure 1.1 ▲ A telephone system.

signals from tens, hundreds, or thousands of other telephones. The form of the signals can be radio waves traveling through air, electrical signals traveling in underground coaxial cable, light pulses traveling in fiber-optic cable, or microwave signals that travel through space. The combined signals are broadcast from a transmission antenna to a receiving antenna. There the combined signals are separated at an exchange, and each is routed to the appropriate telephone, where an earphone acts as a speaker to convert the received electric signals back into sound waves. At each stage of the process, electric circuits operate on the signals. Imagine the challenge involved in designing, building, and operating each circuit in a way that guarantees that all of the hundreds of thousands of simultaneous calls have high-quality connections.

Computer systems use electric signals to process information ranging from word processing to mathematical computations. Systems range in size and power from simple calculators to personal computers to supercomputers that perform such complex tasks as processing weather data and modeling chemical interactions of complex organic molecules. These systems include networks of integrated circuits—miniature assemblies of hundreds, thousands, or millions of electrical components that often operate at speeds and power levels close to fundamental physical limits, including the speed of light and the thermodynamic laws.

Control systems use electric signals to regulate processes. Examples include the control of temperatures, pressures, and flow rates in an oil refinery; the fuel–air mixture in a fuel-injected automobile engine; mechanisms such as the motors, doors, and lights in elevators; and the locks in the Panama Canal. The autopilot and autolanding systems that help to fly and land airplanes are also familiar control systems.

Power systems generate and distribute electric power. Electric power, which is the foundation of our technology-based society, usually is generated in large quantities by nuclear, hydroelectric, solar, and thermal (coal-, oil-, or gas-fired) generators. Power is distributed by a grid of conductors that crisscross the country. A major challenge in designing and operating such a system is to provide sufficient redundancy and control so that failure of any piece of equipment does not leave a city, state, or region completely without power.

Signal-processing systems act on electric signals that represent information. They transform the signals and the information contained in them into a more suitable form. There are many different ways to process the signals and their information. For example, image-processing systems gather massive quantities of data from orbiting weather satellites, reduce the amount of data to a manageable level, and transform the remaining data into a video image for the evening news broadcast. A magnetic resonance imaging (MRI) scan is another example of an image-processing system. It takes signals generated by powerful magnetic fields and radio waves and transforms them into a detailed, three-dimensional image such as the one shown in Fig. 1.2, which can be used to diagnose disease and injury.

Considerable interaction takes place among the engineering disciplines involved in designing and operating these five classes of systems. Thus, communications engineers use digital computers to control the flow of information. Computers contain control systems, and control systems contain computers. Power systems require extensive communications systems to coordinate safely and reliably the operation of components, which may be spread across a continent. A signal-processing system may involve a communications link, a computer, and a control system.

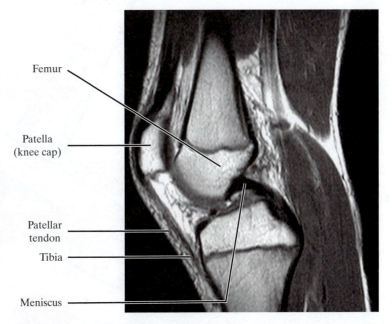

Figure 1.2 ▲ An MRI scan of an adult knee joint.
Neil Borden/Science Source/Getty Images

A good example of the interaction among systems is a commercial airplane, such as the one shown in Fig. 1.3. A sophisticated communications system enables the pilot and the air traffic controller to monitor the plane's location, permitting the air traffic controller to design a safe flight path for all of the nearby aircraft and enabling the pilot to keep the plane on its designated path. An onboard computer system manages engine functions, implements the navigation and flight control systems, and generates video information screens in the cockpit. A complex control system uses cockpit commands to adjust the position and speed of the airplane, producing the appropriate signals to the engines and the control surfaces (such as the wing flaps, ailerons, and rudder) to ensure the plane remains safely airborne and on the desired flight path. The plane must have its own power system to stay aloft and to provide and distribute the electric power needed to keep the cabin lights on, make the coffee, and activate the entertainment system. Signal-processing systems reduce the noise in air traffic communications and transform information about the plane's location into the more meaningful form of a video display in the cockpit. Engineering challenges abound in the design of each of these systems and their integration into a coherent whole. For example, these systems must operate in widely varying and unpredictable environmental conditions. Perhaps the most important engineering challenge is to guarantee that sufficient redundancy is incorporated in the designs, ensuring that passengers arrive safely and on time at their desired destinations.

Although electrical engineers may be interested primarily in one area, they must also be knowledgeable in other areas that interact with this area of interest. This interaction is part of what makes electrical engineering a challenging and exciting profession. The emphasis in engineering is on making things work, so an engineer is free to acquire and use any technique from any field that helps to get the job done.

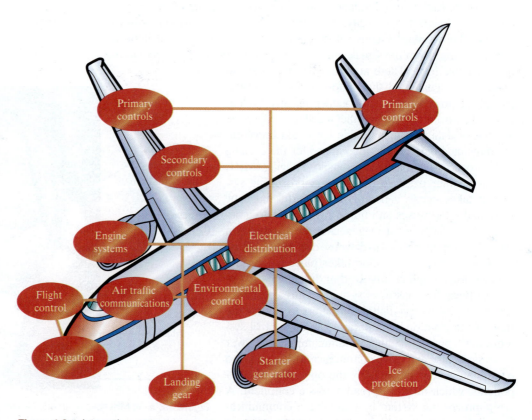

Figure 1.3 ▲ Interacting systems on a commercial aircraft.

Circuit Theory

An **electric circuit** is a mathematical model that approximates the behavior of an actual electrical system. Since electric circuits are found in every branch of electrical engineering, they provide an important foundation for learning how to design and operate systems such as those just described. The models, the mathematical techniques, and the language of circuit theory will form the intellectual framework for your future engineering endeavors.

Note that the term *electric circuit* is commonly used to refer to an actual electrical system as well as to the model that represents it. In this text, when we talk about an electric circuit, we always mean a model, unless otherwise stated. It is the modeling aspect of circuit theory that has broad applications across engineering disciplines.

Circuit theory is a special case of electromagnetic field theory: the study of static and moving electric charges. But applying generalized field theory to the study of electric signals is cumbersome and requires advanced mathematics. Consequently, a course in electromagnetic field theory is not a prerequisite to understanding the material in this book. We do, however, assume that you have had an introductory physics course in which electrical and magnetic phenomena were discussed.

Three basic assumptions permit us to use circuit theory, rather than electromagnetic field theory, to study a physical system represented by an electric circuit.

1. *Electrical effects happen instantaneously throughout a system.* We can make this assumption because we know that electric signals travel at or near the speed of light. Thus, if the system is physically small, electric signals move through it so quickly that we can consider them to affect every point in the system simultaneously. A system that is small enough so that we can make this assumption is called a **lumped-parameter system**.

2. *The net charge on every component in the system is always zero.* Thus, no component can collect a net excess of charge, although some components, as you will learn later, can hold equal but opposite separated charges.

3. *There is no magnetic coupling between the components in a system.* As we demonstrate later, magnetic coupling can occur *within* a component.

That's it; there are no other assumptions. Using circuit theory provides simple solutions (of sufficient accuracy) to problems that would become hopelessly complicated if we were to use electromagnetic field theory. These benefits are so great that engineers sometimes specifically design electrical systems to ensure that these assumptions are met. The importance of assumptions 2 and 3 becomes apparent after we introduce the basic circuit elements and the rules for analyzing interconnected elements.

Let's take a closer look at assumption 1. The question is, "How small does a physical system have to be to qualify as a lumped-parameter system?" To get a quantitative answer to this question, remember that electric signals propagate as waves. If the wavelength of the signal is large compared to the physical dimensions of the system, we have a lumped-parameter system. The wavelength λ is the velocity divided by the repetition rate, or **frequency**, of the signal; that is, $\lambda = c/f$. The frequency f is measured in hertz (Hz). For example, power systems in the United States operate at 60 Hz. If we use the speed of light ($c = 3 \times 10^8$ m/s) as the velocity of propagation, the wavelength is 5×10^6 m. If the power system of interest is physically smaller than this wavelength, we can represent it as a lumped-parameter system and use circuit theory to analyze its behavior.

How do we define *smaller*? A good rule is the *rule of 1/10th*: If the dimension of the system is less than 1/10th the dimension of the wavelength, you have a lumped-parameter system. Thus, as long as the physical dimension of the power system is less than 5×10^5 m (which is about 310 miles), we can treat it as a lumped-parameter system.

Now consider a communication system that sends and receives radio signals. The propagation frequency of radio signals is on the order of 10^9 Hz, so the wavelength is 0.3 m. Using the rule of 1/10th, a communication system qualifies as a lumped-parameter system if its dimension is less than 3 cm. Whenever any of the pertinent physical dimensions of a system under study approaches the wavelength of its signals, we must use electromagnetic field theory to analyze that system. Throughout this book we study circuits derived from lumped-parameter systems.

Problem Solving

As a practicing engineer, you will not be asked to solve problems that have already been solved. Whether you are improving the performance of an existing system or designing a new system, you will be working on unsolved problems. As a student, however, you will read and discuss problems with known solutions. Then, by solving related homework and exam problems on your own, you will begin to develop the skills needed to attack the unsolved problems you'll face as a practicing engineer.

Let's review several general problem-solving strategies. Many of these pertain to thinking about and organizing your solution strategy *before* proceeding with calculations.

1. *Identify what's given and what's to be found.* In problem solving, you need to know your destination before you can select a route for getting there. What is the problem asking you to solve or find? Sometimes the goal of the problem is obvious; other times you may need to paraphrase or make lists or tables of known and unknown information to see your objective.

   On one hand, the problem statement may contain extraneous information that you need to weed out before proceeding. On the other hand, it may offer incomplete information or more complexities than can be handled by the solution methods you know. In that case, you'll need to make assumptions to fill in the missing information or simplify the problem context. Be prepared to circle back and reconsider supposedly extraneous information and/or your assumptions if your calculations get bogged down or produce an answer that doesn't seem to make sense.

2. *Sketch a circuit diagram or other visual model.* Translating a verbal problem description into a visual model is often a useful step in the solution process. If a circuit diagram is already provided, you may need to add information to it, such as labels, values, or reference directions. You may also want to redraw the circuit in a simpler, but equivalent, form. Later in this text you will learn the methods for developing such simplified equivalent circuits.

3. *Think of several solution methods and decide on a way of choosing among them.* This course will help you build a collection of analytical tools, several of which may work on a given problem. But one method may produce fewer equations to be solved than another, or it may require only algebra instead of calculus to reach a solution. Such efficiencies, if you can anticipate them, can streamline your calculations considerably. Having an alternative method in mind also gives you a path to pursue if your first solution attempt bogs down.

4. *Calculate a solution.* Your planning up to this point should have helped you identify a good analytical method and the correct equations for the problem. Now comes the solution of those equations. Paper-and-pencil, calculator, and computer methods are all available for performing the actual calculations of circuit analysis. Efficiency and your instructor's preferences will dictate which tools you should use.

5. *Use your creativity.* If you suspect that your answer is off base or if the calculations seem to go on and on without moving you toward a solution, you should pause and consider alternatives. You may need to revisit your assumptions or select a different solution method. Or you may need to take a less conventional problem-solving approach, such as working backward from a solution. This text provides answers to all of the Assessment Problems and many of the Chapter Problems so that you may work backward when you get stuck. In the real world, you won't be given answers in advance, but you may have a desired problem outcome in mind from which you can work backward. Other creative approaches include allowing yourself to see parallels with other types of problems you've successfully solved, following your intuition or hunches about how to proceed, and simply setting the problem aside temporarily and coming back to it later.

6. *Test your solution.* Ask yourself whether the solution you've obtained makes sense. Does the magnitude of the answer seem reasonable? Is the solution physically realizable? Are the units correct? You may want to rework the problem using an alternative method to validate your original answer and help you develop your intuition about the most efficient solution methods for various kinds of problems. In the real world, safety-critical designs are always checked by several independent means. Getting into the habit of checking your answers will benefit you both as a student and as a practicing engineer.

These problem-solving steps cannot be used as a recipe to solve every problem in this or any other course. You may need to skip, change the order of, or elaborate on certain steps to solve a particular problem. Use these steps as a guideline to develop a problem-solving style that works for you.

1.2 The International System of Units

Engineers use quantitative measures to compare theoretical results to experimental results and compare competing engineering designs. Modern engineering is a multidisciplinary profession in which teams of engineers work together on projects, and they can communicate their results in a meaningful way only if they all use the same units of measure. The International System of Units (abbreviated SI) is used by all the major engineering societies and most engineers throughout the world; hence we use it in this book.

The SI units are based on seven *defined* quantities:

- length
- mass
- time
- electric current
- thermodynamic temperature
- amount of substance
- luminous intensity

These quantities, along with the basic unit and symbol for each, are listed in Table 1.1. Although not strictly SI units, the familiar time units

TABLE 1.1	The International System of Units (SI)	
Quantity	**Basic Unit**	**Symbol**
Length	meter	m
Mass	kilogram	kg
Time	second	s
Electric current	ampere	A
Thermodynamic temperature	degree kelvin	K
Amount of substance	mole	mol
Luminous intensity	candela	cd

National Institute of Standards and Technology Special Publication 330, 2008 Edition, Natl. Inst. Stand. Technol. Spec. Pub. 330, 2008 Ed., 96 pages (March 2008)

of minute (60 s), hour (3600 s), and so on are often used in engineering calculations. In addition, defined quantities are combined to form **derived** units. Some quantities, such as force, energy, power, and electric charge, you already know through previous physics courses. Table 1.2 lists the derived units used in this book.

In many cases, the SI unit is either too small or too large to use conveniently. Standard prefixes corresponding to powers of 10, as listed in Table 1.3, are then applied to the basic unit. Engineers often use only the prefixes for powers divisible by 3; thus centi, deci, deka, and hecto are used rarely. Also, engineers often select the prefix that places the base number in the range between 1 and 1000. Suppose that a time calculation yields a result of 10^{-5} s, that is, 0.00001 s. Most engineers would describe this quantity as 10 μs, that is, 10×10^{-6} s, rather than as 0.01 ms or 10,000 ns.

Example 1.1 illustrates a method for converting from one set of units to another and also uses power-of-10 prefixes.

TABLE 1.2	Derived Units in SI	
Quantity	**Unit Name (Symbol)**	**Formula**
Frequency	hertz (Hz)	s^{-1}
Force	newton (N)	$kg \cdot m/s^2$
Energy or work	joule (J)	$N \cdot m$
Power	watt (W)	J/s
Electric charge	coulomb (C)	$A \cdot s$
Electric potential	volt (V)	J/C
Electric resistance	ohm (Ω)	V/A
Electric conductance	siemens (S)	A/V
Electric capacitance	farad (F)	C/V
Magnetic flux	weber (Wb)	$V \cdot s$
Inductance	henry (H)	Wb/A

National Institute of Standards and Technology Special Publication 330, 2008 Edition, Natl. Inst. Stand. Technol. Spec. Pub. 330, 2008 Ed., 96 pages (March 2008)

TABLE 1.3	Standardized Prefixes to Signify Powers of 10	
Prefix	**Symbol**	**Power**
atto	a	10^{-18}
femto	f	10^{-15}
pico	p	10^{-12}
nano	n	10^{-9}
micro	μ	10^{-6}
milli	m	10^{-3}
centi	c	10^{-2}
deci	d	10^{-1}
deka	da	10
hecto	h	10^2
kilo	k	10^3
mega	M	10^6
giga	G	10^9
tera	T	10^{12}

National Institute of Standards and Technology Special Publication 330, 2008 Edition, Natl. Inst. Stand. Technol. Spec. Pub. 330, 2008 Ed., 96 pages (March 2008)

| EXAMPLE 1.1 | **Using SI Units and Prefixes for Powers of 10** |

If a signal can travel in a cable at 80% of the speed of light, what length of cable, in inches, represents 1 ns?

Solution

First, note that $1\,\text{ns} = 10^{-9}\,\text{s}$. Also, recall that the speed of light $c = 3 \times 10^{8}\,\text{m/s}$. Then, 80% of the speed of light is $0.8c = (0.8)(3 \times 10^{8}) = 2.4 \times 10^{8}\,\text{m/s}$. Using a product of ratios, we can convert 80% of the speed of light from meters per second to inches per nanosecond. The result is the distance in inches traveled in 1 nanosecond:

$$\frac{2.4 \times 10^{8}\ \text{meters}}{1\ \text{second}} \cdot \frac{1\ \text{second}}{10^{9}\ \text{nanoseconds}} \cdot \frac{100\ \text{centimeters}}{1\ \text{meter}} \cdot \frac{1\ \text{inch}}{2.54\ \text{centimeters}}$$

$$= 9.45\ \text{inches/nanosecond.}$$

Therefore, a signal traveling at 80% of the speed of light will cover 9.45 inches of cable in 1 nanosecond.

 ## ASSESSMENT PROBLEMS

Objective 1 — Understand and be able to use SI units and the standard prefixes for powers of 10

1.1 Assume a telephone signal travels through a cable at two-thirds the speed of light. How long does it take the signal to get from New York City to Miami if the distance is approximately 1100 miles?

Answer: 8.85 ms.

1.2 How many dollars per millisecond would the federal government have to collect to retire a deficit of $100 billion in one year?

Answer: $3.17/ms.

SELF-CHECK: Also try Chapter Problems 1.2, 1.3, and 1.6.

1.3 Circuit Analysis: An Overview

We look broadly at engineering design, specifically the design of electric circuits, before becoming involved in the details of circuit analysis. This overview provides you with a perspective on where circuit analysis fits within the whole of circuit design. Even though this book focuses on circuit analysis, we try to provide opportunities for circuit design where appropriate.

All engineering designs begin with a need ①, as shown in Fig. 1.4. This need may come from the desire to improve on an existing design, or it may be something brand new. A careful assessment of the need results in design specifications, which are measurable characteristics of a proposed design. Once a design is proposed, the design specifications ② allow us to assess whether or not the design actually meets the need.

A concept ③ for the design comes next. The concept derives from a complete understanding of the design specifications coupled with an insight into the need, which comes from education and experience. The concept may be realized as a sketch, as a written description, or as some other form. Often the next step is to translate the concept into a mathematical model. A commonly used mathematical model for electrical systems is a circuit model ④.

The elements that make up the circuit model are called ideal circuit components. An **ideal circuit component** is a mathematical model of an actual electrical component, like a battery or a light bulb. The ideal circuit

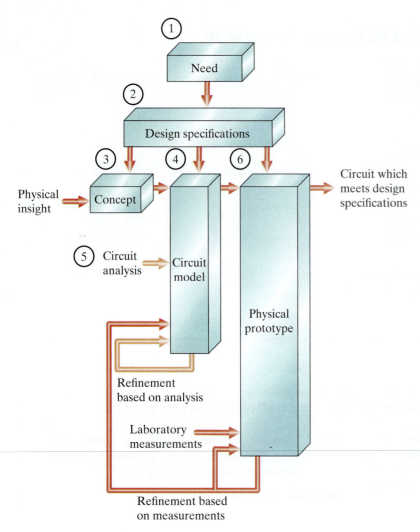

Figure 1.4 ▲ A conceptual model for electrical engineering design.

components used in a circuit model should represent the behavior of the actual electrical components to an acceptable degree of accuracy. The tools of circuit analysis ⑤, the focus of this book, are then applied to the circuit. **Circuit analysis** uses mathematical techniques to predict the behavior of the circuit model and its ideal circuit components. A comparison between the desired behavior, from the design specifications, and the predicted behavior, from circuit analysis, may lead to refinements in the circuit model and its ideal circuit elements. Once the desired and predicted behaviors are in agreement, a physical prototype ⑥ can be constructed.

The **physical prototype** is an actual electrical system, constructed from actual electrical components. Measurements determine the quantitative behavior of the physical system. This actual behavior is compared with the desired behavior from the design specifications and the predicted behavior from circuit analysis. The comparisons may result in refinements to the physical prototype, the circuit model, or both. This iterative process, in which models, components, and systems are continually refined, usually produces a design that accurately satisfies the design specifications and thus meets the need.

Circuit analysis clearly plays a very important role in the design process. Because circuit analysis is applied to circuit models, practicing engineers try to use mature circuit models so that the resulting designs will meet the design specifications in the first iteration. In this book, we use models that have been tested for at least 40 years; you can assume that they are mature. The ability to model actual electrical systems with ideal circuit elements makes circuit theory extremely useful to engineers.

Saying that the interconnection of ideal circuit elements can be used to quantitatively predict the behavior of a system implies that we can describe the interconnection with mathematical equations. For the mathematical equations to be useful, we must write them in terms of measurable quantities. In the case of circuits, these quantities are voltage and current, which we discuss in Section 1.4. The study of circuit analysis involves understanding the behavior of each ideal circuit element in terms of its voltage and current and understanding the constraints imposed on the voltage and current as a result of interconnecting the ideal elements.

1.4 Voltage and Current

The concept of electric charge is the basis for describing all electrical phenomena. Let's review some important characteristics of electric charge.

- Electric charge is bipolar, meaning that electrical effects are described in terms of positive and negative charges.
- Electric charge exists in discrete quantities, which are integer multiples of the electronic charge, 1.6022×10^{-19} C.
- Electrical effects are attributed to both the separation of charge and charges in motion.

In circuit theory, the separation of charge creates an electric force (voltage), and the motion of charge creates an electric fluid (current).

The concepts of voltage and current are useful from an engineering point of view because they can be expressed quantitatively. Whenever positive and negative charges are separated, energy is expended. **Voltage** is the energy per unit charge created by the separation. We express this ratio in differential form as

DEFINITION OF VOLTAGE

$$v = \frac{dw}{dq},$$

(1.1)

where

$$v = \text{the voltage in volts,}$$
$$w = \text{the energy in joules,}$$
$$q = \text{the charge in coulombs.}$$

The electrical effects caused by charges in motion depend on the rate of charge flow. The rate of charge flow is known as the **electric current**, which is expressed as

DEFINITION OF CURRENT

$$i = \frac{dq}{dt},$$

(1.2)

where

$$i = \text{the current in amperes,}$$
$$q = \text{the charge in coulombs,}$$
$$t = \text{the time in seconds.}$$

Equations 1.1 and 1.2 define the magnitude of voltage and current, respectively. The bipolar nature of electric charge requires that we assign polarity references to these variables. We will do so in Section 1.5.

Although current is made up of discrete moving electrons, we do not need to consider them individually because of the enormous number of them. Rather, we can think of electrons and their corresponding charge as one smoothly flowing entity. Thus, i is treated as a continuous variable.

One advantage of using *circuit models* is that we can model a component strictly in terms of the voltage and current at its terminals. Thus, two physically different components could have the same relationship between the terminal voltage and terminal current. If they do, for purposes of circuit analysis, they are identical. Once we know how a component behaves at its terminals, we can analyze its behavior in a circuit. However, when developing *component models*, we are interested in a component's internal behavior. We might want to know, for example, whether charge conduction is taking place because of free electrons moving through the crystal lattice structure of a metal or whether it is because of electrons moving within the covalent bonds of a semiconductor material. These concerns are beyond the realm of circuit theory, so in this book we use component models that have already been developed.

1.5 The Ideal Basic Circuit Element

An **ideal basic circuit element** has three attributes.

1. It has only two terminals, which are points of connection to other circuit components.

2. It is described mathematically in terms of current and/or voltage.

3. It cannot be subdivided into other elements.

Using the word *ideal* implies that a basic circuit element does not exist as a realizable physical component. Ideal elements can be connected in order to model actual devices and systems, as we discussed in Section 1.3. Using the word *basic* implies that the circuit element cannot be further reduced or subdivided into other elements. Thus, the basic circuit elements form the building blocks for constructing circuit models, but they themselves cannot be modeled with any other type of element.

Figure 1.5 represents an ideal basic circuit element. The box is blank because we are making no commitment at this time as to the type of circuit element it is. In Fig. 1.5, the voltage across the terminals of the box is denoted by v, and the current in the circuit element is denoted by i. The plus and minus signs indicate the polarity reference for the voltage, and the arrow placed alongside the current indicates its reference direction. Table 1.4 interprets the voltage polarity and current direction, given positive or negative numerical values of v and i. Note that algebraically the notion of positive charge flowing in one direction is equivalent to the notion of negative charge flowing in the opposite direction.

Assigning the reference polarity for voltage and the reference direction for current is entirely arbitrary. However, once you have assigned the references, you must write all subsequent equations to agree with the chosen references. The most widely used sign convention applied to these references is called the **passive sign convention**, which we use throughout this book.

Figure 1.5 ▲ An ideal basic circuit element.

PASSIVE SIGN CONVENTION

Whenever the reference direction for the current in an element is in the direction of the reference voltage drop across the element (as in Fig. 1.5), use a positive sign in any expression that relates the voltage to the current. Otherwise, use a negative sign.

We apply this sign convention in all the analyses that follow. Our purpose for introducing it even before we have introduced the different types of basic circuit elements is to emphasize that selecting polarity references

TABLE 1.4	Interpretation of Reference Directions in Fig. 1.5	
Positive Value		**Negative Value**
v	voltage drop from terminal 1 to terminal 2	voltage rise from terminal 1 to terminal 2
	or	*or*
	voltage rise from terminal 2 to terminal 1	voltage drop from terminal 2 to terminal 1
i	positive charge flowing from terminal 1 to terminal 2	positive charge flowing from terminal 2 to terminal 1
	or	*or*
	negative charge flowing from terminal 2 to terminal 1	negative charge flowing from terminal 1 to terminal 2

is *not* a function either of the basic elements or the type of interconnections made with the basic elements. We apply and interpret the passive sign convention for power calculations in Section 1.6.

Example 1.2 illustrates one use of the equation defining current.

| **EXAMPLE 1.2** | **Relating Current and Charge** |

No charge exists at the upper terminal of the element in Fig. 1.5 for $t < 0$. At $t = 0$, a 5 A current begins to flow into the upper terminal.

a) Derive the expression for the charge accumulating at the upper terminal of the element for $t > 0$.

b) If the current is stopped after 10 seconds, how much charge has accumulated at the upper terminal?

Solution

a) From the definition of current given in Eq. 1.2, the expression for charge accumulation due to current flow is

$$q(t) = \int_0^t i(x)dx.$$

Therefore,

$$q(t) = \int_0^t 5dx = 5x \Big|_0^t = 5t - 5(0) = 5t \text{ C} \quad \text{for } t > 0.$$

b) The total charge that accumulates at the upper terminal in 10 seconds due to a 5 A current is $q(10) = 5(10) = 50$ C.

ASSESSMENT PROBLEMS

Objective 2—Know and be able to use the definitions of *voltage* and *current*

1.3   The current at the terminals of the element in Fig. 1.5 is

$$i = 0, \qquad t < 0;$$
$$i = 20e^{-5000t} \text{ A}, \qquad t \geq 0.$$

Calculate the total charge (in microcoulombs) entering the element at its upper terminal.

Answer: 4000 μC.

1.4   The expression for the charge entering the upper terminal of Fig. 1.5 is

$$q = \frac{1}{\alpha^2} - \left(\frac{t}{\alpha} + \frac{1}{\alpha^2}\right)e^{-\alpha t} \text{ C}.$$

Find the maximum value of the current entering the terminal if $\alpha = 0.03679 \text{ s}^{-1}$.

Answer: 10 A.

SELF-CHECK: Also try Chapter Problem 1.7.

1.6 Power and Energy

Power and energy calculations are important in circuit analysis. Although voltage and current are useful variables in the analysis and design of electrically based systems, the useful output of the system often is nonelectrical (e.g., sound emitted from a speaker or light from a light bulb), and this output is conveniently expressed in terms of power or energy. Also, all practical devices have limitations on the amount of power that they can handle. In the design process, therefore, voltage and current calculations by themselves are not sufficient to determine whether or not a design meets its specifications.

We now relate power and energy to voltage and current and at the same time use the power calculation to illustrate the passive sign convention. Recall from basic physics that power is the time rate of expending or absorbing energy. (A water pump rated 75 kW can deliver more liters per second than one rated 7.5 kW.) Mathematically, energy per unit time is expressed in the form of a derivative, or

DEFINITION OF POWER

$$p = \frac{dw}{dt},$$ (1.3)

where

$$p = \text{the power in watts,}$$
$$w = \text{the energy in joules,}$$
$$t = \text{the time in seconds.}$$

Thus, 1 W is equivalent to 1 J/s.

The power associated with the flow of charge follows directly from the definition of voltage and current in Eqs. 1.1 and 1.2, or

$$p = \frac{dw}{dt} = \left(\frac{dw}{dq}\right)\left(\frac{dq}{dt}\right),$$

so

POWER EQUATION

$$p = vi,$$ (1.4)

where

$$p = \text{the power in watts,}$$
$$v = \text{the voltage in volts,}$$
$$i = \text{the current in amperes.}$$

Equation 1.4 shows that the **power** associated with a basic circuit element is the product of the current in the element and the voltage across the element. Therefore, power is a quantity associated with a circuit element, and we have to determine from our calculation whether power is being delivered to the circuit element or extracted from it. This information comes from correctly applying and interpreting the passive sign convention (Section 1.5).

If we use the passive sign convention, Eq. 1.4 is correct if the reference direction for the current is in the direction of the reference voltage drop across the terminals. Otherwise, Eq. 1.4 must be written with a minus sign. In other words, if the current reference is in the direction of a reference voltage rise across the terminals, the expression for the power is

$$p = -vi.$$

The algebraic sign of power is based on charge movement through voltage drops and rises. As positive charges move through a drop in voltage, they lose energy, and as they move through a rise in voltage, they gain energy. Figure 1.6 summarizes the relationship between the polarity references for voltage and current and the expression for power.

We can now state the rule for interpreting the algebraic sign of power:

INTERPRETING ALGEBRAIC SIGN OF POWER

- If the power is positive (that is, if $p > 0$), power is being delivered to the circuit element represented by the box.
- If the power is negative (that is, if $p < 0$), power is being extracted from the circuit element.

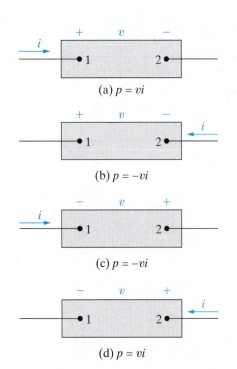

(a) $p = vi$

(b) $p = -vi$

(c) $p = -vi$

(d) $p = vi$

Figure 1.6 ▲ Polarity references and the expression for power.

Example 1.3 shows that the passive sign convention generates the correct sign for power regardless of the voltage polarity and current direction you choose.

EXAMPLE 1.3 | **Using the Passive Sign Convention**

a) Suppose you have selected the polarity references shown in Fig. 1.6(b). Your calculations for the current and voltage yield the following numerical results:

$$i = 4\,\text{A} \quad \text{and} \quad v = -10\,\text{V}.$$

Calculate the power associated with the circuit element and determine whether it is absorbing or supplying power.

b) Your classmate is solving the same problem but has chosen the reference polarities shown in Fig. 1.6(c). Her calculations for the current and voltage show that

$$i = -4\,\text{A} \quad \text{and} \quad v = 10\,\text{V}.$$

What power does she calculate?

Solution

a) The power associated with the circuit element in Fig. 1.6(b) is

$$p = -(-10)(4) = 40\,\text{W}.$$

Thus, the circuit element is absorbing 40 W.

b) Your classmate calculates that the power associated with the circuit element in Fig. 1.6(c) is

$$p = -(10)(-4) = 40\,\text{W}.$$

Using the reference system in Fig. 1.6(c) gives the same conclusion as using the reference system in Fig. 1.6(b)—namely, that the circuit element is absorbing 40 W. In fact, any of the reference systems in Fig. 1.6 yields this same result.

Example 1.4 illustrates the relationship between voltage, current, power, and energy for an ideal basic circuit element and the use of the passive sign convention.

EXAMPLE 1.4 | **Relating Voltage, Current, Power, and Energy**

Assume that the voltage at the terminals of the element in Fig. 1.5, whose current was defined in Assessment Problem 1.3, is

$$v = 0 \qquad\qquad t < 0;$$
$$v = 10e^{-5000t}\,\text{kV}, \qquad t \geq 0.$$

a) Calculate the power supplied to the element at 1 ms.
b) Calculate the total energy (in joules) delivered to the circuit element.

Solution

a) Since the current is entering the + terminal of the voltage drop defined for the element in Fig. 1.5, we use a "+" sign in the power equation.

$$p = vi = (10{,}000e^{-5000t})(20e^{-5000t}) = 200{,}000e^{-10{,}000t}\,\text{W}.$$

$$p(0.001) = 200{,}000e^{-10{,}000(0.001)} = 200{,}000e^{-10}$$

$$= 200{,}000(45.4 \times 10^{-6}) = 9.08\,\text{W}.$$

b) From the definition of power given in Eq. 1.3, the expression for energy is

$$w(t) = \int_0^t p(x)\,dx.$$

To find the total energy delivered, integrate the expresssion for power from zero to infinity. Therefore,

$$w_{\text{total}} = \int_0^\infty 200{,}000e^{-10{,}000x}\,dx = \left.\frac{200{,}000e^{-10{,}000x}}{-10{,}000}\right|_0^\infty$$

$$= -20e^{-\infty} - (-20e^{-0}) = 0 + 20 = 20\,\text{J}.$$

Thus, the total energy supplied to the circuit element is 20 J.

ASSESSMENT PROBLEMS

Objective 3—Know and use the definitions of *power* and *energy*; Objective 4—Be able to use the passive sign convention

1.5 Assume that a 20 V voltage drop occurs across an element from terminal 2 to terminal 1 and that a current of 4 A enters terminal 2.

a) Specify the values of v and i for the polarity references shown in Fig. 1.6(a)–(d).

b) Calculate the power associated with the circuit element.

c) Is the circuit element absorbing or delivering power?

Answer: (a) Circuit 1.6(a): $v = -20$ V, $i = -4$ A;
circuit 1.6(b): $v = -20$ V, $i = 4$ A;
circuit 1.6(c): $v = 20$ V, $i = -4$ A;
circuit 1.6(d): $v = 20$ V, $i = 4$ A;
(b) 80 W;
(c) absorbing.

1.6 The voltage and current at the terminals of the circuit element in Fig. 1.5 are zero for $t < 0$. For $t \geq 0$, they are

$$v = 80{,}000te^{-500t} \text{ V}, \quad t \geq 0;$$

$$i = 15te^{-500t} \text{ A}, \quad t \geq 0.$$

a) Find the time when the power delivered to the circuit element is maximum.

b) Find the maximum value of power.

c) Find the total energy delivered to the circuit element.

Answer: (a) 2 ms;   (b) 649.6 mW;   (c) 2.4 mJ.

1.7 A high-voltage direct-current (dc) transmission line between Celilo, Oregon, and Sylmar, California, is operating at 800 kV and carrying 1800 A, as shown. Calculate the power (in megawatts) at the Oregon end of the line and state the direction of power flow.

Answer: 1440 MW, Celilo to Sylmar

SELF-CHECK: Also try Chapter Problems 1.15, 1.18, and 1.25.

Practical Perspective

Balancing Power

A circuit model for distributing power to a typical home is shown in Fig. 1.7, with voltage polarities and current directions defined for all of the circuit components. Circuit analysis gives values for all of these voltages and currents, as summarized in Table 1.5. To determine whether or not the values given are correct, calculate the power associated with each component. Use the passive sign convention in the power calculations, as shown in the following.

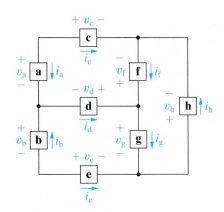

Figure 1.7 ▲ Circuit model for power distribution in a home, with voltages and currents defined.

$$p_a = v_a i_a = (120)(-10) = -1200 \text{ W} \quad p_b = -v_b i_b = -(120)(9) = -1080 \text{ W}$$

$$p_c = v_c i_c = (10)(10) = 100 \text{ W} \quad\quad p_d = -v_d i_d = -(10)(1) = -10 \text{ W}$$

$$p_e = v_e i_e = (-10)(-9) = 90 \text{ W} \quad\quad p_f = -v_f i_f = -(-100)(5) = 500 \text{ W}$$

$$p_g = v_g i_g = (120)(4) = 480 \text{ W} \quad\quad p_h = v_h i_h = (-220)(-5) = 1100 \text{ W}$$

The power calculations show that components a, b, and d are supplying power, since the power values are negative, while components c, e, f, g, and h are absorbing power. Now check to see if the power balances by finding the total power supplied and the total power absorbed.

$$p_{\text{supplied}} = p_a + p_b + p_d = -1200 - 1080 - 10 = -2290 \text{ W}$$

$$p_{\text{absorbed}} = p_c + p_e + p_f + p_g + p_h$$

$$= 100 + 90 + 500 + 480 + 1100 = 2270 \text{ W}$$

$$p_{\text{supplied}} + p_{\text{absorbed}} = -2290 + 2270 = -20 \text{ W}$$

Something is wrong—if the values for voltage and current in this circuit are correct, the total power should be zero! There is an error in the data, and we can find it from the calculated powers if the error exists in the sign of a single component. Note that if we divide the total power by 2, we get -10 W, which is the power calculated for component d. If the power for component d is $+10$ W, the total power would be 0. Circuit analysis techniques from upcoming chapters can be used to show that the current through component d should be -1 A, not $+1$ A as given in Table 1.5.

SELF-CHECK: Assess your understanding of the Practical Perspective by trying Chapter Problems 1.34 and 1.35.

TABLE 1.5	Voltage and Current Values for the Circuit in Fig. 1.7	
Component	v(V)	i(A)
a	120	-10
b	120	9
c	10	10
d	10	1
e	-10	-9
f	-100	5
g	120	4
h	-220	-5

Summary

- The International System of Units (SI) enables engineers to communicate in a meaningful way about quantitative results. Table 1.1 summarizes the SI units; Table 1.2 presents some useful derived SI units. (See page 10.)

- A circuit model is a mathematical representation of an electrical system. Circuit analysis, used to predict the behavior of a circuit model, is based on the variables of voltage and current. (See page 12.)

  - **Voltage** is the energy per unit charge created by charge separation and has the SI unit of volt. (See page 13.)

  $$v = dw/dq$$

  - **Current** is the rate of charge flow and has the SI unit of ampere. (See page 13.)

  $$i = dq/dt$$

- The **ideal basic circuit element** is a two-terminal component that cannot be subdivided; it can be described

mathematically in terms of its terminal voltage and current. (See page 14.)

- The **passive sign convention** uses a positive sign in the expression that relates the voltage and current at the terminals of an element when the reference direction for the current through the element is in the direction of the reference voltage drop across the element. (See page 14.)

- **Power** is energy per unit of time and is equal to the product of the terminal voltage and current; it has the SI unit of watt. (See page 16.)

  $$p = dw/dt = vi$$

The algebraic sign of power is interpreted as follows:

- If $p > 0$, power is being delivered to the circuit or circuit component.

- If $p < 0$, power is being extracted from the circuit or circuit component. (See page 16.)

■ Problems

Section 1.2

1.1 The line described in Assessment Problem 1.7 is 845 mi in length. The line contains four conductors, each weighing 2526 lb per 1000 ft. How many kilograms of conductor are in the line?

1.2 A 32-inch monitor contains 3840×2160 picture elements, or pixels. Each pixel is represented in 24 bits of memory. A byte of memory is 8 bits.

  a) How many megabytes (MB) of memory are required to store the information displayed on the monitor?

  b) To display a video on the monitor, the image must be refreshed 30 times per second. How many terabytes (TB) of memory are required to store a 2 hr video?

  c) For the video described in part (b), how fast must the image data in memory be moved to the monitor? Express your answer in gigabits per second (Gb/s).

1.3 Some species of bamboo can grow (250 mm/day). Assume individual cells in the plant are 10 μm long.

  a) How long, on average, does it take a bamboo stalk to grow 1 cell length?

  b) How many cell lengths are added in one week, on average?

1.4 A hand-held video player displays 480×320 picture elements (pixels) in each frame of the video. Each pixel requires 2 bytes of memory. Videos are displayed at a rate of 30 frames per second. How many hours of video will fit in a 32 gigabyte memory?

1.5 The 16 gigabyte (GB = 2^{30} bytes) flash memory chip for an MP3 player is 11 mm by 15 mm by 1 mm. This memory chip holds 20,000 photos.

  a) How many photos fit into a cube whose sides are 1 mm?

  b) How many bytes of memory are stored in a cube whose sides are 200 μm?

1.6 There are approximately 260 million passenger vehicles registered in the United States. Assume that the battery in the average vehicle stores 540 watt-hours (Wh) of energy. Estimate (in gigawatt-hours) the total energy stored in US passenger vehicles.

Section 1.4

1.7 The current entering the upper terminal of Fig. 1.5 is

$$i = 24 \cos 4000t \text{ A}$$

Assume the charge at the upper terminal is zero at the instant the current is passing through its maximum value. Find the expression for $q(t)$.

1.8 How much energy is imparted to an electron as it flows through a 6 V battery from the positive to the negative terminal? Express your answer in attojoules.

1.9 In electronic circuits it is not unusual to encounter currents in the microampere range. Assume a 35 μA current, due to the flow of electrons. What is the average number of electrons per second that flow past a fixed reference cross section that is perpendicular to the direction of flow?

1.10 There is no charge at the upper terminal of the element in Fig. 1.5 for $t < 0$. At $t = 0$ a current of $125e^{-2500t}$ mA enters the upper terminal.

  a) Derive the expression for the charge that accumulates at the upper terminal for $t > 0$.

  b) Find the total charge that accumulates at the upper terminal.

  c) If the current is stopped at $t = 0.5$ ms, how much charge has accumulated at the upper terminal?

1.11 The current at the terminals of the element in Fig. 1.5 is

$$i = 0, \qquad\qquad t < 0;$$
$$i = 40te^{-500t} \text{ A}, \quad t \geq 0.$$

  a) Find the expression for the charge accumulating at the upper terminal.

  b) Find the charge that has accumulated at $t = 1$ ms.

Sections 1.5–1.6

1.12 When a car has a dead battery, it can often be started by connecting the battery from another car across its terminals. The positive terminals are connected together as are the negative terminals. The connection is illustrated in Fig. P1.12. Assume the current i in Fig. P1.12 is measured and found to be 40 A.

  a) Which car has the dead battery?

  b) If this connection is maintained for 1.5 min, how much energy is transferred to the dead battery?

Figure P1.12

1.13 Two electric circuits, represented by boxes A and B, are connected as shown in Fig. P1.13. The reference direction for the current i in the interconnection and the reference polarity for the voltage v across the interconnection are as shown in the figure. For each of the following sets of numerical values, calculate the power in the interconnection and state whether the power is flowing from A to B or vice versa.

a) $i = 6$ A, $v = 30$ V

b) $i = -8$ A, $v = -20$ V

c) $i = 4$ A $v = -60$ V

d) $i = -9$ A, $v = 40$ V

Figure P1.13

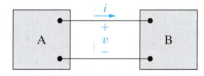

1.14 One 12 V battery supplies 100 mA to a boom box. How much energy does the battery supply in 4 h?

1.15 The references for the voltage and current at the terminals of a circuit element are as shown in Fig. 1.6(d). The numerical values for v and i are -20 V and 5 A.

a) Calculate the power at the terminals and state whether the power is being absorbed or delivered by the element in the box.

b) Given that the current is due to electron flow, state whether the electrons are entering or leaving terminal 2.

c) Do the electrons gain or lose energy as they pass through the element in the box?

1.16 Repeat Problem 1.15 with a current of -5 A.

1.17 The manufacturer of a 6 V dry-cell flashlight battery says that the battery will deliver 15 mA for 60 continuous hours. During that time the voltage will drop from 6 V to 4 V. Assume the drop in voltage is linear with time. How much energy does the battery deliver in this 60 h interval?

1.18 The voltage and current at the terminals of the circuit element in Fig. 1.5 are zero for $t < 0$. For $t \geq 0$ they are

$$v = 75 - 75e^{-1000t} \text{ V},$$

$$i = 50e^{-1000t} \text{mA}.$$

a) Find the maximum value of the power delivered to the circuit.

b) Find the total energy delivered to the element.

1.19 The voltage and current at the terminals of the circuit element in Fig. 1.5 are zero for $t < 0$. For $t \geq 0$ they are

$$v = 15e^{-250t} \text{ V},$$

$$i = 40e^{-250t} \text{ mA}.$$

a) Calculate the power supplied to the element at 10 ms.

b) Calculate the total energy delivered to the circuit element.

1.20 The voltage and current at the terminals of the circuit element in Fig. 1.5 are zero for $t < 0$. For $t \geq 0$ they are

$$v = (1500t + 1)e^{-750t} \text{ V}, \quad t \geq 0;$$

$$i = 40e^{-750t} \text{ mA}, \qquad t \geq 0.$$

a) Find the time when the power delivered to the circuit element is maximum.

b) Find the maximum value of p in milliwatts.

c) Find the total energy delivered to the circuit element in microjoules.

1.21 The voltage and current at the terminals of the circuit element in Fig. 1.5 are zero for $t < 0$. For $t \geq 0$ they are

$$v = 50e^{-1600t} - 50e^{-400t} \text{ V},$$

$$i = 5e^{-1600t} - 5e^{-400t} \text{ mA}.$$

a) Find the power at $t = 625 \ \mu$s.

b) How much energy is delivered to the circuit element between 0 and 625 μs?

c) Find the total energy delivered to the element.

1.22 The voltage and current at the terminals of the circuit element in Fig. 1.5 are zero for $t < 0$. For $t \geq 0$ they are

PSPICE
MULTISIM

$$v = (10{,}000t + 5)e^{-400t} \text{ V},$$

$$i = (40t + 0.05)e^{-400t} \text{ A}.$$

a) At what instant of time is maximum power delivered to the element?

b) Find the maximum power in watts.

c) Find the total energy delivered to the element in microjoules.

1.23 The voltage and current at the terminals of the circuit element in Fig. 1.5 are zero for $t < 0$ and $t > 40$ s. In the interval between 0 and 40 s the expressions are

PSPICE
MULTISIM

$$v = t(1 - 0.025t) \text{ V}, \quad 0 < t < 40 \text{ s};$$

$$i = 4 - 0.2t \text{ A}, \qquad 0 < t < 40 \text{ s}.$$

a) At what instant of time is the power being delivered to the circuit element maximum?

b) What is the power at the time found in part (a)?

c) At what instant of time is the power being extracted from the circuit element maximum?

d) What is the power at the time found in part (c)?

e) Calculate the net energy delivered to the circuit at 0, 10, 20, 30 and 40 s.

1.24 The voltage and current at the terminals of the element in Fig. 1.5 are

PSPICE
MULTISIM

$$v = 250 \cos 800\pi t \text{ V}, \quad i = 8 \sin 800\pi t \text{ A}.$$

a) Find the maximum value of the power being delivered to the element.

b) Find the maximum value of the power being extracted from the element.

c) Find the average value of p in the interval $0 \leq 1 \leq 2.5$ ms.

d) Find the average value of p in the interval $0 \leq t \leq 15.625$ ms.

1.25 The voltage and current at the terminals of the circuit element in Fig. 1.5 are zero for $t < 0$. For $t \geq 0$ they are

PSPICE
MULTISIM

$$v = 100e^{-50t} \sin 150t \text{ V},$$

$$i = 20e^{-50t} \sin 150t \text{ A}.$$

a) Find the power absorbed by the element at $t = 20$ ms.

b) Find the total energy absorbed by the element.

1.26 An industrial battery is charged over a period of several hours at a constant voltage of 120 V. Initially, the current is 10 mA and increases linearly to 15 mA in 10 ks. From 10 ks to 20 ks, the current is constant at 15 mA. From 20 ks to 30 ks the current decreases linearly to 10 mA. At 30 ks the power is disconnected from the battery.

a) Sketch the current from $t = 0$ to $t = 30$ ks.

b) Sketch the power delivered to the battery from $t = 0$ to $t = 30$ ks.

c) Using the sketch of the power, find the total energy delivered to the battery.

1.27 The voltage and current at the terminals of an automobile battery during a charge cycle are shown in Fig. P1.27.

PSPICE
MULTISIM

a) Calculate the total charge transferred to the battery.

b) Calculate the total energy transferred to the battery.

Figure P1.27

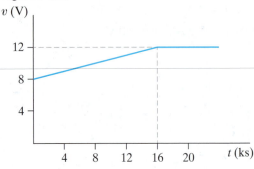

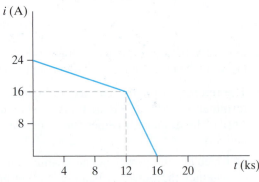

1.28 The voltage and current at the terminals of the circuit element in Fig. 1.5 are shown in Fig. P1.28.

a) Sketch the power versus t plot for $0 \leq t \leq 50$ s.

b) Calculate the energy delivered to the circuit element at $t = 4, 12, 36,$ and 50 s.

Figure P1.28

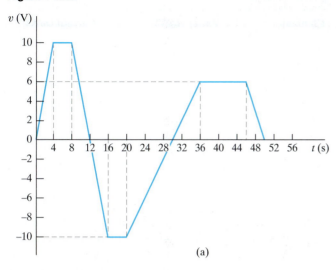

(a)

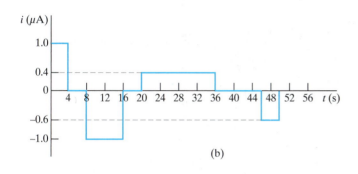

(b)

1.29 The numerical values for the currents and voltages in the circuit in Fig. P1.29 are given in Table P1.29. Find the total power developed in the circuit.

Figure P1.29

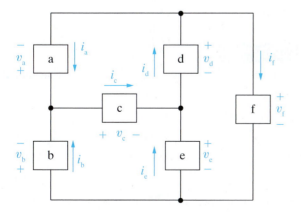

TABLE P1.29

Element	Voltage (V)	Current (mA)
a	−18	−51
b	−18	45
c	2	−6
d	20	−20
e	16	−14
f	36	31

1.30 The voltage and power values for each of the elements shown in Fig. P1.30 are given in Table P1.30.

a) Show that the interconnection of the elements satisfies the power check.

b) Find the value of the current through each of the elements using the values of power and voltage and the current directions shown in the figure.

Figure P1.30

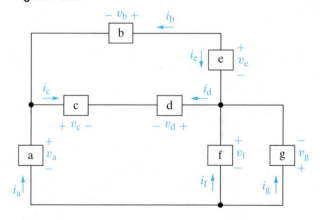

TABLE P1.30

Element	Power (kW)	Voltage (V)
a	0.6 supplied	400
b	0.05 supplied	−100
c	0.4 absorbed	200
d	0.6 supplied	300
e	0.1 absorbed	−200
f	2.0 absorbed	500
g	1.25 supplied	−500

1.31 The numerical values of the voltages and currents in the interconnection seen in Fig. P1.31 are given in Table P1.31. Does the interconnection satisfy the power check?

Figure P1.31

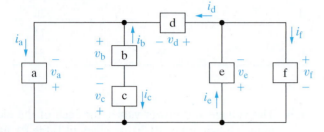

TABLE P1.31

Element	Voltage (kV)	Current (mA)
a	−3	−250
b	4	−400
c	1	400
d	1	150
e	−4	200
f	4	50

1.32 The current and power for each of the interconnected elements in Fig. P1.32 is measured. The values are listed in Table P1.32.

a) Show that the interconnection satisfies the power check.

b) Identify the elements that absorb power.

c) Find the voltage for each of the elements in the interconnection, using the values of power and current and the voltage polarities shown in the figure.

Figure P1.32

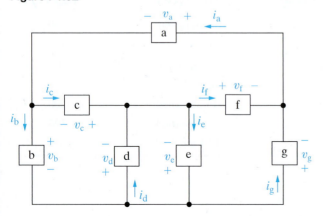

TABLE P1.32

Element	Power (mW)	Current (mA)
a	175	25
b	375	75
c	150	−50
d	−320	40
e	160	20
f	120	−30
g	−660	55

1.33 Assume you are an engineer in charge of a project and one of your subordinate engineers reports that the interconnection in Fig. P1.33 does not pass the power check. The data for the interconnection are given in Table P1.33.

a) Is the subordinate correct? Explain your answer.

b) If the subordinate is correct, can you find the error in the data?

Figure P1.33

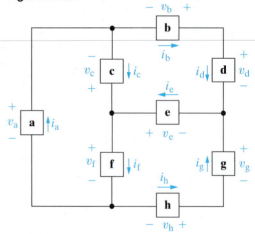

TABLE P1.33

Element	Voltage (V)	Current (A)
a	900	−22.5
b	105	−52.5
c	−600	−30.0
d	585	−52.5
e	−120	30.0
f	300	60.0
g	585	82.5
h	−165	82.5

1.34 Show that the power balances for the circuit shown
PRACTICAL
PERSPECTIVE in Fig. 1.7, using the voltage and current values given
in Table 1.5, with the value of the current for component d changed to −1 A.

1.35 Suppose there is no power lost in the wires used to
PRACTICAL
PERSPECTIVE distribute power in a typical home.

a) Create a new model for the power distribution
circuit by modifying the circuit shown in Fig 1.7.
Use the same names, voltage polarities, and current directions for the components that remain
in this modified model.

b) The following voltages and currents are calculated for the components:

$$v_a = 120 \text{ V} \qquad\qquad i_a = -10 \text{ A}$$

$$v_b = 120 \text{ V} \qquad\qquad i_b = 10 \text{ A}$$

$$v_f = -120 \text{ V} \qquad\qquad i_f = 3 \text{ A}$$

$$v_g = 120 \text{ V}$$

$$v_h = -240 \text{ V} \qquad\qquad i_h = -7 \text{ A}$$

If the power in this modified model balances,
what is the value of the current in component g?

2

Circuit Elements

CHAPTER CONTENTS

2.1 **Voltage and Current Sources** *p. 28*

2.2 **Electrical Resistance (Ohm's Law)** *p. 32*

2.3 **Constructing a Circuit Model** *p. 36*

2.4 **Kirchhoff's Laws** *p. 39*

2.5 **Analyzing a Circuit Containing Dependent Sources** *p. 45*

CHAPTER OBJECTIVES

1 Understand the symbols for and the behavior of the following ideal basic circuit elements: independent voltage and current sources, dependent voltage and current sources, and resistors.

2 Be able to state Ohm's law, Kirchhoff's current law, and Kirchhoff's voltage law, and be able to use these laws to analyze simple circuits.

3 Know how to calculate the power for each element in a simple circuit and be able to determine whether or not the power balances for the whole circuit.

There are five ideal basic circuit elements:

- voltage sources
- current sources
- resistors
- inductors
- capacitors

In this chapter, we discuss the characteristics of the first three circuit elements—voltage sources, current sources, and resistors. Although this may seem like a small number of elements, many practical systems can be modeled with just sources and resistors. They are also a useful starting point because of their relative simplicity; the mathematical relationships between voltage and current in sources and resistors are algebraic. Thus, you will be able to begin learning the basic techniques of circuit analysis with only algebraic manipulations.

We will postpone introducing inductors and capacitors until Chapter 6, because their use requires that you solve integral and differential equations. However, the basic analytical techniques for solving circuits with inductors and capacitors are the same as those introduced in this chapter. So, by the time you need to begin manipulating more difficult equations, you should be very familiar with the methods of writing them.

■ Practical Perspective

Heating with Electric Radiators

You want to heat your small garage using a couple of electric radiators. The power and voltage requirements for each radiator are 1200 W, 240 V. But you are not sure how to wire the radiators to the power supplied to the garage. Should you use the wiring diagram on the left or the one on the right? Does it make any difference?

Once you have studied the material in this chapter, you will be able to answer these questions and determine how to heat the garage. The Practical Perspective at the end of this chapter guides you through the analysis of two circuits based on the two wiring diagrams shown below.

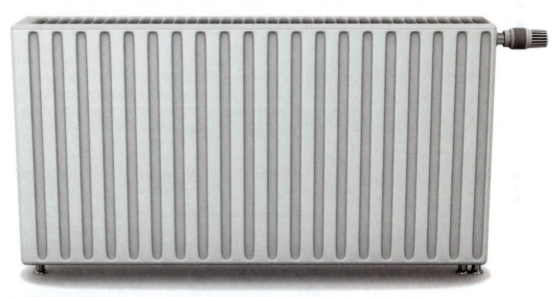

Style-Photography/Fotolia

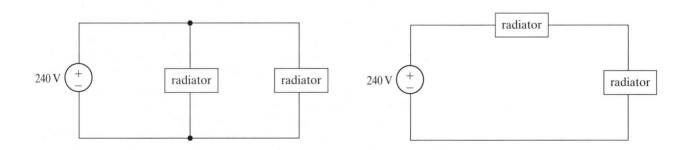

2.1 Voltage and Current Sources

An **electrical source** is a device capable of converting nonelectric energy to electric energy and vice versa. For example, a discharging battery converts chemical energy to electric energy, whereas a charging battery converts electric energy to chemical energy. A *dynamo* is a machine that converts mechanical energy to electric energy and vice versa. For operations in the mechanical-to-electric mode, it is called a *generator*. For transformations from electric to mechanical energy, it is called a *motor*. Electric sources either deliver or absorb electric power while maintaining either voltage or current. This behavior led to the creation of the ideal voltage source and the ideal current source as basic circuit elements.

- An **ideal voltage source** is a circuit element that maintains a prescribed voltage across its terminals regardless of the current flowing in those terminals.
- An **ideal current source** is a circuit element that maintains a prescribed current through its terminals regardless of the voltage across those terminals.

These circuit elements do not exist as practical devices—they are idealized models of actual voltage and current sources.

Using an ideal model for current and voltage sources constrains the mathematical descriptions of these components. For example, because an ideal voltage source provides a steady voltage even if the current in the element changes, it is impossible to specify the current in an ideal voltage source as a function of its voltage. Likewise, if the only information you have about an ideal current source is the value of current supplied, it is impossible to determine the voltage across that current source. We have sacrificed our ability to relate voltage and current in a practical source for the simplicity of using ideal sources in circuit analysis.

Ideal voltage and current sources can be further described as either *independent sources* or *dependent sources*.

- An **independent source** establishes a voltage or current in a circuit without relying on voltages or currents elsewhere in the circuit. The value of the voltage or current supplied is specified by the value of the independent source alone.
- A **dependent source**, in contrast, establishes a voltage or current whose value depends on the value of a voltage or current elsewhere in the circuit. You cannot specify the value of a dependent source unless you know the value of the voltage or current on which it depends.

The circuit symbols for the ideal independent sources are shown in Fig. 2.1. Note that a circle is used to represent an independent source. To completely specify an ideal independent voltage source in a circuit, you must include the value of the supplied voltage and the reference polarity, as shown in Fig. 2.1(a). Similarly, to completely specify an ideal independent current source, you must include the value of the supplied current and its reference direction, as shown in Fig. 2.1(b).

The circuit symbol for an ideal dependent source is a diamond, as shown in Fig. 2.2. There are four possible variations because both dependent current sources and dependent voltage sources can be controlled by either a voltage or a current elsewhere in the circuit. Dependent sources are sometimes called controlled sources.

To completely specify an ideal dependent voltage-controlled voltage source, you must identify the controlling voltage, the equation that

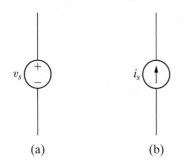

(a)   (b)

Figure 2.1 ▲ The circuit symbols for (a) an ideal independent voltage source and (b) an ideal independent current source.

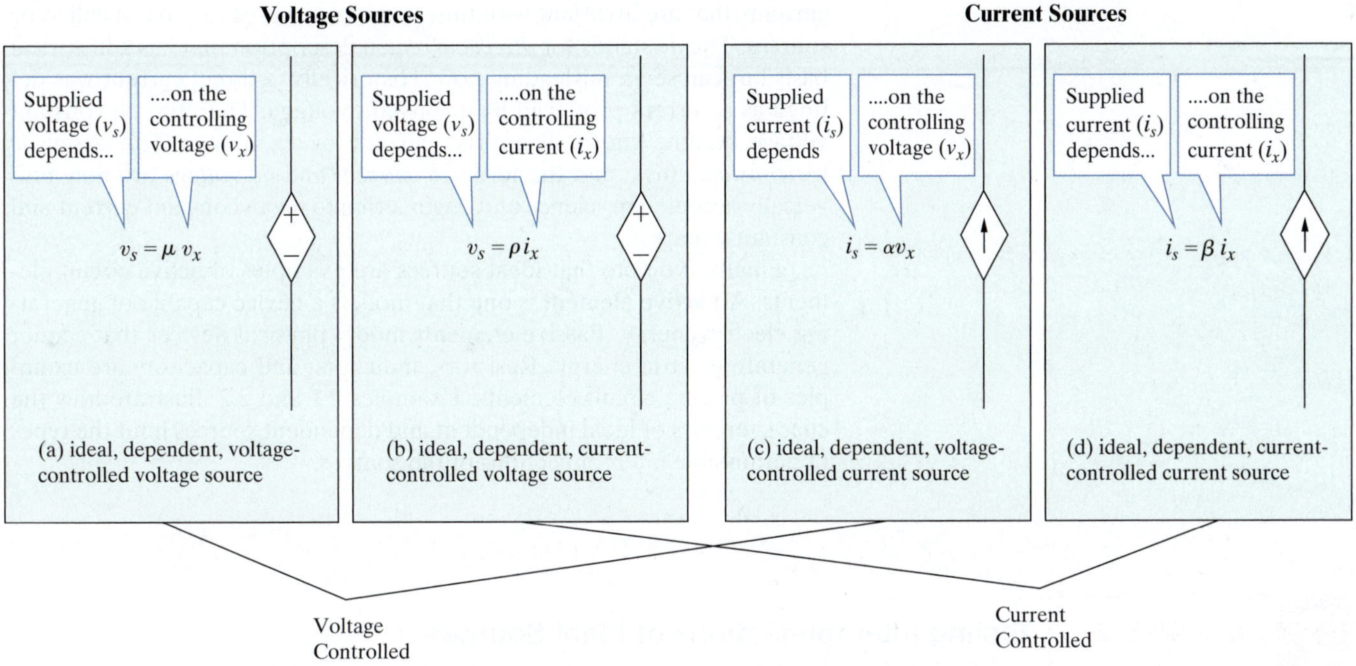

Voltage Sources   **Current Sources**

| Supplied voltage (v_s) depends... | ....on the controlling voltage (v_x) |

$$v_s = \mu\,v_x$$

(a) ideal, dependent, voltage-controlled voltage source

| Supplied voltage (v_s) depends... | ....on the controlling current (i_x) |

$$v_s = \rho\,i_x$$

(b) ideal, dependent, current-controlled voltage source

| Supplied current (i_s) depends | ....on the controlling voltage (v_x) |

$$i_s = \alpha\,v_x$$

(c) ideal, dependent, voltage-controlled current source

| Supplied current (i_s) depends... | ....on the controlling current (i_x) |

$$i_s = \beta\,i_x$$

(d) ideal, dependent, current-controlled current source

Voltage
Controlled

Current
Controlled

Figure 2.2 ▲ (a) (b) Circuit symbols for ideal dependent voltage sources and (c) (d) ideal dependent current sources.

permits you to compute the supplied voltage from the controlling voltage, and the reference polarity for the supplied voltage. For example, in Fig. 2.2(a), the controlling voltage is v_x, the equation that determines the supplied voltage v_s is

$$v_s = \mu v_x,$$

and the reference polarity for v_s is as indicated. Note that μ is a multiplying constant that is dimensionless.

Similar requirements exist for completely specifying the other ideal dependent sources. In Fig. 2.2(b), the controlling current is i_x, the equation for the supplied voltage v_s is

$$v_s = \rho i_x,$$

the reference polarity is as shown, and the multiplying constant ρ has the dimension volts per ampere. In Fig. 2.2(c), the controlling voltage is v_x, the equation for the supplied current i_s is

$$i_s = \alpha v_x,$$

the reference direction is as shown, and the multiplying constant α has the dimension amperes per volt. In Fig. 2.2(d), the controlling current is i_x, the equation for the supplied current i_s is

$$i_s = \beta i_x,$$

the reference direction is as shown, and the multiplying constant β is dimensionless.

Note that the ideal independent and dependent voltage and current sources generate either constant voltages or currents, that is, voltages and currents that are invariant with time. Constant sources are often called **dc sources**. The *dc* stands for *direct current*, a description that has a historical basis but can seem misleading now. Historically, a direct current was defined as a current produced by a constant voltage. Therefore, a constant voltage became known as a direct current, or dc, voltage. The use of *dc* for *constant* stuck, and the terms *dc current* and *dc voltage* are now universally accepted in science and engineering to mean constant current and constant voltage.

Finally, we note that ideal sources are examples of active circuit elements. An **active element** is one that models a device capable of generating electric energy. **Passive elements** model physical devices that cannot generate electric energy. Resistors, inductors, and capacitors are examples of passive circuit elements. Examples 2.1 and 2.2 illustrate how the characteristics of ideal independent and dependent sources limit the types of permissible interconnections of the sources.

EXAMPLE 2.1 | **Testing Interconnections of Ideal Sources**

Use the definitions of the ideal independent voltage and current sources to determine which interconnections in Fig. 2.3 are permitted and which violate the constraints imposed by the ideal sources.

Solution

Connection (a) is permitted. Each source supplies voltage across the same pair of terminals, marked a and b. This requires that each source supply the same voltage with the same polarity, which they do.

Connection (b) is permitted. Each source supplies current through the same pair of terminals, marked a and b. This requires that each source supply the same current in the same direction, which they do.

Connection (c) is not permitted. Each source supplies voltage across the same pair of terminals, marked a and b. This requires that each source supply the same voltage with the same polarity, which they do not.

Connection (d) is not permitted. Each source supplies current through the same pair of terminals, marked a and b. This requires that each source supply the same current in the same direction, which they do not.

Connection (e) is permitted. The voltage source supplies voltage across the pair of terminals marked a and b. The current source supplies current through the same pair of terminals. Because an ideal voltage source supplies the same voltage regardless of the current, and an ideal current source supplies the same current regardless of the voltage, this connection is permitted.

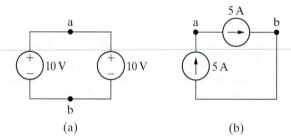

(a)   (b)

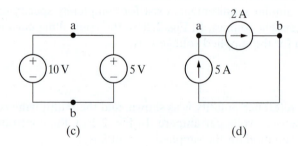

(c)   (d)

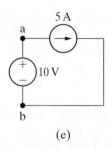

(e)

Figure 2.3 ▲ The circuits for Example 2.1.

| EXAMPLE 2.2 | **Testing Interconnections of Ideal Independent and Dependent Sources** |

State which interconnections in Fig. 2.4 are permitted and which violate the constraints imposed by the ideal sources, using the definitions of the ideal independent and dependent sources.

Solution

Connection (a) is not permitted. Both the independent source and the dependent source supply voltage across the same pair of terminals, labeled a and b. This requires that each source supply the same voltage with the same polarity. The independent source supplies 5 V, but the dependent source supplies 15 V.

Connection (b) is permitted. The independent voltage source supplies voltage across the pair of terminals marked a and b. The dependent current source supplies current through the same pair of terminals. Because an ideal voltage source supplies the same voltage regardless of current, and an ideal current source supplies the same current regardless of voltage, this is a valid connection.

Connection (c) is permitted. The independent current source supplies current through the pair of terminals marked a and b. The dependent voltage source supplies voltage across the same pair of terminals. Because an ideal current source supplies the same current regardless of voltage, and an ideal

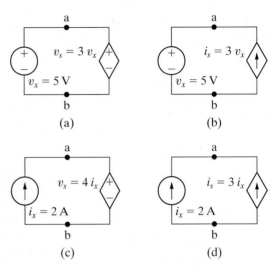

Figure 2.4 ▲ The circuits for Example 2.2.

voltage source supplies the same voltage regardless of current, this is a valid connection.

Connection (d) is not permitted. Both the independent source and the dependent source supply current through the same pair of terminals, labeled a and b. This requires that each source supply the same current in the same direction. The independent source supplies 2 A, but the dependent source supplies 6 A in the opposite direction.

ASSESSMENT PROBLEMS

Objective 1—Understand ideal basic circuit elements

2.1 For the circuit shown,

a) What value of v_g is required in order for the interconnection to be valid?
b) For this value of v_g, find the power associated with the 8 A source.

Answer: (a) −2 V;
(b) −16 W (16 W delivered).

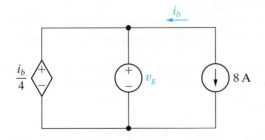

2.2 For the circuit shown,

a) What value of α is required in order for the interconnection to be valid?
b) For the value of α calculated in part (a), find the power associated with the 25 V source.

Answer: (a) 0.6 A/V;
(b) 375 W (375 W absorbed).

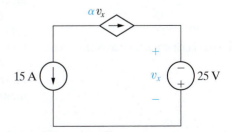

SELF-CHECK: Also try Chapter Problems 2.6 and 2.7.

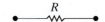

Figure 2.5 ▲ The circuit symbol for a resistor having a resistance R.

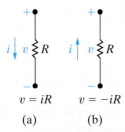

$$v = iR \qquad v = -iR$$

(a)   (b)

Figure 2.6 ▲ Two possible reference choices for the current and voltage at the terminals of a resistor and the resulting equations.

2.2 Electrical Resistance (Ohm's Law)

Resistance is the capacity of materials to impede the flow of current or, more specifically, the flow of electric charge. The circuit element modeling this behavior is the **resistor**. Figure 2.5 shows the resistor's circuit symbol, with R denoting the resistance value of the resistor.

To understand resistance, think about the electrons that make up electric current moving through, interacting with, and being resisted by the atomic structure of some material. The interactions convert some electric energy to thermal energy, dissipated as heat. Many useful electrical devices take advantage of resistance heating, including stoves, toasters, irons, and space heaters.

Most materials resist electric current; the amount of resistance depends on the material. Metals like copper and aluminum have small values of resistance, so they are often used as wires conducting electric current. When represented in a circuit diagram, copper or aluminum wiring isn't usually modeled as a resistor; the wire's resistance is so small compared to the resistance of other circuit elements that we can neglect the wiring resistance to simplify the diagram.

A resistor is an ideal basic circuit element, which is described mathematically using its voltage and current. The relationship between voltage and current for a resistor is known as **Ohm's law**, after Georg Simon Ohm, a German physicist who established its validity early in the nineteenth century. Consider the resistor shown in Fig. 2.6(a), where the current in the resistor is in the direction of the voltage drop across the resistor. For this resistor, Ohm's law is

OHM'S LAW

$$v = iR, \qquad (2.1)$$

where

$$v = \text{the voltage in volts,}$$

$$i = \text{the current in amperes,}$$

$$R = \text{the resistance in ohms.}$$

For the resistor in Fig. 2.6(b), Ohm's law is

$$v = -iR, \qquad (2.2)$$

where v, i, and R are again measured in volts, amperes, and ohms, respectively. We use the passive sign convention (Section 1.5) in determining the algebraic signs in Eqs. 2.1 and 2.2.

Resistance is measured in the SI unit ohms. The Greek letter Omega (Ω) is the standard symbol for an ohm. The circuit diagram symbol for an 8 Ω resistor is shown in Fig. 2.7.

Ohm's law expresses the voltage as a function of the current. However, expressing the current as a function of the voltage also is convenient. Thus, from Eq. 2.1,

Figure 2.7 ▲ The circuit symbol for an 8 Ω resistor.

$$i = \frac{v}{R}, \qquad (2.3)$$

or, from Eq. 2.2,

$$i = -\frac{v}{R}. \tag{2.4}$$

The reciprocal of the resistance is referred to as **conductance**, is symbolized by the letter G, and is measured in siemens (S). Thus,

$$G = \frac{1}{R}. \tag{2.5}$$

An 8 Ω resistor has a conductance value of 0.125 S.

Ideal resistors model the behavior of physical devices. The word *ideal* reminds us that the resistor model makes several simplifying assumptions about the behavior of actual resistive devices. Assuming the resistance of the ideal resistor is constant, so that its value does not vary over time, is the most important of these simplifications. Most actual resistive devices have a time-varying resistance, often because the temperature of the device changes over time. The ideal resistor model represents a physical device whose resistance doesn't vary much from some constant value over the time period of interest in the circuit analysis. In this book, we assume that the simplifying assumptions about resistance devices are valid, and we thus use ideal resistors in circuit analysis.

We can calculate the power at the terminals of a resistor in several ways. The first approach is to use the defining equation (Section 1.6) to calculate the product of the terminal voltage and current. For the resistor shown in Fig. 2.6(a), we write

$$p = vi, \tag{2.6}$$

and for the resistor shown in Fig. 2.6(b), we write

$$p = -vi. \tag{2.7}$$

A second method expresses resistor power in terms of the current and the resistance. Substituting Eq. 2.1 into Eq. 2.6, we obtain

$$p = vi = (iR)i.$$

So

POWER IN A RESISTOR IN TERMS OF CURRENT

$$p = i^2R. \tag{2.8}$$

Likewise, substituting Eq. 2.2 into Eq. 2.7, we have

$$p = -vi = -(-iR)i = i^2R. \tag{2.9}$$

Equations 2.8 and 2.9 are identical, demonstrating that regardless of voltage polarity and current direction, the power at the terminals of a resistor is positive. Therefore, resistors absorb power from the circuit.

A third method expresses resistor power in terms of the voltage and resistance. The expression is independent of the polarity references, so

POWER IN A RESISTOR IN TERMS OF VOLTAGE

$$p = \frac{v^2}{R}. \tag{2.10}$$

Sometimes a resistor's value will be expressed as a conductance rather than as a resistance. Using the relationship between resistance and conductance given in Eq. 2.5, we can also write Eqs. 2.9 and 2.10 in terms of the conductance, or

$$p = \frac{i^2}{G}, \tag{2.11}$$

$$p = v^2 G. \tag{2.12}$$

Equations 2.6–2.12 provide a variety of methods for calculating the power absorbed by a resistor. Each yields the same answer. In analyzing a circuit, look at the information provided and choose the power equation that uses that information directly.

Example 2.3 illustrates Ohm's law for a circuit with an ideal source and a resistor. Power calculations at the terminals of a resistor also are illustrated.

| EXAMPLE 2.3 | Calculating Voltage, Current, and Power for a Simple Resistive Circuit |

In each circuit in Fig. 2.8, either the value of v or i is not known.

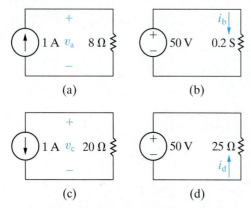

(a)   (b)

(c)   (d)

Figure 2.8 ▲ The circuits for Example 2.3.

a) Calculate the values of v and i.
b) Determine the power dissipated in each resistor.

Solution

a) The voltage v_a in Fig. 2.8(a) is a drop in the direction of the resistor current. The resistor voltage is the product of its current and its resistance, so,

$$v_a = (1)(8) = 8 \text{ V}.$$

The current i_b in the resistor with a conductance of 0.2 S in Fig. 2.8(b) is in the direction of the voltage drop across the resistor. The resistor current is the product of its voltage and its conductance, so

$$i_b = (50)(0.2) = 10 \text{ A}.$$

The voltage v_c in Fig. 2.8(c) is a rise in the direction of the resistor current. The resistor voltage is the product of its current and its resistance, so

$$v_c = -(1)(20) = -20 \text{ V}.$$

The current i_d in the 25 Ω resistor in Fig. 2.8(d) is in the direction of the voltage rise across the resistor. The resistor current is its voltage divided by its resistance, so

$$i_d = -\frac{50}{25} = -2 \text{ A}.$$

b) The power dissipated in each of the four resistors is

$$p_{8\Omega} = \frac{(8)^2}{8} = (1)^2(8) = 8 \text{ W}$$

(using Eq. 2.10 and Eq. 2.9);

$$p_{0.2S} = (50)^2(0.2) = 500 \text{ W}$$

(using Eq. 2.12);

$$p_{20\Omega} = \frac{(-20)^2}{20} = (1)^2(20) = 20 \text{ W}$$

(using Eq. 2.10 and Eq. 2.9);

$$p_{25\Omega} = \frac{(50)^2}{25} = (-2)^2(25) = 100 \text{ W}$$

(using Eq. 2.10 and Eq. 2.9).

◻▪ ASSESSMENT PROBLEMS

Objective 2—Be able to state and use Ohm's law

2.3 For the circuit shown,

a) If $v_g = 1$ kV and $i_g = 5$ mA, find the value of R and the power absorbed by the resistor.

b) If $i_g = 75$ mA and the power delivered by the voltage source is 3 W, find v_g, R, and the power absorbed by the resistor.

c) If $R = 300$ Ω and the power absorbed by R is 480 mW, find i_g and v_g.

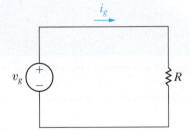

Answer: (a) 200 kΩ, 5 W;
(b) 40 V, 533.33 Ω, 3 W;
(c) 40 mA, 12 V.

2.4 For the circuit shown,

a) If $i_g = 0.5$ A and $G = 50$ mS, find v_g and the power delivered by the current source.

b) If $v_g = 15$ V and the power delivered to the conductor is 9 W, find the conductance G and the source current i_g.

c) If $G = 200$ μS and the power delivered to the conductance is 8 W, find i_g and v_g.

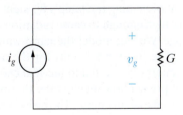

Answer: (a) 10 V, 5 W;
(b) 40 mS, 0.6 A;
(c) 40 mA, 200 V.

SELF-CHECK: Also try Chapter Problems 2.11 and 2.12.

2.3 Constructing a Circuit Model

Let's now move on to using ideal sources and resistors to construct circuit models of real-world systems. Developing a circuit model of a device or system is an important skill. Although this text emphasizes circuit-solving skills, as an electrical engineer you'll need other skills as well, one of the most important of which is modeling.

We develop circuit models in the next two examples. In Example 2.4, we construct a circuit model based on knowing how the system's components behave and how the components are interconnected. In Example 2.5, we create a circuit model by measuring the terminal behavior of a device.

EXAMPLE 2.4 | Constructing a Circuit Model of a Flashlight

Construct a circuit model of a flashlight. Figure 2.9 shows a photograph of a widely available flashlight.

Solution

When a flashlight is regarded as an electrical system, the components of primary interest are the batteries, the lamp, the connector, the case, and the switch. Figure 2.10 shows these components. We now consider the circuit model for each component.

• A dry-cell battery maintains a reasonably constant terminal voltage if the current demand is not excessive. Thus, if the dry-cell battery is operating within its intended limits, we can model it with an ideal voltage source. The prescribed voltage is constant and equal to the sum of two dry-cell values.

• The ultimate output of the lamp is light energy, the result of heating the lamp's filament to a temperature high enough to cause radiation in the visible range. We can model the lamp with an ideal resistor. The resistor accounts for the amount of electric energy converted to thermal energy, but it does not predict how much of the thermal energy is converted to light energy. The resistor representing the lamp also predicts the steady current drain on the batteries, a characteristic of the system that also is of interest. In this model, R_l symbolizes the lamp resistance.

• The connector used in the flashlight serves a dual role. First, it provides an electrical conductive path between the dry cells and the case. Second, it is formed into a springy coil that applies mechanical pressure to the contact between the batteries and the lamp, maintaining contact between the two dry cells and between the dry cells and the lamp. Hence, in choosing the wire for the connector, we may find that its mechanical properties are more important than its electrical

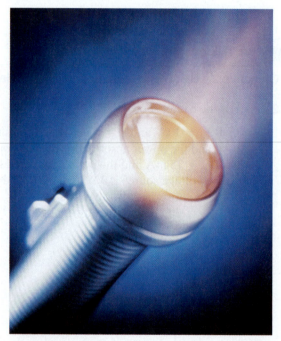

Figure 2.9 ▲ A flashlight can be viewed as an electrical system.
Thom Lang/Corbis/Getty Imagess

properties for the flashlight design. Electrically, we can model the connector with an ideal resistor with resistance R_1.

• The case also serves both a mechanical and an electrical purpose. Mechanically, it contains all the other components and provides a grip for the person using the flashlight. Electrically, it provides a connection between other elements in the flashlight. If the case is metal, it conducts current between the batteries and the lamp. If it is plastic, a metal strip inside the case connects the coiled connector to the switch. An ideal resistor with resistance R_c models the electrical connection provided by the case.

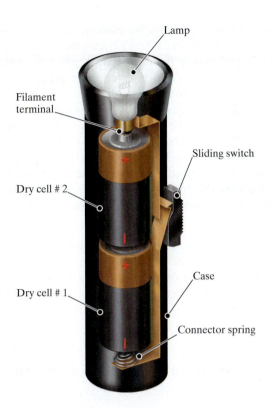

Figure 2.10 ▲ The arrangement of flashlight components.

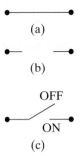

(a)

(b)

OFF

ON

(c)

Figure 2.11 ▲ Circuit symbols. (a) Short circuit. (b) Open circuit. (c) Switch.

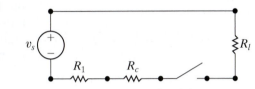

Figure 2.12 ▲ A circuit model for a flashlight.

- The switch has two electrical states: ON or OFF. An ideal switch in the ON state offers no resistance to the current, but it offers infinite resistance to current in the OFF state. These two states represent the limiting values of a resistor; that is, the ON state corresponds to a zero resistance, called a **short circuit** $(R = 0)$, and the OFF state corresponds to an infinite resistance called an **open circuit** $(R = \infty)$. Figures 2.11(a) and (b) depict a short circuit and an open circuit, respectively. The symbol shown in Fig. 2.11(c) represents a switch that can be either a short circuit or an open circuit, depending on the position of its contacts.

We now construct the circuit model of the flashlight shown in Fig. 2.10. Starting with the dry-cell batteries, the positive terminal of the first cell is connected to the negative terminal of the second cell. The positive terminal of the second cell is connected to one terminal of the lamp. The other terminal of the lamp makes contact with one side of the switch, and the other side of the switch is connected to the metal case. The metal spring connects the metal case to the negative terminal of the first dry cell. Note that the connected elements in Fig. 2.10 form a closed path or circuit. Figure 2.12 shows a circuit model for the flashlight.

Our flashlight example provides some general modeling guidelines.

1. **The *electrical* behavior of each physical component is of primary interest in a circuit model.** In the flashlight model, three very different physical components—a lamp, a coiled wire, and a metal case—are all represented by resistors because each circuit component resists the current flowing through the circuit.

2. **Circuit models may need to account for undesired as well as desired electrical effects.** For example, the heat resulting from the lamp resistance produces the light, a desired effect. However, the heat resulting from the case and coil resistance represents an unwanted or *parasitic resistance*. It drains the dry cells and produces no useful output. Such parasitic effects must be considered, or the resulting model may not adequately represent the system.

3. **Modeling requires approximation.** We made several simplifying assumptions in developing the flashlight's circuit model. For example, we assumed an ideal switch, but in practical switches, contact resistance may be large enough to interfere with proper operation of the system. Our model does not predict this behavior. We also assumed that the coiled connector exerts enough pressure to eliminate any contact resistance between the dry cells. Our model does not predict the effect of inadequate pressure. Our use of an ideal voltage source ignores any internal dissipation of energy in the dry cells, which might be due to the parasitic heating just mentioned. We could account for this by adding an ideal resistor between the source and the lamp resistor. Our model assumes the internal loss to be negligible.

We used a basic understanding of the internal components of the flashlight to construct its circuit model. However, sometimes we know only the terminal behavior of a device and must use this information to construct the model. Example 2.5 presents such a modeling problem.

EXAMPLE 2.5 | Constructing a Circuit Model Based on Terminal Measurements

The voltage and current are measured at the terminals of the device illustrated in Fig. 2.13(a), and the values of v_t and i_t are tabulated in Fig. 2.13(b). Construct a circuit model of the device inside the box.

Solution

Plotting the voltage as a function of the current yields the graph shown in Fig. 2.14(a). The equation of the line in this figure is $v_t = 4i_t$, so the terminal voltage is directly proportional to the terminal current. Using Ohm's law, the device inside the box behaves like a 4 Ω resistor. Therefore, the circuit model for the device inside the box is a 4 Ω resistor, as seen in Fig. 2.14(b).

We come back to this technique of using terminal characteristics to construct a circuit model after introducing Kirchhoff's laws and circuit analysis.

v_t (V)	i_t (A)
-40	-10
-20	-5
0	0
20	5
40	10

(a)   (b)

Figure 2.13 ▲ The (a) device and (b) data for Example 2.5.

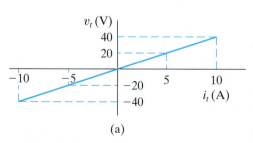

(a)

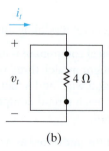

(b)

Figure 2.14 ▲ (a) The values of v_t versus i_t for the device in Fig. 2.13. (b) The circuit model for the device in Fig. 2.13.

SELF-CHECK: Assess your understanding of this example by trying Chapter Problems 2.14 and 2.16.

2.4 Kirchhoff's Laws

A circuit is **solved** when we determine the voltage across and the current in every element. While Ohm's law is an important tool for solving a circuit, it may not be enough to provide a complete solution. Generally, we need two additional algebraic relationships, known as Kirchhoff's laws, to solve most circuits.

Kirchhoff's Current Law

Let's try to solve the flashlight circuit from Example 2.4. We begin by redrawing the circuit as shown in Fig. 2.15, with the switch in the ON state. We have labeled the current and voltage variables associated with each resistor and the current associated with the voltage source, including reference polarities. For convenience, we use the same subscript for the voltage, current, and resistor labels. In Fig. 2.15, we also removed some of the terminal dots of Fig. 2.12 and have inserted nodes. Terminal dots are the start and end points of an individual circuit element. A **node** is a point where two or more circuit elements meet. In Fig. 2.15, the nodes are labeled a, b, c, and d. Node d connects the battery and the lamp and stretches all the way across the top of the diagram, though we label a single point for convenience. The dots on either side of the switch indicate its terminals, but only one is needed to represent a node, labeled node c.

The circuit in Fig. 2.15 has seven unknowns: i_s, i_1, i_c, i_l, v_1, v_c, and v_l. Recall that $v_s = 3$ V, as it represents the sum of the terminal voltages of the two dry cells. To solve the flashlight circuit, we must find values for the seven unknown variables. From algebra, you know that to find n unknown quantities you must solve n simultaneous independent equations. Applying Ohm's law (Section 2.2) to each of the three resistors gives us three of the seven equations we need:

$$v_1 = i_1 R_1, \tag{2.13}$$

$$v_c = i_c R_c, \tag{2.14}$$

$$v_l = i_l R_l. \tag{2.15}$$

What about the other four equations?

Connecting the circuit elements constrains the relationships among the terminal voltages and currents. These constraints are called Kirchhoff's laws, after Gustav Kirchhoff, who first stated them in a paper published in 1848. The two laws that state the constraints in mathematical form are known as Kirchhoff's current law and Kirchhoff's voltage law.

We can now state **Kirchhoff's current law (KCL)**:

KIRCHHOFF'S CURRENT LAW (KCL)

The algebraic sum of all the currents at any node in a circuit equals zero.

To use Kirchhoff's current law at a node, assign an algebraic sign corresponding to the current's reference direction for every current at the node. Assigning a positive sign to a current leaving the node requires assigning a negative sign to a current entering a node. Conversely, giving a negative sign to a current leaving a node requires giving a positive sign to a current entering a node.

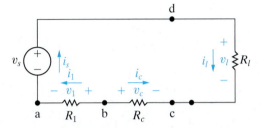

Figure 2.15 ▲ Circuit model of the flashlight with assigned voltage and current variables.

We apply Kirchhoff's current law to the four nodes in the circuit shown in Fig. 2.15, using the convention that currents leaving a node are positive. The four equations are:

$$\text{node a} \qquad i_s - i_1 = 0, \qquad (2.16)$$

$$\text{node b} \qquad i_1 + i_c = 0, \qquad (2.17)$$

$$\text{node c} \qquad -i_c - i_l = 0, \qquad (2.18)$$

$$\text{node d} \qquad i_l - i_s = 0. \qquad (2.19)$$

But Eqs. 2.16–2.19 are not an independent set because any one of the four can be derived from the other three. In any circuit with n nodes, $n - 1$ independent equations can be derived from Kirchhoff's current law.[1] Let's disregard Eq. 2.19 so that we have six independent equations, namely, Eqs. 2.13–2.18. We need one more, which we can derive from Kirchhoff's voltage law.

Kirchhoff's Voltage Law

Before we can state Kirchhoff's voltage law, we must define a **closed path** or **loop**. Starting at an arbitrarily selected node, we trace a closed path in a circuit through selected basic circuit elements and return to the original node without passing through any intermediate node more than once. The circuit shown in Fig. 2.15 has only one closed path or loop. For example, choosing node a as the starting point and tracing the circuit clockwise, we form the closed path by moving through nodes d, c, b, and back to node a. We can now state **Kirchhoff's voltage law (KVL)**:

KIRCHHOFF'S VOLTAGE LAW (KVL)

The algebraic sum of all the voltages around any closed path in a circuit equals zero.

To use Kirchhoff's voltage law, assign an algebraic sign (reference direction) to each voltage in the loop. As we trace a closed path, a voltage will appear either as a rise or a drop in the tracing direction. Assigning a positive sign to a voltage rise requires assigning a negative sign to a voltage drop. Conversely, giving a negative sign to a voltage rise requires giving a positive sign to a voltage drop.

We now apply Kirchhoff's voltage law to the circuit shown in Fig. 2.15, tracing the closed path clockwise and assigning a positive algebraic sign to voltage drops. Starting at node d leads to the expression

$$v_l - v_c + v_1 - v_s = 0. \qquad (2.20)$$

Now we have the seven independent equations needed to find the seven unknown circuit variables in Fig. 2.15.

Solving seven simultaneous equations for the simple flashlight lamp seems excessive. In the coming chapters, we present analytical techniques that solve a simple one-loop circuit like the one shown in Fig. 2.15 using a single equation. Before leaving the flashlight circuit, we observe two analysis details that are important for the techniques presented in subsequent chapters.

---

[1] We say more about this observation in Chapter 4.

1. If you know the current in a resistor, you also know the voltage across the resistor because current and voltage are directly related through Ohm's law. Thus, you can associate one unknown variable with each resistor, either the current or the voltage. For example, choose the current as the unknown variable. Once you solve for the unknown current in the resistor, you can find the voltage across the resistor. In general, if you know the current in a passive element, you can find the voltage across it, greatly reducing the number of simultaneous equations to solve. In the flashlight circuit, choosing the current as the unknown variable eliminates the voltages v_c, v_l, and v_1 as unknowns, and reduces the analytical task to solving four simultaneous equations rather than seven.

2. When only two elements connect to a node, if you know the current in one of the elements, you also know it in the second element by applying Kirchhoff's current law at the node. When just two elements connect at a single node, the elements are said to be **in series**, and you need to define only one unknown current for the two elements. Note that each node in the circuit shown in Fig. 2.15 connects only two elements, so you need to define only one unknown current. Equations 2.16–2.18 lead directly to

$$i_s = i_1 = -i_c = i_l,$$

which states that if you know any one of the element currents, you know them all. For example, choosing i_s as the unknown eliminates i_1, i_c, and i_l. The problem is reduced to determining one unknown, namely, i_s.

Examples 2.6 and 2.7 illustrate how to write circuit equations based on Kirchhoff's laws. Example 2.8 illustrates how to use Kirchhoff's laws and Ohm's law to find an unknown current. Example 2.9 expands on the technique presented in Example 2.5 for constructing a circuit model for a device whose terminal characteristics are known.

EXAMPLE 2.6 | Using Kirchhoff's Current Law

Sum the currents at each node in the circuit shown in Fig. 2.16. Note that there is no connection dot (•) in the center of the diagram, where the 4 Ω branch crosses the branch containing the ideal current source i_a.

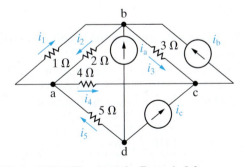

Figure 2.16 ▲ The circuit for Example 2.6.

Solution

In writing the equations, we use a positive sign for a current leaving a node. The four equations are

node a $\quad i_1 + i_4 - i_2 - i_5 = 0,$

node b $\quad i_2 + i_3 - i_1 - i_b - i_a = 0,$

node c $\quad i_b - i_3 - i_4 - i_c = 0,$

node d $\quad i_5 + i_a + i_c = 0.$

EXAMPLE 2.7 | Using Kirchhoff's Voltage Law

Sum the voltages around each designated path in the circuit shown in Fig. 2.17.

Solution

In writing the equations, we use a positive sign for a voltage drop. The four equations are

path a $\qquad -v_1 + v_2 + v_4 - v_b - v_3 = 0,$

path b $\qquad\qquad -v_a + v_3 + v_5 = 0,$

path c $\qquad v_b - v_4 - v_c - v_6 - v_5 = 0,$

path d $\quad -v_a - v_1 + v_2 - v_c + v_7 - v_d = 0.$

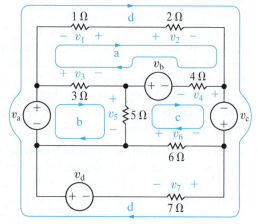

Figure 2.17 ▲ The circuit for Example 2.7.

EXAMPLE 2.8 | Applying Ohm's Law and Kirchhoff's Laws to Find an Unknown Current

a) Use Kirchhoff's laws and Ohm's law to find i_o in the circuit shown in Fig. 2.18.

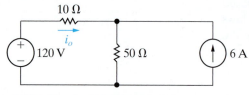

Figure 2.18 ▲ The circuit for Example 2.8.

b) Test the solution for i_o by verifying that the total power generated equals the total power dissipated.

Solution

a) We begin by redrawing the circuit and assigning an unknown current to the 50 Ω resistor and unknown voltages across the 10 Ω and 50 Ω resistors. Figure 2.19 shows the circuit. The nodes are labeled a, b, and c to aid the discussion.

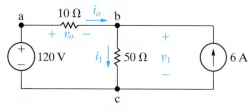

Figure 2.19 ▲ The circuit shown in Fig. 2.18, with the unknowns i_1, v_o, and v_1 defined.

Because i_o also is the current in the 120 V source, we have two unknown currents and therefore must derive two simultaneous equations involving i_o and i_1. One of the equations results from applying Kirchhoff's current law to either node b or c. Summing the currents at node b and assigning a positive sign to the currents leaving the node gives

$$i_1 - i_o - 6 = 0.$$

We obtain the second equation from Kirchhoff's voltage law in combination with Ohm's law. Noting from Ohm's law that $v_o = 10i_o$ and $v_1 = 50i_1$, we sum the voltages clockwise around the closed path c-a-b-c to obtain

$$-120 + 10i_o + 50i_1 = 0.$$

In writing this equation, we assigned a positive sign to voltage drops in the clockwise direction. Solving these two equations (see Appendix A) for i_o and i_1 yields

$$i_o = -3 \text{ A} \quad \text{and} \quad i_1 = 3 \text{ A}.$$

b) The power for the 50 Ω resistor is

$$p_{50\Omega} = (i_1)^2(50) = (3)^2(50) = 450 \text{ W}.$$

The power for the 10 Ω resistor is

$$p_{10\Omega} = (i_o)^2(10) = (-3)^2(10) = 90 \text{ W}.$$

The power for the 120 V source is

$$p_{120V} = -120i_o = -120(-3) = 360 \text{ W}.$$

The power for the 6 A source is

$$p_{6A} = -v_1(6), \text{ and } v_1 = 50i_1 = 50(3) = 150 \text{ V};$$

therefore

$$p_{6A} = -150(6) = -900 \text{ W}.$$

The 6 A source is delivering 900 W, and the 120 V source and the two resistors are absorbing power. The total power absorbed is $p_{6A} + p_{50\Omega} + p_{10\Omega} = 360 + 450 + 90 = 900$ W. Therefore, the solution verifies that the power delivered equals the power absorbed.

EXAMPLE 2.9 Constructing a Circuit Model Based on Terminal Measurements

We measured the terminal voltage and terminal current on the device shown in Fig. 2.20(a) and tabulated the values of v_t and i_t in Fig. 2.20(b).

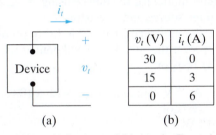

v_t (V)	i_t (A)
30	0
15	3
0	6

(a) (b)

Figure 2.20 ▲ (a) Device and (b) data for Example 2.9.

a) Construct a circuit model of the device inside the box.
b) Using this circuit model, predict the power this device will deliver to a 10 Ω resistor.

Solution

a) Plotting the voltage as a function of the current yields the graph shown in Fig. 2.21(a). The equation of the line plotted is

$$v_t = 30 - 5i_t.$$

What circuit model components produce this relationship between voltage and current? Kirchhoff's voltage law tells us that the voltage drops across two components in series add. From the equation, one of those components produces a 30 V drop regardless of the current, so this component's model is an ideal independent voltage source. The other component produces a

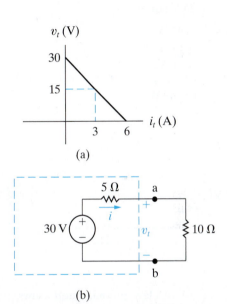

(a)

(b)

Figure 2.21 ▲ (a) The graph of v_t versus i_t for the device in Fig. 2.20(a). (b) The resulting circuit model for the device in Fig. 2.20(a), connected to a 10 Ω resistor.

positive voltage drop in the direction of the current i_t. Because the voltage drop is proportional to the current, Ohm's law tells us that this component's model is an ideal resistor with a value of 5 Ω. The resulting circuit model is depicted in the dashed box in Fig. 2.21(b).

b) Now we attach a 10 Ω resistor to the device in Fig. 2.21(b) to complete the circuit. Kirchhoff's current law tells us that the current in the 10 Ω resistor equals the current in the 5 Ω resistor. Using Kirchhoff's voltage law and Ohm's law, we can write the equation for the voltage drops around the circuit, starting at the voltage source and proceeding clockwise:

$$-30 + 5i + 10i = 0.$$

Solving for i, we get

$$i = 2 \text{ A}.$$

This is the value of current flowing in the $10 \, \Omega$ resistor, so compute the resistor's power using the equation $p = i^2 R$:

$$P_{10\Omega} = (2)^2(10) = 40 \text{ W}.$$

◫▪ ASSESSMENT PROBLEMS

Objective 3—Be able to state and use Ohm's law and Kirchhoff's current and voltage laws

2.5 For the circuit shown, calculate (a) i_5; (b) v_1; (c) v_2; (d) v_5; and (e) the power delivered by the 24 V source.

Answer: (a) 2 A;
(b) −4 V;
(c) 6 V;
(d) 14 V;
(e) 48 W.

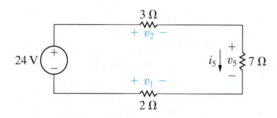

2.6 Use Ohm's law and Kirchhoff's laws to find the value of R in the circuit shown.

Answer: $R = 4 \, \Omega$.

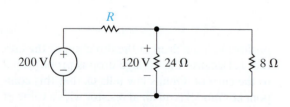

2.7 a) The terminal voltage and terminal current were measured on the device shown. The values of v_t and i_t are provided in the table. Using these values, create the straight-line plot of v_t versus i_t. Compute the equation of the line and use the equation to construct a circuit model for the device using an ideal voltage source and a resistor.

b) Use the model constructed in (a) to predict the power that the device will deliver to a $25 \, \Omega$ resistor.

Answer: (a) A 25 V source in series with a $100 \, \Omega$ resistor;
(b) 1 W.

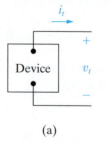

v_t (V)	i_t (A)
25	0
15	0.1
5	0.2
0	0.25

(a)   (b)

2.8 Repeat Assessment Problem 2.7, but use the equation of the graphed line to construct a circuit model containing an ideal current source and a resistor.

Answer: (a) A 0.25 A current source connected between the terminals of a $100 \, \Omega$ resistor;
(b) 1 W.

SELF-CHECK: Also try Chapter Problems 2.17, 2.18, 2.29, and 2.30.

2.5 Analyzing a Circuit Containing Dependent Sources

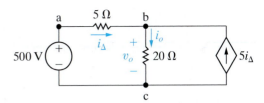

Figure 2.22 ▲ A circuit with a dependent source.

We conclude this introduction to elementary circuit analysis by considering circuits with dependent sources. One such circuit is shown in Fig. 2.22.

We want to use Kirchhoff's laws and Ohm's law to find v_o in this circuit. Before writing equations, it is good practice to examine the circuit diagram closely. This will help us identify the information that is known and the information we must calculate. It may also help us devise a strategy for solving the circuit using only a few calculations.

A look at the circuit in Fig. 2.22 reveals that:

- Once we know i_o, we can calculate v_o using Ohm's law.
- Once we know i_Δ, we also know the current supplied by the dependent source $5i_\Delta$.
- The current in the 500 V source is i_Δ, using Kirchhoff's current law at node a.

There are thus two unknown currents, i_Δ and i_o. We need to construct and solve two independent equations involving these two currents to produce a value for v_o. This is the approach used in Example 2.10.

EXAMPLE 2.10	**Analyzing a Circuit with a Dependent Source**

Find the voltage v_o for the circuit in Fig. 2.22.

Solution

The closed path consisting of the voltage source, the 5 Ω resistor, and the 20 Ω resistor contains the two unknown currents. Apply Kirchhoff's voltage law around this closed path, using Ohm's law to express the voltage across the resistors in terms of the currents in those resistors. Starting at node c and traversing the path clockwise gives:

$$-500 + 5i_\Delta + 20i_o = 0.$$

Now we need a second equation containing these two currents. We can't apply Kirchhoff's voltage law to the closed path formed by the 20 Ω resistor and the dependent current source because we don't know the value of the voltage across the dependent current source. For this same reason, we cannot apply Kirchhoff's voltage law to the closed path containing the voltage source, the 5 Ω resistor, and the dependent source.

We turn to Kirchhoff's current law to generate the second equation. Either node b or node c can be used to construct the second equation from Kirchhoff's current law, since we have already used node a to determine that the current in the voltage source and the 5 Ω resistor is the same. We select node b and produce the following equation, summing the currents leaving the node:

$$-i_\Delta + i_o - 5i_\Delta = 0.$$

Solve the KCL equation for i_o in terms of i_Δ ($i_o = 6i_\Delta$), and then substitute this expression for i_o into the KVL equation to give

$$500 = 5i_\Delta + 20(6i_\Delta) = 125i_\Delta.$$

Therefore,

$$i_\Delta = 500/125 = 4 \text{ A} \quad \text{and} \quad i_o = 6(4) = 24 \text{ A}.$$

Using i_o and Ohm's law for the 20 Ω resistor, we can solve for the voltage v_o:

$$v_o = 20i_o = 480 \text{ V}.$$

Think about a circuit analysis strategy before beginning to write equations because not every closed path yields a useful Kirchhoff's voltage law equation and not every node yields a useful Kirchhoff's current law equation. Think about the problem and select a fruitful approach and useful analysis tools to reduce the number and complexity of equations to be solved. Example 2.11 applies Ohm's law and Kirchhoff's laws to another circuit with a dependent source. Example 2.12 involves a much more complicated circuit, but with a careful choice of analysis tools, the analysis is relatively uncomplicated.

EXAMPLE 2.11 — Applying Ohm's Law and Kirchhoff's Laws to Find an Unknown Voltage

a) Use Kirchhoff's laws and Ohm's law to find the voltage v_o as shown in Fig. 2.23.
b) Show that your solution is consistent with the requirement that the total power developed in the circuit equals the total power dissipated.

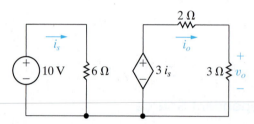

Figure 2.23 ▲ The circuit for Example 2.11.

Solution

a) A close look at the circuit in Fig. 2.23 reveals that:
- There are two closed paths, the one on the left with the current i_s and the one on the right with the current i_o.
- Once i_o is known, we can compute v_o using Ohm's law.

We need two equations for the two currents. Because there are two closed paths and both have voltage sources, we can apply Kirchhoff's voltage law to each, using Ohm's law to express the voltage across the resistors in terms of the current in those resistors. The resulting equations are:

$$-10 + 6i_s = 0 \quad \text{and} \quad -3i_s + 2i_o + 3i_o = 0.$$

Solving for the currents yields

$$i_s = 1.67 \text{ A} \quad \text{and} \quad i_o = 1 \text{ A}.$$

Applying Ohm's law to the 3 Ω resistor gives the desired voltage:

$$v_o = 3i_o = 3 \text{ V}.$$

b) To compute the power delivered to the voltage sources, we use the power equation, $p = vi$, together with the passive sign convention. The power for the independent voltage source is

$$p = -10i_s = -10(1.67) = -16.7 \text{ W}.$$

The power for the dependent voltage source is

$$p = -(3i_s)i_o = -(5)(1) = -5 \text{ W}.$$

Both sources are supplying power, and the total power supplied is 21.7 W.

To compute the power for the resistors, we use the power equation, $p = i^2R$. The power for the 6 Ω resistor is

$$p = (1.67)^2(6) = 16.7 \text{ W}.$$

The power for the 2 Ω resistor is

$$p = (1)^2(2) = 2 \text{ W}.$$

The power for the 3 Ω resistor is

$$p = (1)^2(3) = 3 \text{ W}.$$

The resistors all absorb power, and the total power absorbed is 21.7 W, equal to the total power supplied by the sources.

 EXAMPLE 2.12 | **Applying Ohm's Law and Kirchhoff's Law in an Amplifier Circuit**

The circuit in Fig. 2.24 represents a common configuration encountered in the analysis and design of transistor amplifiers. Assume that the values of all the circuit elements—R_1, R_2, R_C, R_E, V_{CC}, and V_0—are known.

a) Develop the equations needed to determine the current in each element of this circuit.
b) From these equations, devise a formula for computing i_B in terms of the circuit element values.

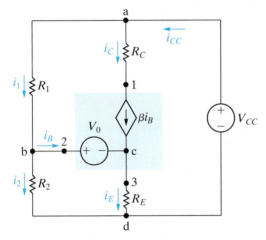

Figure 2.24 ▲ The circuit for Example 2.12.

Solution

Carefully examine the circuit to identify six unknown currents, designated i_1, i_2, i_B, i_C, i_E, and i_{CC}. In defining these six unknown currents, we observed that the resistor R_C is in series with the dependent current source βi_B, so these two components have the same current. We now must derive six independent equations involving these six unknowns.

a) We can derive three equations by applying Kirchhoff's current law to any three of the nodes a, b, c, and d. Let's use nodes a, b, and c and label the currents away from the nodes as positive:

$$(1) \quad i_1 + i_C - i_{CC} = 0,$$

$$(2) \quad i_B + i_2 - i_1 = 0,$$

$$(3) \quad i_E - i_B - i_C = 0.$$

A fourth equation results from imposing the constraint presented by the series connection of R_C and the dependent source:

$$(4) \quad i_C = \beta i_B.$$

We use Kirchhoff's voltage law to derive the remaining two equations. We must select two closed paths, one for each Kirchhoff's voltage law equation. The voltage across the dependent current source is unknown and cannot be determined from the source current βi_B, so select two closed paths that do not contain this dependent current source.

We choose the paths b-c-d-b and b-a-d-b, then use Ohm's law to express resistor voltage in terms of resistor current. Traverse the paths in the clockwise direction and specify voltage drops as positive to yield

$$(5) \quad V_0 + i_E R_E - i_2 R_2 = 0,$$

$$(6) \quad -i_1 R_1 + V_{CC} - i_2 R_2 = 0.$$

b) To get a single equation for i_B in terms of the known circuit variables, you can follow these steps:

- Solve Eq. (6) for i_1, and substitute this solution for i_1 into Eq. (2).
- Solve the transformed Eq. (2) for i_2, and substitute this solution for i_2 into Eq. (5).
- Solve the transformed Eq. (5) for i_E, and substitute this solution for i_E into Eq. (3). Use Eq. (4) to eliminate i_C in Eq. (3).
- Solve the transformed Eq. (3) for i_B, and rearrange the terms to yield

$$i_B = \frac{(V_{CC} R_2)/(R_1 + R_2) - V_0}{(R_1 R_2)/(R_1 + R_2) + (1 + \beta) R_E}. \qquad (2.21)$$

Problem 2.38 asks you to verify these steps. Note that once we know i_B, we can easily obtain the remaining currents.

◼ ASSESSMENT PROBLEMS

Objective 4—Know how to calculate power for each element in a simple circuit

2.9 For the circuit shown, find (a) the current i_1 in microamperes, (b) the voltage v in volts, (c) the total power generated, and (d) the total power absorbed.

Answer: (a) 25 μA;
(b) -2 V;
(c) 6150 μW;
(d) 6150 μW.

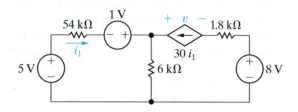

2.10 The current i_ϕ in the circuit shown is 2 A. Calculate

a) v_s,
b) the power absorbed by the independent voltage source,

c) the power delivered by the independent current source,
d) the power delivered by the controlled current source,
e) the total power dissipated in the two resistors.

Answer: (a) 70 V;
(b) 210 W;
(c) 300 W;
(d) 40 W;
(e) 130 W.

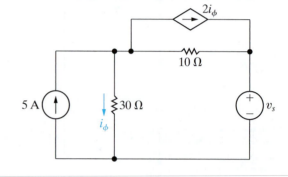

SELF-CHECK: Also try Chapter Problems 2.33 and 2.34.

◼ Practical Perspective

Heating with Electric Radiators

Let's determine which of the two wiring diagrams introduced at the beginning of this chapter should be used to wire the electric radiators to the power supplied to the garage. We begin with the diagram shown in Fig. 2.25. We can turn this into a circuit by modeling the radiators as resistors. The resulting circuit is shown in Fig. 2.26. Note that each radiator has the same resistance, R, and is labeled with a voltage and current value.

To find the unknown voltages and currents for the circuit in Fig. 2.26, begin by using Kirchhoff's voltage law to sum the voltage drops around the path on the circuit's left side:

$$-240 + v_1 = 0 \quad \Rightarrow \quad v_1 = 240 \text{ V}.$$

Now use Kirchhoff's voltage law to sum the voltage drops around the path on the circuit's right side:

$$-v_1 + v_2 = 0 \quad \Rightarrow \quad v_2 = v_1 = 240 \text{ V}.$$

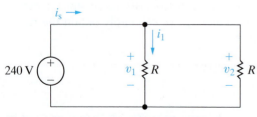

Figure 2.25 ▲ A wiring diagram for two radiators.

Figure 2.26 ▲ A circuit based on Fig. 2.25.

Remember that the power and voltage specifications for each radiator are 1200 W, 240 V. Therefore, the configuration shown in Fig. 2.25 satisfies the voltage specification, since each radiator would have a supplied voltage of 240 V.

Next, calculate the value of resistance R that correctly models each radiator. We want the power associated with each radiator to be 1200 W. Use the equation for resistor power that involves the resistance and the voltage:

$$p_1 = \frac{v_1^2}{R} = \frac{v_2^2}{R} = p_2 \quad \Rightarrow \quad R = \frac{v_1^2}{p_1} = \frac{240^2}{1200} = 48\ \Omega.$$

Each radiator can be modeled as a 48 Ω resistor, with a voltage drop of 240 V and power of 1200 W. The total power for two radiators is thus 2400 W.

Finally, calculate the power for the 240 V source. To do this, calculate the current in the voltage source, i_s, using Kirchhoff's current law to sum the currents leaving the top node in Fig. 2.26. Then use i_s to find the power for the voltage source.

$$-i_s + i_1 + i_2 = 0 \quad \Rightarrow \quad i_s = i_1 + i_2 = \frac{v_1}{R} + \frac{v_2}{R} = \frac{240}{48} + \frac{240}{48} = 10\ \text{A}.$$

$$p_s = -(240)(i_s) = -(240)(10) = -2400\ \text{W}.$$

Thus, the total power in the circuit is $-2400 + 2400 = 0$, and the power balances.

Now look at the other wiring diagram for the radiators, shown in Fig. 2.27. We know that the radiators can be modeled using 48 Ω resistors, which are used to turn the wiring diagram into the circuit in Fig. 2.28.

Start analyzing the circuit in Fig. 2.28 by using Kirchhoff's voltage law to sum the voltage drops around the closed path:

$$-240 + v_x + v_y = 0 \quad \Rightarrow \quad v_x + v_y = 240.$$

Next, use Kirchhoff's current law to sum to currents leaving the node labeled a:

$$-i_x + i_y = 0 \quad \Rightarrow \quad i_x = i_y = i.$$

The current in the two resistors is the same, and we can use that current in Ohm's law equations to replace the two unknown voltages in the Kirchhoff's voltage law equation:

$$48i + 48i = 240 = 96i \quad \Rightarrow \quad i = \frac{240}{96} = 2.5\ \text{A}.$$

Use the current in the two resistors to calculate the power for the two radiators.

$$p_x = p_y = Ri^2 = (48)(2.5)^2 = 300\ \text{W}.$$

Thus, if the radiators are wired as shown in Fig. 2.27, their total power is 600 W. This is insufficient to heat the garage.

Therefore, the way the radiators are wired has a big impact on the amount of heat that will be supplied. When they are wired using the diagram in Fig. 2.25, 2400 W of power will be available, but when they are wired using the diagram in Fig. 2.27, only 600 W of power will be available.

SELF-CHECK: Assess your understanding of the Practical Perspective by solving Chapter Problems 2.41–2.43.

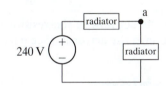

Figure 2.27 ▲ Another way to wire two radiators.

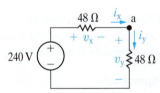

Figure 2.28 ▲ A circuit based on Fig. 2.27.

■ Summary

- The circuit elements introduced in this chapter are voltage sources, current sources, and resistors:

  - An **ideal voltage source** maintains a prescribed voltage regardless of the current in the source. An **ideal current source** maintains a prescribed current regardless of the voltage across the source. Voltage and current sources are either **independent**, that is, not influenced by any other current or voltage in the circuit, or **dependent**, that is, determined by some other current or voltage in the circuit. (See pages 28 and 29.)

  - A **resistor** constrains its voltage and current to be proportional to each other. The value of the proportional constant relating voltage and current in a resistor is called its **resistance** and is measured in ohms. (See page 32.)

- **Ohm's law** establishes the proportionality of voltage and current in a resistor. Specifically,

$$v = iR$$

if the current flow in the resistor is in the direction of the voltage drop across it, or

$$v = -iR$$

if the current flow in the resistor is in the direction of the voltage rise across it. (See page 32.)

- By combining the equation for power, $p = vi$, with Ohm's law, we can determine the power absorbed by a resistor:

$$p = i^2R = v^2/R.$$

(See pages 33–34.)

- Circuits have nodes and closed paths. A **node** is a point where two or more circuit elements join. When just two elements connect to form a node, they are said to be **in series**. A **closed path** is a loop traced through connecting elements, starting and ending at the same node and encountering intermediate nodes only once each. (See pages 39–40.)

- The voltages and currents of interconnected circuit elements obey Kirchhoff's laws:

  - **Kirchhoff's current law** states that the algebraic sum of all the currents at any node in a circuit equals zero. (See page 39.)

  - **Kirchhoff's voltage law** states that the algebraic sum of all the voltages around any closed path in a circuit equals zero. (See page 40.)

- A circuit is solved when the voltage across and the current in every element have been determined. By combining an understanding of independent and dependent sources, Ohm's law, and Kirchhoff's laws, we can solve many simple circuits.

■ Problems

Section 2.1

2.1 If the interconnection in Fig. P2.1 is valid, find the power developed by the current sources. If the interconnection is not valid, explain why.

Figure P2.1

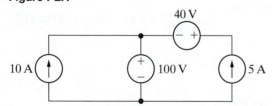

2.2 a) Is the interconnection of ideal sources in the circuit in Fig. P2.2 valid? Explain.

b) Identify which sources are developing power and which sources are absorbing power.

c) Verify that the total power developed in the circuit equals the total power absorbed.

d) Repeat (a)–(c), reversing the polarity of the 18 V source.

Figure P2.2

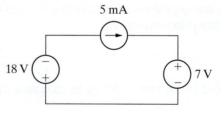

2.3 If the interconnection in Fig. P2.3 is valid, find the total power developed by the voltage sources. If the interconnection is not valid, explain why.

Figure P2.3

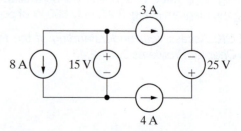

2.4 If the interconnection in Fig. P2.4 is valid, find the total power developed in the circuit. If the interconnection is not valid, explain why.

Figure P2.4

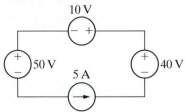

2.5 The interconnection of ideal sources can lead to an indeterminate solution. With this thought in mind, explain why the solutions for v_1 and v_2 in the circuit in Fig. P2.5 are not unique.

Figure P2.5

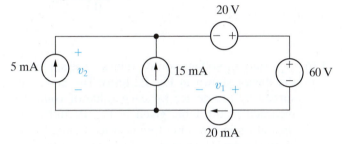

2.6 Consider the interconnection shown in Fig. P2.6.

a) What value of α is required to make this a valid interconnection?

b) For this value of α, find the power associated with the current source.

c) Is the current source supplying or absorbing power?

Figure P2.6

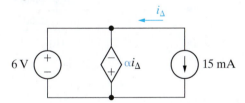

2.7 Consider the interconnection shown in Fig. P2.7.

a) What value of v_1 is required to make this a valid interconnection?

b) For this value of v_1, find the power associated with the voltage source.

Figure P2.7

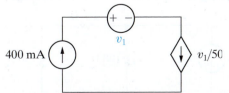

2.8 If the interconnection in Fig. P2.8 is valid, find the total power developed in the circuit. If the interconnection is not valid, explain why.

Figure P2.8

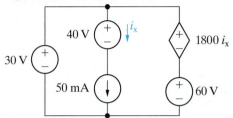

2.9 Find the total power developed in the circuit in Fig. P2.9 if $v_o = 5$ V.

Figure P2.9

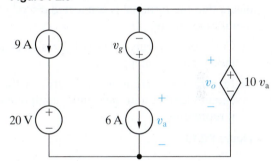

2.10 a) Is the interconnection in Fig. P2.10 valid? Explain.

b) Can you find the total energy developed in the circuit? Explain.

Figure P2.10

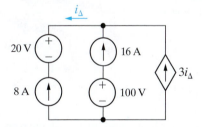

Sections 2.2–2.3

2.11 For the circuit shown in Fig. P2.11

a) Find i.

b) Find the power supplied by the voltage source.

c) Reverse the polarity of the voltage source and repeat parts (a) and (b).

Figure P2.11

2.12 For the circuit shown in Fig. P2.12

a) Find v.

b) Find the power absorbed by the resistor.

c) Reverse the direction of the current source and repeat parts (a) and (b).

Figure P2.12

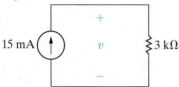

15 mA, v, 3 kΩ

2.13 A pair of automotive headlamps is connected to a 12 V battery via the arrangement shown in Fig. P2.13. In the figure, the triangular symbol ▼ is used to indicate that the terminal is connected directly to the metal frame of the car.

a) Construct a circuit model using resistors and an independent voltage source.

b) Identify the correspondence between the ideal circuit element and the symbol component that it represents.

Figure P2.13

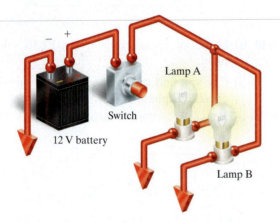

Lamp A
Switch
12 V battery
Lamp B

2.14 The terminal voltage and terminal current were measured on the device shown in Fig. P2.14(a). The values of v and i are given in the table of Fig. P2.14(b). Use the values in the table to construct a circuit model for the device consisting of a single resistor from Appendix H.

Figure P2.14

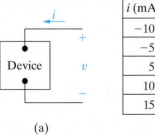

Device, v

i (mA)	v (V)
-10	-120
-5	-60
5	60
10	120
15	180

(a)   (b)

2.15 A variety of current source values were applied to the device shown in Fig. P2.15(a). The power absorbed by the device for each value of current is recorded in the table given in Fig. P2.15(b). Use the values in the table to construct a circuit model for the device consisting of a single resistor from Appendix H.

Figure P2.15

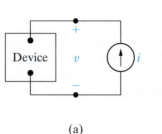

Device, v, i

i (mA)	p (mW)
0.5	8.25
1.0	33.00
1.5	74.25
2.0	132.00
2.5	206.25
3.0	297.00

(a)   (b)

2.16 A variety of voltage source values were applied to the device shown in Fig. P2.16(a). The power absorbed by the device for each value of voltage is recorded in the table given in Fig. P2.16(b). Use the values in the table to construct a circuit model for the device consisting of a single resistor from Appendix H.

Figure P2.16

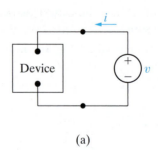

Device, i, v

v (V)	p (mW)
-8	640
-4	160
4	160
8	640
12	1440
16	2560

(a)   (b)

Section 2.4

2.17 a) Find the currents i_1 and i_2 in the circuit in Fig. P2.17.

b) Find the voltage v_o.

c) Verify that the total power developed equals the total power dissipated.

Figure P2.17

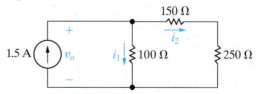

150 Ω, i_2
1.5 A, v_o, i_1, 100 Ω, 250 Ω

2.18 Given the circuit shown in Fig. P2.18, find

PSPICE
MULTISIM

a) the value of i_a,

b) the value of i_b,

c) the value of v_o,

d) the power dissipated in each resistor,

e) the power delivered by the 200 V source.

Figure P2.18

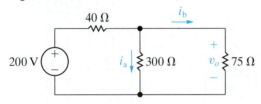

2.19 The current i_a in the circuit shown in Fig. P2.19 is

PSPICE 20 A. Find (a) i_o; (b) i_g; and (c) the power delivered
MULTISIM by the independent current source.

Figure P2.19

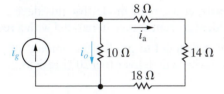

2.20 Consider the circuit shown in Fig. P2.20.

a) Find v_o using Kirchoff's laws and Ohm's law.

b) Test the solution for v_o by verifying that the total power supplied equals the total power absorbed.

Figure P2.20

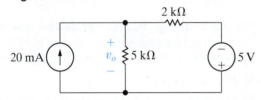

2.21 The current i_x in the circuit shown in Fig. P2.21 is 50 mA, and the voltage v_x is 3.5 V. Find (a) i_1; (b) v_1; (c) v_g; and (d) the power supplied by the voltage source.

Figure P2.21

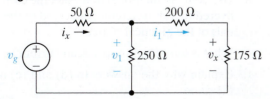

2.22 The current i_o in the circuit in Fig. P2.22 is 2 A.

PSPICE
MULTISIM

a) Find i_1.

b) Find the power dissipated in each resistor.

c) Verify that the total power dissipated in the circuit equals the power developed by the 80 V source.

Figure P2.22

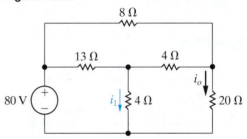

2.23 The voltage across the 22.5 Ω resistor in the circuit in

PSPICE Fig. P2.23 is 90 V, positive at the upper terminal.
MULTISIM

a) Find the power dissipated in each resistor.

b) Find the power supplied by the 240 V ideal voltage source.

c) Verify that the power supplied equals the total power dissipated.

Figure P2.23

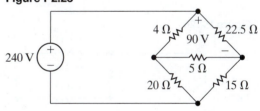

2.24 The currents i_1 and i_2 in the circuit in Fig. P2.24 are 20 A and 15 A, respectively.

a) Find the power supplied by each voltage source.

b) Show that the total power supplied equals the total power dissipated in the resistors.

Figure P2.24

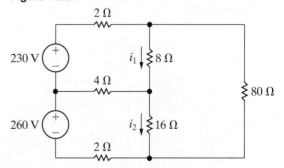

2.25 The currents i_a and i_b in the circuit in Fig. P2.25 are
PSPICE   4 A and -2 A, respectively.
MULTISIM

a) Find i_g.

b) Find the power dissipated in each resistor.

c) Find v_g.

d) Show that the power delivered by the current source is equal to the power absorbed by all the other elements.

Figure P2.25

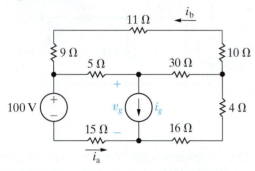

2.26 For the circuit shown in Fig. P2.26, find (a) R and (b)
PSPICE   the power supplied by the 240 V source.
MULTISIM

Figure P2.26

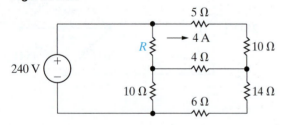

2.27 The variable resistor R in the circuit in Fig. P2.27 is
PSPICE   adjusted until i_o equals 10 mA. Find the value of R.
MULTISIM

Figure P2.27

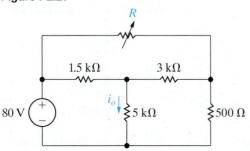

2.28 The voltage and current were measured at the terminals of the device shown in Fig. P2.28(a). The results are tabulated in Fig. P2.28(b).

a) Construct a circuit model for this device using an ideal voltage source in series with a resistor.

b) Use the model to predict the i_t the amount of power the device will deliver to a 40 Ω resistor.

Figure P2.28

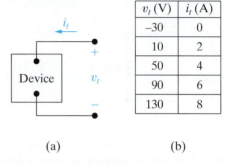

v_t (V)	i_t (A)
-30	0
10	2
50	4
90	6
130	8

(a)   (b)

2.29 The voltage and current were measured at the terminals of the device shown in Fig. P2.29(a). The results are tabulated in Fig. P2.29(b).

a) Construct a circuit model for this device using an ideal current source in parallel with a resistor.

b) Use the model to predict the amount of power the device will deliver to a 20 Ω resistor.

Figure P2.29

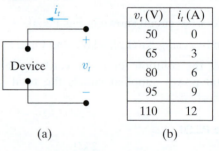

v_t (V)	i_t (A)
50	0
65	3
80	6
95	9
110	12

(a)   (b)

2.30 The table in Fig. P2.30(a) gives the relationship between the terminal voltage and current of the practical constant voltage source shown in Fig. P2.30(b).

a) Plot v_s versus i_s.

b) Construct a circuit model of the practical source that is valid for $0 \leq i_s \leq 24$ mA, based on the equation of the line plotted in (a). (Use an ideal voltage source in series with an ideal resistor.)

c) Use your circuit model to predict the current delivered to a 1 kΩ resistor connected to the terminals of the practical source.

d) Use your circuit model to predict the current delivered to a short circuit connected to the terminals of the practical source.

e) What is the actual short-circuit current?

f) Explain why the answers to (d) and (e) are not the same.

Figure P2.30

v_s (V)	i_s (mA)
24	0
22	8
20	16
18	24
15	32
10	40
0	48

(a)   (b)

2.31 The table in Fig. P2.31(a) gives the relationship between the terminal current and voltage of the practical constant current source shown in Fig. P2.31(b).

a) Plot i_s versus v_s.

b) Construct a circuit model of this current source
that is valid for $0 \le v_s \le 30$ V, based on the
equation of the line plotted in (a).

c) Use your circuit model to predict the current delivered to a 3 kΩ resistor.

d) Use your circuit model to predict the open-
circuit voltage of the current source.

e) What is the actual open-circuit voltage?

f) Explain why the answers to (d) and (e) are not
the same.

Figure P2.31

i_s (mA)	v_s (V)
40	0
35	10
30	20
25	30
18	40
8	50
0	55

(a)   (b)

2.32 Consider the circuit shown in Fig. P2.32.

a) Find i_o.

b) Verify the value of i_o by showing that the power generated in the circuit equals the power absorbed in the circuit.

Figure P2.32

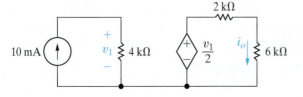

2.33 For the circuit shown in Fig. P2.33, find v_o and the
total power absorbed in the circuit.

Figure P2.33

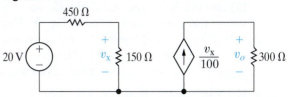

2.34 For the circuit shown in Fig. P2.34, find v_o and the
total power supplied in the circuit.

Figure P2.34

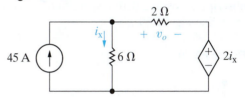

2.35 Find (a) i_o, (b) i_1, and (c) i_2 in the circuit in
Fig. P2.35.

PSPICE
MULTISIM

Figure P2.35

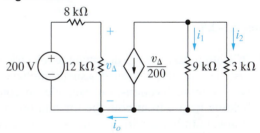

2.36 For the circuit shown in Fig. P2.36, calculate (a)
i_Δ and v_o, and (b) show that the power developed
equals the power absorbed.

PSPICE
MULTISIM

Figure P2.36

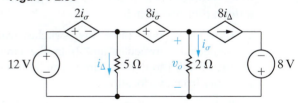

2.37 Find v_1 and v_g in the circuit shown in Fig. P2.37
when v_o equals 5 V. (*Hint:* Start at the right end of
the circuit and work back toward v_g.)

Figure P2.37

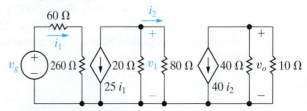

2.38 Derive Eq. 2.21. *Hint:* Use Eqs. (3) and (4) from Example 2.12 to express i_E as a function of i_B. Solve Eq. (2) for i_2 and substitute the result into both Eqs. (5) and (6). Solve the "new" Eq. (6) for i_1 and substitute this result into the "new" Eq. (5). Replace i_E in the "new" Eq. (5) and solve for i_B. Note that because i_{CC} appears only in Eq. (1), the solution for i_B involves the manipulation of only five equations.

2.39 For the circuit shown in Fig. 2.24, $R_1 = 40\,\text{k}\Omega$, $R_2 = 60\,\text{k}\Omega$, $R_C = 750\,\Omega$, $R_E = 120\,\Omega$, $V_{CC} = 10\,\text{V}$, $V_0 = 600\,\text{mV}$, and $\beta = 49$. Calculate i_B, i_C, i_E, v_{3d}, v_{bd}, i_2, i_1, v_{ab}, i_{CC}, and v_{13}. (*Note:* In the double subscript notation on voltage variables, the first subscript is positive with respect to the second subscript. See Fig. P2.39.)

Figure P2.39

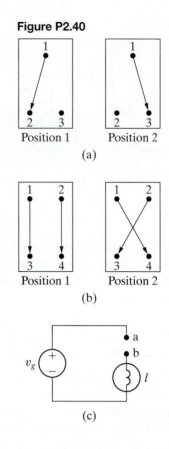

Figure P2.40

(a)

(b)

(c)

Sections 2.1–2.5

2.40 It is often desirable in designing an electric wiring system to be able to control a single appliance from two or more locations, for example, to control a lighting fixture from both the top and bottom of a stairwell. In home wiring systems, this type of control is implemented with three-way and four-way switches. A three-way switch is a three-terminal, two-position switch, and a four-way switch is a four-terminal, two-position switch. The switches are shown schematically in Fig. P2.40(a), which illustrates a three-way switch, and P2.40(b), which illustrates a four-way switch.

a) Show how two three-way switches can be connected between a and b in the circuit in Fig. P2.40(c) so that the lamp *l* can be turned ON or OFF from two locations.

b) If the lamp (appliance) is to be controlled from more than two locations, four-way switches are used in conjunction with two three-way switches. One four-way switch is required for each location in excess of two. Show how one four-way switch plus two three-way switches can be connected between a and b in Fig. P2.40(c) to control the lamp from three locations. (*Hint:* The four-way switch is placed between the three-way switches.)

2.41 Suppose you want to add a third radiator to your garage that is identical to the two radiators you have already installed. All three radiators can be modeled by 48 Ω resistors. Using the wiring diagram shown in Fig. P2.41, calculate the total power for the three radiators.

Figure P2.41

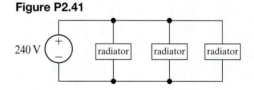

2.42 Repeat Problem 2.41 using the wiring diagram shown in Fig. P2.42. Compare the total radiator power in this configuration with the total radiator power in the configuration shown in Fig. P2.41.

Figure P2.42

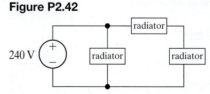

2.43 Repeat Problem 2.41 using the wiring diagram
PRACTICAL
PERSPECTIVE
shown in Fig. P2.43. Compare the total radiator
power in this configuration with the total radiator
power in the configuration shown in Fig. P2.41.

2.44 Repeat Problem 2.41 using the wiring diagram
PRACTICAL
PERSPECTIVE
shown in Fig. P2.44. Compare the total radiator
power in this configuration with the total radiator
power in the configuration shown in Fig. P2.41.

Figure P2.43

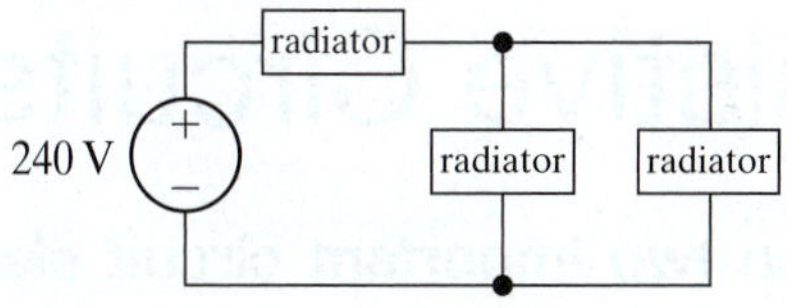

Figure P2.44

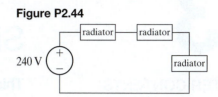

CHAPTER

3

Simple Resistive Circuits

CHAPTER CONTENTS

3.1 **Resistors in Series** *p. 60*

3.2 **Resistors in Parallel** *p. 61*

3.3 **The Voltage-Divider and Current-Divider Circuits** *p. 64*

3.4 **Voltage Division and Current Division** *p. 68*

3.5 **Measuring Voltage and Current** *p. 70*

3.6 **Measuring Resistance—The Wheatstone Bridge** *p. 73*

3.7 **Delta-to-Wye (Pi-to-Tee) Equivalent Circuits** *p. 75*

CHAPTER OBJECTIVES

1 Be able to recognize resistors connected in series and in parallel and use the rules for combining series-connected resistors and parallel-connected resistors to yield equivalent resistance.

2 Know how to design simple voltage-divider and current-divider circuits.

3 Be able to use voltage division and current division appropriately to solve simple circuits.

4 Be able to determine the reading of an ammeter when added to a circuit to measure current; be able to determine the reading of a voltmeter when added to a circuit to measure voltage.

5 Understand how a Wheatstone bridge is used to measure resistance.

6 Know when and how to use delta-to-wye equivalent circuits to solve simple circuits.

This chapter focuses on two important circuit element interconnections: series connections and parallel connections.

- When resistors are connected in series and in parallel, we can combine these resistors into equivalent resistors, reducing the number of circuit elements and the circuit's complexity, and simplifying the circuit analysis.

- When a voltage source is connected to two or more *resistors in series*, the supplied *voltage* divides across the resistors. This common configuration is used as a voltage divider that outputs a specific (desired) voltage, smaller than the supplied voltage. We introduce voltage division, a new tool that simplifies the analysis of these circuits.

- When a current source is connected to two or more *resistors in parallel*, the supplied *current* divides among the resistors. This common configuration is used as a current divider that outputs a specific (desired) current, smaller than the supplied current. We introduce current division, a new tool that simplifies the analysis of these circuits.

Finally, we look at three important measurement instruments.

- The ammeter, which is a practical application of the current divider, is used to measure current.

- The voltmeter, which is a practical application of the voltage divider, is used to measure voltage.

- The Wheatstone bridge, which introduces two new interconnections known as delta and wye connections, is used to measure resistance.

■ Practical Perspective

Resistive Touch Screens

Many mobile phones and tablet computers use touch screens created by applying a transparent resistive material to the glass or acrylic screens. Two screens are typically used, separated by a transparent insulating layer. We can model a touch screen as a grid of resistors in the x-direction and a grid of resistors in the y-direction. The figure on the right depicts one row of the grid in the x-direction, with terminals x_1 and x_2, and one column of the grid in the y-direction, with terminals y_1 and y_2.

A separate electronic circuit repeatedly applies a voltage drop across the grid in the x-direction (between the points x_1 and x_2), then removes that voltage and applies a voltage drop across the grid in the y-direction (between points y_1 and y_2). When the screen is touched, the two resistive layers are pressed together, creating a voltage that is sensed in the x-grid and another voltage that is sensed in the y-grid. These two voltages precisely locate the point where the screen was touched.

How is the voltage created by touching the screen related to the position where the screen was touched? How are the properties of the grids used to calculate the touch position? We will answer these questions in the Practical Perspective at the end of this chapter. The circuit analysis required to answer these questions uses some of the new tools developed in this chapter.

Denis Semenchenko/Shutterstock

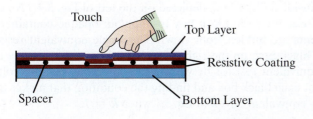

Touch
Top Layer
Resistive Coating
Spacer
Bottom Layer

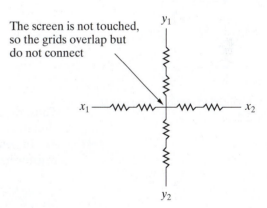

The screen is not touched, so the grids overlap but do not connect

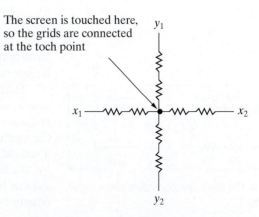

The screen is touched here, so the grids are connected at the toch point

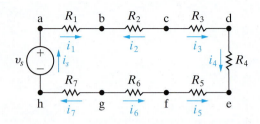

Figure 3.1 ▲ Resistors connected in series.

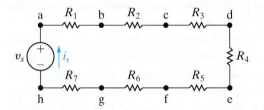

Figure 3.2 ▲ Series resistors with a single unknown current i_s.

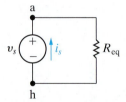

Figure 3.3 ▲ A simplified version of the circuit shown in Fig. 3.2.

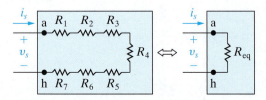

Figure 3.4 ▲ The black box equivalent of the circuit shown in Fig. 3.2.

3.1 Resistors in Series

In Chapter 2, we learned that two elements connected at a single node are said to be *in series*. For example, the seven resistors in Fig. 3.1 are connected in series. **Series-connected circuit elements** carry the same current. Applying Kirchhoff's current law to each node in the circuit, we can show that these resistors carry the same current. The series interconnection in Fig. 3.1 requires that

$$i_s = i_1 = -i_2 = i_3 = i_4 = -i_5 = -i_6 = i_7,$$

so if we know any one of the seven currents, we know them all. Thus, we can redraw Fig. 3.1 as shown in Fig. 3.2, using the single current i_s.

To find i_s, we apply Kirchhoff's voltage law around the single closed loop in the clockwise direction. Defining the voltage across each resistor as a drop in the direction of i_s (Ohm's law) gives

$$-v_s + i_s R_1 + i_s R_2 + i_s R_3 + i_s R_4 + i_s R_5 + i_s R_6 + i_s R_7 = 0,$$

or

$$v_s = i_s(R_1 + R_2 + R_3 + R_4 + R_5 + R_6 + R_7).$$

This equation tells us we can simplify the circuit in Fig. 3.2 by replacing the seven resistors with a single equivalent resistor, R_{eq}, whose numerical value is the sum of the individual resistors; that is,

$$R_{eq} = R_1 + R_2 + R_3 + R_4 + R_5 + R_6 + R_7$$

and

$$v_s = i_s R_{eq}.$$

Thus, we can redraw Fig. 3.2 as shown in Fig. 3.3, which is a much simpler circuit.

In general, if k resistors are connected in series, the equivalent single resistor has a resistance equal to the sum of the k resistances, or

COMBINING RESISTORS IN SERIES

$$R_{eq} = \sum_{i=1}^{k} R_i = R_1 + R_2 + \cdots + R_k. \qquad (3.1)$$

Note that the resistance of the equivalent resistor is always larger than the largest resistor in the series connection.

Think about equivalent resistance by visualizing the series-connected resistors inside a black box, depicted on the left of Fig. 3.4. (An electrical engineer uses the term **black box** to imply an opaque container; that is, the contents are hidden from view.) The single equivalent resistor is in a second black box, on the right of Fig. 3.4. We can derive the equation for the equivalent resistor by writing the equations relating voltage and current for each black box and finding the condition that makes these two equations equivalent, given in Eq. 3.1 when $k = 7$.

3.2 Resistors in Parallel

Two elements connected at both of their nodes are said to be *in parallel*. For example, the four resistors in the circuit in Fig. 3.5 are in parallel. **Parallel-connected circuit elements** have the same voltage across their terminals. Don't assume that two elements are parallel connected merely because they are lined up in parallel in a circuit diagram. The defining characteristic of parallel-connected elements is that they have the same voltage across their terminals. In Fig. 3.6, you can see that R_1 and R_3 are not parallel connected because, between their respective terminals, another resistor dissipates some of the voltage.

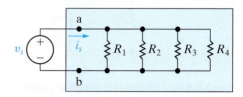

Figure 3.5 ▲ Resistors in parallel.

We can reduce resistors in parallel to a single equivalent resistor using Kirchhoff's current law and Ohm's law. In Fig. 3.5, we let the currents i_1, i_2, i_3, and i_4 be the currents in the resistors R_1 through R_4, respectively. Note that the positive reference direction for each resistor current is through the resistor from node a to node b. From Kirchhoff's current law,

$$i_s = i_1 + i_2 + i_3 + i_4.$$

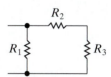

Figure 3.6 ▲ Nonparallel resistors.

The parallel connection of the resistors means that the voltage across each resistor must be the same. Hence, from Ohm's law,

$$i_1 R_1 = i_2 R_2 = i_3 R_3 = i_4 R_4 = v_s.$$

Therefore,

$$i_1 = \frac{v_s}{R_1}, \quad i_2 = \frac{v_s}{R_2}, \quad i_3 = \frac{v_s}{R_3}, \quad \text{and} \quad i_4 = \frac{v_s}{R_4}.$$

Substituting the expressions for the four branch currents into the KCL equation and simplifying yields

$$i_s = v_s \left(\frac{1}{R_1} + \frac{1}{R_2} + \frac{1}{R_3} + \frac{1}{R_4} \right),$$

from which

$$\frac{i_s}{v_s} = \frac{1}{R_{eq}} = \frac{1}{R_1} + \frac{1}{R_2} + \frac{1}{R_3} + \frac{1}{R_4}.$$

This equation shows that the four resistors in Fig. 3.5 can be replaced by a single equivalent resistor, thereby simplifying the circuit. The circuit in Fig. 3.7 illustrates the substitution. The equivalent resistance of k resistors connected in parallel is

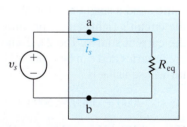

Figure 3.7 ▲ Replacing the four parallel resistors shown in Fig. 3.5 with a single equivalent resistor.

COMBINING RESISTORS IN PARALLEL

$$\frac{1}{R_{eq}} = \sum_{i=1}^{k} \frac{1}{R_i} = \frac{1}{R_1} + \frac{1}{R_2} + \cdots + \frac{1}{R_k}. \tag{3.2}$$

Note that the resistance of the equivalent resistor is always smaller than the resistance of the smallest resistor in the parallel connection. Using conductance when dealing with resistors connected in parallel is sometimes more convenient. In that case, Eq. 3.2 becomes

$$G_{eq} = \sum_{i=1}^{k} G_i = G_1 + G_2 + \cdots + G_k.$$

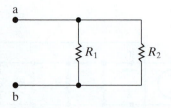

Figure 3.8 ▲ Two resistors connected in parallel.

Many times only two resistors are connected in parallel. Figure 3.8 illustrates this special case. We calculate the equivalent resistance from Eq. 3.2:

$$\frac{1}{R_{eq}} = \frac{1}{R_1} + \frac{1}{R_2} = \frac{R_2 + R_1}{R_1 R_2},$$

or

COMBINING TWO RESISTORS IN PARALLEL

$$R_{eq} = \frac{R_1 R_2}{R_1 + R_2}. \tag{3.3}$$

Thus, for just two resistors in parallel, the equivalent resistance equals the product of the resistances divided by the sum of the resistances. Remember that you can only use this result in the special case of just two resistors in parallel.

Work through Examples 3.1and 3.2 to practice using series and parallel simplifications in circuit analysis.

EXAMPLE 3.1 | Applying Series-Parallel Simplification

a) Find the equivalent resistance seen by the current source in Fig. 3.9, using series and parallel simplifications.

b) Use your results in part (a) to find the power delivered by the current source.

Solution

a) Our goal is a circuit with the 50 mA current source and a single resistor. We start simplifying the circuit's right-hand side, moving left toward the current source. The 2 kΩ and 3 kΩ resistors are in series and can be replaced by a single resistor whose value is

$$2000 + 3000 = 5000 = 5 \text{ k}\Omega.$$

Figure 3.10(a) shows this simplified circuit where the 20 kΩ and the 5 kΩ resistors are now in parallel. We replace these parallel-connected resistors with a single equivalent resistor, calculating its value using the "product over the sum" equation (Eq. 3.3):

$$\frac{(20{,}000)(5000)}{20{,}000 + 5000} = 4000 = 4 \text{ k}\Omega.$$

Figure 3.10(b) shows this result, and now we see the two 4 kΩ resistors are in series. They can be replaced with a single resistor whose value is

$$4000 + 4000 = 8000 = 8 \text{ k}\Omega.$$

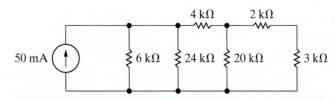

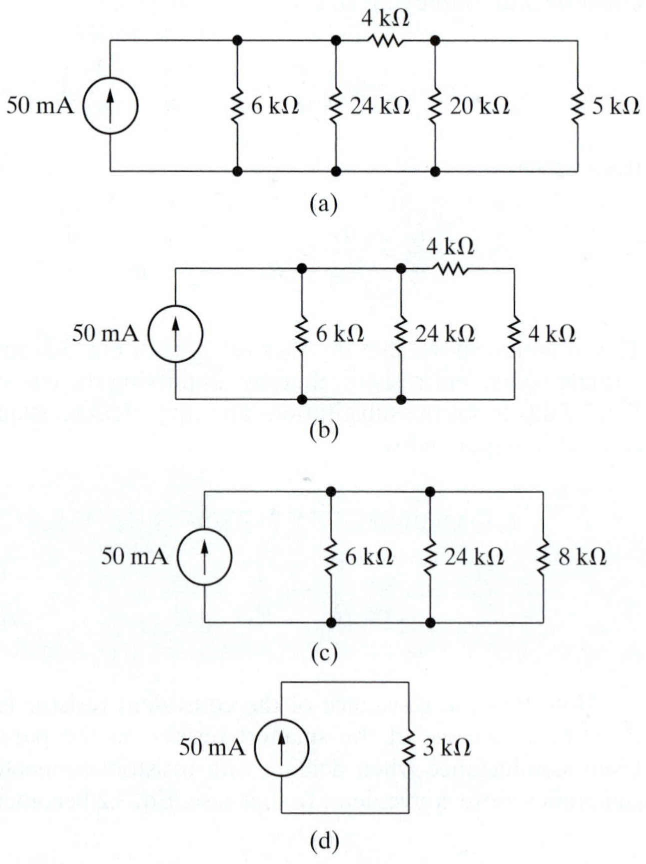

Figure 3.9 ▲ The circuit for Example 3.1.

Figure 3.10 ▲ Simplifying the circuit in Fig. 3.9.

Figure 3.10(c) shows this simplification. Now the $24\,k\Omega$, $6\,k\Omega$, and $8\,k\Omega$ resistors are in parallel. To find their equivalent, we add their inverses and invert the result (Eq. 3.2):

$$\left(\frac{1}{24{,}000} + \frac{1}{6000} + \frac{1}{8000}\right)^{-1}$$

$$= \left(\frac{1}{24{,}000} + \frac{4}{24{,}000} + \frac{3}{24{,}000}\right)^{-1}$$

$$= \left(\frac{8}{24{,}000}\right)^{-1} = \frac{24{,}000}{8} = 3000 = 3\,k\Omega.$$

The equivalent resistance seen by the current source is $3\,k\Omega$, as shown in Fig. 3.10(d).

b) The power of the source and the power of the equivalent $3\,k\Omega$ resistor must sum to zero. Using Fig. 3.10(d), we can easily calculate the resistor's power using its current and resistance to give

$$p = (0.05)^2(3000) = 7.5\text{ W}.$$

The equivalent resistor is absorbing 7.5 W, so the current source must be delivering 7.5 W.

EXAMPLE 3.2   Solving a Circuit Using Series-Parallel Simplification

Find i_s, i_1, and i_2 in the circuit shown in Fig. 3.11.

Solution

Using series-parallel simplifications, we reduce the resistors to the right of the x-y terminals to a single equivalent resistor. On the circuit's right-hand side, the $3\,\Omega$ and $6\,\Omega$ resistors are in series. We replace this series combination with a $9\,\Omega$ resistor, reducing the circuit to the one shown in Fig. 3.12(a). Then we replace the parallel combination of the $9\,\Omega$ and $18\,\Omega$ resistors with a single equivalent resistance of $(18 \times 9)/(18 + 9)$, or $6\,\Omega$. Figure 3.12(b) shows the resulting circuit. The nodes x and y marked on all diagrams should help you trace through the circuit simplification.

From Fig. 3.12(b) you can use Ohm's law to verify that

$$i_s = \frac{120}{(6 + 4)} = 12\text{ A}.$$

Figure 3.13 shows this result and includes the voltage v_1 to help clarify the subsequent discussion. Using Ohm's law, we compute the value of v_1:

$$v_1 = (12)(6) = 72\text{ V}.$$

Since v_1 is the voltage drop from node x to node y, we can return to the circuit shown in Fig. 3.12(a) and again use Ohm's law to calculate i_1 and i_2. Thus,

$$i_1 = \frac{v_1}{18} = \frac{72}{18} = 4\text{ A},$$

$$i_2 = \frac{v_1}{9} = \frac{72}{9} = 8\text{ A}.$$

We have found the three specified currents by using series-parallel reductions in combination with Ohm's law.

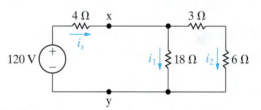

Figure 3.11 ▲ The circuit for Example 3.2.

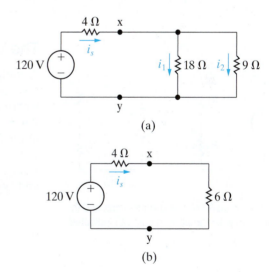

(a)

(b)

Figure 3.12 ▲ A simplification of the circuit shown in Fig. 3.11.

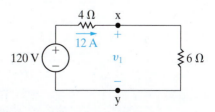

Figure 3.13 ▲ The circuit of Fig. 3.12(b) showing the numerical value of i_s.

Before leaving Example 3.2, you should verify that the solution satisfies Kirchhoff's current law at every node and Kirchhoff's voltage law around every closed path. (There are three closed paths that can be tested.) You can also show that the power delivered by the voltage source equals the total power dissipated in the resistors, and thus the power in the circuit balances. (See Problems 3.8 and 3.9.)

◻ ASSESSMENT PROBLEM

Objective 1—Be able to recognize resistors connected in series and in parallel

3.1   For the circuit shown, find (a) the voltage v, (b) the power delivered to the circuit by the current source, and (c) the power dissipated in the 10 Ω resistor.

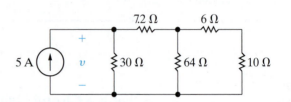

Answer: a) 60 V;

   b) 300 W;

   c) 57.6 W.

SELF-CHECK: Also try Chapter Problems 3.1–3.4.

3.3 The Voltage-Divider and Current-Divider Circuits

The Voltage-Divider Circuit

A **voltage-divider circuit** produces two or more smaller voltages from a single voltage supply. This is especially useful in electronic circuits, where a single circuit may require voltages of +15 V, −15 V, and +5 V. An example of a voltage-divider circuit that creates two voltages is shown in Fig 3.14. We introduce the current i, as shown in Fig. 3.14(b), and recognize (from Kirchhoff's current law) that R_1 and R_2 carry this current. Using Ohm's law to determine the resistor voltages from the current i and applying Kirchhoff's voltage law around the closed loop yields

$$v_s = iR_1 + iR_2,$$

or

$$i = \frac{v_s}{R_1 + R_2}.$$

Using Ohm's law and the expression for i, we calculate v_1 and v_2:

$$v_1 = iR_1 = v_s\frac{R_1}{R_1 + R_2}, \tag{3.4}$$

$$v_2 = iR_2 = v_s\frac{R_2}{R_1 + R_2}. \tag{3.5}$$

Equations 3.4 and 3.5 show that v_1 and v_2 are fractions of v_s. Expressed in words, each fraction is

$$\frac{\text{the resistance across which the divided voltage is defined}}{\text{the sum of the two resistances}}.$$

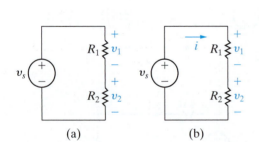

Figure 3.14 ▲ (a) A voltage-divider circuit and (b) the voltage-divider circuit with current i indicated.

Because this ratio is always less than 1.0, the divided voltages v_1 and v_2 are always less than the source voltage v_s.

Work through Example 3.3 to design a simple voltage divider.

EXAMPLE 3.3 | Designing a Simple Voltage Divider

The voltage divider in Fig. 3.14 has a source voltage of 20 V. Determine the values of the resistors R_1 and R_2 to give $v_1 = 15$ V and $v_2 = 5$ V.

Solution

From Eqs. 3.4 and 3.5,

$$15 = \frac{R_1}{R_1 + R_2}(20) \quad \text{and} \quad 5 = \frac{R_2}{R_1 + R_2}(20).$$

Unfortunately, these two equations are not independent. If you solve each equation for R_1, you get $R_1 = 3R_2$. An infinite number of combinations of R_1 and R_2 yield the correct values for v_1 and v_2. For example, if you choose $R_2 = 10\,k\Omega$, then $R_1 = 30\,k\Omega$ gives the correct voltages, but if you choose $R_2 = 400\,\Omega$, then $R_1 = 1200\,\Omega$ gives the correct voltages.

When selecting values for R_1 and R_2, you should consider the power the resistors must dissipate and the effects of connecting the voltage-divider circuit to other circuit components. Example 3.4 uses the voltage divider designed in Example 3.3 to supply 5 V to a 10 kΩ resistor.

EXAMPLE 3.4 | Adding a Resistive Load to a Voltage Divider

a) For the voltage divider designed in Example 3.3, suppose resistors $R_2 = 10\,k\Omega$ and $R_1 = 30\,k\Omega$. Connect a resistor $R_L = 10\,k\Omega$ in parallel with R_2 and determine the voltage across R_L.

b) Repeat part (a) using resistors $R_2 = 400\,\Omega$ and $R_1 = 1200\,\Omega$, but the same value of R_L.

Solution

a) The voltage divider with the resistor R_L is shown in Fig. 3.15. The resistor R_L acts as a load on the voltage-divider circuit. A **load** on any circuit consists of one or more circuit elements that draw power from the circuit. The parallel combination of the

two 10 kΩ resistors, one from the voltage divider and the other the load resistor R_L, gives an equivalent resistance of 5 kΩ. Therefore, from Eq. 3.5,

$$v_o = \frac{5000}{30{,}000 + 5000}(20) = 2.86 \text{ V.}$$

This is certainly not the 5 V we were expecting the voltage divider to deliver to the load, because adding the load resistor changed the voltage-divider circuit.

b) The voltage divider with a different set of resistors and the same load resistor is shown in Fig. 3.16. Again, we expect the load resistor to change the voltage-divider circuit. The parallel combination of the 400 Ω and 10 kΩ resistors

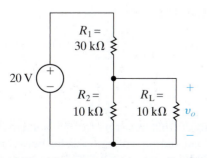

Figure 3.15 ▲ The voltage divider from Example 3.3 with a resistive load.

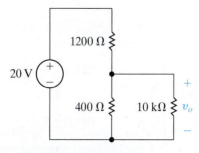

Figure 3.16 ▲ The voltage divider from Example 3.3 with a different choice of R_1 and R_2 resistors and a resistive load.

gives an equivalent resistance of 384.615 Ω. Therefore, from Eq. 3.5,

$$v_o = \frac{384.615}{1200 + 384.615}(20) = 4.85 \text{ V}.$$

This is much closer to the 5 V we expected the voltage divider to deliver to the load. The effect of the load resistor is minimal because the load resistor value is much larger than the value of R_2 in the voltage divider.

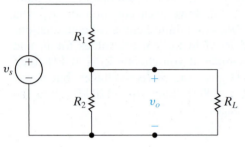

Figure 3.17 ▲ A voltage divider connected to a load R_L.

Figure 3.17 shows a general voltage divider with a load R_L connected. The expression for the output voltage is

$$v_o = \frac{R_{eq}}{R_1 + R_{eq}}v_s,$$

where

$$R_{eq} = \frac{R_2 R_L}{R_2 + R_L}.$$

Substituting the expression for R_{eq} into the equation for v_o and simplifying yields

$$v_o = \frac{R_2}{R_1\left[1 + (R_2/R_L)\right] + R_2} v_s. \qquad (3.6)$$

Note that Eq. 3.6 reduces to Eq. 3.5 as $R_L \rightarrow \infty$, as it should. Equation 3.6 shows that, as long as $R_L \gg R_2$, the voltage ratio v_o/v_s is essentially undisturbed by adding a load to the divider, as we saw in Example 3.4.

Another characteristic of the voltage-divider circuit is its sensitivity to the tolerances of the resistors. By *tolerance* we mean a range of possible values. The resistances of commercially available resistors always vary within some percentage of their stated value. Example 3.5 illustrates the effect of resistor tolerances in a voltage-divider circuit.

EXAMPLE 3.5 | **The Effect of Resistor Tolerance on the Voltage-Divider Circuit**

The resistors used in the voltage-divider circuit shown in Fig. 3.18 have a tolerance of ±10%. Find the maximum and minimum value of v_o.

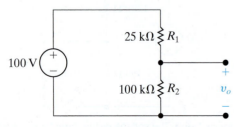

Figure 3.18 ▲ The circuit for Example 3.5.

Solution

From Eq. 3.5, the maximum value of v_o occurs when $R_2 = 110\text{ k}\Omega$ (10% high) and $R_1 = 22.5\text{ k}\Omega$ (10% low), and the minimum value of v_o occurs when $R_2 = 90\text{ k}\Omega$ (10% low) and $R_1 = 27.5\text{ k}\Omega$ (10% high). Therefore

$$v_o(\text{max}) = \frac{110\text{ k}}{110\text{ k} + 22.5\text{ k}}(100) = 83.02 \text{ V}$$

and

$$v_o(\text{min}) = \frac{90\text{ k}}{90\text{ k} + 27.5\text{ k}}(100) = 76.60 \text{ V}.$$

If we choose 10% resistors for this voltage divider, the no-load output voltage will lie between 76.60 and 83.02 V.

The Current-Divider Circuit

The **current-divider circuit** shown in Fig. 3.19 consists of two resistors connected in parallel across a current source. It divides the current i_s between R_1 and R_2. What is the relationship between the current i_s and the current in each resistor (i_1 and i_2)? The voltage across the parallel resistors can be expressed in three ways: as the product of the R_1 resistor and its current i_1, as the product of the R_2 resistor and its current i_2, and as the product of the equivalent resistance seen by the source and the source current. These three expressions for the voltage are given as

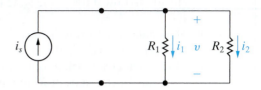

Figure 3.19 ▲ The current-divider circuit.

$$v = i_1 R_1 = i_2 R_2 = \frac{R_1 R_2}{R_1 + R_2} i_s.$$

Therefore,

$$i_1 = \frac{R_2}{R_1 + R_2} i_s, \tag{3.7}$$

$$i_2 = \frac{R_1}{R_1 + R_2} i_s. \tag{3.8}$$

Equations 3.7 and 3.8 show that when the current divides between two resistors in parallel, the current in one resistor equals the current entering the parallel pair multiplied by the other resistance and divided by the sum of the resistors. See how to design a current divider by working through Example 3.6.

EXAMPLE 3.6 | **Designing a Current-Divider Circuit**

Suppose the current source for the current divider shown in Fig. 3.19 is 100 mA. Assuming you have 0.25 W resistors available, what is the largest R_2 resistor you can use to get $i_2 = 50$ mA?

Solution

While you can use Eq. 3.8 to find the ratio of resistors, it should be clear that if the current in one resistor is 50 mA, the current in the other resistor must also be 50 mA, so both resistors must have the same value. Therefore, there are an infinite number of different resistor values which, when used for R_1

and R_2 will give $i_2 = 50$ mA. If the resistors have the same value and the same current, they absorb the same amount of power, which cannot exceed 0.25 W. From the power equation for resistors,

$$p = i^2 R = (0.05)^2 R = 0.25$$

so

$$R = \frac{0.25}{(0.05)^2} = 100 \ \Omega.$$

The largest 0.25 W resistors that can be used to create a current $i_2 = 50$ mA are 100 Ω resistors.

ASSESSMENT PROBLEMS

Objective 2—Know how to design simple voltage-divider and current-divider circuits

3.2   a) Find the no-load value of v_o in the circuit shown.
  b) Find v_o when R_L is 150 kΩ.
  c) How much power is dissipated in the 25 kΩ resistor if the load terminals are accidentally short-circuited?
  d) What is the maximum power dissipated in the 75 kΩ resistor?

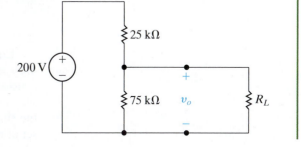

Answer: (a) 150 V;   (c) 1.6 W;
          (b) 133.33 V;   (d) 0.3 W.

3.3 a) Find the value of R that will cause 4 A of current to flow through the 80 Ω resistor in the circuit shown.
   b) How much power will the resistor R from part (a) need to dissipate?
   c) How much power will the current source generate for the value of R from part (a)?

Answer: (a) 30 Ω;
   (b) 7680 W;
   (c) 33,600 W.

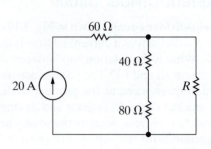

SELF-CHECK: Also try Chapter Problems 3.13, 3.14, and 3.20.

3.4 Voltage Division and Current Division

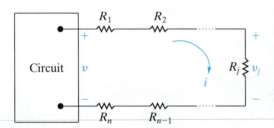

Figure 3.20 ▲ Circuit used to illustrate voltage division.

We now introduce two additional and very useful circuit analysis techniques known as **voltage division** and **current division**. These techniques generalize the results from analyzing the voltage-divider circuit in Fig. 3.14 and the current-divider circuit in Fig. 3.19. We begin with voltage division.

Voltage Division

Consider the circuit shown in Fig. 3.20, where the box on the left contains a single voltage source or any other combination of basic circuit elements that results in the voltage v shown in the figure. To the right of the box are n resistors connected in series. We are interested in finding the voltage drop v_j across an arbitrary resistor R_j in terms of the voltage v. We start by using Ohm's law to calculate i, the current through all of the resistors in series, in terms of the current v and the n resistors:

$$i = \frac{v}{R_1 + R_2 + \cdots + R_n} = \frac{v}{R_{eq}}.$$

The equivalent resistance, R_{eq}, is the sum of the n resistor values because the resistors are in series, as shown in Eq. 3.1. We apply Ohm's law a second time to calculate the voltage drop v_j across the resistor R_j, substituting v/R_{eq} for i:

VOLTAGE DIVISION EQUATION

$$v_j = iR_j = \frac{R_j}{R_{eq}} v. \qquad (3.9)$$

Equation 3.9 is the voltage division equation. It says that the voltage drop v_j across a single resistor R_j from a collection of series-connected resistors is proportional to the total voltage drop v across the set of series-connected resistors. The constant of proportionality is the ratio of the single resistance to the equivalent resistance of the series-connected set of resistors, or R_j/R_{eq}.

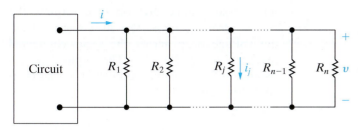

Figure 3.21 ▲ Circuit used to illustrate current division.

Current Division

Now consider the circuit shown in Fig. 3.21, where the box on the left contains a single current source or any other combination of basic circuit elements that results in the current i shown in the figure. To the right of the box are n resistors connected in parallel. We are interested in finding the current i_j through an arbitrary resistor R_j in terms of the current i. We start by using Ohm's law to calculate v, the voltage drop across each of the resistors in parallel, in terms of the current i and the n resistors:

$$v = i(R_1\|R_2\| \cdots \|R_n) = iR_{eq}.$$

The equivalent resistance of n resistors in parallel, R_{eq}, can be calculated using Eq. 3.2. We apply Ohm's law a second time to calculate the current i_j through the resistor R_j, replacing v with iR_{eq}:

CURRENT DIVISION EQUATION

$$i_j = \frac{v}{R_j} = \frac{R_{eq}}{R_j} i. \qquad (3.10)$$

Equation 3.10 is the current division equation. It says that the current i_j through a single resistor R_j from a collection of parallel-connected resistors is proportional to the total current i supplied to the set of parallel-connected resistors. The constant of proportionality is the ratio of the equivalent resistance of the parallel-connected set of resistors to the single resistance, or R_{eq}/R_j. Note that the constant of proportionality in the current division equation is the inverse of the constant of proportionality in the voltage division equation!

Example 3.7 uses voltage division and current division to solve for voltages and currents in a circuit.

EXAMPLE 3.7 | **Using Voltage Division and Current Division to Solve a Circuit**

Use current division to find the current i_o and use voltage division to find the voltage v_o for the circuit in Fig. 3.22.

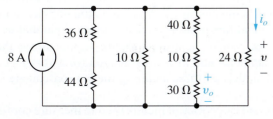

Figure 3.22 ▲ The circuit for Example 3.7.

Solution

We can use Eq. 3.10 if we can find the equivalent resistance of the four parallel branches containing resistors. Using "+" to represent series-connected resistors and "‖" to represent parallel-connected resistors, the equivalent resistance is

$$R_{eq} = (36 + 44)\|10\|(40 + 10 + 30)\|24$$

$$= 80\|10\|80\|24 = \frac{1}{\dfrac{1}{80} + \dfrac{1}{10} + \dfrac{1}{80} + \dfrac{1}{24}} = 6\ \Omega.$$

Applying Eq. 3.10,

$$i_o = \frac{6}{24}(8 \text{ A}) = 2 \text{ A}.$$

We can use Ohm's law to find the voltage drop across the 24 Ω resistor:

$$v = (24)(2) = 48 \text{ V}.$$

This is also the voltage drop across the branch containing the 40 Ω, the 10 Ω, and the 30 Ω resistors in series. Use voltage division to determine the voltage drop v_o across the 30 Ω resistor from the voltage drop across the series-connected resistors, using Eq. 3.9. The equivalent resistance of the series-connected resistors is $40 + 10 + 30 = 80 \text{ Ω}$, so

$$v_o = \frac{30}{80}(48 \text{ V}) = 18 \text{ V}.$$

ASSESSMENT PROBLEM

Objective 3—Be able to use voltage and current division to solve simple circuits

3.4 a) Use voltage division to determine the voltage v_o across the 40 Ω resistor in the circuit shown.

  b) Use v_o from part (a) to determine the current through the 40 Ω resistor, and use this current and current division to calculate the current in the 30 Ω resistor.

  c) How much power is absorbed by the 50 Ω resistor?

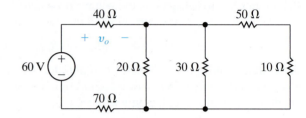

Answer: a) 20 V;
  b) 166.67 mA;
  c) 347.22 mW.

SELF-CHECK: Also try Chapter Problems 3.26 and 3.27.

3.5 Measuring Voltage and Current

Working with actual circuits often requires making voltage and current measurements. The next two sections explore several common devices used to make these measurements. The devices are relatively simple to analyze and offer practical examples of the current- and voltage-divider configurations. We begin by looking at ammeters and voltmeters.

- An **ammeter** is an instrument designed to measure current; it is placed in series with the circuit element whose current is being measured.
- A **voltmeter** is an instrument designed to measure voltage; it is placed in parallel with the element whose voltage is being measured.

Ideal ammeters and voltmeters have no effect on the circuit variable they are designed to measure. That is, an ideal ammeter has an equivalent resistance of 0 Ω and functions as a short circuit in series with the element whose current is being measured. An ideal voltmeter has an infinite equivalent resistance and functions as an open circuit in parallel with the element whose voltage is being measured. Figure 3.23 measures the current in R_1 using an ammeter and measures the voltage across R_2 using a voltmeter. The ideal models for these meters in the same circuit are shown in Fig. 3.24.

There are two broad categories of meters used to measure continuous voltages and currents: analog meters and digital meters.

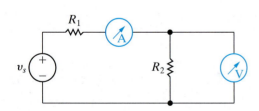

Figure 3.23 ▲ An ammeter connected to measure the current in R_1, and a voltmeter connected to measure the voltage across R_2.

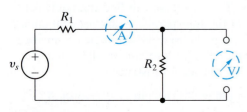

Figure 3.24 ▲ A short-circuit model for the ideal ammeter, and an open-circuit model for the ideal voltmeter.

Analog Meters

Analog meters are based on the d'Arsonval meter movement, which includes a dial readout pointer as shown in Fig. 3.25. The d'Arsonval meter movement consists of a movable coil placed in the field of a permanent magnet. When current flows in the coil, it creates a torque on the coil, causing it to rotate and move a pointer across a calibrated scale. By design, the deflection of the pointer is directly proportional to the current in the movable coil. The coil is characterized by both a voltage rating and a current rating. For example, one commercially available meter movement is rated at 50 mV and 1 mA. This means that when the coil is carrying 1 mA, the voltage drop across the coil is 50 mV and the pointer is deflected to its full-scale position.

An analog ammeter consists of a d'Arsonval movement in parallel with a resistor, as shown in Fig. 3.26. The parallel resistor limits the amount of current in the movement's coil by shunting some of it through R_A. In contrast, an analog voltmeter consists of a d'Arsonval movement in series with a resistor, as shown in Fig. 3.27. Here, the resistor limits the voltage drop across the meter's coil. In both meters, the added resistor determines the full-scale reading of the meter movement.

From these descriptions we see that an analog meter is nonideal; both the added resistor and the meter movement introduce resistance in the circuit where the meter is attached. In fact, any instrument used to make physical measurements extracts energy from the system while making measurements. The more energy extracted by the instruments, the more severely the measurement is disturbed. The equivalent resistance of a real ammeter is not zero, so it adds resistance to the circuit in series with the element whose current is being read. The equivalent resistance of a real voltmeter is not infinite, so it adds resistance to the circuit in parallel with the element whose voltage is being read.

How much these meters disturb the circuit being measured depends on the effective resistance of the meters compared with the resistance in the circuit. For example, using the rule of 1/10th, the effective resistance of an ammeter should be no more than 1/10th of the value of the smallest resistance in the circuit to be sure that the current being measured is nearly the same with or without the ammeter. But in an analog meter, the value of resistance is determined by the desired full-scale reading we wish to make, and it cannot be arbitrarily selected. Examples 3.8 and 3.9 illustrate how to calculate the resistance needed in an analog ammeter or voltmeter. The examples also determine the effective resistance of the meter when it is inserted in a circuit.

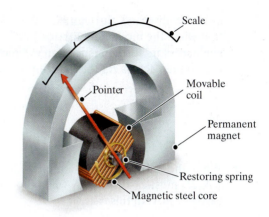

Figure 3.25 ▲ A schematic diagram of a d'Arsonval meter movement.

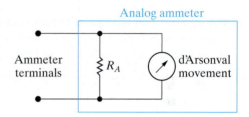

Figure 3.26 ▲ An analog ammeter circuit.

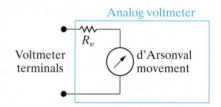

Figure 3.27 ▲ An analog voltmeter circuit.

EXAMPLE 3.8 | **Using a d'Arsonval Ammeter**

a) A 50 mV, 1 mA d'Arsonval movement is to be used in an ammeter with a full-scale reading of 10 mA. Determine R_A.

b) Repeat (a) for a full-scale reading of 1 A.

c) How much resistance is added to the circuit when the 10 mA ammeter is inserted to measure current?

d) Repeat (c) for the 1 A ammeter.

Solution

a) Look at the analog ammeter circuit in Fig. 3.26. The current in the ammeter must divide between the branch with the resistor R_A and the branch with the meter movement. From the problem statement we know that when the current in the ammeter is 10 mA, 1 mA is flowing through the meter coil, which means that 9 mA must be diverted through R_A. We also know that when the movement carries 1 mA, the voltage across its terminals is 50 mV, which is also the voltage across R_A. Using Ohm's law,

$$9 \times 10^{-3} R_A = 50 \times 10^{-3},$$

or

$$R_A = 50/9 = 5.555 \ \Omega.$$

b) When the full-scale deflection of the ammeter is 1 A, R_A must carry 999 mA when the movement carries 1 mA. In this case,

$$999 \times 10^{-3} R_A = 50 \times 10^{-3},$$

or

$$R_A = 50/999 \approx 50.05 \ \text{m}\Omega.$$

c) Let R_m represent the equivalent resistance of the ammeter. When the ammeter current is 10 mA, its voltage drop is 50 mV, so from Ohm's law,

$$R_m = \frac{0.05}{0.01} = 5 \ \Omega.$$

Alternatively, the resistance of the ammeter is the equivalent resistance of the meter movement in parallel with R_A. The resistance of the meter movement is the ratio of its voltage to its current, or $0.05/0.001 = 50 \ \Omega$. Therefore,

$$R_m = 50 \| (50/9) = \frac{(50)(50/9)}{50 + (50/9)} = 5 \ \Omega.$$

d) For the 1 A ammeter

$$R_m = \frac{0.05}{1} = 0.05 \ \Omega,$$

or, alternatively,

$$R_m = 50 \| (50/999) = \frac{(50)(50/999)}{50 + (50/999)} = 0.05 \ \Omega.$$

EXAMPLE 3.9   Using a d'Arsonval Voltmeter

a) A 50 mV, 1 mA d'Arsonval movement is to be used in a voltmeter in which the full-scale reading is 150 V. Determine R_v.

b) Repeat (a) for a full-scale reading of 5 V.

c) How much resistance does the 150 V meter insert into the circuit?

d) Repeat (c) for the 5 V meter.

Solution

a) Look at the analog voltmeter circuit in Fig. 3.27. The voltage across the voltmeter must divide between the resistor R_v and the meter movement. From the problem statement we know that when the voltage across the voltmeter is 150 V, the voltage across the meter coil must be 50 mV. The remaining 149.95 V must be the voltage across R_v. We also know that when the movement's voltage drop is 50 mV, its current is 1 mA, which is also the current in R_v. Using Ohm's law,

$$R_v = \frac{149.95}{0.001} = 149,950 \ \Omega.$$

b) For a full-scale reading of 5 V, the voltage across R_v is 4.95 V and the current in R_v is still 1 mA, so from Ohm's law,

$$R_v = \frac{4.95}{0.001} = 4950 \ \Omega.$$

c) Let R_m represent the equivalent resistance of the voltmeter. When the voltage across the voltmeter is 150 V, its current is 1 mA, so from Ohm's law,

$$R_m = \frac{150}{10^{-3}} = 150,000 \ \Omega.$$

Alternatively, the resistance of the voltmeter is the equivalent resistance of R_v in series with the meter movement. The resistance of the meter movement is the ratio of its voltage to its current, or $50 \ \text{m V}/1 \ \text{mA} = 50 \ \Omega$. Therefore,

$$R_m = 149,950 + 50 = 150,000 \ \Omega.$$

d) For the 5 V voltmeter,

$$R_m = \frac{5}{10^{-3}} = 5000 \ \Omega,$$

or, alternatively,

$$R_m = 4950 + 50 = 5000 \ \Omega.$$

ASSESSMENT PROBLEMS

Objective 4—Be able to determine the reading of ammeters and voltmeters

3.5   a) Find the current in the circuit shown.
    b) If the ammeter in Example 3.8(a) is used to measure the current, what will it read?

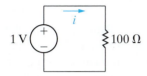

3.6   a) Find the voltage v across the 75 kΩ resistor in the circuit shown.
    b) If the 150 V voltmeter of Example 3.9(a) is used to measure the voltage, what will be the reading?

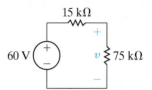

Answer: (a) 10 mA;
        (b) 9.524 mA.

Answer: (a) 50 V;
        (b) 46.15 V.

SELF-CHECK: Also try Chapter Problems 3.33 and 3.37.

Digital Meters

Digital meters measure the continuous voltage or current signal at discrete points in time, called the sampling times. The signal is thus converted from an analog signal, which is continuous in time, to a digital signal, which exists only at discrete instants in time. A more detailed explanation of the workings of digital meters is beyond the scope of this text and course. However, you are likely to see and use digital meters in lab settings because they offer several advantages over analog meters. They introduce less resistance into the circuit to which they are connected (though they are still nonideal), are easier to connect, and take more precise measurements owing to the nature of their readout mechanism.

3.6 Measuring Resistance— The Wheatstone Bridge

While many different circuit configurations are used to measure resistance, here we will focus on just one, the Wheatstone bridge. The Wheatstone bridge circuit is used to precisely measure resistances of medium values, that is, in the range of 1 Ω to 1 MΩ. In commercial models of the Wheatstone bridge, accuracies on the order of ±0.1% are possible. The bridge circuit, shown in Fig. 3.28, consists of four resistors, a dc voltage source, and a detector. The resistance of one of the four resistors can be varied, which is indicated by the arrow through R_3. The dc voltage source is usually a battery, which is indicated by the battery symbol for the voltage source v in Fig. 3.28. The detector is generally a d'Arsonval movement in the microamp range and is called a *galvanometer*. In Fig. 3.28, R_1, R_2, and R_3 are known resistors and R_x is the unknown resistor.

To find the value of R_x, we adjust the variable resistor R_3 until there is no current in the galvanometer. We then calculate the unknown resistor from the simple expression

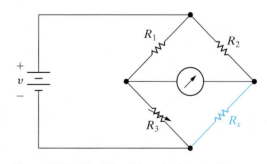

Figure 3.28 ▲ The Wheatstone bridge circuit.

$$R_x = \frac{R_2}{R_1} R_3. \qquad (3.11)$$

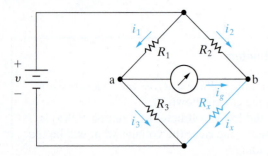

Figure 3.29 ▲ A balanced Wheatstone bridge ($i_g = 0$).

We derive Eq. 3.11 by applying Kirchhoff's laws to the bridge circuit. We redraw the bridge circuit as Fig. 3.29 to show the branch currents in the bridge. When i_g is zero, we say the bridge is *balanced*. At node a, Kirchhoff's current law requires that

$$i_1 = i_3,$$

while at node b, Kirchhoff's current law requires that

$$i_2 = i_x.$$

Now, because i_g is zero, the voltage drop across the detector is also zero, so nodes a and b are at the same potential. Thus, when the bridge is balanced, Kirchhoff's voltage law for the clockwise path containing the galvanometer and resistors R_3 and R_x gives

$$0 + R_x i_x - R_3 i_3 = 0$$

so

$$i_3 R_3 = i_x R_x.$$

Using Kirchhoff's voltage law for the path containing the galvanometer and resistors R_1 and R_2 gives

$$i_1 R_1 = i_2 R_2.$$

Divide the first KVL equation by the second KVL equation to give

$$\frac{i_3 R_3}{i_1 R_1} = \frac{i_x R_x}{i_2 R_2}.$$

Eliminate the currents (because $i_1 = i_3$ and $i_2 = i_x$) and solve for R_x to get Eq. 3.11.

Now that we have verified the validity of Eq. 3.11, some comments about the result are in order. First, note that if $R_2/R_1 = 1$, the unknown resistor R_x equals R_3, so R_3 must vary over a range that includes the value R_x. For example, if the unknown resistance were 1000 Ω and R_3 could be varied from 0 to 100 Ω, the bridge could never be balanced. Thus, to cover a wide range of unknown resistors, we must be able to vary the ratio R_2/R_1. In a commercial Wheatstone bridge, R_1 and R_2 consist of decimal values of resistances that can be switched into the bridge circuit. Normally, the decimal values are 1, 10, 100, and 1000 Ω, so that the ratio R_2/R_1 can be varied from 0.001 to 1000 in decimal steps. The variable resistor R_3 is usually adjustable in integral values of resistance from 1 to 11,000 Ω.

Second, although Eq. 3.11 implies that R_x can vary from zero to infinity, the practical range of R_x is approximately 1 Ω to 1 MΩ. Resistances smaller than 1 Ω are difficult to measure on a standard Wheatstone bridge because of thermoelectric voltages generated at the junctions of dissimilar metals and because of thermal heating effects—that is, $i^2 R$ effects. Resistances larger than 1 Ω are difficult to measure accurately because of leakage currents. In other words, if R_x is large, the current leakage in the electrical insulation may be comparable to the current in the branches of the bridge circuit.

Example 3.10 uses a Wheatstone bridge to measure a range of unknown resistors.

| **EXAMPLE 3.10** | **Using a Wheatstone Bridge to Measure Resistance** |

For the Wheatstone bridge in Fig. 3.30, R_3 can be varied from 10 Ω to 2 kΩ. What range of resistor values can this bridge measure?

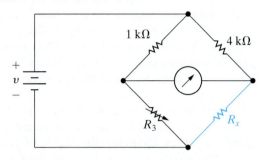

Figure 3.30 ▲ The circuit for Example 3.10.

Solution

When $R_3 = 10\ \Omega$, the bridge is balanced when

$$R_x = \frac{4000}{1000}(10) = 40\ \Omega.$$

When $R_3 = 2$ kΩ, the bridge is balanced when

$$R_x = \frac{4000}{1000}(2000) = 8\ \text{k}\Omega.$$

Therefore, the range of resistor values the bridge can measure is 40 Ω to 8 kΩ.

⌐⌐ ASSESSMENT PROBLEM

Objective 5—Understand how a Wheatstone bridge is used to measure resistance

3.7 The bridge circuit shown is balanced when $R_1 = 100\ \Omega$, $R_2 = 1000\ \Omega$, and $R_3 = 150\ \Omega$. The bridge is energized from a 5 V dc source.

a) What is the value of R_x?

b) Suppose each bridge resistor is capable of dissipating 250 mW. Can the bridge be left in the balanced state without exceeding the power-dissipating capacity of the resistors, thereby damaging the bridge?

SELF-CHECK: Also try Chapter Problem 3.51.

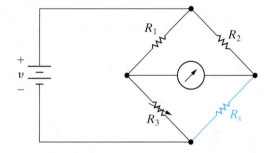

Answer: a) 1500 Ω;
b) yes.

3.7 Delta-to-Wye (Pi-to-Tee) Equivalent Circuits

Look again at the Wheatstone bridge in Fig. 3.28; if we replace the galvanometer with its equivalent resistance R_m, we get the circuit shown in Fig. 3.31(p. 76). Because this circuit does not have any series-connected or parallel-connected resistors, we cannot simplify it using the simple series or parallel equivalent circuits introduced earlier in this chapter. But the five interconnected resistors can be reduced to a single equivalent resistor using a delta-to-wye (Δ-to-Y) or pi-to-tee (π-to-T) equivalent circuit.[1]

---
[1] Δ and Y structures are present in a variety of useful circuits, not just resistive networks. Hence, the Δ-to-Y transformation is a helpful tool in circuit analysis.

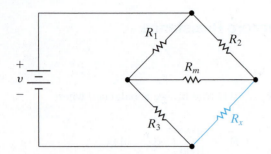

Figure 3.31 ▲ A resistive network generated by a Wheatstone bridge circuit.

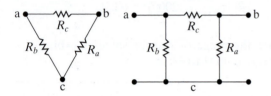

Figure 3.32 ▲ A Δ configuration viewed as a π configuration.

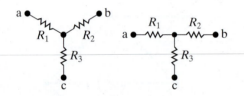

Figure 3.33 ▲ A Y structure viewed as a T structure.

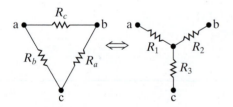

Figure 3.34 ▲ The Δ-to-Y transformation.

The resistors R_1, R_2, and R_m (or R_3, R_m and R_x) in the circuit shown in Fig. 3.31 form a **delta (Δ) interconnection** because the interconnection looks like the Greek letter Δ. It is also called a **pi interconnection** because the Δ can be reshaped like the Greek letter π without disturbing the electrical equivalence of the two configurations, as shown in Fig. 3.32.

The resistors R_1, R_m, and R_3 (or R_2, R_m and R_x) in the circuit shown in Fig. 3.31 form a **wye (Y) interconnection** because the interconnection can be shaped to look like the letter Y. It is easier to see the Y shape of the interconnection in Fig. 3.33. The Y configuration is also called a **tee (T) interconnection** because the Y structure can be reshaped into a T structure without disturbing the electrical equivalence of the two structures, as shown in Fig. 3.33.

Figure 3.34 illustrates the Δ-to-Y (or π-to-T) equivalent circuit transformation. But we cannot transform the Δ interconnection into the Y interconnection simply by changing the shape of the interconnections. Saying the Δ-connected circuit is equivalent to the Y-connected circuit means that the Δ configuration can be replaced with a Y configuration without changing the terminal behavior. Thus, if each circuit is placed in a black box, we can't tell whether the box contains a set of Δ-connected resistors or a set of Y-connected resistors by making external measurements. This condition is true only if the resistance between corresponding terminal pairs is the same for each box. For example, the resistance between terminals a and b must be the same whether we use the Δ-connected set or the Y-connected set. For each pair of terminals in the Δ-connected circuit, compute the equivalent resistance using series and parallel simplifications. For each pair of terminals in the Y-connected circuit, compute the equivalent resistance using only series simplification. The three equivalent resistance equations are

$$R_{ab} = \frac{R_c(R_a + R_b)}{R_a + R_b + R_c} = R_1 + R_2, \tag{3.12}$$

$$R_{bc} = \frac{R_a(R_b + R_c)}{R_a + R_b + R_c} = R_2 + R_3, \tag{3.13}$$

$$R_{ca} = \frac{R_b(R_c + R_a)}{R_a + R_b + R_c} = R_1 + R_3. \tag{3.14}$$

Straightforward algebraic manipulation of Eqs. 3.12–3.14 gives values for the Y-connected resistors in terms of the Δ-connected resistors. Use these equations when transforming three Δ-connected resistors into an equivalent Y connection:

$$R_1 = \frac{R_b R_c}{R_a + R_b + R_c}, \tag{3.15}$$

$$R_2 = \frac{R_c R_a}{R_a + R_b + R_c}, \tag{3.16}$$

$$R_3 = \frac{R_a R_b}{R_a + R_b + R_c}. \tag{3.17}$$

Reversing the Δ-to-Y transformation also is possible, so we can start with a Y structure and replace it with an equivalent Δ structure. The expressions for the three Δ-connected resistors as functions of the three Y-connected resistors are

$$R_a = \frac{R_1R_2 + R_2R_3 + R_3R_1}{R_1}, \quad (3.18)$$

$$R_b = \frac{R_1R_2 + R_2R_3 + R_3R_1}{R_2}, \quad (3.19)$$

$$R_c = \frac{R_1R_2 + R_2R_3 + R_3R_1}{R_3}. \quad (3.20)$$

Example 3.11 uses a Δ-to-Y transformation to simplify a circuit and its analysis.

| **EXAMPLE 3.11** | **Applying a Delta-to-Wye Transform** |

Find the current and power supplied by the 40 V source in the circuit shown in Fig. 3.35.

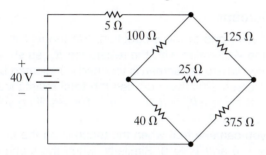

Figure 3.35 ▲ The circuit for Example 3.11.

Solution

This problem is easy to solve if we can find the equivalent resistance seen by the source. Begin this simplification by replacing either the upper Δ (100, 125, 25 Ω) or the lower Δ (40, 25, 37.5 Ω) with its equivalent Y. We choose to replace the upper Δ by computing the three Y resistances, defined in Fig. 3.36, using Eqs. 3.15 to 3.17. Thus,

$$R_1 = \frac{100 \times 125}{250} = 50\ \Omega,$$

$$R_2 = \frac{125 \times 25}{250} = 12.5\ \Omega,$$

$$R_3 = \frac{100 \times 25}{250} = 10\ \Omega.$$

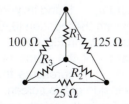

Figure 3.36 ▲ The equivalent Y resistors.

Substituting the Y resistors into the circuit shown in Fig. 3.35 produces the circuit shown in Fig. 3.37. From Fig. 3.37 we can easily calculate the resistance seen by the 40 V source using series-parallel simplifications:

$$R_{eq} = 5 + 50 + (10 + 40) \| (12.5 + 37.5)$$

$$= 55 + \frac{(50)(50)}{50 + 50} = 80\ \Omega.$$

The circuit simplifies to an 80 Ω resistor across a 40 V source, as shown in Fig. 3.38, so the 40 V source delivers current $i = 40/80 = 0.5$ A and power $p = 40(0.5) = 20$ W is delivered to the circuit.

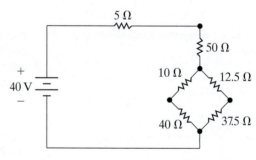

Figure 3.37 ▲ A transformed version of the circuit shown in Fig. 3.35.

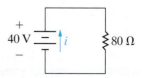

Figure 3.38 ▲ The final step in the simplification of the circuit shown in Fig. 3.35.

■ ASSESSMENT PROBLEM

Objective 6—Know when and how to use delta-to-wye equivalent circuits

3.8   Use a Y-to-Δ transformation to find the voltage v in the circuit shown.

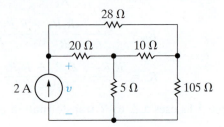

Answer: 35 V.

SELF-CHECK: Also try Chapter Problems 3.59, 3.60, and 3.63.

■ Practical Perspective

Resistive Touch Screens

Let's analyze the resistive grid in the *x*-direction. We model the resistance of the grid in the *x*-direction with the resistance R_x, as shown in Fig. 3.39. The *x*-location where the screen is touched is indicated by the arrow. Touching the screen effectively divides the total resistance, R_x, into two separate resistances αR_x and $(1 - \alpha)R_x$. The resulting voltage drop across the resistance αR_x is V_x.

From the figure you can see that when the touch is on the far right side of the screen, $\alpha = 0$ and $V_x = 0$. Similarly, when the touch is on the far left side of the screen, $\alpha = 1$ and $V_x = V_s$. If the touch is inbetween the two edges of the screen, the value of α is between 0 and 1 and the two parts of the resistance R_x form a voltage divider. We can calculate the voltage V_x using the equation for voltage division:

$$V_x = \frac{\alpha R_x}{\alpha R_x + (1 - \alpha)R_x} V_s = \frac{\alpha R_x}{R_x} V_s = \alpha V_s.$$

We can find the value of α, which represents the location of the touch point with respect to the far right side of the screen, by dividing the voltage across the grid resistance starting at the touch point, V_x, by the voltage applied across the entire resistive grid in the *x*-direction, V_s:

$$\alpha = \frac{V_x}{V_s}.$$

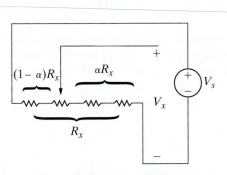

Figure 3.39 ▲ The resistive touch screen grid in the *x*-direction.

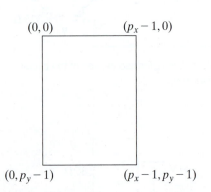

Figure 3.40 ▲ The pixel coordinates of a screen with p_x pixels in the *x*-direction and p_y pixels in the *y*-direction.

Now we want to use the value of α to determine the *x*-coordinate of the touch location on the screen. Typically, the screen coordinates are specified in terms of pixels (short for "picture elements"). For example, the screen of a mobile phone is a grid of pixels, with p_x pixels in the *x*-direction and p_y pixels in the *y*-direction. Each pixel is identified by its *x*-location (a number between 0 and $p_x - 1$) and its *y*-location (a number between 0 and $p_y - 1$). The pixel with the location (0, 0) is in the upper-left-hand corner of the screen, as shown in Fig. 3.40.

Since α represents the location of the touch point with respect to the right side of the screen, $(1 - \alpha)$ represents the location of the touch point with respect to the left side of the screen. Therefore, the *x*-coordinate of the pixel corresponding to the touch point is

$$x = (1 - \alpha)p_x.$$

Note that the value of x is capped at $(p_x - 1)$.

Using the model of the resistive screen grid in the y-direction shown in Fig. 3.41, it is easy to show that the voltage created by a touch at the arrow is given by

$$V_y = \beta V_s.$$

Therefore, the y-coordinate of the pixel corresponding to the touch point is

$$y = (1 - \beta)p_y,$$

where the value of y is capped at $(p_y - 1)$. (See Problem 3.72.)

SELF-CHECK: Assess your understanding of the Practical Perspective by solving Chapter Problems 3.72–3.75.

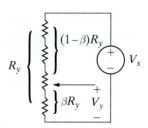

Figure 3.41 ▲ The resistive touch screen grid in the y-direction.

Summary

- **Series resistors** can be combined to obtain a single equivalent resistance according to the equation

$$R_{eq} = \sum_{i=1}^{k} R_i = R_1 + R_2 + \cdots + R_k.$$

(See page 60.)

- **Parallel resistors** can be combined to obtain a single equivalent resistance according to the equation

$$\frac{1}{R_{eq}} = \sum_{i=1}^{k} \frac{1}{R_i} = \frac{1}{R_1} + \frac{1}{R_2} + \cdots + \frac{1}{R_k}.$$

When just two resistors are in parallel, the equation for equivalent resistance can be simplified to give

$$R_{eq} = \frac{R_1 R_2}{R_1 + R_2}.$$

(See pages 61–62.)

- When voltage is divided between series-connected resistors, as shown in the figure, the voltage across each resistor can be found according to the equations

$$v_1 = \frac{R_1}{R_1 + R_2} v_s,$$

$$v_2 = \frac{R_2}{R_1 + R_2} v_s.$$

(See page 64.)

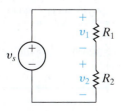

- When current is divided between parallel-connected resistors, as shown in the figure, the current in each resistor can be found according to the equations

$$i_1 = \frac{R_2}{R_1 + R_2} i_s.$$

$$i_2 = \frac{R_1}{R_1 + R_2} i_s.$$

(See page 67.)

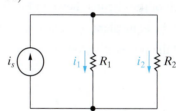

- **Voltage division** is a circuit analysis tool used to find the voltage drop across a single resistance from a collection of series-connected resistances when the voltage drop across the collection is known:

$$v_j = \frac{R_j}{R_{eq}} v,$$

where v_j is the voltage drop across the resistance R_j and v is the voltage drop across the series-connected resistances whose equivalent resistance is R_{eq}. (See page 68.)

- **Current division** is a circuit analysis tool used to find the current through a single resistance from a collection of parallel-connected resistances when the current into the collection is known:

$$i_j = \frac{R_{eq}}{R_j} i,$$

where i_j is the current through the resistance R_j and i is the current into the parallel-connected resistances whose equivalent resistance is R_{eq}. (See page 69.)

- A **voltmeter** measures voltage and must be placed in parallel with the voltage being measured. An ideal voltmeter has infinite internal resistance and thus does not alter the voltage being measured. (See page 70.)

- An **ammeter** measures current and must be placed in series with the current being measured. An ideal ammeter has zero internal resistance and thus does not alter the current being measured. (See page 70.)

- **Digital meters** and **analog meters** have internal resistance, which influences the value of the circuit variable being measured. Meters based on the d'Arsonval meter

movement deliberately include internal resistance as a way to limit the current in the movement's coil. (See pages 71–73.)

- The **Wheatstone bridge** circuit is used to make precise measurements of a resistor's value using four resistors, a dc voltage source, and a galvanometer. A Wheatstone bridge is balanced when the resistors obey Eq. 3.11, resulting in a galvanometer reading of 0 A. (See pages 73–74.)

- A circuit with three resistors connected in a Δ configuration (or a π configuration) can be transformed into an equivalent circuit in which the three resistors are Y connected (or T connected). The Δ-to-Y transformation is given by Eqs. 3.15–3.17; the Y-to-Δ transformation is given by Eqs. 3.18–3.20. (See pages 75–77.)

Problems

Sections 3.1–3.2

3.1 For each of the circuits shown in Fig. P3.1,

  a) identify the resistors connected in series,

  b) simplify the circuit by replacing the series-connected resistors with equivalent resistors.

3.2 For each of the circuits shown in Fig. P3.1,

  a) find the equivalent resistance seen by the source,

  b) find the power developed by the source.

3.3 For each of the circuits shown in Fig. P3.3,

  a) identify the resistors connected in parallel,

  b) simplify the circuit by replacing the parallel-connected resistors with equivalent resistors.

3.4 For each of the circuits shown in Fig. P3.3,

  a) find the equivalent resistance seen by the source,

  b) find the power developed by the source.

Figure P3.1

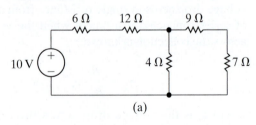

(a)

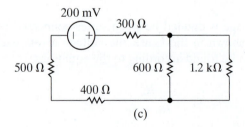

(c)

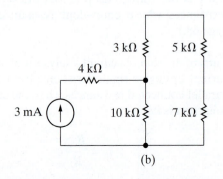

(b)

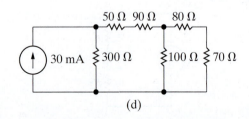

(d)

Figure P3.3

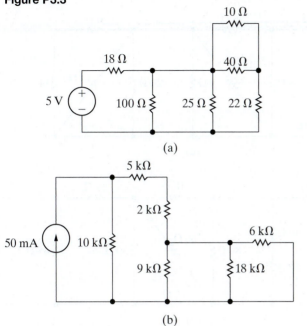

(a)

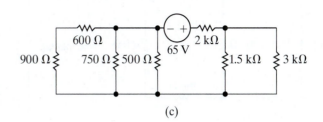

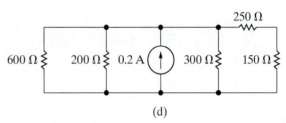

(c)

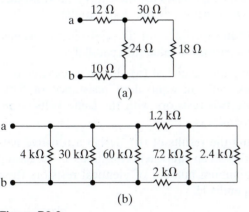

(b)

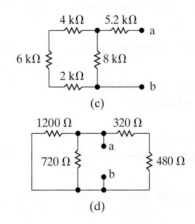

(d)

3.5 Find the equivalent resistance R_{ab} for each of the circuits in Fig. P3.5.

PSPICE
MULTISIM

3.6 Find the equivalent resistance R_{ab} for each of the circuits in Fig. P3.6.

PSPICE
MULTISIM

Figure P3.5

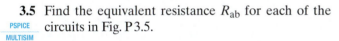

(a)   (c)

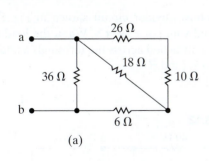

(b)   (d)

Figure P3.6

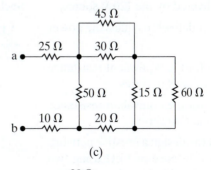

(a)   (c)

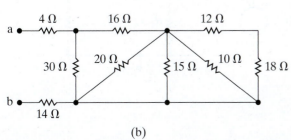

(b)   (d)

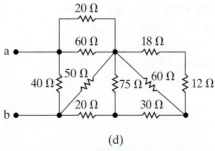

Figure P3.7

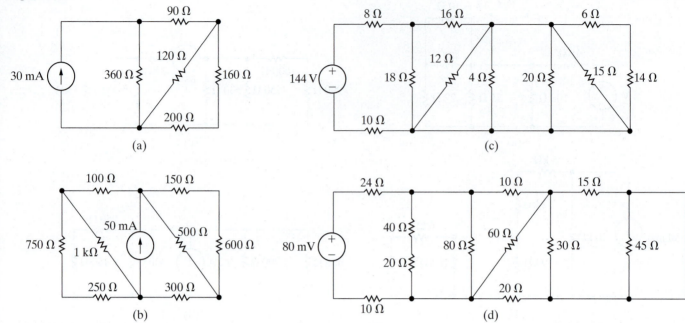

(a)

(b)

(c)

(d)

3.7 a) In the circuits in Fig. P3.7(a)–(d), find the equivalent resistance seen by the source.

   b) For each circuit find the power delivered by the source.

3.8 a) Show that the solution of the circuit in Fig. 3.11 (see Example 3.1) satisfies Kirchhoff's current law at junctions x and y.

   b) Show that the solution of the circuit in Fig. 3.11 satisfies Kirchhoff's voltage law around every closed loop.

3.9 a) Find the power dissipated in each resistor in the circuit shown in Fig. 3.11.

   b) Find the power delivered by the 120 V source.

   c) Show that the power delivered equals the power dissipated.

3.10 a) Find an expression for the equivalent resistance of two resistors of value R in series.

   b) Find an expression for the equivalent resistance of n resistors of value R in series.

   c) Using the results of (a), design a resistive network with an equivalent resistance of $3\,\text{k}\Omega$ using two resistors with the same value from Appendix H.

   d) Using the results of (b), design a resistive network with an equivalent resistance of $4\,\text{k}\Omega$ using a minimum number of identical resistors from Appendix H.

3.11 a) Find an expression for the equivalent resistance of two resistors of value R in parallel.

   b) Find an expression for the equivalent resistance of n resistors of value R in parallel.

   c) Using the results of (a), design a resistive network with an equivalent resistance of $5\,\text{k}\Omega$ using two resistors with the same value from Appendix H.

   d) Using the results of (b), design a resistive network with an equivalent resistance of $4\,\text{k}\Omega$ using a minimum number of identical resistors from Appendix H.

Section 3.3

3.12 In the voltage-divider circuit shown in Fig. P3.12, the no-load value of v_o is 4 V. When the load resistance R_L is attached across the terminals a and b, v_o drops to 3 V. Find R_L.

Figure P3.12

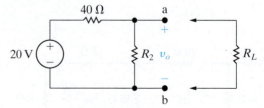

3.13 a) Calculate the no-load voltage v_o for the voltage-divider circuit shown in Fig. P3.13.

DESIGN PROBLEM
PSPICE
MULTISIM

b) Calculate the power dissipated in R_1 and R_2.

c) Assume that only 1 W resistors are available. The no-load voltage is to be the same as in (a). Specify the smallest ohmic values of R_1 and R_2.

Figure P3.13

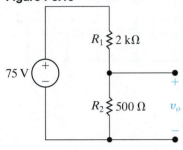

3.14 The no-load voltage in the voltage-divider circuit shown in Fig. P3.14 is 8 V. The smallest load resistor that is ever connected to the divider is 3.6 kΩ. When the divider is loaded, v_o is not to drop below 7.5 V.

DESIGN PROBLEM
PSPICE
MULTISIM

a) Design the divider circuit to meet the specifications just mentioned. Specify the numerical values of R_1 and R_2.

b) Assume the power ratings of commercially available resistors are $1/16, 1/8, 1/4, 1,$ and 2 W. What power rating would you specify?

Figure P3.14

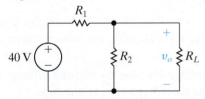

3.15 Assume the voltage divider in Fig. P3.14 has been constructed from 1 W resistors. What is the smallest resistor from Appendix H that can be used as R_L before one of the resistors in the divider is operating at its dissipation limit?

3.16 a) The voltage divider in Fig. P3.16(a) is loaded with the voltage divider shown in Fig. P3.16(b); that is, a is connected to a', and b is connected to b'. Find v_o.

PSPICE
MULTISIM

b) Now assume the voltage divider in Fig. P3.16(b) is connected to the voltage divider in Fig. P3.16(a) by means of a current-controlled voltage source as shown in Fig. P3.16(c). Find v_o.

c) What effect does adding the dependent-voltage source have on the operation of the voltage divider that is connected to the 240 V source?

Figure P3.16

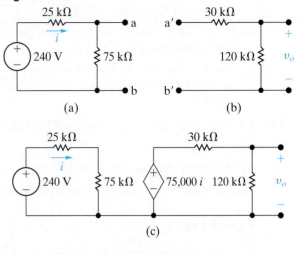

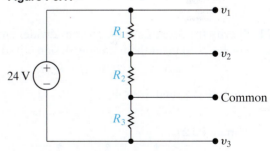

3.17 There is often a need to produce more than one voltage using a voltage divider. For example, the memory components of many personal computers require voltages of −12 V, 5 V, and +12 V, all with respect to a common reference terminal. Select the values of $R_1, R_2,$ and R_3 in the circuit in Fig. P3.17 to meet the following design requirements:

DESIGN PROBLEM

a) The total power supplied to the divider circuit by the 24 V source is 80 W when the divider is unloaded.

b) The three voltages, all measured with respect to the common reference terminal, are $v_1 = 12$ V, $v_2 = 5$ V, and $v_3 = -12$ V.

Figure P3.17

3.18 A voltage divider like that in Fig. 3.17 is to be designed so that $v_o = kv_s$ at no load ($R_L = \infty$) and $v_o = \alpha v_s$ at full load ($R_L = R_o$). Note that by definition $\alpha < k < 1$.

DESIGN PROBLEM

a) Show that

$$R_1 = \frac{k - \alpha}{\alpha k} R_o$$

and

$$R_2 = \frac{k - \alpha}{\alpha(1 - k)} R_o.$$

b) Specify the numerical values of R_1 and R_2 if $k = 0.85$, $\alpha = 0.80$, and $R_o = 34\ \text{k}\Omega$.

c) If $v_s = 60\ \text{V}$, specify the maximum power that will be dissipated in R_1 and R_2.

d) Assume the load resistor is accidentally short circuited. How much power is dissipated in R_1 and R_2?

3.19 For the current-divider circuit in Fig. P3.19 calculate

PSPICE
MULTISIM

    a) i_o and v_o.

    b) the power dissipated in the 6 Ω resistor.

    c) the power developed by the current source.

Figure P3.19

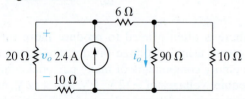

3.20 Find the power dissipated in the 30 Ω resistor in the

PSPICE
MULTISIM

current-divider circuit in Fig. P3.20.

Figure P3.20

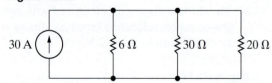

3.21 Specify the resistors in the current-divider circuit in

DESIGN PROBLEM

Fig. P3.21 to meet the following design criteria:

$$i_g = 50\ \text{mA};\ v_g = 25\ \text{V};\ i_1 = 0.6i_2;$$

$$i_3 = 2i_2;\ \text{and}\ i_4 = 4i_1.$$

Figure P3.21

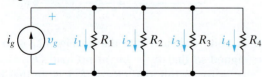

3.22 a) Show that the current in the kth branch of the

PSPICE
MULTISIM

circuit in Fig. P3.22(a) is equal to the source current i_g times the conductance of the kth branch divided by the sum of the conductances, that is,

$$i_k = \frac{i_g G_k}{G_1 + G_2 + G_3 + \cdots + G_k + \cdots + G_N}.$$

    b) Use the result derived in (a) to calculate the current in the 5 Ω resistor in the circuit in Fig. P3.22(b).

Figure P3.22

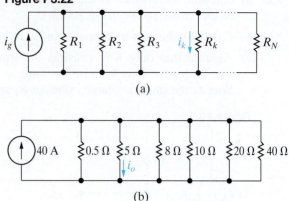

(a)

(b)

Section 3.4

3.23 Look at the circuit in Fig. P3.1(a).

    a) Use voltage division to find the voltage across the 4 Ω resistor, positive at the top.

    b) Use the result from part (a) and voltage division to find the voltage across the 9 Ω resistor, positive on the left.

3.24 Look at the circuit in Fig. P3.1(d).

    a) Use current division to find the current in the 50 Ω resistor from left to right.

    b) Use the result from part (a) and current division to find the current in the 70 Ω resistor from top to bottom.

3.25 Attach a 6 V voltage source between the terminals a–b in Fig. P3.6(b), with the positive terminal at the top.

    a) Use voltage division to find the voltage across the 4 Ω resistor, positive at the top.

    b) Use the result from part (a) to find the current in the 4 Ω resistor from left to right.

    c) Use the result from part (b) and current division to find the current in the 16 Ω resistor from left to right.

    d) Use the result from part (c) and current division to find the current in the 10 Ω resistor from top to bottom.

    e) Use the result from part (d) to find the voltage across the 10 Ω resistor, positive at the top.

    f) Use the result from part (e) and voltage division to find the voltage across the 18 Ω resistor, positive at the top.

3.26 Look at the circuit in Fig. P3.7(a).

    a) Use current division to find the current in the 360 Ω resistor from top to bottom.

    b) Using your result from (a), find the voltage drop across the 360 Ω resistor, positive at the top.

c) Starting with your result from (b), use voltage division to find the voltage across the 90 Ω resistor, positive on the left.

d) Using your result from (c), find the current in the 90 Ω resistor from left to right.

e) Starting with your result from (d), use current division to find the current in the 160 Ω resistor from top to bottom.

3.27 Attach a 450 mA current source between the terminals a–b in Fig. P3.6(a), with the current arrow pointing up.

a) Use current division to find the current in the 36 Ω resistor from top to bottom.

b) Use the result from part (a) to find the voltage across the 36 Ω resistor, positive at the top.

c) Use the result from part (b) and voltage division to find the voltage across the 18 Ω resistor, positive at the top.

d) Use the result from part (c) and voltage division to find the voltage across the 10 Ω resistor, positive at the top.

3.28 For the circuit in Fig. P3.28, find i_g and then use current division to find i_o.

Figure P3.28

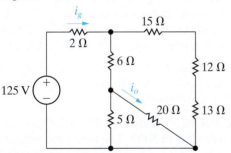

3.29 For the circuit in Fig. P3.29, calculate i_1 and i_2 using current division.

PSPICE
MULTISIM

Figure P3.29

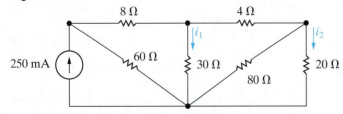

3.30 Find v_1 and v_2 in the circuit in Fig. P3.30 using voltage and/or current division.

PSPICE
MULTISIM

Figure P3.30

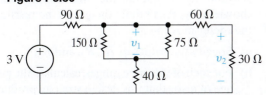

3.31 Find v_o in the circuit in Fig. P3.31 using voltage and/or current division.

PSPICE
MULTISIM

Figure P3.31

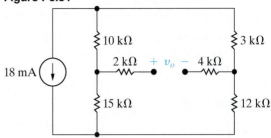

3.32 a) Find the voltage v_x in the circuit in Fig. P3.32 using voltage and/or current division.

PSPICE
MULTISIM

b) Replace the 45 V source with a general voltage source equal to V_s. Assume V_s is positive at the upper terminal. Find v_x as a function of V_s.

Figure P3.32

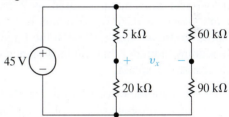

Section 3.5

3.33 A shunt resistor and a 50 mV, 1 mA d'Arsonval movement are used to build a 5 A ammeter. A resistance of 20 mΩ is placed across the terminals of the ammeter. What is the new full-scale range of the ammeter?

3.34 a) Show for the ammeter circuit in Fig. P3.34 that the current in the d'Arsonval movement is always 1/100th of the current being measured.

b) What would the fraction be if the 100 μV, 10 μA movement were used in a 1 A ammeter?

c) Would you expect a uniform scale on a dc d'Arsonval ammeter?

Figure P3.34

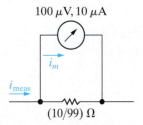

3.35 A d'Arsonval ammeter is shown in Fig. P3.35.

a) Calculate the value of the shunt resistor, R_A, to give a full-scale current reading of 5 A.

b) How much resistance is added to a circuit when the 5 A ammeter in part (a) is inserted to measure current?

c) Calculate the value of the shunt resistor, R_A, to give a full-scale current reading of 100 mA.

d) How much resistance is added to a circuit when the 100 mA ammeter in part (c) is inserted to measure current?

Figure P3.35

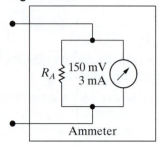

3.36 A d'Arsonval movement is rated at 2 mA and 100 mV. Assume 0.25 W precision resistors are available to use as shunts. What is the largest full-scale-reading ammeter that can be designed using a single resistor? Explain.

DESIGN PROBLEM

3.37 A d'Arsonval voltmeter is shown in Fig. P3.37. Find the value of R_v for each of the following full-scale readings: (a) 50 V, (b) 5 V, (c) 250 mV, and (d) 25 mV.

Figure P3.37

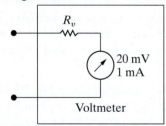

3.38 Suppose the d'Arsonval voltmeter described in Problem 3.37(b) is used to measure the voltage across the 45 Ω resistor in Fig. P3.38.

a) What will the voltmeter read?

b) Find the percentage of error in the voltmeter reading if

$$\% \text{ error} = \left(\frac{\text{measured value}}{\text{true value}} - 1 \right) \times 100.$$

Figure P3.38

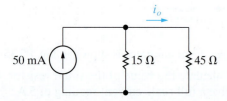

3.39 The ammeter in the circuit in Fig. P3.39 has a resistance of 0.1 Ω. Using the definition of the percentage error in a meter reading found in Problem 3.38, what is the percentage of error in the reading of this ammeter?

Figure P3.39

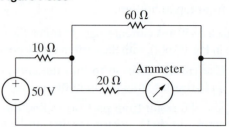

3.40 The ammeter described in Problem 3.39 is used to measure the current i_o in the circuit in Fig. P3.38. What is the percentage of error in the measured value?

3.41 The elements in the circuit in Fig. 2.24 have the following values: $R_1 = 20 \text{ k}\Omega$, $R_2 = 80 \text{ k}\Omega$, $R_C = 0.82 \text{ k}\Omega$, $R_E = 0.2 \text{ k}\Omega$, $V_{CC} = 7.5 \text{ V}$, $V_0 = 0.6 \text{ V}$, and $\beta = 39$.

PSPICE
MULTISIM

a) Calculate the value of i_B in microamperes.

b) Assume that a digital multimeter, when used as a dc ammeter, has a resistance of 1 kΩ. If the meter is inserted between terminals b and 2 to measure the current i_B, what will the meter read?

c) Using the calculated value of i_B in (a) as the correct value, what is the percentage of error in the measurement?

3.42 The voltmeter shown in Fig. P3.42(a) has a full-scale reading of 500 V. The meter movement is rated 100 mV and 0.5 mA. What is the percentage of error in the meter reading if it is used to measure the voltage v in the circuit of Fig. P3.42(b)?

Figure P3.42

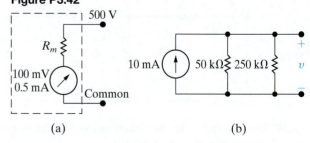

(a)   (b)

3.43 Assume in designing the multirange voltmeter shown in Fig. P3.43 that you ignore the resistance of the meter movement.

DESIGN PROBLEM

a) Specify the values of R_1, R_2, and R_3.

b) For each of the three ranges, calculate the percentage of error that this design strategy produces.

Figure P3.43

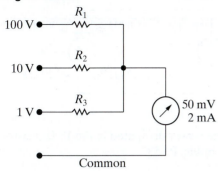

Common

3.44 The voltage-divider circuit shown in Fig. P3.44 is designed so that the no-load output voltage is 7/9ths of the input voltage. A d'Arsonval voltmeter having a sensitivity of $100\ \Omega/V$ and a full-scale rating of 200 V is used to check the operation of the circuit.

a) What will the voltmeter read if it is placed across the 180 V source?

b) What will the voltmeter read if it is placed across the $70\ k\Omega$ resistor?

c) What will the voltmeter read if it is placed across the $20\ k\Omega$ resistor?

d) Will the voltmeter readings obtained in parts (b) and (c) add to the reading recorded in part (a)? Explain why or why not.

Figure P3.44

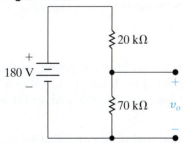

3.45 The circuit model of a dc voltage source is shown in Fig. P3.45. The following voltage measurements are made at the terminals of the source: (1) With the terminals of the source open, the voltage is measured at 50 mV, and (2) with a 15 MΩ resistor connected to the terminals, the voltage is measured at 48.75 mV. All measurements are made with a digital voltmeter that has a meter resistance of 10 MΩ.

Figure P3.45

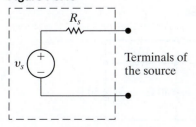

Terminals of the source

a) What is the internal voltage of the source (v_s) in millivolts?

b) What is the internal resistance of the source (R_s) in kiloohms?

3.46 You have been told that the dc voltage of a power supply is about 400 V. When you go to the instrument room to get a dc voltmeter to measure the power supply voltage, you find that there are only two dc voltmeters available. One voltmeter is rated 300 V full scale and has a sensitivity of $1000\ \Omega/V$. The other voltmeter is rated 150 V full scale and has a sensitivity of $800\ \Omega/V$. (*Hint:* you can find the effective resistance of a voltmeter by multiplying its rated full-scale voltage and its sensitivity.)

a) How can you use the two voltmeters to check the power supply voltage?

b) What is the maximum voltage that can be measured?

c) If the power supply voltage is 399 V, what will each voltmeter read?

3.47 Assume that in addition to the two voltmeters described in Problem 3.46, an $80\ k\Omega$ precision resistor is also available. The $80\ k\Omega$ resistor is connected in series with the series-connected voltmeters. This circuit is then connected across the terminals of the power supply. The reading on the 300 V meter is 288 V and the reading on the 150 V meter is 115.2 V. What is the voltage of the power supply?

3.48 Design a d'Arsonval voltmeter that will have the three voltage ranges shown in Fig. P3.48.

DESIGN PROBLEM

a) Specify the values of R_1, R_2, and R_3.

b) Assume that a $750\ k\Omega$ resistor is connected between the 150 V terminal and the common terminal. The voltmeter is then connected to an unknown voltage using the common terminal and the 300 V terminal. The voltmeter reads 288 V. What is the unknown voltage?

c) What is the maximum voltage the voltmeter in (b) can measure?

Figure P3.48

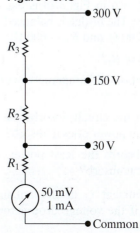

3.49 A 200 kΩ resistor is connected from the 100 V terminal to the common terminal of a dual-scale voltmeter, as shown in Fig. P3.49(a). This modified voltmeter is then used to measure the voltage across the 600 kΩ resistor in the circuit in Fig. P3.49(b).

a) What is the reading on the 820 V scale of the meter?

b) What is the percentage of error in the measured voltage?

Figure P3.49

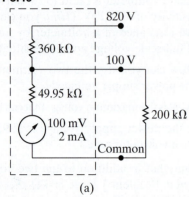

(a)

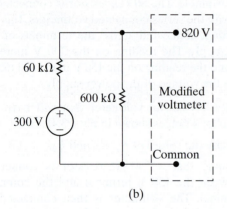

(b)

Section 3.6

3.50 Assume the ideal voltage source in Fig. 3.28 is replaced by an ideal current source. Show that Eq. 3.11 is still valid.

PSPICE
MULTISIM

3.51 The bridge circuit shown in Fig. 3.28 is energized from a 21 V dc source. The bridge is balanced when $R_1 = 800\ \Omega$, $R_2 = 1200\ \Omega$, and $R_3 = 600\ \Omega$.

PSPICE
MULTISIM

a) What is the value of R_x?

b) How much current (in milliamperes) does the dc source supply?

c) Which resistor in the circuit absorbs the most power? How much power does it absorb?

d) Which resistor absorbs the least power? How much power does it absorb?

3.52 Find the detector current i_d in the unbalanced bridge in Fig. P3.52 if the voltage drop across the detector is negligible.

Figure P3.52

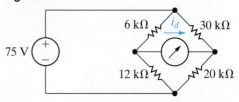

3.53 Find the power dissipated in the 18 Ω resistor in the circuit in Fig. P3.53.

PSPICE
MULTISIM

Figure P3.53

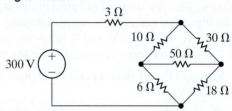

3.54 In the Wheatstone bridge circuit shown in Fig. 3.28, the ratio R_2/R_1 can be set to the following values: 0.001, 0.01, 0.1, 1, 10, 100, and 1000. The resistor R_3 can be varied from 1 to 11,110 Ω, in increments of 1 Ω. An unknown resistor is known to lie between 4 and 5 Ω. What should be the setting of the R_2/R_1 ratio so that the unknown resistor can be measured to four significant figures?

PSPICE
MULTISIM

Section 3.7

3.55 Find the current and power supplied by the 40 V source in the circuit for Example 3.11 (Fig. 3.35) by replacing the lower Δ (25, 37.5, and 40 Ω) with its equivalent Y.

3.56 Find the current and power supplied by the 40 V source in the circuit for Example 3.11 (Fig. 3.35) by replacing the Y on the left (25, 40, and 100 Ω) with its equivalent Δ.

3.57 Find the current and power supplied by the 40 V source in the circuit for Example 3.11 (Fig. 3.35) by replacing the Y on the right (25, 37.5, and 125 Ω) with its equivalent Δ.

3.58 Use a Δ-to-Y transformation to find the voltages v_1 and v_2 in the circuit in Fig. P3.58.

PSPICE
MULTISIM

Figure P3.58

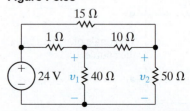

3.59 For the circuit shown in Fig. P3.59, find (a) i_1, (b) v, (c) i_2, and (d) the power supplied by the current source.

PSPICE
MULTISIM

Figure P3.59

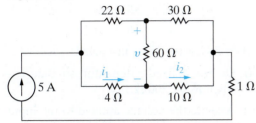

3.60 Find i_o and the power dissipated in the 140 Ω resistor in the circuit in Fig. P3.60.

PSPICE
MULTISIM

Figure P3.60

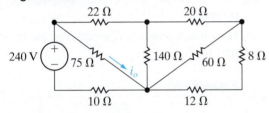

3.61 a) Find the equivalent resistance R_{ab} in the circuit in Fig. P3.61 by using a Y-to-Δ transformation involving resistors R_2, R_3, and R_5.

PSPICE
MULTISIM

    b) Repeat (a) using a Δ-to-Y transformation involving resistors R_3, R_4, and R_5.

    c) Give two additional Δ-to-Y or Y-to-Δ transformations that could be used to find R_{ab}.

Figure P3.61

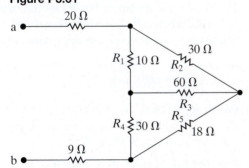

3.62 Use a Y-to-Δ transformation to find (a) i_o; (b) i_1; (c) i_2; and (d) the power delivered by the ideal voltage source in the circuit in Fig. P3.62.

PSPICE
MULTISIM

Figure P3.62

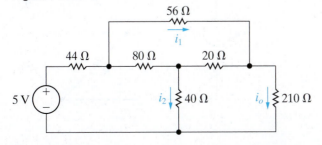

3.63 a) Find the resistance seen by the ideal voltage source in the circuit in Fig. P3.63.

PSPICE
MULTISIM

    b) If v_{ab} equals 400 V, how much power is dissipated in the 31 Ω resistor?

Figure P3.63

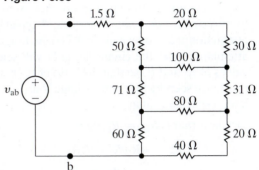

3.64 Show that the expressions for Δ conductances as functions of the three Y conductances are

$$G_a = \frac{G_2 G_3}{G_1 + G_2 + G_3},$$

$$G_b = \frac{G_1 G_3}{G_1 + G_2 + G_3},$$

$$G_c = \frac{G_1 G_2}{G_1 + G_2 + G_3},$$

where

$$G_a = \frac{1}{R_a}, \quad G_1 = \frac{1}{R_1}, \quad \text{etc.}$$

3.65 Derive Eqs. 3.15–3.20 from Eqs. 3.12–3.14. The following two hints should help you get started in the right direction:

1) To find R_1 as a function of R_a, R_b, and R_c, first subtract Eq. 3.13 from Eq. 3.14 and then add this result to Eq. 3.12. Use similar manipulations to find R_2 and R_3 as functions of R_a, R_b, and R_c.

2) To find R_b as a function of R_1, R_2, and R_3, take advantage of the derivations obtained by hint (1), namely, Eqs. 3.15–3.17. Note that these equations can be divided to obtain

$$\frac{R_2}{R_3} = \frac{R_c}{R_b}, \quad \text{or} \quad R_c = \frac{R_2}{R_3} R_b,$$

and

$$\frac{R_1}{R_2} = \frac{R_b}{R_a}, \quad \text{or} \quad R_a = \frac{R_2}{R_1} R_b.$$

Now use these ratios in Eq. 3.14 to eliminate R_a and R_c. Use similar manipulations to find R_a and R_c as functions of R_1, R_2, and R_3.

Sections 3.1–3.7

3.66 Resistor networks are sometimes used as volume-control circuits. In this application, they are referred to as *resistance attenuators* or *pads*. A typical fixed-attenuator pad is shown in Fig. P3.66. In designing an attenuation pad, the circuit designer will select the values of R_1 and R_2 so that the ratio of v_o/v_i and the resistance seen by the input voltage source R_{ab} both have a specified value.

DESIGN PROBLEM

a) Show that if $R_{ab} = R_L$, then

$$R_L^2 = 4R_1(R_1 + R_2),$$

$$\frac{v_o}{v_i} = \frac{R_2}{2R_1 + R_2 + R_L}.$$

b) Select the values of R_1 and R_2 so that $R_{ab} = R_L = 300\,\Omega$ and $v_o/v_i = 0.5$.

c) Choose values from Appendix H that are closest to R_1 and R_2 from part (b). Calculate the percent error in the resulting values for R_{ab} and v_0/v_1 if these new resistor values are used.

Figure P3.66

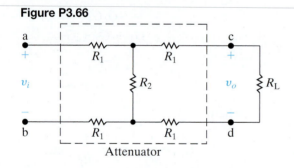

Attenuator

3.67 a) The fixed-attenuator pad shown in Fig. P3.67 is called a *bridged tee*. Use a Y-to-Δ transformation to show that $R_{ab} = R_L$ if $R = R_L$.

DESIGN PROBLEM

b) Show that when $R = R_L$, the voltage ratio v_o/v_i equals 0.50.

Figure P3.67

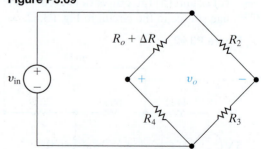

Fixed-attenuator pad

3.68 The design equations for the bridged-tee attenuator circuit in Fig. P3.68 are

PSPICE MULTISIM

$$R_2 = \frac{2RR_L^2}{3R^2 - R_L^2},$$

$$\frac{v_o}{v_i} = \frac{3R - R_L}{3R + R_L},$$

when R_2 has the value just given.

a) Design a fixed attenuator so that $v_i = 3.5v_o$ when $R_L = 300\,\Omega$.

b) Assume the voltage applied to the input of the pad designed in (a) is 42 V. Which resistor in the pad dissipates the most power?

c) How much power is dissipated in the resistor in part (b)?

d) Which resistor in the pad dissipates the least power?

e) How much power is dissipated in the resistor in part (d)?

Figure P3.68

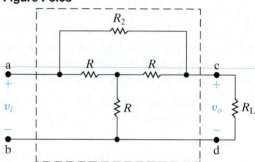

3.69 a) For the circuit shown in Fig. P3.69 the bridge is balanced when $\Delta R = 0$. Show that if $\Delta R \ll R_o$ the bridge output voltage is approximately

PSPICE MULTISIM

$$v_o \approx \frac{-\Delta R R_4}{(R_o + R_4)^2} v_{in}$$

b) Given $R_2 = 1\,k\Omega$, $R_3 = 500\,\Omega$, $R_4 = 5\,k\Omega$, and $v_{in} = 6\,V$, what is the approximate bridge output voltage if ΔR is 3% of R_o?

c) Find the actual value of v_o in part (b).

Figure P3.69

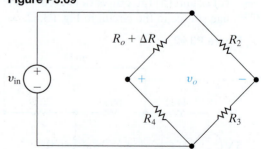

3.70 a) If percent error is defined as

$$\% \text{ error} = \left[\frac{\text{approximate value}}{\text{true value}} - 1 \right] \times 100,$$

show that the percent error in the approximation of v_o in Problem 3.69 is

$$\% \text{ error} = \frac{-(\Delta R)R_3}{(R_2 + R_3)R_4} \times 100.$$

b) Calculate the percent error in v_o, using the values in Problem 3.69(b).

3.71 Assume the error in v_o in the bridge circuit in

DESIGN
PROBLEM
Fig. P3.69 is not to exceed 0.5%. What is the largest percent change in R_o that can be tolerated?

3.72 a) Using Fig. 3.41 derive the expression for the

PRACTICAL
PERSPECTIVE
voltage V_y.

b) Assuming that there are p_y pixels in the y-direction, derive the expression for the y-coordinate of the touch point, using the result from part (a).

3.73 A resistive touch screen has 5 V applied to the grid

PRACTICAL
PERSPECTIVE
in the x-direction and in the y-direction. The screen

PSPICE
has 480 pixels in the x-direction and 800 pixels in

MULTISIM
the y-direction. When the screen is touched, the voltage in the x-grid is 1 V and the voltage in the y-grid is 3.75 V.

a) Calculate the values of α and β.

b) Calculate the x- and y-coordinates of the pixel at the point where the screen was touched.

3.74 A resistive touch screen has 640 pixels in the

PRACTICAL
PERSPECTIVE
x-direction and 1024 pixels in the y-direction.

DESIGN
PROBLEM
The resistive grid has 8 V applied in both the
x- and y-directions. The pixel coordinates at the

PSPICE
touch point are (480, 192). Calculate the voltages

MULTISIM
V_x and V_y.

3.75 Suppose the resistive touch screen described in

PRACTICAL
PERSPECTIVE
Problem 3.74 is simultaneously touched at two points, one with coordinates (480, 192) and the other with coordinates (240, 384).

a) Calculate the voltage measured in the x- and y-grids.

b) Which touch point has your calculation in (a) identified?

4

Techniques of Circuit Analysis

CHAPTER CONTENTS

4.1 **Terminology** *p. 94*

4.2 **Introduction to the Node-Voltage Method** *p. 96*

4.3 **The Node-Voltage Method and Dependent Sources** *p. 98*

4.4 **The Node-Voltage Method: Some Special Cases** *p. 100*

4.5 **Introduction to the Mesh-Current Method** *p. 104*

4.6 **The Mesh-Current Method and Dependent Sources** *p. 107*

4.7 **The Mesh-Current Method: Some Special Cases** *p. 108*

4.8 **The Node-Voltage Method Versus the Mesh-Current Method** *p. 112*

4.9 **Source Transformations** *p. 115*

4.10 **Thévenin and Norton Equivalents** *p. 118*

4.11 **More on Deriving the Thévenin Equivalent** *p. 123*

4.12 **Maximum Power Transfer** *p. 126*

4.13 **Superposition** *p. 129*

CHAPTER OBJECTIVES

1 Understand and be able to use the node-voltage method to solve a circuit.

2 Understand and be able to use the mesh-current method to solve a circuit.

3 Be able to decide whether the node-voltage method or the mesh-current method is the preferred approach to solving a particular circuit.

4 Understand source transformation and be able to use it to solve a circuit.

5 Understand the concept of the Thévenin and Norton equivalent circuits and be able to construct a Thévenin or Norton equivalent for a circuit.

6 Know the condition for maximum power transfer to a resistive load and be able to calculate the value of the load resistor that satisfies this condition.

This chapter introduces two powerful circuit analysis techniques: the *node-voltage method* and the *mesh-current method*. The power of these methods comes from their ability to describe complex circuits using a minimum number of simultaneous equations. The alternative—applying Kirchhoff's laws and Ohm's law—becomes cumbersome for complex circuits.

We also introduce new techniques for simplifying circuits: *source transformations* and *Thévenin* and *Norton equivalent circuits*. Adding these skills to your existing knowledge of series-parallel reductions and Δ-to-Y transformations greatly expands your ability to simplify and solve complex circuits.

Finally, we explore the concepts of *maximum power transfer* and *superposition*. You will learn to use Thévenin equivalent circuits to establish the conditions needed to ensure maximum power is delivered to a resistive load. Superposition helps us analyze circuits that have more than one independent source.

■ Practical Perspective

Circuits with Realistic Resistors

In the last chapter, we examined the effect of imprecise resistor values on the performance of a voltage divider. Resistors are manufactured for only a small number of discrete values, and any given resistor from a batch of resistors will vary from its stated value within some tolerance. Resistors with a tolerance of 1% are more expensive than resistors with a tolerance of 10%. If we know that a particular resistor must be very close to its stated value for the circuit to function correctly, we can then decide to spend the extra money necessary to achieve a tighter tolerance on that resistor's value.

Therefore, we want to predict the effect of varying the value of each resistor in a circuit on the output of that circuit, a technique known as **sensitivity analysis**. Once we have presented additional circuit analysis methods, the topic of sensitivity analysis will be examined.

Resistor color code	
Band color	Value
Black	0
Brown	1
Red	2
Orange	3
Yellow	4
Green	5
Blue	6
Violet	7
Grey	8
White	9
Gold	0.1
Silver	0.01

Tolerance color code	
Band color	±%
Brown	1
Red	2
Gold	5
Silver	10
None	20

Band 1 First figure of value
Band 2 Second figure of value
Band 3 Number of zeros/multiplier
Band 4 Tolerance (±%) See below

Note that the bands are closer to one end than the other

Roomanald/Shutterstock

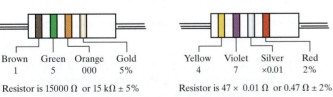

Brown	Green	Orange	Gold
1	5	000	5%

Resistor is 15000 Ω or 15 kΩ ± 5%

Yellow	Violet	Silver	Red
4	7	×0.01	2%

Resistor is 47 × 0.01 Ω or 0.47 Ω ± 2%

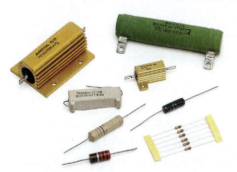

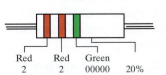

Red	Red	Green	
2	2	00000	20%

Resistor is 2200000 Ω or 2.2 MΩ ± 20%

Brown	Green	Red	Gold
1	5	00	5%

Resistor is 1500 Ω or 1.5 kΩ ± 5%

David J. Green - electrical/Alamy Stock Photo

HL Studios/Pearson Education Ltd

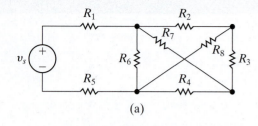

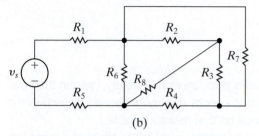

Figure 4.1 ▲ (a) A planar circuit. (b) The same circuit redrawn to verify that it is planar.

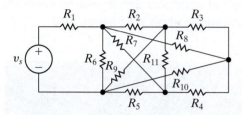

Figure 4.2 ▲ A nonplanar circuit.

4.1 Terminology

Before discussing the node-voltage and mesh-current methods of circuit analysis, we must define a few basic terms. So far, all the circuits presented have been **planar circuits**—that is, those circuits that can be drawn on a plane with no crossing branches. A circuit that is drawn with crossing branches still is considered planar if it can be redrawn with no crossing branches. For example, the circuit shown in Fig. 4.1(a) can be redrawn as Fig. 4.1(b); the circuits are equivalent because all the node connections have been maintained. Therefore, Fig. 4.1(a) is a planar circuit because it can be redrawn as one. Figure 4.2 shows a nonplanar circuit—it cannot be redrawn in such a way that all the node connections are maintained and no branches overlap. Identifying a circuit as planar or nonplanar is important, because

- the node-voltage method is applicable to both planar and nonplanar circuits;
- the mesh-current method is limited to planar circuits.

Describing a Circuit—The Vocabulary

When ideal basic circuit elements (Section 1.5) are interconnected to form a circuit, the resulting interconnection is described in terms of nodes, paths, loops, branches, and meshes. Two of these terms, nodes and loops, were defined in Section 2.4. Now we'll define the rest of these terms. For your convenience, all of these definitions are presented in Table 4.1. The table also includes examples of each definition taken from the circuit in Fig. 4.3, which are developed in Example 4.1.

EXAMPLE 4.1 | **Identifying Node, Branch, Mesh, and Loop in a Circuit**

For the circuit in Fig. 4.3, identify

a) all nodes.

b) all essential nodes.

c) all branches.

d) all essential branches.

e) all meshes.

f) two paths that are not loops or essential branches.

g) two loops that are not meshes.

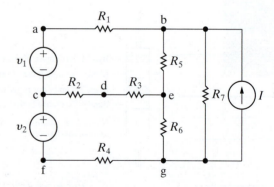

Figure 4.3 ▲ A circuit illustrating nodes, branches, meshes, paths, and loops.

Solution

a) The nodes are a, b, c, d, e, f, and g.

b) The essential nodes are b, c, e, and g.

c) The branches are v_1, v_2, R_1, R_2, R_3, R_4, R_5, R_6, R_7, and I.

d) The essential branches are $v_1 - R_1$, $R_2 - R_3$, $v_2 - R_4$, R_5, R_6, R_7, and I.

e) The meshes are $v_1 - R_1 - R_5 - R_3 - R_2$, $v_2 - R_2 - R_3 - R_6 - R_4$, $R_5 - R_7 - R_6$, and $R_7 - I$.

f) $R_1 - R_5 - R_6$ is a path, but it is not a loop (because it does not have the same starting and ending nodes), nor is it an essential branch (because it does not connect two essential nodes). $v_2 - R_2$ is also a path but is neither a loop nor an essential branch, for the same reasons.

g) $v_1 - R_1 - R_5 - R_6 - R_4 - v_2$ is a loop but is not a mesh because there are two loops within it. $I - R_5 - R_6$ is also a loop but not a mesh.

SELF-CHECK: Assess your understanding of this material by trying Chapter Problem 4.1.

TABLE 4.1	Terms for Describing Circuits	
Name	**Definition**	**Example from Fig. 4.3**
node	A point where two or more circuit elements join	a
essential node	A node where three or more circuit elements join	b
path	A trace of adjoining basic elements with no elements included more than once	$v_1 - R_1 - R_5 - R_6$
branch	A path that connects two nodes	R_1
essential branch	A path that connects two essential nodes without passing through an essential node	$v_1 - R_1$
loop	A path whose last node is the same as the starting node	$v_1 - R_1 - R_5 - R_6 - R_4 - v_2$
mesh	A loop that does not enclose any other loops	$v_1 - R_1 - R_5 - R_3 - R_2$
planar circuit	A circuit that can be drawn on a plane with no crossing branches	Fig. 4.3 is a planar circuit. Fig. 4.2 is a nonplanar circuit.

Simultaneous Equations—How Many?

Recall that we need b independent equations to solve a circuit with b unknown currents. In Fig. 4.3, for example, the circuit has nine branches with unknown currents, so $b = 9$; we need nine independent equations to solve for the unknown currents. Some of the equations can be written by applying Kirchhoff's current law (KCL) to a set of the circuit's nodes. In fact, if the circuit has n nodes, we can derive $n - 1$ independent equations by applying Kirchhoff's current law to *any* set of $n - 1$ nodes.[1] To obtain the rest of the needed $b - (n - 1)$ equations, we apply Kirchhoff's voltage law (KVL) to circuit loops or meshes.

To reduce the number of independent equations needed, we can use essential nodes and essential branches instead of nodes and branches. This is because the number of essential nodes in a circuit is less than or equal to the number of nodes, and the number of essential branches is less than or equal to the number of branches. Thus, our systematic method for writing the necessary equations to solve for the circuit's unknown currents is as follows:

- Count the number of essential nodes, n_e.
- Count the number of essential branches, b_e, where the current is unknown.
- Write $n_e - 1$ equations by applying KCL to any set of $n_e - 1$ nodes.
- Write $b_e - (n_e - 1)$ equations by applying KVL around a set of $b_e - (n_e - 1)$ loops or meshes.

Remember that the voltage for each component in every loop or mesh must be known or must be described in terms of the component's current using Ohm's law.

Let's illustrate our systematic approach by applying it to the circuit from Example 4.1, as seen in Example 4.2.

EXAMPLE 4.2	Using Essential Nodes and Essential Branches to Write Simultaneous Equations

The circuit in Fig. 4.4 has six essential branches, denoted $i_1 - i_6$, where the current is unknown. Use the systematic approach to write the six equations needed to solve for the six unknown currents.

Solution

The essential nodes in the circuit are labeled b, c, e, and g, so $n_e = 4$. From the problem statement we know that the number of essential branches where

---

[1]Applying KCL to the last unused node (the nth node) does *not* generate an independent equation. See Problem 4.4.

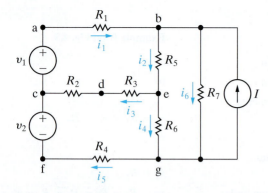

Figure 4.4 ▲ The circuit shown in Fig. 4.3 with six unknown branch currents defined.

the current is unknown is $b_e = 6$. Note that there are seven essential branches in the circuit, but the current in the essential branch containing the current source is known. We need to write six independent equations because there are six unknown currents.

We derive three of the six independent equations by applying Kirchhoff's current law to any three of the four essential nodes. We use the nodes b, c, and e to get

$$-i_1 + i_2 + i_6 - I = 0,$$

$$i_1 - i_3 - i_5 = 0,$$

$$i_3 + i_4 - i_2 = 0.$$

We derive the remaining three equations by applying Kirchhoff's voltage law around three meshes. Remember that the voltage across every component in each mesh must be known or must be expressed as the product of the component's resistance and its current using Ohm's law. Because the circuit has four meshes, we need to dismiss one mesh. We eliminate the $R_7 - I$ mesh because we don't know the voltage across I.[2] Using the other three meshes gives

$$R_1 i_1 + R_5 i_2 + i_3(R_2 + R_3) - v_1 = 0,$$

$$-i_3(R_2 + R_3) + i_4 R_6 + i_5 R_4 - v_2 = 0,$$

$$-i_2 R_5 + i_6 R_7 - i_4 R_6 = 0.$$

Rearranging the six equations to facilitate their solution yields the set

$$-i_1 + i_2 + 0i_3 + 0i_4 + 0i_5 + i_6 = I,$$

$$i_1 + 0i_2 - i_3 + 0i_4 - i_5 + 0i_6 = 0,$$

$$0i_1 - i_2 + i_3 + i_4 + 0i_5 + 0i_6 = 0,$$

$$R_1 i_1 + R_5 i_2 + (R_2 + R_3)i_3 + 0i_4 + 0i_5 + 0i_6 = v_1,$$

$$0i_1 + 0i_2 - (R_2 + R_3)i_3 + R_6 i_4 + R_4 i_5 + 0i_6 = v_2,$$

$$0i_1 - R_5 i_2 + 0i_3 - R_6 i_4 + 0i_5 + R_7 i_6 = 0.$$

SELF-CHECK: Assess your understanding of this material by trying Chapter Problems 4.2, 4.3, and 4.5.

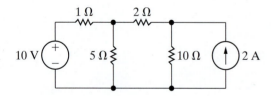

Figure 4.5 ▲ A circuit used to illustrate the node-voltage method of circuit analysis.

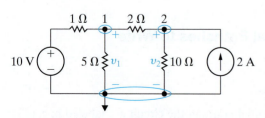

Figure 4.6 ▲ The circuit shown in Fig. 4.5 with a reference node and the node voltages.

4.2 Introduction to the Node-Voltage Method

The **node-voltage method** generates independent simultaneous equations by applying Kirchoff's current law at the essential nodes of the circuit. Solving the simultaneous equations yields the voltage drops between all but one of the essential nodes and a reference essential node. You can use these voltages to calculate voltages and currents for every component in the circuit, thereby solving the circuit. We illustrate the step-by-step procedure using the circuit in Fig. 4.5.

Step 1 is to make a neat layout of the circuit so that no branches cross and to mark the essential nodes on the circuit diagram, as in Fig. 4.6. This circuit has three essential nodes ($n_e = 3$); therefore, we need two ($n_e - 1$) KCL equations to describe the circuit.

Step 2 is to select one of the three essential nodes as a reference node. Although in theory the choice is arbitrary, there is often an obvious and practical choice. Choosing the reference node becomes easier with practice. For example, the node with the most branches is usually a good choice, so we select the lower node in Fig. 4.5 as the reference node. The reference node is identified by the symbol ▼, as in Fig. 4.6. Complete this

---

[2]We say more about this decision in Section 4.7.

step by labeling the remaining essential node voltages on the circuit diagram. A **node voltage** is defined as the voltage rise from the reference node to a nonreference essential node. For this circuit, we must define two node voltages, which are denoted v_1 and v_2 in Fig. 4.6.

In **Step 3** we generate the KCL equations. To do this, write the current leaving each branch connected to a nonreference node as a function of the node voltages and sum the currents to zero in accordance with Kirchhoff's current law. Let's look at node 1. Ohm's law tells us that the current leaving node 1 through the 1 Ω resistor equals the voltage across the resistor ($v_1 - 10$) divided by its resistance (1 Ω). That is, the current equals ($v_1 - 10$)/1. Figure 4.7 depicts these observations. Repeating this reasoning, the current leaving node 1 through the 5 Ω resistor is $v_1/5$, and the current leaving node 1 through the 2 Ω resistor is ($v_1 - v_2$)/2. Because the sum of the three currents leaving node 1 must equal zero, we can write the KCL equation at node 1 as

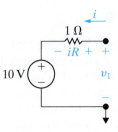

Figure 4.7 ▲ Computation of the branch current *i*.

$$\frac{v_1 - 10}{1} + \frac{v_1}{5} + \frac{v_1 - v_2}{2} = 0. \tag{4.1}$$

Repeating the process for node 2 gives

$$\frac{v_2 - v_1}{2} + \frac{v_2}{10} - 2 = 0. \tag{4.2}$$

Note that the first term in Eq. 4.2 is the current leaving node 2 through the 2 Ω resistor, the second term is the current leaving node 2 through the 10 Ω resistor, and the third term is the current leaving node 2 through the current source.

In **Step 4**, solve the simultaneous equations (see Appendix A). Equations 4.1 and 4.2 are the two simultaneous equations that describe the circuit in terms of the node voltages v_1 and v_2. Solving for v_1 and v_2 yields

$$v_1 = \frac{100}{11} = 9.09 \text{ V},$$

$$v_2 = \frac{120}{11} = 10.91 \text{ V}.$$

Step 5 uses the node voltages to solve for the remaining unknowns in the circuit. Once the node voltages are known, all branch currents can be calculated. Once these are known, the component voltages and powers can be calculated.

A condensed version of the node-voltage method is shown in Analysis Method 4.1. To practice the node-voltage method, work through Example 4.3.

NODE-VOLTAGE METHOD

1. Identify each essential node.
2. Pick and label a reference node; then label the node voltages at the remaining essential nodes.
3. Write a KCL equation for every nonreference essential node.
4. Solve the equations to find the node-voltage values.
5. Solve the circuit using node voltages from Step 4 to find component currents, voltages, and power values.

Analysis Method 4.1 The basic version of the node-voltage method.

EXAMPLE 4.3 | **Using the Node-Voltage Method**

a) Use the node-voltage method of circuit analysis to find the branch currents i_a, i_b, and i_c in the circuit shown in Fig. 4.8.

b) Find the power associated with each source, and state whether the source is delivering or absorbing power.

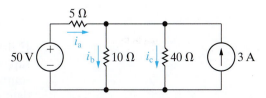

Figure 4.8 ▲ The circuit for Example 4.3.

Solution

a) We begin by noting that the circuit has two essential nodes; thus, we need to write a single KCL equation. **Step 1**: Identify the two essential nodes. **Step 2**: Select the lower node as the reference node and define the unknown node voltage as v_1. Figure 4.9 illustrates these decisions. **Step 3**: Write a KCL equation at the nonreference essential node by summing the currents leaving node 1:

$$\frac{v_1 - 50}{5} + \frac{v_1}{10} + \frac{v_1}{40} - 3 = 0.$$

Step 4: Solve the equation for v_1, giving

$$v_1 = 40 \text{ V}.$$

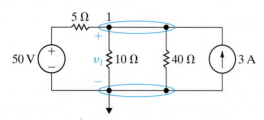

Figure 4.9 ▲ The circuit shown in Fig. 4.8 with a reference node and the unknown node voltage v_1.

Step 5: Use the node voltage v_1 and Ohm's law to find the requested branch currents:

$$i_a = \frac{50 - v_1}{5} = \frac{50 - 40}{5} = 2 \text{ A},$$

$$i_b = \frac{v_1}{10} = \frac{40}{10} = 4 \text{ A},$$

$$i_c = \frac{v_1}{40} = \frac{40}{40} = 1 \text{ A}.$$

b) The power associated with the 50 V source is

$$p_{50V} = -50i_a = -100 \text{ W (delivering)}.$$

The power associated with the 3 A source is

$$p_{3A} = -3v_1 = -3(40) = -120 \text{ W (delivering)}.$$

We check these calculations by noting that the total delivered power is 220 W. The total power absorbed by the three resistors is $4(5) + 16(10) + 1(40)$ or 220 W, which equals the total delivered power.

■ ASSESSMENT PROBLEMS

Objective 1 — Understand and be able to use the node-voltage method

4.1
  a) For the circuit shown, use the node-voltage method to find v_1, v_2, and i_1.
  b) How much power is delivered to the circuit by the 15 A source?
  c) Repeat (b) for the 5 A source.

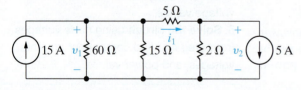

Answer: (a) 60 V, 10 V, 10 A;
  (b) 900 W;
  (c) −50 W.

4.2 Use the node-voltage method to find v in the circuit shown.

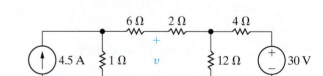

Answer: 15 V.

SELF-CHECK: Also try Chapter Problems 4.6, 4.10, and 4.12.

4.3 The Node-Voltage Method and Dependent Sources

If the circuit contains dependent sources, the KCL equations must be supplemented with the constraint equations imposed by the dependent sources. We will modify Step 3 in the node-voltage method to accommodate dependent sources. Example 4.4 illustrates the application of the node-voltage method to a circuit containing a dependent source.

Using the Node-Voltage Method with Dependent Sources

Use the node-voltage method to find the power dissipated in the 5 Ω resistor in the circuit shown in Fig. 4.10.

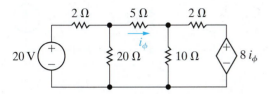

Figure 4.10 ▲ The circuit for Example 4.4.

Solution

Step 1: Identify the circuit's three essential nodes. We will need two KCL equations to describe the circuit.

Step 2: Since four branches terminate on the lower node, we select it as the reference node and label the node voltages at the remaining essential nodes. The results of the first two steps are shown in Fig. 4.11.

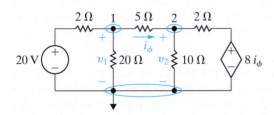

Figure 4.11 ▲ The circuit shown in Fig. 4.10, with a reference node and the node voltages.

Step 3: Generate the simultaneous equations by applying KCL at the nonreference essential nodes. Summing the currents leaving node 1 gives the equation

$$\frac{v_1 - 20}{2} + \frac{v_1}{20} + \frac{v_1 - v_2}{5} = 0.$$

Summing the currents leaving node 2 yields

$$\frac{v_2 - v_1}{5} + \frac{v_2}{10} + \frac{v_2 - 8i_\phi}{2} = 0.$$

As written, these two node-voltage equations contain three unknowns, namely, v_1, v_2, and i_ϕ. We need a third equation, which comes from the constraint imposed by the dependent source. This equation expresses the controlling current of the dependent source, i_ϕ, in terms of the node voltages, or

$$i_\phi = \frac{v_1 - v_2}{5}.$$

As you can see, we need to modify Step 3 in the node-voltage procedure to remind us to write a constraint equation whenever a dependent source is present.

Step 3: Write a KCL equation at each nonreference essential node. If the circuit contains dependent sources, write a dependent source constraint equation that defines the controlling voltage or current of the dependent source in terms of the node voltages.

The condensed form for Step 3 is shown in Analysis Method 4.2.

Step 4: Solve for v_1, v_2, and i_ϕ, giving

$$v_1 = 16 \text{ V}, v_2 = 10 \text{ V}, \text{ and } i_\phi = 1.2 \text{ A}.$$

Step 5: Use the node voltage values to find the current in the 5 Ω resistor and the power dissipated in that resistor:

$$i_\phi = \frac{v_1 - v_2}{5} = \frac{16 - 10}{5} = 1.2 \text{ A},$$

$$p_{5\Omega} = 5i_\phi^2 = 5(1.2)^2 = 7.2 \text{ W}.$$

A good exercise to build your problem-solving intuition is to reconsider this example, using node 2 as the reference node. Does it make the analysis easier or harder?

NODE-VOLTAGE METHOD

3. Write a KCL equation for every nonreference essential node.
- If there are dependent sources, write a constraint equation for each one.

Analysis Method 4.2 Modified Step 3 for the node-voltage method.

▪ ASSESSMENT PROBLEM

Objective 1—Understand and be able to use the node-voltage method

4.3   a) Use the node-voltage method to find the power associated with each source in the circuit shown.

   b) State whether the source is delivering power to the circuit or extracting power from the circuit.

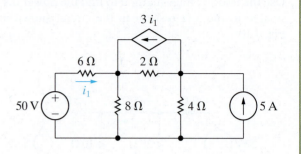

Answer: (a) $p_{50V} = -150$ W, $p_{3i_1} = -144$ W, $p_{5A} = -80$ W;

   (b) all sources are delivering power to the circuit.

SELF-CHECK: Also try Chapter Problems 4.17 and 4.19.

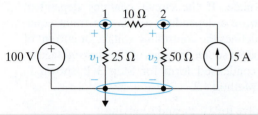

Figure 4.12 ▲ A circuit with a known node voltage.

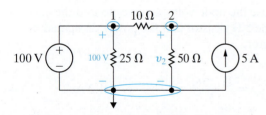

Figure 4.13 ▲ The circuit in Fig. 4.12 with the node voltage v_1 replaced with its value, 100 V.

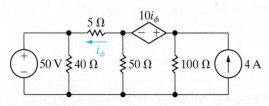

Figure 4.14 ▲ A circuit with a dependent voltage source connected between nodes.

4.4 The Node-Voltage Method: Some Special Cases

Let's explore the special case in which a voltage source is the only element between two essential nodes. The five-step node-voltage method still applies as long as we modify Step 2.

As an example, let's look at the circuit in Fig. 4.12. There are three essential nodes in this circuit, which means two simultaneous equations are needed. Apply Steps 1 and 2 to identify the essential nodes, choose a reference node, and label the remaining nodes. Notice that the 100 V source constrains the voltage between node 1 and the reference node to 100 V. We now modify Step 2 in order to take advantage of this simplification.

Step 2: Pick and label a reference node, then label the node voltages at the remaining essential nodes. If a voltage source is the only element between an essential node and the reference node, replace the node voltage label with the value of the voltage source.

Thus, we can replace the voltage v_1 in the circuit with its value, 100 V, as shown in Fig. 4.13.

Now Step 3 requires only a single KCL equation at node 2:

$$\frac{v_2 - 100}{10} + \frac{v_2}{50} - 5 = 0. \tag{4.3}$$

In Step 4, solve Eq. 4.3 for v_2:

$$v_2 = 125 \text{ V}.$$

Knowing v_2, we can calculate the current in every branch. Use Step 5 to verify that the current into node 1 in the branch containing the independent voltage source is 1.5 A.

As another example, consider the circuit shown in Fig. 4.14, which has four essential nodes and requires three simultaneous equations. However, two essential nodes are connected by an independent voltage source, and two other essential nodes are connected by a current-controlled dependent voltage source. Making simple modifications to the node-voltage method allows us to take advantage of these observations.

There are several possibilities for the reference node. The nodes on each side of the dependent voltage source look attractive because, if chosen, one of the node voltages would be either $+10i_\phi$ (left node is the reference) or $-10i_\phi$ (right node is the reference). The lower node looks even better because one node voltage is immediately known (50 V) and five branches terminate there. We therefore opt for the lower node as the reference.

Figure 4.15 shows the redrawn circuit, after Steps 1 and 2. Notice that we introduced the current i because we cannot express the current in the dependent voltage source branch as a function of the node voltages v_2 and v_3. We write a KCL equation at node 2 to give

$$\frac{v_2 - 50}{5} + \frac{v_2}{50} + i = 0, \qquad (4.4)$$

and at node 3 to give

$$\frac{v_3}{100} - i - 4 = 0. \qquad (4.5)$$

We eliminate i simply by adding Eqs. 4.4 and 4.5 to get

$$\frac{v_2 - 50}{5} + \frac{v_2}{50} + \frac{v_3}{100} - 4 = 0. \qquad (4.6)$$

We will continue the steps of the node-voltage method after we introduce a new concept, the supernode.

The Concept of a Supernode

When a voltage source is between two essential nodes, we can combine those nodes and the source to form a **supernode**. Let's apply the supernode concept to our circuit from Fig. 4.15. Figure 4.16 shows the circuit redrawn with a supernode created by combining nodes 2 and 3. We can remember to look for supernodes by modifying Step 2 one final time:

Step 2: Pick and label a reference node, then label the node voltages at the remaining essential nodes. If a voltage source is the only element between an essential node and the reference node, replace the node voltage label with the value of the voltage source. If a voltage source is the only element between two nonreference essential nodes, combine the two essential nodes and the voltage source into a single supernode.

Step 3 also needs to be modified. Obviously, Kirchhoff's current law must hold for supernodes, so we can write a single KCL equation for the supernode. The supernode also constrains the difference between the node voltages used to create the supernode to the value of the voltage source within the supernode. We therefore need to write a supernode constraint equation. Thus, we arrive at the final version of Step 3.

Step 3: Write a KCL equation at each supernode, then write a KCL equation at each remaining nonreference essential node where the voltage is unknown. If there are any dependent sources, write a dependent source constraint equation for each one, and if there are any supernodes, write a supernode constraint equation for each one.

The final version of the node-voltage method can be found in the end-of-chapter Summary; a condensed version is given in Analysis Method 4.3.

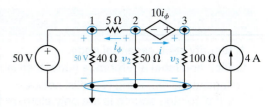

Figure 4.15. ▲ The circuit shown in Fig. 4.14 with the selected node voltages defined.

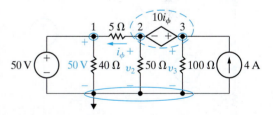

Figure 4.16 ▲ Combining nodes 2 and 3 to form a supernode.

NODE-VOLTAGE METHOD

1. Identify each essential node.

2. Pick and label a reference node; then label the node voltage at the remaining essential nodes.

- If a voltage source is the only element between an essential node and the reference node, replace the node voltage label with the value of the voltage source.
- If a voltage source is the only element between two nonreference essential nodes, combine the nodes and the source into a single supernode.

3. Write a KCL equation for each supernode and every remaining nonreference essential node where the voltage is unknown.

- If there are dependent sources, write a constraint equation for each one.
- If there are supernodes, write a constraint equation for each one.

4. Solve the equations to find the node voltage values and any other unknowns.

5. Solve the circuit using the values from Step 4 to find component currents, voltages, and power values.

Analysis Method 4.3 Complete form of the node-voltage method.

We can now use Step 3 to create the equations for the circuit in Fig. 4.16. First, write the KCL equation for the supernode. Starting with the 5 Ω branch and moving counterclockwise around the supernode, we generate the equation

$$\frac{v_2 - 50}{5} + \frac{v_2}{50} + \frac{v_3}{100} - 4 = 0, \tag{4.7}$$

which is identical to Eq. 4.6. Creating a supernode at nodes 2 and 3 has made writing this equation much easier. Next, write the supernode constraint equation by setting the value of the voltage source in the supernode to the difference between the two node voltages in the supernode:

$$v_3 - v_2 = 10i_\phi. \tag{4.8}$$

Finally, express the current controlling the dependent voltage source as a function of the node voltages:

$$i_\phi = \frac{v_2 - 50}{5}. \tag{4.9}$$

Use Step 5 to solve Eqs. 4.7, 4.8, and 4.9 for the three unknowns v_2, v_3, and i_ϕ to give

$$v_2 = 60 \text{ V}, \quad v_3 = 80 \text{ V}, \quad \text{and} \quad i_\phi = 2 \text{ A}.$$

Let's use the node-voltage method to analyze the amplifier circuit first introduced in Section 2.5 and shown again in Fig. 4.17. When we analyzed this circuit using Ohm's law, KVL, and KCL in Section 2.5, we faced the task of writing and solving six simultaneous equations. Work through Example 4.5 to see how the node-voltage method can simplify the circuit analysis.

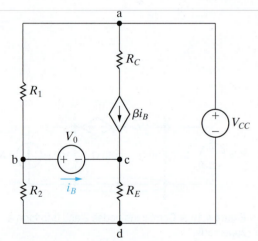

Figure 4.17 ▲ The transistor amplifier circuit shown in Fig 2.24.

EXAMPLE 4.5   **Node-Voltage Analysis of the Amplifier Circuit**

Use the node-voltage method to find i_B in the amplifier circuit shown in Fig. 4.17.

Solution

Step 1: We identify the four essential nodes, which are labeled a, b, c, and d. **Step 2**: Choose node d as the reference node. Then label the voltages at the remaining three essential nodes. **Step 3**: Before writing equations, we notice two special cases. The voltage source V_{CC} in the branch connecting node a and the reference node constrains the voltage between those nodes, so $v_a = V_{CC}$, and the voltage source V_0 in the branch between nodes b and c constrains the voltage between those nodes and creates a supernode. The results of Steps 1 and 2 and the modifications prompted by the special cases are depicted in Fig. 4.18.

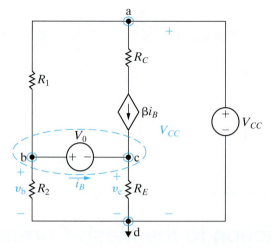

Figure 4.18 ▲ The circuit shown in Fig. 4.17, with voltages and the supernode identified.

Continuing Step 3, write the supernode KCL equation to give

$$\frac{v_b}{R_2} + \frac{v_b - V_{CC}}{R_1} + \frac{v_c}{R_E} - \beta i_B = 0. \quad (4.10)$$

Now write the dependent source constraint equation, which defines the controlling current i_B in terms of the node voltages. Since i_B is the current in a voltage source, we cannot use Ohm's law, so instead, write a KCL equation at node c:

$$i_B = \frac{v_c}{R_E} - \beta i_B. \quad (4.11)$$

The last part of Step 3 is the supernode constraint equation

$$v_b - v_c = V_0. \quad (4.12)$$

Step 3 gave us three equations with three unknowns. To solve these equations, we use back-substitution to eliminate the variables v_b and v_c. Begin by rearranging Eq. 4.11 to give

$$i_B = \frac{v_c}{R_E(1 + \beta)}. \quad (4.13)$$

Next, solve Eq. 4.12 for v_c to give

$$v_c = v_b - V_0. \quad (4.14)$$

Substituting Eqs. 4.13 and 4.14 into Eq. 4.10 and rearranging yields

$$v_b\left[\frac{1}{R_1} + \frac{1}{R_2} + \frac{1}{(1 + \beta)R_E}\right] = \frac{V_{CC}}{R_1} + \frac{V_0}{(1 + \beta)R_E}. \quad (4.15)$$

Solving Eq. 4.15 for v_b yields

$$v_b = \frac{V_{CC}R_2(1 + \beta)R_E + V_0 R_1 R_2}{R_1 R_2 + (1 + \beta)R_E(R_1 + R_2)}. \quad (4.16)$$

You should verify that, when Eq. 4.16 is combined with Eqs. 4.13 and 4.14, the solution for i_B is

$$i_B = \frac{(V_{CC}R_2)/(R_1 + R_2) - V_0}{(R_1 R_2)/(R_1 + R_2) + (1 + \beta)R_E}, \quad (4.17)$$

which is identical to Eq. 2.21. (See Problem 4.31.) Using the node-voltage method to analyze this circuit reduces the problem from manipulating six simultaneous equations (see Problem 2.38) to manipulating three simultaneous equations.

ASSESSMENT PROBLEMS

Objective 1—Understand and be able to use the node-voltage method

4.4 Use the node-voltage method to find v_o in the circuit shown.

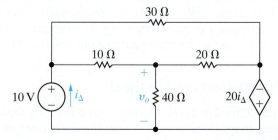

Answer: 24 V.

4.5 Use the node-voltage method to find v in the circuit shown.

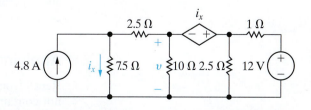

Answer: 8 V.

4.6 Use the node-voltage method to find v_1 in the circuit shown.

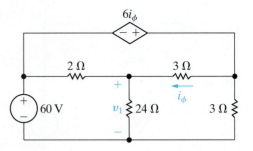

Answer: 48 V.

SELF-CHECK: Also try Chapter Problems 4.22, 4.27, and 4.28.

4.5 Introduction to the Mesh-Current Method

Before we learn about the mesh-current method for solving circuits, let's summarize the node-voltage method. The node-voltage method presented in Sections 4.2–4.4 is used to solve a circuit by writing simultaneous KCL equations at essential nodes. If a dependent source is present, a dependent source constraint equation is required. Special cases exist when a voltage source is the only component in a branch connecting two essential nodes. If one of the essential nodes is the reference node, the node voltage at the other node is the value of the voltage source and no KCL equation is required at that node. If neither of the essential nodes is the reference node, the two nodes and the voltage source are combined into a supernode. A KCL equation is written for the supernode as well as a supernode constraint equation. The simultaneous equations are solved to find the node voltages, and the node voltages can be used to find the voltage, current, and power for every circuit component.

We now turn to the mesh-current method, presented in Sections 4.5–4.7. The mesh-current method is used to solve a circuit by writing simultaneous KVL equations for the circuit's meshes. If a dependent source is present, a dependent source constraint equation is required. Special cases exist when a current source is a component of a mesh. If a current source is a component in only one mesh, the mesh current must have the same value as the current source and no KVL equation is needed for that mesh. If a current source is shared by two adjacent meshes, the two meshes are combined to form a supermesh. A KVL equation is written for the supermesh as well as a supermesh constraint equation. The simultaneous equations are solved to find the mesh currents, and the mesh currents can be used to find the voltage, current, and power for every circuit component.

Did you notice the symmetries in the descriptions of the node-voltage method and the mesh-current method? In engineering, this symmetry is called **duality**, and we will encounter it throughout this text. In comparing the two circuit analysis techniques, we see that essential nodes and meshes are duals, KVL and KCL are duals, voltages and currents are duals, supernodes and supermeshes are duals, and so on. Recognizing the existence of duality can help you master the techniques for circuit analysis presented in this book.

Applying the Mesh-Current Method

Recall from the terms defined in Table 4.1 that a mesh is a loop that does not contain any other loops. You should review the definitions of loop and path in Table 4.1, too. A **mesh current** is the current that exists on the perimeter of a mesh. We represent a mesh current on a circuit diagram using a curved arrow that follows the mesh perimeter, where the arrow-head indicates the current's direction.

Just like the node-voltage method, the mesh-current method is a step-by-step procedure. Let's use the mesh-current method for the circuit shown in Fig. 4.19 to solve for the currents i_1, i_2, and i_3. Because the circuit has two meshes, we expect to write two simultaneous equations.

In **Step 1**, we identify the meshes using a directed curved arrow that follows the perimeter of the mesh.

Step 2 labels the mesh currents using the labels i_a and i_b. The results of Steps 1 and 2 are shown in Fig. 4.20. We can see from this figure that the branch current i_1 equals the mesh current i_a and that the branch current i_2 equals the mesh current i_b. Note that to avoid confusion we use different names for the branch currents and the mesh currents in this circuit.

In **Step 3**, we write the KVL equation for each mesh. Let's start with the mesh whose mesh current is i_a. Pick a starting point anywhere on the perimeter of the mesh and sum the voltage drops for each component in the mesh in the direction of the mesh current until you return to the starting point. Start below the 100 V source and travel in the clockwise direction of the mesh current. The first voltage is due to the 100 V source and has the value −100. The next voltage is across the 4 Ω resistor, whose current is i_a, so from Ohm's law the voltage is $4i_a$. The next voltage is across the 10 Ω resistor, whose current is $(i_a - i_b)$, so from Ohm's law the voltage is $10(i_a - i_b)$. We are back to the starting point, so KVL tells us that the sum of these three voltages is zero:

$$-100 + 4i_a + 10(i_a - i_b) = 0.$$

Repeat this process to get the KVL equation for the i_b mesh. Remember that you can start anywhere on the perimeter of this mesh, so let's start at the left of the 5 Ω resistor. Again we determine the voltage drops for each component in the direction of the i_b mesh. The voltage drop for the 5 Ω resistor is $5i_b$, the voltage drop for the 40 V source is 40, and the voltage drop for the 10 Ω resistor is $10(i_b - i_a)$. We have returned to the starting point, so KVL tells us that the sum of these three voltages is zero:

$$5i_b + 40 + 10(i_b - i_a) = 0.$$

In **Step 4**, we solve these simultaneous mesh current equations (see Appendix A) to find that the mesh current values are

$$i_a = 10 \text{ A}$$

and

$$i_b = 4 \text{ A}.$$

Step 5 uses the mesh currents to solve for the currents, voltages, and power for all components in the circuit. Let's calculate the three branch currents to illustrate:

$$i_1 = i_a = 10 \text{ A},$$

$$i_2 = i_b = 4 \text{ A},$$

$$i_3 = i_a - i_b = 6 \text{ A}.$$

The ability to calculate the branch currents using the mesh currents is crucial to the mesh-current method of circuit analysis. Once you know the mesh currents, you also know the branch currents. And once you know the branch currents, you can compute any voltages or powers of interest. A condensed version of the mesh-current method is given in Analysis Method 4.4.

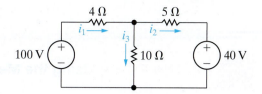

Figure 4.19 ▲ A circuit used to illustrate development of the mesh-current method of circuit analysis.

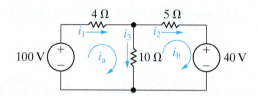

Figure 4.20 ▲ Mesh currents i_a and i_b.

MESH-CURRENT METHOD

1. Identify the meshes with curved directed arrows that follow the perimeter of each mesh.

2. Label the mesh currents for each mesh.

3. Write the KVL equations for each mesh.

4. Solve the KVL equations to find the mesh current values.

5. Solve the circuit using mesh currents from Step 4 to find component currents, voltages, and power values.

Analysis Method 4.4 The basic version of the mesh-current method.

Example 4.6 illustrates how the mesh-current method is used to find source powers and a branch voltage.

EXAMPLE 4.6 | Using the Mesh-Current Method

a) Use the mesh-current method to determine the power associated with each voltage source in the circuit shown in Fig. 4.21.

b) Calculate the voltage v_o across the 8 Ω resistor.

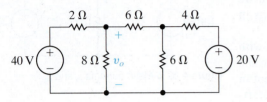

Figure 4.21 ▲ The circuit for Example 4.6.

Solution

a) **Step 1**: We identify the three meshes in the circuit and draw the mesh currents as directed curved arrows following the perimeter of each mesh. It is best to define all mesh currents in the same direction.

Step 2: Label the mesh currents; the results of the first two steps are depicted in Fig. 4.22.

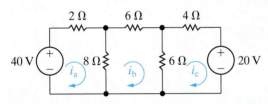

Figure 4.22 ▲ The three mesh currents used to analyze the circuit shown in Fig. 4.21.

Step 3: We use KVL to generate an equation for each mesh by summing the voltages in the direction of the mesh current. In the i_a mesh, start just below the 40 V source and sum the voltages in the clockwise direction to give

$$-40 + 2i_a + 8(i_a - i_b) = 0.$$

In the i_b mesh, start below the 8 Ω resistor and sum the voltages in the clockwise direction to give

$$8(i_b - i_a) + 6i_b + 6(i_b - i_c) = 0.$$

In the i_c mesh, start below the 6 Ω resistor and sum the voltages in the clockwise direction to give

$$6(i_c - i_b) + 4i_c + 20 = 0.$$

Step 4: Solve the three simultaneous mesh current equations from Step 3 to give

$$i_a = 5.6 \text{ A},$$
$$i_b = 2.0 \text{ A},$$
$$i_c = -0.80 \text{ A}.$$

Step 5: Use the mesh currents to find the power for each source. The mesh current i_a equals the branch current in the 40 V source, so the power associated with this source is

$$p_{40V} = -40i_a = -224 \text{ W}.$$

The minus sign means that this source is delivering power to the network. The current in the 20 V source equals the mesh current i_c; therefore

$$p_{20V} = 20i_c = -16 \text{ W}.$$

The 20 V source also is delivering power to the network.

b) The branch current in the 8 Ω resistor in the direction of the voltage drop v_o is $i_a - i_b$. Therefore

$$v_o = 8(i_a - i_b) = 8(3.6) = 28.8 \text{ V}.$$

☐ ASSESSMENT PROBLEM

Objective 2—Understand and be able to use the mesh-current method

4.7   Use the mesh-current method to find (a) the power delivered by the 80 V source to the circuit shown and (b) the power dissipated in the 8 Ω resistor.

Answer: (a) 400 W;
(b) 50 W.

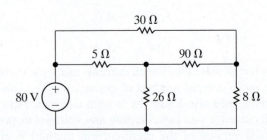

SELF-CHECK: Also try Chapter Problems 4.36 and 4.37.

4.6 The Mesh-Current Method and Dependent Sources

If the circuit contains dependent sources, we must modify the equation-writing step in the mesh-current analysis method, just as we did in the node-voltage analysis method.

Step 3: Write the KVL equation for each mesh; if the circuit contains a dependent source, write a dependent source constraint equation that defines the controlling variable for the dependent source in terms of the mesh currents.

The condensed form for Step 3 is shown in Analysis Method 4.5. Example 4.7 applies the mesh-current method to a circuit with a dependent source.

EXAMPLE 4.7   **Using the Mesh-Current Method with Dependent Sources**

Use the mesh-current method to find the power dissipated in the 4 Ω resistor in the circuit shown in Fig. 4.23.

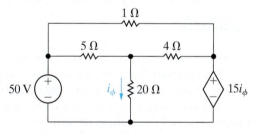

Figure 4.23 ▲ The circuit for Example 4.7.

Solution

Step 1: Begin by drawing the mesh currents in each of the three meshes.

Step 2: Label each mesh current. The resulting circuit is shown in Fig. 4.24.

Step 3: Write a KVL equation for each mesh by picking a starting point anywhere in the mesh and summing the voltages around the mesh in the

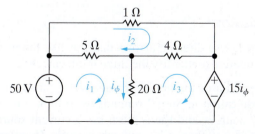

Figure 4.24 ▲ The circuit shown in Fig. 4.23 with the three mesh currents.

direction of the mesh current. When you return to the starting point, set the sum equal to zero. The three mesh-current equations are

$$5(i_1 - i_2) + 20(i_1 - i_3) - 50 = 0,$$
$$5(i_2 - i_1) + 1i_2 + 4(i_2 - i_3) = 0,$$
$$20(i_3 - i_1) + 4(i_3 - i_2) + 15i_\phi = 0.$$

To complete Step 3, express the branch current controlling the dependent voltage source in terms of the mesh currents as

$$i_\phi = i_1 - i_3.$$

Step 4: Solve the four equations generated in Step 3 to find the four unknown currents:

$$i_1 = 29.6 \text{ A}, \quad i_2 = 26 \text{ A}, \quad i_3 = 28 \text{ A}, \quad i_\phi = 1.6 \text{ A}.$$

Step 5: Use the mesh currents to find the power for the 4 Ω resistor. The current in the 4 Ω resistor oriented from left to right is $i_3 - i_2$, or 2 A. Therefore, the power dissipated is

$$p_{4\Omega} = (i_3 - i_2)^2(4) = (2)^2(4) = 16 \text{ W}.$$

What if you had not been told to use the mesh-current method? Would you have chosen the node-voltage method? It reduces the problem to finding one unknown node voltage because of the presence of two voltage sources between essential nodes. We say more about making such choices in Section 4.8.

ASSESSMENT PROBLEMS

Objective 2—Understand and be able to use the mesh-current method

4.8 a) Determine the number of mesh-current
equations needed to solve the circuit shown.
  b) Use the mesh-current method to find how
much power is being delivered by the dependent voltage source.

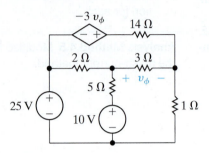

4.9 Use the mesh-current method to find v_o in the
circuit shown.

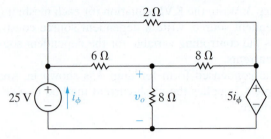

Answer: 16V.

Answer: (a) 3;
  (b) 36 W.

SELF-CHECK: Also try Chapter Problems 4.39 and 4.41.

4.7 The Mesh-Current Method: Some Special Cases

Recall that *voltage* sources present special cases when using the node-*voltage*
method (Section 4.4). So, it is no surprise that *current* sources present special cases when using the mesh-*current* method. There are two special cases,
one that occurs when a current source is in a single mesh, and the other that
occurs when a current source is shared by two adjacent meshes.

When a current source is in a single mesh, the value of the mesh current is known, since it must equal the current of the source. Therefore, we
label the mesh current with its value, and we do not need to write a KVL
equation for that mesh. This leads to the following modification in Step 2
of the mesh-current method.

Step 2: Label the mesh current for each mesh; if there is a current source in
a single mesh, label the mesh current with the value of the current source.

This special case is illustrated in Example 4.8.

EXAMPLE 4.8   **A Special Case in the Mesh-Current Method**

Use the mesh-current method to find branch cur-
rents i_a, i_b, and i_c in the circuit for Example 4.3, repeated here as Fig. 4.25.

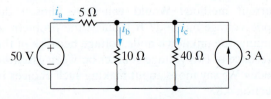

Figure 4.25 ▲ The circuit for Example 4.8.

Solution

Step 1: Use directed arrows that traverse the mesh
perimeters to identify the three mesh currents.

Step 2: Label the mesh currents as i_1, i_2, and i_3. The
modification in Step 2 reminds us to look for cur-
rent sources, and the i_3 mesh has a current source
that is not shared by any other mesh. Therefore,
the i_3 mesh current equals the current supplied by
the source. Note that i_3 and the current source are

in opposite directions, so the current in this mesh should be labeled -3 A. The results of Steps 1 and 2 are shown in Fig. 4.26.

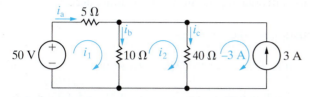

Figure 4.26 ▲ The circuit shown in Fig. 4.25 with the mesh currents identified and labeled.

Step 3: Write the KVL equations for the meshes whose mesh currents are unknown, which in this example are the i_1 and i_2 meshes. Remember to pick a starting point anywhere along the mesh, sum the voltages in the direction of the mesh current, and set the sum equal to zero when you return to the starting point. The resulting simultaneous mesh current equations are

$$-50 + 5i_1 + 10(i_1 - i_2) = 0 \quad \text{and}$$

$$10(i_2 - i_1) + 40(i_2 - (-3)) = 0.$$

Step 4: Solving the simultaneous mesh current equations gives

$$i_1 = 2 \text{ A} \quad \text{and} \quad i_2 = -2 \text{ A}.$$

Step 5: Finally, we use the mesh currents to calculate the branch currents in the circuit, i_a, i_b, and i_c.

$$i_a = i_1 = 2 \text{ A},$$

$$i_b = i_1 - i_2 = 4 \text{ A},$$

$$i_c = i_2 + 3 = 1 \text{ A}.$$

These are the same branch current values as those calculated in Example 4.3. Which of the two circuit analysis methods is better when calculating the branch currents? Which method is better when calculating the power associated with the sources?

Now we turn our attention to the other special case, created by a current source that is shared between two adjacent meshes. The circuit shown in Fig. 4.27 depicts this situation. Applying Steps 1 and 2, we define the mesh currents i_a, i_b, and i_c, as well as the voltage across the 5 A current source. In Step 3, we write the KVL equations for each mesh; let's start with mesh a and pick the starting point just below the 100 V source. The first two voltages we encounter when we traverse the mesh in the direction of the mesh current are -100 V and $3(i_a - i_b)$. But when we get to the current source, we must label the voltage drop across it as v and use this variable in the equation. Thus, for mesh a:

$$-100 + 3(i_a - i_b) + v + 6i_a = 0. \tag{4.18}$$

The same situation arises in mesh c, to give

$$50 + 4i_c - v + 2(i_c - i_b) = 0. \tag{4.19}$$

We now add Eqs. 4.18 and 4.19 to eliminate v; when simplified, the result is

$$-50 + 9i_a - 5i_b + 6i_c = 0. \tag{4.20}$$

We will complete Steps 4 and 5 in the mesh-current method after introducing a new concept, the supermesh.

The Concept of a Supermesh

When a current source is shared between two meshes, we can combine these meshes to form a **supermesh**, which traverses the perimeters of the two meshes and avoids the branch containing the shared current source. Figure 4.28 illustrates the supermesh concept. Using a modified Step 3, we write a KVL equation around the supermesh (denoted by the dashed line), using the original mesh currents to give

$$-100 + 3(i_a - i_b) + 2(i_c - i_b) + 50 + 4i_c + 6i_a = 0,$$

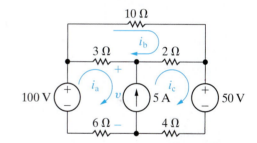

Figure 4.27 ▲ A circuit illustrating mesh analysis when a branch contains an independent current source.

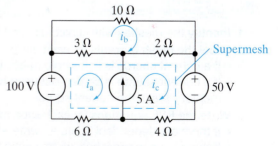

Figure 4.28 ▲ The circuit shown in Fig. 4.27, illustrating the concept of a supermesh.

which simplifies to

$$-50 + 9i_a - 5i_b + 6i_c = 0. \qquad (4.21)$$

Note that Eqs. 4.20 and 4.21 are identical. Thus, the supermesh has eliminated the need for introducing the unknown voltage across the current source.

The KVL equation for the b mesh is

$$10i_b + 2(i_b - i_c) + 3(i_b - i_a) = 0. \qquad (4.22)$$

We have two simultaneous equations, Eqs. 4.21 and 4.22, but three unknowns. Remember that the presence of a supernode in the node-voltage method requires a KCL equation at the supernode and a supernode constraint equation that defines the difference between the node voltages in the supernode as the value of the voltage source in the supernode. In a like manner, the presence of a supermesh in the mesh-current method requires a KVL equation around the supermesh and a supermesh constraint equation that defines the difference between the mesh currents in the supermesh as the value of the shared current source. From Fig. 4.28, the supermesh constraint equation is

$$i_c - i_a = 5. \qquad (4.23)$$

The final version of Steps 2 and 3 in the mesh-current method reminds us how to handle current sources in our circuits:

- **Step 2**: Label the mesh current for each mesh; if there is a current source in a single mesh, label the mesh current with the value of the source. If there is a current source shared between two meshes, create a supermesh by combining the two meshes and mentally erasing the current source.
- **Step 3**: Write a KVL equation around each supermesh and each single mesh where the mesh current is unknown. If there is a dependent source, write a constraint equation defining the controlling quantity for the dependent source in terms of the mesh currents. If there is a supermesh, write a supermesh constraint equation that defines the difference between the two mesh currents in the supermesh as the value of the shared current source.

The final version of the mesh-current method can be found in the end-of-chapter Summary; a condensed version is given in Analysis Method 4.6.

MESH-CURRENT METHOD

1. **Identify the meshes** with curved directed arrows that follow the perimeter of each mesh.
2. **Label the mesh currents** for each mesh.
   - If a current source is in a single mesh, label the mesh current with the value of the current source.
   - If a current source is shared between two meshes, combine the meshes to create a supermesh and mentally erase the current source.
3. **Write the KVL equations** for each supermesh and each single mesh where the current is unknown.
   - If there are dependent sources, write a dependent source constraint equation for each.
   - If there are supermeshes, write a supermesh constraint equation for each.
4. **Solve the KVL equations** and any constraint equations to find the mesh-current values and other unknowns.
5. **Solve the circuit** using mesh currents from Step 4 to find component currents, voltages, and power values.

Analysis Method 4.6 Complete form of the mesh-current method.

Use Step 4 to solve Eqs. 4.21–4.23 and confirm that the solutions for the three mesh currents are

$$i_a = 1.75 \text{ A}, \quad i_b = 1.25 \text{ A}, \quad \text{and} \quad i_c = 6.75 \text{ A}.$$

Work through Example 4.9 to see how the mesh-current method can be used to solve the amplifier circuit from Example 4.5.

EXAMPLE 4.9   Mesh-Current Analysis of the Amplifier Circuit

Use the mesh-current method to find i_B for the amplifier circuit in Fig. 4.29.

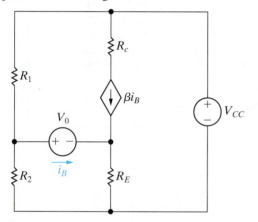

Figure 4.29 ▲ The circuit shown in Fig 2.24.

Solution

Step 1: Use directed arrows that traverse the mesh perimeters to identify the three mesh currents.

Step 2: Label the mesh currents as i_a, i_b, and i_c. Then recognize the current source that is shared between the i_a and i_c meshes. Combine these meshes, bypassing the branch with the shared current source, to create a supermesh. The result of the first two steps is the circuit shown in Fig. 4.30.

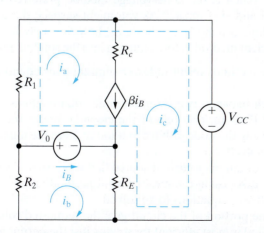

Figure 4.30 ▲ The circuit shown in Fig. 4.29, depicting the supermesh created by the presence of the dependent current source.

Step 3: Using KVL, sum the voltages around the supermesh in terms of the mesh currents i_a, i_b, and i_c to obtain

$$R_1 i_a + v_{CC} + R_E(i_c - i_b) - V_0 = 0. \quad (4.24)$$

The KVL equation for mesh b is

$$R_2 i_b + V_0 + R_E(i_b - i_c) = 0. \quad (4.25)$$

The constraint imposed by the dependent current source is

$$i_B = i_b - i_a. \quad (4.26)$$

The supermesh constraint equation is

$$\beta i_B = i_a - i_c. \quad (4.27)$$

Step 4: Use back-substitution to solve Eqs. 4.24–4.27. Start by combining Eqs. 4.26 and 4.27 to eliminate i_B and solve for i_c to give

$$i_c = (1 + \beta)i_a - \beta i_b. \quad (4.28)$$

We now use Eq. 4.28 to eliminate i_c from Eqs. 4.24 and 4.25:

$$[R_1 + (1 + \beta)R_E]i_a - (1 + \beta)R_E i_b = V_0 - V_{CC}, \quad (4.29)$$

$$-(1 + \beta)R_E i_a + [R_2 + (1 + \beta)R_E]i_b = -V_0. \quad (4.30)$$

You should verify that the solution of Eqs. 4.29 and 4.30 for i_a and i_b gives

$$i_a = \frac{V_0 R_2 - V_{CC} R_2 - V_{CC}(1 + \beta)R_E}{R_1 R_2 + (1 + \beta)R_E(R_1 + R_2)}, \quad (4.31)$$

$$i_b = \frac{-V_0 R_1 - (1 + \beta)R_E V_{CC}}{R_1 R_2 + (1 + \beta)R_E(R_1 + R_2)}. \quad (4.32)$$

Step 5: Use the two mesh currents from Eqs. 4.31 and 4.32, together with the definition for i_B in Eq. 4.26, to find i_B. You should verify that the result is the same as that given by Eq. 2.21.

ASSESSMENT PROBLEMS

Objective 2—Understand and be able to use the mesh-current method

4.10 Use the mesh-current method to find the power dissipated in the 2 Ω resistor in the circuit shown.

Answer: 72 W.

4.11 Use the mesh-current method to find the mesh current i_a in the circuit shown.

Answer: 15 A.

4.12 Use the mesh-current method to find the power dissipated in the 1 Ω resistor in the circuit shown.

Answer: 36 W.

SELF-CHECK: Also try Chapter Problems 4.45, 4.46, 4.50, and 4.52.

4.8 The Node-Voltage Method Versus the Mesh-Current Method

It is natural to ask, "When is the node-voltage method preferred to the mesh-current method and vice versa?" As you might suspect, there is no clear-cut answer. Asking a number of questions, however, may help you identify the more efficient method before plunging into the solution process:

- Does one of the methods result in fewer simultaneous equations to solve?
- Is there a branch between two essential nodes that contains only a voltage source? If so, making one of the essential nodes the reference node and using the node-voltage method reduces the number of equations to be solved.
- Is there a mesh containing a current source that is not shared with an adjacent mesh? If so, using the mesh-current method allows you to reduce the number of equations to be solved.
- Will solving some portion of the circuit give the requested solution? If so, which method is most efficient for solving just the pertinent portion of the circuit?

Perhaps the most important observation is that, for any situation, some time spent thinking about the problem in relation to the various analytical approaches available is time well spent. Examples 4.10 and 4.11 illustrate the process of deciding between the node-voltage and mesh-current methods.

EXAMPLE 4.10 | **Understanding the Node-Voltage Method Versus Mesh-Current Method**

Find the power dissipated in the 300 Ω resistor in the circuit shown in Fig. 4.31.

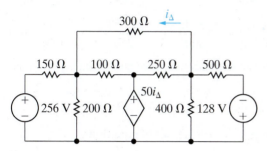

Figure 4.31 ▲ The circuit for Example 4.10.

Solution

To find the power dissipated in the 300 Ω resistor, we need to find either the current in the resistor or the voltage across it. The mesh-current method yields the current in the resistor; this approach requires solving five mesh equations, as depicted in Fig. 4.32, and a dependent source constraint equation, for a total of six simultaneous equations.

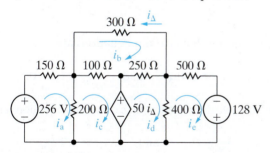

Figure 4.32 ▲ The circuit shown in Fig. 4.31, with the five mesh currents.

Let's now consider using the node-voltage method. The circuit has four essential nodes, and therefore only three node-voltage equations are required to describe the circuit. The dependent voltage source between two essential nodes forms a supernode, requiring a KCL equation and a supernode constraint equation. We have to sum the currents at the remaining essential node, and we need to write the dependent source constraint equation, for a total of four simultaneous equations. Thus, the node-voltage method is the more attractive approach.

Step 1: We begin by identifying the four essential nodes in the circuit of Fig. 4.31. The three black dots at the bottom of the circuit identify a single essential node, and the three black dots in the middle of the circuit are the three remaining essential nodes.

Step 2: Select a reference node. Two essential nodes in the circuit in Fig. 4.31 merit consideration. The first is the reference node in Fig. 4.33, where we also defined the three node voltages v_1, v_2, and v_3, and indicated that nodes 1 and 3 form a supernode because they are connected by a dependent voltage source. If the reference node in Fig. 4.33 is selected, one of the unknown node voltages is the voltage across the 300 Ω resistor, namely, v_2 in Fig. 4.33. Once we know this voltage, we calculate the power in the 300 Ω resistor by using the expression

$$p_{300\Omega} = v_2^2/300.$$

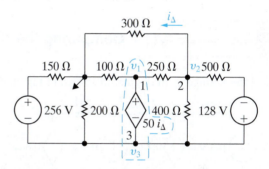

Figure 4.33 ▲ The circuit shown in Fig. 4.31, with a reference node.

The second node worth considering as the reference node is the bottom node in the circuit, as shown in Fig. 4.34. If this reference node is chosen, one of the unknown node voltages is eliminated because $v_b = 50i_\Delta$. We would need to write two KCL equations and a dependent source constraint equation, and solve these three simultaneous equations. However, to find either the current in the 300 Ω resistor or the voltage across it requires an additional calculation once we know the node voltages v_a and v_c.

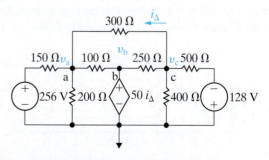

Figure 4.34 ▲ The circuit shown in Fig. 4.31 with an alternative reference node.

Step 3: We compare these two possible reference nodes by generating two sets of KCL equations and constraint equations. The first set pertains to the circuit shown in Fig. 4.33, and the second set is based on the circuit shown in Fig. 4.34.

- Set 1 (Fig. 4.33)
  At the supernode,

$$\frac{v_1}{100} + \frac{v_1 - v_2}{250} + \frac{v_3}{200} + \frac{v_3 - v_2}{400} + \frac{v_3 - (v_2 + 128)}{500}$$

$$+ \frac{v_3 + 256}{150} = 0.$$

At v_2,

$$\frac{v_2}{300} + \frac{v_2 - v_1}{250} + \frac{v_2 - v_3}{400} + \frac{v_2 + 128 - v_3}{500} = 0.$$

The dependent source constraint equation is

$$i_\Delta = \frac{v_2}{300}.$$

The supernode constraint equation is

$$v_1 - v_3 = 50i_\Delta.$$

- Set 2 (Fig. 4.34); remember that $v_b = 50i_\Delta$.
  At v_a,

$$\frac{v_a}{200} + \frac{v_a - 256}{150} + \frac{v_a - 50i_\Delta}{100} + \frac{v_a - v_c}{300} = 0.$$

At v_c,

$$\frac{v_c}{400} + \frac{v_c + 128}{500} + \frac{v_c - 50i_\Delta}{250} + \frac{v_c - v_a}{300} = 0.$$

The dependent source constraint equation is

$$i_\Delta = \frac{v_c - v_a}{300}.$$

Step 4: Solve each set of equations.

Step 5: Verify that both solutions lead to a power calculation of 16.57 W dissipated in the 300 Ω resistor.

EXAMPLE 4.11 **Comparing the Node-Voltage and Mesh-Current Methods**

Find the voltage v_o in the circuit shown in Fig. 4.35.

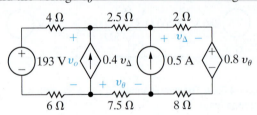

Figure 4.35 ▲ The circuit for Example 4.11.

Solution

We first consider using the mesh-current method.

Step 1: Identify the three mesh currents in the circuit using directed arrows that follow the mesh perimeters.

Step 2: Label the three mesh currents. Because there are two currents sources, each shared by two meshes, we can combine all three meshes into a single supermesh that traverses the perimeter of the entire circuit and avoids the two branches with current sources. The result of these two steps is shown in Fig. 4.36.

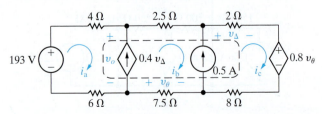

Figure 4.36 ▲ The circuit shown in Fig. 4.35 with the three mesh currents.

Step 3: Write the KCL equation for the supermesh:

$$-193 + 4i_a + 2.5i_b + 2i_c + 0.8v_\theta + 8i_c + 7.5i_b$$
$$+ 6i_a = 0.$$

The supermesh constraint equations are

$$i_b - i_a = 0.4v_\Delta \quad \text{and} \quad i_c - i_b = 0.5,$$

and the two dependent source constraint equations are

$$v_\Delta = 2i_c \quad \text{and} \quad v_\theta = -7.5i_b.$$

Step 4: We must solve the five simultaneous equations generated in Step 3.

Step 5: We need one additional equation to find v_o from the mesh current i_a:

$$v_o = 193 - 10i_a.$$

Now let's consider using the node-voltage method.

Step 1: There are four essential nodes in the circuit of Fig. 4.35, identified by the four black dots in the figure.

Step 2: We can make the unknown voltage v_o one of the three node voltages by choosing the bottom left node as the reference node. After labeling the remaining two node voltages, we have the circuit in Fig. 4.37.

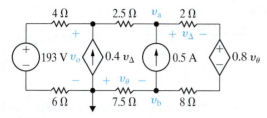

Figure 4.37 ▲ The circuit shown in Fig. 4.35 with node voltages.

Step 3: The KCL equations are

$$\frac{v_o - 193}{10} - 0.4v_\Delta + \frac{v_o - v_a}{2.5} = 0,$$

$$\frac{v_a - v_o}{2.5} - 0.5 + \frac{v_a - (v_b + 0.8v_\theta)}{10} = 0,$$

$$\frac{v_b}{7.5} + 0.5 + \frac{v_b + 0.8v_\theta - v_a}{10} = 0.$$

The dependent source constraint equations are

$$v_\theta = -v_b \quad \text{and} \quad v_\Delta = 2\left[\frac{v_a - (v_b + 0.8v_\theta)}{10}\right].$$

Step 4: Once we solve these five simultaneous equations, we have the value of v_o without writing an additional equation, so Step 5 is not needed.

Based on the comparison of the two methods, the node-voltage method involves a bit less work. You should verify that both approaches give $v_o = 173$ V.

ASSESSMENT PROBLEMS

Objective 3—Deciding between the node-voltage and mesh-current methods

4.13 Find the power delivered by the 2 A current source in the circuit shown.

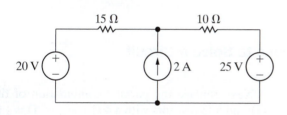

Answer: 70 W.

4.14 Find the power delivered by the 4 A current source in the circuit shown.

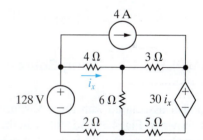

Answer: 40 W.

SELF-CHECK: Also try Chapter Problems 4.54 and 4.56.

4.9 Source Transformations

We are always interested in methods that simplify circuits. Series-parallel reductions and Δ-to-Y transformations are already on our list of simplifying techniques. We now expand the list with source transformations. A **source transformation** allows a voltage source in series with a resistor to be replaced by a current source in parallel with the same resistor or vice versa. Figure 4.38 shows a source transformation. The double-headed arrow emphasizes that a source transformation is bilateral; that is, we can start with either configuration and derive the other.

We need to find the relationship between v_s and i_s that guarantees the two configurations in Fig. 4.38 are equivalent with respect to nodes a and b. Equivalence is achieved if any resistor R_L has the same current and thus the same voltage drop, whether connected between nodes a and b in Fig. 4.38(a) or Fig. 4.38(b).

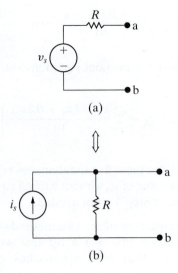

(a)

⇕

(b)

Figure 4.38 ▲ Source transformations.

Suppose R_L is connected between nodes a and b in Fig. 4.38(a). Using Ohm's law, we find that the current in R_L is

$$i_L = \frac{v_s}{R + R_L}. \tag{4.33}$$

Now suppose the same resistor R_L is connected between nodes a and b in Fig. 4.38(b). Using current division, we see that the current in R_L is

$$i_L = \frac{R}{R + R_L} i_s. \tag{4.34}$$

If the two circuits in Fig. 4.38 are equivalent, these resistor currents must be the same. Equating the right-hand sides of Eqs. 4.33 and 4.34 and simplifying gives the **condition of equivalence**:

$$i_s = \frac{v_s}{R}. \tag{4.35}$$

When Eq. 4.35 is satisfied for the circuits in Fig. 4.38, the current in R_L connected between nodes a and b is the same for both circuits for all values of R_L. If the current in R_L is the same for both circuits, then the voltage drop across R_L is the same for both circuits, and the circuits are equivalent at nodes a and b. If the polarity of v_s is reversed, the orientation of i_s must be reversed to maintain equivalence.

Example 4.12 uses source transformations to simplify a circuit-analysis problem.

EXAMPLE 4.12   **Using Source Transformations to Solve a Circuit**

Find the power associated with the 6 V source for the circuit shown in Fig. 4.39 and state whether the 6 V source is absorbing or delivering the power.

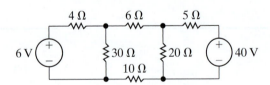

Figure 4.39 ▲ The circuit for Example 4.12.

Solution

If we study the circuit shown in Fig. 4.39, we see ways to simplify the circuit by using source transformations. But we must simplify the circuit in a way that preserves the branch containing the 6 V source. Therefore, begin on the right side of the circuit with the branch containing the 40 V source. We can transform the 40 V source in series with the 5 Ω resistor into an 8 A current source in parallel with a 5 Ω resistor, as shown in Fig. 4.40(a).

Next, replace the parallel combination of the 20 Ω and 5 Ω resistors with a 4 Ω resistor. This 4 Ω resistor is in parallel with the 8 A source and therefore can be replaced with a 32 V source in series with a 4 Ω resistor, as shown in Fig. 4.40(b). The 32 V source is in series with 20 Ω of resistance and, hence, can be replaced by a current source of 1.6 A in parallel with 20 Ω, as shown in Fig. 4.40(c). The 20 Ω and 30 Ω parallel resistors can be reduced to a single 12 Ω resistor. The parallel combination of the 1.6 A current source and the 12 Ω resistor transforms into a voltage source of 19.2 V in series with 12 Ω. Figure 4.40(d) shows the result of this last transformation. The current in the direction of the voltage drop across the 6 V source is $(19.2 - 6)/16$, or 0.825 A. Therefore, the power associated with the 6 V source is

$$p_{6V} = (6)(0.825) = 4.95 \text{ W}$$

and the voltage source is absorbing power.

Practice your circuit analysis skills by using either the node-voltage method or the mesh-current method to solve this circuit and verify that you get the same answer.

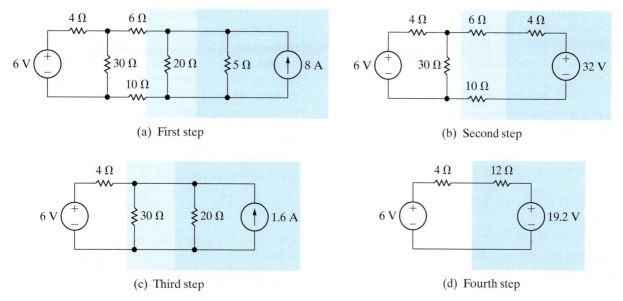

(a) First step

(b) Second step

(c) Third step

(d) Fourth step

Figure 4.40 ▲ Step-by-step simplification of the circuit shown in Fig. 4.39.

A couple of questions arise from the source transformation depicted in Fig. 4.40.

- What happens if there is a resistance R_p in parallel with the voltage source?
- What happens if there is a resistance R_s in series with the current source?

In both cases, the resistance can be removed to create a simpler equivalent circuit with respect to terminals a and b. Figure 4.41 summarizes this observation. The two circuits depicted in Fig. 4.41(a) are equivalent with respect to terminals a and b because they produce the same voltage and current in any resistor R_L inserted between nodes a and b. The same can be said for the circuits in Fig. 4.41(b). Example 4.13 illustrates an application of the equivalent circuits depicted in Fig. 4.41.

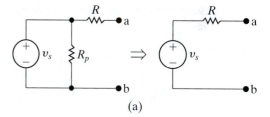

(a)

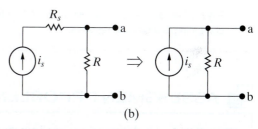

(b)

Figure 4.41 ▲ (a) Generating a simplified equivalent circuit from a circuit with a resistor in parallel with a voltage source; (b) generating a simplified circuit from a circuit with a resistor in series with a current source.

EXAMPLE 4.13 | **Using Special Source Transformation Techniques**

a) Use source transformations to find the voltage v_o in the circuit shown in Fig. 4.42.

b) Find the power developed by the 250 V voltage source.

c) Find the power developed by the 8 A current source.

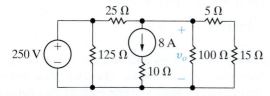

Figure 4.42 ▲ The circuit for Example 4.13.

Solution

a) We begin by removing the 125 Ω and 10 Ω resistors because the 125 Ω resistor is connected in parallel with the 250 V voltage source and the 10 Ω resistor is connected in series with the 8 A current source. We also combine the series-connected resistors in the rightmost branch into a single resistance of 20 Ω. Figure 4.43 shows the simplified circuit.

Now use a source transformation to replace the series-connected 250 V source and 25 Ω resistor with a 10 A source in parallel with the 25 Ω resistor, as shown in Fig. 4.44. We can then use Kirchhoff's current law to combine the parallel

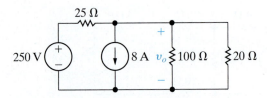

Figure 4.43 ▲ A simplified version of the circuit shown in Fig. 4.42.

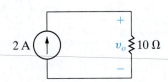

Figure 4.44 ▲ The circuit shown in Fig. 4.43 after a source transformation.

current sources into a single source. The parallel resistors combine into a single resistor. Figure 4.45 shows the result. Hence $v_o = 20$ V.

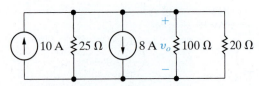

Figure 4.45 ▲ The circuit shown in Fig. 4.44 after combining sources and resistors.

b) We need to return to the original circuit in Fig. 4.42 to calculate the power associated with the sources. While a resistor connected in parallel with a voltage source or a resistor connected in series with a current source can be removed without affecting the terminal behavior of the circuit, these resistors play an important role in how the power is dissipated throughout the circuit. The current supplied by the 250 V source, represented as i_s, equals the current in the 125 Ω resistor plus the current in the 25 Ω resistor. Thus,

$$i_s = \frac{250}{125} + \frac{250 - 20}{25} = 11.2 \text{ A}.$$

Therefore, the power developed by the voltage source is

$$p_{250V}(\text{developed}) = (250)(11.2) = 2800 \text{ W}.$$

c) To find the power developed by the 8 A current source, we first find the voltage across the source. If we let v_s represent the voltage across the source, positive at the upper terminal of the source, we obtain

$$v_s + 8(10) = v_o = 20, \quad \text{or} \quad v_s = -60 \text{ V},$$

and the power developed by the 8 A source is 480 W. Note that the 125 Ω and 10 Ω resistors do not affect the value of v_o but do affect the power calculations. Check your power calculations by determining the power absorbed by all of the resistors in the circuit.

ASSESSMENT PROBLEM

Objective 4—Understand source transformation

4.15   a) Use a series of source transformations to find the voltage v in the circuit shown.
     b) How much power does the 120 V source deliver to the circuit?

Answer: (a) 48 V;
          (b) 374.4 W.

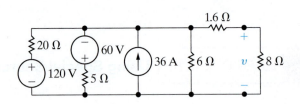

SELF-CHECK: Also try Chapter Problems 4.59 and 4.62.

4.10 Thévenin and Norton Equivalents

At times in circuit analysis, we want to concentrate on what happens at a specific pair of terminals. For example, when we plug a toaster into an outlet, we are interested primarily in the voltage and current at the terminals of the toaster. We have little or no interest in the effect that connecting the toaster has on voltages or currents elsewhere in the circuit supplying the outlet. We can expand this interest in terminal behavior to a set of appliances, each requiring a different amount of power. We then are interested in how the voltage and current delivered at the outlet change as we change appliances. In other words, we want to focus on the behavior of the circuit supplying the outlet, but only at the outlet terminals.

Thévenin and Norton equivalents are circuit simplification techniques. These equivalent circuits retain no information about the internal behavior of the original circuit and focus only on terminal behavior. They are extremely valuable when analyzing complex circuits where one portion of the circuit is fixed, so it can be replaced by a simple Thévenin or Norton equivalent, and another portion of the circuit is changing. Although here we discuss them as they pertain to resistive circuits, Thévenin and Norton equivalent circuits may be used to represent any circuit made up of linear elements.

The Thévenin equivalent circuit is the simplest equivalent for a given circuit and consists of a single voltage source in series with a single resistor. The Norton equivalent circuit is the source transform of the Thévenin equivalent circuit. To better grasp the concept of a Thévenin equivalent circuit, imagine a circuit with a complex interconnection of resistors, independent sources, and dependent sources, as shown in Fig. 4.46(a). We are interested in simplifying this complex circuit with respect to the terminals a and b. The simplified equivalent circuit, shown in Fig. 4.46(b), is the series combination of a voltage source V_{Th} and a resistor R_{Th} called the **Thévenin equivalent circuit**. It is equivalent to the original circuit in the sense that, if we connect the same load across the terminals a and b of each circuit, we get the same voltage and current at the terminals of the load. This equivalence holds for all possible values of load resistance.

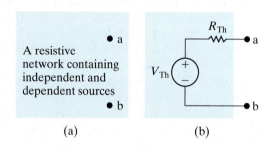

Figure 4.46 ▲ (a) A general circuit. (b) The Thévenin equivalent circuit.

The Thévenin Equivalent

To represent the original circuit by its Thévenin equivalent, we must calculate the Thévenin voltage V_{Th} and the Thévenin resistance R_{Th}. First, we note that if the load resistance is infinitely large, we have an open-circuit condition. The open-circuit voltage at the terminals a and b in the circuit shown in Fig. 4.46(b) is V_{Th}. By hypothesis, this must be the same as the open-circuit voltage at the terminals a and b in the original circuit. Therefore, to find the Thévenin voltage V_{Th}, calculate the open-circuit voltage in the original circuit.

Reducing the load resistance to zero gives us a short-circuit condition. If we place a short circuit across the terminals a and b of the Thévenin equivalent circuit, the short-circuit current directed from a to b is

$$i_{sc} = \frac{V_{Th}}{R_{Th}}. \tag{4.36}$$

By hypothesis, this short-circuit current must be identical to the short-circuit current that exists in a short circuit placed across the terminals a and b of the original network. From Eq. 4.36,

$$R_{Th} = \frac{V_{Th}}{i_{sc}}. \tag{4.37}$$

Thus, the Thévenin resistance is the ratio of the open-circuit voltage to the short-circuit current. Work through Example 4.14 to see how to find the Thévenin equivalent of a circuit.

EXAMPLE 4.14   **Finding a Thévenin Equivalent**

Find the Thévenin equivalent of the circuit in Fig. 4.47.

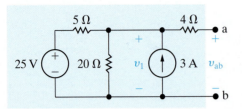

Figure 4.47 ▲ A circuit used to illustrate a Thévenin equivalent.

Solution

To find the Thévenin equivalent of the circuit shown in Fig. 4.47, we first calculate the open-circuit voltage v_{ab}. Note that when the terminals a, b are open, there is no current in the 4 Ω resistor. Therefore the open-circuit voltage v_{ab} is identical to the voltage across the 3 A current source, labeled v_1. We find the voltage by solving a single KCL equation. Choosing the lower node as the reference node, we get

$$\frac{v_1 - 25}{5} + \frac{v_1}{20} - 3 = 0.$$

Solving for v_1 yields

$$v_1 = 32 \text{ V} = V_{Th}.$$

Hence, the Thévenin voltage for the circuit is 32 V.

The next step is to place a short circuit across the terminals a and b and calculate the resulting short-circuit current. Figure 4.48 shows the circuit with the short in place. Note that the short-circuit current is

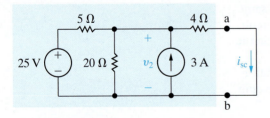

Figure 4.48 ▲ The circuit shown in Fig. 4.47 with terminals a and b short-circuited.

in the direction of the open-circuit voltage drop across the terminals a and b. If the short-circuit current is in the direction of the open-circuit voltage rise across the terminals, a minus sign must be inserted in Eq. 4.37.

The short-circuit current (i_{sc}) is found easily once v_2 is known. Therefore, the problem reduces to finding v_2 with the short in place. Again, if we use the lower node as the reference node, the KCL equation at the node labeled v_2 is

$$\frac{v_2 - 25}{5} + \frac{v_2}{20} - 3 + \frac{v_2}{4} = 0.$$

Solving for v_2 gives

$$v_2 = 16 \text{ V}.$$

Hence, the short-circuit current is

$$i_{sc} = \frac{16}{4} = 4 \text{ A}.$$

We now find the Thévenin resistance by substituting the numerical values for the Thévenin voltage, V_{Th}, and the short-circuit current, i_{sc}, into Eq. 4.37:

$$R_{Th} = \frac{V_{Th}}{i_{sc}} = \frac{32}{4} = 8 \text{ Ω}.$$

Figure 4.49 shows the Thévenin equivalent for the circuit shown in Fig. 4.45.

You should verify that, if a 24 Ω resistor is connected across the terminals a and b in Fig. 4.47, the voltage across the resistor will be 24 V and the current in the resistor will be 1 A, as would be the case with the Thévenin circuit in Fig. 4.49. This same equivalence between the circuits in Figs. 4.47 and 4.49 holds for any resistor value connected between nodes a and b.

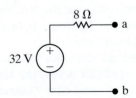

Figure 4.49 ▲ The Thévenin equivalent of the circuit shown in Fig. 4.47.

The Norton Equivalent

A **Norton equivalent circuit** consists of an independent current source in parallel with the Norton equivalent resistance, as shown in Fig. 4.50. We can derive it directly from the original circuit by calculating the open-circuit voltage and the short-circuit current, just as we did when calculating the Thévenin equivalent. The Norton current equals the short-circuit current at the terminals of interest, and the Norton resistance is the ratio of the open-circuit voltage to the short-circuit current, so it is identical to the Thévenin resistance:

$$I_N = i_{sc};$$

$$R_N = \frac{v_{oc}}{i_{sc}} = R_{Th}. \qquad (4.38)$$

If we already have a Thévenin equivalent circuit, we can derive the Norton equivalent circuit from it simply by making a source transformation.

Using Source Transformations

Sometimes we can use source transformations to derive a Thévenin or Norton equivalent circuit. This technique works best when the network contains only independent sources. Dependent sources require us to retain the identity of the controlling voltages and/or currents, and this constraint usually prohibits simplification of the circuit by source transformations. Work through Example 4.15 to see how a series of source transformations leads to a Norton equivalent circuit.

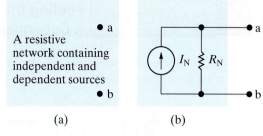

Figure 4.50 ▲ (a) A general circuit. (b) The Norton equivalent circuit.

| **EXAMPLE 4.15** | **Finding a Norton Equivalent** |

Find the Norton equivalent of the circuit in Fig. 4.47 by making a series of source transformations.

Solution

We start on the left side of the circuit and transform the series-connected 25 V source and 5 Ω resistor to a parallel-connected 5 A source and 5 Ω resistor, as shown in Step 1 of Fig. 4.51. Use KCL to combine the parallel-connected 5 A and 3 A sources into a single 8 A source, and combine the parallel-connected 5 Ω and 25 Ω resistors

into a single 4 Ω resistor, as shown in Step 2 of Fig. 4.51. Now transform the parallel-connected 8 A source and 4 Ω resistor to a series-connected 32 V source and 4 Ω resistor, and combine the two series-connected 4 Ω resistors into a single 8 Ω resistor, as shown in Step 3 of Fig. 4.51. Note that the result of Step 3 is the Thévenin equivalent circuit we derived in Example 4.12. Finally, transform the series-connected 32 V source and 8 Ω resistor into a parallel-connected 4 A source and 8 Ω resistor, which is the Norton equivalent circuit, as shown in Step 4 of Fig. 4.51.

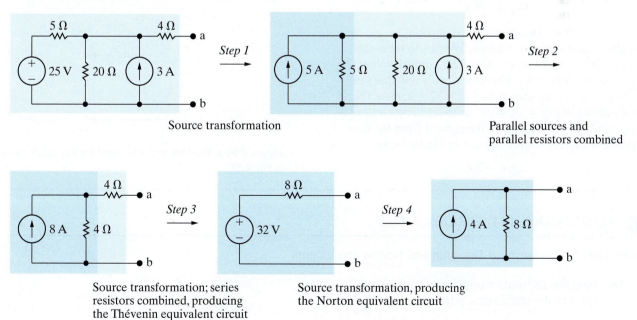

Figure 4.51 ▲ Step-by-step derivation of the Thévenin and Norton equivalents of the circuit shown in Fig. 4.47.

Gain additional practice with Thévenin equivalent circuits and see how to cope with the presence of a dependent source in the original circuit by working through Example 4.16.

EXAMPLE 4.16 Finding the Thévenin Equivalent of a Circuit with a Dependent Source

Find the Thévenin equivalent for the circuit containing dependent sources shown in Fig. 4.52.

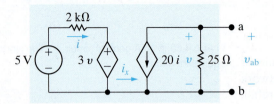

Figure 4.52 ▲ A circuit used to illustrate a Thévenin equivalent when the circuit contains dependent sources.

Solution

The first step in analyzing the circuit in Fig. 4.52 is to recognize that the current labeled i_x must be zero. (Note the absence of a return path for i_x to enter the left-hand portion of the circuit.) The open-circuit, or Thévenin, voltage is the voltage across the 25 Ω resistor. Since $i_x = 0$,

$$V_{Th} = v_{ab} = (-20i)(25) = -500i.$$

The current i is

$$i = \frac{5 - 3v}{2000} = \frac{5 - 3V_{Th}}{2000}.$$

In writing the equation for i, we recognize that the Thévenin voltage is identical to v. When we combine these two equations, we obtain

$$V_{Th} = -5 \text{ V}.$$

To calculate the short-circuit current, we place a short circuit across a and b. When the terminals a and b are shorted together, the control voltage v is reduced to zero. Therefore, with the short in place, the circuit shown in Fig. 4.52 becomes the one shown in Fig. 4.53. With the short circuit shunting the 25 Ω resistor, all of the current from the dependent current source appears in the short, so

$$i_{sc} = -20i.$$

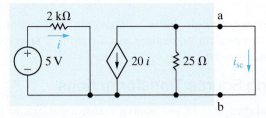

Figure 4.53 ▲ The circuit shown in Fig. 4.52 with terminals a and b short-circuited.

As the voltage controlling the dependent voltage source has been reduced to zero, the current controlling the dependent current source is

$$i = \frac{5}{2000} = 2.5 \text{ mA}.$$

Combining these two equations yields a short-circuit current of

$$i_{sc} = -20(2.5) = -50 \text{ mA}.$$

From i_{sc} and V_{Th} we get

$$R_{Th} = \frac{V_{Th}}{i_{sc}} = \frac{-5}{-0.05} = 100 \text{ Ω}.$$

Figure 4.54 illustrates the Thévenin equivalent for the circuit shown in Fig. 4.52. Note that the reference polarity marks on the Thévenin voltage source in Fig. 4.54 agree with the preceding equation for V_{Th}.

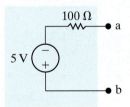

Figure 4.54 ▲ The Thévenin equivalent for the circuit shown in Fig. 4.52.

◨ ASSESSMENT PROBLEMS

Objective 5—Understand Thévenin and Norton equivalents

4.16 Find the Thévenin equivalent circuit with respect to the terminals a, b for the circuit shown.

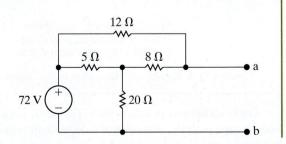

Answer: $V_{ab} = V_{Th} = 64.8 \text{ V}, R_{Th} = 6 \text{ Ω}.$

4.17 Find the Norton equivalent circuit with respect to the terminals a, b for the circuit shown.

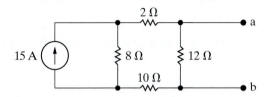

4.18 A voltmeter with an internal resistance of 100 kΩ is used to measure the voltage v_{AB} in the circuit shown. What is the voltmeter reading?

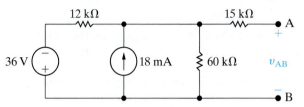

Answer: $I_N = 6$ A (directed toward a), $R_N = 7.5\ \Omega$.

Answer: 120 V.

SELF-CHECK: Also try Chapter Problems 4.64, 4.68, and 4.72.

4.11 More on Deriving the Thévenin Equivalent

We can calculate the Thévenin resistance, R_{Th}, directly from the circuit rather than calculating it as the ratio of the open-circuit voltage to the short-circuit current (Eq. 4.37). If the circuit contains only independent sources and resistors, we can determine R_{Th} by deactivating all independent sources and then calculating the resistance seen looking into the network at the designated terminal pair. A voltage source is deactivated by replacing it with a short circuit, while a current source is deactivated by replacing it with an open circuit. Example 4.17 illustrates this direct method for determining R_{Th}.

EXAMPLE 4.17 | **Finding the Thévenin Equivalent Resistance Directly from the Circuit**

Find R_{Th} the circuit shown in Fig. 4.55.

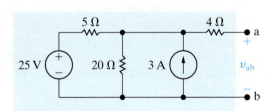

Figure 4.55 ▲ A circuit used to illustrate a Thévenin equivalent.

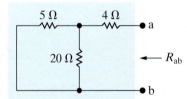

Figure 4.56 ▲ The circuit shown in Fig. 4.55 after deactivation of the independent sources.

Solution

Deactivating the independent sources simplifies the circuit to the one shown in Fig. 4.56. The resistance seen looking into the terminals a and b is denoted R_{ab}, which consists of the 4 Ω resistor in series with the parallel combination of the 5 and 20 Ω resistors. Thus,

$$R_{ab} = R_{Th} = 4 + \frac{(5)(20)}{5 + 20} = 8\ \Omega.$$

Note that deriving R_{Th} directly from the circuit is much simpler than finding R_{Th} from Eq. 4.37, as we did in Example 4.14.

If the circuit or network contains dependent sources, a direct method for finding the Thévenin resistance R_{Th} is as follows. We first deactivate all independent sources, and then we apply either a test voltage source or a test current source to the Thévenin terminals a and b. The Thévenin resistance equals the ratio of the voltage across the test source to the current delivered by the test source. Example 4.18 illustrates this alternative procedure for finding R_{Th}, using the same circuit as Example 4.16.

EXAMPLE 4.18   Finding the Thévenin Equivalent Resistance Using a Test Source

Find the Thévenin resistance R_{Th} for the circuit in Fig. 4.52, using the test source method.

Solution

Begin by deactivating the independent voltage source and exciting the circuit from the terminals a and b with either a test voltage source or a test current source. If we apply a test voltage source, we will know the voltage of the dependent voltage source and hence the controlling current i. Therefore, we opt for the test voltage source. Figure 4.57 shows the circuit for computing the Thévenin resistance.

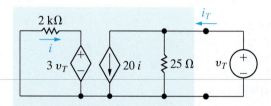

Figure 4.57 ▲ An alternative method for computing the Thévenin resistance.

The test voltage source is denoted v_T, and the current that it delivers to the circuit is labeled i_T. To find the Thévenin resistance, we solve the circuit for the ratio of the voltage to the current at the test source; that is, $R_{\text{Th}} = v_T/i_T$. From Fig. 4.57,

$$i_T = \frac{v_T}{25} + 20i,$$

$$i = \frac{-3v_T}{2000}.$$

We then substitute the expression for i into the equation for i_T and solve the resulting equation for the ratio i_T/v_T:

$$i_T = \frac{v_T}{25} - \frac{60v_T}{2000},$$

$$\frac{i_T}{v_T} = \frac{1}{25} - \frac{6}{200} = \frac{50}{5000} = \frac{1}{100}.$$

The Thévenin resistance is the inverse of the ratio i_T/v_T, so

$$R_{\text{Th}} = \frac{v_T}{i_T} = 100\ \Omega.$$

In a network containing only resistors and dependent sources, the Thévenin equivalent voltage $V_{\text{Th}} = 0$ and the Norton equivalent current $I_N = 0$. It should be clear that if the circuit you start with has no independent sources, its equivalent circuit cannot have any independent sources either. Therefore, the Thévenin and Norton equivalents for a circuit with only dependent sources and resistors is a single equivalent resistance, whose value must be determined using the test source method. The procedure is illustrated in Example 4.19.

EXAMPLE 4.19   Finding the Thévenin Equivalent of a Circuit with Dependent Sources and Resistors

Find the Norton equivalent for the circuit in Fig. 4.58.

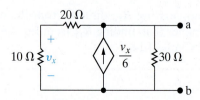

Figure 4.58 ▲ A circuit used to determine a Thévenin equivalent when the circuit contains only dependent sources and resistors.

Solution

The circuit in Fig. 4.58 has no independent sources. Therefore, the Norton equivalent current is zero, and the Norton equivalent circuit consists only of the Norton resistance, R_N. Applying a test source to the terminals a and b is the only way to determine R_N. We have applied a test current source, whose value is i_T, as shown in Fig. 4.59. Analyze this circuit to calculate the voltage across the test source, v_T, and then calculate the Norton equivalent resistance using the ratio of v_T to i_T.

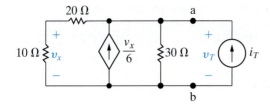

Figure 4.59 ▲ The circuit in Fig. 4.58 with a test current source.

Write a KCL equation at the top essential node to give

$$i_T = \frac{v_T}{20 + 10} - \frac{v_x}{6} + \frac{v_T}{30}.$$

Use voltage division to find the voltage across the 10 Ω resistor:

$$v_x = \frac{10}{20 + 10} v_T = \frac{v_T}{3}.$$

Then,

$$i_T = \frac{v_T}{30} - \frac{v_T}{18} + \frac{v_T}{30} = \frac{v_T}{90}.$$

Therefore, the Norton equivalent of the circuit in Fig. 4.58 is a single resistor whose resistance $R_N = v_T/i_T = 90\ \Omega$.

We conclude this discussion of Thévenin and Norton equivalents with one final example of their application in circuit analysis. Sometimes we can use a Thévenin equivalent to simplify one portion of a circuit, thereby greatly simplifying analysis of the larger network. Let's return to the amplifier circuit first introduced in Section 2.5 and subsequently analyzed in Sections 4.4 and 4.7. Study Example 4.20 to see how a Thévenin equivalent of one portion of this circuit helps us in the analysis of the whole circuit.

EXAMPLE 4.20 | Using a Thévenin Equivalent to Analyze the Amplifier Circuit

Use a Thévenin equivalent of the left side of the amplifier circuit, shown in Fig. 4.60, to find the current i_B.

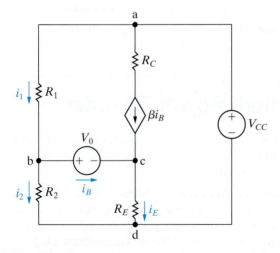

Figure 4.60 ▲ The application of a Thévenin equivalent in circuit analysis.

Solution

We redraw the circuit as shown in Fig. 4.61 to prepare to replace the subcircuit to the left of V_0 with its Thévenin equivalent. You should be able to determine that this modification has no effect on the

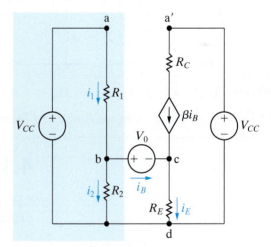

Figure 4.61 ▲ A modified version of the circuit shown in Fig. 4.60.

branch currents i_1, i_2, i_B, and i_E. Then replace the circuit made up of V_{CC}, R_1, and R_2 with a Thévenin equivalent, with respect to the terminals b and d. The Thévenin voltage and resistance are

$$V_{\text{Th}} = \frac{V_{CC} R_2}{R_1 + R_2}, \tag{4.39}$$

$$R_{\text{Th}} = \frac{R_1 R_2}{R_1 + R_2}. \tag{4.40}$$

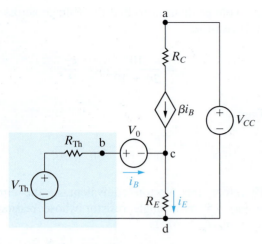

Figure 4.62 ▲ The circuit shown in Fig. 4.61 modified by a Thévenin equivalent.

With the Thévenin equivalent, the circuit in Fig. 4.61 becomes the one shown in Fig. 4.62.

We now derive an equation for i_B by summing the voltages around the left mesh. In writing this mesh equation, we recognize that $i_E = (1 + \beta)i_B$. Thus,

$$V_{Th} = R_{Th}i_B + V_0 + R_E(1 + \beta)i_B,$$

from which

$$i_B = \frac{V_{Th} - V_0}{R_{Th} + (1 + \beta)R_E}. \qquad (4.41)$$

When we substitute Eqs. 4.39 and 4.40 into Eq. 4.41, we get the same expression obtained in Eq. 2.25. Note that once we have incorporated the Thévenin equivalent into the original circuit, we can obtain the solution for i_B by writing a single equation.

ASSESSMENT PROBLEMS

Objective 5—Understand Thévenin and Norton equivalents

4.19 Find the Thévenin equivalent circuit with respect to the terminals a, b for the circuit shown.

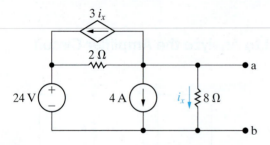

Answer: $V_{Th} = v_{ab} = 8$ V, $R_{Th} = 1\ \Omega$.

4.20 Find the Thévenin equivalent circuit with respect to the terminals a, b for the circuit shown. (*Hint:* Define the voltage at the left-most node as v, and write two nodal equations with V_{Th} as the right node voltage.)

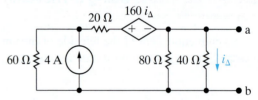

Answer: $V_{Th} = v_{ab} = 30$ V, $R_{Th} = 10\ \Omega$.

SELF-CHECK: Also try Chapter Problems 4.74 and 4.79.

4.12 Maximum Power Transfer

Circuit analysis plays an important role in the analysis of systems designed to transfer power from a source to a load. This text discusses power transfer in two basic types of systems.

- **Systems Optimized for Maximum Efficiency** Power utility systems are a good example of this type because they generate, transmit, and distribute large quantities of electric power. If a power utility system is inefficient, a large percentage of the power generated is lost in the transmission and distribution processes, and thus wasted. We will look at these types of systems in Sections 9.10 and 9.11.
- **Systems Optimized for Maximum Power** Communication and instrumentation systems are good examples because when information (data) is transmitted via electric signals, the power available at the transmitter or detector is limited. Thus, transmitting as much of this power as possible to the receiver (load) is desirable. In such applications, the amount of power being transferred is small, so the transfer efficiency is not a primary concern. Here we consider maximum power transfer in systems that can be modeled by a purely resistive circuit.

Maximum power transfer can best be described with the aid of the circuit shown in Fig. 4.63. We assume a resistive network containing independent and dependent sources and a designated pair of terminals, a and b, to which a load, R_L, is connected. The problem is to determine the value of R_L that permits maximum power delivery to R_L. The first step in this process is to recognize that a resistive network can always be replaced by its Thévenin equivalent. Therefore, we redraw the circuit shown in Fig. 4.63 as the one shown in Fig. 4.64. Replacing the original network by its Thévenin equivalent greatly simplifies the task of finding R_L. To derive R_L, begin by expressing the power dissipated in R_L as a function of the three circuit parameters V_{Th}, R_{Th}, and R_L. Thus

$$p = i^2 R_L = \left(\frac{V_{Th}}{R_{Th} + R_L} \right)^2 R_L. \tag{4.42}$$

Next, we recognize that for a given circuit, V_{Th} and R_{Th} will be fixed. Therefore, the power dissipated is a function of the single variable R_L. To find the value of R_L that maximizes the power, we use elementary calculus to write an equation for the derivative of p with respect to R_L:

$$\frac{dp}{dR_L} = V_{Th}^2 \left[\frac{(R_{Th} + R_L)^2 - R_L \cdot 2(R_{Th} + R_L)}{(R_{Th} + R_L)^4} \right].$$

The derivative is zero and p is maximized when

$$(R_{Th} + R_L)^2 = 2R_L(R_{Th} + R_L).$$

Solving for the load resistance R_L yields

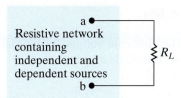

Figure 4.63 ▲ A circuit describing maximum power transfer.

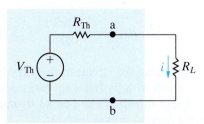

Figure 4.64 ▲ A circuit used to determine the value of R_L for maximum power transfer.

CONDITION FOR MAXIMUM POWER TRANSFERRED TO A RESISTIVE LOAD

$$R_L = R_{Th}. \tag{4.43}$$

Thus, **maximum power transfer** occurs when the load resistance R_L equals the Thévenin resistance R_{Th}. To find the maximum power delivered to R_L, substitute Eq. 4.43 into Eq. 4.42:

MAXIMUM POWER TRANSFERRED TO A RESISTIVE LOAD

$$p_{max} = \frac{V_{Th}^2 R_L}{(2R_L)^2} = \frac{V_{Th}^2}{4R_L}. \tag{4.44}$$

Analyzing a circuit to calculate the load resistor required for maximum power transfer is illustrated in Example 4.21.

EXAMPLE 4.21   **Calculating the Condition for Maximum Power Transfer**

a) For the circuit shown in Fig. 4.65, find the value of R_L that results in maximum power being transferred to R_L.

b) Calculate the maximum power that can be delivered to R_L.

c) When R_L is adjusted for maximum power transfer, what percentage of the power delivered by the 360 V source reaches R_L?

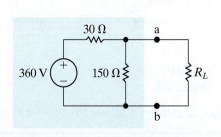

Figure 4.65 ▲ The circuit for Example 4.21.

Solution

a) The Thévenin voltage for the circuit to the left of the terminals a and b is

$$V_{\text{Th}} = \frac{150}{180}(360) = 300 \text{ V}.$$

The Thévenin resistance is

$$R_{\text{Th}} = \frac{(150)(30)}{180} = 25 \text{ }\Omega.$$

Replacing the circuit to the left of the terminals a and b with its Thévenin equivalent gives us the circuit shown in Fig. 4.66, so R_L must equal 25 Ω for maximum power transfer.

b) The maximum power that can be delivered to R_L is

$$p_{\text{max}} = \left(\frac{300}{50}\right)^2(25) = 900 \text{ W}.$$

c) When R_L equals 25 Ω, the voltage v_{ab} is

$$v_{\text{ab}} = \left(\frac{300}{50}\right)(25) = 150 \text{ V}.$$

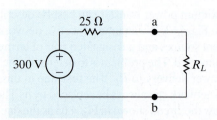

Figure 4.66 ▲ Reduction of the circuit shown in Fig. 4.66 by means of a Thévenin equivalent.

From Fig. 4.65, when v_{ab} equals 150 V, the current in the voltage source in the direction of the voltage rise across the source is

$$i_s = \frac{360 - 150}{30} = \frac{210}{30} = 7 \text{ A}.$$

Therefore, the source is delivering 2520 W to the circuit, or

$$p_s = -i_s(360) = -2520 \text{ W}.$$

The percentage of the source power delivered to the load is

$$\frac{900}{2520} \times 100 = 35.71\%.$$

◻ ASSESSMENT PROBLEMS

Objective 6—Know the condition for and calculate maximum power transfer to resistive load

4.21 a) Find the value of R that enables the circuit shown to deliver maximum power to the terminals a, b.
 b) Find the maximum power delivered to R.

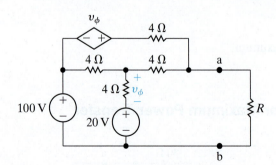

Answer: (a) 3 Ω;
 (b) 1.2 kW.

SELF-CHECK: Also try Chapter Problems 4.87 and 4.89.

4.22 Assume that the circuit in Assessment Problem 4.21 is delivering maximum power to the load resistor R.
 a) How much power is the 100 V source delivering to the network?
 b) Repeat (a) for the dependent voltage source.
 c) What percentage of the total power generated by these two sources is delivered to the load resistor R?

Answer:
 (a) 3000 W;
 (b) 800 W;
 (c) 31.58%.

4.13 Superposition

A linear system obeys the principle of **superposition**, which states that whenever a linear system is excited, or driven, by more than one independent source of energy, the total response is the sum of the individual responses. An individual response is the result of an independent source acting alone. Because we are dealing with circuits made up of interconnected linear-circuit elements, we can apply the principle of superposition directly to the analysis of such circuits when they are driven by more than one independent energy source. At present, we restrict the discussion to simple resistive networks; however, the principle is applicable to any linear system.

Superposition is applied in both the analysis and design of circuits. In analyzing a complex circuit with multiple independent voltage and current sources, there are often fewer, simpler equations to solve when the effects of the independent sources are considered one at a time. Applying superposition can thus simplify circuit analysis. Be aware, though, that sometimes applying superposition actually complicates the analysis, producing more equations to solve than with an alternative method. Superposition is required only if the independent sources in a circuit are fundamentally different. In these early chapters, all independent sources are dc sources, so superposition is not required. We introduce superposition here in anticipation of later chapters in which circuits will require it.

Superposition is applied in design to synthesize a desired circuit response that could not be achieved in a circuit with a single source. If the desired circuit response can be written as a sum of two or more terms, the response can be realized by including one independent source for each term of the response. This approach to the design of circuits with complex responses allows a designer to consider several simple designs instead of one complex design.

We demonstrate the superposition principle in Example 4.22.

EXAMPLE 4.22 | **Using Superposition to Solve a Circuit**

Use the superposition principle to find the branch currents in the circuit shown in Fig. 4.67.

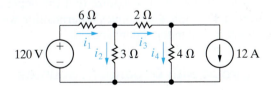

Figure 4.67 ▲ A circuit used to illustrate superposition.

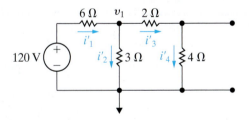

Figure 4.68 ▲ The circuit shown in Fig. 4.67 with the current source deactivated.

Solution

We begin by finding the branch currents resulting from the 120 V voltage source. We denote those currents with a prime. Replacing the ideal current source with an open circuit deactivates it; Fig. 4.68 shows this. The branch currents in this circuit are the result of only the voltage source.

We can easily find the branch currents in the circuit in Fig. 4.68 once we know the node voltage across the 3 Ω resistor. Denoting this voltage v_1, we write

$$\frac{v_1 - 120}{6} + \frac{v_1}{3} + \frac{v_1}{2 + 4} = 0,$$

from which

$$v_1 = 30 \text{ V}.$$

Now we can write the expressions for the branch currents $i'_1 - i'_4$ directly:

$$i'_1 = \frac{120 - 30}{6} = 15 \text{ A},$$

$$i'_2 = \frac{30}{3} = 10 \text{ A},$$

$$i'_3 = i'_4 = \frac{30}{6} = 5 \text{ A}.$$

To find the component of the branch currents resulting from the current source, we deactivate the ideal voltage source and solve the circuit shown in Fig. 4.69. The double-prime notation for the currents indicates they are the components of the total current resulting from the ideal current source.

We determine the branch currents in the circuit shown in Fig. 4.69 by first solving for the node voltages across the 3 and 4 Ω resistors, respectively. Figure 4.70 shows the two node voltages.

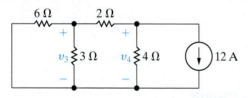

Figure 4.69 ▲ The circuit shown in Fig. 4.67 with the voltage source deactivated.

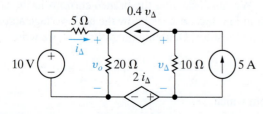

Figure 4.70 ▲ The circuit shown in Fig. 4.69 showing the node voltages v_3 and v_4.

The two KCL equations that describe the circuit are

$$\frac{v_3}{3} + \frac{v_3}{6} + \frac{v_3 - v_4}{2} = 0,$$

$$\frac{v_4 - v_3}{2} + \frac{v_4}{4} + 12 = 0.$$

Solving the simultaneous KCL equations for v_3 and v_4, we get

$$v_3 = -12 \text{ V},$$

$$v_4 = -24 \text{ V}.$$

Now we can write the branch currents i''_1 through i''_4 directly in terms of the node voltages v_3 and v_4:

$$i''_1 = \frac{-v_3}{6} = \frac{12}{6} = 2 \text{ A},$$

$$i''_2 = \frac{v_3}{3} = \frac{-12}{3} = -4 \text{ A},$$

$$i''_3 = \frac{v_3 - v_4}{2} = \frac{-12 + 24}{2} = 6 \text{ A},$$

$$i''_4 = \frac{v_4}{4} = \frac{-24}{4} = -6 \text{ A}.$$

To find the branch currents in the original circuit, that is, the currents i_1, i_2, i_3, and i_4 in Fig. 4.67, we simply add the single-primed currents to the double-primed currents:

$$i_1 = i'_1 + i''_1 = 15 + 2 = 17 \text{ A},$$

$$i_2 = i'_2 + i''_2 = 10 - 4 = 6 \text{ A},$$

$$i_3 = i'_3 + i''_3 = 5 + 6 = 11 \text{ A},$$

$$i_4 = i'_4 + i''_4 = 5 - 6 = -1 \text{ A}.$$

You should verify that the currents i_1, i_2, i_3, and i_4 have the correct values for the branch currents in the circuit shown in Fig. 4.67.

When applying superposition to linear circuits containing both independent and dependent sources, you must recognize that the dependent sources are never deactivated. Example 4.23 applies superposition when a circuit contains both dependent and independent sources.

EXAMPLE 4.23   Using Superposition to Solve a Circuit with Dependent Sources

Use the principle of superposition to find v_o in the circuit shown in Fig. 4.71.

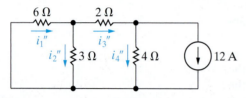

Figure 4.71 ▲ The circuit for Example 4.23.

Solution

We begin by finding the component of v_o resulting from the 10 V source. Figure 4.72 shows the circuit. With the 5 A source deactivated, v'_Δ must equal $(-0.4v'_\Delta)(10)$. Hence, v'_Δ must be zero, the branch containing the two dependent sources is open, and

$$v'_o = \frac{20}{25}(10) = 8 \text{ V}.$$

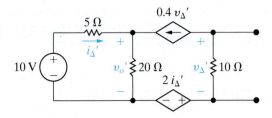

Figure 4.72 ▲ The circuit shown in Fig. 4.71 with the 5 A source deactivated.

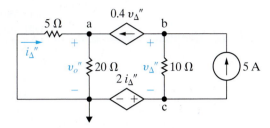

Figure 4.73 ▲ The circuit shown in Fig. 4.71 with the 10 V source deactivated.

When the 10 V source is deactivated, the circuit reduces to the one shown in Fig. 4.73. We have added a reference node and the node designations a, b, and c to aid the discussion. Summing the currents away from node a yields

$$\frac{v_o''}{20} + \frac{v_o''}{5} - 0.4v_\Delta'' = 0, \quad \text{or} \quad 5v_o'' - 8v_\Delta'' = 0.$$

Summing the currents away from node b gives

$$0.4v_\Delta'' + \frac{v_b - 2i_\Delta''}{10} - 5 = 0, \text{ or}$$

$$4v_\Delta'' + v_b - 2i_\Delta'' = 50.$$

We now use

$$v_b = 2i_\Delta'' + v_\Delta''$$

to find the value for v_Δ''. Thus,

$$5v_\Delta'' = 50, \quad \text{or} \quad v_\Delta'' = 10 \text{ V}.$$

From the node a equation,

$$5v_o'' = 80, \quad \text{or} \quad v_o'' = 16 \text{ V}.$$

The value of v_o is the sum of v_o' and v_o'', or 24 V.

SELF-CHECK: Assess your understanding of this material by trying Chapter Problems 4.92 and 4.97.

■ Practical Perspective

Circuits with Realistic Resistors

It is not possible to fabricate identical electrical components. For example, resistors produced from the same manufacturing process can vary in value by as much as 20%. Therefore, in creating an electrical system, the designer must consider the impact that component variation will have on the performance of the system. Sensitivity analysis permits the designer to calculate the impact of variations in the component values on the output of the system. We will see how this information enables a designer to specify an acceptable component value tolerance for each of the system's components.

Consider the circuit shown in Fig. 4.74. We will use sensitivity analysis to determine the sensitivity of the node voltages v_1 and v_2 to changes in the resistor R_1. Using the node-voltage method, we derive the expressions for v_1 and v_2 as functions of the circuit resistors and source currents. The results are given in Eqs. 4.45 and 4.46:

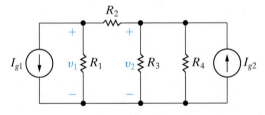

Figure 4.74 ▲ Circuit used to introduce sensitivity analysis.

$$v_1 = \frac{R_1\{R_3R_4I_{g2} - [R_2(R_3 + R_4) + R_3R_4]I_{g1}\}}{(R_1 + R_2)(R_3 + R_4) + R_3R_4}, \quad (4.45)$$

$$v_2 = \frac{R_3R_4[(R_1 + R_2)I_{g2} - R_1I_{g1}]}{(R_1 + R_2)(R_3 + R_4) + R_3R_4}. \quad (4.46)$$

The sensitivity of v_1 with respect to R_1 is found by differentiating Eq. 4.45 with respect to R_1, and similarly the sensitivity of v_2 with respect to R_1 is found by differentiating Eq. 4.46 with respect to R_1. We get

$$\frac{dv_1}{dR_1} = \frac{[R_3R_4 + R_2(R_3 + R_4)]\{R_3R_4I_{g2} - [R_3R_4 + R_2(R_3 + R_4)]I_{g1}\}}{[(R_1 + R_2)(R_3 + R_4) + R_3R_4]^2},$$

(4.47)

$$\frac{dv_2}{dR_1} = \frac{R_3R_4\{R_3R_4I_{g2} - [R_2(R_3 + R_4) + R_3R_4]I_{g1}\}}{[(R_1 + R_2)(R_3 + R_4) + R_3R_4]^2}.$$

(4.48)

We now consider an example with actual component values to illustrate the use of Eqs. 4.47 and 4.48.

EXAMPLE

Assume the nominal values of the components in the circuit in Fig. 4.74 are: $R_1 = 25\ \Omega$; $R_2 = 5\ \Omega$; $R_3 = 50\ \Omega$; $R_4 = 75\ \Omega$; $I_{g1} = 12$ A; and $I_{g2} = 16$ A. Use sensitivity analysis to predict the values of v_1 and v_2 if the value of R_1 is different by 10% from its nominal value.

Solution

From Eqs. 4.45 and 4.46 we find the nominal values of v_1 and v_2. Thus

$$v_1 = \frac{25\{3750(16) - [5(125) + 3750]12\}}{30(125) + 3750} = 25\ \text{V},$$

(4.49)

and

$$v_2 = \frac{3750[30(16) - 25(12)]}{30(125) + 3750} = 90\ \text{V}.$$

(4.50)

Now from Eqs. 4.47 and 4.48 we can find the sensitivity of v_1 and v_2 to changes in R_1. Hence,

$$\frac{dv_1}{dR_1} = \frac{[3750 + 5(125)] - \{3750(16) - [3750 + 5(125)]12\}}{[(30)(125) + 3750]^2}$$

$$= \frac{7}{12}\ \text{V}/\Omega,$$

(4.51)

and

$$\frac{dv_2}{dR_1} = \frac{3750\{3750(16) - [5(125) + 3750]12\}]}{(7500)^2}$$

$$= 0.5\ \text{V}/\Omega.$$

(4.52)

How do we use the results given by Eqs. 4.51 and 4.52? Assume that R_1 is 10% less than its nominal value, that is, $R_1 = 22.5\ \Omega$. Then $\Delta R_1 = -2.5\ \Omega$ and Eq. 4.51 predicts that Δv_1 will be

$$\Delta v_1 = \left(\frac{7}{12}\right)(-2.5) = -1.4583\ \text{V}.$$

Therefore, if R_1 is 10% less than its nominal value, our analysis predicts that v_1 will be

$$v_1 = 25 - 1.4583 = 23.5417\ \text{V}.$$

(4.53)

Similarly, for Eq. 4.52 we have

$$\Delta v_2 = 0.5(-2.5) = -1.25 \text{ V},$$

$$v_2 = 90 - 1.25 = 88.75 \text{ V}. \qquad (4.54)$$

We attempt to confirm the results in Eqs. 4.53 and 4.54 by substituting the value $R_1 = 22.5 \ \Omega$ into Eqs. 4.45 and 4.46. When we do, the results are

$$v_1 = 23.4780 \text{ V},$$

$$v_2 = 88.6960 \text{ V}.$$

Why is there a difference between the values predicted from the sensitivity analysis and the exact values computed by substituting for R_1 in the equations for v_1 and v_2? We can see from Eqs. 4.47 and 4.48 that the sensitivity of v_1 and v_2 with respect to R_1 is a function of R_1 because R_1 appears in the denominator of both Eqs. 4.47 and 4.48. This means that as R_1 changes, the sensitivities change; hence, we cannot expect Eqs. 4.47 and 4.48 to give exact results for large changes in R_1. Note that for a 10% change in R_1, the percent error between the predicted and exact values of v_1 and v_2 is small. Specifically, the percent error in $v_1 = 0.2713\%$ and the percent error in $v_2 = 0.0676\%$.

From this example, we can see that a tremendous amount of work is involved if we are to determine the sensitivity of v_1 and v_2 to changes in the remaining component values, namely, R_2, R_3, R_4, I_{g1}, and I_{g2}. Fortunately, PSpice[2] has a sensitivity function that will perform sensitivity analysis for us. The sensitivity function in PSpice calculates two types of sensitivity. The first is known as the one-unit sensitivity, and the second is known as the 1% sensitivity. In the example circuit, a one-unit change in a resistor would change its value by 1 Ω and a one-unit change in a current source would change its value by 1 A. In contrast, 1% sensitivity analysis determines the effect of changing resistors or sources by 1% of their nominal values.

The result of PSpice sensitivity analysis of the circuit in Fig. 4.69 is shown in Table 4.2. Because we are analyzing a linear circuit, we can

TABLE 4.2 **PSpice Sensitivity Analysis Results**

Element Name	Element Value	Element Sensitivity (Volts/Unit)	Normalized Sensitivity (Volts/Percent)
(a) DC Sensitivities of Node Voltage V1			
R1	25	0.5833	0.1458
R2	5	−5.417	−0.2708
R3	50	0.45	0.225
R4	75	0.2	0.15
IG1	12	−14.58	−1.75
IG2	16	12.5	2
(b) Sensitivities of Output V2			
R1	25	0.5	0.125
R2	5	6.5	0.325
R3	50	0.54	0.27
R4	75	0.24	0.18
IG1	12	−12.5	−1.5
IG2	16	15	2.4

[2]See the PSpice supplement that accompanies this text.

use superposition to predict values of v_1 and v_2 if more than one component's value changes. For example, let us assume R_1 decreases to 24 Ω and R_2 decreases to 4 Ω. From Table 4.2 we can combine the unit sensitivity of v_1 to changes in R_1 and R_2 to get

$$\frac{\Delta v_1}{\Delta R_1} + \frac{\Delta v_1}{\Delta R_2} = 0.5833 - 5.417 = -4.8337 \text{ V}/\Omega.$$

Similarly,

$$\frac{\Delta v_2}{\Delta R_1} + \frac{\Delta v_2}{\Delta R_2} = 0.5 + 6.5 = 7.0 \text{ V}/\Omega.$$

Thus, if both R_1 and R_2 decreased by 1 Ω, we would predict

$$v_1 = 25 + 4.8227 = 29.8337 \text{ V},$$

$$v_2 = 90 - 7 = 83 \text{ V}.$$

If we substitute $R_1 = 24 \ \Omega$ and $R_2 = 4 \ \Omega$ into Eqs. 4.45 and 4.46, we get

$$v_1 = 29.793 \text{ V},$$

$$v_2 = 82.759 \text{ V}.$$

In both cases, our predictions are within a fraction of a volt of the actual node voltage values.

Circuit designers use the results of sensitivity analysis to determine which component value variation has the greatest impact on the output of the circuit. As we can see from the PSpice sensitivity analysis in Table 4.2, the node voltages v_1 and v_2 are much more sensitive to changes in R_2 than to changes in R_1. Specifically, v_1 is (5.417/0.5833) or approximately 9 times more sensitive to changes in R_2 than to changes in R_1, and v_2 is (6.5/0.5) or 13 times more sensitive to changes in R_2 than to changes in R_1. Hence, in the example circuit, the tolerance on R_2 must be more stringent than the tolerance on R_1 if it is important to keep v_1 and v_2 close to their nominal values.

SELF-CHECK: Assess your understanding of this Practical Perspective by trying Chapter Problems 4.105–4.107.

■ Summary

- For the topics in this chapter, mastery of some basic terms, and the concepts they represent, is necessary. Those terms are **node**, **essential node**, **path**, **branch**, **essential branch**, **mesh**, and **planar circuit**. Table 4.1 provides definitions and examples of these terms. (See page 95.)

- Two new circuit analysis techniques were introduced in this chapter:

  - The **node-voltage method** works with both planar and nonplanar circuits. The steps in the node-voltage method are in Table 4.3. (See page 135.)

  - The **mesh-current method** works only with planar circuits. The steps in the mesh-current method are in Table 4.3. (See page 135.)

- Several new circuit simplification techniques were introduced in this chapter:

  - **Source transformations** allow us to exchange a voltage source (v_s) and a series resistor (R) for a current source (i_s) and a parallel resistor (R) and vice versa. The combinations must be equivalent in terms of their terminal voltage and current. Terminal equivalence holds, provided that

    $$i_s = \frac{v_s}{R}.$$

    (See page 115–116.)

TABLE 4.3	**Steps in the Node-Voltage Method and the Mesh-Current Method**	
	Node-Voltage Method	**Mesh-Current Method**
Step 1 Identify nodes/meshes	Identify the essential nodes by circling them on the circuit diagram	Identify the meshes by drawing directed arrows inside each mesh
Step 2 Label node voltages/mesh currents Recognize special cases	Pick and label a reference node; then label the remaining essential node voltages • If a voltage source is the only component in a branch connecting the reference node and another essential node, label the essential node with the value of the voltage source • If a voltage source is the only component in a branch connecting two nonreference essential nodes, create a supernode that includes the voltage source and the two nodes on either side	Label each mesh current • If a current source is in a single mesh, label the mesh current with the value of the current source • If a current source is shared by two adjacent meshes, create a supermesh by combining the two adjacent meshes and temporarily eliminating the branch that contains the current source
Step 3 Write the equations	Write the following equations: • A KCL equation for any supernodes • A KCL equation for any remaining essential nodes where the voltage is unknown • A constraint equation for each dependent source that defines the controlling variable for the dependent source in terms of the node voltages • A constraint equation for each supernode that equates the difference between the two node voltages in the supernode to the voltage source in the supernode	Write the following equations: • A KVL equation for any supermeshes • A KVL equation for any remaining meshes where the current is unknown • A constraint equation for each dependent source that defines the controlling variable for the dependent source in terms of the mesh currents • A constraint equation for each supermesh that equates the difference between the two mesh currents in the supermesh to the current source eliminated to form the supermesh
Step 4 Solve the equations	Solve the equations to find the node voltages	Solve the equations to find the mesh currents
Step 5 Solve for other unknowns	Use the node voltage values to find any unknown voltages, currents, or powers	Use the mesh current values to find any unknown voltages, currents, or powers

- **Thévenin equivalents** and **Norton equivalents** allow us to simplify a circuit composed of sources and resistors into an equivalent circuit consisting of a voltage source and a series resistor (Thévenin) or a current source and a parallel resistor (Norton). The simplified circuit and the original circuit must be equivalent in terms of their terminal voltage and current. Note that

  - The Thévenin voltage (V_{Th}) is the open-circuit voltage across the terminals of the original circuit;

  - The Thévenin resistance (R_{Th}) is the ratio of the Thévenin voltage to the current in a short circuit connecting the terminals of the original circuit;

  - The Norton equivalent is obtained by performing a source transformation on a Thévenin equivalent.

  (See page 118–121.)

- **Maximum power transfer** is a technique for calculating the maximum value of p that can be delivered to a load, R_L. Maximum power transfer occurs when $R_L = R_{\text{Th}}$, the Thévenin resistance as seen from the resistor R_L. The equation for the maximum power transferred is

$$p = \frac{V_{\text{Th}}^2}{4R_L}.$$

  (See page 126–127.)

- In a circuit with multiple independent sources, **superposition** allows us to activate one source at a time and calculate voltages and currents due to each source. To determine the voltages and currents that exist when all independent sources are active, sum the voltages and currents that resulted from each of the sources. Dependent sources are never deactivated when applying superposition. (See page 129.)

■ Problems

Section 4.1

4.1 For the circuit shown in Fig. P4.1, state the numerical value of the number of (a) branches, (b) branches where the current is unknown, (c) essential branches, (d) essential branches where the current is unknown, (e) nodes, (f) essential nodes, and (g) meshes.

Figure P4.1

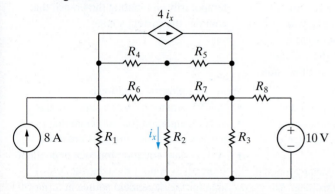

4.2 a) If only the essential nodes and branches are identified in the circuit in Fig. P4.1, how many simultaneous equations are needed to describe the circuit?

b) How many of these equations can be derived using Kirchhoff's current law?

c) How many must be derived using Kirchhoff's voltage law?

d) What two meshes should be avoided in applying the voltage law?

4.3 Assume the voltage v_s in the circuit in Fig. P4.3 is known. The resistors $R_1 - R_7$ are also known.

a) How many unknown currents are there?

b) How many independent equations can be written using Kirchhoff's current law (KCL)?

c) Write an independent set of KCL equations.

d) How many independent equations can be derived from Kirchhoff's voltage law (KVL)?

e) Write a set of independent KVL equations.

Figure P4.3

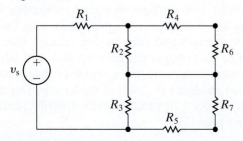

4.4 A current leaving a node is defined as positive.

a) Sum the currents at each essential node in the circuit shown in Fig. P4.3.

b) Show that any one of the equations in (a) can be derived from the remaining three equations.

4.5 Look at the circuit in Fig. 4.4.

a) Write the KCL equation at the essential node labeled g.

b) Show that the KCL equation in part (a) can be derived from the KCL equations at nodes b, c, and e (see Example 4.2).

Section 4.2

4.6 Use the node-voltage method to find v_o in the circuit in Fig. P4.6.

PSPICE
MULTISIM

Figure P4.6

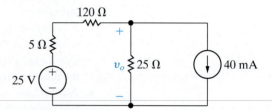

4.7 a) Find the power developed by the 40 mA current source in the circuit in Fig. P4.6.

PSPICE
MULTISIM

b) Find the power developed by the 25 V voltage source in the circuit in Fig. P4.6.

c) Verify that the total power developed equals the total power dissipated.

4.8 A 100 Ω resistor is connected in series with the 40 mA current source in the circuit in Fig. P4.6.

PSPICE
MULTISIM

a) Find v_o.

b) Find the power developed by the 40 mA current source.

c) Find the power developed by the 25 V voltage source.

d) Verify that the total power developed equals the total power dissipated.

e) What effect will any finite resistance connected in series with the 40 mA current source have on the value of v_o?

4.9 Use the node-voltage method to find how much power the 2 A source extracts from the circuit in Fig. P4.9.

PSPICE
MULTISIM

Figure P4.9

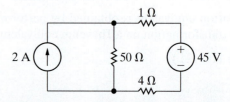

4.10 Use the node-voltage method to find v_1 and v_2 in
the circuit shown in Fig. P4.10.

Figure P4.10

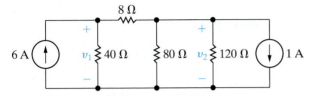

4.11 Use the node-voltage method to find v_1 and v_2 in
the circuit in Fig. P4.11.

Figure P4.11

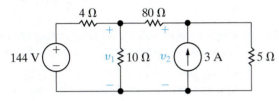

4.12 a) Use the node-voltage method to find the branch
currents $i_a - i_e$ in the circuit shown in Fig. P4.12.

b) Find the total power developed in the circuit.

Figure P4.12

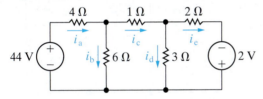

4.13 a) Use the node-voltage method to find v_1, v_2, and
v_3 in the circuit in Fig. P4.13.

b) How much power does the 40 V voltage source
deliver to the circuit?

Figure P4.13

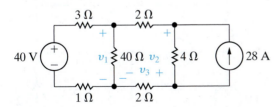

4.14 The circuit shown in Fig. P4.14 is a dc model of a
residential power distribution circuit.

a) Use the node-voltage method to find the branch
currents $i_1 - i_6$.

b) Test your solution for the branch currents by
showing that the total power dissipated equals
the total power developed.

Figure P4.14

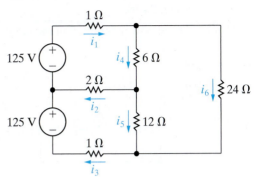

4.15 Use the node-voltage method to find the total
power dissipated in the circuit in Fig. P4.15.

Figure P4.15

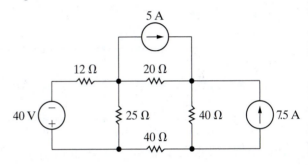

4.16 a) Use the node-voltage method to show that the
output voltage v_o in the circuit in Fig. P4.16 is
equal to the average value of the source voltages.

b) Find v_o if $v_1 = 120$ V, $v_2 = 60$ V, and
$v_3 = -30$ V.

Figure P4.16

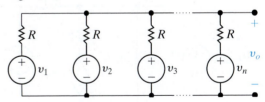

Section 4.3

4.17 Use the node-voltage method to calculate the
power delivered by the dependent voltage source in
the circuit in Fig. P4.17.

Figure P4.17

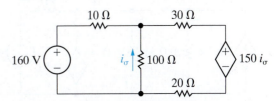

4.18 a) Use the node voltage method to find v_o for the circuit in Fig. P4.18.

b) Find the total power supplied in the circuit.

Figure P4.18

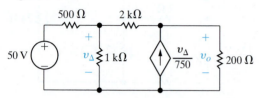

4.19 a) Use the node-voltage method to find the total power developed in the circuit in Fig. P4.19.

PSPICE
MULTISIM

b) Check your answer by finding the total power absorbed in the circuit.

Figure P4.19

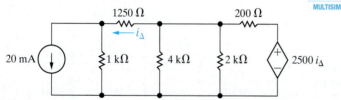

4.20 a) Use the node-voltage method to find v_o in the circuit in Fig. P4.20.

PSPICE
MULTISIM

b) Find the power absorbed by the dependent source.

c) Find the total power developed by the independent sources.

Figure P4.20

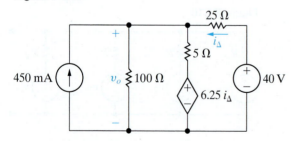

4.21 a) Find the node voltages v_1, v_2, and v_3 in the circuit in Fig. P4.21.

PSPICE
MULTISIM

b) Find the total power dissipated in the circuit.

Figure P4.21

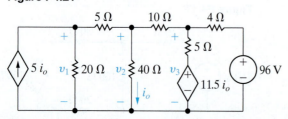

Section 4.4

4.22 Use the node-voltage method to find the value of v_o in the circuit in Fig. P4.22.

PSPICE
MULTISIM

Figure P4.22

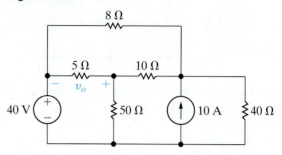

4.23 a) Use the node-voltage method to find the branch currents i_1, i_2, and i_3 in the circuit in Fig. P4.23.

PSPICE
MULTISIM

b) Check your solution for i_1, i_2, and i_3 by showing that the power dissipated in the circuit equals the power developed.

Figure P4.23

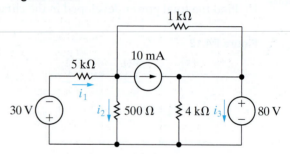

4.24 Use the node-voltage method to find the value of v_o in the circuit in Fig. P4.24.

PSPICE
MULTISIM

Figure P4.24

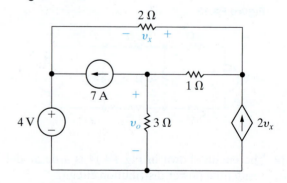

4.25 a) Use the node-voltage method to find the power dissipated in the 2 Ω resistor in the circuit in Fig. P4.25.

b) Find the power supplied by the 230 V source.

Figure P4.25

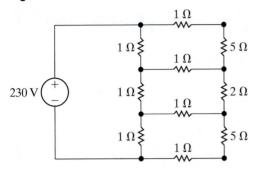

4.26 Use the node-voltage method to find i_o in the circuit in Fig. P4.26.

PSPICE
MULTISIM

Figure P4.26

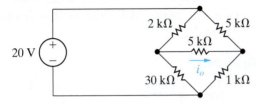

4.27 a) Use the node-voltage method to find v_o and the power delivered by the 2 A current source in the circuit in Fig. P4.27. Use node a as the reference node.

PSPICE
MULTISIM

   b) Repeat part (a), but use node b as the reference node.

   c) Compare the choice of reference node in (a) and (b) Which is better, and why?

Figure P4.27

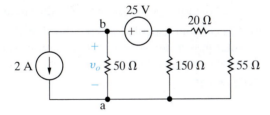

4.28 Use the node-voltage method to find v_o in the circuit in Fig. P4.28.

PSPICE
MULTISIM

Figure P4.28

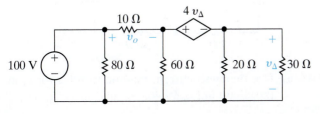

4.29 Use the node-voltage method to find the power developed by the 20 V source in the circuit in Fig. P4.29.

PSPICE
MULTISIM

Figure P4.29

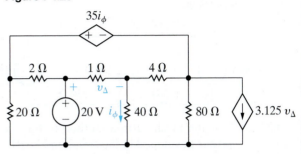

4.30 Assume you are a project engineer and one of your staff is assigned to analyze the circuit shown in Fig. P4.30. The reference node and node numbers given on the figure were assigned by the analyst. Her solution gives the values of v_1 and v_2 as 105 V and 85 V, respectively.

   a) What values did the analyst use for the left-most and right-most node voltages when writing KCL equations at nodes 1 and 2?

   b) Use the values supplied by the analyst to calculate the total power developed in the circuit and the total power dissipated in the circuit.

   c) Do you agree with the solution submitted by the analyst?

Figure P4.30

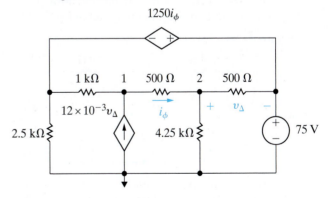

4.31 Show that when Eqs. 4.13, 4.14, and 4.16 are solved for i_B, the result is identical to Eq. 2.21.

Section 4.5

4.32 Solve Problem 4.12 using the mesh-current method.

4.33 Solve Problem 4.14 using the mesh-current method.

4.34 Solve Problem 4.25 using the mesh-current method.

4.35 Solve Problem 4.26 using the mesh-current method.

4.36 a) Use the mesh-current method to find the branch currents i_a, i_b, and i_c in the circuit in Fig. P4.36.

PSPICE
MULTISIM

b) Repeat (a) if the polarity of the 140 V source is reversed.

Figure P4.36

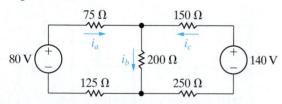

4.37 a) Use the mesh-current method to find the total power developed in the circuit in Fig. P4.37.

PSPICE
MULTISIM

b) Check your answer by showing that the total power developed equals the total power dissipated.

Figure P4.37

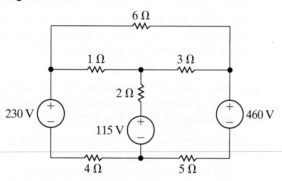

Section 4.6

4.38 Solve Problem 4.17 using the mesh-current method.

4.39 Use the mesh-current method to find the power dissipated in the 15 Ω resistor in the circuit in Fig. P4.39.

PSPICE
MULTISIM

Figure P4.39

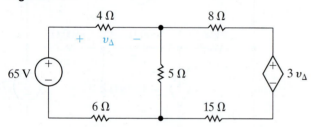

4.40 Use the mesh-current method to find the power developed in the dependent voltage source in the circuit in Fig. P4.40.

PSPICE
MULTISIM

Figure P4.40

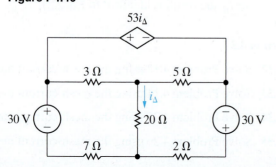

4.41 Use the mesh-current method to find the power delivered by the dependent voltage source in the circuit seen in Fig. P4.41.

PSPICE
MULTISIM

Figure P4.41

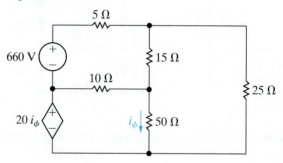

4.42 a) Use the mesh-current method to find v_o in the circuit in Fig. P4.42.

PSPICE
MULTISIM

b) Find the power delivered by the dependent source.

Figure P4.42

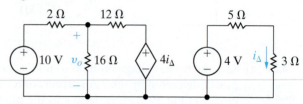

Section 4.7

4.43 Solve Problem 4.10 using the mesh-current method.

4.44 Solve Problem 4.21 using the mesh-current method.

4.45 a) Use the mesh-current method to find how much power the 5 A current source delivers to the circuit in Fig. P4.45.

b) Find the total power delivered to the circuit.

c) Check your calculations by showing that the total power developed in the circuit equals the total power dissipated

Figure P4.45

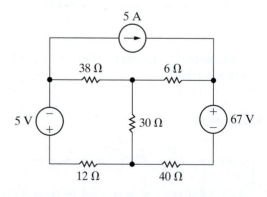

4.46 a) Use the mesh-current method to solve for i_Δ in the circuit in Fig. P4.46.

PSPICE
MULTISIM

b) Find the power delivered by the independent current source.

c) Find the power delivered by the dependent voltage source.

Figure P4.46

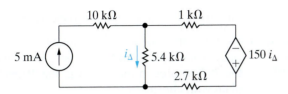

4.47 a) Use the mesh-current method to determine which sources in the circuit in Fig. P4.47 are generating power.

b) Find the total power dissipated in the circuit.

Figure P4.47

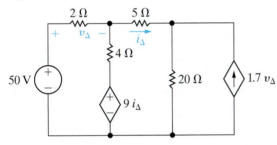

4.48 Use the mesh-current method to find the total power developed in the circuit in Fig. P4.48.

Figure P4.48

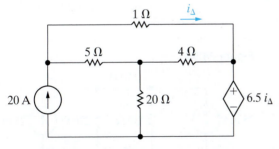

4.49 Solve Problem 4.23 using the mesh-current method.

4.50 Use the mesh-current method to find the total power dissipated in the circuit in Fig. P4.50.

Figure P4.50

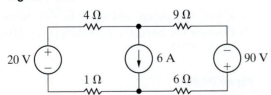

4.51 a) Assume the 20 V source in the circuit in Fig. P4.50 is changed to 60 V. Find the total power dissipated in the circuit.

b) Repeat (a) with the 6 A current source replaced by a short circuit.

c) Explain why the answers to (a) and (b) are the same.

d) Now assume you wish to change the value of the 90 V source, instead of the 20 V source, in the circuit in Fig. P4.50 to get the same power dissipated by the current source that you found in (a) and (b). Use the results in part (c) to calculate the new value of this voltage source.

4.52 a) Use the mesh-current method to find the branch currents in $i_a - i_e$ in the circuit in Fig. P4.52.

b) Check your solution by showing that the total power developed in the circuit equals the total power dissipated.

Figure P4.52

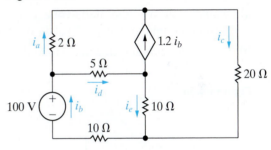

4.53 a) Find the branch currents $i_a - i_e$ for the circuit shown in Fig. P4.53.

b) Check your answers by showing that the total power generated equals the total power dissipated.

Figure P4.53

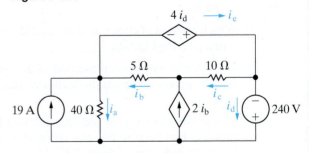

Section 4.8

4.54 The variable dc voltage source in the circuit in Fig. P4.54 is adjusted so that i_o is zero.

a) Would you use the node-voltage or mesh-current method to find v_{dc}? Explain your choice.

b) Find the value of v_{dc}, using the method selected in (a).

c) Check your solution by showing the power developed equals the power dissipated.

Figure P4.54

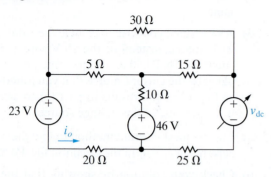

4.55 The variable dc current source in the circuit in Fig. P4.55 is adjusted so that the power developed by the 4 A current source is zero. You want to find the value of i_{dc}.

a) Would you use the node-voltage or mesh-current method to find i_{dc}? Explain your choice.

b) Use the method selected in (a) to find i_{dc}.

Figure P4.55

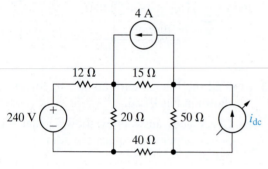

4.56 Assume you have been asked to find the power dissipated in the horizontal 1 kΩ resistor in the circuit in Fig. P4.56.

a) Which method of circuit analysis would you recommend? Explain why.

b) Use your recommended method of analysis to find the power dissipated in the horizontal 1 kΩ resistor.

c) Would you change your recommendation if the problem had been to find the power developed by the 10 mA current source? Explain.

d) Find the power delivered by the 10 mA current source.

Figure P4.56

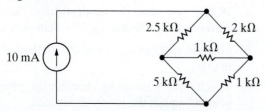

4.57 A 4 kΩ resistor is placed in parallel with the 10 mA current source in the circuit in Fig. P4.56. Assume you have been asked to calculate the power developed by the current source.

a) Which method of circuit analysis would you recommend? Explain why.

b) Find the power developed by the current source.

4.58 a) Would you use the node-voltage or mesh-current method to find the power absorbed by the 20 V source in the circuit in Fig. P4.58? Explain your choice.

b) Use the method you selected in (a) to find the power.

Figure P4.58

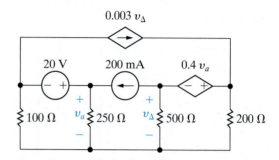

Section 4.9

4.59 a) Use source transformations to find the current i_o in the circuit in Fig. P4.59.

b) Verify your solution by using the node-voltage method to find i_o.

Figure P4.59

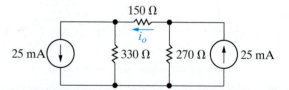

4.60 a) Find the current i_o in the circuit in Fig. P4.60 by making a succession of appropriate source transformations.

b) Using the result obtained in (a), work back through the circuit to find the power developed by the 120 V source.

Figure P4.60

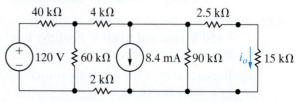

4.61 a) Make a series of source transformations to find the voltage v_0 in the circuit in Fig. P4.61.

b) Verify your solution using the mesh-current method.

Figure P4.61

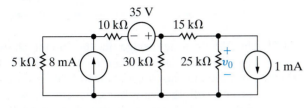

4.62 a) Use a series of source transformations to find i_o in the circuit in Fig. P4.62.

PSPICE
MULTISIM

b) Verify your solution by using the mesh-current method to find i_o.

Figure P4.62

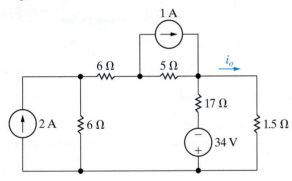

4.63 a) Use source transformations to find v_o in the circuit in Fig. P4.63.

PSPICE
MULTISIM

b) Find the power developed by the 520 V source.

c) Find the power developed by the 1 A current source.

d) Verify that the total power developed equals the total power dissipated.

Figure P4.63

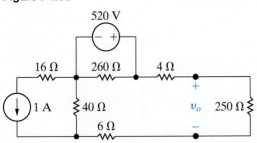

Section 4.10

4.64 Find the Thévenin equivalent with respect to the terminals a, b for the circuit in Fig. P4.64.

PSPICE
MULTISIM

Figure P4.64

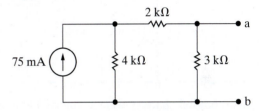

4.65 Find the Norton equivalent with respect to the terminals a, b for the circuit in Fig. P4.65.

Figure P4.65

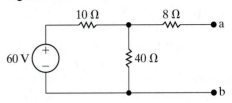

4.66 Find the Norton equivalent with respect to the terminals a, b for the circuit in Fig. P4.66.

PSPICE
MULTISIM

Figure P4.66

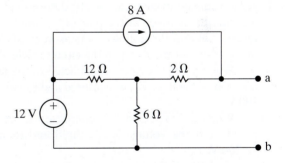

4.67 Find the Thévenin equivalent with respect to the terminals a, b for the circuit in Fig. P4.67.

PSPICE
MULTISIM

Figure P4.67

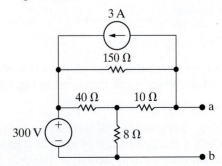

4.68 Find the Norton equivalent with respect to the terminals a, b in the circuit in Fig. P4.68.

PSPICE
MULTISIM

Figure P4.68

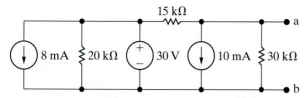

4.69 Determine i_o and v_o in the circuit shown in Fig. P4.69 when R_o is a resistor from Appendix H such that $100 \ \Omega \le R_o < 200 \ \Omega$.

PSPICE
MULTISIM

Figure P4.69

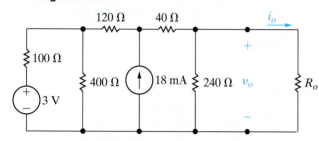

4.70 An automobile battery, when connected to a car radio, provides 12.72 V to the radio. When connected to a set of headlights, it provides 12 V to the headlights. Assume the radio can be modeled as a 6.36 Ω resistor and the headlights can be modeled as a 0.6 Ω resistor. What are the Thévenin and Norton equivalents for the battery?

4.71 A Thévenin equivalent can also be determined from measurements made at the pair of terminals of interest. Assume the following measurements were made at the terminals a,b in the circuit in Fig. P4.71.

When a 20 Ω resistor is connected to the terminals a, b, the voltage v_{ab} is measured and found to be 100 V.

When a 50 Ω resistor is connected to the terminals a, b, the voltage is measured and found to be 200 V.

Find the Thévenin equivalent of the network with respect to the terminals a, b.

Figure P4.71

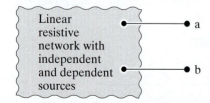

4.72 A voltmeter with a resistance of 85.5 kΩ is used to measure the voltage v_{ab} in the circuit in Fig. P4.72.

PSPICE
MULTISIM

a) What is the voltmeter reading?

b) What is the percentage of error in the voltmeter reading if the percentage of error is defined as [(measured − actual)/actual] × 100?

Figure P4.72

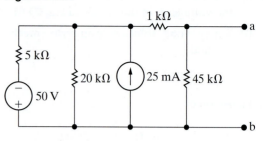

4.73 The Wheatstone bridge in the circuit shown in Fig. P4.73 is balanced when R_3 equals 500 Ω. If the galvanometer has a resistance of 50 Ω, how much current will the galvanometer detect, when the bridge is unbalanced by setting R_3 to 501 Ω? (*Hint:* Find the Thévenin equivalent with respect to the galvanometer terminals when $R_3 = 501 \ \Omega$. Note that once we have found this Thévenin equivalent, it is easy to find the amount of unbalanced current in the galvanometer branch for different galvanometer movements.)

PSPICE
MULTISIM

Figure P4.73

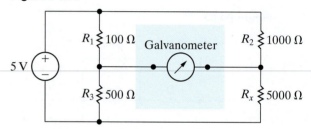

4.74 Determine the Thévenin equivalent with respect to the terminals a, b for the circuit shown in Fig. P4.74.

PSPICE
MULTISIM

Figure P4.74

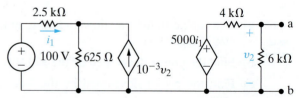

4.75 Find the Norton equivalent with respect to the terminals a, b for the circuit seen in Fig. P4.75.

PSPICE
MULTISIM

Figure P4.75

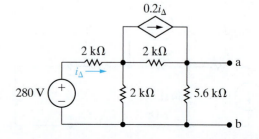

Figure P4.91

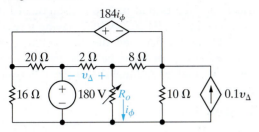

Figure P4.95

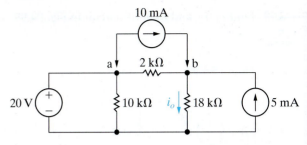

Section 4.13

4.92 a) Use the principle of superposition to find the voltage v in the circuit of Fig. P4.92.

PSPICE
MULTISIM

b) Find the power dissipated in the 10 Ω resistor.

Figure P4.92

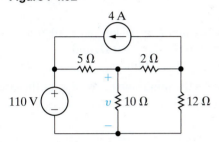

4.93 Use superposition to solve for i_o and v_o in the circuit in Fig. P4.93.

Figure P4.93

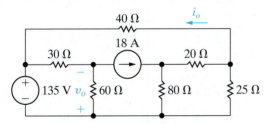

4.94 Use the principle of superposition to find the voltage v_o in the circuit in Fig. P4.94.

PSPICE
MULTISIM

Figure P4.94

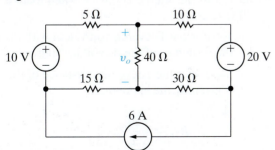

4.95 a) In the circuit in Fig. P4.95, before the 10 mA current source is attached to the terminals a, b, the current i_o is calculated and found to be 1.5 mA. Use superposition to find the value of i_o after the current source is attached.

PSPICE
MULTISIM

b) Verify your solution by finding i_o when all three sources are acting simultaneously.

4.96 Use the principle of superposition to find the current i_o in the circuit shown in Fig. P4.96.

PSPICE
MULTISIM

Figure P4.96

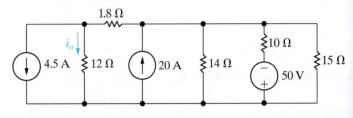

4.97 Use the principle of superposition to find the current i in the circuit of Fig. P4.97.

PSPICE
MULTISIM

Figure P4.97

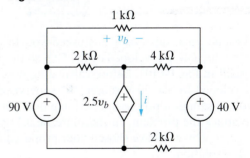

4.98 Use the principle of superposition to find v_o in the circuit in Fig. P4.98.

PSPICE
MULTISIM

Figure P4.98

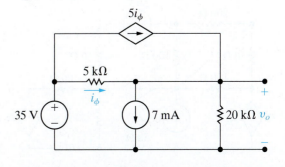

Sections 4.1–4.13

4.99 Find v_1, v_2, and v_3 in the circuit in Fig. P4.99.

Figure P4.99

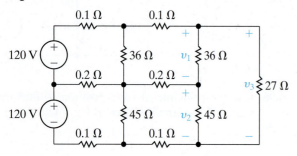

4.100 Find i_1 and i_2 in the circuit in Fig. P4.100.

Figure P4.100

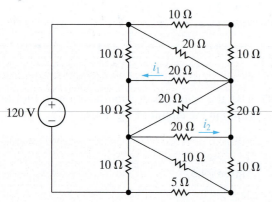

4.101 Assume your supervisor has asked you to determine the power developed by the 50 V source in the circuit in Fig. P4.101. Before calculating the power developed by the 50 V source, the supervisor asks you to submit a proposal describing how you plan to attack the problem. Furthermore, he asks you to explain why you have chosen your proposed method of solution.

a) Describe your plan of attack, explaining your reasoning.

b) Use the method you have outlined in (a) to find the power developed by the 50 V source.

Figure P4.101

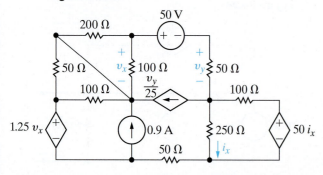

4.102 Two ideal dc voltage sources are connected by electrical conductors that have a resistance of r Ω/m, as shown in Fig. P4.102. A load having a resistance of R Ω moves between the two voltage sources. Let x equal the distance between the load and the source v_1, and let L equal the distance between the sources.

a) Show that

$$v = \frac{v_1 RL + R(v_2 - v_1)x}{RL + 2rLx - 2rx^2}.$$

b) Show that the voltage v will be minimum when

$$x = \frac{L}{v_2 - v_1}\left[-v_1 \pm \sqrt{v_1 v_2 - \frac{R}{2rL}(v_1 - v_2)^2}\right].$$

c) Find x when $L = 16$ km, $v_1 = 1000$ V, $v_2 = 1200$ V, $R = 3.9$ Ω, and $r = 5 \times 10^{-5}$ Ω/m.

d) What is the minimum value of v for the circuit of part (c)?

Figure P4.102

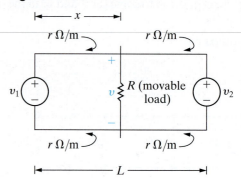

4.103 Laboratory measurements on a dc voltage source yield a terminal voltage of 75 V with no load connected to the source and 60 V when loaded with a 20 Ω resistor.

a) What is the Thévenin equivalent with respect to the terminals of the dc voltage source?

b) Show that the Thévenin resistance of the source is given by the expression

$$R_{Th} = \left(\frac{v_{Th}}{v_o} - 1\right)R_L,$$

where
v_{Th} = the Thévenin voltage,
v_o = the terminal voltage corresponding to the load resistance R_L.

4.104 For the circuit in Fig. 4.74 derive the expressions for the sensitivity of v_1 and v_2 to changes in the source currents I_{g1} and I_{g2}.

4.105 Assume the nominal values for the components in
PRACTICAL the circuit in Fig. 4.74 are: $R_1 = 25\ \Omega$; $R_2 = 5\ \Omega$;
PERSPECTIVE $R_3 = 50\ \Omega$; $R_4 = 75\ \Omega$; $I_{g1} = 12$ A; and $I_{g2} = 16$ A.
PSPICE Predict the values of v_1 and v_2 if I_{g1} decreases to
MULTISIM 11 A and all other components stay at their nominal values. Check your predictions using a tool like PSpice or MATLAB.

4.106 Repeat Problem 4.105 if I_{g2} increases to 17 A, and
PRACTICAL all other components stay at their nominal values.
PERSPECTIVE Check your predictions using a tool like PSpice or MATLAB.

4.107 Repeat Problem 4.105 if I_{g1} decreases to 11 A and
PRACTICAL I_{g2} increases to 17 A. Check your predictions using a
PERSPECTIVE tool like PSpice or MATLAB.
PSPICE
MULTISIM

4.108 Use the results given in Table 4.2 to predict the
PRACTICAL values of v_1 and v_2 if R_1 and R_3 increase to 10%
PERSPECTIVE above their nominal values and R_2 and R_4 decrease to 10% below their nominal values. I_{g1} and I_{g2} remain at their nominal values. Compare your predicted values of v_1 and v_2 with their actual values.

CHAPTER

5

The Operational Amplifier

CHAPTER CONTENTS

5.1 **Operational Amplifier Terminals** *p. 152*

5.2 **Terminal Voltages and Currents** *p. 152*

5.3 **The Inverting-Amplifier Circuit** *p. 156*

5.4 **The Summing-Amplifier Circuit** *p. 158*

5.5 **The Noninverting-Amplifier Circuit** *p. 160*

5.6 **The Difference-Amplifier Circuit** *p. 162*

5.7 **A More Realistic Model for the Operational Amplifier** *p. 167*

CHAPTER OBJECTIVES

1 Be able to name the five op amp terminals and describe and use the voltage and current constraints and the resulting simplifications they lead to in an ideal op amp.

2 Be able to analyze simple circuits containing ideal op amps and recognize the following op amp circuits: inverting amplifier, summing amplifier, noninverting amplifier, and difference amplifier.

3 Understand the more realistic model for an op amp and be able to use this model to analyze simple circuits containing op amps.

This chapter analyzes circuits containing sources, resistors, and a new component, the operational amplifier (op amp). Unlike sources and resistors, the op amp is not an ideal basic circuit element. Instead, it is a complicated integrated circuit consisting of many electronic components such as transistors and diodes that are beyond the scope of this text. We can use op amps in introductory circuits, however, by taking a black box approach that focuses solely on the terminal behavior of the op amp (not on its internal structure or the currents and voltages that exist in this structure).

Operational amplifier circuits were first used as basic building blocks in analog computers. The term *operational* refers to op amp circuits that implement mathematical operations such as integration, differentiation, addition, sign changing, and scaling. While the range of applications has broadened beyond implementing mathematical operations, the original name for the circuit persists.

We do not introduce new circuit analysis techniques in this chapter. Instead, we apply tools we have already introduced to analyze and design interesting and useful op amp circuits that perform scaling, addition, and subtraction. Once we introduce inductors and capacitors in Chapter 6, we will present op amp circuits that integrate and differentiate electric signals.

Initially, we employ an ideal model of the op amp's terminal behavior. At the conclusion of this chapter, we consider a more realistic op amp model that employs a dependent source. This provides additional opportunities to practice analyzing circuits with these sources.

■ Practical Perspective

Strain Gages

How could you measure the amount of bending in metal cables, such as the ones shown in the figure, without physically contacting the cables? One method uses a strain gage, which is a type of **transducer**. A transducer is a device that measures a quantity by converting it into a more convenient form. The quantity we wish to measure in the metal bar is the bending angle, but measuring the angle directly is quite difficult and could even be dangerous. Instead, we attach a strain gage (shown in the line drawing here) to the metal bar. A strain gage is a grid of thin wires whose resistance changes when the wires are lengthened or shortened, according to the equation

$$\Delta R = 2R \frac{\Delta L}{L},$$

where R is the resistance of the gage at rest, $\Delta L / L$ is the fractional lengthening of the gage (which is the definition of "strain"), 2 is a constant typical of the manufacturer's gage factor, and ΔR is the change in resistance due to the bending of the bar. Typically, pairs of strain gages are attached to opposite sides of a bar. When the bar is bent, the wires in one pair of gages get longer and thinner, increasing the resistance, while the wires in the other pair of gages get shorter and thicker, decreasing the resistance.

How can the change in resistance be measured? One way would be to use an ohmmeter. However, the change in resistance experienced by the strain gage is typically much smaller than an ohmmeter can measure accurately. Usually, the pairs of strain gages are connected to form a Wheatstone bridge, and the voltage difference between two legs of the bridge is measured. We use an op amp circuit to amplify, or increase, the voltage difference to make an accurate measurement. After we introduce the operational amplifier and some of the important circuits that employ them, we conclude with an analysis of a strain gage circuit that measures the bending in the metal bar.

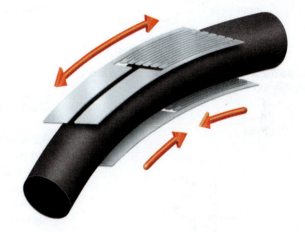

Horiyan/Shutterstock

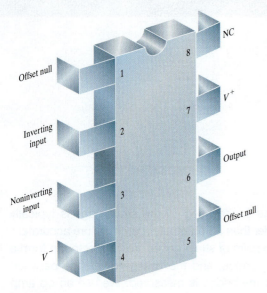

Figure 5.1 ▲ The eight-lead DIP package (top view).

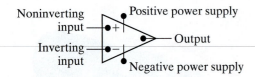

Figure 5.2 ▲ The circuit symbol for an operational amplifier (op amp).

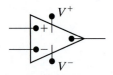

Figure 5.3 ▲ A simplified circuit symbol for an op amp.

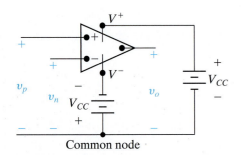

Figure 5.4 ▲ Terminal voltage variables.

5.1 Operational Amplifier Terminals

We begin by looking at the terminal behavior of a commercially available op amp, the μA 741. Fairchild Semiconductor introduced this widely used device in 1968. This op amp is available in several different packages. For our discussion, we assume an eight-lead DIP.[1] Figure 5.1 shows a top view of the package, with the terminals, their names, and numbers. We focus on the following terminals:

- inverting input,
- noninverting input,
- output,
- positive power supply (V^+),
- negative power supply (V^-).

The remaining three terminals are of little or no concern. The two offset null terminals may be used in an auxiliary circuit that compensates for performance degradation owing to aging and imperfections. These terminals are seldom used because degradation is usually negligible. Terminal 8 is an unused terminal; NC stands for no connection, which means that the terminal is not connected to the op amp circuit.

Figure 5.2 shows a common circuit symbol for an op amp that contains the five terminals of primary interest. Because word labels are inconvenient in circuit diagrams, we simplify the terminal designations in the following way. The noninverting input terminal is labeled plus (+), and the inverting input terminal is labeled minus (−). The power supply terminals, which are always drawn outside the triangle, are marked V^+ and V^-. The terminal at the apex of the triangle is always understood to be the output terminal. Figure 5.3 summarizes these simplifications.

5.2 Terminal Voltages and Currents

We now describe the behavior of the op amp using the terminal voltages and currents. The voltage variables are measured from a common reference node.[2] Figure 5.4 shows the five voltage variables with their reference polarities.

All voltages are considered as voltage rises from the common node, a convention we also used in the node-voltage analysis method. A positive supply voltage (V_{CC}) is connected between V^+ and the common node. A negative supply voltage ($-V_{CC}$) is connected between V^- and the common node. The voltage between the inverting input terminal and the common node is denoted v_n. The voltage between the noninverting input terminal and the common node is designated as v_p. The voltage between the output terminal and the common node is denoted v_o.

Figure 5.5 shows the current variables with their reference directions, all of which are into the terminals of the operational amplifier: i_n is the current into the inverting input terminal; i_p is the current into the noninverting input terminal; i_o is the current into the output terminal; i_{c^+} is the current into the positive power supply terminal; and i_{c^-} is the current into the negative power supply terminal.

---

[1] DIP is an abbreviation for *dual in-line package*. This means that the terminals on each side of the package are in line and that the terminals on opposite sides of the package also line up.

[2] The common node is external to the op amp. It is the reference terminal of the circuit in which the op amp is embedded.

The terminal behavior of the op amp as a linear circuit element is characterized by constraints on the input voltages and the input currents. The voltage constraint is derived from the voltage transfer characteristic of the op amp integrated circuit, pictured in Fig. 5.6.

The voltage transfer characteristic describes how the output voltage varies as a function of the input voltages—that is, how voltage is transferred from the input to the output. Note that the op amp's output voltage is a function of the difference between its input voltages, $v_p - v_n$. The equation for the voltage transfer characteristic is

$$v_o = \begin{cases} -V_{CC} & A(v_p - v_n) < -V_{CC}, \\ A(v_p - v_n) & -V_{CC} \leq A(v_p - v_n) \leq +V_{CC}, \\ +V_{CC} & A(v_p - v_n) > +V_{CC}. \end{cases} \tag{5.1}$$

We see from Fig. 5.6 and Eq. 5.1 that the op amp has three distinct regions of operation: negative saturation, linear region, and positive saturation. When the magnitude of the input voltage difference ($|v_p - v_n|$) is small, the op amp behaves as a linear device, so the output voltage is a linear function of the input voltages. More specifically, the output voltage is equal to the difference between the input voltages times the multiplying constant, or **gain**, A. Outside this linear region are two saturation regions. When the output of the op amp saturates, the op amp behaves as a nonlinear device; its output voltage is no longer a linear function of the input voltages.

Op Amp Input Voltage Constraints

When we confine the op amp to its linear operating region, a constraint is imposed on the input voltages, v_p and v_n. The constraint is based on typical numerical values for V_{CC} and A in Eq. 5.1. For most op amps, the recommended dc power supply voltages seldom exceed 20 V, and the gain, A, is rarely less than 10,000, or 10^4. We see from both Fig. 5.6 and Eq. 5.1 that in the linear region, the magnitude of the input voltage difference ($|v_p - v_n|$) must be less than $20/10^4$, or 2 mV.

Node voltages in the circuits we study are typically much larger than 2 mV, so a voltage difference of less than 2 mV means the two voltages are essentially equal. Thus, when an op amp is constrained to its linear operating region and the node voltages are much larger than 2 mV, the constraint on the input voltages of the op amp is

INPUT VOLTAGE CONSTRAINT FOR AN OP AMP

$$v_p = v_n. \tag{5.2}$$

Note that Eq. 5.2 characterizes the relationship between the input voltages for an ideal op amp—that is, an op amp whose value of A is infinite.

We can use Eq. 5.2 only if the op amp is confined to its linear operating region. The op amp stays in its linear region if the op amp circuit includes a signal path from the op amp's output terminal to its inverting input terminal. This configuration is known as **negative feedback** because the signal is fed back from the output and is subtracted from the input signal. The negative feedback causes the input voltage difference to decrease. Because the output voltage is proportional to the input voltage difference, the output voltage is also decreased, and the op amp operates in its linear region.

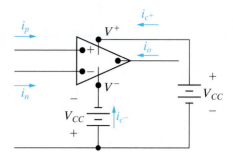

Figure 5.5 ▲ Terminal current variables.

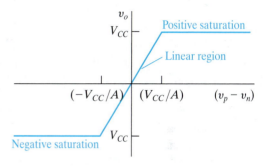

Figure 5.6 ▲ The voltage transfer characteristic of an op amp.

If a circuit containing an op amp does not provide a negative feedback path from the op amp output to the inverting input, then the op amp will normally saturate. But even if the circuit provides a negative feedback path for the op amp, linear operation is not guaranteed. So how do we know whether the op amp is operating in its linear region?

The answer is, we don't! We deal with this dilemma by assuming linear operation, performing the circuit analysis, and then checking our results for contradictions. For example, suppose we assume that an op amp in a circuit is operating in its linear region, and we compute the output voltage of the op amp to be 10 V. On examining the circuit, we discover that V_{CC} is 6 V, resulting in a contradiction, because the op amp's output voltage can be no larger than V_{CC}. Thus, our assumption of linear operation was invalid, and the op amp output must be saturated at 6 V.

We have identified a constraint on the input voltages that is based on the voltage transfer characteristic of the op amp, the assumption that the op amp is restricted to its linear operating region, and typical values for V_{CC}[3] and A. Equation 5.2 represents the voltage constraint for an ideal op amp, that is, with a value of A that is infinite.

Op Amp Input Current Constraints

We now turn our attention to the constraint on the input currents. Analysis of the op amp integrated circuit reveals that the equivalent resistance seen by the input terminals of the op amp is very large, typically 1 MΩ or more. Ideally, the equivalent input resistance is infinite, resulting in the current constraint

INPUT CURRENT CONSTRAINT FOR AN IDEAL OP AMP

$$i_p = i_n = 0. \tag{5.3}$$

Note that the current constraint is not based on assuming that the op amp is confined to its linear operating region, as was the voltage constraint. Together, Eqs. 5.2 and 5.3 form the constraints on terminal behavior that define our ideal op amp model.

From Kirchhoff's current law we know that the sum of the currents entering the operational amplifier is zero, or

$$i_p + i_n + i_o + i_{c^+} + i_{c^-} = 0.$$

Substituting the constraint given by Eq. 5.3 into this KCL equation gives

$$i_o = -(i_{c^+} + i_{c^-}).$$

The equation for i_o tells us that, even though the current at the input terminals is negligible, there may still be appreciable current at the output terminal.

When we use Eqs. 5.2 and 5.3 in analyzing a circuit with an op amp, we are effectively using an ideal model of that op amp. Let's summarize the circuit analysis steps:

Step 1: Check for the presence of a negative feedback path; if it exists, we can assume the op amp is operating in its linear region.

ANALYZING A CIRCUIT WITH AN IDEAL OP AMP

1. Check for a negative feedback path. If it exists, assume the op amp operates in its linear region.
2. Write a KCL equation at the inverting input terminal.
3. Solve the KCL equation and use the solution to find the op amp's output voltage.
4. Compare the op amp's output voltage to the power supply voltages to determine if the op amp is operating in its linear region or if it is saturated.

Analysis Method 5.1 Analyzing an ideal op amp circuit with a negative feedback path.

---

[3] The positive and negative power supply voltages do not have to be equal in magnitude. In the linear operating region, v_o must lie between the two supply voltages. For example, if $V^+ = 15$ V and $V^- = -10$ V, then -10 V $\leq v_o \leq 15$ V.

Step 2: Write a KCL equation at the inverting input terminal, using the input current constraint (Eq. 5.3), the value of v_n (Eq. 5.2), and Ohm's law to find the currents. This equation will usually contain the unknown voltage at the op amp's output terminal.

Step 3: Solve the KCL equation and calculate the voltage at the op amp's output terminal.

Step 4: Compare the voltage at the op amp's output terminal to the power supply voltages to determine whether the op amp is actually in its linear region or whether it has saturated.

These steps are summarized in Analysis Method 5.1. Example 5.1 analyzes an op amp circuit using this analysis method.

EXAMPLE 5.1 | **Analyzing an Op Amp Circuit**

The op amp in the circuit shown in Fig. 5.7 is ideal.

a) Calculate v_o if $v_a = 1$ V and $v_b = 0$ V.

b) Repeat (a) for $v_a = 1$ V and $v_b = 2$ V.

c) If $v_a = 1.5$ V, specify the range of v_b that avoids amplifier saturation.

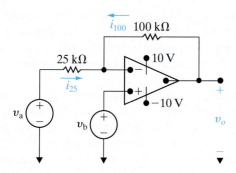

Figure 5.7 ▲ The circuit for Example 5.1.

Solution

a) **Step 1:** A negative feedback path exists from the op amp's output to its inverting input through the 100 kΩ resistor, so we assume the op amp is confined to its linear operating region.

Step 2: The voltage at the inverting input terminal is 0 because $v_p = v_b = 0$ from the connected voltage source, and $v_n = v_p$ from the voltage constraint in Eq. 5.2.

Step 3: Use KCL to sum the currents entering the node labeled v_n to get

$$i_{25} + i_{100} - i_n = 0.$$

Remember that i_n is the current entering the inverting op amp terminal. From Ohm's law,

$$i_{25} = \frac{v_a - v_n}{25,000} = \frac{1 - 0}{25,000} = \frac{1}{25,000};$$

$$i_{100} = \frac{v_o - v_n}{100,000} = \frac{v_o - 0}{100,000} = \frac{v_o}{100,000}.$$

The current constraint requires $i_n = 0$. Substituting the values for the three currents into the KCL equation, we obtain

$$\frac{1}{25} + \frac{v_o}{100} = 0.$$

Hence, v_o is -4 V.

Step 4: Because v_o lies between ± 10 V, our assumption that the op amp is in its linear region of operation is confirmed.

b) Using the same steps as in (a), we get

$$v_p = v_b = v_n = 2 \text{ V},$$

$$i_{25} = -i_{100}.$$

$$i_{25} = \frac{v_a - v_n}{25,000} = \frac{1 - 2}{25,000} = -\frac{1}{25,000};$$

$$i_{100} = \frac{v_o - v_n}{100,000} = \frac{v_o - 2}{100,000}.$$

Therefore, $v_o = 6$ V. Again, v_o lies within ± 10 V.

c) As before, $v_n = v_p = v_b$, and $i_{25} = -i_{100}$. Because $v_a = 1.5$ V,

$$\frac{1.5 - v_b}{25,000} = \frac{v_o - v_b}{100,000}.$$

Solving for v_b as a function of v_o gives

$$v_b = \frac{1}{5}(6 + v_o).$$

Now, if the amplifier operates within its linear region, $-10 \text{ V} \le v_o \le 10$ V. Substituting these limits on v_o into the expression for v_b, we find the range for v_b is

$$-0.8 \text{ V} \le v_b \le 3.2 \text{ V}.$$

ASSESSMENT PROBLEM

Objective 1—Use voltage and current constraints in an ideal op amp

5.1 Assume that the op amp in the circuit shown is ideal.
a) Calculate v_o for the following values of v_s: 0.4, 2.0, 3.5, −0.6, −1.6, and −2.4 V.
b) Specify the range of v_s required to avoid amplifier saturation.

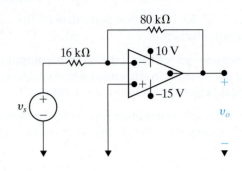

Answer: (a) −2, −10, −15, 3, 8, and 10 V;
(b) −2 V ≤ v_s ≤ 3 V.

SELF-CHECK: Also try Chapter Problems 5.1, 5.3, and 5.4.

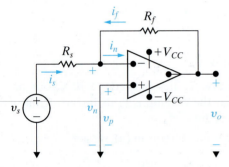

Figure 5.8 ▲ An inverting-amplifier circuit.

5.3 The Inverting-Amplifier Circuit

This section and the three that follow present some important op amp circuits. We begin with the inverting-amplifier circuit, shown in Fig. 5.8. This circuit contains an ideal op amp, two resistors (R_f and R_s), a voltage source (v_s), and a short circuit connecting the noninverting input terminal and the common node.

We can analyze this circuit to obtain an expression for the output voltage, v_o, as a function of the source voltage, v_s. Starting with Step 1, we note the circuit's negative feedback path, so assume the op amp is in its linear operating region. In Step 2, the voltage constraint of Eq. 5.2 sets the voltage at $v_n = 0$, because the voltage at $v_p = 0$. Step 3 generates a single KCL equation at the inverting terminal of the op amp, given as

$$i_s + i_f = i_n.$$

From Ohm's law,

$$i_s = \frac{v_s}{R_s},$$

$$i_f = \frac{v_o}{R_f}.$$

Now we invoke the constraint stated in Eq. 5.3, namely,

$$i_n = 0.$$

Substituting the expressions for i_s, i_f, and i_n into the KCL equation and solving for v_o yields

INVERTING AMPLIFIER EQUATION

$$v_o = \frac{-R_f}{R_s} v_s. \qquad (5.4)$$

Note that the output voltage is an inverted, scaled replica of the input. The sign reverses, or inverts, from input to output. The scaling factor, or gain, is the ratio R_f/R_s, which is usually greater than 1, so $|v_o| > |v_s|$. Hence, we call this circuit an inverting amplifier.

Using Step 4, we see that the result given by Eq. 5.4 is valid only if the op amp shown in the circuit in Fig. 5.8 is operating in its linear region. Even if the op amp is not ideal, Eq. 5.4 is a good approximation. (We demonstrate this in Section 5.7.) Equation 5.4 specifies the gain of the inverting amplifier with the external resistors R_f and R_s. In Step 4, the upper limit on the gain, R_f/R_s, is determined by the power supply voltages and the value of the signal voltage v_s. If we assume equal power supply voltages, that is, $V^+ = -V^- = V_{CC}$, we get

$$|v_o| \leq V_{CC}, \qquad \left|\frac{R_f}{R_s}v_s\right| \leq V_{CC}, \qquad \frac{R_f}{R_s} \leq \left|\frac{V_{CC}}{v_s}\right|.$$

For example, if $V_{CC} = 15$ V and $v_s = 10$ mV, the ratio R_f/R_s must be less than 1500.

In the inverting amplifier circuit shown in Fig. 5.8, the resistor R_f provides the negative feedback connection. That is, it connects the output terminal to the inverting input terminal. If R_f is removed, the feedback path is opened and the amplifier is said to be operating *open loop*. Figure 5.9 shows the open-loop operation.

Opening the feedback path drastically changes the behavior of the circuit. To understand the open-loop circuit, we do not replace the op amp with its ideal model; although A and the input resistance are both large, they are not infinite. Now the output voltage is

$$v_o = -Av_n, \tag{5.5}$$

assuming as before that $V^+ = -V^- = V_{CC}$; then $|v_n| < V_{CC}/A$ for linear operation. Because the inverting input current is almost zero, the voltage drop across R_s is almost zero, and the inverting input voltage nearly equals the signal voltage, v_s; that is, $v_n \approx v_s$. Hence, the op amp can operate open loop in the linear mode only if $|v_s| < V_{CC}/A$. If $|v_s| > V_{CC}/A$, the op amp simply saturates. In particular, if $v_s < -V_{CC}/A$, the op amp saturates at $+V_{CC}$, and if $v_s > V_{CC}/A$, the op amp saturates at $-V_{CC}$. Because the relationship shown in Eq. 5.5 occurs when there is no feedback path, the value of A is often called the **open-loop gain** of the op amp.

Example 5.2 uses the inverting-amplifier equation to design an inverting amplifier using realistic resistor values.

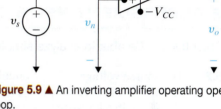

Figure 5.9 ▲ An inverting amplifier operating open loop.

EXAMPLE 5.2 | **Designing an Inverting Amplifier**

a) Design an inverting amplifier (see Fig. 5.8) with a gain of 12. Use ± 15 V power supplies and an ideal op amp.

b) What range of input voltages, v_s, allows the op amp in this design to remain in its linear operating region?

Solution

a) We need to find two resistors whose ratio is 12 from the realistic resistor values listed in Appendix H. There are lots of different possibilities, but let's choose $R_s = 1$ kΩ and $R_f = 12$ kΩ. Use the inverting-amplifier equation (Eq. 5.4) to verify the design:

$$v_o = -\frac{R_f}{R_s}v_s = -\frac{12,000}{1000}v_s = -12v_s.$$

Thus, we have an inverting amplifier with a gain of 12, as shown in Fig. 5.10.

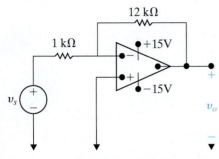

Figure 5.10 ▲ Inverting amplifier for Example 5.2.

b) Solve two different versions of the inverting-amplifier equation for v_s, first using $v_o = +15$ V and then using $v_o = -15$ V:

$$15 = -12v_s \text{ so } v_s = -1.25 \text{ V};$$

$$-15 = -12v_s \text{ so } v_s = 1.25 \text{ V}.$$

Thus, if the input voltage is greater than or equal to -1.25 V and less than or equal to $+1.25$ V, the op amp in the inverting amplifier will remain in its linear operating region.

ASSESSMENT PROBLEM

Objective 2—Be able to analyze simple circuits containing ideal op amps

5.2 The source voltage v_s in the circuit in Assessment Problem 5.1 is -640 mV. The 80 kΩ feedback resistor is replaced by a variable resistor R_x. What range of R_x allows the inverting amplifier to operate in its linear region?

Answer: $0 \leq R_x \leq 250$ kΩ.

SELF-CHECK: Also try Chapter Problems 5.8 and 5.10.

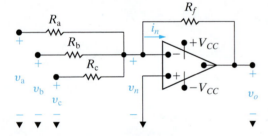

Figure 5.11 ▲ A summing amplifier.

5.4 The Summing-Amplifier Circuit

Figure 5.11 shows a summing amplifier with three input voltages. The output voltage of a summing amplifier is an inverted, scaled sum of the voltages applied to the input of the amplifier. We can find the relationship between the output voltage v_o and the three input voltages, v_a, v_b, and v_c, using Analysis Method 5.1.

The summing amplifier has a negative feedback path that includes the resistor R_f, so in Step 1 we assume the op amp is in its linear region. We then use the ideal op amp voltage constraint in Step 2 together with the ground imposed by the circuit at v_p to determine that $v_n = v_p = 0$. In Step 3, we write a KCL equation at the inverting input terminal, using Ohm's law to specify the current in each resistor in terms of the voltage across that resistor, to get

$$\frac{v_n - v_a}{R_a} + \frac{v_n - v_b}{R_b} + \frac{v_n - v_c}{R_c} + \frac{v_n - v_o}{R_f} + i_n = 0.$$

Apply the voltage constraint from Step 2 and the current constraint $i_n = 0$ to the KCL equation, then solve for v_o to get

INVERTING SUMMING-AMPLIFIER EQUATION

$$v_o = -\left(\frac{R_f}{R_a}v_a + \frac{R_f}{R_b}v_b + \frac{R_f}{R_c}v_c\right). \tag{5.6}$$

Equation 5.6 shows that the output voltage is an inverted, scaled sum of the three input voltages. According to Step 4, this equation is valid only if the value of v_o is between the two power supply voltages.

If $R_a = R_b = R_c = R_s$, then Eq. 5.6 reduces to

$$v_o = -\frac{R_f}{R_s}(v_a + v_b + v_c).$$

Finally, if we make $R_f = R_s$, the output voltage is just the inverted sum of the input voltages. That is,

$$v_o = -(v_a + v_b + v_c).$$

Although we analyzed the summing amplifier with three input signals, the number of input voltages can be increased or decreased as needed. For example, you might wish to sum 16 individually recorded audio signals to form a single audio signal. The summing-amplifier configuration in Fig. 5.11 could include 16 different input resistors whose values specify different amplification factors for each of the input audio tracks. The summing amplifier thus plays the role of an audio mixer. As with inverting-amplifier circuits, the scaling factors in summing-amplifier circuits are determined by the external resistors $R_f, R_a, R_b, R_c, \ldots, R_n$.

Example 5.3 uses the summing-amplifier equation to design an inverting summing amplifier.

EXAMPLE 5.3 | Designing a Summing Amplifier

a) Design a summing amplifier (see Fig. 5.11) whose output voltage is

$$v_o = -4v_a - v_b - 5v_c.$$

Use an ideal op amp with ± 12 V power supplies and a 20 kΩ feedback resistor.

b) Suppose $v_a = 2$ V and $v_c = -1$ V. What range of input voltages for v_b allows the op amp in this design to remain in its linear operating region?

c) Suppose $v_a = 2$ V, $v_b = 3$ V, and $v_c = -1$ V. Using the input resistor values found in part (a), how large can the feedback resistor be before the op amp saturates?

Solution

a) Use the summing-amplifier equation (Eq. 5.6) and the feedback resistor value to find the three input resistor values:

$$-\frac{R_f}{R_a} = -4 \quad \text{so} \quad R_a = \frac{20\text{ k}}{4} = 5\text{ k}\Omega;$$

$$-\frac{R_f}{R_b} = -1 \quad \text{so} \quad R_b = \frac{20\text{ k}}{1} = 20\text{ k}\Omega;$$

$$-\frac{R_f}{R_c} = -5 \quad \text{so} \quad R_c = \frac{20\text{ k}}{5} = 4\text{ k}\Omega.$$

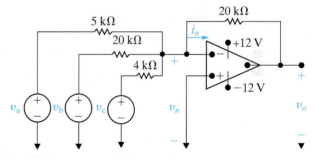

Figure 5.12 ▲ The summing amplifier for Example 3.3(a).

The resulting circuit is shown in Fig. 5.12.

b) Substitute the values for v_a and v_c into the equation for v_o given in the problem statement to get

$$v_o = -4(2) - v_b - 5(-1) = -3 - v_b.$$

Solving this equation for v_b in terms of v_o gives

$$v_b = -3 - v_o.$$

Now substitute the two power supply voltages for the output voltage to find the range of v_b values that keeps the op amp in its linear region:

$$-15\text{ V} \le v_b \le 9\text{ V}.$$

c) Starting with the summing-amplifier equation, Eq. 5.6, substitute the input resistor values found in part (a) and the specified input voltage values.

Remember that the feedback resistor is an unknown in this equation:

$$v_o = -\frac{R_f}{5000}(2) - \frac{R_f}{20,000}(3) - \frac{R_f}{4000}(-1) = -\frac{6R_f}{20,000}.$$

From this equation, it should be clear that if the op amp saturates, it will do so at its negative power supply value, -12 V. Using this voltage

for v_o in the above equation and solving for the feedback resistance gives

$$R_f = 40 \text{ k}\Omega.$$

Given the specified input voltages, this is the largest value of feedback resistance that keeps the op amp in its linear region.

ASSESSMENT PROBLEM

Objective 2—Be able to analyze simple circuits containing ideal op amps

5.3   a) Find v_o in the circuit shown if $v_a = 0.1$ V and $v_b = 0.25$ V.
   b) If $v_b = 0.25$ V, how large can v_a be before the op amp saturates?
   c) If $v_a = 0.10$ V, how large can v_b be before the op amp saturates?
   d) Repeat (a), (b), and (c) with the polarity of v_b reversed.

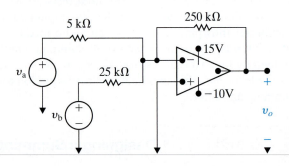

Answer: (a) -7.5 V;
   (b) 0.15 V;
   (c) 0.5 V;
   (d) -2.5, 0.25, and 2 V.

SELF-CHECK: Also try Chapter Problems 5.12, 5.14, and 5.16.

5.5 The Noninverting-Amplifier Circuit

Figure 5.13 depicts a noninverting-amplifier circuit. We use Analysis Method 5.1 to find the expression for the output voltage as a function of the source voltage v_g. Starting with Step 1, we note that the noninverting-amplifier circuit has a negative feedback path with the resistor R_f, so we assume that the op amp is in its linear region. In Step 2, the voltage constraint equation (Eq. 5.2) tells us that $v_n = v_p$. However, the voltage at the noninverting terminal is not 0 because the terminal is not connected to the common node. The current constraint equation (Eq. 5.3) tells us the noninverting input current is zero. Since this current equals the current in the resistor R_s, there is no voltage drop across R_s and $v_p = v_g$. Therefore, $v_n = v_g$ as well.

In Step 3, we write a KCL equation at the noninverting terminal, using the result of Step 2 and the current constraint equation to give

$$\frac{v_g}{R_s} + \frac{v_g - v_o}{R_f} = 0.$$

Solve the KCL equation for v_o to get

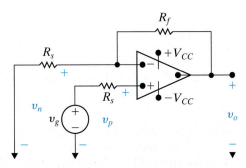

Figure 5.13 ▲ A noninverting amplifier.

NONINVERTING-AMPLIFIER EQUATION

$$v_o = \frac{R_s + R_f}{R_s} v_g. \tag{5.7}$$

From Step 4 we know that keeping the op amp in its linear region requires that

$$\frac{R_s + R_f}{R_s} < \left| \frac{V_{CC}}{v_g} \right|.$$

Once again, the ideal op amp assumption allows us to express the output voltage as a function of the input voltage and the external resistors R_s and R_f.

Example 5.4 illustrates the design of a noninverting amplifier using realistic resistor values.

EXAMPLE 5.4 | **Designing a Noninverting Amplifier**

a) Design a noninverting amplifier (see Fig. 5.13) with a gain of 6. Assume the op amp is ideal.

b) Suppose we wish to amplify a voltage v_g, where $-1.5\,\text{V} \leq v_g \leq +1.5\,\text{V}$. What are the smallest power supply voltages that could be used with the resistors selected in part (a) to ensure that the op amp remains in its linear region?

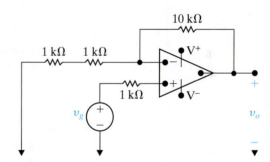

Figure 5.14 ▲ The noninverting amplifier design of Example 5.3.

Solution

a) Using the noninverting-amplifier equation (Eq. 5.7),

$$v_o = \frac{R_s + R_f}{R_s} v_g = 6v_g \quad \text{so} \quad \frac{R_s + R_f}{R_s} = 6.$$

Therefore,

$$R_s + R_f = 6R_s, \quad \text{so} \quad R_f = 5R_s.$$

Look at the realistic resistor values listed in Appendix H. Let's choose $R_f = 10\,\text{k}\Omega$, so $R_s = 2\,\text{k}\Omega$. But there is not a $2\,\text{k}\Omega$ resistor in Appendix H. We can create an equivalent $2\,\text{k}\Omega$ resistor by combining two $1\,\text{k}\Omega$ resistors in series. We can use a third $1\,\text{k}\Omega$ resistor for R_g. The resulting circuit is shown in Fig. 5.14.

b) Solve two different versions of the noninverting-amplifier equation for v_o, first using $v_g = +1.5\,\text{V}$ and then using $v_g = -1.5\,\text{V}$:

$$v_o = 6(1.5) = 9\,\text{V};$$

$$v_o = 6(-1.5) = -9\,\text{V}.$$

Thus, if we use $\pm 9\,\text{V}$ power supplies for the noninverting amplifier designed in part (a) and $-1.5\,\text{V} \leq v_g \leq +1.5\,\text{V}$, the op amp will remain in its linear operating region. The circuit resulting from the analysis in parts (a) and (b) is shown in Fig. 5.14, with $V^+ = 9\,\text{V}$ and $V^- = -9\,\text{V}$.

ASSESSMENT PROBLEM

Objective 2—Be able to analyze simple circuits containing ideal op amps

5.4 Assume that the op amp in the circuit shown is ideal.
   a) Find the output voltage when the variable resistor is set to 60 kΩ.
   b) How large can R_x be before the amplifier saturates?

Answer: (a) 4.8 V;
(b) 75 kΩ.

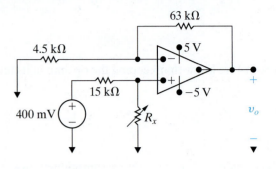

SELF-CHECK: Also try Chapter Problems 5.18 and 5.20.

5.6 The Difference-Amplifier Circuit

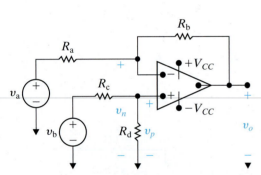

Figure 5.15 ▲ A difference amplifier.

The output voltage of a difference amplifier is proportional to the difference between the two input voltages. To demonstrate, we analyze the difference-amplifier circuit shown in Fig. 5.15, using Analysis Method 5.1. In Step 1, we note the negative feedback path that includes the resistor R_b and thereby assume the op amp is in its linear region. In Step 2, we use the voltage constraint equation (Eq. 5.2) to recognize that $v_n = v_p$. Now we need an expression for v_p. Let's focus on the subcircuit at the noninverting input; employing the current constraint equation (Eq. 5.3), we note that there is no current into the noninverting input. Therefore, the current in the v_b source remains in the loop containing that source and the resistors R_c and R_d. The voltage at the noninverting terminal is the voltage across R_d, which we can find using voltage division:

$$v_p = \frac{R_d}{R_c + R_d} v_b = v_n.$$

At Step 3, write a KCL equation at the inverting terminal, employing the current constraint equation again to see that the current into the inverting terminal is zero:

$$\frac{v_n - v_a}{R_a} + \frac{v_n - v_o}{R_b} = 0.$$

Solving the KCL equation for v_o as a function of both v_a and v_n, we get

$$v_o = \left(\frac{R_a + R_b}{R_a}\right) v_n - \left(\frac{R_b}{R_a}\right) v_a.$$

Substituting the equation for v_n into the equation for v_o gives the desired relationship:

DIFFERENCE-AMPLIFIER EQUATION

$$v_o = \frac{R_d(R_a + R_b)}{R_a(R_c + R_d)} v_b - \frac{R_b}{R_a} v_a. \tag{5.8}$$

Equation 5.8 shows that the output voltage is proportional to the difference between a scaled replica of v_b and a scaled replica of v_a. In general, the scaling factor applied to v_b is not the same as that applied to v_a. However, the scaling factor applied to each input voltage can be made equal by setting

$$\frac{R_a}{R_b} = \frac{R_c}{R_d}. \tag{5.9}$$

When Eq. 5.9 is satisfied, the expression for the output voltage reduces to

SIMPLIFIED DIFFERENCE-AMPLIFIER EQUATION

$$v_o = \frac{R_b}{R_a}(v_b - v_a). \tag{5.10}$$

If Eq. 5.9 is satisfied, the output voltage is a scaled replica of the difference between the input voltages v_b and v_a. As in the previous ideal amplifier circuits, the scaling is controlled by the external resistors. Furthermore, the relationship between the output voltage and the input voltages is not affected by connecting a nonzero load resistance across the output of the amplifier.

Follow Example 5.5 to design a difference amplifier using realistic resistor values.

EXAMPLE 5.5 | Designing a Difference Amplifier

a) Design a difference amplifier (see Fig. 5.15) that amplifies the difference between two input voltages by a gain of 8, using an ideal op amp and ± 8 V power supplies.

b) Suppose $v_a = 1$ V in the difference amplifier designed in part (a). What range of input voltages for v_b will allow the op amp to remain in its linear operating region?

Solution

a) Using the simplified difference-amplifier equation (Eq. 5.10),

$$v_o = \frac{R_b}{R_a}(v_b - v_a) = 8(v_b - v_a) \quad \text{so} \quad \frac{R_b}{R_a} = 8.$$

We want two resistors whose ratio is 8. Look at the realistic resistor values listed in Appendix H. Let's choose $R_b = 12$ kΩ, so $R_a = 1.5$ kΩ, although there are many other possibilities. Note that the simplified difference-amplifier equation requires that

$$\frac{R_a}{R_b} = \frac{R_c}{R_d}.$$

A simple choice for R_c and R_d is $R_c = R_a = 1.5$ kΩ and $R_d = R_b = 12$ kΩ. The resulting circuit is shown in Fig. 5.16.

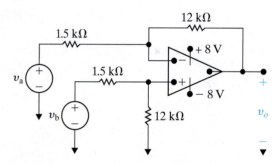

Figure 5.16 ▲ The difference amplifier designed in Example 5.4.

b) Using $v_a = 1$, solve two different versions of the simplified difference-amplifier equation (Eq. 5.10) for v_b in terms of v_o. Then substitute the two limiting values for the output voltage, $v_o = +8$ V and $v_o = -8$ V:

$$v_b = \frac{v_o}{8} + 1 = \frac{8}{8} + 1 = 2 \text{ V};$$

$$v_b = \frac{v_o}{8} + 1 = \frac{-8}{8} + 1 = 0 \text{ V}.$$

Thus, if $v_a = 1$ V in the difference amplifier from part (a), the op amp will remain in its linear region if $0 \text{ V} \leq v_b \leq 2 \text{ V}$.

ASSESSMENT PROBLEM

Objective 2—Be able to analyze simple circuits containing ideal op amps

5.5   a) In the difference amplifier shown, $v_b = 4.0$ V. What range of values for v_a will result in linear operation?

b) Repeat (a) with the 20 kΩ resistor decreased to 8 kΩ.

Answer: (a) 2 V $\leq v_a \leq 6$ V;
(b) 1.2 V $\leq v_a \leq 5.2$ V.

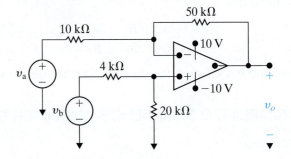

SELF-CHECK: Also try Chapter Problems 5.25, 5.27, and 5.31.

The Difference Amplifier—Another Perspective

Let's examine the difference-amplifier behavior more closely by redefining its inputs in terms of two other voltages. The first is the **differential mode** input, which is the difference between the two input voltages in Fig. 5.15:

$$v_{dm} = v_b - v_a. \tag{5.11}$$

The second is the **common mode** input, which is the average of the two input voltages in Fig. 5.15:

$$v_{cm} = (v_a + v_b)/2. \tag{5.12}$$

Using Eqs. 5.11 and 5.12, we can now represent the original input voltages, v_a and v_b, in terms of the differential mode and common mode voltages, v_{dm} and v_{cm}:

$$v_a = v_{cm} - \frac{1}{2} v_{dm}, \tag{5.13}$$

$$v_b = v_{cm} + \frac{1}{2} v_{dm}. \tag{5.14}$$

Substituting Eqs. 5.13 and 5.14 into Eq. 5.8 gives the output of the difference amplifier in terms of the differential mode and common mode voltages:

$$v_o = \left[\frac{R_a R_d - R_b R_c}{R_a (R_c + R_d)} \right] v_{cm} + \left[\frac{R_d (R_a + R_b) + R_b (R_c + R_d)}{2 R_a (R_c + R_d)} \right] v_{dm}$$

$$= A_{cm} v_{cm} + A_{dm} v_{dm}, \tag{5.15}$$

where A_{cm} is the common mode gain and A_{dm} is the differential mode gain. Now, substitute $R_c = R_a$ and $R_d = R_b$, which are possible values for R_c and R_d that satisfy Eq. 5.9, into Eq. 5.15:

$$v_o = (0) v_{cm} + \left(\frac{R_b}{R_a} \right) v_{dm}. \tag{5.16}$$

Thus, an ideal difference amplifier has $A_{cm} = 0$, amplifies only the differential mode portion of the input voltage, and eliminates the common mode portion of the input voltage. Figure 5.17 shows a difference-amplifier circuit with differential mode and common mode input voltages in place of v_a and v_b.

Equation 5.16 provides an important perspective on the function of the difference amplifier. In many applications it is the differential mode signal that contains the information of interest, while the common mode signal represents the noise found in all electric signals. For example, an electrocardiograph electrode measures the voltages produced by your body that regulate your heartbeat. These voltages have very small magnitudes compared with the electrical noise that the electrode picks up from lights and other electrical equipment in the room. The noise appears as the common mode portion of the measured voltage, whereas the heart rate voltages comprise the differential mode portion. Thus, an ideal difference amplifier (one whose resistors satisfy Eq. 5.9) amplifies only the voltage of interest and suppresses the noise.

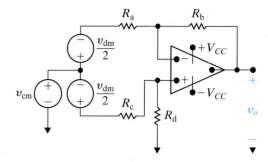

Figure 5.17 ▲ A difference amplifier with common mode and differential mode input voltages.

Measuring Difference-Amplifier Performance— The Common Mode Rejection Ratio

An ideal difference amplifier has zero common mode gain and nonzero (and usually large) differential mode gain. Two factors influence the ideal common mode gain—resistance mismatches (that is, Eq. 5.9 is not satisfied) or a nonideal op amp (that is, Eqs. 5.2 and 5.3 are not satisfied). We focus first on how resistance mismatches affect the performance of a difference amplifier.

Suppose that resistor values are chosen that do not precisely satisfy Eq. 5.9. Instead, the relationship among the resistors R_a, R_b, R_c, and R_d is

$$\frac{R_a}{R_b} = (1 - \epsilon)\frac{R_c}{R_d},$$

so

$$R_a = (1 - \epsilon)R_c \quad \text{and} \quad R_b = R_d,$$

or

$$R_d = (1 - \epsilon)R_b \quad \text{and} \quad R_a = R_c, \tag{5.17}$$

where ϵ is a very small number. We can see the effect of this resistance mismatch on the common mode gain of the difference amplifier by substituting Eq. 5.17 into Eq. 5.15 and simplifying the expression for A_{cm}:

$$\begin{aligned}
A_{cm} &= \frac{R_a(1 - \epsilon)R_b - R_aR_b}{R_a[R_a + (1 - \epsilon)R_b]} \\
&= \frac{-\epsilon R_b}{R_a + (1 - \epsilon)R_b} \\
&\approx \frac{-\epsilon R_b}{R_a + R_b}.
\end{aligned} \tag{5.18}$$

The approximation in Eq. 5.18 is valid because ϵ is very small, and therefore $(1 - \epsilon)$ is approximately 1. Note that when the resistors in the difference amplifier satisfy Eq. 5.9, $\epsilon = 0$ and Eq. 5.18 gives $A_{cm} = 0$.

Now calculate the effect of the resistance mismatch on the differential mode gain by substituting Eq. 5.17 into Eq. 5.15 and simplifying the expression for A_{dm}:

$$
\begin{aligned}
A_{dm} &= \frac{(1 - \epsilon)R_b(R_a + R_b) + R_b[R_a + (1 - \epsilon)R_b]}{2R_a[R_a + (1 - \epsilon)R_b]} \\
&= \frac{R_b}{R_a}\left[1 - \frac{(\epsilon/2)R_a}{R_a + (1 - \epsilon)R_b}\right] \\
&\approx \frac{R_b}{R_a}\left[1 - \frac{(\epsilon/2)R_a}{R_a + R_b}\right].
\end{aligned}
\tag{5.19}
$$

We use the same rationale for the approximation in Eq. 5.19 as in the computation of A_{cm}. When the resistors in the difference amplifier satisfy Eq. 5.9, $\epsilon = 0$ and Eq. 5.19 gives $A_{dm} = R_b/R_a$.

The **common mode rejection ratio (CMRR)** measures a difference amplifier's performance. It is defined as the ratio of the differential mode gain to the common mode gain:

$$
\text{CMRR} = \left|\frac{A_{dm}}{A_{cm}}\right|.
\tag{5.20}
$$

The larger the CMRR, the closer the difference amplifier's behavior is to ideal. We can see the effect of resistance mismatch on the CMRR by substituting Eqs. 5.18 and 5.19 into Eq. 5.20:

$$
\begin{aligned}
\text{CMRR} &\approx \left|\frac{\dfrac{R_b}{R_a}[1 - (R_a\epsilon/2)/(R_a + R_b)]}{-\epsilon R_b/(R_a + R_b)}\right| \\
&\approx \left|\frac{R_a(1 - \epsilon/2) + R_b}{-\epsilon R_a}\right| \\
&\approx \left|\frac{1 + R_b/R_a}{-\epsilon}\right|.
\end{aligned}
\tag{5.21}
$$

From Eq. 5.21, if the resistors in the difference amplifier are matched, $\epsilon = 0$ and CMRR $= \infty$. Even if the resistors are mismatched, we can minimize the impact of the mismatch by making the differential mode gain (R_b/R_a) very large, thereby making the CMRR large.

The second reason for nonzero common mode gain is a nonideal op amp. Remember that the op amp is itself a difference amplifier because in the linear operating region, its output is proportional to the difference of its inputs; that is, $v_o = A(v_p - v_n)$. The output of a nonideal op amp is not strictly proportional to the difference between the inputs (the differential mode input) but also includes a common mode signal. Internal mismatches in the integrated circuit components make the behavior of the op amp nonideal, in the same way that the resistor mismatches in the difference-amplifier circuit make its behavior nonideal. Even though a discussion of nonideal op amps is beyond the scope of this text, note that the CMRR is used to rate op amps in practice by assessing how nearly ideal an op amp's behavior is.

Example 5.6 examines how resistor mismatches affect the CMRR of a difference amplifier.

EXAMPLE 5.6   **Calculating the CMRR**

a) Suppose the R_c resistor in the difference amplifier designed in Example 5.5, shown in Fig. 5.16, is 10% larger than its nominal value. All other resistor values are unchanged. Calculate the common mode gain, the difference mode gain, and the CMRR for the difference amplifier.

b) Repeat part (a) assuming the R_d resistor value is 10% larger than its nominal value and all other resistor values are unchanged.

Solution

a) Use the common mode gain equation in Eq. 5.15 with $R_c = 1500(1.1) = 1650\ \Omega$ to get

$$A_{cm} = \frac{(1500)(12{,}000) - (12{,}000)(1650)}{1500(1650 + 12{,}000)}$$

$$= -0.0879.$$

Then use the difference mode gain equation in Eq. 5.15 with $R_c = 1500(1.1) = 1650\ \Omega$ to get

$$A_{dm} = \frac{12{,}000(1500 + 12{,}000) + 12{,}000(1650 + 12{,}000)}{2(1500)(1650 + 12{,}000)}$$

$$= 7.956.$$

The CMRR (Eq. 5.20) is thus

$$\text{CMRR} = \left|\frac{7.956}{-0.0879}\right| = 90.5.$$

b) Use the common mode gain equation in Eq. 5.15 with $R_d = 12{,}000(1.1) = 13{,}200\ \Omega$ to get

$$A_{cm} = \frac{(1500)(13{,}200) - (12{,}000)(1500)}{1500(1500 + 13{,}200)}$$

$$= 0.08163.$$

Then use the difference mode gain equation in Eq. 5.15 with $R_d = 12{,}000(1.1) = 13{,}200\ \Omega$ to get

$$A_{dm} = \frac{13{,}200(1500 + 12{,}000) + 12{,}000(1500 + 13{,}200)}{2(1500)(1500 + 13{,}200)}$$

$$= 8.0408.$$

The CMRR (Eq. 5.20) is thus

$$\text{CMRR} = \left|\frac{8.0408}{0.08163}\right| = 98.5.$$

ASSESSMENT PROBLEM

Objective 2—Be able to analyze simple circuits containing ideal op amps

5.6   Suppose the 12 kΩ resistor R_d in the difference amplifier in Fig. 5.16 is replaced by a variable resistor.
What range of R_d values will ensure the difference amplifier has a CMRR $\geq |100|$?

Answer: $11\ \text{k}\Omega \leq R_d \leq 13.18\ \text{k}\Omega$.

SELF-CHECK: Assess your understanding of this material by trying Chapter Problems 5.32 and 5.33.

5.7 A More Realistic Model for the Operational Amplifier

We now consider a more realistic model, shown in Fig. 5.18, that predicts the performance of an op amp in its linear region of operation. This model includes three modifications to the ideal op amp: (1) a finite input resistance, R_i; (2) a finite open-loop gain, A; and (3) a nonzero output resistance, R_o.

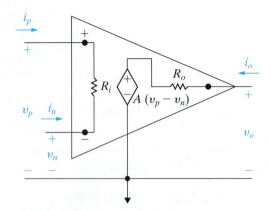

Figure 5.18 ▲ An equivalent circuit for an operational amplifier.

When using the equivalent circuit shown in Fig. 5.18, the assumptions that $v_n = v_p$ (Eq. 5.2) and $i_n = i_p = 0$ (Eq. 5.3) are invalid. Equation 5.1 is also invalid because the output resistance, R_o, is not zero.

Although the presence of A, R_i and R_o makes op amp circuit analysis more cumbersome, it remains straightforward. To illustrate, we use the equivalent circuit shown in Fig. 5.18 when analyzing both an inverting and a noninverting amplifier. We begin with the inverting amplifier.

Analyzing an Inverting-Amplifier Circuit Using a More Realistic Op Amp Model

The circuit for the inverting amplifier, using the op amp circuit shown in Fig. 5.18, is depicted in Fig. 5.19. We can find the output voltage, v_o, as a function of the source voltage, v_s, by writing two KCL equations at the nodes labeled a and b in Fig. 5.19. Note that $v_p = 0$ due to the external short-circuit connection at the noninverting input terminal. The equations are:

$$\text{node a:} \quad \frac{v_n - v_s}{R_s} + \frac{v_n}{R_i} + \frac{v_n - v_o}{R_f} = 0,$$

$$\text{node b:} \quad \frac{v_o - v_n}{R_f} + \frac{v_o - A(-v_n)}{R_o} = 0.$$

We rearrange the KCL equations, preparing to use either back-substitution or Cramer's method to solve for v_o:

$$\left(\frac{1}{R_s} + \frac{1}{R_i} + \frac{1}{R_f} \right) v_n - \frac{1}{R_f} v_o = \frac{1}{R_s} v_s,$$

$$\left(\frac{A}{R_o} - \frac{1}{R_f} \right) v_n + \left(\frac{1}{R_f} + \frac{1}{R_o} \right) v_o = 0.$$

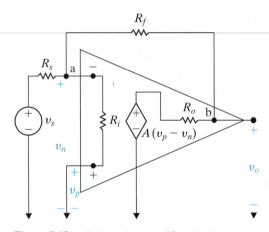

Figure 5.19 ▲ An inverting-amplifier circuit.

Solving for v_o yields

$$v_o = \frac{-A + (R_o/R_f)}{\dfrac{R_s}{R_f} \left(1 + A + \dfrac{R_o}{R_i} \right) + \left(\dfrac{R_s}{R_i} + 1 \right) + \dfrac{R_o}{R_f}} v_s. \tag{5.22}$$

Note that Eq. 5.22 reduces to Eq. 5.4 as $R_o \to 0$, $R_i \to \infty$, and $A \to \infty$.

If the inverting amplifier shown in Fig. 5.19 has a load resistance, R_L, at its output terminal, the relationship between v_o and v_s is

$$v_o = \frac{-A + (R_o/R_f)}{\dfrac{R_s}{R_f} \left(1 + A + \dfrac{R_o}{R_i} + \dfrac{R_o}{R_L} \right) + \left(1 + \dfrac{R_o}{R_L} \right) \left(1 + \dfrac{R_s}{R_i} \right) + \dfrac{R_o}{R_f}} v_s.$$

Analyzing a Noninverting-Amplifier Circuit Using a More Realistic Op Amp Model

Using the equivalent circuit shown in Fig. 5.18 to analyze a noninverting amplifier, we obtain the circuit depicted in Fig. 5.20. Here, the voltage source v_g in series with the resistance R_g represents the signal source.

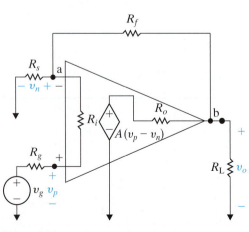

Figure 5.20 ▲ A noninverting-amplifier circuit.

The resistor R_L denotes the load on the amplifier. We derive an expression for v_o as a function of v_g by writing KCL equations at nodes a and b:

$$\text{node a:} \quad \frac{v_n}{R_s} + \frac{v_n - v_g}{R_g + R_i} + \frac{v_n - v_o}{R_f} = 0, \tag{5.23}$$

$$\text{node b:} \quad \frac{v_o - v_n}{R_f} + \frac{v_o}{R_L} + \frac{v_o - A(v_p - v_n)}{R_o} = 0. \tag{5.24}$$

The current in R_g is the same as in R_i, so

$$\frac{v_p - v_g}{R_g} = \frac{v_n - v_g}{R_i + R_g}. \tag{5.25}$$

Use Eq. 5.25 to eliminate v_p from Eq. 5.24, giving

$$v_n\left(\frac{1}{R_s} + \frac{1}{R_g + R_i} + \frac{1}{R_f}\right) - v_o\left(\frac{1}{R_f}\right) = v_g\left(\frac{1}{R_g + R_i}\right),$$

$$v_n\left[\frac{AR_i}{R_o(R_i + R_g)} - \frac{1}{R_f}\right] + v_o\left(\frac{1}{R_f} + \frac{1}{R_o} + \frac{1}{R_L}\right) = v_g\left[\frac{AR_i}{R_o(R_i + R_g)}\right].$$

Solving for v_o yields

$$v_o = \frac{[(R_f + R_s) + (R_s R_o/AR_i)]v_g}{R_s + \dfrac{R_o}{A}(1 + K_r) + \dfrac{R_f R_s + (R_f + R_s)(R_i + R_g)}{AR_i}}, \tag{5.26}$$

where

$$K_r = \frac{R_s + R_g}{R_i} + \frac{R_f + R_s}{R_L} + \frac{R_f R_s + R_f R_g + R_g R_s}{R_i R_L}.$$

Note that Eq. 5.26 reduces to Eq. 5.7 when $R_o \to 0$, $A \to \infty$, and $R_i \to \infty$. For the unloaded ($R_L = \infty$) noninverting amplifier, K_r reduces to $(R_s + R_g)/R_i$ and the expression for v_o becomes

$$v_o = \frac{[(R_f + R_s) + R_s R_o/AR_i]v_g}{R_s + \dfrac{R_o}{A}\left(1 + \dfrac{R_s + R_g}{R_i}\right) + \dfrac{1}{AR_i}[R_f R_s + (R_f + R_s)(R_i + R_g)]}. \tag{5.27}$$

Example 5.7 analyzes a noninverting-amplifier circuit that employs the more realistic op amp model.

EXAMPLE 5.7 | **Analyzing a Noninverting-Amplifier Circuit using a Realistic Op Amp Model**

Here we analyze the noninverting amplifier designed in Example 5.5 using the realistic op amp model in Fig. 5.18. Assume that the open-loop gain $A = 50{,}000$, the input resistance $R_i = 100\ \text{k}\Omega$, and the output resistance $R_o = 7.5\ \text{k}\Omega$. The circuit is shown in Fig. 5.21; note that there is no load resistance at the output.

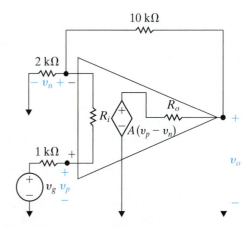

Figure 5.21 ▲ The difference amplifier from Example 5.5, using a realistic op amp model with $A = 50{,}000$, $R_i = 100\ \text{k}\Omega$, and $R_o = 7.5\ \text{k}\Omega$.

a) Calculate the ratio of the output voltage to the source voltage, v_o/v_g.

b) Find the voltages at the op amp input terminals op amp, v_n and v_p, with respect to the common node, when $v_g = 1$ V.

c) Find the voltage difference at the op amp input terminals, $(v_p - v_n)$, when $v_g = 1$ V.

d) Find the current in the signal source, i_g, when the voltage of the source $v_g = 1$ V.

Solution

a) Using Eq. 5.27,

$$\frac{v_o}{v_g} = \frac{10 \text{ k} + 2 \text{ k} + \dfrac{(2 \text{ k})(7.5 \text{ k})}{(100 \text{ k})(50{,}000)}}{2 \text{ k} + \dfrac{7.5 \text{ k}}{50{,}000}\left(1 + \dfrac{2 \text{ k} + 1 \text{ k}}{100 \text{ k}}\right) + \dfrac{1}{50{,}000(100 \text{ k})}\left[(10 \text{ k})(2 \text{ k}) + (10 \text{ k} + 2 \text{ k})(100 \text{ k} + 1 \text{ k})\right]} = 5.9988.$$

Note how close this value is to the gain of 6 specified and achieved in Example 5.5 using the ideal op amp model.

b) From part (a), when $v_g = 1$ V, $v_o = 5.9988$ V. Now use Eq. 5.23 to solve for v_n in terms of v_o and v_g:

$$v_n\left(\frac{1}{2 \text{ k}} + \frac{1}{1 \text{ k} + 100 \text{ k}} + \frac{1}{10 \text{ k}}\right)$$

$$= \frac{1}{100 \text{ k} + 1 \text{ k}} + \frac{5.9988}{10 \text{ k}};$$

$$v_n = 0.999803 \text{ V}.$$

Use Eq. 5.25 to solve for v_p:

$$v_p = \frac{R_g(v_n - v_g)}{R_i + R_g} + v_g = \frac{1 \text{ k}(0.999803 - 1)}{100 \text{ k} + 1 \text{ k}} + 1$$

$$= 0.999996 \text{ V}.$$

c) Using the results from part (b), we find that the voltage difference at the op amp input terminals is

$$v_p - v_n = 192.895 \ \mu\text{V}.$$

While this voltage difference is very small, it is not zero, as we assume when using the ideal op amp model.

d) The current in the signal source is the current in the resistor R_g. Using Ohm's law,

$$i_g = \frac{v_g - v_p}{R_g} = \frac{1 - 0.999996}{1000} = 3.86 \text{ nA}.$$

This is also the current into the noninverting op amp terminal. It is very small but is not zero, as we assume when using the ideal op amp model.

◪ ASSESSMENT PROBLEM

Objective 3—Understand the more realistic model for an op amp

5.7   The inverting amplifier in the circuit shown has an input resistance of 500 kΩ, an output resistance of 5 kΩ, and an open-loop gain of 300,000. Assume that the amplifier is operating in its linear region.

a) Calculate the voltage gain (v_o/v_g) of the amplifier.

b) Calculate the value of v_n in microvolts when $v_g = 1$ V.

c) Calculate the resistance seen by the signal source (v_g).

d) Repeat (a)–(c) using the ideal model for the op amp.

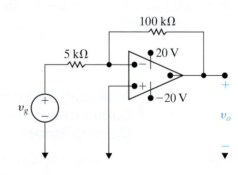

Answer: (a) -19.9985;
(b) $69.995 \ \mu\text{V}$;
(c) $5000.35 \ \Omega$;
(d) $-20, 0 \ \mu\text{V}, 5 \text{ k}\Omega$.

SELF-CHECK: Also try Chapter Problems 5.44 and 5.48.

Practical Perspective

Strain Gages

Changes in the shape of elastic solids are of great importance to engineers who design structures such as aircraft frames that twist, stretch, or bend when subjected to external forces. To use strain gages, you need information about the physical structure of the gage, methods of bonding the gage to the surface of the structure, and the orientation of the gage relative to the forces exerted on the structure. Strain gage measurements are important in many engineering applications, and knowledge of electric circuits is germane to their proper use.

The circuit shown in Fig. 5.22 provides one way to measure the change in resistance experienced by strain gages in applications like the one in the chapter opener. As we will see, this circuit is the familiar difference amplifier, with the strain gage bridge providing the two voltages whose difference is amplified. The pair of strain gages that are lengthened once the bar is bent have the values $R + \Delta R$ in the bridge feeding the difference amplifier, whereas the pair of strain gages that are shortened have the values $R - \Delta R$. We will analyze this circuit to discover the relationship between the output voltage, v_o and the change in resistance, ΔR experienced by the strain gages.

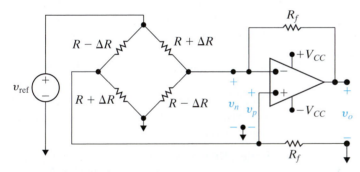

Figure 5.22 ▲ An op amp circuit used for measuring the change in strain gage resistance.

To begin, assume that the op amp is ideal. Writing the KCL equations at the inverting and noninverting input terminals of the op amp, we see

$$\frac{v_{\text{ref}} - v_n}{R + \Delta R} = \frac{v_n}{R - \Delta R} + \frac{v_n - v_o}{R_f}, \tag{5.28}$$

$$\frac{v_{\text{ref}} - v_p}{R - \Delta R} = \frac{v_p}{R + \Delta R} + \frac{v_p}{R_f}. \tag{5.29}$$

Now rearrange Eq. 5.29 to get an expression for the voltage at the noninverting terminal of the op amp:

$$v_p = \frac{v_{\text{ref}}}{(R - \Delta R)\left(\dfrac{1}{R + \Delta R} + \dfrac{1}{R - \Delta R} + \dfrac{1}{R_f}\right)}. \tag{5.30}$$

As usual, we will assume that the op amp is operating in its linear region, so $v_p = v_n$, and the expression for v_p in Eq. 5.30 must also be the expression for v_n. We can thus substitute the right-hand side of Eq. 5.30 for v_n in Eq. 5.28 and solve for v_o. After some algebraic manipulation,

$$v_o = \frac{R_f(2\Delta R)}{R^2 - (\Delta R)^2} v_{\text{ref}}. \tag{5.31}$$

Because the change in resistance experienced by strain gages is very small, $(\Delta R) << R^2$ so $R^2 - (\Delta R)^2 \approx R^2$ and Eq. 5.31 becomes

$$v_o \approx \frac{R_f}{R} 2\delta v_{\text{ref}},$$

where $\delta = \Delta R/R$. By adjusting the value of R_f, we can amplify the small change in resistance to get a large, measurable output voltage, as long as the output voltage value is between the two power supply values.

SELF-CHECK: Assess your understanding of this Practical Perspective by trying Chapter Problem 5.49.

Summary

- The operational amplifier (op amp) is a complex electronic circuit with two input terminals, two power supply terminals and one output terminal. The voltage at the inverting input terminal is v_n, the voltage at the noninverting input terminal is v_p, and the voltage at the output terminal is v_o, all with respect to a common node. The current into the inverting input terminal is i_n, while the current into the noninverting input terminal is i_p. (See page 152.)

- The equation that defines the voltage transfer characteristic of an ideal op amp is

$$v_o = \begin{cases} -V_{CC}, & A(v_p - v_n) < -V_{CC}, \\ A(v_p - v_n), & -V_{CC} \le A(v_p - v_n) \le +V_{CC}, \\ +V_{CC}, & A(v_p - v_n) > +V_{CC}, \end{cases}$$

where A is a proportionality constant known as the open-loop gain, and V_{CC} represents the power supply voltages. (See page 153.)

- A feedback path between an op amp's output and its inverting input can constrain the op amp to its linear operating region where $v_o = A(v_p - v_n)$. (See page 153.)

- A voltage constraint exists when the op amp is confined to its linear operating region due to typical values of V_{CC} and A. If the ideal modeling assumptions are made—meaning A is assumed to be infinite—the ideal op amp model is characterized by the voltage constraint

$$v_p = v_n.$$

(See page 153.)

- A current constraint further characterizes the ideal op amp model, assuming the ideal input resistance of the op amp integrated circuit is infinite. This current constraint is given by

$$i_p = i_n = 0.$$

(See page 154.)

- To analyze an ideal op amp circuit, follow these steps:

  - Check for the presence of a negative feedback path; if it exists, we can assume the op amp is operating in its linear region.

  - Write a KCL equation at the inverting input terminal, using the input current constraint (Eq. 5.3), the value of v_n, and Ohm's law to find the currents. This equation will usually contain the unknown voltage at the op amp's output terminal.

  - Solve the KCL equation and determine the voltage at the op amp's output terminal.

  - Compare the voltage at the op amp's output terminal to the power supply voltages to determine whether the op amp is actually in its linear region or whether it has saturated.

- An inverting amplifier is an op amp circuit producing an output voltage that is an inverted, scaled replica of the input. (See page 156.)

- A summing amplifier is an op amp circuit producing an output voltage that is a scaled sum of the input voltages. (See page 158.)

- A noninverting amplifier is an op amp circuit producing an output voltage that is a scaled replica of the input voltage. (See page 160.)

- A difference amplifier is an op amp circuit producing an output voltage that is a scaled replica of the input voltage difference. (See page 162.)

- The two voltage inputs to a difference amplifier can be used to calculate the common mode and difference mode voltage inputs, v_{cm} and v_{dm}. The output from the difference amplifier can be written in the form

$$v_o = A_{cm}v_{cm} + A_{dm}v_{dm},$$

where A_{cm} is the common mode gain and A_{dm} is the differential mode gain. (See page 164.)

- In an ideal difference amplifier, $A_{cm} = 0$. To measure how nearly ideal a difference amplifier is, we use the common mode rejection ratio:

$$\text{CMRR} = \left| \frac{A_{dm}}{A_{cm}} \right|.$$

An ideal difference amplifier has an infinite CMRR. (See page 166.)

- We considered both a simple, ideal op amp model and a more realistic model in this chapter. The differences between the two models are as follows:

Simplified Model	More Realistic Model
Infinite input resistance	Finite input resistance
Infinite open-loop gain	Finite open-loop gain
Zero output resistance	Nonzero output resistance

(See page 167.)

■ Problems

Sections 5.1–5.2

5.1 The op amp in the circuit in Fig. P5.1 is ideal.

PSPICE
MULTISIM

 a) Label the five op amp terminals with their names.

 b) What ideal op amp constraint determines the value of i_n? What is this value?

 c) What ideal op amp constraint determines the value of $(v_p - v_n)$? What is this value?

 d) Calculate v_o.

Figure P5.1

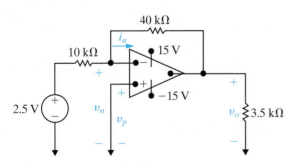

5.2 a) Replace the 2.5 V source in the circuit in Fig. P5.1 and calculate v_o for each of the following source values: -6 V, -3.5 V, -1.25 V, 1 V, 2.4 V, 5.4 V.

 b) Specify the range of voltage source values that will not cause the op amp to saturate.

5.3 Find i_o in the circuit in Fig. P5.3 if the op amp is ideal.

PSPICE
MULTISIM **Figure P5.3**

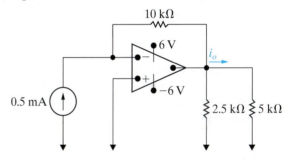

5.4 The op amp in the circuit in Fig. P5.4 is ideal.

PSPICE
MULTISIM

 a) Calculate v_o if $v_a = 4$ V and $v_b = 0$ V.

 b) Calculate v_o if $v_a = 2$ V and $v_b = 0$ V.

 c) Calculate v_o if $v_a = 2$ V and $v_b = 1$ V.

 d) Calculate v_o if $v_a = 1$ V and $v_b = 2$ V.

 e) If $v_b = 1.6$ V, specify the range of v_a such that the amplifier does not saturate.

Figure P5.4

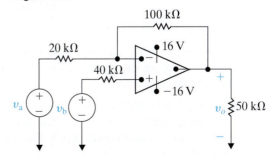

5.5 The op amp in the circuit in Fig. P5.5 is ideal. Calculate the following: a) i_a; b) v_a; c) v_o; d) i_o.

PSPICE
MULTISIM

Figure P5.5

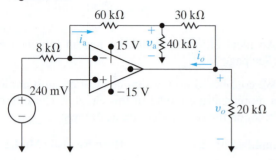

5.6 Find i_L (in milliamperes) in the circuit in Fig. P5.6.

PSPICE
MULTISIM

Figure P5.6

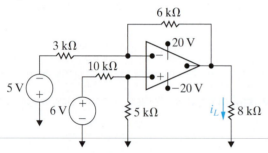

5.7 A voltmeter with a full-scale reading of 10 V is used to measure the output voltage in the circuit in Fig. P5.7. What is the reading of the voltmeter? Assume the op amp is ideal.

PSPICE
MULTISIM

Figure P5.7

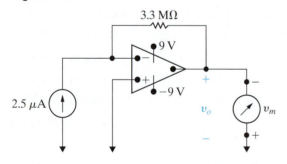

Section 5.3

5.8 a) Design an inverting amplifier with a gain of 2.5, using an ideal op amp. Use a set of identical resistors from Appendix H.

DESIGN
PROBLEM

 b) If you wish to amplify signals between −2 V and +3 V using the circuit you designed in part (a), what are the smallest power supply voltages you can use?

5.9 a) Design an inverting amplifier with a gain of 4. Use an ideal op amp, a 30 kΩ resistor in the feedback path, and ±12 V power supplies.

 b) Using your design from part (a), determine the range of input voltages that will keep the op amp in its linear operating region.

 c) Suppose you wish to amplify a 2 V signal, using your design from part (a) with a variable feedback resistor. What is the largest value of feedback resistance that keeps the op amp in its linear operation region? Using this resistor value, what is the new gain of the inverting amplifier?

5.10 The op amp in the circuit in Fig. P5.10 is ideal.

PSPICE
MULTISIM

 a) Find the range of values for σ in which the op amp does not saturate.

 b) Find i_o (in microamperes) when $\sigma = 0.272$.

Figure P5.10

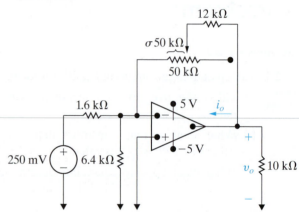

5.11 a) The op amp in the circuit shown in Fig. P5.11 is ideal. The adjustable resistor R_Δ has a maximum value of 100 kΩ, and α is restricted to the range of $0.2 \le \alpha \le 1$. Calculate the range of v_o if $v_g = 40$ mV.

PSPICE
MULTISIM

 b) If α is not restricted, at what value of α will the op amp saturate?

Figure P5.11

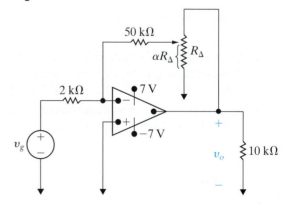

Section 5.4

5.12 The op amp in Fig. P5.12 is ideal.

PSPICE
MULTISIM

a) What circuit configuration is shown in this figure?

b) Find v_o if $v_a = 1.2$ V, $v_b = -1.5$ V, and $v_c = 4$ V.

c) The voltages v_a and v_c remain at 1.2 V and 4 V, respectively. What are the limits on v_b if the op amp operates within its linear region?

Figure P5.12

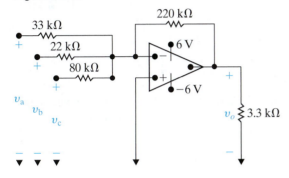

5.13 a) Design an inverting-summing amplifier using a 120 kΩ resistor in the feedback path so that

$$v_o = -(8v_a + 5v_b + 12v_c).$$

Use ±15 V power supplies.

b) Suppose $v_a = 2$ V and $v_c = -1$ V. What range of values for v_b will keep the op amp in its linear operating region?

5.14 Refer to the circuit in Fig. 5.11, where the op amp

PSPICE
MULTISIM

is assumed to be ideal. Given that $R_a = 3$ kΩ, $R_b = 5$ kΩ, $R_c = 25$ kΩ, $v_a = 150$ mV, $v_b = 100$ mV, $v_c = 250$ mV, and $V_{CC} = \pm6$ V, specify the range of R_f for which the op amp operates within its linear region.

5.15 Design an inverting-summing amplifier so that

DESIGN
PROBLEM

PSPICE
MULTISIM

$$v_o = -(8v_a + 4v_b + 10v_c + 6v_d).$$

Start by choosing a feedback resistor (R_f) from Appendix H. Then choose single resistors or construct resistor networks using resistor values in Appendix H to satisfy the design values for R_a, R_b, R_c, and R_d. Draw your final circuit diagram.

5.16 a) The op amp in Fig. P5.16 is ideal. Find v_o if

PSPICE
MULTISIM

$v_a = 4$ V, $v_b = 9$ V, $v_c = 13$ V, and $v_d = 8$ V.

b) Assume v_b, v_c, and v_d retain their values as given in (a). Specify the range of v_a such that the op amp operates within its linear region.

Figure P5.16

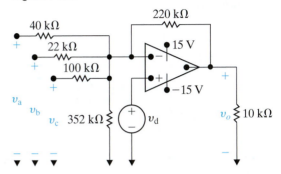

5.17 The 220 kΩ feedback resistor in the circuit in

PSPICE
MULTISIM

Fig. P5.16 is replaced by a variable resistor R_f. The voltages $v_a - v_d$ have the same values as given in Problem 5.16(a).

a) What value of R_f will cause the op amp to saturate? Note that $0 \le R_f \le \infty$.

b) When R_f has the value found in (a), what is the current (in microamperes) into the output terminal of the op amp?

Section 5.5

5.18 The op amp in the circuit of Fig. P5.18 is ideal.

a) What op amp circuit configuration is this?

b) Find v_o in terms of v_s.

c) Find the range of values for v_s such that v_o does not saturate and the op amp remains in its linear region of operation.

Figure P5.18

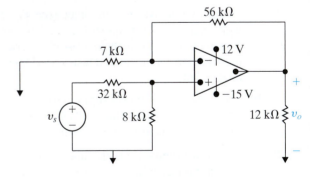

5.19 The op amp in the circuit of Fig. P5.19 is ideal.

a) What op amp circuit configuration is this?

b) Calculate v_o.

Figure P5.19

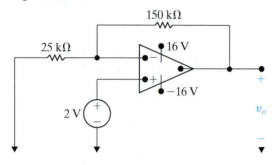

5.20 The op amp in the circuit shown in Fig. P5.20 is ideal,

a) Calculate v_o when v_g equals 3 V.

b) Specify the range of values of v_g so that the op amp operates in a linear mode.

c) Assume that v_g equals 5 V and that the 48 kΩ resistor is replaced with a variable resistor. What value of the variable resistor will cause the op amp to saturate?

Figure P5.20

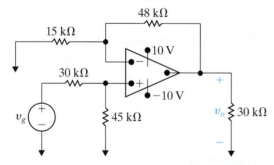

5.21 a) Design a noninverting amplifier (see Fig. 5.13) with a gain of 2.5. Use resistors from Appendix H. You might need to combine resistors in series and in parallel to get the desired resistance. Draw your final circuit.

b) If you use ± 16 V power supplies for the op amp, what range of input values will allow the op amp to stay in its linear operating region?

5.22 a) Design a noninverting amplifier (see Fig. 5.13) with a gain of 6, using a 75 kΩ resistor in the feedback path. Draw your final circuit diagram.

b) Suppose you wish to amplify input signals in the range -2.5 V $\leq v_g \leq 1.5$ V. What are the minimum values of the power supplies that will keep the op amp in its linear operating region?

5.23 The op amp in the circuit of Fig. P5.23 is ideal.

a) What op amp circuit configuration is this?

b) Find v_o in terms of v_s.

c) Find the range of values for v_s such that v_o does not saturate and the op amp remains in its linear region of operation.

Figure P5.23

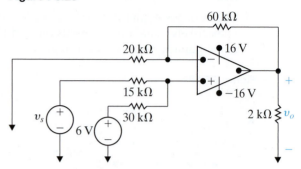

5.24 The circuit in Fig. P5.24 is a noninverting summing amplifier. Assume the op amp is ideal. Design the circuit so that

$$v_o = v_a + 2v_b + 3v_c.$$

a) Specify the numerical values of R_a and R_c.

b) Calculate i_a, i_b, and i_c (in microamperes) when $v_a = 0.7$ V, $v_b = 0.4$ V, and $v_c = 1.1$ V.

Figure P5.24

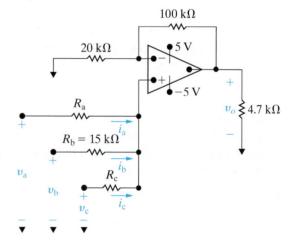

Section 5.6

5.25 The op amp in the circuit of Fig. P5.25 is ideal.

a) What op amp circuit configuration is this?

b) Find an expression for the output voltage v_o in terms of the input voltage v_a.

c) Suppose $v_a = 2$ V. What value of R_f will cause the op amp to saturate?

Figure P5.25

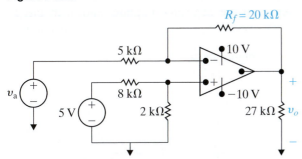

5.26 The resistor R_f in the circuit in Fig. P5.26 is adjusted until the ideal op amp saturates. Specify R_f in kilohms.

Figure P5.26

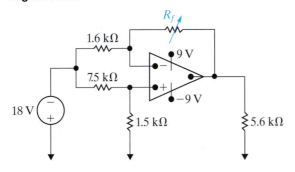

5.27 The resistors in the difference amplifier shown in Fig. 5.15 are $R_a = 10\text{ k}\Omega$, $R_b = 100\text{ k}\Omega$, $R_c = 33\text{ k}\Omega$ and $R_d = 47\text{ k}\Omega$. The signal voltages v_a and v_b are 0.67 and 0.8 V, respectively, and $V_{CC} = \pm 5$ V.

   a) Find v_o.

   b) What is the resistance seen by the signal source v_a?

   c) What is the resistance seen by the signal source v_b?

5.28 Design a difference amplifier (Fig. 5.15) to meet the following criteria: $v_o = 3v_b - 4v_a$. The resistance seen by the signal source v_b is 470 kΩ, and the resistance seen by the signal source v_a is 22 kΩ when the output voltage v_o is zero. Specify the values of R_a, R_b, R_c, and R_d using single resistors or combinations of resistors from Appendix H.

5.29 a) Use the principle of superposition to derive Eq. 5.8.

   b) Derive Eqs. 5.9 and 5.10.

5.30 Select the values of R_b and R_f in the circuit in Fig. P5.30 so that

$$v_o = 8000(i_b - i_a).$$

Use single resistors or combinations of resistors from Appendix H. The op amp is ideal.

Figure P5.30

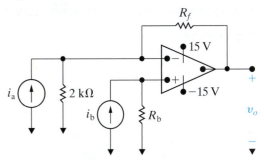

5.31 The op amp in the adder-subtracter circuit shown in Fig. P5.31 is ideal.

   a) Find v_o when $v_a = 1$ V, $v_b = 2$ V, $v_c = 3$ V, and $v_d = 4$ V.

   b) If v_a, v_b, and v_d are held constant, what values of v_c will not saturate the op amp?

Figure P5.31

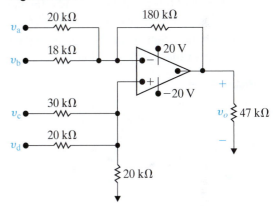

5.32 In the difference amplifier shown in Fig. P5.32, compute (a) the differential mode gain, (b) the common mode gain, and (c) the CMRR.

Figure P5.32

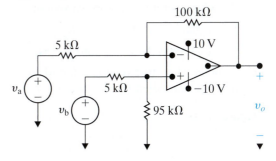

5.33 In the difference amplifier shown in Fig. P5.33, what range of values of R_x yields a CMRR ≥ 1000?

Figure P5.33

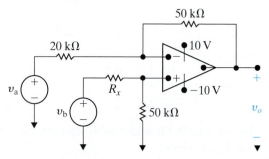

5.34 The op amp in the circuit of Fig. P5.34 is ideal.

a) Plot v_o versus α when $R_f = 4R_1$ and $v_g = 2$ V. Use increments of 0.1 and note by hypothesis that $0 \leq \alpha \leq 1.0$.

b) Write an equation for the straight line you plotted in (a). How are the slope and intercept of the line related to v_g and the ratio R_f/R_1?

c) Using the results from (b), choose values for v_g and the ratio R_f/R_1 such that $v_o = -6\alpha + 4$.

Figure P5.34

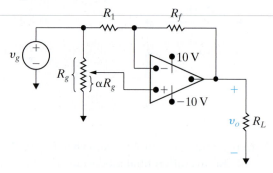

Sections 5.1–5.6

5.35 Assume that the ideal op amp in the circuit seen in Fig. P5.35 is operating in its linear region.

a) Show that $v_o = [(R_1 + R_2)/R_1]v_s$.

b) What happens if $R_1 \to \infty$ and $R_2 \to 0$?

c) Explain why this circuit is referred to as a voltage follower when $R_1 = \infty$ and $R_2 = 0$.

Figure P5.35

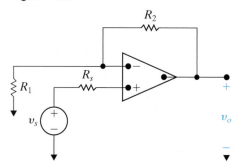

5.36 The voltage v_g shown in Fig. P5.36(a) is applied to the inverting amplifier shown in Fig. P5.36(b). Sketch v_o versus t, assuming the op amp is ideal.

Figure P5.36

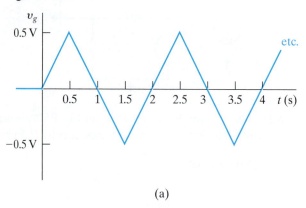

(a)

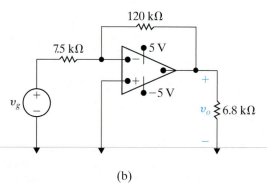

(b)

5.37 The signal voltage v_g in the circuit shown in Fig. P5.37 is described by the following equations:

$$v_g = 0, \qquad\qquad t \leq 0,$$

$$v_g = 4\cos\left(\frac{\pi}{4}\right)t \text{ V}, \quad 0 \leq t \leq \infty.$$

Sketch v_o versus t, assuming the op amp is ideal.

Figure P5.37

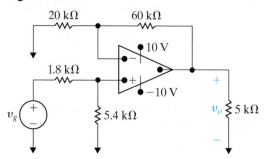

5.38 a) Show that when the ideal op amp in Fig. P5.38 is operating in its linear region,

$$i_a = \frac{3v_g}{R}.$$

b) Show that the ideal op amp will saturate when

$$R_a = \frac{R(\pm V_{CC} - 2v_g)}{3v_g}.$$

Figure P5.38

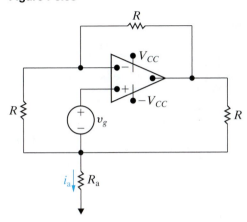

5.39 The op amps in the circuit in Fig. P5.39 are ideal.

a) Find i_a.

b) Find the value of the left source voltage for which $i_a = 0$.

Figure P5.39

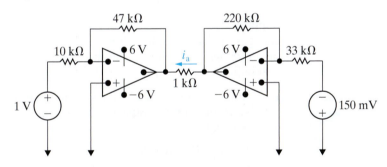

5.40 The two op amps in the circuit in Fig. P5.40 are ideal. Calculate v_{o1} and v_{o2}.

Figure P5.40

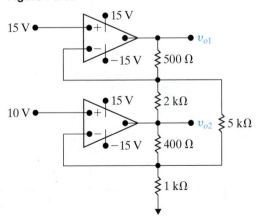

5.41 The circuit inside the shaded area in Fig. P5.41 is a constant current source for a limited range of values of R_L.

a) Find the value of i_L for $R_L = 4\ \text{k}\Omega$.

b) Find the maximum value for R_L for which i_L will have the value in (a).

c) Assume that $R_L = 16\ \text{k}\Omega$. Explain the operation of the circuit. You can assume that $i_n = i_p \approx 0$ under all operating conditions.

d) Sketch i_L versus R_L for $0 \le R_L \le 16\ \text{k}\Omega$.

Figure P5.41

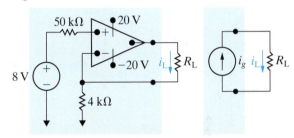

5.42 Assume that the ideal op amp in the circuit in Fig. P5.42 is operating in its linear region.

a) Calculate the power delivered to the 16 kΩ resistor.

b) Repeat (a) with the op amp removed from the circuit, that is, with the 16 kΩ resistor connected in the series with the voltage source and the 48 kΩ resistor.

c) Find the ratio of the power found in (a) to that found in (b).

d) Does the insertion of the op amp between the source and the load serve a useful purpose? Explain.

Figure P5.42

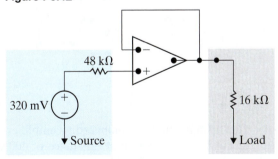

Section 5.7

5.43 Derive Eq. 5.31.

5.44 Repeat Assessment Problem 5.7, given that the inverting amplifier is loaded with a 500 Ω resistor.

5.45 a) Find the Thévenin equivalent circuit with respect to the output terminals a, b for the inverting amplifier of Fig. P5.45. The dc signal source has a value of 880 mV. The op amp has an input resistance of 500 kΩ, an output resistance of 2 kΩ, and an open-loop gain of 100,000.

b) What is the output resistance of the inverting amplifier?

c) What is the resistance (in ohms) seen by the signal source v_s when the load at the terminals a, b is 330 Ω?

Figure P5.45

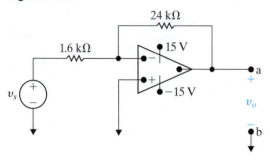

5.46 Repeat Problem 5.45 assuming an ideal op amp.

5.47 Assume the input resistance of the op amp in Fig. P5.47 is infinite and its output resistance is zero.

a) Find v_o as a function of v_g and the open-loop gain A.

b) What is the value of v_o if $v_g = 0.4$ V and $A = 90$?

c) What is the value of v_o if $v_g = 0.4$ V and $A = \infty$?

d) How large does A have to be so that v_o is 95% of its value in (c)?

Figure P5.47

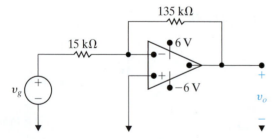

5.48 The op amp in the noninverting amplifier circuit of Fig. P5.48 has an input resistance of 400 kΩ, an output resistance of 5 kΩ, and an open-loop gain of 20,000. Assume that the op amp is operating in its linear region.

a) Calculate the voltage gain (v_o/v_g).

b) Find the inverting and noninverting input voltages v_n and v_p (in millivolts) if $v_g = 1$ V.

c) Calculate the difference $(v_p - v_n)$ in microvolts when $v_g = 1$ V.

d) Find the current drain in picoamperes on the signal source v_g when $v_g = 1$ V.

e) Repeat (a)–(d) assuming an ideal op amp.

Figure P5.48

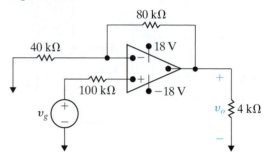

Sections 5.1–5.7

5.49 Suppose the strain gages in the bridge in Fig. 5.22 have the value 120 Ω ± 1%. The power supplies to the op amp are ±15 V, and the reference voltage, v_{ref}, is taken from the positive power supply.

a) Calculate the value of R_f so that when the strain gage that is lengthening reaches its maximum length, the output voltage is 5 V.

b) Suppose that we can accurately measure 50 mV changes in the output voltage. What change in strain gage resistance can be detected in milliohms?

5.50 a) For the circuit shown in Fig. P5.50, show that if $\Delta R \ll R$, the output voltage of the op amp is approximately

$$v_o \approx \frac{R_f\,(R + R_f)}{R^2(R + 2R_f)}\,(-\Delta R)v_{in}.$$

b) Find v_o if $R_f = 470$ kΩ, $R = 10$ kΩ, $\Delta R = 95$ Ω, and $v_{in} = 15$ V.

c) Find the actual value of v_o in (b).

Figure P5.50

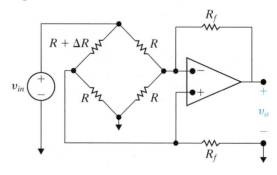

5.51

a) If percent error is defined as

$$\% \text{ error} = \left[\frac{\text{approximate value}}{\text{true value}} - 1\right] \times 100,$$

show that the percent error in the approximation of v_o in Problem 5.50 is

$$\% \text{ error} = \frac{\Delta R (R + R_f)}{R (R + 2R_f)} \times 100.$$

b) Calculate the percent error in v_o for Problem 5.50(b).

5.52 Assume the percent error in the approximation of v_o in the circuit in Fig. P5.50 is not to exceed 1%. What is the largest percent change in R that can be tolerated?

5.53 Assume the resistor in the variable branch of the bridge circuit in Fig. P5.50 is $R - \Delta R$ instead of $R + \Delta R$.

a) What is the expression for v_o if $\Delta R \ll R$?

b) What is the expression for the percent error in v_o as a function of R, R_f, and ΔR?

c) Assume the resistance in the variable arm of the bridge circuit in Fig. P5.50 is 9810 Ω and the values of R, R_f, and v_{in} are the same as in Problem 5.50(b). What is the approximate value of v_o?

d) What is the percent error in the approximation of v_o when the variable arm resistance is 9810 Ω?

Inductance, Capacitance, and Mutual Inductance

CHAPTER CONTENTS

6.1 **The Inductor** *p. 184*

6.2 **The Capacitor** *p. 189*

6.3 **Series-Parallel Combinations of Inductance and Capacitance** *p. 194*

6.4 **Mutual Inductance** *p. 199*

6.5 **A Closer Look at Mutual Inductance** *p. 203*

CHAPTER OBJECTIVES

1 Know and be able to use the equations for voltage, current, power, and energy in an inductor; understand how an inductor behaves in the presence of constant current; and understand the requirement that the current be continuous in an inductor.

2 Know and be able to use the equations for voltage, current, power, and energy in a capacitor; understand how a capacitor behaves in the presence of constant voltage; and understand the requirement that the voltage be continuous in a capacitor.

3 Be able to combine inductors with initial conditions in series and in parallel to form a single equivalent inductor with an initial condition; be able to combine capacitors with initial conditions in series and in parallel to form a single equivalent capacitor with an initial condition.

4 Understand the basic concept of mutual inductance and be able to write mesh-current equations for a circuit containing magnetically coupled coils using the dot convention correctly.

Here we introduce inductors and capacitors, the last two ideal circuit elements mentioned in Chapter 2. Fortunately, the circuit analysis techniques you learned in Chapters 3 and 4 apply to circuits containing inductors and capacitors as well. Once you understand the terminal behavior of these elements in terms of current and voltage, you can use Kirchhoff's laws to describe any interconnections with the other basic elements.

Like other components, inductors and capacitors are easier to describe in terms of circuit variables rather than electromagnetic field variables. Hence, Sections 6.1 and 6.2 briefly review the field concepts underlying inductors and capacitors before focusing on the circuit descriptions. These sections also examine energy in inductors and capacitors. Energy can be stored in both magnetic and electric fields, so inductors and capacitors can store energy. For example, energy can be stored in an inductor and then released to fire a spark plug, or stored in a capacitor and then released to fire a strobe light. In ideal inductors and capacitors, you can extract only as much energy as you have stored. Because inductors and capacitors cannot generate energy, they are classified as **passive elements**.

Section 6.3 describes circuit simplification using series and parallel combinations of capacitors or inductors.

In Sections 6.4 and 6.5, we consider two circuits linked by a magnetic field and thus magnetically coupled. The voltage induced in one circuit is related to the time-varying current in the other circuit by a parameter known as **mutual inductance**. The practical significance of magnetic coupling unfolds as we study the relationships between current, voltage, power, and several new parameters specific to mutual inductance. We introduce these relationships here and then describe their utility in a device called a transformer in Chapters 9 and 10.

■ Practical Perspective

Capacitive Touch Screens

The Practical Perspective in Chapter 3 used a grid of resistors to create a touch screen for a phone or computer monitor. But resistive touch screens have some limitations, the most important of which is that the screen can only process a single touch at any instant in time (see Problem 3.75). This means a resistive touch screen cannot process the "pinch" gesture used by many devices to enlarge or shrink the image on the screen.

Multi-touch screens use a different component within a grid below the screen—capacitors. When you touch a capacitive touch screen, the capacitor's value changes, causing a voltage change. Once you have learned the basic behavior of capacitors and know how they combine in series and in parallel, we will present two possible designs for a multi-touch screen using a grid of capacitors.

cobalt88 /Shutterstock

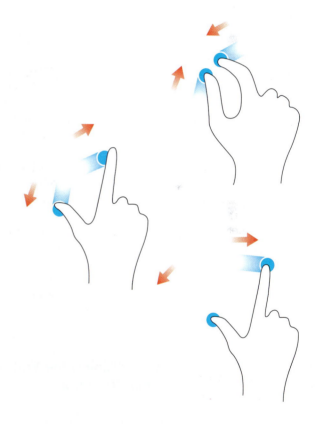

6.1 The Inductor

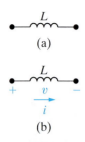

(a)

(b)

Figure 6.1 ▲ (a) The graphic symbol for an inductor with an inductance of L henrys. (b) Assigning reference voltage and current to the inductor, following the passive sign convention.

An inductor is an electrical component that opposes any change in electrical current. It is composed of a coil of wire wound around a supporting core whose material can be magnetic or nonmagnetic. The behavior of inductors is based on phenomena associated with magnetic fields. The source of the magnetic field is charge in motion, or current. If the current is varying with time, the magnetic field is varying with time. A time-varying magnetic field induces a voltage in any conductor linked by the field. The circuit parameter of **inductance** relates the induced voltage to the current.

Figure 6.1(a) shows an inductor, represented graphically as a coiled wire. Its inductance is symbolized by the letter L and is measured in henrys (H). Assigning the reference direction of the current in the direction of the voltage drop across the terminals of the inductor, as shown in Fig. 6.1(b), and using the passive sign convention yields

---

THE INDUCTOR $v - i$ EQUATION

$$v = L\frac{di}{dt},$$ (6.1)

---

where v is measured in volts, L in henrys, i in amperes, and t in seconds. If the current reference is in the direction of the voltage rise, Eq. 6.1 is written with a minus sign.

Note from Eq. 6.1 that the voltage across the terminals of an inductor is proportional to the time rate of change of the current in the inductor. We can make two important observations here. First, if the current is constant, the voltage across the ideal inductor is zero. Thus, the inductor behaves as a short circuit in the presence of a constant, or dc, current. Second, current cannot change instantaneously in an inductor; that is, the current cannot change by a finite amount in zero time. Equation 6.1 tells us that this change would require an infinite voltage, and infinite voltages are not possible. For example, when someone opens the switch on an inductive circuit in an actual system, the current initially continues to flow in the air across the switch, a phenomenon called *arcing*. The arc across the switch prevents the current from dropping to zero instantaneously.

Example 6.1 illustrates the application of Eq. 6.1 to a simple circuit.

EXAMPLE 6.1 | **Determining the Voltage, Given the Current, at the Terminals of an Inductor**

The independent current source in the circuit shown in Fig. 6.2 generates zero current for $t < 0$ and the pulse $10te^{-5t}$ A for $t > 0$.

a) Sketch the current waveform.

b) At what instant of time is the current maximum?

c) Express the voltage across the terminals of the 100 mH inductor as a function of time.

d) Sketch the voltage waveform.

e) Are the voltage and the current at a maximum at the same time?

f) At what instant of time does the voltage change polarity?

g) Is there ever an instantaneous change in voltage across the inductor? If so, at what time?

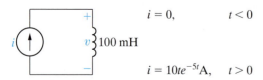

Figure 6.2 ▲ The circuit for Example 6.1.

Solution

a) Figure 6.3 shows the current waveform.

b) $di/dt = 10(-5te^{-5t} + e^{-5t}) = 10e^{-5t}(1 - 5t)$ A/s; $di/dt = 0$ when $t = 0.2$ s. (See Fig. 6.3.)

c) $v = L\,di/dt = (0.1)10e^{-5t}(1 - 5t)$
$= e^{-5t}(1 - 5t)$ V, $t > 0$; $v = 0, t < 0$.

d) Figure 6.4 shows the voltage waveform.

e) No; the voltage is proportional to di/dt, not i.

f) At 0.2 s, which corresponds to the moment when di/dt is passing through zero and changing sign.

g) Yes, at $t = 0$. Note that the voltage can change instantaneously across the terminals of an inductor, even though the current in the inductor cannot change instantaneously.

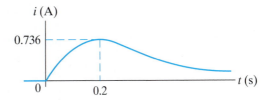

Figure 6.3 ▲ The current waveform for Example 6.1.

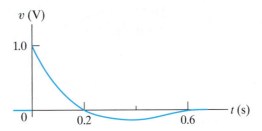

Figure 6.4 ▲ The voltage waveform for Example 6.1.

Current in an Inductor in Terms of the Voltage Across the Inductor

Equation 6.1 expresses the voltage across the terminals of an inductor as a function of the current in the inductor. Now we express the current as a function of the voltage. To find i as a function of v, start by multiplying both sides of Eq. 6.1 by a differential time dt:

$$v\,dt = L\left(\frac{di}{dt}\right)dt.$$

Multiplying the rate at which i varies with t by a differential change in time generates a differential change in i, so the expression simplifies to

$$v\,dt = L\,di.$$

We next integrate both sides of the simplified expression. For convenience, we interchange the two sides of the equation and write

$$L\int_{i(t_0)}^{i(t)} dx = \int_{t_0}^{t} v\,d\tau.$$

Note that we use x and τ as the variables of integration, so i and t become limits on the integrals. Then, divide both sides of the integral equation by L and solve for the inductor current to get

THE INDUCTOR $i - v$ EQUATION

$$i(t) = \frac{1}{L}\int_{t_0}^{t} v\,d\tau + i(t_0), \qquad (6.2)$$

where $i(t_0)$ is the value of the inductor current at the time when we initiate the integration, namely, t_0. In many practical applications, t_0 is zero and Eq. 6.2 becomes

$$i(t) = \frac{1}{L}\int_{0}^{t} v\,d\tau + i(0).$$

Equations 6.1 and 6.2 both give the relationship between the voltage and current at the terminals of an inductor. In both equations, the reference direction for the current is in the direction of the voltage drop across the terminals. Pay attention to the algebraic sign of $i(t_0)$. If the initial current direction and the reference direction for i are the same, the initial current is positive. If the initial current is in the opposite direction, it is negative. Example 6.2 illustrates the application of Eq. 6.2.

EXAMPLE 6.2   Determining the Current, Given the Voltage, at the Terminals of an Inductor

The voltage pulse applied to the 100 mH inductor shown in Fig. 6.5 is 0 for $t < 0$ and is given by the expression

$$v(t) = 20te^{-10t} \text{ V}$$

for $t > 0$. Also assume $i = 0$ for $t \le 0$.

a) Sketch the voltage as a function of time.

b) Find the inductor current as a function of time.

c) Sketch the current as a function of time.

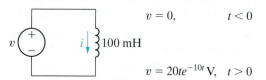

$$v = 0, \qquad t < 0$$
$$v = 20te^{-10t} \text{ V}, \quad t > 0$$

Figure 6.5 ▲ The circuit for Example 6.2.

Solution

a) The voltage as a function of time is shown in Fig. 6.6.

b) The current in the inductor is 0 at $t = 0$. Therefore, the current for $t > 0$ is

$$i = \frac{1}{0.1} \int_0^t 20\tau e^{-10\tau} d\tau + 0$$

$$= 200 \left[\frac{-e^{-10\tau}}{100} (10\tau + 1) \right]\Big|_0^t,$$

$$= 2(1 - 10te^{-10t} - e^{-10t}) \text{ A}, \quad t > 0.$$

c) Figure 6.7 shows the current as a function of time.

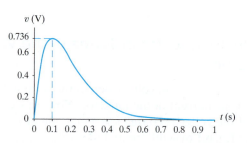

Figure 6.6 ▲ The voltage waveform for Example 6.2.

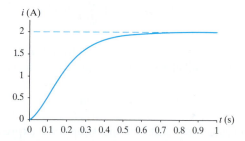

Figure 6.7 ▲ The current waveform for Example 6.2.

In Example 6.2, i approaches a constant value of 2 A as t increases. We say more about this result after discussing the energy stored in an inductor.

Power and Energy in the Inductor

The power and energy relationships for an inductor can be derived directly from the current and voltage relationships. If the current reference is in the direction of the voltage drop across the terminals of the inductor, the power is

$$p = vi. \tag{6.3}$$

Remember that power is in watts, voltage is in volts, and current is in amperes. If we express the inductor voltage as a function of the inductor current, the expression for inductor power becomes

POWER IN AN INDUCTOR

$$p = Li\frac{di}{dt}. \tag{6.4}$$

We can also express the current in terms of the voltage:

$$p = v\left[\frac{1}{L}\int_{t_0}^{t} v \, d\tau + i(t_0)\right].$$

We can use Eq. 6.4 to find the energy stored in the inductor. Power is the time rate of expending energy, so

$$p = \frac{dw}{dt} = Li\frac{di}{dt}.$$

Multiplying both sides by a differential time gives the differential relationship

$$dw = Li \, di.$$

Integrate both sides of the differential relationship, recognizing that the reference for zero energy corresponds to zero current in the inductor. Thus

$$\int_{0}^{w} dx = L\int_{0}^{i} y \, dy,$$

so

ENERGY IN AN INDUCTOR

$$w = \frac{1}{2}Li^2. \tag{6.5}$$

As before, we use different symbols of integration to avoid confusion with the limits placed on the integrals. In Eq. 6.5, the energy is in joules, inductance is in henrys, and current is in amperes. Example 6.3 applies Eqs. 6.3 and 6.5 to the circuits in Examples 6.1 and 6.2 to examine power and energy in these circuits.

EXAMPLE 6.3 | **Determining the Current, Voltage, Power, and Energy for an Inductor**

a) For Example 6.1, plot i, v, p, and w versus time. Line up the plots vertically to allow easy assessment of each variable's behavior.

b) In what time interval is energy being stored in the inductor?

c) In what time interval is energy being extracted from the inductor?

d) What is the maximum energy stored in the inductor?

e) Evaluate the integrals

$$\int_{0}^{0.2} p \, dt \quad \text{and} \quad \int_{0.2}^{\infty} p \, dt,$$

and comment on their significance.

f) Repeat (a)–(c) for Example 6.2.

g) In Example 6.2, why is there a sustained current in the inductor as the voltage approaches zero?

Solution

a) The plots of i and v follow directly from the expressions for i and v obtained in Example 6.1. Applying Eq. 6.3,

$$p = vi = \left[e^{-5t}(1 - 5t)\right](10te^{-5t})$$

$$= 10te^{-10t}(1 - 5t) \text{ W}.$$

Applying Eq. 6.5,

$$w = \frac{1}{2}Li = \frac{1}{2}(0.1)(10te^{-5t})^2 = 5t^2e^{-10t} \text{ J}.$$

The plots of i, v, p, and w are shown in Fig. 6.8.

b) When the energy curve increases, energy is being stored. Thus, from Fig. 6.8, energy is being stored in the time interval 0 to 0.2 s. This corresponds to the interval when $p > 0$.

c) When the energy curve decreases, energy is being extracted. Thus, from Fig. 6.8, energy is being extracted in the time interval 0.2 s to ∞. This corresponds to the interval when $p < 0$.

d) Equation 6.5 tells us that energy is at a maximum when current is at a maximum; the graphs in Fig. 6.8 confirm this. From Example 6.1, $i_{max} = 0.736$ A. Therefore, $w_{max} = 27.07$ mJ.

e) From part (a),

$$p = 10te^{-10t}(1 - 5t) = 10te^{-10t} = 50t^2e^{-10t}.$$

Thus

$$\int_0^{0.2} p\, dt = 10\left[\frac{e^{-10t}}{100}(-10t - 1)\right]_0^{0.2}$$

$$-50\left\{\frac{t^2e^{-10t}}{-10} + \frac{2}{10}\left[\frac{e^{-10t}}{100}(-10t - 1)\right]\right\}_0^{0.2}$$

$$= 0.2e^{-2} = 27.07 \text{ mJ},$$

$$\int_{0.2}^{\infty} p\, dt = 10\left[\frac{e^{-10t}}{100}(-10t - 1)\right]_{0.2}^{\infty}$$

$$-50\left\{\frac{t^2e^{-10t}}{-10} + \frac{2}{10}\left[\frac{e^{-10t}}{100}(-10t - 1)\right]\right\}_{0.2}^{\infty}$$

$$= -0.2e^{-2} = -27.07 \text{ mJ}.$$

Based on the definition of p, the area under the plot of p versus t represents the energy expended over the interval of integration. Hence, integrating the power between 0 and 0.2 s represents the energy

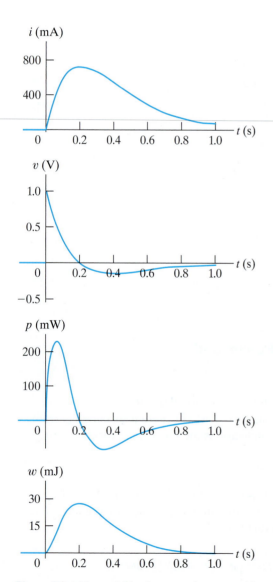

Figure 6.8 ▲ The variables i, v, p, and w versus t for Example 6.1.

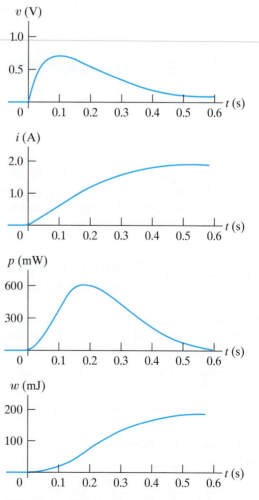

Figure 6.9 ▲ The variables v, i, p, and w versus t for Example 6.2.

stored in the inductor during this time interval. Integrating p between 0.2 s and infinity gives the energy extracted. Note that in this time interval, all the energy originally stored is removed, so after the current peak, no energy is stored in the inductor.

f) The plots of v, i, p, and w follow directly from the expressions for v and i given in Example 6.2 and are shown in Fig. 6.9. Note that in this case

the power is always positive, and hence energy is always being stored during the voltage pulse.

g) The inductor stores energy when the voltage pulse is applied. Because the inductor is ideal, the energy cannot dissipate after the voltage subsides to zero, and thus current continues to circulate in the circuit. Practical inductors require a resistor in the circuit model, which we will examine later in this text.

ASSESSMENT PROBLEM

Objective 1—Know and be able to use the equations for voltage, current, power, and energy in an inductor

6.1 The current source in the circuit shown generates the current pulse

$$i_g(t) = 0, \qquad\qquad t < 0,$$

$$i_g(t) = 8e^{-300t} - 8e^{-1200t}\ \text{A}, \quad t \geq 0.$$

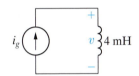

Find (a) $v(0)$; (b) the instant of time, greater than zero, when the voltage v passes through zero; (c) the expression for the power delivered to the inductor; (d) the instant when the power delivered to the inductor is maximum; (e) the maximum power; (f) the instant of time when the stored energy is maximum; and (g) the maximum energy stored in the inductor.

Answer: (a) 28.8 V;
(b) 1.54 ms;
(c) $-76.8e^{-600t} + 384e^{-1500t}$
$\quad -307.2e^{-2400t}$ W, $t \geq 0$;
(d) 411.05 μs;
(e) 32.72 W;
(f) 1.54 ms;
(g) 28.57 mJ.

SELF-CHECK: Also try Chapter Problems 6.3 and 6.11.

6.2 The Capacitor

A capacitor is an electrical component consisting of two conductors separated by an insulator or dielectric material. The capacitor is the only device other than a battery that can store electrical charge. The behavior of capacitors is based on phenomena associated with electric fields. The source of the electric field is separation of charge, or voltage. If the voltage is varying with time, the electric field is varying with time. A time-varying electric field produces a displacement current in the space occupied by the field. The circuit parameter of **capacitance** relates the displacement current to the voltage, where the displacement current is equal to the conduction current at the terminals of the capacitor.

The circuit symbol for a capacitor is two short parallel conductive plates, as shown in Fig. 6.10(a). The capacitance is represented by the letter C and is measured in farads (F). Because the farad is an extremely large quantity of capacitance, practical capacitor values usually lie in the picofarad (pF) to microfarad (μF) range.

The capacitor's symbol reminds us that capacitance occurs whenever electrical conductors are separated by a dielectric, or insulating, material. This condition implies that electric charge is not transported through the capacitor. Although applying a voltage to the terminals of the capacitor cannot move a charge through the dielectric, it can displace a charge within

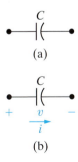

Figure 6.10 ▲ (a) The circuit symbol for a capacitor. (b) Assigning reference voltage and current to the capacitor, following the passive sign convention.

the dielectric. As the voltage varies with time, the displacement of charge also varies with time, causing what is known as the **displacement current**.

The displacement current is indistinguishable from the conduction current at the capacitor's terminals. The current is proportional to the rate at which the voltage across the capacitor varies with time, so

CAPACITOR $i - v$ EQUATION

$$i = C\frac{dv}{dt},$$

(6.6)

where i is measured in amperes, C in farads, v in volts, and t in seconds. In Fig. 6.10(b), the current reference is in the direction of the voltage drop across the capacitor. Using the passive sign convention, we write Eq. 6.6 with a positive sign. If the current reference is in the direction of the voltage rise, Eq. 6.6 is written with a minus sign.

Two important observations follow from Eq. 6.6. First, if the voltage across the terminals is constant, the capacitor current is zero because a conduction current cannot be established in the dielectric material of the capacitor. Only a time-varying voltage can produce a displacement current. Thus, a capacitor behaves as an open circuit in the presence of a constant voltage. Second, voltage cannot change instantaneously across the terminals of a capacitor. Equation 6.6 indicates that such a change would produce infinite current, a physical impossibility.

Equation 6.6 gives the capacitor current as a function of the capacitor voltage. To express the voltage as a function of the current, we multiply both sides of Eq. 6.6 by a differential time dt and then integrate the resulting differentials:

$$i\, dt = C\, dv \quad \text{or} \quad \int_{v(t_0)}^{v(t)} dx = \frac{1}{C}\int_{t_0}^{t} i\, d\tau.$$

Carrying out the integration of the left-hand side of the second equation and rearranging gives

CAPACITOR $v - i$ EQUATION

$$v(t) = \frac{1}{C}\int_{t_0}^{t} i\, d\tau + v(t_0).$$

(6.7)

In many practical applications of Eq. 6.7, the initial time is zero; that is, $t_0 = 0$. Thus, Eq. 6.7 becomes

$$v(t) = \frac{1}{C}\int_{0}^{t} i\, d\tau + v(0).$$

We can easily derive the power and energy relationships for the capacitor. From the definition of power,

CAPACITOR POWER EQUATION

$$p = vi = Cv\frac{dv}{dt},$$

(6.8)

or

$$p = i\left[\frac{1}{C}\int_{t_0}^{t} i\,d\tau + v(t_0)\right].$$

Combining the definition of energy with Eq. 6.8 yields

$$dw = Cv\,dv,$$

from which

$$\int_0^w dx = C\int_0^v y\,dy,$$

or

CAPACITOR ENERGY EQUATION

$$w = \frac{1}{2}Cv^2. \qquad (6.9)$$

In deriving Eq. 6.9, the reference for zero energy corresponds to zero voltage.

Examples 6.4 and 6.5 illustrate the current, voltage, power, and energy relationships for a capacitor.

EXAMPLE 6.4   **Determining Current, Voltage, Power, and Energy for a Capacitor**

The voltage pulse across the terminals of a $0.5\,\mu\text{F}$ capacitor is:

$$v(t) = \begin{cases} 0, & t \le 0\text{ s}; \\ 4t\text{ V}, & 0\text{ s} \le t \le 1\text{ s}; \\ 4e^{-(t-1)}\text{ V}, & t \ge 1\text{ s}. \end{cases}$$

a) Derive the expressions for the capacitor current, power, and energy.

b) Sketch the voltage, current, power, and energy as functions of time. Line up the plots vertically.

c) Specify the time interval when energy is being stored in the capacitor.

d) Specify the time interval when energy is being delivered by the capacitor.

e) Evaluate the integrals

$$\int_0^1 p\,dt \quad\text{and}\quad \int_1^\infty p\,dt$$

and comment on their significance.

Solution

a) From Eq. 6.6,

$$i = \begin{cases} (0.5\,\mu)(0) = 0, & t < 0\text{ s}; \\ (0.5\,\mu)(4) = 2\,\mu\text{A}, & 0\text{ s} < t < 1\text{ s}; \\ (0.5\,\mu)(-4e^{-(t-1)}) = -2e^{-(t-1)}\,\mu\text{A}, & t > 1\text{ s}. \end{cases}$$

The expression for the power is derived from Eq. 6.8:

$$p = \begin{cases} 0, & t \le 0\text{ s}; \\ (4t)(2\mu) = 8t\,\mu\text{W}, & 0\text{ s} \le t < 1\text{ s}; \\ (4e^{-(t-1)})(-2\mu e^{-(t-1)}) = -8e^{-2(t-1)}\,\mu\text{W}, & t > 1\text{ s}. \end{cases}$$

The energy expression follows directly from Eq. 6.9:

$$w = \begin{cases} 0, & t \le 0\text{ s}; \\ \frac{1}{2}(0.5\mu)16t^2 = 4t^2\mu\text{J}, & 0\text{ s} \le t < 1\text{ s}; \\ \frac{1}{2}(0.5\mu)16e^{-2(t-1)} = 4e^{-2(t-1)}\,\mu\text{J}, & t \ge 1\text{ s}. \end{cases}$$

b) Figure 6.11 shows the voltage, current, power, and energy as functions of time.

c) Energy is being stored in the capacitor whenever the power is positive. Hence, energy is being stored in the interval from 0 to 1 s.

d) Energy is being delivered by the capacitor whenever the power is negative. Thus, energy is being delivered for all t greater than 1 s.

e) The integral of $p\,dt$ is the energy associated with the time interval corresponding to the integral's limits. Thus, the first integral represents the energy stored in the capacitor between 0 and 1 s, whereas the second integral represents the energy returned, or delivered, by the capacitor in the interval 1 s to ∞:

$$\int_0^1 p\,dt = \int_0^1 8t\,dt = 4t^2 \Big|_0^1 = 4\ \mu J,$$

$$\int_1^\infty p\,dt = \int_1^\infty (-8e^{-2(t-1)})dt = (-8)\frac{e^{-2(t-1)}}{-2}\Big|_1^\infty = -4\ \mu J.$$

The voltage applied to the capacitor returns to zero as time increases, so the energy returned by this ideal capacitor must equal the energy stored.

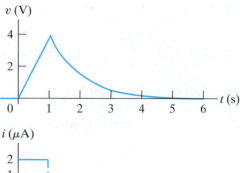
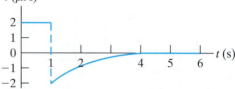
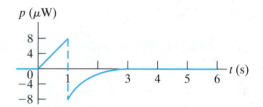
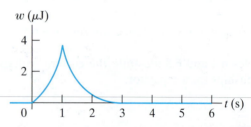

Figure 6.11 ▲ The variables v, i, p, and w versus t for Example 6.4.

EXAMPLE 6.5   **Finding v, p, and w Induced by a Triangular Current Pulse for a Capacitor**

An uncharged $0.2\ \mu F$ capacitor is driven by a triangular current pulse. The current pulse is described by

$$i(t) = \begin{cases} 0, & t \le 0; \\ 5000t\ A, & 0 \le t \le 20\ \mu s; \\ 0.2 - 5000t\ A, & 20 \le t \le 40\ \mu s; \\ 0, & t \ge 40\ \mu s. \end{cases}$$

a) Derive the expressions for the capacitor voltage, power, and energy for each of the four time intervals needed to describe the current.

b) Plot i, v, p, and w versus t. Align the plots as specified in the previous examples.

c) Why does a voltage remain on the capacitor after the current returns to zero?

Solution

a) For $t \le 0$, v, p, and w all are zero.
  For $0 \le t \le 20\ \mu s$,

$$v = \frac{1}{0.2 \times 10^{-6}} \int_0^t (5000\tau)\,d\tau + 0 = 12.5 \times 10^9 t^2\ V,$$

$$p = vi = 62.5 \times 10^{12} t^3\ W,$$

$$w = \frac{1}{2} Cv^2 = 15.625 \times 10^{12} t^4\ J.$$

For $20\ \mu s \le t \le 40\ \mu s$,

$$v(20 \times 10^{-6}) = 12.5 \times 10^9 (20 \times 10^{-6})^2 = 5\ V.$$

Then,

$$v = \frac{1}{0.2 \times 10^{-6}} \int_{20\mu s}^{t} (0.2 - 5000\tau)\,d\tau + 5$$

$$= (10^6 t - 12.5 \times 10^9 t^2 - 10)\ \text{V},$$

$$p = vi,$$

$$= (62.5 \times 10^{12} t^3 - 7.5 \times 10^9 t^2 + 2.5 \times 10^5 t - 2)\ \text{W},$$

$$w = \frac{1}{2}Cv^2,$$

$$= (15.625 \times 10^{12} t^4 - 2.5 \times 10^9 t^3 + 0.125 \times 10^6 t^2$$

$$-2t + 10^{-5})\ \text{J}.$$

For $t \geq 40\ \mu s$,

$$v = 10\ \text{V},$$

$$p = vi = 0,$$

$$w = \frac{1}{2}Cv^2 = 10\ \mu\text{J}.$$

b) The excitation current and the resulting voltage, power, and energy are plotted in Fig. 6.12.

c) Note that the power is always positive for the duration of the current pulse, which means that energy is continuously being stored in the capacitor. When the current returns to zero, the stored energy is trapped because the ideal capacitor cannot dissipate energy. Thus, a voltage remains on the capacitor after its current returns to zero.

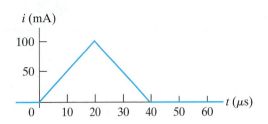

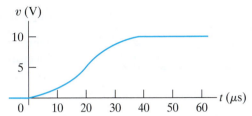

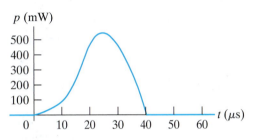

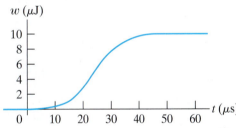

Figure 6.12 ▲ The variables i, v, p, and w versus t for Example 6.5.

ASSESSMENT PROBLEMS

Objective 2—Know and be able to use the equations for voltage, current, power, and energy in a capacitor

6.2   The voltage at the terminals of the 0.6 μF capacitor shown in the figure is 0 for $t < 0$ and $40e^{-15,000t} \sin 30,000t$ V for $t \geq 0$. Find (a) $i(0)$; (b) the power delivered to the capacitor at $t = \pi/80$ ms; and (c) the energy stored in the capacitor at $t = \pi/80$ ms.

0.6 μF

Answer: (a) 0.72 A;
(b) −649.2 mW;
(c) 126.13 μJ.

6.3   The current in the capacitor of Assessment Problem 6.2 is 0 for $t < 0$ and 3 cos 50,000t A for $t \geq 0$. Find (a) $v(t)$; (b) the maximum power delivered to the capacitor at any one instant of time; and (c) the maximum energy stored in the capacitor at any one instant of time.

Answer: (a) 100 sin 50,000t V, $t \geq 0$;
(b) 150 W; (c) 3 mJ.

SELF-CHECK: Also try Chapter Problems 6.17 and 6.21.

6.3 Series-Parallel Combinations of Inductance and Capacitance

Just as series-parallel combinations of resistors can be reduced to a single equivalent resistor, series-parallel combinations of inductors or capacitors can be reduced to a single inductor or capacitor.

Inductors in Series and Parallel

Figure 6.13 shows inductors in series. The series connection means the inductors all have the same current, so we define only one current for the series combination. The voltage drops across the individual inductors are

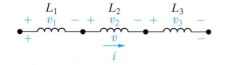

Figure 6.13 ▲ Inductors in series.

$$v_1 = L_1 \frac{di}{dt}, \quad v_2 = L_2 \frac{di}{dt}, \quad \text{and} \quad v_3 = L_3 \frac{di}{dt}.$$

The voltage across the series connection is

$$v = v_1 + v_2 + v_3 = (L_1 + L_2 + L_3) \frac{di}{dt}.$$

Thus, the equivalent inductance of series-connected inductors is the sum of the individual inductances. For n inductors in series,

COMBINING INDUCTORS IN SERIES

$$L_{eq} = \sum_{i=1}^{n} L_i. \tag{6.10}$$

If the original inductors carry an initial current of $i(t_0)$, the equivalent inductor carries the same initial current. Figure 6.14 shows the equivalent circuit for series inductors carrying an initial current.

Inductors in parallel have the same terminal voltage, so the current in each inductor is a function of the terminal voltage and the initial current in that inductor. For the three inductors in parallel shown in Fig. 6.15, the currents for the individual inductors are

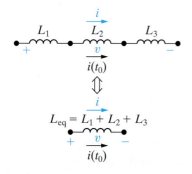

Figure 6.14 ▲ An equivalent circuit for inductors in series carrying an initial current $i(t_0)$.

$$i_1 = \frac{1}{L_1} \int_{t_0}^{t} v \, d\tau + i_1(t_0),$$

$$i_2 = \frac{1}{L_2} \int_{t_0}^{t} v \, d\tau + i_2(t_0),$$

$$i_3 = \frac{1}{L_3} \int_{t_0}^{t} v \, d\tau + i_3(t_0).$$

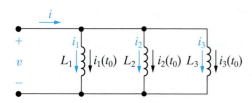

Figure 6.15 ▲ Three inductors in parallel.

The current entering the top node shared by the three parallel inductors is the sum of the inductor currents:

$$i = i_1 + i_2 + i_3.$$

Substituting the expressions for i_1, i_2, and i_3 into the sum yields

$$i = \left(\frac{1}{L_1} + \frac{1}{L_2} + \frac{1}{L_3} \right) \int_{t_0}^{t} v \, d\tau + i_1(t_0) + i_2(t_0) + i_3(t_0). \tag{6.11}$$

The expression for the current as a function of the voltage for a single equivalent inductor is

$$i = \frac{1}{L_{eq}} \int_{t_0}^{t} v \, d\tau + i(t_0).$$ (6.12)

Comparing Eq. 6.12 with Eq. 6.11 yields

$$\frac{1}{L_{eq}} = \frac{1}{L_1} + \frac{1}{L_2} + \frac{1}{L_3},$$

$$i(t_0) = i_1(t_0) + i_2(t_0) + i_3(t_0).$$

Figure 6.16 shows the equivalent circuit for the three parallel inductors in Fig. 6.15.

The results for three inductors in parallel can be extended to n inductors in parallel:

> ## COMBINING INDUCTORS AND THEIR INITIAL CURRENTS IN PARALLEL
>
> $$\frac{1}{L_{eq}} = \sum_{i=1}^{n} \frac{1}{L_i},$$ (6.13)
>
> $$i(t_0) = \sum_{j=1}^{n} i_j(t_0).$$ (6.14)

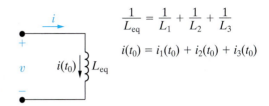

Capacitors in Series and Parallel

Capacitors connected in series can be reduced to a single equivalent capacitor. The reciprocal of the equivalent capacitance is equal to the sum of the reciprocals of the individual capacitances. The initial voltage on the equivalent capacitor is the algebraic sum of the initial voltages on the individual capacitors. Figure 6.17 and the following equations summarize these observations for n series-connected capacitors:

> ## COMBINING CAPACITORS AND THEIR INITIAL VOLTAGES IN SERIES
>
> $$\frac{1}{C_{eq}} = \sum_{i=1}^{n} \frac{1}{C_i},$$ (6.15)
>
> $$v(t_0) = \sum_{j=1}^{n} v_j(t_0).$$ (6.16)

We leave the derivation of the equivalent circuit for series-connected capacitors as an exercise. (See Problem 6.27.)

The equivalent capacitance of capacitors connected in parallel is the sum of the individual capacitances, as Fig. 6.18 page 196 and the following equation show:

> ## COMBINING CAPACITORS IN PARALLEL
>
> $$C_{eq} = \sum_{i=1}^{n} C_i.$$ (6.17)

Capacitors connected in parallel must carry the same voltage. Therefore, the initial voltage across the original parallel capacitors equals the

Figure 6.16 ▲ An equivalent circuit for three inductors in parallel.

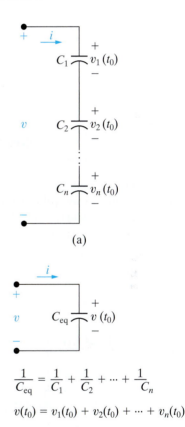

Figure 6.17 ▲ An equivalent circuit for capacitors connected in series. (a) The series capacitors. (b) The equivalent circuit.

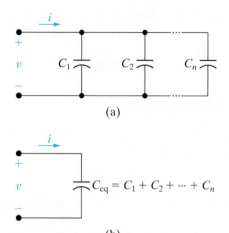

(a)

(b)

Figure 6.18 ▲ An equivalent circuit for capacitors connected in parallel. (a) Capacitors in parallel. (b) The equivalent circuit.

initial voltage across the equivalent capacitance C_{eq}. The derivation of the equivalent circuit for parallel capacitors is left as an exercise. (See Problem 6.28.)

Examples 6.6 and 6.7 use series and parallel combinations to simplify a circuit with multiple inductors and a circuit with multiple capacitors.

EXAMPLE 6.6   Finding the Equivalent Inductance

Figure 6.19 shows four interconnected inductors. The initial currents for two of the inductors are also shown in Fig. 6.19. A single equivalent inductor, together with its initial current, is shown in Fig. 6.20.

a) Find the equivalent inductance, L_{eq}.

b) Find the initial current in the equivalent inductor.

Solution

a) Begin by replacing the parallel-connected 12 mH and 24 mH inductors with a single equivalent inductor whose inductance is

$$\left(\frac{1}{0.012} + \frac{1}{0.024}\right)^{-1} = 0.008 = 8 \text{ mH}.$$

Now the 8 mH, 6 mH, and 10 mH inductors are in series. Combining them gives

$$L_{eq} = 0.008 + 0.006 + 0.010 = 0.024 = 24 \text{ mH}.$$

b) The initial current in the equivalent inductor, i, is the same as the current entering the node to the left of the 24 mH inductor. The KCL equation at that node, summing the currents entering the node, is

$$i - 10 + 6 = 0.$$

Therefore, the initial current in the equivalent inductor is $i = 4$ A.

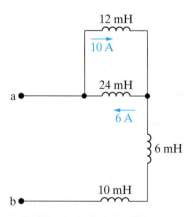

Figure 6.19 ▲ Interconnected inductors for Example 6.6.

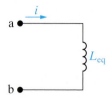

Figure 6.20 ▲ The equivalent inductor for the inductors in Fig. 6.19

EXAMPLE 6.7 | **Finding the Equivalent Capacitance**

Figure 6.21 shows four interconnected capacitors. The initial voltages for three of the capacitors are also shown in Fig. 6.21. A single equivalent capacitor, together with its initial voltage, is shown in Fig. 6.22.

a) Find the equivalent capacitance, C_{eq}.

b) Find the initial voltage across the equivalent capacitor.

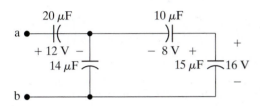

a •

$20\ \mu F$ $10\ \mu F$

$+\ 12\ V\ -$ $-\ 8\ V\ +$

$14\ \mu F$ $15\ \mu F$ $16\ V$

b •

Figure 6.21 ▲ Interconnected capacitors for Example 6.7.

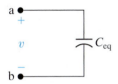

a •

v C_{eq}

b •

Figure 6.22 ▲ The equivalent capacitor for the capacitors in Fig. 6.21.

Solution

a) Begin by replacing the $10\ \mu F$ and $15\ \mu F$ capacitors with a single equivalent capacitor whose capacitance is

$$\left(\frac{1}{10 \times 10^{-6}} + \frac{1}{15 \times 10^{-6}}\right)^{-1} = 6 \times 10^{-6} = 6\ \mu F.$$

Next, combine the $6\ \mu F$ capacitor from the first simplification with the $14\ \mu F$ capacitor to give

$$6 \times 10^{-6} + 14 \times 10^{-6} = 20 \times 10^{-6} = 20\ \mu F.$$

Finally, combine the $20\ \mu F$ from the previous simplification with the $20\ \mu F$ on the left side of the circuit to give

$$C_{eq} = \left(\frac{1}{20 \times 10^{-6}} + \frac{1}{20 \times 10^{-6}}\right)^{-1} = 10 \times 10^{-6} = 10\ \mu F.$$

b) To find the initial voltage from a to b, use KVL to sum the initial voltages for the capacitors on the perimeter of the circuit. This gives

$$12 - 8 + 16 = 20\ V.$$

Therefore, the initial voltage across the equivalent capacitor is 20 V.

⌂ ASSESSMENT PROBLEMS

Objective 3—Be able to combine inductors or capacitors in series and in parallel to form a single equivalent inductor

6.4 The initial values of i_1 and i_2 in the circuit shown are $+3$ A and -5 A, respectively. The voltage at the terminals of the parallel inductors for $t \geq 0$ is $-30e^{-5t}$ mV.
   a) If the parallel inductors are replaced by a single inductor, what is its inductance?
   b) Find the initial current and its reference direction in the equivalent inductor.
   c) Use the equivalent inductor to find $i(t)$.
   d) Find $i_1(t)$ and $i_2(t)$. Verify that the solutions for $i_1(t), i_2(t)$, and $i(t)$ satisfy Kirchhoff's current law.

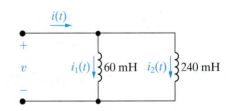

Answer: (a) 48 mH;
(b) 2 A, up;
(c) $0.125e^{-5t} - 2.125$ A, $t \geq 0$;
(d) $i_1(t) = 0.1e^{-5t} + 2.9$ A, $t \geq 0$,
   $i_2(t) = 0.025e^{-5t} - 5.025$ A, $t \geq 0$.

6.5 The current at the terminals of the two capacitors shown is $240e^{-10t} \mu A$ for $t \geq 0$. The initial values of v_1 and v_2 are -10 V and -5 V, respectively. Calculate the total energy trapped in the capacitors as $t \rightarrow \infty$. (*Hint:* Don't combine the capacitors in series—find the energy trapped in each, and then add.)

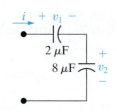

Answer: $20 \mu J$.

SELF-CHECK: Also try Chapter Problems 6.22, 6.24, 6.27, and 6.31.

Inductor and Capacitor Symmetry

We introduced the concept of symmetry, or duality, in Chapter 4. In that chapter, we recognized several examples of duality, including

- essential nodes and meshes;
- KCL and KVL;
- the node voltage method and the mesh current method.

There are also examples of duality in Sections 6.1 through 6.3; these are summarized in Table 6.1. From this table we note many dual relationships, including

- voltage and current;
- open circuits and short circuits;
- inductance and capacitance;
- series connections and parallel connections.

TABLE 6.1 Inductor and Capacitor Duality

	Inductors	**Capacitors**
Primary v-i equation	$v(t) = L\dfrac{di(t)}{dt}$	$i(t) = C\dfrac{dv(t)}{dt}$
Alternate v-i equation	$i(t) = \dfrac{1}{L}\displaystyle\int_{t_0}^{t} v(\tau)d\tau + i(t_0)$	$v(t) = \dfrac{1}{C}\displaystyle\int_{t_0}^{t} i(\tau)d\tau + v(t_0)$
Initial condition	$i(t_0)$	$v(t_0)$
Behavior with a constant source	If $i(t) = I$, $v(t) = 0$ and the inductor behaves like a short circuit	If $v(t) = V$, $i(t) = 0$ and the capacitor behaves like an open circuit
Continuity requirement	$i(t)$ is continuous for all time so $v(t)$ is finite	$v(t)$ is continuous for all time so $i(t)$ is finite
Power equation	$p(t) = v(t)i(t) = Li(t)\dfrac{di(t)}{dt}$	$p(t) = v(t)i(t) = Cv(t)\dfrac{dv(t)}{dt}$
Energy equation	$w(t) = \dfrac{1}{2}Li(t)^2$	$w(t) = \dfrac{1}{2}Cv(t)^2$
Series-connected equivalent	$L_{eq} = \displaystyle\sum_{j=1}^{n} L_j$ $i_{eq}(t_0) = i_j(t_0)$ for all j	$\dfrac{1}{C_{eq}} = \displaystyle\sum_{j=1}^{n} \dfrac{1}{C_j}$ $v_{eq}(t_0) = \displaystyle\sum_{j=1}^{n} v_j(t_0)$
Parallel-connected equivalent	$\dfrac{1}{L_{eq}} = \displaystyle\sum_{j=1}^{n} \dfrac{1}{L_j}$ $i_{eq}(t_0) = \displaystyle\sum_{j=1}^{n} i_j(t_0)$	$C_{eq} = \displaystyle\sum_{j=1}^{n} C_j$ $v_{eq}(t_0) = v_j(t_0)$ for all j

Recognizing the symmetry in the characteristics of inductors and capacitors makes it easier to understand and remember these characteristics.

6.4 Mutual Inductance

A time-varying current in an inductor coil creates a time-varying magnetic field that surrounds that coil, as we saw in Section 6.1. Inductance relates the voltage drop across an inductor to the time-varying current in that inductor. As you will see, it makes sense to rename this parameter self-inductance.

Now we consider circuits containing two inductors with time-varying currents, each creating a time-varying magnetic field. In these circuits, the magnetic field generated by one coil's current envelops the other coil and vice versa, thereby magnetically coupling the two inductors and the circuits that contain them. Figure 6.23 shows an example of such a circuit, where L_1 and L_2 are the self-inductances of the individual inductors and M is the mutual inductance associated with the magnetic coupling. Note the double-headed arrow adjacent to M; the arrows indicate the pair of coils with this value of mutual inductance. We need this notation to accommodate circuits with more than one pair of magnetically coupled coils.

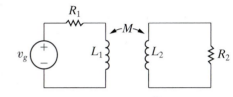

Figure 6.23 ▲ Two magnetically coupled coils.

The mesh-current method is the easiest way to analyze circuits with magnetically coupled inductors. Following Analysis Method 4.6 (page 110), in Step 1 we identify the meshes, and in Step 2 we label the mesh currents. The result is shown in Fig. 6.24. In Step 3, we write a KVL equation around each mesh, summing the voltages across each mesh component. There will be two voltages across each inductor coil because the coils are magnetically coupled. One voltage is the self-induced voltage, which is the product of the self-inductance of the coil and the first derivative of that coil's current. The other voltage is the mutually induced voltage, which is the product of the mutual inductance of the coils and the first derivative of the current in the other coil. Consider the inductor on the left in Fig. 6.24 with self-inductance L_1. The self-induced voltage across this coil is $L_1(di_1/dt)$, and the mutually induced voltage across this coil is $M(di_2/dt)$. But what about the polarities of these two voltages?

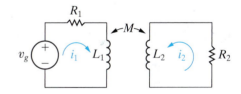

Figure 6.24 ▲ Coil currents i_1 and i_2 used to describe the circuit shown in Fig. 6.23.

Using the passive sign convention, we find that the self-induced voltage is a voltage drop in the direction of the current producing the voltage. But the polarity of the mutually induced voltage depends on the way the inductor coils are wound in relation to the reference direction of coil currents. Showing the details of mutually coupled windings is very cumbersome. Instead, we use the **dot convention** to determine the mutually induced voltage polarities. A dot is placed on one terminal of each winding, as shown in Fig. 6.25. These dots carry the sign information, so we can draw the inductor coils schematically rather than showing how they wrap around a core structure.

The dot convention can be summarized as follows:

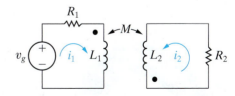

Figure 6.25 ▲ The circuit of Fig. 6.24 with dots added to the coils indicating the polarity of the mutually induced voltages.

DOT CONVENTION FOR MAGNETICALLY COUPLED INDUCTOR COILS

When the reference direction for a current enters the dotted terminal of a coil, the reference polarity of the voltage that it induces in the other coil is positive at its dotted terminal.

Or, stated alternatively,

ALTERNATE DOT CONVENTION

When the reference direction for a current leaves the dotted terminal of a coil, the reference polarity of the voltage that it induces in the other coil is negative at its dotted terminal.

Usually, dot markings will be provided for you in the circuit diagrams in this text, as they are in Fig. 6.25. Let's use the dot convention to complete Step 3 in the mesh current method. Start with the mesh on the left in Fig. 6.25 and sum the voltages around the mesh in the direction of the i_1 mesh current to get

$$-v_g + i_1 R_1 + L_1 \frac{di_1}{dt} - M \frac{di_2}{dt} = 0,$$

The first three terms in the sum, including their signs, should be familiar to you. Look at the fourth term: note that the i_2 current enters the undotted terminal of the L_2 inductor and creates a mutually induced voltage across the L_1 inductor that is positive at its undotted terminal. This is a voltage rise with respect to the direction of i_1, so the sign of $M \, di_2/dt$ in the sum is negative.

Now consider the mesh on the right in Fig. 6.25, and sum the voltages around the mesh in the direction of the i_2 mesh current to get

$$i_2 R_2 + L_2 \frac{di_2}{dt} - M \frac{di_1}{dt} = 0.$$

Look at the third term in the sum: the i_1 current enters the dotted terminal of the L_1 inductor and creates a mutually induced voltage across the L_2 inductor that is positive at its dotted terminal. This is a voltage rise with respect to the direction of i_2, so the sign of $M \, di_1/dt$ in the sum is negative. Figure 6.26 shows the self- and mutually induced voltages across both inductor coils, together with their polarity marks.

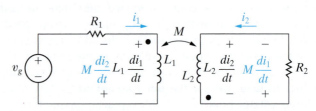

Figure 6.26 ▲ The self- and mutually induced voltages appearing across the coils shown in Fig. 6.25.

The Procedure for Determining Dot Markings

If the polarity dots are not given, you can often determine their locations by examining the physical configuration of the actual circuit or by testing it in the laboratory. We present both of these procedures.

The first procedure assumes that we know the physical arrangement of the two coils and the mode of each winding in a magnetically coupled circuit. Use the following six steps, applied here to Fig. 6.27, to create the dot markings:

1. Arbitrarily select one terminal—say, the D terminal—of one coil and mark it with a dot.

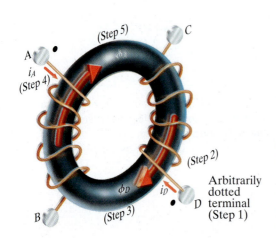

Figure 6.27 ▲ A set of coils showing a method for determining a set of dot markings.

2. Assign a current into the dotted terminal and label it i_D.

3. Use the right-hand rule[1] to determine the direction of the magnetic field established by i_D *inside* the coupled coils and label this field ϕ_D.

4. Arbitrarily pick one terminal of the second coil—say, terminal A—and assign a current into this terminal, showing the current as i_A.

5. Use the right-hand rule to determine the direction of the flux established by i_A *inside* the coupled coils and label this flux ϕ_A.

6. Compare the directions of the two fluxes ϕ_D and ϕ_A. If the fluxes have the same reference direction, place a dot on the terminal of the second coil where the test current (i_A) enters. If the fluxes have different reference directions, place a dot on the terminal of the second coil where the test current leaves. In Fig. 6.27, the fluxes ϕ_D and ϕ_A have the same reference direction, and therefore a dot goes on terminal A.

The second procedure determines the relative polarities of magnetically coupled coils experimentally. This method is used when it is impossible to determine how the coils are wound on the core. One experimental method is to connect a dc voltage source, a resistor, a switch, and a dc voltmeter to the pair of coils, as shown in Fig. 6.28. The shaded box covering the coils implies that physical inspection of the coils is not possible. The resistor R limits the magnitude of the current supplied by the dc voltage source.

Begin by marking the coil terminal connected to the positive terminal of the dc source, via the switch and limiting resistor, with a dot, as shown in Fig. 6.28. When the switch is closed, observe the voltmeter deflection. If the momentary deflection is upscale, place a dot on the coil terminal connected to the voltmeter's positive terminal. If the deflection is downscale, place a dot on the coil terminal connected to the voltmeter's negative terminal.

Example 6.8 constructs the equations for a circuit with magnetically coupled coils, using the dot convention.

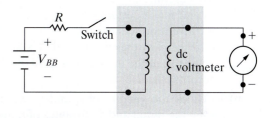

Figure 6.28 ▲ An experimental setup for determining polarity marks.

EXAMPLE 6.8	**Finding Mesh-Current Equations for a Circuit with Magnetically Coupled Coils**

a) Use the mesh-current method to write equations for the circuit in Fig. 6.29 in terms of the currents i_1 and i_2.

b) Verify that if there is no energy stored in the circuit at $t = 0$ and if $i_g = 16 - 16e^{-5t}$ A, the solutions for i_1 and i_2 are

$$i_1 = 4 + 64e^{-5t} - 68e^{-4t} \text{ A},$$

$$i_2 = 1 - 52e^{-5t} + 51e^{-4t} \text{ A}.$$

Solution

a) Follow the steps in Analysis Method 4.6. Steps 1 and 2 identify the meshes and label the mesh currents, as shown in Fig. 6.29. In Step 3, we write

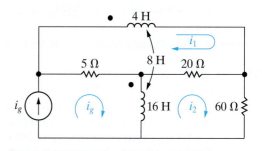

Figure 6.29 ▲ The circuit for Example 6.8.

a KVL equation for each mesh where the current is unknown. Summing the voltages around the i_1 mesh yields

$$4\frac{di_1}{dt} + 8\frac{d}{dt}(i_g - i_2) + 20(i_1 - i_2) + 5(i_1 - i_g) = 0.$$

[1]See discussion of Faraday's law on page 203.

Look carefully at the second term in this equation and make certain you understand how the dot convention was used. Note that the voltage across the 4 H coil due to the current $(i_g - i_2)$, that is, $8d(i_g - i_2)/dt$, is a voltage drop in the direction of i_1, so this term has a positive sign. The KVL equation for the i_2 mesh is

$$20(i_2 - i_1) + 60i_2 + 16\frac{d}{dt}(i_2 - i_g) - 8\frac{di_1}{dt} = 0.$$

Look carefully at the fourth term in this equation and make certain you understand how the dot convention was used. The voltage induced in the 16 H coil by the current i_1, that is, $8\,di_1/dt$, is a voltage rise in the direction of i_2, so this term has a negative sign.

b) To check the validity of i_1 and i_2, we begin by testing the initial and final values of i_1 and i_2. We know by hypothesis that $i_1(0) = i_2(0) = 0$. From the given solutions we have

$$i_1(0) = 4 + 64 - 68 = 0,$$

$$i_2(0) = 1 - 52 + 51 = 0.$$

Now we observe that as t approaches infinity, the source current (i_g) approaches a constant value of 16 A, and therefore the magnetically coupled coils behave as short circuits. Hence, at $t = \infty$ the circuit reduces to that shown in Fig. 6.30. From Fig. 6.30 we see that at $t = \infty$ the three resistors are in parallel across the 16 A source. The equivalent resistance is 3.75 Ω, and thus the

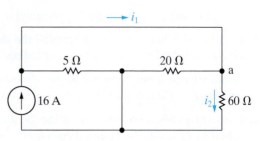

Figure 6.30 ▲ The circuit of Example 6.8 when $t = \infty$.

voltage across the 16 A current source is 60 V. Write a KCL equation at node a, using Ohm's law to find the currents in the 20 Ω and 60 Ω resistors to give

$$i_1(\infty) = \frac{60}{20} + \frac{60}{60} = 4 \text{ A.}$$

Using Ohm's law,

$$i_2(\infty) = \frac{60}{60} = 1 \text{ A.}$$

These values agree with the final values predicted by the solutions for i_1 and i_2:

$$i_1(\infty) = 4 + 64(0) - 68(0) = 4 \text{ A,}$$

$$i_2(\infty) = 1 - 52(0) + 51(0) = 1 \text{ A.}$$

Finally, we check the solutions to see if they satisfy the differential equations derived in (a). We will leave this final check to the reader via Problem 6.36.

◻■ ASSESSMENT PROBLEM

Objective 4—Use the dot convention to write mesh-current equations for mutually coupled coils

6.6   a) Write a set of mesh-current equations for the circuit in Example 6.8 if the dot on the 4 H inductor is at the right-hand terminal, the reference direction of i_g is reversed, and the 60 Ω resistor is increased to 780 Ω.
b) Verify that if there is no energy stored in the circuit at $t = 0$, and if $i_g = 1.96 - 1.96e^{-4t}$ A, the solutions to the differential equations derived in (a) of this Assessment Problem are

$$i_1 = -0.4 - 11.6e^{-4t} + 12e^{-5t} \text{ A,}$$

$$i_2 = -0.01 - 0.99e^{-4t} + e^{-5t} \text{ A.}$$

Answer: (a) $4(di_1/dt) + 25i_1 + 8(di_2/dt) - 20i_2$
$= -5i_g - 8(di_g/dt)$
and

$$8(di_1/dt) - 20i_1 + 16(di_2/dt) + 800i_2$$
$= -16(di_g/dt);$

(b) verification.

SELF-CHECK: Also try Chapter Problem 6.41.

6.5 A Closer Look at Mutual Inductance

Here we take a closer look at self-inductance, and then we turn to a deeper inspection of mutual inductance, examining the limitations and assumptions made in Section 6.4.

A Review of Self-Inductance

Michael Faraday, who studied inductance in the early 1800s, envisioned a magnetic field consisting of lines of force surrounding the current-carrying conductor. Picture these lines of force as energy-storing elastic bands that close on themselves. As the current increases and decreases, the elastic bands (that is, the lines of force) spread and collapse about the conductor. The voltage induced in the conductor is proportional to the number of lines that collapse into, or cut, the conductor. This image of induced voltage is expressed by Faraday's law:

$$v = \frac{d\lambda}{dt}, \tag{6.18}$$

where λ is the flux linkage, measured in weber-turns.

So, how is Faraday's law related to inductance as defined in Section 6.1? Let's look at the coil depicted in Fig. 6.31. The lines threading the N turns, labeled ϕ, represent the magnetic lines of force that make up the magnetic field, which has a spatial orientation and a strength. Use the right-hand rule to determine the spatial orientation: When the fingers of the right hand are wrapped around the coil and point in the direction of the current, the thumb points in the direction of that portion of the magnetic field inside the coil.

To determine the magnetic field strength, begin by defining flux linkage, introduced in Faraday's law (Eq. 6.18). The flux linkage is the product of the magnetic field, ϕ, measured in webers (Wb), and the number of turns linked by the field, N:

$$\lambda = N\phi.$$

The magnitude of the flux, ϕ, is related to the magnitude of the coil current by the relationship

$$\phi = \mathcal{P}Ni$$

where N is the number of turns on the coil, and $\mathcal{P}$ is the permeance of the space occupied by the flux. When the space containing the flux is made up of magnetic materials (such as iron, nickel, and cobalt), the permeance varies with the flux, giving a nonlinear relationship between ϕ and i. But when the space containing the flux is composed of nonmagnetic materials, the permeance is constant, giving a linear relationship between ϕ and i. Note that the flux is also proportional to the number of turns on the coil.

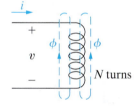

Figure 6.31 ▲ Representation of a magnetic field linking an *N*-turn coil.

Here, we assume that the core material—the space containing the flux—is nonmagnetic. Then, substituting the expressions for flux (ϕ) and flux linkage (λ) into Eq. 6.18 takes us from Faraday's law to the inductor equation:

$$v = \frac{d\lambda}{dt} = \frac{d(N\phi)}{dt}$$

$$= N\frac{d\phi}{dt} = N\frac{d}{dt}(\mathscr{P}Ni)$$

$$= N^2\mathscr{P}\frac{di}{dt} = L\frac{di}{dt}. \tag{6.19}$$

The polarity of the induced voltage in the circuit in Fig. 6.31 depends on the current creating the magnetic field. For example, when i is increasing, di/dt is positive and v is positive, so energy is required to establish the magnetic field. The product vi gives the rate at which energy is stored in the field. When i is decreasing, di/dt is negative, v is negative, and the field collapses about the coil, returning energy to the circuit.

Given this deeper look at self-inductance, we now reexamine mutual inductance.

The Concept of Mutual Inductance

Figure 6.32 shows two magnetically coupled coils. Use the procedure presented in Section 6.4 to verify that the dot markings on the two coils agree with the direction of the coil windings and currents shown. The number of turns on each coil are N_1 and N_2, respectively. Coil 1 is energized by a time-varying current source that establishes the current i_1 in the N_1 turns. Coil 2 is not energized and is open. The coils are wound on a nonmagnetic core. The flux produced by the current i_1 can be divided into two components, labeled ϕ_{11} and ϕ_{21}. The flux component ϕ_{11} is the flux produced by i_1 that links only the N_1 turns. The component ϕ_{21} is the flux produced by i_1 that links both the N_2 and N_1 turns. The first digit in the flux subscript gives the coil number, and the second digit refers to the coil current. Thus, ϕ_{11} is a flux linking coil 1 and produced by a current in coil 1, whereas ϕ_{21} is a flux linking coil 2 and produced by a current in coil 1.

The total flux linking coil 1 is ϕ_1, the sum of ϕ_{11} and ϕ_{21}:

$$\phi_1 = \phi_{11} + \phi_{21}.$$

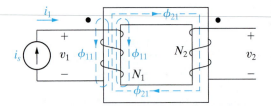

Figure 6.32 ▲ Two magnetically coupled coils.

The flux ϕ_1 and its components ϕ_{11} and ϕ_{21} are related to the coil current i_1 as follows:

$$\phi_1 = \mathscr{P}_1 N_1 i_1,$$

$$\phi_{11} = \mathscr{P}_{11} N_1 i_1,$$

$$\phi_{21} = \mathscr{P}_{21} N_1 i_1,$$

where $\mathscr{P}_1$ is the permeance of the space occupied by the flux ϕ_1, $\mathscr{P}_{11}$ is the permeance of the space occupied by the flux ϕ_{11}, and $\mathscr{P}_{21}$ is the permeance of the space occupied by the flux ϕ_{21}. Combining these four equations and simplifying yields the relationship between the permeance of the space occupied by the total flux ϕ_1 and the permeances of the spaces occupied by its components ϕ_{11} and ϕ_{21}:

$$\mathscr{P}_1 = \mathscr{P}_{11} + \mathscr{P}_{21}. \tag{6.20}$$

We use Faraday's law to derive expressions for v_1 and v_2:

$$v_1 = \frac{d\lambda_1}{dt} = \frac{d(N_1\phi_1)}{dt} = N_1\frac{d}{dt}(\phi_{11} + \phi_{21})$$

$$= N_1^2(\mathscr{P}_{11} + \mathscr{P}_{21})\frac{di_1}{dt} = N_1^2\mathscr{P}_1\frac{di_1}{dt} = L_1\frac{di_1}{dt}, \tag{6.21}$$

and

$$v_2 = \frac{d\lambda_2}{dt} = \frac{d(N_2\phi_{21})}{dt} = N_2\frac{d}{dt}(\mathscr{P}_{21}N_1i_1)$$

$$= N_2N_1\mathscr{P}_{21}\frac{di_1}{dt} = M_{21}\frac{di_1}{dt}.$$

The coefficient of di_1/dt in the equation for v_1 is the self-inductance of coil 1. The coefficient of di_1/dt in the equation for v_2 is the mutual inductance between coils 1 and 2. Thus

$$M_{21} = N_2N_1\mathscr{P}_{21}. \tag{6.22}$$

The subscript on M specifies an inductance that relates the voltage induced in coil 2 to the current in coil 1.

Figure 6.33 again shows two magnetically coupled coils, but now coil 2 is energized by a time-varying current source (i_2) and coil 1 is open. The total flux linking coil 2 is

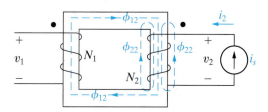

Figure 6.33 ▲ The magnetically coupled coils of Fig. 6.32, with coil 2 excited and coil 1 open.

$$\phi_2 = \phi_{22} + \phi_{12}.$$

The flux ϕ_2 and its components ϕ_{22} and ϕ_{12} are related to the coil current i_2 as follows:

$$\phi_2 = \mathscr{P}_2N_2i_2,$$

$$\phi_{22} = \mathscr{P}_{22}N_2i_2,$$

$$\phi_{12} = \mathscr{P}_{12}N_2i_2.$$

The voltages v_2 and v_1 are

$$v_2 = \frac{d\lambda_2}{dt} = N_2^2\mathscr{P}_2\frac{di_2}{dt} = L_2\frac{di_2}{dt}, \tag{6.23}$$

$$v_1 = \frac{d\lambda_1}{dt} = \frac{d}{dt}(N_1\phi_{12}) = N_1N_2\mathscr{P}_{12}\frac{di_2}{dt} = M_{12}\frac{di_2}{dt}.$$

The coefficient of mutual inductance that relates the voltage induced in coil 1 to the time-varying current in coil 2 is

$$M_{12} = N_1N_2\mathscr{P}_{12}. \tag{6.24}$$

For nonmagnetic materials, the permeances $\mathscr{P}_{12}$ and $\mathscr{P}_{21}$ are equal, so from Eqs. 6.22 and 6.24,

$$M_{12} = M_{21} = M.$$

Hence, for linear circuits with just two magnetically coupled coils, attaching subscripts to the coefficient of mutual inductance is not necessary.

Mutual Inductance in Terms of Self-Inductance

Here we derive the relationship between mutual inductance and self-inductance. From Eqs. 6.21 and 6.23,

$$L_1 = N_1^2 \mathscr{P}_1,$$

$$L_2 = N_2^2 \mathscr{P}_2,$$

so

$$L_1 L_2 = N_1^2 N_2^2 \mathscr{P}_1 \mathscr{P}_2.$$

Use Eq. 6.20 and the corresponding expression for $\mathscr{P}_2$ to write

$$L_1 L_2 = N_1^2 N_2^2 (\mathscr{P}_{11} + \mathscr{P}_{21})(\mathscr{P}_{22} + \mathscr{P}_{12}).$$

But for a linear system, $\mathscr{P}_{21} = \mathscr{P}_{12}$, so the expression for $L_1 L_2$ becomes

$$L_1 L_2 = (N_1 N_2 \mathscr{P}_{12})^2 \left(1 + \frac{\mathscr{P}_{11}}{\mathscr{P}_{12}}\right)\left(1 + \frac{\mathscr{P}_{22}}{\mathscr{P}_{12}}\right)$$

$$= M^2 \left(1 + \frac{\mathscr{P}_{11}}{\mathscr{P}_{12}}\right)\left(1 + \frac{\mathscr{P}_{22}}{\mathscr{P}_{12}}\right). \tag{6.25}$$

Now replace the two terms involving permeances by a single constant, defined as

$$\frac{1}{k^2} = \left(1 + \frac{\mathscr{P}_{11}}{\mathscr{P}_{12}}\right)\left(1 + \frac{\mathscr{P}_{22}}{\mathscr{P}_{12}}\right). \tag{6.26}$$

Substituting Eq. 6.26 into Eq. 6.25 and rearranging yields

$$M^2 = k^2 L_1 L_2$$

or

RELATING SELF-INDUCTANCES AND MUTUAL INDUCTANCE

$$M = k\sqrt{L_1 L_2}, \tag{6.27}$$

where the constant k is called the **coefficient of coupling**. From Eq. 6.26,

$$\frac{1}{k^2} \geq 1 \quad \text{so} \quad k \leq 1.$$

In fact, the coefficient of coupling must lie between 0 and 1, or

$$0 \le k \le 1. \tag{6.28}$$

The coefficient of coupling is 0 when the two coils have no common flux; that is, when $\phi_{12} = \phi_{21} = 0$. This condition implies that $\mathscr{P}_{12} = 0$, and Eq. 6.26 indicates that $1/k^2 = \infty$, or $k = 0$. If there is no flux linkage between the coils, obviously M is 0.

The coefficient of coupling is equal to 1 when ϕ_{11} and ϕ_{22} are 0. This condition implies that all the flux that links coil 1 also links coil 2, so $\mathscr{P}_{11} = \mathscr{P}_{22} = 0$, which represents an ideal state. In reality, winding two coils so that they share precisely the same flux is physically impossible. Magnetic materials (such as alloys of iron, cobalt, and nickel) create a space with high permeance and are used to establish coefficients of coupling that approach unity. (We say more about this important quality of magnetic materials in Chapter 9.)

Energy Calculations

We conclude by calculating the total energy stored in magnetically coupled coils. Along the way, we confirm two observations made earlier: For linear magnetic coupling, (1) $M_{12} = M_{21} = M$, and (2) $M = k\sqrt{L_1 L_2}$, where $0 \le k \le 1$.

Look at the circuit shown in Fig. 6.34. Initially, assume that the currents i_1 and i_2 are zero and that this zero-current state corresponds to zero energy stored in the coils. Then let i_1 increase from zero to some arbitrary value I_1 and compute the energy stored when $i_1 = I_1$. Because $i_2 = 0$, the total power input into the pair of coils is $v_1 i_1$, and the energy stored is

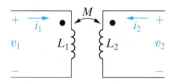

Figure 6.34 ▲ The circuit used to derive the basic energy relationships.

$$\int_0^{W_1} dw = L_1 \int_0^{I_1} i_1 di_1,$$

$$W_1 = \frac{1}{2} L_1 I_1^2.$$

Now we hold i_1 constant at I_1 and increase i_2 from zero to some arbitrary value I_2. During this time interval, the voltage induced in coil 2 by i_1 is zero because I_1 is constant. The voltage induced in coil 1 by i_2 is $M_{12} di_2/dt$. Therefore, the power input to the pair of coils is

$$p = I_1 M_{12} \frac{di_2}{dt} + i_2 v_2.$$

The total energy stored in the pair of coils when $i_2 = I_2$ is

$$\int_{W_1}^{W} dw = \int_0^{I_2} I_1 M_{12} di_2 + \int_0^{I_2} L_2 i_2 di_2,$$

or

$$W = W_1 + I_1 I_2 M_{12} + \frac{1}{2} L_2 I_2^2,$$

$$= \frac{1}{2} L_1 I_1^2 + \frac{1}{2} L_2 I_2^2 + I_1 I_2 M_{12}. \tag{6.29}$$

If we reverse the procedure—that is, if we first increase i_2 from zero to I_2 and then increase i_1 from zero to I_1—the total energy stored is

$$W = \frac{1}{2} L_1 I_1^2 + \frac{1}{2} L_2 I_2^2 + I_1 I_2 M_{21}. \tag{6.30}$$

Equations 6.29 and 6.30 express the total energy stored in a pair of linearly coupled coils as a function of the coil currents, the self-inductances, and the mutual inductance. Note that the only difference between these equations is the coefficient of the current product I_1I_2. We use Eq. 6.29 if i_1 is established first and Eq. 6.30 if i_2 is established first.

When the coupling medium is linear, the total energy stored is the same regardless of the order used to establish I_1 and I_2 because the resultant magnetic flux depends only on the final values of i_1 and i_2, not on how the currents reached their final values. If the resultant flux is the same, the stored energy is the same. Therefore, for linear coupling, $M_{12} = M_{21}$. Also, because I_1 and I_2 are arbitrary values of i_1 and i_2, respectively, we represent the coil currents by their instantaneous values i_1 and i_2. Thus, at any instant of time, the total energy stored in the coupled coils is

$$w(t) = \frac{1}{2} L_1 i_1^2 + \frac{1}{2} L_2 i_2^2 + M i_1 i_2. \qquad (6.31)$$

We derived Eq. 6.31 by assuming that both coil currents entered dotted terminals. We leave it to you to verify that, if one current enters a dotted terminal while the other leaves such a terminal, the algebraic sign of the term $M i_1 i_2$ reverses. Thus, in general,

ENERGY STORED IN MAGNETICALLY COUPLED COILS

$$w(t) = \frac{1}{2} L_1 i_1^2 + \frac{1}{2} L_2 i_2^2 \pm M i_1 i_2. \qquad (6.32)$$

We can use Eq. 6.32 to show that M cannot exceed $\sqrt{L_1 L_2}$. The magnetically coupled coils are passive elements, so the total energy stored can never be negative. If $w(t)$ can never be negative, Eq. 6.32 indicates that

$$\frac{1}{2} L_1 i_1^2 + \frac{1}{2} L_2 i_2^2 - M i_1 i_2 \geq 0$$

when i_1 and i_2 are either both positive or both negative. The limiting value of M occurs when

$$\frac{1}{2} L_1 i_1^2 + \frac{1}{2} L_2 i_2^2 - M i_1 i_2 = 0. \qquad (6.33)$$

To find the limiting value of M, we add and subtract the term $i_1 i_2 \sqrt{L_1 L_2}$ to the left-hand side of Eq. 6.33. Doing so generates a term that is a perfect square:

$$\left(\sqrt{\frac{L_1}{2}} i_1 - \sqrt{\frac{L_2}{2}} i_2 \right)^2 + i_1 i_2 (\sqrt{L_1 L_2} - M) = 0. \qquad (6.34)$$

The squared term in Eq. 6.34 can never be negative, but it can be zero. Therefore, $w(t) \geq 0$ only if

$$\sqrt{L_1 L_2} \geq M, \qquad (6.35)$$

which is another way of saying that

$$M = k \sqrt{L_1 L_2} \qquad (0 \leq k \leq 1).$$

We derived Eq. 6.35 by assuming that i_1 and i_2 are either both positive or both negative. However, we get the same result if i_1 and i_2 have opposite signs because in this case we obtain the limiting value of M by selecting the plus sign in Eq. 6.32.

Work through Example 6.9 to practice calculating the coupling coefficient and the stored energy for magnetically coupled coils.

EXAMPLE 6.9 | Calculating the Coupling Coefficient and Stored Energy for Magnetically Coupled Coils

The mutual inductance and self-inductances of the coils in Fig. 6.34 are $M = 40$ mH, $L_1 = 25$ mH, and $L_2 = 100$ mH.

a) Calculate the coupling coefficient.

b) Calculate the energy stored in the coupled coils when $i_1 = 10$ A and $i_2 = 15$ A.

c) If the coupling coefficient is increased to 1 and $i_1 = 10$ A, what value of i_2 results in zero stored energy?

Solution

a) $k = \dfrac{M}{\sqrt{L_1 L_2}} = \dfrac{0.04}{\sqrt{(0.025)(0.1)}} = 0.8.$

b) $w = \dfrac{1}{2}(0.025)(10)^2 + \dfrac{1}{2}(0.1)(15)^2$

$+ (0.04)(10)(15) = 18.5$ J.

c) When $k = 1, M = \sqrt{(0.025)(0.1)} = 0.05 = 50$ mH. The energy in the coils is now

$$\frac{1}{2}(0.025)(10)^2 + \frac{1}{2}(0.1)(i_2)^2 + (0.05)(10)(i_2) = 0$$

so i_2 must satisfy the quadratic equation

$$0.05i_2^2 + 0.5i_2 + 1.25 = 0.$$

Use the quadratic formula to find i_2:

$$i_2 = \frac{-0.5 \pm \sqrt{0.5^2 - 4(0.05)(1.25)}}{2(0.05)} = -5 \text{ A}.$$

You should verify that the energy is zero for this value of i_2, when $k = 1$.

ASSESSMENT PROBLEM

Objective 4—Understand the concept of mutual inductance

6.7 Consider the magnetically coupled coils described in Example 6.9. Assume that the physical arrangement of the coils results in $\mathcal{P}_1 = \mathcal{P}_2$. If coil 1 has 500 turns, how many turns does coil 2 have?

Answer:

1000 turns

SELF-CHECK: Assess your understanding of this material by trying Chapter Problems 6.45 and 6.48.

■ Practical Perspective

Capacitive Touch Screens

Capacitive touch screens are often used in applications where two or more simultaneous touch points must be detected. We will discuss two designs for a multi-touch screen. The first design employs a grid of electrodes, as shown in Fig. 6.35. When energized, a small parasitic capacitance, C_p, exists between each electrode strip and ground, as

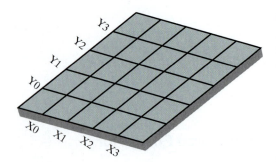

Figure 6.35 ▲ Multi-touch screen with grid of electrodes.

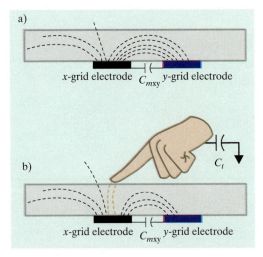

Figure 6.36 ▲ (a) Parasitic capacitance between electrode and ground with no touch; (b) Additional capacitance introduced by a touch.

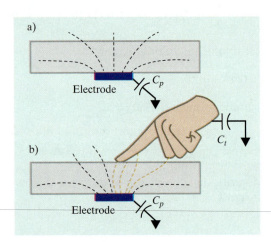

Figure 6.37 ▲ (a) Mutual capacitance between an x-grid and a y-grid electrode; (b) Additional capacitance introduced by a touch.

shown in Fig. 6.36(a). When the screen is touched, say at position x, y on the screen, a second capacitance exists due to the transfer of a small amount of charge from the screen to the human body, which acts like a conductor. This introduces a second capacitance at the point of touch with respect to ground, as shown in Fig. 6.36(b).

A touch-screen controller is continually monitoring the capacitance between the electrodes in the grid and ground. If the screen is untouched, the capacitance between every electrode in the x-grid and ground is C_p; the same is true for the capacitance between every electrode in the y-grid and ground.

When the screen is touched at a single point, C_t and C_p combine in parallel. The equivalent capacitance between the x-grid electrode closest to the touch point and ground is now

$$C_{tx} = C_t + C_p.$$

Likewise, the equivalent capacitance between the y-grid electrode closest to the touch point and ground is now

$$C_{ty} = C_t + C_p.$$

Thus, a screen touch increases the capacitance between the electrodes and ground for the x- and y-grid electrodes closest to the touch point.

What happens when there are two simultaneous points where the screen is touched? Assume that the first touch point has coordinates x_1, y_1 and the second touch point has coordinates x_2, y_2. Now there are four screen locations that correspond to an increase in capacitance: x_1, y_1; x_1, y_2; x_2, y_1; and x_2, y_2. Two of those screen locations match the two touch points, and the other two points are called "ghost" points because the screen was not touched at those points. Therefore, this method for implementing a capacitive touch screen cannot accurately identify more than a single touch point.

Most modern capacitive touch screens do not use the "self-capacitance" design. Instead of measuring the capacitance between each x-grid electrode and ground, and each y-grid electrode and ground, the capacitance between each x-grid electrode and each y-grid electrode is measured. This capacitance is known as "mutual" capacitance and is shown in Fig. 6.37(a).

When the screen is touched, say at position x, y on the screen, a second capacitance again exists due to the transfer of a small amount of charge from the screen to the human body. The second capacitance exists at the point of touch with respect to ground, as shown in Fig. 6.37(b). Therefore, whenever there is a change in the mutual capacitance, C_{mxy}, the screen touch point can be uniquely identified as x, y. If the screen is touched at the points x_1, y_1 and x_2, y_2 then precisely two mutual capacitances change: C_{mx1y1} and C_{mx2y2}. There are no "ghost" points identified, as there were in the self-capacitance design. The mutual capacitance design produces a multi-touch screen capable of identifying two or more touch points uniquely and accurately.

SELF-CHECK: Assess your understanding of the Practical Perspective by solving Chapter Problems 6.51–6.53.

■ Summary

- **Inductance** is a linear circuit parameter that relates the voltage induced by a time-varying magnetic field to the current producing the field. (See page 184.)

- **Capacitance** is a linear circuit parameter that relates the current induced by a time-varying electric field to the voltage producing the field. (See page 189.)

- Inductors and capacitors are passive elements; they can store and release energy, but they cannot generate or dissipate energy. (See page 182.)

- The instantaneous power at the terminals of an inductor or capacitor can be positive or negative, depending on whether energy is being delivered to or extracted from the element.

- An inductor:

  - does not permit an instantaneous change in its terminal current,

  - does permit an instantaneous change in its teminal voltage, and

  - behaves as a short circuit in the presence of a constant terminal current. (See page 184.)

- A capacitor:

  - does not permit an instantaneous change in its terminal voltage,

  - does permit an instantaneous change in its terminal current, and

  - behaves as an open circuit in the presence of a constant terminal voltage. (See page 190.)

- Equations for voltage, current, power, and energy in ideal inductors and capacitors are given in Table 6.1. (See page 198.)

- Inductors in series or in parallel can be replaced by an equivalent inductor. Capacitors in series or in parallel can be replaced by an equivalent capacitor. The equations are summarized in Table 6.1. The table includes the initial conditions for series and parallel equivalent circuits involving inductors and capacitors.

- **Mutual inductance**, M, is the circuit parameter relating the voltage induced in one circuit to a time-varying current in another circuit. Specifically,

$$v_1 = L_1 \frac{di_1}{dt} + M_{12} \frac{di_2}{dt}$$

$$v_2 = M_{21} \frac{di_1}{dt} + L_2 \frac{di_2}{dt},$$

where v_1 and i_1 are the voltage and current in circuit 1, and v_2 and i_2 are the voltage and current in circuit 2. For coils wound on nonmagnetic cores, $M_{12} = M_{21} = M$. (See page 199.)

- The **dot convention** establishes the polarity of mutually induced voltages:

  When the reference direction for a current enters the dotted terminal of a coil, the reference polarity of the voltage that the current induces in the other coil is positive at its dotted terminal.

  Or, alternatively,

  When the reference direction for a current leaves the dotted terminal of a coil, the reference polarity of the voltage that the current induces in the other coil is negative at its dotted terminal.

(See page 199.)

- The relationship between the self-inductance of each winding and the mutual inductance between windings is

$$M = k\sqrt{L_1 L_2}.$$

The **coefficient of coupling**, k, is a measure of the degree of magnetic coupling. By definition, $0 \le k \le 1$. (See page 206.)

- The energy stored in magnetically coupled coils in a linear medium is related to the coil currents and inductances by the relationship

$$w = \frac{1}{2} L_1 i_1^2 + \frac{1}{2} L_2 i_2^2 \pm M i_1 i_2.$$

(See page 208.)

■ Problems

Section 6.1

6.1 The current in a 50 μH inductor is known to be

$$i_L = 18te^{-10t}\text{A} \quad \text{for } t \geq 0.$$

a) Find the voltage across the inductor for $t > 0$. (Assume the passive sign convention.)

b) Find the power (in microwatts) at the terminals of the inductor when $t = 200$ ms.

c) Is the inductor absorbing or delivering power at 200 ms?

d) Find the energy (in microjoules) stored in the inductor at 200 ms.

e) Find the maximum energy (in microjoules) stored in the inductor and the time (in milliseconds) when it occurs.

6.2 The voltage at the terminals of the 200 μH inductor in Fig. P6.2(a) is shown in Fig. P6.2(b). The inductor current i is known to be zero for $t \leq 0$.

a) Derive the expressions for i for $t \geq 0$.

b) Sketch i versus t for $0 \leq t \leq \infty$.

Figure P6.2

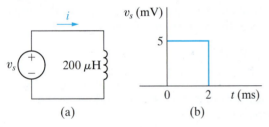

(a)   (b)

6.3 The triangular current pulse shown in Fig. P6.3 is applied to a 20 mH inductor.

a) Write the expressions that describe $i(t)$ in the four intervals $t < 0, 0 \leq t \leq 5$ ms, 5 ms $\leq t \leq 10$ ms, and $t > 10$ ms.

b) Derive the expressions for the inductor voltage, power, and energy. Use the passive sign convention.

Figure P6.3

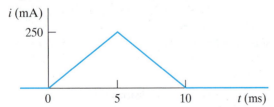

6.4 The current in a 200 mH inductor is

$$i = 75 \text{ mA}, \qquad\qquad t \leq 0;$$

$$i = (B_1 \cos 200t + B_2 \sin 200t)e^{-50t} \text{ A}, \quad t \geq 0.$$

The voltage across the inductor (passive sign convention) is 4.25 V at $t = 0$. Calculate the power at the terminals of the inductor at $t = 25$ ms. State whether the inductor is absorbing or delivering power.

6.5 The current in a 20 mH inductor is known to be

$$i = 40 \text{ mA}, \qquad\qquad t \leq 0;$$

$$i = A_1 e^{-10,000t} + A_2 e^{-40,000t} \text{A}, \quad t \geq 0.$$

The voltage across the inductor (passive sign convention) is 28 V at $t = 0$.

a) Find the expression for the voltage across the inductor for $t > 0$.

b) Find the time, greater than zero, when the power at the terminals of the inductor is zero.

6.6 Assume in Problem 6.5 that the value of the voltage across the inductor at $t = 0$ is -68 V instead of 28 V.

a) Find the numerical expressions for i and v for $t \geq 0$.

b) Specify the time intervals when the inductor is storing energy and the time intervals when the inductor is delivering energy.

c) Show that the total energy extracted from the inductor is equal to the total energy stored.

6.7 Evaluate the integral

$$\int_0^\infty p \, dt$$

for Example 6.2. Comment on the significance of the result.

6.8 a) Find the inductor current in the circuit in Fig. P6.8 if $v = 30 \cos 500t$ V, $L = 15$ mH, and $i(0) = -4$ A.

b) Sketch v, i, p, and w versus t. In making these sketches, use the format used in Fig. 6.8. Plot over one complete cycle of the voltage waveform.

c) Describe the subintervals in the time interval between 0 and 4π ms when power is being absorbed by the inductor. Repeat for the subintervals when power is being delivered by the inductor.

Figure P6.8

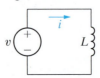

6.9 The current in a 25 mH inductor is known to be -10 A for $t \leq 0$ and $(-10 \cos 400t - 5 \sin 400t) e^{-200t}$ A for $t \geq 0$. Assume the passive sign convention.

  a) At what instant of time is the voltage across the inductor maximum?

  b) What is the maximum voltage?

6.10 The current in and the voltage across a 5 H inductor are known to be zero for $t \leq 0$. The voltage across the inductor is given by the graph in Fig. P6.10 for $t \geq 0$.

  a) Derive the expression for the current as a function of time in the intervals $0 \leq t \leq 1$ s, 1 s $\leq t \leq 3$ s, 3 s $\leq t \leq 5$ s, 5 s $\leq t \leq 6$ s, and 6 s $\leq t < \infty$.

  b) For $t > 0$, what is the current in the inductor when the voltage is zero?

  c) Sketch i versus t for $0 \leq t < \infty$.

Figure P6.10

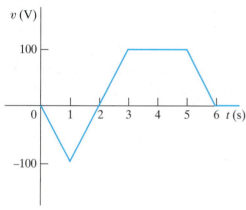

6.11 The current in the 2.5 mH inductor in Fig. P6.11 is known to be 1 A for $t < 0$. The inductor voltage for $t \geq 0$ is given by the expression

$$v_L(t) = 3e^{-4t} \text{ mV}, \quad 0^+ \leq t \leq 2 \text{ s};$$

$$v_L(t) = -3e^{-4(t-2)} \text{ mV}, \quad 2 \text{ s} \leq t < \infty.$$

Sketch $v_L(t)$ and $i_L(t)$ for $0 \leq t < \infty$.

Figure P6.11

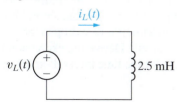

6.12 Initially there was no energy stored in the 5 H inductor in the circuit in Fig. P6.12 when it was placed across the terminals of the voltmeter. At $t = 0$ the inductor was switched instantaneously to position b where it remained for 1.6 s before returning instantaneously to position a. The d'Arsonval voltmeter

has a full-scale reading of 20 V and a sensitivity of 1000 Ω/V. What will the reading of the voltmeter be at the instant the switch returns to position a if the inertia of the d'Arsonval movement is negligible?

Figure P6.12

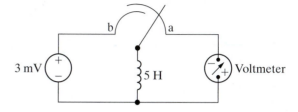

Section 6.2

6.13 The voltage across a 5 μF capacitor is known to be

$$v_c = 500te^{-2500t} \text{ V} \quad \text{for} \quad t \geq 0.$$

  a) Find the current through the capacitor for $t > 0$. Assume the passive sign convention.

  b) Find the power at the terminals of the capacitor when $t = 100 \mu$s.

  c) Is the capacitor absorbing or delivering power at $t = 100 \mu$s?

  d) Find the energy stored in the capacitor at $t = 100 \mu$s.

  e) Find the maximum energy stored in the capacitors and the time when the maximum occurs.

6.14 The voltage at the terminals of the capacitor in Fig. 6.10 is known to be

$$v = \begin{cases} -10 \text{ V}, & t \leq 0; \\ 40 - 1e^{-1000t}(50 \cos 500t + 20 \sin 500 \, t) \text{ V}, & t \geq 0. \end{cases}$$

Assume $C = 0.8 \mu$F.

  a) Find the current in the capacitor for $t < 0$.

  b) Find the current in the capacitor for $t > 0$.

  c) Is there an instantaneous change in the voltage across the capacitor at $t = 0$?

  d) Is there an instantaneous change in the current in the capacitor at $t = 0$?

  e) How much energy (in millijoules) is stored in the capacitor at $t = \infty$?

6.15 The triangular voltage pulse shown in Fig. P6.15 is applied to a 200 μF capacitor.

  a) Write the expressions that describe $v(t)$ in the five time intervals $t < 0$, $0 \leq t \leq 2$ s, 2 s $\leq t \leq 6$ s, 6 s $\leq t \leq 8$ s, and $t > 8$ s.

  b) Derive the expressions for the capacitor current, power, and energy for the time intervals in part (a). Use the passive sign convention.

  c) Identify the time intervals between 0 and 8 s when power is being delivered by the capacitor.

Repeat for the time intervals when power is being absorbed by the capacitor.

Figure P6.15

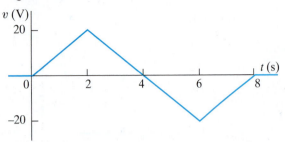

6.16 The expressions for voltage, power, and energy derived in Example 6.5 involved both integration and manipulation of algebraic expressions. As an engineer, you cannot accept such results on faith alone. That is, you should develop the habit of asking yourself, "Do these results make sense in terms of the known behavior of the circuit they purport to describe?" With these thoughts in mind, test the expressions of Example 6.5 by performing the following checks:

a) Check the expressions to see whether the voltage is continuous in passing from one time interval to the next.

b) Check the power expression in each interval by selecting a time within the interval and seeing whether it gives the same result as the corresponding product of v and i. For example, test at 10 and 30 μs.

c) Check the energy expression within each interval by selecting a time within the interval and seeing whether the energy equation gives the same result as $\frac{1}{2}Cv^2$. Use 10 and 30 μs as test points.

6.17 A 20 μF capacitor is subjected to a voltage pulse having a duration of 1 s. The pulse is described by the following equations:

$$v_c(t) = \begin{cases} 30t^2 \text{ V}, & 0 \le t \le 0.5 \text{ s}; \\ 30(t-1)^2 \text{ V}, & 0.5 \text{ s} \le t \le 1 \text{ s}; \\ 0 & \text{elsewhere.} \end{cases}$$

Sketch the current pulse that exists in the capacitor during the 1 s interval.

6.18 The voltage across the terminals of a 250 nF capacitor is

PSPICE
MULTISIM

$$v = \begin{cases} 50 \text{ V}, & t \le 0; \\ (A_1 e^{-4000t} + A_2 t e^{-4000t}) \text{ V}, & t \ge 0. \end{cases}$$

The initial current in the capacitor is 400 mA. Assume the passive sign convention.

a) What is the initial energy stored in the capacitor?

b) Evaluate the coefficients A_1 and A_2.

c) What is the expression for the capacitor current?

6.19 The initial voltage on the 0.5 μF capacitor shown in Fig. P6.19(a) is -20 V. The capacitor current has the waveform shown in Fig. P6.19(b).

PSPICE
MULTISIM

a) How much energy, in microjoules, is stored in the capacitor at $t = 500 \mu$s?

b) Repeat (a) for $t = \infty$.

Figure P6.19

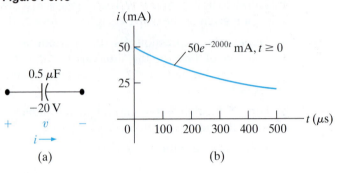

6.20 The current shown in Fig. P6.20 is applied to a 250 nF capacitor. The initial voltage on the capacitor is zero.

PSPICE
MULTISIM

a) Find the charge on the capacitor at $t = 30 \mu$s.

b) Find the voltage on the capacitor at $t = 50 \mu$s.

c) How much energy is stored in the capacitor by this current?

Figure P6.20

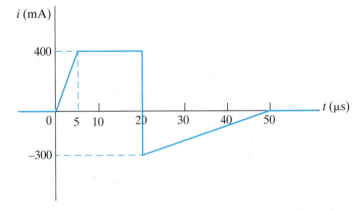

6.21 The rectangular-shaped current pulse shown in Fig. P6.21 is applied to a 5 μF capacitor. The initial voltage on the capacitor is a 12 V drop in the reference direction of the current. Derive the expression for the capacitor voltage for the time intervals in (a)–(e).

PSPICE
MULTISIM

a) $0 \le t \le 5 \mu$s;

b) 5μs $\le t \le 20 \mu$s;

c) 20μs $\le t \le 25 \mu$s;

d) 25μs $\le t \le 35 \mu$s;

e) 35μs $\le t < \infty$.

f) Sketch $v(t)$ over the interval -20μs $\le t \le 40 \mu$s.

Figure P6.21

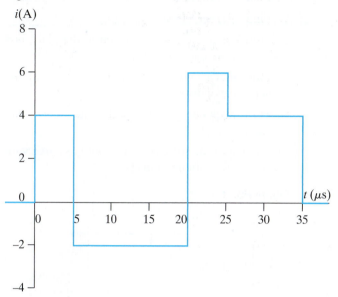

i(A)

Section 6.3

6.22 Use realistic inductor values from Appendix H to construct series and parallel combinations of inductors to yield the equivalent inductances specified below. Try to minimize the number of inductors used. Assume that no initial energy is stored in any of the inductors.

a) 8 mH;

b) 45 μH;

c) 180 μH.

6.23 Assume that the initial energy stored in the induc-
PSPICE   tors of Figs. P6.23(a) and (b) is zero. Find the equiv-
MULTISIM alent inductance with respect to the terminals a, b.

Figure P6.23

(a)

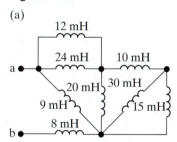

(b)

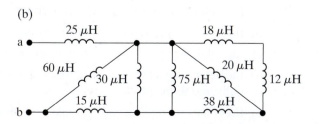

6.24 The three inductors in the circuit in Fig. P6.24 are
PSPICE   connected across the terminals of a black box at
MULTISIM $t = 0$. The resulting voltage for $t > 0$ is known to be

$$v_o = 2000e^{-100t} \text{ V}.$$

If $i_1(0) = -6$ A and $i_2(0) = 1$ A, find

a) $i_o(0)$;

b) $i_o(t), t \geq 0$;

c) $i_1(t), t \geq 0$;

d) $i_2(t), t \geq 0$;

e) the initial energy stored in the three inductors;

f) the total energy delivered to the black box; and

g) the energy trapped in the ideal inductors.

Figure P6.24

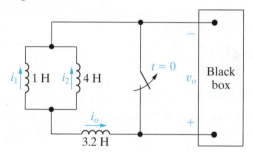

6.25 For the circuit shown in Fig. P6.24, how many milli-seconds after the switch is opened is the energy delivered to the black box 80% of the total energy delivered?

6.26 The two parallel inductors in Fig. P6.26 are connect-ed across the terminals of a black box at $t = 0$. The resulting voltage v for $t > 0$ is known to be $12e^{-t}$ V. It is also known that $i_1(0) = 2$ A and $i_2(0) = 4$ A.

a) Replace the original inductors with an equiva-lent inductor and find $i(t)$ for $t \geq 0$.

b) Find $i_1(t)$ for $t \geq 0$.

c) Find $i_2(t)$ for $t \geq 0$.

d) How much energy is delivered to the black box in the time interval $0 \leq t < \infty$?

e) How much energy was initially stored in the par-allel inductors?

f) How much energy is trapped in the ideal inductors?

g) Show that your solutions for i_1 and i_2 agree with the answer obtained in (f).

Figure P6.26

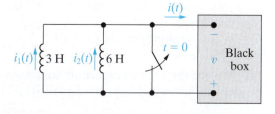

6.27 Derive the equivalent circuit for a series connection of ideal capacitors. Assume that each capacitor has its own initial voltage. Denote these initial voltages as $v_1(t_0)$, $v_2(t_0)$, and so on. (*Hint*: Sum the voltages across the string of capacitors, recognizing that the series connection forces the current in each capacitor to be the same.)

6.28 Derive the equivalent circuit for a parallel connection of ideal capacitors. Assume that the initial voltage across the paralleled capacitors is $v(t_0)$. (*Hint*: Sum the currents into the string of capacitors, recognizing that the parallel connection forces the voltage across each capacitor to be the same.)

6.29 Use realistic capacitor values from Appendix H to construct series and parallel combinations of capacitors to yield the equivalent capacitances specified below. Try to minimize the number of capacitors used. Assume that no initial energy is stored in any of the capacitors.

  a) 480 pF;

  b) 600 nF;

  c) 120 μF.

6.30 Find the equivalent capacitance with respect to the terminals a, b for the circuits shown in Fig. P6.30.

Figure P6.30

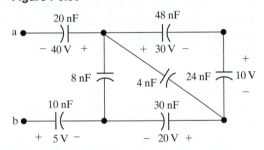

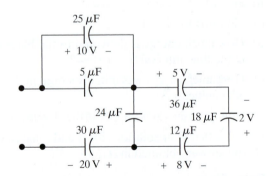

6.31 The two series-connected capacitors in Fig. P6.31 are connected to the terminals of a black box at $t = 0$. The resulting current $i(t)$ for $t > 0$ is known to be $20e^{-t}$ μA.

  a) Replace the original capacitors with an equivalent capacitor and find $v_o(t)$ for $t \geq 0$.

  b) Find $v_1(t)$ for $t \geq 0$.

  c) Find $v_2(t)$ for $t \geq 0$.

  d) How much energy is delivered to the black box in the time interval $0 \leq t < \infty$?

  e) How much energy was initially stored in the series capacitors?

  f) How much energy is trapped in the ideal capacitors?

  g) Show that the solutions for v_1 and v_2 agree with the answer obtained in (f).

Figure P6.31

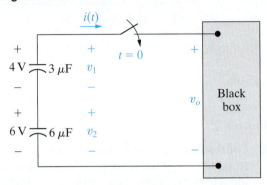

6.32 The four capacitors in the circuit in Fig. P6.32 are connected across the terminals of a black box at $t = 0$. The resulting current i_b for $t > 0$ is known to be

$$i_b = -5e^{-50t} \text{ mA.}$$

If $v_a(0) = -20$ V, $v_c(0) = -30$ V, and $v_d(0) = 250$ V, find the following for $t \geq 0$: (a) $v_b(t)$, (b) $v_a(t)$, (c) $v_c(t)$, (d) $v_d(t)$, (e) $i_1(t)$, and (f) $i_2(t)$.

Figure P6.32

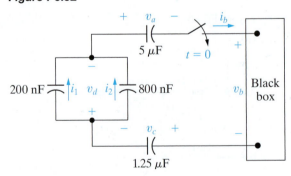

6.33 For the circuit in Fig. P6.32, calculate

  a) the initial energy stored in the capacitors;

  b) the final energy stored in the capacitors;

  c) the total energy delivered to the black box;

  d) the percentage of the initial energy stored that is delivered to the black box; and

  e) the time, in milliseconds, it takes to deliver 7.5 mJ to the black box.

6.34 At $t = 0$, a series-connected capacitor and inductor are placed across the terminals of a black box, as shown in Fig. P6.34. For $t > 0$, it is known that

$$i_o = 1.5e^{-16,000t} - 0.5e^{-16,000t} \text{ A}.$$

If $v_c(0) = -50$ V find v_o for $t \geq 0$.

Figure P6.34

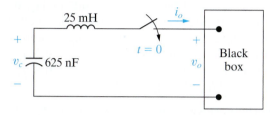

6.35 The current in the circuit in Fig. P6.35 is known to be

$$i_o = 5e^{-2000t}(2 \cos 4000t + \sin 4000t) \text{ A}$$

for $t \geq 0^+$. Find $v_1(0^+)$ and $v_2(0^+)$.

Figure P6.35

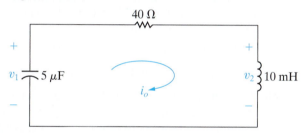

Section 6.4

6.36 a) Show that the differential equations derived in (a) of Example 6.8 can be rearranged as follows:

$$4\frac{di_1}{dt} + 25i - 8\frac{di_2}{dt} - 20i_2 = 5i_g - 8\frac{di_g}{dt};$$

$$-8\frac{di_1}{dt} - 20i_1 + 16\frac{di_2}{dt} + 80i_2 = 16\frac{di_g}{dt}.$$

  b) Show that the solutions for i_1, and i_2 given in (b) of Example 6.8 satisfy the differential equations given in part (a) of this problem.

6.37 a) Show that the two coupled coils in Fig. P6.37 can be replaced by a single coil having an inductance of $L_{ab} = L_1 + L_2 + 2M$. (*Hint:* Express v_{ab} as a function of i_{ab}.)

  b) Show that if the connections to the terminals of the coil labeled L_2 are reversed, $L_{ab} = L_1 + L_2 - 2M$.

Figure P6.37

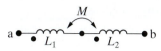

6.38 a) Show that the two magnetically coupled coils in Fig. P6.38 can be replaced by a single coil having an inductance of

$$L_{ab} = \frac{L_1 L_2 - M^2}{L_1 + L_2 - 2M}.$$

  (*Hint:* Let i_1 and i_2 be clockwise mesh currents in the left and right "windows" of Fig. P6.38, respectively. Sum the voltages around the two meshes. In mesh 1 let v_{ab} be the unspecified applied voltage. Solve for di_1/dt as a function of v_{ab}.)

  b) Show that if the magnetic polarity of coil 2 is reversed, then

$$L_{ab} = \frac{L_1 L_2 - M^2}{L_1 + L_2 + 2M}.$$

Figure P6.38

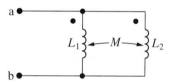

6.39 Let v_g represent the voltage across the current source in the circuit in Fig. 6.29. The reference for v_g is positive at the upper terminal of the current source.

  a) Find v_g as a function of time when $i_g = 16 - 16e^{-5t}$ A.

  b) What is the initial value of v_g?

  c) Find the expression for the power developed by the current source.

  d) How much power is the current source developing when t is infinite?

  e) Calculate the power dissipated in each resistor when t is infinite.

6.40 Let v_o represent the voltage across the 16 H inductor in the circuit in Fig. 6.29. Assume v_o is positive at the dot. As in Example 6.8, $i_g = 16 - 16e^{-5t}$ A.

  a) Can you find v_o without having to differentiate the expressions for the currents? Explain.

  b) Derive the expression for v_o.

  c) Check your answer in (b) using the appropriate current derivatives and inductances.

6.41 There is no energy stored in the circuit in Fig. P6.41 at the time the switch is opened.

a) Derive the differential equation that governs the behavior of i_2 if $L_1 = 5\,H$, $L_2 = 0.2\,H$, $M = 0.5\,H$, and $R_o = 10\,\Omega$.

b) Show that when $i_g = e^{-10t} - 10\,A$, $t \geq 0$, the differential equation derived in (a) is satisfied when $i_2 = 625e^{-10t} - 250e^{-50t}\,mA$, $t \geq 0$.

c) Find the expression for the voltage v_1 across the current source.

d) What is the initial value of v_1? Does this make sense in terms of known circuit behavior?

Figure P6.41

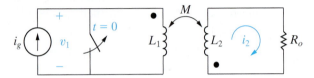

6.42 The polarity markings on two coils are to be determined experimentally. The experimental setup is shown in Fig. P6.42. Assume that the terminal connected to the positive terminal of the battery has been given a polarity mark as shown. When the switch is *opened*, the dc voltmeter kicks downscale. Where should the polarity mark be placed on the coil connected to the voltmeter?

Figure P6.42

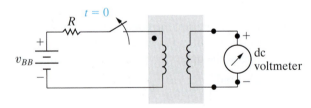

6.43 The physical construction of two pairs of magnetically coupled coils is shown in Fig. P6.43. Assume that the magnetic flux is confined to the core material in each structure. Show two possible locations for the dot markings on each pair of coils.

Figure P6.43

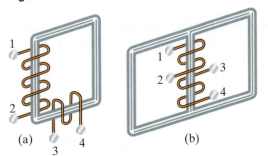

(a)

(b)

Section 6.5

6.44 a) Starting with Eq. 6.26, show that the coefficient of coupling can also be expressed as

$$k = \sqrt{\left(\frac{\phi_{21}}{\phi_1}\right)\left(\frac{\phi_{12}}{\phi_2}\right)}.$$

b) On the basis of the fractions ϕ_{21}/ϕ_1 and ϕ_{12}/ϕ_2, explain why k is less than 1.0.

6.45 The self-inductances of the coils in Fig. 6.34 are $L_1 = 18\,mH$ and $L_2 = 32\,mH$. If the coefficient of coupling is 0.85, calculate the energy stored in the system in millijoules when (a) $i_1 = 6\,A$, $i_2 = 9\,A$; (b) $i_1 = -6\,A$, $i_2 = -9\,A$; (c) $i_1 = -6\,A$, $i_2 = 9\,A$; and (d) $i_1 = 6\,A$, $i_2 = -9A$.

6.46 The coefficient of coupling in Problem 6.45 is increased to 1.0.

a) If i_1 equals 6 A, what value of i_2 results in zero stored energy?

b) Is there any physically realizable value of i_2 that can make the stored energy negative?

6.47 Two magnetically coupled coils are wound on a nonmagnetic core. The self-inductance of coil 1 is 288 mH, the mutual inductance is 90 mH, the coefficient of coupling is 0.75, and the physical structure of the coils is such that $\mathcal{P}_{11} = \mathcal{P}_{22}$.

a) Find L_2 and the turns ratio N_1/N_2.

b) If $N_1 = 1200$, what is the value of $\mathcal{P}_1$ and $\mathcal{P}_2$?

6.48 Two magnetically coupled coils have self-inductances of 60 mH and 9.6 mH, respectively. The mutual inductance between the coils is 22.8 mH.

a) What is the coefficient of coupling?

b) For these two coils, what is the largest value that M can have?

c) Assume that the physical structure of these coupled coils is such that $\mathcal{P}_1 = \mathcal{P}_2$. What is the turns ratio N_1/N_2 if N_1 is the number of turns on the 60 mH coil?

6.49 The self-inductances of two magnetically coupled coils are $L_1 = 180\,\mu H$ and $L_2 = 500\,\mu H$. The coupling medium is nonmagnetic. If coil 1 has 300 turns and coil 2 has 500 turns, find $\mathcal{P}_{11}$ and $\mathcal{P}_{21}$ (in nanowebers per ampere) when the coefficient of coupling is 0.6.

6.50 The self-inductances of two magnetically coupled coils are 72 mH and 40.5 mH, respectively. The 72 mH coil has 250 turns, and the coefficient of coupling between the coils is $^2/_3$. The coupling medium is nonmagnetic. When coil 1 is excited with coil 2 open, the flux linking only coil 1 is 0.2 as large as the flux linking coil 2.

a) How many turns does coil 2 have?

b) What is the value of $\mathscr{P}_2$ in nanowebers per ampere?

c) What is the value of $\mathscr{P}_{11}$ in nanowebers per ampere?

d) What is the ratio (ϕ_{22}/ϕ_{12})?

Sections 6.1–6.5

6.51 Suppose a capacitive touch screen that uses the mutual-capacitance design, as shown in Fig. 6.37, is touched at the point x, y. Determine the mutual capacitance at that point, C'_{mxy}, in terms of the mutual capacitance at the point without a touch, C_{mxy}, and the capacitance introduced by the touch, C_t.

6.52 a) Assume the parasitic capacitance in the self-capacitance design, $C_p = 30 \, \text{pF}$, and the capacitance introduced by a touch is 15 pF (see Fig. 6.36[b]). What is the capacitance at the touch point with respect to ground for the x-grid and y-grid electrodes closest to the touch point?

b) Assume the mutual capacitance in the mutual-capacitance design, $C_{mxy} = 30 \, \text{pF}$, and the capacitance introduced by a touch is 15 pF (see Fig. 6.37[b]). What is the mutual capacitance between the x- and y-grid electrodes closest to the touch point?

c) Compare your results in parts (a) and (b) — does touching the screen increase or decrease the capacitance in these two different capacitive touch screen designs?

6.53 a) As shown in the Practical Perspective, the self-capacitance design does not permit a true multi-touch screen — if the screen is touched at two difference points, a total of four touch points are identified, the two actual touch points and two ghost points. If a self-capacitance touch screen is touched at the x, y coordinates (2.1, 4.3) and (3.2, 2.5), what are the four touch locations that will be identified? (Assume the touch coordinates are measured in inches from the upper left corner of the screen.)

b) A self-capacitance touch screen can still function as a multi-touch screen for several common gestures. For example, suppose at time t_1 the two touch points are those identified in part (a), and at time t_2 four touch points associated with the x, y coordinates (1.8, 4.9) and (3.9, 1.8) are identified. Comparing the four points at t_1 with the four points at t_2, software can recognize a pinch gesture — should the screen be zoomed in or zoomed out?

c) Repeat part (b), assuming that at time t_2 four touch points associated with the x, y coordinates (2.8, 3.9) and (3.0, 2.8) are identified.

7

Response of First-Order *RL* and *RC* Circuits

CHAPTER CONTENTS

7.1 **The Natural Response of an *RL* Circuit** *p. 222*

7.2 **The Natural Response of an *RC* Circuit** *p. 228*

7.3 **The Step Response of *RL* and *RC* Circuits** *p. 233*

7.4 **A General Solution for Step and Natural Responses** *p. 241*

7.5 **Sequential Switching** *p. 246*

7.6 **Unbounded Response** *p. 250*

7.7 **The Integrating Amplifier** *p. 252*

CHAPTER OBJECTIVES

1 Be able to determine the natural response of both *RL* and *RC* circuits.

2 Be able to determine the step response of both *RL* and *RC* circuits.

3 Know how to analyze circuits with sequential switching.

4 Be able to analyze op amp circuits containing resistors and a single capacitor.

In this chapter, we focus on circuits that consist only of sources, resistors, and either (but not both) inductors or capacitors. We call these circuits **RL** (resistor-inductor) and **RC** (resistor-capacitor) **circuits**. In Chapter 6, we saw that inductors and capacitors can store energy. We analyze *RL* and *RC* circuits to determine the currents and voltages that arise when energy is either released or acquired by an inductor or capacitor in response to an abrupt change in a dc voltage or current source.

We divide our analysis of *RL* and *RC* circuits into three phases.

- First Phase: Find the currents and voltages that arise when stored energy in an inductor or capacitor is suddenly released to a resistive network. This happens when the inductor or capacitor is abruptly disconnected from its dc source. Thus, we can reduce the circuit to one of the two equivalent forms shown in Fig. 7.1 on page 222. These currents and voltages characterize the **natural response** of the circuit because the nature of the circuit itself, not external sources of excitation, determines its behavior.

- Second Phase: Find the currents and voltages that arise when energy is being acquired by an inductor or capacitor when a dc voltage or current source is suddenly applied. This response is called the **step response**.

- Third Phase: Develop a general method for finding the response of *RL* and *RC* circuits to any abrupt change in a dc voltage or current source. A general method exists because the process for finding both the natural and step responses is the same.

Figure 7.2 (page 222) shows the four general configurations of *RL* and *RC* circuits. Note that when there are no independent sources in the circuit, the Thévenin voltage or Norton current is zero, and the circuit reduces to one of those shown in Fig. 7.1; that is, we have a natural-response problem.

RL and *RC* circuits are also known as **first-order circuits** because their voltages and currents are described by first-order differential equations. No matter how complex a circuit may appear, if it can be reduced to a Thévenin or Norton equivalent connected to the

■ Practical Perspective

Artificial Pacemaker

The muscle that makes up the heart contracts due to rhythmical electrical impulses. Pacemaker cells control the impulse frequency. In adults, the pacemaker cells establish a resting heart rate of about 72 beats per minutes. Sometimes, however, damaged pacemaker cells produce a very low resting heart rate (a condition known as bradycardia) or a very high resting heart rate (a condition known as tachycardia). When either happens, a normal heart rhythm can be restored by implanting an artificial pacemaker that mimics the pacemaker cells by delivering electrical impulses to the heart. Examples of internal and external artificial pacemakers are shown in the figures below.

Artificial pacemakers are very small and lightweight. They have a programmable microprocessor that adjusts the heart rate based on several parameters, an efficient battery with a life of up to 15 years, and a circuit that generates the pulse. The simplest circuit consists of a resistor and a capacitor. After we introduce and study the *RC* circuit, we will look at an *RC* circuit design for an artificial pacemaker.

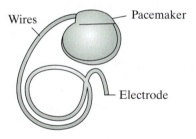

Swapan Photography/Shutterstock

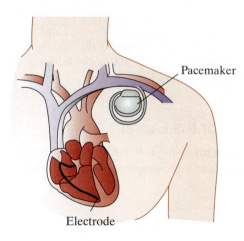

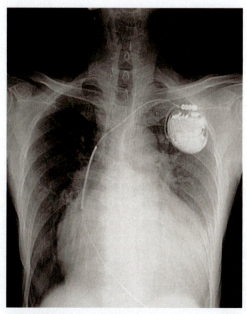

Tewan Banditrukkanka/Shutterstock

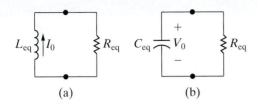

Figure 7.1 ▲ The two forms of the circuits for natural response. (a) *RL* circuit. (b) *RC* circuit.

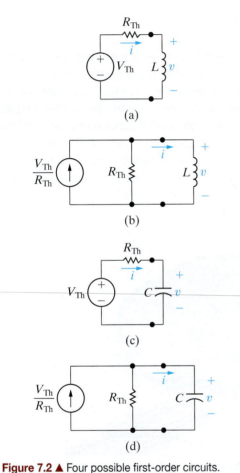

Figure 7.2 ▲ Four possible first-order circuits.
(a) An inductor connected to a Thévenin equivalent.
(b) An inductor connected to a Norton equivalent.
(c) A capacitor connected to a Thévenin equivalent.
(d) A capacitor connected to a Norton equivalent.

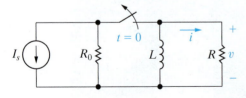

Figure 7.3 ▲ An *RL* circuit.

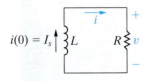

Figure 7.4 ▲ The circuit shown in Fig. 7.3, for $t \geq 0$.

terminals of an equivalent inductor or capacitor, it is a first-order circuit. If the original circuit has two or more inductors or capacitors, they must be interconnected so that they can be replaced by a single equivalent element.

After introducing the techniques for analyzing the natural and step responses of first-order circuits, we will discuss three special cases:

- Sequential switching (circuits in which switching occurs at two or more instants in time);
- Circuits with an unbounded response;
- The integrating amplifier circuit (containing an ideal op amp).

7.1 The Natural Response of an *RL* Circuit

We can describe the natural response of an *RL* circuit using the circuit shown in Fig. 7.3. We assume that the independent current source generates a constant current I_s and that the switch has been in a closed position for a long time. We define the phrase *a long time* more accurately later in this section. For now it means that all currents and voltages have reached a constant value. Thus only constant, or dc, currents exist in the circuit just before the switch opens, and the voltage across the inductor is zero ($L \, di/dt = 0$). Therefore, before the stored energy is released,

- The inductor behaves like a short circuit;
- The entire source current I_s appears in the inductive branch; and
- There is no current in either R_0 or R.

To find the natural response, we find the voltage and current at the terminals of the resistor R after the switch has been opened—that is, after the source and its parallel resistor R_0 have been disconnected and the inductor begins releasing energy. If we let $t = 0$ denote the instant when the switch is opened, we find $v(t)$ and $i(t)$ for $t \geq 0$. For $t \geq 0$, the circuit shown in Fig. 7.3 reduces to the one shown in Fig. 7.4.

Deriving the Expression for the Current

To find $i(t)$, we write an expression involving i, R, and L for the circuit in Fig. 7.4 using Kirchhoff's voltage law. Summing the voltages around the closed loop gives

$$L\frac{di}{dt} + Ri = 0, \qquad (7.1)$$

where we used the passive sign convention. Equation 7.1 is a **first-order** ordinary differential equation because it involves the ordinary derivative of the unknown, di/dt, and the highest order derivative appearing in the equation is 1.

We can go one step further in describing this equation. The coefficients in the equation, R and L, are constants; that is, they are not functions of either the dependent variable i or the independent variable t. Thus, the equation can also be described as an ordinary differential equation with constant coefficients.

To solve Eq. 7.1, we divide by L, move the term involving i to the right-hand side, and then multiply both sides by a differential time dt. The result is

$$\frac{di}{dt}\,dt = -\frac{R}{L}\,i\,dt.$$

Next, we recognize the left-hand side of this equation simplifies to a differential change in the current i, that is, di. We now divide through by i, getting

$$\frac{di}{i} = -\frac{R}{L}\,dt.$$

We obtain an explicit expression for i as a function of t by integrating both sides. Using x and y as variables of integration yields

$$\int_{i(t_0)}^{i(t)} \frac{dx}{x} = -\frac{R}{L}\int_{t_0}^{t} dy,$$

where $i(t_0)$ is the current at time t_0 and $i(t)$ is the current at time t. Here, $t_0 = 0$. Therefore, carrying out the indicated integration gives

$$\ln\frac{i(t)}{i(0)} = -\frac{R}{L}\,t.$$

Based on the definition of the natural logarithm, we can solve for the current to get

$$i(t) = i(0)e^{-(R/L)t}.$$

Recall from Chapter 6 that the inductor current cannot change instantaneously. Therefore, in the first instant after the switch has been opened, the current in the inductor remains unchanged. If 0^- is the time just prior to switching and 0^+ is the time immediately following switching, then

INITIAL INDUCTOR CURRENT

$$i(0^-) = i(0^+) = I_0. \tag{7.2}$$

I_0 is the inductor's initial current, as in Fig. 7.1(a), and has the same direction as the reference direction of i. Hence, the equation for the current becomes

$$i(t) = I_0 e^{-(R/L)t}, \quad t \geq 0.$$

Figure 7.5 shows this response, where the current has an initial value I_0 and decreases exponentially toward zero as t increases.

Note that the expression for $i(t)$ includes the term $e^{-(R/L)t}$. The coefficient of t—namely, R/L—determines the rate at which the current approaches zero. The reciprocal of this coefficient is the **time constant** of the circuit, denoted

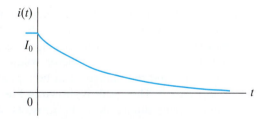

Figure 7.5 ▲ The current response for the circuit shown in Fig. 7.4.

TIME CONSTANT FOR *RL* CIRCUIT

$$\tau = \frac{L}{R}. \tag{7.3}$$

Using the time constant, we write the expression for current as

NATURAL RESPONSE OF AN *RL* CIRCUIT

$$i(t) = I_0 e^{-t/\tau}, \quad t \geq 0. \tag{7.4}$$

RL NATURAL-RESPONSE METHOD

1. **Determine the initial inductor current,** I_0, by analyzing the circuit for $t < 0$.
2. **Calculate the time constant,** $\tau = L/R$, where R is the equivalent resistance connected to the inductor for $t \geq 0$.
3. **Write the equation for the inductor current,** $i(t) = I_0 e^{-t/\tau}$, for $t \geq 0$.
4. **Calculate other quantities of interest** using the inductor current.

Analysis Method 7.1 Finding the *RL* natural response.

Now we have a step-by-step method for finding the natural response of an *RL* circuit.

Step 1: Determine the initial current, I_0, in the inductor. This usually involves analyzing the circuit for $t < 0$.

Step 2: Calculate the time constant, τ. To do this, you need to find the equivalent resistance attached to the inductor for $t \geq 0$.

Step 3: Write the equation for the inductor current for $t \geq 0$ by substituting the values for the initial current and the time constant into Eq. 7.4.

Step 4: Calculate any other quantities of interest, such as resistor current and voltage, using resistive circuit analysis techniques.

This method is summarized in Analysis Method 7.1 and is illustrated in Example 7.1.

EXAMPLE 7.1 | **Determining the Natural Response of an *RL* Circuit**

The switch in the circuit shown in Fig. 7.6 has been closed for a long time before it is opened at $t = 0$. Find

a) $i_L(t)$ for $t \geq 0$,

b) $i_o(t)$ for $t \geq 0^+$,

c) $v_o(t)$ for $t \geq 0^+$,

d) the percentage of the total energy stored in the 2 H inductor that is dissipated in the 10 Ω resistor.

Solution

Use Analysis Method 7.1.

a) **Step 1:** To determine the initial current in the inductor, draw the circuit in Fig. 7.6 for $t < 0$. The switch has been closed for a long time prior to $t = 0$, so we know the inductor voltage is zero at $t = 0^-$ and the inductor can be replaced by a short circuit. The result is shown in Fig. 7.7. The short circuit shunts all of the resistors, so it has

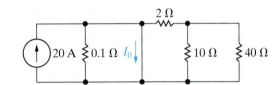

Figure 7.7 ▲ The circuit for Example 7.1 when $t < 0$.

all of the current from the source. Therefore, the current in the inductor at $t = 0^-$ is 20 A and

$$I_0 = i_L(0^-) = i_L(0^+) = 20 \text{ A}.$$

Step 2: To calculate the time constant, τ, we need to find the equivalent resistance attached to the inductor when $t \geq 0$. To do this, draw the circuit in Fig. 7.6 for $t \geq 0$. Since the switch is now open, the current source and its parallel resistor are removed from the circuit, as shown in Fig. 7.8.

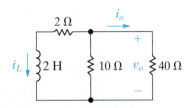

Figure 7.8 ▲ The circuit for Example 7.1 when $t \geq 0$.

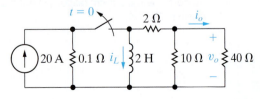

Figure 7.6 ▲ The circuit for Example 7.1.

From this circuit you can see that the equivalent resistance attached to the inductor is

$$R_{eq} = 2 + (40 \| 10) = 10 \ \Omega$$

and the time constant of the circuit is

$$\tau = \frac{L}{R_{eq}} = \frac{2}{10} = 0.2 \text{ s.}$$

Step 3: Write the equation for the inductor current by substituting the values for the initial current and the time constant into Eq. 7.4 to give

$$i_L(t) = I_0 e^{-t/\tau} = 20e^{-t/0.2} = 20e^{-5t} \text{ A,} \quad t \geq 0.$$

Step 4: We use resistive circuit analysis in the remaining parts of this problem to find additional currents and voltages.

b) We find the current in the 40 Ω resistor in Fig. 7.8 using current division; that is,

$$i_o = -i_L \frac{10}{10 + 40}.$$

Note that this expression is valid for $t \geq 0^+$ because $i_o = 0$ at $t = 0^-$, so the resistor current i_o changes instantaneously. Thus,

$$i_o(t) = -0.25 i_L(t) = -4e^{-5t} \text{ A}, t \geq 0^+.$$

c) We find the voltage v_o in Fig. 7.8 by applying Ohm's law:

$$v_o(t) = 40 i_o = -160e^{-5t} \text{ V,} \quad t \geq 0^+.$$

d) The power dissipated in the 10 Ω resistor in Fig. 7.8 is

$$p_{10\Omega}(t) = \frac{v_o^2}{10} = 2560e^{-10t} \text{ W,} \quad t \geq 0^+.$$

The total energy dissipated in the 10 Ω resistor is

$$w_{10\Omega}(t) = \int_0^\infty 2560e^{-10t} \, dt = 256 \text{ J.}$$

The initial energy stored in the 2 H inductor is

$$w(0) = \frac{1}{2} L i_L^2(0) = \frac{1}{2}(2)(20)^2 = 400 \text{ J.}$$

Therefore, the percentage of energy dissipated in the 10 Ω resistor is

$$\frac{256}{400}(100) = 64\%.$$

ASSESSMENT PROBLEMS

Objective 1 — Be able to determine the natural response of both *RL* and *RC* circuits

7.1 The switch in the circuit shown has been closed for a long time and is opened at $t = 0$.
a) Calculate the initial value of i.
b) Calculate the initial energy stored in the inductor.
c) What is the time constant of the circuit for $t > 0$?
d) What is the numerical expression for $i(t)$ for $t \geq 0$?
e) What percentage of the initial energy stored has been dissipated in the 2 Ω resistor 5 ms after the switch has been opened?

Answer: a) -12.5 A;
b) 625 mJ;
c) 4 ms;
d) $-12.5e^{-250t}$ A, $\quad t \geq 0$;
e) 91.8%.

7.2 At $t = 0$, the switch in the circuit shown moves instantaneously from position a to position b.
a) Calculate v_o for $t \geq 0^+$.
b) What percentage of the initial energy stored in the inductor is eventually dissipated in the 4 Ω resistor?

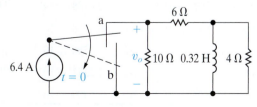

Answer: a) $-8e^{-10t}$ V, $t \geq 0$;
b) 80%.

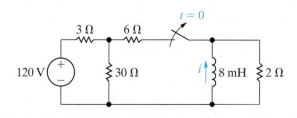

SELF-CHECK: Also try Chapter Problems 7.1, 7.2, and 7.4.

Deriving the Expressions for Voltage, Power, and Energy

We derive the voltage across the resistor in Fig. 7.4 using Ohm's law:

$$v = iR = I_0 Re^{-t/\tau}, \quad t \geq 0^+. \tag{7.5}$$

Note that while the current is defined for $t \geq 0$ (Eq. 7.4), the voltage is defined only for $t \geq 0^+$, not at $t = 0$. At $t = 0$ a step change occurs in the voltage. For $t < 0$, the derivative of the current is zero, so the voltage is also zero ($v = L \, di/dt = 0$). Thus

$$v(0^-) = 0,$$

$$v(0^+) = I_0 R,$$

where $v(0^+)$ is obtained from Eq. 7.5 with $t = 0^{+1}$. The value of the voltage at $t = 0$ is undefined owing to the step change at $t = 0$. Thus, we use $t \geq 0^+$ when defining the region of validity for the voltage in Eq. 7.5.

We derive the power dissipated in the resistor from any of the following expressions:

$$p = vi, \quad p = i^2 R, \quad \text{or} \quad p = \frac{v^2}{R}.$$

Whichever form is used, the resulting expression can be reduced to

$$p = I_0^2 Re^{-2t/\tau}, \quad t \geq 0^+.$$

The energy delivered to the resistor during any interval of time after the switch has been opened is

$$w = \int_0^t p \, dx = \int_0^t I_0^2 Re^{-2x/\tau} dx$$

$$= \frac{\tau}{2} I_0^2 R(1 - e^{-2t/\tau})$$

$$= \frac{L}{2R} I_0^2 R(1 - e^{-2t/\tau})$$

$$= \frac{1}{2} L I_0^2 (1 - e^{-2t/\tau}), \quad t \geq 0^+.$$

Note from the energy equation that as t becomes infinite, the energy dissipated in the resistor approaches the initial energy stored in the inductor.

The Significance of the Time Constant

The time constant is an important parameter for first-order circuits. You can express the time elapsed after switching as an integer multiple of τ. For example, one time constant after the inductor begins releasing its stored energy to the resistor, the current has been reduced to e^{-1}, or approximately 0.37 of its initial value.

Table 7.1 gives the value of $e^{-t/\tau}$ for integer multiples of τ from 1 to 10. Note that the current is less than 1% of its initial value when the elapsed

TABLE 7.1 Value of $e^{-t/\tau}$ For t Equal to Integral Multiples of τ

t	$e^{-t/\tau}$	t	$e^{-t/\tau}$
τ	3.6788×10^{-1}	6τ	2.4788×10^{-3}
2τ	1.3534×10^{-1}	7τ	9.1188×10^{-4}
3τ	4.9787×10^{-2}	8τ	3.3546×10^{-4}
4τ	1.8316×10^{-2}	9τ	1.2341×10^{-4}
5τ	6.7379×10^{-3}	10τ	4.5400×10^{-5}

---

[1] We can define the expressions 0^- and 0^+ more formally. The expression $x(0^-)$ refers to the limit of the variable x as $t \to 0$ from the left, or from negative time. The expression $x(0^+)$ refers to the limit of the variable x as $t \to 0$ from the right, or from positive time.

time exceeds five time constants. So, five time constants after switching has occurred, the currents and voltages have essentially reached their final values. After switching, the changes to the currents and voltages are momentary events and represent the **transient response** of the circuit.

By contrast, the phrase *a long time* implies that five or more time constants have elapsed, for first-order circuits. The response that exists a long time after switching is called the **steady-state response**. The phrase *a long time* then also means the time it takes the circuit to reach its steady-state value.

The time constant also represents the time required for the current to reach its final value if the current continues to change at its initial rate. To illustrate, we evaluate di/dt at 0^+ and assume that the current continues to change at this rate:

$$\frac{di}{dt}(0^+) = -\frac{I_0}{\tau} e^{-0^+/\tau} = -\frac{I_0}{\tau}.$$

Now, if i starts at I_0 and decreases at a constant rate of I_0/τ amperes per second, the expression for i becomes

$$i = I_0 - \frac{I_0}{\tau} t.$$

This expression indicates that i would reach its final value of zero in τ seconds. Figure 7.9 shows how this graphic interpretation can be used to estimate a circuit's time constant from a plot of its natural response. Such a plot could be generated on an oscilloscope measuring output current. Drawing the tangent to the natural-response plot at $t = 0$ and reading the value at which the tangent intersects the time axis gives the value of τ. This allows you to determine the time constant of a circuit even if you don't know its component values.

Up to this point, we have dealt with circuits having a single inductor. But the techniques presented apply to circuits with multiple inductors if the inductors can be combined into a single equivalent inductor. Example 7.2 finds the natural response of an *RL* circuit that contains two inductors.

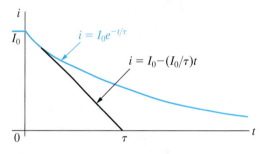

Figure 7.9 ▲ A graphic interpretation of the time constant of the *RL* circuit shown in Fig. 7.4.

EXAMPLE 7.2   **Determining the Natural Response of an *RL* Circuit with Parallel Inductors**

In the circuit shown in Fig. 7.10, the initial currents in inductors L_1 and L_2 have been established by sources not shown. The switch is opened at $t = 0$.

a) Find i_1, i_2, and i_3 for $t \geq 0$.

b) Calculate the initial energy stored in the parallel inductors.

c) Determine how much energy is stored in the inductors as $t \rightarrow \infty$.

d) Show that the total energy delivered to the resistive network equals the difference between the results obtained in (b) and (c).

Solution

a) The key to finding currents i_1, i_2, and i_3 lies in knowing the voltage $v(t)$. We can easily find $v(t)$ if we simplify the circuit shown in Fig. 7.10 to the equivalent form shown in Fig. 7.11. The parallel inductors combine to give an equivalent inductance of 4 H, carrying an initial current of 12 A. The resistive network reduces to a single resistance of $40 \parallel [4 + (15 \parallel 10)] = 8\ \Omega$. We can now use Analysis Method 7.1.

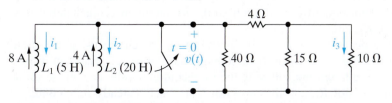

Figure 7.10 ▲ The circuit for Example 7.2.

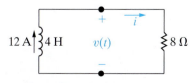

Figure 7.11 ▲ A simplification of the circuit shown in Fig. 7.10.

Step 1: The initial current in the inductor in Fig. 7.11 is $I_0 = 12$ A.

Step 2: The equivalent resistance attached to the inductor in Fig. 7.11 is 8 Ω. Therefore, the time constant is

$$\tau = \frac{L}{R} = \frac{4}{8} = 0.5 \text{ s.}$$

Step 3: The inductor current in Fig. 7.11 is

$$i(t) = I_0 e^{-t/\tau} = 12e^{-t/0.5} = 12e^{-2t} \text{ A}, \quad t \geq 0.$$

Step 4: We will use additional circuit analysis techniques to find the currents i_1, i_2, and i_3. To begin, note that in Fig. 7.11, $v(t) = 8i(t)$, so

$$v(t) = 96e^{-2t} \text{ V}, \quad t \geq 0^+.$$

From the circuit in Fig. 7.10, we see that that $v(t) = 0$ at $t = 0^-$, so the expression for $v(t)$ is valid for $t \geq 0^+$. After obtaining $v(t)$, we can calculate i_1 and i_2 using the relationship between current and voltage in inductors:

$$i_1 = \frac{1}{5} \int_0^t 96e^{-2x} \, dx - 8$$

$$= 1.6 - 9.6e^{-2t} \text{ A}, \quad t \geq 0,$$

$$i_2 = \frac{1}{20} \int_0^t 96e^{-2x} \, dx - 4$$

$$= -1.6 - 2.4e^{-2t} \text{ A}, \quad t \geq 0.$$

We will use two steps to find i_3; in the first step, calculate the voltage across the parallel 15 Ω and 10 Ω resistors using voltage division. Calling that voltage $v_{15\|10}$, positive at the top of the circuit, we get

$$v_{15\|10} = \frac{15\|10}{4 + 15\|10} v = \frac{6}{10}(96e^{-2t}) = 57.6e^{-2t} \text{ V}, t \geq 0^+.$$

SELF-CHECK: Also try Chapter Problem 7.18.

Now use Ohm's law to calculate i_3, giving

$$i_3 = \frac{v_{15\|10}}{10} = 5.76e^{-2t} \text{ A}, \quad t \geq 0^+.$$

Note that the expressions for the inductor currents i_1 and i_2 are valid for $t \geq 0$, whereas the expression for the resistor current i_3 is valid for $t \geq 0^+$.

b) The initial energy stored in the inductors is

$$w = \frac{1}{2}(5)(8)^2 + \frac{1}{2}(20)(4)^2 = 320 \text{ J.}$$

c) As $t \to \infty$, $i_1 \to 1.6$ A and $i_2 \to -1.6$ A. Therefore, a long time after the switch opens, the energy stored in the two inductors is

$$w = \frac{1}{2}(5)(1.6)^2 + \frac{1}{2}(20)(-1.6)^2 = 32 \text{ J.}$$

d) We obtain the total energy delivered to the resistive network by integrating the expression for the instantaneous power from zero to infinity:

$$w = \int_0^\infty p \, dt = \int_0^\infty (96e^{-2t})(12e^{-2t}) \, dt$$

$$= 1152 \frac{e^{-4t}}{-4} \Big|_0^\infty = 288 \text{ J.}$$

This result is the difference between the initially stored energy (320 J) and the energy trapped in the parallel inductors (32 J). Also, note that the equivalent inductor for the parallel inductors (which predicts the terminal behavior of the parallel combination) has an initial energy of $\frac{1}{2}(4)(12)^2 = 288$ J; that is, the energy stored in the equivalent inductor represents the amount of energy that will be delivered to the resistive network at the terminals of the original inductors.

7.2 The Natural Response of an *RC* Circuit

The form of an *RC* circuit's natural response is analogous to that of an *RL* circuit. Consequently, we don't treat the *RC* circuit in as much detail as we did the *RL* circuit.

We develop the natural response of an *RC* circuit using the circuit shown in Fig. 7.12. Begin by assuming that the switch has been in position a for a long time, allowing the loop containing the dc voltage source V_g, the resistor R_1, and the capacitor C to reach a steady-state condition. Recall from Chapter 6 that a capacitor behaves as an open circuit in the presence of a constant voltage, so the source voltage appears across the capacitor terminals. In Section 7.3, we will discuss how the capacitor voltage builds to the steady-state value of the dc voltage source. Here it is important to remember that when the switch is moved from position a to position b (at $t = 0$), the voltage on the capacitor is V_g. Because there can be no instantaneous change in the voltage at the terminals of a capacitor, the problem reduces to solving the circuit shown in Fig. 7.13.

Figure 7.12 ▲ An *RC* circuit.

Deriving the Expression for the Voltage

We can easily find the voltage $v(t)$ by writing a KCL equation. Using the lower node between R and C as the reference node and summing the currents away from the upper node between R and C gives

$$C\frac{dv}{dt} + \frac{v}{R} = 0. \tag{7.6}$$

Comparing Eq. 7.6 with Eq. 7.1, you should see that the mathematical techniques used to find $i(t)$ in the *RL* circuit can be used to find $v(t)$ in the *RC* circuit. We leave it to you to show that

$$v(t) = v(0)e^{-t/RC}, \quad t \geq 0.$$

As we have already noted, the initial voltage on the capacitor equals the voltage source voltage V_g, or

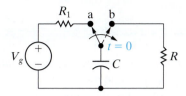

Figure 7.13 ▲ The circuit shown in Fig. 7.12, after switching.

CAPACITOR VOLTAGE

$$v(0^-) = v(0) = v(0^+) = V_g = V_0, \tag{7.7}$$

where V_0 denotes the initial voltage on the capacitor. The time constant for the *RC* circuit equals the product of the resistance and capacitance, namely,

TIME CONSTANT FOR *RC* CIRCUIT

$$\tau = RC. \tag{7.8}$$

Therefore, the general expression for the voltage becomes

NATURAL RESPONSE OF AN *RC* CIRCUIT

$$v(t) = V_0 e^{-t/\tau}, \quad t \geq 0, \tag{7.9}$$

which indicates that the natural response of an *RC* circuit is an exponential decay of the initial voltage. The time constant RC governs the rate of decay. Figure 7.14 shows the plot of Eq. 7.9 and the graphic interpretation of the time constant.

Now we have a step-by-step method for finding the natural response of an *RC* circuit.

Step 1: Determine the initial voltage, V_0, across the capacitor. This usually involves analyzing the circuit for $t < 0$.

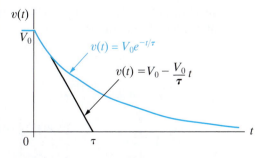

Figure 7.14 ▲ The natural response of an *RC* circuit.

RC NATURAL-RESPONSE METHOD

1. **Determine the initial capacitor voltage,** V_0, by analyzing the circuit for $t < 0$.
2. **Calculate the time constant,** $\tau = RC$, where R is the equivalent resistance connected to the capacitor for $t \geq 0$.
3. **Write the equation for capacitor voltage,** $v(t) = V_0 e^{-t/\tau}$, for $t \geq 0$.
4. **Calculate other quantities of interest** using the capacitor voltage.

Analysis Method 7.2 Finding the *RC* natural response.

Step 2: Calculate the time constant, τ. To do this, you need to find the equivalent resistance attached to the capacitor for $t \geq 0$.

Step 3: Write the equation for the capacitor voltage for $t \geq 0$ by substituting the values for the initial voltage and the time constant into Eq. 7.9.

Step 4: Calculate any other quantities of interest, such as resistor current and voltage, using resistive circuit analysis techniques.

This method is summarized in Analysis Method 7.2.

After determining $v(t)$, we can easily derive the expressions for i, p, and w:

$$i(t) = \frac{v(t)}{R} = \frac{V_0}{R} e^{-t/\tau}, \quad t \geq 0^+,$$

$$p = vi = \frac{V_0^2}{R} e^{-2t/\tau}, \quad t \geq 0^+,$$

$$w = \int_0^t p \, dx = \int_0^t \frac{V_0^2}{R} e^{-2x/\tau} \, dx$$

$$= \frac{1}{2} CV_0^2 (1 - e^{-2t/\tau}), \quad t \geq 0.$$

Example 7.3 uses Analysis Method 7.2 to determine the natural response of an *RC* circuit. Analysis Method 7.2 applies to circuits with a single capacitor, but it can also be used to analyze circuits with multiple capacitors if they can all be combined into a single equivalent capacitor. Example 7.4 considers a circuit with two capacitors.

EXAMPLE 7.3 | **Determining the Natural Response of an *RC* Circuit**

The switch in the circuit shown in Fig. 7.15 has been in position x for a long time. At $t = 0$, the switch moves instantaneously to position y. Find

a) $v_C(t)$ for $t \geq 0$,

b) $v_o(t)$ for $t \geq 0^+$,

c) $i_o(t)$ for $t \geq 0^+$, and

d) the total energy dissipated in the 60 kΩ resistor.

Solution

Use Analysis Method 7.2.

a) **Step 1:** Determine the initial capacitor voltage V_0 by drawing the circuit in Fig. 7.15 for $t < 0$. The result is shown in Fig. 7.16, and since the capacitor behaves like an open circuit, its initial voltage equals the source voltage:

$$V_0 = 100 \text{ V}.$$

Step 2: Calculate the time constant. To do this, draw the circuit in Fig. 7.15 for $t \geq 0$, as shown in Fig. 7.17, and find the equivalent resistance attached to the capacitor:

$$R_{eq} = 32 \times 10^3 + (240 \times 10^3 \| 60 \times 10^3)$$

$$= 80 \text{ k}\Omega,$$

$$\tau = R_{eq}C = (80 \times 10^3)(0.5 \times 10^{-6})$$

$$= 40 \text{ ms}.$$

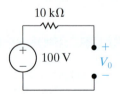

Figure 7.16 ▲ The circuit in Fig. 7.15 for $t < 0$.

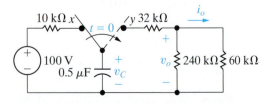

Figure 7.15 ▲ The circuit for Example 7.3.

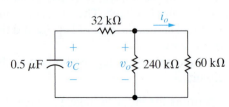

Figure 7.17 ▲ The circuit in Fig. 7.15 for $t \geq 0$.

Step 3: Write the equation for the capacitor voltage by substituting the values for V_0 and τ into Eq. 7.9:

$$v_C(t) = 100e^{-t/0.04} = 100e^{-25t} \text{ V}, \quad t \geq 0.$$

Step 4: Determine the remaining quantities using resistive circuit analysis techniques for the circuit in Fig. 7.17.

b) To find $v_o(t)$ in Fig. 7.17, note that the resistive circuit forms a voltage divider across the terminals of the capacitor. Thus

$$v_o(t) = \frac{240 \times 10^3 \| 60 \times 10^3}{32 \times 10^3 + (240 \times 10^3 \| 60 \times 10^3)} v_C(t)$$

$$= 0.6(100e^{-25t}) = 60e^{-25t} \text{ V}, \quad t \geq 0^+.$$

This expression for $v_o(t)$ is valid for $t \geq 0^+$ because $v_o(0^-)$ is zero. Thus, we have an instantaneous change in the voltage across the 240 kΩ resistor.

c) We find the current $i_o(t)$ from Ohm's law:

$$i_o(t) = \frac{v_o(t)}{60 \times 10^3} = e^{-25t} \text{ mA}, \quad t \geq 0^+.$$

d) The power dissipated in the 60 kΩ resistor is

$$p_{60k\Omega}(t) = i_o^2(t)(60 \times 10^3) = 60e^{-50t} \text{ mW}, \quad t \geq 0^+.$$

The total energy dissipated is

$$w_{60k\Omega} = \int_0^\infty i_o^2(t)(60 \times 10^3)dt = 1.2 \text{ mJ}.$$

EXAMPLE 7.4	**Determining the Natural Response of an *RC* Circuit with Series Capacitors**

The initial voltages on capacitors C_1 and C_2 in the circuit shown in Fig. 7.18 have been established by sources not shown. The switch is closed at $t = 0$.

a) Find $v_1(t)$, $v_2(t)$, and $v(t)$ for $t \geq 0$ and $i(t)$ for $t \geq 0^+$.

b) Calculate the initial energy stored in the capacitors C_1 and C_2.

c) Determine how much energy is stored in the capacitors as $t \to \infty$.

d) Show that the total energy delivered to the 250 kΩ resistor is the difference between the results obtained in (b) and (c).

Solution

a) Once we know $v(t)$, we can obtain the current $i(t)$ from Ohm's law. After determining $i(t)$, we can calculate $v_1(t)$ and $v_2(t)$ because the voltage across a capacitor is a function of the capacitor current. To find $v(t)$, we replace the series-connected capacitors with an equivalent capacitor. It has a

capacitance of 4μF and is charged to a voltage of 20 V. Therefore, the circuit shown in Fig. 7.18 reduces to the one shown in Fig. 7.19. We can now use Analysis Method 7.2 to determine $v(t)$.

Step 1: The initial voltage across the capacitor in Fig. 7.19 is $V_0 = 20$ V.

Step 2: The resistance attached to the capacitor in Fig. 7.19 is 250 kΩ. Therefore, the time constant is

$$\tau = (250 \times 10^3)(4 \times 10^{-6}) = 1 \text{ s}.$$

Step 3: Write the equation for the capacitor voltage by substituting the values for V_0 and τ in Eq. 7.9 to give

$$v(t) = 20e^{-t} \text{ V}, \quad t \geq 0.$$

Step 4: Determine the currents and voltages requested using the techniques described at the start of the Solution.

For the circuit in Fig. 7.19, use Ohm's law to find the current $i(t)$:

$$i(t) = \frac{v(t)}{250,000} = 80e^{-t} \, \mu\text{A}, \quad t \geq 0^+.$$

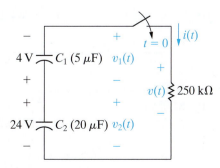

Figure 7.18 ▲ The circuit for Example 7.4.

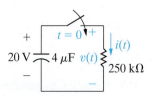

Figure 7.19 ▲ A simplification of the circuit shown in Fig. 7.18.

Knowing $i(t)$, we calculate the expressions for $v_1(t)$ and $v_2(t)$ for the circuit in Fig. 7.18:

$$v_1(t) = -\frac{1}{5 \times 10^{-6}} \int_0^t 80 \times 10^{-6} e^{-x} dx - 4$$

$$= (16e^{-t} - 20) \text{ V}, \quad t \geq 0,$$

$$v_2(t) = -\frac{1}{20 \times 10^{-6}} \int_0^t 80 \times 10^{-6} e^{-x} dx + 24$$

$$= (4e^{-t} + 20) \text{ V}, \quad t \geq 0.$$

b) The initial energy stored in C_1 is

$$w_1 = \frac{1}{2}(5 \times 10^{-6})(4)^2 = 40 \ \mu\text{J}.$$

The initial energy stored in C_2 is

$$w_2 = \frac{1}{2}(20 \times 10^{-6})(24)^2 = 5760 \ \mu\text{J}.$$

The total energy stored in the two capacitors is

$$w_o = 40 + 5760 = 5800 \ \mu\text{J}.$$

c) As $t \to \infty$,

$$v_1 \to -20 \text{ V and } v_2 \to +20 \text{ V}.$$

Therefore, the energy stored in the two capacitors is

$$w_\infty = \frac{1}{2}(5 \times 10^{-6})(-20)^2 + \frac{1}{2}(20 \times 10^{-6})(20)^2$$

$$= 5000 \ \mu\text{J}.$$

d) The total energy delivered to the 250 kΩ resistor is

$$w = \int_0^\infty p \, dt = \int_0^\infty (20e^{-t})(80 \times 10^{-6}e^{-t}) dt = 800 \mu\text{J}.$$

Comparing the results obtained in (b) and (c) shows that

$$800 \ \mu\text{J} = (5800 - 5000) \ \mu\text{J}.$$

The energy stored in the equivalent capacitor in Fig. 7.19 is $\frac{1}{2}(4 \times 10^{-6})(20)^2 = 800 \ \mu\text{J}$. Because this capacitor predicts the terminal behavior of the original series-connected capacitors, the energy stored in the equivalent capacitor is the energy delivered to the 250 kΩ resistor.

ASSESSMENT PROBLEMS

Objective 1—Be able to determine the natural response of both *RL* and *RC* circuits

7.3 The switch in the circuit shown has been closed for a long time and is opened at $t = 0$. Find
a) the initial value of $v(t)$,
b) the time constant for $t > 0$,
c) the numerical expression for $v(t)$ after the switch has been opened,
d) the initial energy stored in the capacitor, and
e) the length of time required to dissipate 75% of the initially stored energy.

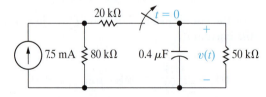

Answer: a) 200 V;
  b) 20 ms;
  c) $200e^{-50t}$ V, $t \geq 0$;
  d) 8 mJ;
  e) 13.86 ms.

7.4 The switch in the circuit shown has been closed for a long time before being opened at $t = 0$.
a) Find $v_o(t)$ for $t \geq 0$.
b) What percentage of the initial energy stored in the circuit has been dissipated after the switch has been open for 60 ms?

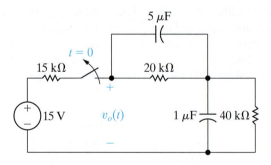

Answer: a) $8e^{-25t} + 4e^{-10t}$ V, $t \geq 0$;
  b) 81.05%.

SELF-CHECK: Also try Chapter Problems 7.22 and 7.33.

7.3 The Step Response of *RL* and *RC* Circuits

The response of a circuit to the sudden application of a constant voltage or current source is called the **step response** of the circuit. In this section, we find the step response of first-order *RL* and *RC* circuits by describing the currents and voltages generated when either dc voltage or current sources are suddenly applied. We also show how the circuit responds when energy is being stored in the inductor or capacitor. We begin with the step response of an *RL* circuit.

The Step Response of an *RL* Circuit

We modify the first-order circuit shown in Fig. 7.2(a) by adding a switch and develop the step response of an *RL* circuit using the resulting circuit, shown in Fig. 7.20. We assume a nonzero initial current $i(0)$, so the inductor has stored energy at the time the switch is closed. We want to find the expressions for the current in the circuit and for the voltage across the inductor after the switch has been closed. We use circuit analysis to derive the differential equation that describes the circuit in terms of the variable of interest, and then we use elementary calculus to solve the equation, just as we did in Section 7.1.

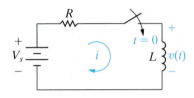

Figure 7.20 ▲ A circuit used to illustrate the step response of a first-order *RL* circuit.

After the switch in Fig. 7.20 has been closed, KVL requires that

$$V_s = Ri + L\frac{di}{dt}, \tag{7.10}$$

which can be solved for the current by separating the variables i and t and then integrating. We begin by solving Eq. 7.10 for the derivative di/dt:

$$\frac{di}{dt} = \frac{-Ri + V_s}{L} = \frac{-R}{L}\left(i - \frac{V_s}{R}\right).$$

Next, we multiply both sides of the equation for di/dt by a differential time dt. This step reduces the left-hand side of the equation to a differential change in the current. Thus

$$di = \frac{-R}{L}\left(i - \frac{V_s}{R}\right)dt.$$

We now separate the variables in the equation for di to get

$$\frac{di}{i - (V_s/R)} = \frac{-R}{L}dt$$

and then integrate both sides. Using x and y as variables for the integration, we obtain

$$\int_{I_0}^{i(t)} \frac{dx}{x - (V_s/R)} = \frac{-R}{L}\int_0^t dy,$$

where I_0 is the current at $t = 0$ and $i(t)$ is the current at any $t > 0$. Evaluating the integrals gives

$$\ln\frac{i(t) - (V_s/R)}{I_0 - (V_s/R)} = \frac{-R}{L}t,$$

from which

$$\frac{i(t) - (V_s/R)}{I_0 - (V_s/R)} = e^{-(R/L)t},$$

or

$$i(t) = \frac{V_s}{R} + \left(I_0 - \frac{V_s}{R}\right)e^{-(R/L)t}. \tag{7.11}$$

Equation 7.11 indicates that after the switch is closed, the current changes exponentially from its initial value I_0 to a final value $I_f = V_s/R$. The time constant of the circuit, $\tau = L/R$, determines the rate of change. We can determine the final value of the current by analyzing the circuit in Fig. 7.20 as $t \to \infty$, and we can calculate the time constant by finding the equivalent resistance attached to the inductor in the circuit shown in Fig. 7.20 for $t \geq 0$. Now express the equation for inductor current (Eq. 7.11) in terms of the initial value of the current, the final value of the current, and the time constant of the circuit to give

STEP RESPONSE OF AN *RL* CIRCUIT

$$i(t) = I_f + (I_0 - I_f)e^{-t/\tau}. \tag{7.12}$$

RL STEP-RESPONSE METHOD

1. **Determine the initial inductor current,** I_0, by analyzing the circuit for $t < 0$.
2. **Calculate the time constant,** $\tau = L/R$, where R is the equivalent resistance connected to the inductor for $t \geq 0$.
3. **Find the final value for the inductor current,** I_f, by analyzing the circuit as $t \to \infty$.
4. **Write the equation for inductor current,** $i(t) = I_f + (I_0 - I_f)e^{-t/\tau}$, for $t \geq 0$.
5. **Calculate other quantities of interest** using the inductor current.

Analysis Method 7.3   Finding the *RL* step response.

Using Eq. 7.12, we can construct a step-by-step procedure to calculate the step response of an *RL* circuit.

Step 1: Determine the initial current, I_0, in the inductor. This usually involves analyzing the circuit for $t < 0$.

Step 2: Calculate the time constant, τ. To do this, you need to find the equivalent resistance attached to the inductor for $t \geq 0$.

Step 3: Calculate the final value of the inductor current, I_f, by analyzing the circuit as t approaches infinity.

Step 4: Write the equation for the inductor current when $t \geq 0$ by substituting the values for the initial current, the time constant, and the final current into Eq. 7.12.

Step 5: Calculate any other quantities of interest, such as resistor current and voltage, using resistive circuit analysis techniques.

This method is summarized in Analysis Method 7.3 and is applied to a specific circuit in Example 7.5.

EXAMPLE 7.5 | **Determining the Step Response of an *RL* Circuit**

The switch in the circuit shown in Fig. 7.21 has been in position a for a long time. At $t = 0$, the switch moves from position a to position b. The switch is a make-before-break type; that is, the connection at position b is established before the connection at position a is broken, so the inductor current is continuous.

a) Find the expression for $i(t)$ for $t \geq 0$.

b) What is the initial voltage across the inductor just after the switch has been moved to position b?

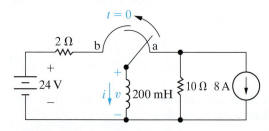

Figure 7.21 ▲ The circuit for Example 7.5.

c) Does this initial voltage make sense in terms of circuit behavior?

d) How many milliseconds after the switch has been moved does the inductor voltage equal 24 V?

e) Plot both $i(t)$ and $v(t)$ versus t.

Solution

a) Use Analysis Method 7.3 to find the inductor current.

Step 1: Determine the initial current in the inductor. To do this, draw the circuit in Fig. 7.21 when $t < 0$ and the switch is in position a, as shown in Fig. 7.22. Note that since the switch has been in position a for a long time, the inductor behaves like a short circuit that carries all of the current from the 8 A current source. Therefore, $I_0 = -8$ A because the inductor current and the source current are in opposite directions.

Step 2: Calculate the time constant for the circuit. Start by drawing the circuit in Fig. 7.21 when $t \geq 0$ and the switch is in position b, as shown in Fig. 7.23. Then determine the Thévenin equivalent resistance for the circuit attached to the inductor. Since the circuit attached to the inductor is already a Thévenin equivalent circuit, the Thévenin equivalent resistance is 2 Ω and $\tau = 0.2/2 = 0.1$ s.

Step 3: Calculate the final value for the inductor current. To do this, draw the circuit in Fig. 7.21 as $t \to \infty$, when the switch is in position b, as shown in Fig. 7.24. Since the switch has been in position b for a long time, the inductor behaves like a short circuit, as seen in Fig. 7.24, and the current can be found from Ohm's law. Therefore, $I_f = 24/2 = 12$ A.

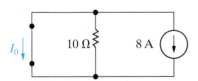

Figure 7.22 ▲ The circuit in Fig. 7.21 for $t < 0$.

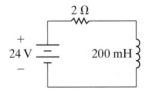

Figure 7.23 ▲ The circuit in Fig. 7.21 for $t \geq 0$.

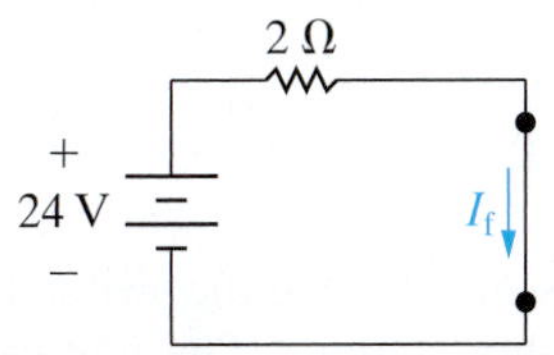

Figure 7.24 ▲ The circuit in Fig. 7.21 as $t \to \infty$.

Step 4: Write the equation for the inductor current when $t \geq 0$ by substituting the values for the initial current, the time constant, and the final current into Eq. 7.12 to give

$$i = I_f + (I_0 - I_f)e^{-t/\tau}$$
$$= 12 + (-8 - 12)e^{-t/0.1}$$
$$= 12 - 20e^{-10t} \text{ A}, t \geq 0.$$

Step 5: Calculate any other quantities of interest, which we do in the remainder of this example.

b) The voltage across the inductor is

$$v = L\frac{di}{dt}$$
$$= 0.2(200e^{-10t})$$
$$= 40e^{-10t} \text{ V}, \quad t \geq 0^+.$$

The initial inductor voltage is

$$v(0^+) = 40 \text{ V}.$$

c) Yes. In the instant after the switch has been moved to position b, the inductor current is 8 A counterclockwise around the newly formed closed path. This current causes a 16 V drop across the 2 Ω resistor. This voltage drop adds to the 24 V drop across the source, producing a 40 V drop across the inductor.

d) We find the time at which the inductor voltage equals 24 V by solving the expression

$$24 = 40e^{-10t}$$

for t:

$$t = \frac{1}{10} \ln \frac{40}{24}$$
$$= 51.08 \text{ ms}.$$

e) Figure 7.25 shows the graphs of $i(t)$ and $v(t)$ versus t. Note that at the instant of time when the current equals zero, the inductor voltage equals the source voltage of 24 V, as predicted by Kirchhoff's voltage law.

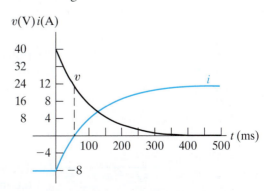

Figure 7.25 ▲ The current and voltage waveforms for Example 7.5.

ASSESSMENT PROBLEM

Objective 2—Be able to determine the step response of both *RL* and *RC* circuits

7.5 Assume that the switch in the circuit shown in Fig. 7.21 has been in position b for a long time and that at $t = 0$ it moves to position a. Find (a) $i(0^+)$; (b) $v(0^+)$; (c) $\tau, t > 0$; (d) $i(t), t \geq 0$; and (e) $v(t), t \geq 0^+$.

Answer: a) 12 A;
b) −200 V;
c) 20 ms;
d) $-8 + 20e^{-50t}$ A, $\quad t \geq 0$;
e) $-200e^{-50t}$ V, $\quad t \geq 0^+$.

SELF-CHECK: Also try Chapter Problems 7.35, 7.36, and 7.38.

Observations on the Step Response of an *RL* Circuit

Let's take a closer look at the *RL* step response of the circuit shown in Fig. 7.20. If the initial energy in the inductor is zero, I_0 is zero and Eq. 7.11 reduces to

$$i(t) = \frac{V_s}{R} - \frac{V_s}{R} e^{-(R/L)t}. \tag{7.13}$$

One time constant after the switch has been closed, the current will have reached approximately 63% of its final value, or

$$i(\tau) = \frac{V_s}{R} - \frac{V_s}{R} e^{-1} \approx 0.6321 \frac{V_s}{R}.$$

If the current were to continue to increase at its initial rate, it would reach its final value at $t = \tau$; that is, because

$$\frac{di}{dt} = \frac{-V_s}{R}\left(\frac{-1}{\tau}\right)e^{-t/\tau} = \frac{V_s}{L} e^{-t/\tau},$$

the initial rate at which $i(t)$ increases is

$$\frac{di}{dt}(0) = \frac{V_s}{L}.$$

If the current were to continue to increase at this rate, the expression for i would be

$$i = \frac{V_s}{L} t, \tag{7.14}$$

so at $t = \tau$,

$$i = \frac{V_s}{L}\frac{L}{R} = \frac{V_s}{R}.$$

Equations 7.13 and 7.14 are plotted in Fig. 7.26. The values for $i(\tau)$ and I_f are also shown in this figure.

The voltage across an inductor is $L\,di/dt$, so from Eq. 7.11, for $t \geq 0^+$,

$$v = L\left(\frac{-R}{L}\right)\left(I_0 - \frac{V_s}{R}\right)e^{-(R/L)t} = (V_s - I_0R)e^{-(R/L)t}. \tag{7.15}$$

The voltage across the inductor is zero before the switch is closed because the inductor is behaving like a short circuit. The voltage equation indicates that the inductor voltage jumps to $V_s - I_0R$ at the instant the switch is closed and then decays exponentially to zero.

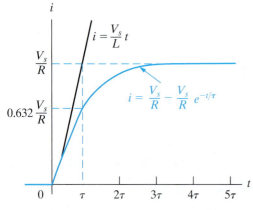

Figure 7.26 ▲ The step response of the *RL* circuit shown in Fig. 7.20 when $I_0 = 0$.

Does the value of v at $t = 0^+$ make sense? Because the initial current is I_0 and the inductor prevents an instantaneous change in current, the current is I_0 in the instant after the switch has been closed. The voltage drop across the resistor is I_0R, so the voltage across the inductor is the source voltage minus the resistor voltage, that is, $V_s - I_0R$.

When the initial inductor current is zero, Eq. 7.15 simplifies to

$$v = V_se^{-(R/L)t}.$$

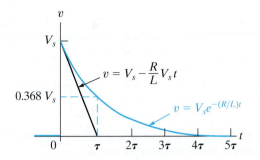

Figure 7.27 ▲ Inductor voltage versus time.

If the initial current is zero, the voltage across the inductor jumps to V_s when the switch closes. We also expect the inductor voltage to approach zero as t increases because the current in the circuit is approaching the constant value of V_s/R. Figure 7.27 shows the plot of the simplified voltage equation and the relationship between the time constant and the initial rate at which the inductor voltage is decreasing.

We can also describe the voltage $v(t)$ across the inductor in Fig. 7.20 directly, without first calculating the circuit current. We begin by noting that the voltage across the resistor is the difference between the source voltage and the inductor voltage. Using Ohm's law, we write

$$i = \frac{V_s}{R} - \frac{v}{R},$$

where V_s is a constant. Differentiating both sides with respect to time yields

$$\frac{di}{dt} = -\frac{1}{R}\frac{dv}{dt}.$$

Then, if we multiply each side of this equation by the inductance L, we get an expression for the voltage across the inductor on the left-hand side, or

$$v = -\frac{L}{R}\frac{dv}{dt}.$$

Putting this differential equation into standard form yields

$$\frac{dv}{dt} + \frac{R}{L}v = 0. \tag{7.16}$$

You should verify (in Problem 7.39) that the solution to Eq. 7.16 is identical to that given in Eq. 7.15.

At this point, a general observation about the step response of an *RL* circuit is pertinent. (This observation will prove helpful later.) When we derived the differential equation for the inductor current, we obtained Eq. 7.10, which we can rewrite as

$$\frac{di}{dt} + \frac{R}{L}i = \frac{V_s}{L}. \tag{7.17}$$

Observe that Eqs. 7.16 and 7.17 have the same form. Specifically, each equates the sum of the first derivative of the variable and a constant times the variable to a constant value. In Eq. 7.16, the constant on the right-hand side happens to be zero; hence, this equation takes on the same form as the natural-response equations in Section 7.1. In both Eq. 7.16 and Eq. 7.17, the constant multiplying the dependent variable is the reciprocal of the time constant; that is, $R/L = 1/\tau$. We will encounter a similar situation in the derivations for the step response of an *RC* circuit. In Section 7.4, we will use these observations to develop a general approach to finding the natural and step responses of *RL* and *RC* circuits.

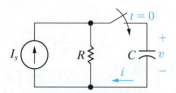

Figure 7.28 ▲ A circuit used to illustrate the step response of a first-order *RC* circuit.

The Step Response of an *RC* Circuit

We can find the step response of a first-order *RC* circuit by analyzing the circuit shown in Fig. 7.28. For mathematical convenience, we choose the Norton equivalent of the network connected to the equivalent capacitor. Summing the currents away from the top node in Fig. 7.28 generates the differential equation

$$C\frac{dv}{dt} + \frac{v}{R} = I_s.$$

Dividing both sides of the differential equation by C gives

$$\frac{dv}{dt} + \frac{v}{RC} = \frac{I_s}{C}. \qquad (7.18)$$

Comparing Eq. 7.18 with Eq. 7.17 reveals that the form of the solution for v is the same as that for the current in the inductive circuit, namely, Eq. 7.11. Therefore, by substituting the appropriate variables and coefficients, we can write the solution for v directly. The translation requires that I_s replace V_s, C replace L, $1/R$ replace R, and V_0 replace I_0. We get

$$v = I_s R + (V_0 - I_s R)e^{-t/RC}, \quad t \geq 0. \qquad (7.19)$$

Equation 7.19 indicates that after the switch has been closed, the voltage changes exponentially from its initial value V_0 to a final value $V_f = I_s R$. The time constant of the circuit, $\tau = RC$, determines the rate of change. We can determine the final value of the voltage by analyzing the circuit in Fig. 7.28 as $t \to 0$, and we can calculate the time constant by finding the equivalent resistance attached to the capacitor in the circuit shown in Fig. 7.28 for $t \geq 0$. Now express the equation for capacitor voltage (Eq. 7.19) in terms of the initial value of the voltage, the final value of the voltage, and the time constant of the circuit to give

STEP RESPONSE OF AN *RC* CIRCUIT

$$v(t) = V_f + (V_0 - V_f)e^{-t/\tau}. \qquad (7.20)$$

Using Eq. 7.20, we can construct a step-by-step procedure to calculate the step response of an *RC* circuit.

Step 1: Determine the initial voltage, V_0, across the capacitor. This usually involves analyzing the circuit for $t < 0$.

Step 2: Calculate the time constant, τ. To do this, you need to find the equivalent resistance attached to the capacitor for $t \geq 0$.

Step 3: Calculate the final value of the capacitor voltage, V_f, by analyzing the circuit as t approaches infinity.

Step 4: Write the equation for the capacitor voltage when $t \geq 0$ by substituting the values for the initial voltage, the time constant, and the final voltage into Eq. 7.20.

Step 5: Calculate any other quantities of interest, such as resistor current and voltage, using resistive circuit analysis techniques.

This method is summarized in Analysis Method 7.4.

RC STEP-RESPONSE METHOD

1. **Determine the initial capacitor voltage,** V_0, by analyzing the circuit for $t < 0$.
2. **Calculate the time constant,** $\tau = RC$, where R is the equivalent resistance connected to the inductor for $t \geq 0$.
3. **Calculate the final value for the capacitor voltage,** V_f, by analyzing the circuit as $t \to \infty$.
4. **Write the equation for the capacitor voltage,** $v(t) = V_f + (V_0 - V_f)e^{-t/\tau}$, for $t \geq 0$.
5. **Calculate other quantities of interest** using the capacitor voltage.

Analysis Method 7.4   Finding the *RC* step response.

A similar derivation for the current in the capacitor yields the differential equation

$$\frac{di}{dt} + \frac{1}{RC} i = 0. \tag{7.21}$$

Equation 7.21 has the same form as Eq. 7.16, so the solution for i is obtained by using the same translations used for the solution of Eq. 7.18. Thus

$$i = \left(I_s - \frac{V_0}{R} \right) e^{-t/RC}, \quad t \geq 0^+, \tag{7.22}$$

where V_0 is the initial voltage across the capacitor.

We obtained Eqs. 7.19 and 7.22 by applying a mathematical analogy to the solution for the step response of the *RL* circuit. Let's see whether these solutions for the *RC* circuit make sense in terms of known circuit behavior. From Eq. 7.19, we have already observed that the initial capacitor voltage is V_0, the final capacitor voltage is $I_s R$, and the time constant of the circuit is RC. Also note that the solution for v is valid for $t \geq 0$. These observations are consistent with the behavior of a capacitor in parallel with a resistor when driven by a constant current source.

Equation 7.22 predicts that the current in the capacitor at $t = 0^+$ is $I_s - V_0/R$. This prediction makes sense because the capacitor voltage cannot change instantaneously, and therefore the initial current in the resistor is V_0/R. The capacitor branch current changes instantaneously from zero at $t = 0^-$ to $I_s - V_0/R$ at $t = 0^+$. The capacitor current is zero at $t = \infty$.

Example 7.6 illustrates how to use Analysis Method 7.4 to find the step response of a first-order *RC* circuit.

EXAMPLE 7.6 | **Determining the Step Response of an *RC* Circuit**

The switch in the circuit shown in Fig. 7.29 has been in position 1 for a long time. At $t = 0$, the switch moves to position 2. Find

a) $v_o(t)$ for $t \geq 0$ and

b) $i_o(t)$ for $t \geq 0^+$.

Solution

Use Analysis Method 7.4.

a) **Step 1:** Determine the initial voltage across the capacitor by analyzing the circuit in Fig. 7.29 for $t < 0$. Do this by redrawing the circuit with the switch in position 1, as shown in Fig. 7.30. Note that the capacitor behaves like an open circuit because the switch has been in position 1 for a

long time. The capacitor's initial voltage is the same as the voltage across the 60 kΩ resistor, which we can find using voltage division:

$$V_0 = \frac{60{,}000}{60{,}000 + 20{,}000} (40) = 30 \text{ V}.$$

Step 2: Calculate the time constant by finding the equivalent resistance attached to the capacitor for $t \geq 0$ in the circuit of Fig. 7.29. Begin by drawing the circuit in Fig. 7.29 with the switch in position 2, as shown in Fig. 7.31(a). Then find the Norton equivalent with respect to the terminals of the capacitor. Begin by computing the open-circuit voltage, which is given by the -75 V

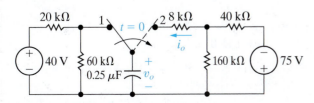

Figure 7.29 ▲ The circuit for Example 7.6.

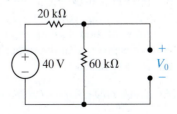

Figure 7.30 ▲ The circuit in Fig. 7.29 when $t < 0$.

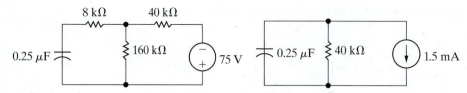

Figure 7.31 ▲ (a) The circuit in Fig. 7.29 when $t \geq 0$; (b) replacing the circuit to the right of the capacitor in part (a) with its Norton equivalent.

source divided across the $40\,\text{k}\Omega$ and $160\,\text{k}\Omega$ resistors:

$$V_{\text{oc}} = \frac{160 \times 10^3}{(40 + 160) \times 10^3}(-75) = -60\ \text{V}.$$

Next, calculate the Thévenin resistance, as seen to the right of the capacitor, by shorting the 75 V source and making series and parallel combinations of the resistors:

$$R_{\text{Th}} = 8000 + 40{,}000 \| 160{,}000 = 40\,\text{k}\Omega.$$

The value of the Norton current source is the ratio of the open-circuit voltage to the Thévenin resistance, or $-60/(40 \times 10^3) = -1.5\ \text{mA}$. The resulting Norton equivalent circuit is shown in Fig. 7.31(b). From Fig. 7.31(b) we see that the equivalent resistance attached to the capacitor is $40\,\text{k}\Omega$, so the time constant is

$$\tau = RC = (40 \times 10^3)(0.25 \times 10^{-6}) = 10\ \text{ms}.$$

Step 3: Calculate the final value of the capacitor voltage by analyzing the circuit in Fig. 7.29 as $t \to \infty$. The circuit is shown in Fig. 7.32, and since

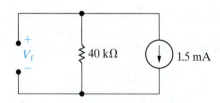

Figure 7.32 ▲ The circuit in Fig. 7.29 as $t \to \infty$.

the switch has been in position 2 for a long time, the capacitor behaves like an open circuit. The final capacitor voltage equals the voltage across the $40\,\text{k}\Omega$ resistor, so

$$V_{\text{f}} = -(40 \times 10^3)(1.5 \times 10^{-3}) = -60\ \text{V}.$$

Step 4: Write the equation for capacitor voltage by substituting the values for initial capacitor voltage, time constant, and final capacitor voltage into Eq. 7.20 to give

$$\begin{aligned}
v_o &= V_{\text{f}} + (V_0 - V_{\text{f}})e^{-t/\tau} \\
&= -60 + [30 - (-60)]e^{-t/0.01} \\
&= -60 + 90e^{-100t}\ \text{V}, \quad t \geq 0.
\end{aligned}$$

Step 5: We calculate the other quantity of interest, i_o, in part (b).

b) Write the solution for i_o using the relationship between current and voltage in a capacitor to give

$$\begin{aligned}
i_o &= C\frac{dv_o}{dt} = (0.25 \times 10^{-6})(-9000e^{-100t}) \\
&= -2.25e^{-100t}\ \text{mA}.
\end{aligned}$$

Because $dv_o(0^-)/dt = 0$, the expression for i_o clearly is valid only for $t \geq 0^+$.

◼ ASSESSMENT PROBLEM

Objective 2—Be able to determine the step response of both *RL* and *RC* circuits

7.6 a) Find the expression for the voltage across the $160\,\text{k}\Omega$ resistor in the circuit shown in Fig. 7.29. Let this voltage be denoted v_A, and assume that the reference polarity for the voltage is positive at the upper terminal of the $160\,\text{k}\Omega$ resistor.

b) Specify the interval of time for which the expression obtained in (a) is valid.

Answer: a) $-60 + 72e^{-100t}\ \text{V}$;
b) $t \geq 0^+$.

SELF-CHECK: Also try Chapter Problems 7.52 and 7.55.

7.4 A General Solution for Step and Natural Responses

We can construct a general approach to finding either the natural response or the step response of the first-order RL and RC circuits because their differential equations all have the same form. To generalize the solution of these four possible circuits, we let $x(t)$ represent the unknown quantity, where $x(t)$ represents the circuit variable that is required to be continuous for all time. Thus, $x(t)$ is the inductor current for RL natural- and step-response circuits and is the capacitor voltage for RC natural- and step-response circuits. From Eqs. 7.16, 7.17, 7.18, and 7.21, we know that the differential equation describing both the natural and step responses of the RL and RC circuits takes the form

$$\frac{dx}{dt} + \frac{x}{\tau} = K,$$

where the value of the constant K is zero for the natural response and nonzero for the step response.

How are the natural and step responses of RL circuits different, and how are they the same? Compare the circuits we analyzed to determine the natural and step responses of RL circuits using Figs. 7.3 and 7.20. The step-response circuit contains an independent source for $t \geq 0$, but the natural-response circuit does not. Now compare the analysis of the RL natural response and the RL step response (Analysis Methods 7.1 and 7.3). There is one extra calculation in the step-response analysis that computes the final value of the inductor current. We can make Analysis Methods 7.1 and 7.3 exactly the same by recognizing that the final value of the inductor current in the natural-response circuit is 0. These same observations hold when comparing both the natural and step responses of RC circuits (Figs. 7.12 and 7.28) and the natural-response analysis and step-response analysis of RC circuits (Analysis Methods 7.2 and 7.4). So we can make Analysis Methods 7.2 and 7.4 exactly the same by recognizing that the final value of the capacitor voltage in the natural-response circuit is 0.

Next, let's look at how the step response of an RL circuit is similar to and different from the step response of an RC circuit. Comparing Analysis Methods 7.2 and 7.4, we identify four important differences:

- In the RL circuit, we find the inductor current for $t \geq 0$; in the RC circuit, we find the capacitor voltage for $t \geq 0$.
- We analyze the RL circuit when $t < 0$ to find the initial inductor current; we analyze the RC circuit when $t < 0$ to find the initial capacitor voltage.
- We analyze the RL circuit when $t \geq 0$ to find the equivalent resistance attached to the inductor and use it to calculate the circuit's time constant, $\tau = L/R$; we analyze the RC circuit when $t \geq 0$ to find the equivalent resistance attached to the capacitor and use it to calculate the circuit's time constant, $\tau = RC$.
- We analyze the RL circuit as $t \to \infty$ to find the final inductor current; we analyze the RC circuit as $t \to \infty$ to find the final capacitor voltage. If the circuit exhibits a natural response instead of a step response, we know that the final values are zero without performing circuit analysis.

Based on these comparisons, we can create a general step-by-step method to calculate the natural and step responses of both RL and RC circuits.

Step 1: Identify the variable $x(t)$, which is the quantity that is required to be continuous for all time. This is the inductor current in RL circuits and the capacitor voltage in RC circuits.

GENERAL METHOD FOR NATURAL AND STEP RESPONSE OF *RL* AND *RC* CIRCUITS

1. **Identify the variable x(t),** which is the inductor current for *RL* circuits and capacitor voltage for *RC* circuits.
2. **Calculate the initial value X_0,** by analyzing the circuit to find $x(t)$ for $t < 0$.
3. **Calculate the time constant τ**; for *RL* circuits $\tau = L/R$ and for *RC* circuits $\tau = RC$, where R is the equivalent resistance connected to the inductor or capacitor for $t \geq 0$.
4. **Calculate the final value X_f,** by analyzing the circuit to find $x(t)$ as $t \to \infty$; for the natural response, $X_f = 0$.
5. **Write the equation for x(t),** $x(t) = X_f + (X_0 - X_f)e^{-t/\tau}$, for $t \geq 0$.
6. **Calculate other quantities of interest** using $x(t)$.

Analysis Method 7.5   Finding the *RL* and *RC* natural and step response.

Step 2: Calculate the initial value X_0, by analyzing the circuit to find $x(t)$ for $t < 0$.

Step 3: Calculate the time constant, τ, for the circuit by analyzing the circuit for $t \geq 0$ to find the equivalent resistance attached to the inductor or capacitor. For *RL* circuits, $\tau = L/R$, and for *RC* circuits, $\tau = RC$.

Step 4: Calculate the final value X_f, by analyzing the circuit to find $x(t)$ as $t \to \infty$. If the circuit exhibits a natural response, $X_f = 0$, so no calculation is needed.

Step 5: Write the equation for $x(t)$ by substituting the initial value X_0, the time constant τ, and the final value X_f, into the expression

GENERAL SOLUTION FOR NATURAL AND STEP RESPONSES OF *RL* AND *RC* CIRCUITS

$$x(t) = X_f + (X_0 - X_f)e^{-t/\tau}, t \geq 0. \qquad (7.23)$$

Step 6: Use $x(t)$ to find any other quantities of interest in the circuit.

These general steps are summarized in Analysis Method 7.5. Examples 7.7–7.9 illustrate how to use Analysis Method 7.5 to find the natural or step responses of *RC* or *RL* circuits.

EXAMPLE 7.7 | Using the General Solution Method to Find an *RL* Circuit's Natural Response

The switch in the circuit shown in Fig. 7.33 has been closed for a long time. At $t = 0$ the switch opens and remains open.

a) What is the initial value of i_o?

b) What is the time constant of the circuit when the switch is open?

c) What is the final value of i_o?

d) What is the expression for $i_o(t)$ when $t \geq 0$?

e) What is the expression for $v_o(t)$ when $t \geq 0$?

f) Find $v_o(0^-)$ and $v_o(0^+)$.

Solution

Use Analysis Method 7.5.

a) **Step 1**: Identify the inductor current, i_o, as the variable of interest, because this is an *RL* circuit.
Step 2: Calculate the initial value of i_o. The switch has been closed for a long time, so the inductor behaves like a short circuit. Therefore, the current through the inductor is the current in the 25 Ω

resistor. Using current division, the current in the 25 Ω resistor is $[(100\|25)/25](0.075) = 60$ mA, so $I_0 = i_o(0) = 60$ mA.

b) **Step 3**: Calculate the time constant $\tau = L/R$. When $t \geq 0$, the equivalent resistance attached to the inductor is the series combination of the 100 Ω and 25 Ω resistors, or 125 Ω and. Therefore,

$$\tau = \frac{0.05}{125} = 0.4 \text{ ms}.$$

c) **Step 4**: Calculate the final value for the inductor current, I_f. This is a natural-response problem because for $t \geq 0$ there is no source in the circuit. Eventually, all of the energy stored in the inductor before the switch opens is dissipated by the resistors and the inductor current is zero, so $I_f = 0$.

d) **Step 5**: Write the equation for the inductor current by substituting the values for I_0, τ, and I_f into Eq. 7.23 to give

$$i_o(t) = I_f + (I_0 - I_f)e^{-t/\tau} = 0 + (0.06 - 0)e^{-t/0.4 \times 10^{-3}}$$

$$= 60e^{-2500t} \text{ mA}, \quad t \geq 0.$$

e) **Step 6**: Use the inductor current to find the voltage across the 100 Ω, using Ohm's law. The result is

$$v_o(t) = -100i_o = -6e^{-2500t} \text{ V}, t \geq 0^+.$$

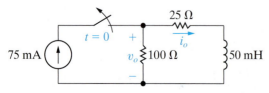

Figure 7.33 ▲ The circuit for Example 7.7.

f) From part (a), when $t < 0$ the switch is closed, and the current divides between the 100 Ω and 25 Ω resistors. We know that the current in the 25 Ω is 60 mA, so the current in the 100 Ω must be $75 - 60 = 15$ mA. Using Ohm's law,

$$v_o(0^-) = 100(0.015) = 1.5 \text{ V}.$$

From part (e)

$$v_o(0^+) = -6e^{-2500(0^+)} = -6 \text{ V}.$$

There is a discontinuity in the voltage across the 100 Ω resistor at $t = 0$.

EXAMPLE 7.8   Using the General Solution Method to Find an *RC* Circuit's Step Response

The switch in the circuit shown in Fig. 7.34 has been in position a for a long time. At $t = 0$ the switch is moved to position b.

a) What is the expression for $v_C(t)$ when $t \geq 0$?

b) What is the expression for $i(t)$ when $t \geq 0^+$?

c) How long after the switch is in position b does the capacitor voltage equal zero?

d) Plot $v_C(t)$ and $i(t)$ versus t.

Solution

Use Analysis Method 7.5.

a) **Step 1:** Identify the capacitor voltage, v_C, as the variable of interest, because this is an *RC* circuit.

Step 2: Calculate the initial value of v_C. The switch has been in position a for a long time, so the capacitor looks like an open circuit. Therefore, the voltage across the capacitor is the voltage across the 60 Ω resistor. Using voltage division, the voltage across the 60 Ω resistor is $[60/(60 + 20)](40) = 30$ V, positive at the lower terminal of the resistor. But v_C is positive at the upper terminal of the capacitor, so $V_0 = v_C(0) = -30$ V.

Step 3: Calculate the time constant $\tau = RC$. When $t \geq 0$, the equivalent resistance attached to the capacitor has the value 400 kΩ. Therefore,

$$\tau = (400 \times 10^3)(0.5 \times 10^{-6}) = 0.2 \text{ s}.$$

Step 4: Calculate the final value for the capacitor voltage, V_f. As $t \to \infty$, the switch has been in position b for a long time, and the capacitor behaves like an open circuit in the presence of the

90 V source. Because of the open circuit, there is no current in the 400 kΩ resistor, so $V_f = 90$ V.

Step 5: Write the equation for capacitor voltage by substituting the values for V_o, τ, and V_f into Eq. 7.23 to give

$$v_C(t) = V_f + (V_0 - V_f)e^{-t/\tau} = 90 + (-30 - 90)e^{-t/0.2}$$

$$= 90 - 120e^{-5t} \text{ V}, \quad t \geq 0.$$

b) **Step 6:** Use the relationship between voltage and current for capacitors to find the capacitor voltage. The result is

$$i(t) = C\frac{dv_C}{dt} = (0.5 \times 10^{-6})[-5(-120e^{-5t})]$$

$$= 300e^{-5t} \text{ } \mu A, \quad t \geq 0^+.$$

c) To find how long the switch must be in position b before the capacitor voltage becomes zero, we solve the equation derived in (a) for the time when $v_C(t) = 0$:

$$120e^{-5t} = 90 \text{ or } e^{5t} = \frac{120}{90},$$

so

$$t = \frac{1}{5}\ln\left(\frac{4}{3}\right) = 57.54 \text{ ms}.$$

Note that when $v_C = 0$, the voltage drop across the 400 kΩ resistor is 90 V so $i = 225$ μA.

d) Figure 7.35 shows the graphs of $v_C(t)$ and $i(t)$ versus t.

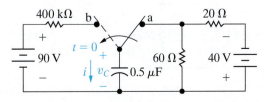

Figure 7.34 ▲ The circuit for Example 7.8.

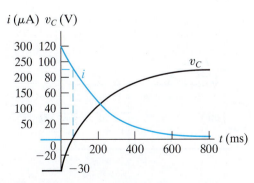

Figure 7.35 ▲ The current and voltage waveforms for Example 7.8.

| EXAMPLE 7.9 | Using the General Solution Method to Find an *RL* Circuit's Step Response |

The switch in the circuit shown in Fig. 7.36 has been open for a long time. At $t = 0$ the switch is closed. Find the expression for

a) $i(t)$ when $t \geq 0$ and

b) $v(t)$ when $t \geq 0^+$.

Solution

Use Analysis Method 7.5.

a) **Step 1:** Identify the inductor current, i, as the variable of interest, because this is an *RL* circuit.

Step 2: Calculate the initial value of i. The switch has been open for a long time, so from Ohm's law, the initial current in the inductor is $20/(1 + 3) = 5$ A. Thus, $I_0 = i(0) = 5$ A.

Step 3: Calculate the time constant $\tau = L/R$. When $t \geq 0$, the switch is closed, shunting the

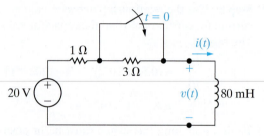

Figure 7.36 ▲ The circuit for Example 7.9.

3 Ω resistor. The remaining resistance attached to the inductor has the value 1 Ω. Therefore,

$$\tau = \frac{80 \times 10^{-3}}{1} = 80 \text{ ms.}$$

Step 4: Calculate the final value for the inductor current. As $t \to \infty$, the switch has been closed for a long time, and the inductor behaves like a short circuit in the presence of the 20 V source. Using Ohm's law, the current in the inductor is $20/1 = 20$ A, so $I_f = 20$ A.

Step 5: Write the equation for inductor current by substituting the values for I_o, τ, and I_f into Eq. 7.23 to give

$$i(t) = I_f + (I_0 - I_f)e^{-t/\tau} = 20 + (5 - 20)e^{-t/0.08}$$

$$= 20 - 15e^{-12.5t} \text{ A}, \quad t \geq 0.$$

b) **Step 6:** Use the relationship between voltage and current for inductors to find the inductor voltage. The result is

$$v(t) = L\frac{di}{dt} = (80 \times 10^{-3})[-12.5(-15e^{-12.5t})]$$

$$= 15e^{-12.5t} \text{ V}, \quad t \geq 0^+.$$

ASSESSMENT PROBLEMS

Objectives 1 and 2—Be able to determine the natural and step response of both *RL* and *RC* circuits

7.7 The switch in the circuit has been closed for a long time. At $t = 0$ the switch opens and stays open. Find the expression for $v_o(t)$ for $t \geq 0$.

Answer: $15e^{-8000t}$ V, $t \geq 0$.

7.8 The switch in the circuit has been open for a long time. The initial charge on the capacitor is zero. At $t = 0$ the switch is closed. Find the expression for

a) $i(t)$ for $t \geq 0^+$ and

b) $v(t)$ for $t \geq 0^+$.

Answer: a) $3e^{-200t}$ mA, $t \geq 0^+$;

b) $150 - 60e^{-200t}$ V, $t \geq 0^+$.

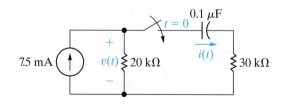

SELF-CHECK: Also try Chapter Problems 7.56 and 7.59.

Example 7.10 shows that Eq. 7.23 can even be used to find the step response of some circuits containing magnetically coupled coils.

| **EXAMPLE 7.10** | **Determining the Step Response of a Circuit with Magnetically Coupled Coils** |

There is no energy stored in the circuit in Fig. 7.37 at the time the switch is closed.

a) Find the solutions for i_o, v_o, i_1, and i_2.

b) Show that the solutions obtained in (a) make sense in terms of known circuit behavior.

Solution

a) For the circuit in Fig. 7.37, the magnetically coupled coils can be replaced by a single inductor having an inductance of

$$L_{eq} = \frac{L_1 L_2 - M^2}{L_1 + L_2 - 2M} = \frac{45 - 36}{18 - 12} = 1.5 \text{ H}.$$

(See Problem 6.38.) It follows that the circuit in Fig. 7.37 can be simplified, as shown in Fig. 7.38. We can apply Analysis Method 7.5 to the circuit in Fig. 7.38.

Step 1: Identify the inductor current, i_o, as the variable of interest, because this is an RL circuit.

Step 2: Calculate the initial value of i. By hypothesis, there is no initial energy stored in the coils, so there is no initial current in the equivalent 1.5 H inductor. Thus, $I_0 = i(0) = 0$.

Step 3: Calculate the time constant $\tau = L/R$. When $t \geq 0$, the switch is closed and the resistance attached to the inductor has the value 7.5 Ω. Therefore,

$$\tau = \frac{1.5}{7.5} = 0.2 \text{ s}.$$

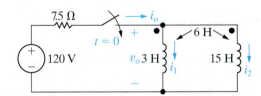

Figure 7.37 ▲ The circuit for Example 7.10.

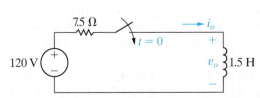

Figure 7.38 ▲ The circuit in Fig. 7.37 with the magnetically coupled coils replaced by an equivalent coil.

Step 4: Calculate the final value for the inductor current. As $t \rightarrow \infty$, the switch has been closed for a long time, and the inductor behaves like a short circuit in the presence of the 120 V source. Using Ohm's law, the current in the inductor is $120/7.5 = 16$ A, so $I_f = 16$ A.

Step 5: Write the equation for inductor current by substituting the values for I_0, τ, and I_f into Eq. 7.23 to give

$$i_o(t) = I_f + (I_0 - I_f)e^{-t/\tau} = 16 + (0 - 16)e^{-t/0.2}$$
$$= 16 - 16e^{-5t} \text{ A}, t \geq 0.$$

Step 6: Use the inductor current and the relationship between voltage and current for inductors to find the inductor voltage. The result is

$$v_o(t) = L\frac{di}{dt} = (1.5)[-5(-16e^{-5t})]$$
$$= 120e^{-5t} \text{ V}, t \geq 0^+.$$

To find i_1 and i_2 we use KVL for the mesh containing the coupled coils in Fig. 7.37 to see that

$$3\frac{di_1}{dt} + 6\frac{di_2}{dt} = 6\frac{di_1}{dt} + 15\frac{di_2}{dt}$$

or

$$\frac{di_1}{dt} = -3\frac{di_2}{dt}.$$

It also follows from Fig. 7.37 and KCL that $i_o = i_1 + i_2$, so

$$\frac{di_o}{dt} = \frac{di_1}{dt} + \frac{di_2}{dt} = -3\frac{di_2}{dt} + \frac{di_2}{dt} = -2\frac{di_2}{dt}.$$

Therefore

$$80e^{-5t} = -2\frac{di_2}{dt}.$$

Because $i_2(0)$ is zero, we have

$$i_2 = \int_0^t -40e^{-5x}\, dx$$
$$= -8 + 8e^{-5t} \text{ A}, \quad t \geq 0.$$

Using $i_o = i_1 + i_2$, we get

$$i_1 = (16 - 16e^{-5t}) - (-8 + 8e^{-5t})$$
$$= 24 - 24e^{-5t} \text{ A}, \quad t \geq 0.$$

b) First, we observe that $i_o(0)$, $i_1(0)$, and $i_2(0)$ are all zero, which is consistent with the statement that no energy is stored in the circuit at the instant the switch is closed. Then we observe that $v_o(0^+) = 120$ V, which is consistent with the fact that $i_o(0) = 0$. Now we see that the solutions for i_1 and i_2 are consistent with the solution for v_o by observing

$$v_o = 3\frac{di_1}{dt} + 6\frac{di_2}{dt}$$

$$= 360e^{-5t} - 240e^{-5t}$$

$$= 120e^{-5t} \text{ V}, \quad t \geq 0^+,$$

or

$$v_o = 6\frac{di_1}{dt} + 15\frac{di_2}{dt}$$

$$= 720e^{-5t} - 600e^{-5t}$$

$$= 120e^{-5t} \text{ V}, \quad t \geq 0^+.$$

The final values of i_1 and i_2 can be checked using flux linkages. The flux linking the 3 H coil (λ_1) must be equal to the flux linking the 15 H coil (λ_2) because

$$v_o = \frac{d\lambda_1}{dt} = \frac{d\lambda_2}{dt}.$$

Now

$$\lambda_1 = 3i_1 + 6i_2 \text{ Wb-turns}$$

and

$$\lambda_2 = 6i_1 + 15i_2 \text{ Wb-turns}.$$

Regardless of which expression we use, we obtain

$$\lambda_1 = \lambda_2 = 24 - 24e^{-5t} \text{ Wb-turns.}$$

Note the solution for λ_1 or λ_2 is consistent with the solution for v_o.

The final value of the flux linking either coil 1 or coil 2 is 24 Wb-turns; that is,

$$\lambda_1(\infty) = \lambda_2(\infty) = 24 \text{ Wb-turns.}$$

The final value of i_1 is

$$i_1(\infty) = 24 \text{ A}$$

and the final value of i_2 is

$$i_2(\infty) = -8 \text{ A.}$$

The consistency between these final values for i_1 and i_2 and the final value of the flux linkage can be seen from the expressions:

$$\lambda_1(\infty) = 3i_1(\infty) + 6i_2(\infty)$$

$$= 3(24) + 6(-8) = 24 \text{ Wb-turns,}$$

$$\lambda_2(\infty) = 6i_1(\infty) + 15i_2(\infty)$$

$$= 6(24) + 15(-8) = 24 \text{ Wb-turns.}$$

The final values of i_1 and i_2 can only be checked via flux linkage because at $t = \infty$ the two coils are ideal short circuits. We cannot use current division when the two branches have no resistance

SELF-CHECK: Assess your understanding of this material by using the general solution method to solve Chapter Problems 7.68 and 7.69.

7.5 Sequential Switching

Whenever switching occurs at two or more distinct times in a circuit, we have **sequential switching**. For example, a single, two-position switch may be in position 1 at t_1 and in position 2 at t_2, or multiple switches may be opened or closed in sequence. We determine the voltages and currents generated by a switching sequence using the techniques described previously in this chapter, primarily Analysis Method 7.5. We derive the expressions for $v(t)$ and $i(t)$ for a given position of the switch or switches and then use these solutions to determine the initial conditions for the next position of the switch or switches.

With sequential switching problems, a premium is placed on obtaining the initial value $x(t_0)$. Recall that anything but inductive currents and capacitive voltages can change instantaneously at the time of switching. Thus, solving first for inductive currents and capacitive voltages is even more pertinent in sequential switching problems. Drawing the circuit that pertains to each time interval in such a problem is often helpful in the solution process.

Examples 7.11 and 7.12 illustrate the analysis techniques for circuits with sequential switching. The first example is a natural-response problem with two switching times, and the second is a step-response problem.

EXAMPLE 7.11 | Analyzing an *RL* Circuit that has Sequential Switching

The two switches in the circuit shown in Fig. 7.39 have been closed for a long time. At $t = 0$, switch 1 is opened. Then, 35 ms later, switch 2 is opened.

a) Find $i_L(t)$ for $0 \le t \le 35$ ms.

b) Find i_L for $t \ge 35$ ms.

c) What percentage of the initial energy stored in the 150 mH inductor is dissipated in the 18 Ω resistor?

d) Repeat (c) for the 3 Ω resistor.

e) Repeat (c) for the 6 Ω resistor.

Solution

We use Analysis Method 7.5 to solve this problem.

a) **Step 1:** Identify the inductor current, i_L, as the variable of interest, because this is an *RL* circuit.

Step 2: Calculate the initial value of i. For $t < 0$ both switches are closed, causing the 150 mH inductor to short-circuit the 18 Ω resistor. The equivalent circuit is shown in Fig. 7.40. We determine the initial current in the inductor by solving for $i_L(0^-)$ in the circuit shown in Fig. 7.40. After making several source transformations, we find $i_L(0^-)$ to be 6 A, so $I_0 = 6$ A.

Step 3: Calculate the time constant $\tau = L/R$. For $0 \le t \le 35$ ms, switch 1 is open (switch 2 is closed), which disconnects the 60 V voltage

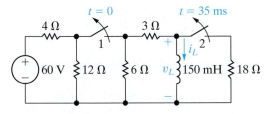

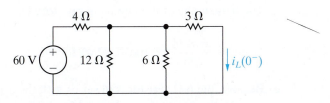

Figure 7.39 ▲ The circuit for Example 7.11.

Figure 7.40 ▲ The circuit shown in Fig. 7.39, for $t < 0$.

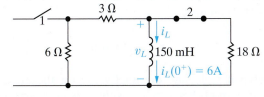

Figure 7.41 ▲ The circuit shown in Fig. 7.39, for $0 \le t \le 35$ ms.

source and the 4 Ω and 12 Ω resistors from the circuit. The inductor is no longer behaving as a short circuit (because the dc source is no longer in the circuit), so the 18 Ω resistor is no longer short-circuited. The equivalent circuit is shown in Fig. 7.41. Note that the equivalent resistance across the terminals of the inductor is the parallel combination of 9 Ω and 18 Ω, or 6 Ω. Therefore,

$$\tau = \frac{0.15}{6} = 25 \text{ ms.}$$

Step 4: Calculate the final value for the inductor current. For $0 \le t \le 35$ ms there is no source in the circuit, so during this time period we have a natural-response problem and the final value of the inductor current is zero. Thus, $I_f = 0$.

Step 5: Write the equation for the inductor current for $0 \le t \le 35$ ms by substituting the values for I_o, τ, and I_f into Eq. 7.23 to give

$$i_L(t) = I_f + (I_0 - I_f)e^{-t/\tau} = 0 + (6 - 0)e^{-t/0.025}$$
$$= 6e^{-40t} \text{ A}, 0 \le t \le 35 \text{ ms.}$$

b) Now we repeat Steps 2–5 for $t \ge 35$ ms.

Step 2: Calculate the initial value of the inductor current for this time segment. When $t = 35$ ms, the value of the inductor current is determined from the inductor current equation for the previous time segment because the inductor current must be continuous for all time. So,

$$i_L(35 \times 10^{-3}) = 6e^{-40(35 \times 10^{-3})} = 6e^{-1.4} = 1.48 \text{ A.}$$

Thus, for $t \ge 35$ ms, $I_0 = 1.48$ A.

Step 3: Calculate the time constant $\tau = L/R$. For $t \ge 35$ ms, both switches are open, and the circuit reduces to the one shown in Fig. 7.42. Note that the

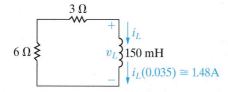

Figure 7.42 ▲ The circuit shown in Fig. 7.39, for $t \geq 35$ ms.

equivalent resistance across the terminals of the inductor is the series combination of 3 Ω and 6 Ω or 9 Ω. Therefore,

$$\tau = \frac{0.15}{9} = 16.67 \text{ ms.}$$

Step 4: Calculate the final value for the inductor current. For $t \geq 35$ ms there is no source in the circuit, so during this time period we have a natural-response problem, and the final value of the inductor current is zero. Thus, $I_f = 0$.

Step 5: Write the equation for inductor current when $t \geq 35$ ms by substituting the values for I_o, τ, and I_f into Eq. 7.23 to give

$$i_L(t) = I_f + (I_0 - I_f)e^{-(t - t_0)/\tau}$$

$$= 0 + (1.48 - 0)e^{-(t - 0.035)/0.01667}$$

$$= 1.48e^{-60(t - 0.035)} \text{ A, } t \geq 35 \text{ ms.}$$

Note that when switch 2 is opened, the time constant changes and the exponential function is shifted in time by 35 ms.

Step 6: Use the inductor current to solve the remaining parts of this problem.

c) The 18 Ω resistor is in the circuit only during the first 35 ms of the switching sequence. During this interval, the voltage across the resistor is

$$v_L = 0.15\frac{d}{dt}(6e^{-40t})$$

$$= -36e^{-40t} \text{ V, } \quad 0 < t < 35 \text{ ms.}$$

The power dissipated in the 18 Ω resistor is

$$p = \frac{v_L^2}{18} = 72e^{-80t} \text{ W, } \quad 0 < t < 35 \text{ ms.}$$

Hence, the energy dissipated is

$$w = \int_0^{0.035} 72e^{-80t} \, dt$$

$$= \frac{72}{-80}e^{-80t}\Big|_0^{0.035}$$

$$= 0.9(1 - e^{-2.8}) = 845.27 \text{ mJ.}$$

The initial energy stored in the 150 mH inductor is

$$w_0 = \frac{1}{2}(0.15)(6)^2 = 2.7 \text{ J} = 2700 \text{ mJ.}$$

Therefore, $(845.27/2700) \times 100$, or 31.31% of the initial energy stored in the 150 mH inductor is dissipated in the 18 Ω resistor.

d) For $0 < t < 35$ ms, the voltage across the 3 Ω resistor is

$$v_{3\Omega} = \left(\frac{v_L}{9}\right)(3)$$

$$= \frac{1}{3}v_L$$

$$= -12e^{-40t} \text{ V.}$$

Therefore, the energy dissipated in the 3 Ω resistor in the first 35 ms is

$$w_{3\Omega} = \int_0^{0.035} \frac{(-12e^{-40t})^2}{3} \, dt$$

$$= 0.6(1 - e^{-2.8})$$

$$= 563.51 \text{ mJ.}$$

For $t > 35$ ms, the current in the 3 Ω resistor is

$$i_{3\Omega} = i_L = (6e^{-1.4})e^{-60(t - 0.035)} \text{ A.}$$

Hence, the energy dissipated in the 3 Ω resistor for $t > 35$ ms is

$$w_{3\Omega} = \int_{0.035}^{\infty} 3i_{3\Omega}^2 \, dt$$

$$= \int_{0.035}^{\infty} 3(6e^{-1.4})^2(e^{-60(t - 0.035)})^2 \, dt$$

$$= 108e^{-2.8} \times \frac{e^{-120(t - 0.035)}}{-120}\Big|_{0.035}^{\infty}$$

$$= \frac{108}{120}e^{-2.8} = 54.73 \text{ mJ.}$$

The total energy dissipated in the 3 Ω resistor is

$$w_{3\Omega}(\text{total}) = 563.51 + 54.73$$

$$= 618.24 \text{ mJ.}$$

The percentage of the initial energy stored is

$$\frac{618.24}{2700} \times 100 = 22.90\%.$$

e) Because the 6 Ω resistor is in series with the 3 Ω resistor, the energy dissipated and the percentage of the initial energy stored will be twice that of the 3 Ω resistor:

$$w_{6\Omega}(\text{total}) = 1236.48 \text{ mJ},$$

and the percentage of the initial energy stored is 45.80%. We check these calculations by observing that

$$1236.48 + 618.24 + 845.27 = 2699.99 \text{ mJ}$$

and

$$31.31 + 22.90 + 45.80 = 100.01\%.$$

The small discrepancies in the summations are the result of roundoff errors.

EXAMPLE 7.12 Analyzing an *RC* Circuit that Has Sequential Switching

The uncharged capacitor in the circuit shown in Fig. 7.43 is initially switched to terminal a of the three-position switch. At $t = 0$, the switch is moved to position b, where it remains for 15 ms. After the 15 ms delay, the switch is moved to position c, where it remains indefinitely.

a) Derive the numerical expression for the voltage across the capacitor.

b) Plot the capacitor voltage versus time.

c) When will the voltage on the capacitor equal 200 V?

Solution

We use Analysis Method 7.5 to solve this problem.

a) **Step 1:** Identify the capacitor voltage, v, as the variable of interest, because this is an *RC* circuit.

Step 2: Calculate the initial value of v. For $t < 0$, the capacitor is initially uncharged, so $V_0 = v(0) = 0$ V.

Step 3: Calculate the time constant $\tau = RC$. When $0 \leq t \leq 15$ ms, the equivalent resistance attached to the capacitor has the value 100 kΩ. Therefore,

$$\tau = (100 \times 10^3)(0.1 \times 10^{-6}) = 10 \text{ ms}.$$

Step 4: Calculate the final value for the capacitor voltage, V_f. If the switch were to remain in position b for a long time, the capacitor would eventually behave like an open circuit in the presence of the 400 V source. Because of the open

circuit, there would be no current in the 100 kΩ resistor, so $V_f = 400$ V.

Step 5: Write the equation for capacitor voltage by substituting the values for V_0, τ, and V_f into Eq. 7.23 to give

$$v(t) = V_f + (V_0 - V_f)e^{-t/\tau} = 400 + (0 - 400)e^{-t/0.01}$$

$$= 400 - 400e^{-100t} \text{ V}, \quad 0 \leq t \leq 15 \text{ ms}.$$

Now we repeat Steps 2–5 for the next time interval, $t \geq 15$ ms.

Step 2: Calculate the initial value of v. At $t = 15$ ms, the capacitor voltage is determined by the equation we derived for the previous time interval. So,

$$v(0.015) = 400 - 400e^{-100(0.015)} = 400 - 400e^{-1.5}$$

$$= 310.75 \text{ V}.$$

Thus, $V_0 = v(0.015) = 310.75$ V.

Step 3: Calculate the time constant $\tau = RC$. When $t \geq 15$ ms, the equivalent resistance attached to the capacitor has the value 50 kΩ. Therefore,

$$\tau = (50 \times 10^3)(0.1 \times 10^{-6}) = 5 \text{ ms}.$$

Step 4: Calculate the final value for the capacitor voltage, V_f. For $t \geq 15$ ms, the switch remains in position c for a long time, and there is no source in the circuit. During this time interval, the circuit exhibits a natural response, so $V_f = 0$.

Step 5: Write the equation for capacitor voltage by substituting the values for V_0, τ, and V_f into Eq. 7.23 to give

$$v(t) = V_f + (V_0 - V_f)e^{-(t-t_0)/\tau}$$

$$= 0 + (310.75 - 0)e^{-(t-0.015)/0.005}$$

$$= 310.75e^{-200(t-0.015)} \text{ V}, \quad t \geq 15 \text{ ms}.$$

Step 6: Use the capacitor voltage to solve the remaining parts of this problem.

b) Figure 7.44 shows the plot of v versus t.

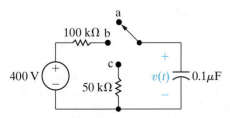

Figure 7.43 ▲ The circuit for Example 7.12.

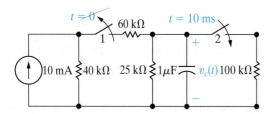

$$v = 400 - 400e^{-100t}$$

$$v = 310.75e^{-200(t-0.015)}$$

Figure 7.44 ▲ The capacitor voltage for Example 7.12.

c) The plot in Fig. 7.44 reveals that the capacitor voltage will equal 200 V at two different times: once in the interval between 0 and 15 ms and

once after 15 ms. We find the first time by solving the expression

$$200 = 400 - 400e^{-100t_1},$$

which yields $t_1 = 6.93$ ms. We find the second time by solving the expression

$$200 = 310.75e^{-200(t_2-0.015)}.$$

In this case, $t_2 = 17.20$ ms.

ASSESSMENT PROBLEMS

Objective 3—Know how to analyze circuits with sequential switching

7.9   In the circuit shown, switch 1 has been closed, and switch 2 has been open for a long time. At $t = 0$, switch 1 is opened. Then 10 ms later, switch 2 is closed. Find

a) $v_c(t)$ for $0 \leq t \leq 0.01$ s,
b) $v_c(t)$ for $t \geq 0.01$ s,
c) the total energy dissipated in the 25 kΩ resistor, and
d) the total energy dissipated in the 100 kΩ resistor.

Answer: a) $80e^{-40t}$ V;
b) $53.63e^{-50(t-0.01)}$ V;
c) 2.91 mJ;
d) 0.29 mJ.

7.10   Switch a in the circuit shown has been open for a long time, and switch b has been closed for a long time. Switch a is closed at $t = 0$ and, after remaining closed for 1 s, is opened again. Switch b is opened simultaneously, and both switches remain open indefinitely. Determine the expression for the inductor current i that is valid when
(a) $0 \leq t \leq 1$ s and (b) $t \geq 1$ s.

Answer: a) $3 - 3e^{-0.5t}$ A;
b) $-4.8 + 5.98e^{-1.25(t-1)}$ A.

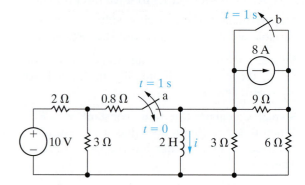

SELF-CHECK: Also try Chapter Problems 7.73 and 7.78.

7.6 Unbounded Response

A circuit response may grow, rather than decay, exponentially with time. This type of response, called an **unbounded response**, is possible if the circuit contains dependent sources. In circuits with an unbounded response, the Thévenin equivalent resistance with respect to the terminals of either an inductor or a capacitor is negative. This negative resistance generates a negative time constant, and the resulting currents and voltages increase without limit. In an actual circuit, a component eventually breaks down or saturates, halting the unbounded response.

We cannot use Analysis Method 7.5 to analyze a circuit with an unbounded response because calculating a final value of voltage or current is not possible. Instead, we must derive the differential equation describing the circuit's response and then solve it using the separation of variables technique. Example 7.13 analyzes a circuit with an unbounded response to illustrate this technique.

EXAMPLE 7.13 Finding the Unbounded Response in an *RC* Circuit

a) When the switch is closed in the circuit shown in Fig. 7.45, the voltage on the capacitor is 10 V. Find the expression for v_o for $t \geq 0$.

b) Assume that the capacitor short-circuits when its terminal voltage reaches 150 V. How many milliseconds elapse before the capacitor short-circuits?

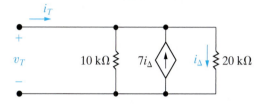

Figure 7.46 ▲ The test-source method used to find R_{Th}.

Solution

a) We need to write the differential equation that describes the capacitor voltage, v_o. To make this task easier, let's simplify the circuit attached to the capacitor by replacing it with its Thévenin equivalent. This subcircuit is shown in Fig. 7.46, and as you can see, it does not contain an independent source. Thus, the Thévenin equivalent consists of a single resistor. To find the Thévenin equivalent resistance for the circuit in Fig. 7.46, we use the test-source method (see Example 4.18, p. 124), where v_T is the test voltage and i_T is the test current. Writing a KCL equation at the top node, we get

$$i_T = \frac{v_T}{10 \times 10^3} - 7\left(\frac{v_T}{20 \times 10^3}\right) + \frac{v_T}{20 \times 10^3}$$

Solving for the ratio v_T/i_T yields the Thévenin resistance:

$$R_{Th} = \frac{v_T}{i_T} = -5 \, k\Omega.$$

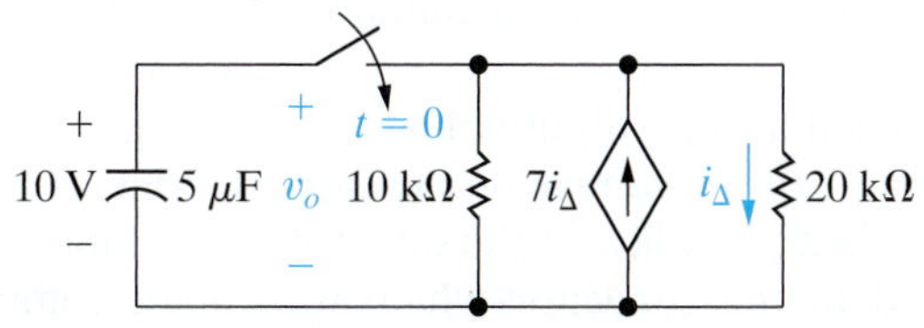

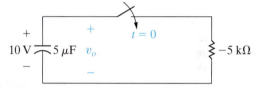

Figure 7.45 ▲ The circuit for Example 7.13.

Figure 7.47 ▲ A simplification of the circuit shown in Fig. 7.45.

We replace the two resistors and the dependent source in Fig. 7.45 with R_{Th} to get the circuit shown in Fig. 7.47. For $t \geq 0$, write a KCL equation at the top node to construct the differential equation describing this circuit:

$$(5 \times 10^{-6})\frac{dv_o}{dt} + \frac{v_o}{-5000} = 0$$

Dividing by the coefficient of the first derivative yields

$$\frac{dv_o}{dt} - 40v_o = 0.$$

This equation has the same form as Eq. 7.1, so we can find $v_o(t)$ using the same separation of variables technique applied to Eq. 7.1 (see p. 223). Thus, the capacitor voltage is

$$v_o(t) = 10e^{40t} \, V, \quad t \geq 0.$$

b) $v_o = 150$ V when $e^{40t} = 15$.
Therefore, $40t = \ln 15$, and $t = 67.70$ ms.

SELF-CHECK: Assess your understanding of this material by trying Chapter Problems 7.87 and 7.88.

As an engineer, you should be aware that interconnected circuit elements may create unbounded responses. If such interconnections are unintended, the resulting circuit may experience unexpected, and potentially dangerous, component failures.

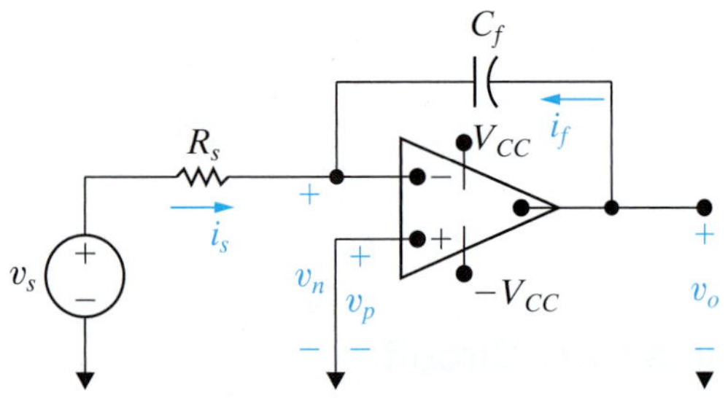

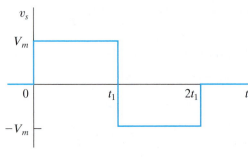

Figure 7.48 ▲ An integrating amplifier.

7.7 The Integrating Amplifier

We are now ready to analyze an integrating-amplifier circuit, shown in Fig. 7.48, which is based on the op amp presented in Chapter 5. This circuit generates an output voltage proportional to the integral of the input voltage. In Fig. 7.48, we added the branch currents i_f and i_s, along with the node voltages v_n and v_p, to aid our analysis. We assume that the op amp is ideal. Write a KCL equation at the inverting input node, and remember that the current into the ideal op amp at its input terminals is zero, to get

$$i_f + i_s = 0.$$

Also, the ideal op amp constrains the voltages at its two input terminals to give

$$v_n = v_p.$$

In the integrating amplifier circuit, $v_p = 0$, so using Ohm's law

$$i_s = \frac{v_s}{R_s},$$

and using the relationship between voltage and current for a capacitor,

$$i_f = C_f \frac{dv_o}{dt}.$$

Substituting the expressions for i_s and i_f into the KCL equation and solving for dv_o/dt, we get

$$\frac{dv_o}{dt} = -\frac{1}{R_sC_f}v_s.$$

Multiplying both sides of this equation by a differential time dt and then integrating from t_0 to t generates the equation

$$v_o(t) = -\frac{1}{R_sC_f}\int_{t_0}^{t} v_s\, dy + v_o(t_0). \tag{7.24}$$

In Eq. 7.24, t_0 represents the instant in time when we begin the integration. Thus, $v_o(t_0)$ is the value of the output voltage at that time. Also, because $v_n = v_p = 0$, $v_o(t_0)$ is identical to the initial voltage on the feedback capacitor C_f.

Equation 7.24 states that the output voltage of an integrating amplifier equals the initial value of the voltage on the capacitor plus an inverted (minus sign), scaled $(1/R_sC_f)$ replica of the integral of the input voltage. If no energy is stored in the capacitor when integration commences, Eq. 7.24 reduces to

$$v_o(t) = -\frac{1}{R_sC_f}\int_{t_0}^{t} v_s\, dy.$$

For example, assume that the input voltage is the rectangular voltage pulse shown in Fig. 7.49. Assume also that the initial value of $v_o(t)$ is zero at the instant v_s steps from 0 to V_m. Using Eq. 7.24, we see that

$$v_o = -\frac{1}{R_sC_f}V_mt + 0, \quad 0 \le t \le t_1.$$

Figure 7.49 ▲ An input voltage signal.

When t lies between t_1 and $2t_1$,

$$v_o = -\frac{1}{R_s C_f} \int_{t_1}^{t} (-V_m) \, dy - \frac{1}{R_s C_f} V_m t_1$$

$$= \frac{V_m}{R_s C_f} t - \frac{2V_m}{R_s C_f} t_1, \quad t_1 \le t \le 2t_1.$$

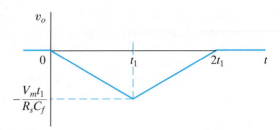

Figure 7.50 ▲ The output voltage of an integrating amplifier.

Figure 7.50 shows a plot of $v_o(t)$ versus t. Clearly, the output voltage is an inverted, scaled replica of the integral of the input voltage.

The output voltage is proportional to the integral of the input voltage only if the op amp operates within its linear range—that is, if it doesn't saturate. Examples 7.14 and 7.15 further illustrate the analysis of the integrating amplifier.

EXAMPLE 7.14 | **Analyzing an Integrating Amplifier**

Assume that the numerical values for the voltage shown in Fig. 7.49 are $V_m = 50$ mV and $t_1 = 1$ s. We apply this voltage to the integrating-amplifier circuit shown in Fig. 7.48. The circuit parameters of the amplifier are $R_s = 100$ kΩ, $C_f = 0.1 \, \mu$F, and $V_{CC} = 6$ V. The capacitor's initial voltage is zero.

a) Calculate $v_o(t)$.

b) Plot $v_o(t)$ versus t.

Solution

a) For $0 \le t \le 1$ s,

$$v_o = \frac{-1}{(100 \times 10^3)(0.1 \times 10^{-6})} 50 \times 10^{-3} t + 0$$

$$= -5t \text{ V}, \quad 0 \le t \le 1 \text{ s}.$$

For $1 \le t \le 2$ s,

$$v_o = (5t - 10) \text{ V}.$$

b) Figure 7.51 shows a plot of $v_o(t)$ versus t.

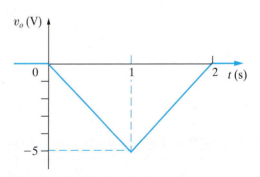

Figure 7.51 ▲ The output voltage for Example 7.14.

EXAMPLE 7.15 | **Analyzing an Integrating Amplifier that Has Sequential Switching**

At the instant the switch makes contact with terminal a in the circuit shown in Fig. 7.52, the voltage on the $0.1 \, \mu$F capacitor is 5 V. The switch

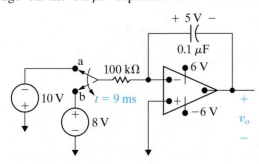

Figure 7.52 ▲ The circuit for Example 7.15.

remains at terminal a for 9 ms and then moves instantaneously to terminal b. How many milliseconds after making contact with terminal b does the op amp saturate?

Solution

The expression for the output voltage during the time the switch is at terminal a is

$$v_o = \frac{-1}{(100 \times 10^3)(0.1 \times 10^{-6})} \int_{0}^{t} (-10) \, dy + (-5)$$

$$= (1000t - 5) \text{ V}.$$

Thus, 9 ms after the switch makes contact with terminal a, the output voltage is $1000(9 \times 10^{-3}) - 5 = 4$ V. Note that the op amp does not saturate during its first 9 ms of operation.

The expression for the output voltage after the switch moves to terminal b is

$$v_o = \frac{-1}{(100 \times 10^3)(0.1 \times 10^{-6})} \int_{9 \times 10^{-3}}^{t} 8 dy + 4$$

$$= -800(t - 9 \times 10^{-3}) + 4$$

$$= (11.2 - 800t) \text{ V}.$$

When the switch is at terminal b, the voltage is decreasing, and the op amp eventually saturates at -6 V. Therefore, we set the expression for v_o equal to -6 V to obtain the saturation time t_s:

$$11.2 - 800t_s = -6,$$

or

$$t_s = 21.5 \text{ ms.}$$

Thus, the integrating amplifier saturates 21.5 ms after making contact with terminal b.

From the examples, we see that the integrating amplifier can perform the integration function very well but only within specified limits that avoid saturating the op amp. The op amp saturates because charge accumulates on the feedback capacitor. We can prevent saturation by placing a resistor in parallel with the feedback capacitor. We examine such a circuit in Chapter 8.

Note that we can convert the integrating amplifier to a differentiating amplifier by interchanging the input resistance R_s and the feedback capacitor C_f. Then

$$v_o = -R_s C_f \frac{dv_s}{dt}.$$

We leave the derivation of this expression as an exercise for you. The differentiating amplifier is seldom used because in practice it is a source of unwanted or noisy signals.

Finally, we can use an inductor instead of a capacitor to create both integrating- and differentiating-amplifier circuits. Since it is easier to fabricate capacitors for integrated-circuit devices, inductors are rarely used in integrating amplifiers.

ASSESSMENT PROBLEMS

Objective 4—Be able to analyze op amp circuits containing resistors and a single capacitor

7.11 There is no energy stored in the capacitor at the time the switch in the circuit makes contact with terminal a. The switch remains at position a for 32 ms and then moves instantaneously to position b. How many milliseconds after making contact with terminal a does the op amp saturate?

Answer: 262 ms.

7.12 a) When the switch closes in the circuit shown, there is no energy stored in the capacitor. How long does it take to saturate the op amp?
b) Repeat (a) with an initial voltage on the capacitor of 1 V, positive at the upper terminal.

Answer: a) 1.11 ms;
b) 1.76 ms.

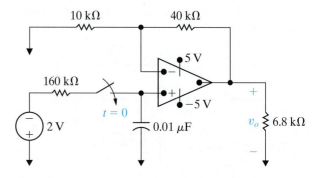

SELF-CHECK: Also try Chapter Problems 7.94 and 7.95.

■ Practical Perspective

Artificial Pacemaker

The *RC* circuit shown in Fig. 7.53 can be used in an artificial pacemaker to establish a normal heart rhythm by generating periodic electrical impulses. The box labeled "controller" behaves as an open circuit until the voltage drop across the capacitor reaches a preset limit. Once that limit is reached, the capacitor discharges its stored energy in the form of an electrical impulse to the heart, starts to recharge, and then the process repeats.

Before we develop the analytical expressions that describe the circuit's behavior, let's get a feel for how the circuit works. First, when the controller behaves like an open circuit, the dc voltage source will charge the capacitor via the resistor R, toward a value of V_s volts. But once the capacitor voltage reaches V_{max}, the controller behaves like a short circuit, enabling the capacitor to discharge. Once the capacitor discharge is complete, the controller again acts like an open circuit and the capacitor starts to recharge. This cycle of charging and discharging the capacitor establishes the desired heart rhythm, as shown in Fig. 7.54.

In drawing Fig. 7.54, we have chosen $t = 0$ at the instant the capacitor starts to charge. This plot also assumes that the circuit has reached the repetitive stage of its operation and that the time to discharge the capacitor is negligible when compared to the recharge time. We need an equation for $v_C(t)$ as a function of V_s, R, and C to design the artificial pacemaker circuit.

To begin the analysis, we assume that the circuit has been in operation for a long time. Let $t = 0$ at the instant when the capacitor has completely discharged and the controller is acting as an open circuit. From the circuit we find the initial and final values of the capacitor voltage and the circuit's time constant:

$$V_0 = v_C(0) = 0;$$

$$V_f = v_C(\infty) = V_s; \text{ and } \tau = RC.$$

To find the capacitor voltage while the capacitor is charging, substitute the initial and final values of the capacitor voltage and the circuit's time constant into Eq. 7.23 and simplify to get

$$v_C(t) = V_s (1 - e^{-t/RC}).$$

Suppose the controller is programmed to generate an electrical pulse that stimulates the heart when $v_C = 0.75 V_s = V_{max}$. Given values of R and C, we can determine the resulting heart rate, H, in beats per minute as follows:

$$H = \frac{60}{-RC \ln\ 0.25} \quad \text{[beats per minute]}$$

A more realistic design problem requires you to calculate the value of resistance, R, given V_{max} as a percentage of V_s, C, and the desired heart rate in beats per minute. Developing an equation for resistance, R, is the focus of Problem 7.106.

SELF-CHECK: Assess your understanding of the Practical Perspective by solving Chapter Problems 7.104–7.107.

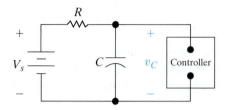

Figure 7.53 ▲ An artificial pacemaker circuit.

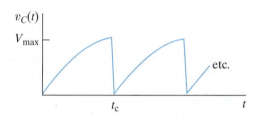

Figure 7.54 ▲ Capacitor voltage versus time for the circuit in Fig. 7.55.

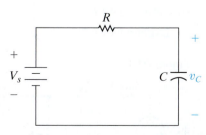

Figure 7.55 ▲ The artificial pacemaker circuit at $t = 0$, when the capacitor is charging.

Summary

- A first-order circuit may be reduced to a Thévenin (or Norton) equivalent connected to either a single equivalent inductor or capacitor. (See page 220.)

- The **natural response** is the currents and voltages that exist when stored energy is released to a circuit that contains no independent sources. (See page 220.)

- The **time constant** of an *RL* circuit equals the equivalent inductance divided by the Thévenin resistance as viewed from the terminals of the equivalent inductor. (See page 223.)

- The **time constant** of an *RC* circuit equals the equivalent capacitance times the Thévenin resistance as viewed from the terminals of the equivalent capacitor. (See page 229.)

- The **step response** is the currents and voltages that result from abrupt changes in dc sources connected to a circuit. Stored energy may or may not be present at the time the abrupt changes take place. (See page 233.)

- Analysis Method 7.5 can be used to find the solution for the natural and step responses of both *RL* and *RC* circuits:

  Step 1: Identify the variable $x(t)$, which is the quantity that is required to be continuous for all time. This is the inductor current in *RL* circuits and the capacitor voltage in *RC* circuits.

  Step 2: Calculate the initial value X_0, by analyzing the circuit to find $x(t)$ for $t < 0$.

Step 3: Calculate the time constant, τ, for the circuit by analyzing the circuit for $t \geq 0$ to find the equivalent resistance attached to the inductor or capacitor. For *RL* circuits, $\tau = L/R$, and for *RC* circuits, $\tau = RC$.

Step 4: Calculate the final value X_f, by analyzing the circuit to find $x(t)$ as $t \to \infty$. If the circuit exhibits a natural response, $X_f = 0$, so no calculation is needed.

Step 5: Write the equation for $x(t)$ by substituting the initial value X_0, the time constant τ, and the final value X_f, into the expression $x(t) = X_f + (X_0 - X_f)e^{-t/\tau}, t \geq 0$.

Step 6: Use $x(t)$ to find any other quantities of interest in the circuit. (See page 242.)

- **Sequential switching** in first-order circuits is analyzed by dividing the analysis into time intervals corresponding to specific switch positions. Initial values for a particular interval are determined from the solution corresponding to the immediately preceding interval. (See page 246.)

- An **unbounded response** occurs when the Thévenin resistance is negative, which is possible when the first-order circuit contains dependent sources. (See page 250.)

- An **integrating amplifier** consists of an ideal op amp, a capacitor in the negative feedback branch, and a resistor in series with the signal source. It outputs the integral of the signal source, within specified limits that avoid saturating the op amp. (See page 252.)

Problems

Section 7.1

7.1 The switch in the circuit in Fig. P7.1 has been closed for a long time before opening at $t = 0$.

PSPICE
MULTISIM

    a) Find $i_1(0^-)$ and $i_2(0^-)$.

    b) Find $i_1(0^+)$ and $i_2(0^+)$.

    c) Find $i_1(t)$ for $t \geq 0$.

    d) Find $i_2(t)$ for $t \geq 0^+$.

    e) Explain why $i_2(0^-) \neq i_2(0^+)$.

Figure P7.1

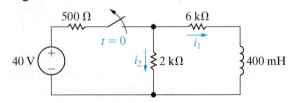

7.2 In the circuit shown in Fig. P7.2, the switch makes contact with position b just before breaking contact with position a. This is known as a make-before-break switch and it ensures that the inductor current is continuous. The interval of time between "making" and "breaking" is assumed to be negligible. The switch has been in the a position for a long time. At $t = 0$ the switch is thrown from position a to position b.

    a) Determine the initial current in the inductor.

    b) Determine the time constant of the circuit for $t > 0$.

    c) Find i, v_1, and v_2 for $t \geq 0$.

    d) What percentage of the initial energy stored in the inductor is dissipated in the 90 Ω resistor 1 ms after the switch is thrown from position a to position b?

Figure P7.2

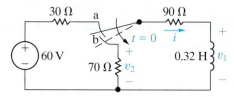

7.3 The switch in the circuit in Fig. P7.3 has been open
PSPICE for a long time. At $t = 0$ the switch is closed.
MULTISIM

a) Determine $i_o(0)$ and $i_o(\infty)$.

b) Determine $i_o(t)$ for $t \geq 0$.

c) How many milliseconds after the switch has
    been closed will i_o equal 5 A?

Figure P7.3

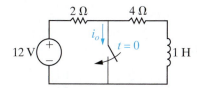

7.4 The switch shown in Fig. P7.4 has been open for a
long time before closing at $t = 0$.

a) Find $i_o(0^-)$, $i_L(0^-)$, and $v_L(0^-)$.

b) Find $i_o(0^+)$, $i_L(0^+)$, and $v_L(0^+)$.

c) Find $i_o(\infty)$, $i_L(\infty)$, and $v_L(\infty)$.

d) Write the expression for $i_L(t)$ for $t \geq 0$.

e) Write the expression for $i_o(t)$ for $t \geq 0^+$.

f) Write the expression for $v_L(t)$ for $t \geq 0^+$.

Figure P7.4

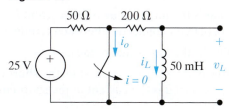

7.5 The switch in the circuit in Fig. P7.5 has been in po-
PSPICE sition 1 for a long time. At $t = 0$, the switch moves
MULTISIM instantaneously to position 2. Find $v_o(t)$ for $t \geq 0^+$.

Figure P7.5

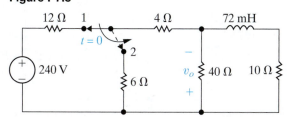

7.6 For the circuit of Fig. P7.5, what percentage of the
initial energy stored in the inductor is eventually
dissipated in the 40 Ω resistor?

7.7 The switch in the circuit in Fig. P7.7 has been closed
PSPICE for a long time. At $t = 0$ it is opened.
MULTISIM

a) Write the expression for $i_o(t)$ for $t \geq 0$.

b) Write the expression for $v_o(t)$ for $t \geq 0^+$.

Figure P7.7

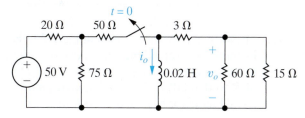

7.8 In the circuit in Fig. P7.8, the voltage and current
expressions are

$$v = 400e^{-5t} \text{ V}, \quad t \geq 0^+;$$

$$i = 10e^{-5t} \text{ A}, \quad t \geq 0.$$

Find

a) R.

b) τ (in milliseconds).

c) L.

d) the initial energy stored in the inductor.

e) the time (in milliseconds) it takes to dissipate
    80% of the initial stored energy.

Figure P7.8

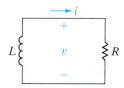

7.9 a) Use component values from Appendix H to cre-
         ate a first-order RL circuit (see Fig. 7.4) with a
         time constant of 1 ms. Use a single inductor and
         a network of resistors, if necessary. Draw your
         circuit.

b) Suppose the inductor you chose in part (a) has
    an initial current of 10 mA. Write an expression
    for the current through the inductor for $t \geq 0$.

c) Using your result from part (b), calculate the
    time at which half of the initial energy stored
    in the inductor has been dissipated by the re-
    sistor.

7.10 In the circuit in Fig. P7.10, the switch has been
closed for a long time before opening at $t = 0$.

a) Find the value of L so that $v_o(t)$ equals 0.5 $v_o(0^+)$
    when $t = 1$ ms.

b) Find the percentage of the stored energy that
    has been dissipated in the 10 Ω resistor when
    $t = 1$ ms.

Figure P7.10

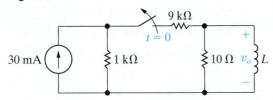

7.11 The switch in the circuit seen in Fig. P7.11 has been in position 1 for a long time. At $t = 0$, the switch moves instantaneously to position 2. Find the value of R so that 20% of the initial energy stored in the 30 mH inductor is dissipated in R in 15 μs.

Figure P7.11

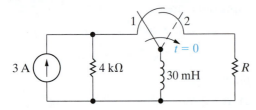

7.12 In the circuit in Fig. P7.11, let I_g represent the dc current source, σ represent the fraction of initial energy stored in the inductor that is dissipated in t_o seconds, and L represent the inductance.

a) Show that

$$R = \frac{L \ln [1/(1 - \sigma)]}{2t_o}.$$

b) Test the expression derived in (a) by using it to find the value of R in Problem 7.11.

7.13 The two switches in the circuit seen in Fig. P7.13 are synchronized. The switches have been closed for a long time before opening at $t = 0$.

a) How many microseconds after the switches are open is the energy dissipated in the 4 kΩ resistor 10% of the initial energy stored in the 6 H inductor?

b) At the time calculated in (a), what percentage of the total energy stored in the inductor has been dissipated?

Figure P7.13

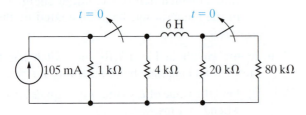

7.14 The switch in the circuit in Fig. P7.14 has been closed for a long time before opening at $t = 0$. Find $v_o(t)$ for $t \geq 0^+$.

PSPICE
MULTISIM

Figure P7.14

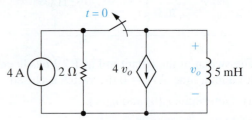

7.15 The switch in Fig. P7.15 has been closed for a long time before opening at $t = 0$. Find

a) $i_L(t)$, $t \geq 0$.

b) $v_L(t)$, $t \geq 0^+$.

c) $i_\Delta(t)$, $t \geq 0^+$.

Figure P7.15

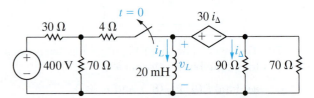

7.16 What percentage of the initial energy stored in the inductor in the circuit in Fig. P7.15 is dissipated by the current-controlled voltage source?

7.17 The 165 V, 5 Ω source in the circuit in Fig. P7.17 is inadvertently short-circuited at its terminals a, b. At the time the fault occurs, the circuit has been in operation for a long time.

PSPICE
MULTISIM

a) What is the initial value of the current i_{ab} in the short-circuit connection between terminals a, b?

b) What is the final value of the current i_{ab}?

c) How many microseconds after the short circuit has occurred is the current in the short equal to 19 A?

Figure P7.17

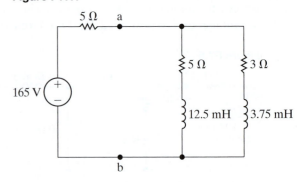

7.18 In the circuit shown in Fig. P7.18, the switch has
PSPICE   been in position a for a long time. At $t = 0$, it moves
MULTISIM   instantaneously from a to b.

a) Find $i_o(t)$ for $t \geq 0$.

b) What is the total energy delivered to the 8 Ω resistor?

c) How many time constants does it take to deliver 95% of the energy found in (b)?

Figure P7.18

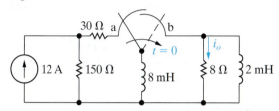

7.19 The two switches shown in the circuit in Fig. P7.19
PSPICE   operate simultaneously. Prior to $t = 0$ each switch
MULTISIM   has been in its indicated position for a long time.
At $t = 0$ the two switches move instantaneously to
their new positions. Find

a) $v_o(t), t \geq 0^+$.

b) $i_o(t), t \geq 0$.

Figure P7.19

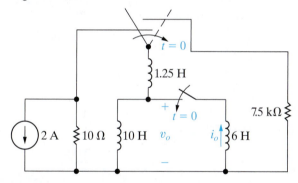

7.20 For the circuit seen in Fig. P7.19, find

a) the total energy dissipated in the 7.5 kΩ resistor.

b) the energy trapped in the ideal inductors.

Section 7.2

7.21 The switch in the circuit in Fig. P7.21 has been in the
left position for a long time. At $t = 0$ it moves to the
right position and stays there.

a) Find the initial voltage drop across the capacitor.

b) Find the initial energy stored by the capacitor.

c) Find the time constant of this circuit for $t > 0$.

d) Write the expression for the capacitor voltage $v(t)$ for $t \geq 0$.

Figure P7.21

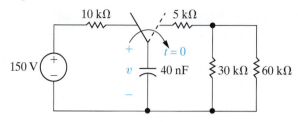

7.22 The switch in the circuit in Fig. P7.22 has been in the
left position for a long time. At $t = 0$ it moves to the
right position and stays there.

a) Write the expression for the capacitor voltage, $v(t)$, for $t \geq 0$.

b) Write the expression for the current through the 2.4 kΩ resistor, $i(t)$, for $t \geq 0^+$.

Figure P7.22

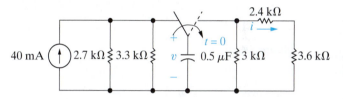

7.23 What percentage of the initial energy stored in the
capacitor in Fig. P7.22 is dissipated by the 3 kΩ resistor 500 μs after the switch is thrown?

7.24 The switch shown in Fig. P7.24 has been open for a
long time before closing at $t = 0$. Write the expression for the capacitor voltage, $v(t)$, for $t \geq 0$.

Figure P7.24

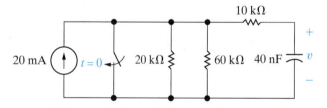

7.25 The switch in the circuit in Fig. P7.25 is closed at
PSPICE $t = 0$ after being open for a long time.
MULTISIM
a) Find $i_1(0^-)$ and $i_2(0^-)$.

b) Find $i_1(0^+)$ and $i_2(0^+)$.

c) Explain why $i_1(0^-) = i_1(0^+)$.

d) Explain why $i_2(0^-) \neq i_2(0^+)$.

e) Find $i_1(t)$ for $t \geq 0$.

f) Find $i_2(t)$ for $t \geq 0^+$.

Figure P7.25

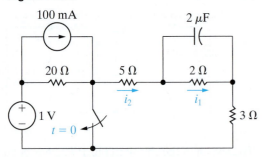

7.26 In the circuit shown in Fig. P7.26, both switches operate together; that is, they either open or close at the same time. The switches are closed a long time before opening at $t = 0$.

a) How many microjoules of energy have been dissipated in the 12 kΩ resistor 12 ms after the switches open?

b) How long does it take to dissipate 75% of the initially stored energy?

Figure P7.26

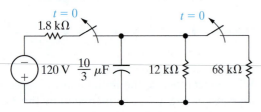

7.27 In the circuit in Fig. P7.27 the voltage and current expressions are

$$v = 48e^{-25t} \text{ V}, \quad t \geq 0;$$
$$i = 12e^{-25t} \text{ mA}, \quad t \geq 0^+.$$

Find

a) R.

b) C.

c) τ (in milliseconds).

d) the initial energy stored in the capacitor.

e) the amount of energy that has been dissipated by the resistor 60 ms after the voltage begins to decay.

Figure P7.27

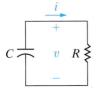

7.28 a) Use component values from Appendix H to create a first-order *RC* circuit (see Fig. 7.13) with a time constant of 50 ms. Use a single capacitor and a network of resistors, if necessary. Draw your circuit.

b) Suppose the capacitor you chose in part (a) has an initial voltage drop of 50 V. Write an expression for the voltage drop across the capacitor for $t \geq 0$.

c) Using you result from part (b), calculate the time at which the voltage drop across the capacitor has reached 10 V.

7.29 The switch in the circuit in Fig. P7.29 has been in position 1 for a long time before moving to position 2 at $t = 0$. Find $i_o(t)$ for $t \geq 0^+$.

PSPICE
MULTISIM

Figure P7.29

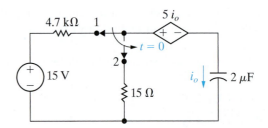

7.30 The switch in the circuit seen in Fig. P7.30 has been in position x for a long time. At $t = 0$, the switch moves instantaneously to position y.

a) Find α so that the time constant for $t > 0$ is 40 ms.

b) For the α found in (a), find v_Δ.

Figure P7.30

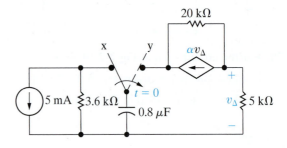

7.31 a) In Problem 7.30, how many microjoules of energy are generated by the dependent current source during the time the capacitor discharges to 0 V?

b) Show that for $t \geq 0$ the total energy stored and generated in the capacitive circuit equals the total energy dissipated.

7.32 At the time the switch is closed in the circuit in Fig. P7.32, the voltage across the parallel capacitors is 50 V and the voltage on the 250 nF capacitor is 40 V.

a) What percentage of the initial energy stored in the three capacitors is dissipated in the 24 kΩ resistor?

b) Repeat (a) for the 400 Ω and 16 kΩ resistors.

c) What percentage of the initial energy is trapped in the capacitors?

Figure P7.32

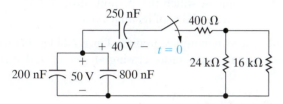

7.33 The switch in the circuit in Fig. P7.33 has been in position a for a long time and $v_2 = 0$ V. At $t = 0$, the switch is thrown to position b. Calculate

a) i, v_1, and v_2 for $t \geq 0^+$,

b) the energy stored in the 1 μF capacitor at $t = 0$, and

c) the energy trapped in the circuit and the total energy dissipated in the 25 kΩ resistor if the switch remains in position b indefinitely.

Figure P7.33

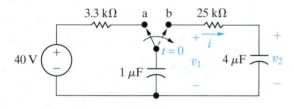

7.34 After the circuit in Fig. P7.34 has been in operation for a long time, a screwdriver is inadvertently connected across the terminals a, b. Assume the resistance of the screwdriver is negligible.

a) Find the current in the screwdriver at $t = 0^+$ and $t = \infty$.

b) Derive the expression for the current in the screwdriver for $t \geq 0^+$.

Figure P7.34

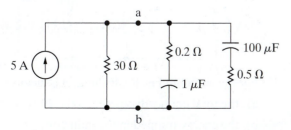

Section 7.3

7.35 After the switch in the circuit of Fig. P7.35 has been open for a long time, it is closed at $t = 0$. Calculate (a) the initial value of i; (b) the final value of i; (c) the time constant for $t \geq 0$; and (d) the numerical expression for $i(t)$ when $t \geq 0$.

Figure P7.35

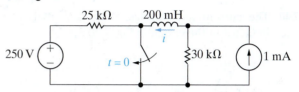

7.36 The switch in the circuit shown in Fig. P7.36 has been in position a for a long time. At $t = 0$, the switch moves instantaneously to position b.

a) Find the numerical expression for $i_o(t)$ when $t \geq 0$.

b) Find the numerical expression for $v_o(t)$ for $t \geq 0^+$.

Figure P7.36

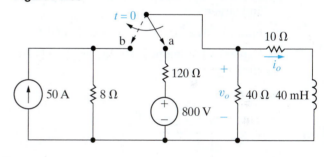

7.37 Repeat Problem 7.36 assuming that the switch in the circuit in Fig. P7.36 has been in position b for a long time and then moves to position a at $t = 0$ and stays there.

7.38 The switch in the circuit shown in Fig. P7.38 has been in position a for a long time before moving to position b at $t = 0$.

a) Find the numerical expressions for $i_L(t)$ and $v_o(t)$ for $t \geq 0$.

b) Find the numerical values of $v_L(0^+)$ and $v_o(0^+)$.

Figure P7.38

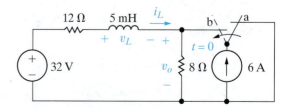

7.39 a) Derive Eq. 7.16 by first converting the Thévenin equivalent in Fig. 7.20 to a Norton equivalent and then summing the currents away from the upper node, using the inductor voltage v as the variable of interest.

b) Use the separation of variables technique to find the solution to Eq. 7.16. Verify that your solution agrees with the solution given in Eq. 7.15.

7.40 The current and voltage at the terminals of the inductor in the circuit in Fig. 7.20 are

$$i(t) = (4 + 4e^{-40t}) \text{ A}, \quad t \geq 0;$$
$$v(t) = -80e^{-40t} \text{ V}, \quad t \geq 0^+.$$

a) Specify the numerical values of V_s, R, I_o, and L.

b) How many milliseconds after the switch has been closed does the energy stored in the inductor reach 9 J?

7.41 The switch in the circuit shown in Fig. P7.41 has been closed for a long time. The switch opens at $t = 0$. For $t \geq 0^+$:

a) Find $v_o(t)$ as a function of I_g, R_1, R_2, and L.

b) Explain what happens to $v_o(t)$ as R_2 gets larger and larger.

c) Find v_{SW} as a function of I_g, R_1, R_2, and L.

d) Explain what happens to v_{SW} as R_2 gets larger and larger.

Figure P7.41

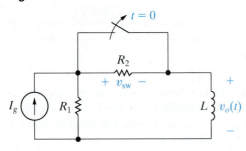

7.42 The switch in the circuit in Fig. P7.42 has been closed for a long time. A student abruptly opens the switch and reports to her instructor that when the switch opened, an electric arc with noticeable persistence was established across the switch, and at the same time the voltmeter placed across the coil was damaged. On the basis of your analysis of the circuit in Problem 7.41, can you explain to the student why this happened?

Figure P7.42

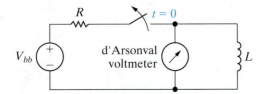

7.43 a) Use component values from Appendix H to create a first-order *RL* circuit (see Fig. 7.20) with a time constant of 8 μs. Use a single inductor and a network of resistors, if necessary. Draw your circuit.

b) Suppose the inductor you chose in part (a) has no initial stored energy. At $t = 0$, a switch connects a voltage source with a value of 25 V in series with the inductor and equivalent resistance. Write an expression for the current through the inductor for $t \geq 0$.

c) Using your result from part (b), calculate the time at which the current through the inductor reaches 75% of its final value.

7.44 The switch in the circuit in Fig. P7.44 has been open a long time before closing at $t = 0$. Find $v_o(t)$ for $t \geq 0^+$.

PSPICE
MULTISIM

Figure P7.44

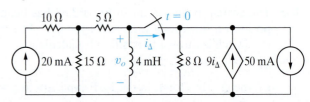

7.45 The switch in the circuit in Fig. P7.45 has been open a long time before closing at $t = 0$. Find $i_o(t)$ for $t \geq 0$.

PSPICE
MULTISIM

Figure P7.45

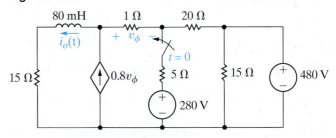

7.46 The switch in the circuit in Fig. P7.46 has been in position 1 for a long time. At $t = 0$ it moves instantaneously to position 2. How many milliseconds after the switch moves does v_o equal 100 V?

PSPICE
MULTISIM

Figure P7.46

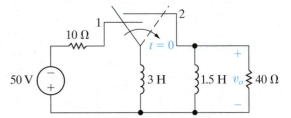

7.47 For the circuit in Fig. P7.46, find (in joules):

a) the total energy dissipated in the 40 Ω resistor,

b) the energy trapped in the inductors, and

c) the initial energy stored in the inductors.

7.48 The switch in the circuit in Fig. P7.48 has been open
a long time before closing at $t = 0$. Find $v_o(t)$ for
$t \geq 0^+$.

Figure P7.48

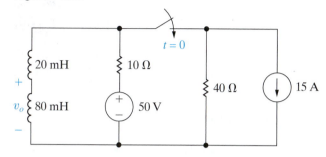

7.49 The make-before-break switch in the circuit of
Fig. P7.49 has been in position a for a long time.
At $t = 0$, the switch moves instantaneously to po-
sition b. Find

a) $v_o(t), t \geq 0^+$.

b) $i_1(t), t \geq 0$.

c) $i_2(t), t \geq 0$.

Figure P7.49

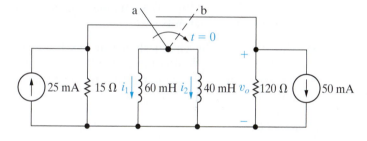

7.50 There is no energy stored in the inductors L_1 and L_2
at the time the switch is opened in the circuit shown
in Fig. P7.50.

a) Derive the expressions for the currents $i_1(t)$ and
$i_2(t)$ for $t \geq 0$.

b) Use the expressions derived in (a) to find $i_1(\infty)$
and $i_2(\infty)$.

Figure P7.50

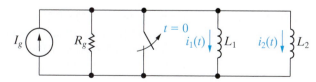

7.51 Assume that the switch in the circuit of Fig. P7.51
has been in position a for a long time and that at
$t = 0$ it is moved to position b. Find (a) $v_C(0^+)$;
(b) τ for $t > 0$; (c) $v_C(\infty)$; (d) $i(0^+)$; (e) $v_C, t \geq 0$;
and (f) $i, t \geq 0^+$.

Figure P7.51

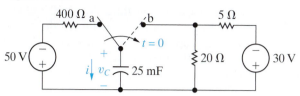

7.52 The switch in the circuit seen in Fig. P7.52 has been
in position a for a long time. At $t = 0$, the switch
moves instantaneously to position b. For $t \geq 0^+$,
find

a) $v_o(t)$.

b) $i_o(t)$.

Figure P7.52

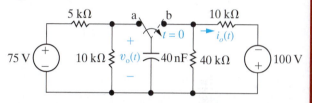

7.53 a) The switch in the circuit in Fig. P7.53 has been
in position a for a long time. At $t = 0$, the switch
moves instantaneously to position b and stays
there. Find the initial and final values of the capacitor voltage, the time constant for $t \geq 0$, and
the expression for the capacitor voltage for $t \geq 0$.

b) Now suppose the switch in the circuit in Fig.
P7.53 has been in position b for a long time. At
$t = 0$, the switch moves instantaneously to position a and stays there. Find the initial and final
values of the capacitor voltage, the time constant
for $t \geq 0$, and the expression for the capacitor
voltage for $t \geq 0$.

Figure P7.53

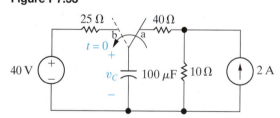

7.54 The circuit in Fig. P7.54 has been in operation for
a long time. At $t = 0$, the voltage source reverses
polarity and the current source drops from 3 mA to
2 mA. Find $v_o(t)$ for $t \geq 0$.

Figure P7.54

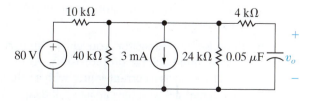

7.55 The switch in the circuit of Fig. P7.55 has been in position a for a long time. At $t = 0$ the switch is moved to position b. Calculate (a) the initial voltage on the capacitor; (b) the final voltage on the capacitor; (c) the time constant (in microseconds) for $t > 0$; and (d) the length of time (in microseconds) required for the capacitor voltage to reach zero after the switch is moved to position b.

Figure P7.55

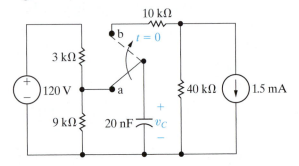

7.56 The switch in the circuit seen in Fig. P7.56 has been in position a for a long time. At $t = 0$, the switch moves instantaneously to position b. Find $v_o(t)$ and $i_o(t)$ for $t \geq 0^+$.

PSPICE
MULTISIM

Figure P7.56

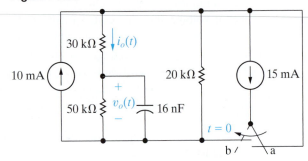

7.57 The current and voltage at the terminals of the capacitor in the circuit in Fig. 7.28 are

$$i(t) = 3e^{-2500t} \text{ mA}, \qquad t \geq 0^+;$$

$$v(t) = (40 - 24e^{-2500t}) \text{ V}, \quad t \geq 0.$$

a) Specify the numerical values of I_s, V_o, R, C, and τ.

b) How many microseconds after the switch has been closed does the energy stored in the capacitor reach 81% of its final value?

7.58 a) Use component values from Appendix H to create a first-order *RC* circuit (see Fig. 7.28) with a time constant of 250 ms. Use a single capacitor and a network of resistors, if necessary. Draw your circuit.

b) Suppose the capacitor you chose in part (a) has an initial voltage drop of 100 V. At $t = 0$, a switch connects a current source with a value of 1 mA in parallel with the capacitor and equivalent

resistance. Write an expression for the voltage drop across the capacitor for $t \geq 0$.

c) Using your result from part (b), calculate the time at which the voltage drop across the capicitor reaches 50 V.

7.59 The switch in the circuit in Fig. P7.59 has been in position a for a long time. At $t = 0$, the switch moves instantaneously to position b. At the instant the switch makes contact with terminal b, switch 2 opens. Find $v_o(t)$ for $t \geq 0$.

PSPICE
MULTISIM

Figure P7.59

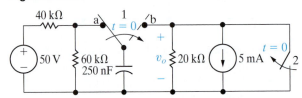

7.60 a) Derive Eq. 7.17 by first converting the Norton equivalent circuit shown in Fig. 7.28 to a Thévenin equivalent and then summing the voltages around the closed loop, using the capacitor current i as the relevant variable.

b) Use the separation of variables technique to find the solution to Eq. 7.17. Verify that your solution agrees with that of Eq. 7.22.

7.61 The switch in the circuit shown in Fig. P7.61 has been in the OFF position for a long time. At $t = 0$, the switch moves instantaneously to the ON position. Find $v_o(t)$ for $t \geq 0$.

PSPICE
MULTISIM

Figure P7.61

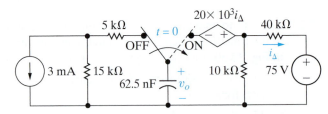

7.62 Assume that the switch in the circuit of Fig. P7.61 has been in the ON position for a long time before switching instantaneously to the OFF position at $t = 0$. Find $v_o(t)$ for $t \geq 0$.

PSPICE
MULTISIM

7.63 The switch in the circuit shown in Fig. P7.63 opens at $t = 0$ after being closed for a long time. How many milliseconds after the switch opens is the energy stored in the capacitor 36% of its final value?

PSPICE
MULTISIM

Figure P7.63

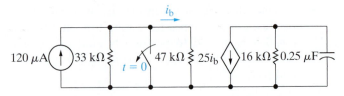

7.64 The switch in the circuit of Fig. P7.64 has been in
PSPICE position a for a long time. At $t = 0$, it moves instan-
MULTISIM taneously to position b. For $t \geq 0^+$, find

a) $v_o(t)$.

b) $i_o(t)$.

c) $v_1(t)$.

d) $v_2(t)$.

e) the energy trapped in the capacitors as $t \to \infty$.

Figure P7.64

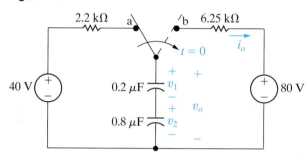

7.65 The switch in the circuit in Fig. P7.65 has been in
position x for a long time. The initial charge on the
60 nF capacitor is zero. At $t = 0$, the switch moves
instantaneously to position y.

a) Find $v_o(t)$ for $t \geq 0^+$.

b) Find $v_1(t)$ for $t \geq 0$.

Figure P7.65

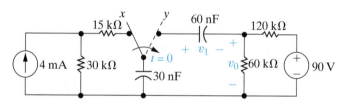

7.66 There is no energy stored in the capacitors C_1 and
C_2 at the time the switch is closed in the circuit seen
in Fig. P7.66.

a) Derive the expressions for $v_1(t)$ and $v_2(t)$ for
$t \geq 0$.

b) Use the expressions derived in (a) to find $v_1(\infty)$
and $v_2(\infty)$.

Figure P7.66

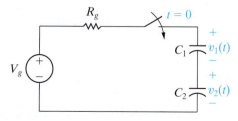

Section 7.4

7.67 Repeat (a) and (b) in Example 7.10 if the mutual
inductance is reduced to zero.

7.68 There is no energy stored in the circuit of Fig. P7.68
at the time the switch is closed.

a) Find $i_o(t)$ for $t \geq 0$.

b) Find $v_o(t)$ for $t \geq 0^+$.

c) Find $i_1(t)$ for $t \geq 0$.

d) Find $i_2(t)$ for $t \geq 0$.

e) Do your answers make sense in terms of known
circuit behavior?

Figure P7.68

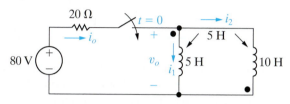

7.69 There is no energy stored in the circuit in Fig. P7.69
PSPICE at the time the switch is closed.
MULTISIM

a) Find $i_o(t)$ for $t \geq 0$.

b) Find $v_o(t)$ for $t \geq 0^+$.

c) Find $i_1(t)$ for $t \geq 0$.

d) Find $i_2(t)$ for $t \geq 0$.

e) Do your answers make sense in terms of known
circuit behavior?

Figure P7.69

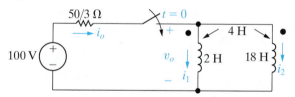

7.70 There is no energy stored in the circuit in Fig. P7.70
PSPICE at the time the switch is closed.
MULTISIM

a) Find $i(t)$ for $t \geq 0$.

b) Find $v_1(t)$ for $t \geq 0^+$.

c) Find $v_2(t)$ for $t \geq 0$.

d) Do your answers make sense in terms of known
circuit behavior?

Figure P7.70

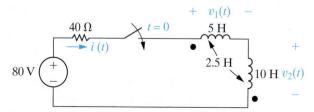

7.71 Repeat Problem 70 if the dot on the 10 H coil is at
PSPICE the top of the coil.
MULTISIM

Section 7.5

7.72 In the circuit in Fig. P7.72, switch A has been open and switch B has been closed for a long time. At $t = 0$, switch A closes. Five seconds after switch A closes, switch B opens. Find $i_L(t)$ for $t \geq 0$.

PSPICE
MULTISIM

Figure P7.72

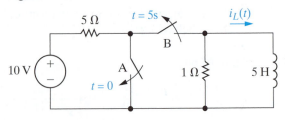

7.73 The action of the two switches in the circuit seen in Fig. P7.73 is as follows. For $t < 0$, switch 1 is in position a and switch 2 is open. This state has existed for a long time. At $t = 0$, switch 1 moves instantaneously from position a to position b, while switch 2 remains open. Ten milliseconds after switch 1 operates, switch 2 closes, remains closed for 10 ms and then opens. Find v_o 25 ms after switch 1 moves to position b.

PSPICE
MULTISIM

Figure P7.73

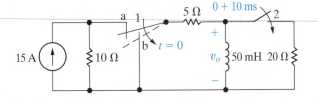

7.74 For the circuit in Fig. P7.73, how many milliseconds after switch 1 moves to position b is the energy stored in the inductor 4% of its initial value?

7.75 The switch in the circuit shown in Fig. P7.75 has been in position a for a long time. At $t = 0$, the switch is moved to position b, where it remains for 1 ms. The switch is then moved to position c, where it remains indefinitely. Find

PSPICE
MULTISIM

a) $i(0^+)$.

b) $i(200 \ \mu s)$.

c) $i(6 \ \text{ms})$.

d) $v(1^{-1} \ \text{ms})$.

e) $v(1^+ \ \text{ms})$.

Figure P7.75

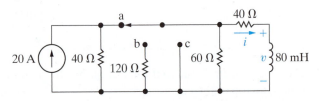

7.76 There is no energy stored in the capacitor in the circuit in Fig. P7.76 when switch 1 closes at $t = 0$. Switch 2 closes 10 microseconds later. Find $v_o(t)$ for $t \geq 0$.

PSPICE
MULTISIM

Figure P7.76

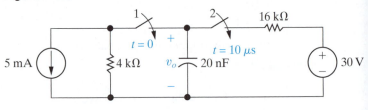

7.77 The capacitor in the circuit seen in Fig. P7.77 has been charged to 300 V. At $t = 0$, switch 1 closes, causing the capacitor to discharge into the resistive network. Switch 2 closes 200 μs after switch 1 closes. Find the magnitude and direction of the current in the second switch 300 μs after switch 1 closes.

PSPICE
MULTISIM

Figure P7.77

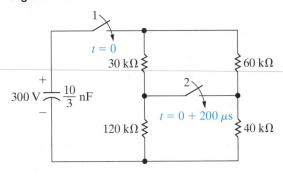

7.78 The switch in the circuit in Fig. P7.78 has been in position a for a long time. At $t = 0$, it moves instantaneously to position b, where it remains for 50 ms before moving instantaneously to position c. Find v_o for $t \geq 0$.

PSPICE
MULTISIM

Figure P7.78

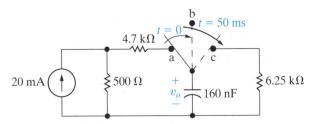

7.79 In the circuit in Fig. P7.79, switch 1 has been in position a and switch 2 has been closed for a long time. At $t = 0$, switch 1 moves instantaneously to position b. Eight hundred microseconds later, switch 2 opens, remains open for 300 μs, and then recloses. Find v_o 1.5 ms after switch 1 makes contact with terminal b.

Figure P7.79

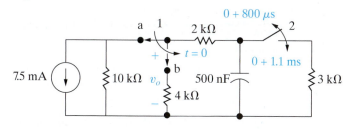

7.80 For the circuit in Fig. P7.79, what percentage of the
initial energy stored in the 500 nF capacitor is dissipated in the 3 kΩ resistor?

PSPICE
MULTISIM

7.81 The current source in the circuit in Fig. P7.81(a)
generates the current pulse shown in Fig. P7.81(b).
There is no energy stored at $t = 0$.

PSPICE
MULTISIM

   a) Derive the numerical expressions for $v_o(t)$ for
     the time intervals $t < 0, 0 \le t \le 75 \ \mu s$, and
     $75 \ \mu s \le t < \infty$.

   b) Calculate $v_o(75^- \ \mu s)$ and $v_o(75^+ \ \mu s)$.

   c) Calculate $i_o(75^- \ \mu s)$ and $i_o(75^+ \ \mu s)$.

Figure P7.81

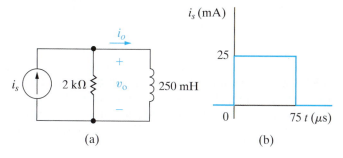

(a)   (b)

7.82 The voltage waveform shown in Fig. P7.82(a) is
applied to the circuit of Fig. P7.82(b). The initial
current in the inductor is zero.

PSPICE
MULTISIM

   a) Calculate $v_o(t)$.

   b) Make a sketch of $v_o(t)$ versus t.

   c) Find i_o at $t = 5$ ms.

Figure P7.82

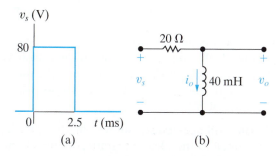

(a)   (b)

7.83 The voltage signal source in the circuit in
Fig. P7.83(a) is generating the signal shown in
Fig. P7.83(b). There is no stored energy at $t = 0$.

PSPICE
MULTISIM

   a) Derive the expressions for $v_o(t)$ that apply in the
     intervals $t < 0; 0 \le t \le 4$ ms; 4 ms $\le t \le 8$ ms;
     and 8 ms $\le t < \infty$.

   b) Sketch v_o and v_s on the same coordinate axes.

   c) Repeat (a) and (b) with R reduced to 50 kΩ.

Figure P7.83

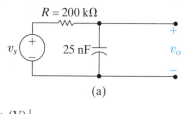

(a)

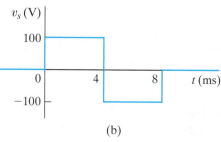

(b)

7.84 The voltage waveform shown in Fig. P7.84(a) is
applied to the circuit of Fig. P7.84(b). The initial
voltage on the capacitor is zero.

PSPICE
MULTISIM

   a) Calculate $v_o(t)$.

   b) Make a sketch of $v_o(t)$ versus t.

Figure P7.84

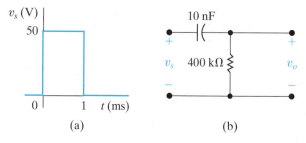

(a)   (b)

Section 7.6

7.85 The inductor current in the circuit in Fig. P7.85 is
25 mA at the instant the switch is opened. The
inductor will malfunction whenever the magnitude
of the inductor current equals or exceeds 5 A. How
long after the switch is opened does the inductor
malfunction?

PSPICE
MULTISIM

Figure P7.85

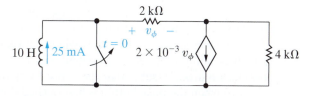

7.86 The capacitor in the circuit shown in Fig. P7.86 is charged to 20 V at the time the switch is closed. If the capacitor ruptures when its terminal voltage equals or exceeds 20 kV, how long does it take to rupture the capacitor?

Figure P7.86

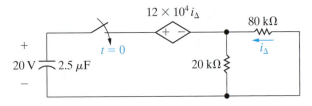

7.87 The switch in the circuit in Fig. P7.87 has been closed for a long time. The maximum voltage rating of the 1.6 μF capacitor is 14.4 kV. How long after the switch is opened does the voltage across the capacitor reach the maximum voltage rating?

Figure P7.87

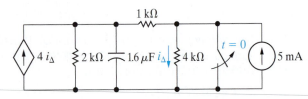

7.88 The gap in the circuit seen in Fig. P7.88 will arc over whenever the voltage across the gap reaches 45 kV. The initial current in the inductor is zero. The value of β is adjusted so the Thévenin resistance with respect to the terminals of the inductor is -5 kΩ.

a) What is the value of β?

b) How many microseconds after the switch has been closed will the gap arc over?

Figure P7.88

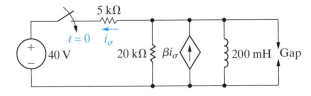

7.89 The circuit shown in Fig. P7.89 is used to close the switch between a and b for a predetermined length of time. The electric relay holds its contact arms down as long as the voltage across the relay coil exceeds 5 V. When the coil voltage equals 5 V, the relay contacts return to their initial position by a mechanical spring action. The switch between a and b is initially closed by momentarily pressing the push button. Assume that the capacitor is fully charged when the push button is first pushed down.

The resistance of the relay coil is 25 kΩ, and the inductance of the coil is negligible.

a) How long will the switch between a and b remain closed?

b) Write the numerical expression for *i* from the time the relay contacts first open to the time the capacitor is completely charged.

c) How many milliseconds (after the circuit between a and b is interrupted) does it take the capacitor to reach 85% of its final value?

Figure P7.89

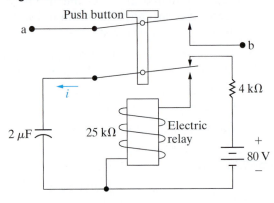

Section 7.7

7.90 The energy stored in the capacitor in the circuit shown in Fig. P7.90 is zero at the instant the switch is closed. The ideal operational amplifier reaches saturation in 15 ms. What is the numerical value of *R*?

Figure P7.90

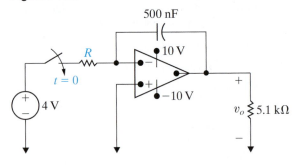

7.91 At the instant the switch is closed in the circuit of Fig. P7.90, the capacitor is charged to 6 V, positive at the right-hand terminal. If the ideal operational amplifier saturates in 40 ms, what is the value of *R*?

7.92 The voltage pulse shown in Fig. P7.92(a) is applied to the ideal integrating amplifier shown in Fig. P7.92(b). Derive the numerical expression for $v_o(t)$ when $v_o(0) = 0$ for the time intervals

a) $t < 0$.

b) $0 \le t \le 250$ ms.

c) 250 ms $\le t \le 500$ ms.

d) 500 ms $\le t$.

Figure P7.92

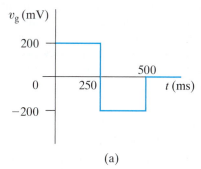

(a)

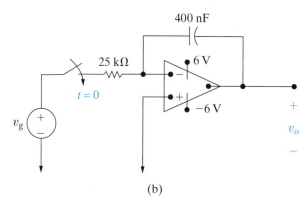

(b)

7.93 Repeat Problem 7.92 with a 5 MΩ resistor placed
PSPICE across the 400 nF feedback capacitor.
MULTISIM

7.94 There is no energy stored in the capacitors in the
PSPICE circuit shown in Fig. P7.94 at the instant the two
MULTISIM switches close. Assume the op amp is ideal.

a) Find v_o as a function of v_a, v_b, R, and C.

b) On the basis of the result obtained in (a), describe the operation of the circuit.

c) How long will it take to saturate the amplifier if $v_a = 40$ mV; $v_b = 15$ mV; $R = 50$ kΩ;
$C = 10$ nF; and $V_{CC} = 6$ V?

Figure P7.94

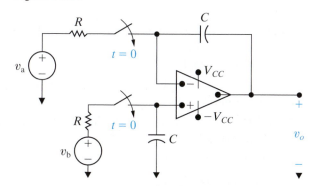

7.95 At the instant the switch of Fig. P7.95 is closed, the
PSPICE voltage on the capacitor is 56 V. Assume an ideal
MULTISIM op amp. How many milliseconds after the switch is
closed will the output voltage v_o equal zero?

Figure P7.95

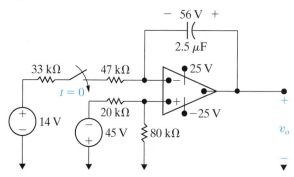

7.96 The voltage source in the circuit in Fig. P7.96(a)
PSPICE is generating the triangular waveform shown in
MULTISIM Fig. P7.96(b). Assume the energy stored in the
capacitor is zero at $t = 0$ and the op amp is ideal.

a) Derive the numerical expressions for $v_o(t)$
for the following time intervals: $0 \le t \le 1$ μs;
1 μs $\le t \le 3$ μs; and 3 μs $\le t \le 4$ μs.

b) Sketch the output waveform between 0 and 4 μs.

c) If the triangular input voltage continues to repeat itself for $t > 4$ μs, what would you expect
the output voltage to be? Explain.

Figure P7.96

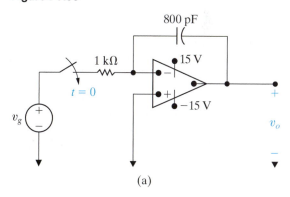

(a)

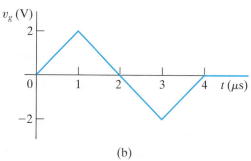

(b)

Sections 7.1–7.7

7.97 The circuit shown in Fig. P7.97 is known as a *monostable multivibrator*. The adjective *monostable* is used to describe the fact that the circuit has one stable state. That is, if left alone, the electronic switch T_2 will be on, and T_1 will be off. (The operation of the ideal transistor switch is described in detail in Problem 7.99.) T_2 can be turned off by momentarily closing the switch S. After S returns to its open position, T_2 will return to its on state.

a) Show that if T_2 is on, T_1 is off and will stay off.

b) Explain why T_2 is turned off when S is momentarily closed.

c) Show that T_2 will stay off for $RC \ln 2$ s.

Figure P7.97

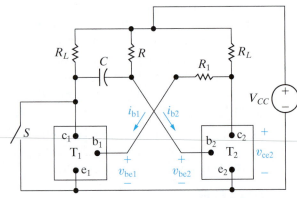

7.98 The parameter values in the circuit in Fig. P7.97 are $V_{CC} = 6$ V; $R_1 = 5.0$ kΩ; $R_L = 20$ kΩ; $C = 250$ pF; and $R = 23,083$ Ω.

a) Sketch v_{ce2} versus t, assuming that after S is momentarily closed, it remains open until the circuit has reached its stable state. Assume S is closed at $t = 0$. Make your sketch for the interval $-5 \le t \le 10$ μs.

b) Repeat (a) for i_{b2} versus t.

7.99 The circuit shown in Fig. P7.99 is known as an *astable multivibrator* and finds wide application in pulse circuits. The purpose of this problem is to relate the charging and discharging of the capacitors to the operation of the circuit. The key to analyzing the circuit is to understand the behavior of the ideal transistor switches T_1 and T_2. The circuit is designed so that the switches automatically alternate between on and off. When T_1 is off, T_2 is on and vice versa. Thus in the analysis of this circuit, we assume a switch is either on or off. We also assume that the ideal transistor switch can change its state instantaneously. In other words, it can snap from off to on and vice versa. When a transistor switch is on, (1) the base current i_b is greater than

zero, (2) the terminal voltage v_{be} is zero, and (3) the terminal voltage v_{ce} is zero. Thus, when a transistor switch is on, it presents a short circuit between the terminals b,e and c,e. When a transistor switch is off, (1) the terminal voltage v_{be} is negative, (2) the base current is zero, and (3) there is an open circuit between the terminals c,e. Thus when a transistor switch is off, it presents an open circuit between the terminals b,e and c,e. Assume that T_2 has been on and has just snapped off, while T_1 has been off and has just snapped on. You may assume that at this instance, C_2 is charged to the supply voltage V_{CC}, and the charge on C_1 is zero. Also assume $C_1 = C_2$ and $R_1 = R_2 = 10R_L$.

a) Derive the expression for v_{be2} during the interval that T_2 is off.

b) Derive the expression for v_{ce2} during the interval that T_2 is off.

c) Find the length of time T_2 is off.

d) Find the value of v_{ce2} at the end of the interval that T_2 is off.

e) Derive the expression for i_{b1} during the interval that T_2 is off.

f) Find the value of i_{b1} at the end of the interval that T_2 is off.

g) Sketch v_{ce2} versus t during the interval that T_2 is off.

h) Sketch i_{b1} versus t during the interval that T_2 is off.

Figure P7.99

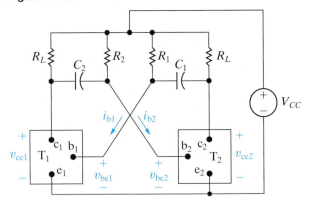

7.100 The component values in the circuit of Fig. P7.99 are $V_{CC} = 10$ V; $R_L = 1$ kΩ; $C_1 = C_2 = 1$ nF; and $R_1 = R_2 = 14.43$ kΩ.

a) How long is T_2 in the off state during one cycle of operation?

b) How long is T_2 in the on state during one cycle of operation?

c) Repeat (a) for T_1.

d) Repeat (b) for T_1.

e) At the first instant after T_1 turns ON, what is the value of i_{b1}?

f) At the instant just before T_1 turns OFF, what is the value of i_{b1}?

g) What is the value of v_{ce2} at the instant just before T_2 turns ON?

7.101 Repeat Problem 7.100 with $C_2 = 0.8$ nF. All other component values are unchanged.

7.102 The astable multivibrator circuit in Fig. P7.99 is to satisfy the following criteria: (1) One transistor switch is to be ON for 48 μs and OFF for 36 μs for each cycle; (2) $R_L = 2$ kΩ; (3) $V_{CC} = 5$ V; (4) $R_1 = R_2$; and (5) $6R_L \leq R_1 \leq 50 R_L$. What are the limiting values for the capacitors C_1 and C_2?

7.103 The relay shown in Fig. P7.103 connects the 30 V dc generator to the dc bus as long as the relay current is greater than 0.4 A. If the relay current drops to 0.4 A or less, the spring-loaded relay immediately connects the dc bus to the 30 V standby battery. The resistance of the relay winding is 60 Ω. The inductance of the relay winding is to be determined.

a) Assume the prime motor driving the 30 V dc generator abruptly slows down, causing the generated voltage to drop suddenly to 21 V. What value of L will assure that the standby battery will be connected to the dc bus in 0.5 seconds?

b) Using the value of L determined in (a), state how long it will take the relay to operate if the generated voltage suddenly drops to zero.

Figure P7.103

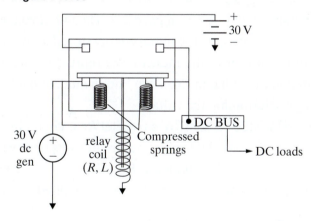

7.104 Derive the expression for heart rate in beats per minute given the values of R and C and assuming that the capacitor discharges when its voltage reaches 75% of the source voltage V_s. The expression, given in the Practical Perspective, is repeated here for convenience:

$$H = \frac{60}{-RC \ln 0.25} \quad \text{[beats per minute]}.$$

7.105 Use an expression similar to the one derived in Problem 7.104 to calculate the heart rate in beats per minute for $R = 150$ kΩ, $C = 6$ μF, if the capacitor discharges when its voltage reaches 60% of the source voltage V_s.

7.106 Show that the resistance required to achieve a heart rate H, in beats per minute, is given by the equation

$$R = \frac{-60}{HC \ln\left(1 - \dfrac{V_{max}}{V_s}\right)},$$

where C is the capacitance, V_s is the source voltage, and V_{max} is the capacitor voltage at which discharge occurs.

7.107 Use the expression derived in Problem 7.106 to calculate the resistance required to achieve a heart rate of 70 beats per minute using a capacitance of 2.5 μF and assuming that the capacitor discharges when its voltage reaches 68% of the source voltage.

Natural and Step Responses of *RLC* Circuits

◻️ CHAPTER CONTENTS

8.1 **Introduction to the Natural Response of a Parallel *RLC* Circuit** *p. 274*

8.2 **The Forms of the Natural Response of a Parallel *RLC* Circuit** *p. 278*

8.3 **The Step Response of a Parallel *RLC* Circuit** *p. 289*

8.4 **The Natural and Step Response of a Series *RLC* Circuit** *p. 296*

8.5 **A Circuit with Two Integrating Amplifiers** *p. 303*

◻️ CHAPTER OBJECTIVES

1 Be able to determine the natural response and the step response of parallel *RLC* circuits.

2 Be able to determine the natural response and the step response of series *RLC* circuits.

In this chapter, we discuss the natural response and step response of circuits containing a resistor, an inductor, and a capacitor, known as *RLC* circuits. We limit our analysis to two simple structures: the parallel *RLC* circuit and the series *RLC* circuit.

We begin with the natural response of a parallel *RLC* circuit and cover this material in two sections: one section discusses the solution of the second-order differential equation that describes the circuit, and the other presents the three distinct forms that the solution can take. After introducing these three forms, we show that the same forms apply to the step response of a parallel *RLC* circuit as well as to the natural and step responses of series *RLC* circuits. The chapter concludes with an introduction to an op-amp-based circuit whose output is also characterized by a second-order differential equation.

Parallel *RLC* Circuits

We characterize the natural response of a parallel *RLC* circuit by finding the voltage across the parallel branches created by the release of energy stored in the inductor or capacitor, or both. The circuit is shown in Fig. 8.1 on page 274. The initial voltage on the capacitor, V_0, represents the initial energy stored in the capacitor. The initial current in the inductor, I_0, represents the initial energy stored in the inductor. You can find the individual branch currents after determining the voltage.

We derive the step response of a parallel *RLC* circuit by using Fig. 8.2 on page 274. We determine the circuit's response when a dc current source is applied suddenly. Energy may or may not be stored in the circuit when the current source is applied.

Series *RLC* Circuits

We characterize the natural response of a series *RLC* circuit by finding the current generated in the series-connected elements by the release of initially stored energy in the inductor, capacitor, or both. The circuit is shown in Fig. 8.3 on page 274. As before, the initial inductor current, I_0, and the initial capacitor voltage, V_0, represent the initially stored energy. You can find the individual element voltages after determining the current.

■ Practical Perspective

Clock for Computer Timing

The digital circuits found in most computers require a timing signal that synchronizes the operation of the circuits. Consider a laptop computer whose processor speed is 2 GHz. This means that the central processing unit for this computer can perform about 2×10^9 simple operations every second.

The timing signal, produced by a clock generator chip, is typically a square wave with the required clock frequency. The square wave is obtained from a sinusoidal wave with the required clock frequency. Typically, the sinusoidal wave is generated by a precisely cut quartz crystal with an applied voltage. The crystal produces a stable frequency suitable for synchronizing digital circuits.

We can also generate a sinusoidal wave using a circuit with an inductor and a capacitor. By choosing the values of inductance and capacitance, we can create a sinusoid with a specific frequency. We will examine such a design once we have presented the fundamental concepts of second-order circuits.

Scanrail/fotolia

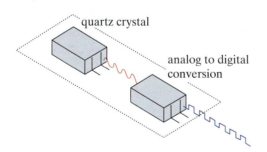

David J. Green/Alamy Stock Photo

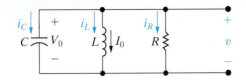

Figure 8.1 ▲ A circuit used to illustrate the natural response of a parallel *RLC* circuit.

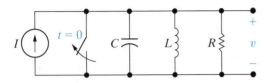

Figure 8.2 ▲ A circuit used to illustrate the step response of a parallel *RLC* circuit.

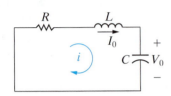

Figure 8.3 ▲ A circuit used to illustrate the natural response of a series *RLC* circuit.

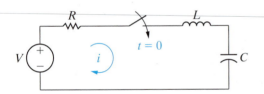

Figure 8.4 ▲ A circuit used to illustrate the step response of a series *RLC* circuit.

We describe the step response of a series *RLC* circuit using the circuit shown in Fig. 8.4. We determine the circuit's response to the sudden application of the dc voltage source. Energy may or may not be stored in the circuit when the switch is closed.

8.1 Introduction to the Natural Response of a Parallel *RLC* Circuit

To find the natural response of the circuit shown in Fig. 8.1, we begin by deriving the differential equation that the voltage v satisfies. We choose to find the voltage because it is the same for each component. Once we know the voltage, we can find every branch current by using the current–voltage relationship for the branch component. We write the differential equation for the voltage using KCL to sum the currents leaving the top node, where each current is expressed as a function of the unknown voltage v:

$$\frac{v}{R} + \frac{1}{L} \int_0^t v \, d\tau + I_0 + C\frac{dv}{dt} = 0.$$

We eliminate the integral in the KCL equation by differentiating once with respect to t, and because I_0 is a constant, we get

$$\frac{1}{R}\frac{dv}{dt} + \frac{v}{L} + C\frac{d^2v}{dt^2} = 0.$$

We now divide the equation by the capacitance C and arrange the derivatives in descending order:

$$\frac{d^2v}{dt^2} + \frac{1}{RC}\frac{dv}{dt} + \frac{v}{LC} = 0. \tag{8.1}$$

Equation 8.1 is an ordinary, second-order differential equation with constant coefficients because it describes a circuit with both an inductor and a capacitor. Therefore, we also call *RLC* circuits **second-order circuits**.

The General Solution of the Second-Order Differential Equation

We can't solve Eq. 8.1 by separating the variables and integrating, as we were able to do with the first-order equations in Chapter 7. Instead, we solve Eq. 8.1 by assuming that the voltage is of the form

$$v = Ae^{st}, \tag{8.2}$$

where A and s are unknown constants.

Why did we choose an exponential form for v, given in Eq. 8.2? The reason is that Eq. 8.1 requires the sum of the second derivative of v, and the first derivative of v times a constant, and v times a constant equal zero for all values of t. This can occur only if higher-order derivatives of v have the same form as v. The exponential function satisfies this criterion. Furthermore, note that the solutions of the first-order equations we derived in Chapter 7 were all exponential; thus, it seems reasonable to assume that the solution of the second-order equation is also exponential.

If Eq. 8.2 is a solution of Eq. 8.1, it must satisfy Eq. 8.1 for all values of t. Substituting Eq. 8.2 into Eq. 8.1 generates the expression

$$As^2 e^{st} + \frac{As}{RC} e^{st} + \frac{Ae^{st}}{LC} = 0,$$

or

$$Ae^{st}\left(s^2 + \frac{s}{RC} + \frac{1}{LC}\right) = 0,$$

which can be satisfied for all values of t only if A is zero or the parenthetical term is zero because $e^{st} \neq 0$ for any finite values of st. We cannot use $A = 0$ as a general solution because to do so implies that the voltage is zero for all time—a physical impossibility if energy is stored in either the inductor or capacitor. Therefore, in order for Eq. 8.2 to be a solution of Eq. 8.1, the parenthetical term must be zero, or

CHARACTERISTIC EQUATION, PARALLEL *RLC* CIRCUIT

$$s^2 + \frac{s}{RC} + \frac{1}{LC} = 0. \tag{8.3}$$

Equation 8.3 is called the **characteristic equation** of the differential equation because the roots of this quadratic equation determine the mathematical character of $v(t)$.

The two roots of Eq. 8.3 are

$$s_1 = -\frac{1}{2RC} + \sqrt{\left(\frac{1}{2RC}\right)^2 - \frac{1}{LC}},$$

$$s_2 = -\frac{1}{2RC} - \sqrt{\left(\frac{1}{2RC}\right)^2 - \frac{1}{LC}}.$$

If either root is substituted into Eq. 8.2, v satisfies the differential equation in Eq. 8.1, regardless of the value of A. Therefore, both

$$v = A_1 e^{s_1 t} \text{ and}$$

$$v = A_2 e^{s_2 t}$$

satisfy Eq. 8.1. Denoting these two solutions v_1 and v_2, respectively, we can show that their sum also is a solution. Specifically, if we let

$$v = v_1 + v_2 = A_1 e^{s_1 t} + A_2 e^{s_2 t},$$

then

$$\frac{dv}{dt} = A_1 s_1 e^{s_1 t} + A_2 s_2 e^{s_2 t},$$

$$\frac{d^2 v}{dt^2} = A_1 s_1^2 e^{s_1 t} + A_2 s_2^2 e^{s_2 t}.$$

Substituting these expressions for v and its first and second derivatives into Eq. 8.1 gives

$$A_1 e^{s_1 t}\left(s_1^2 + \frac{1}{RC}s_1 + \frac{1}{LC}\right) + A_2 e^{s_2 t}\left(s_2^2 + \frac{1}{RC}s_2 + \frac{1}{LC}\right) = 0.$$

But each parenthetical term is zero because by definition s_1 and s_2 are roots of the characteristic equation. Hence, the natural response of the parallel *RLC* circuit shown in Fig. 8.1 is

$$v = A_1 e^{s_1 t} + A_2 e^{s_2 t}. \tag{8.4}$$

In Eq. 8.4, the constants s_1 and s_2, which are the roots of the characteristic equation, are determined by the circuit parameters R, L, and C. The constants A_1 and A_2 are determined by the initial conditions for the inductor and the capacitor.

To find the natural response (Eq. 8.4), we begin by finding the roots of the characteristic equation, s_1 and s_2, which we first wrote in terms of the circuit parameters. We now rewrite them as follows:

$$s_1 = -\alpha + \sqrt{\alpha^2 - \omega_0^2}, \tag{8.5}$$

$$s_2 = -\alpha - \sqrt{\alpha^2 - \omega_0^2}, \tag{8.6}$$

where

NEPER FREQUENCY, PARALLEL *RLC* CIRCUIT

$$\alpha = \frac{1}{2RC}, \tag{8.7}$$

RESONANT RADIAN FREQUENCY, PARALLEL *RLC* CIRCUIT

$$\omega_0 = \frac{1}{\sqrt{LC}}. \tag{8.8}$$

These results are summarized in Table 8.1.

TABLE 8.1   Natural-Response Parameters of the Parallel *RLC* Circuit

Parameter	Terminology	Value in Natural Response
s_1, s_2	Characteristic roots	$s_1 = -\alpha + \sqrt{\alpha^2 - \omega_0^2}$ $s_2 = -\alpha - \sqrt{\alpha^2 - \omega_0^2}$
α	Neper frequency	$\alpha = \dfrac{1}{2RC}$
ω_0	Resonant radian frequency	$\omega_0 = \dfrac{1}{\sqrt{LC}}$

The exponent of e must be dimensionless, so both s_1 and s_2 (and hence α and ω_0) must have the dimension of the inverse of time, or frequency. To distinguish among the frequencies s_1, s_2, α, and ω_0, we use the following terminology: s_1 and s_2 are the *complex frequencies*, α is the *neper frequency*, and ω_0 is the *resonant radian frequency*. The full significance of this terminology unfolds as we move through the remaining chapters of this book. These frequencies all have the dimension of angular frequency per time. For the complex frequencies, the neper frequency, and the resonant radian frequency, we specify values using the unit *radians per second* (rad/s).

The form of the roots s_1 and s_2 depends on the values of α and ω_0. There are three possibilities.

- If $\omega_0^2 < \alpha^2$, both roots will be real and distinct. For reasons to be discussed later, we call the voltage response **overdamped**.
- If $\omega_0^2 > \alpha^2$, both s_1 and s_2 will be complex and, in addition, will be conjugates of each other. In this situation, we call the voltage response **underdamped**.
- If $\omega_0^2 = \alpha^2$, s_1 and s_2 will be real and equal. Here, we call the voltage response **critically damped**.

As we shall see, damping affects the way the voltage response reaches its final (or steady-state) value. We discuss each case separately in Section 8.2.

Example 8.1 illustrates how the values of R, L, and C determine the numerical values of s_1 and s_2.

EXAMPLE 8.1   **Finding the Roots of the Characteristic Equation of a Parallel *RLC* Circuit**

a) Find the roots of the characteristic equation that governs the transient behavior of the voltage shown in Fig. 8.5 if $R = 200\ \Omega$, $L = 50$ mH, and $C = 0.2\ \mu$F.

b) Will the response be overdamped, underdamped, or critically damped?

c) Repeat (a) and (b) for $R = 312.5\ \Omega$.

d) What value of R causes the response to be critically damped?

Solution

a) For the given values of R, L, and C,

$$\alpha = \frac{1}{2RC} = \frac{1}{2(200)(0.2 \times 10^{-6})} = 12{,}500\ \text{rad/s},$$

$$\omega_0^2 = \frac{1}{LC} = \frac{1}{(50 \times 10^{-3})(0.2 \times 10^{-6})} = 10^8\ \text{rad}^2/\text{s}^2.$$

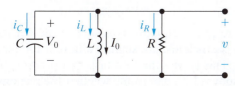

Figure 8.5 ▲ A circuit used to illustrate the natural response of a parallel *RLC* circuit.

From Eqs. 8.5 and 8.6,

$$s_1 = -12{,}500 + \sqrt{(12{,}500)^2 - 10^8}$$

$$= -12{,}500 + 7500 = -5000\ \text{rad/s},$$

$$s_1 = -12{,}500 - \sqrt{(12{,}500)^2 - 10^8}$$

$$= -12{,}500 - 7500 = -20{,}000\ \text{rad/s}.$$

b) The voltage response is overdamped because $\omega_0^2 < \alpha^2$.

c) For $R = 312.5\ \Omega$,

$$\alpha = \frac{1}{2RC} = \frac{1}{2(312.5)(0.2 \times 10^{-6})} = 8000\ \text{rad/s}.$$

Since ω_0^2 remains at $10^8\ \text{rad}^2/\text{s}^2$,

$$s_1 = -8000 + \sqrt{(8000)^2 - 10^8}$$

$$= -8000 + j6000\ \text{rad/s},$$

$$s_2 = -8000 - \sqrt{(8000)^2 - 10^8}$$

$$= -8000 - j6000\ \text{rad/s}.$$

(In electrical engineering, the imaginary number $\sqrt{-1}$ is represented by the letter j because the letter i represents current.)

In this case, the voltage response is underdamped since $\omega_0^2 > \alpha^2$.

d) For critical damping, $\alpha^2 = \omega_0^2$, so

$$\left(\frac{1}{2RC}\right)^2 = \frac{1}{LC} = 10^8,$$

or

$$\frac{1}{2RC} = 10^4,$$

and

$$R = \frac{1}{2(10^4)(0.2 \times 10^{-6})} = 250 \ \Omega.$$

ASSESSMENT PROBLEM

Objective 1—Be able to determine the natural response and the step response of parallel *RLC* circuits

8.1 The resistance and inductance of the circuit in Fig. 8.5 are 100 Ω and 20 mH, respectively.

a) Find the value of C that makes the voltage response critically damped.

b) If C is adjusted to give a neper frequency of 5 krad/s, find the value of C and the roots of the characteristic equation.

c) If C is adjusted to give a resonant frequency of 20 krad/s, find the value of C and the roots of the characteristic equation.

Answer: (a) 500 nF;

(b) $C = 1 \ \mu\text{F}$,
$s_1 = -5000 + j5000 \ \text{rad/s}$,
$s_2 = -5000 - j5000 \ \text{rad/s}$;

(c) $C = 125 \ \text{nF}$,
$s_1 = -5359 \ \text{rad/s}$,
$s_2 = -74{,}641 \ \text{rad/s}$.

SELF-CHECK: Also try Chapter Problem 8.3.

8.2 The Forms of the Natural Response of a Parallel *RLC* Circuit

In this section, we find the natural response for each of the three types of damping: overdamped, underdamped, and critically damped. As we have already seen, the values of s_1 and s_2 determine the type of damping. We need to find values for the two coefficients A_1 and A_2 so that we can completely characterize the natural response given in Eq. 8.4. This requires two equations based on the following observations:

- The initial value of the voltage in Eq. 8.4 must be the same as the initial value of the voltage in the circuit.
- The initial value of the first derivative of the voltage in Eq. 8.4 must be the same as the initial value of the first derivative of the voltage in the circuit.

As we will see, the natural-response equations, as well as the equations for evaluating the unknown coefficients, are slightly different for each of the three types of damping. This is why the first task that presents itself when finding the natural response is to determine whether the response is overdamped, underdamped, or critically damped.

The Overdamped Response

When the roots of the characteristic equation are real and distinct, the response of a parallel *RLC* circuit is overdamped. The solution for the voltage is

> **PARALLEL *RLC* NATURAL RESPONSE: OVERDAMPED**
>
> $$v = A_1 e^{s_1 t} + A_2 e^{s_2 t}, \tag{8.9}$$

where s_1 and s_2 are the roots of the characteristic equation. The constants A_1 and A_2 are determined by the initial conditions, specifically from the values of $v(0^+)$ and $dv(0^+)/dt$, which in turn are determined from the initial voltage on the capacitor, V_0, and the initial current in the inductor, I_0.

To determine the values of A_1 and A_2, we need two independent equations. The first equation sets the initial value v from Eq. 8.9 equal to the initial value of v in the circuit, which is the initial voltage for the capacitor, V_0. The resulting equation is

$$v(0^+) = A_1 + A_2 = V_0. \tag{8.10}$$

The second equation sets the initial value of dv/dt from Eq. 8.9 equal to the initial value of dv/dt in the circuit. The initial value of dv/dt from Eq. 8.9 is

$$\frac{dv(0^+)}{dt} = s_1 A_1 + s_2 A_2.$$

But how do we find the initial value of dv/dt from the circuit? Remember that dv/dt appears in the equation relating voltage and current for a capacitor,

$$i_C = C \frac{dv}{dt}.$$

We can solve the capacitor equation for dv/dt and find its initial value in terms of the initial current in the capacitor:

$$\frac{dv(0^+)}{dt} = \frac{i_C(0^+)}{C}.$$

Now we use KCL to find the initial current in the capacitor. We know that the sum of the three branch currents at $t = 0^+$ must be zero. The initial current in the resistive branch is the initial voltage V_0 divided by the resistance, and the initial current in the inductive branch is I_0. Using the reference system depicted in Fig. 8.5, we obtain

$$i_C(0^+) = \frac{-V_0}{R} - I_0.$$

Now we have the second equation needed to find the values of A_1 and A_2 in Eq. 8.9:

$$\frac{dv(0^+)}{dt} = A_1 s_1 + A_2 s_2 = \frac{1}{C}\left(\frac{-V_0}{R} - I_0\right). \tag{8.11}$$

The method for finding the voltage for the parallel *RLC* circuit in Fig. 8.5 is as follows:

Step 1: Determine the initial values of capacitor voltage, V_0, and inductor current, I_0, by analyzing the parallel *RLC* circuit for $t < 0$.

Step 2: Determine the values of the neper frequency, α, and the resonant radian frequency, ω_0, using the values of R, L, and C and the equations in Table 8.1.

Step 3: Compare α^2 and ω_0^2. If $\alpha^2 > \omega_0^2$, the response is overdamped:

$$v(t) = A_1 e^{s_1 t} + A_2 e^{s_2 t}, \, t \ge 0.$$

Step 4: If the response is overdamped, calculate the values of s_1 and s_2 from α and ω_0, using the equations in Table 8.1.

Step 5: If the response is overdamped, calculate the values of A_1 and A_2 by solving Eqs. 8.10 and 8.11 simultaneously:

$$A_1 + A_2 = V_0;$$

$$A_1 s_1 + A_2 s_2 = \frac{1}{C}\left(\frac{-V_0}{R} - I_0\right).$$

Step 6: Write the equation for $v(t)$ from Step 3 using the results from Steps 4 and 5. Find any desired branch current using the relationship between voltage and current for the component in the branch.

A condensed version of this method is given in Analysis Method 8.1. Examples 8.2 and 8.3 use this method to find the overdamped response of a parallel *RLC* circuit.

NATURAL RESPONSE OF A PARALLEL *RLC* CIRCUIT

1. Determine the initial capacitor voltage (V_0) and inductor current (I_0) from the circuit.
2. Determine the values of α and ω_0 using the equations in Table 8.1.
3. If $\alpha^2 > \omega_0^2$, the response is overdamped and $v(t) = A_1 e^{s_1 t} + A_2 e^{s_2 t}, t \ge 0$.
4. If the response is overdamped, calculate s_1 and s_2 using the equations in Table 8.1.
5. If the response is overdamped, calculate A_1 and A_2 by simultaneously solving Eqs. 8.10 and 8.11.
6. Write the equation for $v(t)$ from Step 3 using the results from Steps 4 and 5; find any desired branch currents.

Analysis Method 8.1 The natural response of an overdamped parallel *RLC* circuit.

<table>
<tr><td>**EXAMPLE 8.2**</td><td>**Finding the Overdamped Natural Response of a Parallel *RLC* Circuit**</td></tr>
</table>

For the circuit in Fig. 8.6, $v(0^+) = 12$ V, and $i_L(0^+) = 30$ mA.

a) Find the expression for $v(t)$.
b) Sketch $v(t)$ in the interval $0 \le t \le 250 \,\mu s$.

Solution

a) We use Analysis Method 8.1 to find the voltage.

Step 1: Determine the initial values of capacitor voltage, V_0, and inductor current, I_0; since these values are given in the problem statement, no circuit analysis is required.

Step 2: Determine the values of α and ω_0 using the equations in Table 8.1:

$$\alpha = \frac{1}{2RC} = \frac{1}{2(200)(0.2 \times 10^{-6})} = 12{,}500 \text{ rad/s},$$

$$\omega_0^2 = \frac{1}{LC} = \frac{1}{(50 \times 10^{-3})(0.2 \times 10^{-6})} = 10^8 \text{ rad}^2/\text{s}^2.$$

Step 3: Compare α^2 and ω_0^2; since $\alpha^2 > \omega_0^2$, the response is overdamped and

$$v(t) = A_1 e^{s_1 t} + A_2 e^{s_2 t}, \quad t \ge 0.$$

Step 4: Since the response is overdamped, calculate the values of s_1 and s_2:

$$s_1 = -\alpha + \sqrt{\alpha^2 - \omega_0^2} = -12{,}500 + \sqrt{(12{,}500)^2 - 10^8}$$
$$= -12{,}500 + 7500 = -5000 \text{ rad/s};$$

$$s_2 = -\alpha - \sqrt{\alpha^2 - \omega_0^2} = -12{,}500 - \sqrt{(12{,}500)^2 - 10^8}$$
$$= -12{,}500 - 7500 = -20{,}000 \text{ rad/s}.$$

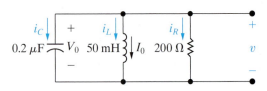

Figure 8.6 ▲ The circuit for Example 8.2.

Step 5: Since the response is overdamped, calculate the values of A_1 and A_2 by simultaneously solving

$$A_1 + A_2 = V_0 = 12;$$

$$A_1 s_1 + A_2 s_2 = \frac{1}{C}\left(\frac{-V_0}{R} - I_0\right) \quad \text{so}$$

$$-5000A_1 - 20,000A_2 = \frac{1}{0.2 \times 10^{-6}}\left(\frac{-12}{200} - 0.03\right)$$

$$= -450,000.$$

Solving,

$$A_1 = -14 \text{ V} \quad \text{and} \quad A_2 = 26 \text{ V}.$$

Step 6: Write the equation for $v(t)$ using the results from Steps 4 and 5:

$$v(t) = (-14e^{-5000t} + 26e^{-20,000t}) \text{ V}, \quad t \ge 0.$$

b) Figure 8.7 shows a plot of $v(t)$ versus t over the interval $0 \le t \le 250 \ \mu s$.

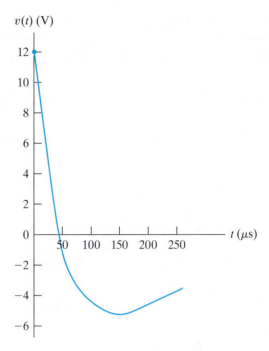

Figure 8.7 ▲ The voltage response for Example 8.2.

EXAMPLE 8.3	**Calculating Branch Currents in the Natural Response of a Parallel *RLC* Circuit**

Derive the expressions for the three branch currents i_R, i_L, and i_C in Example 8.2 (Fig. 8.6) during the time the stored energy is being released.

Solution

We know the voltage across the three branches from the solution in Example 8.2 is

$$v(t) = (-14e^{-5000t} + 26e^{-20,000t}) \text{ V}, \quad t \ge 0.$$

The current in the resistive branch is then

$$i_R(t) = \frac{v(t)}{200} = (-70e^{-5000t} + 130e^{-20,000t}) \text{ mA}, \quad t \ge 0.$$

There are two ways to find the current in the inductive branch. One way is to use the integral relationship that exists between the current and the voltage at the terminals of an inductor:

$$i_L(t) = \frac{1}{L}\int_0^t v_L(x)\, dx + I_0.$$

A second approach is to find the current in the capacitive branch first and then use the fact that

$i_R + i_L + i_C = 0$. Let's use this approach. The current in the capacitive branch is

$$i_C(t) = C\frac{dv}{dt}$$

$$= 0.2 \times 10^{-6}(70,000e^{-5000t} - 520,000e^{-20,000t})$$

$$= (14e^{-5000t} - 104e^{-20,000t}) \text{ mA}, \quad t \ge 0^+.$$

Now find the inductive branch current from the relationship

$$i_L(t) = -i_R(t) - i_C(t)$$

$$= (56e^{-5000t} - 26e^{-20,000t}) \text{ mA}, \quad t \ge 0.$$

We leave it to you, in Assessment Problem 8.2, to show that the integral relationship between voltage and current in an inductor leads to the same result. Note that the expression for i_L agrees with the initial inductor current, as it must.

ASSESSMENT PROBLEMS

8.2 Use the integral relationship between i_L and v to find the expression for i_L in Fig. 8.6.

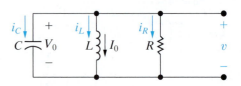

Answer: $i_L(t) = (56e^{-5000t} - 26e^{-20,000t})$ mA, $t \geq 0$.

8.3 The element values in the circuit shown are $R = 2\text{ k}\Omega$, $L = 250\text{ mH}$, and $C = 10\text{ nF}$. The initial current I_0 in the inductor is -4 A, and the initial voltage on the capacitor is 0 V. The output signal is the voltage v. Find (a) $i_R(0^+)$; (b) $i_C(0^+)$; (c) $dv(0^+)/dt$; (d) A_1; (e) A_2; and (f) $v(t)$ when $t \geq 0$.

Answer: (a) 0;
(b) 4 A;
(c) 4×10^8 V/s;
(d) 13,333 V;
(e) $-13,333$ V;
(f) $13,333(e^{-10,000t} - e^{-40,000t})$ V.

SELF-CHECK: Also try Chapter Problems 8.6 and 8.16.

The Underdamped Voltage Response

When $\omega_0^2 > \alpha^2$, the roots of the characteristic equation are complex numbers, and the response is underdamped. For convenience, we express the roots s_1 and s_2 as

$$s_1 = -\alpha + \sqrt{-(\omega_0^2 - \alpha^2)}$$

$$= -\alpha + j\sqrt{\omega_0^2 - \alpha^2}$$

$$= -\alpha + j\omega_d \tag{8.12}$$

$$s_2 = -\alpha - j\omega_d, \tag{8.13}$$

where

DAMPED RADIAN FREQUENCY

$$\omega_d = \sqrt{\omega_0^2 - \alpha^2}. \tag{8.14}$$

The term ω_d is called the **damped radian frequency**. Later we explain the reason for this terminology.

The underdamped voltage response of a parallel *RLC* circuit is

PARALLEL *RLC* NATURAL RESPONSE

$$v(t) = B_1 e^{-\alpha t} \cos \omega_d t + B_2 e^{-\alpha t} \sin \omega_d t, \tag{8.15}$$

which follows from Eq. 8.9. In making the transition from Eq. 8.9 to Eq. 8.15, we use the Euler identity:

$$e^{\pm j\theta} = \cos \theta \pm j \sin \theta.$$

Thus,

$$v(t) = A_1 e^{(-\alpha + j\omega_d)t} + A_2 e^{-(\alpha + j\omega_d)t}$$

$$= A_1 e^{-\alpha t} e^{j\omega_d t} + A_2 e^{-\alpha t} e^{-j\omega_d t}$$

$$= e^{-\alpha t}(A_1 \cos \omega_d t + j A_1 \sin \omega_d t + A_2 \cos \omega_d t - j A_2 \sin \omega_d t)$$

$$= e^{-\alpha t}[\,(A_1 + A_2) \cos \omega_d t + j(A_1 - A_2) \sin \omega_d t\,].$$

At this point in the transition from Eq. 8.9 to Eq. 8.15, replace the arbitrary constants $A_1 + A_2$ and $j(A_1 - A_2)$ with new arbitrary constants denoted B_1 and B_2 to get

$$v = e^{-\alpha t}(B_1 \cos \omega_d t + B_2 \sin \omega_d t)$$

$$= B_1 e^{-\alpha t} \cos \omega_d t + B_2 e^{-\alpha t} \sin \omega_d t.$$

The constants B_1 and B_2 are real, not complex, because the voltage is a real function. Don't be misled by the fact that $B_2 = j(A_1 - A_2)$. In this underdamped case, A_1 and A_2 are complex conjugates, and thus B_1 and B_2 are real. (See Problems 8.22 and 8.23.) Defining the underdamped response in terms of the coefficients B_1 and B_2 yields a simple expression for the voltage, v. We determine B_1 and B_2 in the same way that we found A_1 and A_2 for the overdamped response—by solving two simultaneous equations. The first equation sets the initial value v from Eq. 8.15 equal to the initial value of v in the circuit. The second equation sets the initial value of dv/dt from Eq. 8.15 equal to the initial value of dv/dt in the circuit. Note that the initial values of v and dv/dt in the circuit are the same in both the underdamped and overdamped cases. For the underdamped response, the two simultaneous equations that determine B_1 and B_2 are

$$v(0^+) = B_1 = V_0, \tag{8.16}$$

$$\frac{dv(0^+)}{dt} = -\alpha B_1 + \omega_d B_2 = \frac{1}{C}\left(\frac{-V_0}{R} - I_0\right). \tag{8.17}$$

The overall process for finding the underdamped response is the same as that for the overdamped response, although the response equations and the simultaneous equations used to find the constants are slightly different. We can modify Steps 3, 4, and 5 in the method for finding the voltage for the parallel *RLC* circuit to accommodate the differences.

Step 3: Compare α^2 and ω_0^2. If $\alpha^2 > \omega_0^2$, the response is overdamped:

$$v(t) = A_1 e^{s_1 t} + A_2 e^{s_2 t}, \quad t \geq 0.$$

If $\alpha^2 < \omega_0^2$, the response is overdamped:

$$v(t) = B_1 e^{-\alpha t} \cos \omega_d t + B_2 e^{-\alpha t} \sin \omega_d t, \quad t \geq 0.$$

Step 4: If the response is overdamped, calculate the values of s_1 and s_2 from α and ω_0, using the equations in Table 8.1. If the response is underdamped, calculate the value of ω_d from

$$\omega_d = \sqrt{\omega_0^2 - \alpha^2}.$$

NATURAL RESPONSE OF A PARALLEL *RLC* CIRCUIT

1. Determine the initial capacitor voltage (V_0) and inductor current (I_0) from the circuit.

2. Determine the values of α and ω_0 using the equations in Table 8.1.

3. If $\alpha^2 > \omega_0^2$, the response is overdamped and $v(t) = A_1 e^{s_1 t} + A_2 e^{s_2 t}$, $t \geq 0$; **If $\alpha^2 < \omega_0^2$, the response is underdamped** and $v(t) = B_1 e^{-\alpha t} \cos \omega_d t + B_2 e^{-\alpha t} \sin \omega_d t$, $t \geq 0$.

4. If the response is overdamped, calculate s_1 and s_2 using the equations in Table 8.1; **If the response is underdamped, calculate ω_d using $\omega_d = \sqrt{\omega_0^2 - \alpha^2}$.**

5. If the response is overdamped, calculate A_1 and A_2 by simultaneously solving Eqs. 8.10 and 8.11; **If the response is underdamped, calculate B_1 and B_2 by simultaneously solving** Eqs. 8.16 and 8.17.

6. Write the equation for $v(t)$ from Step 3 using the results from Steps 4 and 5; find any desired branch currents.

Analysis Method 8.2 The natural response of an overdamped or underdamped parallel *RLC* circuit.

Step 5: If the response is overdamped, calculate the values of A_1 and A_2 by solving Eqs. 8.10 and 8.11 simultaneously:

$$A_1 + A_2 = V_0;$$

$$A_1 s_1 + A_2 s_2 = \frac{1}{C}\left(\frac{-V_0}{R} - I_0\right).$$

If the response is underdamped, calculate the values of B_1 and B_2 by solving Eqs. 8.16 and 8.17 simultaneously:

$$B_1 = V_0;$$

$$-\alpha B_1 + \omega_d B_2 = \frac{1}{C}\left(\frac{-V_0}{R} - I_0\right).$$

These modified steps are condensed in Analysis Method 8.2. We examine the characteristics of the underdamped response following Example 8.4, which analyzes a circuit whose response is underdamped.

EXAMPLE 8.4	**Finding the Underdamped Natural Response of a Parallel *RLC* Circuit**

In the circuit shown in Fig. 8.8, $V_0 = 0$, and $I_0 = -12.25$ mA.

a) Calculate the voltage response for $t \geq 0$.

b) Plot $v(t)$ versus t for the time interval $0 \leq t \leq 11$ ms.

Solution

Use Analysis Method 8.2.

a) **Step 1:** Determine the initial values of capacitor voltage, V_0, and inductor current, I_0; since these values are given in the problem statement, no circuit analysis is required.

Step 2: Determine the values of α and ω_0 using the equations in Table 8.1:

$$\alpha = \frac{1}{2RC} = \frac{1}{2(20{,}000)(125 \times 10^{-9})} = 200 \text{ rad/s},$$

$$\omega_0^2 = \frac{1}{LC} = \frac{1}{(8)(125 \times 10^{-9})} = 10^6 \text{ rad}^2/\text{s}^2.$$

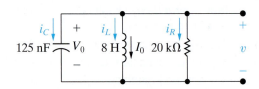

Figure 8.8 ▲ The circuit for Example 8.4.

Step 3: Compare α^2 and ω_0^2; since $\alpha^2 < \omega_0^2$, the response is underdamped and

$$v(t) = B_1 e^{-\alpha t} \cos \omega_d t + B_2 e^{-\alpha t} \sin \omega_d t, \quad t \geq 0.$$

Step 4: Since the response is underdamped, calculate the value of ω_d:

$$\omega_d = \sqrt{\omega_0^2 - \alpha^2} = \sqrt{10^6 - (200)^2} = 979.80 \text{ rad/s}.$$

Step 5: Since the response is underdamped, calculate the values of B_1 and B_2 by simultaneously solving

$$B_1 = V_0 = 0;$$

$$-\alpha B_1 + \omega_d B_2 = \frac{1}{C}\left(\frac{-V_0}{R} - I_0\right) \quad \text{so}$$

$$-200 B_1 - 979.80 B_2 = \frac{1}{125 \times 10^{-9}}\left(\frac{-0}{20{,}000} - (-12.25 \times 10^{-3})\right)$$

$$= 98{,}000.$$

Solving,

$$B_1 = 0 \text{ V} \quad \text{and} \quad B_2 = 100 \text{ V}.$$

Step 6: Write the equation for $v(t)$ using the results from Steps 4 and 5:

$$v(t) = 100 e^{-200t} \sin 979.80t \text{ V}, \quad t \geq 0.$$

b) Figure 8.9 shows the plot of $v(t)$ versus t for the first 11 ms after the stored energy is released. It clearly indicates that the underdamped response is a damped oscillation. The voltage $v(t)$ approaches its final value, alternating between values that are greater than and less than the final value. Furthermore, these swings about the final value decrease exponentially with time.

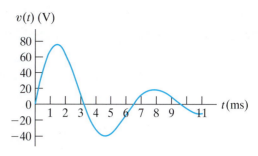

Figure 8.9 ▲ The voltage response for Example 8.4.

ASSESSMENT PROBLEM

Objective 1—Be able to determine the natural and the step response of parallel *RLC* circuits

8.4 A 10 mH inductor, a 1 μF capacitor, and a variable resistor are connected in parallel in the circuit shown. The resistor is adjusted so that the roots of the characteristic equation are $-8000 \pm j6000$ rad/s. The initial voltage on the capacitor is 10 V, and the initial current in the inductor is 80 mA. Find

a) R;

b) $dv(0^+)/dt$;

c) B_1 and B_2 in the solution for v; and

d) $i_L(t)$.

Answer: (a) $62.5\ \Omega$;
(b) $-240{,}000$ V/s;
(c) $B_1 = 10$ V, $B_2 = -80/3$ V;
(d) $10e^{-8000t}[8 \cos 6000t + (82/3) \sin 6000t]$ mA when $t \ge 0$.

SELF-CHECK: Also try Chapter Problems 8.8 and 8.14.

Characteristics of the Underdamped Response

Let's look at the general nature of the underdamped response. From Eq. 8.15 and the plot in Fig. 8.9 we know that the voltage alternates, or oscillates between positive and negative values. The voltage oscillates because there are two types of energy-storage elements in the circuit: the inductor and the capacitor. (A mechanical analogy of this electric circuit is that of a mass suspended on a spring, where oscillation is possible because energy can be stored in both the spring and the moving mass.) The oscillation rate is fixed by ω_d and the oscillation amplitude decreases exponentially at a rate determined by α, so α is also called the **damping factor** or **damping coefficient**. That explains why ω_d is called the damped radian frequency.

If there is no damping, $\alpha = 0$ and the frequency of oscillation is ω_0. Whenever there is a dissipative element, R, in the circuit, α is not zero and the frequency of oscillation, ω_d, is less than ω_0. Thus, when α is not zero, the frequency of oscillation is said to be damped. As the dissipative losses in the circuit decrease, the persistence of the oscillations increases, and the frequency of the oscillations approaches ω_0. In other words, as $R \to \infty$, the energy dissipation in the circuit in Fig. 8.8 approaches zero because $p = v^2/R \to 0$. As $R \to \infty$, $\alpha \to 0$ and $\omega_d \to \omega_0$; thus, the voltage oscillates and its amplitude does not decay.

In Example 8.4, if R is increased to infinity, the solution for $v(t)$ becomes

$$v(t) = 98 \sin 1000t \text{ V}, \quad t \ge 0.$$

In this case, the oscillation is sustained at a frequency of 1000 rad/s and the maximum amplitude of the voltage is 98 V.

Let's examine the qualitative differences between an underdamped and an overdamped response. In an underdamped system, the response oscillates, or "bounces," about its final value. This oscillation is also called *ringing*. In an overdamped system, the response approaches its final value without ringing or in what is sometimes described as a "sluggish" manner. When specifying the desired response of a second-order system, you have two options:

- Reach the final value in the shortest time possible, without concern for the small oscillations about that final value that will eventually cease. Therefore, design the system components to achieve an underdamped response.
- Do not allow the response to exceed its final value, perhaps to ensure that components are not damaged. Therefore, design the system components to achieve an overdamped response, and accept a relatively slow rise to the final value.

The Critically Damped Voltage Response

The response of a parallel *RLC* circuit is critically damped when $\omega_0^2 = \alpha^2$, or $\omega_0 = \alpha$. When a circuit is critically damped, the response is on the verge of oscillating, and the roots of the characteristic equation are real and equal:

$$s_1 = s_2 = -\alpha = -\frac{1}{2RC}. \tag{8.18}$$

If we substitute $s_1 = s_2 = -\alpha$, into the voltage equation (Eq. 8.9), the equation becomes

$$v = (A_1 + A_2)e^{-\alpha t} = A_0 e^{-\alpha t},$$

where A_0 is an arbitrary constant. But this expression for v cannot satisfy two independent initial conditions (V_0, I_0) with only one constant, A_0.

Thus, when the roots of the characteristic equation are equal, the solution for the differential equation (Eq. 8.1) must take a different form, namely,

> **PARALLEL *RLC* NATURAL RESPONSE–CRITICALLY DAMPED**
>
> $$v(t) = D_1 t e^{-\alpha t} + D_2 e^{-\alpha t}. \tag{8.19}$$

This solution involves a simple exponential term plus the product of a linear term and an exponential term. The justification of Eq. 8.19 is left for an introductory course in differential equations.

There are only two unknowns in Eq. 8.19, D_1 and D_2. We find their values in the same way we found A_1 and A_2 for the overdamped response and B_1 and B_2 for the underdamped response—by solving two simultaneous equations.

One equation sets the initial value v from Eq. 8.19 equal to the initial value of v in the circuit. The second equation sets the initial value of dv/dt from Eq. 8.19 equal to the initial value of dv/dt in the circuit. Note that the initial values of v and dv/dt in the circuit are the same in the underdamped, overdamped, and critically damped cases. For the critically damped response, the two simultaneous equations that determine D_1 and D_2 are

$$v(0^+) = D_2 = V_0, \tag{8.20}$$

$$\frac{dv(0^+)}{dt} = D_1 - \alpha D_2 = \frac{1}{C}\left(\frac{-V_0}{R} - I_0\right). \tag{8.21}$$

The overall process for finding the critically damped response is the same as that for the overdamped and underdamped responses, but again, the response equation and the simultaneous equations used to find the constants are slightly different. We can modify Steps 3, 4, and 5 in the method for finding the voltage for the parallel *RLC* circuit to accommodate the differences.

Step 3: Compare α^2 and ω_0^2. If $\alpha^2 > \omega_0^2$, the response is overdamped:

$$v(t) = A_1 e^{s_1 t} + A_2 e^{s_2 t}, \quad t \geq 0.$$

If $\alpha^2 < \omega_0^2$, the response is overdamped:

$$v(t) = B_1 e^{-\alpha t}\cos\omega_d t + B_2 e^{-\alpha t}\sin\omega_d t, \quad t \geq 0.$$

If $\alpha^2 = \omega_0^2$, the response is critically damped:

$$v(t) = D_1 t e^{-\alpha t} + D_2 e^{-\alpha t}, \quad t \geq 0.$$

Step 4: If the response is overdamped, calculate the values of s_1 and s_2 from α and ω_0, using the equations in Table 8.1. If the response is underdamped, calculate the value of ω_d from

$$\omega_d = \sqrt{\omega_0^2 - \alpha^2}.$$

If the response is critically damped, you can skip this step.

Step 5: If the response is overdamped, calculate the values of A_1 and A_2 by solving Eqs. 8.10 and 8.11 simultaneously:

$$A_1 + A_2 = V_0;$$

$$A_1 s_1 + A_2 s_2 = \frac{1}{C}\left(\frac{-V_0}{R} - I_0\right).$$

If the response is underdamped, calculate the values of B_1 and B_2 by solving Eqs. 8.16 and 8.17 simultaneously:

$$B_1 = V_0;$$

$$-\alpha B_1 + \omega_d B_2 = \frac{1}{C}\left(\frac{-V_0}{R} - I_0\right).$$

If the response is critically damped, calculate the values of D_1 and D_2 by solving Eqs. 8.20 and 8.21 simultaneously:

$$D_2 = V_0;$$

$$D_1 - \alpha D_2 = \frac{1}{C}\left(\frac{-V_0}{R} - I_0\right).$$

These modified steps are condensed in Analysis Method 8.3, while Table 8.2 collects all of the equations needed to find the natural response of any parallel *RLC* circuit.

NATURAL RESPONSE OF A PARALLEL *RLC* CIRCUIT

1. Determine the initial capacitor voltage (V_0) and inductor current (I_0) from the circuit.

2. Determine the values of α and ω_0 using the equations in Table 8.2.

3. If $\alpha^2 > \omega_0^2$, the response is overdamped and $v(t) = A_1 e^{s_1 t} + A_2 e^{s_2 t}, \quad t \geq 0$;

If $\alpha^2 < \omega_0^2$, the response is underdamped and $v(t) = B_1 e^{-\alpha t}\cos\omega_d t + B_2 e^{-\alpha t}\sin\omega_d t, \quad t \geq 0$;

If $\alpha^2 = \omega_0^2$ the response is critically damped and $v(t) = D_1 t e^{-\alpha t} + D_2 e^{-\alpha t}, \quad t \geq 0$.

4. If the response is overdamped, calculate s_1 and s_2 using the equations in Table 8.2;

If the response is underdamped, calculate ω_d using the equation in Table 8.2.

5. If the response is overdamped, calculate A_1 and A_2 by simultaneously solving the equations in Table 8.2;

If the response is underdamped, calculate B_1 and B_2 by simultaneously solving the equations in Table 8.2;

If the response is critically damped, calculate D_1 and D_2 by simultaneously solving the equations in Table 8.2.

6. Write the equation for $v(t)$ from Step 3 using the results from Steps 4 and 5; find any desired branch currents.

Analysis Method 8.3 The natural response of parallel *RLC* circuits.

TABLE 8.2   Equations for analyzing the natural response of parallel *RLC* circuits

Characteristic equation	$s^2 + \dfrac{1}{RC}s + \dfrac{1}{LC} = 0$
Neper, resonant, and damped frequencies	$\alpha = \dfrac{1}{2RC} \quad \omega_0 = \sqrt{\dfrac{1}{LC}} \quad \omega_d = \sqrt{\omega_0^2 - \alpha^2}$
Roots of the characteristic equation	$s_1 = -\alpha + \sqrt{\alpha^2 - \omega_0^2}, \quad s_2 = -\alpha - \sqrt{\alpha^2 - \omega_0^2}$
$\alpha^2 > \omega_0^2$: overdamped	$v(t) = A_1 e^{s_1 t} + A_2 e^{s_2 t}, \ t \geq 0$
	$v(0^+) = A_1 + A_2 = V_0$
	$\dfrac{dv(0^+)}{dt} = s_1 A_1 + s_2 A_2 = \dfrac{1}{C}\left(\dfrac{-V_0}{R} - I_0\right)$
$\alpha^2 < \omega_0^2$: underdamped	$v(t) = B_1 e^{-\alpha t}\cos\omega_d t + B_2 e^{-\alpha t}\sin\omega_d t, \ t \geq 0$
	$v(0^+) = B_1 = V_0$
	$\dfrac{dv(0^+)}{dt} = -\alpha B_1 + \omega_d B_2 = \dfrac{1}{C}\left(\dfrac{-V_0}{R} - I_0\right)$
$\alpha^2 = \omega_0^2$: critically damped	$v(t) = D_1 t e^{-\alpha t} + D_2 e^{-\alpha t}, \ t \geq 0$
	$v(0^+) = D_2 = V_0$
	$\dfrac{dv(0^+)}{dt} = D_1 - \alpha D_2 = \dfrac{1}{C}\left(\dfrac{-V_0}{R} - I_0\right)$

(Note that the equations in the last three rows assume that the reference direction for the current in every component is in the direction of the reference voltage drop across that component.)

You will rarely encounter critically damped systems in practice, largely because ω_0 must equal α exactly. Both of these quantities depend on circuit parameters, and in a real circuit it is very difficult to choose component values that satisfy an exact equality relationship. Even so, Example 8.5 illustrates the approach for finding the critically damped response of a parallel *RLC* circuit.

EXAMPLE 8.5   Finding the Critically Damped Natural Response of a Parallel *RLC* Circuit

a) For the circuit in Example 8.4 (Fig. 8.8), find the value of R that results in a critically damped voltage response.

b) Calculate $v(t)$ for $t \geq 0$.

c) Plot $v(t)$ versus t for $0 \leq t \leq 7$ ms.

Solution

a) From Example 8.4, we know that $\omega_0^2 = 10^6$. Therefore, for critical damping,

$$\alpha = 10^3 = \frac{1}{2RC},$$

or

$$R = \frac{1}{2(1000)(125 \times 10^{-9})} = 4000 \ \Omega.$$

b) Follow the steps in Analysis Method 8.3 to find the voltage v.

Step 1: Determine the initial values of capacitor voltage, V_0, and inductor current, I_0; since these

values are given in Example 8.4, no circuit analysis is required.

Step 2: From part (a), we know that

$$\alpha = \omega_0 = 1000 \text{ rad/s}.$$

Step 3: Compare α^2 and ω_0^2; since $\alpha^2 = \omega_0^2$, the response is critically damped and

$$v(t) = D_1 t e^{-\alpha t} + D_2 e^{-\alpha t}, \quad t \geq 0.$$

Step 4: Since the response is critically damped, this step is not needed.

Step 5: Since the response is critically damped, calculate the values of D_1 and D_2:

$$D_2 = V_0 = 0;$$

$$D_1 - \alpha D_2 = \frac{1}{C}\left(\frac{-V_0}{R} - I_0\right) \text{ so}$$

$$D_1 - 1000 D_2 = \frac{1}{125 \times 10^{-9}}\left(\frac{-0}{4000} - (-12.25 \times 10^{-3})\right)$$

$$= 98{,}000.$$

Therefore,

$$D_1 = 98,000 \text{ V/s} \quad \text{and} \quad D_2 = 0 \text{ V}.$$

Step 6: Write the equation for $v(t)$ using the results from Steps 4 and 5:

$$v(t) = 98,000te^{-1000t} \text{ V}, \quad t \geq 0.$$

c) Figure 8.10 shows a plot of $v(t)$ versus t in the interval $0 \leq t \leq 7$ ms.

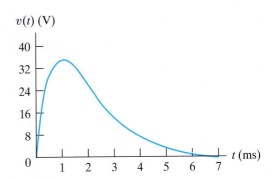

Figure 8.10 ▲ The voltage response for Example 8.5.

ASSESSMENT PROBLEM

Objective 1—Be able to determine the natural and the step response of parallel *RLC* circuits

8.5   The resistor in the circuit in Assessment Problem 8.4 is adjusted for critical damping. The inductance and capacitance values are 0.4 H and 10 μF, respectively. The initial energy stored in the circuit is 25 mJ and is distributed equally between the inductor and capacitor. Find (a) R; (b) V_0; (c) I_0; (d) D_1 and D_2 in the solution for v; and (e) i_R, $t \geq 0^+$.

Answer: (a) 100 Ω;

(b) 50 V;

(c) 250 mA;

(d) $-50,000$ V/s, 50 V;

(e) $i_R(t) = (-500te^{-500t} + 0.50e^{-500t})$ A, $t \geq 0^+$.

SELF-CHECK: Also try Chapter Problems 8.9 and 8.15.

8.3 The Step Response of a Parallel *RLC* Circuit

Now we find the step response of a parallel *RLC* circuit, represented by the circuit shown in Fig. 8.11. The step response results from the sudden application of a dc current source. Energy may or may not be stored in the circuit when the current source is applied. We develop a general approach to the step response by finding the current in the inductive branch (i_L).

Why do we find the inductor current? Remember that for the natural response, we found the voltage because it was the same for all of the parallel-connected components, and we could use the voltage to find the current in any branch. But to find the step response, we need to satisfy three constraints. Two of the constraints are established by the initial values of the capacitor voltage and the inductor current, just as in the natural-response problem. For the step-response problem, a third constraint arises from a nonzero final value that exists because there is a source in the circuit for $t \geq 0$.

We draw the circuit in Fig. 8.11 as $t \rightarrow \infty$. In the presence of the dc current source, the capacitor behaves like an open circuit, and the inductor behaves like a short circuit, which shunts the resistor. The resulting circuit is shown in Fig. 8.12, where we see that the only nonzero final value is the inductor current. This explains why we find the inductor current and not the voltage in the parallel *RLC* step response.

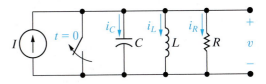

Figure 8.11 ▲ A circuit used to describe the step response of a parallel *RLC* circuit.

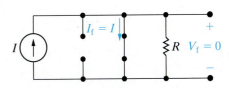

Figure 8.12 ▲ The circuit in Fig. 8.11 as $t \rightarrow \infty$.

To find the inductor current i_L in the circuit in Fig. 8.11, we begin with a KCL equation for the circuit's top node:

$$i_L + i_R + i_C = I,$$

or

$$i_L + \frac{v}{R} + C\frac{dv}{dt} = I. \tag{8.22}$$

Because

$$v = L\frac{di_L}{dt},$$

we get

$$\frac{dv}{dt} = L\frac{d^2i_L}{dt^2}.$$

Now we can write Eq. 8.22 using only the inductor current and its first and second derivatives, to give

$$i_L + \frac{L}{R}\frac{di_L}{dt} + LC\frac{d^2i_L}{dt^2} = I.$$

For convenience, we divide through by LC and rearrange terms:

$$\frac{d^2i_L}{dt^2} + \frac{1}{RC}\frac{di_L}{dt} + \frac{i_L}{LC} = \frac{I}{LC}. \tag{8.23}$$

Compare Eq. 8.23 with Eq. 8.1—they have the same form, but note the nonzero constant on the right-hand side of Eq. 8.23. Before showing how to solve Eq. 8.23 directly, we find its solution indirectly. When we know the solution of Eq. 8.23, explaining the direct approach will be easier.

The Indirect Approach

We can solve for i_L indirectly by first finding the voltage v. We use the techniques introduced in Section 8.2 because the differential equation that v must satisfy is identical to Eq. 8.1. To see this, we simply return to Eq. 8.22 and express i_L as a function of v; thus

$$\frac{1}{L}\int_0^t v d\tau + I_0 + \frac{v}{R} + C\frac{dv}{dt} = I,$$

where I_0 is the initial current in the inductor. Differentiating once with respect to t reduces the right-hand side to zero because I is a constant and eliminates I_0 from the left-hand side for the same reason. Thus

$$\frac{v}{L} + \frac{1}{R}\frac{dv}{dt} + C\frac{d^2v}{dt^2} = 0,$$

or

$$\frac{d^2v}{dt^2} + \frac{1}{RC}\frac{dv}{dt} + \frac{v}{LC} = 0.$$

As discussed in Section 8.2, the solution for v depends on the roots of the characteristic equation. Thus, the three possible solutions are

$$v = A_1 e^{s_1 t} + A_2 e^{s_2 t},$$

$$v = B_1 e^{-\alpha t} \cos \omega_d t + B_2 e^{-\alpha t} \sin \omega_d t,$$

$$v = D_1 t e^{-\alpha t} + D_2 e^{-\alpha t}.$$

A word of caution: Because there is a source in the circuit for $t > 0$, you must take into account the value of the source current at $t = 0^+$ when you evaluate the coefficients in the three expressions for v.

To find the three possible solutions for i_L, we substitute the three expressions for v into Eq. 8.22. When this has been done, you should be able to verify that the three solutions for i_L will be

---

PARALLEL *RLC* STEP-RESPONSE FORMS

$$i_L = I + A'_1 e^{s_1 t} + A'_2 e^{s_2 t} \text{ (overdamped)} \tag{8.24}$$

$$i_L = I + B'_1 e^{-\alpha t} \cos \omega_d t + B'_2 e^{-\alpha t} \sin \omega_d t \text{ (underdamped)} \tag{8.25}$$

$$i_L = I + D'_1 t e^{-\alpha t} + D'_2 e^{-\alpha t} \text{ (critically damped)} \tag{8.26}$$

---

where A'_1, A'_2, B'_1, B'_2, D'_1, and D'_2, are arbitrary constants. In each case, the primed constants can be found indirectly in terms of the arbitrary constants associated with the voltage solution. However, this approach is cumbersome.

The Direct Approach

As we have just seen, the solution for a second-order differential equation with a constant forcing function equals the forced response, plus a response function identical in form to the natural response. Thus, we can always write the solution for the step response in the form

$$i = I_f + \left\{ \begin{matrix} \text{function of the same form} \\ \text{as the natural response} \end{matrix} \right\},$$

or

$$v = V_f + \left\{ \begin{matrix} \text{function of the same form} \\ \text{as the natural response} \end{matrix} \right\},$$

where I_f and V_f represent the final value of the response function. The final value may be zero, as we saw for the voltage v in the circuit in Fig. 8.12.

As we have already noted, the only quantity with a nonzero final value in the circuit of Fig. 8.11 is the inductor current. Let's look at how to alter the parallel *RLC* natural-response method to construct a method for finding the parallel *RLC* step response for the inductor current.

Step 1: Determine the initial values of capacitor voltage, V_0, and inductor current, I_0, by analyzing the parallel *RLC* circuit for $t < 0$. In this step, we also need to find the final value of the inductor current, I_f, by analyzing the circuit as $t \to \infty$.

Step 2: Determine the values of the neper frequency, α, and the resonant radian frequency, ω_0, using the values of R, L, and C. No modifications are needed for this step.

Step 3: Compare α^2 and ω_0^2. Here we replace the natural response for the voltage with the step response for the inductor current, given in Eqs. 8.24–8.26.

Step 4: If the response is overdamped, calculate the values of s_1 and s_2. If the response is underdamped, calculate the value of ω_d. If the response is critically damped, you can skip this step. No changes are needed in this step.

Step 5: Calculate the values of the A', B', and D' coefficients by simultaneously solving two equations. To construct the first equation, we evaluate the expression for $i_L(t)$ from Step 3 at $t = 0^+$ and set it equal to the initial inductor current, I_0. For example, in the overdamped case the first equation is

$$I_f + A_1' + A_2' = I_0;$$

in the underdamped case, the first equation is

$$I_f + B_1' = I_0;$$

and in the critically damped case, the first equation is

$$I_f + D_2' = I_0.$$

To construct the second equation, we find the di_L/dt from the inductor current in Step 3, evaluate it at $t = 0^+$, and set it equal to the initial value of di_L/dt from the circuit. How do we find di_L/dt from the circuit? We use the relationship between voltage and current in an inductor to get

$$\frac{di_L(0^+)}{dt} = \frac{v_L(0^+)}{L} = \frac{V_0}{L}.$$

In the overdamped case, the second equation is

$$A_1' s_1 + A_2' s_2 = \frac{V_0}{L};$$

in the underdamped case, the second equation is

$$-\alpha B_1' + \omega_d B_2' = \frac{V_0}{L};$$

and in the critically damped case, the second equation is

$$D_1' - \alpha D_2' = \frac{V_0}{L}.$$

Step 6: Write the equation for $i_L(t)$ from Step 3 using the results from Steps 4 and 5. Find the voltage v and the remaining branch currents using the relationship between voltage and current for the component in each branch.

The steps for finding the step response of a parallel *RLC* circuit are condensed into Analysis Method 8.4. All of the equations you will need are collected in Table 8.3. Examples 8.6–8.10 illustrate how to use Table 8.3 and Analysis Method 8.4 when finding the step response of a parallel *RLC* circuit.

STEP RESPONSE OF A PARALLEL *RLC* CIRCUIT

1. Determine the initial capacitor voltage (V_0), the initial inductor current (I_0), and the final inductor current (I_f) from the circuit.

2. Determine the values of α and ω_0 using the equations in Table 8.3.

3. If $\alpha^2 > \omega_0^2$, the response is overdamped and $i_L(t) = I_f + A_1' e^{s_1 t} + A_2' e^{s_2 t}$, $t \geq 0^+$; If $\alpha^2 > \omega_0^2$ the response is underdamped and $i_L(t) = I_f + B_1' e^{-\alpha t} \cos \omega_d t + B_2' e^{-\alpha t} \sin \omega_d t$, $t \geq 0^+$; If $\alpha^2 = \omega_0^2$, the response is critically damped and $i_L(t) = I_f + D_1' t e^{-\alpha t} + D_2' e^{-\alpha t}$, $t \geq 0^+$.

4. If the response is overdamped, calculate s_1 and s_2 using the equations in Table 8.3; If the response is underdamped, calculate ω_d using the equation in Table 8.3.

5. If the response is overdamped, calculate A_1' and A_2' by simultaneously solving the equations in Table 8.3; If the response is underdamped, calculate B_1' and B_2' by simultaneously solving the equations in Table 8.3; If the response is critically damped, calculate D_1' and D_2' by simultaneously solving the equations in Table 8.3.

6. Write the equation for $i_L(t)$ from Step 3 using the results from Steps 4 and 5; find the inductor voltage and any desired branch currents.

Analysis Method 8.4 The step response of parallel *RLC* circuits.

TABLE 8.3   **Equations for analyzing the step response of parallel *RLC* circuits**

Characteristic equation	$s^2 + \dfrac{1}{RC}s + \dfrac{1}{LC} = \dfrac{I}{LC}$
Neper, resonant, and damped frequencies	$\alpha = \dfrac{1}{2RC} \qquad \omega_0 = \sqrt{\dfrac{1}{LC}} \qquad \omega_d = \sqrt{\omega_0^2 - \alpha^2}$
Roots of the characteristic equation	$s_1 = -\alpha + \sqrt{\alpha^2 - \omega_0^2}, \quad s_2 = -\alpha - \sqrt{\alpha^2 - \omega_0^2}$
$\alpha^2 > \omega_0^2$: overdamped	$i_L(t) = I_f + A_1' e^{s_1 t} + A_2' e^{s_2 t}, \quad t \geq 0$
	$i_L(0^+) = I_f + A_1' + A_2' = I_0$
	$\dfrac{di_L(0^+)}{dt} = s_1 A_1' + s_2 A_2' = \dfrac{V_0}{L}$
$\alpha^2 < \omega_0^2$: underdamped	$i_L(t) = I_f + B_1' e^{-\alpha t}\cos \omega_d t + B_2' e^{-\alpha t}\sin \omega_d t, \quad t \geq 0$
	$i_L(0^+) = I_f + B_1' = I_0$
	$\dfrac{di_L(0^+)}{dt} = -\alpha B_1' + \omega_d B_2' = \dfrac{V_0}{L}$
$\alpha^2 = \omega_0^2$: critically damped	$i_L(t) = I_f + D_1' t e^{-\alpha t} + D_2' e^{-\alpha t}, \quad t \geq 0$
	$i_L(0^+) = I_f + D_2' = I_0$
	$\dfrac{di_L(0^+)}{dt} = D_1' - \alpha D_2' = \dfrac{V_0}{L}$

(Note that the equations in the last three rows assume that the reference direction for the current in every component is in the direction of the reference voltage drop across that component.)

EXAMPLE 8.6   **Finding the Overdamped Step Response of a Parallel *RLC* Circuit**

The initial energy stored in the circuit in Fig. 8.13 is zero. At $t = 0$, a dc current source of 24 mA is applied to the circuit. The value of the resistor is 400 Ω. Find $i_L(t)$ for $t \geq 0$.

Solution

Follow the steps in Analysis Method 8.4.

Step 1: The initial values of capacitor voltage, V_0, and inductor current, I_0 are both zero because the initial stored energy is zero. As $t \to \infty$, the capacitor behaves like an open circuit and the inductor behaves like a short circuit that shunts the resistor, so all of the current from the source is in the inductor. Thus, $I_f = 24$ mA.

Step 2: Using the equations in the second row of Table 8.3,

$$\alpha = \frac{1}{2RC} = \frac{1}{2(400)(25 \times 10^{-9})} = 50{,}000 \text{ rad/s};$$

$$\omega_0 = \sqrt{\frac{1}{LC}} = \sqrt{\frac{1}{(0.025)(25 \times 10^{-9})}} = 40{,}000 \text{ rad/s}.$$

Step 3: Compare α^2 and ω_0^2; since $\alpha^2 > \omega_0^2$, the response is overdamped and from Table 8.3,

$$i_L(t) = I_f + A_1' e^{s_1 t} + A_2' e^{s_2 t}, \, t \geq 0.$$

Step 4: Calculate s_1 and s_2 using the equations in the third row of Table 8.3:

$$s_1 = -\alpha + \sqrt{\alpha^2 - \omega_0^2} = -50{,}000 + \sqrt{50{,}000^2 - 40{,}000^2}$$
$$= -50{,}000 + 30{,}000 = -20{,}000 \text{ rad/s};$$

$$s_2 = -\alpha - \sqrt{\alpha^2 - \omega_0^2} = -50{,}000 - \sqrt{50{,}000^2 - 40{,}000^2}$$
$$= -50{,}000 - 30{,}000 = -80{,}000 \text{ rad/s}.$$

Step 5: Calculate the values of A_1' and A_2' by simultaneously solving the equations from row 4 in Table 8.3:

$$I_f + A_1' + A_2' = I_0 \quad \text{so} \quad 0.024 + A_1' + A_2' = 0;$$

$$s_1 A_1' + s_2 A_2' = \frac{V_0}{L} \quad \text{so} \quad -20{,}000 A_1' - 80{,}000 A_2' = 0.$$

Solving,

$$A_1' = -32 \text{ mA} \quad \text{and} \quad A_2' = 8 \text{ mA}.$$

Step 6: Write the equation from Step 3 using the results from Steps 4 and 5 to give

$$i_L(t) = (24 - 32^{-20{,}000t} + 8e^{-80{,}000t}) \text{ mA}, \quad t \geq 0.$$

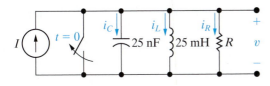

Figure 8.13 ▲ The circuit for Example 8.6.

| EXAMPLE 8.7 | Finding the Underdamped Step Response of a Parallel *RLC* Circuit |

The resistor in the circuit in Example 8.6 (Fig. 8.13) is increased to 625 Ω. Find $i_L(t)$ for $t \geq 0$.

Solution

Follow the steps in Analysis Method 8.4.

Step 1: From Step 1 of Example 8.6, $V_0 = 0$, $I_0 = 0$, $I_f = 24$ mA.

Step 2: Since only R has changed, $\omega_0 = 40{,}000$ rad/s from Example 8.6 and

$$\alpha = \frac{1}{2RC} = \frac{1}{2(625)(25 \times 10^{-9})} = 32{,}000 \text{ rad/s.}$$

Step 3: Compare α^2 and ω_0^2; since $\alpha^2 < \omega_0^2$, the response is underdamped and from Table 8.3,

$$i_L(t) = I_f + B_1' e^{-\alpha t} \cos \omega_d t + B_2' e^{-\alpha t} \sin \omega_d t, \ t \geq 0.$$

Step 4: Calculate ω_d using the equation in the second row of Table 8.3:

$$\omega_d = \sqrt{\omega_0^2 - \alpha^2} = \sqrt{40{,}000^2 - 32{,}000^2} = 24{,}000 \text{ rad/s.}$$

Step 5: Calculate the values of B_1' and B_2' by simultaneously solving the equations from row 5 in Table 8.3:

$$I_f + B_1' = I_0 \quad \text{so} \quad 0.024 + B_1' = 0;$$

$$-\alpha B_1' + \omega_d B_2' = \frac{V_0}{L} \quad \text{so} \quad -32{,}000 B_1' + 24{,}000 B_2' = 0.$$

Solving,

$$B_1' = -24 \text{ mA} \quad \text{and} \quad B_2' = -32 \text{ mA.}$$

Step 6: Write the equation from Step 3 using the results from Steps 4 and 5 to give

$$i_L(t) = (24 - 24e^{-32{,}000t} \cos 24{,}000t$$
$$- 32e^{-32{,}000t} \sin 24{,}000t) \text{ mA}, \quad t \geq 0.$$

| EXAMPLE 8.8 | Finding the Critically Damped Step Response of a Parallel *RLC* Circuit |

The resistor in the circuit in Example 8.6 (Fig. 8.13) is set at 500 Ω. Find i_L for $t \geq 0$.

Solution

Follow the steps in Analysis Method 8.4.

Step 1: From Step 1 of Example 8.6,
$$V_0 = 0, I_0 = 0, I_f = 24 \text{ mA.}$$

Step 2: Since only R has changed,
$$\omega_0 = 40{,}000 \text{ rad/s,}$$
and from the second row of Table 8.3

$$\alpha = \frac{1}{2RC} = \frac{1}{2(500)(25 \times 10^{-9})} = 40{,}000 \text{ rad/s.}$$

Step 3: Compare α^2 and ω_0^2; since $\alpha^2 = \omega_0^2$, the response is critically damped and from Table 8.3,
$$i_L(t) = I_f + D_1' t e^{-\alpha t} + D_2' e^{-\alpha t}, t \geq 0.$$

Step 4: The response is critically damped, so this step is not needed.

Step 5: Calculate the values of D_1' and D_2' by simultaneously solving the equations from row 6 in Table 8.3:

$$I_f + D_2' = I_0 \quad \text{so} \quad 0.024 + D_2' = 0;$$

$$D_1' - \alpha D_2' = \frac{V_0}{L} \quad \text{so} \quad D_1' - 40{,}000 D_2' = 0.$$

Solving,

$$D_1' = -960{,}000 \text{ mA/s} \quad \text{and} \quad D_2' = -24 \text{ mA.}$$

Step 6: Write the equation from Step 3 using the results from Steps 4 and 5 to give

$$i_L(t) = (24 - 960{,}000t e^{-40{,}000t} - 24e^{-40{,}000t}) \text{ mA}, t \geq 0.$$

EXAMPLE 8.9 | **Comparing the Three-Step Response Forms**

a) Plot on a single graph, over a range from 0 to 220 μs, the overdamped, underdamped, and critically damped responses derived in Examples 8.6–8.8.

b) Use the plots of (a) to find the time required for i_L to reach 90% of its final value.

c) On the basis of the results obtained in (b), which response would you specify in a design that puts a premium on reaching 90% of the final value of the output in the shortest time?

d) Which response would you specify in a design that must ensure that the final value of the current is never exceeded?

Solution

a) See Fig. 8.14.

b) The final value of i_L is 24 mA, so we can read the times off the plots corresponding to $i_L = 21.6$ mA. Thus, $t_{od} = 130\ \mu$s, $t_{cd} = 97\ \mu$s, and $t_{ud} = 74\ \mu$s.

c) The underdamped response reaches 90% of the final value in the fastest time, so it is the desired response type when speed is the most important design specification.

d) From the plot, you can see that the underdamped response overshoots the final value of current, whereas neither the critically damped nor the overdamped response produces currents in excess of 24 mA. Although specifying either of the latter two responses would meet the design specification, it is best to use the overdamped response. It would be impractical to require a design to achieve the exact component values that ensure a critically damped response.

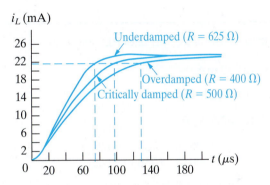

Figure 8.14 ▲ The current plots for Example 8.9.

EXAMPLE 8.10 | **Finding Step Response of a Parallel *RLC* Circuit with Initial Stored Energy**

Energy is stored in the circuit in Example 8.8 (Fig. 8.13, with $R = 500\ \Omega$) at the instant the dc current source is applied. The initial current in the inductor is 29 mA, and the initial voltage across the capacitor is 50 V. Find $i_L(t)$ for $t \geq 0$ and $v(t)$ for $t \geq 0$.

Solution

Follow Analysis Method 8.4.

Step 1: From the problem statement, $V_0 = 50$ V and $I_0 = 29$ mA. The final value of the inductor current is unchanged from the problem in Example 8.8, so $I_f = 24$ mA.

Step 2: From Example 8.8, $\alpha = 40,000$ rad/s and $\omega_0 = 40,000$ rad/s.

Step 3: Compare α^2 and ω_0^2; since $\alpha^2 = \omega_0^2$, the response is critically damped and from Table 8.3,

$$i_L(t) = I_f + D_1' t e^{-\alpha t} + D_2' e^{-\alpha t}, \quad t \geq 0.$$

Step 4: The response is critically damped, so this step is not needed.

Step 5: Calculate the values of D_1' and D_2' by simultaneously solving the equations from row 6 in Table 8.3:

$$I_f + D_2' = I_0 \quad \text{so} \quad 0.024 + D_2' = 0.029;$$

$$D_1' - \alpha D_2' = \frac{V_0}{L} \quad \text{so} \quad D_1' - 40,000 D_2' = \frac{50}{0.025} = 2000.$$

Solving,

$$D_1' = 2.2 \times 10^6 \,\text{mA/s} \quad \text{and} \quad D_2' = 5 \,\text{mA}.$$

Step 6: Write the equation from Step 3 using the results from Steps 4 and 5 to give

$$i_L(t) = (24 - 2.2 \times 10^6 te^{-40,000t} + 5e^{-40,000t})\,\text{mA}, \quad t \geq 0.$$

We can get the expression for $v(t), t \geq 0$ by using the relationship between the voltage and current in an inductor:

$$v(t) = L\frac{di_L}{dt}$$

$$= (25 \times 10^{-3})[(2.2 \times 10^6)\,(-40,000)te^{-40,000t}$$
$$+ 2.2 \times 10^6 e^{-40,000t}$$
$$+ (5)\,(-40,000)e^{-40,000t}] \times 10^{-3}$$
$$= (-2.2 \times 10^6 te^{-40,000t} + 50e^{-40,000t})\,\text{V}, \quad t \geq 0.$$

To check this result, let's verify that the initial voltage across the inductor is 50 V:

$$v(0) = -2.2 \times 10^6 (0)\,(1) + 50(1) = 50\,\text{V}.$$

ASSESSMENT PROBLEM

Objective 1—Be able to determine the natural response and the step response of parallel *RLC* circuits

8.6 In the circuit shown, $R = 500\,\Omega$, $L = 0.64\,\text{H}$, $C = 1\,\mu\text{F}$, and $I = -1\,\text{A}$. The initial voltage drop across the capacitor is 40 V and the initial inductor current is 0.5 A. Find (a) $i_R(0^+)$; (b) $i_C(0^+)$; (c) $di_L(0^+)/dt$; (d) s_1, s_2; (e) $i_L(t)$ for $t \geq 0$; and (f) $v(t)$ for $t \geq 0^+$.

Answer: (a) 80 mA;

(b) −1.58 A;

(c) 62.5 A/s;

(d) $(-1000 + j750)\,\text{rad/s}$, $(-1000 - j750)\,\text{rad/s}$;

(e) $-1 + e^{-1000t}[1.5\cos 750t + 2.0833\sin 750t]\,\text{A}$, for $t \geq 0$;

(f) $e^{-1000t}(40\cos 750t - 2053.33\sin 750t)\,\text{V}$, for $t \geq 0^+$.

SELF-CHECK: Also try Chapter Problems 8.32–8.34.

8.4 The Natural and Step Response of a Series *RLC* Circuit

The procedures for finding the natural or step responses of a series *RLC* circuit are the same as those used to find the natural or step responses of a parallel *RLC* circuit because both circuits are described by differential equations that have the same form. For the natural-response problem, we solve for the current because it is the same for all circuit components. We begin by summing the voltages, expressed in terms of the current, around the closed path in the circuit shown in Fig. 8.15. Thus

$$Ri + L\frac{di}{dt} + \frac{1}{C}\int_0^t i\,d\tau + V_0 = 0.$$

We now differentiate once with respect to t to get

$$R\frac{di}{dt} + L\frac{d^2i}{dt^2} + \frac{i}{C} = 0,$$

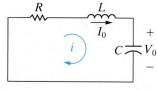

Figure 8.15 ▲ A circuit used to illustrate the natural response of a series *RLC* circuit.

which we can rearrange as

$$\frac{d^2i}{dt^2} + \frac{R}{L}\frac{di}{dt} + \frac{i}{LC} = 0. \tag{8.27}$$

Comparing Eq. 8.27 with Eq. 8.1 reveals that they have the same form. Therefore, to find the solution of Eq. 8.27, we follow the same process that led us to the solution of Eq. 8.1.

From Eq. 8.27, the characteristic equation for the series *RLC* circuit is

CHARACTERISTIC EQUATION, SERIES *RLC* CIRCUIT

$$s^2 + \frac{R}{L}s + \frac{1}{LC} = 0. \tag{8.28}$$

The roots of the characteristic equation are

$$s_{1,2} = -\frac{R}{2L} \pm \sqrt{\left(\frac{R}{2L}\right)^2 - \frac{1}{LC}},$$

or

$$s_{1,2} = -\alpha \pm \sqrt{\alpha^2 - \omega_0^2}.$$

The neper frequency (α) for the series *RLC* circuit is

NEPER FREQUENCY, SERIES *RLC* CIRCUIT

$$\alpha = \frac{R}{2L} \text{ rad/s}, \tag{8.29}$$

and the expression for the resonant radian frequency is

RESONANT RADIAN FREQUENCY, SERIES *RLC* CIRCUIT

$$\omega_0 = \frac{1}{\sqrt{LC}} \text{ rad/s}. \tag{8.30}$$

Note that the equation for the neper frequency of the series *RLC* circuit differs from that of the parallel *RLC* circuit, but the equations for the resonant radian frequencies are the same.

The current response will be overdamped, underdamped, or critically damped according to whether $\omega_0^2 < \alpha^2$, $\omega_0^2 > \alpha^2$, or $\omega_0^2 = \alpha^2$, respectively. Thus, the three possible solutions for the current are as follows:

SERIES *RLC* NATURAL-RESPONSE FORMS

$$i(t) = A_1 e^{s_1 t} + A_2 e^{s_2 t} \text{ (overdamped)}, \tag{8.31}$$

$$i(t) = B_1 e^{-\alpha t}\cos\omega_d t$$
$$+ B_2 e^{-\alpha t}\sin\omega_d t \text{ (underdamped)}, \tag{8.32}$$

$$i(t) = D_1 t e^{-\alpha t} + D_2 e^{-\alpha t} \text{ (critically damped)}. \tag{8.33}$$

Once you know the current you can find the voltage across any circuit element.

Let's look at how to alter the parallel *RLC* natural-response method to construct a method for finding the series *RLC* natural response for the circuit's current.

Step 1: Determine the initial values of capacitor voltage, V_0, and inductor current, I_0, by analyzing the *RLC* circuit. This step is unchanged.

Step 2: Determine the values of the neper frequency, α, and the resonant radian frequency, ω_0, using the values of R, L, and C. While the equation for ω_0 is the same for both circuits, the equation for α is different: for the parallel *RLC*, $\alpha = 1/(2RC)$, and for the series *RLC*, $\alpha = R/(2L)$.

Step 3: Compare α^2 and ω_0^2 to determine the response form. Use the appropriate equation for the circuit current from Eqs. 8.31–8.33.

Step 4: If the response is overdamped, calculate the values of s_1 and s_2. If the response is underdamped, calculate the value of ω_d. If the response is critically damped, you can skip this step. No changes are needed in this step.

Step 5: Calculate the values of the A, B, and D coefficients by simultaneously solving two equations. To construct the first equation, we evaluate the expression for $i(t)$ from Step 3 at $t = 0^+$ and set it equal to the initial inductor current, I_0. For example, in the overdamped case the first equation is

$$A_1' + A_2' = I_0;$$

in the underdamped case, the first equation is

$$B_1' = I_0;$$

and in the critically damped case, the first equation is

$$D_2' = I_0.$$

To construct the second equation, we find di/dt from the circuit current in Step 3, evaluate it at $t = 0^+$ and set it equal to the initial value of di/dt from the circuit. How do we find the initial value of di/dt from the circuit? We use the relationship between voltage and current in an inductor to get

$$\frac{di(0^+)}{dt} = \frac{v_L(0^+)}{L}.$$

But we don't know the initial voltage across the inductor, so we use KVL to find it. We know that the sum of the three component voltages at $t = 0^+$ must be zero. The voltage across the resistor at $t = 0^+$ is the product of the initial current (I_0) and the resistance, and the voltage across the capacitor at $t = 0^+$ is V_0. Using the reference system in Fig. 8.15, we obtain

$$v_L(0^+) = -RI_0 - V_0.$$

So the initial value of di/dt from the circuit is

$$\frac{di(0^+)}{dt} = \frac{1}{L}(-RI_0 - V_0).$$

Thus, in the overdamped case, the second equation is

$$A_1's_1 + A_2's_2 = \frac{1}{L}(-RI_0 - V_0)$$

in the underdamped case, the second equation is

$$-\alpha B_1' + \omega_d B_2' = \frac{1}{L}(-RI_0 - V_0);$$

and in the critically damped case, the second equation is

$$D_1' - \alpha D_2' = \frac{1}{L}(-RI_0 - V_0).$$

Step 6: Write the equation for $i(t)$ from Step 3 using the results from Steps 4 and 5. Find the voltage for any component using its relationship between voltage and current.

The steps for finding the natural response for a parallel RLC circuit are condensed into Analysis Method 8.5. All of the equations you will need are collected in Table 8.4.

To verify that the procedure for finding the step response of a series *RLC* circuit is similar to that for a parallel *RLC* circuit, we show that the differential equation that describes the capacitor voltage in Fig. 8.16 has the same form as the differential equation that describes the inductor current in Fig. 8.11. Applying KVL to the circuit shown in Fig. 8.16 gives

$$V = Ri + L\frac{di}{dt} + v_C.$$

The current (i) is related to the capacitor voltage (v_C) by the expression

$$i = C\frac{dv_C}{dt},$$

from which

$$\frac{di}{dt} = C\frac{d^2 v_C}{dt^2}.$$

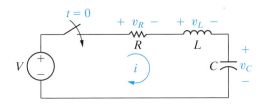

Figure 8.16 ▲ A circuit used to illustrate the step response of a series *RLC* circuit.

NATURAL RESPONSE OF A SERIES *RLC* CIRCUITS

1. Determine the initial capacitor voltage (V_0) and inductor current (I_0) from the circuit.

2. Determine the values of α and ω_0 using the equations in Table 8.4.

3. If $\alpha^2 > \omega_0^2$, the response is overdamped and $i(t) = A_1 e^{s_1 t} + A_2 e^{s_2 t}, \quad t \geq 0$;
If $\alpha^2 < \omega_0^2$ the response is underdamped and $i(t) = B_1 e^{-\alpha t}\cos \omega_d t + B_2 e^{-\alpha t}\sin \omega_d t, \quad t \geq 0$;
If $\alpha^2 = \omega_0^2$, the response is critically damped and $i(t) = D_1 t e^{-\alpha t} + D_2 e^{-\alpha t}, \quad t \geq 0$.

**4. If the response is overdamped, calculate s_1 and s_2 using the equations in Table 8.4;
If the response is underdamped, calculate ω_d using the equation in Table 8.4.**

**5. If the response is overdamped, calculate A_1 and A_2 by simultaneously solving the equations in Table 8.4;
If the response is underdamped, calculate B_1 and B_2 by simultaneously solving the equations in Table 8.4;
If the response is critically damped, calculate D_1 and D_2 by simultaneously solving the equations in Table 8.4.**

6. Write the equation for $i(t)$ from Step 3 using the results from Steps 4 and 5; find any desired component voltages.

Analysis Method 8.5 The natural response of series *RLC* circuits.

TABLE 8.4	Equations for analyzing the natural response of series *RLC* circuits
Characteristic equation	$s^2 + \dfrac{R}{L}s + \dfrac{1}{LC} = 0$
Neper, resonant, and damped frequencies	$\alpha = \dfrac{R}{2L} \qquad \omega_0 = \sqrt{\dfrac{1}{LC}} \qquad \omega_d = \sqrt{\omega_0^2 - \alpha^2}$
Roots of the characteristic equation	$s_1 = -\alpha + \sqrt{\alpha^2 - \omega_0^2}, \quad s_2 = -\alpha - \sqrt{\alpha^2 - \omega_0^2}$
$\alpha^2 > \omega_0^2$: overdamped	$i(t) = A_1 e^{s_1 t} + A_2 e^{s_2 t}, t \geq 0$ $i(0^+) = A_1 + A_2 = I_0$ $\dfrac{di(0^+)}{dt} = s_1 A_1 + s_2 A_2 = \dfrac{1}{L}(-RI_0 - V_0)$
$\alpha^2 < \omega_0^2$: underdamped	$i(t) = B_1 e^{-\alpha t}\cos \omega_d t + B_2 e^{-\alpha t}\sin \omega_d t, \quad t \geq 0$ $i(0^+) = B_1 = I_0$ $\dfrac{di(0^+)}{dt} = -\alpha B_1 + \omega_d B_2 = \dfrac{1}{L}(-RI_0 - V_0)$
$\alpha^2 = \omega_0^2$: critically damped	$i(t) = D_1 t e^{-\alpha t} + D_2 e^{-\alpha t}, \quad t \geq 0$ $i(0^+) = D_2 = I_0$ $\dfrac{di(0^+)}{dt} = D_1 - \alpha D_2 = \dfrac{1}{L}(-RI_0 - V_0)$

(Note that the equations in the last three rows assume that the reference direction for the current in every component is in the direction of the reference voltage drop across that component.)

Substitute the expressions for the current and its first derivative into the KVL equation to get

$$\frac{d^2 v_C}{dt^2} + \frac{R}{L}\frac{dv_C}{dt} + \frac{v_C}{LC} = \frac{V}{LC}. \qquad (8.34)$$

Equation 8.34 has the same form as Eq. 8.23; therefore, the procedure for finding v_C parallels that for finding i_L. The three possible solutions for v_C are as follows:

PARALLEL *RLC* STEP-RESPONSE FORMS

$$v_C = V_f + A_1' e^{s_1 t} + A_2' e^{s_2 t} \quad (\text{overdamped}), \qquad (8.35)$$

$$v_C = V_f + B_1' e^{-\alpha t} \cos \omega_d t$$
$$\quad + B_2' e^{-\alpha t} \sin \omega_d t \quad (\text{underdamped}), \qquad (8.36)$$

$$v_C = V_f + D_1' t e^{-\alpha t} + D_2' e^{-\alpha t} \quad (\text{critically damped}), \qquad (8.37)$$

where V_f is the final value of v_C. Hence, from the circuit shown in Fig. 8.15, the final value of v_C is the dc source voltage V.

Let's look at how to alter the series *RLC* natural-response method (Analysis Method 8.5) to find the series *RLC* step response for the capacitor voltage.

Step 1: Determine the initial values of capacitor voltage, V_0, and inductor current, I_0, by analyzing the *RLC* circuit. In this step, we also need to find the final value of the capacitor voltage, V_f, by analyzing the circuit as $t \to \infty$.

Step 2: Determine the values of the neper frequency, α, and the resonant radian frequency, ω_0, using the values of R, L, and C. No modifications are needed for this step.

Step 3: Compare α^2 and ω_0^2. Here we replace the natural response for the current with the step response for the capacitor voltage, given in Eqs. 8.35–8.37.

Step 4: If the response is overdamped, calculate the values of s_1 and s_2. If the response is underdamped, calculate the value of ω_d. If the response is critically damped, you can skip this step. No changes are needed in this step.

Step 5: Here we calculate the values of the A', B', and D' coefficients by simultaneously solving two equations. To construct the first equation, we evaluate the expression for $v_C(t)$ from Step 3 at $t = 0^+$ and set it equal to the initial capacitor voltage, V_0. For example, in the overdamped case the first equation is

$$V_f + A_1' + A_2' = V_0;$$

in the underdamped case, the first equation is

$$V_f + B_1' = V_0;$$

and in the critically damped case, the first equation is

$$V_f + D_2' = V_0.$$

To construct the second equation, we find dv_C/dt from the capacitor voltage in Step 3, evaluate it at $t = 0^+$ and set it equal to the initial value of dv_C/dt from the circuit. How do we find dv_C/dt at $t = 0^+$ from the circuit? We use the relationship between voltage and current in a capacitor to get

$$\frac{dv_C(0^+)}{dt} = \frac{i_C(0^+)}{C} = \frac{I_0}{C}.$$

In the overdamped case, the second equation is

$$A_1's_1 + A_2's_2 = \frac{I_0}{C};$$

in the underdamped case, the second equation is

$$-\alpha B_1' + \omega_d B_2' = \frac{I_0}{C};$$

and in the critically damped case, the second equation is

$$D_1' - \alpha D_2' = \frac{I_0}{C}.$$

Step 6: Here we must write the equation for $v_C(t)$ from Step 3 using the results from Steps 4 and 5. Find the current i and the remaining component voltages using the relationships between voltage and current.

The steps for finding the step response for a series *RLC* circuit are condensed into Analysis Method 8.6. All of the equations you will need are collected in Table 8.5.

Examples 8.11 and 8.12 find the natural and step responses of a series *RLC* circuit.

TABLE 8.5	Equations for analyzing the step response of series *RLC* circuits
Characteristic equation	$s^2 + \dfrac{R}{L}s + \dfrac{1}{LC} = \dfrac{V}{LC}$
Neper, resonant, and damped frequencies	$\alpha = \dfrac{R}{2L} \qquad \omega_0 = \sqrt{\dfrac{1}{LC}} \qquad \omega_d = \sqrt{\omega_0^2 - \alpha^2}$
Roots of the characteristic equation	$s_1 = -\alpha + \sqrt{\alpha^2 - \omega_0^2}, \quad s_2 = -\alpha - \sqrt{\alpha^2 - \omega_0^2}$
$\alpha^2 > \omega_0^2$: overdamped	$v_C(t) = V_f + A_1'e^{s_1t} + A_2'e^{s_2t}, t \geq 0$ $v_C(0^+) = V_f + A_1' + A_2' = V_0$ $\dfrac{dv_C(0^+)}{dt} = s_1A_1' + s_2A_2' = \dfrac{I_0}{C}$
$\alpha^2 < \omega_0^2$: underdamped	$v_C(t) = V_f + B_1'e^{-\alpha t}\cos\omega_d t + B_2'e^{-\alpha t}\sin\omega_d t, t \geq 0$ $v_C(0^+) = V_f + B_1' = V_0$ $\dfrac{dv_C(0^+)}{dt} = -\alpha B_1' + \omega_d B_2' = \dfrac{I_0}{C}$
$\alpha^2 = \omega_0^2$: critically damped	$v_C(t) = V_f + D_1'te^{-\alpha t} + D_2'e^{-\alpha t}, t \geq 0$ $v_C(0^+) = V_f + D_2' = V_0$ $\dfrac{dv_C(0^+)}{dt} = D_1' - \alpha D_2' = \dfrac{I_0}{C}$

(Note that the equations in the last three rows assume that the reference direction for the current in every component is in the direction of the reference voltage drop across that component.)

STEP RESPONSE OF A SERIES *RLC* CIRCUITS

1. Determine the initial capacitor voltage (V_0), the initial inductor current (I_0), and the final capacitor voltage (V_f) from the circuit.

2. Determine the values of α and ω_0 using the equations in Table 8.5.

3. If $\alpha^2 > \omega_0^2$, the response is **overdamped** and $v_C(t) = V_f + A_1'e^{s_1t} + A_2'e^{s_2t}$, $t \geq 0^+$;

If $\alpha^2 < \omega_0^2$, the response is **underdamped** and $v_C(t) = V_f + B_1'e^{-\alpha t}\cos\omega_d t + B_2'e^{-\alpha t}\sin\omega_d t$, $t \geq 0^+$;

If $\alpha^2 = \omega_0^2$, the response is **critically damped** and $v_C(t) = V_f + D_1'te^{-\alpha t} + D_2'e^{-\alpha t}$, $t \geq 0^+$.

4. If the response is overdamped, calculate s_1 and s_2 using the equations in Table 8.5;

If the response is underdamped, calculate ω_d using the equation in Table 8.5.

5. If the response is overdamped, calculate A_1' and A_2' by simultaneously solving the equations in Table 8.5;

If the response is underdamped, calculate B_1' and B_2' by simultaneously solving the equations in Table 8.5;

If the response is critically damped, calculate D_1' and D_2' by simultaneously solving the equations in Table 8.5.

6. Write the equation for $v_C(t)$ from Step 3 using the results from Steps 4 and 5; find the inductor voltage and any desired branch currents.

Analysis Method 8.6 The step response of series *RLC* circuits.

EXAMPLE 8.11   Finding the Natural Response of a Series *RLC* Circuit

The 0.1 μF capacitor in the circuit shown in Fig. 8.17 is charged to 100 V. At $t = 0$ the capacitor is discharged through a series combination of a 100 mH inductor and a 560 Ω resistor.

a) Find $i(t)$ for $t \geq 0$.

b) Find $v_C(t)$ for $t \geq 0$.

Solution

a) This is a natural-response problem because there is no source in the circuit for $t \geq 0$; follow Analysis Method 8.5, which uses Table 8.4.

Step 1: Determine the initial values of capacitor voltage, V_0, and inductor current, I_0. From the problem statement and the circuit configuration, $V_0 = 100$ V and $I_0 = 0$.

Step 2: Determine the values of α and ω_0 using the equations in Table 8.4:

$$\alpha = \frac{R}{2L} = \frac{560}{2(0.1)} = 2800 \text{ rad/s},$$

$$\omega_0^2 = \frac{1}{LC} = \frac{1}{(0.1)(0.1 \times 10^{-6})} = 10^8 \text{ rad}^2/\text{s}^2.$$

Step 3: Compare α^2 and ω_0^2; since $\alpha^2 < \omega_0^2$, the response is underdamped and from row 5 in Table 8.4,

$$i(t) = B_1 e^{-\alpha t} \cos \omega_d t + B_2 e^{-\alpha t} \sin \omega_d t, \ t \geq 0.$$

Step 4: Since the response is underdamped, calculate the value of ω_d from row 2 in Table 8.4:

$$\omega_d = \sqrt{\omega_0^2 - \alpha^2} = \sqrt{10^8 - 2800^2} = 9600 \text{ rad/s}.$$

Step 5: Since the response is underdamped, calculate the values of B_1 and B_2 by simultaneously solving the equations from row 5 in Table 8.4:

$$B_1 = I_0 = 0;$$

$$-\alpha B_1 + \omega_d B_2 = \frac{1}{L}(-RI_0 - V_0) \quad \text{so}$$

$$-2800 B_1 + 9600 B_2 = \frac{1}{0.1}(-(560)(0) - 100) = -1000.$$

Solving,

$$B_1 = 0 \quad \text{and} \quad B_2 = -0.1042 \text{ A}.$$

Step 6: Write the equation for $i(t)$ using the results from Steps 4 and 5:

$$i(t) = -0.1042 e^{-2800t} \sin 9600t \text{ A}, \quad t \geq 0.$$

b) To find $v_C(t)$, we can use either of the following relationships:

$$v_C = \frac{1}{C} \int_0^t i d\tau + 100 \quad \text{or} \quad v_C = -\left(Ri + L\frac{di}{dt}\right).$$

Whichever expression is used (the second is recommended), the result is

$$v_C(t) = (100 \cos 9600t + 29.17 \sin 9600t) e^{-2800t} \text{ V}, \quad t \geq 0.$$

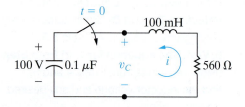

Figure 8.17 ▲ The circuit for Example 8.11.

EXAMPLE 8.12   Finding the Step Response of a Series *RLC* Circuit

No energy is stored in the 100 mH inductor or the 0.4 μF capacitor when the switch in the circuit shown in Fig. 8.18 is closed. Find $v_C(t)$ for $t \geq 0$.

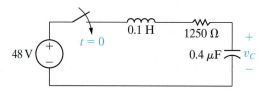

Figure 8.18 ▲ The circuit for Example 8.12.

Solution

This is a step-response problem because there is a source in the circuit for $t \geq 0$; follow Analysis Method 8.6, which uses Table 8.5.

Step 1: Determine the initial values of capacitor voltage, V_0, and inductor current, I_0, and the final value of the capacitor voltage, V_f. From the problem statement and the circuit configuration, $V_0 = 0$, $I_0 = 0$, and $V_f = 48$ V.

Step 2: Determine the values of α and ω_0 using the equations in Table 8.5:

$$\alpha = \frac{R}{2L} = \frac{1250}{2(0.1)} = 6250 \text{ rad/s};$$

$$\omega_0^2 = \frac{1}{LC} = \frac{1}{(0.1)(0.4 \times 10^{-6})} = 25 \times 10^6 \text{ rad}^2/\text{s}^2.$$

Step 3: Compare α^2 and ω_0^2; since $\alpha^2 > \omega_0^2$, the response is overdamped and from row 4 in Table 8.5:

$$v_C(t) = V_f + A_1'e^{s_1 t} + A_2'e^{s_2 t}, \quad t \ge 0.$$

Step 4: Since the response is overdamped, calculate the values of s_1 and s_2 from row 2 in Table 8.5:

$$s_1 = -\alpha + \sqrt{\alpha^2 - \omega_0^2} = -6250 + \sqrt{6250^2 - 5000^2}$$

$$= -6250 + 3750 = -2500 \text{ rad/s};$$

$$s_2 = -\alpha - \sqrt{\alpha^2 - \omega_0^2} = -6250 - \sqrt{6250^2 - 5000^2}$$

$$= -6250 - 3750 = -10,000 \text{ rad/s}.$$

Step 5: Since the response is overdamped, calculate the values of A_1' and A_2' by simultaneously solving the equations from row 4 in Table 8.5:

$$V_f + A_1' + A_2' = V_0 \quad \text{so} \quad 48 + A_1' + A_2' = 0;$$

$$s_1 A_1' + s_2 A_2' = \frac{I_0}{C} \quad \text{so} \quad -2500 A_1' - 10,000 A_2' = 0.$$

Solving,

$$A_1' = -64 \text{ V} \quad \text{and} \quad A_2' = 16 \text{ V}.$$

Step 6: Write the equation for $v_C(t)$ in Step 3 using the results from Steps 4 and 5:

$$v_C(t) = (48 - 64e^{-2500t} + 16e^{-10,000t}) \text{ V}, \quad t \ge 0.$$

ASSESSMENT PROBLEMS

Objective 2—Be able to determine the natural response and the step response of series *RLC* circuits

8.7 The switch in the circuit shown has been in position a for a long time. At $t = 0$, it moves to position b. Find $v_C(t)$ for $t \ge 0$.

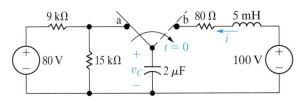

Answer: $[100 - e^{-8000t}(50 \cos 6000t + 66.67 \sin 6000t)]$ V for $t \ge 0$.

8.8 Find $i(t)$ for $t \ge 0$ for the circuit in Assessment Problem 8.7.

Answer: $(1.67e^{-8000t} \sin 6000t)$ A for $t \ge 0$.

8.9 Repeat Assessment Problems 8.7 and 8.8 if the 80 Ω resistor is replaced by a 100 Ω resistor.

Answer:

$$v_C(t) = (100 - 500,000te^{-10,000t} - 50e^{-10,000t}) \text{ V}, \quad t \ge 0;$$

$$i(t) = 10,000te^{-10,000t} \text{ A}, \quad t \ge 0.$$

SELF-CHECK: Also try Chapter Problems 8.50–8.52.

8.5 A Circuit with Two Integrating Amplifiers

A circuit containing two integrating amplifiers connected in cascade[1] is also a second-order circuit; that is, the output voltage of the second integrator is related to the input voltage of the first by a second-order differential equation. We begin our analysis of a circuit containing two cascaded amplifiers with the circuit shown in Fig. 8.19.

---

[1] In a cascade connection, the output signal of the first amplifier (v_{o1} in Fig. 8.19) is the input signal for the second amplifier.

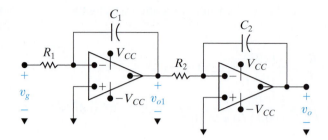

Figure 8.19 ▲ Two integrating amplifiers connected in cascade.

We derive the differential equation that defines the relationship between v_o and v_g, assuming that the op amps are ideal. Begin the derivation by summing the currents at the inverting input terminal of the first integrator:

$$\frac{0 - v_g}{R_1} + C_1 \frac{d}{dt}(0 - v_{o1}) = 0.$$

Simplifying and rearranging, we get

$$\frac{dv_{o1}}{dt} = -\frac{1}{R_1 C_1} v_g. \tag{8.38}$$

Now, sum the currents away from the inverting input terminal of the second integrating amplifier:

$$\frac{0 - v_{o1}}{R_2} + C_2 \frac{d}{dt}(0 - v_o) = 0,$$

or

$$\frac{dv_o}{dt} = -\frac{1}{R_2 C_2} v_{o1}.$$

Differentiating both sides of this equation gives

$$\frac{d^2 v_o}{dt^2} = -\frac{1}{R_2 C_2} \frac{dv_{o1}}{dt}.$$

We find the differential equation that governs the relationship between v_o and v_g by substituting for dv_{o1}/dt, using Eq. 8.38:

$$\frac{d^2 v_o}{dt^2} = \frac{1}{R_1 C_1} \frac{1}{R_2 C_2} v_g. \tag{8.39}$$

Example 8.13 illustrates the step response of a circuit containing two cascaded integrating amplifiers.

EXAMPLE 8.13 | Analyzing Two Cascaded Integrating Amplifiers

No energy is stored in the circuit shown in Fig. 8.20 when the input voltage v_g jumps instantaneously from 0 to 25 mV.

a) Derive the expression for $v_o(t)$ for $0 \le t \le t_{sat}$.

b) Find the time t_{sat} when the circuit saturates.

Solution

a) Figure 8.20 indicates that the amplifier scaling factors are

$$\frac{1}{R_1C_1} = \frac{1000}{(250)(0.1)} = 40,$$

$$\frac{1}{R_2C_2} = \frac{1000}{(500)(1)} = 2.$$

Now, because $v_g = 25$ mV for $t > 0$, Eq. 8.39 becomes

$$\frac{d^2v_o}{dt^2} = (40)(2)(25 \times 10^{-3}) = 2.$$

To solve for v_o, we let

$$g(t) = \frac{dv_o}{dt},$$

then,

$$\frac{dg(t)}{dt} = 2, \quad \text{and} \quad dg(t) = 2dt.$$

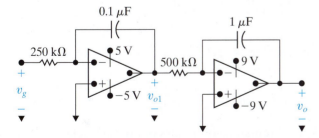

Figure 8.20 ▲ The circuit for Example 8.13.

Hence

$$\int_{g(0)}^{g(t)} dy = 2\int_0^t dx,$$

from which

$$g(t) - g(0) = 2t.$$

However,

$$g(0) = \frac{dv_o(0)}{dt} = 0,$$

because the energy stored in the circuit initially is zero, and the op amps are ideal. (See Problem 8.59.) Then,

$$\frac{dv_o}{dt} = 2t, \text{ and } v_o = t^2 + v_o(0).$$

But $v_o(0) = 0$, so the expression for v_o becomes

$$v_o = t^2, \quad 0 \le t \le t_{sat}.$$

b) The second integrating amplifier saturates when v_o reaches 9 V or $t = 3$ s. But it is possible that the first integrating amplifier saturates before $t = 3$ s. To explore this possibility, use Eq. 8.38 to find dv_{o1}/dt:

$$\frac{dv_{o1}}{dt} = -40(25) \times 10^{-3} = -1.$$

Solving for v_{o1} yields

$$v_{o1} = -t.$$

Thus, at $t = 3$ s, $v_{o1} = -3$ V, and, because the power supply voltage on the first integrating amplifier is ± 5 V, the circuit reaches saturation when the second amplifier saturates. When one of the op amps saturates, we no longer can use the linear model to predict the behavior of the circuit.

SELF-CHECK: Assess your understanding of this material by trying Chapter Problem 8.63.

Two Integrating Amplifiers with Feedback Resistors

Figure 8.21 depicts a variation of the circuit shown in Fig. 8.19. Recall from Section 7.7 that the op amp in the integrating amplifier saturates due to the feedback capacitor's accumulation of charge. Here, a resistor is

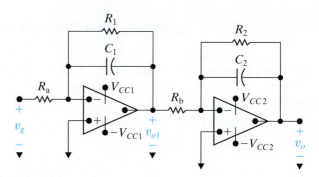

Figure 8.21 ▲ Cascaded integrating amplifiers with feedback resistors.

placed in parallel with each feedback capacitor (C_1 and C_2) to overcome this problem. We derive the equation for the output voltage, v_o, and determine the impact of these feedback resistors on the integrating amplifiers from Example 8.13.

We begin the derivation of the second-order differential equation that relates v_{o1} to v_g by summing the currents at the inverting input node of the first integrator:

$$\frac{0 - v_g}{R_a} + \frac{0 - v_{o1}}{R_1} + C_1 \frac{d}{dt}(0 - v_{o1}) = 0.$$

Simplifying and rearranging, we get

$$\frac{dv_{o1}}{dt} + \frac{1}{R_1 C_1} v_{o1} = \frac{-v_g}{R_a C_1}.$$

For convenience, we let $\tau_1 = R_1 C_1$ and write Eq. 8.41

$$\frac{dv_{o1}}{dt} + \frac{v_{o1}}{\tau_1} = \frac{-v_g}{R_a C_1}. \tag{8.40}$$

Next, sum the currents at the inverting input terminal of the second integrator:

$$\frac{0 - v_{o1}}{R_b} + \frac{0 - v_o}{R_2} + C_2 \frac{d}{dt}(0 - v_o) = 0.$$

Simplifying and rearranging, we get

$$\frac{dv_o}{dt} + \frac{v_o}{\tau_2} = \frac{-v_{o1}}{R_b C_2}, \tag{8.41}$$

where $\tau_2 = R_2 C_2$. Differentiating both sides of Eq. 8.41 yields

$$\frac{d^2 v_o}{dt^2} + \frac{1}{\tau_2} \frac{dv_o}{dt} = -\frac{1}{R_b C_2} \frac{dv_{o1}}{dt}.$$

From Eq. 8.40,

$$\frac{dv_{o1}}{dt} = \frac{-v_{o1}}{\tau_1} - \frac{v_g}{R_a C_1},$$

and from Eq. 8.41,

$$v_{o1} = -R_bC_2\frac{dv_o}{dt} - \frac{R_bC_2}{\tau_2}v_o.$$

Finally, eliminate v_{o1} and dv_{o1}/dt from the second-order differential equation and obtain the desired relationship:

$$\frac{d^2v_o}{dt^2} + \left(\frac{1}{\tau_1} + \frac{1}{\tau_2}\right)\frac{dv_o}{dt} + \left(\frac{1}{\tau_1\tau_2}\right)v_o = \frac{v_g}{R_aC_1R_bC_2}. \qquad (8.42)$$

From Eq. 8.42, the characteristic equation is

$$s^2 + \left(\frac{1}{\tau_1} + \frac{1}{\tau_2}\right)s + \frac{1}{\tau_1\tau_2} = 0.$$

The roots of the characteristic equation are real, namely,

$$s_1 = \frac{-1}{\tau_1},$$

$$s_2 = \frac{-1}{\tau_2}.$$

Example 8.14 determines the step response of two cascaded integrating amplifiers when the feedback capacitors are shunted with feedback resistors.

EXAMPLE 8.14 | **Analyzing Two Cascaded Integrating Amplifiers with Feedback Resistors**

The parameters for the circuit shown in Fig. 8.21 are $R_a = 100\ \text{k}\Omega$, $R_1 = 500\ \text{k}\Omega$, $C_1 = 0.1\ \mu\text{F}$, $R_b = 25\ \text{k}\Omega$, $R_2 = 100\ \text{k}\Omega$, and $C_2 = 1\ \mu\text{F}$. The power supply voltage for each op amp is ±6 V. The signal voltage (v_g) for the cascaded integrating amplifiers jumps from 0 to 250 mV at $t = 0$. No energy is stored in the feedback capacitors at the instant the signal is applied.

a) Find differential equation that governs v_o.

b) Find $v_o(t)$ for $t \geq 0$.

c) Find the differential equation that governs v_{o1}.

d) Find $v_{o1}(t)$ for $t \geq 0$.

Solution

a) From the numerical values of the circuit parameters, we have $\tau_1 = R_1C_1 = 0.05$ s; $\tau_2 = R_2C_2 = 0.10$ s, and $v_g/R_aC_1R_bC_2 = 1000\ \text{V/s}^2$. Substituting these values into Eq. 8.42 gives

$$\frac{d^2v_o}{dt^2} + 30\frac{dv_o}{dt} + 200v_o = 1000.$$

b) The roots of the characteristic equation are $s_1 = -20$ rad/s and $s_2 = -10$ rad/s. The final value

of v_o is the product of the input voltage and the gain of each stage because the capacitors behave as open circuits as $t \to \infty$. Thus,

$$v_o(\infty) = (250 \times 10^{-3})\frac{(-500)}{100}\frac{(-100)}{25} = 5\ \text{V}.$$

The solution for v_o thus takes the form:

$$v_o = 5 + A_1'e^{-10t} + A_2'e^{-20t}.$$

With $v_o(0) = 0$ and $dv_o(0)/dt = 0$, the numerical values of A_1' and A_2' are $A_1' = -10$ V and $A_2' = 5$ V. Therefore, the solution for v_o is

$$v_o(t) = (5 - 10e^{-10t} + 5e^{-20t})\ \text{V}, \quad t \geq 0.$$

The solution assumes that neither op amp saturates. We have already noted that the final value of v_o is 5 V, which is less than 6 V; hence, the second op amp does not saturate. The final value of v_{o1} is $(250 \times 10^{-3})(-500/100)$, or -1.25 V. Therefore, the first op amp does not saturate, and our assumption and solution are correct.

c) Substituting the numerical values of the parameters into Eq. 8.40 generates the desired differential equation:

$$\frac{dv_{o1}}{dt} + 20v_{o1} = -25.$$

d) We have already noted the initial and final values of v_{o1}, along with the time constant τ_1. Thus, we write the solution in accordance with Analysis Method 7.5, developed in Section 7.4:

$$v_{o1} = -1.25 + [\,0 - (-1.25)\,]e^{-20t}$$

$$= -1.25 + 1.25e^{-20t} \text{ V}, \quad t \geq 0.$$

SELF-CHECK: Assess your understanding of this material by trying Chapter Problem 8.64.

■ Practical Perspective

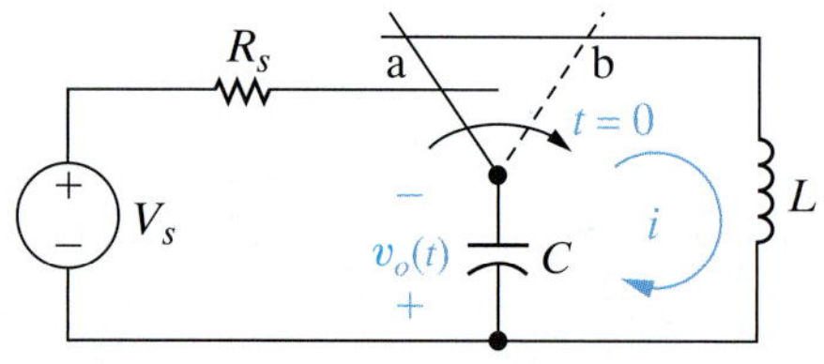

Figure 8.22 ▲ An *LC* natural response circuit.

Clock for Computer Timing

Consider the circuit in Fig. 8.22, where the output is the voltage drop across the capacitor. For $t \geq 0$ this circuit looks like the series *RLC* natural-response circuit of Fig. 8.3 without its resistor. When we analyze this *LC* circuit, we will discover that its output is an undamped sinusoid, which a computer's clock generator could use instead of the typical quartz crystal oscillator. We will be able to specify the frequency of the clock by selecting appropriate values for the inductor and capacitor.

Since this is a series *RLC* natural-response problem, we follow Analysis Method 8.5, which uses Table 8.4. Remember that $R = 0$ for $t \geq 0$.

Step 1: Determine the initial values of capacitor voltage, V_0, and inductor current, I_0. How does the circuit behave when $t < 0$? When the switch is in the a position, all of the components are connected. The capacitor acts like an open circuit, whose voltage is V_0, and the inductor acts like a short circuit, whose current is I_0. The capacitor and inductor have the same voltage, and since the inductor's voltage is 0, $V_0 = 0$. The current in the inductor is the current in the loop containing the voltage source, the resistor, and the inductor, which is $V_0 = V_s/R_s$.

Step 2: Determine the values of α and ω_0 using the equations in Table 8.4:

$$\alpha = \frac{R}{2L} = \frac{0}{2L} = 0,$$

$$\omega_0^2 = \frac{1}{LC}.$$

Step 3: Compare α^2 and ω_0^2; since $\alpha^2 < \omega_0^2$, the response is underdamped and from row 5 in Table 8.4:

$$i(t) = B_1 e^{-\alpha t} \cos \omega_d t + B_2 e^{-\alpha t} \sin \omega_d t, \, t \geq 0.$$

Step 4: Since the response is underdamped, calculate the value of ω_d from row 2 in Table 8.4:

$$\omega_d = \sqrt{\omega_0^2 - \alpha^2} = \sqrt{\omega_0^2 - 0^2} = \omega_0.$$

Step 5: Since the response is underdamped, calculate the values of B_1 and B_2 by simultaneously solving the equations from row 5 in Table 8.4:

$$B_1 = I_0 = \frac{V_s}{R_s};$$

$$-\alpha B_1 + \omega_d B_2 = \frac{1}{L}(-R_s I_0 - V_0) \quad \text{so}$$

$$-(0)B_1 + \omega_0 B_2 = \frac{1}{L}\left(-(0)\left(\frac{V_s}{R_s}\right) - 0\right) = 0.$$

Solving,

$$B_1 = \frac{V_s}{R_s} \quad \text{and} \quad B_2 = 0.$$

Step 6: Write the equation for $i(t)$ using the results from Steps 4 and 5:

$$i(t) = \frac{V_s}{R_s}e^{-(0)t}\cos\omega_0 t, \quad t \geq 0.$$

We can now use the expression for the current in the circuit to find the voltage output by the capacitor:

$$v_o(t) = \frac{1}{C}\int_0^t i(x)dx = \frac{1}{C}\int_0^t \frac{V_s}{R_s}\cos\omega_0 x\,dx = \frac{V_s}{\omega_0 R_s C}\sin\omega_0 t, \quad t \geq 0.$$

By choosing values for L and C, we can use the circuit in Fig. 8.22 to generate an undamped sinusoid when $t \geq 0$ for a computer's clock generator.

So why is a quartz crystal used to generate the sinusoid for the clock generator instead of the LC circuit of Fig. 8.22? Remember that our analysis of the LC circuit assumed that the inductor and capacitor are ideal. But ideal inductors and capacitors do not exist—real inductors and capacitors have a small amount of resistance. We leave it to you to examine the effect of this small amount of resistance on the performance of an LC oscillator in the Chapter Problems.

SELF-CHECK: Assess your understanding of the Practical Perspective by solving Chapter Problems 8.66–8.68.

◼ Summary

- The **characteristic equation** for both the parallel and series RLC circuits has the form

$$s^2 + 2\alpha s + \omega_0^2 = 0,$$

  where $\alpha = 1/2RC$ for the parallel circuit, $\alpha = R/2L$ for the series circuit, and $\omega_0^2 = 1/LC$ for both the parallel and series circuits. (See pages 275 and 297.)

- The roots of the characteristic equation are

$$s_{1,2} = -\alpha \pm \sqrt{\alpha^2 - \omega_0^2}.$$

  (See page 276.)

- The form of the natural and step responses of series and parallel RLC circuits depends on the values of α^2 and ω_0^2; such responses can be **overdamped, underdamped,** or **critically damped.** These terms describe the impact of the dissipative element (R) on the response. The **neper frequency,** α, reflects the effect of R. (See pages 276 and 277.)

- To determine the natural response of a parallel RLC circuit, follow the steps in Analysis Method 8.3, using the equations in Table 8.2. (See page 287.)

- To determine the step response of a parallel RLC circuit, follow the steps in Analysis Method 8.4, using the equations in Table 8.3. (See page 292.)

- To determine the natural response of a series *RLC* circuit, follow the steps in Analysis Method 8.5, using the equations in Table 8.4. (See page 299.)

- To determine the step response of a series *RLC* circuit, follow the steps in Analysis Method 8.6, using the equations in Table 8.5. (See page 301.)

- When two integrating amplifiers with ideal op amps are connected in cascade, the output voltage of the second integrator is related to the input voltage of the first by an ordinary, second-order differential equation. Therefore, the techniques developed in this chapter may be used to analyze the behavior of a cascaded integrator. (See pages 303 and 304.)

- We can overcome the limitation of a simple integrating amplifier—the saturation of the op amp due to charge accumulating in the feedback capacitor—by placing a resistor in parallel with the capacitor in the feedback path. (See page 305.)

■ Problems

Sections 8.1–8.2

8.1 The resistance, inductance, and capacitance in a parallel *RLC* circuit are 1 kΩ, 12.5 H, and 2 μF, respectively.

a) Calculate the roots of the characteristic equation that describe the voltage response of the circuit.

b) Will the response be over-, under-, or critically damped?

c) What value of *R* will yield a damped frequency of 120 rad/s?

d) What are the roots of the characteristic equation for the value of *R* found in (c)?

e) What value of *R* will result in a critically damped response?

8.2 The circuit elements in the circuit in Fig. 8.1 are
PSPICE $R = 200\ \Omega$, $L = 50\ \text{mH}$, and $C = 0.2\ \mu\text{F}$. The
MULTISIM initial inductor current is -45 mA and the initial capacitor voltage is 15 V.

a) Calculate the initial current in each branch of the circuit.

b) Find $v(t)$ for $t \geq 0$.

c) Find $i_L(t)$ for $t \geq 0$.

8.3 The resistance in Problem 8.2 is increased to 250 Ω.
PSPICE Find the expression for $v(t)$ for $t \geq 0$.
MULTISIM

8.4 The resistance in Problem 8.2 is increased to 312.5 Ω.
PSPICE Find the expression for $v(t)$ for $t \geq 0$.
MULTISIM

8.5 The resistor in the circuit in Example 8.4 is changed
PSPICE to 3200 Ω.
MULTISIM

a) Find the numerical expression for $v(t)$ when $t \geq 0$.

b) Plot $v(t)$ versus t for the time interval $0 \leq t \leq 7$ ms. Compare this response with the one in Example 8.4 ($R = 20$ kΩ) and Example 8.5 ($R = 4$ kΩ). In particular, compare peak values of $v(t)$ and the times when these peak values occur.

8.6 Suppose the inductor in the circuit shown in Fig. 8.1 has a value of 10 mH. The voltage response for $t \geq 0$ is

$$v(t) = 40e^{-1000t} - 90e^{-4000t}\ \text{V}.$$

a) Determine the numerical values of ω_0, α, C, and R.

b) Calculate $i_R(t), i_L(t)$, and $i_C(t)$ for $t \geq 0^+$.

8.7 The natural response for the circuit shown in Fig. 8.1 is known to be

$$v(t) = 3e^{-100t} + 3e^{-900t}\ \text{V}, \quad t \geq 0.$$

If $C = 2.5\ \mu$F and $L = (40/9)$ H, find $i_L(0^+)$ in milliamperes.

8.8 The natural voltage response of the circuit in Fig. 8.1 is

$$v(t) = 120e^{-400t}\cos 300t + 80e^{-400t}\sin 300t\ \text{V},$$

when the capacitor is 250 μF. Find (a) L; (b) R; (c) V_0; (d) I_0; and (e) $i_L(t)$.

8.9 The voltage response for the circuit in Fig. 8.1 is known to be

$$v(t) = D_1 t e^{-500t} + D_2 e^{-500t}, \quad t \geq 0.$$

The initial current in the inductor (I_0) is -10 mA, and the initial voltage on the capacitor (V_0) is 8 V. The inductance is 4 H.

a) Find the values of R, C, D_1, and D_2.

b) Find $i_C(t)$ for $t \geq 0^+$.

8.10 In the circuit shown in Fig. 8.1, a 20 mH inductor is
PSPICE shunted by a 500 nF capacitor, the resistor *R* is adjust-
MULTISIM ed for critical damping, $V_0 = 40$ V, and $I_0 = 120$ mA.

a) Calculate the numerical value of *R*.

b) Calculate $v(t)$ for $t \geq 0$.

c) Find $v(t)$ when $i_C(t) = 0$.

d) What percentage of the initially stored energy remains stored in the circuit at the instant $i_C(t)$ is 0?

8.11 a) Design a parallel *RLC* circuit (see Fig. 8.1) using component values from Appendix H, with a resonant radian frequency of 5000 rad/s. Choose a resistor or create a resistor network so that the response is critically damped. Draw your circuit.

b) Calculate the roots of the characteristic equation for the resistance in part (a).

8.12 a) Change the resistance for the circuit you designed in Problem 8.11(a) so that the response is underdamped. Continue to use components from Appendix H. Calculate the roots of the characteristic equation for this new resistance.

b) Change the resistance for the circuit you designed in Problem 8.11(a) so that the response is overdamped. Continue to use components from Appendix H. Calculate the roots of the characteristic equation for this new resistance.

8.13 The initial value of the voltage v in the circuit in Fig. 8.1 is zero, and the initial value of the capacitor current, $i_C(0^+)$, is 45 mA. The expression for the capacitor current is known to be

$$i_C(t) = A_1 e^{-200t} + A_2 e^{-800t}, \quad t \geq 0^+,$$

when R is 250 Ω. Find

a) the values of $\alpha, \omega_0, L, C, A_1,$ and A_2

$$\left(\text{Hint: } \frac{di_C(0^+)}{dt} = -\frac{di_L(0^+)}{dt} - \frac{di_R(0^+)}{dt} = \frac{-v(0)}{L} - \frac{1}{R}\frac{i_C(0^+)}{C}\right),$$

b) the expression for $v(t), \quad t \geq 0$,

c) the expression for $i_R(t) \geq 0$,

d) the expression for $i_L(t) \geq 0$.

8.14 The two switches in the circuit seen in Fig. P8.14 operate synchronously. When switch 1 is in position a, switch 2 is in position d. When switch 1 moves to position b, switch 2 moves to position c. Switch 1 has been in position a for a long time. At $t = 0$, the switches move to their alternate positions. Find $v_o(t)$ for $t \geq 0$.

PSPICE
MULTISIM

8.15 The resistor in the circuit of Fig. P8.14 is increased from 100 Ω to 125 Ω. Find $v_o(t)$ for $t \geq 0$.

PSPICE
MULTISIM

8.16 The resistor in the circuit of Fig. P8.14 is increased from 100 Ω to 200 Ω. Find $v_o(t)$ for $t \geq 0$.

PSPICE
MULTISIM

8.17 The switch in the circuit of Fig. P8.17 has been in position a for a long time. At $t = 0$ the switch moves instantaneously to position b. Find $v_o(t)$ for $t \geq 0$.

PSPICE
MULTISIM

Figure P8.17

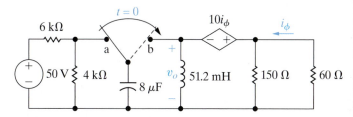

8.18 The inductor in the circuit of Fig. P8.17 is increased to 80 mH. Find $v_o(t)$ for $t \geq 0$.

8.19 The inductor in the circuit of Fig. P8.17 is increased to 125 mH. Find $v_o(t)$ for $t \geq 0$.

8.20 In the circuit in Fig. 8.1, $R = 12.5$ Ω, $C = 80$ mF, $L = (50/101)$ H, $V_0 = 0$ V, and $I_0 = -4$ A.

PSPICE
MULTISIM

a) Find $v(t)$ for $t \geq 0$.

b) Find the first three values of t for which dv/dt is zero. Let these values of t be denoted $t_1, t_2,$ and t_3.

c) Show that $t_3 - t_1 = T_d$.

d) Show that $t_2 - t_1 = T_d/2$.

e) Calculate $v(t_1), v(t_2),$ and $v(t_3)$.

f) Sketch $v(t)$ versus t for $0 \leq t \leq t_2$.

8.21 a) Find $v(t)$ for $t \geq 0$ in the circuit in Problem 8.20 if the 12.5 Ω resistor is removed from the circuit.

PSPICE
MULTISIM

b) Calculate the frequency of $v(t)$ in hertz.

c) Calculate the maximum amplitude of $v(t)$ in volts.

8.22 Assume the underdamped voltage response of the circuit in Fig. 8.1 is written as

$$v(t) = (A_1 + A_2)e^{-\alpha t} \cos \omega_d t + j(A_1 - A_2)e^{-\alpha t} \sin \omega_d t.$$

The initial value of the inductor current is I_0, and the initial value of the capacitor voltage is V_0. Show that A_2 is the conjugate of A_1. (Hint: Use the same process as outlined in the chapter to find A_1 and A_2.)

8.23 Show that the results obtained from Problem 8.22 — that is, the expressions for A_1 and A_2 — are consistent with Eqs. 8.16 and 8.17 in the text.

Section 8.3

8.24 For the circuit in Example 8.6, find, for $t \geq 0$, (a) $v(t)$; (b) $i_R(t)$; and (c) $i_C(t)$.

PSPICE
MULTISIM

Figure P8.14

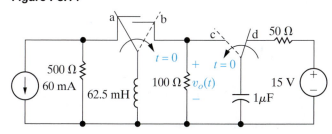

8.25 For the circuit in Example 8.7, find, for $t \geq 0$, (a) $v(t)$ and (b) $i_C(t)$.

8.26 For the circuit in Example 8.8, find $v(t)$ for $t \geq 0$.

8.27 The switch in the circuit in Fig. P8.27 has been open for a long time before closing at $t = 0$. Find $i_o(t)$ for $t \geq 0$.

Figure P8.27

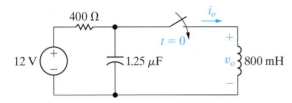

8.28 a) For the circuit in Fig. P8.27, find v_o for $t \geq 0$.

b) Show that your solution for v_o is consistent with the solution for i_o in Problem 8.27.

8.29 The switch in the circuit in Fig. P8.29 has been open a long time before closing at $t = 0$. At the time the switch closes, the capacitor has no stored energy. Find v_o for $t \geq 0$.

Figure P8.29

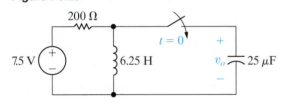

8.30 There is no energy stored in the circuit in Fig. P8.30 when the switch is closed at $t = 0$. Find $i_o(t)$ for $t \geq 0$.

Figure P8.30

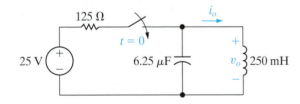

8.31 a) For the circuit in Fig. P8.30, find v_o for $t \geq 0$.

b) Show that your solution for v_o is consistent with the solution for i_o in Problem 8.30.

8.32 Assume that at the instant the 2 A current source is applied to the circuit in Fig. P8.32, the initial current in the 25 mH inductor is 1 A, and the initial voltage on the capacitor is 50 V (positive at the upper terminal). Find the expression for $i_L(t)$ for $t \geq 0$ if R equals 12.5 Ω.

Figure P8.32

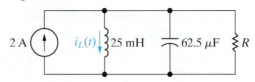

8.33 The resistance in the circuit in Fig. P8.32 is changed to 8 Ω. Find $i_L(t)$ for $t \geq 0$.

8.34 The resistance in the circuit in Fig. P8.32 is changed to 10 Ω. Find $i_L(t)$ for $t \geq 0$.

8.35 Switches 1 and 2 in the circuit in Fig. P8.35 are synchronized. When switch 1 is opened, switch 2 closes and vice versa. Switch 1 has been open a long time before closing at $t = 0$. Find $i_L(t)$ for $t \geq 0$.

Figure P8.35

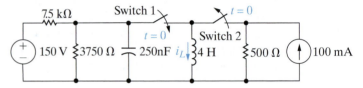

8.36 The switch in the circuit in Fig. P8.36 has been in the left position for a long time before moving to the right position at $t = 0$. Find

a) $i_L(t)$ for $t \geq 0$,

b) $v_C(t)$ for $t \geq 0$.

Figure P8.36

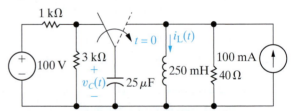

8.37 Consider the circuit in Fig. P8.36.

a) Find the total energy delivered to the inductor.

b) Find the total energy delivered to the 40 Ω resistor.

c) Find the total energy delivered to the capacitor.

d) Find the total energy delivered by the current source.

e) Check the results of parts (a) through (d) against the conservation of energy principle.

8.38 The switch in the circuit in Fig. P8.38 has been open for a long time before closing at $t = 0$. Find $i_L(t)$ for $t \geq 0$.

Figure P8.38

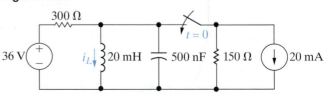

Section 8.4

8.39 In the circuit in Fig. P8.39, the resistor is adjusted for critical damping. The initial capacitor voltage is 20 V, and the initial inductor current is 30 mA.

a) Find the numerical value of R.

b) Find the numerical values of i and di/dt immediately after the switch is closed.

c) Find $v_C(t)$ for $t \geq 0$.

Figure P8.39

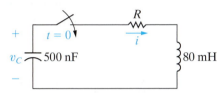

8.40 The initial energy stored in the 50 nF capacitor in the circuit in Fig. P8.40 is 90 μJ. The initial energy stored in the inductor is zero. The roots of the characteristic equation that describes the natural behavior of the current i are -1000 s^{-1} and $-4,000$ s^{-1}.

a) Find the numerical values of R and L.

b) Find the numerical values of $i(0)$ and $di(0)/dt$ immediately after the switch has been closed.

c) Find $i(t)$ for $t \geq 0$.

d) How many microseconds after the switch closes does the current reach its maximum value?

e) What is the maximum value of i in milliamperes?

f) Find $v_L(t)$ for $t \geq 0$.

Figure P8.40

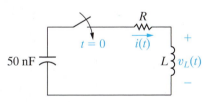

8.41 The current in the circuit in Fig. 8.3 is known to be

$$i = B_1 e^{-2000t} \cos 1500t + B_2 e^{-2000t} \sin 1500t, \quad t \geq 0.$$

The capacitor has a value of 80 nF; the initial value of the current is 7.5 mA; and the initial voltage on the capacitor is -30 V. Find the values of R, L, B_1, and B_2.

8.42 Find the voltage across the 80 nF capacitor for the circuit described in Problem 8.41. Assume the reference polarity for the capacitor voltage is positive at the upper terminal.

8.43 a) Design a series RLC circuit (see Fig. 8.3) using component values from Appendix H, with a resonant radian frequency of 20 krad/s. Choose a resistor or create a resistor network so that the response is critically damped. Draw your circuit.

b) Calculate the roots of the characteristic equation for the resistance in part (a).

8.44 a) Change the resistance for the circuit you designed in Problem 8.43(a) so that the response is underdamped. Continue to use components from Appendix H. Calculate the roots of the characteristic equation for this new resistance.

b) Change the resistance for the circuit you designed in Problem 8.43(a) so that the response is overdamped. Continue to use components from Appendix H. Calculate the roots of the characteristic equation for this new resistance.

8.45 The switch in the circuit shown in Fig. P8.45 has been in position a for a long time. At $t = 0$, the switch is moved instantaneously to position b. Find $i(t)$ for $t \geq 0$.

Figure P8.45

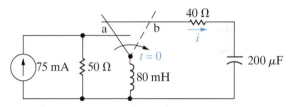

8.46 The switch in the circuit in Fig. P8.46 has been in position a for a long time. At $t = 0$, the switch moves instantaneously to position b.

a) What is the initial value of v_a?

b) What is the initial value of dv_a/dt?

c) What is the numerical expression for $v_a(t)$ for $t \geq 0$?

Figure P8.46

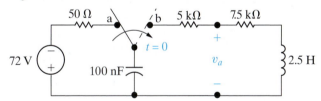

8.47 The circuit shown in Fig. P8.47 has been in operation for a long time. At $t = 0$, the two switches move to the new positions shown in the figure. Find

a) $i_o(t)$ for $t \geq 0$,

b) $v_o(t)$ for $t \geq 0$.

Figure P8.47

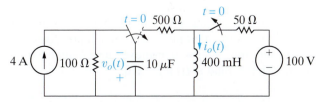

8.48 The switch in the circuit shown in Fig. P8.48 has been closed for a long time. The switch opens at $t = 0$. Find $v_o(t)$ for $t \geq 0^+$.

Figure P8.48

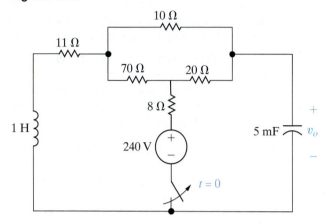

8.49 The circuit shown in Fig. P8.49 has been in operation for a long time. At $t = 0$, the source voltage suddenly increases to 250 V. Find $v_o(t)$ for $t \geq 0$.

PSPICE
MULTISIM

Figure P8.49

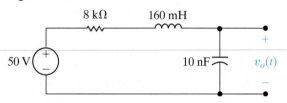

8.50 The initial energy stored in the circuit in Fig. P8.50 is zero. Find $v_o(t)$ for $t \geq 0$.

PSPICE
MULTISIM

Figure P8.50

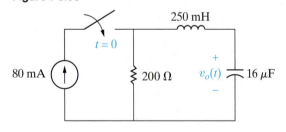

8.51 The resistor in the circuit shown in Fig. P8.50 is changed to 250 Ω. The initial energy stored is still zero. Find $v_o(t)$ for $t \geq 0$.

PSPICE
MULTISIM

8.52 The resistor in the circuit shown in Fig. P8.50 is changed to 312.5 Ω. The initial energy stored is still zero. Find $v_o(t)$ for $t \geq 0$.

8.53 The two switches in the circuit seen in Fig. P8.53 operate synchronously. When switch 1 is in position a, switch 2 is closed. When switch 1 is in position b, switch 2 is open. Switch 1 has been in position a for a long time. At $t = 0$, it moves instantaneously to position b. Find $v_C(t)$ for $t \geq 0$.

PSPICE
MULTISIM

Figure P8.53

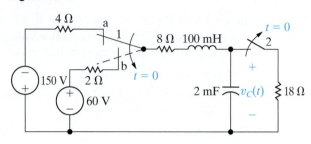

8.54 The switch in the circuit of Fig. P8.54 has been in position a for a long time. At $t = 0$ the switch moves instantaneously to position b. Find $v_o(t)$ for $t \geq 0$.

PSPICE
MULTISIM

Figure P8.54

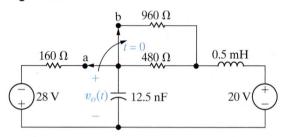

8.55 The switch in the circuit shown in Fig. P8.55 has been closed for a long time before it is opened at $t = 0$. Assume that the circuit parameters are such that the response is underdamped.

a) Derive the expression for $v_o(t)$ as a function of V_g, α, ω_d, C, and R for $t \geq 0$.

b) Derive the expression for the value of t when the magnitude of v_o is maximum.

Figure P8.55

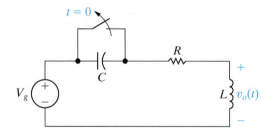

8.56 The circuit parameters in the circuit of Fig. P8.55 are $R = 4800$ Ω, $L = 64$ mH, $C = 4$ nF, and $v_g = -72$ V.

PSPICE
MULTISIM

a) Express $v_o(t)$ numerically for $t \geq 0$.

b) How many microseconds after the switch opens is the inductor voltage maximum?

c) What is the maximum value of the inductor voltage?

d) Repeat (a)–(c) with R reduced to 480 Ω.

8.57 Assume that the capacitor voltage in the circuit of Fig. 8.16 is underdamped. Also assume that no energy is stored in the circuit elements when the switch is closed.

a) Show that $dv_C/dt = (\omega_0^2/\omega_d)Ve^{-\alpha t} \sin \omega_d t$.

b) Show that $dv_C/dt = 0$ when $t = n\pi/\omega_d$, where $n = 0, 1, 2, \ldots$.

c) Let $t_n = n\pi/\omega_d$, and show that

$$v_C(t_n) = V - V(-1)^n e^{-\alpha n\pi/\omega_d}.$$

d) Show that

$$\alpha = \frac{1}{T_d} \ln \frac{v_C(t_1) - V}{v_C(t_3) - V},$$

where $T_d = t_3 - t_1$.

8.58 The voltage across a 100 nF capacitor in the circuit of Fig. 8.16 is described as follows: After the switch has been closed for several seconds, the voltage is constant at 100 V. The first time the voltage exceeds 100 V, it reaches a peak of 163.84 V. This occurs $\pi/7$ ms after the switch has been closed. The second time the voltage exceeds 100 V, it reaches a peak of 126.02 V. This second peak occurs $3\pi/7$ after the switch has been closed. At the time when the switch is closed, there is no energy stored in either the capacitor or the inductor. Find the numerical values of R and L. (Hint: Work Problem 8.57 first.)

Section 8.5

8.59 Show that, if no energy is stored in the circuit shown in Fig. 8.20 at the instant v_g jumps in value, then dv_o/dt equals zero at $t = 0$.

8.60 a) Find the equation for $v_o(t)$ for $0 \le t \le t_{sat}$ in the circuit shown in Fig. 8.20 if $v_{o1}(0) = 5$ V and $v_o(0) = 8$ V.

PSPICE
MULTISIM

b) How long does the circuit take to reach saturation?

8.61 a) Rework Example 8.14 with feedback resistors R_1 and R_2 removed.

b) Rework Example 8.14 with $v_{o1}(0) = -2$ V and $v_o(0) = 4$ V.

8.62 a) Derive the differential equation that relates the output voltage to the input voltage for the circuit shown in Fig. P8.62.

b) Compare the result with Eq. 8.39 when $R_1C_1 = R_2C_2 = RC$ in Fig. 8.19.

c) What is the advantage of the circuit shown in Fig. P8.62?

Figure P8.62

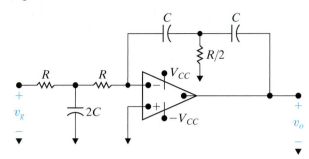

8.63 The voltage signal of Fig. P8.63(a) is applied to the cascaded integrating amplifiers shown in Fig. P8.63(b). There is no energy stored in the capacitors at the instant the signal is applied.

PSPICE
MULTISIM

a) Derive the numerical expressions for $v_o(t)$ and $v_{o1}(t)$ for the time intervals $0 \le t \le 0.5$ s and 0.5 s $\le t \le t_{sat}$.

b) Compute the value of t_{sat}.

Figure P8.63

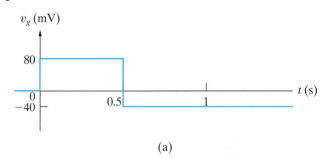

(a)

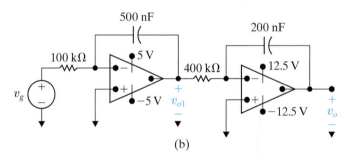

(b)

8.64 The circuit in Fig. P8.63(b) is modified by adding a 1 MΩ resistor in parallel with the 500 nF capacitor and a 5 MΩ resistor in parallel with the 200 nF capacitor. As in Problem 8.63, there is no energy stored in the capacitors at the time the signal is applied. Derive the numerical expressions for $v_o(t)$ and $v_{o1}(t)$ for the time intervals $0 \le t \le 0.5$ s and $t \ge 0.5$ s.

PSPICE
MULTISIM

8.65 We now wish to illustrate how several op amp circuits can be interconnected to solve a differential equation.

a) Derive the differential equation for the spring-mass system shown in Fig. P8.65(a). Assume that the force exerted by the spring is directly proportional to the spring displacement, that the mass is constant, and that the frictional force is directly proportional to the velocity of the moving mass.

b) Rewrite the differential equation derived in (a) so that the highest order derivative is expressed as a function of all the other terms in the equation.

Now assume that a voltage equal to d^2x/dt^2 is available and by successive integrations generates dx/dt and x. We can synthesize the coefficients in the equations by scaling amplifiers, and we use a summing amplifier to combine the terms required to generate d^2x/dt^2. With these ideas in mind, analyze the interconnection shown in Fig. P8.65(b). In particular, describe the purpose of each shaded area in the circuit and describe the signal at the points labeled B, C, D, E, and F, assuming the signal at A represents d^2x/dt^2. Also discuss the parameters $R; R_1, C_1; R_2, C_2; R_3, R_4; R_5, R_6;$ and R_7, R_8 in terms of the coefficients in the differential equation.

Figure P8.65

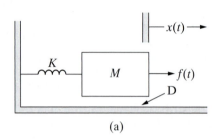

(a)

(b)

Sections 8.1–8.5

8.66 a) Suppose the circuit in Fig. 8.22 has a 5 nH induc-
PRACTICAL PERSPECTIVE
tor and a 2 pF capacitor. Calculate the frequency, in GHz, of the sinusoidal output for $t \geq 0$.

b) The dc voltage source and series-connected resistor in Fig. 8.22 are used to establish the initial energy in the inductor. If $V = 10$ V and $R_s = 25 \, \Omega$, calculate the initial energy stored in the inductor.

c) What is the total energy stored in the LC circuit for any time $t \geq 0$?

8.67 Consider the LC oscillator circuit in Fig. 8.22. As-
PRACTICAL PERSPECTIVE
sume that $V = 4$ V, $R_s = 10 \, \Omega$, and $L = 1$ nH.

a) Calculate the value of capacitance, C, that will produce a sinusoidal output with a frequency of 2 GHz for $t \geq 0$.

b) Write the expression for the output voltage, $v_o(t)$, for $t \geq 0$.

8.68 Suppose the inductor and capacitor in the LC oscil-
PRACTICAL PERSPECTIVE
lator circuit in Fig. 8.22 are not ideal, but instead have some small resistance that can be lumped together. Assume that $V = 10$ V, $R_s = 25 \, \Omega$, $L = 5$ nH, and $C = 2$ pF, just as in Problem 8.66. Suppose the resistance associated with the inductor and capacitor is 10 mΩ.

a) Calculate the values of the neper frequency, α, and the resonant radian frequency, ω_0.

b) Is the response of this circuit overdamped, underdamped, or critically damped?

c) What is the actual frequency of oscillation, in GHz?

d) Approximately how long will the circuit oscillate?

Sinusoidal
Steady-State Analysis

CHAPTER CONTENTS

9.1 **The Sinusoidal Source** *p. 320*

9.2 **The Sinusoidal Response** *p. 323*

9.3 **The Phasor** *p. 324*

9.4 **The Passive Circuit Elements in the Frequency Domain** *p. 327*

9.5 **Kirchhoff's Laws in the Frequency Domain** *p. 332*

9.6 **Series, Parallel, and Delta-to-Wye Simplifications** *p. 333*

9.7 **Source Transformations and Thévenin–Norton Equivalent Circuits** *p. 340*

9.8 **The Node-Voltage Method** *p. 344*

9.9 **The Mesh-Current Method** *p. 345*

9.10 **The Transformer** *p. 347*

9.11 **The Ideal Transformer** *p. 351*

9.12 **Phasor Diagrams** *p. 357*

CHAPTER OBJECTIVES

1 Understand phasor concepts and be able to perform a phasor transform and an inverse phasor transform.

2 Be able to transform a circuit with a sinusoidal source into the frequency domain using phasor concepts.

3 Know how to use the following circuit analysis techniques to solve a circuit in the frequency domain:
   - Kirchhoff's laws;
   - Series, parallel, and delta-to-wye simplifications;
   - Voltage and current division;
   - Thévenin and Norton equivalents;
   - Node-voltage method; and
   - Mesh-current method.

4 Be able to analyze circuits containing linear transformers using phasor methods.

5 Understand the ideal transformer constraints and be able to analyze circuits containing ideal transformers using phasor methods.

Thus far, we have focused on circuits with constant sources; in this chapter we are now ready to consider circuits energized by sinusoidal voltage or current sources. For these circuits, we will calculate the values of the specified output voltages and currents in the steady state. This means we will not know the complete response of the circuits, which in general is the sum of the transient (or natural) response and the steady-state response. Our analysis will only characterize a circuit's response once the transient component has decayed to zero.

Sinusoidal sources and their effect on circuit behavior form an important area of study for several reasons.

- Generating, transmitting, distributing, and consuming electric energy occurs under essentially sinusoidal steady-state conditions.
- Understanding sinusoidal behavior makes it possible to predict the behavior of circuits with nonsinusoidal sources.
- Specifying the behavior of an electrical system in terms of its steady-state sinusoidal response simplifies the design. If the system satisfies the specifications, the designer knows that the circuit will respond satisfactorily to nonsinusoidal inputs.

The remaining chapters of this book are largely based on the techniques used when analyzing circuits with sinusoidal sources. Fortunately, the circuit analysis and simplification techniques from Chapters 1–4 work for circuits with sinusoidal as well as dc sources, so some of the material in this chapter will be very familiar to you. The challenges of sinusoidal analysis include developing the appropriate component models, writing the equations that describe the resulting circuit, and working with complex numbers.

Practical Perspective

A Household Distribution Circuit

Power systems that generate, transmit, and distribute electrical power are designed to operate in the sinusoidal steady state. The standard household distribution circuit used in the United States supplies both 120 V and 240 V.

Consider the following situation. At the end of a day of fieldwork, a farmer returns to his farmstead, checks his hog confinement building, and finds to his dismay that the hogs are dead. The problem is traced to a blown fuse that caused a 240 V fan motor to stop. The loss of ventilation led to the suffocation of the livestock. The interrupted fuse is located in the main switch that connects the farmstead to the electrical service.

Before the insurance company settles the claim, it wants to know if the electric circuit supplying the farmstead functioned properly. The lawyers for the insurance company are puzzled because the farmer's wife, who was in the house on the day of the accident convalescing from minor surgery, was able to watch TV during the afternoon. Furthermore, when she went to the kitchen to start preparing the evening meal, the electric clock indicated the correct time. The lawyers have hired you to explain why the electric clock in the kitchen and the television set in the living room continued to operate after the fuse in the main switch blew.

We will explore this situation and answer the question after learning how to calculate the steady-state response of circuits with sinusoidal sources.

Yulia Grigoryeva/Shutterstock

Björn Erlandsson /123RF

Tetra Images/Alamy Stock Photo

9.1 The Sinusoidal Source

A **sinusoidal voltage source** (independent or dependent) produces a voltage that varies sinusoidally with time. A **sinusoidal current source** (independent or dependent) produces a current that varies sinusoidally with time. We begin by reviewing the sinusoidal function, using a voltage source as an example, but our observations also apply to current sources.

We can express a sinusoidally varying function with either the sine function or the cosine function. Although they work equally well, we cannot use both functional forms simultaneously. We will use the cosine function throughout our discussion. Hence, we write a sinusoidally varying voltage as

$$v = V_m \cos(\omega t + \phi). \tag{9.1}$$

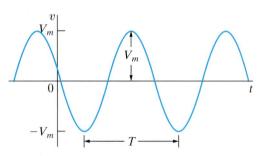

Figure 9.1 ▲ A sinusoidal voltage.

To aid discussion of the parameters in Eq. 9.1, we show the voltage versus time plot in Fig. 9.1. The coefficient V_m gives the maximum **amplitude** of the sinusoidal voltage. Because ± 1 bounds the cosine function, $\pm V_m$ bounds the amplitude, as seen in Fig. 9.1. You can also see that the sinusoidal function repeats at regular intervals; therefore, it is a periodic function. A periodic function is characterized by the time required for the function to pass through all its possible values. This time is the **period** of the function, T. and is measured in seconds. The reciprocal of T gives the number of cycles per second, or the frequency, of the periodic function, and is denoted f, so

$$f = \frac{1}{T}. \tag{9.2}$$

A cycle per second is called a hertz, abbreviated Hz. (The term *cycles per second* rarely is used in contemporary technical literature.)

Now look at the coefficient of t in Eq. 9.1. Omega (ω) represents the **angular frequency** of the sinusoidal function and is related to both T and f:

$$\omega = 2\pi f = 2\pi/T \, (\text{radians/second}). \tag{9.3}$$

Equation 9.3 tells us that the cosine (or sine) function passes through a complete set of values each time its argument, ωt, passes through 2π rad (360°). From Eq. 9.3, we see that whenever t is an integral multiple of T, the argument ωt increases by an integral multiple of 2π rad.

The angle ϕ in Eq. 9.1 is the **phase angle** of the sinusoidal voltage. It determines the value of the sinusoidal function at $t = 0$; therefore, it fixes the point on the periodic wave where we start measuring time. Changing the phase angle ϕ shifts the sinusoidal function along the time axis but has no effect on either the amplitude (V_m) or the angular frequency (ω). Note, for example, that reducing ϕ to zero shifts the sinusoidal function shown in Fig. 9.1 ϕ/ω time units to the right, as shown in Fig. 9.2. When compared with a sinusoidal function with $\phi = 0$, a sinusoidal function with a positive ϕ is shifted to the left, while a sinusoidal function with a negative ϕ is shifted to the right. (See Problem 9.4.)

Remember that ωt and ϕ must carry the same units because the argument of the sinusoidal function is ($\omega t + \phi$). With ωt expressed in radians, you would expect ϕ to also be in radians. However, ϕ normally is given in degrees, and ωt is converted from radians to degrees before the two quantities are added. The conversion from radians to degrees is given by

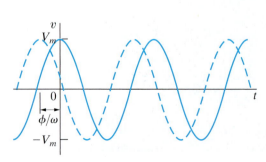

Figure 9.2 ▲ The sinusoidal voltage from Fig. 9.1 shifted to the right when $\phi = 0$.

$$(\text{number of degrees}) = \frac{180°}{\pi}(\text{number of radians}).$$

Another important characteristic of the sinusoidal voltage (or current) is its **rms value**. The rms value of a periodic function is defined as the square **r**oot of the **m**ean value of the **s**quared function. Hence, if $v = V_m \cos(\omega t + \phi)$, the rms value of v is

$$V_{rms} = \sqrt{\frac{1}{T} \int_{t_0}^{t_0 + T} V_m^2 \cos^2(\omega t + \phi) \, dt}. \tag{9.4}$$

Note from Eq. 9.4 that we obtain the mean value of the squared voltage by integrating v^2 over one period (that is, from t_0 to $t_0 + T$) and then dividing by the range of integration, T. Note further that the starting point for the integration t_0 is arbitrary.

The quantity under the square root sign in Eq. 9.4 reduces to $V_m^2/2$. (See Problem 9.7.) Hence, the rms value of v is

RMS VALUE OF A SINUSOIDAL VOLTAGE SOURCE

$$V_{rms} = \frac{V_m}{\sqrt{2}}. \tag{9.5}$$

The rms value of the sinusoidal voltage depends only on the maximum amplitude of v, namely, V_m. The rms value is not a function of either the frequency or the phase angle. In Chapter 10, we explain the importance of the rms value and use it extensively to calculate power in circuits with sinusoidal sources.

We can completely describe a specific sinusoidal signal if we know its frequency, phase angle, and amplitude. Examples 9.1, 9.2, and 9.3 illustrate these basic properties of the sinusoidal function. In Example 9.4, we calculate the rms value of a periodic function, and in so doing we clarify the meaning of *root mean square*.

EXAMPLE 9.1   Finding the Characteristics of a Sinusoidal Current

A sinusoidal current has a maximum amplitude of 20 A. The current passes through one complete cycle in 1 ms. The magnitude of the current at $t = 0$ is 10 A.

a) What is the frequency of the current in hertz?

b) What is the frequency in radians per second?

c) Write the expression for $i(t)$ using the cosine function. Express ϕ in degrees.

d) What is the rms value of the current?

Solution

a) From the statement of the problem, $T = 1$ ms; hence, $f = 1/T = 1/0.001 = 1000$ Hz.

b) $\omega = 2\pi f = 2\pi(1000) = 2000\pi$ rad/sec.

c) We have

$$i(t) = I_m \cos(\omega t + \phi)$$
$$= 20 \cos(2000\pi t + \phi) \, \text{A},$$

but $i(0) = 10$ A. Therefore, $10 = 20 \cos \phi$, so $\phi = 60°$. Thus, the expression for $i(t)$ becomes

$$i(t) = 20 \cos(2000\pi t + 60°) \, \text{A}.$$

d) From Eq. 9.5, the rms value of a sinusoidal current is $I_m/\sqrt{2}$. Therefore, the rms value is $20/\sqrt{2}$, or 14.14 A.

EXAMPLE 9.2 | Finding the Characteristics of a Sinusoidal Voltage

A sinusoidal voltage is given by the expression $v = 300 \cos(120\pi t + 30°)$ V.

a) What is the period of the voltage in milliseconds?

b) What is the frequency in hertz?

c) What is the magnitude of v at $t = 2.778$ ms?

d) What is the rms value of v?

Solution

a) From the expression for v, $\omega = 120\pi$ rad/s. Because $\omega = 2\pi/T$, $T = 2\pi/\omega = 1/60$ s, or 16.667 ms.

b) The frequency is $1/T$, or 60 Hz.

c) From (a), $\omega = 2\pi/16.667$; thus, at $t = 2.778$ ms, ωt is nearly 1.047 rad, or 60°. Therefore, $v(2.778 \text{ ms}) = 300 \cos(60° + 30°) = 0$ V.

d) $V_{rms} = 300/\sqrt{2} = 212.13$ V.

EXAMPLE 9.3 | Translating a Sine Expression to a Cosine Expression

We can translate the sine function to the cosine function by subtracting 90° ($\pi/2$ rad) from the argument of the sine function.

a) Verify this translation by showing that

$$\sin(\omega t + \theta) = \cos(\omega t + \theta - 90°).$$

b) Use the result in (a) to express $\sin(\omega t + 30°)$ as a cosine function.

Solution

a) Verification involves direct application of the trigonometric identity

$$\cos(\alpha - \beta) = \cos\alpha \cos\beta + \sin\alpha \sin\beta.$$

We let $\alpha = \omega t + \theta$ and $\beta = 90°$. From the trigonometric identity,

$$\cos(\omega t + \theta - 90°) = \cos(\omega t + \theta)\cos(90°) + \sin(\omega t + \theta)\sin(90°).$$

Since $\cos 90° = 0$ and $\sin 90° = 1$, we have

$$\cos(\omega t + \theta - 90°) = \sin(\omega t + \theta).$$

b) From (a) we have

$$\sin(\omega t + 30°) = \cos(\omega t + 30° - 90°) = \cos(\omega t - 60°).$$

EXAMPLE 9.4 | Calculating the rms Value of a Triangular Waveform

Calculate the rms value of the periodic triangular current shown in Fig. 9.3. Express your answer in terms of the peak current I_p.

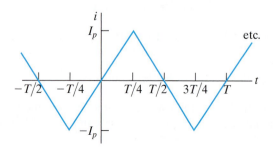

Figure 9.3 ▲ Periodic triangular current.

Solution

From the definition of rms, the rms value of i is

$$I_{rms} = \sqrt{\frac{1}{T}\int_{t_0}^{t_0+T} i^2 dt}.$$

Interpreting the integral under the square root sign as the area under the squared function for an interval of one period helps us find the rms value. The squared function, with the area between 0 and T shaded, is shown in Fig. 9.4. Notice that the area under the squared current for an interval of one period is equal to four times the area under the squared current for the interval 0 to $T/4$ seconds; that is,

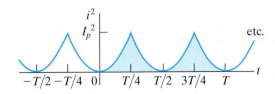

Figure 9.4 ▲ i^2 versus t.

$$\int_{t_0}^{t_0+T} i^2 dt = 4\int_0^{T/4} i^2 dt.$$

The analytical expression for i in the interval 0 to $T/4$ is

$$i = \frac{4I_p}{T}t, \quad 0 < t < T/4.$$

The area under the squared function for one period is

$$\int_{t_0}^{t_0+T} i^2 dt = 4\int_0^{T/4} \frac{16I_p^2}{T^2}t^2 dt = \frac{I_p^2 T}{3}.$$

The mean, or average, value of the function is simply the area for one period divided by the period. Thus

$$i_{mean} = \frac{1}{T}\frac{I_p^2 T}{3} = \frac{1}{3}I_p^2.$$

The rms value of the current is the square root of this mean value. Hence

$$I_{rms} = \frac{I_p}{\sqrt{3}}.$$

SELF-CHECK: Assess your understanding of this material by trying Chapter Problems 9.1, 9.2, and 9.8.

9.2 The Sinusoidal Response

As stated in the Introduction, this chapter focuses on the steady-state response to sinusoidal sources. But we begin by characterizing the total response, which will help you keep the steady-state solution in perspective.

The circuit shown in Fig. 9.5 describes the general problem, where v_s is a sinusoidal voltage described by

$$v_s = V_m \cos(\omega t + \phi).$$

For convenience, we assume the circuit's initial current is zero, and we measure time from the moment the switch is closed. We want to find $i(t)$ for $t \geq 0$, using a method similar to the one used when finding the step response of an RL circuit (Chapter 7). But here, the voltage source is time-varying sinusoidal voltage rather than a constant voltage. Applying KVL to the circuit in Fig. 9.5 gives us the ordinary differential equation

$$L\frac{di}{dt} + Ri = V_m \cos(\omega t + \phi). \tag{9.6}$$

The solution for Eq. 9.6 is discussed in an introductory course in differential equations. We ask those of you who have not yet studied differential equations to accept that the solution for i is

$$i = \frac{-V_m}{\sqrt{R^2 + \omega^2 L^2}}\cos(\phi - \theta)e^{-(R/L)t} + \frac{V_m}{\sqrt{R^2 + \omega^2 L^2}}\cos(\omega t + \phi - \theta), \tag{9.7}$$

where

$$\theta = \tan^{-1}\left(\frac{\omega L}{R}\right).$$

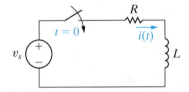

Figure 9.5 ▲ An RL circuit excited by a sinusoidal voltage source.

Thus, we can easily determine θ for a circuit driven by a sinusoidal source of known frequency. We can check that Eq. 9.7 is valid by showing that it satisfies Eq. 9.6 for all values of $t \geq 0$; this exercise is left for your exploration in Problem 9.9.

Look carefully at the two terms on the right-hand side of Eq. 9.7. The first term is a decaying exponential function whose time constant is $\tau = L/R$. This term is the **transient component** of the current because it decays to zero as $t \to 0$. Remember from Chapter 7 that this transient component has less than 1% of its initial value when $t = 5\tau$.

The second term is a cosine whose frequency is ω, the same as the frequency of the voltage source. This is the **steady-state component** of the current because it persists as long as the switch remains closed and the source continues to supply the sinusoidal voltage. In this chapter, we find only the steady-state response of circuits with sinusoidal sources; that is, we find the response once its transient component has decayed to zero. We develop a technique for calculating the steady-state response directly, thus avoiding the problem of solving the differential equation. However, when we use this technique, we cannot find either the transient component or the total response.

Using the steady-state component of Eq. 9.7, we identify four important characteristics of the steady-state solution:

1. The steady-state solution is a cosine function, just like the circuit's source.

2. The frequency of the solution is identical to the frequency of the source. This condition is always true in a linear circuit when the circuit parameters, R, L, and C, are constant. (If frequencies in the solution are not present in the source, there is a nonlinear element in the circuit.)

3. The maximum amplitude of the steady-state response, in general, differs from the maximum amplitude of the source. For the circuit in Fig. 9.5, the maximum amplitude of the current is $V_m / \sqrt{R^2 + \omega^2 L^2}$, while the maximum amplitude of the source is V_m.

4. The phase angle of the steady-state response, in general, differs from the phase angle of the source. For the circuit being discussed, the phase angle of the current is $\phi - \theta$, and that of the voltage source is ϕ.

These characteristics motivate the phasor method, which we introduce in Section 9.3. Note that finding only the steady-state response means finding only its maximum amplitude and phase angle. The waveform and frequency of the steady-state response are already known because they are the same as the circuit's source.

SELF-CHECK: Assess your understanding of this material by trying Chapter Problem 9.10.

9.3 The Phasor

A **phasor** is a complex number that carries the amplitude and phase angle information of a sinusoidal function.[1] The phasor concept is rooted in Euler's identity, which relates the exponential function to the trigonometric function:

$$e^{\pm j\theta} = \cos\theta \pm j\sin\theta.$$

Euler's identity gives us another way of representing the cosine and sine functions. We can think of the cosine function as the real part of the

---

[1] You can review complex numbers by reading Appendix B.

exponential function and the sine function as the imaginary part of the exponential function; that is,

$$\cos \theta = \mathcal{R}\{e^{j\theta}\}$$

and

$$\sin \theta = \mathcal{I}\{e^{j\theta}\},$$

where $\mathcal{R}$ means "the real part of" and $\mathcal{I}$ means "the imaginary part of."

Because we chose to use the cosine function to represent sinusoidal signals (see Section 9.1), we can apply Euler's identity directly. In particular, we write the sinusoidal voltage function given in Eq. 9.1 by replacing the cosine function with the real part of the complex exponential:

$$v = V_m \cos(\omega t + \phi)$$

$$= V_m \mathcal{R}\{e^{j(\omega t + \phi)}\}$$

$$= V_m \mathcal{R}\{e^{j\omega t}e^{j\phi}\}.$$

We can move the constant V_m inside the argument of the $\mathcal{R}$ function without altering the equation. We can also reverse the order of the two exponential functions inside the argument and write the voltage as

$$v = \mathcal{R}\{V_m e^{j\phi}e^{j\omega t}\}.$$

In this expression for the voltage, note that the quantity $V_m e^{j\phi}$ is a complex number that carries the amplitude and phase angle of the cosine function we started with (Eq. 9.1). We define this complex number as the **phasor representation**, or **phasor transform**, of the given sinusoidal function. Thus

PHASOR TRANSFORM

$$\mathbf{V} = V_m e^{j\phi} = \mathcal{P}\{V_m \cos(\omega t + \phi)\}, \qquad (9.8)$$

where the notation $\mathcal{P}\{V_m \cos(\omega t + \phi)\}$ is read as "the phasor transform of $V_m \cos(\omega t + \phi)$." Thus, the phasor transform transfers the sinusoidal function from the time domain to the complex-number domain, which is also called the **frequency domain**, since the response depends, in general, on ω. As in Eq. 9.8, throughout this book we represent a phasor quantity by using a boldface capital letter.

Equation 9.8 is the polar form of a phasor, but we also can express a phasor in rectangular form. Thus, we rewrite Eq. 9.8 as

$$\mathbf{V} = V_m \cos \phi + jV_m \sin \phi.$$

Both polar and rectangular forms are useful in circuit applications of the phasor concept.

We see from Eq. 9.8 that phasors always have the form $Ae^{j\phi}$, where A is the amplitude of the underlying voltage or current. It is common to abbreviate phasors using the angle notation $A\underline{/\phi^\circ}$, where

$$A\underline{/\phi^\circ} \equiv Ae^{j\phi}.$$

We use this angle notation extensively in the material that follows.

Inverse Phasor Transform

Using Eq. 9.8, we can transform a sinusoidal function to a phasor. We can also reverse the process; that is, we can transform a phasor back to the original sinusoidal function. If $\mathbf{V} = 100\underline{/-26°}$ V, the expression for v is $100 \cos(\omega t - 26°)$ V because we have decided to use the cosine function for all sinusoids. Notice that the phasor cannot give us the value of ω because it carries only amplitude and phase information. When we transform a phasor to the corresponding time-domain expression, we use the *inverse phasor transform* function, as shown in the equation

> ### INVERSE PHASOR TRANSFORM
>
> $$\mathcal{P}^{-1}\{V_m e^{j\theta}\} = \mathcal{R}\{V_m e^{j\theta} e^{j\omega t}\} = V_m \cos(\omega t + \theta°) \quad (9.9)$$

where the notation $\mathcal{P}^{-1}\{V_m e^{j\phi}\}$ is read as "the inverse phasor transform of $V_m e^{j\phi}$." Using Eq. 9.9, we find the inverse phasor transform by multiplying the phasor by $e^{j\omega t}$ and extracting the real part of the product.

Before applying the phasor transform to circuit analysis, we use it to solve a problem with which you are already familiar: adding sinusoidal functions. Example 9.5 shows how the phasor transform greatly simplifies this type of problem.

EXAMPLE 9.5 | Adding Cosines Using Phasors

If $y_1 = 20 \cos(\omega t - 30°)$ and $y_2 = 40 \cos(\omega t + 60°)$, express $y = y_1 + y_2$ as a single sinusoidal function.

a) Solve by using trigonometric identities.

b) Solve by using the phasor concept.

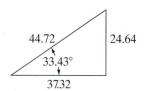

Figure 9.6 ▲ A right triangle used in the solution for y.

Solution

a) First, we expand both y_1 and y_2, using the cosine of the sum of two angles, to get

$$y_1 = 20 \cos \omega t \cos 30° + 20 \sin \omega t \sin 30°;$$

$$y_2 = 40 \cos \omega t \cos 60° - 40 \sin \omega t \sin 60°.$$

Adding y_1 and y_2, we obtain

$$y = (20 \cos 30° + 40 \cos 60°) \cos \omega t$$

$$+ (20 \sin 30° - 40 \sin 60°) \sin \omega t$$

$$= 37.32 \cos \omega t - 24.64 \sin \omega t.$$

To combine these two terms, we treat the coefficients of the cosine and sine as sides of a right triangle (Fig. 9.6) and then multiply and divide the right-hand side by the hypotenuse. Our expression for y becomes

$$y = 44.72\left(\frac{37.32}{44.72} \cos \omega t - \frac{24.64}{44.72} \sin \omega t\right)$$

$$= 44.72 (\cos 33.43° \cos \omega t - \sin 33.43° \sin \omega t).$$

Again, we invoke the identity involving the cosine of the sum of two angles and write

$$y = 44.72 \cos(\omega t + 33.43°).$$

b) The sum of the two cosines is

$$y = 20 \cos(\omega t - 30°) + 40 \cos(\omega t + 60°).$$

Use Euler's identity to rewrite the righthand side of this equation as

$$y = \mathcal{R}\{20 e^{-j30°} e^{j\omega}\} + \mathcal{R}\{40 e^{j60°} e^{j\omega}\}$$

$$= \mathcal{R}\{20 e^{-j30°} e^{j\omega} + 40 e^{j60°} e^{j\omega}\}.$$

Factoring out the term $e^{j\omega}$ from each term gives

$$y = \Re\{(20e^{-j30^\circ} + 40e^{j60^\circ})e^{j\omega}\}.$$

We can calculate the sum of the two phasors using the angle notation:

$$\begin{aligned} 20\angle{-30^\circ} + 40\angle{60^\circ} &= (17.32 - j10) + (20 + j34.64) \\ &= 37.32 + j24.64 \\ &= 44.72\angle 33.43^\circ. \end{aligned}$$

Therefore,

$$\begin{aligned} y &= \Re\{44.72e^{j33.43^\circ}e^{j\omega}\} \\ &= 44.72\cos(\omega t + 33.43^\circ). \end{aligned}$$

Adding sinusoidal functions using phasors is clearly easier than using trigonometric identities. Note that it requires the ability to move back and forth between the polar and rectangular forms of complex numbers.

ASSESSMENT PROBLEMS

Objective 1—Understand phasor concepts and be able to perform a phasor transform and an inverse phasor transform

9.1 Find the phasor transform of each trigonometric function:

a) $v = 170\cos(377t - 40^\circ)$ V.

b) $i = 10\sin(1000t + 20^\circ)$ A.

c) $i = [5\cos(\omega t + 36.87^\circ) + 10\cos(\omega t - 53.13^\circ)]$ A.

d) $v = [300\cos(20{,}000\pi t + 45^\circ) - 100\sin(20{,}000\pi t + 30^\circ)]$ mV.

Answer: (a) $170\angle{-40^\circ}$ V;

(b) $10\angle{-70^\circ}$ A;

(c) $11.18\angle{-26.57^\circ}$ A;

(d) $339.90\angle{61.51^\circ}$ mV.

9.2 Find the time-domain expression corresponding to each phasor:

a) $\mathbf{V} = 18.6\angle{-54^\circ}$ V.

b) $\mathbf{I} = (20\angle{45^\circ} - 50\angle{-30^\circ})$ mA.

c) $\mathbf{V} = (20 + j80 - 30\angle{15^\circ})$ V.

Answer: (a) $18.6\cos(\omega t - 54^\circ)$ V;

(b) $48.81\cos(\omega t + 126.68^\circ)$ mA;

(c) $72.79\cos(\omega t + 97.08^\circ)$ V.

SELF-CHECK: Also try Chapter Problem 9.11.

9.4 The Passive Circuit Elements in the Frequency Domain

Applying the phasor transform in circuit analysis is a two-step process.

1. Establish the relationship between the phasor current and the phasor voltage at the terminals of the passive circuit elements. We complete this step in this section, analyzing the resistor, inductor, and capacitor in the phasor domain.
2. Develop the phasor-domain version of Kirchhoff's laws, which we discuss in Section 9.5.

The V-I Relationship for a Resistor

From Ohm's law, if the current in a resistor is $i = I_m\cos(\omega t + \theta_i)$, the voltage at the terminals of the resistor, as shown in Fig. 9.7, is

$$\begin{aligned} v &= R[I_m\cos(\omega t + \theta_i)] \\ &= RI_m[\cos(\omega t + \theta_i)], \end{aligned}$$

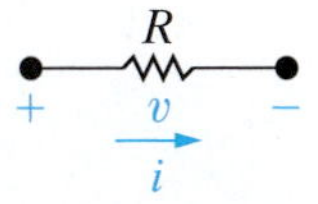

Figure 9.7 ▲ A resistive element carrying a sinusoidal current.

where I_m is the maximum amplitude of the current in amperes and θ_i is the phase angle of the current.

The phasor transform of this voltage is

$$\mathbf{V} = RI_m e^{j\theta_i} = RI_m\underline{/\theta_i}.$$

But $I_m\underline{/\theta_i}$ is the phasor representation of the sinusoidal current, so we can write the voltage phasor as

> ### RELATIONSHIP BETWEEN PHASOR VOLTAGE AND PHASOR CURRENT FOR A RESISTOR
>
> $$\mathbf{V} = R\mathbf{I}, \qquad (9.10)$$

which states that the phasor voltage at the terminals of a resistor is the resistance times the phasor current—the phasor version of Ohm's law. Figure 9.8 shows the circuit diagram for a resistor in the frequency domain.

Equation 9.10 contains an important piece of information—namely, that at the terminals of a resistor, there is no phase shift between the current and voltage. Figure 9.9 depicts this phase relationship, where the phase angle of both the voltage and the current waveforms is 60°. The signals are said to be **in phase** because they both reach corresponding values on their respective curves at the same time (for example, they are at their positive maxima at the same instant).

Figure 9.8 ▲ The frequency-domain equivalent circuit of a resistor.

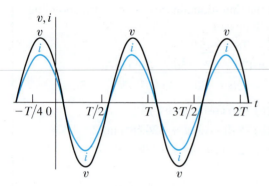

Figure 9.9 ▲ A plot showing that the voltage and current at the terminals of a resistor are in phase.

The V-I Relationship for an Inductor

We derive the relationship between the phasor current and phasor voltage at the terminals of an inductor by assuming a sinusoidal current and using $L\,di/dt$ to establish the corresponding voltage. Thus, for $i = I_m \cos(\omega t + \theta_i)$, the expression for the voltage is

$$v = L\frac{di}{dt} = -\omega L I_m \sin(\omega t + \theta_i).$$

We now replace the sine function with the cosine function:

$$v = -\omega L I_m \cos(\omega t + \theta_i - 90°).$$

The phasor representation of the voltage is then

$$\mathbf{V} = -\omega L I_m e^{j(\theta_i - 90°)}$$

$$= -\omega L I_m e^{j\theta_i} e^{-j90°}$$

$$= j\omega L I_m e^{j\theta_i}$$

$$= j\omega L I_m\underline{/\theta_i}.$$

Note that, in deriving the expression for the phasor voltage, we used the identity

$$e^{-j90°} = \cos 90° - j\sin 90° = -j.$$

Also, $I_m\underline{/\theta_i}$ is the phasor representation of the sinusoidal current, so we can express the phasor voltage in terms of the phasor current:

RELATIONSHIP BETWEEN PHASOR VOLTAGE AND PHASOR CURRENT FOR AN INDUCTOR

$$\mathbf{V} = j\omega L \mathbf{I}. \tag{9.11}$$

Equation 9.11 states that the phasor voltage at the terminals of an inductor equals $j\omega L$ times the phasor current. Figure 9.10 shows the frequency-domain equivalent circuit for the inductor. Note that the relationship between phasor voltage and phasor current for an inductor also applies for the mutual inductance in one coil due to current flowing in another mutually coupled coil. That is, the phasor voltage at the terminals of one coil in a mutually coupled pair of coils equals $j\omega M$ times the phasor current in the other coil.

We can rewrite Eq. 9.11 as

Figure 9.10 ▲ The frequency-domain equivalent circuit for an inductor.

$$\mathbf{V} = (\omega L \underline{/90°}) I_m \underline{/\theta_i}$$

$$= \omega L I_m \underline{/(\theta_i + 90)°},$$

which indicates that the voltage and current are out of phase by exactly 90°. Specifically, the voltage leads the current by 90°, or, equivalently, the current lags the voltage by 90°. Figure 9.11 illustrates the concept of *voltage leading current* or *current lagging voltage*. For example, the voltage reaches its negative peak exactly 90° before the current reaches its negative peak. The same observation can be made with respect to the zero-going-positive crossing or the positive peak.

We can also express the phase shift in seconds. A phase shift of 90° corresponds to one-fourth of a period; hence, the voltage leads the current by $T/4$, or $\frac{1}{4f}$ second.

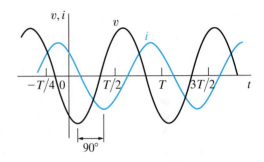

Figure 9.11 ▲ A plot showing the phase relationship between the current and voltage at the terminals of an inductor ($\theta_i = 60°$).

The V-I Relationship for a Capacitor

To determine the relationship between the phasor current and phasor voltage at the terminals of a capacitor, we start with the relationship between current and voltage for a capacitor in the time domain,

$$i = C\frac{dv}{dt},$$

and assume that

$$v = V_m \cos(\omega t + \theta_v).$$

Therefore,

$$i = C\frac{dv}{dt} = -\omega C V_m \sin(\omega t + \theta_v).$$

We now rewrite the expression for the current using the cosine function:

$$i = -\omega C V_m \cos(\omega t + \theta_v - 90°).$$

The phasor representation of the current is

$$\mathbf{I} = -\omega C V_m e^{j(\theta_v - 90°)}$$

$$= -\omega C V_m e^{j\theta_v} e^{-j90°}$$

$$= j\omega C V_m e^{j\theta_v}$$

$$= j\omega C V_m \underline{/\theta_v}.$$

Since $V_m \underline{/\theta_v}$ is the phasor representation of the sinusoidal voltage, we can express the current phasor in terms of the voltage phasor as

$$\mathbf{I} = j\omega C \mathbf{V}.$$

Now express the voltage phasor in terms of the current phasor, to conform to the phasor equations for resistors and inductors:

> ### RELATIONSHIP BETWEEN PHASOR VOLTAGE AND PHASOR CURRENT FOR A CAPACITOR
>
> $$\mathbf{V} = \frac{1}{j\omega C} \mathbf{I}. \tag{9.12}$$

Equation 9.12 demonstrates that the equivalent circuit for the capacitor in the phasor domain is as shown in Fig. 9.12.

The voltage across the terminals of a capacitor lags behind the current by 90°. We can show this by rewriting Eq. 9.12 as

$$\mathbf{V} = \left(\frac{1}{\omega C} \underline{/-90°}\right) I_m \underline{/\theta_i}$$

$$= \frac{I_m}{\omega C} \underline{/(\theta_i - 90)°}.$$

Thus, we can also say that the current leads the voltage by 90°. Figure 9.13 shows the phase relationship between the current and voltage at the terminals of a capacitor.

Impedance and Reactance

We conclude this discussion of passive circuit elements in the frequency domain with an important observation. When we compare Eqs. 9.10, 9.11, and 9.12, we note that they are all of the form

> ### DEFINITION OF IMPEDANCE
>
> $$\mathbf{V} = Z\mathbf{I}, \tag{9.13}$$

where Z represents the **impedance** of the circuit element. Solving for Z in Eq. 9.13, you can see that impedance is the ratio of a circuit element's voltage phasor to its current phasor. Thus, the impedance of a resistor is R, the impedance of an inductor is $j\omega L$, the impedance of mutual inductance is $j\omega M$, and the impedance of a capacitor is $1/j\omega C$. In all cases, impedance is measured in ohms. Note that, although impedance is a complex number, it is not a phasor. Remember, a phasor is a complex number that

Figure 9.12 ▲ The frequency domain equivalent circuit of a capacitor.

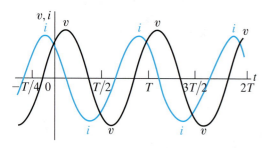

Figure 9.13 ▲ A plot showing the phase relationship between the current and voltage at the terminals of a capacitor ($\theta_i = 60°$).

results from the phasor transform of a cosine waveform. Thus, although all phasors are complex numbers, not all complex numbers are phasors.

Impedance in the frequency domain is the quantity analogous to resistance, inductance, and capacitance in the time domain. The imaginary part of the impedance is called **reactance**. The values of impedance and reactance for each of the component values are summarized in Table 9.1.

And, finally, a reminder. The passive sign convention holds in the frequency domain. If the reference direction for the current phasor in a circuit element is in the direction of the voltage phasor rise across the element, you must insert a minus sign into the equation that relates the voltage phasor to the current phasor.

Work through Example 9.6 to practice transforming circuit components from the time domain to the phasor domain.

TABLE 9.1	Impedance and Reactance Values	
Circuit Element	**Impedance**	**Reactance**
Resistor	R	—
Inductor	$j\omega L$	ωL
Capacitor	$j(-1/\omega C)$	$-1/\omega C$

EXAMPLE 9.6	**Calculating Component Voltages Using Phasor Techniques**

Figure 9.14 shows a resistor and an inductor connected in series. The current in these components is

$$i = 50\cos(1000t + 45°)\ \text{mA}.$$

The phasor transform of these components is shown in Fig. 9.15. Find

a) Z_R;

b) Z_L;

c) $\mathbf{I}$;

d) $\mathbf{V}_R$;

e) $\mathbf{V}_L$.

Figure 9.15 ▲ The phasor transform of the components in Fig. 9.14.

Solution

a) $Z_R = R = 100\ \Omega$.

b) $Z_L = j\omega L = j(1000)(0.05) = j50\ \Omega$.

c) $\mathbf{I} = \mathscr{P}\{0.05\cos(1000t + 45°\} = 0.05\underline{/45°}$
   $= 50\underline{/45°}\ \text{mA}$.

d) $\mathbf{V}_R = Z_R\mathbf{I} = (100)(0.05\underline{/45°}) = 5\underline{/45°}\ \text{V}.$

e) $\mathbf{V}_L = Z_L\mathbf{I} = (j50)(0.05\underline{/45°}) = 2.5\underline{/135°}\ \text{V}.$

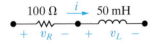

Figure 9.14 ▲ The components for Example 9.6.

ASSESSMENT PROBLEMS

Objective 2—Be able to transform a circuit with a sinusoidal source into the frequency domain using phasor concepts

9.3 The current in the 20 mH inductor is $10\cos(10{,}000t + 30°)$ mA. Calculate (a) the inductive reactance; (b) the impedance of the inductor; (c) the phasor voltage V; and (d) the steady-state expression for $v(t)$.

Answer: (a) 200 Ω;
(b) $j200$ Ω;
(c) $2\underline{/120°}$ V;
(d) $2\cos(10{,}000t + 120°)$ V.

9.4 The voltage across the terminals of the 5 μF capacitor is $30\cos(4000t + 25°)$ V. Calculate (a) the capacitive reactance; (b) the impedance of the capacitor; (c) the phasor current $\mathbf{I}$; and (d) the steady-state expression for $i(t)$.

$5\ \mu\text{F}$

Answer: (a) -50 Ω;
(b) $-j50$ Ω;
(c) $0.6\underline{/115°}$ A;
(d) $0.6\cos(4000t + 115°)$ A.

SELF-CHECK: Also try Chapter Problems 9.13 and 9.14.

9.5 Kirchhoff's Laws in the Frequency Domain

Kirchhoff's Voltage Law in the Frequency Domain

We begin by assuming that $v_1, v_2, \ldots, v_n$ represent voltages around a closed path in a circuit. We also assume that the circuit is operating in a sinusoidal steady state. Thus, Kirchhoff's voltage law requires that

$$v_1 + v_2 + \cdots + v_n = 0,$$

which in the sinusoidal steady state becomes

$$V_{m_1} \cos(\omega t + \theta_1) + V_{m_2} \cos(\omega t + \theta_2) + \cdots + V_{m_n} \cos(\omega t + \theta_n) = 0.$$

We now use Euler's identity to write the KVL equation as

$$\mathcal{R}\{V_{m_1}e^{j\theta_1}e^{j\omega t}\} + \mathcal{R}\{V_{m_2}e^{j\theta_2}e^{j\omega t}\} + \cdots + \mathcal{R}\{V_{m_n}e^{j\theta_n}e^{j\omega t}\} = 0,$$

which we simplify as

$$\mathcal{R}\{V_{m_1}e^{j\theta_1}e^{j\omega t} + V_{m_2}e^{j\theta_2}e^{j\omega t} + \cdots + V_{m_n}e^{j\theta_n}e^{j\omega t}\} = 0.$$

Factoring the term $e^{j\omega t}$ from each term yields

$$\mathcal{R}\{(V_{m_1}e^{j\theta_1} + V_{m_2}e^{j\theta_2} + \cdots + V_{m_n}e^{j\theta_n})e^{j\omega t}\} = 0,$$

or

$$\mathcal{R}\{(\mathbf{V}_1 + \mathbf{V}_2 + \cdots + \mathbf{V}_n)e^{j\omega t}\} = 0.$$

But $e^{j\omega t} \neq 0$, so

KVL IN THE FREQUENCY DOMAIN

$$\mathbf{V}_1 + \mathbf{V}_2 + \cdots + \mathbf{V}_n = 0, \tag{9.14}$$

which is the statement of Kirchhoff's voltage law as it applies to phasor voltages.

Kirchhoff's Current Law in the Frequency Domain

A similar derivation applies to a set of sinusoidal currents. Thus, if

$$i_1 + i_2 + \cdots + i_n = 0,$$

then

KCL IN THE FREQUENCY DOMAIN

$$\mathbf{I}_1 + \mathbf{I}_2 + \cdots + \mathbf{I}_n = 0, \tag{9.15}$$

where $\mathbf{I}_1, \mathbf{I}_2, \ldots, \mathbf{I}_n$ are the phasor representations of the individual currents $i_1, i_2, \ldots, i_n$. Thus, Eq. 9.15 states Kirchhoff's current law as it applies to phasor currents.

Equations 9.13, 9.14, and 9.15 form the basis for circuit analysis in the frequency domain. Note that Eq. 9.13 has the same algebraic form as Ohm's law and that Eqs. 9.14 and 9.15 state Kirchhoff's laws for phasor quantities. Therefore, you can use all the techniques developed for analyzing resistive circuits to find phasor currents and voltages. No new analytical techniques are needed; the basic circuit analysis and simplification tools covered in Chapters 2–4 can all be used to analyze circuits in the frequency domain. Phasor-circuit analysis consists of two fundamental tasks: (1) You must be able to construct the frequency-domain model of a circuit; and (2) you must be able to manipulate complex numbers and/or quantities algebraically.

Example 9.7 illustrates the use of KVL in the frequency domain.

EXAMPLE 9.7	**Using KVL in the Frequency Domain**

a) Use the results from Example 9.6 to calculate the phasor voltage drop, from left to right, across the series combination of the resistive and inductive impedances in Fig. 9.15.

b) Use the phasor voltage found in (a) to calculate the steady-state voltage drop, from left to right, across the series combination of resistor and inductor in Fig. 9.14.

Solution

a) Using KVL, the phasor voltage drop from left to right in Fig. 9.15 is

$$\mathbf{V} = \mathbf{V}_1 + \mathbf{V}_2 = 5\underline{/45^\circ} + 2.5\underline{/135^\circ}$$
$$= 5.59\underline{/71.565^\circ}\ \text{V}.$$

b) To find the steady-state voltage drop across the resistor and inductor in Fig. 9.14, we need to apply the inverse phasor transform to the phasor $\mathbf{V}$ from part (a). We need the frequency of the current defined in Example 9.6, which is $\omega = 1000\ \text{rad/s}$:

$$v_{ss}(t) = \mathscr{P}^{-1}\{V\} = \mathscr{P}^{-1}\{\ 5.59\underline{/71.565^\circ}\}$$
$$= 5.59\cos(1000t + 71.565^\circ)\ \text{V}.$$

 ASSESSMENT PROBLEM

Objective 3—Know how to use circuit analysis techniques to solve a circuit in the frequency domain

9.5 Four branches terminate at a common node. The reference direction of each branch current (i_1, i_2, i_3, and i_4) is toward the node. If

$i_1 = 100\cos(\omega t + 25^\circ)\ \text{A}$,
$i_2 = 100\cos(\omega t + 145^\circ)\ \text{A}$, and
$i_3 = 100\cos(\omega t - 95^\circ)\ \text{A}$, find i_4.

Answer: $i_4 = 0$.

SELF-CHECK: Also try Chapter Problem 9.15.

9.6 Series, Parallel, and Delta-to-Wye Simplifications

The rules for combining impedances in series or parallel and for making delta-to-wye transformations are the same as those for resistors. The only difference is that combining impedances involves the algebraic manipulation of complex numbers.

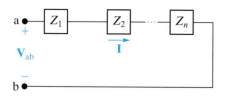

Figure 9.16 ▲ Impedances in series.

Combining Impedances in Series

Impedances in series can be combined into a single equivalent impedance whose value is the sum of the individual impedances. The circuit shown in Fig. 9.16 defines the problem in general terms. The impedances $Z_1, Z_2, \ldots, Z_n$ are connected in series between terminals a,b. When impedances are in series, they carry the same phasor current $\mathbf{I}$. From Eq. 9.13, the voltage drop across each impedance is $Z_1\mathbf{I}, Z_2\mathbf{I}, \ldots, Z_n\mathbf{I}$, and from Kirchhoff's voltage law,

$$\mathbf{V}_{ab} = Z_1\mathbf{I} + Z_2\mathbf{I} + \cdots + Z_n\mathbf{I}$$

$$= (Z_1 + Z_2 + \cdots + Z_n)\mathbf{I}.$$

The equivalent impedance between terminals a,b is

COMBINING IMPEDANCES IN SERIES

$$Z_{ab} = \frac{\mathbf{V}_{ab}}{\mathbf{I}} = Z_1 + Z_2 + \cdots + Z_n. \tag{9.16}$$

Remember from Chapter 3 that we can use voltage division to find the voltage across a single component from a collection of series-connected components whose total voltage is known (Eq. 3.9). We derived the voltage division equation using the equation for the equivalent resistance of series-connected resistors. Using the same process, we can derive the voltage division equation for frequency-domain circuits, where $\mathbf{V}_s$ is the voltage applied to a collection of series-connected impedances, $\mathbf{V}_j$ is the voltage across the impedance Z_j, and Z_{eq} is the equivalent impedance of the series-connected impedances:

VOLTAGE DIVISION IN THE FREQUENCY DOMAIN

$$\mathbf{V}_j = \frac{Z_j}{Z_{eq}} \mathbf{V}_s. \tag{9.17}$$

Example 9.8 illustrates the following frequency-domain circuit analysis techniques: combining impedances in series, Ohm's law for phasors, and voltage division.

EXAMPLE 9.8 | Combining Impedances in Series

A 90 Ω resistor, a 32 mH inductor, and a 5 μF capacitor are connected in series across the terminals of a sinusoidal voltage source, as shown in Fig. 9.17. The steady-state expression for the source voltage v_s is $750 \cos(5000t + 30°)$ V.

a) Construct the frequency-domain equivalent circuit.

b) Calculate the phasor voltage $\mathbf{V}$ using voltage division for the circuit from part (a).

c) Find the steady-state voltage v using the inverse phasor transform.

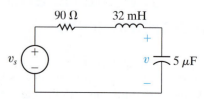

Figure 9.17 ▲ The circuit for Example 9.8.

Solution

a) From the expression for v_s, we have $\omega = 5000$ rad/s. Therefore, the impedance of the inductor is

$$Z_L = j\omega L = j(5000)(32 \times 10^{-3}) = j160 \ \Omega,$$

and the impedance of the capacitor is

$$Z_C = j\frac{-1}{\omega C} = -j\frac{1}{(5000)(5 \times 10^{-6})} = -j40 \ \Omega.$$

The phasor transform of v_s is

$$\mathbf{V}_s = 750\underline{/30°} \text{ V.}$$

Figure 9.18 illustrates the frequency-domain equivalent circuit of the circuit shown in Fig. 9.17.

b) Using voltage division, we see that the phasor voltage $\mathbf{V}$ is proportional to the source voltage; from Eq. 9.17,

$$\mathbf{V} = \frac{-j40}{90 + j160 - j40}(750\underline{/30°}) = 200\underline{/-113.13°} \text{ V.}$$

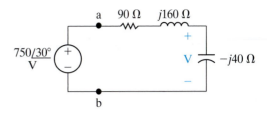

Figure 9.18 ▲ The frequency-domain equivalent circuit of the circuit shown in Fig. 9.17.

Note that we used Eq. 9.16 to find the equivalent impedance of the series-connected impedances in the circuit.

c) Find the steady-state voltage v using the inverse phasor transform of $\mathbf{V}$ from part (b). Remember that the source frequency is 5000 rad/s:

$$v(t) = 200 \cos(5000t - 113.13°) \text{ V.}$$

This voltage is the steady-state component of the complete response, which is what remains once the transient component has decayed to zero.

ASSESSMENT PROBLEM

Objective 3—Know how to use circuit analysis techniques to solve a circuit in the frequency domain

9.6   Using the values of resistance and inductance in the circuit of Fig. 9.15, let $\mathbf{V}_s = 125\underline{/-60°}$ V and $\omega = 5000$ rad/s. Find

a) the value of capacitance that yields a steady-state output current i with a phase angle of $-105°$;

b) the magnitude of the steady-state output current i.

Answer: (a) 2.86 μF;
  (b) 0.982 A.

SELF-CHECK: Also try Chapter Problem 9.17.

Combining Impedances in Parallel

Impedances connected in parallel can be reduced to an equivalent impedance using the reciprocal relationship

COMBINING IMPEDANCES IN PARALLEL

$$\frac{1}{Z_{ab}} = \frac{1}{Z_1} + \frac{1}{Z_2} + \cdots + \frac{1}{Z_n}. \tag{9.18}$$

Figure 9.19 depicts the parallel connection of impedances. Note that when impedances are in parallel, they have the same voltage across their terminals. We derive Eq. 9.18 directly from Fig. 9.19 by combining

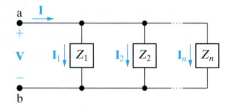

Figure 9.19 ▲ Impedances in parallel.

Kirchhoff's current law with the phasor-domain version of Ohm's law, that is, Eq. 9.13. From Fig. 9.19,

$$\mathbf{I} = \mathbf{I}_1 + \mathbf{I}_2 + \cdots + \mathbf{I}_n,$$

or

$$\frac{\mathbf{V}}{Z_{ab}} = \frac{\mathbf{V}}{Z_1} + \frac{\mathbf{V}}{Z_2} + \cdots + \frac{\mathbf{V}}{Z_n}.$$

Canceling the common voltage term from both sides gives us Eq. 9.18.
From Eq. 9.18, for the special case of just two impedances in parallel,

$$Z_{ab} = \frac{Z_1 Z_2}{Z_1 + Z_2}. \qquad (9.19)$$

We can also express Eq. 9.18 in terms of **admittance**, defined as the reciprocal of impedance and denoted Y. Thus

$$Y = \frac{1}{Z} = G + jB \text{ (siemens)}.$$

Admittance is a complex number whose real part, G, is called **conductance** and whose imaginary part, B, is called **susceptance**. Like admittance, conductance and susceptance are measured in siemens (S). Replacing impedances with admittances in Eq. 9.18, we get

$$Y_{ab} = Y_1 + Y_2 + \cdots + Y_n.$$

TABLE 9.2	Admittance and Susceptance Values	
Circuit Element	**Admittance** (Y)	**Susceptance**
Resistor	G (conductance)	—
Inductor	$j(-1/\omega L)$	$-1/\omega L$
Capacitor	$j\omega C$	ωC

The admittance of each of the ideal passive circuit elements also is worth noting and is summarized in Table 9.2.

Finally, remember from Chapter 3 that we can use current division to find the current in a single branch from a collection of parallel-connected branches whose total current is known (Eq. 3.10). We derived the current division equation using the equation for the equivalent resistance of parallel-connected resistors. Using the same process, we can derive the current division equation for frequency-domain circuits, where $\mathbf{I}_s$ is the current supplied to a collection of parallel-connected impedances, $\mathbf{I}_j$ is the current in the branch containing impedance Z_j, and Z_{eq} is the equivalent impedance of the parallel-connected impedances:

CURRENT DIVISION IN THE FREQUENCY DOMAIN

$$\mathbf{I}_j = \frac{Z_{eq}}{Z_j} \mathbf{I}_s. \qquad (9.20)$$

We use Eq. 9.18 to calculate the equivalent impedance in Eq. 9.20.
Example 9.9 analyzes a circuit in the frequency domain using series and parallel combinations of impedances and current division.

EXAMPLE 9.9 Combining Impedances in Series and in Parallel

The sinusoidal current source in the circuit shown in Fig. 9.20 produces the current $i_s = 8 \cos 200{,}000t$ A.

a) Construct the frequency-domain equivalent circuit.

b) Find the equivalent admittance to the right of the current source.

c) Use the equivalent admittance from part (b) to find the phasor voltage $\mathbf{V}$.

d) Find the phasor current $\mathbf{I}$, using current division.

e) Find the steady-state expressions for v and i.

Solution

a) The phasor transform of the current source is $8\,\underline{/0°}$; the resistors transform directly to the frequency domain as 10 and 6 Ω; the 40 μH inductor has an impedance of $j8$ Ω at the given frequency of 200,000 rad/s; and at this frequency the 1 μF capacitor has an impedance of $-j5$ Ω. Figure 9.21 shows the frequency-domain equivalent circuit and symbols representing the phasor transforms of the unknowns.

b) We first find the equivalent admittance to the right of the current source by adding the admittances of each branch. The admittance of the first branch is

$$Y_1 = \frac{1}{10} = 0.1 \text{ S},$$

the admittance of the second branch is

$$Y_2 = \frac{1}{6 + j8} = 0.06 - j0.08 \text{ S},$$

and the admittance of the third branch is

$$Y_3 = \frac{1}{-j5} = j0.2 \text{ S}.$$

The admittance of the three branches is

$$Y_{eq} = Y_1 + Y_2 + Y_3$$

$$= 0.16 + j0.12 \text{ S}$$

$$= 0.2\,\underline{/36.87°} \text{ S}.$$

c) Use the equivalent admittance from part (b) to find the phasor voltage $\mathbf{V}$.
The impedance seen by the current source is

$$Z_{eq} = \frac{1}{Y_{eq}} = 5\,\underline{/-36.87°} \ \Omega.$$

The phasor voltage $\mathbf{V}$ is

$$\mathbf{V} = Z_{eq}\mathbf{I} = 40\,\underline{/-36.87°} \text{ V}.$$

d) Using Eq. 9.20, together with the equivalent impedance found in part (c), we get

$$\mathbf{I} = \frac{5\,\underline{/-36.87°}}{6 + j8}(8\,\underline{/0°}) = 4\,\underline{/-90°} \text{ A}.$$

You can verify this answer using the phasor voltage across the branch, $\mathbf{V}$, and the impedance of the branch, $(6 + j8)$ Ω.

e) From the phasors found in parts (c) and (d), the steady-state time-domain expressions are

$$v(t) = 40 \cos(200{,}000t - 36.87°) \text{ V},$$

$$i(t) = 4 \cos(200{,}000t - 90°) \text{ A}.$$

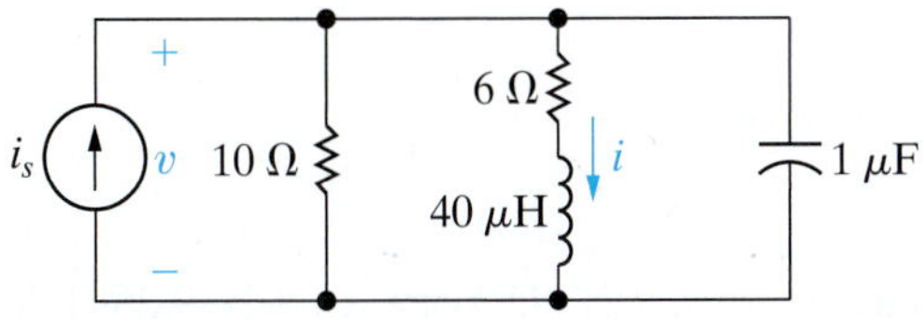

Figure 9.20 ▲ The circuit for Example 9.9.

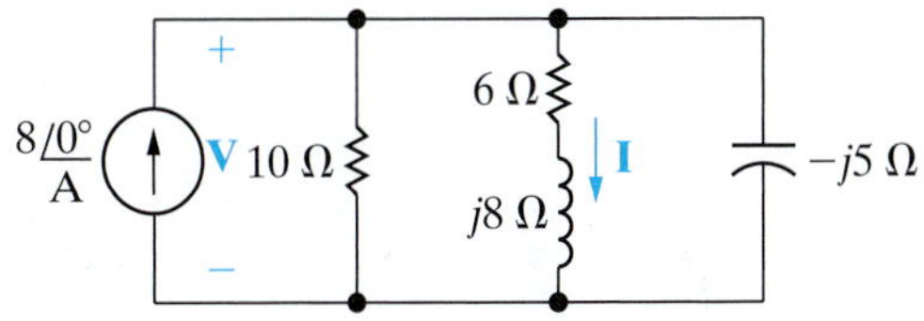

Figure 9.21 ▲ The frequency-domain equivalent circuit.

ASSESSMENT PROBLEMS

Objective 3—Know how to use circuit analysis techniques to solve a circuit in the frequency domain

9.7 A 20 Ω resistor is connected in parallel with a 5 mH inductor. This parallel combination is connected in series with a 5 Ω resistor and a 25 μF capacitor.
a) Calculate the impedance of this interconnection if the frequency is 2 krad/s.
b) Repeat (a) for a frequency of 8 krad/s.
c) At what finite frequency does the impedance of the interconnection become purely resistive?
d) What is the impedance at the frequency found in (c)?

Answer: (a) $9 - j12$ Ω;
(b) $21 + j3$ Ω;
(c) 4 krad/s;
(d) 15 Ω.

9.8 The interconnection described in Assessment Problem 9.7 is connected across the terminals of a voltage source that is generating $v = 150 \cos 4000t$ V. What is the maximum amplitude of the current in the 5 mH inductor?

Answer: 7.07 A.

SELF-CHECK: Also try Chapter Problems 9.28, 9.31, and 9.36.

Delta-to-Wye Transformations

The Δ-to-Y transformation for resistive circuits, discussed in Section 3.7, also applies to impedances. Figure 9.22 defines the Δ-connected impedances along with the Y-equivalent circuit. The Y impedances as functions of the Δ impedances are

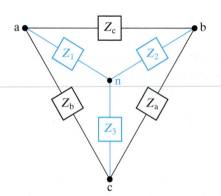

Figure 9.22 ▲ The delta-to-wye transformation.

$$Z_1 = \frac{Z_b Z_c}{Z_a + Z_b + Z_c}, \tag{9.21}$$

$$Z_2 = \frac{Z_c Z_a}{Z_a + Z_b + Z_c}, \tag{9.22}$$

$$Z_3 = \frac{Z_a Z_b}{Z_a + Z_b + Z_c}. \tag{9.23}$$

The Δ-to-Y transformation also may be reversed; that is, we can start with the Y structure and replace it with an equivalent Δ structure. The Δ impedances as functions of the Y impedances are

$$Z_a = \frac{Z_1 Z_2 + Z_2 Z_3 + Z_3 Z_1}{Z_1}, \tag{9.24}$$

$$Z_b = \frac{Z_1 Z_2 + Z_2 Z_3 + Z_3 Z_1}{Z_2}, \tag{9.25}$$

$$Z_c = \frac{Z_1 Z_2 + Z_2 Z_3 + Z_3 Z_1}{Z_3}. \tag{9.26}$$

The process used to derive Eqs. 9.21–9.23 or Eqs. 9.24–9.26 is the same as that used to derive the corresponding equations for resistive circuits. In fact, comparing Eqs. 3.15–3.17 with Eqs. 9.21–9.23 and Eqs. 3.18–3.20 with Eqs. 9.24–9.26 reveals that the symbol Z has replaced the symbol R. You may want to review Problem 3.65 concerning the derivation of the Δ-to-Y transformation.

Example 9.10 uses the Δ-to-Y transformation in phasor-circuit analysis.

EXAMPLE 9.10 **Using a Delta-to-Wye Transform in the Frequency Domain**

Use a Δ-to-Y impedance transformation to find $\mathbf{I}_0$, $\mathbf{I}_1$, $\mathbf{I}_2$, $\mathbf{I}_3$, $\mathbf{I}_4$, $\mathbf{I}_5$, $\mathbf{V}_1$, and $\mathbf{V}_2$ in the circuit in Fig. 9.23.

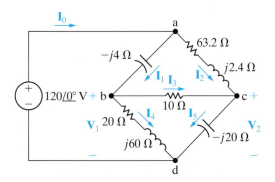

Figure 9.23 ▲ The circuit for Example 9.10.

Solution

It is not possible to simplify the circuit in Fig. 9.23 using series and parallel combinations of impedances. But if we replace either the upper delta (abc) or the lower delta (bcd) with its Y equivalent, we can simplify the resulting circuit by series-parallel combinations. To decide which delta to replace, find the sum of the impedances around each delta. This quantity forms the denominator for the equivalent Y impedances. The sum around the lower delta is $30 + j40$, so we choose to transform the lower delta to its equivalent Y. The Y impedance connecting to terminal b is

$$Z_1 = \frac{(20 + j60)(10)}{30 + j40} = 12 + j4\,\Omega,$$

the Y impedance connecting to terminal c is

$$Z_2 = \frac{10(-j20)}{30 + j40} = -3.2 - j2.4\ \Omega,$$

and the Y impedance connecting to terminal d is

$$Z_3 = \frac{(20 + j60)(-j20)}{30 + j40} = 8 - j24\ \Omega.$$

Inserting the Y-equivalent impedances into the circuit results in the circuit shown in Fig. 9.24.

We can simplify the circuit in Fig. 9.24 by making series-parallel combinations. The impedence of the abn branch is

$$Z_{abn} = 12 + j4 - j4 = 12\ \Omega,$$

and the impedance of the acn branch is

$$Z_{acn} = 63.2 + j2.4 - j2.4 - 3.2 = 60\ \Omega.$$

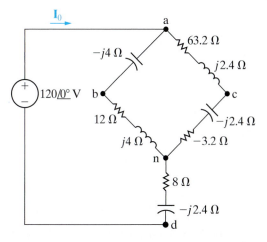

Figure 9.24 ▲ The circuit shown in Fig. 9.23, with the lower delta replaced by its equivalent wye.

Note that the abn branch is in parallel with the acn branch. Therefore, we may replace these two branches with a single branch having an impedance of

$$Z_{an} = \frac{(60)(12)}{72} = 10\ \Omega.$$

Combining this $10\ \Omega$ resistor with the impedance between n and d reduces the circuit to the one shown in Fig. 9.25. From that circuit,

$$\mathbf{I}_0 = \frac{120\ \underline{/0°}}{18 - j24} = 4\ \underline{/53.13°} = 2.4 + j3.2\ \text{A}.$$

Once we know $\mathbf{I}_0$, we can work back through the equivalent circuits to find the branch currents in the original circuit. We begin by noting that $\mathbf{I}_0$ is the current in the branch nd of Fig. 9.24. Therefore,

$$\mathbf{V}_{nd} = (8 - j24)\,\mathbf{I}_0 = 96 - j32\ \text{V}.$$

We can now calculate the voltage $\mathbf{V}_{an}$ because

$$\mathbf{V} = \mathbf{V}_{an} + \mathbf{V}_{nd}$$

where $\mathbf{V}$ is the phasor voltage of the source and $\mathbf{V}_{nd}$ is known. Thus

$$\mathbf{V}_{an} = 120 - (96 - j32) = 24 + j32\ \text{V}.$$

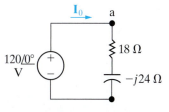

Figure 9.25 ▲ A simplified version of the circuit shown in Fig. 9.24.

We now compute the branch currents $\mathbf{I}_{abn}$ and $\mathbf{I}_{acn}$ using $\mathbf{V}_{an}$ and the equivalent impedance of each branch:

$$\mathbf{I}_{abn} = \frac{24 + j32}{12} = 2 + j\frac{8}{3} \text{ A,}$$

$$\mathbf{I}_{acn} = \frac{24 + j32}{60} = \frac{4}{10} + j\frac{8}{15} \text{ A.}$$

In terms of the branch currents defined in Fig. 9.23,

$$\mathbf{I}_1 = \mathbf{I}_{abn} = 2 + j\frac{8}{3} \text{ A,}$$

$$\mathbf{I}_2 = \mathbf{I}_{acn} = \frac{4}{10} + j\frac{8}{15} \text{ A.}$$

We check the calculations of $\mathbf{I}_1$ and $\mathbf{I}_2$ by noting that

$$\mathbf{I}_1 + \mathbf{I}_2 = 2.4 + j3.2 = \mathbf{I}_0.$$

To find the branch currents $\mathbf{I}_3$, $\mathbf{I}_4$, and $\mathbf{I}_5$, we must first calculate the voltages $\mathbf{V}_1$ and $\mathbf{V}_2$. Referring to Fig. 9.23, we note that

$$\mathbf{V}_1 = 120 \underline{/0°} - (-j4)\mathbf{I}_1 = \frac{328}{3} + j8 \text{ V,}$$

$$\mathbf{V}_2 = 120 \underline{/0°} - (63.2 + j2.4)\mathbf{I}_2 = 96 - j\frac{104}{3} \text{ V.}$$

We now calculate the branch currents $\mathbf{I}_3$, $\mathbf{I}_4$, and $\mathbf{I}_5$:

$$\mathbf{I}_3 = \frac{\mathbf{V}_1 - \mathbf{V}_2}{10} = \frac{4}{3} + j\frac{12.8}{3} \text{ A,}$$

$$\mathbf{I}_4 = \frac{\mathbf{V}_1}{20 + j60} = \frac{2}{3} - j1.6 \text{ A,}$$

$$\mathbf{I}_5 = \frac{\mathbf{V}_2}{-j20} = \frac{26}{15} + j4.8 \text{ A.}$$

We check the calculations by noting that

$$\mathbf{I}_4 + \mathbf{I}_5 = \frac{2}{3} + \frac{26}{15} - j1.6 + j4.8 = 2.4 + j3.2 = \mathbf{I}_0,$$

$$\mathbf{I}_3 + \mathbf{I}_4 = \frac{4}{3} + \frac{2}{3} + j\frac{12.8}{3} - j1.6 = 2 + j\frac{8}{3} = \mathbf{I}_1,$$

$$\mathbf{I}_3 + \mathbf{I}_2 = \frac{4}{3} + \frac{4}{10} + j\frac{12.8}{3} + j\frac{8}{15} = \frac{26}{15} + j4.8 = \mathbf{I}_5.$$

ASSESSMENT PROBLEM

Objective 3—Know how to use circuit analysis techniques to solve a circuit in the frequency domain

9.9 Use a Y-to-Δ transformation to find the current I in the circuit shown.

Answer: $\mathbf{I} = 4 \underline{/28.07°}$ A.

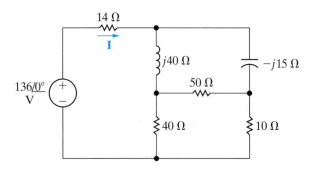

SELF-CHECK: Also try Chapter Problem 9.42.

9.7 Source Transformations and Thévenin–Norton Equivalent Circuits

The source transformations introduced in Section 4.9 and the Thévenin–Norton equivalent circuits discussed in Section 4.10 are analytical techniques that also can be applied to frequency-domain circuits. We prove these techniques are valid by following the same process used in Sections 4.9 and 4.10, except that we substitute impedance (Z) for resistance (R). Figure 9.26 shows a source-transformation equivalent circuit in the frequency domain.

Figure 9.27 illustrates the frequency-domain version of a Thévenin equivalent circuit, and Figure 9.28 shows the frequency-domain equivalent of a Norton equivalent circuit. The techniques for finding the Thévenin equivalent voltage and impedance are identical to those used for resistive circuits, except that the frequency-domain equivalent circuit involves phasors and complex numbers. The same holds for finding the Norton equivalent current and impedance.

Example 9.11 demonstrates the application of the source-transformation equivalent circuit to frequency-domain analysis. Example 9.12 illustrates the details of finding a Thévenin equivalent circuit in the frequency domain.

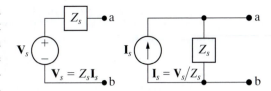

Figure 9.26 ▲ A source transformation in the frequency domain.

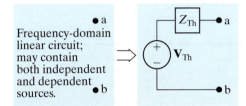

Figure 9.27 ▲ The frequency-domain version of a Thévenin equivalent circuit.

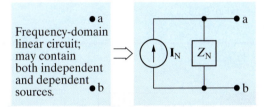

Figure 9.28 ▲ The frequency-domain version of a Norton equivalent circuit.

EXAMPLE 9.11 | **Performing Source Transformations in the Frequency Domain**

Use a series of source transformations to find the phasor voltage V_0 in the circuit shown in Fig. 9.29.

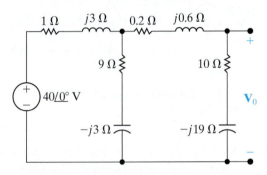

Figure 9.29 ▲ The circuit for Example 9.11.

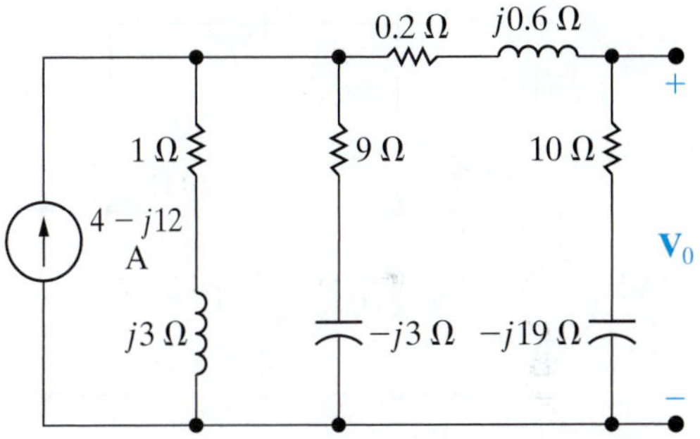

Figure 9.30 ▲ The first step in reducing the circuit shown in Fig. 9.29.

Solution

Begin by replacing the series combination of the voltage source $(40 \underline{/0°})$ and the impedance of $1 + j3 \ \Omega$ with the parallel combination of a current source and the $1 + j3 \ \Omega$ impedance. The current source is

$$I = \frac{40}{1 + j3} = 4 - j12 \text{ A}.$$

The resulting circuit is shown in Fig. 9.30. We used the polarity of the 40 V source to determine the direction for **I**.

Next, we combine the two parallel branches into a single impedance,

$$Z = \frac{(1 + j3)(9 - j3)}{10} = 1.8 + j2.4 \ \Omega,$$

which is in parallel with the current source of $4 - j12$ A. Another source transformation converts this parallel combination to a series combination of a voltage source and the impedance of $1.8 + j2.4 \ \Omega$. The voltage of the voltage source is

$$V = (4 - j12)(1.8 + j2.4) = 36 - j12 \text{ V}.$$

The resulting circuit is shown in Fig. 9.31. We added the current I_0 to this circuit to assist us in finding V_0.

We have now reduced the circuit to a simple series connection. We calculate the current I_0 by dividing the voltage of the source by the total series impedance:

$$I_0 = \frac{36 - j12}{12 - j16} = 1.56 + j1.08 \text{ A}.$$

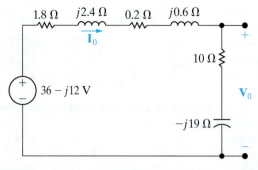

Figure 9.31 ▲ The second step in reducing the circuit shown in Fig. 9.29.

Now multiply $\mathbf{I}_0$ by the impedance $10 - j19$ to get $\mathbf{V}_0$:

$$\mathbf{V}_0 = (1.56 + j1.08)(10 - j19) = 36.12 - j18.84 \text{ V}.$$

You can verify this result by using voltage division to calculate $\mathbf{V}_0$.

EXAMPLE 9.12 Finding a Thévenin Equivalent in the Frequency Domain

Find the Thévenin equivalent circuit with respect to terminals a,b for the circuit shown in Fig. 9.32.

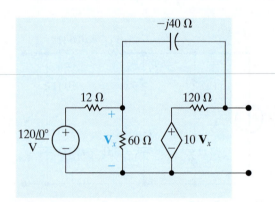

Figure 9.32 ▲ The circuit for Example 9.12.

Solution

We first determine the Thévenin equivalent voltage. This voltage is the open-circuit voltage appearing at terminals a,b. We choose the reference for the Thévenin voltage as positive at terminal a. We can make two source transformations using the 120 V, 12 Ω, and 60 Ω circuit elements to simplify the left-hand side of the circuit. These transformations must preserve the identity of the controlling voltage $\mathbf{V}_x$ because of the dependent voltage source.

The first source transformation replaces the series combination of the 120 V source and 12 Ω resistor with a 10 A current source in parallel with 12 Ω. Next, we replace the parallel combination of the 12 and 60 Ω resistors with a single 10 Ω resistor. Finally, we replace the parallel-connected 10 A source and 10 Ω resistor with a 100 V source in series with 10 Ω. Figure 9.33 shows the resulting circuit.

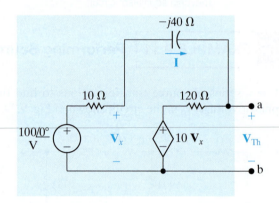

Figure 9.33 ▲ A simplified version of the circuit shown in Fig. 9.32.

We added the current $\mathbf{I}$ to Fig. 9.33; note that once we know its value, we can compute the Thévenin voltage. Use KVL to find $\mathbf{I}$ by summing the voltages around the closed path in the circuit. Hence

$$100 = 10\mathbf{I} - j40\mathbf{I} + 120\mathbf{I} + 10\mathbf{V}_x = (130 - j40)\mathbf{I} + 10\mathbf{V}_x.$$

We relate the controlling voltage $\mathbf{V}_x$ to the current $\mathbf{I}$ by noting from Fig. 9.33 that

$$\mathbf{V}_x = 100 - 10\mathbf{I}.$$

Then,

$$\mathbf{I} = \frac{-900}{30 - j40} = -10.8 - j14.4 \text{ A}.$$

Finally, we note from Fig. 9.33 that

$$\begin{aligned}
\mathbf{V}_{\text{Th}} &= 10\mathbf{V}_x + 120\mathbf{I} \\
&= 10(100 - 10\mathbf{I}) + 120\mathbf{I} \\
&= 1000 + 20(-10.8 - j14.4) \\
&= 784 - j288 \text{ V}.
\end{aligned}$$

We can find the Thévenin impedance using any of the techniques in Sections 4.10–4.11 for finding Thévenin resistance. We use the test-source method in this example. We begin by deactivating all independent sources in the circuit, and then we apply either a test-voltage source or a test-current source to the terminals of interest. The ratio of the voltage to the current at the test source is the Thévenin impedance. Figure 9.34 presents the result of applying this technique to the circuit shown in Fig. 9.32 while preserving the identity of $\mathbf{V}_x$.

We added branch currents $\mathbf{I}_a$ and $\mathbf{I}_b$ to simplify the calculation of $\mathbf{I}_T$. You should verify the following relationships by applying Ohm's law, KVL, and KCL for phasors:

$$\mathbf{I}_a = \frac{\mathbf{V}_T}{10 - j40}, \quad \mathbf{V}_x = 10\mathbf{I}_a,$$

$$\mathbf{I}_b = \frac{\mathbf{V}_T - 10\mathbf{V}_x}{120}$$

$$= \frac{-\mathbf{V}_T(9 + j4)}{120(1 - j4)},$$

$$\mathbf{I}_T = \mathbf{I}_a + \mathbf{I}_b$$

$$= \frac{\mathbf{V}_T}{10 - j40}\left(1 - \frac{9 + j4}{12}\right)$$

$$= \frac{\mathbf{V}_T(3 - j4)}{12(10 - j40)},$$

$$Z_{\text{Th}} = \frac{\mathbf{V}_T}{\mathbf{I}_T} = 91.2 - j38.4 \ \Omega.$$

Figure 9.35 depicts the Thévenin equivalent circuit.

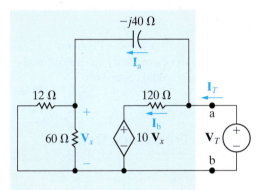

Figure 9.34 ▲ A circuit for calculating the Thévenin equivalent impedance.

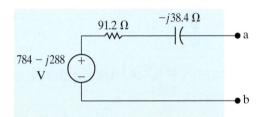

Figure 9.35 ▲ The Thévenin equivalent for the circuit shown in Fig. 9.32.

ASSESSMENT PROBLEMS

Objective 3—Know how to use circuit analysis techniques to solve a circuit in the frequency domain

9.10 Find the steady-state expression for $v_o(t)$ in the circuit shown by using source transformations. The sinusoidal voltage sources are

$$v_1 = 240 \cos(4000t + 53.13°) \text{ V},$$

$$v_2 = 96 \sin 4000t \text{ V}.$$

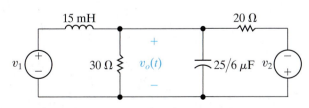

Answer: $48 \cos(4000t + 36.87°)$ V.

9.11 Find the Thévenin equivalent with respect to terminals a,b in the circuit shown.

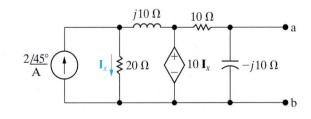

Answer: $\mathbf{V}_{\text{Th}} = \mathbf{V}_{ab} = 10 \underline{/45°}$ V;
$Z_{\text{Th}} = 5 - j5 \ \Omega$.

SELF-CHECK: Also try Chapter Problems 9.43, 9.44, and 9.48.

9.8 The Node-Voltage Method

In Sections 4.2–4.4, we introduced the node-voltage method of circuit analysis, culminating in Analysis Method 4.3 (p. 102). We can use this analysis method to find the steady-state response for circuits with sinusoidal sources. We need to make a few modifications:

- If the circuit is in the time domain, it must be transformed to the appropriate frequency domain. To do this, transform known voltages and currents to phasors, replace unknown voltages and currents with phasor symbols, and replace the component values for resistors, inductors, mutually coupled coils, and capacitors with their impedance values.
- Follow the steps in Analysis Method 4.3 to find the values of the unknown voltage and current phasors of interest.
- Apply the inverse phasor transform to the voltage and current phasors to find the steady-state values of the corresponding voltages and currents in the time domain.

Example 9.13 illustrates these steps. Assessment Problem 9.12 and many of the chapter problems give you an opportunity to use the node-voltage method to solve for steady-state sinusoidal responses.

EXAMPLE 9.13 | **Using the Node-Voltage Method in the Frequency Domain**

Use the node-voltage method to find the branch currents i_a, i_b, and i_c in the steady-state, for the circuit shown in Fig. 9.36. The value of the current source in this circuit is $i_s = 10.6 \cos(500t)$ A.

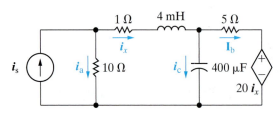

Figure 9.36 ▲ The circuit for Example 9.13.

Solution

We begin by transforming the circuit into the frequency domain. To do this, we replace the value of the current source with its phasor transform, $10.6\underline{/0°}$ A. We also replace the currents i_a, i_b, i_c, and i_x with corresponding phasor symbols $\mathbf{I}_a$, $\mathbf{I}_b$, $\mathbf{I}_c$, and $\mathbf{I}_x$. Then we replace the inductor and capacitor values with their impedances, using the frequency of the source:

$$Z_L = j(500)(4 \times 10^{-3}) = j2 \ \Omega;$$

$$Z_C = \frac{-j}{(500)(400 \times 10^{-6})} = -j5 \ \Omega.$$

The resulting frequency-domain circuit is shown in Fig. 9.37.

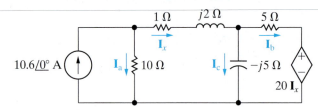

Figure 9.37 ▲ The circuit in Fig. 9.36, transformed into the frequency domain.

Now we can employ Analysis Method 4.3.

Step 1: The circuit has three essential nodes, two at the top and one on the bottom. We will need two KCL equations to describe the circuit.

Step 2: Four branches terminate on the bottom node, so we select it as the reference node and label the node voltages at the remaining essential nodes. The results of the first two steps are shown in Fig. 9.38.

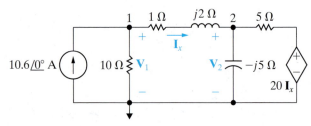

Figure 9.38 ▲ The circuit shown in Fig. 9.37, with the node voltages defined.

Step 3: Apply KCL at the nonreference essential nodes to give

$$-10.6 + \frac{\mathbf{V}_1}{10} + \frac{\mathbf{V}_1 - \mathbf{V}_2}{1 + j2} = 0,$$

and

$$\frac{\mathbf{V}_2 - \mathbf{V}_1}{1 + j2} + \frac{\mathbf{V}_2}{-j5} + \frac{\mathbf{V}_2 - 20\mathbf{I}_x}{5} = 0.$$

The circuit has a dependent source, so we need a dependent source constraint equation that defines $\mathbf{I}_x$ in terms of the node voltages:

$$\mathbf{I}_x = \frac{\mathbf{V}_1 - \mathbf{V}_2}{1 + j2}.$$

Step 4: Solve the three equations from Step 3 for $\mathbf{V}_1, \mathbf{V}_2,$ and $\mathbf{I}_x$:

$$\mathbf{V}_1 = 68.4 - j16.8 \text{ V},$$

$$\mathbf{V}_2 = 68 - j26 \text{ V},$$

$$\mathbf{I}_x = 3.76 + j1.68 \text{ A}.$$

Step 5: Use the phasor values from Step 4 to find the three branch currents from Fig. 9.37:

$$\mathbf{I}_a = \frac{\mathbf{V}_1}{10} = 6.84 - j1.68 \text{ A} = 7.04\underline{/-13.8°} \text{ A},$$

$$\mathbf{I}_b = \frac{\mathbf{V}_2 - 20\mathbf{I}_x}{5} = -1.44 - j11.92 \text{ A} = 12\underline{/-96.89°} \text{ A},$$

$$\mathbf{I}_c = \frac{\mathbf{V}_2}{-j5} = 5.2 + j13.6 \text{ A} = 14.56\underline{/69.08°} \text{ A}.$$

We find the steady-state values of the branch currents in the time-domain circuit of Fig. 9.37 by applying the inverse phasor transform to the results of Step 5. Remember that the frequency of the current source in the circuit is 500 rad/s. The results are

$$i_a = 7.04 \cos(500t - 13.8°) \text{ A},$$

$$i_b = 12 \cos(500t - 96.89°) \text{ A},$$

$$i_c = 14.56 \cos(500t + 69.08°) \text{ A}.$$

◾ ASSESSMENT PROBLEM

Objective 3—Know how to use circuit analysis techniques to solve a circuit in the frequency domain

9.12 Use the node-voltage method to find the steady-state expression for $v(t)$ in the circuit shown. The sinusoidal sources are $i_s = 10 \cos \omega t$ A and $v_s = 100 \sin \omega t$ V, where $\omega = 50$ krad/s.

Answer: $v(t) = 31.62 \cos(50{,}000t - 71.57°)$ V.

SELF-CHECK: Also try Chapter 9.55 and 9.59.

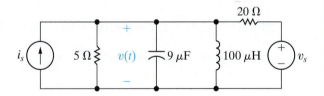

9.9 The Mesh-Current Method

We can also use the mesh-current method to analyze frequency-domain circuits. If a problem begins with a circuit in the time domain, it needs to be transformed into the frequency domain. Then, Analysis Method 4.6 (p. 110) can be used to find the mesh-current phasors, just as it was used to find mesh currents in resistive circuits. Finally, apply the inverse phasor transform to the phasor voltages and currents to find the steady-state voltages and currents in the time domain. We use the mesh current method to analyze a frequency-domain circuit in Example 9.14.

EXAMPLE 9.14   **Using the Mesh-Current Method in the Frequency Domain**

Use the mesh-current method to find the voltages $\mathbf{V}_1$, $\mathbf{V}_2$, and $\mathbf{V}_3$ in the circuit shown in Fig. 9.39.

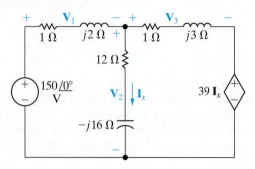

Figure 9.39 ▲ The circuit for Example 9.14.

Solution

The circuit is already in the frequency domain, so we apply Analysis Method 4.6.

Step 1: Use directed arrows that traverse the mesh perimeters to identify the two mesh current phasors.

Step 2: Label the mesh current phasors as $\mathbf{I}_1$ and $\mathbf{I}_2$, as shown in Fig. 9.40.

Step 3: Write the KVL equations for the meshes:

$$150 = (1 + j2)\mathbf{I}_1 + (12 - j16)(\mathbf{I}_1 - \mathbf{I}_2),$$

$$0 = (12 - j16)(\mathbf{I}_2 - \mathbf{I}_1) + (1 + j3)\mathbf{I}_2 + 39\mathbf{I}_x.$$

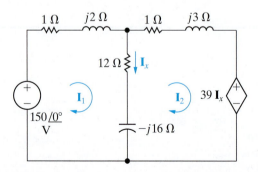

Figure 9.40 ▲ Mesh currents used to solve the circuit shown in Fig. 9.39.

The circuit in Fig. 9.40 has a dependent source, so we need a constraint equation that defines $\mathbf{I}_x$ in terms of the mesh currents. The resulting equation is

$$\mathbf{I}_x = \mathbf{I}_1 - \mathbf{I}_2.$$

Step 4: Solving the simultaneous equations in Step 3 gives

$$\mathbf{I}_1 = -26 - j52 \text{ A},$$

$$\mathbf{I}_2 = -24 - j58 \text{ A},$$

$$\mathbf{I}_x = -2 + j6 \text{ A}.$$

Step 5: Finally, we use the mesh-current phasors from Step 4 to find the phasor voltages identified in the circuit of Fig. 9.39:

$$\mathbf{V}_1 = (1 + j2)\mathbf{I}_1 = 78 - j104 \text{ V},$$

$$\mathbf{V}_2 = (12 - j16)\mathbf{I}_x = 72 + j104 \text{ V},$$

$$\mathbf{V}_3 = (1 + j3)\mathbf{I}_2 = 150 - j130 \text{ V}.$$

Also

$$39\mathbf{I}_x = -78 + j234 \text{ V}.$$

We check these calculations by summing the voltages around closed paths:

$$-150 + \mathbf{V}_1 + \mathbf{V}_2 = -150 + 78 - j104 + 72$$
$$+ j104 = 0,$$

$$-\mathbf{V}_2 + \mathbf{V}_3 + 39\mathbf{I}_x = -72 - j104 + 150 - j130$$
$$- 78 + j234 = 0,$$

$$-150 + \mathbf{V}_1 + \mathbf{V}_3 + 39\mathbf{I}_x = -150 + 78 - j104 + 150$$
$$- j130 - 78 + j234 = 0.$$

ASSESSMENT PROBLEM

Objective 3—Know how to use circuit analysis techniques to solve a circuit in the frequency domain

9.13 Use the mesh-current method to find the phasor current **I** in the circuit shown.

Answer: $29 + j2$ A $= 29.07 \underline{/3.95°}$ A.

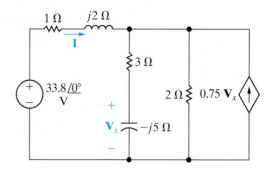

SELF-CHECK: Also try Chapter Problems 9.61 and 9.64.

9.10 The Transformer

A transformer is a device based on the magnetic coupling that characterizes mutually coupled inductor coils. Transformers are used in both communication and power circuits. In communication circuits, the transformer is used to match impedances and eliminate dc signals from portions of the system. In power circuits, transformers are used to establish ac voltage levels that facilitate the transmission, distribution, and consumption of electrical power. We need to know how a transformer behaves in the sinusoidal steady state when analyzing both communication and power systems. In this section, we will discuss the sinusoidal steady-state behavior of the **linear transformer**, which is found primarily in communication circuits. In Section 9.11, we will present the **ideal transformer**, which is used to model the ferromagnetic transformers found in power systems.

When analyzing circuits containing mutually coupled inductor coils, we use the mesh-current method. The node-voltage method is hard to use when mutual inductance is present because the currents in the coupled coils cannot be written by inspection as functions of the node voltages.

The Analysis of a Linear Transformer Circuit

A simple **transformer** is formed when two coils are wound on a single core to ensure magnetic coupling. Figure 9.41 shows the frequency-domain circuit model of a system that uses a transformer to connect a load to a source. The transformer winding connected to the source is called the **primary winding**, and the winding connected to the load is called the **secondary winding**. The transformer circuit parameters are

R_1 = the resistance of the primary winding,

R_2 = the resistance of the secondary winding,

L_1 = the self-inductance of the primary winding,

L_2 = the self-inductance of the secondary winding,

M = the mutual inductance.

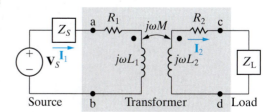

Figure 9.41 ▲ The frequency domain circuit model for a transformer used to connect a load to a source.

The internal voltage of the sinusoidal source is $\mathbf{V}_S$, and the internal impedance of the source is Z_S. The impedance Z_L represents the load connected to the secondary winding of the transformer. The phasor currents $\mathbf{I}_1$ and $\mathbf{I}_2$ are the primary and secondary currents of the transformer, respectively.

We analyze the circuit in Fig. 9.41 to find $\mathbf{I}_1$ and $\mathbf{I}_2$ as functions of the circuit parameters $\mathbf{V}_S$, Z_S, R_1, L_1, L_2, R_2, M, Z_L, and ω. Let's write the two KVL equations that describe the circuit:

$$\mathbf{V}_S = (Z_S + R_1 + j\omega L_1)\mathbf{I}_1 - j\omega M \mathbf{I}_2,$$

$$0 = -j\omega M \mathbf{I}_1 + (R_2 + j\omega L_2 + Z_L)\,\mathbf{I}_2.$$

We define

$$Z_{11} = Z_S + R_1 + j\omega L_1, \tag{9.27}$$

$$Z_{22} = R_2 + j\omega L_2 + Z_L, \tag{9.28}$$

where Z_{11} is the total self-impedance of the mesh containing the primary winding of the transformer and Z_{22} is the total self-impedance of the mesh containing the secondary winding. Using the impedances defined in Eqs. 9.27 and 9.28, we solve the mesh-current equations for $\mathbf{I}_1$ and $\mathbf{I}_2$ to give

$$\mathbf{I}_1 = \frac{Z_{22}}{Z_{11}Z_{22} + \omega^2 M^2}\,\mathbf{V}_S, \tag{9.29}$$

$$\mathbf{I}_2 = \frac{j\omega M}{Z_{11}Z_{22} + \omega^2 M^2}\,\mathbf{V}_S = \frac{j\omega M}{Z_{22}}\,\mathbf{I}_1. \tag{9.30}$$

We are also interested in finding the impedance seen when we look into the transformer from the terminals a and b. The internal source voltage $\mathbf{V}_S$ is attached to an equivalent impedance whose value is the ratio of the source-voltage phasor to the primary current phasor, or

$$\frac{\mathbf{V}_S}{\mathbf{I}_1} = Z_{\text{int}} = \frac{Z_{11}Z_{22} + \omega^2 M^2}{Z_{22}} = Z_{11} + \frac{\omega^2 M^2}{Z_{22}}.$$

The impedance at the terminals of the source is $Z_{\text{int}} - Z_S$, so

$$Z_{ab} = Z_{11} + \frac{\omega^2 M^2}{Z_{22}} - Z_S = R_1 + j\omega L_1 + \frac{\omega^2 M^2}{(R_2 + j\omega L_2 + Z_L)}. \tag{9.31}$$

Note that the impedance Z_{ab} is independent of the magnetic polarity of the transformer because the mutual inductance M appears in Eq. 9.31 as a squared quantity. The impedance Z_{ab} is interesting because it describes how the transformer affects the impedance of the load as seen from the source. Without the transformer, the load would be connected directly to the source, and the source would see a load impedance of Z_L; with the transformer, the load is connected to the source through the transformer, and the source sees a load impedance that is a modified version of Z_L, as seen in the third term of Eq. 9.31.

Reflected Impedance

The third term in Eq. 9.31 is called the **reflected impedance** (Z_r) because it is the equivalent impedance of the secondary coil and load impedance transmitted, or reflected, to the primary side of the transformer. Note that the reflected impedance exists because of the mutual inductance. If the two coils are not coupled, M is zero, Z_r is zero, and Z_{ab} is the self-impedance of the primary coil.

To consider reflected impedance in more detail, we first express the load impedance in rectangular form:

$$Z_L = R_L + jX_L,$$

where the load reactance X_L carries its own algebraic sign. That is, X_L is a positive number if the load is inductive and a negative number if the load is capacitive. We can now write the reflected impedance in rectangular form:

$$
\begin{aligned}
Z_r &= \frac{\omega^2 M^2}{R_2 + R_L + j(\omega L_2 + X_L)} \\[1em]
&= \frac{\omega^2 M^2[(R_2 + R_L) - j(\omega L_2 + X_L)]}{(R_2 + R_L)^2 + (\omega L_2 + X_L)^2} \\[1em]
&= \frac{\omega^2 M^2}{|Z_{22}|^2}[(R_2 + R_L) - j(\omega L_2 + X_L)].
\end{aligned}
\tag{9.32}
$$

The derivation of Eq. 9.32 uses the fact that, when Z_L is written in rectangular form, the self-impedance of the mesh containing the secondary winding is

$$Z_{22} = R_2 + R_L + j(\omega L_2 + X_L).$$

In Eq. 9.32 we see that the self-impedance of the secondary circuit is reflected into the primary circuit by a scaling factor of $(\omega M/|Z_{22}|)^2$ and that the sign of the reactive component $(\omega L_2 + X_L)$ is reversed. Thus, the linear transformer reflects the complex conjugate of the self-impedance of the secondary circuit (Z_{22}^*) into the primary winding with a scalar multiplier. Example 9.15 analyzes a circuit with a linear transformer.

EXAMPLE 9.15 | **Analyzing a Linear Transformer in the Frequency Domain**

The parameters of a linear transformer are $R_1 = 200\ \Omega$, $R_2 = 100\ \Omega$, $L_1 = 9$ H, $L_2 = 4$ H, and $k = 0.5$. The transformer couples a load impedance with an 800 Ω resistor in series with a 1 μF capacitor to a sinusoidal voltage source. The 300 V (rms) source has an internal impedance of $500 + j100\ \Omega$ and a frequency of 400 rad/s.

a) Construct a frequency-domain equivalent circuit of the system.

b) Calculate the self-impedance of the primary circuit.

c) Calculate the self-impedance of the secondary circuit.

d) Calculate the impedance reflected into the primary winding.

e) Calculate the scaling factor for the reflected impedance.

f) Calculate the impedance seen looking into the primary terminals of the transformer.

g) Calculate the Thévenin equivalent with respect to the terminals of the load impedance.

Solution

a) Figure 9.42 shows the frequency-domain equivalent circuit. Note that the internal voltage of the source serves as the reference phasor because it is assigned a phase angle of $0°$, and that $\mathbf{V}_1$ and $\mathbf{V}_2$ represent the terminal voltages of the transformer.

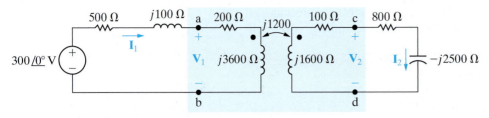

Figure 9.42 ▲ The frequency-domain equivalent circuit for Example 9.15.

In constructing the circuit in Fig. 9.42, we made the following calculations:

$$j\omega L_1 = j(400)(9) = j3600 \ \Omega,$$

$$j\omega L_2 = j(400)(4) = j1600 \ \Omega,$$

$$M = 0.5\sqrt{(9)(4)} = 3 \ \text{H},$$

$$j\omega M = j(400)(3) = j1200 \ \Omega,$$

$$\frac{-j}{\omega C} = \frac{-j}{(400)(1 \times 10^{-6})} = -j2500 \ \Omega.$$

b) From Eq. 9.27, the self-impedance of the primary circuit is

$$Z_{11} = 500 + j100 + 200 + j3600 = 700 + j3700 \ \Omega.$$

c) From Eq. 9.28, the self-impedance of the secondary circuit is

$$Z_{22} = 100 + j1600 + 800 - j2500 = 900 - j900 \ \Omega.$$

d) From Eq. 9.32, the impedance reflected into the primary winding is

$$Z_r = \left(\frac{1200}{|900 - j900|}\right)^2 (900 + j900)$$

$$= \frac{8}{9}(900 + j900) = 800 + j800 \ \Omega.$$

e) The scaling factor by which Z_{22}^* is reflected is 8/9.

f) The impedance seen looking into the primary terminals of the transformer is the impedance of the primary winding, Z_{11}, plus the reflected impedance, Z_r; thus

$$Z_{ab} = 200 + j3600 + 800 + j800 = 1000 + j4400 \ \Omega.$$

g) The Thévenin voltage is the open-circuit value of $\mathbf{V}_{cd}$, which equals $j1200$ times the open-circuit value of $\mathbf{I}_1$. The open-circuit value of $\mathbf{I}_1$ is

$$\mathbf{I}_1 = \frac{300 \ \underline{/0^\circ}}{700 + j3700}$$

$$= 79.67 \ \underline{/-79.29^\circ} \ \text{mA}.$$

Therefore

$$\mathbf{V}_{\text{Th}} = j1200(79.67 \ \underline{/-79.29^\circ}) \times 10^{-3}$$

$$= 95.60 \ \underline{/10.71^\circ} \ \text{V}.$$

The Thévenin impedance equals the impedance of the secondary winding, plus the impedance reflected from the primary when the voltage source is replaced by a short circuit. Thus

$$Z_{\text{Th}} = 100 + j1600 + \left(\frac{1200}{|700 + j3700|}\right)^2 (700 - j3700)$$

$$= 171.09 + j1224.26 \ \Omega.$$

The Thévenin equivalent is shown in Fig. 9.43.

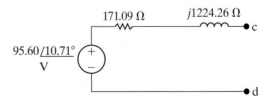

Figure 9.43 ▲ The Thévenin equivalent circuit for Example 9.15.

ASSESSMENT PROBLEM

Objective 4—Be able to analyze circuits containing linear transformers using phasor methods

9.14 A linear transformer couples a load consisting of a 360 Ω resistor in series with a 0.25 H inductor to a sinusoidal voltage source, as shown. The voltage source has an internal impedance of $184 + j0\ \Omega$ and a maximum voltage of 245.20 V, and it is operating at 800 rad/s. The transformer parameters are $R_1 = 100\ \Omega$, $L_1 = 0.5$ H, $R_2 = 40\ \Omega$, $L_2 = 0.125$ H, and $k = 0.4$. Calculate (a) the reflected impedance; (b) the primary current; and (c) the secondary current.

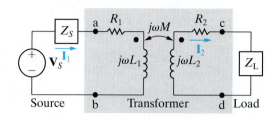

Answer: (a) $10.24 - j7.68\ \Omega$;
(b) $0.5 \cos(800t - 53.13°)$ A;
(c) $80 \cos 800t$ mA.

SELF-CHECK: Also try Chapter Problems 9.74 and 9.77.

9.11 The Ideal Transformer

An **ideal transformer** consists of two magnetically coupled coils having N_1 and N_2 turns, respectively, and exhibiting these three properties:

1. The coefficient of coupling is unity ($k = 1$).

2. The self-inductance of each coil is infinite ($L_1 = L_2 = \infty$).

3. The coil losses, due to parasitic resistance, are negligible.

Understanding the behavior of ideal transformers begins with Eq. 9.31, which describes the impedance at the terminals of a source connected to a linear transformer. We repeat this equation in the following discussion and examine it further.

Exploring Limiting Values

Equation 9.31, repeated here as Eq. 9.33, defines the relationship between the input impedance (Z_{ab}) and load impedance (Z_L) for a linear transformer:

$$Z_{ab} = Z_{11} + \frac{\omega^2 M^2}{Z_{22}} - Z_s$$

$$= R_1 + j\omega L_1 + \frac{\omega^2 M^2}{(R_2 + j\omega L_2 + Z_L)}. \tag{9.33}$$

Let's consider what happens to Eq. 9.33 as L_1 and L_2 each become infinitely large and, at the same time, the coefficient of coupling approaches unity. Transformers wound on ferromagnetic cores can approach these conditions. Even though such transformers are nonlinear, we can obtain some useful information using an ideal model that ignores the nonlinearities.

To show how Z_{ab} changes when $k = 1$ and L_1 and L_2 approach infinity, we first introduce the notation

$$Z_{22} = R_2 + R_L + j(\omega L_2 + X_L) = R_{22} + jX_{22}$$

and then rearrange Eq. 9.33:

$$Z_{ab} = R_1 + \frac{\omega^2 M^2 R_{22}}{R_{22}^2 + X_{22}^2} + j\left(\omega L_1 - \frac{\omega^2 M^2 X_{22}}{R_{22}^2 + X_{22}^2}\right) \quad (9.34)$$

$$= R_{ab} + jX_{ab}.$$

At this point, we must be careful with the imaginary part of Z_{ab} because, as L_1 and L_2 approach infinity, X_{ab} is the difference between two large quantities. Thus, before letting L_1 and L_2 increase, we write the imaginary part of Z_{ab} as

$$X_{ab} = \omega L_1 - \frac{(\omega L_1)(\omega L_2) X_{22}}{R_{22}^2 + X_{22}^2} = \omega L_1\left(1 - \frac{\omega L_2 X_{22}}{R_{22}^2 + X_{22}^2}\right),$$

where we recognize that, when $k = 1$, $M^2 = L_1 L_2$. Putting the term multiplying ωL_1 over a common denominator gives

$$X_{ab} = \omega L_1\left(\frac{R_{22}^2 + \omega L_2 X_L + X_L^2}{R_{22}^2 + X_{22}^2}\right).$$

Factoring ωL_2 out of the numerator and $(\omega L_2)^2$ out of the denominator, then simplifying, yields

$$X_{ab} = \frac{L_1}{L_2}\frac{X_L + (R_{22}^2 + X_L^2)/\omega L_2}{(R_{22}/\omega L_2)^2 + [1 + (X_L/\omega L_2)]^2}.$$

As k approaches 1.0, the ratio L_1/L_2 approaches the constant value of $(N_1/N_2)^2$. This follows from the relationship between L_1 and N_1 (Eq. 6.21), the relationship between L_2 and N_2 (Eq. 6.23), and the fact that, as the coupling becomes extremely tight, the two permeances $\mathcal{P}_1$ and $\mathcal{P}_2$ become equal. The expression for X_{ab} simplifies to

$$X_{ab} = \left(\frac{N_1}{N_2}\right)^2 X_L,$$

as $L_1 \to \infty$, $L_2 \to \infty$, and $k \to 1.0$.

The same reasoning leads to simplification of the reflected resistance in Eq. 9.34:

$$\frac{\omega^2 M^2 R_{22}}{R_{22}^2 + X_{22}^2} = \frac{L_1}{L_2}R_{22} = \left(\frac{N_1}{N_2}\right)^2 R_{22}.$$

Substituting the simplified forms for X_{ab} and the reflected resistance in Eq. 9.34 yields

$$Z_{ab} = R_1 + \left(\frac{N_1}{N_2}\right)^2 R_2 + \left(\frac{N_1}{N_2}\right)^2 (R_L + jX_L).$$

Compare this expression for Z_{ab} with the one given in Eq. 9.33. We see that when the coefficient of coupling approaches unity and the self-inductances of the coupled coils approach infinity, the transformer reflects the secondary winding resistance and the load impedance to the primary side by a scaling factor equal to the turns ratio (N_1/N_2) squared. Hence, we may describe the terminal behavior of the ideal transformer in terms of two characteristics. First, the magnitude of the volts per turn is the same for each coil, or

$$\left|\frac{\mathbf{V}_1}{N_1}\right| = \left|\frac{\mathbf{V}_2}{N_2}\right|. \quad (9.35)$$

Second, the magnitude of the ampere-turns is the same for each coil, or

$$|\mathbf{I}_1 N_1| = |\mathbf{I}_2 N_2|. \qquad (9.36)$$

We use magnitude signs in Eqs. 9.35 and 9.36 because we have not yet established reference polarities for the currents and voltages; we discuss the removal of the magnitude signs shortly.

Figure 9.44 shows two lossless ($R_1 = R_2 = 0$) magnetically coupled coils. We use Fig. 9.44 to validate Eqs. 9.35 and 9.36. In Fig. 9.44(a), coil 2 is open; in Fig. 9.44(b), coil 2 is shorted. Although we carry out the following analysis in terms of sinusoidal steady-state operation, the results also apply to instantaneous values of v and i.

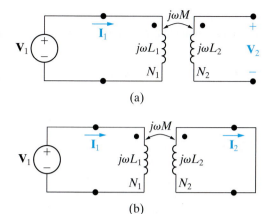

Figure 9.44 ▲ The circuits used to verify Eqs. 9.35 and 9.36 for an ideal transformer.

Determining the Voltage and Current Ratios

Note in Fig. 9.44(a) that the voltage at the terminals of the open-circuit coil is entirely the result of the current in coil 1; therefore,

$$\mathbf{V}_2 = j\omega M \mathbf{I}_1.$$

The current in coil 1 is

$$\mathbf{I}_1 = \frac{\mathbf{V}_1}{j\omega L_1}.$$

Thus,

$$\mathbf{V}_2 = \frac{M}{L_1} \mathbf{V}_1.$$

For unity coupling ($k = 1$), the mutual inductance equals $\sqrt{L_1 L_2}$, so the expression for $\mathbf{V}_2$ becomes

$$\mathbf{V}_2 = \sqrt{\frac{L_2}{L_1}} \mathbf{V}_1.$$

For unity coupling, the flux linking coil 1 is the same as the flux linking coil 2, so we need only one permeance to describe the self-inductance of each coil. Thus,

$$\mathbf{V}_2 = \sqrt{\frac{N_2^2 \mathscr{P}}{N_1^2 \mathscr{P}}} \mathbf{V}_1 = \frac{N_2}{N_1} \mathbf{V}_1$$

or

> ### VOLTAGE RELATIONSHIP FOR AN IDEAL TRANSFORMER
>
> $$\frac{\mathbf{V}_1}{N_1} = \frac{\mathbf{V}_2}{N_2}. \qquad (9.37)$$

Summing the voltages around the shorted coil of Fig. 9.44(b) yields

$$0 = -j\omega M \mathbf{I}_1 + j\omega L_2 \mathbf{I}_2,$$

which, when $k = 1$, gives

$$\frac{\mathbf{I}_1}{\mathbf{I}_2} = \frac{L_2}{M} = \frac{L_2}{\sqrt{L_1 L_2}} = \sqrt{\frac{L_2}{L_1}} = \frac{N_2}{N_1}.$$

Therefore,

> ### CURRENT RELATIONSHIP FOR AN IDEAL TRANSFORMER
>
> $$\mathbf{I}_1 N_1 = \mathbf{I}_2 N_2. \tag{9.38}$$

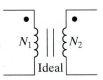

Figure 9.45 ▲ The graphic symbol for an ideal transformer.

Figure 9.45 shows the graphic symbol for an ideal transformer. The vertical lines in the symbol represent the layers of magnetic material from which ferromagnetic cores are often made. Coils wound on a ferromagnetic core behave very much like an ideal transformer, for several reasons. The ferromagnetic material creates a space with high permeance. Thus, most of the magnetic flux is trapped inside the core material, establishing tight magnetic coupling between coils that share the same core. High permeance also means high self-inductance because $L = N^2 \mathscr{P}$. Finally, ferromagnetically coupled coils efficiently transfer power from one coil to the other. Efficiencies in excess of 95% are common, so neglecting losses is a valid approximation for many applications.

Determining the Polarity of the Voltage and Current Ratios

We now turn to the removal of the magnitude signs from Eqs. 9.35 and 9.36. Note that magnitude signs do not appear in Eqs. 9.37 and 9.38 because we established reference polarities for voltages and reference directions for currents in Fig. 9.44. In addition, we specified the magnetic polarity dots of the two coupled coils.

The rules for assigning the proper algebraic sign to Eqs. 9.35 and 9.36 are as follows:

> ### DOT CONVENTION FOR IDEAL TRANSFORMERS
>
> If the coil voltages $\mathbf{V}_1$ and $\mathbf{V}_2$ are both positive or negative at the dot-marked terminals, use a plus sign in Eq. 9.35. Otherwise, use a negative sign.
>
> If the coil currents $\mathbf{I}_1$ and $\mathbf{I}_2$ are both directed into or out of the dot-marked terminals, use a minus sign in Eq. 9.36. Otherwise, use a plus sign.

The four circuits shown in Fig. 9.46 illustrate these rules.

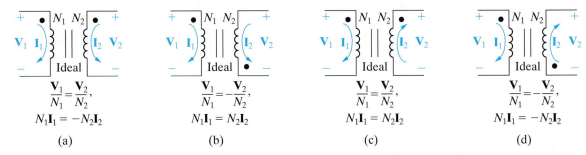

$\dfrac{\mathbf{V}_1}{N_1} = \dfrac{\mathbf{V}_2}{N_2},$	$\dfrac{\mathbf{V}_1}{N_1} = -\dfrac{\mathbf{V}_2}{N_2},$	$\dfrac{\mathbf{V}_1}{N_1} = \dfrac{\mathbf{V}_2}{N_2},$	$\dfrac{\mathbf{V}_1}{N_1} = -\dfrac{\mathbf{V}_2}{N_2},$
$N_1\mathbf{I}_1 = -N_2\mathbf{I}_2$	$N_1\mathbf{I}_1 = N_2\mathbf{I}_2$	$N_1\mathbf{I}_1 = N_2\mathbf{I}_2$	$N_1\mathbf{I}_1 = -N_2\mathbf{I}_2$
(a)	(b)	(c)	(d)

Figure 9.46 ▲ Circuits that show the proper algebraic signs for relating the terminal voltages and currents of an ideal transformer.

The turns ratio for the two windings is an important parameter of the ideal transformer. In this text, we use a to denote the ratio N_2/N_1, so

$$a = \frac{N_2}{N_1}.$$ (9.39)

Figure 9.47 shows three ways to represent the turns ratio of an ideal transformer. Figure 9.47(a) shows the number of turns in each coil explicitly. Figure 9.47(b) shows that the ratio N_2/N_1 is 5 to 1, and Fig. 9.47(c) shows that the ratio N_2/N_1 is 1 to $\frac{1}{5}$.

Example 9.16 analyzes a circuit containing an ideal transformer.

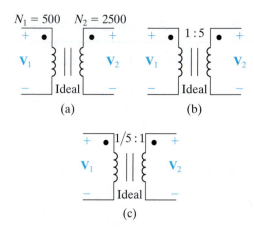

Figure 9.47 ▲ Three ways to show that the turns ratio of an ideal transformer is 5.

EXAMPLE 9.16 | Analyzing an Ideal Transformer Circuit in the Frequency Domain

The load impedance connected to the secondary winding of the ideal transformer in Fig. 9.48 is a 237.5 mΩ resistor in series with a 125 μH inductor.

If the sinusoidal voltage source (v_g) is generating the voltage 2500 cos 400t V, find the steady-state expressions for: (a) i_1; (b) v_1; (c) i_2; and (d) v_2.

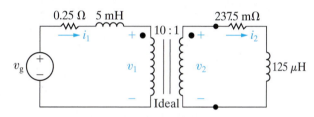

Figure 9.48 ▲ The circuit for Example 9.16.

Solution

a) We begin by transforming the circuit to the frequency domain. The voltage source has the phasor value 2500 $\underline{/0°}$ V; the 5 mH inductor has an impedance of $j2$ Ω; and the 125 μH inductor has an impedance of $j0.05$ Ω. The resulting frequency domain circuit is shown in Fig. 9.49.

Writing a KCL equation for the left-hand mesh in Fig. 9.49 gives

$$2500 \underline{/0°} = (0.25 + j2)\mathbf{I}_1 + \mathbf{V}_1.$$

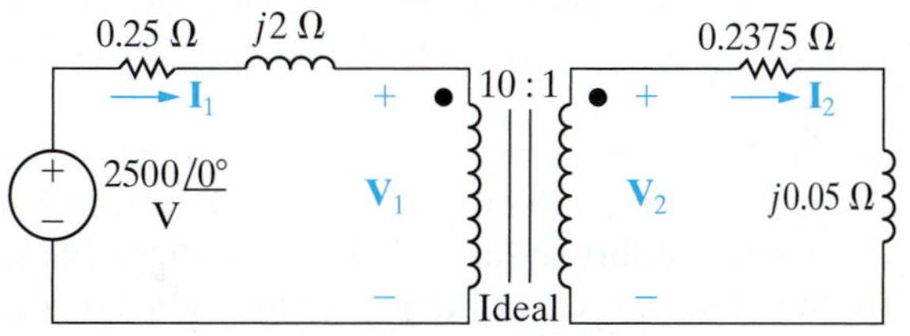

Figure 9.49 ▲ Phasor domain circuit for Example 9.16.

Using the relationship between $\mathbf{V}_1$ and $\mathbf{V}_2$ for the ideal transformer and using Ohm's law to express $\mathbf{V}_2$ in terms of $\mathbf{I}_2$, we have

$$\mathbf{V}_1 = 10\mathbf{V}_2 = 10[(0.2375 + j0.05)\mathbf{I}_2].$$

Because

$$\mathbf{I}_2 = 10\mathbf{I}_1$$

we have

$$\mathbf{V}_1 = 10(0.2375 + j0.05)(10\mathbf{I}_1)$$

$$= (23.75 + j5)\mathbf{I}_1.$$

Therefore

$$2500 \underline{/0°} = (24 + j7)\mathbf{I}_1,$$

or

$$\mathbf{I}_1 = 100 \underline{/-16.26°} \text{ A}.$$

Thus, the steady-state expression for i_1 is

$$i_1 = 100 \cos(400t - 16.26°) \text{ A}.$$

b) $\mathbf{V}_1 = 2500\underline{/0°} - (100 \underline{/-16.26°})(0.25 + j2)$

$$= 2420 - j185 = 2427.06 \underline{/-4.37°} \text{ V}.$$

Hence, in the steady state

$$v_1 = 2427.06 \cos(400t - 4.37°) \text{ V}.$$

c) $\mathbf{I}_2 = 10\mathbf{I}_1 = 1000 \underline{/-16.26°}$ A.

Therefore, in the steady state

$$i_2 = 1000 \cos(400t - 16.26°) \text{ A}.$$

d) $\mathbf{V}_2 = 0.1\mathbf{V}_1 = 242.71 \underline{/-4.37°}$ V,

so in the steady state,

$$v_2 = 242.71 \cos(400t - 4.37°) \text{ V}.$$

ASSESSMENT PROBLEM

Objective 5—Be able to analyze circuits with ideal transformers

9.15 The source voltage in the phasor domain circuit in the accompanying figure is $25 \underline{/0°}$ kV. Find the amplitude and phase angle of $\mathbf{V}_2$ and $\mathbf{I}_2$.

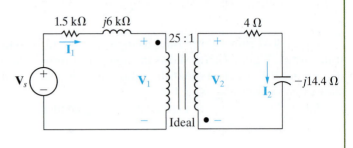

Answer: $\mathbf{V}_2 = 1868.15 \underline{/142.39°}$ V;
$\mathbf{I}_2 = 125 \underline{/216.87°}$ A.

SELF-CHECK: Also try Chapter Problem 9.79.

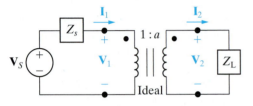

Figure 9.50 ▲ Using an ideal transformer to couple a load to a source.

Using an Ideal Transformer for Impedance Matching

Ideal transformers can be used to increase or decrease the impedance level of a load, as illustrated by the circuit shown in Fig. 9.50. The impedance seen by the practical voltage source ($\mathbf{V}_s$ in series with Z_s) is $\mathbf{V}_1/\mathbf{I}_1$. The voltage and current at the terminals of the load impedance ($\mathbf{V}_2$ and $\mathbf{I}_2$) are related to $\mathbf{V}_1$ and $\mathbf{I}_1$ by the transformer turns ratio; thus

$$\mathbf{V}_1 = \frac{\mathbf{V}_2}{a},$$

and

$$\mathbf{I}_1 = a\mathbf{I}_2.$$

Therefore, the impedance seen by the practical source is

$$Z_{IN} = \frac{\mathbf{V}_1}{\mathbf{I}_1} = \frac{1}{a^2}\frac{\mathbf{V}_2}{\mathbf{I}_2},$$

but the ratio $\mathbf{V}_2/\mathbf{I}_2$ is the load impedance Z_L, so the expression for Z_{IN} becomes

$$Z_{IN} = \frac{1}{a^2}Z_L. \tag{9.40}$$

Thus, the ideal transformer's secondary coil reflects the load impedance back to the primary coil, with the scaling factor $1/a^2$.

Note that the ideal transformer changes the magnitude of Z_L but does not affect its phase angle. Whether Z_{IN} is greater or less than Z_L depends on the turns ratio a. The ideal transformer—or its practical counterpart, the ferromagnetic core transformer—can be used to match the magnitude of Z_L to the magnitude of Z_s. We will discuss why this may be desirable in Chapter 10.

Ideal transformers are also used to increase or decrease voltages from a source to a load, as we will see in Chapter 10. Thus, ideal transformers are used widely in the electric utility industry, where it is desirable to decrease, or step down, the voltage level at the power line to safer residential voltage levels.

9.12 Phasor Diagrams

When we analyze the sinusoidal steady-state operation of a circuit using phasors, a diagram of the phasor currents and voltages can give us additional insight into the circuit's behavior. A phasor diagram shows the magnitude and phase angle of each phasor quantity in the complex-number plane. Phase angles are measured counterclockwise from the positive real axis, and magnitudes are measured from the origin of the axes. For example, Fig. 9.51 shows the phasor quantities $10\,\underline{/30°}$, $12\,\underline{/150°}$, $5\,\underline{/-45°}$, and $8\,\underline{/-170°}$.

Because constructing phasor diagrams for circuits usually involves both currents and voltages, we use two different magnitude scales, one for currents and one for voltages. Visualizing a phasor on the complex-number plane is a good way to check your calculations. For example, suppose you convert the phasor $-7 - j3$ to polar form. Before making your calculation, you should anticipate a magnitude greater than 7 and an angle in the third quadrant that is more negative than $-135°$ or less positive than $225°$, as illustrated in Fig. 9.52.

Examples 9.17 and 9.18 construct and use phasor diagrams. You can use phasor diagrams to get additional insight into the steady-state sinusoidal operation of a circuit. For example, Problem 9.85 uses a phasor diagram to explain the operation of a phase-shifting circuit.

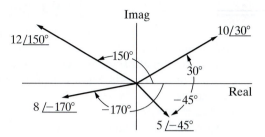

Figure 9.51 ▲ A graphic representation of phasors.

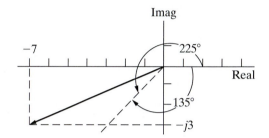

Figure 9.52 ▲ The complex number $-7 - j3 = 7.62\,\underline{/-156.80°}$.

EXAMPLE 9.17	**Using Phasor Diagrams to Analyze a Circuit**

Use a phasor diagram for the circuit in Fig. 9.53 to find the value of R that causes the current through that resistor, i_R, to lag the source current, i_s, by 45° when $\omega = 5$ krad/s.

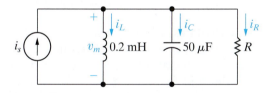

Figure 9.53 ▲ The circuit for Example 9.17.

Solution

According to KCL, the sum of the currents $\mathbf{I}_R$, $\mathbf{I}_L$, and $\mathbf{I}_C$ must equal the source current $\mathbf{I}_s$. If we assume that the phase angle of the voltage $\mathbf{V}_m$ is zero, we can draw the current phasors for each of the components. The current phasor for the inductor is

$$\mathbf{I}_L = \frac{V_m\,\underline{/0°}}{j\,(5000)(0.2 \times 10^{-3})} = V_m\,\underline{/-90°},$$

whereas the current phasor for the capacitor is

$$\mathbf{I}_C = \frac{V_m\,\underline{/0°}}{-j(5000)(50 \times 10^{-6})} = 4V_m\,\underline{/90°},$$

and the current phasor for the resistor is

$$\mathbf{I}_R = \frac{V_m\,\underline{/0°}}{R} = \frac{V_m}{R}\,\underline{/0°}.$$

These phasors are shown in Fig. 9.54. The phasor diagram also shows the source current phasor, sketched as a dashed line, which must be the sum of the current phasors of the three circuit components and must be at an angle that is 45° more positive than the current phasor for the resistor. As you can see, summing the phasors makes an isosceles triangle, so the length of the current phasor for the resistor must equal $3V_m$. Therefore, the value of the resistor is $\frac{1}{3}\,\Omega$.

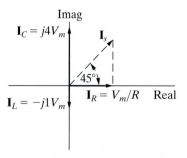

Figure 9.54 ▲ The phasor diagram for the currents in Fig. 9.53.

EXAMPLE 9.18 | **Using Phasor Diagrams to Analyze Capacitive Loading Effects**

The circuit in Fig. 9.55 has a load consisting of the parallel combination of the resistor and inductor. Use phasor diagrams to explore the effect of adding a capacitor across the terminals of the load on the amplitude of $\mathbf{V}_s$ if we adjust $\mathbf{V}_s$ so that the amplitude of $\mathbf{V}_L$ remains constant. Utility companies use this technique to control the voltage drop on their lines.

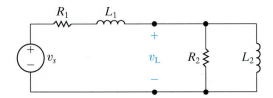

Figure 9.55 ▲ The circuit for Example 9.18.

Solution

We begin by assuming zero capacitance across the load. After constructing the phasor diagram for the zero-capacitance case, we can add the capacitor and study its effect on the amplitude of $\mathbf{V}_s$, holding the amplitude of $\mathbf{V}_L$ constant. Figure 9.56 shows the frequency-domain equivalent of the circuit in Fig. 9.55. We added the phasor branch currents $\mathbf{I}$, $\mathbf{I}_a$, and $\mathbf{I}_b$ to Fig. 9.56 to assist our analysis.

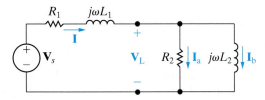

Figure 9.56 ▲ The frequency-domain equivalent of the circuit in Fig. 9.55.

Figure 9.57 shows the stepwise evolution of the phasor diagram. Keep in mind that in this example we are not interested in specific phasor values and positions, but rather in the general effect of adding a capacitor across the terminals of the load. Thus, we want to develop the relative positions of the phasors before and after the capacitor is added.

Relating the phasor diagram to the circuit shown in Fig. 9.56, we make the following observations:

- We choose $\mathbf{V}_L$ as our reference because we are holding its amplitude constant. For convenience, we place this phasor on the positive real axis.
- We know that $\mathbf{I}_a$ is in phase with $\mathbf{V}_L$ and that its magnitude is $|\mathbf{V}_L|/R_2$. (On the phasor diagram, the magnitude scale for the current phasors is independent of the magnitude scale for the voltage phasors.)

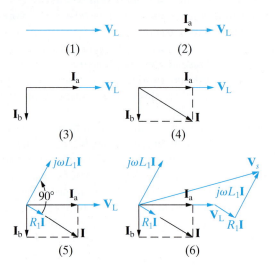

Figure 9.57 ▲ The step-by-step evolution of the phasor diagram for the circuit in Fig. 9.56.

- We know that $\mathbf{I}_b$ lags behind $\mathbf{V}_L$ by 90° and that its magnitude is $|\mathbf{V}_L|/\omega L_2$.
- The current $\mathbf{I}$ is equal to the sum of $\mathbf{I}_a$ and $\mathbf{I}_b$.
- The voltage drop across R_1 is in phase with the current $\mathbf{I}$, and the voltage drop across $j\omega L_1$ leads $\mathbf{I}$ by 90°.
- The source voltage is given by
  $\mathbf{V}_s = \mathbf{V}_L + (R_1 + j\omega L_1)\mathbf{I}$.

The completed phasor diagram shown in Step 6 of Fig. 9.57 shows the amplitude and phase angle relationships among all the currents and voltages in Fig. 9.56.

Now add the capacitor branch shown in Fig. 9.58. We are holding $\mathbf{V}_L$ constant, so we construct the phasor diagram for the circuit in Fig. 9.58 following the same steps as those in Fig. 9.57, except that, in Step 4, we add the capacitor current $\mathbf{I}_c$ to the diagram, where $\mathbf{I}_c$ leads $\mathbf{V}_L$ by 90°, with $|\mathbf{I}_c| = |\mathbf{V}_L\omega C|$. Figure 9.59 shows the effect of $\mathbf{I}_c$ on the current $\mathbf{I}$: Both the magnitude and phase angle of $\mathbf{I}$ change as the magnitude of $\mathbf{I}_c$ changes. As $\mathbf{I}$ changes, so do the magnitude and phase angle of the voltage drop across the impedance $(R_1 + j\omega L_1)$, resulting in changes to the magnitude and phase angle of $\mathbf{V}_s$. The phasor diagram shown in Fig. 9.60 depicts these observations. The dashed phasors represent the pertinent currents and voltages before adding the capacitor.

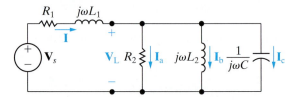

Figure 9.58 ▲ The addition of a capacitor to the circuit shown in Fig. 9.56.

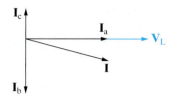

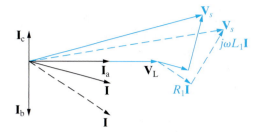

Figure 9.60 ▲ The effect of adding a load-shunting capacitor to the circuit shown in Fig. 9.53 if $\mathbf{V}_L$ is held constant.

Figure 9.59 ▲ The effect of the capacitor current $\mathbf{I}_c$ on the line current $\mathbf{I}$.

Thus, comparing the dashed phasors of $\mathbf{I}$, $R_1\mathbf{I}$, $j\omega L_1\mathbf{I}$, and $\mathbf{V}_s$ with their solid counterparts clearly shows the effect of adding C to the circuit. In particular, adding the capacitor reduces the source voltage amplitude while maintaining the load voltage amplitude. Practically, this result means that, as the load increases (i.e., as $\mathbf{I}_a$ and $\mathbf{I}_b$ increase), we can add capacitors to the system (i.e., increase $\mathbf{I}_c$), so

that under heavy load conditions we can maintain $\mathbf{V}_L$ without increasing the amplitude of the source voltage.

SELF-CHECK: Assess your understanding of this material by trying Chapter Problems 9.83 and 9.84.

■ Practical Perspective

A Household Distribution Circuit

After determining the loads on the three-wire distribution circuit prior to the interruption of fuse A, you are able to construct the frequency-domain circuit model shown in Fig. 9.61. The impedances of the wires connecting the source to the loads are assumed negligible.

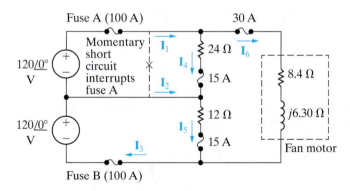

Figure 9.61 ▲ The three-wire household distribution circuit model.

Let's begin by calculating all of the branch current phasors in the figure, prior to the interruption of Fuse A. We calculate $\mathbf{I}_4$, $\mathbf{I}_5$, and $\mathbf{I}_6$ using Ohm's law:

$$\mathbf{I}_4 = \frac{120}{24} = 5\underline{/0°} \text{ A};$$

$$\mathbf{I}_5 = \frac{120}{12} = 10\underline{/0°} \text{ A};$$

$$\mathbf{I}_6 = \frac{240}{8.4 + j6.3} = 18.29 - j13.71 \text{ A} = 22.86\underline{/-36.87°} \text{ A}.$$

We calculate $\mathbf{I}_1$, $\mathbf{I}_2$, and $\mathbf{I}_3$ using KCL and the other branch currents:

$$\mathbf{I}_1 = \mathbf{I}_4 + \mathbf{I}_6 = 23.29 - j13.71 \text{ A} = 27.02\underline{/-30.5°} \text{ A};$$

$$\mathbf{I}_2 = \mathbf{I}_5 - \mathbf{I}_4 = 5\underline{/0°} \text{ A};$$

$$\mathbf{I}_3 = \mathbf{I}_5 + \mathbf{I}_6 = 28.29 - j13.71 \text{ A} = 31.44\underline{/-25.87°} \text{ A}.$$

Now calculate those same branch current phasors after Fuse A is interrupted. We assume that the fan motor behaves like a short circuit when it stalls, and the interrupted fuse behaves like an open circuit. The circuit model now looks like Fig. 9.62. To analyze this circuit, we write two mesh current equations using the mesh current phasors shown in Fig. 9.62:

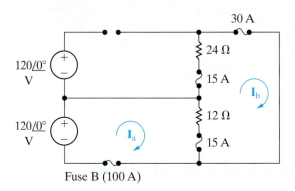

Figure 9.62 ▲ The circuit in Fig. 9.61 after Fuse A is interrupted and the fan motor stalls.

$$12(\mathbf{I}_a - \mathbf{I}_b) = 120;$$

$$24\mathbf{I}_b + 12(\mathbf{I}_b - \mathbf{I}_a) = 0.$$

Solving the mesh current equations, we get

$$\mathbf{I}_a = 15 \text{ A};$$

$$\mathbf{I}_b = 5 \text{ A}.$$

Using these mesh current phasors to calculate the new branch current phasors from Fig. 9.61, we get

$$\mathbf{I}_1 = 0 \text{ A};$$

$$\mathbf{I}_2 = \mathbf{I}_3 = \mathbf{I}_a = 15 \text{ A};$$

$$\mathbf{I}_6 = \mathbf{I}_b = 5 \text{ A};$$

$$\mathbf{I}_4 = \mathbf{I}_1 - \mathbf{I}_6 = -5 \text{ A};$$

$$\mathbf{I}_5 = \mathbf{I}_2 + \mathbf{I}_4 = 10 \text{ A}.$$

We can see that even though Fuse A is interrupted and the $\mathbf{I}_1$ branch current is zero, all of the other branch currents are nonzero. The television and clock continued to operate because they are represented by the 12 Ω load that still has an ample supply of current.

SELF-CHECK: Assess your understanding of this Practical Perspective by trying Chapter Problems 9.86 and 9.87.

Summary

- The general equation for a **sinusoidal source** is

$$v = V_m \cos(\omega t + \phi) \text{ (voltage source)},$$

or

$$i = I_m \cos(\omega t + \phi) \text{ (current source)},$$

where V_m (or I_m) is the maximum amplitude, ω is the frequency, and ϕ is the phase angle. (See page 320.)

- The frequency, ω, of a sinusoidal response is the same as the frequency of the sinusoidal source driving the circuit. The amplitude and phase angle of the response are usually different from those of the source. (See page 323.)

- The best way to find the steady-state voltages and currents in a circuit driven by sinusoidal sources is to perform the analysis in the frequency domain. The following mathematical transforms allow us to move between the time and frequency domains.

  - The phasor transform (from the time domain to the frequency domain):
  $$\mathbf{V} = V_m e^{j\phi} = \mathscr{P}\{V_m \cos(\omega t + \phi)\}.$$

  - The inverse phasor transform (from the frequency domain to the time domain):
  $$\mathscr{P}^{-1}\{V_m e^{j\phi} = \mathscr{R}\{V_m e^{j\phi}e^{j\omega t}\}.$$

  (See pages 324–325.)

- In a circuit with a sinusoidal source, the voltage leads the current by 90° at the terminals of an inductor, and the current leads the voltage by 90° at the terminals of a capacitor. (See pages 327–331.)

- **Impedance** (Z) relates the phasor current and phasor voltage for resistors, inductors, and capacitors in an equation that has the same form as Ohm's law,

$$\mathbf{V} = Z\mathbf{I},$$

where the reference direction for $\mathbf{I}$ obeys the passive sign convention. The reciprocal of impedance is **admittance** (Y), so another way to express the current-voltage relationship for resistors, inductors, and capacitors in the frequency domain is

$$\mathbf{V} = \mathbf{I}/Y.$$

(See pages 330 and 336.)

- The equations for impedance and admittance for resistors, inductors, and capacitors are summarized in Table 9.3.

- All of the circuit analysis techniques developed in Chapters 2–4 for resistive circuits also apply to sinusoidal steady-state circuits in the frequency domain. These techniques include KVL, KCL, series, and parallel combinations of impedances, voltage and current division, node-voltage and mesh-current methods, Thévenin and Norton equivalents, and source transformation.

- The two-winding **linear transformer** is a coupling device made up of two coils wound on the same nonmagnetic core. **Reflected impedance** is the impedance of the secondary circuit as seen from the terminals of the primary circuit, or vice versa. The reflected impedance of a linear transformer seen from the primary side is the complex conjugate of the self-impedance of the secondary circuit scaled by the factor $(\omega M/|Z_{22}|)^2$. (See pages 347–349.)

- The two-winding **ideal transformer** is a linear transformer with the following special properties: perfect coupling ($k = 1$), infinite self-inductance in each coil ($L_1 = L_2 = \infty$), and lossless coils ($R_1 = R_2 = 0$). The circuit behavior is governed by the turns ratio $a = N_2/N_1$. In particular, the volts per turn is the same for each winding, or

$$\frac{\mathbf{V}_1}{N_1} = \pm\frac{\mathbf{V}_2}{N_2},$$

and the ampere turns are the same for each winding, or

$$N_1\mathbf{I}_1 = \pm N_2\mathbf{I}_2.$$

(See page 356.)

TABLE 9.3	Impedance and Related Values			
Element	**Impedance** (Z)	**Reactance**	**Admittance** (Y)	**Susceptance**
Resistor	R (resistance)	—	G (conductance)	—
Capacitor	$j(-1/\omega C)$	$-1/\omega C$	$j\omega C$	ωC
Inductor	$j\omega L$	ωL	$j(-1/\omega L)$	$-1/\omega L$

Problems

Section 9.1

9.1 Consider the sinusoidal voltage

$$v(t) = 40 \cos(100\pi t + 60°) \text{ V.}$$

a) What is the maximum amplitude of the voltage?

b) What is the frequency in hertz?

c) What is the frequency in radians per second?

d) What is the phase angle in radians?

e) What is the phase angle in degrees?

f) What is the period in milliseconds?

g) What is the first time after $t = 0$ that $v = -40$ V?

h) The sinusoidal function is shifted $10/3$ ms to the right along the time axis. What is the expression for $v(t)$?

i) What is the minimum number of milliseconds that the function must be shifted to the left if the expression for $v(t)$ is $40 \sin 100\pi t$ V?

9.2 A sinusoidal voltage is given by the expression

$$v = 10 \cos(3769.91t - 53.13°) \text{ V.}$$

Find (a) f in hertz; (b) T in milliseconds; (c) V_m; (d) $v(0)$; (e) ϕ in degrees and radians; (f) the smallest positive value of t at which $v = 0$; and (g) the smallest positive value of t at which $dv/dt = 0$.

9.3 At $t = -2$ ms, a sinusoidal voltage is known to be zero and going positive. The voltage is next zero at $t = 8$ ms. It is also known that the voltage is 80.9 V at $t = 0$.

a) What is the frequency of v in hertz?

b) What is the expression for v?

9.4 In a single graph, sketch $v = 100 \cos(\omega t + \phi)$ versus ωt for $\phi = -60°, -30°, 0°, 30°,$ and $60°$.

a) State whether the voltage function is shifting to the right or left as ϕ becomes more positive.

b) What is the direction of shift if ϕ changes from 0 to $-30°$?

9.5 A sinusoidal voltage is zero at $t = (-2\pi/3)$ ms and increasing at a rate of $80,000$ V/s. The maximum amplitude of the voltage is 80 V.

a) What is the frequency of v in radians per second?

b) What is the expression for v?

9.6 The rms value of the sinusoidal voltage supplied to the convenience outlet of a home in the United States is 120 V. What is the maximum value of the voltage at the outlet?

9.7 Show that

$$\int_{t_0}^{t_0+T} V_m^2 \cos^2(\omega t + \phi)dt = \frac{V_m^2 T}{2}.$$

9.8 Find the rms value of the half-wave rectified sinusoidal voltage shown in Fig. P9.8.

Figure P9.8

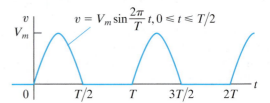

$$v = V_m \sin\frac{2\pi}{T} t, 0 \leqslant t \leqslant T/2$$

Section 9.2

9.9 a) Verify that Eq. 9.7 is the solution of Eq. 9.6. This can be done by substituting Eq. 9.7 into the left-hand side of Eq. 9.6 and then noting that it equals the right-hand side for all values of $t > 0$. At $t = 0$, Eq. 9.7 should reduce to the initial value of the current.

b) Because the transient component vanishes as time elapses and because our solution must satisfy the differential equation for all values of t, the steady-state component, by itself, must also satisfy the differential equation. Verify this observation by showing that the steady-state component of Eq. 9.7 satisfies Eq. 9.6.

9.10 The voltage applied to the circuit shown in Fig. 9.5 at $t = 0$ is $20 \cos(800t + 25°)$ V. The circuit resistance is 80Ω and the initial current in the 75 mH inductor is zero.

a) Find $i(t)$ for $t \geq 0$.

b) Write the expressions for the transient and steady-state components of $i(t)$.

c) Find the numerical value of i after the switch has been closed for 1.875 ms.

d) What are the maximum amplitude, frequency (in radians per second), and phase angle of the steady-state current?

e) By how many degrees are the voltage and the steady-state current out of phase?

Sections 9.3–9.4

9.11 Use the concept of the phasor to combine the following sinusoidal functions into a single trigonometric expression:

a) $y = 50 \cos(500t + 60°) + 100 \cos(500t - 30°)$,

b) $y = 200 \cos(377t + 50°) - 100 \sin(377t + 150°)$,

c) $y = 80 \cos(100t + 30°) - 100 \sin(100t - 135°) + 50 \cos(100t - 90°)$,

d) $y = 250 \cos \omega t + 250 \cos(\omega t + 120°) + 250 \cos(\omega t - 120°)$.

9.12 The expressions for the steady-state voltage and current at the terminals of the circuit seen in Fig. P9.12 are

$$v_g = 300 \cos(5000\pi t + 78°) \text{ V},$$

$$i_g = 6 \sin(5000\pi t + 123°) \text{ A}.$$

a) What is the impedance seen by the source?

b) By how many microseconds is the current out of phase with the voltage?

Figure P9.12

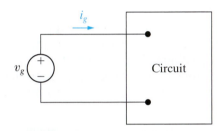

9.13 A 400 Hz sinusoidal voltage with a maximum amplitude of 100 V at $t = 0$ is applied across the terminals of an inductor. The maximum amplitude of the steady-state current in the inductor is 20 A.

a) What is the frequency of the inductor current?

b) If the phase angle of the voltage is zero, what is the phase angle of the current?

c) What is the inductive reactance of the inductor?

d) What is the inductance of the inductor in millihenrys?

e) What is the impedance of the inductor?

9.14 A 50 kHz sinusoidal voltage has zero phase angle and a maximum amplitude of 10 mV. When this voltage is applied across the terminals of a capacitor, the resulting steady-state current has a maximum amplitude of 628.32 μA.

a) What is the frequency of the current in radians per second?

b) What is the phase angle of the current?

c) What is the capacitive reactance of the capacitor?

d) What is the capacitance of the capacitor in microfarads?

e) What is the impedance of the capacitor?

Sections 9.5 and 9.6

9.15 A 40 Ω resistor, a 5 mH inductor, and a 1.25 μF capacitor are connected in series. The series-connected elements are energized by a sinusoidal voltage source whose voltage is $600 \cos(8000t + 20°)$ V.

PSPICE
MULTISIM

a) Draw the frequency-domain equivalent circuit.

b) Reference the current in the direction of the voltage rise across the source, and find the phasor current.

c) Find the steady-state expression for $i(t)$.

9.16 A 10 Ω resistor and a 5 μF capacitor are connected in parallel. This parallel combination is also in parallel with the series combination of a 8 Ω resistor and a 300 μH inductor. These three parallel branches are driven by a sinusoidal current source whose current is $922 \cos(20{,}000t + 30°)$ A.

PSPICE
MULTISIM

a) Draw the frequency-domain equivalent circuit.

b) Reference the voltage across the current source as a rise in the direction of the source current, and find the phasor voltage.

c) Find the steady-state expression for $v(t)$.

9.17 Three branches having impedances of $3 + j4$ Ω, $16 - j12$ Ω, and $-j4$ Ω, respectively, are connected in parallel. What are the equivalent (a) admittance, (b) conductance, and (c) susceptance of the parallel connection in millisiemens? (d) If the parallel branches are excited from a sinusoidal current source where $i = 8 \cos \omega t$ A, what is the maximum amplitude of the current in the purely capacitive branch?

9.18 a) Show that, at a given frequency ω, the circuits in Fig. P9.18(a) and (b) will have the same impedance between the terminals a,b if

$$R_1 = \frac{\omega^2 L_2^2 R_2}{R_2^2 + \omega^2 L_2^2}, \qquad L_1 = \frac{R_2^2 L_2}{R_2^2 + \omega^2 L_2^2}.$$

b) Find the values of resistance and inductance that when connected in series will have the same impedance at 4 krad/s as that of a 5 kΩ resistor connected in parallel with a 1.25 H inductor.

Figure P9.18

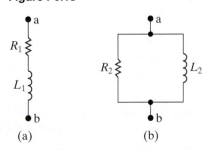

(a)   (b)

9.19 a) Show that at a given frequency ω, the circuits in Fig. P9.18(a) and (b) will have the same impedance between the terminals a,b if

$$R_2 = \frac{R_1^2 + \omega^2 L_1^2}{R_1}, \qquad L_2 = \frac{R_1^2 + \omega^2 L_1^2}{\omega^2 L_1}.$$

(*Hint:* The two circuits will have the same impedance if they have the same admittance.)

b) Find the values of resistance and inductance that when connected in parallel will have the same impedance at 1 krad/s as an 8 kΩ resistor connected in series with a 4 H inductor.

9.20 a) Show that at a given frequency ω, the circuits in Fig. P9.20(a) and (b) will have the same impedance between the terminals a,b if

$$R_1 = \frac{R_2}{1 + \omega^2 R_2^2 C_2^2},$$

$$C_1 = \frac{1 + \omega^2 R_2^2 C_2^2}{\omega^2 R_2^2 C_2}.$$

b) Find the values of resistance and capacitance that when connected in series will have the same impedance at 40 krad/s as that of a 1000 Ω resistor connected in parallel with a 50 nF capacitor.

Figure P9.20

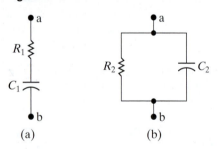

(a)   (b)

9.21 a) Show that at a given frequency ω, the circuits in Fig 9.20(a) and (b) will have the same impedance between the terminals a,b if

$$R_2 = \frac{1 + \omega^2 R_1^2 C_1^2}{\omega^2 R_1 C_1^2},$$

$$C_2 = \frac{C_1}{1 + \omega^2 R_1^2 C_1^2}.$$

(*Hint:* The two circuits will have the same impedance if they have the same admittance.)

b) Find the values of resistance and capacitance that when connected in parallel will give the same impedance at 50 krad/s as that of a 1 kΩ resistor connected in series with a capacitance of 40 nF.

9.22 a) Using component values from Appendix H, combine at least one resistor, inductor, and

capacitor in series to create an impedance of $300 - j400\ \Omega$ at a frequency of 10,000 rad/s.

b) At what frequency does the circuit from part (a) have an impedance that is purely resistive?

9.23 a) Using component values from Appendix H, combine at least one resistor and one inductor in parallel to create an impedance of $40 + j20\ \Omega$ at a frequency of 5000 rad/s. (*Hint:* Use the results of Problem 9.19.)

b) Using component values from Appendix H, combine at least one resistor and one capacitor in parallel to create an impedance of $40 - j20\ \Omega$ at a frequency of 5000 rad/s. (*Hint:* Use the result of Problem 9.21.)

9.24 a) Using component values from Appendix H, find a single capacitor or a network of capacitors that, when combined in parallel with the *RL* circuit from Problem 9.23(a), gives an equivalent impedance that is purely resistive at a frequency of 5000 rad/s.

b) Using component values from Appendix H, find a single inductor or a network of inductors that, when combined in parallel with the *RC* circuit from Problem 9.23(b), gives an equivalent impedance that is purely resistive at a frequency of 5000 rad/s.

9.25 Find the admittance Y_{ab} in the circuit seen in Fig. P9.25. Express Y_{ab} in both polar and rectangular form. Give the value of Y_{ab} in millisiemens.

Figure P9.25

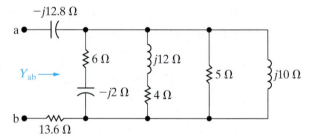

9.26 Find the impedance Z_{ab} in the circuit seen in Fig. P9.26. Express Z_{ab} in both polar and rectangular form.

Figure P9.26

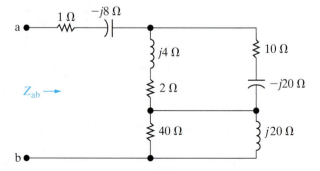

9.27 a) For the circuit shown in Fig. P9.27, find the frequency (in radians per second) at which the impedance Z_{ab} is purely resistive.

  b) Find the value of Z_{ab} at the frequency of (a).

Figure P9.27

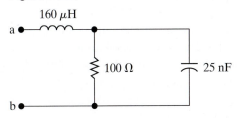

9.28 The circuit in Fig. P9.28 is operating in the sinusoidal steady state. Find the steady-state expression for $v_o(t)$ if $v_g = 40 \sin 50{,}000t$ V.

Figure P9.28

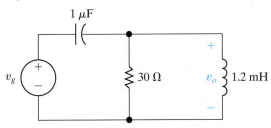

9.29 Find the steady-state expression for $i_o(t)$ in the circuit in Fig. P9.29 if $v_s = 100 \sin 50t$ mV.

Figure P9.29

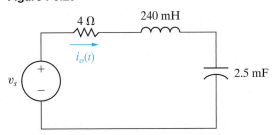

9.30 The circuit in Fig. P9.30 is operating in the sinusoidal steady state. Find $i_o(t)$ if $v_s(t) = 25 \sin 4000t$ V.

Figure P9.30

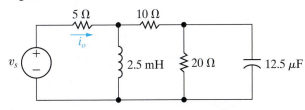

9.31 Find the steady-state expression for v_o in the circuit of Fig. P9.31 if $i_g = 0.5 \cos 2000t$ A.

Figure P9.31

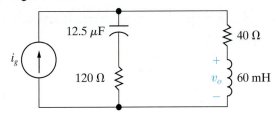

9.32 The phasor current $\mathbf{I}_b$ in the circuit shown in Fig. P9.32 is $25 \, \underline{/0°}$ mA.

  a) Find $\mathbf{I}_a$, $\mathbf{I}_c$, and $\mathbf{I}_g$.

  b) If $\omega = 1500$ rad/s, write expressions for $i_a(t)$, $i_c(t)$, and $i_g(t)$.

Figure P9.32

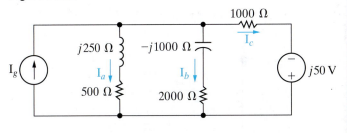

9.33 a) For the circuit shown in Fig. P9.33, find the steady-state expression for v_o if $i_g = 25 \cos 50{,}000t$ mA.

  b) By how many microseconds does v_o lead i_g?

Figure P9.33

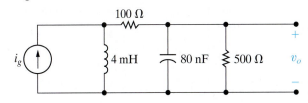

9.34 Find the value of Z in the circuit seen in Fig. P9.34 if $\mathbf{V}_g = 100 - j50$ V, $\mathbf{I}_g = 30 + j20$ A, and $\mathbf{V}_1 = 140 + j30$ V.

Figure P9.34

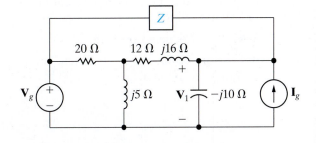

9.35 Find $\mathbf{I}_b$ and Z in the circuit shown in Fig. P9.35 if $\mathbf{V}_g = 25\underline{/0°}$ V and $\mathbf{I}_a = 5\underline{/90°}$ A.

Figure P9.35

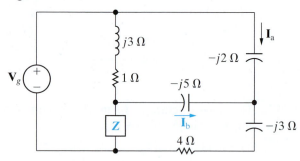

9.36 The circuit shown in Fig. P9.36 is operating in the sinusoidal steady state. Find the value of ω if

$$i_o = 100 \sin(\omega t + 173.13°) \text{ mA},$$

$$v_g = 500 \cos(\omega t + 30°) \text{ V}.$$

Figure P9.36

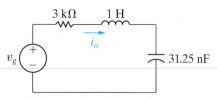

9.37 The frequency of the sinusoidal voltage source in the circuit in Fig. P9.37 is adjusted until i_g is in phase with v_g.

PSPICE
MULTISIM

a) What is the value of ω in radians per second?

b) If $v_g = 15 \cos \omega t$ V (where ω is the frequency found in [a]), what is the steady-state expression for i_g?

Figure P9.37

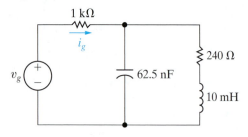

9.38 The frequency of the sinusoidal voltage source in the circuit in Fig. P9.38 is adjusted until the current i_o is in phase with v_g.

PSPICE
MULTISIM

a) Find the frequency in hertz.

b) Find the steady-state expression for i_g (at the frequency found in [a]) if $v_g = 90 \cos \omega t$ V.

Figure P9.38

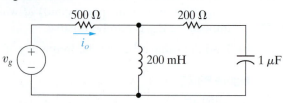

9.39 a) The frequency of the source voltage in the circuit in Fig. P9.39 is adjusted until v_g is in phase with i_g. What is the value of ω in radians per second?

PSPICE
MULTISIM

b) If $i_g = 60 \cos \omega t$ mA (where ω is the frequency found in [a]), what is the steady-state expression for v_g?

Figure P9.39

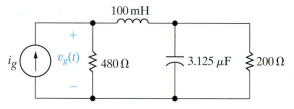

9.40 The circuit shown in Fig. P9.40 is operating in the sinusoidal steady state. The capacitor is adjusted until the current i_g is in phase with the sinusoidal voltage v_g.

PSPICE
MULTISIM

a) Specify the capacitance in microfarads if $v_g = 80 \cos 5000t$ V.

b) Give the steady-state expression for i_g when C has the value found in (a).

Figure P9.40

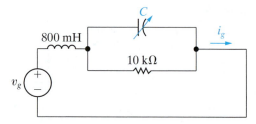

9.41 a) The source voltage in the circuit in Fig. P9.41 is $v_g = 200 \cos 500t$ V. Find the values of L such that i_g is in phase with v_g when the circuit is operating in the steady state.

PSPICE
MULTISIM

b) For the values of L found in (a), find the steady-state expressions for i_g.

Figure P9.41

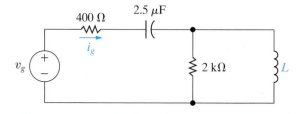

9.42 Find Z_{ab} for the circuit shown in Fig P9.42.

Figure P9.42

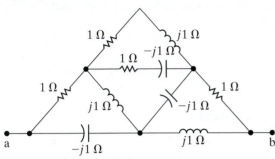

Section 9.7

9.43 Use source transformations to find the Thévenin equivalent circuit with respect to the terminals a,b for the circuit shown in Fig. P9.43.

Figure P9.43

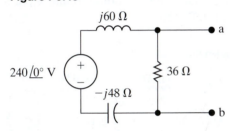

9.44 Use source transformations to find the Norton equivalent circuit with respect to the terminals a,b for the circuit shown in Fig. P9.44.

Figure P9.44

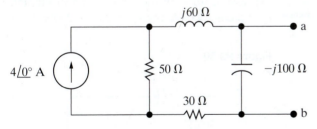

9.45 The sinusoidal voltage source in the circuit in Fig. P9.45 is developing a voltage equal to $50 \sin 400t$ V.

a) Find the Thévenin voltage with respect to the terminals a,b.

b) Find the Thévenin impedance with respect to the terminals a,b.

c) Draw the Thévenin equivalent.

Figure P9.45

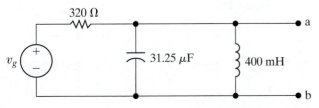

9.46 Find the Norton equivalent circuit with respect to the terminals a,b for the circuit shown in Fig. P9.46.

Figure P9.46

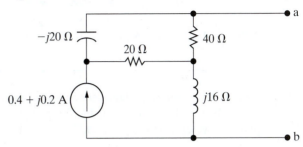

9.47 The device in Fig. P9.47 is represented in the frequency domain by a Norton equivalent. When a resistor having an impedance of 5 kΩ is connected across the device, the value of V_0 is $(5 - j15)$ V. When a capacitor having an impedance of $-j3$ kΩ is connected across the device, the value of I_0 is $(4.5 - j6)$ mA. Find the Norton current I_N and the Norton impedance Z_N.

Figure P9.47

9.48 Find the Norton equivalent with respect to terminals a,b in the circuit of Fig. P9.48.

Figure P9.48

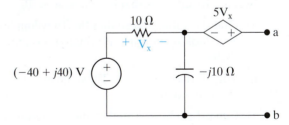

9.49 Find the Norton equivalent circuit with respect to the terminals a,b for the circuit shown in Fig. P9.49 when $V_s = 5\underline{/0°}$ V.

Figure P9.49

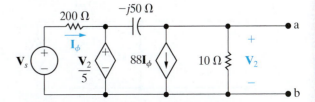

9.50 Find the Thévenin equivalent circuit with respect to the terminals a,b of the circuit shown in Fig. P9.50.

Figure P9.50

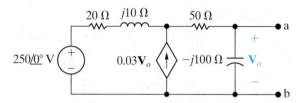

9.51 Find the Thévenin impedance seen looking into the terminals a,b of the circuit in Fig. P9.51 if the frequency of operation is $(200/\pi)$ kHz.

Figure P9.51

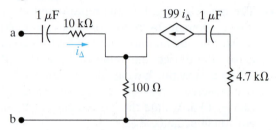

9.52 Find Z_{ab} in the circuit shown in Fig. P9.52 when the circuit is operating at a frequency of 100 krad/s.

Figure P9.52

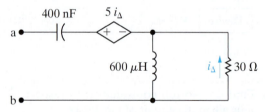

9.53 The circuit shown in Fig. P9.53 is operating at a frequency of 10 rad/s. Assume α is real and lies between -10 and $+10$, that is, $-10 \le \alpha \le 10$.

a) Find the value of α so that the Thévenin impedance looking into the terminals a,b is purely resistive.

b) What is the value of the Thévenin impedance for the α found in (a)?

c) Can α be adjusted so that the Thévenin impedance equals $500 - j500\ \Omega$? If so, what is the value of α?

d) For what values of α will the Thévenin impedance be inductive?

Figure P9.53

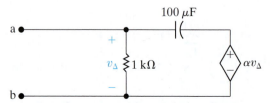

Section 9.8

9.54 Use the node-voltage method to find the steady-state expression for $v_o(t)$ in the circuit in Fig. P9.54 if

$$v_{g1} = 25 \sin(400t + 143.13°) \text{ V},$$

$$v_{g2} = 18.03 \cos(400t + 33.69°) \text{ V}.$$

Figure P9.54

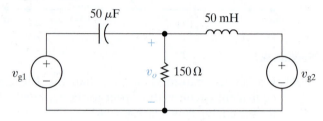

9.55 Use the node-voltage method to find V_o in the circuit in Fig. P9.55.

Figure P9.55

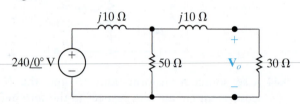

9.56 Use the node voltage method to find the steady-state expression for i_o in the circuit seen in Fig. P9.56 if $i_g = 5 \cos 2500t$ A and $v_g = 20 \cos(2500t + 90°)$ V.

Figure P9.56

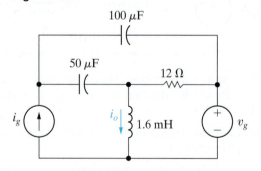

9.57 Use the node-voltage method to find $\mathbf{V}_o$ and $\mathbf{I}_o$ in the circuit seen in Fig. P9.57.

Figure P9.57

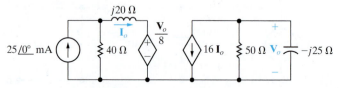

9.58 Use the node-voltage method to find the phasor voltage $\mathbf{V}_0$ in the circuit shown in Fig. P9.58. Express the voltage in both polar and rectangular form.

Figure P9.58

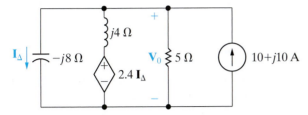

Section 9.9

9.59 Use the mesh-current method to find the steady-state expression for the branch currents i_a and i_b in the circuit seen in Fig. P9.59 if $v_a = 50 \sin 10^6 t$ V and $v_b = 25 \cos(10^6 t + 90°)$ V.

PSPICE
MULTISIM

Figure P9.59

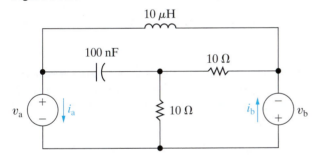

9.60 Use the mesh-current method to find the steady-state expression for $i_o(t)$ in the circuit in Fig. P9.60 if

$$v_a = 100 \cos 50{,}000t \text{ V},$$

$$v_b = 100 \sin(50{,}000t + 180°) \text{ V}.$$

Figure P9.60

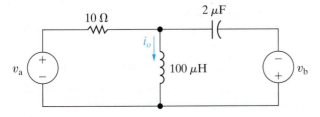

9.61 Use the mesh-current method to find the steady-state expression for $v_o(t)$ in the circuit in Fig. P9.54.

9.62 Use the mesh-current method to find the steady-state expression for $i_o(t)$ in the circuit in Fig. P9.56.

9.63 Use the mesh-current method to find the branch currents $\mathbf{I}_a$, $\mathbf{I}_b$, $\mathbf{I}_c$, and $\mathbf{I}_d$ in the circuit shown in Fig. P9.63.

Figure P9.63

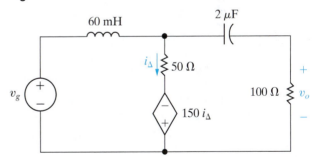

9.64 Use the mesh-current method to find the steady-state expression for v_o in the circuit seen in Fig. P9.64 if v_g equals $400 \cos 5000t$ V.

PSPICE
MULTISIM

Figure P9.64

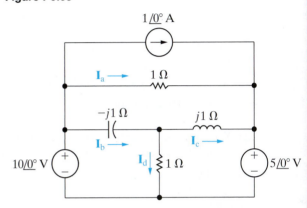

Sections 9.5–9.9

9.65 Use voltage division to find the steady-state expression for $v_o(t)$ in the circuit in Fig. P9.65 if $v_g = 100 \cos 8000t$ V.

PSPICE
MULTISIM

Figure P9.65

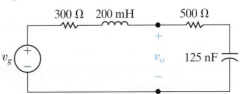

9.66 Use current division to find the steady-state expression for i_o in the circuit in Fig. P9.66 if $i_g = 400 \cos 20{,}000t$ mA.

Figure P9.66

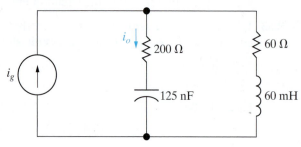

9.67 For the circuit in Fig. P9.67, suppose

$$v_1 = 20 \cos(2000t - 36.87°) \text{ V},$$

$$v_2 = 10 \cos(5000t + 16.26°) \text{ V}.$$

a) What circuit analysis technique must be used to find the steady-state expression for $v_o(t)$?

b) Find the steady-state expression for $v_o(t)$.

Figure P9.67

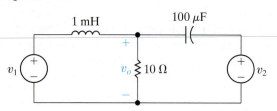

9.68 For the circuit in Fig. P9.68, suppose

$$v_a = 10 \cos 16{,}000t \text{ V},$$

$$v_b = 20 \cos 4000t \text{ V}.$$

a) What circuit analysis technique must be used to find the steady-state expression for $v_o(t)$?

b) Find the steady-state expression for $v_o(t)$.

Figure P9.68

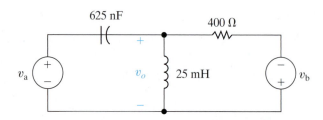

9.69 The op amp in the circuit in Fig. P9.69 is ideal.

a) Find the steady-state expression for $v_o(t)$.

b) How large can the amplitude of v_g be before the amplifier saturates?

Figure P9.69

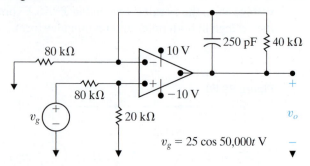

$$v_g = 25 \cos 50{,}000t \text{ V}$$

9.70 The op amp in the circuit seen in Fig. P9.70 is ideal. Find the steady-state expression for $v_o(t)$ when $v_g = 2 \cos 10^6 t$ V.

Figure P9.70

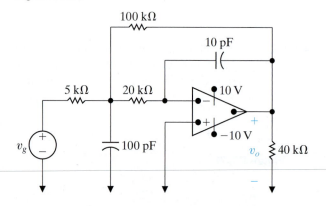

9.71 The sinusoidal voltage source in the circuit shown in Fig. P9.71 is generating the voltage $v_g = 4 \cos 200t$ V. If the op amp is ideal, what is the steady-state expression for $v_o(t)$?

Figure P9.71

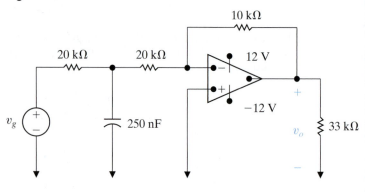

9.72 The 250 nF capacitor in the circuit seen in Fig. P9.71 is replaced with a variable capacitor. The capacitor is adjusted until the output voltage leads the input voltage by 135°.

a) Find the value of C in microfarads.

b) Write the steady-state expression for $v_o(t)$ when C has the value found in (a).

9.73 The op amp in the circuit shown in Fig. P9.73 is ideal. The voltage of the ideal sinusoidal source is $v_g = 6 \cos 10^5 t$ V.

PSPICE
MULTISIM

  a) How small can C_o be before the steady-state output voltage no longer has a pure sinusoidal waveform?

  b) For the value of C_o found in (a), write the steady-state expression for v_o.

Figure P9.73

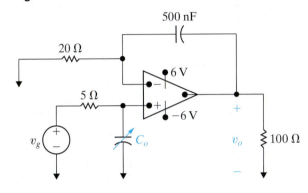

Section 9.10

9.74 a) Find the steady-state expressions for the currents i_g and i_L in the circuit in Fig. P9.74 when $v_g = 70 \cos 5000t$ V.

PSPICE
MULTISIM

  b) Find the coefficient of coupling.

  c) Find the energy stored in the magnetically coupled coils at $t = 100\pi$ μs and $t = 200\pi$ μs.

Figure P9.74

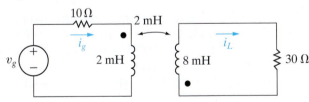

9.75 For the circuit in Fig. P9.75, find the Thévenin equivalent with respect to the terminals c,d.

Figure P9.75

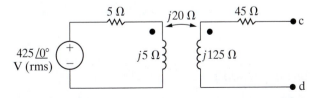

9.76 The value of k in the circuit in Fig. P9.76 is adjusted so that Z_{ab} is purely resistive when $\omega = 4$ krad/s. Find Z_{ab}.

Figure P9.76

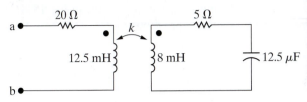

9.77 The sinusoidal voltage source in the circuit seen in Fig. P9.77 is operating at a frequency of 200 krad/s. The coefficient of coupling is adjusted until the peak amplitude of i_1 is maximum.

PSPICE
MULTISIM

  a) What is the value of k?

  b) What is the peak amplitude of i_1 if $v_g = 560 \cos(2 \times 10^5 t)$ V?

Figure P9.77

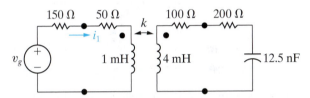

9.78 A series combination of a 300 Ω resistor and a 100 mH inductor is connected to a sinusoidal voltage source by a linear transformer. The source is operating at a frequency of 1000 rad/s. At this frequency, the internal impedance of the source is $(100 + j13.74)$ Ω. The rms voltage at the terminals of the source is 50 V when it is not loaded. The parameters of the linear transformer are $R_1 = 41.68$ Ω, $L_1 = 180$ mH, $R_2 = 500$ Ω, $L_2 = 500$ mH, and $M = 270$ mH.

  a) What is the value of the impedance reflected into the primary?

  b) What is the value of the impedance seen from the terminals of the practical source?

Section 9.11

9.79 Find the impedance Z_{ab} in the circuit in Fig. P9.79 if $Z_L = 200 \underline{/-45^\circ}$ Ω.

Figure P9.79

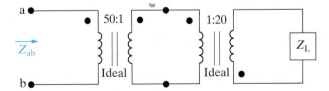

9.80 At first glance, it may appear from Eq. 9.34 that an inductive load could make the reactance seen looking into the primary terminals (i.e., X_{ab}) look capacitive. Intuitively, we know this is impossible. Show that X_{ab} can never be negative if X_L is an inductive reactance.

9.81 a) Show that the impedance seen looking into the terminals a,b in the circuit in Fig. P9.81 is given by the expression

$$Z_{ab} = \left(1 + \frac{N_1}{N_2}\right)^2 Z_L.$$

b) Show that if the polarity terminals of either one of the coils is reversed,

$$Z_{ab} = \left(1 - \frac{N_1}{N_2}\right)^2 Z_L.$$

Figure P9.81

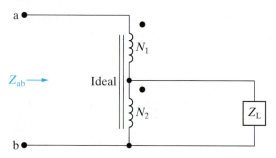

9.82 a) Show that the impedance seen looking into the terminals a,b in the circuit in Fig. P9.82 is given by the expression

$$Z_{ab} = \frac{Z_L}{\left(1 + \dfrac{N_1}{N_2}\right)^2}.$$

b) Show that if the polarity terminal of either one of the coils is reversed that

$$Z_{ab} = \frac{Z_L}{\left(1 - \dfrac{N_1}{N_2}\right)^2}.$$

Figure P9.82

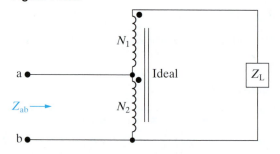

Section 9.12

9.83 The parameters in the circuit shown in Fig. 9.56 are $R_1 = 0.1\ \Omega$, $\omega L_1 = 0.8\ \Omega$, $R_2 = 24\ \Omega$, $\omega L_2 = 32\ \Omega$, and $V_L = 240 + j0$ V.

a) Calculate the phasor voltage V_s.

b) Connect a capacitor in parallel with the inductor, hold $\mathbf{V}_L$ constant, and adjust the capacitor until the magnitude of $\mathbf{I}$ is a minimum. What is the capacitive reactance? What is the value of $\mathbf{V}_s$?

c) Find the value of the capacitive reactance that keeps the magnitude of $\mathbf{I}$ as small as possible and that at the same time makes

$$|\mathbf{V}_s| = |\mathbf{V}_L| = 240 \text{ V}.$$

9.84 a) For the circuit shown in Fig. P9.84, compute $\mathbf{V}_s$ and $\mathbf{V}_l$.

b) Construct a phasor diagram showing the relationship between $\mathbf{V}_s$, $\mathbf{V}_l$, and the load voltage of $120\ \underline{/0°}$ V.

c) Repeat parts (a) and (b), given that the load resistance changes from 7.5 Ω to 2.5 Ω and the load reactance changes from 12 Ω to 4 Ω. Assume that the load voltage remains constant at $120\underline{/0°}$ V. How much must the amplitude of $\mathbf{V}_s$ be increased in order to maintain the load voltage at 120 V?

d) Repeat part (c), given that at the same time the load resistance and reactance change, a capacitive reactance of $-2\ \Omega$ is connected across the load terminals.

Figure P9.84

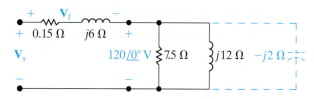

9.85 Show by using a phasor diagram what happens to the magnitude and phase angle of the voltage v_o in the circuit in Fig. P9.85 as R_x is varied from zero to infinity. The amplitude and phase angle of the source voltage are held constant as R_x varies.

PSPICE
MULTISIM

Figure P9.85

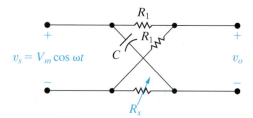

Sections 9.1–9.12

9.86 a) Calculate the branch currents $\mathbf{I}_1 - \mathbf{I}_6$ in the circuit in Fig. P9.86.

PRACTICAL PERSPECTIVE

b) Find the primary current $\mathbf{I}_p$.

Figure P9.86

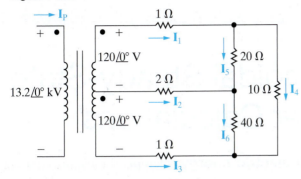

b) Show that $\mathbf{V}_1 = \mathbf{V}_2$ if $R_1 = R_2$.

c) Open the neutral branch and calculate $\mathbf{V}_1$ and $\mathbf{V}_2$ if $R_1 = 40\ \Omega, R_2 = 400\ \Omega$, and $R_3 = 8\ \Omega$.

d) Close the neutral branch and repeat (c).

e) On the basis of your calculations, explain why the neutral conductor is never fused in such a manner that it could open while the hot conductors are energized.

9.87 Suppose the 40 Ω resistance in the distribution cir-
PRACTICAL   cuit in Fig. P9.86 is replaced by a 20 Ω resistance.
PERSPECTIVE

a) Recalculate the branch current in the 2 Ω resistor, $\mathbf{I}_2$.

b) Recalculate the primary current, $\mathbf{I}_p$.

c) On the basis of your answers, is it desirable to have the resistance of the two 120 V loads be equal?

9.88 A residential wiring circuit is shown in Fig. P9.88. In
PRACTICAL   this model, the resistor R_3 is used to model a 250 V
PERSPECTIVE appliance (such as an electric range), and the resistors R_1 and R_2 are used to model 125 V appliances
(such as a lamp, toaster, and iron). The branches
carrying $\mathbf{I}_1$ and $\mathbf{I}_2$ are modeling what electricians refer to as the hot conductors in the circuit, and the
branch carrying $\mathbf{I}_n$ is modeling the neutral conductor. Our purpose in analyzing the circuit is to show
the importance of the neutral conductor in the satisfactory operation of the circuit. You are to choose
the method for analyzing the circuit.

a) Show that $\mathbf{I}_n$ is zero if $R_1 = R_2$.

Figure P9.88

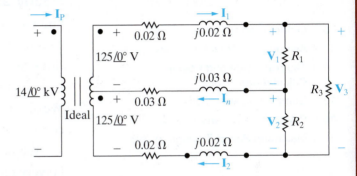

9.89 a) Find the primary current $\mathbf{I}_p$ for (c) and (d) in
PRACTICAL   Problem 9.88.
PERSPECTIVE

b) Do your answers make sense in terms of known circuit behavior?

9.90 Assume the fan motor in Fig. 9.61 is equipped with
PRACTICAL   a thermal cutout designed to interrupt the motor
PERSPECTIVE circuit if the motor current becomes excessive.
Would you expect the thermal cutout to operate?
Explain.

9.91 Explain why fuse B in Fig. 9.61 is not interrupted
PRACTICAL   when the fan motor stalls.
PERSPECTIVE

CHAPTER
10

CHAPTER CONTENTS

10.1 **Instantaneous Power** *p. 376*

10.2 **Average and Reactive Power** *p. 377*

10.3 **The rms Value and Power Calculations** *p. 382*

10.4 **Complex Power** *p. 384*

10.5 **Power Calculations** *p. 386*

10.6 **Maximum Power Transfer** *p. 393*

CHAPTER OBJECTIVES

1 Understand the following ac power concepts, their relationships to one another, and how to calculate them in a circuit:

- Instantaneous power;
- Average (real) power;
- Reactive power;
- Complex power; and
- Power factor.

2 Understand the condition for maximum real power delivered to a load in an ac circuit and be able to calculate the load impedance required to deliver maximum real power to the load.

3 Be able to calculate all forms of ac power in ac circuits with linear transformers and in ac circuits with ideal transformers.

Sinusoidal Steady-State Power Calculations

Nearly all electric energy is supplied by sinusoidal voltages and currents. Thus, after our Chapter 9 discussion of sinusoidal circuits, we now consider sinusoidal steady-state power calculations. We are primarily interested in the average power delivered to or supplied by a pair of terminals in the sinusoidal steady state. We also present other power quantities, including reactive power, complex power, and apparent power.

We begin and end this chapter with two concepts that should be very familiar to you from previous chapters: the basic equation for power (Section 10.1) and maximum power transfer (Section 10.6). In between, we discuss the general techniques for calculating power, which will be familiar from your studies in Chapters 1 and 4, although some additional mathematical techniques are required here to deal with sinusoidal, rather than dc, signals. We also revisit the rms value of a sinusoid, briefly introduced in Chapter 9, because it is used extensively in power calculations.

A wide variety of problems deal with the delivery of energy to do work, ranging from determining the power rating for safely and efficiently operating an appliance to designing the vast array of generators, transformers, and wires that provide electric energy to household and industrial consumers. Thus, power engineering is an important and exciting subdiscipline in electrical engineering.

■ Practical Perspective

Vampire Power

Even when we are not using many of the common electrical devices found in our homes, schools, and businesses, they can still be consuming power. This "standby power" can run an internal clock, charge batteries, display time or other quantities, monitor temperature or other environmental measures, or search for signals to receive. Devices such as microwave ovens, DVRs, televisions, remote controls, and computers all consume power when not in use.

The ac adapters used to charge many portable devices are a common source of standby power. Even when the device is unplugged from the adapter, the adapter can continue to consume power if it is plugged into the wall outlet. Because the plug on the adapter looks somewhat like vampire fangs, standby power has become known as "vampire power." It is power that is used even while we sleep.

How much vampire power do electrical devices in our home use over the course of a year? Is there a way to reduce or eliminate vampire power? These questions are explored in the Practical Perspective example at the end of the chapter and in the chapter problems.

Pung/Shutterstock

katalinks/Fotolia

Route55/Shutterstock

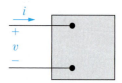

Figure 10.1 ▲ The black box representation of a circuit used for calculating power.

10.1 Instantaneous Power

Consider the familiar circuit in Fig. 10.1. Here, v and i are steady-state sinusoidal signals, given by

$$v = V_m \cos(\omega t + \theta_v),$$

$$i = I_m \cos(\omega t + \theta_i),$$

where θ_v is the voltage phase angle and θ_i is the current phase angle. Using the passive sign convention, we find that the power at any instant of time is

$$p = vi.$$

This is **instantaneous power**. Instantaneous power is measured in watts when the voltage is in volts and the current is in amperes.

Because the circuit operates in the sinusoidal steady state, we can choose any convenient reference for zero time. It is convenient to define zero time at the instant the current passes through a positive maximum. This reference system requires a shift of both the voltage and current by θ_i. Thus, the equations for voltage and current become

$$v = V_m \cos(\omega t + \theta_v - \theta_i), \tag{10.1}$$

$$i = I_m \cos \omega t. \tag{10.2}$$

When we substitute Eqs. 10.1 and 10.2 into the power equation, the expression for the instantaneous power becomes

$$p = V_m I_m \cos(\omega t + \theta_v - \theta_i) \cos \omega t.$$

We could use this equation directly to find the average power. Instead, we use a couple of trigonometric identities to construct a much more informative expression. We begin with the trigonometric identity[1]

$$\cos \alpha \cos \beta = \frac{1}{2} \cos(\alpha - \beta) + \frac{1}{2} \cos(\alpha + \beta)$$

and let $\alpha = \omega t + \theta_v - \theta_i$ and $\beta = \omega t$ to give

$$p = \frac{V_m I_m}{2} \cos(\theta_v - \theta_i) + \frac{V_m I_m}{2} \cos(2\omega t + \theta_v - \theta_i).$$

Now use the trigonometric identity

$$\cos(\alpha + \beta) = \cos \alpha \cos \beta - \sin \alpha \sin \beta$$

to expand the second term on the right-hand side of the expression for p, which gives

> ### INSTANTANEOUS POWER
>
> $$p = \frac{V_m I_m}{2} \cos(\theta_v - \theta_i) + \frac{V_m I_m}{2} \cos(\theta_v - \theta_i) \cos 2\omega t$$
>
> $$- \frac{V_m I_m}{2} \sin(\theta_v - \theta_i) \sin 2\omega t. \tag{10.3}$$

---

[1] See Appendix F.

Examine the three terms on the right-hand side of Eq. 10.3. The first term is a constant; it is not a function of time. The other two terms are sinusoids, each with a frequency that is double the frequency of the voltage and current in Eqs. 10.1 and 10.2. You can make these same observations in the plot of Fig. 10.2, which depicts v, i, and p, assuming $\theta_v = 60°$ and $\theta_i = 0°$. You can see that the frequency of the instantaneous power is twice the frequency of the voltage or current. Therefore, the instantaneous power goes through two complete cycles for every cycle of either the voltage or the current.

Also note that the instantaneous power may be negative for a portion of each cycle. When the power is negative, the energy stored in the inductors or capacitors is being extracted. The instantaneous power varies with time when a circuit operates in the sinusoidal steady state. As a result, some motor-driven appliances (such as refrigerators) experience vibration and require resilient motor mountings to prevent excessive vibration.

In the next section, we use Eq. 10.3 to find the average power at the terminals of the circuit in Fig. 10.1 and also introduce the concept of reactive power.

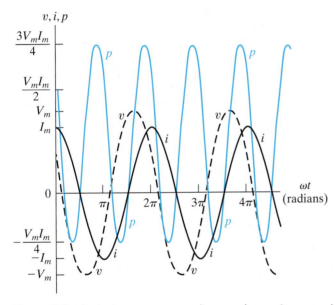

Figure 10.2 ▲ Instantaneous power, voltage, and current versus ωt for steady-state sinusoidal operation.

10.2 Average and Reactive Power

As we have already noted, Eq. 10.3 has three terms, which we can rewrite as follows:

$$p = P + P\cos 2\omega t - Q\sin 2\omega t, \qquad (10.4)$$

where

AVERAGE (REAL) POWER

$$P = \frac{V_m I_m}{2}\cos(\theta_v - \theta_i), \qquad (10.5)$$

REACTIVE POWER

$$Q = \frac{V_m I_m}{2} \sin(\theta_v - \theta_i). \qquad (10.6)$$

P is the **average power**, and Q is the **reactive power**. Average power is sometimes called **real power** because it describes the power in a circuit that is transformed from electric to nonelectric energy. Although the two terms are interchangeable, we primarily use the term *average power* in this text.

It is easy to see why P is called the average power: it is the average of the instantaneous power over one period. In equation form,

$$P = \frac{1}{T} \int_{t_o}^{t_0 + T} p \, dt, \qquad (10.7)$$

where T is the period of the sinusoidal function. The limits on the integral imply that we can initiate the integration process at any convenient time t_0 but that we must terminate the integration exactly one period later. (We could integrate over nT periods, where n is an integer, provided we multiply the integral by $1/nT$.)

We could find the average power by substituting Eq. 10.3 directly into Eq. 10.7 and integrating. But the average value of p is given by the first term on the right-hand side of Eq. 10.3 because the integral of both $\cos 2\omega t$ and $\sin 2\omega t$ over one period is zero. Thus, the average power is given in Eq. 10.5.

We can develop a better understanding of all the terms in Eq. 10.4 and the relationships among them by examining the power in circuits that are purely resistive, purely inductive, or purely capacitive.

Power for Purely Resistive Circuits

If the circuit between the terminals in Fig. 10.1 is purely resistive, the voltage and current are in phase, which means that $\theta_v = \theta_i$. Equation 10.4 then reduces to

$$p = P + P \cos 2\omega t.$$

The instantaneous power for a resistor is called the **instantaneous real power**. Figure 10.3 shows a graph of p for a purely resistive circuit with $\omega = 377$ rad/s. By definition, the average power, P, is the average of p over one period. Looking at the graph, we see that $P = 1$ for this circuit. Note that the instantaneous real power can never be negative, which is seen in its equation and is also shown in Fig. 10.3. In other words, power cannot be extracted from a purely resistive network. Resistors dissipate electric energy in the form of thermal energy.

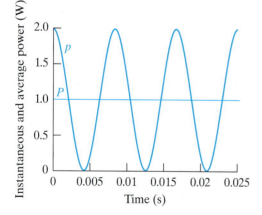

Figure 10.3 ▲ Instantaneous power and average power for a purely resistive circuit.

Power for Purely Inductive Circuits

If the circuit between the terminals in Fig. 10.1 is purely inductive, the current lags the voltage by 90° (that is, $\theta_i = \theta_v - 90°$); therefore, $\theta_v - \theta_i = +90°$. The expression for the instantaneous power then reduces to

$$p = -Q \sin 2\omega t.$$

In a purely inductive circuit, the average power is zero, and energy is not transformed from electric to nonelectric form. Instead, the instantaneous power in a purely inductive circuit is continually exchanged between the circuit and the source driving the circuit, at a frequency of 2ω. When p is positive, energy is stored in the magnetic fields associated with the inductive elements, and when p is negative, energy is extracted from the magnetic fields.

We measure the power of purely inductive circuits using the reactive power Q. The name *reactive power* recognizes an inductor as a reactive element; its impedance is purely reactive. Note that average power P and reactive power Q carry the same dimension. To distinguish between average and reactive power, we use the units *watt* (W) for average power and **var** (*volt-amp reactive*, or VAR) for reactive power. Figure 10.4 plots the instantaneous power for a purely inductive circuit, assuming $\omega = 377$ rad/s and $Q = 1$ VAR.

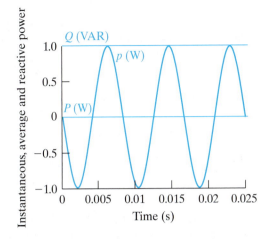

Figure 10.4 ▲ Instantaneous power, average power, and reactive power for a purely inductive circuit.

Power for Purely Capacitive Circuits

If the circuit between the terminals in Fig. 10.1 is purely capacitive, the current leads the voltage by 90° (that is, $\theta_i = \theta_v + 90°$); thus, $\theta_v - \theta_i = -90°$. The expression for the instantaneous power then becomes

$$p = -Q \sin 2\omega t.$$

Again, the average power is zero, and energy is not transformed from electric to nonelectric form. Instead, the power is continually exchanged between the source driving the circuit and the electric field associated with the capacitive elements. Figure 10.5 plots the instantaneous power for a purely capacitive circuit, assuming $\omega = 377$ rad/s and $Q = -1$ VAR.

Note that the decision to use the current as the reference (see Eq. 10.2) means that Q is positive for inductors (because $\theta_v - \theta_i = 90°$) and negative for capacitors (because $\theta_v - \theta_i = -90°$). Power engineers recognize this difference in the algebraic sign of Q by saying that inductors demand (or absorb) magnetizing vars and capacitors furnish (or deliver) magnetizing vars. We say more about this convention later.

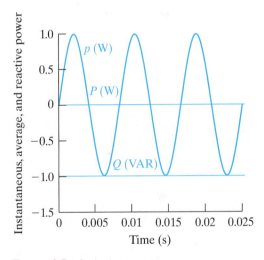

Figure 10.5 ▲ Instantaneous power, average power, and reactive power for a purely capacitive circuit.

The Power Factor

The angle $\theta_v - \theta_i$ is used when computing both average and reactive power and is referred to as the **power factor angle**. The cosine of this angle is called the **power factor**, abbreviated pf, and the sine of this angle is called the **reactive factor**, abbreviated rf. Thus

> ### POWER FACTOR
>
> $$\text{pf} = \cos(\theta_v - \theta_i), \tag{10.8}$$

$$\text{rf} = \sin(\theta_v - \theta_i).$$

Even if you know the value of the power factor, you cannot determine the power factor angle because $\cos(\theta_v - \theta_i) = \cos(\theta_i - \theta_v)$. To completely describe this angle, we use the phrases **lagging power factor and leading power factor**. Lagging power factor means that current lags voltage—hence, an inductive load. Leading power factor means that current leads voltage—hence, a capacitive load. Both the power factor and the reactive factor are convenient quantities to use in describing electrical loads.

Example 10.1 illustrates the interpretation of P and Q using numerical calculations.

| EXAMPLE 10.1 | Calculating Average and Reactive Power |

a) Calculate the average power and the reactive power at the terminals of the network shown in Fig. 10.6 if

$$v = 100 \cos(\omega t + 15°) \text{ V},$$

$$i = 4 \sin(\omega t - 15°) \text{ A}.$$

b) State whether the network inside the box is absorbing or delivering average power.

c) State whether the network inside the box is absorbing or supplying magnetizing vars.

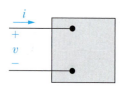

Figure 10.6 ▲ A pair of terminals used for calculating power.

Solution

a) Because i is expressed in terms of the sine function, the first step in calculating P and Q is to rewrite i as a cosine function:

$$i = 4 \cos(\omega t - 105°) \text{ A}.$$

We now calculate P and Q directly from Eqs. 10.5 and 10.6, using the passive sign convention. Thus

$$P = \frac{1}{2} (100)(4) \cos[15° - (-105°)] = -100 \text{ W},$$

$$Q = \frac{1}{2} 100(4) \sin[15° - (-105°)] = 173.21 \text{ VAR}.$$

b) The value of P is negative, so the network inside the box is delivering average power to the terminals.

c) The value of Q is positive, so the network inside the box is absorbing magnetizing vars at its terminals.

ASSESSMENT PROBLEMS

Objective 1—Understand ac power concepts, their relationships to one another, and how to calcuate them in a circuit

10.1 For each of the following sets of voltage and current, calculate the real and reactive power in the line between networks A and B in the circuit shown. In each case, state whether the power flow is from A to B or vice versa. Also state whether magnetizing vars are being transferred from A to B or vice versa.

a) $v = 100 \cos(\omega t - 45°) \text{ V}$;
   $i = 20 \cos(\omega t + 15°) \text{ A}$.

b) $v = 100 \cos(\omega t - 45°) \text{ V}$;
   $i = 20 \cos(\omega t + 165°) \text{ A}$.

c) $v = 100 \cos(\omega t - 45°) \text{ V}$;
   $i = 20 \cos(\omega t - 105°) \text{ A}$.

d) $v = 100 \cos \omega t \text{ V}$;
   $i = 20 \cos(\omega t + 120°) \text{ A}$.

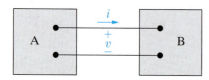

Answer: (a) $P = 500$ W (A to B);
$Q = -866.03$ VAR (B to A).

(b) $P = -866.03$ W (B to A);
$Q = 500$ VAR (A to B).

(c) $P = 500$ W (A to B);
$Q = 866.03$ VAR (A to B).

(d) $P = -500$ W (B to A);
$Q = -866.03$ VAR (B to A).

10.2 Compute the power factor and the reactive factor for the network inside the box in Fig. 10.6, whose voltage and current are described in Example 10.1.

Hint: Use $-i$ to calculate the power factor and reactive factor.

Answer: pf = 0.5 leading; rf = −0.866.

SELF-CHECK: Also try Chapter Problem 10.1.

Appliance Ratings

Average power is used to quantify the power needs of household appliances. Your monthly electric bill is based on the number of kilowatt-hours used by the household. Table 10.1 presents data for some common appliances, including the average hours per month and the number of months the appliance is used, the annual kilowatt-hour (kwh) consumption, and the annual cost of operation. For example, a coffee maker has a monthly use of 30 hours, or one hour per day, is used every month of the year, and consumes 60 kwh per year, at a cost of about $10. Therefore, the coffee maker consumes 60 kwh/360 hours = 0.167 kW = 167 W of power during the hour it operates each day.

Example 10.2 uses Table 10.1 to determine whether four common appliances can all be in operation without exceeding the current-carrying capacity of the household.

TABLE 10.1   Annual Energy Requirements of Electric Household Appliances

Appliance	Hours in Use per Month	Months Used	Annual kWH	Annual Cost
A/C—central	120	3	1080	$173
Clothes dryer—electric	24	12	901	$144
Clothes washer (does not include cost of hot water)	28	12	108	$17
Coffee maker (residential)	30	12	60	$10
Computer—desktop with monitor	*	*	127	$20
Computer—laptop	*	*	23	$4
Dishwasher—heat dry (does not include hot water)	30	12	293	$47
DVD player	60	12	18	$3
Fan—ceiling (does not include lights)	150	6	72	$12
Fan—table/box/floor	60	3	28	$4
Game console (includes standby/phantom load)	*	*	65	$10
Heating system—electric heat—baseboard, 10 ft	240	5	1500	$240
Humidifier	240	12	360	$58
Lighting—incandescent, 75 Watt	60	12	54	$9
Lighting—CFL, 20 Watt (75W incandescent equivalent)	90	12	22	$3
Lighting—LED, 10 Watt (75W incandescent equivalent)	90	12	12	$2
Microwave	9	12	101	$16
Oven—electric	9	12	284	$45
Refrigerator—19–21.4 cu ft —2001–2008	720	12	533	$85
Refrigerator—19–21.4 cu ft (new ENERGY STAR)	720	12	336	$54
Set-top box, cable/satellite receiver	720	12	249	$40
Television—50"+ non-ENERGY STAR TV	*	*	215	$34
Water heater—electric (newer base model .95 energy factor)	n/a	12	4559	$729

*Draws power in standby mode

Notes:
- Hours in Use per Month is based on a typical four-person household in a northern US state.
- Annual kWh may vary considerably depending on model, age, and use.
- Annual Cost is based on 16 cents per kilowatt hour (kWH).
- Data used with permission from https://www.efficiencyvermont.com/tips-tools/tools/electric-usage-chart-tool.

EXAMPLE 10.2 **Making Power Calculations Involving Household Appliances**

The branch circuit supplying the outlets in a typical home kitchen is wired with #12 conductor and is protected by either a 20 A fuse or a 20 A circuit breaker. Assume that the following 120 V appliances are in operation at the same time: a coffee maker, microwave, dishwasher, and older refrigerator. Will the circuit be interrupted by the protective device?

Solution

We have already estimated that the average power used by the coffee maker is 167 W. Using Table 10.1, we find that the average power used by the other three appliances is

$$P_{\text{microwave}} = \frac{101}{(9)(12)} = 0.935 \text{ kW} = 935 \text{ W},$$

$$P_{\text{dishwasher}} = \frac{293}{(30)(12)} = 0.814 \text{ kW} = 814 \text{ W},$$

$$P_{\text{refrigerator}} = \frac{533}{(720)(12)} = 0.062 \text{ kW} = 62 \text{ W}.$$

The total average power used by the four appliances is

$$P_{\text{total}} = 167 + 935 + 814 + 62 = 1978 \text{ W}.$$

The total current in the protective device is

$$I = \frac{P_{\text{total}}}{V} = \frac{1978}{120} = 16.5 \text{ A}.$$

Since the current is less than 20 A, the protective device will not interrupt the circuit.

SELF-CHECK: Assess your understanding of this material by trying Chapter Problem 10.2.

10.3 The rms Value and Power Calculations

When we introduced the rms value of a sinusoidal voltage (or current) in Section 9.1, we mentioned that it would play an important role in power calculations. We now discuss this role.

Assume a sinusoidal voltage is applied to the terminals of a resistor, as shown in Fig. 10.7, and that we want to determine the average power delivered to the resistor. From Eq. 10.7,

$$P = \frac{1}{T} \int_{t_0}^{t_0+T} \frac{V_m^2 \cos^2(\omega t + \phi_v)}{R} \, dt$$

$$= \frac{1}{R} \left[\frac{1}{T} \int_{t_0}^{t_0+T} V_m^2 \cos^2(\omega t + \phi_v) \, dt \right].$$

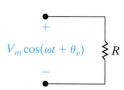

$V_m \cos(\omega t + \theta_v)$ R

Figure 10.7 ▲ A sinusoidal voltage applied to the terminals of a resistor.

From Eq. 9.4, we see that the bracketed term is the rms value of the voltage squared. Therefore, the average power delivered to R is

$$P = \frac{V_{rms}^2}{R}.$$

If the resistor has a sinusoidal current, say, $I_m \cos(\omega t + \phi_i)$, the average power delivered to the resistor is

$$P = I_{rms}^2 R.$$

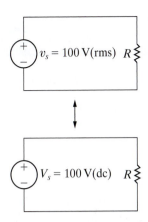

Figure 10.8 ▲ The effective value of v_s [100 V(rms)] delivers the same power to R as the dc voltage v_s [100 V(dc)].

The rms value is also referred to as the **effective value** of the sinusoidal voltage (or current). The rms value has an interesting property: Given an equivalent resistive load, R, and an equivalent time period, T, the rms value of a sinusoidal source delivers the same energy to R as does a dc source of the same value. For example, a dc source of 100 V delivers the same energy in T seconds that a sinusoidal source of 100 V(rms) delivers, assuming equivalent load resistances (see Problem 10.16). Figure 10.8 demonstrates this equivalence. The effect of the two sources is identical with respect to energy delivery. So we use the terms *effective value* and *rms value* interchangeably.

The average power given by Eq. 10.5 and the reactive power given by Eq. 10.6 can be written in terms of effective values:

$$P = \frac{V_m I_m}{2} \cos(\theta_v - \theta_i)$$

$$= \frac{V_m}{\sqrt{2}} \frac{I_m}{\sqrt{2}} \cos(\theta_v - \theta_i)$$

$$= V_{rms} I_{rms} \cos(\theta_v - \theta_i) \tag{10.9}$$

and, by similar manipulation,

$$Q = V_{rms} I_{rms} \sin(\theta_v - \theta_i). \tag{10.10}$$

Using the effective values of sinusoidal signals in power calculations is so widespread that we specify the voltage and current ratings of circuits and equipment using rms values. For example, the voltage rating of residential electric wiring is often 240 V/120 V service. These voltages are the rms values of the sinusoidal voltages supplied by the utility company, which provides power at two voltage levels, accommodating low-voltage appliances (such as televisions) and higher-voltage appliances (such as electric ranges). Appliances such as electric lamps, irons, and toasters all carry rms ratings on their nameplates. For example, a 120 V, 100 W lamp has a resistance of $120^2/100$, or 144 Ω, and draws an rms current of 120/144, or 0.833 A. The peak value of the lamp current is $0.833\sqrt{2}$, or 1.18 A.

The phasor transform of a sinusoidal function may also be expressed as an rms value. The magnitude of the rms phasor is equal to the rms value of the sinusoidal function. We indicate that a phasor is based on an rms value using either an explicit statement, a parenthetical "rms" adjacent to the phasor's units, or the subscript "rms."

In Example 10.3, we use rms values to calculate power.

EXAMPLE 10.3   Determining Average Power Delivered to a Resistor by a Sinusoidal Voltage

a) A sinusoidal voltage having a maximum amplitude of 625 V is applied to the terminals of a 50 Ω resistor. Find the average power delivered to the resistor.

b) Repeat (a) by first finding the current in the resistor.

Solution

a) The rms value of the sinusoidal voltage is $625/\sqrt{2}$, or approximately 441.94 V. The average power delivered to the 50 Ω resistor is

$$P = \frac{V_{rms}^2}{R} = \frac{(441.94)^2}{50} = 3906.25 \text{ W}.$$

b) The maximum amplitude of the current in the resistor is $625/50$, or 12.5 A. The rms value of the current is $12.5/\sqrt{2}$, or approximately 8.84 A. Hence, the average power delivered to the resistor is

$$P = (8.84)^2 50 = 3906.25 \text{ W}.$$

◾ ASSESSMENT PROBLEM

Objective 1—Understand ac power concepts, their relationships to one another, and how to calculate them in a circuit

10.3 The periodic triangular current in Example 9.4, repeated here, has a peak value of 180 mA. Find the average power that this current delivers to a 5 kΩ resistor.

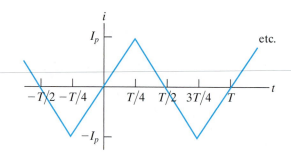

Answer: 54 W.

SELF-CHECK: Also try Chapter Problem 10.14.

10.4 Complex Power

Before discussing the methods for calculating real and reactive power in circuits operating in the sinusoidal steady state, we introduce and define complex power. **Complex power** is the complex sum of real power and reactive power, or

> **COMPLEX POWER**
>
> $$S = P + jQ. \tag{10.11}$$

As you will see, we can compute the complex power using the voltage and current phasors for a circuit. Equation 10.11 can then be used to determine the average and reactive power, because $P = \mathcal{R}\{S\}$ and $Q = \mathcal{I}\{S\}$.

Complex power has the same units as average or reactive power. However, to distinguish complex power from both average and reactive power, we use the units **volt-amps** (VA). Thus, we use volt-amps for complex power, watts for average power, and vars for reactive power, as summarized in Table 10.2.

Complex power provides a geometric relationship among several different power quantities. In Eq. 10.11, envision P, Q, and $|S|$ as the sides of a right triangle, as shown in Fig. 10.9. We can show that the angle θ in the power triangle is the power factor angle $\theta_v - \theta_i$. For the right triangle shown in Fig. 10.9,

$$\tan \theta = \frac{Q}{P}.$$

But from the definitions of P and Q (Eqs. 10.5 and 10.6, respectively),

$$\frac{Q}{P} = \frac{(V_m I_m/2)\sin(\theta_v - \theta_i)}{(V_m I_m/2)\cos(\theta_v - \theta_i)}$$

$$= \tan(\theta_v - \theta_i).$$

Therefore, $\theta = \theta_v - \theta_i$. The geometric relationships for a right triangle mean that the four power triangle dimensions (the three sides and the power factor angle) can be determined if any two of the four are known.

The magnitude of complex power is referred to as **apparent power**. Specifically,

APPARENT POWER

$$|S| = \sqrt{P^2 + Q^2}. \tag{10.12}$$

Apparent power, like complex power, is measured in volt-amps. The apparent power, or volt-amp, requirement of a device designed to convert electric energy to a nonelectric form is more useful than the average power requirement. The apparent power represents the volt-amp capacity required to supply the average power used by the device. As you can see from the power triangle in Fig. 10.9, unless the power factor angle is 0° (that is, the device is purely resistive, pf = 1, and $Q = 0$), the volt-amp capacity required by the device is larger than the average power used by the device.

Many appliances (including refrigerators, fans, air conditioners, fluorescent lighting fixtures, and washing machines) and most industrial loads operate at a lagging power factor. The power factor of these loads can be *corrected* either by adding a capacitor to the device itself or by connecting capacitors across the line feeding the load; the latter method is often used for large industrial loads. Many of the problems at the end of the chapter explore methods for correcting a lagging power factor load and improving the operation of a circuit.

Example 10.4 uses a power triangle to calculate several quantities associated with power in an electrical load.

TABLE 10.2	Three Power Quantities and Their Units
Quantity	**Units**
Complex power	volt-amps
Average power	watts
Reactive power	var

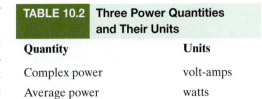

Figure 10.9 ▲ A power triangle.

EXAMPLE 10.4 | Calculating Complex Power

An electrical load operates at 240 V(rms). The load absorbs an average power of 8 kW at a lagging power factor of 0.8.

a) Calculate the complex power of the load.

b) Calculate the impedance of the load, Z.

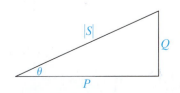

Figure 10.10 ▲ A power triangle.

Solution

a) Because the power factor is lagging, we know that the load is inductive and that the algebraic sign of the reactive power is positive. From the power triangle shown in Fig. 10.10,

$$P = |S| \cos \theta,$$

$$Q = |S| \sin \theta.$$

Since $\cos \theta = 0.8$, we know that $\sin \theta = 0.6$. Therefore

$$|S| = \frac{P}{\cos \theta} = \frac{8000}{0.8} = 10{,}000 = 10 \text{ kVA},$$

$$Q = |S| \sin \theta = (10{,}000)(0.6) = 6 \text{ kvar},$$

and

$$S = P + jQ = 8 + j6 \text{ kVA}.$$

b) From the problem statement, we know that $P = 8$ kW for the load. Using Eq. 10.9,

$$P = V_{\text{rms}} I_{\text{rms}} \cos(\theta_v - \theta_i)$$

$$= (240) I_{\text{rms}} (0.8)$$

$$= 8000 \text{ W}.$$

Solving for I_{rms},

$$I_{\text{rms}} = 41.67 \text{ A(rms)}.$$

We already know the angle of the load impedance because it is the power factor angle:

$$\theta = \cos^{-1}(0.8) = 36.87°.$$

We also know that θ is positive because the power factor is lagging, indicating an inductive load. Compute the load impedance magnitude using its definition as the ratio of the magnitude of the load voltage to the magnitude of the load current:

$$|Z| = \frac{|V_{\text{rms}}|}{|I_{\text{rms}}|} = \frac{240}{41.67} = 5.76.$$

Hence,

$$Z = 5.76 \underline{/36.87°} \ \Omega = 4.608 + j3.456 \ \Omega.$$

ASSESSMENT PROBLEM

Objective 1—Understand ac power concepts, their relationships to one another, and how to calculate them in a circuit

10.4 A load consisting of a 1.35 kΩ resistor in parallel with a 405 mH inductor is connected across the terminals of a sinusoidal voltage source v_g, where $v_g = 90 \cos 2500t$ V. Find

a) the average power delivered to the load,

b) the reactive power for the load,

c) the apparent power for the load, and

d) the power factor of the load.

Answer: (a) 3 W;
(b) 4 VAR;
(c) 5 VA;
(d) 0.6 lagging.

SELF-CHECK: Also try Chapter Problem 10.17.

10.5 Power Calculations

We now develop additional equations for calculating real, reactive, and complex power. We begin by combining Eqs. 10.5, 10.6, and 10.11 to get

$$S = \frac{V_m I_m}{2} \cos(\theta_v - \theta_i) + j \frac{V_m I_m}{2} \sin(\theta_v - \theta_i)$$

$$= \frac{V_m I_m}{2} [\cos(\theta_v - \theta_i) + j \sin(\theta_v - \theta_i)]$$

$$= \frac{V_m I_m}{2} e^{j(\theta_v - \theta_i)} = \frac{1}{2} V_m I_m \underline{/(\theta_v - \theta_i)}.$$

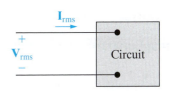

If we use the rms values of the sinusoidal voltage and current, the expression for the complex power becomes

$$S = V_{rms}I_{rms}\underline{/(\theta_v - \theta_i)}.$$

Therefore, if the phasor current and voltage are known at a pair of terminals, the complex power associated with that pair of terminals is either one half the product of the phasor voltage and the conjugate of the phasor current, or the product of the rms phasor voltage and the conjugate of the rms phasor current. We can show this for the rms phasor voltage and current in Fig. 10.11 as follows:

Figure 10.11 ▲ The phasor voltage and current associated with a pair of terminals.

$$S = V_{rms}I_{rms}\underline{/(\theta_v - \theta_i)}$$
$$= V_{rms}I_{rms}e^{j(\theta_v - \theta_i)}$$
$$= V_{rms}e^{j\theta_v}I_{rms}e^{-j\theta_i}$$

so

> ## COMPLEX POWER, ALTERNATE FORM
>
> $$S = \mathbf{V}_{rms}\mathbf{I}_{rms}^*. \tag{10.13}$$

Note that $\mathbf{I}_{rms}^* = I_{rms}e^{-j\theta_i}$ follows from Euler's identity and the trigonometric identities $\cos(-\theta) = \cos(\theta)$ and $\sin(-\theta) = -\sin(\theta)$:

$$I_{rms}e^{-j\theta_i} = I_{rms}\cos(-\theta_i) + jI_{rms}\sin(-\theta_i)$$
$$= I_{rms}\cos(\theta_i) - jI_{rms}\sin(\theta_i)$$
$$= \mathbf{I}_{rms}^*.$$

If the voltage and current phasors are not specified as rms values, the derivation technique used for Eq. 10.13 yields

$$S = \frac{1}{2}\mathbf{V}\mathbf{I}^*. \tag{10.14}$$

Both Eqs. 10.13 and 10.14 use the passive sign convention. If the current reference is in the direction of the voltage rise across the terminals, we insert a minus sign on the right-hand side of each equation.

Example 10.5 uses Eq. 10.14 in a power calculation, with the phasor representation of the voltage and current from Example 10.1

EXAMPLE 10.5 **Calculating Power Using Phasor Voltage and Current**

a) Calculate the average power and the reactive power at the terminals of the network shown in Fig. 10.12 if

$$\mathbf{V} = 100\underline{/15°}\text{ V,}$$

$$\mathbf{I} = 4\underline{/-105°}\text{ A.}$$

b) State whether the network inside the box is absorbing or delivering average power.

c) State whether the network inside the box is absorbing or supplying magnetizing vars.

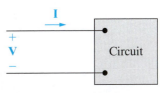

Figure 10.12 ▲ The circuit for Example 10.5.

Solution

a) From Eq. 10.14,

$$S = \frac{1}{2}(100\,\underline{/\,15°})(4\,\underline{/+105°}) = 200\,\underline{/\,120°}$$

$$= -100 + j173.21 \text{ VA.}$$

Once we calculate the complex power, we can read off both the real and reactive powers, because $S = P + jQ$. Thus

$$P = -100 \text{ W,}$$
$$Q = 173.21 \text{ var.}$$

b) The value of P is negative, so the network inside the box is delivering average power to the terminals.

c) The value of Q is positive, so the network inside the box is absorbing magnetizing vars at its terminals.

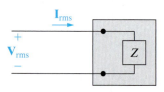

Figure 10.13 ▲ The general circuit of Fig. 10.11 replaced with an equivalent impedance.

Alternate Forms for Complex Power

Equations 10.13 and 10.14 have several useful variations. We use the rms form of the equations throughout because voltages and currents are most often given as rms values in power computations.

The first variation of Eq. 10.13 replaces the voltage with the product of the current times the impedance. We can always represent the circuit inside the box of Fig. 10.11 by an equivalent impedance, as shown in Fig. 10.13. Then,

$$\mathbf{V}_{\text{rms}} = Z\mathbf{I}_{\text{rms}}.$$

Replacing the rms voltage phasor in Eq. 10.13 yields

$$
\begin{aligned}
S &= Z\,\mathbf{I}_{\text{rms}}\mathbf{I}^*_{\text{rms}} \\
&= |\mathbf{I}_{\text{rms}}|^2\,Z \\
&= |\mathbf{I}_{\text{rms}}|^2(R + jX) \\
&= |\mathbf{I}_{\text{rms}}|^2 R + j|\mathbf{I}_{\text{rms}}|^2 X = P + jQ,
\end{aligned}
\tag{10.15}
$$

from which

$$P = |\mathbf{I}_{\text{rms}}|^2 R = \frac{1}{2}I_m^2 R, \tag{10.16}$$

$$Q = |\mathbf{I}_{\text{rms}}|^2 X = \frac{1}{2}I_m^2 X. \tag{10.17}$$

In Eqs. 10.15 and 10.17, X is the reactance of either the equivalent inductance or the equivalent capacitance of the circuit. Remember that reactance is positive for inductive circuits and negative for capacitive circuits.

A second variation of Eq. 10.13 replaces the current with the voltage divided by the impedance:

$$S = \mathbf{V}_{\text{rms}}\left(\frac{\mathbf{V}_{\text{rms}}}{Z}\right)^* = \frac{|\mathbf{V}_{\text{rms}}|^2}{Z^*} = P + jQ. \tag{10.18}$$

Note that if Z is a pure resistive element,

$$P = \frac{|\mathbf{V}_{\text{rms}}|^2}{R}, \tag{10.19}$$

and if Z is a pure reactive element,

$$Q = \frac{|\mathbf{V}_{\text{rms}}|^2}{X}. \tag{10.20}$$

In Eq. 10.20, X is positive for an inductor and negative for a capacitor.

Examples 10.6–10.8 demonstrate various power calculations in circuits operating in the sinusoidal steady state.

| **EXAMPLE 10.6** | **Calculating Average and Reactive Power** |

In the circuit shown in Fig. 10.14, a load having an impedance of $39 + j26 \ \Omega$ is fed from a voltage source through a line having an impedance of $1 + j4 \ \Omega$. The source voltage is 250 V(rms).

a) Calculate the load current phasor $\mathbf{I_L}$ and voltage phasor $\mathbf{V_L}$.

b) Calculate the average and reactive power delivered to the load.

c) Calculate the average and reactive power delivered to the line.

d) Calculate the average and reactive power supplied by the source.

Solution

a) The line and load impedances are in series across the voltage source, so the load current equals the voltage divided by the total impedance, or

$$\mathbf{I_L} = \frac{250 \ \underline{/0°}}{40 + j30} = 4 - j3 = 5 \ \underline{/-36.87°} \ \text{A (rms)}.$$

Because the voltage is given as an rms value, the current value is also rms. The load voltage is the product of the load current and load impedance:

$$\mathbf{V_L} = (39 + j26) \mathbf{I_L} = 234 - j13$$

$$= 234.36 \ \underline{/-3.18°} \ \text{V (rms)}.$$

b) Use Eq. 10.13 to find the average and reactive power delivered to the load. Therefore

$$S = \mathbf{V_L I_L^*} = (234 - j13)(4 + j3)$$

$$= 975 + j650 \ \text{VA}.$$

Thus, the load is absorbing an average power of 975 W and a reactive power of 650 var.

c) Because the line current is known, the average and reactive power delivered to the line are most easily calculated using Eqs. 10.16 and 10.17. Thus

$$P = (5)^2(1) = 25 \ \text{W},$$

$$Q = (5)^2(4) = 100 \ \text{var}.$$

Note that the reactive power associated with the line is positive because the line reactance is inductive.

d) We can calculate the average and reactive power delivered by the source by adding the complex power delivered to the line to that delivered to the load, or

$$S = 25 + j100 + 975 + j650$$

$$= 1000 + j750 \ \text{VA}.$$

The complex power at the source can also be calculated from Eq. 10.13:

$$S_s = -250\mathbf{I_L^*}.$$

The minus sign is inserted in Eq. 10.13 whenever the current reference is in the direction of a voltage rise. Thus

$$S_s = -250(4 + j3) = -(1000 + j750) \ \text{VA}.$$

The minus sign implies that both average power and magnetizing reactive power are being delivered by the source. This result agrees with the previous calculation of S, as it must, because the source supplies all the average and reactive power absorbed by the line and load.

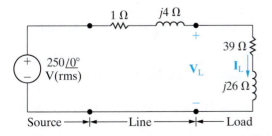

Figure 10.14 ▲ The circuit for Example 10.6.

EXAMPLE 10.7 | Calculating Power in Parallel Loads

The two loads in the circuit shown in Fig. 10.15 can be described as follows: Load 1 absorbs 8 kW at a leading power factor of 0.8. Load 2 absorbs 20 kVA at a lagging power factor of 0.6.

a) Determine the power factor of the two loads in parallel.

b) Determine the apparent power required to supply the loads, the magnitude of the current, $\mathbf{I}_s$, and the average power loss in the transmission line.

c) Given that the frequency of the source is 60 Hz, compute the value of the capacitor that would correct the power factor to 1 if placed in parallel with the two loads. Recalculate the values in (b) for the load with the corrected power factor.

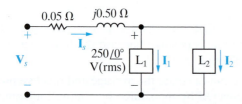

Figure 10.15 ▲ The circuit for Example 10.7.

Solution

a) All voltage and current phasors in this problem are rms values. Note from the circuit diagram in Fig. 10.15 that $\mathbf{I}_s = \mathbf{I}_1 + \mathbf{I}_2$. The total complex power absorbed by the two loads is

$$S = (250)\mathbf{I}_s^*$$

$$= (250)(\mathbf{I}_1 + \mathbf{I}_2)^*$$

$$= (250)\mathbf{I}_1^* + (250)\mathbf{I}_2^*$$

$$= S_1 + S_2.$$

We can sum the complex powers geometrically, using the power triangles for each load, as shown in Fig. 10.16. By hypothesis,

$$S_1 = 8000 - j\frac{8000(.6)}{(.8)}$$

$$= 8000 - j6000 \text{ VA},$$

$$S_2 = 20,000(.6) + j\,20,000(.8)$$

$$= 12,000 + j16,000 \text{ VA}.$$

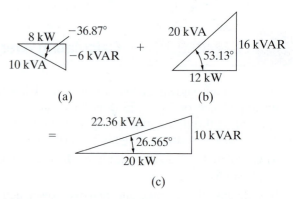

Figure 10.16 ▲ (a) The power triangle for load 1; (b) The power triangle for load 2; (c) The sum of the power triangles.

It follows that

$$S = 20,000 + j10,000 \text{ VA},$$

and

$$\mathbf{I}_s^* = \frac{20,000 + j10,000}{250} = 80 + j40 \text{ A(rms)}.$$

Therefore

$$\mathbf{I}_s = 80 - j40 = 89.44 \underline{/-26.57°} \text{ A(rms)}.$$

Thus, the power factor of the combined load is

$$\text{pf} = \cos(0 + 26.57°) = 0.8944 \text{ lagging}.$$

The power factor of the two loads in parallel is lagging because the net reactive power is positive.

b) The apparent power supplied to the two loads is

$$|S| = |20,000 + j10,000| = 22.36 \text{ kVA}.$$

The magnitude of the current supplying this apparent power is

$$|\mathbf{I}_s| = |80 - j40| = 89.44 \text{ A(rms)}.$$

The average power lost in the line results from the current in the line resistance:

$$P_{\text{line}} = |\mathbf{I}_s|^2 R = (89.44)^2(0.05) = 400 \text{ W}.$$

Note that the power supplied totals $20,000 + 400 = 20,400$ W, even though the loads require a total of only 20,000 W.

c) As we can see from the power triangle in Fig. 10.16(c), we can correct the power factor to 1 if we place a capacitor in parallel with the existing loads that supplies 10 kVAR of magnetizing reactive power. The value of the capacitor is calculated as follows. First, find the capacitive reactance from Eq. 10.20:

$$X = \frac{|V_{rms}|^2}{Q}$$

$$= \frac{(250)^2}{-10,000}$$

$$= -6.25 \ \Omega.$$

Recall that the reactive impedance of a capacitor is $-1/\omega C$, and $\omega = 2\pi(60) = 376.99$ rad/s, because the source frequency is 60 Hz. Thus,

$$C = \frac{-1}{\omega X} = \frac{-1}{(376.99)(-6.25)} = 424.4 \ \mu F.$$

Adding the capacitor as the third load is represented in geometric form as the sum of the two power triangles shown in Fig. 10.17. When the power factor is 1, the apparent power and the average power are the same, as seen from the power triangle in Fig. 10.17(c). Therefore, once the power factor has been corrected, the apparent power is

$$|S| = P = 20 \text{ kVA}.$$

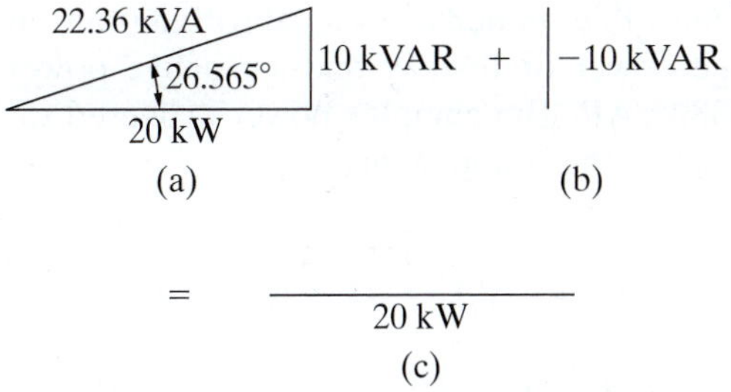

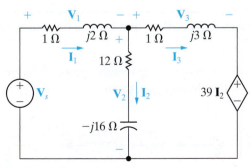

22.36 kVA — 26.565° — 20 kW — 10 kVAR + | −10 kVAR

(a)   (b)

$$= \frac{}{20 \text{ kW}}$$

(c)

Figure 10.17 ▲ (a) The sum of the power triangles for loads 1 and 2; (b) The power triangle for a 424.4 μF capacitor at 60 Hz; (c) The sum of the power triangles in (a) and (b).

The magnitude of the current that supplies this apparent power is

$$|I_s| = \frac{20,000}{250} = 80 \text{ A(rms)}.$$

The average power lost in the line is thus reduced to

$$P_{line} = |I_s|^2 R = (80)^2(0.05) = 320 \text{ W}.$$

Now, the power supplied totals
$$20,000 + 320 = 20,320 \text{ W}.$$

Note that the addition of the capacitor has reduced the line loss by 20%, from 400 W to 320 W.

EXAMPLE 10.8 | **Balancing Power Delivered with Power Absorbed in an AC Circuit**

a) Calculate the total average and reactive power delivered to each impedance in the circuit shown in Fig. 10.18.

b) Calculate the average and reactive powers associated with each source in the circuit.

c) Verify that the average power delivered equals the average power absorbed and that the magnetizing reactive power delivered equals the magnetizing reactive power absorbed.

Solution

a) The complex power delivered to the $(1 + j2) \ \Omega$ impedance is

$$S_1 = \frac{1}{2} \mathbf{V}_1 \mathbf{I}_1^* = P_1 + jQ_1$$

$$= \frac{1}{2}(78 - j104)(-26 + j52)$$

$$= \frac{1}{2}(3380 + j6760)$$

$$= 1690 + j3380 \text{ VA}.$$

$\mathbf{V}_s = 150 \underline{/0°} \text{ V}$

$\mathbf{V}_1 = (78 - j104) \text{ V}$ $\mathbf{I}_1 = (-26 - j52) \text{ A}$

$\mathbf{V}_2 = (72 + j104) \text{ V}$ $\mathbf{I}_2 = (-2 + j6) \text{ A}$

$\mathbf{V}_3 = (150 - j130) \text{ V}$ $\mathbf{I}_3 = (-24 - j58) \text{ A}$

Figure 10.18 ▲ The circuit, with solution, for Example 10.8.

Thus, this impedance is absorbing an average power of 1690 W and a reactive power of 3380 VAR. The complex power delivered to the $(12 - j16)\,\Omega$ impedance is

$$S_2 = \frac{1}{2}\mathbf{V}_2\mathbf{I}_2^* = P_2 + jQ_2$$

$$= \frac{1}{2}(72 + j104)(-2 - j6)$$

$$= 240 - j320 \text{ VA}.$$

Therefore, the impedance in the vertical branch is absorbing 240 W and delivering 320 VAR. The complex power delivered to the $(1 + j3)\,\Omega$ impedance is

$$S_3 = \frac{1}{2}\mathbf{V}_3\mathbf{I}_3^* = P_3 + jQ_3$$

$$= \frac{1}{2}(150 - j130)(-24 + j58)$$

$$= 1970 + j5910 \text{ VA}.$$

This impedance is absorbing 1970 W and 5910 VAR.

b) The complex power associated with the independent voltage source is

$$S_s = -\frac{1}{2}\mathbf{V}_s\mathbf{I}_1^* = P_s + jQ_s$$

$$= -\frac{1}{2}(150)(-26 + j52)$$

$$= 1950 - j3900 \text{ VA}.$$

Note that the independent voltage source is absorbing an average power of 1950 W and delivering 3900 VAR. The complex power associated with the current-controlled voltage source is

$$S_x = \frac{1}{2}(39\mathbf{I}_2)(\mathbf{I}_2^*) = P_x + jQ_x$$

$$= \frac{1}{2}(-78 + j234)(-24 + j58)$$

$$= -5850 - j5070 \text{ VA}.$$

Both average power and magnetizing reactive power are being delivered by the dependent source.

c) The total power absorbed by the passive impedances and the independent voltage source is

$$P_{\text{absorbed}} = P_1 + P_2 + P_3 + P_s = 5850 \text{ W}.$$

The dependent voltage source is the only circuit element delivering average power. Thus

$$P_{\text{delivered}} = 5850 \text{ W}.$$

Magnetizing reactive power is being absorbed by the two horizontal branches. Thus

$$Q_{\text{absorbed}} = Q_1 + Q_3 = 9290 \text{ VAR}.$$

Magnetizing reactive power is being delivered by the independent voltage source, the capacitor in the vertical impedance branch, and the dependent voltage source. Therefore

$$Q_{\text{delivered}} = 9290 \text{ VAR}.$$

ASSESSMENT PROBLEMS

Objective 1—Understand ac power concepts, their relationships to one another, and how to calculate them in a circuit

10.5   The load impedance in the circuit shown is shunted by a capacitor having a capacitive reactance of $-52\,\Omega$. Calculate:
   a) the rms phasors $\mathbf{V}_L$ and $\mathbf{I}_L$,
   b) the average power and magnetizing reactive power absorbed by the $(39 + j26)\,\Omega$ load impedance,
   c) the average power and magnetizing reactive power absorbed by the $(1 + j4)\,\Omega$ line impedance,
   d) the average power and magnetizing reactive power delivered by the source, and
   e) the magnetizing reactive power delivered by the shunting capacitor.

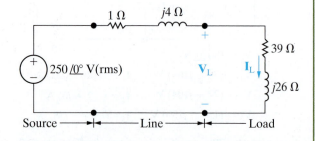

Answer: (a) $252.20 \underline{/-4.54°}$ V(rms),
$5.38 \underline{/-38.23°}$ A(rms);
(b) 1129.09 W, 752.73 VAR;
(c) 23.52 W, 94.09 VAR;
(d) 1152.62 W, −376.36 VAR;
(e) 1223.18 VAR.

10.6 The voltage at the terminals of a load is 250 V(rms). The load is absorbing an average power of 40 kW and delivering a magnetizing reactive power of 30 kVAR. Derive two equivalent impedance models of the load.

Answer: 1 Ω in series with 0.75 Ω of capacitive reactance; 1.5625 Ω in parallel with 2.083 Ω of capacitive reactance.

SELF-CHECK: Also try Chapter Problems 10.18, 10.24, and 10.27.

10.7 Find the phasor voltage $\mathbf{V}_s$ in the circuit shown if loads L_1 and L_2 are absorbing 15 kVA at 0.6 pf lagging and 6 kVA at 0.8 pf leading, respectively. Express $\mathbf{V}_s$ in polar form.

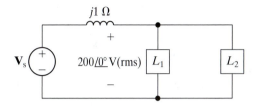

Answer: $251.64 \underline{/15.91°}$ V(rms).

10.6 Maximum Power Transfer

Recall from Chapter 4 that certain systems—for example, those that transmit information via electric signals—need to transfer a maximum amount of power from the source to the load. We now determine the condition for maximum power transfer in sinusoidal steady-state networks, beginning with Fig. 10.19. We must determine the load impedance Z_L that maximizes the average power delivered to terminals a and b. Any linear network can be replaced by a Thévenin equivalent circuit, so we will use the circuit in Fig. 10.20 to find the value of Z_L that results in maximum average power delivered to Z_L.

For maximum average power transfer, Z_L must equal the conjugate of the Thévenin impedance; that is,

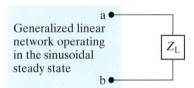

Figure 10.19 ▲ A circuit describing maximum power transfer.

Figure 10.20 ▲ The circuit shown in Fig. 10.19, with the network replaced by its Thévenin equivalent.

CONDITION FOR MAXIMUM AVERAGE POWER TRANSFER	
$$Z_L = Z_{Th}^*.$$	(10.21)

We derive Eq. 10.21 by a straightforward application of elementary calculus. We begin by expressing Z_{Th} and Z_L in rectangular form:

$$Z_{Th} = R_{Th} + jX_{Th},$$

$$Z_L = R_L + jX_L.$$

In these impedance equations, the reactance term carries its own algebraic sign—positive for inductance and negative for capacitance. We assume that the Thévenin voltage amplitude is an rms value. We also use the Thévenin voltage as the reference phasor, that is, the phasor whose phase angle is 0°. Then, from Fig. 10.20, the rms value of the load current $\mathbf{I}$ is

$$\mathbf{I} = \frac{\mathbf{V}_{Th}}{(R_{Th} + R_L) + j(X_{Th} + X_L)}.$$

The average power delivered to the load is

$$P = |\mathbf{I}|^2 R_L.$$

Therefore,

$$P = \frac{|\mathbf{V}_{Th}|^2 R_L}{(R_L + R_{Th})^2 + (X_L + X_{Th})^2}. \tag{10.22}$$

In Eq. 10.22, remember that V_{Th}, R_{Th}, and X_{Th} are fixed quantities, whereas R_L and X_L are independent variables. Therefore, to maximize P, we must find the values of R_L and X_L that make both $\partial P/\partial R_L$ and $\partial P/\partial X_L$ zero. From Eq. 10.22,

$$\frac{\partial P}{\partial X_L} = \frac{-|\mathbf{V}_{Th}|^2 2 R_L (X_L + X_{Th})}{[(R_L + R_{Th})^2 + (X_L + X_{Th})^2]^2},$$

$$\frac{\partial P}{\partial R_L} = \frac{|\mathbf{V}_{Th}|^2 [(R_L + R_{Th})^2 + (X_L + X_{Th})^2 - 2 R_L (R_L + R_{Th})]}{[(R_L + R_{Th})^2 + (X_L + X_{Th})^2]^2}.$$

From its equation, $\partial P/\partial X_L$ is zero when

$$X_L = -X_{Th}.$$

From its equation, $\partial P/\partial R_L$ is zero when

$$R_L = \sqrt{R_{Th}^2 + (X_L + X_{Th})^2}.$$

Note that when we combine the expressions for X_L and R_L, both partial derivatives are zero when $Z_L = Z_{Th}^*$.

The Maximum Average Power Absorbed

When $Z_L = Z_{Th}^*$, we can use the circuit in Fig. 10.20 to calculate the maximum average power that is delivered to Z_L. The rms load current is $\mathbf{V}_{Th}/2R_L$ because $Z_L = Z_{Th}^*$, and the maximum average power delivered to the load is

$$P_{max} = \frac{|\mathbf{V}_{Th}|^2 R_L}{4 R_L^2} = \frac{1}{4} \frac{|\mathbf{V}_{Th}|^2}{R_L}. \tag{10.23}$$

If the Thévenin voltage phasor is expressed using its maximum amplitude rather than its rms amplitude, Eq. 10.23 becomes

$$P_{max} = \frac{1}{8} \frac{|\mathbf{V}_m|^2}{R_L}. \tag{10.24}$$

Maximum Power Transfer When Z_L is Restricted

Maximum average power can be delivered to Z_L only if Z_L can be set equal to the conjugate of Z_{Th}. In some situations, this is not possible. First, R_L and X_L may be restricted to a limited range of values. To maximize power in this situation, set X_L as close to $-X_{Th}$ as possible and then adjust R_L as close to $\sqrt{R_{Th}^2 + (X_L + X_{Th})^2}$ as possible (see Example 10.10).

A second type of restriction occurs when the magnitude of Z_L can be varied, but its phase angle cannot. Under this restriction, maximum power is delivered to the load when the magnitude of Z_L is set equal to the magnitude of Z_{Th}; that is, when

$$|Z_L| = |Z_{Th}|.$$

The proof of this is left to you as Problem 10.45.

For purely resistive networks, maximum power transfer occurs when the load resistance equals the Thévenin resistance. Note that we first derived this result in the introduction to maximum power transfer in Chapter 4.

Examples 10.9–10.11 calculate the load impedance Z_L that produces maximum average power transfer to the load, for several different situations. Example 10.12 finds the condition for maximum power transfer to a load for a circuit with an ideal transformer.

EXAMPLE 10.9	**Determining Maximum Power Transfer without Load Restrictions**

a) For the circuit shown in Fig. 10.21, determine the impedance Z_L that results in maximum average power transferred to Z_L.

b) What is the maximum average power transferred to the load impedance determined in (a)?

Solution

a) To begin, determine the Thévenin equivalent with respect to the load terminals a, b. After two source transformations involving the 20 V source, the 5 Ω resistor, and the 20 Ω resistor, we simplify the circuit shown in Fig. 10.21 to the one shown in Fig. 10.22. Use voltage division in the simplified circuit to get

$$\mathbf{V}_{Th} = \frac{-j6}{4 + j3 - j6}\,(16\underline{/0°})$$

$$= 19.2\underline{/-53.13°} = 11.52 - j15.36 \text{ V}.$$

To find the Thévenin impedance, deactivate the source in Fig. 10.22 and calculate the impedance seen looking into the terminals a and b. Thus,

$$Z_{Th} = -j6 \,\|\, (4 + j3) = \frac{(-j6)(4 + j3)}{4 + j3 - j6}$$

$$= 5.76 - j1.68 \ \Omega.$$

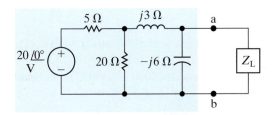

Figure 10.21 ▲ The circuit for Example 10.9.

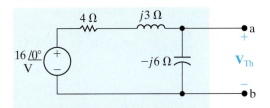

Figure 10.22 ▲ A simplification of Fig. 10.21 by source transformations.

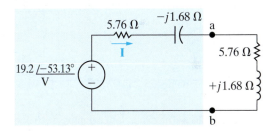

Figure 10.23 ▲ The circuit shown in Fig. 10.21, with the original network replaced by its Thévenin equivalent.

For maximum average power transfer, the load impedance must be the conjugate of Z_{Th}, so

$$Z_L = 5.76 + j1.68 \ \Omega.$$

b) We calculate the maximum average power delivered to Z_L using the circuit in Fig. 10.23, which has the Thévenin equivalent of the original network attached to the load impedance calculated in part (a). From Fig. 10.23, the rms magnitude of the load current $\mathbf{I}$ is

$$I_{rms} = \frac{19.2/\sqrt{2}}{2(5.76)} = 1.1785 \text{ A(rms)}.$$

The average power delivered to the load is

$$P = I_{rms}^2(5.76) = 8 \text{ W}.$$

EXAMPLE 10.10 **Determining Maximum Power Transfer with Load Impedance Restriction**

a) For the circuit shown in Fig. 10.24, what value of Z_L results in maximum average power transfer to Z_L? What is the maximum power in milliwatts?

b) Assume that the load resistance can be varied between 0 and 4000 Ω and that the capacitive reactance of the load can be varied between 0 and $-2000\ \Omega$. What settings of R_L and X_L transfer the most average power to the load? What is the maximum average power that can be transferred under these restrictions?

Solution

a) If there are no restrictions on R_L and X_L, maximum average power is delivered to the load if the load impedance equals the conjugate of the Thévenin impedance. Therefore we set

$$R_L = 3000\ \Omega \quad \text{and} \quad X_L = -4000\ \Omega,$$

or

$$Z_L = 3000 - j4000\ \Omega.$$

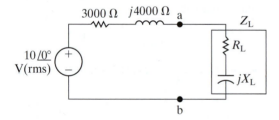

Figure 10.24 ▲ The circuit for Examples 10.10 and 10.11.

Because the source voltage is an rms value, the average power delivered to Z_L is

$$P = \frac{1}{4}\frac{10^2}{3000} = \frac{25}{3}\,\text{mW} = 8.33\ \text{mW}.$$

b) Now R_L and X_L are restricted, so first we set X_L as close to $-4000\ \Omega$ as possible; thus, $X_L = -2000\ \Omega$. Next, we set R_L as close to $\sqrt{R_{Th}^2 + (X_L + X_{Th})^2}$ as possible. Thus

$$R_L = \sqrt{3000^2 + (-2000 + 4000)^2} = 3605.55\ \Omega.$$

Since R_L can be varied from 0 to 4000 Ω, we can set R_L to 3605.55 Ω. Therefore, the load impedance value is

$$Z_L = 3605.55 - j2000\ \Omega.$$

For this value of Z_L, the value of the load current is

$$\mathbf{I}_{rms} = \frac{10\ \underline{/0^\circ}}{6605.55 + j2000} = 1.4489\ \underline{/-16.85^\circ}\ \text{mA(rms)}.$$

The average power delivered to the load is

$$P = (1.4489 \times 10^{-3})^2(3605.55) = 7.57\ \text{mW}.$$

This is the maximum power delivered to a load with the specified restrictions on R_L and X_L. Note that this is less than the 8.33 mW that can be delivered if there are no restrictions, as we found in part (a).

EXAMPLE 10.11 **Finding Maximum Power Transfer with Impedance Angle Restrictions**

A load impedance having a constant phase angle of -36.87° is connected across the terminals a and b in the circuit shown in Fig. 10.24. The magnitude of Z_L is varied until the average power delivered is maximized under the given restriction.

a) Specify Z_L in rectangular form.

b) Calculate the average power delivered to Z_L.

Solution

a) When only the magnitude of Z_L can be varied, maximum power is delivered to the load when

the magnitude of Z_L equals the magnitude of Z_{Th}. So,

$$|Z_L| = |Z_{Th}| = |3000 + j4000| = 5000\ \Omega.$$

Therefore,

$$Z_L = 5000\ \underline{/-36.87^\circ} = 4000 - j3000\ \Omega.$$

b) When Z_L equals $4000 - j3000\ \Omega$, the load current is

$$\mathbf{I}_{rms} = \frac{10}{7000 + j1000} = 1.4142\ \underline{/-8.13^\circ}\ \text{mA(rms)},$$

and the average power delivered to the load is

$$P = (1.4142 \times 10^{-3})^2(4000) = 8 \text{ mW}.$$

This quantity is the maximum power that can be delivered by this circuit to a load impedance

whose angle is constant at $-36.87°$. Again, this quantity is less than the maximum power that can be delivered if there are no restrictions on Z_L.

ASSESSMENT PROBLEM

Objective 2—Understand the condition for maximum real power delivered to a load in an ac circuit

10.8 The source current in the circuit shown is $3 \cos 5000t$ A.

a) What impedance should be connected across terminals a and b for maximum average power transfer?

b) What is the average power transferred to the impedance in (a)?

c) Assume that the load is restricted to pure resistance. What size resistor connected across a and b will result in the maximum average power transferred?

d) What is the average power transferred to the resistor in (c)?

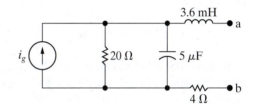

Answer: (a) $20 - j10$ Ω;

(b) 18 W;

(c) 22.36 Ω;

(d) 17 W.

SELF-CHECK: Also try Chapter Problems 10.43, 10.49, and 10.50.

| EXAMPLE 10.12 | **Finding Maximum Power Transfer in a Circuit with an Ideal Transformer** |

The variable resistor in the circuit in Fig. 10.25 is adjusted until maximum average power is delivered to R_L.

a) What is the value of R_L in ohms?

b) What is the maximum average power (in watts) delivered to R_L?

Solution

a) We first find the Thévenin equivalent with respect to the terminals of R_L. We determine the open-circuit voltage using the circuit in Fig. 10.26. The variables $\mathbf{V}_1$, $\mathbf{V}_2$, $\mathbf{I}_1$, and $\mathbf{I}_2$ have been added to aid the discussion.

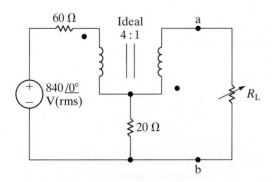

Figure 10.25 ▲ The circuit for Example 10.12.

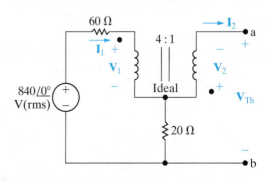

Figure 10.26 ▲ The circuit used to find the Thévenin voltage.

The ideal transformer imposes the following constraints on the variables V_1, V_2, I_1, and I_2:

$$V_2 = \frac{1}{4}V_1, \qquad I_1 = -\frac{1}{4}I_2.$$

The open-circuit value of I_2 is zero; hence, I_1 is zero. It follows that

$$V_1 = 840\ \underline{/0°}\ V(rms), \qquad V_2 = 210\ \underline{/0°}\ V(rms).$$

From Fig. 10.26 we note that V_{Th} is the negative of V_2, because there is no current in the 20 Ω resistor. Hence

$$V_{Th} = -210\ \underline{/0°}\ V(rms).$$

We determine the short-circuit current using the circuit in Fig. 10.27. Since I_1 and I_2 are mesh currents, write a KVL equation for each mesh:

$$840\ \underline{/0°} = 80I_1 - 20I_2 + V_1,$$

$$0 = 20I_2 - 20I_1 + V_2.$$

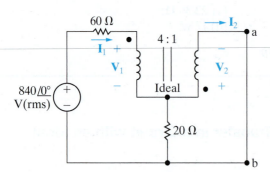

Figure 10.27 ▲ The circuit used to calculate the short circuit current.

Combine these two KVL equations with the constraint equations to get

$$840\ \underline{/0°} = -40I_2 + V_1,$$

$$0 = 25I_2 + \frac{V_1}{4}.$$

Solving for the short circuit value of I_2 yields

$$I_2 = -6\ A(rms).$$

Therefore, the Thévenin resistance is

$$R_{Th} = \frac{-210}{-6} = 35\ \Omega.$$

Maximum power will be delivered to R_L when R_L equals 35 Ω.

b) We determine the maximum power delivered to R_L using the Thévenin equivalent circuit in Fig. 10.28. From this circuit, the rms current in the load resistor is $(-210/70)$ A(rms). Therefore,

$$P_{max} = \left(\frac{-210}{70}\right)^2 (35) = 315\ W.$$

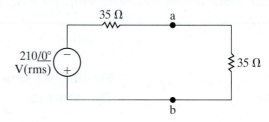

Figure 10.28 ▲ The Thévenin equivalent loaded for maximum power transfer.

ASSESSMENT PROBLEMS

Objective 3—Be able to calculate all forms of ac power in ac circuits with linear transformers and ideal transformers

10.9 Find the average power delivered to the 100 Ω resistor in the circuit shown if $v_g = 660\cos 5000t$ V.

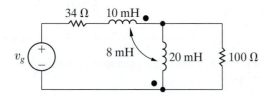

Answer: 612.5 W.

10.10 a) Find the average power delivered to the 400 Ω resistor in the circuit shown if $v_g = 248\cos 10{,}000t$ V.
b) Find the average power delivered to the 375 Ω resistor.
c) Find the power developed by the ideal voltage source. Check your result by showing that the power absorbed equals the power developed.

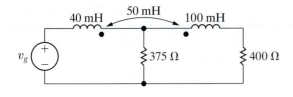

Answer: (a) 50 W;
(b) 49.2 W;
(c) 99.2 W, 50 + 49.2 = 99.2 W.

SELF-CHECK: Also try Chapter Problems 10.56, 10.59, and 10.60.

10.11 Solve Example 10.12 if the polarity dot on the coil connected to terminal a is at the top.

Answer: (a) 15 Ω;
(b) 735 W.

10.12 Solve Example 10.12 if the voltage source is reduced to 146 $\underline{/0°}$ V(rms) and the turns ratio is reversed to 1:4.

Answer: (a) 1460 Ω;
(b) 58.4 W.

■ Practical Perspective

Vampire Power

Vampire power, or standby power, may cost you more than you think. The average household has about 40 electrical products that draw power, even when they are turned off. Approximately 5% of typical residential power consumption can be attributed to standby power. Table 10.3 provides the power consumption for several different devices. Notice that when a device is "off" it is often still consuming power.

Consider a typical mobile phone charger. According to the values given in Table 10.3, when the charger is detached from the phone it consumes only a fraction of the power required when the charger is

TABLE 10.3 Average power consumption of common electrical devices

Electrical device[+]	Power [W][*]
Mobile phone charger	
Attached to phone, phone charging	3.68
Plugged into wall outlet but not into phone	0.26
Notebook computer AC adapter	
Attached to computer, computer charging	44.28
Attached to computer, computer sleeping	15.77
Attached to computer, computer off	8.9
Plugged into wall outlet but not into computer	4.42
DVD player	
On and playing	9.91
On and not playing	7.54
Off	1.55
Microwave oven	
Ready with door closed	3.08
Ready with door open	25.79
Cooking	1433.0
Inkjet multifunction printer	
On	9.16
Off	5.26

[*]Data in this table from Lawrence Berkeley National Laboratory report (http://standby.lbl.gov/standby.html).
[+]This value is the average of the power measured for many types of each device.

attached to the phone and the phone is charging. Suppose you charge your phone for three hours each day but leave the charger plugged into the wall outlet 24 hours a day. Recall that the electric company charges you based on the number of kilowatt-hours (kWh) you use in a given month. A device that uses 1000 W of power continuously over one hour has consumed 1 kWh. Let's calculate the number of kilowatt-hours used by the phone charger in one month:

$$P[\text{kWh}] = \frac{30[3(3.68) + 21(0.26)]}{1000} = 1.8 \text{ kWh}.$$

Now do the calculation again, this time assuming that you unplug the charger when it is not being used to charge the phone:

$$P[\text{kWh}] = \frac{30[3(3.68) + 21(0)]}{1000} = 0.33 \text{ kWh}.$$

Keeping the charger plugged in when you are not using it consumes more than 5 times the power needed to charge your phone every day. You can therefore minimize the cost of vampire power by unplugging electrical devices if they are not being used.

Why does the phone charger consume power when not plugged into the phone? The electronic circuitry in your phone uses 5 V(dc) sources to supply power. The phone charger must transform the 120 V(rms) signal supplied by the wall outlet into a signal that can be used to charge the phone. Phone chargers can use linear transformers, together with other circuitry, to output the voltage needed by the phone.

Consider the circuit in Fig. 10.29. The linear transformer is part of the circuitry used to reduce the voltage supplied by the source to the level required by the phone. The additional components needed to complete this task are not shown in the circuit. When the phone is un-plugged from the circuit in Fig. 10.29, but the circuit is still connected to the 120 V(rms) source, there is still a path for the current, as shown in Fig. 10.30. The current is

$$\mathbf{I} = \frac{120}{R_s + R_1 + j\omega L_1}.$$

The real power, delivered by the voltage source and supplied to the resistors, is

$$P = (R_s + R_1)|\mathbf{I}|^2.$$

This is the vampire power being consumed by the phone charger even when it is not connected to the phone.

SELF-CHECK: Assess you understanding of this Practical Perspective by trying Chapter Problems 10.67–10.69.

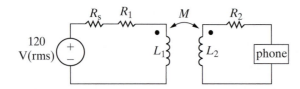

Figure 10.29 ▲ A linear transformer used in a phone charger.

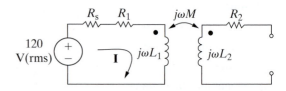

Figure 10.30 ▲ The phone charger circuit when the phone is not connected.

Summary

- **Instantaneous power** is the product of the instantaneous terminal voltage and current, or $p = \pm vi$. The positive sign is used when the reference direction for the current is from the positive to the negative reference polarity of the voltage. The frequency of the instantaneous power is twice the frequency of the voltage (or current). (See page 376.)

- **Average power** is the average value of the instantaneous power over one period. It is the power converted from electric to nonelectric form and vice versa, so it is also called real power. Average power is given by

$$P = \frac{1}{2} V_m I_m \cos(\theta_v - \theta_i)$$

$$= V_{rms} I_{rms} \cos(\theta_v - \theta_i).$$

(See page 377.)

- **Reactive power** is the electric power exchanged between the magnetic field of an inductor and the source that drives it or between the electric field of a capacitor and the source that drives it. Reactive power is never converted to nonelectric power. Reactive power is given by

$$Q = \frac{1}{2} V_m I_m \sin(\theta_v - \theta_i)$$

$$= V_{rms} I_{rms} \sin(\theta_v - \theta_i).$$

Both average power and reactive power can be expressed in terms of either peak (V_m, I_m) or rms (V_{rms}, I_{rms}) current and voltage. RMS values, also called *effective values*, are widely used in both household and industrial applications. (See page 378.)

- The **power factor** is the cosine of the phase angle between the voltage and the current:

$$pf = \cos(\theta_v - \theta_i).$$

The terms *lagging* and *leading*, added to the description of the power factor, indicate whether the current is lagging or leading the voltage and thus whether the load is inductive or capacitive. (See page 379.)

- The **reactive factor** is the sine of the phase angle between the voltage and the current:

$$rf = \sin(\theta_v - \theta_i).$$

(See page 379.)

- **Complex power** is the complex sum of the real and reactive powers, or

$$S = P + jQ$$

$$= \frac{1}{2} \mathbf{VI}^* = \mathbf{V}_{rms} \mathbf{I}_{rms}^*$$

$$= |\mathbf{I}_{rms}|^2 Z = \frac{|\mathbf{V}_{rms}|^2}{Z^*}.$$

(See page 384.)

- **Apparent power** is the magnitude of the complex power:

$$|S| = \sqrt{P^2 + Q^2}.$$

(See page 385.)

- The **watt** is used as the unit for both instantaneous and real power. The **var** (volt amp reactive, or VAR) is used as the unit for reactive power. The **volt-amp** (VA) is used as the unit for complex and apparent power. (See page 385.)

- **Maximum power transfer** occurs in circuits operating in the sinusoidal steady state when the load impedance is the conjugate of the Thévenin impedance as viewed from the terminals of the load impedance. (See page 393.)

Problems

Sections 10.1–10.2

10.1 The following sets of values for v and i pertain to the circuit seen in Fig. 10.1. For each set of values, calculate P and Q and state whether the circuit inside the box is absorbing or delivering (1) average power and (2) magnetizing vars.

a) $v = 100 \cos(\omega t + 50°)$ V,
   $i = 10 \cos(\omega t + 15°)$ A.

b) $v = 40 \cos(\omega t - 15°)$ V,
   $i = 5 \cos(\omega t + 60°)$ A.

c) $v = 400 \cos(\omega t + 30°)$ V,
   $i = 10 \sin(\omega t + 240°)$ A.

d) $v = 200 \sin(\omega t + 250°)$ V,
   $i = 5 \cos(\omega t + 40°)$ A.

10.2 a) A college student wakes up on a warm day. The central air conditioning is on, and the room feels comfortable. She turns on the dishwasher, takes some milk out of the old refrigerator, and puts some oatmeal in the microwave oven to cook. If all of these appliances in her dorm room are supplied by a 120 V(rms) branch circuit protected by a 50 A(rms) circuit breaker, will the breaker interrupt her morning?

b) Her roommate wakes up and moves wet clothes from the washer to the dryer. Now does the circuit breaker interrupt their morning?

10.3 Show that the maximum value of the instantaneous power given by Eq. 10.3 is $P + \sqrt{P^2 + Q^2}$ and that the minimum value is $P - \sqrt{P^2 + Q^2}$.

10.4 A load consisting of a 480 Ω resistor in parallel with a $(5/9)$ μF capacitor is connected across the terminals of a sinusoidal voltage source v_g, where $v_g = 240 \cos 5000t$ V.

a) What is the peak value of the instantaneous power delivered by the source?

b) What is the peak value of the instantaneous power absorbed by the source?

c) What is the average power delivered to the load?

d) What is the reactive power delivered to the load?

e) Does the load absorb or generate magnetizing vars?

f) What is the power factor of the load?

g) What is the reactive factor of the load?

10.5 a) Calculate the real and reactive power associated with each circuit element in the circuit in Fig. P9.60.

b) Verify that the average power generated equals the average power absorbed.

c) Verify that the magnetizing vars generated equal the magnetizing vars absorbed.

10.6 Repeat Problem 10.5 for the circuit shown in Fig. P9.64.

10.7 Find the average power delivered by the ideal current source in the circuit in Fig. P10.7 if $i_g = 40 \cos 1250t$ mA.

Figure P10.7

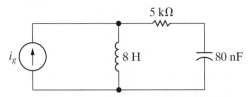

10.8 The op amp in the circuit shown in Fig. P10.8 is ideal. Calculate the average power delivered to the 1 kΩ resistor when $v_g = \cos 1000t$ V.

Figure P10.8

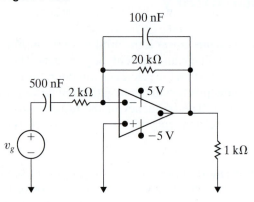

10.9 Find the average power dissipated in the 40 Ω resistor in the circuit seen in Fig. P10.9 if $i_g = 4 \cos 10^5 t$ A.

Figure P10.9

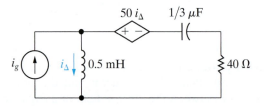

10.10 The load impedance in Fig. P10.10 absorbs 2.5 kW and generates 5 kVAR. The sinusoidal voltage source develops 7.5 kW.

a) Find the values of inductive line reactance that will satisfy these constraints.

b) For each value of line reactance found in (a), show that the magnetizing vars developed equals the magnetizing vars absorbed.

Figure P10.10

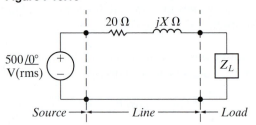

Section 10.3

10.11 Find the rms value of the periodic current shown in Fig. P10.11.

Figure P10.11

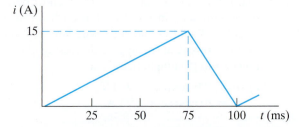

10.12 The periodic current shown in Fig. P10.11 dissipates an average power of 3 kW in a resistor. What is the value of the resistor?

10.13 a) Find the rms value of the periodic voltage shown in Fig. P10.13.

b) Suppose the voltage in part (a) is applied to the terminals of a 40 Ω resistor. Calculate the average power dissipated by the resistor.

c) When the voltage in part (a) is applied to a different resistor, that resistor dissipates 10 mW of average power. What is the value of the resistor?

Figure P10.13

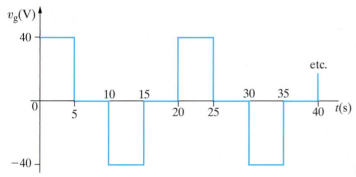

10.14 a) Find the rms value of the periodic voltage shown in Fig. P10.14.

b) If this voltage is applied to the terminals of a 2.25 Ω resistor, what is the average power dissipated in the resistor?

Figure P10.14

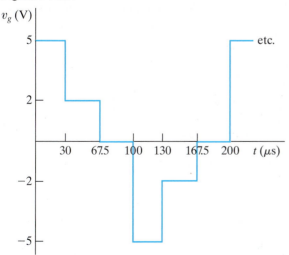

10.15 a) A personal computer with a monitor and keyboard requires 40 W at 115 V(rms). Calculate the rms value of the current carried by its power cord.

b) A laser printer for the personal computer in (a) is rated at 90 W at 115 V(rms). If this printer is plugged into the same wall outlet as the computer, what is the rms value of the current drawn from the outlet?

10.16 A dc voltage equal to V_{dc} V is applied to a resistor of R Ω. A sinusoidal voltage equal to v_s V is also applied to a resistor of R Ω. Show that the dc voltage will deliver the same amount of energy in T seconds (where T is the period of the sinusoidal voltage) as the sinusoidal voltage provided V_{dc} equals the rms value of v_s. (Hint: Equate the two expressions for the energy delivered to the resistor.)

Sections 10.4–10.5

10.17 a) Find V_L (rms) and θ for the circuit in Fig. P10.17 if the load absorbs 2500 VA at a lagging power factor of 0.8.

b) Construct a phasor diagram of each solution obtained in (a).

Figure P10.17

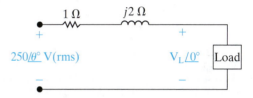

10.18 a) Find the average power, the reactive power, and the apparent power supplied by the voltage source in the circuit in Fig. P10.18 if $v_g = 40 \cos 10^6 t$ V.

b) Check your answer in (a) by showing $P_{dev} = \Sigma P_{abs}$.

c) Check your answer in (a) by showing $Q_{dev} = \Sigma Q_{abs}$.

Figure P10.18

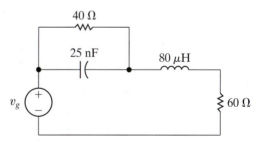

10.19 The voltage $\mathbf{V}_g$ in the frequency-domain circuit shown in Fig. P10.19 is $170\underline{/0^\circ}$ V(rms).

a) Find the average and reactive power for the voltage source.

b) Is the voltage source absorbing or delivering average power?

c) Is the voltage source absorbing or delivering magnetizing vars?

d) Find the average and reactive powers associated with each impedance branch in the circuit.

e) Check the balance between delivered and absorbed average power.

f) Check the balance between delivered and absorbed magnetizing vars.

Figure P10.19

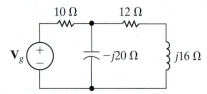

10.20 Find the average power, the reactive power, and the apparent power absorbed by the load in the circuit in Fig. P10.20 if v_g equals 150 cos 250t V.

PSPICE
MULTISIM

Figure P10.20

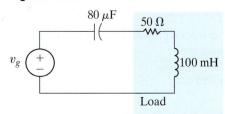

Load

10.21 The two loads shown in Fig. P10.21 can be described as follows: Load 1 absorbs an average power of 60 kW and delivers 70 kVAR of reactive power; Load 2 has an impedance of $(24 + j7)$ Ω. The voltage at the terminals of the loads is $2500\sqrt{2}$ cos 120π t V.

a) Find the rms value of the source voltage.

b) By how many microseconds is the load voltage out of phase with the source voltage?

c) Does the load voltage lead or lag the source voltage?

Figure P10.21

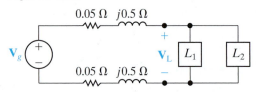

10.22 Two 125 V(rms) loads are connected in parallel. The two loads draw a total average power of 4500 W at a power factor of 0.96 leading. One of the loads draws 2700 W at a power factor of 0.8 lagging. What is the power factor of the other load?

10.23 The three parallel loads in the circuit shown in Fig. 10.23 can be described as follows: Load 1 is absorbing an average power of 6 kW and delivering reactive power of 8 kvars; Load 2 is absorbing an average power of 9 kW and reactive power of 3 kvars; Load 3 is a 25 Ω resistor in parallel with a capacitor whose reactance is −5 Ω. Find the rms magnitude and the phase angle of $\mathbf{V}_g$ if $\mathbf{V}_o = 250\underline{/0°}$ V(rms).

Figure P10.23

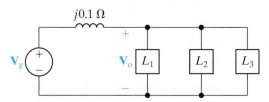

10.24 Three loads are connected in parallel across a 250 V(rms) line, as shown in Fig. P10.24. Load 1 absorbs 16 kW and 18 kVAR; Load 2 absorbs 10 kVA at 0.6 leading; Load 3 absorbs 8 kW at unity power factor.

a) Find the impedance that is equivalent to the three parallel loads.

b) Find the power factor of the equivalent load as seen from the line's input terminals.

Figure P10.24

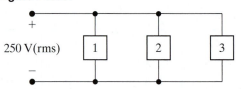

10.25 The three loads in Problem 10.24 are fed from a line having a series impedance 0.01 + $j0.08$ Ω, as shown in Fig. P10.25.

a) Calculate the rms value of the voltage ($\mathbf{V}_s$) at the sending end of the line.

b) Calculate the average and reactive powers associated with the line impedance.

c) Calculate the average and reactive powers at the sending end of the line.

d) Calculate the efficiency (η) of the line if the efficiency is defined as

$$\eta = (P_{\text{load}}/P_{\text{sending end}}) \times 100.$$

Figure P10.25

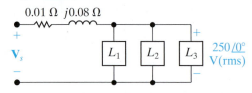

10.26 Consider the circuit described in Problem 9.78.

a) What is the rms magnitude of the voltage across the load impedance?

b) What percentage of the average power developed by the practical source is delivered to the load impedance?

10.27 The three loads in the circuit in Fig. P10.27 can be described as follows: Load 1 is a 240 Ω resistor in series with an inductive reactance of 70 Ω; load 2 is a capacitive reactance of 120 Ω in series with a 160 Ω resistor; and load 3 is a 30 Ω resistor in series with a capacitive reactance of 40 Ω. The frequency of the voltage source is 60 Hz.

a) Give the power factor and reactive factor of each load.

b) Give the power factor and reactive factor of the composite load seen by the voltage source.

Figure P10.27

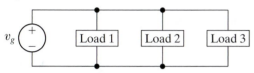

10.28 The three loads in the circuit seen in Fig. P10.28 are described as follows: Load 1 is absorbing 7.5 kW and 2.5 kVAR; Load 2 is absorbing 10 kVA at a power factor of 0.28 leading; Load 3 is a 12.5 Ω resistor in parallel with an inductance whose reactance is 50 Ω.

a) Calculate the average power and the magnetizing reactive power delivered by each source if $\mathbf{V}_{g1} = \mathbf{V}_{g2} = 250\underline{/0°}$ V(rms).

b) Check your calculations by showing your results are consistent with the requirements

$$\sum P_{\text{dev}} = \sum P_{\text{abs}};$$

$$\sum P_{\text{dev}} = \sum P_{\text{abs}}.$$

Figure P10.28

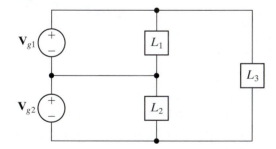

10.29 Suppose the circuit shown in Fig. P10.28 represents a residential distribution circuit in which the impedances of the service conductors are negligible and $\mathbf{V}_{g1} = \mathbf{V}_{g2} = 120\underline{/0°}$ V(rms). The three loads in the circuit are L_1 (a new refrigerator, a coffee maker, and a microwave oven); L_2 (2 incandescent lamps, a humidifier, and a ceiling fan); and L_3 (a clothes washer and a clothes dryer). Assume that all of these appliances are in operation at the same time. The service conductors are protected with 20 A circuit breakers. Will the service to this residence be interrupted? Why or why not?

10.30 The three loads in the circuit seen in Fig. P10.30 are

$$S_1 = 4 + j1 \text{ kVA,}$$

$$S_2 = 5 + j2 \text{ kVA,}$$

$$S_3 = 10 + j0 \text{ kVA.}$$

a) Calculate the complex power associated with each voltage source, $\mathbf{V}_{g1}$ and $\mathbf{V}_{g2}$.

b) Verify that the total real and reactive power delivered by the sources equals the total real and reactive power absorbed by the network.

Figure P10.30

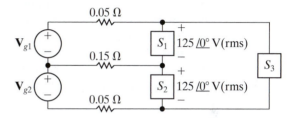

10.31 A group of small appliances on a 60 Hz system requires 20 kVA at 0.85 pf lagging when operated at 125 V(rms). The impedance of the feeder supplying the appliances is 0.01 + j0.08 Ω. The voltage at the load end of the feeder is 125 V(rms).

a) What is the rms magnitude of the voltage at the source end of the feeder?

b) What is the average power loss in the feeder?

c) What size capacitor (in microfarads) across the load end of the feeder is needed to improve the load power factor to unity?

d) After the capacitor is installed, what is the rms magnitude of the voltage at the source end of the feeder if the load voltage is maintained at 125 V(rms)?

e) What is the average power loss in the feeder for (d)?

10.32 a) Find the average power dissipated in the line in Fig. P10.32.

b) Find the capacitive reactance that when connected in parallel with the load will make the load look purely resistive.

c) What is the equivalent impedance of the load in (b)?

d) Find the average power dissipated in the line when the capacitive reactance is connected across the load.

e) Express the power loss in (d) as a percentage of the power loss found in (a).

Figure P10.32

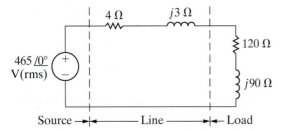

Source →|←————— Line ————→|← Load

10.33 The steady-state voltage drop between the load and the sending end of the line seen in Fig. P10.33 is excessive. A capacitor is placed in parallel with the 192 kVA load and is adjusted until the steady-state voltage at the sending end of the line has the same magnitude as the voltage at the load end, that is, 4800 V(rms). The 192 kVA load is operating at a power factor of 0.8 lag. Calculate the size of the capacitor in microfarads if the circuit is operating at 60 Hz. In selecting the capacitor, keep in mind the need to keep the power loss in the line at a reasonable level.

Figure P10.33

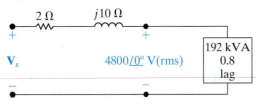

10.34 A factory has an electrical load of 1600 kW at a lagging power factor of 0.8. An additional variable power factor load is to be added to the factory. The new load will add 320 kW to the real power load of the factory. The power factor of the added load is to be adjusted so that the overall power factor of the factory is 0.96 lagging.

a) Specify the reactive power associated with the added load.

b) Does the added load absorb or deliver magnetizing vars?

c) What is the power factor of the additional load?

d) Assume that the voltage at the input to the factory is 2400 V(rms). What is the rms magnitude of the current into the factory before the variable power factor load is added?

e) What is the rms magnitude of the current into the factory after the variable power factor load has been added?

10.35 Assume the factory described in Problem 10.34 is fed from a line having an impedance of 0.05 + j0.4 Ω. The voltage at the factory is maintained at 2400 V(rms).

a) Find the average power loss in the line before and after the load is added.

b) Find the magnitude of the voltage at the sending end of the line before and after the load is added.

10.36 a) Find the six branch currents $I_a - I_f$ in the circuit in Fig. P10.36.

b) Find the complex power in each branch of the circuit.

c) Check your calculations by verifying that the average power developed equals the average power dissipated.

d) Check your calculations by verifying that the magnetizing vars generated equal the magnetizing vars absorbed.

Figure P10.36

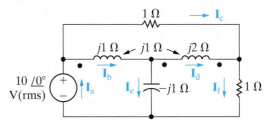

10.37 a) Find the average power delivered to the 8 Ω resistor in the circuit in Fig. P10.37.

b) Find the average power developed by the ideal sinusoidal voltage source.

c) Find Z_{ab}.

d) Show that the average power developed equals the average power dissipated.

Figure P10.37

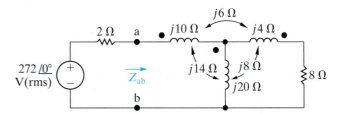

10.38 a) Find the average power delivered by the sinusoidal current source in the circuit of Fig. P10.38.

b) Find the average power delivered to the 20 Ω resistor.

Figure P10.38

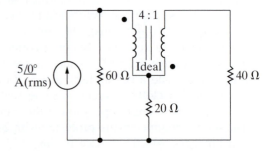

10.39 a) Find the average power dissipated in each resistor in the circuit in Fig. P10.39.

b) Check your answer by showing that the total power developed equals the total power absorbed.

Figure P10.39

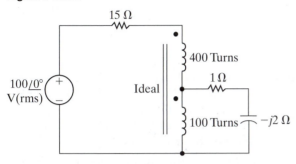

10.40 The sinusoidal voltage source in the circuit in Fig. P10.40 is developing an rms voltage of 2000 V. The 4 Ω load in the circuit is absorbing four times as much average power as the 25 Ω load. The two loads are matched to the sinusoidal source that has an internal impedance of $500 \underline{/0°}$ Ω.

a) Specify the numerical values of a_1 and a_2.

b) Calculate the power delivered to the 25 Ω load.

c) Calculate the rms value of the voltage across the 4 Ω resistor.

Figure P10.40

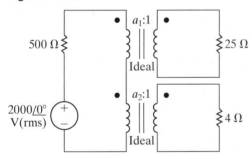

Section 10.6

10.41 Suppose an impedance equal to the conjugate of the Thévenin impedance is connected to the terminals c, d of the circuit shown in Fig. P9.75.

a) Find the average power developed by the sinusoidal voltage source.

b) What percentage of the power developed by the source is lost in the linear transformer?

10.42 The phasor voltage $\mathbf{V}_{ab}$ in the circuit shown in Fig. P10.42 is $240 \underline{/0°}$ V(rms) when no external load is connected to the terminals a, b. When a load having an impedance of $80 - j60$ Ω is connected across a, b, the value of $\mathbf{V}_{ab}$ is $115.2 + j33.6$ V(rms).

a) Find the impedance that should be connected across a, b for maximum average power transfer.

b) Find the maximum average power transferred to the load of (a).

c) Construct the impedance of part (a) using components from Appendix H if the source frequency is 1000 Hz.

Figure P10.42

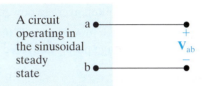

10.43 a) Determine the load impedance for the circuit shown in Fig. P10.43 that will result in maximum average power being transferred to the load if $\omega = 10$ krad/s.

b) Determine the maximum average power delivered to the load from part (a) if $v_g = 120 \cos 10{,}000t$ V.

c) Repeat part (b) when Z_L consists of two components from Appendix H whose values yield a maximum average power closest to the value calculated in part (a).

Figure P10.43

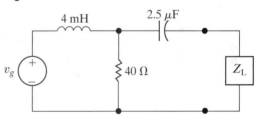

10.44 The load impedance Z_L for the circuit shown in Fig. P10.44 is adjusted until maximum average power is delivered to Z_L.

a) Find the maximum average power delivered to Z_L.

b) What percentage of the total power developed in the circuit is delivered to Z_L?

Figure P10.44

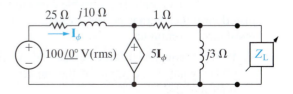

10.45 Prove that if only the magnitude of the load impedance can be varied, the most average power is transferred to the load when $|Z_L| = |Z_{Th}|$. (*Hint:* In deriving the expression for the load's average power, write the load impedance (Z_L) in the form $Z_L = |Z_L|\cos\theta + j|Z_L|\sin\theta$, and note that only $|Z_L|$ is variable.)

10.46 The variable resistor in the circuit shown in Fig. P10.46 is adjusted until the average power it absorbs is maximum.

a) Find R.

b) Find the maximum average power.

c) Find the resistor in Appendix H that would have the most average power delivered to it.

Figure P10.46

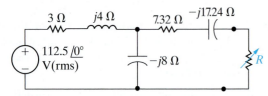

10.47 The variable resistor R_o in the circuit shown in Fig. P10.47 is adjusted until maximum average power is delivered to R_o.

a) What is the value of R_o in ohms?

b) Calculate the average power delivered to R_o.

c) If R_o is replaced with a variable impedance Z_o, what is the maximum average power that can be delivered to Z_o?

d) In (c), what percentage of the circuit's developed power is delivered to the load Z_o?

Figure P10.47

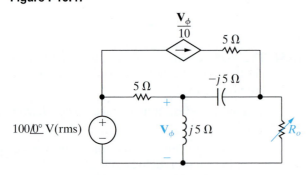

10.48 The sending-end voltage in the circuit seen in Fig. P10.48 is adjusted so that the load voltage is always 13,800 V(rms). The variable capacitor is adjusted until the average power dissipated in the line resistance is minimum.

a) If the frequency of the sinusoidal source is 60 Hz, what is the value of the capacitance in microfarads?

b) If the capacitor is removed from the circuit, what percentage increase in the magnitude of $\mathbf{V}_s$ is necessary to maintain 13,800 V(rms) at the load?

c) If the capacitor is removed from the circuit, what is the percentage increase in line loss?

Figure P10.48

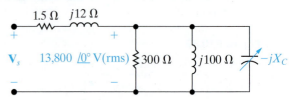

10.49 The peak amplitude of the sinusoidal voltage source in the circuit shown in Fig. P10.49 is 180 V, and its frequency is 5000 rad/s. The load resistor can be varied from 0 to 4000 Ω, and the load capacitor can be varied from 0.1 μF to 0.5 μF.

a) Calculate the average power delivered to the load when $R_o = 2000\ \Omega$ and $C_o = 0.2\ \mu$F.

b) Determine the settings of R_o and C_o that will result in the most average power being transferred to R_o.

c) What is the average power in (b)? Is it greater than the power in (a)?

d) If there are no constraints on R_o and C_o, what is the maximum average power that can be delivered to a load?

e) What are the values of R_o and C_o for the condition of (d)?

f) Is the average power calculated in (d) larger than that calculated in (c)?

Figure P10.49

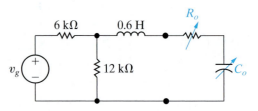

10.50 a) Assume that R_o in Fig. P10.49 can be varied between 0 and 10 kΩ. Repeat (b) and (c) of Problem 10.49.

b) Is the new average power calculated in (a) greater than that found in Problem 10.49(a)?

c) Is the new average power calculated in (a) less than that found in 10.49(d)?

10.51 a) Find the steady-state expression for the currents i_g and i_L in the circuit in Fig. P10.51 when $v_g = 400\cos 400t$ V.

b) Find the coefficient of coupling.

c) Find the energy stored in the magnetically coupled coils at $t = 1.25\pi$ ms and $t = 2.5\pi$ ms.

d) Find the power delivered to the 375 Ω resistor.

e) If the 375 Ω resistor is replaced by a variable resistor R_L, what value of R_L will yield maximum average power transfer to R_L?

f) What is the maximum average power in (e)?

g) Assume the 375 Ω resistor is replaced by a variable impedance Z_L. What value of Z_L will result in maximum average power transfer to Z_L?

h) What is the maximum average power in (g)?

Figure P10.51

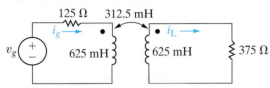

10.52 The values of the parameters in the circuit shown in Fig. P10.52 are $L_1 = 8$ mH; $L_2 = 2$ mH; $k = 0.75$; $R_g = 1\ \Omega$; and $R_L = 7\ \Omega$. If $v_g = 54\sqrt{2}\ \cos 1000t$ V, find

a) the rms magnitude of v_o;

b) the average power delivered to R_L;

c) the percentage of the average power generated by the ideal voltage source that is delivered to R_L.

Figure P10.52

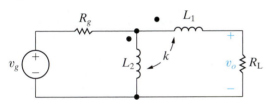

10.53 Assume the coefficient of coupling in the circuit in Fig. P10.52 is adjustable.

a) Find the value of k that makes v_o equal to zero.

b) Find the power developed by the source when k has the value found in (a).

10.54 Assume the load resistor (R_L) in the circuit in Fig. P10.52 is adjustable.

a) What value of R_L will result in the maximum average power being transferred to R_L?

b) What is the value of the maximum power transferred?

10.55 Find the impedance seen by the ideal voltage source in the circuit in Fig. P10.55 when Z_o is adjusted for maximum average power transfer to Z_o.

Figure P10.55

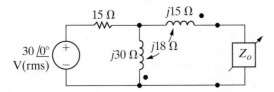

10.56 For the frequency-domain circuit in Fig. P10.56, calculate:

a) the rms magnitude of $\mathbf{V}_o$;

b) the average power dissipated in the 9 Ω resistor;

c) the percentage of the average power generated by the ideal voltage source that is delivered to the 9 Ω load resistor.

Figure P10.56

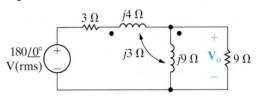

10.57 The 9 Ω resistor in the circuit in Fig. P10.56 is replaced with a variable impedance Z_o. Assume Z_o is adjusted for maximum average power transfer to Z_o.

a) What is the maximum average power that can be delivered to Z_o?

b) What is the average power developed by the ideal voltage source when maximum average power is delivered to Z_o?

c) Choose single components from Appendix H to form an impedance that dissipates average power closest to the value in part (a). Assume the source frequency is 60 Hz.

10.58 The impedance Z_L in the circuit in Fig. P10.58 is adjusted for maximum average power transfer to Z_L. The internal impedance of the sinusoidal voltage source is $4 + j7\ \Omega$.

a) What is the maximum average power delivered to Z_L?

b) What percentage of the average power delivered to the linear transformer is delivered to Z_L?

Figure P10.58

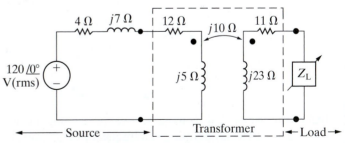

10.59 Find the average power delivered to the 5 kΩ resistor in the circuit of Fig. P10.59.

Figure P10.59

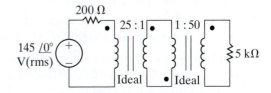

10.60 The ideal transformer connected to the 5 kΩ load in Problem 10.59 is replaced with an ideal transformer that has a turns ratio of 1:a.

a) What value of a results in maximum average power being delivered to the 5 kΩ resistor?

b) What is the maximum average power?

10.61 The load impedance Z_L in the circuit in Fig. P10.61 is adjusted until maximum average power is transferred to Z_L.

a) Specify the value of Z_L if $N_1 = 3600$ turns and $N_2 = 600$ turns.

b) Specify the values of $\mathbf{I}_L$ and $\mathbf{V}_L$ when Z_L is absorbing maximum average power.

Figure P10.61

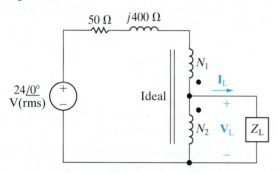

10.62 a) Find the turns ratio N_1/N_2 for the ideal transformer in the circuit in Fig. P10.62 so that maximum average power is delivered to the 400 Ω load.

b) Find the average power delivered to the 400 Ω load.

c) Find the voltage $\mathbf{V}_1$.

d) What percentage of the power developed by the ideal current source is delivered to the 400 Ω resistor?

Figure P10.62

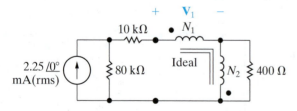

10.63 The sinusoidal voltage source in the circuit in Fig. P10.63 is operating at a frequency of 20 krad/s. The variable capacitive reactance in the circuit is adjusted until the average power delivered to the 100 Ω resistor is as large as possible.

a) Find the value of C in microfarads.

b) When C has the value found in (a), what is the average power delivered to the 100 Ω resistor?

c) Replace the 100 Ω resistor with a variable resistor R_o. Specify the value of R_o so that maximum average power is delivered to R_o.

d) What is the maximum average power that can be delivered to R_o?

Figure P10.63

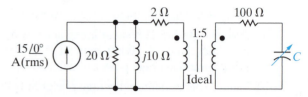

10.64 The variable load resistor R_L in the circuit shown in Fig. P10.64 is adjusted for maximum average power transfer to R_L.

PSPICE
MULTISIM

a) Find the maximum average power.

b) What percentage of the average power developed by the ideal voltage source is delivered to R_L when R_L is absorbing maximum average power?

c) Test your solution by showing that the power developed by the ideal voltage source equals the power dissipated in the circuit.

Figure P10.64

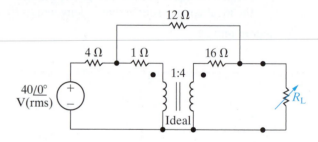

10.65 Repeat Problem 10.64 for the circuit shown in Fig. P10.65.

PSPICE
MULTISIM

Figure P10.65

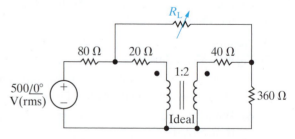

10.66 a) If N_1 equals 1000 turns, how many turns should be placed on the N_2 winding of the ideal transformer in the circuit seen in Fig. P10.66 so that maximum average power is delivered to the 6800 Ω load?

b) Find the average power delivered to the 6800 Ω resistor.

c) What percentage of the average power delivered by the ideal voltage source is dissipated in the linear transformer?

Figure P10.66

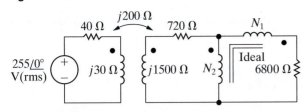

Sections 10.1–10.6

10.67 a) Use the values in Table 10.3 to calculate the
PRACTICAL   number of kilowatt-hours consumed in one
PERSPECTIVE   month by a notebook computer AC adapter if
every day the computer is charging for 5 hours
and sleeping for 19 hours..

b) Repeat the calculation in part (a) assuming that
the computer is charging for 5 hours and off for
19 hours.

c) Repeat the calculation in part (a) assuming that
the computer is charging for 5 hours and disconnected from the AC adapter for 19 hours, but the
AC adapter remains plugged into the wall outlet.

d) Repeat the calculation in part (a) assuming that
the computer is charging for 5 hours and the AC
adapter is unplugged from the wall outlet for 19
hours.

10.68 a) Suppose you use your microwave oven for
PRACTICAL   12 minutes each day. The remaining time, the
PERSPECTIVE   oven is ready with the door closed. Use the values in Table 10.3 to calculate the total number
of kilowatt-hours used by the microwave oven
in one month.

b) What percentage of the power used by the microwave oven in one month is consumed when
the oven is ready with the door closed?

10.69 Determine the amount of power, in watts, con-
PRACTICAL   sumed by the transformer in Fig. 10.30. Assume that
PERSPECTIVE   the voltage source is ideal ($R_s = 0\ \Omega$), $R_1 = 5\ \Omega$,
and $L_1 = 250$ mH. The frequency of the 120 V(rms)
source is 60 Hz.

10.70 Repeat Problem 10.69, but assume that the linear
PRACTICAL   transformer has been improved so that $R_1 = 50$ mΩ.
PERSPECTIVE   All other values are unchanged.

10.71 Repeat Problem 10.69 assuming that the linear
PRACTICAL   transformer in Fig. 10.30 has been replaced by an
PERSPECTIVE   ideal transformer with a turns ratio of 30:1. (Hint:
you shouldn't need to make any calculations to determine the amount of power consumed.)

CHAPTER CONTENTS

11.1 **Balanced Three-Phase Voltages** *p. 414*

11.2 **Three-Phase Voltage Sources** *p. 415*

11.3 **Analysis of the Wye-Wye Circuit** *p. 416*

11.4 **Analysis of the Wye-Delta Circuit** *p. 422*

11.5 **Power Calculations in Balanced Three-Phase Circuits** *p. 425*

11.6 **Measuring Average Power in Three-Phase Circuits** *p. 430*

CHAPTER OBJECTIVES

1 Know how to analyze a balanced, three-phase wye-wye connected circuit.

2 Know how to analyze a balanced, three-phase wye-delta connected circuit.

3 Be able to calculate power (average, reactive, and complex) in any three-phase circuit.

Balanced Three-Phase Circuits

We use three-phase circuits to generate, transmit, distribute, and consume large blocks of electric power. The comprehensive analysis of such systems is a field of study in its own right; we cannot cover it in a single chapter. Fortunately, an understanding of the steady-state sinusoidal behavior of balanced three-phase circuits is sufficient for engineers who do not specialize in power systems. We analyze balanced three-phase circuits using several shortcuts based on circuit-analysis techniques discussed in earlier chapters.

The basic structure of a three-phase system consists of voltage sources connected to loads by means of transformers and transmission lines. To analyze such a circuit, we can reduce it to a voltage source connected to a load via a line. The omission of the transformer simplifies the discussion without jeopardizing a basic understanding of the three-phase system. Figure 11.1 on page 414 shows a basic circuit. A defining characteristic of a balanced three-phase circuit is that it contains a set of balanced three-phase voltages at its source. We begin by considering these voltages, and then we move to the voltage and current relationships for the Y-Y and Y-Δ circuits. After considering voltage and current in such circuits, we conclude with sections on power and power measurement.

■ Practical Perspective

Transmission and Distribution of Electric Power

In this chapter, we introduce circuits that are designed to handle large blocks of electric power. These are the circuits used to transport electric power from generating plants to both industrial and residential customers. We introduced a very basic residential customer circuit in the design perspective for Chapter 1. Now we introduce the type of circuit that delivers electric power to an entire residential subdivision.

One of the requirements imposed on electric utilities is to maintain the rms voltage level at the customer's premises. Whether lightly loaded, as at 3:00 a.m., or heavily loaded, as at midafternoon on a hot, humid day, the utility must supply the same rms voltage. To satisfy this requirement, utility systems place capacitors at strategic locations in the distribution network. The capacitors supply magnetizing vars close to the loads

requiring them, adjusting the power factor of the load. We illustrate this concept after we have analyzed balanced three-phase circuits.

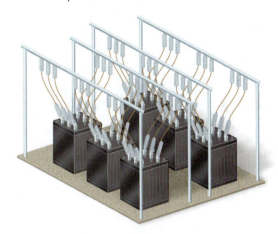

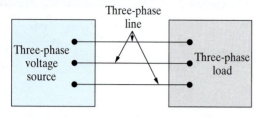

Figure 11.1 ▲ A basic three-phase circuit.

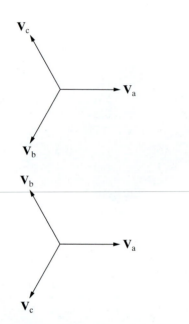

Figure 11.2 ▲ Phasor diagrams of a balanced set of three-phase voltages. (a) The abc (positive) sequence. (b) The acb (negative) sequence.

11.1 Balanced Three-Phase Voltages

A set of balanced three-phase voltages consists of three sinusoidal voltages that have identical amplitudes and frequencies but are out of phase with each other by exactly 120°. We refer to the three phases as a, b, and c, and use the a-phase as the reference phase. The three voltages are referred to as the **a-phase voltage**, the **b-phase voltage**, and the **c-phase voltage**.

Only two possible phase relationships can exist between the a-phase voltage and the b- and c-phase voltages. One possibility is for the b-phase voltage to lag the a-phase voltage by 120°, in which case the c-phase voltage must lead the a-phase voltage by 120°. This phase relationship is known as the **abc (or positive) phase sequence**. Using phasor notation, we see that the abc phase sequence is

$$\mathbf{V}_a = V_m \, \underline{/0°},$$
$$\mathbf{V}_b = V_m \, \underline{/-120°}, \tag{11.1}$$
$$\mathbf{V}_c = V_m \, \underline{/+120°}.$$

The only other possibility is for the b-phase voltage to lead the a-phase voltage by 120°, in which case the c-phase voltage must lag the a-phase voltage by 120°. This phase relationship is known as the **acb (or negative) phase sequence**. In phasor notation, the acb phase sequence is

$$\mathbf{V}_a = V_m \, \underline{/0°},$$
$$\mathbf{V}_b = V_m \, \underline{/+120°}, \tag{11.2}$$
$$\mathbf{V}_c = V_m \, \underline{/-120°}.$$

Figure 11.2 shows the phasor diagrams of the voltages in Eqs. 11.1 and 11.2. The phase sequence is the clockwise order of the subscripts around the diagram, starting from $\mathbf{V}_a$. You can use the phasor diagrams to show that the sum of the three phasor voltages in a balanced set is zero. This important characteristic can also be derived from Eq. 11.1 or 11.2 to give

$$\mathbf{V}_a + \mathbf{V}_b + \mathbf{V}_c = 0.$$

Because the sum of the phasor voltages is zero, the sum of the instantaneous voltages also is zero; that is,

$$v_a + v_b + v_c = 0.$$

Now that we know the nature of a balanced set of three-phase voltages, we can state the first of the analytical shortcuts alluded to in the introduction to this chapter: if we know the phase sequence and one voltage in the set, we know the entire set. Thus, for a balanced three-phase system, once we determine the voltages (or currents) in one phase, we can use the 120° phase angle difference and the phase sequence to find the voltages or currents in the remaining two phases.

SELF-CHECK: Assess your understanding of three-phase voltages by trying Chapter Problems 11.1 and 11.2.

11.2 Three-Phase Voltage Sources

A three-phase voltage source is a generator with three separate windings distributed around the periphery of the stator. Each winding comprises one phase of the generator. The rotor of the generator is an electromagnet driven at synchronous speed by a prime mover, such as a steam or gas turbine. Rotation of the electromagnet induces a sinusoidal voltage in each winding. The phase windings are designed so that the sinusoidal voltages induced in them are equal in amplitude and out of phase with each other by 120°. The phase windings are stationary with respect to the rotating electromagnet, so the frequency of the voltage induced in each winding is the same. Figure 11.3 shows a sketch of a two-pole three-phase source.

There are two ways of interconnecting the separate phase windings to form a three-phase source: as a wye (Y) or as a delta (Δ). Figure 11.4 shows both, with ideal voltage sources modeling the phase windings of the three-phase generator. The common terminal in the Y-connected source, labeled n in Fig. 11.4(a), is called the **neutral terminal** of the source, which may or may not be available for external connections.

When constructing a circuit model of a three-phase generator, we need to consider the impedance of each phase winding. Sometimes it is so small (compared with other impedances in the circuit) that we don't include it in the generator model; the model then consists solely of ideal voltage sources, as in Fig. 11.4. If the impedance of each phase winding is not negligible, we connect an inductive winding impedance in series with an ideal sinusoidal voltage source in each phase. Because the winding construction in each phase is the same, we make the winding impedances in each phase identical. Figure 11.5 shows two models of a three-phase generator, one using a Y connection and the other using a Δ connection. In both models, R_w is the winding resistance, and X_w is the inductive reactance of the winding.

Because both three-phase sources and three-phase loads can be either Y-connected or Δ-connected, the basic circuit in Fig. 11.1 represents four different configurations:

Source	Load
Y	Y
Y	Δ
Δ	Y
Δ	Δ

We begin by analyzing the Y-Y circuit. The remaining three configurations can be reduced to a Y-Y equivalent circuit, so analysis of the Y-Y circuit is the key to solving all balanced three-phase arrangements. We then illustrate how to transform a Y-Δ circuit to a Y-Y circuit and leave the analysis of the Δ-Y and Δ-Δ circuits to you in the Problems.

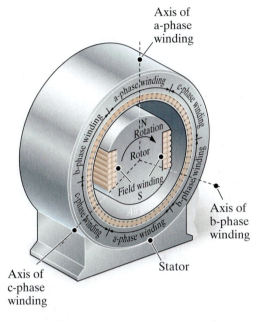

Figure 11.3 ▲ A sketch of a three-phase voltage source.

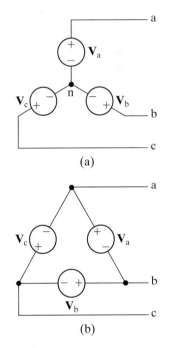

Figure 11.4 ▲ The two basic connections of an ideal three-phase source. (a) A Y-connected source. (b) A Δ-connected source.

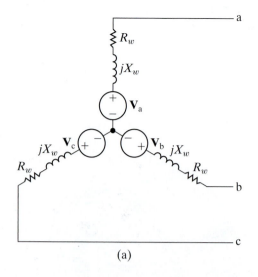

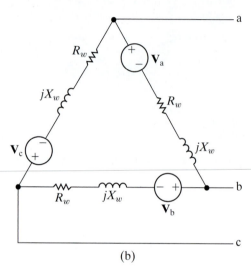

Figure 11.5 ▲ A model of a three-phase source with winding impedance: (a) a Y-connected source; and (b) a Δ-connected source.

11.3 Analysis of the Wye-Wye Circuit

Figure 11.6 illustrates a general Y-Y circuit, in which we included a fourth conductor that connects the source neutral to the load neutral. A fourth conductor is possible only in the Y-Y arrangement. For convenience, we transformed the Y connections into "tipped-over tees." In Fig. 11.6, Z_{ga}, Z_{gb}, and Z_{gc} represent the internal impedance associated with each phase winding of the voltage generator; Z_{la}, Z_{lb} and Z_{lc} represent the impedance of the lines connecting a phase of the source to a phase of the load; Z_0 is the impedance of the neutral conductor connecting the source neutral to the load neutral; and Z_A, Z_B, and Z_C represent the impedance of each phase of the load.

We can describe this circuit with a single KCL equation. Using the source neutral as the reference node and letting $\mathbf{V}_N$ denote the node voltage between the nodes N and n, we get

$$\frac{\mathbf{V}_N}{Z_0} + \frac{\mathbf{V}_N - \mathbf{V}_{a'n}}{Z_A + Z_{la} + Z_{ga}} + \frac{\mathbf{V}_N - \mathbf{V}_{b'n}}{Z_B + Z_{lb} + Z_{gb}} + \frac{\mathbf{V}_N - \mathbf{V}_{c'n}}{Z_C + Z_{lc} + Z_{gc}} = 0.$$

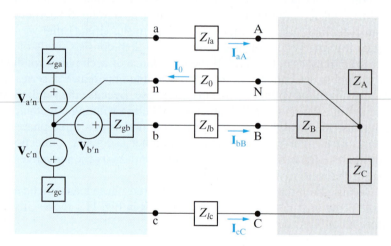

Figure 11.6 ▲ A three-phase Y-Y system.

What happens to the KCL equation if the three-phase circuit in Fig. 11.6 is balanced? To answer this question, we must formally define a balanced three-phase circuit. A three-phase circuit is balanced if it satisfies the following criteria:

CONDITIONS FOR A BALANCED THREE-PHASE CIRCUIT

1. The voltage sources form a balanced three-phase set.

2. The impedance of each phase of the voltage source is the same.

3. The impedance of each line is the same.

4. The impedance of each phase of the load is the same.

There is no restriction on the impedance of the neutral conductor; its value has no effect on whether the system is balanced.

The circuit in Fig. 11.6 is balanced if

1. $\mathbf{V}_{a'n}$, $\mathbf{V}_{b'n}$, and $\mathbf{V}_{c'n}$ are a set of balanced three-phase voltages.

2. $Z_{ga} = Z_{gb} = Z_{gc}$.

3. $Z_{la} = Z_{lb} = Z_{lc}$.

4. $Z_A = Z_B = Z_C$.

If the circuit in Fig. 11.6 is balanced, we can rewrite the KCL equation at $\mathbf{V}_N$ as

$$\mathbf{V}_N\left(\frac{1}{Z_0} + \frac{3}{Z_\phi}\right) = \frac{\mathbf{V}_{a'n} + \mathbf{V}_{b'n} + \mathbf{V}_{c'n}}{Z_\phi}, \tag{11.3}$$

where

$$Z_\phi = Z_A + Z_{la} + Z_{ga} = Z_B + Z_{lb} + Z_{gb} = Z_C + Z_{lc} + Z_{gc}.$$

The right-hand side of Eq. 11.3 is zero because by hypothesis the numerator is a set of balanced three-phase voltages and Z_ϕ is not zero. The only value of $\mathbf{V}_N$ that satisfies Eq. 11.3 is zero. Therefore, for a balanced three-phase circuit,

$$\mathbf{V}_N = 0. \tag{11.4}$$

When a three-phase circuit is balanced, $\mathbf{V}_N$ is zero, so the voltage between the source neutral, n, and the load neutral, N, is zero. Consequently, the current in the neutral conductor is zero, so we can either remove the neutral conductor from a balanced Y-Y configuration ($\mathbf{I}_0 = 0$) or replace it with a short circuit between the nodes n and N ($\mathbf{V}_N = 0$). Both equivalents are convenient to use when modeling balanced three-phase circuits.

If the three-phase circuit in Fig. 11.6 is balanced, the three line currents are

$$\mathbf{I}_{aA} = \frac{\mathbf{V}_{a'n} - \mathbf{V}_N}{Z_A + Z_{la} + Z_{ga}} = \frac{\mathbf{V}_{a'n}}{Z_\phi}, \tag{11.5}$$

$$\mathbf{I}_{bB} = \frac{\mathbf{V}_{b'n} - \mathbf{V}_N}{Z_B + Z_{lb} + Z_{gb}} = \frac{\mathbf{V}_{b'n}}{Z_\phi},$$

$$\mathbf{I}_{cC} = \frac{\mathbf{V}_{c'n} - \mathbf{V}_N}{Z_C + Z_{lc} + Z_{gc}} = \frac{\mathbf{V}_{c'n}}{Z_\phi}.$$

From these equations, we see that the three line currents form a balanced set; that is, the current in each line is equal in amplitude and frequency and is 120° out of phase with the other two line currents. Thus, if we calculate the current $\mathbf{I}_{aA}$ and we know the phase sequence, we have a shortcut for finding $\mathbf{I}_{bB}$ and $\mathbf{I}_{cC}$. This is the same shortcut used to find the b- and c-phase source voltages from the a-phase source voltage.

Terminology

We use Fig. 11.6 to define some important terms.

- **Line voltage** is the voltage across any pair of lines. In Fig. 11.6, the three line voltages at the load are $\mathbf{V}_{AB}$, $\mathbf{V}_{BC}$, and $\mathbf{V}_{CA}$.
- **Phase voltage** is the voltage across a single phase. In Fig. 11.6, the three phase voltages at the load are $\mathbf{V}_{AN}$, $\mathbf{V}_{BN}$, and $\mathbf{V}_{CN}$.
- **Line current** is the current in a single line. In Fig. 11.6, the three line currents are $\mathbf{I}_{aA}$, $\mathbf{I}_{bB}$, and $\mathbf{I}_{cC}$.
- **Phase current** is the current in a single phase. In Fig. 11.6, the three phase currents for the load are $\mathbf{I}_{AN}$, $\mathbf{I}_{BN}$, and $\mathbf{I}_{CN}$.

From Fig. 11.6 you can see that when the load is Y-connected; the line current and phase current in each phase are identical, but the line voltage and phase voltage in each phase are different.

The Greek letter phi (ϕ) is widely used in the literature to denote a per-phase quantity. Thus, $\mathbf{V}_\phi$, $\mathbf{I}_\phi$, Z_ϕ, P_ϕ, and Q_ϕ are interpreted as voltage-per-phase, current-per-phase, impedance-per-phase, real power-per-phase, and reactive power-per-phase, respectively.

Since three-phase systems are designed to handle large blocks of electric power, all voltage and current specifications are rms values. When voltage ratings are given, they refer specifically to the rating of the line voltage. Thus, when a three-phase transmission line is rated at 345 kV, the value of the line-to-line voltage is 345,000 V(rms). In this chapter, we express all voltages and currents as rms values.

Constructing a Single-Phase Equivalent Circuit

We can use Eq. 11.5 to construct an equivalent circuit for the a-phase of the balanced Y-Y circuit. From this equation, the line current in the a-phase is the voltage generated in the a-phase winding of the generator divided by the total impedance in the a-phase of the circuit. Thus, Eq. 11.5 describes the circuit shown in Fig. 11.7, in which the neutral conductor has been replaced by a short circuit. The circuit in Fig. 11.7 is the **single-phase equivalent circuit** for a balanced three-phase circuit. Once we solve this circuit, we can write down the voltages and currents in the other two phases, using the relationships among the phases. Thus, drawing a single-phase equivalent circuit is an important first step in analyzing a balanced three-phase circuit.

Note that the current in the neutral conductor in Fig. 11.7 is $\mathbf{I}_{aA}$. This is not the same as the current in the neutral conductor of the balanced three-phase circuit, which is

$$\mathbf{I}_0 = \mathbf{I}_{aA} + \mathbf{I}_{bB} + \mathbf{I}_{cC}.$$

Thus, the circuit shown in Fig. 11.7 gives the correct value of the line current but only the a-phase component of the neutral current. In a balanced three-phase circuit, the line currents form a balanced three-phase set, and $\mathbf{I}_0 = 0$.

Once we know the line current ($\mathbf{I}_{aA}$) in Fig. 11.7, we know the phase current ($\mathbf{I}_{AN}$) because they are equal. We can also use Ohm's law to calculate the phase voltage ($\mathbf{V}_{AN}$) from the phase current and

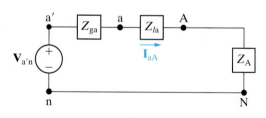

Figure 11.7 ▲ A single-phase equivalent circuit.

the load impedance. But how do we calculate the line voltage, $\mathbf{V}_{AB}$? The line voltage is not even shown in the single-phase equivalent circuit in Fig. 11.7. In a Y-connected load, the line voltages and the phase voltages are related. Using Fig. 11.8, we can describe those relationships using KVL:

$$\mathbf{V}_{AB} = \mathbf{V}_{AN} - \mathbf{V}_{BN},$$

$$\mathbf{V}_{BC} = \mathbf{V}_{BN} - \mathbf{V}_{CN},$$

$$\mathbf{V}_{CA} = \mathbf{V}_{CN} - \mathbf{V}_{AN}.$$

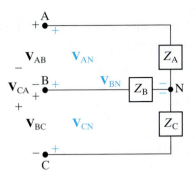

Figure 11.8 ▲ Line-to-line and line-to-neutral voltages.

To derive the relationship between the line voltages and the phase voltages, we assume a positive, or abc, sequence. Using the phase voltage in the a-phase as the reference,

$$\mathbf{V}_{AN} = V_\phi \,\underline{/0°},$$

$$\mathbf{V}_{BN} = V_\phi \,\underline{/-120°},$$

$$\mathbf{V}_{CN} = V_\phi \,\underline{/+120°},$$

where V_ϕ represents the magnitude of the phase voltage. Substituting the phase voltages into the equations for the line voltages yields

$$\mathbf{V}_{AB} = V_\phi \,\underline{/0°} - V_\phi \,\underline{/-120°} = \sqrt{3}V_\phi \,\underline{/30°}, \tag{11.6}$$

$$\mathbf{V}_{BC} = V_\phi \,\underline{/-120°} - V_\phi \,\underline{/120°} = \sqrt{3}V_\phi \,\underline{/-90°}, \tag{11.7}$$

$$\mathbf{V}_{CA} = V_\phi \,\underline{/120°} - V_\phi \,\underline{/0°} = \sqrt{3}V_\phi \,\underline{/150°}. \tag{11.8}$$

Equations 11.6–11.8 reveal that

1. The magnitude of the line voltage is $\sqrt{3}$ times the magnitude of the phase voltage.

2. The line voltages form a balanced three-phase set.

3. The set of line voltages leads the set of phase voltages by 30°.

We leave it to you to demonstrate that for a negative sequence, the only change is that the set of line voltages lags the set of phase voltages by 30°. The phasor diagrams shown in Fig. 11.9 summarize these observations. Here, again, is a shortcut in the analysis of a balanced system: If you know any phase voltage in the circuit, say $\mathbf{V}_{BN}$, you can determine the corresponding line voltage, which is $\mathbf{V}_{BC}$, and vice versa.

Example 11.1 shows how to use a single-phase equivalent circuit to solve a balanced three-phase Y-Y circuit.

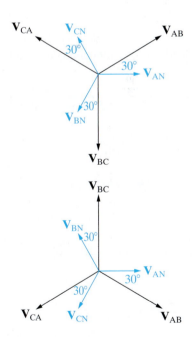

Figure 11.9 ▲ Phasor diagrams showing the relationship between line-to-line and line-to-neutral voltages in a balanced system. (a) The abc sequence. (b) The acb sequence.

EXAMPLE 11.1 | Analyzing a Wye-Wye Circuit

A balanced three-phase Y-connected generator with positive sequence has an impedance of $0.2 + j0.5 \ \Omega/\phi$ and an internal voltage of $120 \ \text{V}/\phi$. The generator feeds a balanced three-phase Y-connected load having an impedance of $39 + j28 \ \Omega/\phi$. The impedance of the line connecting the generator to the load is $0.8 + j1.5 \ \Omega/\phi$. The internal voltage of the generator in the a-phase is the reference phasor.

a) Construct the a-phase equivalent circuit of the system.

b) Calculate the three line currents $\mathbf{I}_{aA}$, $\mathbf{I}_{bB}$, and $\mathbf{I}_{cC}$.

c) Calculate the three phase voltages at the load, $\mathbf{V}_{AN}$, $\mathbf{V}_{BN}$, and $\mathbf{V}_{CN}$.

d) Calculate the line voltages $\mathbf{V}_{AB}$, $\mathbf{V}_{BC}$, and $\mathbf{V}_{CA}$ at the terminals of the load.

e) Calculate the phase voltages at the terminals of the generator, $\mathbf{V}_{an}$, $\mathbf{V}_{bn}$, and $\mathbf{V}_{cn}$.

f) Calculate the line voltages $\mathbf{V}_{ab}$, $\mathbf{V}_{bc}$, and $\mathbf{V}_{ca}$ at the terminals of the generator.

g) Repeat (a)–(f) for a negative phase sequence.

Solution

a) Figure 11.10 shows the single-phase equivalent circuit.

b) The a-phase line current is the voltage in the a-phase divided by the total impedance in the a-phase:

$$\mathbf{I}_{aA} = \frac{120 \ \underline{/0°}}{(0.2 + 0.8 + 39) + j(0.5 + 1.5 + 28)}$$

$$= \frac{120 \ \underline{/0°}}{40 + j30}$$

$$= 2.4 \ \underline{/-36.87°} \ \text{A}.$$

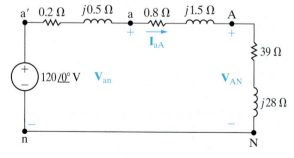

Figure 11.10 ▲ The single-phase equivalent circuit for Example 11.1.

For a positive phase sequence,

$$\mathbf{I}_{bB} = 2.4 \ \underline{/-156.87°} \ \text{A},$$

$$\mathbf{I}_{cC} = 2.4 \ \underline{/83.13°} \ \text{A}.$$

c) The line current and the phase current in the a-phase are equal. The phase voltage across the load in the a-phase is the product of the single-phase impedance of the load and the phase current in the load:

$$\mathbf{V}_{AN} = (39 + j28)(2.4 \ \underline{/-36.87°})$$

$$= 115.22 \ \underline{/-1.19°} \ \text{V}.$$

For a positive phase sequence,

$$\mathbf{V}_{BN} = 115.22 \ \underline{/-121.19°} \ \text{V},$$

$$\mathbf{V}_{CN} = 115.22 \ \underline{/118.81°} \ \text{V}.$$

d) For a positive phase sequence, the magnitude of the line voltages is $\sqrt{3}$ times the magnitude of the phase voltages, and the line voltages lead the phase voltages by 30°. Thus

$$\mathbf{V}_{AB} = (\sqrt{3} \ \underline{/30°})\mathbf{V}_{AN}$$

$$= 199.58 \ \underline{/28.81°} \ \text{V},$$

$$\mathbf{V}_{BC} = 199.58 \ \underline{/-91.19°} \ \text{V},$$

$$\mathbf{V}_{CA} = 199.58 \ \underline{/148.81°} \ \text{V}.$$

e) The phase voltage of the source in the a-phase is the voltage of the ideal source minus the voltage across the source impedance. Therefore,

$$\mathbf{V}_{an} = 120 - (0.2 + j0.5)(2.4 \ \underline{/-36.87°})$$

$$= 120 - 1.29 \ \underline{/31.33°}$$

$$= 118.90 - j0.67$$

$$= 118.90 \ \underline{/-0.32°} \ \text{V}.$$

For a positive phase sequence,

$$\mathbf{V}_{bn} = 118.90 \ \underline{/-120.32°} \ \text{V},$$

$$\mathbf{V}_{cn} = 118.90 \ \underline{/119.68°} \ \text{V}.$$

f) The line voltages at the source terminals are

$$\mathbf{V}_{ab} = (\sqrt{3}\,\underline{/30°})\mathbf{V}_{an}$$

$$= 205.94\,\underline{/29.68°}\text{ V,}$$

$$\mathbf{V}_{bc} = 205.94\,\underline{/-90.32°}\text{ V,}$$

$$\mathbf{V}_{ca} = 205.94\,\underline{/149.68°}\text{ V.}$$

g) Changing the phase sequence has no effect on the single-phase equivalent circuit. The three line currents are

$$\mathbf{I}_{aA} = 2.4\,\underline{/-36.87°}\text{ A,}$$

$$\mathbf{I}_{bB} = 2.4\,\underline{/83.13°}\text{ A,}$$

$$\mathbf{I}_{cC} = 2.4\,\underline{/-156.87°}\text{ A.}$$

The phase voltages at the load are

$$\mathbf{V}_{AN} = 115.22\,\underline{/-1.19°}\text{ V,}$$

$$\mathbf{V}_{BN} = 115.22\,\underline{/118.81°}\text{ V,}$$

$$\mathbf{V}_{CN} = 115.22\,\underline{/-121.19°}\text{ V.}$$

For a negative phase sequence, the line voltages lag the phase voltages by 30°:

$$\mathbf{V}_{AB} = (\sqrt{3}\,\underline{/-30°})\mathbf{V}_{AN}$$

$$= 199.58\,\underline{/-31.19°}\text{ V,}$$

$$\mathbf{V}_{BC} = 199.58\,\underline{/88.81°}\text{ V,}$$

$$\mathbf{V}_{CA} = 199.58\,\underline{/-151.19°}\text{ V.}$$

The phase voltages at the terminals of the generator are

$$\mathbf{V}_{an} = 118.90\,\underline{/-0.32°}\text{ V,}$$

$$\mathbf{V}_{bn} = 118.90\,\underline{/119.68°}\text{ V,}$$

$$\mathbf{V}_{cn} = 118.90\,\underline{/-120.32°}\text{ V.}$$

The line voltages at the terminals of the generator are

$$\mathbf{V}_{ab} = (\sqrt{3}\,\underline{/-30°})\mathbf{V}_{an}$$

$$= 205.94\,\underline{/-30.32°}\text{ V,}$$

$$\mathbf{V}_{bc} = 205.94\,\underline{/89.68°}\text{ V,}$$

$$\mathbf{V}_{ca} = 205.94\,\underline{/-150.32°}\text{ V.}$$

ASSESSMENT PROBLEMS

Objective 1—Know how to analyze a balanced, three-phase wye-wye circuit

11.1 The voltage from A to N in a balanced three-phase circuit is $240\,\underline{/-30°}$ V. If the phase sequence is positive, what is the value of $\mathbf{V}_{BC}$?

Answer: $415.69\,\underline{/-120°}$ V.

11.2 The c-phase voltage of a balanced three-phase Y-connected system is $450\,\underline{/-25°}$ V. If the phase sequence is negative, what is the value of $\mathbf{V}_{AB}$?

Answer: $779.42\,\underline{/65°}$ V.

11.3 The phase voltage at the terminals of a balanced three-phase Y-connected load is 2400 V. The load has an impedance of $16 + j12\ \Omega/\phi$ and is fed from a line having an impedance of $0.10 + j0.80\ \Omega/\phi$. The Y-connected source at the sending end of the line has a phase sequence of acb and an internal impedance of $0.02 + j0.16\ \Omega/\phi$. Use the a-phase voltage at the load as the reference and calculate (a) the line currents $\mathbf{I}_{aA}$, $\mathbf{I}_{bB}$, and $\mathbf{I}_{cC}$; (b) the line voltages at the source, $\mathbf{V}_{ab}$, $\mathbf{V}_{bc}$, and $\mathbf{V}_{ca}$; and (c) the internal phase-to-neutral voltages at the source, $\mathbf{V}_{a'n}$, $\mathbf{V}_{b'n}$, and $\mathbf{V}_{c'n}$.

Answer: (a) $\mathbf{I}_{aA} = 120\,\underline{/-36.87°}$ A,

$\qquad\qquad \mathbf{I}_{bB} = 120\,\underline{/83.13°}$ A, and

$\qquad\qquad \mathbf{I}_{cC} = 120\,\underline{/-156.87°}$ A;

$\qquad$ (b) $\mathbf{V}_{ab} = 4275.02\,\underline{/-28.38°}$ V,

$\qquad\qquad \mathbf{V}_{bc} = 4275.02\,\underline{/91.62°}$ V, and

$\qquad\qquad \mathbf{V}_{ca} = 4275.02\,\underline{/-148.38°}$ V;

$\qquad$ (c) $\mathbf{V}_{a'n} = 2482.05\,\underline{/1.93°}$ V,

$\qquad\qquad \mathbf{V}_{b'n} = 2482.05\,\underline{/121.93°}$ V, and

$\qquad\qquad \mathbf{V}_{c'n} = 2482.05\,\underline{/-118.07°}$ V.

SELF-CHECK: Also try Chapter Problems 11.9, 11.11, and 11.12.

11.4 Analysis of the Wye-Delta Circuit

A balanced, three-phase, Y-Δ circuit is shown in Fig. 11.11. From this circuit you can see that the line voltage and the phase voltage in each phase of the load are the same. For example, $\mathbf{V}_{AN} = \mathbf{V}_{AB}$. But the line current and the phase current in each phase of the load are not the same. This is an important difference when comparing the Y-Δ circuit with the Y-Y circuit, where the line and phase *currents* are the same in each phase of the load but the line and phase *voltages* are different.

When the load in a three-phase circuit is connected in a delta, it can be transformed into a wye by using the delta-to-wye transformation discussed in Section 9.6. When the load is balanced, the impedance of each leg of the wye is one-third the impedance of each leg of the delta, or

> **RELATIONSHIP BETWEEN DELTA-CONNECTED AND WYE-CONNECTED IMPEDANCES**
>
> $$Z_Y = \frac{Z_\Delta}{3},$$ (11.9)

which follows directly from Eqs. 9.21–9.23. After the Δ load has been replaced by its Y equivalent, the a-phase can be modeled by the single-phase equivalent circuit shown in Fig. 11.12.

We use this circuit to calculate the line currents, and we then use the line currents to find the currents in each phase of the original Δ-connected load. The relationship between the line currents and the phase currents in each phase of the delta can be derived using the circuit shown in Fig. 11.11. We assume a positive phase sequence and let I_ϕ represent the magnitude of the phase current. Then

$$\mathbf{I}_{AB} = I_\phi \underline{/0°},$$

$$\mathbf{I}_{BC} = I_\phi \underline{/-120°},$$

$$\mathbf{I}_{CA} = I_\phi \underline{/120°}.$$

In writing these equations, we arbitrarily selected $\mathbf{I}_{AB}$ as the reference phasor.

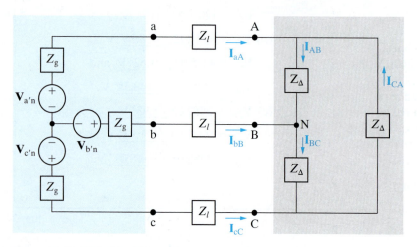

Figure 11.11 ▲ A balanced three-phase Y-Δ system.

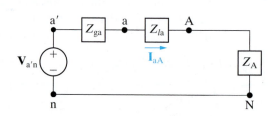

Figure 11.12 ▲ A single-phase equivalent circuit.

We can write the line currents in terms of the phase currents by applying KCL at the nodes labeled A, B, and C in Fig. 11.11:

$$\mathbf{I}_{aA} = \mathbf{I}_{AB} - \mathbf{I}_{CA} \tag{11.10}$$

$$= I_\phi\,\underline{/0°} - I_\phi\,\underline{/120°}$$

$$= \sqrt{3}I_\phi\,\underline{/-30°},$$

$$\mathbf{I}_{bB} = \mathbf{I}_{BC} - \mathbf{I}_{AB} \tag{11.11}$$

$$= I_\phi\,\underline{/-120°} - I_\phi\,\underline{/0°}$$

$$= \sqrt{3}I_\phi\,\underline{/-150°},$$

$$\mathbf{I}_{cC} = \mathbf{I}_{CA} - \mathbf{I}_{BC} \tag{11.12}$$

$$= I_\phi\,\underline{/120°} - I_\phi\,\underline{/-120°}$$

$$= \sqrt{3}I_\phi\,\underline{/90°}.$$

Comparing the line currents and the phase currents reveals that the magnitude of the line currents is $\sqrt{3}$ times the magnitude of the phase currents and that the set of line currents lags the set of phase currents by 30°.

We leave it to you to verify that, for a negative phase sequence, the line currents are still $\sqrt{3}$ times larger than the phase currents, but they lead the phase currents by 30°. Thus, we have a shortcut for calculating line currents from phase currents (or vice versa) for a balanced three-phase Δ-connected load. Figure 11.13 summarizes this shortcut graphically. Example 11.2 analyzes a balanced three-phase circuit having a Y-connected source and a Δ-connected load, using a single-phase equivalent circuit.

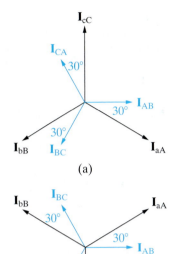

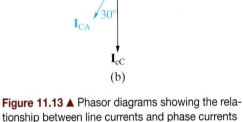

Figure 11.13 ▲ Phasor diagrams showing the relationship between line currents and phase currents in a Δ-connected load. (a) The positive sequence. (b) The negative sequence.

EXAMPLE 11.2   Analyzing a Wye-Delta Circuit

The Y-connected source in Example 11.1 feeds a Δ-connected load through a distribution line having an impedance of $0.3 + j0.9\ \Omega/\phi$. The load impedance is $118.5 + j85.8\ \Omega/\phi$. Use the internal voltage of the generator in the a-phase as the reference.

a) Construct a single-phase equivalent circuit of the three-phase system.

b) Calculate the line currents $\mathbf{I}_{aA}$, $\mathbf{I}_{bB}$, and $\mathbf{I}_{cC}$.

c) Calculate the phase voltages at the load terminals.

d) Calculate the phase currents of the load.

e) Calculate the line voltages at the source terminals.

Solution

a) Figure 11.14 shows the single-phase equivalent circuit. The load impedance of the Y equivalent is

$$\frac{118.5 + j85.8}{3} = 39.5 + j28.6\ \Omega/\phi.$$

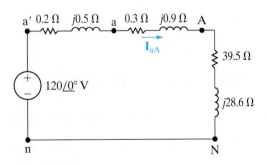

Figure 11.14 ▲ The single-phase equivalent circuit for Example 11.2.

b) The a-phase line current is the source voltage in the a-phase divided by the sum of the impedances in the a-phase:

$$\mathbf{I}_{aA} = \frac{120\,\underline{/0°}}{(0.2 + 0.3 + 39.5) + j(0.5 + 0.9 + 28.6)}$$

$$= \frac{120\,\underline{/0°}}{40 + j30} = 2.4\,\underline{/-36.87°}\ \text{A}.$$

Hence

$$\mathbf{I}_{bB} = 2.4 \; \underline{/-156.87°} \; \text{A},$$

$$\mathbf{I}_{cC} = 2.4 \; \underline{/83.13°} \; \text{A}.$$

c) Because the load is Δ connected, the phase voltages are the same as the line voltages. To calculate the line voltages, we first calculate $\mathbf{V}_{AN}$ using the single-phase equivalent circuit in Fig. 11.14:

$$\mathbf{V}_{AN} = (39.5 + j28.6)(2.4\underline{/-36.87°})$$

$$= 117.04 \; \underline{/-0.96°} \; \text{V}.$$

Because the phase sequence is positive, the line voltage $\mathbf{V}_{AB}$ is

$$\mathbf{V}_{AB} = (\sqrt{3} \; \underline{/30°}) \; \mathbf{V}_{AN}$$

$$= 202.72 \; \underline{/29.04°} \; \text{V}.$$

Therefore

$$\mathbf{V}_{BC} = 202.72 \; \underline{/-90.96°} \; \text{V},$$

$$\mathbf{V}_{CA} = 202.72 \; \underline{/149.04°} \; \text{V}.$$

d) The phase currents of the load can be calculated directly from the line currents:

$$\mathbf{I}_{AB} = \left(\frac{1}{\sqrt{3}} \; \underline{/30°} \right) \mathbf{I}_{aA}$$

$$= 1.39 \; \underline{/-6.87°} \; \text{A}.$$

Once we know $\mathbf{I}_{AB}$, we also know the other load phase currents:

$$\mathbf{I}_{BC} = 1.39 \; \underline{/-126.87°} \; \text{A},$$

$$\mathbf{I}_{CA} = 1.39 \; \underline{/113.13°} \; \text{A}.$$

Note that we can check the calculation of $\mathbf{I}_{AB}$ by using the previously calculated $\mathbf{V}_{AB}$ and the impedance of the Δ-connected load; that is,

$$\mathbf{I}_{AB} = \frac{\mathbf{V}_{AB}}{Z_\phi} = \frac{202.72\underline{/29.04°}}{118.5 + j85.8}$$

$$= 1.39 \; \underline{/-6.87°} \; \text{A}.$$

e) To calculate the line voltage at the terminals of the source, we first calculate $\mathbf{V}_{an}$. Figure 11.14 shows that $\mathbf{V}_{an}$ is the voltage drop across the line impedance plus the load impedance, so

$$\mathbf{V}_{an} = (39.8 + j29.5)(2.4 \; \underline{/-36.87°})$$

$$= 118.90 \; \underline{/-0.32°} \; \text{V}.$$

The line voltage $\mathbf{V}_{ab}$ is

$$\mathbf{V}_{ab} = (\sqrt{3} \; \underline{/30°})\mathbf{V}_{an}$$

$$= 205.94 \; \underline{/29.68°} \; \text{V}.$$

Therefore

$$\mathbf{V}_{bc} = 205.94 \; \underline{/-90.32°} \; \text{V},$$

$$\mathbf{V}_{ca} = 205.94 \; \underline{/149.68°} \; \text{V}.$$

ASSESSMENT PROBLEMS

Objective 2—Know how to analyze a balanced, three-phase wye-delta connected circuit

11.4 The current $\mathbf{I}_{CA}$ in a balanced three-phase Δ-connected load is $8 \; \underline{/-15°}$ A. If the phase sequence is positive, what is the value of $\mathbf{I}_{cC}$?

Answer: $13.86 \; \underline{/-45°}$ A.

11.5 A balanced three-phase Δ-connected load is fed from a balanced three-phase circuit. The reference for the b-phase line current is toward the load. The value of the current in the b-phase is $12 \; \underline{/65°}$ A. If the phase sequence is negative, what is the value of $\mathbf{I}_{AB}$?

Answer: $6.93 \; \underline{/-85°}$ A.

11.6 The line voltage $\mathbf{V}_{AB}$ at the terminals of a balanced three-phase Δ-connected load is $4160 \; \underline{/0°}$ V. The line current $\mathbf{I}_{aA}$ is $69.28 \; \underline{/-10°}$ A.
a) Calculate the per-phase impedance of the load if the phase sequence is positive.
b) Repeat (a) for a negative phase sequence.

Answer: (a) $104 \; \underline{/-20°}$ Ω;

(b) $104 \; \underline{/40°}$ Ω.

11.7 The line voltage at the terminals of a balanced Δ-connected load is 110 V. Each phase of the load consists of a 3.667 Ω resistor in parallel with a 2.75 Ω inductive impedance. What is the magnitude of the current in the line feeding the load?

Answer: 86.60 A.

SELF-CHECK: Also try Chapter Problems 11.14–11.16.

11.5 Power Calculations in Balanced Three-Phase Circuits

So far, we have analyzed balanced three-phase circuits using a single-phase equivalent circuit to determine line currents, phase currents, line voltages, and phase voltages. We now calculate power for balanced three-phase circuits. We begin by considering the average power delivered to a balanced Y-connected load.

Average Power in a Balanced Wye Load

Figure 11.15 shows a Y-connected load, along with its pertinent currents and voltages. We calculate the average power associated with any one phase by using the techniques introduced in Chapter 10. Using Eq. 10.9, we express the average power associated with the a-phase as

$$P_A = |\mathbf{V}_{AN}||\mathbf{I}_{aA}| \cos(\theta_{vA} - \theta_{iA}), \tag{11.13}$$

where θ_{vA} and θ_{iA} denote the phase angles of $\mathbf{V}_{AN}$ and $\mathbf{I}_{aA}$, respectively. Using the notation introduced in Eq. 11.13, we can find the power associated with the b- and c-phases:

$$P_B = |\mathbf{V}_{BN}||\mathbf{I}_{bB}| \cos(\theta_{vB} - \theta_{iB}), \tag{11.14}$$

$$P_C = |\mathbf{V}_{CN}||\mathbf{I}_{cC}| \cos(\theta_{vC} - \theta_{iC}). \tag{11.15}$$

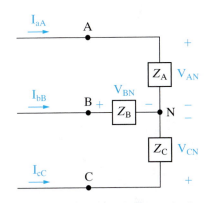

Figure 11.15 ▲ A balanced Y load used to introduce average power calculations in three-phase circuits.

In Eqs. 11.13–11.15, all phasor currents and voltages use the rms value of the sinusoidal function they represent.

In a balanced three-phase system, the magnitude of each line-to-neutral voltage is the same, as is the magnitude of each phase current. The argument of the cosine functions is also the same for all three phases. We introduce the following notation to take advantage of these observations:

$$V_\phi = |\mathbf{V}_{AN}| = |\mathbf{V}_{BN}| = |\mathbf{V}_{CN}|,$$

$$I_\phi = |\mathbf{I}_{aA}| = |\mathbf{I}_{bB}| = |\mathbf{I}_{cC}|,$$

and

$$\theta_\phi = \theta_{vA} - \theta_{iA} = \theta_{vB} - \theta_{iB} = \theta_{vC} - \theta_{iC}.$$

Moreover, for a balanced system, the power delivered to each phase of the load is the same, so

$$P_A = P_B = P_C = P_\phi = V_\phi I_\phi \cos\theta_\phi,$$

where P_ϕ represents the average power per phase. The total average power delivered to the balanced Y-connected load is simply three times the power per phase, or

$$P_T = 3P_\phi = 3V_\phi I_\phi \cos\theta_\phi. \tag{11.16}$$

Expressing the total power in terms of the rms magnitudes of the line voltage and current is also desirable. If we let V_L and I_L represent the rms magnitudes of the line voltage and current, respectively, we can modify Eq. 11.16 as follows:

TOTAL REAL POWER IN A BALANCED THREE-PHASE LOAD

$$P_T = 3\left(\frac{V_L}{\sqrt{3}}\right)I_L \cos\theta_\phi = \sqrt{3}V_L I_L \cos\theta_\phi. \quad (11.17)$$

In deriving Eq. 11.17, we recognized that, for a balanced Y-connected load, the magnitude of the phase voltage is the magnitude of the line voltage divided by $\sqrt{3}$, and that the magnitude of the line current is equal to the magnitude of the phase current. When using Eq. 11.17 to calculate the total power delivered to the load, remember that θ_ϕ is the phase angle between the phase voltage and current.

Reactive and Complex Power in a Balanced Wye Load

We can also calculate the reactive power and complex power associated with any phase of a Y-connected load using the techniques introduced in Chapter 10. For a balanced load, the expressions for the reactive power are

$$Q_\phi = V_\phi I_\phi \sin\theta_\phi,$$

TOTAL REACTIVE POWER IN A BALANCED THREE-PHASE LOAD

$$Q_T = 3Q_\phi = \sqrt{3}V_L I_L \sin\theta_\phi. \quad (11.18)$$

Use Eq. 10.13 to express the complex power of any phase. For a balanced Y-connected load,

$$S_\phi = \mathbf{V}_{AN}\mathbf{I}_{aA}^* = \mathbf{V}_{BN}\mathbf{I}_{bB}^* = \mathbf{V}_{CN}\mathbf{I}_{cC}^* = \mathbf{V}_\phi\mathbf{I}_\phi^*,$$

where $\mathbf{V}_\phi$ and $\mathbf{I}_\phi$ represent a phase voltage and current for the same phase. Thus, in general,

$$S_\phi = P_\phi + jQ_\phi = \mathbf{V}_\phi\mathbf{I}_\phi^*,$$

TOTAL COMPLEX POWER IN A BALANCED THREE-PHASE LOAD

$$S_T = 3S_\phi = \sqrt{3}V_L I_L \underline{/\theta_\phi}. \quad (11.19)$$

Power Calculations in a Balanced Delta Load

If the load is Δ-connected, the calculation of power—reactive or complex—is basically the same as that for a Y-connected load. Figure 11.16 shows a Δ-connected load, along with its pertinent currents and voltages. The power associated with each phase is

$$P_A = |\mathbf{V}_{AB}||\mathbf{I}_{AB}| \cos(\theta_{vAB} - \theta_{iAB}),$$

$$P_B = |\mathbf{V}_{BC}||\mathbf{I}_{BC}| \cos(\theta_{vBC} - \theta_{iBC}),$$

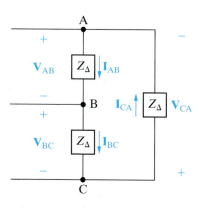

Figure 11.16 ▲ A Δ-connected load used to discuss power calculations.

$$P_C = |\mathbf{V}_{CA}||\mathbf{I}_{CA}| \cos(\theta_{vCA} - \theta_{iCA}).$$

For a balanced load,

$$|\mathbf{V}_{AB}| = |\mathbf{V}_{BC}| = |\mathbf{V}_{CA}| = V_\phi,$$

$$|\mathbf{I}_{AB}| = |\mathbf{I}_{BC}| = |\mathbf{I}_{CA}| = I_\phi,$$

$$\theta_{vAB} - \theta_{iAB} = \theta_{vBC} - \theta_{iBC} = \theta_{vCA} - \theta_{iCA} = \theta_\phi,$$

and

$$P_A = P_B = P_C = P_\phi = V_\phi I_\phi \cos \theta_\phi.$$

Thus, whether you have a balanced Y- or Δ-connected load, the average power per phase is equal to the product of the rms magnitude of the phase voltage, the rms magnitude of the phase current, and the cosine of the angle between the phase voltage and current.

The total average power delivered to a balanced Δ-connected load is

$$P_T = 3P_\phi = 3V_\phi I_\phi \cos \theta_\phi$$

$$= 3V_L\left(\frac{I_L}{\sqrt{3}}\right) \cos \theta_\phi$$

$$= \sqrt{3}V_L I_L \cos \theta_\phi.$$

Note that this equation is the same as Eq. 11.17. The expressions for reactive power and complex power also are the same as those developed for the Y load:

$$Q_\phi = V_\phi I_\phi \sin \theta_\phi;$$

$$Q_T = 3Q_\phi = 3V_\phi I_\phi \sin \theta_\phi;$$

$$S_\phi = P_\phi + jQ_\phi = \mathbf{V}_\phi \mathbf{I}_\phi^*;$$

$$S_T = 3S_\phi = \sqrt{3}V_L I_L \underline{/\theta_\phi}.$$

Instantaneous Power in Three-Phase Circuits

Although we are primarily interested in average, reactive, and complex power calculations, computing the total instantaneous power is also important. In a balanced three-phase circuit, this power has an interesting property: it is invariant with time! Thus, the torque developed at the shaft of a three-phase motor is constant, which in turn means less vibration in machinery powered by three-phase motors.

Let the instantaneous line-to-neutral voltage v_{AN} be the reference, and, as before, θ_ϕ is the phase angle $\theta_{vA} - \theta_{iA}$. Then, for a positive phase sequence, the instantaneous power in each phase is

$$p_A = v_{AN}i_{aA} = V_m I_m \cos \omega t \cos(\omega t - \theta_\phi),$$

$$p_B = v_{BN}i_{bB} = V_m I_m \cos(\omega t - 120°) \cos(\omega t - \theta_\phi - 120°),$$

$$p_C = v_{CN}i_{cC} = V_m I_m \cos(\omega t + 120°) \cos(\omega t - \theta_\phi + 120°),$$

where V_m and I_m represent the maximum amplitude of the phase voltage and line current, respectively. The total instantaneous power is the sum of the instantaneous phase powers, which reduces to $1.5V_m I_m \cos \theta_\phi$; that is,

$$p_T = p_A + p_B + p_C = 1.5V_m I_m \cos \theta_\phi.$$

Note that this result is consistent with Eq. 11.16 since $V_m = \sqrt{2}V_\phi$ and $I_m = \sqrt{2}I_\phi$ (see Problem 11.31).

Examples 11.3–11.5 illustrate power calculations in balanced three-phase circuits.

EXAMPLE 11.3	**Calculating Power in a Three-Phase Wye-Wye Circuit**

a) Calculate the average power per phase delivered to the Y-connected load of Example 11.1.

b) Calculate the total average power delivered to the load.

c) Calculate the total average power lost in the line.

d) Calculate the total average power lost in the generator.

e) Calculate the total number of magnetizing vars absorbed by the load.

f) Calculate the total complex power delivered by the source.

Solution

a) From Example 11.1, $V_\phi = 115.22$ V, $I_\phi = 2.4$ A, and $\theta_\phi = -1.19 - (-36.87) = 35.68°$. Therefore

$$P_\phi = (115.22)(2.4) \cos 35.68°$$

$$= 224.64 \text{ W}.$$

The power per phase may also be calculated from $I_\phi^2 R_\phi$, or

$$P_\phi = (2.4)^2(39) = 224.64 \text{ W}.$$

b) The total average power delivered to the load is $P_T = 3P_\phi = 673.92$ W. We calculated the line voltage in Example 11.1, so we can also use Eq. 11.17:

$$P_T = \sqrt{3}(199.58)(2.4) \cos 35.68°$$

$$= 673.92 \text{ W}.$$

c) The total power lost in the line is

$$P_{\text{line}} = 3(2.4)^2(0.8) = 13.824 \text{ W}.$$

d) The total internal power lost in the generator is

$$P_{\text{gen}} = 3(2.4)^2(0.2) = 3.456 \text{ W}.$$

e) The total number of magnetizing vars absorbed by the load is

$$Q_T = \sqrt{3}(199.58)(2.4) \sin 35.68°$$

$$= 483.84 \text{ VAR}.$$

f) The total complex power associated with the source is

$$S_T = 3S_\phi = -3(120)(2.4) \underline{/36.87°}$$

$$= -691.20 - j518.40 \text{ VA}.$$

The minus sign indicates that the average power and magnetizing reactive power are being delivered to the circuit. We check this result by calculating the total average and reactive power absorbed by the circuit:

$$P = 673.92 + 13.824 + 3.456$$

$$= 691.20 \text{ W (check)},$$

$$Q = 483.84 + 3(2.4)^2(1.5) + 3(2.4)^2(0.5)$$

$$= 483.84 + 25.92 + 8.64$$

$$= 518.40 \text{ VAR (check)}.$$

EXAMPLE 11.4	**Calculating Power in a Three-Phase Wye-Delta Circuit**

a) Calculate the total complex power delivered to the Δ-connected load of Example 11.2.

b) What percentage of the average power at the sending end of the line is delivered to the load?

Solution

a) Using the a-phase values from the solution of Example 11.2, we obtain

$$V_\phi = \mathbf{V}_{AB} = 202.72 \underline{/29.04°} \text{ V},$$

$$I_\phi = \mathbf{I}_{AB} = 1.39 \underline{/-6.87°} \text{ A}.$$

Using Eq. 11.19, we have

$$S_T = 3(202.72 \,\underline{/29.04°})(1.39 \,\underline{/6.87°})$$

$$= 682.56 + j494.21 \text{ VA}.$$

b) The total average power at the sending end of the distribution line equals the total average power delivered to the load plus the total average power lost in the line; therefore

$$P_{input} = 682.56 + 3(2.4)^2(0.3)$$

$$= 687.74 \text{ W}.$$

The fraction of the average power reaching the load is 682.56/687.74, or 99.25%. Nearly 100% of the average power at the input is delivered to the load because the resistance of the line is quite small compared to the load resistance.

EXAMPLE 11.5   Calculating Three-Phase Power with an Unspecified Load

A balanced three-phase load requires 480 kW at a lagging power factor of 0.8. The load is fed from a line having an impedance of $0.005 + j0.025 \ \Omega/\phi$. The line voltage at the terminals of the load is 600 V.

a) Construct a single-phase equivalent circuit of the system.

b) Calculate the magnitude of the line current.

c) Calculate the magnitude of the line voltage at the sending end of the line.

d) Calculate the power factor at the sending end of the line.

Solution

a) Figure 11.17 shows the single-phase equivalent circuit. We arbitrarily selected the line-to-neutral voltage at the load as the reference.

b) The line current $\mathbf{I}_{aA}^*$ appears in the equation for the complex power of the load:

$$\left(\frac{600}{\sqrt{3}}\right)\mathbf{I}_{aA}^* = (160 + j120)10^3 \text{ VA}.$$

Solving for $\mathbf{I}_{aA}^*$ we get

$$\mathbf{I}_{aA}^* = 577.35 \,\underline{/36.87°} \text{ A}.$$

Therefore,

$$\mathbf{I}_{aA} = 577.35 \,\underline{/-36.87°} \text{ A}.$$

The magnitude of the line current is the magnitude of $\mathbf{I}_{aA}$, so $I_L = 577.35$ A.

We obtain an alternative solution for I_L from the expression

$$P_T = \sqrt{3} V_L I_L \cos \theta_\phi$$

$$= \sqrt{3}(600) I_L (0.8)$$

$$= 480,000 \text{ W};$$

$$I_L = \frac{480,000}{\sqrt{3}(600)(0.8)}$$

$$= \frac{1000}{\sqrt{3}}$$

$$= 577.35 \text{ A}.$$

c) To calculate the magnitude of the line voltage at the sending end, we first calculate $\mathbf{V}_{an}$. From Fig. 11.17,

$$\mathbf{V}_{an} = \mathbf{V}_{AN} + Z_\ell \mathbf{I}_{aA}$$

$$= \frac{600}{\sqrt{3}} + (0.005 + j0.025)(577.35 \,\underline{/-36.87°})$$

$$= 357.51 \,\underline{/1.57°} \text{ V}.$$

Thus

$$V_L = \sqrt{3} |\mathbf{V}_{an}|$$

$$= 619.23 \text{ V}.$$

d) The power factor at the sending end of the line is the cosine of the phase angle between $\mathbf{V}_{an}$ and $\mathbf{I}_{aA}$:

$$pf = \cos[1.57° - (-36.87°)]$$

$$= \cos 38.44°$$

$$= 0.783 \text{ lagging}.$$

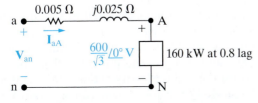

Figure 11.17 ▲ The single-phase equivalent circuit for Example 11.5.

An alternative method for calculating the power factor is to first calculate the complex power at the sending end of the line:

$$S_\phi = (160 + j120)10^3 + (577.35)^2(0.005 + j0.025)$$

$$= 161.67 + j128.33 \text{ kVA}$$

$$= 206.41 \underline{/38.44°} \text{ kVA}.$$

The power factor is

$$pf = \cos 38.44°$$

$$= 0.783 \text{ lagging}.$$

Finally, if we calculate the total complex power at the sending end, after first calculating the magnitude of the line current, we can use this value to calculate V_L. That is,

$$\sqrt{3}V_L I_L = 3(206.41) \times 10^3,$$

$$V_L = \frac{3(206.41) \times 10^3}{\sqrt{3}(577.35)},$$

$$= 619.23 \text{ V}.$$

ASSESSMENT PROBLEMS

Objective 3—Be able to calculate power (average, reactive, and complex) in any three-phase circuit

11.8 The three-phase average power rating of the central processing unit (CPU) on a mainframe digital computer is 22,659 W. The three-phase line supplying the computer has a line voltage rating of 208 V. The line current is 73.8 A. The computer absorbs magnetizing VARs.
   a) Calculate the total magnetizing reactive power absorbed by the CPU.
   b) Calculate the power factor.

Answer: (a) 13,909.50 VAR;
   (b) 0.852 lagging.

11.9 The complex power associated with each phase of a balanced load is 144 + j192 kVA. The line voltage at the terminals of the load is 2450 V.
   a) What is the magnitude of the line current feeding the load?
   b) The load is delta connected, and the impedance of each phase consists of a resistance in parallel with a reactance. Calculate R and X.
   c) The load is wye connected, and the impedance of each phase consists of a resistance in series with a reactance. Calculate R and X.

Answer: (a) 169.67 A;
   (b) $R = 41.68 \ \Omega, X = 31.26 \ \Omega$;
   (c) $R = 5 \ \Omega, X = 6.67 \ \Omega$.

SELF-CHECK: Also try Chapter Problems 11.25 and 11.27.

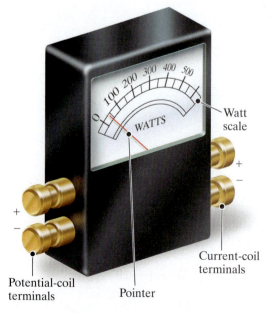

Figure 11.18 ▲ The key features of the electrodynamometer wattmeter.

11.6 Measuring Average Power in Three-Phase Circuits

The instrument used to measure power in three-phase circuits is the electrodynamometer wattmeter. It contains two coils. One coil, called the **current coil**, is stationary and is designed to carry a current proportional to the load current. The second coil, called the **potential coil**, is movable and carries a current proportional to the load voltage. The important features of the wattmeter are shown in Fig. 11.18.

The average deflection of the pointer attached to the movable coil is proportional to the product of the rms current in the current coil, the rms voltage impressed on the potential coil, and the cosine of the phase angle between the voltage and current. The pointer deflects in a direction that depends on the instantaneous polarity of the current-coil current and the potential-coil voltage. Therefore, each coil has one terminal with a polarity mark—usually a plus sign—but sometimes the double polarity mark ± is used. The wattmeter deflects upscale when (1) the polarity-marked terminal of the current coil is toward the source, and (2) the polarity-marked terminal of the potential coil is connected to the same line in which the current coil has been inserted.

The Two-Wattmeter Method

Consider a general network inside a box, with power supplied by n conducting lines. Such a system is shown in Fig. 11.19. If we wish to measure the total power at the terminals of the box, we need to know $n - 1$ currents and voltages. This follows because if we choose one terminal as a reference, there are only $n - 1$ independent voltages. Likewise, only $n - 1$ independent currents can exist in the n conductors entering the box. Thus, the total power is the sum of $n - 1$ product terms; that is, $p = v_1 i_1 + v_2 i_2 + \cdots + v_{n-1} i_{n-1}$.

Applying this general observation, we can see that for a three-conductor circuit, whether balanced or not, we need only two wattmeters to measure the total power. For a four-conductor circuit, we need three wattmeters if the three-phase circuit is unbalanced, but only two wattmeters if it is balanced, because in the latter case there is no current in the neutral line. Thus, only two wattmeters are needed to measure the total average power in any balanced three-phase system.

The two-wattmeter method determines the magnitude and algebraic sign of the average power indicated by each wattmeter. We can describe the basic problem using the circuit shown in Fig. 11.20, where the two wattmeters are indicated by the shaded boxes and labeled W_1 and W_2. The coil notations cc and pc stand for current coil and potential coil, respectively. The current coils of the wattmeters are inserted in lines aA and cC, making line bB the reference line for the two potential coils. The load is Y-connected, and its per-phase impedance is $Z_\phi = |Z| \underline{/\theta}$. This is a general representation, as any Δ-connected load can be represented by its Y equivalent; furthermore, for the balanced case, the impedance angle θ is unaffected by the Δ-to-Y transformation.

We now develop general equations for the readings of the two wattmeters, making the following assumptions.

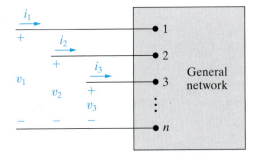

Figure 11.19 ▲ A general circuit whose power is supplied by n conductors.

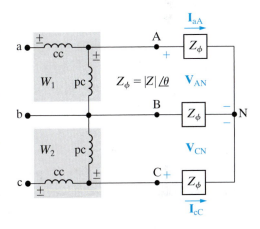

Figure 11.20 ▲ A circuit used to analyze the two-wattmeter method of measuring average power delivered to a balanced load.

- The current drawn by the potential coil of the wattmeter is negligible compared with the line current measured by the current coil.
- The loads can be modeled by passive circuit elements, so the phase angle of the load impedance (θ in Fig. 11.20) lies between $-90°$ (pure capacitance) and $+90°$ (pure inductance).
- The phase sequence is positive.

From our introductory discussion of the average deflection of the wattmeter and the placement of wattmeter 1 in Fig. 11.20, we note that the wattmeter reading, W_1, is

$$W_1 = |\mathbf{V}_{AB}||\mathbf{I}_{aA}| \cos \theta_1 \tag{11.20}$$

$$= V_L I_L \cos \theta_1.$$

It follows that

$$W_2 = |\mathbf{V}_{CB}||\mathbf{I}_{cC}| \cos \theta_2 \tag{11.21}$$

$$= V_L I_L \cos \theta_2.$$

In Eq. 11.20, θ_1 is the phase angle between $\mathbf{V}_{AB}$ and $\mathbf{I}_{aA}$, and in Eq. 11.21, θ_2 is the phase angle between $\mathbf{V}_{CB}$ and $\mathbf{I}_{cC}$.

To calculate W_1 and W_2, we express θ_1 and θ_2 in terms of the impedance angle θ, which is also the same as the phase angle between the phase voltage and current. For a positive phase sequence,

$$\theta_1 = \theta + 30° = \theta_\phi + 30°, \tag{11.22}$$

$$\theta_2 = \theta - 30° = \theta_\phi - 30°. \tag{11.23}$$

The derivation of Eqs. 11.22 and 11.23 is left as an exercise (see Problem 11.46). When we substitute Eqs. 11.22 and 11.23 into Eqs. 11.20 and 11.21, respectively, we get

$$W_1 = V_L I_L \cos(\theta_\phi + 30°),$$

$$W_2 = V_L I_L \cos(\theta_\phi - 30°).$$

To find the total power, we add W_1 and W_2; thus

$$P_T = W_1 + W_2 = 2V_L I_L \cos\theta_\phi \cos 30° \tag{11.24}$$

$$= \sqrt{3}V_L I_L \cos\theta_\phi,$$

which is the expression for the total average power in a three-phase circuit, given in Eq. 11.17. Therefore, we have confirmed that the sum of the two wattmeter readings is the total average power.

A closer look at the expressions for W_1 and W_2 reveals the following about the readings of the two wattmeters:

1. If the power factor is greater than 0.5, both wattmeters read positive.
2. If the power factor equals 0.5, one wattmeter reads zero.
3. If the power factor is less than 0.5, one wattmeter reads negative.
4. Reversing the phase sequence will interchange the readings on the two wattmeters.

Example 11.6 and Problems 11.41–11.52 illustrate these observations.

EXAMPLE 11.6 | Computing Wattmeter Readings in Three-Phase Circuits

Calculate the reading of each wattmeter in the circuit in Fig. 11.20 if the phase voltage at the load is 120 V and

a) $Z_\phi = 8 + j6 \ \Omega$;
b) $Z_\phi = 8 - j6 \ \Omega$;
c) $Z_\phi = 5 + j5 \sqrt{3} \ \Omega$; and
d) $Z_\phi = 10 \ /\!\!-75° \ \Omega$.

e) Verify for (a)–(d) that the sum of the wattmeter readings equals the total power delivered to the load.

Solution

a) $Z_\phi = 10 \ /\underline{36.87°} \ \Omega$, $V_L = 120 \sqrt{3}$ V,

$$I_L = 120/10 = 12 \text{ A}.$$

$$W_1 = (120\sqrt{3})(12) \cos(36.87° + 30°) = 979.75 \text{ W},$$

$$W_2 = (120\sqrt{3})(12) \cos(36.87° - 30°) = 2476.25 \text{ W}.$$

The power factor is $\cos 36.87° = 0.8$, so as expected, both wattmeter readings are positive.

b) $Z_\phi = 10 \underline{/-36.87°}$ Ω, $V_L = 120\sqrt{3}$ V,

   $I_L = 120/10 = 12$ A.

$W_1 = (120\sqrt{3})(12)\cos(-36.87° + 30°) = 2476.25$ W,

$W_2 = (120\sqrt{3})(12)\cos(-36.87° - 30°) = 979.75$ W.

The power factor is $\cos -36.87° = 0.8$, so as expected, both wattmeter readings are positive. But the readings on the wattmeters are interchanged when compared to the results of part (a) because the sign of the power factor angle changed, which has the same impact on the wattmeter readings as a change in phase sequence.

c) $Z_\phi = 5(1 + j\sqrt{3}) = 10 \underline{/60°}$ Ω, $V_L = 120\sqrt{3}$ V, and $I_L = 12$ A.

$W_1 = (120\sqrt{3})(12)\cos(60° + 30°) = 0$,

$W_2 = (120\sqrt{3})(12)\cos(60° - 30°) = 2160$ W.

The power factor is $\cos 60° = 0.5$, so as expected, one of the wattmeter readings is zero.

d) $Z_\phi = 10 \underline{/-75°}$ Ω, $V_L = 120\sqrt{3}$ V, $I_L = 12$ A.

$W_1 = (120\sqrt{3})(12)\cos(-75° + 30°) = 1763.63$ W,

$W_2 = (120\sqrt{3})(12)\cos(-75° - 30°) = -645.53$ W.

The power factor is $\cos -75° = 0.26$, so as expected, one wattmeter reading is negative.

e) For each load impedance value, the real power delivered to a single phase of the load is $I_L^2 R_L$,

where R_L is the resistive impedance of the load. Since the three-phase circuit is balanced, the total real power delivered to the three-phase load is $3I_L^2 R_L$.

For the impedance in part (a),

$$P_T = 3(12)^2(8) = 3456 \text{ W},$$

$$W_1 + W_2 = 979.75 + 2476.25 = 3456 \text{ W}.$$

For the impedance in part (b),

$$P_T = 3(12)^2(8) = 3456 \text{ W},$$

$$W_1 + W_2 = 2476.25 + 979.75 = 3456 \text{ W}.$$

For the impedance in part (c),

$$P_T = 3(12)^2(5) = 2160 \text{ W},$$

$$W_1 + W_2 = 0 + 2160 = 2160 \text{ W}.$$

For the impedance in part (d),

$$Z_\phi = 2.5882 - j9.6593 \text{ Ω so}$$

$$P_T = 3(12)^2(2.5882) = 1118.10 \text{ W},$$

$$W_1 + W_2 = 1763.63 - 645.53 = 1118.10 \text{ W}.$$

SELF-CHECK: Assess your understanding of the two-wattmeter method by trying Chapter Problems 11.41 and 11.45.

■ Practical Perspective

Transmission and Distribution of Electric Power

At the start of this chapter we noted that utilities must maintain the rms voltage level at their customer's premises. Although the acceptable deviation from a nominal level may vary among different utilities, we will assume that the tolerance is ±5%. Thus, a nominal rms voltage of 120 V could range from 114 V to 126 V. We also pointed out that strategically located capacitors can be used to support voltage levels.

The circuit shown in Fig. 11.21 represents a substation in a municipal system. We assume that the system is balanced, the line-to-line voltage at the substation is 13.8 kV, the impedance of the distribution line is $0.6 + j4.8$ Ω/ϕ, and the load at the substation at 3:00 p.m. on a hot, humid day in July is 3.6 MW and 3.6 magnetizing MVAR.

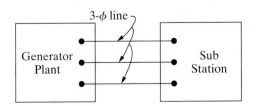

Figure 11.21 ▲ A substation connected to a power plant via a three-phase line.

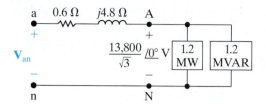

Figure 11.22 ▲ A single phase equivalent circuit for the system in Fig. 11.21.

Using the line-to-neutral voltage at the substation as a reference, the single-phase equivalent circuit for the system in Fig. 11.21 is shown in Fig. 11.22. The line current can be calculated from the expression for the complex power at the substation. Thus,

$$\frac{13,800}{\sqrt{3}} \mathbf{I}_{aA}^* = (1.2 + j1.2)10^6.$$

It follows that

$$\mathbf{I}_{aA}^* = 150.61 + j150.61 \text{ A}$$

or

$$\mathbf{I}_{aA} = 150.61 - j150.61 \text{ A}.$$

The line-to-neutral voltage at the generating plant is the voltage drop across the load, $\mathbf{V}_{AN}$, plus the voltage across the transmission line, so

$$\mathbf{V}_{an} = \frac{13,800}{\sqrt{3}} \underline{/0°} + (0.6 + j4.8)(150.61 - j150.61)$$

$$= 8780.74 + j632.58 = 8803.50 \underline{/4.12°} \text{ V}.$$

Therefore, the magnitude of the line voltage at the generating plant is

$$|\mathbf{V}_{ab}| = \sqrt{3}(8803.50) = 15,248.11 \text{ V}.$$

We are assuming the utility is required to keep the voltage level within ±5% of the nominal value. This means the magnitude of the line-to-line voltage at the power plant should not exceed 14.5 kV nor be less than 13.1 kV. The magnitude of the line voltage at the generating plant exceeds 14.5 kV, so could cause problems for customers.

To address this problem, connect a capacitor bank to the substation bus that supplies the magnetizing vars required by the load. Now the generator does not need to supply this reactive power, the load has a unity power factor, and the line current $\mathbf{I}_{aA}$ becomes

$$\mathbf{I}_{aA} = 150.61 + j0 \text{ A}.$$

Therefore, the voltage at the generating plant necessary to maintain a line-to-line voltage of 13,800 V at the substation is

$$\mathbf{V}_{an} = \frac{13,800}{\sqrt{3}} \underline{/0°} + (0.6 + j4.8)(150.61 + j0)$$

$$= 8057.80 + j722.94 = 8090.17 \underline{/5.13°} \text{ V}.$$

Hence

$$|\mathbf{V}_{ab}| = \sqrt{3}(8090.17) = 14,012.58 \text{ V}.$$

This voltage level falls within the allowable range of 13.1 kV to 14.5 kV.

SELF-CHECK: Assess your understanding of this Practical Perspective by trying Chapter Problems 11.53(a)–(b) and 11.54, 11.57, and 11.58.

■ Summary

- A set of balanced three-phase voltages consists of three sinusoidal voltages that have identical amplitudes and frequencies but are out of phase with each other by exactly 120°. For the abc (or positive) phase sequence,

$$\mathbf{V}_a = V_m \underline{/0°}, \quad \mathbf{V}_b = V_m \underline{/-120°}, \quad \mathbf{V}_c = V_m \underline{/120°}.$$

  For the acb (or negative) phase sequence,

$$\mathbf{V}_a = V_m \underline{/0°}, \quad \mathbf{V}_b = V_m \underline{/120°}, \quad \mathbf{V}_c = V_m \underline{/-120°}.$$

  (See page 414)

- Three-phase systems can be unbalanced or balanced. A three-phase system is balanced when the following conditions are satisfied:

  1. The voltage sources form a balanced three-phase set.

  2. The impedance of each phase of the voltage source is the same.

  3. The impedance of each line is the same.

  4. The impedance of each phase of the load is the same.

  (See page 417.)

- A **single-phase equivalent circuit** is used to calculate the line current and the phase voltage in one phase of the Y-Y structure. The a-phase is normally chosen for this purpose. If the structure is not Y-Y, any Δ connections should be transformed into equivalent Y connections before creating a single-phase equivalent circuit. (See page 418.)

- Once we know the line current and phase voltage in the a-phase equivalent circuit, we can use analytical short-cuts to find any current or voltage in a balanced three-phase circuit, based on the following facts:

  - The b- and c-phase currents and voltages are identical to the a-phase current and voltage except for a 120° shift in phase. The direction of the phase shift depends on the phase sequence.

- The set of line voltages is out of phase with the set of phase voltages by ±30°. The plus sign corresponds to the positive phase sequence, while the minus sign corresponds to the negative phase sequence.

- In a Y-Y circuit, the magnitude of a line voltage is $\sqrt{3}$ times the magnitude of a phase voltage.

- The set of line currents is out of phase with the set of phase currents in Δ-connected sources and loads by ∓30°. The minus sign corresponds to the positive phase sequence, while the plus sign corresponds to the negative phase sequence.

- The magnitude of a line current is $\sqrt{3}$ times the magnitude of a phase current in a Δ-connected source or load.

  (See pages 419 and 423.)

- The techniques for calculating per-phase average power, reactive power, and complex power are identical to those introduced in Chapter 10. (See page 425.)

- The total real, reactive, and complex power can be determined either by multiplying the corresponding per-phase quantity by 3 or by using the expressions based on line current and line voltage, as given by Eqs. 11.17, 11.18, and 11.19. (See page 426.)

- The total instantaneous power in a balanced three-phase circuit is constant and equals $1.5 V_m I_m \text{pf}$. (See page 427.)

- A wattmeter measures the average power delivered to a load by using a current coil connected in series with the load and a potential coil connected in parallel with the load. (See page 430.)

- The total average power in a balanced three-phase circuit can be measured by summing the readings of two wattmeters connected in two different phases of the circuit. (See page 432.)

Problems

All phasor voltages in the following Problems are stated in terms of the rms value.

Section 11.1

11.1 What is the phase sequence of each of the following sets of voltages?

a) $v_a = 180 \cos(\omega t + 27°)$ V,
$v_b = 180 \cos(\omega t + 147°)$ V,
$v_c = 180 \cos(\omega t - 93°)$ V.

b) $v_a = 4160 \cos(\omega t - 18°)$ V,
$v_b = 4160 \cos(\omega t - 138°)$ V,
$v_c = 4160 \cos(\omega t + 102°)$ V.

11.2 For each set of voltages, state whether or not the
voltages form a balanced three-phase set. If the set is balanced, state whether the phase sequence is positive or negative. If the set is not balanced, explain why.

a) $v_a = 180 \cos 377t$ V,
$v_b = 180 \cos(377t - 120°)$ V,
$v_c = 180 \cos(377t - 240°)$ V.

b) $v_a = 180 \sin 377t$ V,
$v_b = 180 \sin(377t + 120°)$ V,
$v_c = 180 \sin(377t - 120°)$ V.

c) $v_a = -400 \sin 377t$ V,
$v_b = 400 \sin(377t + 210°)$ V,
$v_c = 400 \cos(377t - 30°)$ V.

d) $v_a = 200 \cos(2000t + 30°)$ V,
$v_b = 201 \cos(2000t + 150°)$ V,
$v_c = 200 \cos(2000t + 270°)$ V.

e) $v_a = 208 \cos(630t + 42°)$ V,
$v_b = 208 \cos(630t - 78°)$ V,
$v_c = 208 \cos(630t - 201°)$ V.

f) $v_a = 240 \cos 377t$ V,
$v_b = 240 \cos(377t - 120°)$ V,
$v_c = 240 \cos(397t + 120°)$ V.

11.3 Verify that the sum of the three voltage phasors is zero for either Eq. 11.1 or Eq. 11.2.

Section 11.2

11.4 Refer to the circuit in Fig. 11.5(b). Assume that there are no external connections to the terminals a, b, c. Assume further that the three windings are from a three-phase generator whose voltages are those described in Problem 11.2(b). Determine the current circulating in the Δ-connected generator.

11.5 Repeat Problem 11.4 but assume that the three-phase voltages are those described in Problem 11.2(c).

Section 11.3

11.6 a) Is the circuit in Fig. P11.6 a balanced or unbalanced three-phase system? Explain.

b) Find $\mathbf{I}_0$.

Figure P11.6

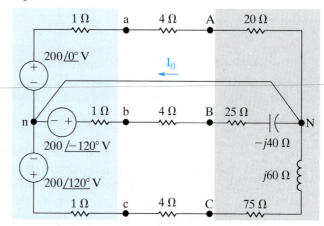

11.7 a) Find $\mathbf{I}_0$ in the circuit in Fig. P11.7.
b) Find $\mathbf{V}_{AN}$.

c) Find $\mathbf{V}_{AB}$.

d) Is the circuit a balanced or unbalanced three-phase system?

Figure P11.7

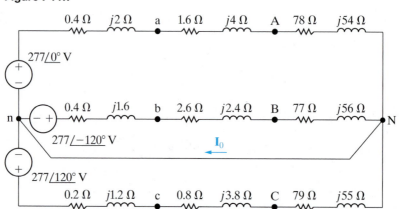

11.8 Find the rms value of $\mathbf{I}_0$ in the unbalanced three-phase circuit seen in Fig. P11.8.

Figure P11.8

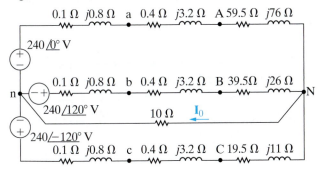

11.9 The time-domain expressions for three line-to-neutral voltages at the terminals of a Y-connected load are

$$v_{AN} = 169.71 \cos(\omega t + 26°) \text{ V},$$

$$v_{BN} = 169.71 \cos(\omega t - 94°) \text{ V},$$

$$v_{CN} = 169.71 \cos(\omega t + 146°) \text{ V}.$$

What are the time-domain expressions for the three line-to-line voltages v_{AB}, v_{BC}, and v_{CA}?

11.10 A balanced three-phase circuit has the following characteristics:

- Y-Y connected;
- The line voltage at the source is
  $\mathbf{V}_{ab} = 110\sqrt{3}\underline{/-60°}$ V;
- The phase sequence is positive;
- The line impedance is $3 + j2 \ \Omega/\phi$;
- The load impedance is $37 + j28 \ \Omega/\phi$.

a) Draw the single phase equivalent circuit for the a-phase.

b) Calculate the line current in the a-phase.

c) Calculate the line voltage at the load in the a-phase.

11.11 The magnitude of the line voltage at the terminals of a balanced Y-connected load is 660 V. The load impedance is $30.48 + j22.86 \ \Omega/\phi$. The load is fed from a line that has an impedance of $0.25 + j2 \ \Omega/\phi$.

a) What is the magnitude of the line current?

b) What is the magnitude of the line voltage at the source?

11.12 The magnitude of the phase voltage of an ideal balanced three-phase Y-connected source is 125 V. The source is connected to a balanced Y-connected load by a distribution line that has an impedance

of $0.1 + j0.8 \ \Omega/\phi$. The load impedance is $19.9 + j14.2 \ \Omega/\phi$. The phase sequence of the source is acb. Use the a-phase voltage of the source as the reference. Specify the magnitude and phase angle of the following quantities: (a) the three line currents, (b) the three line voltages at the source, (c) the three phase voltages at the load, and (d) the three line voltages at the load.

Section 11.4

11.13 A balanced Δ-connected load has an impedance of $60 + j45 \ \Omega/\phi$. The load is fed through a line having an impedance of $0.8 + j0.6 \ \Omega/\phi$. The phase voltage at the terminals of the load is 480 V. The phase sequence is positive. Use $\mathbf{V}_{AB}$ as the reference.

a) Calculate the three phase currents of the load.

b) Calculate the three line currents.

c) Calculate the three line voltages at the sending end of the line.

11.14 A balanced, three-phase circuit is characterized as follows:

- Y-Δ connected;
- Source voltage in the b-phase is $150\underline{/135°}$ V;
- Source phase sequence is acb;
- Line impedance is $2 + j3 \ \Omega/\phi$;
- Load impedance is $129 + j171 \ \Omega/\phi$.

a) Draw the single phase equivalent for the a-phase.

b) Calculate the a-phase line current.

c) Calculate the a-phase line voltage for the three-phase load.

11.15 An acb sequence balanced three-phase Y-connected source supplies power to a balanced, three-phase Δ-connected load with an impedance of $12 + j9 \ \Omega/\phi$. The source voltage in the b-phase is $240\underline{/-50°}$ V. The line impedance is $1 + j1 \ \Omega/\phi$. Draw the single phase equivalent circuit for the a-phase and use it to find the current in the a-phase of the load.

11.16 In a balanced three-phase system, the source is a balanced Y with an abc phase sequence and a line voltage $\mathbf{V}_{ab} = 208\underline{/50°}$ V. The load is a balanced Y in parallel with a balanced Δ. The phase impedance of the Y is $4 + j3 \ \Omega/\phi$ and the phase impedance of the Δ is $3 - j9 \ \Omega/\phi$. The line impedance is $1.4 + j0.8 \ \Omega/\phi$. Draw the single phase equivalent circuit and use it to calculate the line voltage at the load in the a-phase.

11.17 A balanced Y-connected load having an impedance of $72 + j21 \ \Omega/\phi$ is connected in parallel with a balanced Δ-connected load having an impedance of $150\underline{/0°} \ \Omega/\phi$. The parallel loads are fed from a line having an impedance of $j1 \ \Omega/\phi$. The magnitude of the line-to-line voltage of the Y-load is 7650 V.

a) Calculate the magnitude of the current in the line feeding the loads.

b) Calculate the magnitude of the phase current in the Y-connected load.

c) Calculate the magnitude of the phase current in the Δ-connected load.

d) Calculate the magnitude of the line voltage at the sending end of the line.

11.18 A three-phase Δ-connected generator has an internal impedance of $9 + j90 \ \text{m}\Omega/\phi$. When the load is removed from the generator, the magnitude of the terminal voltage is 13,800 V. The generator feeds a Δ-connected load through a transmission line with an impedance of $20 + j180 \ \text{m}\Omega/\phi$. The per-phase impedance of the load is $7.056 + j3.417 \ \Omega$.

a) Construct a single-phase equivalent circuit.

b) Calculate the magnitude of the line current.

c) Calculate the magnitude of the line voltage at the terminals of the load.

d) Calculate the magnitude of the line voltage at the terminals of the source.

e) Calculate the magnitude of the phase current in the load.

f) Calculate the magnitude of the phase current in the source.

11.19 The impedance Z in the balanced three-phase circuit in Fig. P11.19 is $160 + j120 \ \Omega$. Find

a) $\mathbf{I}_{AB}$, $\mathbf{I}_{BC}$, and $\mathbf{I}_{CA}$,

b) $\mathbf{I}_{aA}$, $\mathbf{I}_{bB}$, and $\mathbf{I}_{cC}$,

c) $\mathbf{I}_{ba}$, $\mathbf{I}_{cb}$, and $\mathbf{I}_{ac}$.

Figure P11.19

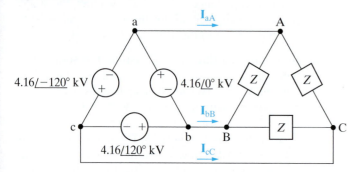

11.20 For the circuit shown in Fig. P11.20, find

a) the phase currents $\mathbf{I}_{AB}$, $\mathbf{I}_{BC}$, and $\mathbf{I}_{CA}$,

b) the line currents $\mathbf{I}_{aA}$, $\mathbf{I}_{bB}$, and $\mathbf{I}_{cC}$

when $Z_1 = 2.4 - j0.7 \ \Omega$, $Z_2 = 8 + j6 \ \Omega$, and $Z_3 = 20 + j0 \ \Omega$.

Figure P11.20

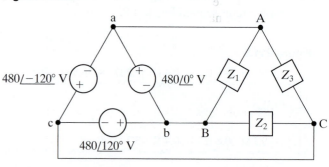

11.21 A balanced three-phase Δ-connected source is shown in Fig. P11.21.

a) Find the Y-connected equivalent circuit.

b) Show that the Y-connected equivalent circuit delivers the same open-circuit voltage as the original Δ-connected source.

c) Apply an external short circuit to the terminals A, B, and C. Use the Δ-connected source to find the three line currents $\mathbf{I}_{aA}$, $\mathbf{I}_{bB}$, and $\mathbf{I}_{cC}$.

d) Repeat (c) but use the Y-equivalent source to find the three line currents.

Figure P11.21

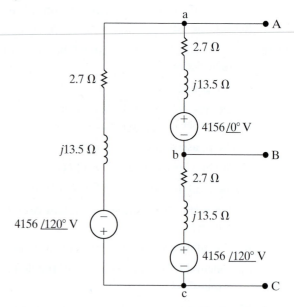

11.22 The Δ-connected source of Problem 11.21 is connected to a Y-connected load by means of a balanced three-phase distribution line. The load impedance is $1910 - j636 \ \Omega/\phi$. and the line impedance is $9.1 + j71.5 \ \Omega/\phi$.

a) Construct a single-phase equivalent circuit for the system.

b) Determine the magnitude of the line voltage at the terminals of the load.

c) Determine the magnitude of the phase current in the Δ-source.

d) Determine the magnitude of the line voltage at the terminals of the source.

Section 11.5

11.23 a) Find the rms magnitude and the phase angle of $\mathbf{I}_{CA}$ in the circuit shown in Fig. P11.23.

b) What percent of the average power delivered by the three-phase source is dissipated in the three-phase load?

Figure P11.23

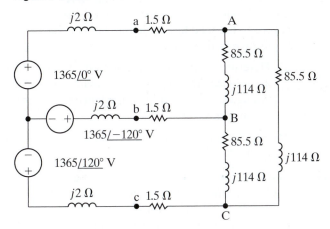

11.24 A balanced three-phase source is supplying 60 kVA at 0.6 lagging to two balanced Y-connected parallel loads. The distribution line connecting the source to the load has negligible impedance. Load 1 is purely resistive and absorbs 30 kW. Find the per-phase impedance of Load 2 if the line voltage is $120\sqrt{3}$ V and the impedance components are in series.

11.25 In a balanced three-phase system, the source has an abc sequence, is Y-connected, and $\mathbf{V}_{an} = 120\underline{/20°}$ V. The source feeds two loads, both of which are Y-connected. The impedance of load 1 is $8 + j6 \ \Omega/\phi$. The complex power for the a-phase of load 2 is $600\underline{/36°}$ VA. Find the total complex power supplied by the source.

11.26 The line-to-neutral voltage at the terminals of the balanced three-phase load in the circuit shown in Fig. P11.26 is 1200 V. At this voltage, the load is absorbing 500 kVA at 0.96 pf lag.

a) Using $\mathbf{V}_{AN}$ as the reference, find $\mathbf{I}_{na}$ in polar form.

b) Calculate the complex power associated with the ideal three-phase source.

c) Check that the total average power delivered equals the total average power absorbed.

d) Check that the total magnetizing reactive power delivered equals the total magnetizing reactive power absorbed.

Figure P11.26

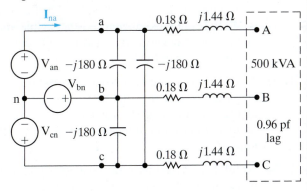

11.27 A three-phase positive sequence Y-connected source supplies 14 kVA with a power factor of 0.75 lagging to a parallel combination of a Y-connected load and a Δ-connected load. The Y-connected load uses 9 kVA at a power factor of 0.6 lagging and has an a-phase current of $10\underline{/-30°}$ A.

a) Find the complex power per phase of the Δ-connected load.

b) Find the magnitude of the line voltage.

11.28 A balanced three-phase distribution line has an impedance of $1 + j8 \ \Omega/\phi$. This line is used to supply three balanced three-phase loads that are connected in parallel. The three loads are $L_1 = 120$ kVA at 0.96 pf leading, $L_2 = 180$ kVA at 0.80 pf lagging, and $L_3 = 100.8$ kW and 15.6 kVAR (magnetizing). The magnitude of the line voltage at the terminals of the loads is $2400\sqrt{3}$ V.

a) What is the magnitude of the line voltage at the sending end of the line?

b) What is the percent efficiency of the distribution line with respect to average power?

11.29 The three tools described below are part of a university's machine shop. Each piece of equipment is a balanced three-phase load rated at 220 V. Calculate (a) the magnitude of the line current supplying these three tools and (b) the power factor of the combined load.

- Drill press: 10.2 kVA at 0.87 pf lag.

- Lathe: 4.2 kW at 0.91 pf lag.

- Band saw: line current 36.8 A, 7.25 kVAR.

11.30 Calculate the complex power in each phase of the unbalanced load in Problem 11.20.

11.31 Show that the total instantaneous power in a balanced three-phase circuit is constant and equal to $1.5V_m I_m \cos\theta_\phi$, where V_m and I_m represent the maximum amplitudes of the phase voltage and phase current, respectively.

11.32 The total apparent power supplied in a balanced, three-phase Y-Δ system is 4800 VA. The line voltage is 240 V. If the line impedance is negligible and the power factor angle of the load is −50°, determine the impedance of the load.

11.33 A balanced three-phase load absorbs 96 kVA at a lagging power factor of 0.8 when the line voltage at the terminals of the load is 480 V. Find four equivalent circuits that can be used to model this load.

11.34 At full load, a commercially available 100 hp, three-phase induction motor operates at an efficiency of 97% and a power factor of 0.88 lag. The motor is supplied from a three-phase outlet with a line-voltage rating of 208 V.

　　a) What is the magnitude of the line current drawn from the 208 V outlet? (1 hp = 746 W.)

　　b) Calculate the reactive power supplied to the motor.

11.35 A three-phase line has an impedance of $0.1 + j0.8 \; \Omega/\phi$. The line feeds two balanced three-phase loads connected in parallel. The first load is absorbing a total of 630 kW and absorbing 840 kVAR magnetizing vars. The second load is Y-connected and has an impedance of $15.36 - j4.48 \; \Omega/\phi$. The line-to-neutral voltage at the load end of the line is 4000 V. What is the magnitude of the line voltage at the source end of the line?

11.36 Three balanced three-phase loads are connected in parallel. Load 1 is Y-connected with an impedance of $400 + j300 \; \Omega/\phi$; load 2 is Δ-connected with an impedance of $2400 + j1800 \; \Omega/\phi$; and load 3 is $172.8 + j2203.2$ kVA. The loads are fed from a distribution line with an impedance of $2 + j16 \; \Omega/\phi$. The magnitude of the line-to-neutral voltage at the load end of the line is $24\sqrt{3}$ kV.

　　a) Calculate the total complex power at the sending end of the line.

　　b) What percentage of the average power at the sending end of the line is delivered to the loads?

11.37 The output of the balanced positive-sequence three-phase source in Fig. P11.37 is 41.6 kVA at a lagging power factor of 0.707. The line voltage at the source is 240 V.

　　a) Find the magnitude of the line voltage at the load.

　　b) Find the total complex power at the terminals of the load.

Figure P11.37

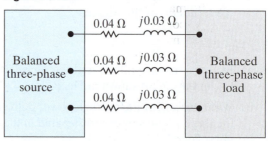

11.38 A balanced three-phase source is supplying 60 kVA at 0.96 pf leading to two balanced Δ-connected parallel loads. The distribution line connecting the source to the load has negligible impedance. Load 1 is purely resistive and absorbs 45 kW.

　　a) Determine the types of components and their impedances in each phase of load 2 if the line voltage is 630 V and the impedance components are in series.

　　b) Repeat (a) with the impedance components in parallel.

11.39 The total power delivered to a balanced three-phase load when operating at a line voltage of $2400\sqrt{3}$ V is 720 kW at a lagging power factor of 0.8. The impedance of the distribution line supplying the load is $0.8 + j6.4 \; \Omega/\phi$. Under these operating conditions, the drop in the magnitude of the line voltage between the sending end and the load end of the line is excessive. To compensate, a bank of Y-connected capacitors is placed in parallel with the load. The capacitor bank is designed to furnish 576 kVAR of magnetizing reactive power when operated at a line voltage of $2400\sqrt{3}$ V.

　　a) What is the magnitude of the voltage at the sending end of the line when the load is operating at a line voltage of $2400\sqrt{3}$ V and the capacitor bank is disconnected?

　　b) Repeat (a) with the capacitor bank connected.

　　c) What is the average power efficiency of the line in (a)?

　　d) What is the average power efficiency in (b)?

　　e) If the system is operating at a frequency of 60 Hz, what is the size of each capacitor in microfarads?

11.40 A balanced bank of delta-connected capacitors is connected in parallel with the load described in Assessment Problem 11.9. The effect is to place a capacitor in parallel with the load in each phase. The line voltage at the terminals of the load thus remains at 2450 V. The circuit is operating at a frequency of 60 Hz. The capacitors are adjusted so that the magnitude of the line current feeding the parallel combination of the load and capacitor bank is at its minimum.

a) What is the size of each capacitor in microfarads?

b) Repeat (a) for wye-connected capacitors.

c) What is the magnitude of the line current?

Section 11.6

11.41 The two-wattmeter method is used to measure the power at the load end of the line in Example 11.1. Calculate the reading of each wattmeter.

11.42 The wattmeters in the circuit in Fig. 11.20 read as follows: $W_1 = 40{,}823.09$ W, and $W_2 = 103{,}176.91$ W. The magnitude of the line voltage is $2400\sqrt{3}$ V. The phase sequence is positive. Find Z_ϕ.

11.43 In the balanced three-phase circuit shown in Fig. P11.43, the current coil of the wattmeter is connected in line aA, and the potential coil of the wattmeter is connected across lines b and c. Show that the wattmeter reading multiplied by $\sqrt{3}$ equals the total reactive power associated with the load. The phase sequence is positive.

Figure P11.43

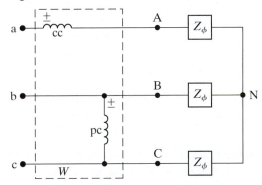

11.44 The line-to-neutral voltage in the circuit in Fig. P11.43 is 680 V, the phase sequence is positive, and the load impedance is $16 + j12 \ \Omega/\phi$.

a) Calculate the wattmeter reading.

b) Calculate the total reactive power associated with the load.

11.45 The two wattmeters in Fig. 11.20 can be used to compute the total reactive power of the load.

a) Prove this statement by showing that $\sqrt{3}(W_2 - W_1) = \sqrt{3}V_L I_L \sin \theta_\phi$.

b) Compute the total reactive power from the wattmeter readings for each of the loads in Example 11.6. Check your computations by calculating the total reactive power directly from the given voltage and impedance.

11.46 Derive Eqs. 11.22 and 11.23.

11.47 a) Calculate the complex power associated with each phase of the balanced load in Problem 11.19.

b) If the two-wattmeter method is used to measure the average power delivered to the load, specify the reading of each meter.

11.48 The two-wattmeter method is used to measure the power delivered to the unbalanced load in Problem 11.20. The current coil of wattmeter 1 is placed in line aA and that of wattmeter 2 is placed in line bB.

a) Calculate the reading of wattmeter 1.

b) Calculate the reading of wattmeter 2.

c) Show that the sum of the two wattmeter readings equals the total power delivered to the unbalanced load.

11.49 The balanced three-phase load shown in Fig. P11.49 is fed from a balanced, positive-sequence, three-phase Y-connected source. The impedance of the line connecting the source to the load is negligible. The line-to-neutral voltage of the source is 7200 V.

a) Find the reading of the wattmeter in watts.

b) Explain how you would connect a second wattmeter in the circuit so that the two wattmeters would measure the total power.

c) Calculate the reading of the second wattmeter.

d) Verify that the sum of the two wattmeter readings equals the total average power delivered to the load.

Figure P11.49

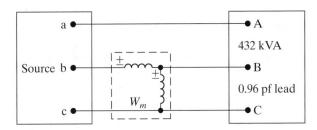

11.50 a) Calculate the reading of each wattmeter in the circuit shown in Fig. P11.50. The value of Z_ϕ is $40 \ \underline{/-30°} \ \Omega$.

b) Verify that the sum of the wattmeter readings equals the total average power delivered to the Δ-connected load.

Figure P11.50

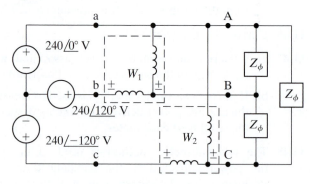

11.51 a) Calculate the reading of each wattmeter in the circuit shown in Fig. P11.51 when $Z = 13.44 + j46.08 \, \Omega$.

b) Check that the sum of the two wattmeter readings equals the total power delivered to the load.

c) Check that $\sqrt{3}(W_1 - W_2)$ equals the total magnetizing vars delivered to the load.

Figure P11.51

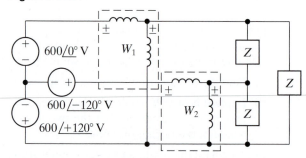

11.52 a) Find the reading of each wattmeter in the circuit shown in Fig. P11.52 if $Z_A = 20 \angle 30° \, \Omega$, $Z_B = 60 \angle 0° \, \Omega$, and $Z_C = 40 \angle -30° \, \Omega$.

b) Show that the sum of the wattmeter readings equals the total average power delivered to the unbalanced three-phase load.

Figure P11.52

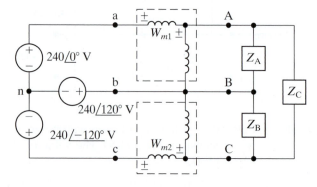

Sections 11.1–11.6

11.53 Refer to the Practical Perspective example:

PRACTICAL PERSPECTIVE a) Construct a power triangle for the substation load before the capacitors are connected to the bus.

b) Repeat (a) after the capacitors are connected to the bus.

c) Using the line-to-neutral voltage at the substation as a reference, construct a phasor diagram that depicts the relationship between $\mathbf{V}_{AN}$ and $\mathbf{V}_{an}$ before the capacitors are added.

d) Assume a positive phase sequence and construct a phasor diagram that depicts the relationship between $\mathbf{V}_{AB}$ and $\mathbf{V}_{ab}$.

11.54 Refer to the Practical Perspective example. Assume the frequency of the utility is 60 Hz.

PRACTICAL PERSPECTIVE

a) What is the μF rating of each capacitor if the capacitors are delta-connected?

b) What is the μF rating of each capacitor if the capacitors are wye-connected?

11.55 Choose a single capacitor from Appendix H that is closest to the μF rating of the wye-connected capacitor from Problem 11.54(b).

PRACTICAL PERSPECTIVE

a) How much reactive power will a capacitor bank using this new value supply?

b) What line-to-line voltage at the generating plant will be required when this new capacitor bank is connected to the substation bus?

11.56 Choose a single capacitor from Appendix H that is closest to the μF rating of the delta-connected capacitor from Problem 11.54(a).

PRACTICAL PERSPECTIVE

a) How much reactive power will a capacitor bank using this new value supply?

b) What line-to-line voltage at the generating plant will be required when this new capacitor bank is connected to the substation bus?

11.57 In the Practical Perspective example, what happens to the voltage level at the generating plant if the substation is maintained at 13.8 kV, the substation load is removed, and the added capacitor bank remains connected?

PRACTICAL PERSPECTIVE

11.58 In the Practical Perspective example, calculate the total line loss in kW before and after the capacitors are connected to the substation bus.

PRACTICAL PERSPECTIVE

11.59 Assume the load on the substation bus in the Practical Perspective example drops to 240 kW and 600 magnetizing kVAR. Also assume the capacitors remain connected to the substation.

PRACTICAL PERSPECTIVE

a) What is the magnitude of the line-to-line voltage at the generating plant that is required to maintain a line-to-line voltage of 13.8 kV at the substation?

b) Will this power plant voltage level cause problems for other customers?

11.60 Assume in Problem 11.59 that when the load drops to 240 kW and 600 magnetizing kVAR the capacitor bank at the substation is disconnected. Also assume that the line-to-line voltage at the substation is maintained at 13.8 kV.

a) What is the magnitude of the line-to-line voltage at the generating plant?

b) Is the voltage level found in (a) within the acceptable range of variation?

c) What is the total line loss in kW when the capacitors stay on line after the load drops to $240 + j600$ kVA?

d) What is the total line loss in kW when the capacitors are removed after the load drops to $240 + j600$ kVA?

e) Based on your calculations, would you recommend disconnecting the capacitors after the load drops to $240 + j600$ kVA? Explain.

CHAPTER

12

CHAPTER CONTENTS

12.1 **Definition of the Laplace Transform** *p. 446*

12.2 **The Step Function** *p. 447*

12.3 **The Impulse Function** *p. 449*

12.4 **Functional Transforms** *p. 452*

12.5 **Operational Transforms** *p. 453*

12.6 **Applying the Laplace Transform** *p. 458*

12.7 **Inverse Transforms** *p. 460*

12.8 **Poles and Zeros of $F(s)$** *p. 470*

12.9 **Initial- and Final-Value Theorems** *p. 472*

CHAPTER OBJECTIVES

1 Be able to calculate the Laplace transform of a function using the definition of Laplace transform, the Laplace transform table, and/or a table of operational transforms.

2 Be able to calculate the inverse Laplace transform using partial fraction expansion and the Laplace transform table.

3 Be able to find and plot the poles and zeros for a rational function in the s-domain.

4 Understand and know how to use the initial-value theorem and the final-value theorem.

Introduction to the Laplace Transform

We now introduce the Laplace transform, which forms the basis of a powerful technique that is widely used to analyze linear, lumped-parameter circuits. We need the Laplace transform analysis technique because we can use it to solve the following types of problems.

- Finding the transient behavior of circuits whose describing equations consist of more than a single node-voltage or mesh-current differential equation. In other words, analyzing multiple-node and multiple-mesh circuits that are described by sets of linear differential equations.

- Determining the transient response of circuits whose signal sources vary in ways more complicated than the simple dc level jumps considered in Chapters 7 and 8.

- Calculating the transfer function for a circuit and using it to find the steady-state sinusoidal response of that circuit when the frequency of the sinusoidal source is varied. We discuss the transfer function in Chapter 13.

- Relating the time-domain behavior of a circuit to its frequency-domain behavior, in a systematic fashion.

In this chapter, we define the Laplace transform, discuss its pertinent characteristics, and present a systematic method for transforming expressions from the frequency domain to the time domain.

Practical Perspective

Transient Effects

As we learned in Chapter 9, power delivered from electrical wall outlets in the United States can be modeled as a sinusoidal voltage or current source, where the frequency of the sinusoid is 60 Hz. We used the phasor transform, introduced in Chapter 9, to find the steady-state response of a circuit to a sinusoidal source.

But in many cases, we need to consider the complete response of a circuit to a sinusoidal source. Remember that the complete response has two parts—the steady-state response that takes the same form as the input to the circuit, and the transient response that decays to zero as time progresses. When a circuit's source is a 60 Hz sinusoid, the steady-state response is also a 60 Hz sinusoid whose magnitude and phase angle can be calculated using phasor circuit analysis. The transient response depends on the components that make up the circuit, the values of those components, and the way the components are interconnected. Once the source is switched into the circuit, the voltage and current for every circuit component are the sum of a transient expression and a steady-state expression.

The transient part of the voltage and current eventually decays to zero. But, initially, the sum of the transient part and the steady-state part might exceed the voltage or current rating of the circuit component. This is why it is important to determine the complete response of a circuit. The Laplace transform techniques introduced in this chapter can be used to find the complete response of a circuit to a sinusoidal source.

Consider the *RLC* circuit shown below, comprised of components from Appendix H and powered by a 60 Hz sinusoidal source. As detailed in Appendix H, the 10 mH inductor has a current rating of 40 mA. The amplitude of the sinusoidal source has been chosen so that this rating is not exceeded in the steady state. Once we have presented the Laplace transform method, we will be able to determine whether this current rating is exceeded when the source is first switched on and both the transient and steady-state components of the inductor current exist.

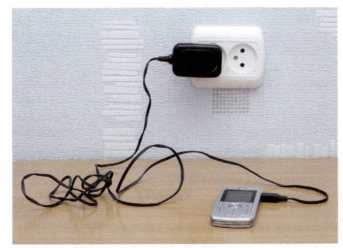

Most popular/Shutterstock

12.1 Definition of the Laplace Transform

The **Laplace transform** of a function is given by the expression

LAPLACE TRANSFORM

$$\mathcal{L}\{f(t)\} = \int_0^\infty f(t)e^{-st}\,dt, \qquad (12.1)$$

where the symbol $\mathcal{L}\{f(t)\}$ is read "the Laplace transform of $f(t)$."

The Laplace transform of $f(t)$ is also denoted $F(s)$; that is,

$$F(s) = \mathcal{L}\{f(t)\}. \qquad (12.2)$$

This notation emphasizes that when the integral in Eq. 12.1 has been evaluated, the resulting expression is a function of s. In our applications, t represents the time domain and s represents the frequency domain. Note that the dimension of s must be reciprocal time, or frequency, because the exponent of e in the integral of Eq. 12.1 must be dimensionless. The Laplace transform transforms the problem from the time domain to the frequency domain. Once we solve the problem in the frequency domain, we inverse-transform the solution back to the time domain.

Recall that the phasor is also a transform. As we know from Chapter 9, it converts a sinusoidal signal into a complex number for easier, algebraic computation of circuit values. After determining the phasor value of a signal, we transform it back to its time-domain expression. Both the Laplace transform and the phasor transform exhibit an essential feature of mathematical transforms: They create a new domain to make the mathematical manipulations easier. After finding the unknown in the new domain, we inverse-transform it back to the original domain.

In circuit analysis, we use the Laplace transform to transform a set of integrodifferential equations in the time domain to a set of algebraic equations in the frequency domain. We can therefore find the solution for an unknown quantity by solving a set of algebraic equations.

Before we illustrate some of the important properties of the Laplace transform, some general comments are in order. First, note that the integral in Eq. 12.1 is improper because the upper limit is infinite. Thus, we are confronted immediately with the question of whether the integral converges. In other words, does a given $f(t)$ have a Laplace transform? Obviously, the functions of primary interest in engineering analysis have Laplace transforms; otherwise we would not be interested in the transform. In linear circuit analysis, we excite circuits with sources that have Laplace transforms. Excitation functions such as t^t or e^{t^2}, which do not have Laplace transforms, are of no interest here.

Second, because the lower limit on the integral is zero, the Laplace transform ignores $f(t)$ for negative values of t. That is, $F(s)$ is determined by the behavior of $f(t)$ only for positive values of t. To emphasize that the lower limit is zero, Eq. 12.1 is frequently referred to as the **one-sided,** or **unilateral, Laplace transform**. In the two-sided, or bilateral, Laplace transform, the lower limit is $-\infty$. We do not use the bilateral form here; hence, $F(s)$ is understood to be the one-sided transform.

The Laplace transform's lower limit creates a concern: what happens when $f(t)$ has a discontinuity at the origin? If $f(t)$ is continuous at the origin, like the function in Fig. 12.1(a), $f(0)$ is not ambiguous. However, if $f(t)$ has a finite discontinuity at the origin, like the function in Fig. 12.1(b), should the Laplace transform integral include or exclude the discontinuity? That is, should we make the lower limit 0^- and include the discontinuity, or choose 0^+ as the lower limit and exclude the discontinuity? (We use the notation 0^- and 0^+ to denote values of t just to the left and right of the origin, respectively.) Actually, we may choose either as long as we are consistent. For reasons we explain later, we choose 0^- as the lower limit.

Because we are using 0^- as the lower limit, we note that the integral from 0^- to 0^+ is zero, except when the discontinuity at the origin is an impulse function. We discuss this situation in Section 12.3. The two functions shown in Fig. 12.1 have the same unilateral Laplace transform because there is no impulse function at the origin for the function in Fig. 12.1(b).

The one-sided Laplace transform ignores $f(t)$ for $t < 0^-$. We use initial conditions to account for what happens prior to 0^-. Thus, the Laplace transform predicts the response to a function that begins after initial conditions have been established.

In the discussion that follows, we divide the Laplace transforms into two types: functional transforms and operational transforms. A **functional transform** is the Laplace transform of a specific function, such as $\sin \omega t$, t, e^{-at}, and so on. An **operational transform** defines a general mathematical property of the Laplace transform, such as finding the transform of the derivative of $f(t)$. Before considering functional and operational transforms, we introduce the step and impulse functions.

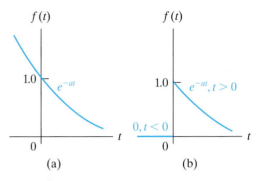

Figure 12.1 ▲ A continuous and discontinuous function at the origin. (a) $f(t)$ is continuous at the origin. (b) $f(t)$ is discontinuous at the origin.

12.2 The Step Function

When a circuit contains a switch, a change in the switch position creates abrupt changes in currents and voltages, as we have seen in previous chapters. An abrupt change is represented mathematically as a discontinuity, which can occur at any instant in time. We represent such discontinuities in the functions that describe the currents and voltages using step and impulse functions.

Figure 12.2 illustrates the step function. The symbol for the step function is $Ku(t)$. The mathematical definition of the **step function** is

$$Ku(t) = 0, \quad t < 0,$$
$$Ku(t) = K, \quad t > 0. \tag{12.3}$$

If K is 1, the function in Eq. 12.3 is the **unit step**, $u(t)$.

The step function is not defined at $t = 0$. If we need to define the transition between 0^- and 0^+, we assume that it is linear and that

$$Ku(0) = 0.5K.$$

As before, 0^- and 0^+ represent symmetric points arbitrarily close to the left and right of the origin. Figure 12.3 illustrates the linear transition from 0^- to 0^+.

A discontinuity may occur at some time other than $t = 0$; for example, it may occur in sequential switching. A step that occurs at $t = a$ is represented by the function $Ku(t - a)$. Thus

$$Ku(t - a) = 0, \quad t < a,$$
$$Ku(t - a) = K, \quad t > a.$$

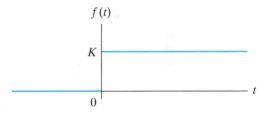

Figure 12.2 ▲ The step function.

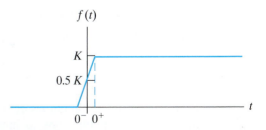

Figure 12.3 ▲ The linear approximation to the step function.

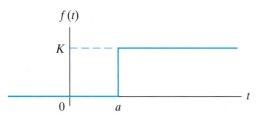

Figure 12.4 ▲ A step function occurring at $t = a$ when $a > 0$.

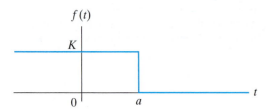

Figure 12.5 ▲ A step function $Ku(a - t)$ for $a > 0$.

If $a > 0$, the step occurs to the right of the origin, and if $a < 0$, the step occurs to the left of the origin. The step function's value is 0 when the argument $t - a$ is negative, and it is K when the argument is positive. Figure 12.4 illustrates a step that occurs at $t = a$.

A step function equal to K for $t < a$ is written as $Ku(a - t)$. Thus

$$Ku(a - t) = K, \quad t < a,$$
$$Ku(a - t) = 0, \quad t > a.$$

The discontinuity is to the left of the origin when $a < 0$. This type of step is shown in Fig. 12.5.

We can add two step functions to create a function that describes a finite-width pulse. For example, the function $K[u(t - 1) - u(t - 3)]$ has the value K for $1 < t < 3$ and the value 0 everywhere else, so it is a finite-width pulse of height K initiated at $t = 1$ and terminated at $t = 3$. In defining this pulse, think of the step function $u(t - 1)$ as "turning on" the constant value K at $t = 1$, and the step function $-u(t - 3)$ as "turning off" the constant value K at $t = 3$. We use step functions to turn on and turn off linear functions at desired times in Example 12.1.

EXAMPLE 12.1 **Using Step Functions to Represent a Function of Finite Duration**

Use step functions to write an expression for the function illustrated in Fig. 12.6.

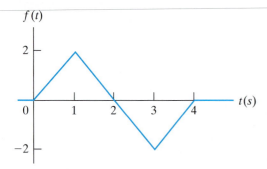

Figure 12.6 ▲ The function for Example 12.1.

Solution

The function shown in Fig. 12.6 is made up of linear segments, so it is a piecewise linear function. You have probably seen such functions defined using a different functional form for each of its time segments. For this function, the piecewise-linear definition is

$$f(t) = \begin{aligned} &= 0, &&t \leq 0; \\ &= 2t, &&0 \leq t \leq 1 \text{ s}; \\ &= -2t + 4, &&1 \text{ s} \leq t \leq 3 \text{ s}; \\ &= 2t - 8, &&3 \text{ s} \leq t \leq 4 \text{ s}; \\ &= 0, &&t \geq 4 \text{ s}. \end{aligned}$$

We can construct a single continuous definition for this function using step functions to initiate and

terminate each linear segment at the proper times. In other words, we use the step function to turn on and turn off the three nonzero pieces of the function:

$$+2t, \text{ on at } t = 0, \text{ off at } t = 1;$$
$$-2t + 4, \text{ on at } t = 1, \text{ off at } t = 3;$$
$$+2t - 8, \text{ on at } t = 3, \text{ off at } t = 4.$$

These straight line segments and their equations are shown in Fig. 12.7. The expression for $f(t)$, valid for all values of t, is

$$f(t) = 2t[u(t) - u(t - 1)] + (-2t + 4)[u(t - 1) - u(t - 3)] + (2t - 8)[u(t - 3) - u(t - 4)].$$

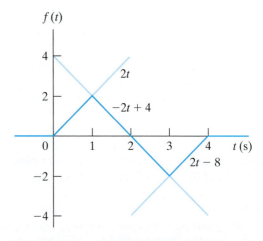

Figure 12.7 ▲ Definition of the three line segments turned on and off with step functions to form the function shown in Fig. 12.6.

SELF-CHECK: Assess your understanding of step functions by trying Chapter Problems 12.1 and 12.2.

12.3 The Impulse Function

An **impulse** is a signal of infinite amplitude and zero duration. Such signals don't exist in nature, but some circuit signals come very close to approximating this definition, so a mathematical model of an impulse is useful. For example, the impulse function[1] enables us to define the derivative at a discontinuity, such as the one in the function of Fig. 12.1(b), and thus to define the Laplace transform of that derivative. Also, voltage and current impulses occur in circuit analysis either because of a switching operation or because the circuit is excited by an impulsive source. We will analyze these situations in Chapter 13, but here we define the impulse function generally.

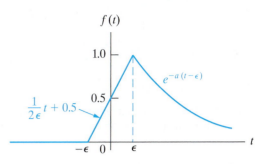

Figure 12.8 ▲ A magnified view of the discontinuity in Fig. 12.1(b), assuming a linear transition between $-\epsilon$ and $+\epsilon$.

Describing the Impulse Function

We describe the impulse function by considering how we would define the derivative of the function in Fig. 12.1(b) at its discontinuity. First, we assume that the function varies linearly across the discontinuity, as shown in Fig. 12.8. In this figure, note that as $\epsilon \to 0$, an abrupt discontinuity occurs at the origin. When we differentiate the function, the derivative between $-\epsilon$ and $+\epsilon$ is constant, with a value of $1/2\epsilon$. For $t > \epsilon$, the derivative is $-ae^{-a(t-\epsilon)}$. Figure 12.9 shows these observations graphically. As ϵ approaches zero, the value of $f'(t)$ between $\pm\epsilon$ approaches infinity. At the same time, the duration of this large value is approaching zero. Furthermore, the area under $f'(t)$ between $\pm\epsilon$ remains constant as $\epsilon \to 0$. In this example, the area is unity. As ϵ approaches zero, we say that the function between $\pm\epsilon$ approaches a **unit impulse function**, denoted $\delta(t)$. Thus, the derivative of $f(t)$ at the origin approaches a unit impulse function as ϵ approaches zero, or

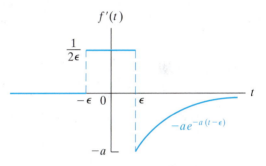

Figure 12.9 ▲ The derivative of the function shown in Fig. 12.8.

$$f'(0) \to \delta(t) \quad \text{as } \epsilon \to 0.$$

If the area under the impulse function curve is other than unity, the impulse function is denoted $K\delta(t)$, where K is the area. K is often referred to as the **strength** of the impulse function.

To summarize, an impulse function is created from a variable-parameter function whose parameter approaches zero. The variable-parameter function must exhibit the following three characteristics as the parameter approaches zero:

1. The amplitude of the function approaches infinity.

2. The duration of the function approaches zero.

3. The area under the variable-parameter function is constant as the parameter changes.

Many different variable-parameter functions have these three characteristics. In Fig. 12.8, we used a linear function $f(t) = 0.5t/\epsilon + 0.5$. Another example of a variable-parameter function with the three characteristics is the exponential function:

$$f(t) = \frac{K}{2\epsilon} e^{-|t|/\epsilon}.$$

As ϵ approaches zero, the function becomes infinite at the origin and at the same time decays to zero in an infinitesimal length of time.

---

[1] The impulse function is also known as the Dirac delta function.

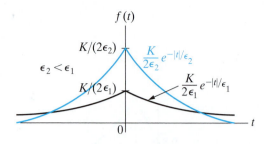

Figure 12.10 ▲ A variable-parameter function used to generate an impulse function.

Figure 12.10 illustrates $f(t)$ as $\epsilon \to 0$. An impulse function is created as $\epsilon \to 0$ if the area under the function is independent of ϵ. Thus,

$$\text{Area} = \int_{-\infty}^{0} \frac{K}{2\epsilon} e^{t/\epsilon} dt + \int_{0}^{\infty} \frac{K}{2\epsilon} e^{-t/\epsilon} dt$$

$$= \frac{K}{2\epsilon} \cdot \frac{e^{t/\epsilon}}{1/\epsilon} \bigg|_{-\infty}^{0} + \frac{K}{2\epsilon} \cdot \frac{e^{-t/\epsilon}}{-1/\epsilon} \bigg|_{0}^{\infty}$$

$$= \frac{K}{2} + \frac{K}{2} = K,$$

so the area under the curve is constant and equal to K units. Therefore, as $\epsilon \to 0$, $f(t) \to K\delta(t)$.

Defining the Impulse Function and Its Sifting Property

Mathematically, the **impulse function** is defined as

$$\int_{-\infty}^{\infty} K\delta(t)dt = K; \tag{12.4}$$

$$\delta(t) = 0, \quad t \neq 0. \tag{12.5}$$

Equation 12.4 states that the area under the impulse function is constant. This area represents the strength of the impulse. Equation 12.5 states that the impulse is zero everywhere except at $t = 0$. An impulse that occurs at $t = a$ is denoted $K\delta(t - a)$. The graphic symbol for the impulse function is an arrow. The strength of the impulse is given parenthetically next to the head of the arrow. Figure 12.11 shows the impulses $K\delta(t)$ and $K\delta(t - a)$.

An important property of the impulse function is the **sifting property**, which is expressed as

$$\int_{-\infty}^{\infty} f(t)\delta(t - a)\,dt = f(a), \tag{12.6}$$

where we assume the function $f(t)$ is continuous at $t = a$, the location of the impulse. Equation 12.6 shows that the impulse function sifts out all values of $f(t)$ except the one at $t = a$. Equation 12.6 follows from Eqs. 12.4 and 12.5, noting that $\delta(t - a)$ is zero everywhere except at $t = a$, and hence the integral can be written

$$\int_{-\infty}^{\infty} f(t)\delta(t - a)\,dt = \int_{a-\epsilon}^{a+\epsilon} f(t)\delta(t - a)\,dt.$$

But because $f(t)$ is continuous at a, it takes on the value $f(a)$ as $t \to a$, so

$$\int_{a-\epsilon}^{a+\epsilon} f(a)\delta(t - a)\,dt = f(a)\int_{a-\epsilon}^{a+\epsilon} \delta(t - a)\,dt$$

$$= f(a).$$

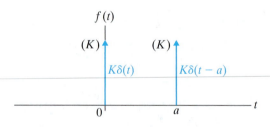

Figure 12.11 ▲ A graphic representation of the impulse $K\delta(t)$ and $K\delta(t - a)$.

Laplace Transform and Derivatives of the Impulse Function

We use the sifting property of the impulse function to find its Laplace transform:

$$\mathcal{L}\{\delta(t)\} = \int_{0^-}^{\infty} \delta(t)e^{-st}\,dt = e^{-s(0)} = 1,$$

which is an important Laplace transform. We can also define the derivatives of the impulse function and the Laplace transform of these derivatives. We discuss the first derivative, along with its transform, and then state the result for the higher-order derivatives.

The function illustrated in Fig. 12.12(a) generates an impulse function as $\epsilon \to 0$. Figure 12.12(b) shows the derivative of this impulse-generating function, which is defined as the derivative of the impulse $[\delta'(t)]$ as $\epsilon \to 0$. The derivative of the impulse function sometimes is referred to as a moment function, or unit doublet.

To find the Laplace transform of $\delta'(t)$, we apply the defining integral to the function shown in Fig. 12.12(b) and, after integrating, let $\epsilon \to 0$. Then

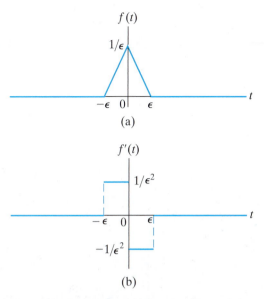

$$\mathcal{L}\{\delta'(t)\} = \lim_{\epsilon \to 0}\left[\int_{-\epsilon}^{0^-}\frac{1}{\epsilon^2}e^{-st}\,dt + \int_{0^+}^{\epsilon}\left(-\frac{1}{\epsilon^2}\right)e^{-st}dt\right]$$

$$= \lim_{\epsilon \to 0}\frac{e^{s\epsilon} + e^{-s\epsilon} - 2}{s\epsilon^2}$$

$$= \lim_{\epsilon \to 0}\frac{se^{s\epsilon} - se^{-s\epsilon}}{2\epsilon s}$$

$$= \lim_{\epsilon \to 0}\frac{s^2 e^{s\epsilon} + s^2 e^{-s\epsilon}}{2s}$$

$$= s.$$

Figure 12.12 ▲ The first derivative of the impulse function. (a) The impulse-generating function used to define the first derivative of the impulse. (b) The first derivative of the impulse-generating function that approaches $\delta'(t)$ as $\epsilon \to 0$.

We used l'Hôpital's rule twice in this derivation, to evaluate the indeterminate form $0/0$.

Higher-order derivatives can be generated in a similar manner (see Problem 12.6), and the defining integral can then be used to find the Laplace transforms. The Laplace transform of the nth derivative of the impulse function is

$$\mathcal{L}\{\delta^{(n)}(t)\} = s^n.$$

Finally, note that the derivative of a step function is an impulse function; that is,

$$\delta(t) = \frac{du(t)}{dt}.$$

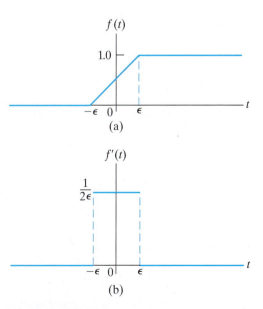

Figure 12.13 depicts the relationship between the impulse function and the step function. The function shown in Fig. 12.13(a) approaches a unit step function as $\epsilon \to 0$. The function shown in Fig. 12.13(b)—the derivative of the function in Fig. 12.13(a)—approaches a unit impulse as $\epsilon \to 0$.

The impulse function is an extremely useful concept in circuit analysis, and we say more about it in the following chapters. We introduced the concept here so that we can include discontinuities at the origin in our definition of the Laplace transform.

SELF-CHECK: Assess your understanding of the impulse function by trying Chapter end-of-text Problems 12.8 and 12.10.

Figure 12.13 ▲ The impulse function as the derivative of the step function: (a) $f(t) \to u(t)$ as $\epsilon \to 0$; and (b) $f'(t) \to \delta(t)$ as $\epsilon \to 0$.

12.4 Functional Transforms

The Laplace transform of a specified function of t is called a functional transform. Because we are using the unilateral, or one-sided, Laplace transform, we define all functions to be zero for $t < 0^-$.

We derived one functional transform pair in Section 12.3, where we showed that the Laplace transform of the unit impulse function equals 1. Next, we find the Laplace transform of the unit step function at the origin, where

$$\mathcal{L}\{u(t)\} = \int_0^\infty f(t)e^{-st}\,dt = \int_{0^+}^\infty 1e^{-st}dt$$

$$= \frac{e^{-st}}{-s}\bigg|_{0^+}^\infty = \frac{1}{s}.$$

Thus, the Laplace transform of the unit step function is $1/s$.

The Laplace transform of the decaying exponential function shown in Fig. 12.14 is

$$\mathcal{L}\{e^{-at}\} = \int_{0^+}^\infty e^{-at}e^{-st}dt = \int_{0^+}^\infty e^{-(a+s)t}dt = \frac{1}{s+a}.$$

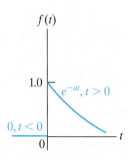

$f(t)$

1.0

$e^{-at}, t > 0$

$0, t < 0$

0

t

Figure 12.14 ▲ A decaying exponential function.

In deriving the Laplace transforms of the unit step function and the decaying exponential function, we used the fact that integration across the discontinuity at the origin is zero.

Let's find the Laplace transform of the sinusoidal function shown in Fig. 12.15. The expression for $f(t)$ for $t > 0^-$ is $\sin \omega t$; hence, the Laplace transform is

$$\mathcal{L}\{\sin \omega t\} = \int_{0^-}^\infty (\sin \omega t)e^{-st}\,dt$$

$$= \int_{0^-}^\infty \left(\frac{e^{j\omega t} - e^{-j\omega t}}{2j}\right)e^{-st}\,dt$$

$$= \int_{0^-}^\infty \frac{e^{-(s-j\omega)t} - e^{-(s+j\omega)t}}{2j}\,dt$$

$$= \frac{1}{2j}\left(\frac{1}{s-j\omega} - \frac{1}{s+j\omega}\right)$$

$$= \frac{\omega}{s^2 + \omega^2}.$$

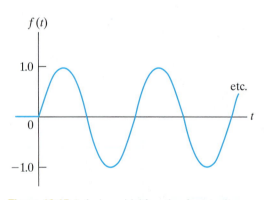

$f(t)$

1.0

etc.

0

t

-1.0

Figure 12.15 ▲ A sinusoidal function for $t > 0$.

Table 12.1 gives an abbreviated list of functional Laplace transform pairs. It includes the functions of most interest in an introductory course on circuit applications.

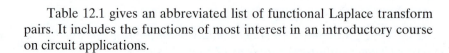

TABLE 12.1   **An Abbreviated List of Laplace Transform Pairs**

Type	$f(t)$ $(t > 0^-)$	$F(s)$
(impulse)	$\delta(t)$	1
(step)	$u(t)$	$\dfrac{1}{s}$
(ramp)	t	$\dfrac{1}{s^2}$
(exponential)	e^{-at}	$\dfrac{1}{s+a}$
(sine)	$\sin \omega t$	$\dfrac{\omega}{s^2+\omega^2}$
(cosine)	$\cos \omega t$	$\dfrac{s}{s^2+\omega^2}$
(damped ramp)	te^{-at}	$\dfrac{1}{(s+a)^2}$
(damped sine)	$e^{-at}\sin \omega t$	$\dfrac{\omega}{(s+a)^2+\omega^2}$
(damped cosine)	$e^{-at}\cos \omega t$	$\dfrac{s+a}{(s+a)^2+\omega^2}$

 ASSESSMENT PROBLEM

Objective 1—Be able to calculate the Laplace transform of a function using the definition of Laplace transform

12.1 Use the defining integral to
 a) find the Laplace transform of $\cosh \beta t$;
 b) find the Laplace transform of $\sinh \beta t$.

Answer: (a) $s/(s^2 - \beta^2)$;

(b) $\beta/(s^2 - \beta^2)$.

SELF-CHECK: Also try Chapter Problem 12.20.

12.5 Operational Transforms

Operational Laplace transforms define how mathematical operations performed on $f(t)$ affect its Laplace transform, $F(s)$. Operational transforms also define how mathematical operations performed on $F(s)$ affect its corresponding time-domain function $f(t)$. The operations we consider include (1) multiplication by a constant; (2) addition and subtraction; (3) differentiation; (4) integration; (5) translation in the time domain; (6) translation in the frequency domain; and (7) scale changing.

Multiplication by a Constant

From the defining integral, if

$$\mathcal{L}\{f(t)\} = F(s),$$

then

$$\mathcal{L}\{Kf(t)\} = KF(s).$$

Thus, multiplication of $f(t)$ by a constant corresponds to multiplying $F(s)$ by the same constant.

Addition and Subtraction

Addition and subtraction in the time domain translates into addition and subtraction in the frequency domain. Thus if

$$\mathcal{L}\{f_1(t)\} = F_1(s),$$

$$\mathcal{L}\{f_2(t)\} = F_2(s),$$

$$\mathcal{L}\{f_3(t)\} = F_3(s),$$

then

$$\mathcal{L}\{f_1(t) + f_2(t) - f_3(t)\} = F_1(s) + F_2(s) - F_3(s).$$

Use the defining integral to derive this operation transform by recognizing that the integral of a sum of functions equals the sum of each function's integral.

Differentiation

We use the definition of the Laplace transform to find the operational transform for differentiation in the time domain:

$$\mathcal{L}\left\{\frac{df(t)}{dt}\right\} = \int_{0^-}^{\infty} \left[\frac{df(t)}{dt}\right] e^{-st}\, dt.$$

We integrate by parts to evaluate this integral. Let $u = e^{-st}$ and $dv = [df(t)/dt]\, dt$ to give

$$\mathcal{L}\left\{\frac{df(t)}{dt}\right\} = e^{-st}f(t)\Big|_{0^-}^{\infty} - \int_{0^-}^{\infty} f(t)(-se^{-st}dt).$$

Because we are assuming that $f(t)$ has a Laplace transform, evaluating $e^{-st}f(t)$ at $t = \infty$ gives zero. We complete the evaluation of the integral to get

$$-f(0^-) + s\int_{0^-}^{\infty} f(t)e^{-st}dt = sF(s) - f(0^-).$$

Thus, the Laplace transform of the derivative of f(t) is

$$\mathcal{L}\left\{\frac{df(t)}{dt}\right\} = sF(s) - f(0^-).$$

This important result shows that differentiation in the time domain transforms to an algebraic operation in the s domain.

Use the Laplace transform of the first derivative of $f(t)$ to find the Laplace transform of higher-order derivatives. For example, to find the Laplace transform of the second derivative of $f(t)$, we first let

$$g(t) = \frac{df(t)}{dt}.$$

Now we find the Laplace transform of $g(t)$:

$$G(s) = sF(s) - f(0^-).$$

But because

$$\frac{dg(t)}{dt} = \frac{d^2f(t)}{dt^2},$$

we know

$$\mathscr{L}\left\{\frac{dg(t)}{dt}\right\} = \mathscr{L}\left\{\frac{d^2f(t)}{dt^2}\right\} = sG(s) - g(0^-).$$

Therefore,

$$\mathscr{L}\left\{\frac{d^2f(t)}{dt^2}\right\} = s^2F(s) - sf(0^-) - \frac{df(0^-)}{dt}.$$

We find the Laplace transform of the nth derivative by successively applying the preceding process, which leads to the general result

$$\mathscr{L}\left\{\frac{d^nf(t)}{dt^n}\right\} = s^nF(s) - s^{n-1}f(0^-) - s^{n-2}\frac{df(0^-)}{dt}$$

$$- s^{n-3}\frac{d^2f(0^-)}{dt^2} - \cdots - \frac{d^{n-1}f(0^-)}{dt^{n-1}}.$$

Integration

We find the Laplace transform of the integral of $f(t)$ by again applying the defining integral:

$$\mathscr{L}\left\{\int_{0^-}^{t} f(x)\, dx\right\} = \int_{0^-}^{\infty}\left[\int_{0^-}^{t} f(x)\, dx\right] e^{-st}\, dt.$$

Integrate by parts to evaluate the integral on the right-hand side of this expression. Let

$$u = \int_{0^-}^{t} f(x)\, dx,$$

$$dv = e^{-st}\, dt.$$

Then

$$du = f(t)\, dt,$$

$$v = -\frac{e^{-st}}{s}.$$

The integration-by-parts formula yields

$$\mathcal{L}\left\{\int_{0^-}^{t} f(x)dx\right\} = -\frac{e^{-st}}{s}\int_{0^-}^{t} f(x)dx\bigg|_{0^-}^{\infty} + \int_{0^-}^{\infty} \frac{e^{-st}}{s} f(t)\,dt.$$

The first term on the right-hand side is zero at both the upper and lower limits. The value at the lower limit is zero because both limits on the integral are the same. The value at the upper limit is zero because $e^{-st} \rightarrow 0$ as $t \rightarrow \infty$. The second term on the right-hand side is $F(s)/s$; therefore

$$\mathcal{L}\left\{\int_{0^-}^{t} f(x)\,dx\right\} = \frac{F(s)}{s},$$

which reveals that integration in the time domain transforms to multiplication by $1/s$ in the s domain. We have therefore demonstrated that the Laplace transform translates a set of integrodifferential equations into a set of algebraic equations.

Translation in the Time Domain

If we start with any function $f(t)u(t)$, we can represent the same function, translated in time by the constant a, as $f(t - a)u(t - a)$.[2] To find the Laplace transform of $f(t - a)u(t - a)$, start with the defining integral:

$$\mathcal{L}\{(t - a)u(t - a)\} = \int_{0^-}^{\infty} u(t - a)f(t - a)e^{-st}\,dt$$

$$= \int_{a}^{\infty} f(t - a)e^{-st}\,dt.$$

In writing this equation, we took advantage of $u(t - a) = 0$ for $t < a$. Now change the variable of integration by letting $x = t - a$. Then $x = 0$ when $t = a$, $x = \infty$ when $t = \infty$ and $dx = dt$. Rewrite the integral as

$$\mathcal{L}\{f(t - a)u(t - a)\} = \int_{0}^{\infty} f(x)e^{-s(x+a)}\,dx$$

$$= e^{-sa}\int_{0}^{\infty} f(x)e^{-sx}\,dx$$

$$= e^{-as}F(s).$$

Thus

$$\mathcal{L}\{f(t - a)u(t - a)\} = e^{-as}F(s), \quad a > 0.$$

Translation in the time domain corresponds to multiplication by an exponential in the frequency domain.

---

[2]Note that throughout we multiply any arbitrary function $f(t)$ by the unit step function $u(t)$ to ensure that the resulting function is defined for all positive time.

For example, knowing that

$$\mathcal{L}\{tu(t)\} = \frac{1}{s^2},$$

we can use this operational transform to find the Laplace transform of $(t - a)u(t - a)$:

$$\mathcal{L}\{(t - a)u(t - a)\} = \frac{e^{-as}}{s^2}.$$

Translation in the Frequency Domain

Translation in the frequency domain corresponds to multiplication by an exponential in the time domain:

$$\mathcal{L}\{e^{-at} f(t)\} = F(s + a),$$

which follows from the defining integral. Problem 12.15 asks you to derive this result.

We can use this operational transform to derive new transform pairs. Thus, knowing that

$$\mathcal{L}\{\cos \omega t\} = \frac{s}{s^2 + \omega^2},$$

we use the effect of translation in the frequency domain to deduce that

$$\mathcal{L}\{e^{-at} \cos \omega t\} = \frac{s + a}{(s + a)^2 + \omega^2}.$$

Scale Changing

The scale-change property gives the relationship between $f(t)$ and $F(s)$ when the time variable is multiplied by a positive constant:

$$\mathcal{L}\{f(at)\} = \frac{1}{a} F\left(\frac{s}{a}\right), \quad a > 0.$$

The derivation is left to Problem 12.16. The scale-change property is particularly useful in experimental work, when time-scale changes are made to assist in building a model of a system.

We use this operational transform to formulate new transform pairs. Thus, knowing that

$$\mathcal{L}\{\cos t\} = \frac{s}{s^2 + 1},$$

we use the effect of scale changing to show that

$$\mathcal{L}\{\cos \omega t\} = \frac{1}{\omega} \frac{s/\omega}{(s/\omega)^2 + 1} = \frac{s}{s^2 + \omega^2}.$$

Table 12.2 lists these operational transforms.

TABLE 12.2 An Abbreviated List of Operational Transforms

Operation	$f(t)$	$F(s)$
Multiplication by a constant	$Kf(t)$	$KF(s)$
Addition/subtraction	$f_1(t) + f_2(t) - f_3(t) + \cdots$	$F_1(s) + F_2(s) - F_3(s) + \cdots$
First derivative (time)	$\dfrac{df(t)}{dt}$	$sF(s) - f(0^-)$
Second derivative (time)	$\dfrac{d^2f(t)}{dt^2}$	$s^2F(s) - sf(0^-) - \dfrac{df(0^-)}{dt}$
nth derivative (time)	$\dfrac{d^nf(t)}{dt^n}$	$s^nF(s) - s^{n-1}f(0^-) - s^{n-2}\dfrac{df(0^-)}{dt}$ $- s^{n-3}\dfrac{df^2(0^-)}{dt^2} - \cdots - \dfrac{d^{n-1}f(0^-)}{dt^{n-1}}$
Time integral	$\displaystyle\int_0^t f(x)\, dx$	$\dfrac{F(s)}{s}$
Translation in time	$f(t - a)u(t - a), a > 0$	$e^{-as}F(s)$
Translation in frequency	$e^{-at}f(t)$	$F(s + a)$
Scale changing	$f(at), a > 0$	$\dfrac{1}{a}F\left(\dfrac{s}{a}\right)$
First derivative (s)	$tf(t)$	$-\dfrac{dF(s)}{ds}$
nth derivative (s)	$t^nf(t)$	$(-1)^n\dfrac{d^nF(s)}{ds^n}$
s integral	$\dfrac{f(t)}{t}$	$\displaystyle\int_s^\infty F(u)\, du$

 ASSESSMENT PROBLEM

Objective 1—Be able to calculate the Laplace transform of a function using the Laplace transform table or a table of operational transforms

12.2 Use the appropriate operational transform from Table 12.2 to find the Laplace transform of each function:

a) t^2e^{-at};

b) $\dfrac{d}{dt}\,(e^{-at}\sinh \beta t)$;

c) $t\cos \omega t$.

Answer: (a) $\dfrac{2}{(s + a)^3}$;

(b) $\dfrac{\beta s}{(s + a)^2 - \beta^2}$;

(c) $\dfrac{s^2 - \omega^2}{(s^2 + \omega^2)^2}$.

SELF-CHECK: Also try Chapter Problems 12.18 and 12.22.

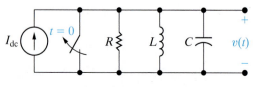

Figure 12.16 ▲ A parallel *RLC* circuit.

12.6 Applying the Laplace Transform

We now use the Laplace transform to solve the ordinary integrodifferential equations that describe the behavior of a lumped-parameter circuit, such as the one shown in Fig. 12.16. We assume that no initial energy is stored in the circuit at the instant the switch is opened. The problem is to find the time-domain expression for $v(t)$ when $t \geq 0$.

We begin by writing the integrodifferential equation that $v(t)$ must satisfy, using a single KCL equation to describe the circuit. Summing the currents away from the top node in the circuit gives:

$$\frac{v(t)}{R} + \frac{1}{L} \int_0^t v(x) \, dx + C\frac{dv(t)}{dt} = I_{dc}u(t).$$

Note that we represented the switch opening at $t = 0$ with the product of the source current and the unit step function, $I_{dc}u(t)$.

We transform the KCL equation into the s-domain using four operational transforms (multiplication by a constant, addition, integration, and differentiation) and one functional transform (unit step) to get

$$\frac{V(s)}{R} + \frac{1}{L}\frac{V(s)}{s} + C[sV(s) - v(0^-)] = I_{dc}\left(\frac{1}{s}\right).$$

The result is an algebraic equation with one unknown variable, $V(s)$. We are assuming that the circuit parameters R, L, and C, as well as the source current I_{dc}, are known; the initial voltage on the capacitor $v(0^-)$ is zero because the initial energy stored in the circuit is zero. Thus, we have reduced the problem to solving an algebraic equation.

Solving for $V(s)$ gives

$$V(s)\left(\frac{1}{R} + \frac{1}{sL} + sC\right) = \frac{I_{dc}}{s},$$

$$V(s) = \frac{I_{dc}/C}{s^2 + (1/RC)s + (1/LC)}.$$

To find $v(t)$, we must inverse-transform the expression for $V(s)$. We denote this inverse operation as

$$v(t) = \mathcal{L}^{-1}\{V(s)\}.$$

The inverse transform, which takes the solution from the s-domain to the time domain, is the subject of Section 12.7. In that section, we also present a final, critical step: checking the validity of the resulting time-domain expression. This final step is not unique to the Laplace transform; it is always a good idea to test any derived solution to be sure it makes sense in terms of known system behavior.

Before continuing, we simplify the notation by dropping the parenthetical t in time-domain expressions and the parenthetical s in frequency-domain expressions. We use lowercase letters for all time-domain variables, and we represent the corresponding s-domain variables with uppercase letters. Thus

$$\mathcal{L}\{v\} = V \quad \text{or} \quad v = \mathcal{L}^{-1}\{V\},$$

$$\mathcal{L}\{i\} = I \quad \text{or} \quad i = \mathcal{L}^{-1}\{I\},$$

$$\mathcal{L}\{f\} = F \quad \text{or} \quad f = \mathcal{L}^{-1}\{F\},$$

and so on.

Example 12.2 supplies component values for the circuit in Fig. 12.16 and uses Laplace transforms to predict the circuit's output voltage.

EXAMPLE 12.2 | Using Laplace Transforms to Predict a Circuit's Response

Suppose for the circuit in Fig. 12.16, $I_{dc} = 24$ mA, $R = 400 \ \Omega$, $L = 25$ mH, and $C = 25$ nF. There is no energy stored in the circuit when the switch opens at $t = 0$. Find the Laplace transform of $v(t)$ and use it to predict the functional form of $v(t)$.

Solution

Using the expression for $V(s)$ found for the circuit in Fig. 12.16,

$$V(s) = \frac{I_{dc}/C}{s^2 + (1/RC)s + (1/LC)}$$

$$= \frac{0.024/(25 \times 10^{-9})}{s^2 + [1/(400)(25 \times 10^{-9})]s + [1/(0.025)(25 \times 10^{-9})]}$$

$$= \frac{96 \times 10^4}{s^2 + 10^5 s + 16 \times 10^8}.$$

The expression for $V(s)$ is not a familiar functional Laplace transform, so we cannot use Tables 12.1 or 12.2 to find its inverse transform. But note that we can rewrite the denominator of $V(s)$ as the product of two factors:

$$V(s) = \frac{96 \times 10^4}{(s + 20{,}000)(s + 80{,}000)}.$$

Now, recognize that we can write $V(s)$ as the sum of two terms, with each factor appearing in the denominator of one term:

$$V(s) = \frac{K_1}{(s + 20{,}000)} + \frac{K_2}{(s + 80{,}000)}.$$

Each of these terms looks like a familiar Laplace transform. Using the transform for the exponential function (Table 12.1) and the operational transforms for multiplication by a constant and addition (Table 12.2) we predict that $v(t)$ is the sum of two exponential terms in the form $e^{-20{,}000t}$ and $e^{-80{,}000t}$.

SELF-CHECK: Assess your understanding of this material by trying Chapter Problem 12.26.

12.7 Inverse Transforms

The expression for $V(s)$ in Example 12.2 is a **rational** function of s. This means we can write $V(s)$ as a ratio of two polynomials in s where only integer powers of s appear in the polynomials. For linear, lumped-parameter circuits with constant component values, the s-domain expressions for the unknown voltages and currents are always rational functions of s. (For verification, work Problems 12.27–12.31.) If we can inverse-transform rational functions of s, we can find the time-domain expressions for the voltages and currents. This section presents a straightforward and systematic technique for finding the inverse transform of a rational function.

In general, we need to find the inverse transform of a function that has the form

$$F(s) = \frac{N(s)}{D(s)} = \frac{a_n s^n + a_{n-1} s^{n-1} + \cdots + a_1 s + a_0}{b_m s^m + b_{m-1} s^{m-1} + \cdots + b_1 s + b_0}. \quad (12.7)$$

The coefficients a and b are real constants, and the exponents m and n are positive integers. The ratio $N(s)/D(s)$ is called a **proper rational function** if $m > n$, and an **improper rational function** if $m \leq n$. Only a proper rational function can be expanded as a sum of partial fractions. This restriction poses no problem, as we show at the end of this section.

Partial Fraction Expansion: Proper Rational Functions

A proper rational function is expanded into a sum of partial fractions by writing a term or a series of terms for each root of $D(s)$. Thus, $D(s)$ must be factored before we construct a partial fraction expansion. For each distinct root of $D(s)$, a single term appears in the sum of partial fractions. For each multiple root of $D(s)$ of multiplicity r, the sum of partial fractions contains r terms. For example, in the rational function

$$\frac{s + 6}{s(s + 3)(s + 1)^2},$$

the denominator has four roots. Two of these roots are distinct—namely, at $s = 0$ and $s = -3$. A multiple root of multiplicity 2 occurs at $s = -1$. Thus, the partial fraction expansion takes the form

$$\frac{s + 6}{s(s + 3)(s + 1)^2} \equiv \frac{K_1}{s} + \frac{K_2}{s + 3} + \frac{K_3}{(s + 1)^2} + \frac{K_4}{s + 1}.$$

To find the inverse transform from the sum of partial fractions, we identify the $f(t)$ corresponding to each term in the sum using the functional and operational transform tables. Use Tables 12.1 and 12.2 to verify that

$$\mathcal{L}^{-1}\left\{\frac{K_1}{s} + \frac{K_2}{s + 3} + \frac{K_3}{(s + 1)^2} + \frac{K_4}{s + 1}\right\}$$

$$= (K_1 + K_2 e^{-3t} + K_3 t e^{-t} + K_4 e^{-t})u(t).$$

Now we need to find the numerator coefficients ($K_1, K_2, K_3, \ldots$) that appear in each partial fraction term. There are only four different types of partial fraction terms because the roots of $D(s)$ can be (1) real and distinct; (2) complex and distinct; (3) real and repeated; or (4) complex and repeated. We develop a technique to determine the numerator coefficient for each type of partial fraction term. Before doing so, a few general comments are in order.

We used the identity sign $\equiv$ in the partial fraction expansion to emphasize that expanding a rational function into a sum of partial fractions establishes an identical equation. This means that both sides of the equation must be the same for all values of the variable s. Also, the identity relationship must hold when the same mathematical operation is applied to both sides. These observations will help us calculate the coefficient values.

Before creating a partial fraction expansion, you should verify that the rational function is proper. This check is important because the procedure for finding the partial fraction coefficients will not prevent you from generating invalid results if the rational function is improper. We present a procedure for checking the coefficients, but you can avoid wasted effort by always asking, "Is $F(s)$ a proper rational function?"

Partial Fraction Expansion: Distinct Real Roots of *D*(s)

Let's find the partial fraction expansion of a proper rational function whose denominator has distinct real roots. For example,

$$F(s) = \frac{N(s)}{(s + p_1)(s + p_2)(s + p_3)} \equiv \frac{K_1}{(s + p_1)} + \frac{K_2}{(s + p_2)} + \frac{K_3}{(s + p_3)}.$$

To find K_1 for the first partial fraction term, we multiply both sides of the identity by the denominator beneath K_1 to get

$$\frac{N(s)}{(s + p_2)(s + p_3)} \equiv K_1 + \frac{K_2(s + p_1)}{(s + p_2)} + \frac{K_3(s + p_1)}{(s + p_3)}.$$

Then when we evaluate both sides of the identity for $s = -p_1$, which is the root of the partial fraction term whose coefficient is K_1:

$$\frac{N(s)}{(s + p_2)(s + p_3)}\bigg|_{s=-p_1}$$

$$\equiv K_1 + \frac{K_2(s + p_1)}{(s + p_2)}\bigg|_{s=-p_1} + \frac{K_3(s + p_1)}{(s + p_3)}\bigg|_{s=-p_1} = K_1.$$

The right-hand side is always the desired K, and the left-hand side is always its numerical value. To find K_2 and K_3, repeat the steps used to find K_1. Example 12.3 illustrates this process.

EXAMPLE 12.3 | **Finding the Inverse Laplace Transform when *F(s)* has Distinct Real Roots**

Use partial fraction expansion to find the inverse Laplace transform of

$$F(s) = \frac{96(s + 5)(s + 12)}{s(s + 8)(s + 6)}.$$

Solution

The partial fraction expansion of $F(s)$ is

$$F(s) = \frac{96(s + 5)(s + 12)}{s(s + 8)(s + 6)} \equiv \frac{K_1}{s} + \frac{K_2}{(s + 8)} + \frac{K_3}{(s + 6)}.$$

To find the value of K_1, we multiply both sides by s and then evaluate both sides at $s = 0$:

$$\frac{96(s + 5)(s + 12)}{(s + 8)(s + 6)}\bigg|_{s=0} \equiv K_1 + \frac{K_2 s}{s + 8}\bigg|_{s=0} + \frac{K_3 s}{s + 6}\bigg|_{s=0},$$

or

$$\frac{96(5)(12)}{8(6)} = K_1 = 120.$$

To find the value of K_2, we multiply both sides by $s + 8$ and then evaluate both sides at $s = -8$:

$$\frac{96(s + 5)(s + 12)}{s(s + 6)}\bigg|_{s=-8}$$

$$\equiv \frac{K_1(s + 8)}{s}\bigg|_{s=-8} + K_2 + \frac{K_3(s + 8)}{(s + 6)}\bigg|_{s=-8},$$

or

$$\frac{96(-3)(4)}{(-8)(-2)} = K_2 = -72.$$

Then K_3 is

$$\frac{96(s + 5)(s + 12)}{s(s + 8)}\bigg|_{s=-6} = K_3 = 48.$$

Therefore,

$$\frac{96(s + 5)(s + 12)}{s(s + 8)(s + 6)} \equiv \frac{120}{s} - \frac{72}{(s + 8)} + \frac{48}{(s + 6)}.$$

It is a good idea to test this result. While the choice of test values is completely open, we choose values that are easy to verify. For example, testing at either -5 or -12 is convenient because in both cases the left-hand side reduces to zero. Choosing -5 yields

$$\frac{120}{-5} - \frac{72}{3} + \frac{48}{1} = -24 - 24 + 48 = 0,$$

whereas testing -12 gives

$$\frac{120}{-12} - \frac{72}{-4} + \frac{48}{-6} = -10 + 18 - 8 = 0.$$

Now find the inverse transform of F(s) using the tables of functional and operational transforms (Tables 12.1 and 12.2):

$$\mathcal{L}^{-1}\left\{\frac{96(s + 5)(s + 12)}{s(s + 8)(s + 6)}\right\} = (120 - 78e^{-8t} + 48e^{-6t})u(t).$$

ASSESSMENT PROBLEMS

Objective 2—Be able to calculate the inverse Laplace transform using partial fraction expansion and the Laplace transform table

12.3 Find $f(t)$ if

$$F(s) = \frac{6s^2 + 26s + 26}{(s+1)(s+2)(s+3)}.$$

Answer: $f(t) = (3e^{-t} + 2e^{-2t} + e^{-3t})u(t)$.

12.4 Find $f(t)$ if

$$F(s) = \frac{7s^2 + 63s + 134}{(s+3)(s+4)(s+5)}.$$

Answer: $f(t) = (4e^{-3t} + 6e^{-4t} - 3e^{-5t})u(t)$.

12.5 Find $v(t)$ for the circuit analyzed in Example 12.2 and shown in Fig. 12.16.

Answer: $v(t) = (16e^{-20,000t} - 16e^{-80,000t})u(t)$ V.

SELF-CHECK: Also try Chapter Problems 12.40(a) and (b).

Partial Fraction Expansion: Distinct Complex Roots of $D(s)$

We begin by noting that if $F(s)$ describes the Laplace transform of a voltage or current in a physically realizable circuit, the factors of $D(s)$ that have complex roots will always come in conjugate pairs. Let's assume that

$$F(s) = \frac{A(s + z_1)}{(s + \alpha - j\beta)(s + \alpha + j\beta)} \equiv \frac{K_1}{(s + \alpha - j\beta)} + \frac{K_2}{(s + \alpha + j\beta)}.$$

Here we assume that z_1 is a real number.

We find the unknown coefficients, K_1 and K_2, using the same technique we employed when the factors of $D(s)$ were real and distinct. As you will see, the only difference is that the algebra involves complex numbers. To find K_1, we multiply both sides of the identity by the denominator beneath K_1 to get

$$\frac{A(s + z_1)}{(s + \alpha + j\beta)} \equiv K_1 + \frac{K_2(s + \alpha - j\beta)}{(s + \alpha + j\beta)}.$$

Then when we evaluate both sides of the identity for $s = -\alpha + j\beta$, which is the root of the partial fraction term whose coefficient is K_1:

$$\left.\frac{A(s + z_1)}{(s + \alpha + j\beta)}\right|_{s=-\alpha+j\beta} = K_1 + \left.\frac{K_2(s + \alpha - j\beta)}{(s + \alpha + j\beta)}\right|_{s=-\alpha+j\beta} = K_1.$$

We evaluate the left-hand side of the identity to get an expression for K_1:

$$K_1 = \frac{A(-\alpha + z_1 + j\beta)}{(j2\beta)} = \frac{A}{2} + j\left(\frac{\alpha - z_1}{2\beta}\right).$$

As expected, K_1 is a complex number. We repeat the steps used to find K_1 when finding the expression for K_2:

$$\frac{A(s + z_1)}{(s + \alpha - j\beta)} = \frac{K_1(s + \alpha + j\beta)}{(s + \alpha - j\beta)} + K_2;$$

$$\left.\frac{A(s + z_1)}{(s + \alpha - j\beta)}\right|_{s=-\alpha-j\beta} = \left.\frac{K_1(s + \alpha + j\beta)}{(s + \alpha - j\beta)}\right|_{s=-\alpha-j\beta} + K_2;$$

$$K_2 = \frac{A(-\alpha + z_1 - j\beta)}{(-j2\beta)} = \frac{A}{2} - j\left(\frac{\alpha - z_1}{2\beta}\right).$$

Compare the expression for K_1 with the expression for K_2; K_1 and K_2 are conjugates. This will always be the case when $D(s)$ has complex roots, so we need only calculate one coefficient, since the other is its conjugate. We can use the following general polar form for K_1 and K_2:

$$K_1 = |K|e^{j\theta} \quad \text{and} \quad K_2 = |K|e^{-j\theta}.$$

Using the polar form for the general coefficients K_1 and K_2, we can find the inverse Laplace transform of F(s), with the help of the functional and operational transform tables (Tables 12.1 and 12.2) and Euler's identity:

$$\mathcal{L}^{-1}\left\{\frac{|K|e^{j\theta}}{s + \alpha - j\beta} + \frac{|K|e^{-j\theta}}{s + \alpha + j\beta}\right\} = |K|e^{j\theta}e^{-(\alpha-j\beta)t} + |K|e^{-j\theta}e^{-(\alpha+j\beta)t}$$

$$= |K|e^{-\alpha t}e^{j(\beta t+\theta)} + |K|e^{-\alpha t}e^{-j(\beta t+\theta)}$$

$$= |K|e^{-\alpha t}[\cos(\beta t + \theta) + j\sin(\beta t + \theta)]$$

$$+ |K|e^{-\alpha t}[\cos(\beta t + \theta) - j\sin(\beta t + \theta)]$$

$$= 2|K|e^{-\alpha t}\cos(\beta t + \theta).$$

Because distinct complex roots appear frequently in lumped-parameter linear circuit analysis, we summarize these results with a new transform pair. Whenever $D(s)$ contains distinct complex roots—that is, factors of the form $(s + \alpha - j\beta)(s + \alpha + j\beta)$—a pair of terms of the form

$$\frac{K}{s + \alpha - j\beta} + \frac{K^*}{s + \alpha + j\beta}$$

appears in the partial fraction expansion, where the partial fraction coefficient is, in general, a complex number. In polar form,

$$K = |K|e^{j\theta} = |K|\underline{/\theta}, \tag{12.8}$$

where $|K|$ denotes the magnitude of the complex coefficient. Then

$$K^* = |K|e^{-j\theta} = |K|\underline{/-\theta}. \tag{12.9}$$

The pair of complex conjugate partial fraction terms always inverse-transforms as

$$\mathcal{L}^{-1}\left\{\frac{K}{s + \alpha - j\beta} + \frac{K^*}{s + \alpha + j\beta}\right\}$$

$$= 2|K|e^{-\alpha t}\cos(\beta t + \theta). \tag{12.10}$$

In applying Eq. 12.10, it is important to note that K is defined as the coefficient associated with the denominator $s + \alpha - j\beta$, and K^* is defined as the coefficient associated with the denominator $s + \alpha + j\beta$.

Example 12.4 finds the inverse Laplace transform of an s-domain function with distinct complex roots.

EXAMPLE 12.4   **Finding the Inverse Laplace Transform when $F(s)$ has Distinct Complex Roots**

Use partial fraction expansion to find the inverse Laplace transform of

$$F(s) = \frac{100(s + 3)}{(s + 6)(s^2 + 6s + 25)}.$$

Solution

We begin by noting that $F(s)$ is a proper rational function. Next we find the roots of the quadratic term $s^2 + 6s + 25$:

$$s^2 + 6s + 25 = (s + 3 - j4)(s + 3 + j4).$$

With the denominator in factored form, we create the partial fraction expansion:

$$\frac{100(s + 3)}{(s + 6)(s^2 + 6s + 25)} \equiv$$

$$\frac{K_1}{s + 6} + \frac{K_2}{s + 3 - j4} + \frac{K_2^*}{s + 3 + j4}.$$

Find K_1 and K_2 using the same process employed in Example 12.3:

$$K_1 = \frac{100(s + 3)}{s^2 + 6s + 25}\bigg|_{s=-6} = \frac{100(-3)}{25} = -12,$$

$$K_2 = \frac{100(s + 3)}{(s + 6)(s + 3 + j4)}\bigg|_{s=-3+j4} = \frac{100(j4)}{(3 + j4)(j8)}$$

$$= 6 - j8 = 10e^{-j53.13°}.$$

Thus

$$\frac{100(s + 3)}{(s + 6)(s^2 + 6s + 25)}$$

$$\equiv \frac{-12}{s + 6} + \frac{10\,\angle{-53.13°}}{s + 3 - j4} + \frac{10\,\angle{53.13°}}{s + 3 + j4}.$$

Before inverse-transforming the terms in the partial fraction expansion, we check the expansion numerically. We test using $s = -3$ because the left-hand side reduces to zero at this value:

$$F(s) = \frac{-12}{3} + \frac{10\,\angle{-53.13°}}{-j4} + \frac{10\,\angle{53.13°}}{j4}$$

$$= -4 + 2.5\,\angle{36.87°} + 2.5\,\angle{-36.87°}$$

$$= -4 + 2.0 + j1.5 + 2.0 - j1.5 = 0.$$

Finally, perform the inverse-transform, using the functional and operational transform tables (Tables 12.1 and 12.2) and the new transform pair in Eq. 12.10:

$$\mathcal{L}^{-1}\left\{ \frac{100(s + 3)}{(s + 6)(s^2 + 6s + 25)} \right\}$$

$$= [-12e^{-6t} + 20e^{-3t}\cos(4t - 53.13°)]u(t).$$

◢ ASSESSMENT PROBLEMS

Objective 2—Be able to calculate the inverse Laplace transform using partial fraction expansion and the Laplace transform table

12.6 Find $f(t)$ if

$$F(s) = \frac{10(s^2 + 119)}{(s + 5)(s^2 + 10s + 169)}.$$

Answer: $f(t) = (10e^{-5t} - 8.33e^{-5t}\sin 12t)u(t)$.

12.7 Suppose for the circuit in Fig. 12.16, $I_{dc} = 24$ mA, $R = 625$ Ω, $L = 25$ mH, and $C = 25$ nF. These are the same values used in Example 12.2 except for the value of R. There is no energy stored in the circuit when the switch opens at $t = 0$.

a) Find $V(s)$, the Laplace transform of $v(t)$.

b) Find $v(t)$ by finding the inverse transform of the partial fraction expansion of $V(s)$.

Answer: (a) $V(s) = \dfrac{96 \times 10^4}{s^2 + 64{,}000s + 16 \times 10^8}$;

(b) $v(t) = 40e^{-32{,}000t} \cos(24{,}000t - 90°)u(t)$ V.

SELF-CHECK: Also try Chapter Problems 12.41(c) and (d).

Partial Fraction Expansion: Repeated Real Roots of $D(s)$

We describe two methods for finding the partial fraction coefficients for terms generated by a multiple root with multiplicity r. Method A requires you to solve several simultaneous algebraic equations. Method B is a modification of the process we have been using to find the coefficients for the partial fraction terms associated with distinct roots. Both methods begin with the identity relating the original s-domain function, which must be a proper rational function, and its partial fraction expansion:

$$\frac{N(s)}{D(s)} \equiv \frac{K_1}{(s + p)^r} + \frac{K_2}{(s + p)^{r-1}} + \cdots + \frac{K_r}{(s + p)}.$$

Method A

1. Combine all terms in the partial fraction expansion over the common denominator, $D(s)$. Call the numerator of this new function $N_1(s)$.

2. Collect all of the terms in the numerator $N_1(s)$ according to their power of s. The coefficient of each power of s will include some of the unknown partial fraction coefficients.

3. Equate the coefficient of each power of s in $N_1(s)$ with the coefficient of the corresponding power of s in $N(s)$. The result is a collection of simultaneous equations whose unknowns are the partial fraction coefficients.

4. Solve the simultaneous equations to find the partial fraction coefficients.

Method B

1. Multiply both sides of the identity defining the partial fraction expansion by the multiple root raised to its rth power. Call the resulting identity I_r.

2. Find K in the numerator of the factor raised to the rth power by evaluating both sides of I_r at the multiple root.

3. To find K in the numerator of the factor raised to the $(r - 1)$ power, differentiate both sides of I_r with respect to s. Call the resulting identity $I_{(r-1)}$. Evaluate both sides of $I_{(r-1)}$ at the multiple root. The right-hand side is always the desired K, and the left-hand side is always its numerical value.

4. Repeat Step 3 to find the remaining partial fraction coefficients by differentiating $I_{(r-1)}$ to get $I_{(r-2)}$ and so on. In total, you will have differentiated $I_r (r - 1)$ times.

Example 12.5 uses Method A to find the inverse Laplace transform for an s-domain function with repeated real roots.

| EXAMPLE 12.5 | Finding the Inverse Laplace Transform when *F(s)* has Repeated Real Roots |

Use partial fraction expansion to find the inverse Laplace transform of

$$F(s) = \frac{100(s + 25)}{s(s + 5)^3}.$$

Solution

We begin by noting that $F(s)$ is a proper rational function. Next we find partial fraction expansion of $F(s)$:

$$\frac{100(s + 25)}{s(s + 5)^3} \equiv \frac{K_1}{s} + \frac{K_2}{(s + 5)^3} + \frac{K_3}{(s + 5)^2} + \frac{K_4}{s + 5}.$$

We will use Method A to find the partial fraction coefficients. Begin by multiplying the numerator and denominator of each term in the partial fraction expansion with an expression that creates a denominator of $s(s + 5)^3$ in each term:

$$\frac{100(s + 25)}{s(s + 5)^3} \equiv \frac{K_1(s + 5)^3}{s(s + 5)^3} + \frac{K_2 s}{s(s + 5)^3}$$

$$+ \frac{K_3 s(s + 5)}{s(s + 5)^3} + \frac{K_4 s(s + 5)^2}{s(s + 5)^3}.$$

Combine the terms on the right-hand side over their common denominator, and expand the resulting numerator by collectiong the coefficients for each power of s. The numerator on the right-hand side is

$$(K_1 + K_4)s^3 + (15K_1 + K_3 + 10K_4)s^2$$
$$+ (35K_1 + K_2 + 5K_3 + 25K_4)s$$
$$+ (125K_1).$$

Equate the coefficients of each power of s in the numerators on the right-hand side and left-hand side to create four simultaneous equations:

$$K_1 + K_4 = 0;$$

$$15K_1 + K_3 + 10K_4 = 0;$$

$$35K_1 + K_2 + 5K_3 + 25K_4 = 100;$$

$$125K_1 = 2500.$$

Solve the simultaneous equations to find

$$K_1 = 20; \quad K_2 = -400; \quad K_3 = -100; \quad K_4 = -20.$$

Therefore, the partial fraction expansion is

$$\frac{100(s + 25)}{s(s + 5)^3} \equiv \frac{20}{s} - \frac{400}{(s + 5)^3} - \frac{100}{(s + 5)^2} - \frac{20}{s + 5}.$$

At this point, we can check our expansion by testing both sides at $s = -25$; for this value of s, both sides should equal zero. The result of evaluating the partial fraction expansion at $s = -25$ is

$$\frac{20}{-25} - \frac{400}{(-20)^3} - \frac{100}{(-20)^2} - \frac{20}{(-20)} = 0.$$

Use the functional and operational transform tables (Tables 12.1 and 12.2) to transform each term in the partial fraction expansion. Thus, the inverse transform of $F(s)$ is

$$\mathcal{L}^{-1}\left\{\frac{100(s + 25)}{s(s + 5)^3}\right\} = \left[20 - 200t^2 e^{-5t} - 100te^{-5t} - 20e^{-5t}\right]u(t).$$

▗ ASSESSMENT PROBLEMS

Objective 2—Be able to calculate the inverse Laplace transform using partial fraction expansion and the Laplace transform table

12.8 Find *f(t)* if

$$F(s) = \frac{(4s^2 + 7s + 1)}{s(s + 1)^2}.$$

Answer: $f(t) = (1 + 2te^{-t} + 3e^{-t})u(t)$.

12.9 Suppose for the circuit in Fig. 12.16, $I_{dc} = 24$ mA, $R = 500\ \Omega, L = 25$ mH, and $C = 25$ nF. These are the same values used in Example 12.2 and

Assessment Problem 12.7 except for the value of R. There is no energy stored in the circuit when the switch opens at $t = 0$.
a) Find $V(s)$, the Laplace transform of $v(t)$.
b) Find $v(t)$ by finding the inverse transform of the partial fraction expansion of $V(s)$.

Answer: (a) $V(s) = \dfrac{96 \times 10^4}{s^2 + 80{,}000s + 16 \times 10^8};$

(b) $v(t) = 96 \times 10^4 te^{-40{,}000t}u(t)$ V.

SELF-CHECK: Also try Chapter Problems 12.42(b) and (c).

Partial Fraction Expansion: Repeated Complex Roots of $D(s)$

We can find the coefficients of the partial fraction terms corresponding to repeated complex roots using either Method A or Method B. The algebra involves complex numbers. Recall that complex roots always appear in conjugate pairs and that the coefficients associated with a conjugate pair are also conjugates, so that only half the Ks need to be evaluated. We illustrate the process using Method B in Example 12.6.

EXAMPLE 12.6	**Finding the Inverse Laplace Transform when $F(s)$ has Repeated Complex Roots**

Use partial fraction expansion to find the inverse Laplace transform of

$$F(s) = \frac{768}{(s^2 + 6s + 25)^2}.$$

Solution

After factoring the denominator polynomial, we write

$$F(s) = \frac{768}{(s + 3 - j4)^2(s + 3 + j4)^2}$$

$$\equiv \frac{K_1}{(s + 3 - j4)^2} + \frac{K_2}{s + 3 - j4}$$

$$+ \frac{K_1^*}{(s + 3 + j4)^2} + \frac{K_2^*}{s + 3 + j4}.$$

Now we need to evaluate only K_1 and K_2, because K_1^* and K_2^* are conjugate values. We use Method B to find these two partial fraction coefficients. The value of K_1 is

$$K_1 = \frac{768}{(s + 3 + j4)^2}\bigg|_{s=-3+j4} = \frac{768}{(j8)^2} = -12.$$

To find the value of K_2, multiply $F(s)$ by $(s + 3 - j4)^2$, find the first derivative of the result with respect to s, and evaluate for $s = -3 + j4$:

$$K_2 = \frac{d}{ds}\left[\frac{768}{(s + 3 + j4)^2}\right]_{s=-3+j4}$$

$$= \frac{-2(768)}{(s + 3 + j4)^3}\bigg|_{s=-3+j4}$$

$$= \frac{-2(768)}{(j8)^3} = -j3 = 3\;\underline{/-90°}.$$

From the values for K_1 and K_2, $K_1^* = -12$, and $K_2^* = j3 = 3\;\underline{/90°}$. Group the partial fraction expansion by conjugate terms to obtain

$$F(s) = \left[\frac{-12}{(s + 3 - j4)^2} + \frac{-12}{(s + 3 + j4)^2}\right]$$

$$+ \left[\frac{3\;\underline{/-90°}}{(s + 3 - j4)} + \frac{3\;\underline{/90°}}{(s + 3 + j4)}\right].$$

Inverse-transform $F(s)$ by applying the functional and operational transforms (Tables 12.1 and 12.2) to the terms in the partial fraction expansion. The result is

$$f(t) = [-24te^{-3t}\cos 4t + 6e^{-3t}\cos(4t - 90°)]u(t).$$

ASSESSMENT PROBLEM

Objective 2—Be able to calculate the inverse Laplace transform using partial fraction expansion and the Laplace transform table

12.10 Find $f(t)$ if

$$F(s) = \frac{40}{(s^2 + 4s + 5)^2}.$$

SELF-CHECK: Also try Chapter Problem 12.43(b).

Answer: $f(t) = (-20te^{-2t}\cos t + 20e^{-2t}\sin t)u(t).$

Note that if $F(s)$ has a real root a of multiplicity r in its denominator, the partial fraction expansion has a term of the form

$$\frac{K}{(s + a)^r}.$$

The inverse transform of this term is

$$\mathcal{L}^{-1}\left\{\frac{K}{(s + a)^r}\right\} = \frac{Kt^{r-1}e^{-at}}{(r - 1)!}u(t). \qquad (12.11)$$

If $F(s)$ has a complex root of $\alpha + j\beta$ of multiplicity r in its denominator, the partial fraction expansion has a conjugate pair of terms in the form

$$\frac{K}{(s + \alpha - j\beta)^r} + \frac{K^*}{(s + \alpha + j\beta)^r}.$$

The inverse transform of this pair is

$$\mathcal{L}^{-1}\left\{\frac{|K|\underline{/\theta}}{(s + \alpha - j\beta)^r} + \frac{|K|\underline{/-\theta}}{(s + \alpha + j\beta)^r}\right\}$$

$$= \left[\frac{2|K|t^{r-1}}{(r - 1)!}e^{-\alpha t}\cos(\beta t + \theta)\right]u(t). \qquad (12.12)$$

Equations 12.11 and 12.12 are the key to finding the inverse transform for any partial fraction expansion with repeated roots. One further note regarding these two equations: In circuit analysis problems, r is seldom greater than 2. Therefore, the inverse transform of a rational function can be handled with four transform pairs. Table 12.3 lists these pairs.

Partial Fraction Expansion: Improper Rational Functions

An improper rational function can always be written as the sum of a polynomial and a proper rational function. The polynomial is then inverse-transformed into impulse functions and derivatives of impulse functions, while the proper rational function is inverse-transformed by the techniques outlined in this section.

We illustrate the procedure in Example 12.7.

Pair Number	Nature of Roots	$F(s)$	$f(t)$
1	Distinct real	$\dfrac{K}{s + a}$	$Ke^{-at}u(t)$
2	Repeated real	$\dfrac{K}{(s + a)^2}$	$Kte^{-at}u(t)$
3	Distinct complex	$\dfrac{K}{s + \alpha - j\beta} + \dfrac{K^*}{s + \alpha + j\beta}$	$2|K|e^{-\alpha t}\cos(\beta t + \theta)u(t)$
4	Repeated complex	$\dfrac{K}{(s + \alpha - j\beta)^2} + \dfrac{K^*}{(s + \alpha + j\beta)^2}$	$2t|K|e^{-\alpha t}\cos(\beta t + \theta)u(t)$

TABLE 12.3 Four Useful Transform Pairs

Note: In pairs 1 and 2, K is a real quantity, whereas in pairs 3 and 4, K is the complex quantity $|K|\underline{/\theta}$.

EXAMPLE 12.7	**Finding the Inverse Laplace Transform of an Improper Rational Function**

Use partial fraction expansion to find the inverse Laplace transform of

$$F(s) = \frac{s^4 + 13s^3 + 66s^2 + 200s + 300}{s^2 + 9s + 20}.$$

Solution

The order of the numerator is 4, while the order of the denominator is 2, so $F(s)$ is an improper rational function. To write it as the sum of a polynomial and a proper rational function, divide the denominator into the numerator until the remainder is a proper rational function. The result is

$$F(s) = (s^2 + 9s + 20)\overline{)s^4 + 13s^3 + 66s^2 + 200s + 300}$$

$$= s^2 + 4s + 10 + \frac{30s + 10}{s^2 + 9s + 20}.$$

Next, expand the proper rational function into a sum of partial fractions:

$$\frac{30s + 100}{s^2 + 9s + 20} = \frac{30s + 100}{(s + 4)(s + 5)} \equiv \frac{-20}{s + 4} + \frac{50}{s + 5}.$$

Replace the proper rational function in $F(s)$ with the partial fraction expansion to get

$$F(s) = s^2 + 4s + 10 - \frac{20}{s + 4} + \frac{50}{s + 5}.$$

Using the tables of functional and operational transforms (Tables 12.1 and 12.2) we can now inverse-transform $F(s)$. Hence

$$f(t) = \frac{d^2\delta(t)}{dt^2} + 4\frac{d\delta(t)}{dt} + 10\delta(t) - (20e^{-4t} - 50e^{-5t})u(t).$$

ASSESSMENT PROBLEMS

Objective 2—Be able to calculate the inverse Laplace transform using partial fraction expansion and the Laplace transform table

12.11 Find $f(t)$ if

$$F(s) = \frac{(5s^2 + 29s + 32)}{(s + 2)(s + 4)}.$$

Answer: $5\delta(t) - (3e^{-2t} - 2e^{-4t})u(t)$.

12.12 Find $f(t)$ if

$$F(s) = \frac{(2s^3 + 8s^2 + 2s - 4)}{(s^2 + 5s + 4)}.$$

Answer: $2\frac{d\delta(t)}{dt} - 2\delta(t) + 4e^{-4t}u(t)$.

SELF-CHECK: Also try Chapter Problem 12.43(d).

12.8 Poles and Zeros of *F*(s)

The rational function of Eq. 12.7 can also be expressed as the ratio of two factored polynomials. In other words, we can write $F(s)$ as

$$F(s) = \frac{K(s + z_1)(s + z_2) \cdots (s + z_n)}{(s + p_1)(s + p_2) \cdots (s + p_m)}, \qquad (12.13)$$

where K is the constant a_n/b_m.

The roots of the denominator polynomial, that is, $-p_1, -p_2, -p_3, \ldots, -p_m$, are called the **poles of *F*(s)**; they are the values of s at which $F(s)$ becomes infinitely large. The roots of the numerator polynomial, that is, $-z_1, -z_2, -z_3, \ldots, -z_n$ are called the **zeros of *F*(s)**; they are the values of s at which $F(s)$ becomes zero. We can visualize the poles and zeros of $F(s)$ as points on a complex s plane. In the complex s plane, we use the horizontal axis to plot the real values of s and the vertical axis to plot the imaginary values of s. Example 12.8 finds the poles and zeros for two different functions of s and plots the locations of the poles and zeros on the complex plane.

| **EXAMPLE 12.8** | **Finding and Plotting the Poles and Zeros of an *s*-Domain Function** |

a) Suppose $F_1(s)$ is given by

$$F_1(s) = \frac{40s^3 + 440s^2 + 2200s + 5000}{4s^4 + 88s^3 + 880s^2 + 4000s}.$$

Find the poles and zeros of $F_1(s)$ and plot them on the complex s-plane.

b) Suppose $F_2(s)$ is given by

$$F_2(s) = \frac{8s^2 + 120s + 400}{2s^4 + 20s^3 + 70s^2 + 100s + 48}.$$

Find the poles and zeros of $F_2(s)$ and plot them on the complex s-plane.

Solution

a) Begin by factoring out a constant in the numerator and denominator of $F_1(s)$. Then factor the numerator and denominator polynomials. The result is

$$F_1(s) = \frac{40(s^3 + 11s^2 + 55s + 125)}{4(s^4 + 22s^3 + 220s^2 + 1000s)}$$

$$= \frac{10(s + 5)(s + 3 - j4)(s + 3 + j4)}{s(s + 10)(s + 6 - j8)(s + 6 + j8)}.$$

The poles of $F_1(s)$ are at 0, -10, $-6 + j8$, and $-6 - j8$. The zeros are at -5, $-3 + j4$, and $-3 - j4$. Figure 12.17 shows the poles and zeros plotted on the s plane, where X's represent poles and O's represent zeros.

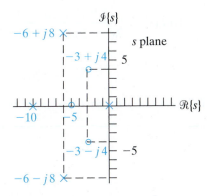

Figure 12.17 ▲ Plotting poles and zeros on the s plane for $F_1(s)$ in Example 12.8.

b) Factor out a constant in the numerator and denominator of $F_2(s)$. Then factor the numerator and denominator polynomials to give

$$F_2(s) = \frac{8(s^2 + 15s + 50)}{2(s^4 + 10s^3 + 35s^2 + 50s + 24)}$$

$$= \frac{4(s + 5)(s + 10)}{(s + 1)(s + 2)(s + 3)(s + 4)}.$$

The zeros of $F_2(s)$ are -5 and -10. The poles of $F_2(s)$ are -1, -2, -3, and -4. They are plotted in the complex s-plane in Fig. 12.18. Note that $F_2(s)$ also has a second-order zero at infinity because for large values of s the function reduces to $4/s^2$, and $F_2(s) = 0$ when $s = \infty$. In general, $F(s)$ can have either an rth-order pole or an rth-order zero at infinity. In this text, we are interested in the poles and zeros located in the finite s plane. Therefore, when we refer to the poles and zeros of a rational function of s, we are referring to the finite poles and zeros.

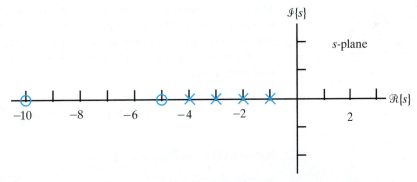

Figure 12.18 ▲ Plotting poles and zeros on the s plane for $F_2(s)$ in Example 10.8.

ASSESSMENT PROBLEM

Objective 3—Be able to find and plot the poles and zeros for a rational function in the s-domain

12.13 Find the poles and zeros for the following rational functions of s.

a) $F(s) = \dfrac{20s^2 + 80s + 100}{s^3 + 10s^2 + 21s}$;

b) $F(s) = \dfrac{10s^2 + 20s + 10}{2s^3 + 28s^2 + 258s + 712}$;

c) $F(s) = \dfrac{125s}{5s^4 + 90s^3 + 670s^2 + 2360s + 3400}$.

Answer: (a) Zeros at $-2 + j$ and $-2 - j$, poles at 0, -3, and -7;
(b) Two zeros at -1, poles at -4, $-5 + j8$, and $-5 - j8$;
(c) Zero at 0, poles at $-5 + j3$, $-5 - j3$, $-4 + j2$, and $-4 - j2$.

SELF-CHECK: Also try Chapter Problem 12.46.

12.9 Initial- and Final-Value Theorems

The initial- and final-value theorems enable us to determine the behavior of $f(t)$ at 0 and ∞ from $F(s)$. Hence, we can check the initial and final values of $f(t)$ to see if they conform to known circuit behavior, before actually finding the inverse transform of $F(s)$.

The initial-value theorem states that

INITIAL-VALUE THEOREM

$$\lim_{t \to 0^+} f(t) = \lim_{s \to \infty} sF(s),$$
(12.14)

and the final-value theorem states that

FINAL-VALUE THEOREM

$$\lim_{t \to \infty} f(t) = \lim_{s \to 0} sF(s).$$
(12.15)

The initial-value theorem assumes that $f(t)$ does not have an impulse function at the origin. The final-value theorem is valid only if the poles of $F(s)$, except for a first-order pole at the origin, lie in the left half of the s plane.

To prove Eq. 12.14, we start with the operational transform of the first derivative:

$$\mathcal{L}\left\{\frac{df}{dt}\right\} = sF(s) - f(0^-) = \int_{0^-}^{\infty} \frac{df}{dt} e^{-st} dt.$$

Now we take the limit as $s \to \infty$:

$$\lim_{s \to \infty} [sF(s) - f(0^-)] = \lim_{s \to \infty} \int_{0^-}^{\infty} \frac{df}{dt} e^{-st} dt.$$
(12.16)

Observe that the right-hand side of Eq. 12.16 can be written as

$$\lim_{s \to \infty} \left(\int_{0^-}^{0^+} \frac{df}{dt} e^0 dt + \int_{0^+}^{\infty} \frac{df}{dt} e^{-st} dt \right).$$

As $s \to \infty$, $(df/dt) e^{-st} \to 0$; hence, the second integral vanishes in the limit. The first integral reduces to $f(0^+) - f(0^-)$, which is independent of s. Thus, the right-hand side of Eq. 12.16 becomes

$$\lim_{s \to \infty} \int_{0^-}^{\infty} \frac{df}{dt} e^{-st} dt = f(0^+) - f(0^-).$$

Because $f(0^-)$ is independent of s, the left-hand side of Eq. 12.16 can be written as

$$\lim_{s \to \infty} [sF(s) - f(0^-)] = \lim_{s \to 0} [sF(s)] - f(0^-).$$

Therefore,

$$\lim_{s \to \infty} sF(s) = f(0^+) = \lim_{t \to 0^+} f(t),$$

which completes the proof of the initial-value theorem.

The proof of the final-value theorem also starts with the operational transform of the first derivative. Here we take the limit as $s \to 0$:

$$\lim_{s \to 0} [sF(s) - f(0^-)] = \lim_{s \to 0} \left(\int_{0^-}^{\infty} \frac{df}{dt} e^{-st} dt \right). \tag{12.17}$$

The integration is with respect to t and the limit operation is with respect to s, so the right-hand side of Eq. 12.17 reduces to

$$\lim_{s \to 0} \left(\int_{0^-}^{\infty} \frac{df}{dt} e^{-st} dt \right) = \int_{0^-}^{\infty} \frac{df}{dt} dt.$$

Because the upper limit on the integral is infinite, this integral may also be written as a limit process:

$$\int_{0^-}^{\infty} \frac{df}{dt} dt = \lim_{t \to \infty} \int_{0^-}^{t} \frac{df}{dy} dy,$$

where we use y as the symbol of integration to avoid confusion with the upper limit on the integral. Carrying out the integration on the right-hand side gives

$$\lim_{t \to \infty} [f(t) - f(0^-)] = \lim_{t \to \infty} [f(t)] - f(0^-).$$

Substituting this expression into the right-hand side of Eq. 12.17 gives

$$\lim_{s \to 0} [sF(s)] - f(0^-) = \lim_{t \to \infty} [f(t)] - f(0^-).$$

Since $f(0^-)$ cancels, we get

$$\lim_{s \to 0} sF(s) = \lim_{t \to \infty} f(t),$$

which completes the proof of the final value theorem.

The final-value theorem is useful only if $f(\infty)$ exists. This condition is true only if all the poles of $F(s)$, except for a single pole at the origin, lie in the left half of the s plane.

Example 12.9 applies the initial- and final-value theorems to the s-domain function from Example 12.2.

EXAMPLE 12.9 **Applying the Initial- and Final-Value Theorems**

Suppose for the circuit in Fig. 12.16, $I_{dc} = 24$ mA, $R = 400\ \Omega$, $L = 25$ mH, and $C = 25$ nF. There is no energy stored in the circuit when the switch opens at $t = 0$. In Example 12.2, we found the Laplace transform of $v(t)$ is

$$V(s) = \frac{96 \times 10^4}{s^2 + 10^5 s + 16 \times 10^8}.$$

Use the initial- and final-value theorems to predict the initial and final values of $v(t)$ and verify that $V(s)$ correctly predicts the values of $v(0^+)$ and $v(\infty)$ from the circuit.

Solution

From the initial-value theorem,

$$\lim_{s \to \infty} sV(s) = \lim_{s \to \infty} \frac{96 \times 10^4 s}{s^2 + 10^5 s + 16 \times 10^8}.$$

To evaluate the limit on the right-hand side, divide numerator and denominator by the highest power of s in the denominator, in this case s^2, and find the limit as $1/s \to 0$:

$$\lim_{s \to \infty} \frac{96 \times 10^4 s}{s^2 + 10^5 s + 16 \times 10^8}$$

$$= \lim_{1/s \to 0} \frac{96 \times 10^4 (1/s)}{1 + 10^5 (1/s) + 16 \times 10^8 (1/s)^2}$$

$$= \frac{0}{1 + 0 + 0} = 0 = \lim_{t \to 0} v(t) = v(0^+).$$

Since the problem states that there is no energy stored in the circuit prior to the switch opening at $t = 0$, we have confirmed that the initial voltage is zero.

Before applying the final-value theorem, find the poles of $V(s)$. They are $-20,000$ and $-80,000$, so both lie in the left-half complex plane, and we can use the final-value theorem to get

$$\lim_{s \to 0} sV(s) = \lim_{s \to 0} \frac{96 \times 10^4 s}{s^2 + 10^5 s + 16 \times 10^8} = 0.$$

As we expected from the circuit, as $t \to \infty$, the final-value theorem gives $v(\infty) = 0$. Thus, $V(s)$ correctly predicts the initial and final values of $v(t)$.

ASSESSMENT PROBLEM

Objective 4—Understand and know how to use the initial-value theorem and the final-value theorem

12.14 Use the initial- and final-value theorems to find the initial and final values of $f(t)$ in Assessment Problems 12.4, 12.8, and 12.10.

Answer: 7, 0; 4, 1; and 0, 0.

SELF-CHECK: Also try Chapter Problem 12.52.

■ **Practical Perspective**

Transient Effects

The circuit introduced in the Practical Perspective at the beginning of the chapter is repeated in Fig. 12.19 with the switch closed and the chosen sinusoidal source.

We use Laplace methods to determine the complete response of the inductor current, i_L. To begin, use KVL to sum the voltage drops around the circuit, in the clockwise direction:

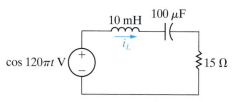

Figure 12.19 ▲ A series *RLC* circuit with a 60 Hz sinusoidal source.

$$15i_L + 0.01 \frac{di_L}{dt} + \frac{1}{100 \times 10^{-6}} \int_0^t i_L(x)dx = \cos 120\pi t.$$

Using Tables 12.1 and 12.2, we find the Laplace transform of the KVL equation:

$$15I_L + 0.01sI_L + 10^4\frac{I_L}{s} = \frac{s}{s^2 + (120\pi)^2}.$$

Solve this equation for I_L:

$$I_L = \frac{100s^2}{[s^2 + 1500s + 10^6][s^2 + (120\pi^2)]}.$$

Note that the expression for I_L has two complex conjugate pairs of poles, so the partial fraction expansion of I_L will have four terms:

$$I_L = \frac{K_1}{(s + 750 - j661.44)} + \frac{K_1^*}{(s + 750 + j661.44)}$$

$$+ \frac{K_2}{(s - j120\pi)} + \frac{K_2^*}{(s + j120\pi)}.$$

Determine the values of K_1 and K_2:

$$K_1 = \frac{100s^2}{[s + 750s + j661.44][s^2 + (120\pi)^2]}\Bigg|_{s=-750+j661.44}$$

$$= 0.07357\underline{/-97.89°},$$

$$K_2 = \frac{100s^2}{[s^2 + 1500s + 10^6][s + j120\pi]}\Bigg|_{s=j120\pi} = 0.018345\underline{/56.61°}.$$

Therefore, the s-domain expression for the inductor current is

$$I_L = \frac{0.07357\underline{/-97.89°}}{(s + 750 - j661.44)} + \frac{0.07357\underline{/97.89°}}{(s + 750 + j661.44)}$$

$$+ \frac{0.018345\underline{/56.61°}}{(s - j120\pi)} + \frac{0.018345\underline{/-56.61°}}{(s + j120\pi)}.$$

Finally, we use Table 12.3 to calculate the inverse Laplace transform and find i_L:

$$i_L = 147.14e^{-750t}\cos(661.44t - 97.89°)$$

$$+ 36.69\cos(120\pi t + 56.61°) \text{ mA}.$$

The first term in the inductor current is the transient response, which will decay to zero in about 7 ms. The second term in the inductor current is the steady-state response, which has the same frequency as the 60 Hz sinusoidal source and will persist as long as this source is connected in the circuit. Note that the amplitude of the steady-state response is 36.69 mA, which is less than the 40 mA current rating of the inductor. But the transient response has an initial amplitude of 147.14 mA, far greater than the 40 mA current rating. Calculate the value of the inductor current at $t = 0$:

$$i_L(0) = 147.14(1)\cos(-97.89°) + 36.69\cos(56.61°) = -6.21\mu A.$$

Clearly, the transient part of the response does not cause the inductor current to exceed its rating initially. But we need a plot of the complete response to determine whether or not the current rating is ever exceeded, as shown in Fig. 12.20.

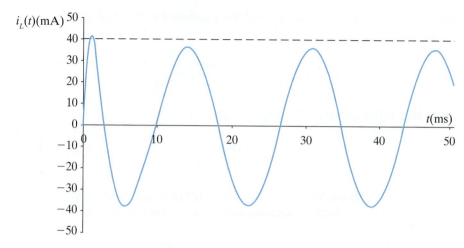

Figure 12.20 ▲ Plot of the inductor current for the circuit in Fig. 12.19.

The plot suggests we check the value of the inductor current at 1 ms:

$$i_L(0.001) = 147.14e^{-0.75}\cos(-59.99°) + 36.69\cos(78.21°) = 42.4 \text{ mA}.$$

Thus, the current rating is exceeded in the inductor, at least momentarily. If we determine that we never want to exceed the current rating, we should reduce the magnitude of the sinusoidal source. This example illustrates the importance of considering the complete response of a circuit to a sinusoidal input, even if we are satisfied with the steady-state response.

SELF-CHECK: Access your understanding of the Practical Perspective by trying Chapter Problems 12.57 and 12.58.

■ Summary

- The **Laplace transform** is a tool for converting time-domain equations into frequency-domain equations, according to the following general definition:

$$\mathcal{L}\{f(t)\} = \int_0^\infty f(t)e^{-st}\,dt = F(s),$$

  where $f(t)$ is the time-domain expression and $F(s)$ is the frequency-domain expression. (See page 446.)

- The **step function** $Ku(t)$ describes a function that experiences a discontinuity from one constant level to another at some point in time. K is the magnitude of the jump; if $K = 1$, $Ku(t)$ is the **unit step function**. (See page 447.)

- The **impulse function** $K\delta(t)$ is defined as

$$\int_{-\infty}^\infty K\delta(t)dt = K,$$

$$\delta(t) = 0, \quad t \neq 0.$$

  K is the strength of the impulse; if $K = 1$, $K\delta(t)$ is the **unit impulse function**. (See page 449.)

- A **functional transform** is the Laplace transform of a specific function. Important functional transform pairs are summarized in Table 12.1. (See page 453.)

- **Operational transforms** define the general mathematical properties of the Laplace transform. Important operational transform pairs are summarized in Table 12.2. (See page 458.)

- In linear lumped-parameter circuits, $F(s)$ is a rational function of s. (See page 460.)

- If $F(s)$ is a proper rational function, the inverse transform is found by a partial fraction expansion. (See page 461.)

- If $F(s)$ is an improper rational function, it can be inverse-transformed by first expanding it into a sum of a polynomial and a proper rational function. (See page 469.)

- $F(s)$ can be expressed as the ratio of two factored polynomials. The roots of the denominator are called **poles** and are plotted as Xs on the complex s plane. The roots of the numerator are called **zeros** and are plotted as 0s on the complex s plane. (See page 470.)

- The initial-value theorem states that

$$\lim_{t \to 0^+} f(t) = \lim_{s \to \infty} sF(s).$$

The theorem assumes that $f(t)$ contains no impulse functions. (See page 472.)

- The final-value theorem states that

$$\lim_{t \to \infty} f(t) = \lim_{s \to 0^+} sF(s).$$

The theorem is valid only if the poles of $F(s)$, except for a first-order pole at the origin, lie in the left half of the s plane. (See page 472.)

- The initial- and final-value theorems allow us to predict the initial and final values of $f(t)$ from an s-domain expression. (See page 474.)

■ Problems

Section 12.2

12.1 Use step functions to write the expression for each of the functions shown in Fig. P12.1.

Figure P12.1

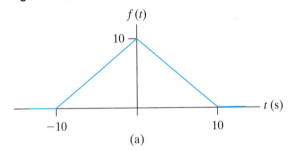

(a)

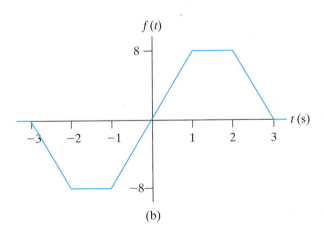

(b)

12.2 Use step functions to write the expression for each function shown in Fig. P12.2.

Figure P12.2

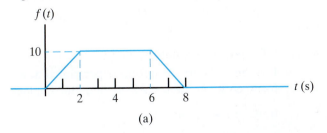

(a)

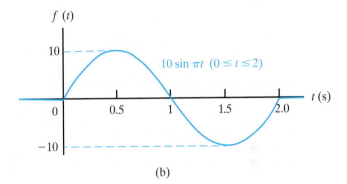

(b)

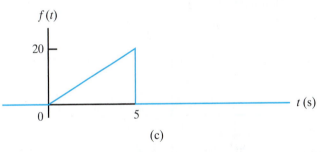

(c)

12.3 Make a sketch of $f(t)$ for -15 s $\le$ t ≤ 35 s when $f(t)$ is given by the following expression:

$$f(t) = (100 + 10t)u(t + 10) - (50 + 10t)u(t + 5)$$
$$+ (50 - 10t)u(t - 5) - (150 - 10t)u(t - 15)$$
$$+ (10t - 250)u(t - 25) - (10t - 300)u(t - 3).$$

12.4 Step functions can be used to define a *window* function. Thus $u(t - 1) - u(t - 4)$ defines a window 1 unit high and 3 units wide located on the time axis between 1 and 4.

A function $f(t)$ is defined as follows:

$$\begin{aligned}
f(t) &= 0, & t \le 0; \\
&= -20t, & 0 \le t \le 1 \text{ s}; \\
&= -20, & 1 \text{ s} \le t \le 2 \text{ s}; \\
&= 2 \cos \frac{\pi}{2} t, & 2 \text{ s} \le t \le 4 \text{ s}; \\
&= 100 - 20t, & 4 \text{ s} \le t \le 5 \text{ s}; \\
&= 0, & 5 \text{ s} \le t < \infty.
\end{aligned}$$

a) Sketch $f(t)$ over the interval $-1\,\text{s} \le t \le 6\,\text{s}$.

b) Use the concept of the window function to write an expression for $f(t)$.

Section 12.3

12.5 Explain why the following function generates an impulse function as $\epsilon \to 0$:

$$f(t) = \frac{\epsilon/\pi}{\epsilon^2 + t^2}, \quad -\infty \le t \le \infty.$$

12.6 a) Find the area under the function shown in Fig. 12.12(a).

b) What is the duration of the function when $\epsilon = 0$?

c) What is the magnitude of $f(0)$ when $\epsilon = 0$?

12.7 The triangular pulses shown in Fig. P12.7 are equivalent to the rectangular pulses in Fig. 12.12(b), because they both enclose the same area $(1/\epsilon)$ and they both approach infinity proportional to $1/\epsilon^2$ as $\epsilon \to 0$. Use this triangular-pulse representation for $\delta'(t)$ to find the Laplace transform of $\delta''(t)$.

Figure P12.7

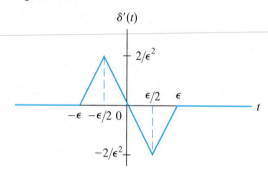

12.8 Evaluate the following integrals:

a) $I = \displaystyle\int_{-1}^{3} (t^3 + 2)[\delta(t) + 8\delta(t - 1)]dt.$

b) $I = \displaystyle\int_{-2}^{2} t^2[\delta(t) + \delta(t + 1.5) + \delta(t - 3)]dt.$

12.9 In Section 12.3, we used the sifting property of the impulse function to show that $\mathcal{L}\{\delta(t)\} = 1$. Show that we can obtain the same result by finding the Laplace transform of the rectangular pulse that exists between $\pm\epsilon$ in Fig. 12.9 and then finding the limit of this transform as $\epsilon \to 0$.

12.10 Find $f(t)$ if

$$f(t) = \frac{1}{2\pi} \int_{-\infty}^{\infty} F(\omega)e^{j\omega t}\,d\omega,$$

and

$$F(\omega) = \frac{4 + j\omega}{9 + j\omega}\,\pi\delta(\omega).$$

12.11 Show that

$$\mathcal{L}\{\delta^{(n)}(t)\} = s^n.$$

12.12 a) Show that

$$\int_{-\infty}^{\infty} f(t)\delta'(t - a)\,dt = -f'(a).$$

(*Hint:* Integrate by parts.)

b) Use the formula in (a) to show that

$$\mathcal{L}\{\delta'(t)\} = s.$$

Sections 12.4–12.5

12.13 Find the Laplace transform of each of the following functions:

a) $f(t) = 20e^{-500(t-10)}u(t - 10).$

b) $f(t) = (5t + 20)[u(t + 4) - u(t + 2)]$
$\qquad - 5t[u(t + 2) - u(t - 2)]$
$\qquad + (5t - 20)[u(t - 2) - u(t - 4)].$

12.14 a) Find the Laplace transform of the function illustrated in Fig. P12.14.

b) Find the Laplace transform of the first derivative of the function illustrated in Fig. P12.14.

c) Find the Laplace transform of the second derivative of the function illustrated in Fig. P12.14.

Figure P12.14

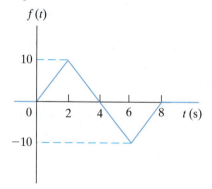

12.15 Show that

$$\mathcal{L}\{e^{-at}f(t)\} = F(s + a).$$

12.16 Show that

$$\mathcal{L}\{f(at)\} = \frac{1}{a}F\left(\frac{s}{a}\right).$$

12.17 a) Find the Laplace transform of te^{-at}.

b) Use the first derivative (time) operational transform given in Table 12.2 to find the Laplace transform of $\dfrac{d}{dt}(te^{-at})$.

c) Check your result in part (b) by first differentiating and then transforming the resulting expression.

12.18 a) Find $\mathcal{L}\left\{\int_{0^-}^{t} e^{-ax}dx\right\}$.

b) Check the results of (a) by first integrating and then transforming.

12.19 a) Find the Laplace transform of

$$\int_{0^-}^{t} x \, dx$$

by first integrating and then transforming.

b) Check the result obtained in (a) by using the time integral operational transform given in Table 12.2.

12.20 Find the Laplace transform of each of the following functions:

a) $f(t) = te^{-at}$;

b) $f(t) = \sin \omega t$;

c) $f(t) = \sin(\omega t + \theta)$;

d) $f(t) = t$;

e) $f(t) = \cosh(t + \theta)$.

(*Hint:* See Assessment Problem 12.1.)

12.21 a) Given that $F(s) = \mathcal{L}\{f(t)\}$, show that

$$-\frac{dF(s)}{ds} = \mathcal{L}\{t f(t)\}.$$

b) Show that

$$(-1)^n \frac{d^n F(s)}{ds^n} = \mathcal{L}\{t^n f(t)\}.$$

c) Use the result of (b) to find $\mathcal{L}\{t^5\}$, $\mathcal{L}\{t \sin \beta t\}$, and $\mathcal{L}\{te^{-1}\cosh t\}$.

12.22 a) Find $\mathcal{L}\left\{\frac{d}{dt}\sin \omega t\right\}$.

b) Find $\mathcal{L}\left\{\frac{d}{dt}\cos \omega t\right\}$.

c) Find $\mathcal{L}\left\{\frac{d^3}{dt^3}t^2 u(t)\right\}$.

d) Check the results of parts (a), (b), and (c) by first differentiating and then transforming.

12.23 Find the Laplace transform (when $\epsilon \to 0$) of the derivative of the exponential function illustrated in Fig. 12.8, using each of the following two methods:

a) First differentiate the function and then find the transform of the resulting function.

b) Use the first derivative (time) operational transform given in Table 12.2.

12.24 Find the Laplace transform for (a) and (b).

a) $f(t) = \frac{d}{dt}(e^{-at}\cos \omega t)$.

b) $f(t) = \int_{0^-}^{t} e^{-ax}\sin \omega x \, dx$.

c) Verify the results obtained in (a) and (b) by first carrying out the indicated mathematical operation and then finding the Laplace transform.

12.25 a) Show that if $F(s) = \mathcal{L}\{f(t)\}$, and $\{f(t)/t\}$ is Laplace-transformable, then

$$\int_{s}^{\infty} F(u)du = \mathcal{L}\left\{\frac{f(t)}{t}\right\}.$$

(*Hint:* Use the defining integral to write

$$\int_{s}^{\infty} F(u)du = \int_{s}^{\infty}\left(\int_{0^-}^{\infty} f(t)e^{-ut}dt\right)du$$

and then reverse the order of integration.)

b) Starting with the Laplace transform of $t \sin \beta t$, from Problem 12.21(c), use the operational transform given in (a) of this problem to find $\mathcal{L}\{\sin \beta t\}$.

Section 12.6

12.26 In the circuit shown in Fig. 12.16, the dc current source is replaced with a sinusoidal source that delivers a current of $1.2 \cos t$ A. The circuit components are $R = 1\ \Omega$, $C = 625$ mF, and $L = 1.6$ H. Find the numerical expression for $V(s)$.

12.27 There is no energy stored in the circuit shown in Fig. P12.27 at the time the switch is opened. In Section 12.6, we derived the integrodifferential equation that governs the behavior of the voltage v_o. We also showed that the Laplace transform of v_o is

$$V_o(s) = \frac{I_{dc}/C}{s^2 + (1/RC)s + (1/LC)}.$$

Use $V_o(s)$ to show that the Laplace transform of i_o is

$$I_o(s) = \frac{sI_{dc}}{s^2 + (1/RC)s + (1/LC)}.$$

Figure P12.27

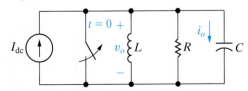

12.28 The switch in the circuit in Fig. P12.28 has been in position a for a long time. At $t = 0$, the switch moves instantaneously to position b.

a) Derive the integrodifferential equation that governs the behavior of the current i_o for $t \geq 0^+$.

b) Show that

$$I_o(s) = \frac{I_{dc}[s + (1/RC)]}{[s^2 + (1/RC)s + (1/LC)]}.$$

Figure P12.28

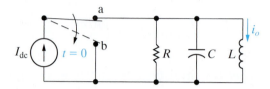

12.29 The switch in the circuit in Fig. P12.29 has been in position a for a long time. At $t = 0$, the switch moves instantaneously to position b.

a) Derive the integrodifferential equation that governs the behavior of the voltage v_o for $t \geq 0^+$.

b) Show that

$$V_o(s) = \frac{V_{dc}[s + (R/L)]}{[s^2 + (R/L)s + (1/LC)]}.$$

Figure P12.29

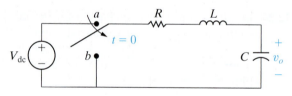

12.30 The switch in the circuit in Fig. P12.30 has been open for a long time. At $t = 0$, the switch closes.

a) Derive the integrodifferential equation that governs the behavior of the voltage v_o for $t \geq 0$.

b) Show that

$$V_o(s) = \frac{V_{dc}/RC}{s^2 + (1/RC)s + (1/LC)}.$$

c) Show that

$$I_o(s) = \frac{V_{dc}/RLC}{s[s^2 + (1/RC)s + (1/LC)]}.$$

Figure P12.30

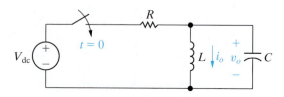

12.31 There is no energy stored in the circuit shown in Fig. P12.31 at the time the switch is opened.

a) Derive the integrodifferential equations that govern the behavior of the node voltages v_1 and v_2.

b) Show that

$$V_2(s) = \frac{sI_g(s)}{C[s^2 + (R/L)s + (1/LC)]}.$$

Figure P12.31

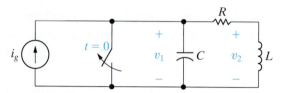

12.32 a) Write the two simultaneous differential equations that describe the circuit shown in Fig. P12.32 in terms of the mesh currents i_1 and i_2.

b) Laplace-transform the equations derived in (a). Assume that the initial energy stored in the circuit is zero.

c) Solve the equations in (b) for $I_1(s)$ and $I_2(s)$.

Figure P12.32

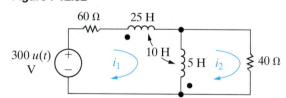

Section 12.7

12.33 Find $v(t)$ in Problem 12.26.

12.34 The parameter values for the circuit in Fig. P12.27 are as follows: $R = 1\,k\Omega$, $L = 12.5\,H$, $C = 2\,\mu F$, and $I_{dc} = 30\,mA$.

a) Find $v_o(t)$ for $t \geq 0$.

b) Find $i_o(t)$ for $t \geq 0$.

c) Does your solution for $i_o(t)$ make sense when $t = 0$? Explain.

12.35 The circuit parameters in the circuit in Fig. P12.28 are $R = 500\,\Omega$, $L = 250\,mH$, and $C = 250\,nF$. If $I_{dc} = 5\,mA$, find $i_o(t)$ for $t \geq 0$.

12.36 The circuit parameters in the circuit in Fig. P12.29 are $R = 4\,k\Omega$, $L = 400\,mH$, and $C = 156.25\,nF$. If $V_{dc} = 120\,V$, find $v_o(t)$ for $t \geq 0$.

12.37 The circuit parameters in the circuit in Fig. P12.30 are $R = 5\,k\Omega$; $L = 200\,mH$; and $C = 100\,nF$. If V_{dc} is 35 V, find

a) $v_o(t)$ for $t \geq 0$

b) $i_o(t)$ for $t \geq 0$

12.38 The circuit parameters in the circuit in Fig. P12.31
PSPICE are $R = 1600\ \Omega$; $L = 200$ mH; and $C = 200$ nF. If
MULTISIM $i_g(t) = 6$ mA, find $v_2(t)$.

12.39 Use the results from Problem 12.32 and the circuit shown in Fig P12.32 to

a) Find $i_1(t)$ and $i_2(t)$.

b) Find $i_1(\infty)$ and $i_2(\infty)$.

c) Do the solutions for i_1 and i_2 make sense? Explain.

12.40 Find $f(t)$ for each of the following functions:

a) $F(s) = \dfrac{8s^2 + 37s + 32}{(s+1)(s+2)(s+4)}$.

b) $F(s) = \dfrac{8s^3 + 89s^2 + 311s + 300}{s(s+2)(s^2 + 8s + 15)}$.

c) $F(s) = \dfrac{22s^2 + 60s + 58}{(s+1)(s^2 + 4s + 5)}$.

d) $F(s) = \dfrac{250(s+7)(s+14)}{s(s^2 + 14s + 50)}$.

12.41 Find $f(t)$ for each of the following functions.

a) $F(s) = \dfrac{280}{s^2 + 14s + 245}$.

b) $F(s) = \dfrac{-s^2 + 52s + 445}{s(s^2 + 10s + 89)}$.

c) $F(s) = \dfrac{14s^2 + 56s + 152}{(s+6)(s^2 + 4s + 20)}$.

d) $F(s) = \dfrac{8(s+1)^2}{(s^2 + 10s + 34)(s^2 + 8s + 20)}$.

12.42 Find $f(t)$ for each of the following functions.

a) $F(s) = \dfrac{320}{s^2(s+8)}$.

b) $F(s) = \dfrac{80(s+3)}{s(s+2)^2}$.

c) $F(s) = \dfrac{60(s+5)}{(s+1)^2(s^2 + 6s + 25)}$.

d) $F(s) = \dfrac{25(s+4)^2}{s^2(s+5)^2}$.

12.43 Find $f(t)$ for each of the following functions.

a) $F(s) = \dfrac{135}{s(s+3)^3}$.

b) $F(s) = \dfrac{10(s+2)^2}{(s^2 + 2s + 2)^2}$.

c) $F(s) = \dfrac{25s^2 + 395s + 1494}{s^2 + 15s + 54}$.

d) $F(s) = \dfrac{5s^3 + 20s^2 - 49s - 108}{s^2 + 7s + 10}$.

12.44 Derive the transform pair given by Eq. 12.10.

12.45 a) Derive the transform pair given by Eq. 12.11.

b) Derive the transform pair given by Eq. 12.12.

Sections 12.8–12.9

12.46 Find the poles and zeros for the s-domain functions in Problems 12.42(b) and 12.42(c).

12.47 Find the poles and zeros for the s-domain functions in Problems 12.43(a) and 12.43(b).

12.48 a) Use the initial-value theorem to find the initial value of v in Problem 12.26.

b) Can the final-value theorem be used to find the steady-state value of v? Why or why not?

12.49 Use the initial- and final-value theorems to check the initial and final values of the current and voltage in Problem 12.27.

12.50 Use the initial- and final-value theorems to check the initial and final values of the current in Problem 12.28.

12.51 Use the initial- and final-value theorems to check the initial and final values of the current and voltage in Problem 12.30.

12.52 Apply the initial- and final-value theorems to each transform pair in Problem 12.40.

12.53 Apply the initial- and final-value theorems to each transform pair in Problem 12.41.

12.54 Apply the initial- and final-value theorems to each transform pair in Problem 12.42.

12.55 Apply the initial- and final-value theorems to each transform pair in Problem 12.43.

Sections 12.1–12.9

12.56 a) Use phasor circuit analysis techniques from
PRACTICAL Chapter 9 to determine the steady-state expres-
PERSPECTIVE sion for the inductor current in Fig. 12.19.

b) How does your result in part (a) compare to the complete response for inductor current calculated in the Practical Perspective?

12.57 Find the maximum magnitude of the sinusoidal
PRACTICAL source in Fig. 12.19 such that the complete response
PERSPECTIVE of the inductor current does not exceed the 40 mA current rating at $t = 1$ ms.

12.58 Suppose the input to the circuit in Fig 12.19 is a
PRACTICAL damped ramp of the form Kte^{-100t} V. Find the larg-
PERSPECTIVE est value of K such that the inductor current does not exceed the 40 mA current rating.

The Laplace Transform in Circuit Analysis

CHAPTER CONTENTS

13.1 **Circuit Elements in the s Domain** *p. 484*

13.2 **Circuit Analysis in the s Domain** *p. 486*

13.3 **Applications** *p. 488*

13.4 **The Transfer Function** *p. 500*

13.5 **The Transfer Function in Partial Fraction Expansions** *p. 502*

13.6 **The Transfer Function and the Convolution Integral** *p. 505*

13.7 **The Transfer Function and the Steady-State Sinusoidal Response** *p. 511*

13.8 **The Impulse Function in Circuit Analysis** *p. 514*

CHAPTER OBJECTIVES

1 Be able to transform a circuit into the *s* domain using Laplace transforms; be sure you understand how to represent the initial conditions on energy-storage elements in the *s* domain.

2 Know how to analyze a circuit in the *s* domain and be able to transform an *s*-domain solution back to the time domain.

3 Understand the definition and significance of the transfer function and be able to calculate the transfer function for a circuit using *s*-domain techniques.

4 Know how to use a circuit's transfer function to calculate the circuit's unit impulse response, its unit step response, and its steady-state response to a sinusoidal input.

The **Laplace transform** has two characteristics that make it an attractive tool for circuit analysis.

- It transforms a set of linear constant-coefficient differential equations into a set of linear polynomial equations, which are easier to manipulate and solve.
- It automatically introduces into the polynomial equations the initial values of the current and voltage variables. Thus, initial conditions are an inherent part of the transform process. (This contrasts with the classical approach to the solution of differential equations, in which initial conditions are considered when the unknown coefficients are evaluated.)

We begin this chapter by showing how to avoid writing time-domain integrodifferential equations and transforming them into the *s* domain. In Section 13.1, we'll develop the *s*-domain circuit models for resistors, inductors, and capacitors. Then we can transform entire circuits into the *s* domain and write the *s*-domain equations directly. Section 13.2 reviews Ohm's and Kirchhoff's laws for *s*-domain circuits. After establishing these fundamentals, we apply the Laplace transform method to a variety of circuit problems in Section 13.3, using the circuit analysis and simplification tools first introduced for resistive circuits.

After solving for the circuit response in the *s* domain, we inverse-transform back to the time domain, using partial fraction expansion (as demonstrated in Chapter 12). As before, checking the final time-domain equations in terms of the initial conditions and final values is an important step in the solution process.

We introduce a new concept, the transfer function, in Section 13.4. The transfer function for a circuit is the ratio of the Laplace transform of its output to the Laplace transform of its input. In Chapters 14 and 15, we'll use the transfer function in circuit design, but here we focus on using it for circuit analysis. We continue this chapter with a look at the role of partial fraction expansion (Section 13.5) and the convolution integral (Section 13.6) when using the transfer function in circuit analysis. We conclude with a discussion of the impulse function in circuit analysis.

■ Practical Perspective

Surge Suppressors

When using personal computers and other sensitive electronic equipment, we need to provide protection from voltage surges. These surges can occur during a lightning storm or just by switching an electrical device on or off. A commercially available surge suppressor is shown in the accompanying figure.

How can flipping a switch to turn on a light or turn off a hair dryer cause a voltage surge? At the end of this chapter, we will answer that question using Laplace transform techniques to analyze a circuit. We will illustrate how a voltage surge can be created by switching off a resistive load in a circuit operating in the sinusoidal steady state.

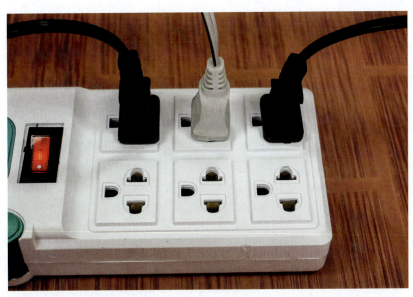

Virote Chuenwiset/Shutterstock

Jhaz Photography/Shutterstock

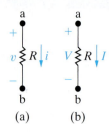

Figure 13.1 ▲ The resistor. (a) Time domain. (b) Frequency domain.

Figure 13.2 ▲ An inductor of L henrys carrying an initial current of I_0 amperes.

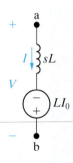

Figure 13.3 ▲ The series equivalent circuit for an inductor of L henrys carrying an initial current of I_0 amperes.

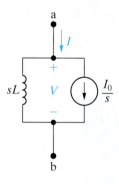

Figure 13.4 ▲ The parallel equivalent circuit for an inductor of L henrys carrying an initial current of I_0 amperes.

Figure 13.5 ▲ The s-domain circuit for an inductor when the initial current is zero.

13.1 Circuit Elements in the s Domain

A three-step procedure transforms each circuit element into an s-domain equivalent circuit.

1. Write the time-domain equation relating the terminal voltage to the terminal current.

2. Laplace transform the time-domain equation to generate an algebraic relationship between the s-domain current and voltage.

3. Construct a circuit model that satisfies the relationship between the s-domain current and voltage.

Note the s-domain dimensions: the s-domain voltage dimension is volt-seconds [V-s], the s-domain current dimension is ampere-seconds [A-s], and thus the s-domain voltage-to-current ratio dimension is volts per ampere, or ohms. An impedance in the s domain is measured in ohms, and an admittance is measured in siemens. We use the passive sign convention in all the derivations.

A Resistor in the s Domain

We begin with the resistor. From Ohm's law,

$$v = Ri.$$

Because R is a constant, the Laplace transform of Ohm's law is

$$V = RI, \tag{13.1}$$

where

$$V = \mathcal{L}\{v\} \quad \text{and} \quad I = \mathcal{L}\{i\}.$$

From Eq. 13.1 we see that the s-domain equivalent circuit of a resistor is a resistance of R ohms that carries a current of I ampere-seconds and has a terminal voltage of V volt-seconds.

Figure 13.1 shows the time- and frequency-domain circuits of the resistor. Note that going from the time domain to the frequency domain does not change the resistance element.

An Inductor in the s Domain

Figure 13.2 shows an inductor carrying an initial current of I_0 amperes. The time-domain equation relating the terminal voltage to the terminal current is

$$v = L\frac{di}{dt}.$$

The Laplace transform of the inductor equation gives

$$V = L[sI - i(0^-)] = sLI - LI_0. \tag{13.2}$$

One circuit configuration that satisfies Eq. 13.2 is an impedance of sL ohms in series with an independent voltage source of LI_0 volt-seconds, as shown in Fig. 13.3. Note that the polarity marks on the voltage source LI_0 agree with the minus sign in Eq. 13.2. Note also that LI_0 carries its own algebraic sign; that is, if the initial value of i is opposite to the reference direction for i, then I_0 has a negative value.

We can also solve Eq. 13.2 for the current I and then construct the circuit to satisfy the resulting equation. The current I is given by

$$I = \frac{V + LI_0}{sL} = \frac{V}{sL} + \frac{I_0}{s}.$$

The s-domain equivalent circuit that satisfies this equation is an impedance of sL ohms in parallel with an independent current source of I_0/s ampere-seconds,

There are two other ways to construct the s domain circuit in Fig. 13.4: (1) find the Norton equivalent of the circuit shown in Fig. 13.3 or (2) start with the inductor current as a function of the inductor voltage in the time domain and then Laplace transform the resulting integral equation. We leave these two approaches to Problems 13.1 and 13.2.

If the initial energy stored in the inductor is zero, that is, if $I_0 = 0$, the s-domain equivalent circuit is an inductor with an impedance of sL ohms. Figure 13.5 shows this circuit.

A Capacitor in the s Domain

A capacitor with initial stored energy also has two s-domain equivalent circuits. Figure 13.6 shows a capacitor initially charged to V_0 volts. The capacitor current is

$$i = C \frac{dv}{dt}.$$

Transforming the capacitor equation yields

$$I = C[sV - v(0^-)] = sCV - CV_0, \tag{13.3}$$

so the s-domain current I is the sum of two branch currents. One branch contains an admittance of sC siemens, and the second branch contains an independent current source of CV_0 ampere-seconds. Figure 13.7 shows this parallel equivalent circuit.

To derive the other equivalent circuit for the charged capacitor, solve Eq. 13.3 for V:

$$V = \left(\frac{1}{sC}\right)I + \frac{V_0}{s}.$$

Figure 13.8 shows the circuit that satisfies the equation for capacitor voltage, which is a series combination of an impedance and an independent voltage source.

In the equivalent circuits shown in Figs. 13.7 and 13.8, V_0 carries its own algebraic sign. In other words, if the polarity of V_0 is opposite to the reference polarity for v, V_0 is a negative quantity. If the initial voltage on the capacitor is zero, both equivalent circuits reduce to an impedance of $1/sC$ ohms, as shown in Fig. 13.9.

In Chapter 9 we used the phasor transform to turn a time-domain circuit into a frequency-domain circuit. In this chapter, we use the s-domain equivalent circuits, summarized in Table 13.1, to transform a time-domain circuit into the s domain. When the time-domain circuit contains inductors and capacitors with initial stored energy, you need to decide whether to use the parallel or series s-domain equivalent circuit. With a little forethought and some experience, the best choice is often evident.

Figure 13.6 ▲ A capacitor of C farads initially charged to V_0 volts.

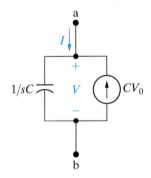

Figure 13.7 ▲ The parallel equivalent circuit for a capacitor initially charged to V_0 volts.

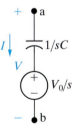

Figure 13.8 ▲ The series equivalent circuit for a capacitor initially charged to V_0 volts.

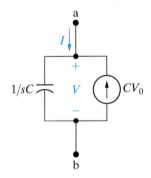

Figure 13.9 ▲ The s-domain circuit for a capacitor when the initial voltage is zero.

TABLE 13.1	Summary of the s-Domain Equivalent Circuits

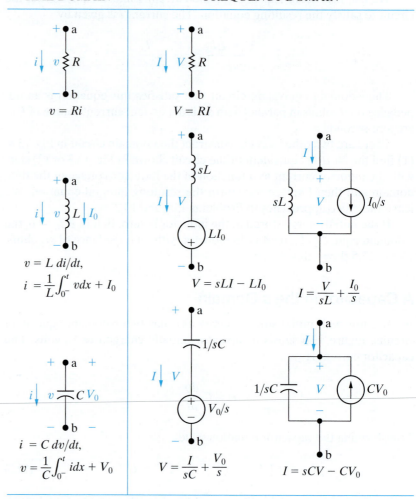

TIME DOMAIN

$v = Ri$

$v = L\,di/dt,$
$i = \dfrac{1}{L}\displaystyle\int_{0^-}^{t} v\,dx + I_0$

$i = C\,dv/dt,$
$v = \dfrac{1}{C}\displaystyle\int_{0^-}^{t} i\,dx + V_0$

FREQUENCY DOMAIN

$V = RI$

$V = sLI - LI_0$

$I = \dfrac{V}{sL} + \dfrac{I_0}{s}$

$V = \dfrac{I}{sC} + \dfrac{V_0}{s}$

$I = sCV - CV_0$

13.2 Circuit Analysis in the s Domain

Before presenting a method for using the s-domain equivalent circuits in analysis, we make some important observations.

- If no energy is stored in the inductor or capacitor, the relationship between the s-domain voltage and current for each passive element is:

OHM'S LAW IN THE S-DOMAIN

$$V = ZI, \tag{13.4}$$

where Z is the s-domain impedance of the element. Thus, a resistor has an impedance of R ohms, an inductor has an impedance of sL ohms, and a capacitor has an impedance of $1/sC$ ohms. The relationship contained in Eq. 13.4 also appears in Figs. 13.1(b), 13.5, and 13.9. Equation 13.4 is known as Ohm's law in the s domain. The reciprocal of the impedance is admittance. Therefore, the s-domain admittance of a resistor is $1/R$ siemens, an inductor has an admittance of $1/sL$ siemens, and a capacitor has an admittance of sC siemens.

- The rules for combining impedances and admittances in the *s* domain are the same as those for frequency-domain circuits. Thus, series-parallel simplifications and Δ-to-Y conversions also are applicable to *s*-domain analysis.
- Kirchhoff's laws apply to *s*-domain currents and voltages because an operational Laplace transform states that the Laplace transform of a sum of time-domain functions equals the sum of the Laplace transforms of the individual functions (see Table 12.2). The algebraic sum of the currents at a node is zero in the time domain, so the algebraic sum of the Laplace-transformed currents is also zero. A similar statement holds for the algebraic sum of the Laplace-transformed voltages around a closed path. The *s*-domain version of Kirchhoff's laws is

$$\text{at every node in a circuit,} \quad \text{alg} \sum I = 0, \qquad (13.5)$$

$$\text{around every closed path in a circuit,} \quad \text{alg} \sum V = 0. \qquad (13.6)$$

Therefore, because Ohm's law, KVL, and KCL hold in the *s* domain, all of the circuit analysis techniques developed in Chapters 2–4 for resistive circuits can be used to analyze circuits in the *s*-domain. These techniques include combining impedances in series and parallel to find equivalent impedances, voltage division and current division, the node-voltage method and the mesh-current method, source transformation, and Thévenin and Norton equivalent circuits. This leads us to the following step-by-step procedure for using Laplace transform techniques to analyze circuits.

Step 1 determines the initial current in each inductor and the initial voltage across each capacitor by analyzing the time-domain circuit for $t < 0$.

Step 2 transforms each independent voltage or current source defined by time-domain functions into the *s* domain using the functional and operational transforms in Tables 12.1 and 12.2.

Step 3 transforms voltages and currents represented by time-domain symbols such as $v(t)$ and $i(t)$ into corresponding *s*-domain symbols such as V and I.

Step 4 transforms any remaining components in the time-domain circuit into the *s* domain using the circuits in Table 13.1. When inductors and capacitors have nonzero initial values, calculated in Step 1, these initial conditions are represented by independent sources in series or parallel with the component impedances.

Step 5 analyzes the resulting *s*-domain circuit using the techniques developed for resistive circuits in Chapters 2–4. The analysis produces *s*-domain voltages and currents that should each be represented as a ratio of two polynomials in *s*.

Step 6 applies the initial- and final-value theorems to the *s*-domain functions from Step 5 to check the values of the corresponding time-domain functions at $t = 0$ and $t = \infty$. Note that it might not be possible to apply one or both of these theorems, depending on the form of the *s*-domain function.

Step 7 represents each *s*-domain voltage and current of interest as a partial fraction expansion and then uses Table 12.3 to inverse-Laplace-transform the *s*-domain voltages and currents back to the time domain.

This analysis method yields the complete response to any circuit whose voltage and current sources have Laplace transforms. This method represents the most comprehensive circuit analysis technique presented in this text and is summarized in Analysis Method 13.1. Example 13.1 applies the first five steps in Analysis Method 13.1 to a time-domain circuit.

LAPLACE TRANSFORM METHOD

1. **Determine the initial conditions** for inductors and capacitors.
2. **Laplace-transform independent voltage and current functions** using Tables 12.1 and 12.2.
3. **Transform symbolic time-domain voltages and currents** into *s*-domain symbols.
4. **Transform remaining circuit components** into the *s* domain using Table 13.1.
5. **Analyze the *s*-domain circuit** using resistive circuit analysis techniques; represent the resulting *s*-domain voltages and currents as ratios of polynomials in *s*.
6. **Use the initial- and final-value theorems** to check the *s*-domain voltages and currents.
7. **Inverse-Laplace-transform the *s*-domain voltages and currents** using partial fraction expansion and Table 12.3.

Analysis Method 13.1 Laplace-transform circuit analysis method.

<table>
<tr><td>

EXAMPLE 13.1

</td><td>

Transforming a Circuit into the s Domain

</td></tr>
</table>

A 500 Ω resistor, a 16 mH inductor, and a 25 nF capacitor are connected in parallel.

a) Express the admittance of this parallel combination of elements as a rational function of s.

b) Compute the numerical values of the zeros and poles.

Solution

a) We use Analysis Method 13.1 to transform the three parallel-connected components from the time domain to the s domain.

Step 1: There are no initial conditions for the inductor and capacitor.

Step 2: There are no voltage or current sources.

Step 3: There are no symbolic voltages or currents.

Step 4: Because there are no initial conditions, each component is represented by its s-domain impedance:

$$Z_R = 500 \ \Omega;$$

$$Z_L = 0.016s \ \Omega;$$

$$Z_C = \frac{1}{25 \times 10^{-9}s} = \frac{40 \times 10^6}{s} \ \Omega.$$

The s-domain circuit is shown in Fig. 13.10.

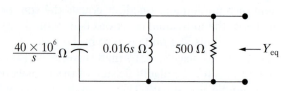

Figure 13.10 ▲ The s domain circuit for Example 13.1.

Step 5: Find the equivalent admittance by adding the inverse of the three impedances:

$$Y_{eq} = \frac{1}{500} + \frac{1}{0.016s} + \frac{s}{40 \times 10^6}$$

$$= \frac{s^2 + 80,000s + 25 \times 10^8}{40 \times 10^6 s} \ \text{S}.$$

b) The numerator factors are $(s + 40,000 + j30,000)$ and $(s + 40,000 - j30,000)$. Therefore the zeros are $-40,000 + j30,000$ and $-40,000 - j30,000$. There is a pole at 0.

ASSESSMENT PROBLEM

Objective 1 — Be able to transform a circuit into the s domain using Laplace transforms

13.1 The parallel circuit in Example 13.1 is placed in series with a 2000 Ω resistor.
a) Express the impedance of this series combination as a rational function of s.
b) Compute the numerical values of the zeros and poles.

Answer: (a) $2000(s + 50,000)^2/(s^2 + 80,000s + 25 \times 10^8)$;

(b) $-z_1 = -z_2 = -50,000$;

$-p_1 = -40,000 - j30,000$,

$-p_2 = -40,000 + j30,000$.

SELF-CHECK: Also try Chapter Problems 13.4 and 13.5.

13.3 Applications

We now use Analysis Method 13.1 to find the complete response of several example circuits. We start with two familiar circuits from Chapters 7 and 8 to show that the Laplace transform approach yields the same results found using the first- and second-order circuit analysis techniques. Example 13.2 solves an *RC* circuit, and Example 13.3 solves an *RLC* circuit.

| EXAMPLE 13.2 | The Natural Response of an *RC* Circuit |

The circuit in Fig. 13.11 was analyzed in Example 7.3 using first-order circuit analysis techniques. Use the Laplace transform method to find $v_o(t)$ for $t \geq 0^+$.

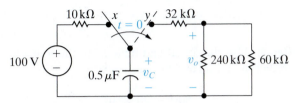

Figure 13.11 ▲ The *RC* circuit for Example 13.2.

Solution

Apply the Laplace transform method using Analysis Method 13.1.

Step 1: Determine the initial voltage across the capacitor by analyzing the circuit in Fig. 13.11 for $t < 0$. Because the switch has been in position x for a long time, the capacitor behaves like an open circuit. The voltage across the open circuit $V_0 = 100$ V.

Step 2: For $t \geq 0$, there are no voltage or current sources in the circuit, so we can skip this step.

Step 3: The voltage across the 240 kΩ resistor is represented in the s domain as V, as shown in Fig. 13.12.

Step 4: The impedance of the three resistors is their resistance. The impedance of the capacitor is

$$Z_C = \frac{1}{sC} = \frac{1}{0.5 \times 10^{-6}s} = \frac{2 \times 10^6}{s} \; \Omega.$$

Because the capacitor has an initial condition, we must decide whether to represent it using a series-connected voltage source or a parallel-connected current source, as shown in Table 13.1. Here we use the series-connected voltage source, whose value is $V_0/s = 100/s$ V-s. The s-domain circuit that results from Steps 1–4 is shown in Fig. 13.12.

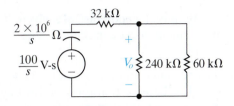

Figure 13.12 ▲ The circuit in Fig. 13.11 for $t \geq 0$, transformed into the s domain.

Step 5: Begin by combining the parallel-connected 240 kΩ and 60 kΩ resistors into a single equivalent 48 kΩ resistor whose voltage is V_o. Now use voltage division to find V_o:

$$V_o = \frac{48,000}{(2 \times 10^6/s) + 32,000 + 48,000}\left(\frac{100}{s}\right)$$

$$= \frac{60}{s + 25}.$$

Step 6: Use the initial- and final-value theorems to show that the initial value of v_o is 60 V and final value of v_o is zero, as we expect from the circuit in Fig. 13.11.

$$\lim_{s \to \infty} sV_o = \lim_{s \to \infty} \frac{60s}{s + 25} = \lim_{(1/s) \to 0} \frac{60}{1 + (25/s)}$$

$$= \frac{60}{1 + 0} = 60 = \lim_{t \to 0} v_o(t);$$

$$\lim_{s \to 0} sV_o = \lim_{s \to 0} \frac{60s}{s + 25} = \frac{60(0)}{0 + 25} = 0 = \lim_{t \to \infty} v_o(t).$$

Step 7: Since V_o is already a partial fraction, we can use the transforms in Table 12.3 to find v_o:

$$v_o(t) = \mathcal{L}^{-1}\left\{\frac{60}{s + 25}\right\} = 60e^{-25t}u(t) \text{ V}.$$

This matches the voltage found in Example 7.3 using first-order circuit analysis methods.

| EXAMPLE 13.3 | The Step Response of an *RLC* Circuit |

Consider the circuit in Fig. 13.13, where the initial current in the inductor is 29 mA and the initial voltage across the capacitor is 50 V. This circuit was analyzed in Example 8.10 using second-order circuit analysis techniques. Use the Laplace transform method to find $v(t)$ for $t \geq 0$.

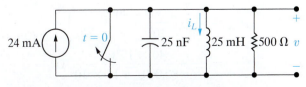

Figure 13.13 ▲ The parallel *RLC* circuit for Example 13.3.

Solution

Apply the Laplace transform method using Analysis Method 13.1.

Step 1: Both initial conditions are given in the problem statement, so we can skip this step.

Step 2: We can describe the parallel combination of the current source and the switch in Fig. 13.13 as $24u(t)$ mA. The Laplace transform of this function is $24/s$ mA-s, which is the current source value in the s-domain circuit in Fig. 13.14.

Step 3: The voltage across the $240\,k\Omega$ resistor is represented in the s domain as V, as shown in Fig. 13.12.

Step 4: The impedance of the resistor is its resistance. The impedance of the inductor is

$$Z_L = sL = 0.025s = \frac{s}{40}\,\Omega$$

and the impedance of the capacitor is

$$Z_C = \frac{1}{sC} = \frac{1}{25 \times 10^{-9}s} = \frac{40 \times 10^6}{s}\,\Omega.$$

Both the inductor and capacitor have nonzero initial conditions. We must decide whether to represent them as series-connected voltage sources or as parallel-connected current sources, as shown in Table 13.1. Here we use parallel-connected current sources because the resulting circuit has only parallel-connected components. The value of the current source in parallel with the inductor is $I_0/s = 0.029/s$ A-s, with the current arrow directed down. The value of the current source in parallel with the capacitor is $CV_0 = (50)(25 \times 10^{-9}) = 1.25\,\mu$ V-s, with the current arrow directed up. The s-domain circuit that results from Steps 1–4 is shown in Fig. 13.14.

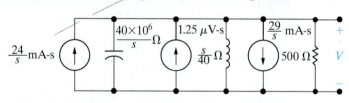

Figure 13.14 ▲ The circuit in Fig. 13.13 for $t \geq 0$, transformed into the s domain.

Step 5: Begin by combining the three parallel-connected impedances into a single equivalent impedance:

$$Z_{eq} = \left(\frac{s}{40 \times 10^6} + \frac{40}{s} + \frac{1}{500}\right)^{-1}$$

$$= \frac{40 \times 10^6 s}{s^2 + 80,000s + 16 \times 10^8}\,\Omega.$$

To find the voltage across this equivalent impedance, multiply by the sum of the three parallel-connected currents:

$$V = Z_{eq}\left(\frac{0.024}{s} + 1.25 \times 10^{-6} - \frac{0.029}{s}\right)$$

$$= \frac{50s - 20 \times 10^4}{s^2 + 80,000s + 16 \times 10^8}\,\text{V-s.}$$

Step 6: Use the initial and final value theorems to predict the initial and final values of v:

$$\lim_{s \to \infty} sV = \lim_{s \to \infty} \frac{50s^2 - 20 \times 10^4 s}{s^2 + 80,000s + 16 \times 10^8}$$

$$= \lim_{(1/s) \to 0} \frac{50 - 20 \times 10^4 (1/s)}{1 + 80,000(1/s) + 16 \times 10^8 (1/s)^2}$$

$$= \frac{50}{1} = 50 = \lim_{t \to 0} v(t);$$

$$\lim_{s \to 0} sV = \lim_{s \to 0} \frac{50s^2 - 20 \times 10^4 s}{s^2 + 80,000s + 16 \times 10^8}$$

$$= \frac{(0)}{16 \times 10^8} = 0 = \lim_{t \to \infty} v(t).$$

The initial-value theorem predicts the correct initial voltage from the problem statement, $V_0 = 50$ V. To confirm the final value of the voltage, envision the circuit in Fig. 13.13 as $t \to \infty$. The inductor is behaving like a short circuit, and the capacitor is behaving like an open circuit in parallel with the short circuit. Therefore, the final value of capacitor voltage in the circuit is zero, as predicted by the final-value theorem.

Step 7: The partial fraction expansion of V is

$$V = \frac{50s - 20 \times 10^4}{(s + 40,000)^2}$$

$$= \frac{-2.2 \times 10^6}{(s + 40,000)^2} + \frac{50}{(s + 40,000)}.$$

Now use the transforms in Table 12.3 to find v:

$$v(t) = \mathcal{L}^{-1}\left\{\frac{-2.2 \times 10^6}{(s + 40,000)^2} + \frac{50}{(s + 40,000)}\right\}$$

$$= (-2.2 \times 10^6\, te^{-40,000t} + 50e^{-40,000t})u(t)\,\text{V.}$$

This matches the voltage found in Example 8.10 using second-order circuit analysis methods.

ASSESSMENT PROBLEMS

Objective 2—Know how to analyze a circuit in the s domain and be able to transform an s domain solution to the time domain

13.2 The switch in the circuit shown has been in position a for a long time. At $t = 0$, the switch is thrown to position b.
a) Find I, V_1, and V_2 as rational functions of s.
b) Find the time-domain expressions for i, v_1, and v_2.

Answer: (a) $I = 0.02/(s + 1250)$,
$$V_1 = 80/(s + 1250),$$
$$V_2 = 20/(s + 1250);$$

(b) $i = 20e^{-1250t}u(t)$ mA,
$$v_1 = 80e^{-1250t}u(t) \text{ V},$$
$$v_2 = 20e^{-1250t}u(t) \text{ V}.$$

13.3 The energy stored in the circuit shown is zero at the time when the switch is closed.
a) Find the s-domain expression for I.
b) Find the time-domain expression for i when $t > 0$.
c) Find the s-domain expression for V.
d) Find the time-domain expression for v when $t > 0$.

Answer: (a) $I = 40/(s^2 + 1.2s + 1)$;
(b) $i = (50e^{-0.6t} \sin 0.8t)u(t)$ A;
(c) $V = 160s/(s^2 + 1.2s + 1)$;
(d) $v = [200e^{-0.6t} \cos (0.8t + 36.87°)]u(t)$ V.

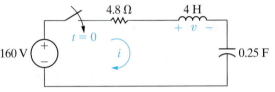

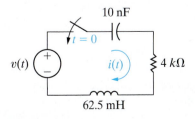

SELF-CHECK: Also try Chapter Problems 13.9, 13.15, and 13.16.

Next, we analyze a circuit with a sinusoidal source using Laplace methods. Phasor methods, presented in Chapter 9, were used to analyze circuits with sinusoidal sources but have an important limitation—they only produce the steady-state response of the circuit, not the complete response. Recall that the complete response consists of both the steady-state and the transient response. Laplace methods do not have this limitation, so they produce a circuit's complete response to a sinusoidal source, as seen in Example 13.4.

EXAMPLE 13.4   **Analyzing a Circuit with a Sinusoidal Source**

The circuit in Fig. 13.15 has no initial stored energy. At $t = 0$, the switch closes and the circuit is driven by a sinusoidal source $v(t) = 15 \cos 40,000t$ V. Use Laplace methods to find $i(t)$ for $t \geq 0$.

Figure 13.15 ▲ The circuit for Example 13.4.

Solution

Apply the Laplace transform method using Analysis Method 13.1.

Step 1: There is no initial stored energy, so both initial conditions are zero.

Step 2: Using the functional and operational transform tables (Tables 12.1 and 12.2), we see that the Laplace transform of $v(t)$ is

$$V = \mathcal{L}\{15 \cos 40,000t\} = \frac{15s}{s^2 + 40,000^2} \text{ V-s.}$$

This is the value of the voltage source for the s-domain circuit in Fig. 13.16.

Step 3: The current is represented in the s domain as I, as shown in Fig. 13.16.

Step 4: The impedance of the resistor is its resistance. The impedance of the inductor is

$$Z_L = sL = 0.0625s = \frac{s}{16} \ \Omega$$

and the impedance of the capacitor is

$$Z_C = \frac{1}{sC} = \frac{1}{10 \times 10^{-9} s} = \frac{10^8}{s} \ \Omega.$$

The s-domain circuit that results from Steps 1–4 is shown in Fig. 13.16.

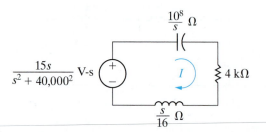

Figure 13.16 ▲ The circuit in Fig 13.16 for $t \geq 0$, transformed into the s domain.

Step 5: To find the current I, divide the source voltage by the sum of the impedances:

$$I = \frac{\dfrac{15s}{s^2 + 40,000^2}}{4000 + \dfrac{10^8}{s} + \dfrac{s}{16}}$$

$$= \frac{240s^2}{(s^2 + 40,000^2)(s^2 + 64,000s + 16 \times 10^8)} \ \text{A-s}.$$

Step 6: Use the initial value theorem to predict the initial value of i:

$$\lim_{s \to \infty} sI = \lim_{s \to \infty} \frac{240s^3}{(s^2 + 40,000^2)(s^2 + 64,000s + 16 \times 10^8)}$$

$$= \lim_{(1/s) \to 0} \frac{240 \left(\dfrac{1}{s} \right)}{\left[1 + \left(\dfrac{40,000}{s} \right)^2 \right] \left[1 + \left(\dfrac{60,000}{s} \right) + \left(\dfrac{16 \times 10^8}{s^2} \right) \right]}$$

$$= \frac{0}{(1)(1)} = 0 = \lim_{t \to 0} i(t).$$

The initial-value theorem predicts correctly that the initial current is zero. We cannot use the final-value theorem to predict the final current because I has two poles on the imaginary axis at $\pm j40,000$ rad/s.

Step 7: The partial fraction expansion of I is

$$I = \frac{240s^2}{(s^2 + 40,000^2)(s^2 + 64,000s + 16 \times 10^8)}$$

$$= \frac{K_1}{s - j40,000} + \frac{K_1^*}{s + j40,000}$$

$$+ \frac{K_2}{s + 32,000 - j24,000} + \frac{K_2^*}{s + 32,000 + j24,000}.$$

Finding K_1 and K_2,

$$K_1 = \frac{240s^2}{(s + j40,000)(s^2 + 64,000s + 16 \times 10^8)} \bigg|_{s = j40,000}$$

$$= 1.875 \times 10^{-3};$$

$$K_2 = \frac{240s^2}{(s^2 + 40,000^2)(s + 32,000 + j24000)} \bigg|_{s = -32,000 + j40,000}$$

$$= 3.125 \times 10^{-3} \underline{/-126.87^\circ}.$$

Now use the transforms in Table 12.3 to find v:

$$i(t) = \mathcal{L}^{-1} \left\{ \frac{1.875 \times 10^{-3}}{s - j40,000} + \frac{1.875 \times 10^{-3}}{s + j40,000} \right.$$

$$+ \frac{3.125 \times 10^{-3} \underline{/-126.87^\circ}}{s + 32,000 - j24,000} + \left. \frac{3.125 \times 10^{-3} \underline{/126.87^\circ}}{s + 32,000 + j24,000} \right\}$$

$$= [\, 3.75 \cos 40,000t$$

$$+ 6.25e^{-32,000t} \cos(24,000t - 126.87^\circ)\,]u(t) \ \text{mA}.$$

The first term in the expression for $i(t)$ is the steady-state response. Its frequency matches the frequency of the source, and this term persists for all time. You should use the phasor methods from Chapter 9 to verify this result. The second term in the expression for $i(t)$ is the transient response, or the natural response. Note that it decays to zero as $t \to \infty$. This part of the response is independent of the voltage source and is based only on the passive component values and their interconnections.

ASSESSMENT PROBLEM

Objective 2—Know how to analyze a circuit in the *s* domain and be able to transform an *s* domain solution to the time domain

13.4   The energy stored in the circuit shown is zero at the time when the switch is opened. The current source is 24 cos 40,000*t* mA. Find the voltage $v(t)$ for $t \geq 0$.

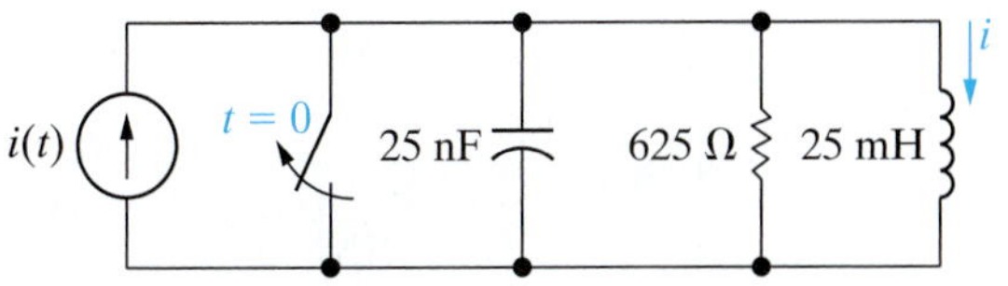

Answer: $(15 \sin 40{,}000t - 25e^{-32{,}000t} \sin 24{,}000t)u(t)$ mA.

SELF-CHECK: Also try Chapter Problem 13.23.

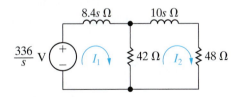

Until now, we avoided analyzing circuits with inductors and capacitors that have two or more meshes, or three or more essential nodes. Such circuits are described by two or more simultaneous differential equations, and the techniques for solving these systems of equations are beyond the scope of this text. However, using Laplace techniques, we can transform a circuit into the *s* domain and write a set of simultaneous algebraic equations, whose solution is much more manageable. Example 13.5 illustrates this by solving a circuit with two meshes.

EXAMPLE 13.5 | **Analyzing a Circuit with Multiple Meshes**

The circuit in Fig. 13.17 has no initial stored energy. At $t = 0$, the switch closes. Use Laplace methods to find $i_1(t)$ and $i_2(t)$ for $t \geq 0$.

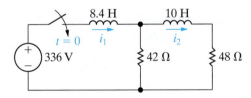

Figure 13.17 ▲ A multiple-mesh *RL* circuit.

Solution

Apply the Laplace transform method using Analysis Method 13.1.

Step 1: There is no initial stored energy, so both initial conditions are zero.

Step 2: The series connection of the 336 V dc voltage source and the switch can be described in the time domain as $336u(t)$ V. Using the functional and operational transform tables (Tables 12.1 and 12.2), the Laplace transform of this voltage is

$$\mathcal{L}\{336u(t)\} = \frac{336}{s} \text{ V-s.}$$

This is the value of the voltage source for the *s*-domain circuit in Fig. 13.18.

Step 3: The currents are represented in the *s* domain as I_1 and I_2, as shown in Fig. 13.18.

Step 4: The impedance of the resistors is their resistance. The impedance of the inductors is

$$Z_{L1} = sL_1 = 8.4s \ \Omega; \quad Z_{L2} = sL_2 = 10s \ \Omega.$$

The *s*-domain circuit that results from Steps 1–4 is shown in Fig. 13.18.

Figure 13.18 ▲ The s-domain equivalent circuit for the circuit shown in Fig. 13.17.

Step 5: The two KVL equations are

$$\frac{336}{s} = (42 + 8.4s)I_1 - 42I_2,$$

$$0 = -42I_1 + (90 + 10s)I_2.$$

Using Cramer's method, we get

$$\Delta = \begin{vmatrix} 42 + 8.4s & -42 \\ -42 & 90 + 10s \end{vmatrix}$$

$$= 84(s^2 + 14s + 24)$$

$$= 84(s + 2)(s + 12),$$

$$N_1 = \begin{vmatrix} 336/s & -42 \\ 0 & 90 + 10s \end{vmatrix}$$

$$= \frac{3360(s + 9)}{s},$$

$$N_2 = \begin{vmatrix} 42 + 8.4s & 336/s \\ -42 & 0 \end{vmatrix}$$

$$= \frac{14{,}112}{s}.$$

Therefore,

$$I_1 = \frac{N_1}{\Delta} = \frac{40(s + 9)}{s(s + 2)(s + 12)},$$

$$I_2 = \frac{N_2}{\Delta} = \frac{168}{s(s + 2)(s + 12)}.$$

Step 6: Use the initial-value theorem to predict the initial values of i_1 and i_2:

$$\lim_{s \to \infty} sI_1 = \lim_{s \to \infty} \frac{40(s + 9)}{(s + 2)(s + 12)}$$

$$= \lim_{(1/s) \to 0} \frac{40[(1/s) + 9(1/s)^2]}{(1 + (2/s))(1 + (12/s))}$$

$$= \frac{0}{(1)(1)} = 0 = \lim_{t \to 0} i_1(t);$$

$$\lim_{s \to \infty} sI_2 = \lim_{s \to \infty} \frac{168}{(s + 2)(s + 12)}$$

$$= \lim_{(1/s) \to 0} \frac{168(1/s)^2}{(1 + (2/s))(1 + (12/s))}$$

$$= \frac{0}{(1)(1)} = 0 = \lim_{t \to 0} i_2(t).$$

The initial-value theorem predicts correctly that the initial currents are zero. Now use the final-value theorem to predict the values of i_1 and i_2 as $t \to \infty$:

$$\lim_{s \to 0} sI_1 = \lim_{s \to 0} \frac{40(s + 9)}{(s + 2)(s + 12)} = \frac{40(9)}{(2)(12)}$$

$$= 15 \text{ A} = \lim_{t \to \infty} i_1(t);$$

$$\lim_{s \to 0} sI_2 = \lim_{s \to 0} \frac{168}{(s + 2)(s + 12)}$$

$$= \frac{168}{(2)(12)} = 7 \text{ A} = \lim_{t \to \infty} i_2(t).$$

To verify these values, consider the circuit in Fig. 13.16 as $t \to \infty$. The inductors now behave like short circuits, as shown in Fig. 13.19.

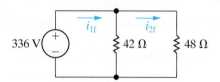

Figure 13.19 ▲ The circuit in Fig 13.17 as $t \to \infty$.

The two KVL equations that describe this circuit are

$$42i_{1f} - 42i_{2f} = 336;$$

$$-42i_{1f} + 90i_{2f} = 0.$$

Solving, we find the final value of i_1 is 15 A and the final value of i_2 is 7 A, as predicted by the final value theorem.

Step 7: Expanding I_1 and I_2 into a sum of partial fractions gives

$$I_1 = \frac{15}{s} - \frac{14}{s + 2} - \frac{1}{s + 12},$$

$$I_2 = \frac{7}{s} - \frac{8.4}{s + 2} + \frac{1.4}{s + 12}.$$

We obtain the expressions for i_1 and i_2 by inverse-transforming I_1 and I_2, using Table 12.3:

$$i_1 = (15 - 14e^{-2t} - e^{-12t})u(t) \text{ A},$$

$$i_2 = (7 - 8.4e^{-2t} + 1.4e^{-12t})u(t) \text{ A}.$$

One final test involves calculating the voltage drop across the 42 Ω resistor using three different methods. From the circuit, the voltage across the 42 Ω resistor (positive at the top) is

$$v = 42(i_1 - i_2) = 336 - 8.4\frac{di_1}{dt} = 48i_2 + 10\frac{di_2}{dt}.$$

You should verify that regardless of which expression is used, the voltage is

$$v = (336 - 235.2e^{-2t} - 100.80e^{-12t})u(t) \text{ V}.$$

We are thus confident that the solutions for i_1 and i_2 are correct.

ASSESSMENT PROBLEM

Objective 2—Know how to analyze a circuit in the *s* domain and be able to transform an *s*-domain solution to the time domain

13.5 The dc current and dc voltage sources are applied simultaneously to the circuit shown. No energy is stored in the circuit at the instant of application.

a) Derive the *s*-domain expressions for V_1 and V_2.
b) For $t > 0$, derive the time-domain expressions for v_1 and v_2.
c) Calculate $v_1(0^+)$ and $v_2(0^+)$.
d) Compute the steady-state values of v_1 and v_2.

Answer: (a) $V_1 = [5(s + 3)]/[s(s + 0.5)(s + 2)]$,
$V_2 = [2.5(s^2 + 6)]/[s(s + 0.5)(s + 2)]$;

(b) $v_1 = \left(15 - \dfrac{50}{3}e^{-0.5t} + \dfrac{5}{3}e^{-2t}\right)u(t)$ V,

$v_2 = \left(15 - \dfrac{125}{6}e^{-0.5t} + \dfrac{25}{3}e^{-2t}\right)u(t)$ V;

(c) $v_1(0^+) = 0$, $v_2(0^+) = 2.5$ V;

(d) $v_1 = v_2 = 15$ V.

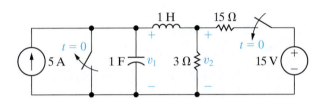

SELF-CHECK: Also try Chapter Problems 13.19 and 13.20.

We can use Thévenin and Norton equivalents to simplify circuits in the *s* domain. When one part of the circuit changes frequently while another part remains constant, we can find a simpler equivalent for the constant part of the circuit that makes analysis of the entire circuit easier. Example 13.6 creates a Thévenin equivalent to simplify part of an *s*-domain circuit.

EXAMPLE 13.6 | **Creating a Thévenin Equivalent in the *s* Domain**

We want to find the capacitor current, i_C, for the circuit in Fig. 13.20. The circuit has no initial stored energy, and at $t = 0$, the switch closes. Find the Thévenin equivalent for the circuit to the left of the terminals a and b in the *s* domain, using Laplace methods. Then analyze the simplified circuit to find $i_C(t)$ for $t \geq 0$.

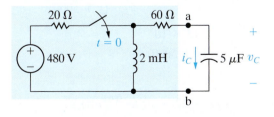

Figure 13.20 ▲ The circuit for Example 13.6.

Solution

Apply the Laplace transform method using Analysis Method 13.1.

Step 1: There is no initial stored energy, so both initial conditions are zero.

Step 2: The series connection of the 480 V dc voltage source and the switch can be described in the time domain as $480u(t)$ V. Using the functional and operational transform tables (Tables 12.1 and 12.2), the Laplace transform of this voltage is

$$\mathscr{L}\{480u(t)\} = \frac{480}{s} \text{ V-s.}$$

This is the value of the voltage source for the *s*-domain circuit in Fig. 13.21.

Step 3: The capacitor current and voltage are represented in the s domain as I_C and V_C, as shown in Fig. 13.21.

Step 4: The impedance of the resistors is their resistance. The impedances of the inductor and capacitor are

$$Z_L = sL = 0.002s = \frac{s}{500} \; \Omega;$$

$$Z_C = \frac{1}{sC} = \frac{1}{5 \times 10^{-6}s} = \frac{2 \times 10^5}{s} \; \Omega.$$

The s-domain circuit that results from Steps 1–4 is shown in Fig. 13.21.

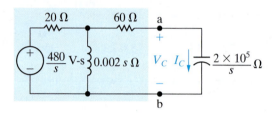

Figure 13.21 ▲ The Laplace transform of the circuit shown in Fig. 13.20.

Step 5: The Thévenin voltage is the open-circuit voltage across terminals a and b. Under open-circuit conditions, there is no voltage across the 60 Ω resistor. Using voltage division,

$$V_{Th} = \frac{(480/s)(0.002s)}{20 + 0.002s} = \frac{480}{s + 10^4}.$$

The Thévenin impedance seen from terminals a and b equals the 60 Ω resistor in series with the parallel combination of the 20 Ω resistor and the inductive impedance. Thus

$$Z_{Th} = 60 + \frac{0.002s(20)}{20 + 0.002s} = \frac{80(s + 7500)}{s + 10^4}.$$

Using the Thévenin equivalent, we reduce the circuit shown in Fig. 13.21 to the one shown in Fig. 13.22.

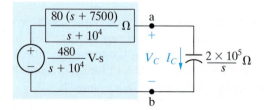

Figure 13.22 ▲ A simplified version of the circuit shown in Fig. 13.21, using a Thévenin equivalent.

In this circuit, the capacitor current I_C equals the Thévenin voltage divided by the total series impedance. Thus,

$$I_C = \frac{\dfrac{480}{s + 10^4}}{\dfrac{80(s + 7500)}{s + 10^4} + \dfrac{2 \times 10^5}{s}} = \frac{6s}{(s + 5000)^2}.$$

Step 6: Use the initial- and final-value theorems to predict the initial and final values of i_C:

$$\lim_{s \to \infty} sI_C = \lim_{s \to \infty} \frac{6s^2}{(s + 5000)^2} = \lim_{(1/s) \to 0} \frac{6}{(1 + (5000/s))^2}$$

$$= \frac{6}{(1)^2} = 6 \text{ A} = \lim_{t \to 0} i_C(t);$$

$$\lim_{s \to 0} sI_C = \lim_{s \to 0} \frac{6s^2}{(s + 5000)^2} = \frac{0}{(5000)^2} = 0 = \lim_{t \to \infty} i_C(t).$$

Let's calculate the initial capacitor current from the circuit in Fig. 13.20. The initial inductor current is zero and the initial capacitor voltage is zero, so the initial capacitor current is $480/(20 + 60)$ or 6 A, which agrees with the prediction of the initial-value theorem. The final value of the capacitor current is zero because as $t \to \infty$ in the circuit shown in Fig. 13.20, the capacitor behaves like an open circuit. This final capacitor current also agrees with the prediction of the final-value theorem.

Step 7: The partial fraction expansion of I_C is

$$I_C = \frac{-30{,}000}{(s + 5000)^2} + \frac{6}{s + 5000},$$

and its inverse transform is

$$i_C = (-30{,}000te^{-5000t} + 6e^{-5000t})u(t) \text{ A}.$$

Suppose we also want to find the voltage drop across the capacitor, v_C. Once we know i_C, we could find v_C by integration in the time domain:

$$v_C = 2 \times 10^5 \int_{0^-}^{t} (6 - 30{,}000x)e^{-5000x} \, dx.$$

Although the integration is not difficult, we can avoid it altogether by first finding the s-domain expression for V_C and then using the inverse transform to find v_C. Thus

$$V_C = \frac{1}{sC} I_C = \frac{2 \times 10^5}{s} \frac{6s}{(s + 5000)^2}$$

$$= \frac{12 \times 10^5}{(s + 5000)^2},$$

and

$$v_C = 12 \times 10^5 te^{-5000t}u(t) \text{ V}.$$

You can explore this circuit's behavior further in Problem 13.35.

⬛ ASSESSMENT PROBLEM

Objective 2—Know how to analyze a circuit in the *s* domain and be able to transform an *s*-domain solution to the time domain

13.6 The initial charge on the capacitor in the circuit shown is zero.
a) Find the *s*-domain Thévenin equivalent circuit with respect to terminals a and b.
b) Find the *s*-domain expression for the current that the circuit delivers to a load consisting of a 1 H inductor in series with a 2 Ω resistor.

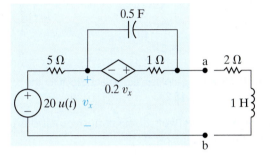

Answer: (a) $V_{Th} = V_{ab} = [20(s + 2.4)]/[s(s + 2)]$,
$Z_{Th} = 5(s + 2.8)/(s + 2)$;
(b) $I_{ab} = [20(s + 2.4)]/[s(s + 3)(s + 6)]$.

SELF-CHECK: Also try Chapter Problem 13.37.

We can also use Laplace transform methods to analyze circuits with mutually coupled coils. Example 13.7 illustrates this process.

EXAMPLE 13.7 | Analyzing a Circuit with Mutual Inductance

Consider the circuit in Fig. 13.23. The make-before-break switch has been in position a for a long time. At $t = 0$, the switch moves instantaneously to position b. Use Laplace methods to find $i_2(t)$ for $t \geq 0$.

Solution

Apply the Laplace transform method using Analysis Method 13.1.

Step 1: Because the switch has been in position a for a long time, both inductors behave like short circuits. The current in the 2 H inductor is $60/(9 + 3) = 5$ A, and the current in the 8 H inductor is zero. Therefore, $i_1(0^-) = 5$ A and $i_2(0^-) = 0$.

Step 2: For $t \geq 0$, there is no independent source in the circuit, so we can skip this step.

Step 3: The currents in the two coils are represented in the *s* domain as I_1 and I_2.

Step 4: Before calculating the impedance of the resistors and inductors, we replace the magnetically coupled coils with a T-equivalent circuit.[1] Figure 13.24 shows the new circuit.

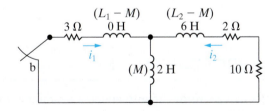

Figure 13.24 ▲ The circuit shown in Fig. 13.23, with the magnetically coupled coils replaced by a T-equivalent circuit.

The impedance of the resistors is their resistance, and the impedance of the inductors is sL. Because we plan to use the mesh-current method in the *s* domain, we use the series-equivalent circuit for inductors carrying initial current. We place a voltage source in the vertical leg of the tee, to represent the initial value of the current in that

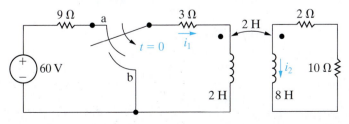

Figure 13.23 ▲ The circuit for Example 13.7, containing magnetically coupled coils.

---

1 See Appendix C.

vertical leg, which is $i_1(0^-) + i_2(0^-)$, or 5 A. The s-domain circuit resulting from Steps 1–4 is shown in Fig. 13.25.

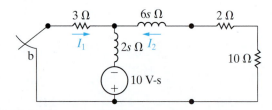

Figure 13.25 ▲ The s-domain equivalent circuit for the circuit shown in Fig. 13.24.

Step 5: Write the two s-domain mesh equations for the circuit in Fig. 13.25:

$$(3 + 2s)I_1 + 2sI_2 = 10;$$

$$2sI_1 + (12 + 8s)I_2 = 10.$$

Solving for I_2 yields

$$I_2 = \frac{2.5}{(s + 1)(s + 3)}.$$

Step 6: Use the initial- and final-value theorems to predict the initial and final values of i_2:

$$\lim_{s \to \infty} sI_2 = \lim_{s \to \infty} \frac{2.5s}{(s + 1)(s + 3)}$$

$$= \lim_{(1/s) \to 0} \frac{2.5(1/s)}{(1 + (1/s))(1 + (3/s))}$$

$$= \frac{0}{(1)(1)} = 0 = \lim_{t \to 0} i_2(t);$$

$$\lim_{s \to 0} sI_2 = \lim_{s \to 0} \frac{2.5s}{(s + 1)(s + 3)} = \frac{0}{(1)(3)}$$

$$= 0 = \lim_{t \to \infty} i_2(t).$$

The initial- and final-value theorems correctly predict the initial and final values of the current i_2 for the circuit in Fig. 13.23.

Step 7: The partial fraction expansion of I_2 is

$$I_2 = \frac{1.25}{s + 1} - \frac{1.25}{s + 3}.$$

Using Table 12.3, the inverse Laplace transform of I_2 is

$$i_2 = (1.25e^{-t} - 1.25e^{-3t})u(t) \text{ A}.$$

Figure 13.26 shows a plot of i_2 versus t. This response makes sense in terms of the known physical behavior of the magnetically coupled coils. A current can exist in the L_2 inductor only if there is a time-varying current in the L_1 inductor. As i_1 decreases from its initial value of 5 A, i_2 increases from zero and then approaches zero as i_1 approaches zero.

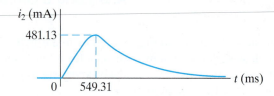

Figure 13.26 ▲ The plot of i_2 versus t for the circuit shown in Fig. 13.23.

ASSESSMENT PROBLEM

Objective 2—Know how to analyze a circuit in the s domain and be able to transform an s-domain solution to the time domain

13.7 a) Using the results from Example 13.7 for the circuit in Fig. 13.23, verify that i_2 reaches a peak value of 481.13 mA at $t = 549.31$ ms.

b) Find i_1, for $t > 0$, for the circuit shown in Fig. 13.23.

c) Compute di_1/dt when i_2 is at its peak value.

d) Express i_2 as a function of di_1/dt when i_2 is at its peak value.

e) Use the results obtained in (c) and (d) to calculate the peak value of i_2.

Answer: (a) $di_2/dt = 0$ when $t = 0.5 \ln 3 = 549.31$ ms;

(b) $i_1 = 2.5(e^{-t} + e^{-3t})u(t)$ A;

(c) -2.89 A/s;

(d) $i_2 = -\left(\frac{M}{12}\right)\frac{di_1}{dt}$;

(e) 481.13 mA.

SELF-CHECK: Also try Chapter Problems 13.38 and 13.39.

Because we are analyzing linear lumped-parameter circuits, we can use superposition to divide the response into components that can be identified with particular sources and initial conditions. We need to identify these components in order to use the transfer function, which we introduce in the next section. Example 13.8 illustrates superposition in the s domain by revisiting the circuit from Example 13.3 and separating its output voltage, v, into three components, two associated with the circuit's initial conditions and one associated with the circuit's independent source.

EXAMPLE 13.8 | **Applying Superposition in the s Domain**

Repeat the analysis of the s-domain circuit for Example 13.3, shown in Fig. 13.14. Use superposition to separate the time-domain voltage v into three components, each one being the response to one of the three sources in Fig. 13.14.

Solution

Since each source in Fig. 13.14 is a current source, we will use Ohm's law to find each component of V, using the equivalent impedance of the three parallel-connected impedances. This equivalent impedance, calculated in Example 13.3, is

$$Z_{eq} = \frac{40 \times 10^6 s}{s^2 + 80{,}000s + 16 \times 10^8} \ \Omega.$$

Define V' as the component of V in Fig. 13.14 due to the initial capacitor energy. Find V' using the circuit in Fig. 13.27, where the only source is the one that represents the initial condition for the capacitor.

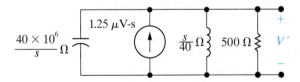

Figure 13.27 ▲ The response of the circuit in Fig. 13.14 due to the initial capacitor voltage.

$$V' = Z_{eq}(1.25 \times 10^{-6}) = \frac{50s}{s^2 + 80{,}000s + 16 \times 10^8}.$$

The partial fraction expansion of V' is

$$V' = \frac{-2 \times 10^6}{(s + 40{,}000)^2} + \frac{50}{(s + 40{,}000)}.$$

Use Table 12.3 to find the inverse Laplace transform of V':

$$v' = \mathcal{L}^{-1}\{V'\} = (50e^{-40{,}000t} - 2 \times 10^6 t e^{-40{,}000t})u(t) \ \text{V}.$$

Therefore, v' is the component of v that is due only to the initial energy of the capacitor in the circuit of Fig. 13.13.

Next, define V'' as the component of V in Fig. 13.14 due to the initial inductor energy. Find V'' using the circuit in Fig. 13.28, where the only source is the one that represents the initial condition for the inductor.

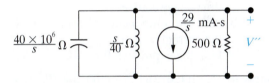

Figure 13.28 ▲ The response of the circuit in Fig. 13.14 due to the initial inductor current.

$$V'' = Z_{eq}\left(\frac{-0.029}{s}\right) = \frac{-1.16 \times 10^6}{s^2 + 80{,}000s + 16 \times 10^8}.$$

The partial fraction expansion of V'' is

$$V'' = \frac{-1.16 \times 10^6}{(s + 40{,}000)^2}.$$

Use Table 12.3 to find the inverse Laplace transform of V'':

$$v'' = \mathcal{L}^{-1}\{V''\} = (-1.16 \times 10^6 t e^{-40{,}000t})u(t) \ \text{V}.$$

Therefore, v'' is the component of v that is due only to the initial energy of the inductor in the circuit of Fig. 13.13.

Finally, define V''' as the component of V in Fig. 13.14 due to the 24 mA dc current source. Find V''' using the circuit in Fig. 13.29, where the only source is the one that represents the 24 mA dc current source.

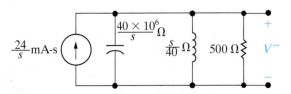

Figure 13.29 ▲ The response of the circuit in Fig. 13.14 due to the independent dc current source.

$$V''' = Z_{eq}\left(\frac{0.024}{s}\right) = \frac{0.96 \times 10^6}{s^2 + 80{,}000s + 16 \times 10^8}.$$

The partial fraction expansion of V''' is

$$V''' = \frac{0.96 \times 10^6}{(s + 40{,}000)^2}.$$

Use Table 12.3 to find the inverse Laplace transform of V''':

$$v''' = \mathcal{L}^{-1}\{V'''\} = (0.96 \times 10^6 t e^{-40{,}000t})u(t)\ \text{V}.$$

Therefore, v''' is the component of v that is due only to the 24 mA current source in the circuit of Fig. 13.13.

The voltage v in the circuit of Fig. 13.13 is the sum of the three component voltages we just found:

$$\begin{aligned}v &= v' + v'' + v'''\\ &= (50e^{-40{,}000t} - 2.2 \times 10^6 t e^{-40{,}000t})u(t)\ \text{V}.\end{aligned}$$

This is the result found in Example 13.3.

ASSESSMENT PROBLEM

Objective 2—Know how to analyze a circuit in the _s_ domain and be able to transform an _s_-domain solution to the time domain

13.8 The energy stored in the circuit shown is zero at the instant the two sources are turned on.
   a) Find the component of v for $t > 0$ owing to the voltage source.
   b) Find the component of v for $t > 0$ owing to the current source.
   c) Find the expression for v when $t > 0$.

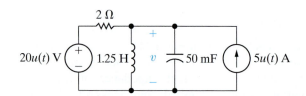

Answer: (a) $[(100/3)e^{-2t} - (100/3)e^{-8t}]u(t)\ \text{V}$;

   (b) $[(50/3)e^{-2t} - (50/3)e^{-8t}]u(t)\ \text{V}$;

   (c) $[50e^{-2t} - 50e^{-8t}]u(t)\ \text{V}$.

SELF-CHECK: Also try Chapter Problem 13.44.

13.4 The Transfer Function

The **transfer function** is defined as the s-domain ratio of the Laplace transform of the output (response) to the Laplace transform of the input (source). As we will see, the transfer function characterizes a circuit's behavior in a single s-domain expression, without revealing what components make up the circuit or how those components are interconnected. In computing the transfer function, we only consider circuits where all initial conditions are zero. If a circuit has multiple independent sources, we can find the transfer function for each source and use superposition to find the response to all sources.

The transfer function is

DEFINITION OF A TRANSFER FUNCTION

$$H(s) = \frac{Y(s)}{X(s)}, \tag{13.7}$$

where $Y(s)$ is the Laplace transform of the output signal, and $X(s)$ is the Laplace transform of the input signal. Note that the transfer function depends on what is defined as the output signal. Consider, for example, the series circuit shown in Fig. 13.30. If the current is defined as the output signal of the circuit,

$$H(s) = \frac{I}{V_g} = \frac{1}{R + sL + 1/sC} = \frac{sC}{s^2LC + RCs + 1}.$$

In deriving $H(s)$, we recognized that I corresponds to the output $Y(s)$ and V_g corresponds to the input $X(s)$.

If, instead, the capacitor voltage is defined as the output signal of the circuit shown in Fig. 13.30, then the transfer function is

$$H(s) = \frac{V}{V_g} = \frac{1/sC}{R + sL + 1/sC} = \frac{1}{s^2 LC + RCs + 1}.$$

Thus, because circuits may have multiple sources and because the definition of the output signal of interest can vary, a single circuit can generate many transfer functions. Remember that when multiple sources are involved, no single transfer function can represent the total output; transfer functions associated with each source must be combined using superposition to yield the total response. Example 13.9 illustrates the computation of a transfer function for known numerical values of R, L, and C.

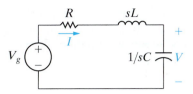

Figure 13.30 ▲ A series *RLC* circuit.

EXAMPLE 13.9 | **Deriving the Transfer Function of a Circuit**

The voltage source v_g drives the circuit shown in Fig. 13.31. The output signal is the voltage across the capacitor, v_o.

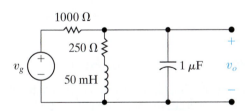

Figure 13.31 ▲ The circuit for Example 13.9.

a) Find the transfer function for this circuit.

b) Calculate the numerical values for the poles and zeros of the transfer function.

Solution

a) Use Analysis Method 13.1 to construct and analyze the s-domain circuit.

Step 1: The circuit in Fig. 13.31 has no initial stored energy, an assumption we always make when calculating a circuit's transfer function. Therefore, we can skip this step.

Step 2: There are no independent sources described by a time-domain function in this circuit, so we can skip this step.

Step 3: Represent the source and output voltages using V_g and V_o, respectively, as shown in Fig. 13.32.

Step 4: The impedance of the resistors is their resistance. The impedances of the inductor and capacitor are

$$Z_L = sL = 0.05s \ \Omega;$$

$$Z_C = \frac{1}{sC} = \frac{1}{s(10^{-6})} = \frac{10^6}{s} \ \Omega.$$

The s-domain circuit resulting from Steps 1–4 is shown in Fig. 13.32.

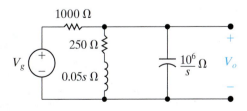

Figure 13.32 ▲ The s-domain equivalent circuit for the circuit shown in Fig. 13.31.

Step 5: From the problem statement, the transfer function is the ratio of V_o/V_g. Writing a node-voltage equation at the upper node and summing the currents leaving the node gives

$$\frac{V_o - V_g}{1000} + \frac{V_o}{250 + 0.05s} + \frac{V_o s}{10^6} = 0.$$

Solving for V_o yields

$$V_o = \frac{1000(s + 5000)V_g}{s^2 + 6000s + 25 \times 10^6}.$$

Hence, the transfer function is

$$H(s) = \frac{V_o}{V_g} = \frac{1000(s + 5000)}{s^2 + 6000s + 25 \times 10^6}.$$

b) The poles of $H(s)$ are the roots of the denominator polynomial. Therefore

$$-p_1 = -3000 - j4000,$$
$$-p_2 = -3000 + j4000.$$

The zeros of $H(s)$ are the roots of the numerator polynomial; thus, $H(s)$ has a zero at

$$-z_1 = -5000.$$

ASSESSMENT PROBLEM

Objective 3—Understand the definition and significance of the transfer function; be able to derive a transfer function

13.9 a) Derive the numerical expression for the transfer function V_o/I_g for the circuit shown.
b) Give the numerical value of each pole and zero of $H(s)$.

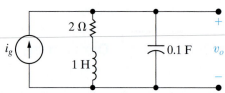

Answer: (a) $H(s) = 10(s + 2)/(s^2 + 2s + 10)$;
(b) $-p_1 = -1 + j3$, $-p_2 = -1 - j3$,
$-z = -2$.

SELF-CHECK: Also try Chapter Problem 13.51.

The Location of Poles and Zeros of H(s)

For linear lumped-parameter circuits, $H(s)$ is always a rational function of s. Complex poles and zeros always occur in conjugate pairs. The poles of $H(s)$ must lie in the left half of the s plane if the response to a bounded source (one whose values lie within some finite bounds) is to be bounded. The zeros of $H(s)$ may lie in either the right half or the left half of the s plane.

With these general characteristics in mind, we next discuss the role that $H(s)$ plays in determining the circuit's output.

13.5 The Transfer Function in Partial Fraction Expansions

Using the definition of a circuit's transfer function (Eq. 13.7), we can find the circuit's output by multiplying the transfer function and the circuit's input:

$$Y(s) = H(s)X(s). \tag{13.8}$$

We have already noted that $H(s)$ is a rational function of s; $X(s)$ also is a rational function of s for the time-domain functions of most interest in circuit analysis (see Table 12.1).

Expanding the right-hand side of Eq. 13.8 into a sum of partial fractions produces a term for each pole of $H(s)$ and $X(s)$. The terms generated by the poles of $H(s)$ correspond to the transient components of the total response. The terms generated by the poles of $X(s)$ correspond to the steady-state components of the response, which exist after the transient components have become negligible. Example 13.10 illustrates these general observations.

EXAMPLE 13.10   Analyzing the Transfer Function of a Circuit

The circuit in Example 13.9 (Fig. 13.31) is driven by a voltage source whose voltage increases linearly with time, namely, $v_g = 50tu(t)$ V.

a) Use the transfer function to find v_o.

b) Identify the transient component of the response.

c) Identify the steady-state component of the response.

d) Sketch v_o versus t for $0 \le t \le 1.5$ ms.

Solution

a) From Example 13.9,

$$H(s) = \frac{1000(s + 5000)}{s^2 + 6000s + 25 \times 10^6}.$$

The Laplace transform of the source voltage is $50/s^2$; therefore, the s-domain expression for the output voltage is

$$V_o = \frac{1000(s + 5000)}{(s^2 + 6000s + 25 \times 10^6)} \frac{50}{s^2}.$$

The partial fraction expansion of V_o is

$$V_o = \frac{K_1}{s + 3000 - j4000}$$

$$+ \frac{K_1^*}{s + 3000 + j4000} + \frac{K_2}{s^2} + \frac{K_3}{s}.$$

We evaluate the coefficients K_1, K_2, and K_3 by using the techniques described in Section 12.7:

$$K_1 = 5\sqrt{5} \times 10^{-4} \underline{/79.70°};$$

$$K_1^* = 5\sqrt{5} \times 10^{-4} \underline{/-79.70°};$$

$$K_2 = 10;$$

$$K_3 = -4 \times 10^{-4}.$$

Using Table 12.3, the time-domain expression for v_o is

$$v_o = [10\sqrt{5} \times 10^{-4}e^{-3000t}\cos(4000t + 79.70°)$$

$$+ 10t - 4 \times 10^{-4}]u(t) \text{ V}.$$

b) The transient component of v_o is

$$10\sqrt{5} \times 10^{-4}e^{-3000t}\cos(4000t + 79.70°) \text{ V}.$$

Note that this term is generated by the poles $(-3000 + j4000)$ and $(-3000 - j4000)$ of the transfer function.

c) The steady-state component of the response is

$$(10t - 4 \times 10^{-4})u(t) \text{ V}.$$

These two terms are generated by the second-order pole (K/s^2) of the input voltage.

d) Figure 13.33 shows a sketch of v_o versus t. Note that the deviation from the steady-state solution $10{,}000t - 0.4$ mV is imperceptible after approximately 1 ms.

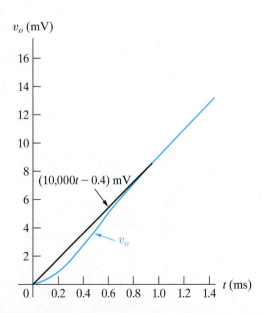

Figure 13.33 ▲ The graph of v_o versus t for Example 13.10.

Observations on the Use of *H(s)* in Circuit Analysis

Example 13.10 related the components that make up a circuit's response to the poles of the transfer function, $H(s)$, and the poles of the circuit's input in the *s* domain, using a partial fraction expansion. However, the example raises questions about driving a circuit with an increasing ramp voltage that generates an increasing ramp response. Eventually, excessive voltage will cause the circuit components to fail, and then our linear model is invalid. But some practical applications have input ramp functions that increase to some maximum value over a finite time interval, so the response to a ramp input is important. If the time it takes for the ramp to reach its maximum value is long compared with the time constants of the circuit, the solution assuming an unbounded ramp is valid for this finite time interval.

Here are some additional observations about a circuit's transfer function, defined in Eq. 13.7.

- If the circuit's input is delayed by *a* seconds,

$$\mathcal{L}\{x(t - a)u(t - a)\} = e^{-as}X(s),$$

then from Eq. 13.8, the circuit's output is

$$Y(s) = H(s)X(s)e^{-as}.$$

If $y(t) = \mathcal{L}^{-1}\{H(s)X(s)\}$, then,

$$y(t - a)u(t - a) = \mathcal{L}^{-1}\{H(s)X(s)e^{-as}\}.$$

Therefore, delaying the input by *a* seconds delays the output by *a* seconds. A circuit that exhibits this characteristic is **time invariant**.

- If the circuit's input is a unit impulse, the circuit's output equals the inverse transform of the transfer function. Thus, if

$$x(t) = \delta(t), \quad \text{then } X(s) = 1$$

and

$$Y(s) = H(s).$$

Hence,

$$y(t) = h(t),$$

so the inverse transform of the transfer function equals the unit impulse response of the circuit.

- A circuit's unit impulse response is also its natural response because applying an impulsive source is equivalent to instantaneously storing energy in the circuit (see Section 13.8). The subsequent release of this stored energy is the circuit's natural response (see Problem 13.84).

- A circuit's unit impulse response, $h(t)$, contains enough information to compute the response to any source that drives the circuit. We can extract a circuit's response to an arbitrary source from the circuit's unit impulse response, using the convolution integral. This technique is demonstrated in the next section.

ASSESSMENT PROBLEMS

Objective 4—Know how to use a circuit's transfer function to calculate the circuit's impulse response, unit step response, and steady-state response to sinusoidal input

13.10 Find (a) the unit step and (b) the unit impulse response of the circuit shown in Assessment Problem 13.9.

Answer: (a) $[2 + (10/3)e^{-t} \cos (3t - 126.87°)]u(t)$ V;
(b) $10.54e^{-t} \cos (3t - 18.43°)u(t)$ V.

13.11 The unit impulse response of a circuit is

$$v_o(t) = 10{,}000e^{-70t} \cos (240t + \theta) \text{ V},$$

where $\tan \theta = \dfrac{7}{24}$.

a) Find the transfer function of the circuit.

b) Find the unit step response of the circuit.

Answer: (a) $9600s/(s^2 + 140s + 62{,}500)$;
(b) $[40e^{-70t} \sin 240t]u(t)$ V.

SELF-CHECK: Also try Chapter Problems 13.55 and 13.56.

13.6 The Transfer Function and the Convolution Integral

The **convolution integral** relates the output $y(t)$ of a linear time-invariant circuit to the input $x(t)$ of the circuit and the circuit's impulse response $h(t)$. The integral relationship can be expressed in two ways:

CONVOLUTION INTEGRAL

$$y(t) = \int_{-\infty}^{\infty} h(\lambda)x(t - \lambda)\, d\lambda = \int_{-\infty}^{\infty} h(t - \lambda)x(\lambda)\, d\lambda. \quad (13.9)$$

We are interested in the convolution integral for several reasons.

- We can find a circuit's output for any input by working entirely in the time domain. This is beneficial when $x(t)$ and $h(t)$ are known only through experimental data. In such cases, using Laplace transform methods may be awkward or even impossible, as we would need to compute the Laplace transform of experimental data.
- The convolution integral introduces the concepts of memory and the weighting function into analysis. The concept of memory enables us to predict, to some degree, how closely the output waveform replicates the input waveform, using the impulse response (or the weighting function) $h(t)$.
- The convolution integral provides a formal procedure for finding the inverse transform of products of Laplace transforms.

To derive Eq. 13.9, we assume that the circuit is linear and time invariant. Because the circuit is linear, the principle of superposition is valid, and because it is time invariant, the response delay and the input delay are the same. Consider Fig. 13.34, in which the block containing $h(t)$ represents any linear time-invariant circuit whose impulse response is known,

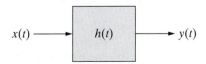

Figure 13.34 ▲ A block diagram of a general circuit.

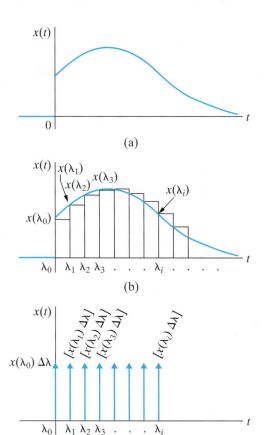

Figure 13.35 ▲ The excitation signal of $x(t)$. (a) A general excitation signal. (b) Approximating $x(t)$ with a series of pulses. (c) Approximating $x(t)$ with a series of impulses.

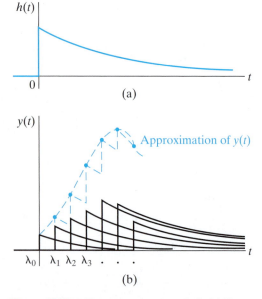

Figure 13.36 ▲ The approximation of $y(t)$. (a) The impulse response of the box shown in Fig. 13.34. (b) Summing the impulse responses.

$x(t)$ represents the input signal and $y(t)$ represents the output signal. We also assume the following:

- $x(t)$ is the general signal shown in Fig. 13.35(a).
- $x(t) = 0$ for $t < 0^-$. Once we derive the convolution integral assuming $x(t) = 0$ for $t < 0^-$, extending the integral to include excitation functions that exist for all time is straightforward.
- A discontinuity in $x(t)$ at the origin (between 0^- and 0^+) is permitted.

Begin by approximating $x(t)$ with a series of rectangular pulses of uniform width $\Delta\lambda$, as shown in Fig. 13.35(b). Thus

$$x(t) = x_0(t) + x_1(t) + \cdots + x_i(t) + \cdots,$$

where $x_i(t)$ is a rectangular pulse that equals $x(\lambda_i)$ between λ_i and λ_{i+1} and is zero elsewhere. Note that the ith pulse can be expressed using step functions; that is,

$$x_i(t) = x(\lambda_i)\{u(t - \lambda_i) - u[t - (\lambda_i + \Delta\lambda)]\}.$$

Continue to approximate $x(t)$ by making the pulse width $\Delta\lambda$ so small that we can approximate the ith component using an impulse function of strength $x(\lambda_i)\Delta\lambda$. Figure 13.35(c) shows this impulse representation, where the brackets beside each arrow represent the impulse strength. The impulse representation of $x(t)$ is

$$x(t) = x(\lambda_0)\Delta\lambda\delta(t - \lambda_0) + x(\lambda_1)\Delta\lambda\delta(t - \lambda_1) + \cdots$$

$$+ x(\lambda_i)\Delta\lambda\delta(t - \lambda_i) + \cdots.$$

When we represent $x(t)$ with a series of impulse functions (which occur at equally spaced intervals of time, that is, at $\lambda_0, \lambda_1, \lambda_2, \dots$), the output function $y(t)$ is the sum of uniformly delayed impulse responses. The strength of each response depends on the strength of the impulse driving the circuit. For example, let's assume that the unit impulse response of the circuit represented by the box in Fig. 13.34 is the exponential decay function shown in Fig. 13.36(a). Then the approximation of $y(t)$ is the sum of the impulse responses shown in Fig. 13.36(b).

Analytically, the expression for $y(t)$ is

$$y(t) = x(\lambda_0)\Delta\lambda h(t - \lambda_0) + x(\lambda_1)\Delta\lambda h(t - \lambda_1)$$

$$+ x(\lambda_2)\Delta\lambda h(t - \lambda_2) + \cdots$$

$$+ x(\lambda_i)\Delta\lambda h(t - \lambda_i) + \cdots.$$

As $\Delta\lambda \to 0$, the sum approaches a continuous integral, or

$$\sum_{i=\infty}^{\infty} x(\lambda_i)h(t - \lambda_i)\Delta\lambda \to \int_0^\infty x(\lambda)h(t - \lambda)\, d\lambda.$$

Therefore,

$$y(t) = \int_0^\infty x(\lambda)h(t - \lambda)\, d\lambda.$$

If $x(t)$ exists over all time, then the lower limit on the integral is $-\infty$. Thus, in general,

$$y(t) = \int_{-\infty}^{\infty} x(\lambda)h(t - \lambda)\,d\lambda,$$

which is the second form of the convolution integral given in Eq. 13.9.

To derive the first form of the integral in Eq. 13.9, we change the variable of integration in the second form of the integral. Let $u = t - \lambda$, and then $du = -d\lambda$, $u = -\infty$ when $\lambda = \infty$, and $u = +\infty$ when $\lambda = -\infty$. Now we can write

$$y(t) = \int_{\infty}^{-\infty} x(t - u)h(u)(-du),$$

or

$$y(t) = \int_{-\infty}^{\infty} x(t - u)h(u)(du).$$

Since u is a symbol of integration, this integral is equivalent to the first form of the convolution integral, Eq. 13.9.

The convolution integral relating $y(t)$, $h(t)$, and $x(t)$, (Eq. 13.9), is often written using a shorthand notation:

$$y(t) = h(t) * x(t) = x(t) * h(t), \tag{13.10}$$

where the asterisk represents the integral relationship between $h(t)$ and $x(t)$. Thus, $h(t) * x(t)$ is read as "$h(t)$ is convolved with $x(t)$" and implies that

$$h(t) * x(t) = \int_{-\infty}^{\infty} h(\lambda)x(t - \lambda)\,d\lambda,$$

whereas $x(t) * h(t)$ is read as "$x(t)$ is convolved with $h(t)$" and implies that

$$x(t) * h(t) = \int_{-\infty}^{\infty} x(\lambda)h(t - \lambda)\,d\lambda.$$

The integrals in Eq. 13.9 give the most general relationship for the convolution of two functions. However, when we apply the convolution integral, we can change the lower limit to zero and the upper limit to t. Then we can write Eq. 13.9 as

$$y(t) = \int_{0}^{t} h(\lambda)x(t - \lambda)\,d\lambda = \int_{0}^{t} x(\lambda)h(t - \lambda)\,d\lambda.$$

We change the limits for two reasons. First, for physically realizable circuits, $h(t)$ is zero for $t < 0$ because there is no impulse response before you apply an impulse. Second, we start measuring time at the instant we turn on the input $x(t)$; therefore $x(t) = 0$ for $t < 0^-$.

A graphic interpretation of the convolution integrals (Eq. 13.9) helps us use convolution as a computational tool. We begin with the first integral and assume that the circuit's impulse response is the exponential decay function shown in Fig. 13.37(a) and its input function has the waveform shown in Fig. 13.37(b). In each of the plots, we replace t with λ, the symbol of integration. Replacing λ with $-\lambda$ folds the excitation function over the vertical axis, as shown in Fig 13.37(d). The folding operation explains why Eq. 13.9 is called the *convolution* integral. Replacing $-\lambda$ with $t - \lambda$ slides the folded function to the right, as shown in Fig. 13.37(d).

At any specified value of t, the output function $y(t)$ is the area under the product function $h(\lambda)x(t - \lambda)$, as shown in Fig. 13.37(e). This plot explains why we can set the lower limit on the convolution integral to zero and the upper limit to t. For $\lambda < 0$, the product $h(\lambda)x(t - \lambda)$ is zero because $h(\lambda)$ is zero. For $\lambda > t$, the product $h(\lambda)x(t - \lambda)$ is zero because $x(t - \lambda)$ is zero.

Figure 13.38 shows the second form of the convolution integral. Once again, the product function in Fig. 13.38(e) explains why we use zero for the lower limit and t for the upper limit.

Example 13.11 illustrates how to use the convolution integral and the unit impulse response to find a circuit's output in response to a given input.

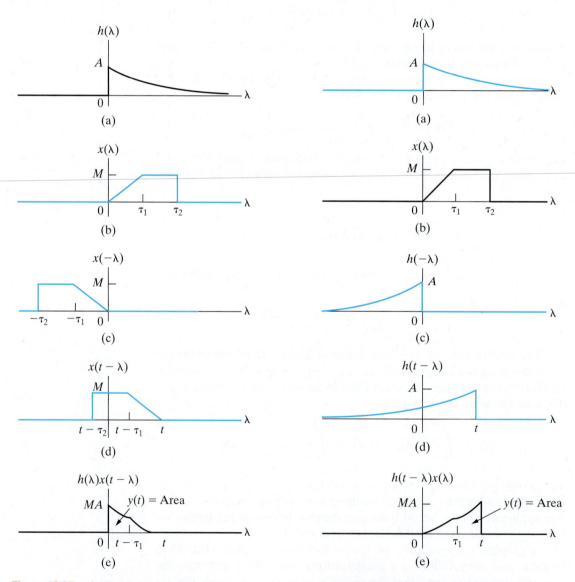

Figure 13.37 ▲ A graphic interpretation of the convolution integral $\int_0^t h(\lambda)x(t - \lambda)\, d\lambda$. (a) The impulse response. (b) The excitation function. (c) The folded excitation function. (d) The folded excitation function displaced t units. (e) The product $h(\lambda)x(t - \lambda)$.

Figure 13.38 ▲ A graphic interpretation of the convolution integral $\int_0^t h(t - \lambda)x(\lambda)\, d\lambda$. (a) The impulse response. (b) The excitation function. (c) The folded impulse response. (d) The folded impulse response displaced t units. (e) The product $h(t - \lambda)x(\lambda)$.

<table><tr><td>**EXAMPLE 13.11**</td><td>**Using the Convolution Integral to Find an Output Signal**</td></tr></table>

The input voltage v_i for the circuit shown in Fig. 13.39(a) is shown in Fig. 13.39(b).

a) Use the convolution integral to find v_o.

b) Plot v_o over the range of $0 \le t \le 15$ s.

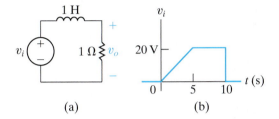

(a)   (b)

Figure 13.39 ▲ The circuit and excitation voltage for Example 13.11. (a) The circuit. (b) The input voltage.

Solution

a) Begin by finding the unit impulse response of the circuit. The s-domain equivalent of the circuit in Fig. 13.39(a) replaces the inductor with an impedance of $s\ \Omega$ and the two voltages with V_o and V_i. Using voltage division,

$$V_o = \frac{1}{s + 1} V_i.$$

When v_i is a unit impulse function $\delta(t)$, $V_i = 1$ and $H(s) = V_o/V_i = 1/(s + 1)$. Then,

$$v_o = h(t) = e^{-t}u(t),$$

from which

$$h(\lambda) = e^{-\lambda}u(\lambda).$$

Using the first form of the convolution integral in Eq. 13.9, we construct the impulse response and folded input function shown in Fig. 13.40, which helps us select the convolution integral limits. As we slide the folded input function to the right, we break the integration into three intervals: $0 \le t \le 5$; $5 \le t \le 10$; and $10 \le t \le \infty$. The integration intervals correspond to points where the input function's definition changes. Figure 13.41 depicts the position of the folded input function for each interval.

The expression for v_i in the time interval $0 \le t \le 5$ is

$$v_i = 4t, \quad 0 \le t \le 5\ \text{s}.$$

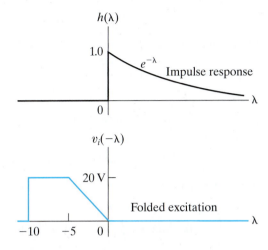

Figure 13.40 ▲ The impulse response and the folded input function for Example 13.11.

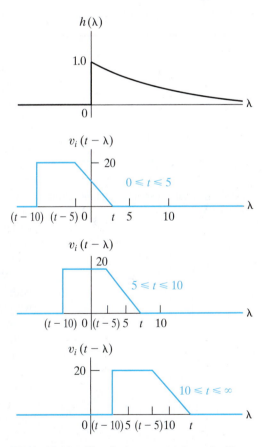

Figure 13.41 ▲ The displacement $v_i(t - \lambda)$ of for three different time intervals.

Hence, the expression for the folded input function in the interval $t - 5 \leq \lambda \leq t$ is

$$v_i(t - \lambda) = 4(t - \lambda), \quad t - 5 \leq \lambda \leq t.$$

In the other two time intervals, v_i is a constant.

We can now set up the three convolution integrals to find v_o. For $0 \leq t \leq 5$ s:

$$v_o = \int_0^t 4(t - \lambda)e^{-\lambda}\, d\lambda$$

$$= 4(e^{-t} + t - 1)\text{ V}.$$

For $5 \leq t \leq 10$ s,

$$v_o = \int_0^{t-5} 20e^{-\lambda}\, d\lambda + \int_{t-5}^t 4(t - \lambda)e^{-\lambda}\, d\lambda$$

$$= 4(5 + e^{-t} - e^{-(t-5)})\text{ V}.$$

For $10 \leq t \leq \infty$ s,

$$v_o = \int_{t-10}^{t-5} 20e^{-\lambda}\, d\lambda + \int_{t-5}^t 4(t - \lambda)e^{-\lambda}\, d\lambda$$

$$= 4(e^{-t} - e^{-(t-5)} + 5e^{-(t-10)})\text{ V}.$$

b) Values for v_o are shown graphically in Fig. 13.42.

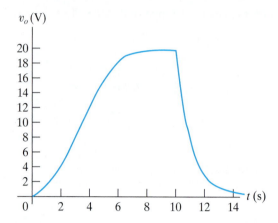

Figure 13.42 ▲ The voltage response versus time for Example 13.11.

SELF-CHECK: Assess your understanding of convolution by trying Chapter Problems 13.61 and 13.62.

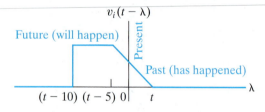

Figure 13.43 ▲ The past, present, and future values of the input function.

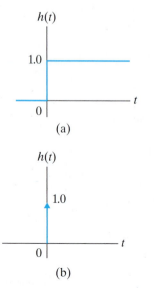

Figure 13.44 ▲ Weighting functions. (a) Perfect memory. (b) No memory.

The Concepts of Memory and the Weighting Function

We mentioned at the beginning of this section that the convolution integral introduces the concepts of memory and the weighting function into circuit analysis. The graphical interpretation of the convolution integral is the easiest way to understand these concepts. We can view the folding and sliding of a circuit's input function on a time scale divided into past, present, and future time intervals. The vertical axis represents the present value of $x(t)$, with past values to the right of the vertical axis and future values to the left. Figure 13.43 shows this description of $x(t)$, using the input function from Example 13.11.

When we combine the past, present, and future views of $x(t - \tau)$ with the impulse response of the circuit, we see that the impulse response weights $x(t)$ according to present and past values. For example, Fig. 13.41 shows that the impulse response in Example 13.11 gives less weight to past values of $x(t)$ than to the present value of $x(t)$. In other words, the circuit "remembers" less and less about past input values. Therefore, in Fig. 13.42, v_o quickly approaches zero when the present value of the input is zero (that is, when $t > 10$ s). Since the present value of the input receives more weight than the past values, the output quickly approaches the present value of the input.

Because the convolution integral uses the product of $x(t - \lambda)$ and $h(\lambda)$, the impulse response is considered the circuit's **weighting function**. The weighting function, in turn, determines how much memory the circuit has. **Memory** represents how accurately the circuit's response matches its input. For example, if the impulse response, or weighting function, is flat, as shown in Fig. 13.44(a), it gives equal weight to all values of $x(t)$, past and present. Such a circuit has a perfect memory. However, if the impulse response is an impulse function, as shown in Fig. 13.44(b), it gives no weight to past values of $x(t)$ and the circuit has no memory.

The more memory a circuit has, the more distortion exists between the circuit's input waveform and its output. We can demonstrate this by assuming that the circuit has no memory, that is, $h(t) = A\delta(t)$, and then noting from the convolution integral that

$$y(t) = \int_0^t h(\lambda)x(t - \lambda)\, d\lambda$$

$$= \int_0^t A\delta(\lambda)x(t - \lambda)\, d\lambda$$

$$= Ax(t).$$

The expression for $y(t)$ shows that, if the circuit has no memory, the output is a scaled replica of the input.

The circuit shown in Example 13.11 illustrates the distortion between input and output for a circuit that has some memory. This distortion is clear when we plot the input and output waveforms on the same graph, as in Fig. 13.45.

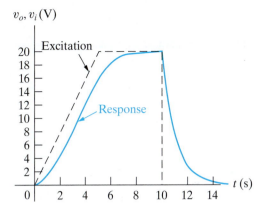

Figure 13.45 ▲ The input and output waveforms for Example 13.11.

13.7 The Transfer Function and the Steady-State Sinusoidal Response

Once we have computed a circuit's transfer function, we can use it to find the steady-state response to a sinusoidal input. To show this, we assume that

$$x(t) = A\cos(\omega t + \phi),$$

and then we use Eq. 13.8 to find the steady-state solution of $y(t)$. To find the Laplace transform of $x(t)$, we first write $x(t)$ as

$$x(t) = A\cos\omega t\cos\phi - A\sin\omega t\sin\phi,$$

from which

$$X(s) = \frac{(A\cos\phi)s}{s^2 + \omega^2} - \frac{(A\sin\phi)\omega}{s^2 + \omega^2}$$

$$= \frac{A(s\cos\phi - \omega\sin\phi)}{s^2 + \omega^2}.$$

Substituting the expression for $X(s)$ into Eq. 13.8 gives the s-domain expression for the response:

$$Y(s) = H(s)\frac{A(s\cos\phi - \omega\sin\phi)}{s^2 + \omega^2}.$$

Think about the partial fraction expansion of $Y(s)$. The number of terms in the expansion depends on the number of poles of $H(s)$. Because $H(s)$ is not specified, we write the expansion of $Y(s)$ as

$$Y(s) = \frac{K_1}{s - j\omega} + \frac{K_1^*}{s + j\omega} + \sum \text{ terms generated by the poles of } H(s).$$

The first two terms in the partial fraction expansion correspond to the complex conjugate poles of the sinusoidal input because $s^2 + \omega^2 = (s - j\omega)(s + j\omega)$. But the terms corresponding to the poles of $H(s)$ do not contribute to the steady-state response of $y(t)$, because all these poles lie in the left half of the s plane. Consequently, the corresponding time-domain terms approach zero as t increases. Thus, only the first two partial fraction terms determine the steady-state response, so we only have to calculate a single partial fraction coefficient, K_1:

$$K_1 = \left.\frac{H(s)A(s \cos \phi - \omega \sin \phi)}{s + j\omega}\right|_{s = j\omega}$$

$$= \frac{H(j\omega)A(j\omega \cos \phi - \omega \sin \phi)}{2j\omega}$$

$$= \frac{H(j\omega)A(\cos \phi + j \sin \phi)}{2} = \frac{1}{2} H(j\omega)Ae^{j\phi}.$$

In general, $H(j\omega)$ is a complex number, which we can write in polar form as

$$H(j\omega) = |H(j\omega)|e^{j\theta(\omega)}.$$

We see that the transfer function's magnitude, $|H(j\omega)|$, and its phase angle, $\theta(\omega)$, vary with the frequency ω. Substituting the polar form for $H(j\omega)$ into the equation for K_1 and simplifying, we see that the expression for K_1 becomes

$$K_1 = \frac{A}{2} |H(j\omega)|e^{j[\theta(\omega) + \phi]}.$$

We find the steady-state component for $y(t)$ by inverse-transforming the first two terms in the partial fraction expansion of $Y(s)$, ignoring the terms generated by the poles of $H(s)$. Thus

SINUSOIDAL STEADY-STATE RESPONSE COMPUTED USING A TRANSFER FUNCTION

$$y_{ss}(t) = A|H(j\omega)| \cos [\omega t + \phi + \theta(\omega)], \qquad (13.11)$$

which tells us how to find a circuit's steady-state response to a sinusoidal input using the circuit's transfer function:

- Determine the input sinusoid's magnitude, A, frequency, ω, and phase angle, ϕ
- Evaluate the circuit's transfer function, $H(s)$, for $s = j\omega$.
- Transform $H(j\omega)$ into polar form, with a magnitude $|H(j\omega)|$ and a phase angle θ.
- Write the steady-state output as a cosine with the amplitude $A|H(j\omega)|$, a phase angle of $\phi + \theta$, and a frequency of ω.

Example 13.12 uses a circuit's transfer function to find its sinusoidal steady-state response.

<table>
<tr><td>**EXAMPLE 13.12**</td><td>**Using the Transfer Function to Find the Steady-State Sinusoidal Response**</td></tr>
</table>

The circuit from Example 13.9 is shown in Fig. 13.46. The sinusoidal source voltage is

$$v_g = 120 \cos(5000t + 30°)\text{V}.$$

Find the steady-state expression for v_o.

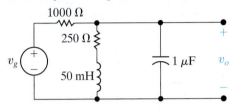

Figure 13.46 ▲ The circuit for Example 13.12.

Solution

From Example 13.9,

$$H(s) = \frac{1000(s + 5000)}{s^2 + 6000s + 25 \times 10^6}.$$

The frequency of the voltage source is 5000 rad/s, so we evaluate $H(j5000)$:

$$H(j5000) = \frac{1000(5000 + j5000)}{-25 \times 10^6 + j5000(6000) + 25 \times 10^6}$$

$$= \frac{1 + j1}{j6} = \frac{1 - j1}{6} = \frac{\sqrt{2}}{6}\angle{-45°}.$$

Then, from Eq. 13.11,

$$v_{o_{ss}} = \frac{(120)\sqrt{2}}{6}\cos(5000t + 30° - 45°)$$

$$= 20\sqrt{2}\cos(5000t - 15°)\text{ V}.$$

The relationship between $H(s)$ and $H(j\omega)$ provides a link between the time domain and the frequency domain. Using Eq. 13.11, we can find the sinusoidal steady-state response of a circuit by evaluating $H(j\omega)$. Theoretically, we can reverse the process; instead of using $H(s)$ to find $H(j\omega)$, we can use $H(j\omega)$ to find $H(s)$. To do so, we determine $H(j\omega)$ experimentally and then construct $H(s)$ from the data. Once we know $H(s)$, we can find the circuit's response to other inputs. While this experimental approach is not always possible, in some cases it does provide a way to find $H(s)$ for a circuit whose components and their values are unknown.

The transfer function is also used to find a circuit's frequency response, a concept we introduce in the next chapter.

ASSESSMENT PROBLEMS

Objective 4—Know how to use a circuit's transfer function to calculate the circuit's impulse response, unit step response, and steady-state response to sinusoidal input

13.12 The current source in the circuit shown is delivering 10 cos 4t A. Use the transfer function to compute the steady-state expression for v_o.

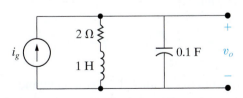

Answer: 44.7cos(4t − 63.43°) V.

13.13 a) For the circuit shown, find the steady-state expression for v_o when $v_g = 10 \cos 50,000t$ V.

b) Replace the 50 kΩ resistor with a variable resistor and compute the value of resistance necessary to cause v_o to lead v_g by 120°.

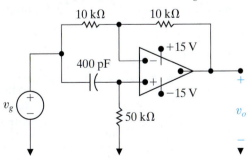

Answer: (a) 10 cos(50,000t + 90°) V;
(b) 28,867.51 Ω.

SELF-CHECK: Also try Chapter Problems 13.79 and 13.82.

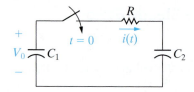

Figure 13.47 ▲ A circuit showing the creation of an impulsive current.

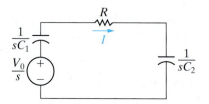

Figure 13.48 ▲ The s-domain equivalent circuit for the circuit shown in Fig. 13.47.

13.8 The Impulse Function in Circuit Analysis

Circuit analysis uses impulse functions to represent a switching operation or because a circuit's input is an impulsive source. We can use Laplace transform methods to determine the impulsive currents and voltages created during switching, and to find a circuit's response to an impulsive source. To begin, we create an impulse function with a switching operation.

Switching Operations

We use two different circuits to create an impulse function with a switching operation: a capacitor circuit, and a series inductor circuit.

Capacitor Circuit

In the circuit shown in Fig. 13.47, the capacitor C_1 is charged to an initial voltage of V_0 at the time the switch is closed. The initial charge on C_2 is zero. We want to find the expression for $i(t)$ as $R \to 0$. Figure 13.48 shows the s-domain equivalent circuit, and we use Ohm's law to find the s-domain current I:

$$I = \frac{V_0/s}{R + (1/sC_1) + (1/sC_2)}$$

$$= \frac{V_0/R}{s + (1/RC_e)},$$

where the equivalent capacitance $C_1 C_2/(C_1 + C_2)$ is replaced by C_e.

We inverse-transform the expression for I using Table 12.3 to get

$$i = \left(\frac{V_0}{R} e^{-t/RC_e}\right) u(t),$$

which indicates that as R decreases, the initial current (V_0/R) increases and the time constant (RC_e) decreases. Thus, as R gets smaller, the current starts from a larger initial value and then drops off more rapidly. Figure 13.49 shows these characteristics of i.

As R approaches zero, the initial value of i approaches infinity and the duration of i approaches zero. If the area under the current function is independent of R, i approaches an impulse function. Physically, the total area under the i versus t curve represents the total charge transferred to C_2 after the switch is closed. Thus

$$\text{Area} = q = \int_{0^-}^{\infty} \frac{V_0}{R} e^{-t/RC_e} dt = V_0 C_e,$$

so the total charge transferred to C_2 equals $V_0 C_e$ coulombs and is independent of R. Thus, as R approaches zero, the current approaches an impulse with strength $V_0 C_e$; that is,

$$i \to V_0 C_e \delta(t).$$

The physical interpretation of this expression says that when $R = 0$, a finite amount of charge is transferred to C_2 instantaneously. Set $R = 0$ in the circuit of Fig. 13.47 to see why we get an instantaneous transfer of charge. With $R = 0$, closing the switch creates a contradiction: we apply a

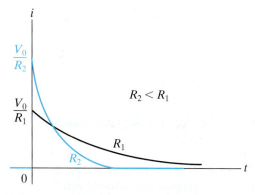

Figure 13.49 ▲ The plot of $i(t)$ versus t for two different values of R.

voltage across a capacitor whose initial voltage is zero. The capacitor voltage can change instantaneously only if there is an instantaneous transfer of charge. When the switch is closed, the voltage across C_2 does not jump to V_0 but to its final value of

$$v_2 = \frac{C_1 V_0}{C_1 + C_2}.$$

We leave the derivation of v_2 to you (see Problem 13.83).

If we set $R = 0$ at the outset, Laplace transform analysis predicts the impulsive current response. Thus,

$$I = \frac{V_0/s}{(1/sC_1) + (1/sC_2)} = \frac{C_1 C_2 V_0}{C_1 + C_2} = C_e V_0.$$

In writing this equation for I, we use the capacitor voltages at $t = 0^-$. The inverse transform of I, which is constant, is the constant times the impulse function; therefore,

$$i = C_e V_0 \delta(t).$$

Example 13.13 uses Laplace methods to analyze an inductor circuit whose output contains an impulse.

EXAMPLE 13.13   A Series Inductor Circuit with an Impulsive Response

The switch in the circuit shown in Fig. 13.50 has been closed for a long time. At $t = 0$ it opens. Use Laplace methods to find the output voltage, v_o, and the current in the 3 H inductor, i_1.

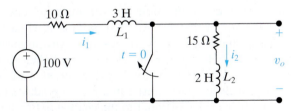

Figure 13.50 ▲ The circuit for Example 13.13.

Solution

Use Analysis Method 13.1 to construct and analyze the s-domain circuit.

Step 1: For $t < 0$, the switch is closed, and the inductors behave like short circuits. The current in the 2 H inductor is zero, since there is no source in that part of the circuit. The current in the 3 H inductor is $100/10 = 10$ A.

Step 2: The combination of the dc voltage source and the switch is defined by the time-domain function $100u(t)$. The Laplace transform of this function is $100/s$ V-s, which labels the voltage source in the s-domain circuit in Fig. 13.51.

Step 3: Represent the output voltage and current using V_o and I_1, respectively, as shown in Fig. 13.51. Note that in the s-domain circuit, the current in both inductors is the same.

Step 4: The impedance of the resistors is their resistance. The impedance of the inductors is sL. We also need to represent the nonzero initial current in the 3 H inductor using either a series-connected voltage source or a parallel-connected current source. Here, we choose the voltage source, so the resulting s-domain circuit has a single mesh. The voltage source has the value $LI_0 = 3(10) = 30$ V-s. The s-domain circuit resulting from Steps 1–4 is shown in Fig. 13.51.

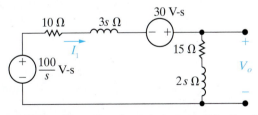

Figure 13.51 ▲ The s-domain equivalent circuit for the circuit shown in Fig. 13.50.

Step 5: From the circuit in Fig. 13.51, we find V_o using voltage division:

$$V_o = \frac{2s + 15}{5s + 25}\left(\frac{100}{s} + 30\right) = \frac{12s^2 + 130s + 300}{s^2 + 5s}.$$

Now use Ohm's law to find I_1:

$$I_1 = \frac{(100/s) + 30}{5s + 25} = \frac{6s + 20}{s(s + 5)}.$$

Step 6: We will discuss initial and final values in this circuit after we find the output voltage and current, so for now we skip this step.

Step 7: The expression for V_o is an improper rational function, so before we find its partial fraction expansion we must divide numerator by denominator to get

$$V_o = \frac{12s^2 + 130s + 300}{s^2 + 5s} = 12 + \frac{70s + 300}{s(s + 5)}.$$

The second term is a proper rational function of s, so we can find the partial fraction expansion:

$$V_o = 12 + \frac{60}{s} + \frac{10}{(s + 5)}.$$

Using Tables 12.1–12.3, we can inverse-Laplace-transform V_o to get

$$v_o = 12\delta(t) + (60 + 10e^{-5t})u(t) \text{ V}.$$

The partial fraction expansion for I_1 is

$$I_1 = \frac{6s + 20}{s(s + 5)} = \frac{4}{s} + \frac{2}{(s + 5)}.$$

Using Table 12.3, we can inverse-Laplace-transform I_1 to get

$$i_1 = (4 + 2e^{-5t})u(t) \text{ A}.$$

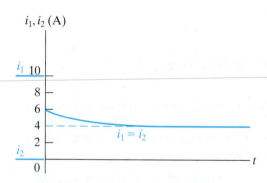

Figure 13.52 ▲ The inductor currents versus t for the circuit shown in Fig. 13.50.

Do the solutions for the output voltage and current in the circuit of Fig. 13.50 make sense? Before the switch is opened, the current in the 3 H inductor is 10 A, and the current in the 2 H inductor is 0 A. From the equation for i_1 derived in Example 13.13, we know that at $t = 0^+$, the current in both inductors is 6 A. This means the current in the 3 H inductor changes instantaneously from 10 to 6 A, while the current in the 2 H inductor changes instantaneously from 0 to 6 A. Then, the current decreases exponentially from 6 A to a final value of 4 A. This final value is easily verified from the circuit; that is, it should equal 100/25, or 4 A. Figure 13.52 shows these characteristics of i_1 and i_2.

Do these instantaneous jumps in the inductor current make sense in terms of the physical behavior of the circuit? First, note that when the switch opens in Fig. 13.50, the two inductors are in series. Any impulsive voltage appearing across the 3 H inductor must be exactly balanced by an impulsive voltage across the 2 H inductor because the sum of the impulsive voltages around a closed path must equal zero. Faraday's law states that the induced voltage is proportional to the change in flux linkage ($v = d\lambda/dt$). Therefore, the change in flux linkage must sum to zero. In other words, the total flux linkage immediately after switching is the same as that before switching. For the circuit here, the flux linkage before switching is

$$\lambda = L_1 i_1 + L_2 i_2 = 3(10) + 2(0) = 30 \text{ Wb-turns}.$$

Immediately after switching, it is

$$\lambda = (L_1 + L_2)i(0^+) = 5i(0^+).$$

Therefore,

$$i_1(0^+) = 30/5 = 6 \text{ A}.$$

Thus, the solution for i_1 in Example 13.13 agrees with the principle of the conservation of flux linkage.

We now test the validity of v_o from Example 13.13. First, we check the impulsive term $12\delta(t)$. The instantaneous jump of i_2 from 0 to 6 A at $t = 0$ means the derivative of i_2 contains the impulse $6\delta(t)$. This current impulse results in a $12\delta(t)$ impulse in the voltage across the 2 H inductor. For $t > 0$, $di_2/dt = di_1/dt = -10e^{-5t}$A/s. The voltage v_o is the sum of the voltage across the 15 Ω resistor and the voltage across the 2 H inductor:

$$v_o = 15(4 + 2e^{-5t}) + 2(-10e^{-5t})$$
$$= (60 + 10e^{-5t})u(t) \text{ V}.$$

This expression agrees with the last two terms of v_o from Example 13.13. Thus, the expression for v_o does make sense in terms of known circuit behavior.

We can also check the instantaneous drop from 10 to 6 A in the current i_1. This drop means the derivative of i_1 contains the impulse $-4\delta(t)$. Therefore, the voltage across L_1 includes the impulse $-12\delta(t)$. This impulse exactly balances the impulse included in the voltage across L_2; that is, the sum of the impulsive voltages around a closed path equals zero.

Impulsive Sources

Impulse functions can appear in circuit sources as well as circuit outputs. These sources are called **impulsive sources**. An impulsive source creates a finite amount of energy in the circuit instantaneously. A mechanical analogy is striking a bell with an impulsive clapper blow. After the energy has been transferred to the bell, the natural response of the bell determines the tone emitted (that is, the frequency of the resulting sound waves) and the tone's duration.

In the circuit shown in Fig. 13.53, an impulsive voltage source with a strength of V_0 volt-seconds is applied to a series connection of a resistor and an inductor. When the voltage source is applied, the initial energy in the inductor is zero, so the initial current is zero. There is no voltage drop across R, so the impulsive voltage source appears directly across L. The impulsive voltage across the inductor creates an instantaneous current. The current is

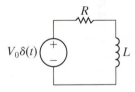

Figure 13.53 ▲ An *RL* circuit excited by an impulsive voltage source.

$$i = \frac{1}{L} \int_{0^-}^{t} V_0\delta(x) \, dx.$$

The integral of $\delta(t)$ over any interval that includes zero is 1, so

$$i(0^+) = \frac{V_0}{L} \text{ A}.$$

Thus, in an infinitesimal moment, the impulsive voltage source has stored energy in the inductor, given by

$$w = \frac{1}{2}L\left(\frac{V_0}{L}\right)^2 = \frac{1}{2}\frac{V_0^2}{L} \text{ J}.$$

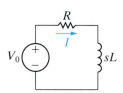

Figure 13.54 ▲ The s-domain equivalent circuit for the circuit shown in Fig. 13.53.

The initial current V_0/L now decays to zero due to the natural response of the circuit; that is,

$$i = \frac{V_0}{L} e^{-t/\tau} u(t),$$

where $\tau = L/R$. Remember from Chapter 7 that the natural response occurs as passive elements release or store energy. When a circuit is driven by an impulsive source, the total response is completely defined by the natural response; the duration of the impulsive source is so infinitesimal that it does not create a forced response.

We can also derive the inductor current using the Laplace transform method. Figure 13.54 shows the s-domain equivalent of the circuit in Fig. 13.53. From this circuit,

$$I = \frac{V_0}{R + sL} = \frac{V_0/L}{s + (R/L)}.$$

The inverse Laplace transform is

$$i = \frac{V_0}{L} e^{-(R/L)t} = \frac{V_0}{L} e^{-t/\tau} u(t).$$

Thus, the Laplace transform method gives the correct solution for $i \geq 0^+$.

Finally, we consider the case in which internally generated impulses and externally applied impulses occur simultaneously. As we see in Example 13.14, the Laplace transform method automatically ensures the correct solution for $t > 0^+$ if inductor currents and capacitor voltages at $t = 0^-$ are used in constructing the s-domain equivalent circuit and if externally applied impulses are represented by their transforms.

EXAMPLE 13.14

A Circuit with Both Internally Generated and Externally Applied Impulses

The switch in the circuit shown in Fig. 13.55 has been closed for a long time. At $t = 0$ it opens. Use Laplace methods to find the output voltage, v_o, and the current in the 3 H inductor, i_1.

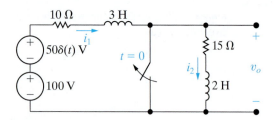

Figure 13.55 ▲ The circuit shown in Fig. 13.50 with an impulsive voltage source added in series with the 100 V source.

Solution

Note that the circuit in Fig. 13.55 was created from the circuit for Example 13.13 (Fig. 13.50) by adding an impulsive voltage source of $50\delta(t)$ in series with the 100 V source. In Example 13.13, we found the s-domain circuit that corresponds to the time-domain circuit in Fig. 13.50. The s-domain

circuit is in Fig. 13.51. To find the s-domain circuit corresponding to the circuit in Fig. 13.55, we add another voltage source in series, whose value is $\mathcal{L}\{50\delta(t)\} = 50$ V-s. The s-domain equivalent circuit is shown in Fig. 13.56.

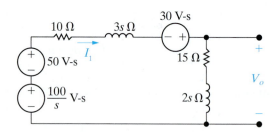

Figure 13.56 ▲ The s-domain equivalent circuit for the circuit shown in Fig. 13.55.

From this circuit, the expression for I_1 is

$$I_1 = \frac{(100/s) + 50 + 30}{5s + 25} = \frac{16s + 20}{s(s + 5)}.$$

The partial fraction expansion of I_1 is

$$I_1 = \frac{4}{s} + \frac{12}{s + 5},$$

and the inverse Laplace transform of I_1 is

$$i_1 = (4 + 12e^{-5t})u(t) \text{ A}.$$

The expression for V_o is

$$V_o = (2s + 15)I_1 = \frac{32s^2 + 280s + 300}{s(s + 5)},$$

which is an improper rational function. Before we can find the partial fraction expansion, we

divide the numerator of V_o by its denominator to get

$$V_o = 32 + \frac{120s + 300}{s(s + 5)}.$$

The partial fraction expansion of V_o is then

$$V_o = 32 + \frac{60}{s} + \frac{60}{s + 5}$$

and the inverse Laplace transform of V_o is

$$v_o = 32\delta(t) + (60 + 60e^{-5t})u(t) \text{ V}.$$

As before, we test the results of Example 13.14 to see whether they make sense. From the expression for i_1, we see that the current in L_1 and L_2 is 16 A at $t = 0^+$. In Example 13.13, at the instant the switch opens, i_1 decreases from 10 to 6 A and i_2 increases from 0 to 6 A. Superimposed on these changes is a 10 A current in L_1 and L_2, due to the impulsive voltage source; that is,

$$i_1 = \frac{1}{3 + 2} \int_{0^-}^{t} 50\delta(x)dx = 10 \text{ A}.$$

Therefore, i_1 increases suddenly from 10 to 16 A, while i_2 increases suddenly from 0 to 16 A. The final value of both currents is 4 A. Figure 13.57 shows i_1 and i_2 graphically.

We can also find the abrupt changes in i_1 and i_2 without using superposition. The sum of the impulsive voltages across the 3 H and 2 H inductors equals $50\delta(t)$. Thus, the change in flux linkage must sum to 50; that is,

$$\Delta\lambda_1 + \Delta\lambda_2 = 50.$$

Because $\lambda = Li$, we get

$$3\Delta i_1 + 2\Delta i_2 = 50.$$

But because i_1 and i_2 must be equal after the switch opens,

$$i_1(0^-) + \Delta i_1 = i_2(0^-) + \Delta i_2.$$

Thus,

$$10 + \Delta i_1 = 0 + \Delta i_2.$$

Solving for Δi_1 and Δi_2 yields

$$\Delta i_1 = 6 \text{ A},$$

$$\Delta i_2 = 16 \text{ A}.$$

These expressions agree with the previous check.

Figure 13.57 also indicates that the derivatives of i_1 and i_2 will contain an impulse at $t = 0$. Specifically, the derivative of i_1 will have an impulse of $6\delta(t)$, and the derivative of i_2 will have an impulse of $16\delta(t)$. Figure 13.58 illustrates the derivatives of i_1 and i_2.

Now let's turn to the expression for the output voltage, v_o, found in Example 13.14. The impulsive component $32\delta(t)$ agrees with the impulse

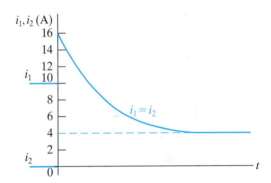

Figure 13.57 ▲ The inductor currents versus t for the circuit shown in Fig. 13.55.

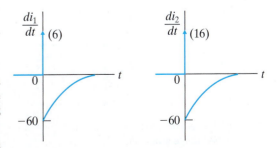

Figure 13.58 ▲ The derivatives of i_1 and i_2.

$16\delta(t)$ that characterizes di_2/dt at the origin. The term $(60e^{-5t} + 60)$ agrees with the fact that for $t > 0^+$,

$$v_o = 15i_2 + 2\frac{di_2}{dt} = 15i_1 + 2\frac{di_1}{dt}.$$

We test the impulsive component of di_1/dt by noting that it produces an impulsive voltage of $(3)6\delta(t)$, or $18\delta(t)$, across L_1. This voltage, added to $32\delta(t)$ across L_2, gives us $50\delta(t)$. Thus, the algebraic sum of the impulsive voltages around the mesh is zero.

To summarize, the Laplace transform will correctly predict the impulsive currents and voltages created by switching. However, the s-domain equivalent circuits must be based on initial conditions at $t = 0^-$, that is, on the initial conditions that exist prior to the switching. The Laplace transform will correctly predict the response to impulsive input sources by representing these sources in the s domain by their Laplace transforms.

SELF-CHECK: Assess your understanding of the impulse function in circuit analysis by trying Chapter Problems 13.90 and 13.91.

■ Practical Perspective

Surge Suppressors

As mentioned at the beginning of this chapter, voltage surges can occur in a circuit that is operating in the sinusoidal steady state. We use Laplace transform methods to see how a voltage surge is created between the line and neutral conductors of a household circuit when a load is switched off during sinusoidal steady-state operation.

Consider the circuit shown in Fig. 13.59, which models a household circuit with three loads, one of which is switched off at $t = 0$. To simplify the analysis, we assume that the line-to-neutral voltage, $\mathbf{V}_o$, is $120\underline{/0°}$ V (rms), a standard household voltage, and that when the load is switched off at $t = 0$, the value of $\mathbf{V}_g$ does not change. After the switch is opened, we can construct the s-domain circuit, as shown in Fig. 13.60. Note that because the phase angle of the voltage across the inductive load is 0°, the initial current through the inductive load is 0. Therefore, only the line inductance has a nonzero initial condition, which is modeled in the s-domain circuit as a voltage source with the value $L_\ell I_0$, as seen in Fig. 13.60.

Just before the switch is opened at $t = 0$, each of the loads has a steady-state sinusoidal voltage with a peak magnitude of $120\sqrt{2} = 169.7$ V. All of the current in the line from the voltage source divides among the three loads. When the switch is opened at $t = 0$, all of the current in the line appears in the remaining resistive load because the inductive load current is 0 at $t = 0$ and the inductor current cannot change instantaneously. Therefore, the voltage drop across the remaining loads can experience a surge as the line current is directed through the resistive load.

For example, if the initial current in the line is 25 A (rms) and the impedance of the resistive load is 12 Ω, the voltage drop across the resistor surges from 169.7 V to $(25)(\sqrt{2})(12) = 424.3$ V when the switch is opened. If the resistive load cannot handle this amount of voltage, it needs to be protected with a surge suppressor such as the one shown at the beginning of the chapter.

SELF-CHECK: Assess your understanding of this Practical Perspective by trying Chapter Problems 13.94 and 13.95.

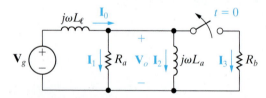

Figure 13.59 ▲ Circuit used to introduce a switching surge voltage.

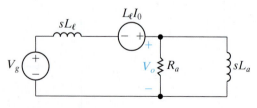

Figure 13.60 ▲ The circuit in Fig. 13.59, transformed to the s-domain.

■ Summary

- We can represent each of the circuit elements as an s-domain equivalent circuit by Laplace-transforming the voltage-current equation for each element:

  - Resistor: $V = RI$

  - Inductor: $V = sLI - LI_0$

  - Capacitor: $V = (1/sC)I + V_o/s$

  In these equations, $V = \mathcal{L}\{v\}$, $I = \mathcal{L}\{i\}$, I_0 is the initial current through the inductor, and V_0 is the initial voltage across the capacitor. (See pages 484–485.)

- Using Steps 1–4 in Analysis Method 13.1, we can transform the time-domain circuit into the s domain. Table 13.1 summarizes the equivalent circuits for resistors, inductors, and capacitors in the s domain. (See page 486.)

- Solve the s-domain equivalent circuit in Analysis Method 13.1, Step 5, by writing algebraic equations using the circuit analysis techniques from resistive circuits. Step 6 checks the resulting s-domain voltages and currents using the initial- and final-value theorems, where possible. Step 7 finds the partial fraction expansion for the s-domain voltages and currents, and uses Table 12.3 to find the inverse Laplace transforms. (See page 487.)

- Circuit analysis in the s domain is particularly advantageous for solving transient response problems in linear lumped parameter circuits when initial conditions are known. It is also useful for problems involving multiple simultaneous mesh-current or node-voltage equations because it reduces problems to algebraic rather than differential equations. (See page 489.)

- The **transfer function** is the s-domain ratio of a circuit's output to its input. It is defined as

$$H(s) = \frac{Y(s)}{X(s)},$$

  where $Y(s)$ is the Laplace transform of the output signal, and $X(s)$ is the Laplace transform of the input signal. (See page 500.)

- The partial fraction expansion of the product $H(s)X(s)$ yields a term for each pole of $H(s)$ and $X(s)$. The $H(s)$ terms correspond to the transient component of the total response, while the $X(s)$ terms correspond to the steady-state component. (See page 503.)

- If a circuit is driven by a unit impulse, $x(t) = \delta(t)$, then the response of the circuit equals the inverse Laplace transform of the transfer function, $y(t) = \mathcal{L}^{-1}\{H(s)\} = h(t)$. (See pages 504–505.)

- When a circuit is **time-invariant**, delaying the input by a seconds delays the output by a seconds. (See page 504.)

- The output of a circuit, $y(t)$, can be computed by convolving the input, $x(t)$, with the impulse response of the circuit, $h(t)$:

$$y(t) = h(t) * x(t) = \int_0^t h(\lambda) x(t - \lambda)\, d\lambda$$

$$= x(t) * h(t) = \int_0^t x(\lambda) h(t - \lambda)\, d\lambda.$$

  A graphical interpretation of the convolution integral usually helps you to compute $y(t)$. (See pages 505 and 508.)

- We can use the transfer function of a circuit to compute its steady-state response to a sinusoidal source. To do so, make the substitution $s = j\omega$ in $H(s)$ and represent the resulting complex number as a magnitude and phase angle. If

$$x(t) = A \cos(\omega t + \phi),$$

$$H(j\omega) = |H(j\omega)| e^{j\theta(\omega)},$$

  then

$$y_{ss}(t) = A|H(j\omega)| \cos[\omega t + \phi + \theta(\omega)].$$

  (See page 512.)

- Laplace transform analysis correctly predicts impulsive currents and voltages arising from switching and impulsive sources. You must ensure that the s-domain equivalent circuits are based on initial conditions prior to the switching, that is, at $t = 0^-$. (See page 520.)

Problems

Section 13.1

13.1 Derive the *s*-domain equivalent circuit shown in Fig. 13.4 by expressing the inductor current *i* as a function of the terminal voltage *v* and then finding the Laplace transform of this time-domain integral equation.

13.2 Find the Norton equivalent of the circuit shown in Fig. 13.3.

13.3 Find the Thévenin equivalent of the circuit shown in Fig. 13.7.

Section 13.2

13.4 A 400 Ω resistor, a 2.5 mH inductor, and a 40 nF capacitor are in series.

a) Express the *s*-domain impedance of this series combination as a rational function.

b) Give the numerical value of the poles and zeros of the impedance.

13.5 An 2 kΩ resistor, a 6.25 H inductor, and a 250 nF capacitor are in parallel.

a) Express the *s*-domain impedance of this parallel combination as a rational function.

b) Give the numerical values of the poles and zeros of the impedance.

13.6 A 250 Ω resistor is in series with an 80 mH inductor. This series combination is in parallel with a 500 nF capacitor.

a) Express the equivalent *s*-domain impedance of these parallel branches as a rational function.

b) Determine the numerical values of the poles and zeros.

13.7 Find the poles and zeros of the impedance seen looking into the terminals a,b of the circuit shown in Fig. P13.7.

Figure P13.7

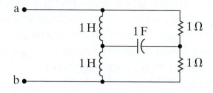

13.8 Find the poles and zeros of the impedance seen looking into the terminals a,b of the circuit shown in Fig. P13.8.

Figure P13.8

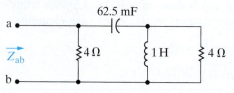

Section 13.3

13.9 The switch in the circuit shown in Fig. P13.9 has
been in position x for a long time. At $t = 0$, the switch moves instantaneously to position y.

a) Construct an *s*-domain circuit for $t > 0$.

b) Find V_o.

c) Find v_o.

Figure P13.9

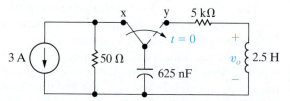

13.10 The switch in the circuit in Fig. P13.10 has been closed for a long time. At $t = 0$, the switch is opened.

a) Find i_o for $t \geq 0$.

b) Find v_o for $t \geq 0$.

Figure P13.10

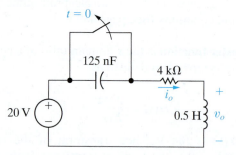

13.11 Find V_o and v_o in the circuit shown in Fig. P13.11 if the initial energy is zero and the switch is closed at $t = 0$.

Figure P13.11

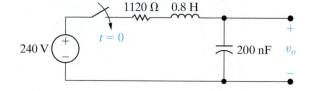

13.12 Repeat Problem 13.11 if the initial voltage on the capacitor is 72 V positive at the lower terminal.

13.13 a) Find the s-domain expression for V_o in the circuit in Fig. P13.13.

b) Use the s-domain expression derived in (a) to predict the initial- and final-values of v_o.

c) Find the time-domain expression for v_o.

Figure P13.13

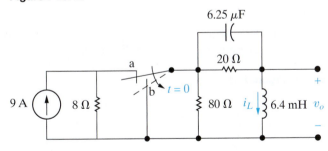

13.14 Find the time-domain expression for the current in the inductor in Fig. P13.13. Assume the reference direction for i_L is down.

13.15 The switch in the circuit in Fig. P13.15 has been in position a for a long time. At $t = 0$, the switch moves instantaneously to position b.

a) Construct the s-domain circuit for $t > 0$.

b) Find V_o.

c) Find I_L.

d) Find v_o for $t > 0$.

e) Find i_L for $t \geq 0$.

Figure P13.15

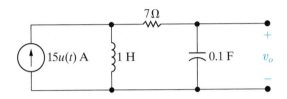

13.16 The switch in the circuit in Fig. P13.16 has been in position a for a long time. At $t = 0$, it moves instantaneously from a to b.

a) Construct the s-domain circuit for $t > 0$.

b) Find $V_o(s)$.

c) Find $v_o(t)$ for $t \geq 0$.

Figure P13.16

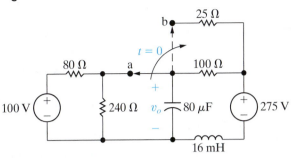

13.17 The make-before-break switch in the circuit in Fig. P13.17 has been in position a for a long time. At $t = 0$, it moves instantaneously to position b. Find i_o for $t \geq 0$.

Figure P13.17

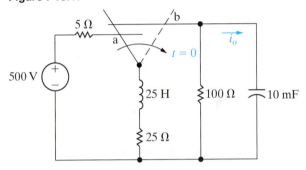

13.18 The switch in the circuit in Fig. P13.18 has been closed for a long time before opening at $t = 0$.

a) Construct the s-domain equivalent circuit for $t > 0$.

b) Find I_o.

c) Find i_o for $t \geq 0$.

Figure P13.18

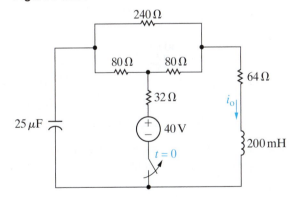

13.19 There is no energy stored in the circuit in Fig. P13.19 at the time the sources are energized.

a) Find $I_1(s)$ and $I_2(s)$.

b) Use the initial- and final-value theorems to check the initial- and final-values of $i_1(t)$ and $i_2(t)$.

c) Find $i_1(t)$ and $i_2(t)$ for $t \geq 0$.

Figure P13.19

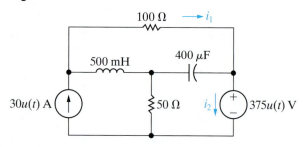

13.20 There is no energy stored in the circuit in Fig. P13.20 at $t = 0^-$.

a) Use the mesh current method to find i_o.

b) Find the time domain expression for v_o.

c) Do your answers in (a) and (b) make sense in terms of known circuit behavior? Explain.

Figure P13.20

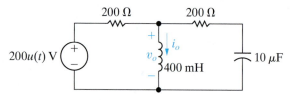

13.21 There is no energy stored in the circuit in Fig. P13.21 at $t = 0^-$.

a) Find V_o.

b) Find v_o.

c) Does your solution for v_o make sense in terms of known circuit behavior? Explain.

Figure P13.21

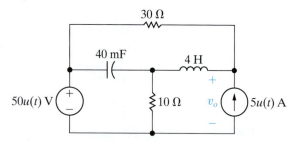

13.22 There is no energy stored in the circuit in Fig. P13.22 at the time the voltage source is turned on, and $v_g = 325u(t)$ V.

a) Find V_o and I_o.

b) Find v_o and i_o.

c) Do the solutions for v_o and i_o make sense in terms of known circuit behavior? Explain.

Figure P13.22

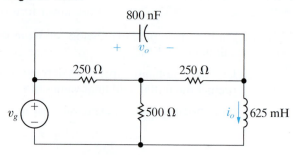

13.23 There is no initial energy in the circuit in Fig. P13.23 before the switch closes at $t = 0$. Find $i_o(t)$ for $t \geq 0$.

Figure P13.23

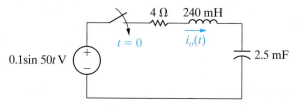

13.24 There is no initial energy in the circuit in Fig. P13.24 before the switch closes at $t = 0$. Find $v_o(t)$ for $t \geq 0$.

Figure P13.24

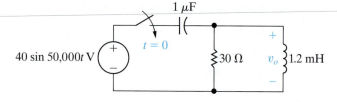

13.25 There is no energy stored in the circuit in Fig. P13.25 at the time the switch is closed.

a) Find v_o for $t \geq 0$.

b) Does your solution make sense in terms of known circuit behavior? Explain.

Figure P13.25

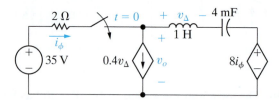

13.26 Find v_o in the circuit shown in Fig. P13.26 if $i_g = 5u(t)$ mA. There is no energy stored in the circuit at $t = 0$.

Figure P13.26

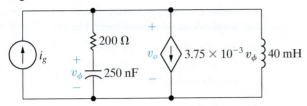

13.27 The initial energy in the circuit in Fig. P13.27 is zero.
The ideal voltage source is $120u(t)$ V.

a) Find $I(s)$.

b) Use the initial- and final-value theorems to find $i(0^+)$ and $i(\infty)$.

c) Do the values obtained in (b) agree with known circuit behavior? Explain.

d) Find $i(t)$.

Figure P13.27

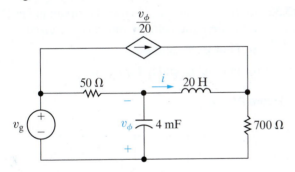

13.28 The switch in the circuit in Fig. P13.28 has been closed for a long time before opening at $t = 0$. Find v_o for $t \geq 0$.

Figure P13.28

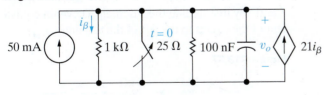

13.29 The switch in the circuit seen in Fig. P13.29 has been in position a for a long time. At $t = 0$, it moves instantaneously to position b.

a) Find V_o.

b) Find v_o.

Figure P13.29

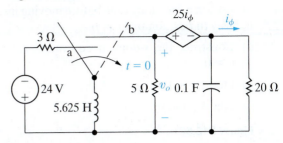

13.30 There is no energy stored in the circuit in Fig. P13.30 at the time the current source turns on. Given that $i_g = 100u(t)$ A:

a) Find $I_o(s)$.

b) Use the initial- and final-value theorems to find $i_o(0^+)$ and $i_o(\infty)$.

c) Determine if the results obtained in (b) agree with known circuit behavior.

d) Find $i_o(t)$.

Figure P13.30

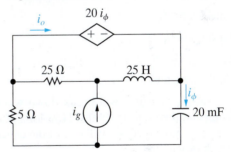

13.31 There is no energy stored in the capacitors in the circuit in Fig. P13.31 at the time the switch is closed.

a) Construct the s-domain circuit for $t > 0$.

b) Find I_1, V_1, and V_2.

c) Find i_1, v_1, and v_2.

d) Do your answers for i_1, v_1, and v_2 make sense in terms of known circuit behavior? Explain.

Figure P13.31

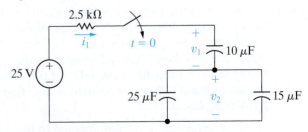

13.32 The switch in the circuit seen in Fig. P13.32 has been in position a for a long time before moving instantaneously to position b at $t = 0$.

PSPICE
MULTISIM

a) Construct the s-domain equivalent circuit for $t > 0$.

b) Find V_1 and v_1.

c) Find V_2 and v_2.

Figure P13.32

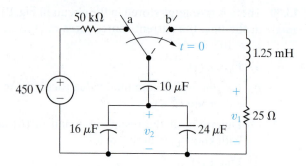

13.33 There is no energy stored in the circuit in Fig. P13.33 at the time the current source is energized.

PSPICE
MULTISIM

a) Find I_a and I_b.

b) Find i_a and i_b.

c) Find V_a, V_b, and I_c.

d) Find v_a, v_b, and v_c.

e) Assume a capacitor will break down whenever its terminal voltage is 1000 V. How long after the current source turns on will one of the capacitors break down?

Figure P13.33

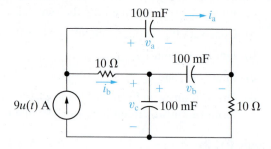

13.34 The switch in the circuit shown in Fig. P13.34 has been open for a long time. The voltage of the sinusoidal source is $v_g = V_m \sin(\omega t + \phi)$. The switch closes at $t = 0$. Note that the angle ϕ in the voltage expression determines the value of the voltage at the moment when the switch closes, that is, $v_g(0) = V_m \sin \phi$.

a) Use the Laplace transform method to find i for $t > 0$.

b) Using the expression derived in (a), write the expression for the current after the switch has been closed for a long time.

c) Using the expression derived in (a), write the expression for the transient component of i.

d) Find the steady-state expression for i using the phasor method. Verify that your expression is equivalent to that obtained in (b).

e) Specify the value of ϕ so that the circuit passes directly into steady-state operation when the switch is closed.

Figure P13.34

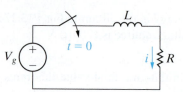

13.35 Using the expression for capacitor voltage found in Example 13.6, show that the capacitor current in Fig. 13.20 is positive for $0 < t < 200 \, \mu s$ and negative for $t > 200 \, \mu s$. Also show that at $200 \, \mu s$, the current is zero and that this corresponds to when dv_C/dt is zero.

13.36 There is no energy stored in the circuit in Fig. P13.36 at the time the voltage source is energized.

PSPICE
MULTISIM

a) Find V_o, I_o, and I_L.

b) Find v_o, i_o, and i_L for $t \geq 0$.

Figure P13.36

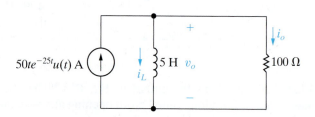

13.37 The two switches in the circuit shown in Fig. P13.37 operate simultaneously. There is no energy stored in the circuit at the instant the switches close. Find $i(t)$ for $t \geq 0^+$ by first finding the s-domain Thévenin equivalent of the circuit to the left of the terminals a, b.

PSPICE
MULTISIM

Figure P13.37

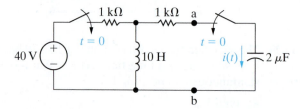

13.38 There is no energy stored in the circuit in Fig. P13.38 at the time the switch is closed.

PSPICE
MULTISIM

a) Find V_o.

b) Use the initial- and final-value theorems to find $v_o(0^+)$ and $v_o(\infty)$.

c) Find v_o.

Figure P13.38

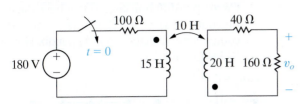

13.39 Find v_o in the circuit in Fig. P13.38 if the polarity dot on the 20 H coil is at the top.

PSPICE
MULTISIM

13.40 The magnetically coupled coils in the circuit seen in Fig. P13.40 carry initial currents of 15 and 10 A, as shown.

PSPICE
MULTISIM

  a) Find the initial energy stored in the circuit.

  b) Find I_1 and I_2.

  c) Find i_1 and i_2.

  d) Find the total energy dissipated in the 120 and 270 Ω resistors.

  e) Repeat (a)–(d), with the dot on the 18 H inductor at the lower terminal.

Figure P13.40

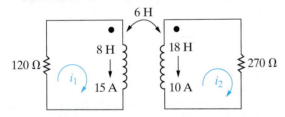

13.41 In the circuit in Fig. P13.41, switch 1 closes at $t = 0$, and the make-before-break switch moves instantaneously from position a to position b.

PSPICE
MULTISIM

  a) Construct the s-domain equivalent circuit for $t > 0$.

  b) Find I_1.

  c) Use the initial- and final-value theorems to check the initial and final values of i_1.

  d) Find i_1 for $t \geq 0^+$.

Figure P13.41

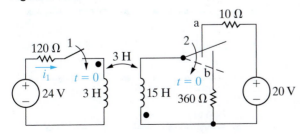

13.42 The make-before-break switch in the circuit seen in Fig. P13.42 has been in position a for a long time. At $t = 0$, it moves instantaneously to position b. Find i_o for $t \geq 0$.

PSPICE
MULTISIM

Figure P13.42

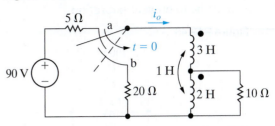

13.43 The switch in the circuit seen in Fig. P13.43 has been closed for a long time before opening at $t = 0$. Use the Laplace transform method of analysis to find v_o.

PSPICE
MULTISIM

Figure P13.43

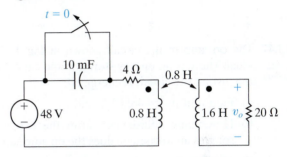

13.44 There is no energy stored in the circuit seen in Fig. P13.44 at the time the two sources are energized.

PSPICE
MULTISIM

  a) Use the principle of superposition to find V_o.

  b) Find v_o for $t > 0$.

Figure P13.44

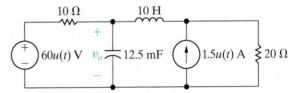

13.45 The op amp in the circuit seen in Fig. P13.45 is ideal. There is no energy stored in the capacitors at the time the circuit is energized. Determine (a) V_o, (b) v_o, and (c) how long it takes to saturate the operational amplifier.

PSPICE
MULTISIM

Figure P13.45

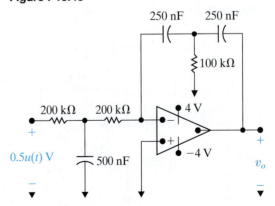

13.46 The op amp in the circuit shown in Fig. P13.46 is ideal. There is no energy stored in the circuit at the time it is energized. If $v_g = 20{,}000tu(t)$ V, find (a) V_o, (b) v_o, (c) how long it takes to saturate the operational

PSPICE
MULTISIM

amplifier, and (d) how small the rate of increase in v_g must be to prevent saturation.

Figure P13.46

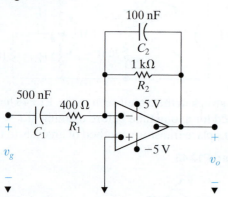

13.47 The op amp in the circuit shown in Fig. P13.47 is
PSPICE   ideal. There is no energy stored in the capacitors at
MULTISIM the instant the circuit is energized.

a) Find v_o if $v_{g1} = 40u(t)$ V and $v_{g2} = 16u(t)$ V.

b) How many milliseconds after the two voltage sources are turned on does the op amp saturate?

Figure P13.47

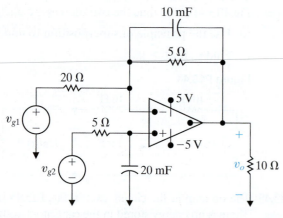

13.48 Find $v_o(t)$ in the circuit shown in Fig. P13.48 if the
PSPICE   ideal op amp operates within its linear range and
MULTISIM $v_g = 4.8u(t)$ V.

Figure P13.48

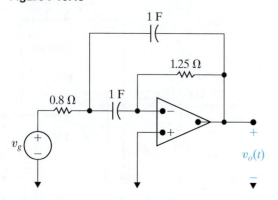

Sections 13.4–13.5

13.49 a) Find the transfer function $H(s) = V_o/V_i$ for the circuit shown in Fig. P13.49(a).

b) Find the transfer function $H(s) = V_o/V_i$ for the circuit shown in Fig. P13.49(b).

c) Create two different circuits that have the transfer function $H(s) = V_o/V_i = s/(s + 10,000)$. Use components selected from Appendix H and Figs. P13.49(a) and (b).

Figure P13.49

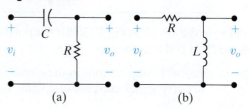

(a)   (b)

13.50 a) Find the transfer function $H(s) = V_o/V_i$ for the
PSPICE   circuit shown in Fig. P13.50(a).
MULTISIM

b) Find the transfer function $H(s) = V_o/V_i$ for the circuit shown in Fig. P13.50(b).

c) Create two different circuits that have the transfer function $H(s) = V_o/V_i = 1000/(s + 1000)$. Use components selected from Appendix H and Figs. P13.50(a) and (b).

Figure P13.50

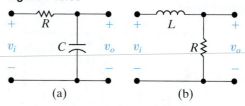

(a)   (b)

13.51 Find the numerical expression for the transfer func-
PSPICE   tion (V_o/V_i) of each circuit in Fig. P13.51 and give the
MULTISIM numerical value of the poles and zeros of each trans-
fer function.

Figure P13.51

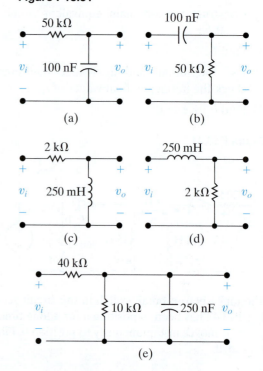

13.52 a) Find the transfer function $H(s) = V_o/V_i$ for the circuit shown in Fig. P13.52. Identify the poles and zeros for this transfer function.

b) Find three components from Appendix H which when used in the circuit of Fig. P13.52 will result in a transfer function with two poles that are distinct real numbers. Calculate the values of the poles.

c) Find three components from Appendix H which when used in the circuit of Fig. P13.52 will result in a transfer function with two poles, both with the same value. Calculate the value of the poles.

d) Find three components from Appendix H which when used in the circuit of Fig. P13.52 will result in a transfer function with two poles that are complex conjugate complex numbers. Calculate the values of the poles.

Figure P13.52

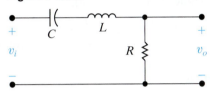

13.53 Repeat Problem 13.52 for the circuit in Fig. P13.53. In parts (b)–(d), let $L = 10$ mH and $C = 1$ μF, and use one or two resistors from Appendix H.

Figure P13.53

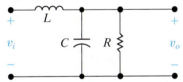

13.54 a) Find the numerical expression for the transfer function $H(s) = V_o/V_i$ for the circuit in Fig. P13.54.

b) Give the numerical value of each pole and zero of $H(s)$.

Figure P13.54

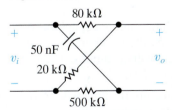

13.55 The operational amplifier in the circuit in Fig. P13.55 is ideal.

a) Derive the numerical expression of the transfer function $H(s) = V_o/V_g$ for the circuit in Fig. P13.55.

b) Give the numerical value of each pole and zero of $H(s)$.

c) Use the transfer function to find the unit impulse response for the circuit in Fig. P13.55.

Figure P13.55

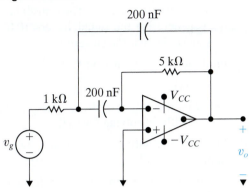

13.56 The operational amplifier in the circuit in Fig. P13.56 is ideal.

a) Find the numerical expression for the transfer function $H(s) = V_o/V_g$.

b) Give the numerical value of each zero and pole of $H(s)$.

c) Use the transfer function to find the unit step response for the circuit in Fig. P13.56.

Figure P13.56

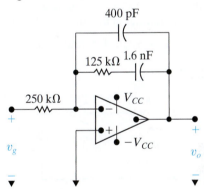

13.57 The operational amplifier in the circuit in Fig. P13.57 is ideal.

a) Find the numerical expression for the transfer function $H(s) = V_o/V_g$.

b) Give the numerical value of each zero and pole of $H(s)$.

c) Use the transfer function to find the unit impulse response and unit step response for the circuit in Fig. P13.57.

Figure P13.57

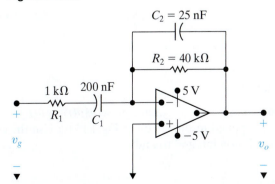

13.58 There is no energy stored in the circuit in Fig. P13.58 at
the time the switch is opened. The sinusoidal current
source is generating the signal $100 \cos 10{,}000t$ mA.
The response signal is the current i_o.

PSPICE
MULTISIM

a) Find the transfer function I_o/I_g.

b) Find $I_o(s)$.

c) Describe the nature of the transient component
of $i_o(t)$ without solving for $i_o(t)$.

d) Describe the nature of the steady-state component of $i_o(t)$ without solving for $i_o(t)$.

e) Verify the observations made in (c) and (d) by
finding $i_o(t)$.

Figure P13.58

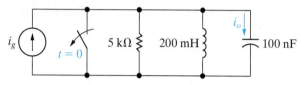

13.59 a) Find the transfer function I_o/I_g as a function of
μ for the circuit seen in Fig. P13.59.

PSPICE
MULTISIM

b) Find the largest value of μ that will produce a
bounded output signal for a bounded input signal.

c) Find i_o for $\mu = -3, 0, 4, 5$, and 6 if $i_g = 5u(t)$ A.

Figure P13.59

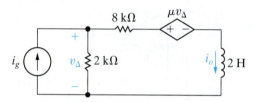

13.60 In the circuit of Fig. P13.60 i_o is the output signal and
v_g is the input signal. Find the poles and zeros of the
transfer function, assuming there is no initial energy
stored in the linear transformer or in the capacitor.

Figure P13.60

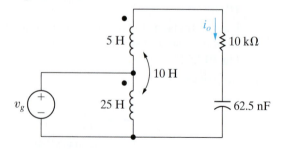

Section 13.6

13.61 A rectangular voltage pulse $v_i[u(t) - u(t - 1)]$ V is
applied to the circuit in Fig. P13.61. Use the convolution integral to find v_o.

Figure P13.61

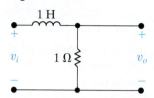

13.62 Interchange the inductor and resistor in
Problem 13.61 and again use the convolution
integral to find v_o.

13.63 a) Given $y(t) = h(t) * x(t)$, find $y(t)$ when $h(t)$
and $x(t)$ are the rectangular pulses shown in
Fig. P13.63(a).

b) Repeat (a) when $h(t)$ changes to the rectangular
pulse shown in Fig. P13.63(b).

c) Repeat (a) when $h(t)$ changes to the rectangular
pulse shown in Fig. P13.63(c).

d) Sketch $y(t)$ versus t for (a)–(c) on a single graph.

e) Do the sketches in (d) make sense? Explain.

Figure P13.63

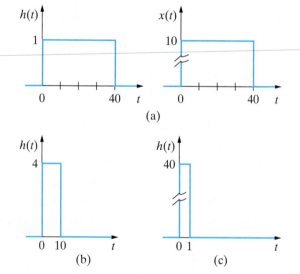

13.64 Assume the voltage impulse response of a circuit
can be modeled by the triangular waveform shown
in Fig. P13.64. The voltage input signal to this circuit
is the step function $10u(t)$ V.

a) Use the convolution integral to derive the expressions for the output voltage.

b) Sketch the output voltage over the interval 0 to
15 s.

c) Repeat parts (a) and (b) if the area under the
voltage impulse response stays the same but the
width of the impulse response narrows to 4 s.

d) Which output waveform is closer to replicating
the input waveform: (b) or (c)? Explain.

Figure P13.64

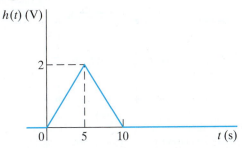

13.65 a) Find $h(t) * x(t)$ when $h(t)$ and $x(t)$ are the rectangular pulses shown in Fig. P13.65(a).

b) Repeat (a) when $x(t)$ changes to the rectangular pulse shown in Fig. P13.65(b).

c) Repeat (a) when $h(t)$ changes to the rectangular pulse shown in Fig. P13.65(c).

Figure P13.65

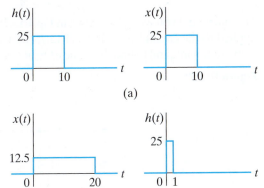

13.66 a) Use the convolution integral to find the output voltage of the circuit in Fig. P13.51(a) if the input voltage is the rectangular pulse shown in Fig. P13.66.

b) Sketch $v_o(t)$ versus t for the time interval $0 \le t \le 15$ ms.

Figure P13.66

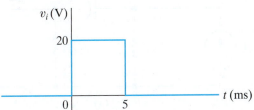

13.67 a) Repeat Problem 13.66, given that the resistor in the circuit in Fig. P13.51(a) is decreased to 5 kΩ.

b) Does decreasing the resistor increase or decrease the memory of the circuit?

c) Which circuit comes closer to transmitting a replica of the input voltage?

13.68 The voltage impulse response of a circuit is shown in Fig. P13.68(a). The input signal to the circuit is the rectangular voltage pulse shown in Fig. P13.68(b).

a) Derive the equations for the output voltage. Note the range of time for which each equation is applicable.

b) Sketch v_o for $-1 \le t \le 34$ s.

Figure P13.68

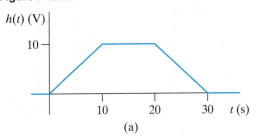

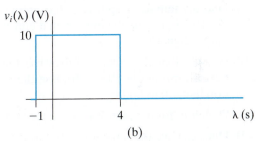

13.69 Use the convolution integral to find v_o in the circuit seen in Fig. P13.69 if $v_i = 75u(t)$ V.

Figure P13.69

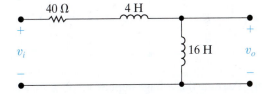

13.70 The input voltage in the circuit seen in Fig. P13.70 is

$$v_i = 5[u(t) - u(t - 0.5)] \text{ V.}$$

a) Use the convolution integral to find v_o.

b) Sketch v_o for $0 \le t \le 1$ s.

Figure P13.70

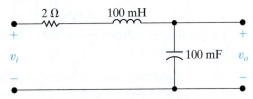

13.71 a) Find the impulse response of the circuit shown in Fig. P13.71(a) if v_g is the input signal and v_o is the output signal.

b) Given that v_g has the waveform shown in Fig. P13.71(b), use the convolution integral to find v_o.

c) Does v_o have the same waveform as v_g? Why or why not?

Figure P13.71

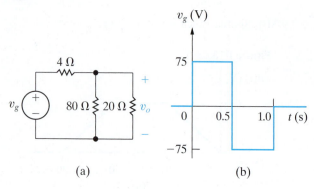

(a)   (b)

13.72 a) Find the impulse response of the circuit seen in Fig. P13.72 if v_g is the input signal and v_o is the output signal.

b) Assume that the voltage source has the waveform shown in Fig. P13.71(b). Use the convolution integral to find v_o.

c) Sketch v_o for $0 \le t \le 2$ s.

d) Does v_o have the same waveform as v_g? Why or why not?

Figure P13.72

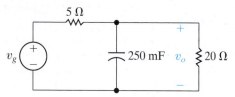

13.73 a) Assume the voltage impulse response of a circuit is

$$h(t) = \begin{cases} 0, & t < 0; \\ 10e^{-4t}, & t \ge 0. \end{cases}$$

Use the convolution integral to find the output voltage if the input signal is $10u(t)$ V.

b) Repeat (a) if the voltage impulse response is

$$h(t) = \begin{cases} 0, & t < 0; \\ 10(1 - 2t), & 0 \le t \le 0.5 \text{ s}; \\ 0, & t \ge 0.5 \text{ s}. \end{cases}$$

c) Plot the output voltage versus time for (a) and (b) for $0 \le t \le 1$ s.

13.74 a) Use the convolution integral to find v_o in the circuit in Fig. P13.74(a) if i_g is the pulse shown in Fig. P13.74(b).

b) Use the convolution integral to find i_o.

c) Show that your solutions for v_o and i_o are consistent by calculating v_o and i_o at 100^- ms, 100^+ ms, 200^- ms, and 200^+ ms.

Figure P13.74

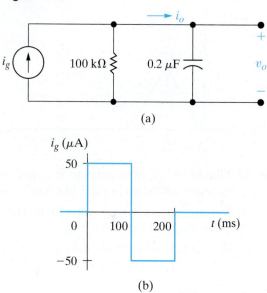

(a)

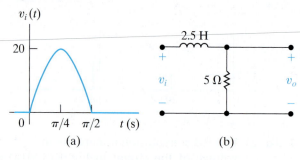

(b)

13.75 The sinusoidal voltage pulse shown in Fig. P13.75(a) is applied to the circuit shown in Fig. P13.75(b). Use the convolution integral to find the value of v_o at $t = 2.2$ s.

Figure P13.75

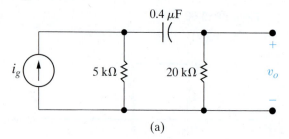

(a)   (b)

13.76 The current source in the circuit shown in Fig. P13.76(a) is generating the waveform shown in Fig. P13.76(b). Use the convolution integral to find v_o at $t = 5$ ms.

Figure P13.76

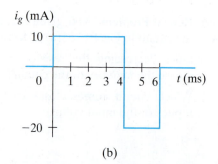

(a)

(b)

13.77 a) Show that if $y(t) = h(t) * x(t)$, then

$$Y(s) = H(x)X(s).$$

b) Use the result given in (a) to find $f(t)$ if

$$F(s) = \frac{a}{s(s + a)^2}.$$

Section 13.7

13.78 The transfer function for a linear time-invariant circuit is

$$H(s) = \frac{I_o}{I_g} = \frac{125(s + 400)}{s(s^2 + 200s + 10^4)}.$$

If $i_g = 80 \cos 500t$ A, what is the steady-state expression for i_o?

13.79 The transfer function for a linear time-invariant circuit is

$$H(s) = \frac{V_o}{V_g} = \frac{4(s + 3)}{s^2 + 8s + 41}.$$

If $v_g = 40 \cos 3t$ V, what is the steady-state expression for v_o?

13.80 When an input voltage of $30u(t)$ V is applied to a circuit, the response is known to be

$$v_o = (50e^{-8000t} - 20e^{-5000t})u(t) \text{ V}.$$

What will the steady-state response be if $v_g = 120 \cos 6000\,t$ V?

13.81 The op amp in the circuit seen in Fig. P13.81 is ideal.

a) Find the transfer function V_o/V_g.

b) Find v_o if $v_g = 0.6\, u(t)$ V.

c) Find the steady-state expression for v_o if $v_g = 2 \cos 10{,}000t$ V.

Figure P13.81

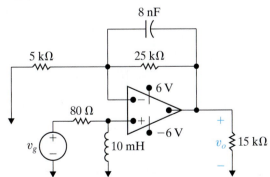

13.82 The operational amplifier in the circuit seen in

Fig. P13.82 is ideal and is operating within its linear region.

a) Calculate the transfer function V_o/V_g.

b) If $v_g = 2 \cos 400t$ V, what is the steady-state expression for v_o?

Figure P13.82

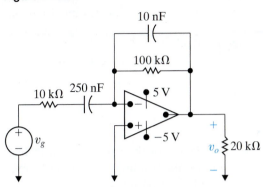

Section 13.8

13.83 Show that after $V_0 C_e$ coulombs are transferred from C_1 to C_2 in the circuit shown in Fig. 13.47 (with $R = 0$), the voltage across each capacitor is $C_1 V_0/(C_1 + C_2)$. (*Hint:* Use the conservation-of-charge principle.)

13.84 The voltage source in the circuit in Example 13.9 is changed to a unit impulse; that is, $v_g = \delta(t)$.

a) How much energy does the impulsive voltage source store in the capacitor?

b) How much energy does it store in the inductor?

c) Use the transfer function to find $v_o(t)$.

d) Show that the response found in (c) is identical to the response generated by first charging the capacitor to 1000 V and then releasing the charge to the circuit, as shown in Fig. P13.84.

Figure P13.84

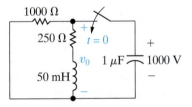

13.85 There is no energy stored in the circuit in Fig. P13.85 at the time the impulsive voltage is applied.

a) Find $v_o(t)$ for $t \geq 0$.

b) Does your solution make sense in terms of known circuit behavior? Explain.

Figure P13.85

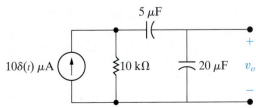

13.86 The inductor L_1 in the circuit shown in Fig. P13.86 is carrying an initial current of ρ A at the instant the switch opens. Find (a) $v(t)$; (b) $i_1(t)$; (c) $i_2(t)$; and (d) $\lambda(t)$, where $\lambda(t)$ is the total flux linkage in the circuit.

Figure P13.86

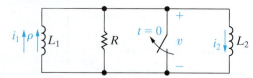

13.87 a) Let $R \to \infty$ in the circuit shown in Fig. P13.86, and use the solutions derived in Problem 13.86 to find $v(t)$, $i_1(t)$, and $i_2(t)$.

b) Let $R = \infty$ in the circuit shown in Fig. P13.86 and use the Laplace transform method to find $v(t)$, $i_1(t)$, and $i_2(t)$.

13.88 There is no energy stored in the circuit in Fig. P13.88 at the time the impulse voltage is applied.

a) Find i_1 for $t \geq 0^+$.

b) Find i_2 for $t \geq 0^+$.

c) Find v_o for $t \geq 0^+$.

d) Do your solutions for i_1, i_2, and v_o make sense in terms of known circuit behavior? Explain.

Figure P13.88

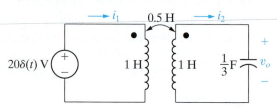

13.89 There is no energy stored in the circuit in Fig. P13.89 at the time the impulsive current is applied.

a) Find v_o for $t \geq 0^+$.

b) Does your solution make sense in terms of known circuit behavior? Explain.

Figure P13.89

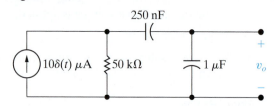

13.90 The switch in the circuit in Fig. P13.90 has been in position a for a long time. At $t = 0$, the switch moves to position b. Compute (a) $v_1(0^-)$; (b) $v_2(0^-)$; (c) $v_3(0^-)$; (d) $i(t)$; (e) $v_1(0^+)$; (f) $v_2(0^+)$; and (g) $v_3(0^+)$.

Figure P13.90

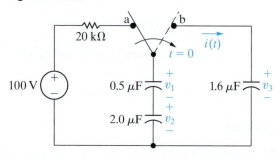

13.91 The switch in the circuit in Fig. P13.91 has been closed for a long time. The switch opens at $t = 0$. Compute (a) $i_1(0^-)$; (b) $i_1(0^+)$; (c) $i_2(0^-)$; (d) $i_2(0^+)$; (e) $i_1(t)$; (f) $i_2(t)$; and (g) $v(t)$.

Figure P13.91

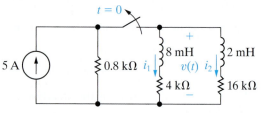

13.92 The parallel combination of R_2 and C_2 in the circuit shown in Fig. P13.92 represents the input circuit to a cathode-ray oscilloscope (CRO). The parallel combination of R_1 and C_1 is a circuit model of a compensating lead that is used to connect the CRO to the source. There is no energy stored in C_1 or C_2 at the time when the 10 V source is connected to the CRO via the compensating lead. The circuit values are $C_1 = 4$ pF, $C_2 = 16$ pF, $R_1 = 1.25$ MΩ, and $R_2 = 5$ MΩ.

a) Find v_o.

b) Find i_o.

c) Repeat (a) and (b) given C_1 is changed to 64 pF.

Figure P13.92

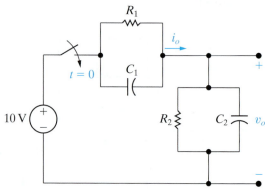

13.93 Show that if $R_1C_1 = R_2C_2$ in the circuit shown in Fig. P13.92, v_o will be a scaled replica of the source voltage.

Sections 13.1–13.8

13.94 Assume the line-to-neutral voltage V_o in the 60 Hz

PRACTICAL
PERSPECTIVE
circuit of Fig. 13.59 is $120\underline{/0°}$ V (rms). Load R_a is absorbing 1200 W; load R_b is absorbing 1800 W; and the inductive load is absorbing 350 magnetizing VAR. The inductive reactance of the line is 1 Ω. Assume V_g does not change after the switch opens.

a) Calculate the initial value of $i_2(t)$ and $i_o(t)$.

b) Find V_o, $v_o(t)$, and $v_o(0^+)$ using the s-domain circuit of Fig. 13.60.

c) Test the steady-state component of v_o using phasor domain analysis.

d) Using a computer program of your choice, plot v_o vs. t for $0 \le t \le 20$ ms.

13.95 Assume the switch in the circuit in Fig. 13.59 opens at

PRACTICAL
PERSPECTIVE
the instant the sinusoidal steady-state voltage v_o is zero and going positive, i.e., $v_o = 120\sqrt{2} \sin 120\pi t$ V.

a) Find $v_o(t)$ for $t \ge 0$.

b) Using a computer program of your choice, plot $v_o(t)$ vs. t for $0 \le t \le 20$ ms.

c) Compare the disturbance in the voltage in part (a) with that obtained in part (c) of Problem 13.94.

13.96 The purpose of this problem is to show that the

PRACTICAL
PERSPECTIVE
line-to-neutral voltage in the circuit in Fig. 13.59 can go directly into steady state if the load R_b is disconnected from the circuit at precisely the right time. Let $v_o = V_m \cos(120\pi t - \theta)$ V, where $V_m = 120\sqrt{2}$. Assume v_g does not change after R_b is disconnected.

a) Find the value of θ (in degrees) so that v_o goes directly into steady-state operation when the load R_b is disconnected.

b) For the value of θ found in part (a), find $v_o(t)$ for $t \ge 0$.

c) Using a computer program of your choice, plot on a single graph, for -10 ms $\le t \le 10$ ms, $v_o(t)$ before and after load R_b is disconnected.

CHAPTER CONTENTS

14.1 **Some Preliminaries** *p. 538*
14.2 **Low-Pass Filters** *p. 539*
14.3 **High-Pass Filters** *p. 545*
14.4 **Bandpass Filters** *p. 550*
14.5 **Bandreject Filters** *p. 560*

CHAPTER OBJECTIVES

1 Know the *RL* and *RC* circuit configurations that act as low-pass filters and be able to design *RL* and *RC* circuit component values to meet a specified cutoff frequency.

2 Know the *RL* and *RC* circuit configurations that act as high-pass filters and be able to design *RL* and *RC* circuit component values to meet a specified cutoff frequency.

3 Know the *RLC* circuit configurations that act as bandpass filters, understand the definition of and relationship among the center frequency, cutoff frequencies, bandwidth, and quality factor of a bandpass filter, and be able to design *RLC* circuit component values to meet design specifications.

4 Know the *RLC* circuit configurations that act as bandreject filters, understand the definition of and relationship among the center frequency, cutoff frequencies, bandwidth, and quality factor of a band reject filter, and be able to design *RLC* circuit component values to meet design specifications.

Introduction to Frequency Selective Circuits

Up to this point in our analysis of circuits with sinusoidal sources, the source frequency has been held constant. In this chapter, we analyze the effect of varying source frequency on circuit voltages and currents. The result of this analysis is the **frequency response** of a circuit.

We've seen in previous chapters that a circuit's response depends on the types of elements in the circuit, the way the elements are connected, and the element impedances. Varying the frequency of a sinusoidal source does not change the element types or their connections, but it does change the capacitor and inductor impedances because these impedances are functions of frequency.

We can design circuits whose output signals reside within a desired frequency range, excluding all other frequencies that appear in the circuit's input. Such circuits are called **frequency-selective circuits**. Many devices that communicate via electric signals, such as telephones, radios, televisions, and satellites, employ frequency-selective circuits.

Frequency-selective circuits are also called **filters** because they filter out certain input signals on the basis of frequency. Note that no practical frequency-selective circuit can perfectly or completely filter out selected frequencies. Rather, filters **attenuate**— that is, weaken or lessen the effect of—any input signals with frequencies outside the desired frequency band. For example, your home stereo system may have a graphic equalizer. Each band in the graphic equalizer is a filter that amplifies sounds (audible frequencies) in the frequency range of the band and attenuates frequencies outside of that band. Thus, the graphic equalizer enables you to change the sound volume in each frequency band.

We begin this chapter by analyzing circuits from each of the four major categories of filters: low pass, high pass, band pass, and band reject. The transfer function of a circuit is the starting point for the frequency response analysis. Pay close attention to the similarities among the transfer functions of circuits that perform the same filtering function. We will employ these similarities when designing filter circuits in Chapter 15.

■ Practical Perspective

Pushbutton Telephone Circuits

A pushbutton telephone produces tones that you hear when you press a button. You may have wondered about these tones. How are they used to tell the telephone system which button was pushed? Why are tones used at all? Why do the tones sound musical? How does the phone system tell the difference between button tones and the normal sounds of people talking or singing?

The telephone system was designed to handle audio signals—those with frequencies between 300 Hz and 3 kHz. Thus, all signals from the system to the user have to be audible—including the dial tone and the busy signal. Similarly, all signals from the user to the system have to be audible, including the signal that the user has pressed a button. It is important to distinguish button signals from the normal audio signal, so a dual-tone-multiple-frequency (DTMF) design is employed. When a number button is pressed, a unique pair of sinusoidal tones with very precise frequencies is sent by the phone to the telephone system. The DTMF frequency and timing specifications make it unlikely that a human voice could produce the exact tone pairs, even if the person were trying. In the central telephone facility, electric circuits monitor the audio signal, listening for the tone pairs that signal a number. In the Practical Perspective example at the end of the chapter, we will examine the design of the DTMF filters used to determine which button has been pushed.

Urbanbuzz/Alamy Stock Photo

Fuse/Getty Images

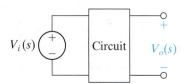

Figure 14.1 ▲ A circuit with voltage input and output.

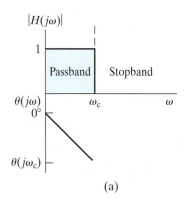

(a)

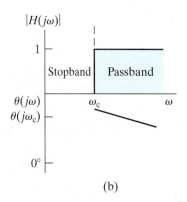

(b)

Figure 14.2 ▲ Ideal frequency response plots for (a) an ideal low-pass filter and (b) an ideal high-pass filter.

14.1 Some Preliminaries

Recall from Section 13.7 that a circuit's transfer function provides an easy way to compute the steady-state response to a sinusoidal input. There, we considered only fixed-frequency sources. To study the frequency response of a circuit, we replace a fixed-frequency sinusoidal source with a varying-frequency sinusoidal source. The transfer function is still an immensely useful tool because the magnitude and phase of the output signal depend only on the magnitude and phase of the transfer function $H(j\omega)$, which vary as a function of the source frequency ω. Note that this method of analyzing the output of a circuit as its input frequency varies assumes that we can vary the input frequency without changing its magnitude or phase angle.

To further simplify this first look at frequency-selective circuits, we will also restrict our attention to cases where both the input and output signals are sinusoidal voltages, as illustrated in Fig. 14.1. Thus, the circuit's transfer function will be the ratio of the Laplace transform of the output voltage to the Laplace transform of the input voltage, or $H(s) = V_o(s)/V_i(s)$. We should keep in mind, however, that for a particular application, a current may be either the input signal or output signal of interest.

The signals passed from the input to the output fall within a band of frequencies called the **passband**. Input voltages outside this band have their magnitudes attenuated by the circuit and are thus effectively prevented from reaching the circuit's output. Frequencies not in a circuit's passband are in its **stopband**. Frequency-selective circuits are categorized by the location of the passband.

We can identify the type of frequency-selective circuit by examining its **frequency response plot**. A frequency response plot shows how a circuit's transfer function (both amplitude and phase) changes as the source frequency changes. A frequency response plot has two parts.

- A graph of $|H(j\omega)|$ versus frequency ω, called the **magnitude plot**.
- A graph of $\theta(j\omega)$ versus frequency ω, called the **phase angle plot**.

The ideal frequency response plots for the four major categories of filters are shown in Figs. 14.2 and 14.3. Figure 14.2 illustrates the ideal plots for a low-pass and a high-pass filter, respectively. Both filters have one passband and one stopband, which are defined by the **cutoff frequency** that separates them. The names *low pass* and *high pass* are derived from the magnitude plots: a **low-pass filter** passes signals at frequencies lower than the cutoff frequency from the input to the output, and a **high-pass filter** passes signals at frequencies higher than the cutoff frequency. Thus, the terms *low* and *high* as used here do not refer to any absolute values of frequency, but rather to relative values with respect to the cutoff frequency.

Note from the graphs for both these filters (as well as those for the bandpass and bandreject filters in Fig. 14.3) that the phase angle plot for an ideal filter varies linearly in the passband. It is of no interest outside the passband because there the magnitude is zero. Linear phase variation is necessary to avoid phase distortion.

The two remaining categories of filters each have two cutoff frequencies. Figure 14.3(a) illustrates the ideal frequency response plot of a **bandpass filter**, which passes a source voltage to the output only when the source frequency is within the band defined by the two cutoff frequencies. Figure 14.3(b) shows the ideal plot of a **bandreject filter**, which passes a source voltage to the output only when the source frequency is outside the band defined by the two cutoff frequencies. The bandreject filter thus rejects, or stops, the source voltage from reaching the output when its frequency is within the band defined by the cutoff frequencies.

In specifying a filter using any of the circuits from this chapter, it is important to note that the magnitude and phase angle characteristics are not independent. In other words, the characteristics of a circuit that result in a particular magnitude plot will also dictate the form of the phase angle plot and vice versa. For example, once we select a desired form for the magnitude response of a circuit, the phase angle response is also determined. Alternatively, if we select a desired form for the phase angle response, the magnitude response is also determined. Although there are some frequency-selective circuits for which the magnitude and phase angle behavior can be independently specified, these circuits are not presented here.

The next sections present examples of circuits from each of the four filter categories. These circuits represent only a few of the many circuits that act as filters. As you read, focus your attention on trying to identify what properties of a circuit determine its behavior as a filter. For example, you should note the various forms of the transfer functions that perform the same filtering function. Identifying the form of a filter's transfer function will ultimately help you in designing filter circuits for particular applications.

All of the filters in this chapter are **passive filters** because their behavior depends only on passive elements: resistors, capacitors, and inductors. Usually, the largest output amplitude a passive filter can achieve equals the input amplitude. The only passive filter described in this chapter that can amplify its output is the series RLC resonant filter. Also, if you place an impedance in series with the source or in parallel with the load, the maximum output amplitude will decrease. Because many practical filter applications require amplification (a ratio of output-to-input amplitude greater than 1), passive filters have some significant disadvantages. Many active filter circuits, introduced in Chapter 15, provide amplification, and thereby overcome this passive filter disadvantage.

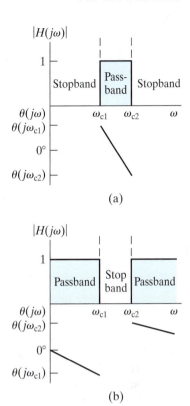

Figure 14.3 ▲ Ideal frequency response plots for (a) an ideal bandpass filter and (b) an ideal bandreject filter.

14.2 Low-Pass Filters

Two circuits that behave as low-pass filters are the series RL circuit and the series RC circuit. Let's investigate the circuit characteristics that affect the cutoff frequency.

The Series *RL* Circuit—Qualitative Analysis

A series RL circuit is shown in Fig. 14.4(a). The circuit's input is a sinusoidal voltage source with varying frequency. The circuit's output is the voltage across the resistor. Suppose the source frequency starts very low and increases gradually. We know that the behavior of the ideal resistor does not change because its impedance is independent of frequency. But consider how the behavior of the inductor changes.

Recall that the impedance of an inductor is $j\omega L$. At low frequencies, the inductor's impedance is very small compared with the resistor's impedance, and the inductor effectively functions as a short circuit. The term *low frequencies* thus refers to any frequencies for which $\omega L \ll R$. The equivalent circuit for $\omega = 0$ is shown in Fig. 14.4(b). In this equivalent circuit, the output voltage and the input voltage are equal both in magnitude and in phase angle.

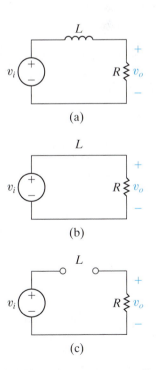

Figure 14.4 ▲ (a) A series RL low-pass filter. (b) The equivalent circuit at $\omega = 0$. and (c) The equivalent circuit at $\omega = \infty$.

As the frequency increases, the impedance of the inductor increases relative to that of the resistor. Increasing the inductor's impedance causes a corresponding increase in the magnitude of the voltage drop across the inductor and a corresponding decrease in the output voltage magnitude. Increasing the inductor's impedance also introduces a shift in phase angle between the inductor's voltage and current, resulting in a phase angle difference between the input and output voltage. The output voltage always lags the input voltage, and as the frequency increases, this phase lag approaches 90°.

At high frequencies, the inductor's impedance is very large compared with the resistor's impedance, and the inductor thus functions as an open circuit, effectively blocking the flow of current in the circuit. The term *high frequencies* thus refers to any frequencies for which $\omega L \gg R$. The equivalent circuit for $\omega = \infty$ is shown in Fig. 14.4(c), where the output voltage magnitude is zero. The phase angle of the output voltage is 90° more negative than that of the input voltage, so the output lags the input by 90°.

Based on the behavior of the output voltage magnitude, this series RL circuit selectively passes low-frequency inputs to the output, and it blocks high-frequency inputs from reaching the output. The circuit's response to varying input frequency is shown in Fig. 14.5, and these plots represent the frequency response of the series RL circuit in Fig. 14.4(a). The upper plot shows how $|H(j\omega)|$ varies with frequency. The lower plot shows how $\theta(j\omega)$ varies with frequency. Appendix E presents a method for constructing these plots.

We have also superimposed the ideal magnitude plot for a low-pass filter from Fig. 14.2(a) on the magnitude plot of the RL filter in Fig. 14.5 (see the dashed line). There is an obvious difference between the magnitude plots of an ideal filter and an actual RL filter. The ideal filter exhibits a discontinuity in magnitude at the cutoff frequency, ω_c, which creates an abrupt transition between the passband and the stopband. While this is, ideally, how we would like our filters to perform, it is not possible to use real components to construct a circuit with an abrupt transition in magnitude. Circuits acting as low-pass filters have a magnitude response that transitions gradually between the passband and the stopband. Hence, the magnitude plot of a real circuit requires us to define what we mean by the cutoff frequency, ω_c.

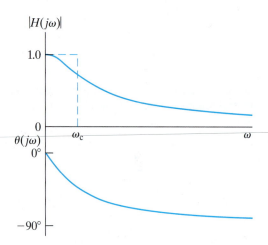

Figure 14.5 ▲ The frequency response plot for the series RL circuit in Fig. 14.4(a).

Defining the Cutoff Frequency

We need to define the cutoff frequency, ω_c, for realistic filter circuits because the magnitude plot does not allow us to identify a single frequency that divides the passband and the stopband. The definition for cutoff frequency widely used by electrical engineers is the frequency for which the transfer function magnitude is decreased by the factor $1/\sqrt{2}$ from its maximum value:

CUTOFF FREQUENCY DEFINITION

$$|H(j\omega_c)| = \frac{1}{\sqrt{2}} H_{\max}, \qquad (14.1)$$

where $H_{\max}$ is the maximum magnitude of the transfer function. It follows from Eq. 14.1 that the passband of a filter is defined as the range of frequencies for which the amplitude of the output voltage is at least 70.7% of the maximum possible amplitude.

Defining the cutoff frequency using the constant $1/\sqrt{2}$ is not an arbitrary choice. Examining another consequence of the cutoff frequency explains this choice. Recall from Section 10.5 that the average power delivered by any circuit to a load is proportional to V_L^2, where V_L is the amplitude of the voltage drop across the load:

$$P = \frac{1}{2}\frac{V_L^2}{R}.$$

If the circuit has a sinusoidal voltage source, $V_i(j\omega)$, then the load voltage is also a sinusoid, and its amplitude is a function of the frequency ω. We define P_{max} as the value of the average power delivered to a load when the magnitude of the load voltage is maximum:

$$P_{max} = \frac{1}{2}\frac{V_{L\,max}^2}{R}.$$

If we vary the frequency of the sinusoidal voltage source, $V_i(j\omega)$, the load voltage is a maximum when the magnitude of the circuit's transfer function is also a maximum:

$$V_{Lmax} = H_{max}|V_i|. \tag{14.2}$$

Now consider what happens to the average power when the frequency of the voltage source is ω_c. Using Eqs. 14.1 and 14.2, the magnitude of the load voltage at ω_c is

$$|V_L(j\omega_c)| = |H(j\omega_c)|\,|V_i|$$

$$= \frac{1}{\sqrt{2}}H_{max}|V_i|$$

$$= \frac{1}{\sqrt{2}}V_{L\,max}.$$

Now, compute the power delivered to the load at the cutoff frequency:

$$P(j\omega_c) = \frac{1}{2}\frac{|V_L^2(j\omega_c)|}{R}$$

$$= \frac{1}{2}\frac{\left(\frac{1}{\sqrt{2}}V_{Lmax}\right)^2}{R}$$

$$= \frac{1}{2}\frac{V_{Lmax}^2/2}{R}$$

$$= \frac{P_{max}}{2}.$$

We see that, at the cutoff frequency ω_c, the average power delivered by the circuit is one half the maximum average power. Thus, ω_c is also called the **half-power frequency**. In the filter's passband, the average power delivered to a load is at least 50% of the maximum average power.

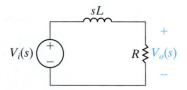

Figure 14.6 ▲ The s-domain equivalent for the circuit in Fig. 14.4(a).

The Series *RL* Circuit—Quantitative Analysis

Now that we have defined the cutoff frequency for filter circuits, we can analyze the series *RL* circuit to find the relationship between the component values and the cutoff frequency for this low-pass filter. We begin by constructing the s-domain equivalent of the circuit in Fig. 14.4(a), assuming the initial conditions are zero. The resulting equivalent circuit is shown in Fig. 14.6. The voltage transfer function for this circuit is

$$H(s) = \frac{V_o(s)}{V_i(s)} = \frac{R/L}{s + R/L}. \tag{14.3}$$

To study the frequency response, we make the substitution $s = j\omega$ in Eq. 14.3:

$$H(j\omega) = \frac{R/L}{j\omega + R/L}.$$

We can now separate $H(j\omega)$ into two equations. The first defines the transfer function magnitude as a function of frequency; the second defines the transfer function phase angle as a function of frequency:

$$|H(j\omega)| = \frac{R/L}{\sqrt{\omega^2 + (R/L)^2}}, \tag{14.4}$$

$$\theta(j\omega) = -\tan^{-1}\left(\frac{\omega L}{R}\right).$$

Close examination of Eq. 14.4 provides the quantitative support for the magnitude plot shown in Fig. 14.5.

- When $\omega = 0$, the denominator and the numerator are equal and $|H(j0)| = 1$. This means that at $\omega = 0$, the input voltage is passed to the output terminals without a change in the voltage magnitude.
- As the frequency increases, the numerator of Eq. 14.4 is unchanged, but the denominator gets larger. Thus, $|H(j\omega)|$ decreases as the frequency increases, as shown in the plot in Fig. 14.5. Likewise, as the frequency increases, the phase angle changes from its dc value of 0°, becoming more negative.
- As $\omega \rightarrow \infty$, the denominator of Eq. 14.4 approches infinity and $|H(j\omega)| \rightarrow 0$, as seen in Fig. 14.5. As $\omega \rightarrow \infty$, the phase angle approaches −90°, as seen from the phase angle plot in Fig. 14.5.

Using Eq. 14.4, we can compute the cutoff frequency, ω_c. Remember that ω_c is defined as the frequency at which $|H(j\omega_c)| = (1/\sqrt{2})H_{max}$. For the low-pass filter, $H_{max} = |H(j0)| = 1$, as seen in Fig. 14.5. Thus, for the circuit in Fig. 14.4(a),

$$|H(j\omega_c)| = \frac{1}{\sqrt{2}}|1| = \frac{R/L}{\sqrt{\omega_c^2 + (R/L)^2}}.$$

Solving for ω_c, we get

CUTOFF FREQUENCY FOR *RL* FILTERS

$$\omega_c = \frac{R}{L}. \tag{14.5}$$

Equation 14.5 provides an important result. The cutoff frequency, ω_c, can be set to any desired value by appropriately selecting values for R and L. We can therefore design a low-pass filter with whatever cutoff frequency is needed. Example 14.1 demonstrates the design potential of Eq. 14.5.

EXAMPLE 14.1 | Designing a Low-Pass Filter

Electrocardiology is the study of the electric signals produced by the heart. These signals maintain the heart's rhythmic beat and are measured by an instrument called an electrocardiograph. This instrument must be capable of detecting periodic signals whose frequency is about 1 Hz (the normal heart rate is 72 beats per minute). The instrument must operate in the presence of sinusoidal noise consisting of signals from the surrounding electrical environment, whose fundamental frequency is 60 Hz—the frequency at which electric power is supplied.

Choose values for R and L in the circuit of Fig. 14.4(a) such that the resulting circuit could be used in an electrocardiograph to filter out any noise above 10 Hz and pass the electric signals from the heart at or near 1 Hz. Then compute the magnitude of V_o at 1 Hz, 10 Hz, and 60 Hz to see how well the filter performs.

Solution

The problem is to select values for R and L that yield a low-pass filter with a cutoff frequency of 10 Hz. From Eq. 14.5, we see that R and L cannot be specified independently to generate a value for ω_c. Therefore, let's choose a commonly available value of L, 100 mH. Before we use Eq. 14.5 to compute the value of R needed to obtain the desired cutoff frequency, we need to convert the cutoff frequency from hertz to radians per second:

$$\omega_c = 2\pi(10) = 20\pi \text{ rad/s.}$$

Now, solve for the value of R that, together with $L = 100$ mH, will yield a low-pass filter, with a cutoff frequency of 10 Hz:

$$R = \omega_c L = (20\pi)(100 \times 10^{-3}) = 6.28 \ \Omega.$$

We can compute the magnitude of V_o using the equation $|V_o| = |H(j\omega)| \cdot |V_i|$:

$$|V_o(\omega)| = \frac{R/L}{\sqrt{\omega^2 + (R/L)^2}} |V_i|$$

$$= \frac{20\pi}{\sqrt{\omega^2 + 400\pi^2}} |V_i|.$$

Table 14.1 summarizes the computed magnitude values for the frequencies 1 Hz, 10 Hz, and 60 Hz. As expected, the input and output voltages have the same magnitudes at the low frequency because the circuit is a low-pass filter. At the cutoff frequency, the output voltage magnitude has been reduced by $1/\sqrt{2}$ from the unity passband magnitude. At 60 Hz, the output voltage magnitude has been reduced by a factor of about 6, achieving the desired attenuation of the noise that could corrupt the signal the electrocardiograph is designed to measure.

TABLE 14.1	Input and Output Voltage Magnitudes for Several Frequencies	
f(Hz)	$|V_i|$(V)	$|V_o|$(V)
1	1.0	0.995
10	1.0	0.707
60	1.0	0.164

A Series *RC* Circuit

The series *RC* circuit shown in Fig. 14.7 also behaves as a low-pass filter. We can verify this via the same qualitative analysis used previously. In fact, such a qualitative examination is an important problem-solving step that you should get in the habit of performing when analyzing filters. Doing so enables you to predict the filtering characteristics (low pass, high pass, etc.) and thus also predict the general form of the transfer function. If the calculated transfer function matches the qualitatively predicted form, you have an important accuracy check.

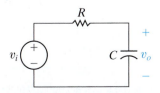

Figure 14.7 ▲ A series *RC* low-pass filter.

Note that the circuit's output is defined as the voltage across the capacitor. We again use three frequency regions to determine how the series RC circuit in Fig. 14.7 behaves:

- *Zero frequency* ($\omega = 0$): The impedance of the capacitor is infinite, and the capacitor acts as an open circuit. The input and output voltages are thus the same.
- *Frequencies increasing from zero*: The impedance of the capacitor decreases relative to the impedance of the resistor, and the source voltage divides between the resistive impedance and the capacitive impedance. The output voltage is thus smaller than the source voltage.
- *Infinite frequency* ($\omega = \infty$): The impedance of the capacitor is zero, and the capacitor acts as a short circuit. The output voltage is thus zero.

Based on this analysis, the series RC circuit functions as a low-pass filter. Example 14.2 explores this circuit quantitatively.

EXAMPLE 14.2 | **Designing a Series *RC* Low-Pass Filter**

For the series RC circuit in Fig. 14.7:

a) Find the transfer function between the source voltage and the output voltage.

b) Determine an equation for the cutoff frequency in the series RC circuit.

c) Choose values for R and C that will yield a low-pass filter with a cutoff frequency of 3 kHz.

Solution

a) To derive an expression for the transfer function, we first construct the s-domain equivalent of the circuit in Fig. 14.7, as shown in Fig. 14.8. Using s-domain voltage division on the equivalent circuit, we find

$$H(s) = \frac{\dfrac{1}{RC}}{s + \dfrac{1}{RC}}.$$

Now, substitute $s = j\omega$ and compute the magnitude of the resulting expression:

$$|H(j\omega)| = \frac{\dfrac{1}{RC}}{\sqrt{\omega^2 + \left(\dfrac{1}{RC}\right)^2}}.$$

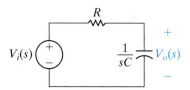

Figure 14.8 ▲ The s-domain equivalent for the circuit in Fig. 14.7.

b) At the cutoff frequency ω_c, $|H(j\omega)|$ is equal to $(1/\sqrt{2})H_{\max}$. For a low-pass filter, $H_{\max} = H(j0)$, and for the circuit in Fig. 14.8, $H(j0) = 1$. We can then describe the relationship among the quantities R, C, and ω_c:

$$|H(j\omega_c)| = \frac{1}{\sqrt{2}}(1) = \frac{\dfrac{1}{RC}}{\sqrt{\omega_c^2 + \left(\dfrac{1}{RC}\right)^2}}.$$

Solving this equation for ω_c, we get

CUTOFF FREQUENCY FOR *RC* FILTERS

$$\omega_c = \frac{1}{RC}.$$

c) From the results in (b), we see that the cutoff frequency is determined by the values of R and C. Because R and C cannot be determined independently, let's choose $C = 1\,\mu F$. Given a choice, we will usually specify a value for C first, rather than for R, because the number of available capacitor values is much smaller than the number of resistor values. Remember that we have to convert the specified cutoff frequency from 3 kHz to $(2\pi)(3)$ krad/s:

$$R = \frac{1}{\omega_c C}$$

$$= \frac{1}{(2\pi)(3 \times 10^3)(1 \times 10^{-6})}$$

$$= 53.05\ \Omega.$$

Figure 14.9 summarizes the two low-pass filter circuits we have analyzed. Look carefully at the transfer functions and notice their similar form—they differ only in the terms that specify the cutoff frequency. We can therefore identify a general form for the transfer functions of these two low-pass filters:

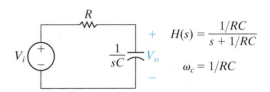

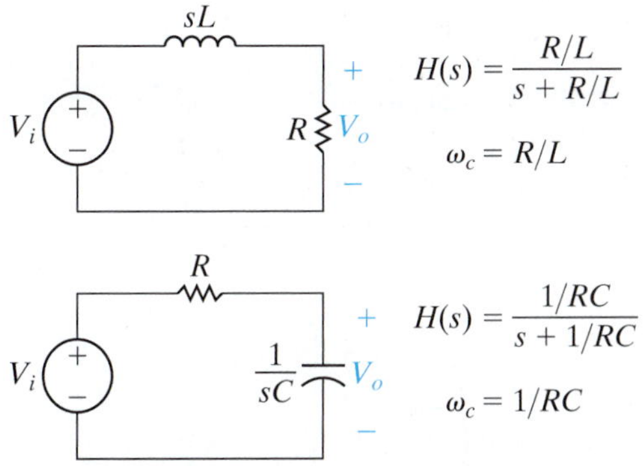

$$H(s) = \frac{R/L}{s + R/L}$$
$$\omega_c = R/L$$

$$H(s) = \frac{1/RC}{s + 1/RC}$$
$$\omega_c = 1/RC$$

> ## TRANSFER FUNCTION FOR LOW-PASS FILTERS
>
> $$H(s) = \frac{\omega_c}{s + \omega_c}. \qquad (14.6)$$

Any circuit with a voltage ratio described by Eq. 14.6 would behave as a low-pass filter with a cutoff frequency of ω_c. The problems at the end of the chapter give you other examples of circuits with this voltage ratio.

Figure 14.9 ▲ Two low-pass filters, the series *RL* and the series *RC*, together with their transfer functions and cutoff frequencies.

Relating the Frequency Domain to the Time Domain

Finally, you might have noticed one other important relationship. Recall our discussion of the natural responses of the first-order *RL* and *RC* circuits in Chapter 7. An important parameter for these circuits is the time constant, τ, which characterizes the shape of the time response. For the *RL* circuit, the time constant has the value L/R (Eq. 7.3); for the *RC* circuit, the time constant is RC (Eq. 7.8). Compare the time constants to the cutoff frequencies for these circuits and notice that

$$\tau = 1/\omega_c. \qquad (14.7)$$

This result is a direct consequence of the relationship between the time response of a circuit and its frequency response, as revealed by the Laplace transform. The discussion of memory and weighting as represented in the convolution integral of Section 13.6 shows that as $\omega_c \to \infty$, the filter has no memory, and the output approaches a scaled replica of the input; that is, no filtering has occurred. As $\omega_c \to 0$, the filter has increased memory, and the output voltage is a distortion of the input because filtering has occurred.

ASSESSMENT PROBLEMS

Objective 1—Know the *RL* and *RC* circuit configurations that act as low-pass filters

14.1 A series *RC* low-pass filter requires a cutoff frequency of 8 kHz. Use $R = 10\ \text{k}\Omega$ and compute the value of C required.

Answer: 1.99 nF.

14.2 A series *RL* low-pass filter with a cutoff frequency of 2 kHz is needed. Using $R = 5\ \text{k}\Omega$,

SELF-CHECK: Also try Chapter Problems 14.1 and 14.3.

compute (a) L; (b) $|H(j\omega)|$ at 50 kHz; and (c) $\theta(j\omega)$ at 50 kHz.

Answer: (a) 0.40 H;

(b) 0.04;

(c) −87.71°.

14.3 High-Pass Filters

Now we examine two circuits that function as high-pass filters. Once again, they are the series *RL* circuit and the series *RC* circuit. We will see that the same series circuit can act as either a low-pass or a high-pass filter, depending on where the output voltage is defined. We will also determine

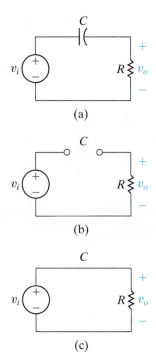

Figure 14.10 ▲ (a) A series RC high-pass filter; (b) the equivalent circuit at $\omega = 0$; and (c) the equivalent circuit at $\omega = \infty$.

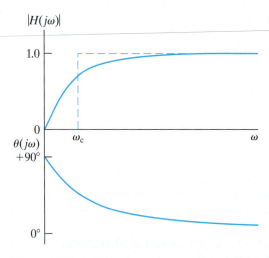

Figure 14.11 ▲ The frequency response plot for the series RC circuit in Fig. 14.10(a).

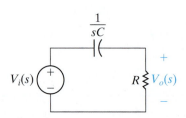

Figure 14.12 ▲ The s-domain equivalent of the circuit in Fig. 14.10(a).

the relationship between the component values and the cutoff frequency of these filters.

The Series RC Circuit—Qualitative Analysis

A series RC circuit is shown in Fig. 14.10(a). In contrast to its low-pass counterpart in Fig. 14.7, the output voltage here is defined across the resistor, not the capacitor. Because of this, the effect of the changing capacitive impedance is different than it was in the low-pass configuration.

At $\omega = 0$, the capacitor behaves like an open circuit, so there is no current in the resistor. This is illustrated in the equivalent circuit in Fig. 14.10(b). In this circuit, there is no voltage across the resistor and $v_o = 0$. The circuit filters out the low-frequency input voltage before it reaches the circuit's output.

As the frequency of the voltage source increases, the impedance of the capacitor decreases relative to the impedance of the resistor, and the source voltage is now divided between the capacitor and the resistor. The output voltage magnitude thus begins to increase.

When the frequency of the source is infinite ($\omega = \infty$), the capacitor behaves as a short circuit, so the capacitor voltage is zero. This is illustrated in the equivalent circuit in Fig. 14.10(c). In this circuit, the input and output voltages are the same.

The phase angle difference between the input and output voltages also varies as the frequency of the source changes. For $\omega = \infty$, the output voltage is the same as the input voltage, so the phase angle difference is zero. As the frequency of the source decreases and the impedance of the capacitor increases, a phase shift is introduced between the voltage and the current in the capacitor. This creates a phase difference between the input and output voltages. The phase angle of the output voltage leads that of the source voltage. When $\omega = 0$, this phase angle difference reaches its maximum of $+90°$.

Based on our qualitative analysis, we see that when the output is defined as the voltage across the resistor, the series RC circuit behaves as a high-pass filter. The components and connections are identical to the low-pass series RC circuit, but the choice of output is different. Thus, we have confirmed the earlier observation that the filtering characteristics of a circuit depend on the definition of the output as well as on circuit components, values, and connections.

Figure 14.11 shows the frequency response plot for the series RC high-pass filter. For reference, the dashed lines indicate the magnitude plot for an ideal high-pass filter. We now turn to a quantitative analysis of this same circuit.

The Series RC Circuit—Quantitative Analysis

To begin, construct the s-domain equivalent of the circuit in Fig. 14.10(a), as shown in Fig. 14.12. We use voltage division to find the transfer function:

$$H(s) = \frac{V_o(s)}{V_i(s)} = \frac{s}{s + 1/RC}.$$

Making the substitution $s = j\omega$ gives

$$H(j\omega) = \frac{j\omega}{j\omega + 1/RC}.$$

Next, we separate $H(j\omega)$ into two equations. The first is the equation describing the magnitude of the transfer function; the second is the equation describing the phase angle of the transfer function:

$$|H(j\omega)| = \frac{\omega}{\sqrt{\omega^2 + (1/RC)^2}}, \qquad (14.8)$$

$$\theta(j\omega) = 90° - \tan^{-1}\omega RC.$$

A close look at the equations for the magnitude and phase angle of the transfer function confirms the shape of the frequency response plot in Fig. 14.11. Using Eq. 14.8, we can calculate the cutoff frequency for the series RC high-pass filter. Recall that at the cutoff frequency, the magnitude of the transfer function is $(1/\sqrt{2})H_{max}$. For a high-pass filter, $H_{max} = |H(j\infty)|$, as seen from Fig. 14.11. We can construct an equation for ω_c by setting the left-hand side of Eq. 14.8 to $(1/\sqrt{2})|H(j\infty)|$, noting that for this series RC circuit, $|H(j\infty)| = 1$:

$$\frac{1}{\sqrt{2}} = \frac{\omega_c}{\sqrt{\omega_c^2 + (1/RC)^2}}.$$

Solving for ω_c, we get

$$\omega_c = \frac{1}{RC}. \qquad (14.9)$$

Equation 14.9 presents a familiar result. The cutoff frequency for the series RC circuit has the value $1/RC$, whether the circuit is configured as a low-pass filter in Fig. 14.7 or as a high-pass filter in Fig. 14.10(a). This is not a surprising result, as we have already discovered a connection between the cutoff frequency, ω_c, and the time constant, τ, of a circuit.

Example 14.3 analyzes a series RL circuit, this time configured as a high-pass filter. Example 14.4 examines the effect of adding a load resistor at the output of the filter.

EXAMPLE 14.3   **Designing a Series _RL_ High-Pass Filter**

Show that the series RL circuit in Fig. 14.13 also acts like a high-pass filter:

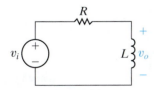

Figure 14.13 ▲ The circuit for Example 14.3.

a) Derive an expression for the circuit's transfer function.

b) Use the result from (a) to determine an equation for the cutoff frequency in the series RL circuit.

c) Choose values for R and L that will yield a high-pass filter with a cutoff frequency of 15 kHz.

Solution

a) Begin by constructing the s-domain equivalent of the series RL circuit, as shown in Fig. 14.14. Then use voltage division to find the transfer function:

$$H(s) = \frac{s}{s + R/L}.$$

Making the substitution $s = j\omega$, we get

$$H(j\omega) = \frac{j\omega}{j\omega + R/L}.$$

Notice that this equation has the same form as the equation for the series RC high-pass filter.

b) To find an equation for the cutoff frequency, first compute the magnitude of $H(j\omega)$:

$$|H(j\omega)| = \frac{\omega}{\sqrt{\omega^2 + (R/L)^2}}.$$

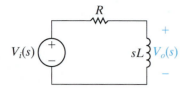

Figure 14.14 ▲ The s-domain equivalent of the circuit in Fig. 14.13.

Then, as before, we set the left-hand side of this equation to $(1/\sqrt{2})H_{max}$, based on the definition of the cutoff frequency ω_c. Remember that $H_{max} = |H(j\infty)|$ for a high-pass filter, and for

the series RL circuit, $|H(j\infty)| = 1$. We solve the resulting equation for the cutoff frequency:

$$\frac{1}{\sqrt{2}} = \frac{\omega_c}{\sqrt{\omega_c^2 + (R/L)^2}}, \quad \omega_c = \frac{R}{L}.$$

This is the same cutoff frequency we computed for the series RL low-pass filter.

c) Using the equation for ω_c computed in (b), we find that it is not possible to specify values for R and L independently. Therefore, let's arbitrarily select a value of 500 Ω for R. Remember to convert the cutoff frequency to radians per second:

$$L = \frac{R}{\omega_c} = \frac{500}{(2\pi)(15,000)} = 5.31 \text{ mH}.$$

EXAMPLE 14.4 | Loading the Series *RL* High-Pass Filter

Examine the effect of placing a load resistor in parallel with the inductor in the RL high-pass filter shown in Fig. 14.15:

a) Determine the transfer function for the circuit in Fig. 14.15.

b) Sketch the magnitude plot for the loaded RL high-pass filter, using the values for R and L from the circuit in Example 14.3(c) and letting $R_L = R$. On the same graph, sketch the magnitude plot for the unloaded RL high-pass filter of Example 14.3(c).

Solution

a) Begin by transforming the circuit in Fig. 14.15 to the s-domain, as shown in Fig. 14.16. Use voltage division across the parallel combination of inductor and load resistor to compute the transfer function:

$$H(s) = \frac{\dfrac{R_L s L}{R_L + sL}}{R + \dfrac{R_L s L}{R_L + sL}} = \frac{\left(\dfrac{R_L}{R + R_L}\right)s}{s + \left(\dfrac{R_L}{R + R_L}\right)\dfrac{R}{L}} = \frac{Ks}{s + \omega_c},$$

where

$$K = \frac{R_L}{R + R_L}, \quad \omega_c = KR/L.$$

Note that ω_c is the cutoff frequency of the loaded filter.

b) For the unloaded RL high-pass filter from Example 14.3(c), the passband magnitude is 1, and the cutoff frequency is 15 kHz. For the loaded RL high-pass filter, $R = R_L = 500\ \Omega$, so $K = 1/2$. Thus, for the loaded filter, the passband

magnitude is $(1)(1/2) = 1/2$, and the cutoff frequency is $(15,000)(1/2) = 7.5$ kHz. A sketch of the magnitude plots of the loaded and unloaded circuits is shown in Fig. 14.17.

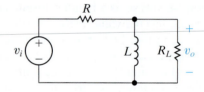

Figure 14.15 ▲ The circuit for Example 14.4.

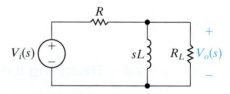

Figure 14.16 ▲ The s-domain equivalent of the circuit in Fig. 14.15.

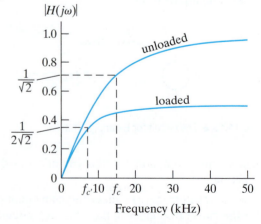

Figure 14.17 ▲ The magnitude plots for the unloaded RL high-pass filter of Fig 14.13 and the loaded RL high-pass filter of Fig. 14.15.

Let's compare the transfer functions of the unloaded filter in Example 14.3 and the loaded filter in Example 14.4. Both transfer functions are in the form:

$$H(s) = \frac{Ks}{s + K(R/L)},$$

with $K = 1$ for the unloaded filter and $K = R_L/(R + R_L)$ for the loaded filter. Note that the value of K for the loaded circuit reduces to the value of K for the unloaded circuit when $R_L = \infty$; that is, when there is no load resistor.

The cutoff frequencies for both filters can be seen directly from their transfer functions. In both cases, $\omega_c = K(R/L)$, where $K = 1$ for the unloaded circuit, and $K = R_L/(R + R_L)$ for the loaded circuit. Again, the cutoff frequency for the loaded circuit reduces to that of the unloaded circuit when $R_L = \infty$. Because $R_L/(R + R_L) < 1$, the effect of the load resistor is to reduce the passband magnitude by the factor K and to lower the cutoff frequency by the same factor.

We predicted these results at the beginning of this chapter. When the output voltage amplitude of a passive high-pass filter is maximum, it equals the amplitude of the filter's input voltage. Placing a load across the filter, as we did in Example 14.4, decreases the output voltage amplitude. If we need to amplify signals in the passband, we must turn to active filters, such as those discussed in Chapter 15.

The effect of a load on a filter's transfer function poses another dilemma in circuit design. We typically begin with a transfer function specification and then design a filter to meet that specification. We may or may not know what the load on the filter will be. Ideally, we want the filter's transfer function to remain the same regardless of the load on it, but this is not possible for the passive filters presented here.

Figure 14.18 summarizes the high-pass filter circuits we have analyzed. Looking at the expressions for $H(s)$, we see that they differ only in the denominator, which includes the cutoff frequency. As we did with the low-pass filters in Eq. 14.6, we state a general form for the transfer function of these two high-pass filters:

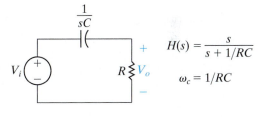

$$H(s) = \frac{s}{s + 1/RC}$$
$$\omega_c = 1/RC$$

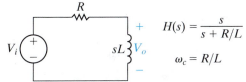

$$H(s) = \frac{s}{s + R/L}$$
$$\omega_c = R/L$$

Figure 14.18 ▲ Two high-pass filters, the series RC and the series RL, together with their transfer functions and cutoff frequencies.

TRANSFER FUNCTION FOR HIGH-PASS FILTERS

$$H(s) = \frac{s}{s + \omega_c}. \qquad (14.10)$$

Any circuit with a voltage ratio described by Eq. 14.10 would behave as a high-pass filter with a cutoff frequency of ω_c. The problems at the end of the chapter give you other examples of circuits with this voltage ratio.

We have drawn attention to another important relationship. We have discovered that a series RC circuit has the same cutoff frequency whether it is configured as a low-pass filter or as a high-pass filter. The same is true of a series RL circuit. Given the connection between the cutoff frequency of a filter circuit and its time constant, we should expect the cutoff frequency to be a characteristic parameter of the circuit whose value depends only on the circuit components, their values, and the way they are connected.

ASSESSMENT PROBLEMS

Objective 2—Know the *RL* and *RC* circuit configurations that act as high-pass filters

14.3 A series *RL* high-pass filter has $R = 5\text{ k}\Omega$ and $L = 3.5\text{ mH}$. What is ω_c for this filter?

Answer: 1.43 Mrad/s.

14.4 A series *RC* high-pass filter has $C = 1\ \mu\text{F}$. Compute the cutoff frequency for the following values of *R*: (a) 100 Ω; (b) 5 kΩ; and (c) 30 kΩ.

Answer: (a) 10 krad/s;

(b) 200 rad/s;

(c) 33.33 rad/s.

14.5 Compute the transfer function of a series *RC* low-pass filter that has a load resistor R_L in parallel with its capacitor.

Answer: $H(s) = \dfrac{\dfrac{1}{RC}}{s + \dfrac{1}{KRC}}$, where $K = \dfrac{R_L}{R + R_L}$.

SELF-CHECK: Also try Chapter Problems 14.13 and 14.17.

14.4 Bandpass Filters

The next filters we examine are those that pass voltages within a band of frequencies to the output while filtering out voltages at frequencies outside this band. These filters are somewhat more complicated than the low-pass and high-pass filters of the previous sections. As we have already seen in Fig. 14.3(a), ideal bandpass filters have two cutoff frequencies, ω_{c1} and ω_{c2}, which identify the passband. For realistic bandpass filters, these cutoff frequencies are again defined as the frequencies for which the magnitude of the transfer function equals $(1/\sqrt{2})H_{\max}$.

Center Frequency, Bandwidth, and Quality Factor

Besides the cutoff frequencies ω_{c1} and ω_{c2}, three other important parameters characterize a bandpass filter. The first is the **center frequency**, ω_o, defined as the frequency for which a circuit's transfer function is purely real. Another name for the center frequency is the **resonant frequency**. This is the same name given to the frequency that characterizes the natural response of the second-order circuits in Chapter 8 because they are the same frequencies! When a circuit is driven at the resonant frequency, we say that the circuit is *in resonance* because the frequency of the forcing function is the same as the natural frequency of the circuit. The center frequency is the geometric center of the passband; that is, $\omega_o = \sqrt{\omega_{c1}\omega_{c2}}$. For bandpass filters, the magnitude of the transfer function is maximum at the center frequency ($H_{\max} = |H(j\omega_o)|$).

The second parameter is the **bandwidth**, β, which is the width of the passband. The final parameter is the **quality factor** Q, which is the ratio of the center frequency to the bandwidth. The quality factor describes the width of the passband, independent of its location on the frequency axis. It also describes the shape of the magnitude plot, independent of frequency.

Although five different parameters characterize bandpass filters—ω_{c1}, ω_{c2}, ω_o, β, and Q—only two of the five can be specified independently. That is, once we specify any two of these parameter values, the other three can be calculated from the dependent relationships among them. We explore these relationships next, as we derive expressions for the five characteristic parameters in terms of the component values for two *RLC* circuits that act as bandpass filters.

The Series *RLC* Circuit—Qualitative Analysis

Figure 14.19(a) depicts a series *RLC* circuit. We want to consider the effect of changing the source frequency on the magnitude of the output voltage. As before, changes to the source frequency result in changes to the impedance of the capacitor and the inductor. This time, the qualitative analysis is somewhat more complicated because the circuit has both an inductor and a capacitor.

At $\omega = 0$, the capacitor behaves like an open circuit, and the inductor behaves like a short circuit. The equivalent circuit is shown in Fig. 14.19(b). The open circuit representing the capacitor impedance prevents current from reaching the resistor, and the resulting output voltage is zero.

At $\omega = \infty$, the capacitor behaves like a short circuit, and the inductor behaves like an open circuit. The equivalent circuit is shown in Fig. 14.19(c). The inductor now prevents current from reaching the resistor, and again the output voltage is zero.

But what happens in the frequency region between $\omega = 0$ and $\omega = \infty$? Between these two extremes, both the capacitor and the inductor have finite impedances. In this region, the source voltage divides among the capacitor, inductor, and resistor. Remember that the capacitor impedance is negative, whereas the inductor impedance is positive. Thus, at some frequency, the capacitor impedance and the inductor impedance have equal magnitudes and opposite signs; the two impedances cancel out, so the output voltage equals the source voltage. This special frequency is the center frequency, ω_o. On either side of ω_o, the output voltage is less than the source voltage. Note that at ω_o, the series combination of the inductor and capacitor behaves like a short circuit.

The frequency response plot for the circuit in Fig. 14.19(a) is shown in Fig. 14.20. The ideal bandpass filter magnitude plot is overlaid (as a dashed line) on the transfer function magnitude plot for comparison.

Now consider what happens to the transfer function phase angle. At the center frequency, ω_o, the phase angles of the input and output voltages are equal, so the phase angle of the transfer function is zero. As the frequency decreases, the capacitor phase angle is larger than the inductor phase angle. Because the capacitor contributes positive phase shift, the transfer function phase angle is positive. At very low frequencies, the transfer function phase angle is $+90°$.

Conversely, if the frequency increases from the center frequency, the inductor phase angle is larger than the capacitor phase angle. The inductor contributes negative phase shift, so the transfer function phase angle is negative. At very high frequencies, the transfer function phase angle is $-90°$. The plot of the transfer function phase angle is shown in Fig. 14.20.

The Series *RLC* Circuit—Quantitative Analysis

We begin by drawing the *s*-domain equivalent for the series *RLC* circuit, as shown in Fig. 14.21. Using voltage division, we find that the transfer function equation is

$$H(s) = \frac{(R/L)s}{s^2 + (R/L)s + (1/LC)}. \tag{14.11}$$

As before, we substitute $s = j\omega$ into Eq. 14.11 and produce the equations for the magnitude and the phase angle of the transfer function:

$$|H(j\omega)| = \frac{\omega(R/L)}{\sqrt{[(1/LC) - \omega^2]^2 + [\omega(R/L)]^2}}, \tag{14.12}$$

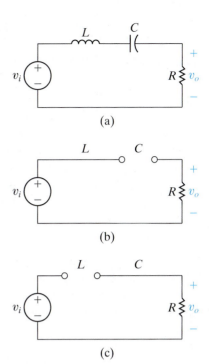

Figure 14.19 ▲ (a) A series *RLC* bandpass filter; (b) the equivalent circuit for $\omega = 0$; and (c) the equivalent circuit for $\omega = \infty$.

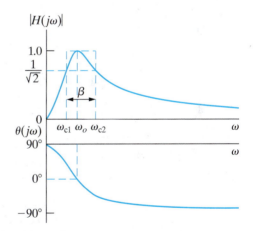

Figure 14.20 ▲ The frequency response plot for the series *RLC* bandpass filter circuit in Fig. 14.19.

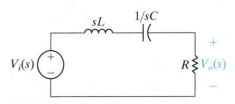

Figure 14.21 ▲ The *s*-domain equivalent for the circuit in Fig. 14.19(a).

$$\theta(j\omega) = 90° - \tan^{-1}\left[\frac{\omega(R/L)}{(1/LC) - \omega^2}\right].$$

Now calculate the five parameters that characterize this RLC band-pass filter. Recall that the center frequency, ω_o, is defined as the frequency for which the circuit's transfer function is purely real. The transfer function for the RLC circuit in Fig. 14.21 will be real when the capacitor and inductor impedances sum to zero:

$$j\omega_o L + \frac{1}{j\omega_o C} = 0.$$

Solving for ω_o, we get

CENTER FREQUENCY

$$\omega_o = \sqrt{\frac{1}{LC}}. \tag{14.13}$$

Next, calculate the cutoff frequencies, ω_{c1} and ω_{c2}. Remember that at the cutoff frequencies, the magnitude of the transfer function is $(1/\sqrt{2})H_{max}$. Because $H_{max} = |H(j\omega_o)|$, we find H_{max} by substituting Eq. 14.13 into Eq. 14.12:

$$H_{max} = |H(j\omega_o)|$$

$$= \frac{\omega_o(R/L)}{\sqrt{[(1/LC) - \omega_o^2]^2 + (\omega_o R/L)^2}}$$

$$= \frac{\sqrt{(1/LC)}(R/L)}{\sqrt{[(1/LC) - (1/LC)]^2 + [\sqrt{(1/LC)}(R/L)]^2}} = 1.$$

Now set the left-hand side of Eq. 14.12 to $(1/\sqrt{2})H_{max}$ (which equals $1/\sqrt{2}$) and prepare to solve for ω_c:

$$\frac{1}{\sqrt{2}} = \frac{\omega_c(R/L)}{\sqrt{[(1/LC) - \omega_c^2]^2 + (\omega_c R/L)^2}}$$

$$= \frac{1}{\sqrt{[(\omega_c L/R) - (1/\omega_c RC)]^2 + 1}}.$$

We can equate the denominators of the two sides of this expression and simplify to get

$$\pm 1 = \omega_c \frac{L}{R} - \frac{1}{\omega_c RC}.$$

Rearrange to get the following quadratic equation:

$$\omega_c^2 \pm \frac{R}{L}\omega_c - \frac{1}{LC} = 0.$$

The solution of the quadratic equation yields four values for the cut-off frequency. Only two of these values are positive and have physical significance; they identify the passband of this filter:

CUTOFF FREQUENCIES, SERIES *RLC* FILTERS

$$\omega_{c1} = -\frac{R}{2L} + \sqrt{\left(\frac{R}{2L}\right)^2 + \left(\frac{1}{LC}\right)}, \qquad (14.14)$$

$$\omega_{c2} = \frac{R}{2L} + \sqrt{\left(\frac{R}{2L}\right)^2 + \left(\frac{1}{LC}\right)}. \qquad (14.15)$$

We can use Eqs. 14.14 and 14.15 to confirm that the center frequency, ω_o, is the geometric mean of the two cutoff frequencies:

RELATIONSHIP BETWEEN CENTER FREQUENCY AND CUTOFF FREQUENCIES

$$\omega_o = \sqrt{\omega_{c1} \cdot \omega_{c2}}$$

$$= \sqrt{\left[-\frac{R}{2L} + \sqrt{\left(\frac{R}{2L}\right)^2 + \left(\frac{1}{LC}\right)}\right]\left[\frac{R}{2L} + \sqrt{\left(\frac{R}{2L}\right)^2 + \left(\frac{1}{LC}\right)}\right]}$$

$$= \sqrt{\frac{1}{LC}}.$$

Recall that the bandwidth of a bandpass filter is defined as the difference between the two cutoff frequencies. Because $\omega_{c2} > \omega_{c1}$ we can compute the bandwidth by subtracting Eq. 14.14 from Eq. 14.15:

RELATIONSHIP BETWEEN BANDWIDTH AND CUTOFF FREQUENCIES

$$\beta = \omega_{c2} - \omega_{c1}$$

$$= \left[\frac{R}{2L} + \sqrt{\left(\frac{R}{2L}\right)^2 + \left(\frac{1}{LC}\right)}\right] - \left[-\frac{R}{2L} + \sqrt{\left(\frac{R}{2L}\right)^2 + \left(\frac{1}{LC}\right)}\right],$$

so

BANDWIDTH, SERIES *RLC* FILTERS

$$\beta = \frac{R}{L}. \qquad (14.16)$$

The quality factor, the last of the five characteristic parameters, is defined as the ratio of center frequency to bandwidth. Using Eqs. 14.13 and 14.16:

RELATIONSHIP AMONG QUALITY FACTOR, CENTER FREQUENCY AND BANDWIDTH

$$Q = \frac{\omega_o}{\beta}$$

$$= \frac{\sqrt{1/LC}}{(R/L)},$$

so

QUALITY FACTOR, SERIES *RLC* FILTERS

$$Q = \sqrt{\frac{L}{R^2 C}}. \tag{14.17}$$

We now have five parameters that characterize the series *RLC* bandpass filter: two cutoff frequencies, ω_{c1} and ω_{c2}, which delimit the passband; the center frequency, ω_o, at which the magnitude of the transfer function is maximum; the bandwidth, β, a measure of the width of the passband; and the quality factor, Q, a second measure of passband width. Remember, only two of these parameters can be specified independently in a design. We have already observed that the quality factor is the ratio of the center frequency to the bandwidth. We can also rewrite the equations for the cutoff frequencies in terms of the center frequency and the bandwidth:

$$\omega_{c1} = -\frac{\beta}{2} + \sqrt{\left(\frac{\beta}{2}\right)^2 + \omega_o^2},$$

$$\omega_{c2} = \frac{\beta}{2} + \sqrt{\left(\frac{\beta}{2}\right)^2 + \omega_o^2}.$$

Alternative forms for these equations express the cutoff frequencies in terms of the quality factor and the center frequency (see Problem 14.18):

$$\omega_{c1} = \omega_o \cdot \left[-\frac{1}{2Q} + \sqrt{1 + \left(\frac{1}{2Q}\right)^2} \right],$$

$$\omega_{c2} = \omega_o \cdot \left[\frac{1}{2Q} + \sqrt{1 + \left(\frac{1}{2Q}\right)^2} \right].$$

Examples 14.5, 14.6, and 14.7 illustrate the design of bandpass filters, introduce another *RLC* circuit that behaves as a bandpass filter, and examine the effects of source resistance on the characteristic parameters of a series *RLC* bandpass filter.

EXAMPLE 14.5 | Designing a Bandpass Filter

A graphic equalizer is an audio amplifier that allows you to select different levels of amplification within different frequency regions. Using the series *RLC* circuit in Fig. 14.19(a), choose values for *R*, *L*, and *C* that yield a bandpass circuit able to select inputs within the 1 kHz–10 kHz frequency band. Such a circuit might be used in a graphic equalizer to select this frequency band from the larger audio band (generally 0–20 kHz) prior to amplification.

Solution

We need to compute values for *R*, *L*, and *C* that produce a bandpass filter with cutoff frequencies of 1 kHz and 10 kHz. There are many possible approaches to a solution. For instance, we could use Eqs. 14.14 and 14.15, which specify ω_{c1} and ω_{c2} in terms of *R*, *L*, and *C*. Because of the form of these equations, the algebraic manipulations might get complicated. Instead, we will calculate the center frequency, ω_o, from the cutoff frequencies and then use Eq. 14.13 to compute *L* and *C* from ω_o. Next we will calculate the bandwidth, β, from the cutoff frequencies and finally, use Eq. 14.16 to compute *R* from β. While this approach involves several computational steps, each calculation is fairly simple.

Any approach we choose will provide only two equations—insufficient to solve for the three unknowns—because of the dependencies among the bandpass filter characteristics. Thus, we need to select a value for either *R*, *L*, or *C* and use the two equations we've chosen to calculate the remaining component values. Here, we arbitrarily choose 1 μF as the capacitor value.

We compute the center frequency as the geometric mean of the cutoff frequencies:

$$f_o = \sqrt{f_{c1}f_{c2}} = \sqrt{(1000)(10{,}000)} = 3162.28 \text{ Hz.}$$

Next, use Eq. 14.13 to find *L* using *C* and the center frequency, which must be converted to radians/sec:

$$L = \frac{1}{\omega_o^2 C} = \frac{1}{[2\pi(3162.28)]^2(10^{-6})}$$

$$= 2.533 \text{ mH.}$$

The bandwidth is the difference between the two cutoff frequency values, so

$$\beta = \omega_{c2} - \omega_{c1} = 10{,}000 - 1000 = 9 \text{ kHz.}$$

Now convert the bandwidth to radians/sec and use Eq. 14.16 to calculate *R*:

$$R = \beta L = [2\pi(9000)](2.533 \times 10^{-3}) = 143.24 \ \Omega.$$

To check whether these component values produce the bandpass filter we want, substitute them into Eqs. 14.14 and 14.15. We find that

$$\omega_{c1} = 6283.19 \text{ rad/s } (1000 \text{ Hz}),$$

$$\omega_{c2} = 62{,}831.85 \text{ rad/s } (10{,}000 \text{ Hz}),$$

which are the cutoff frequencies specified for the filter.

This example reminds us that only two of the five bandpass filter parameters can be specified independently. The other three parameters can be computed from the two that are specified. In turn, these five parameter values depend on the three component values, *R*, *L*, and *C*, of which only two can be specified independently. It is almost always easiest to calculate the center frequency and bandwidth from whatever two parameters are specified, and then use Eqs. 14.13 and 14.16 to calculate the two unknown component values.

EXAMPLE 14.6 | Designing a Parallel *RLC* Bandpass Filter

a) Show that the *RLC* circuit in Fig. 14.22 is also a bandpass filter by deriving an expression for the transfer function *H*(*s*). Note that this circuit is a parallel *RLC* circuit with the parallel-connected current source and resistor source-transformed to a series-connected voltage source and resistor. This permits us to continue defining the filter transfer functions as ratios of output to input voltages.

b) Compute the center frequency, ω_o.

c) Calculate the cutoff frequencies, ω_{c1} and ω_{c2}, the bandwidth, β, and the quality factor, *Q*.

d) Compute values for *R* and *L* to yield a bandpass filter with a center frequency of 5 kHz and a bandwidth of 200 Hz, using a 5 μF capacitor.

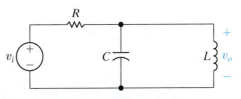

Figure 14.22 ▲ The circuit for Example 14.6.

Solution

a) Begin by transforming the circuit in Fig. 14.22 to the s domain; the result is shown in Fig. 14.23. We can find the transfer function for the s-domain circuit using voltage division if we first compute the equivalent impedance of the parallel combination of L and C, identified as $Z_{eq}(s)$ in Fig. 14.23:

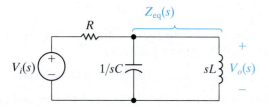

Figure 14.23 ▲ The s-domain equivalent of the circuit in Fig. 14.22.

$$Z_{eq} = sL \| \frac{1}{sC} = \frac{(sL)(1/sC)}{sL + (1/sC)} = \frac{sL}{s^2LC + 1}.$$

Now,

$$H(s) = \frac{V_o(s)}{V_i(s)} = \frac{Z_{eq}}{R + Z_{eq}} = \frac{\dfrac{s}{RC}}{s^2 + \dfrac{s}{RC} + \dfrac{1}{LC}}.$$

b) To find the center frequency, we need to calculate the frequency for which the transfer function magnitude is maximum. Substituting $s = j\omega$ in $H(s)$,

$$|H(j\omega)| = \frac{\dfrac{\omega}{RC}}{\sqrt{\left(\dfrac{1}{LC} - \omega^2\right)^2 + \left(\dfrac{\omega}{RC}\right)^2}}$$

$$= \frac{1}{\sqrt{1 + \left(\omega RC - \dfrac{1}{\omega \dfrac{L}{R}}\right)^2}}.$$

The magnitude of this transfer function is maximum when

$$\left(\frac{1}{LC} - \omega^2\right)^2 = 0.$$

Thus,

$$\omega_o = \sqrt{\frac{1}{LC}}$$

and

$$H_{max} = |H(j\omega_o)| = 1.$$

c) At the cutoff frequencies, the magnitude of the transfer function is $(1/\sqrt{2})H_{max} = 1/\sqrt{2}$. Substituting this constant on the left-hand side of the magnitude equation and simplifying, we get

$$\left[\omega_c RC - \frac{1}{\omega_c \dfrac{L}{R}}\right] = \pm 1.$$

Rearranging this equation once again produces two quadratic equations for the cutoff frequencies, with four solutions. Only two of them are positive and therefore have physical significance:

CUTOFF FREQUENCIES, PARALLEL RLC FILTERS

$$\omega_{c1} = -\frac{1}{2RC} + \sqrt{\left(\frac{1}{2RC}\right)^2 + \frac{1}{LC}},$$

$$\omega_{c2} = \frac{1}{2RC} + \sqrt{\left(\frac{1}{2RC}\right)^2 + \frac{1}{LC}}.$$

We compute the bandwidth from the cutoff frequencies:

BANDWIDTH, PARALLEL RLC FILTERS

$$\beta = \omega_{c2} - \omega_{c1} = \frac{1}{RC}.$$

Finally, use the definition of quality factor to calculate Q:

QUALITY FACTOR, PARALLEL RLC FILTERS

$$Q = \frac{\omega_o}{\beta} = \sqrt{\frac{R^2 C}{L}}.$$

Notice that once again we can specify the cutoff frequencies for this bandpass filter in terms of its center frequency and bandwidth:

$$\omega_{c1} = -\frac{\beta}{2} + \sqrt{\left(\frac{\beta}{2}\right)^2 + \omega_o^2},$$

$$\omega_{c2} = \frac{\beta}{2} + \sqrt{\left(\frac{\beta}{2}\right)^2 + \omega_o^2}.$$

d) Use the equation for bandwidth in (c) to compute a value for R, given $C = 5\,\mu F$. Remember to convert the bandwidth to radians/sec:

$$R = \frac{1}{\beta C}$$

$$= \frac{1}{(2\pi)(200)(5 \times 10^{-6})}$$

$$= 159.15\ \Omega.$$

Using the value of capacitance and the equation for center frequency in (b), compute the inductor value:

$$L = \frac{1}{\omega_o^2 C}$$

$$= \frac{1}{[2\pi(5000)]^2(5 \times 10^{-6})}$$

$$= 202.64\ \mu H.$$

| EXAMPLE 14.7 | Determining Effect of a Nonideal Voltage Source on a *RLC* Bandpass Filter |

For each of the bandpass filters we have constructed, we have always assumed an ideal voltage source, that is, a voltage source with no series resistance. When we design with a filter using values of R, L, and C whose equivalent impedance has a magnitude close to the actual impedance of the voltage source, it is not valid to assume the voltage source is ideal. In this example, we determine the effect of the nonzero source resistance R_i on the series *RLC* bandpass filter characteristics.

a) Determine the transfer function for the circuit in Fig. 14.24.

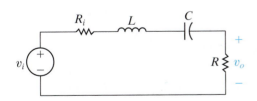

Figure 14.24 ▲ The circuit for Example 14.7.

b) Sketch the magnitude plot for the circuit in Fig. 14.24, using the values for R, L, and C from Example 14.5 and setting $R_i = R$. On the same graph, sketch the magnitude plot for the circuit in Example 14.5, where $R_i = 0$.

Solution

a) Begin by transforming the circuit in Fig. 14.24 into the s domain, as shown in Fig. 14.25. Now

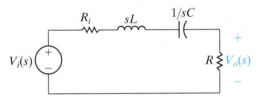

Figure 14.25 ▲ The s-domain equivalent of the circuit in Fig. 14.24.

construct the transfer function using voltage division:

$$H(s) = \frac{\dfrac{R}{L} s}{s^2 + \left(\dfrac{R + R_i}{L}\right)s + \dfrac{1}{LC}}.$$

Substitute $s = j\omega$ and calculate the transfer function magnitude:

$$|H(j\omega)| = \frac{\dfrac{R}{L}\omega}{\sqrt{\left(\dfrac{1}{LC} - \omega^2\right)^2 + \left(\omega\,\dfrac{R + R_i}{L}\right)^2}}.$$

The center frequency, ω_o, is the frequency at which this transfer function magnitude is maximum, which is

$$\omega_o = \sqrt{\frac{1}{LC}}.$$

At the center frequency, the maximum magnitude is

$$H_{\max} = |H(j\omega_o)| = \frac{R}{R_i + R}.$$

The cutoff frequencies can be computed by setting the transfer function magnitude equal to $(1/\sqrt{2})H_{\max}$:

$$\omega_{c1} = -\frac{R + R_i}{2L} + \sqrt{\left(\frac{R + R_i}{2L}\right)^2 + \frac{1}{LC}},$$

$$\omega_{c2} = \frac{R + R_i}{2L} + \sqrt{\left(\frac{R + R_i}{2L}\right)^2 + \frac{1}{LC}}.$$

The bandwidth is calculated from the cutoff frequencies:

$$\beta = \omega_{c2} - \omega_{c1} = \frac{R + R_i}{L}.$$

Finally, the quality factor is computed from the center frequency and the bandwidth:

$$Q = \frac{\omega_o}{\beta} = \frac{\sqrt{L/C}}{R + R_i}.$$

From this analysis, note that we can write the transfer function of the series RLC bandpass filter with non-zero source resistance as

$$H(s) = \frac{K\beta s}{s^2 + \beta s + \omega_o^2},$$

where

$$K = \frac{R}{R + R_i}.$$

Note that when $R_i = 0$, $K = 1$ and the transfer function is

$$H(s) = \frac{\beta s}{s^2 + \beta s + \omega_o^2}.$$

b) The circuit in Example 14.5 has a center frequency of 3162.28 Hz and a bandwidth of 9 kHz, and $H_{\max} = 1$. If we use the same values for R, L, and C in the circuit in Fig. 14.24 and let $R_i = R$, then the center frequency remains at 3162.28 kHz, but $\beta = (R + R_i)/L = 18$ kHz, and $H_{\max} = R/(R + R_i) = 1/2$. The transfer function magnitudes for these two bandpass filters are plotted on the same graph in Fig. 14.26.

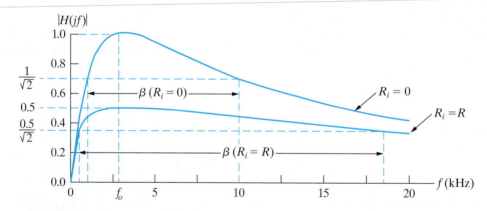

Figure 14.26 ▲ The magnitude plots for a series RLC bandpass filter with a zero source resistance and a nonzero source resistance.

If we compare the characteristic parameter values for the filter with $R_i = 0$ to the values for the filter with $R_i \neq 0$, we see the following:

- The center frequencies are the same.
- The maximum transfer function magnitude for the filter with $R_i \neq 0$ is smaller than that for the filter with $R_i = 0$.
- The bandwidth for the filter with $R_i \neq 0$ is larger than that for the filter with $R_i = 0$. Thus, the cutoff frequencies and the quality factors for the two circuits are also different.

Adding a nonzero source resistance to a series RLC bandpass filter leaves the center frequency unchanged but widens the passband and reduces the passband magnitude.

Here we see the same design challenge we saw when adding a load resistor to the high-pass filter. We would like to design bandpass filters with filtering properties that are unchanged by the internal resistance of the voltage source. Unfortunately, this is not possible for filters constructed from passive elements. In Chapter 15, we will discover that active filters are insensitive to changes in source resistance and thus are better suited to designs in which this is an important issue.

Figure 14.27 summarizes the two *RLC* bandpass filters we have studied. Note that the expressions for the circuit transfer functions have the same form. As we have done previously, we can create a general form for the transfer functions of these two bandpass filters:

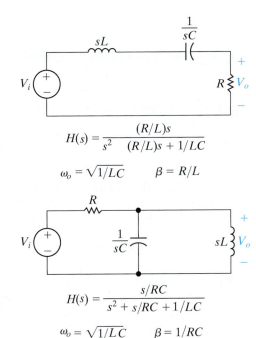

$$H(s) = \frac{(R/L)s}{s^2 + (R/L)s + 1/LC}$$

$$\omega_o = \sqrt{1/LC} \qquad \beta = R/L$$

$$H(s) = \frac{s/RC}{s^2 + s/RC + 1/LC}$$

$$\omega_o = \sqrt{1/LC} \qquad \beta = 1/RC$$

Figure 14.27 ▲ Two *RLC* bandpass filters, together with equations for the transfer function, center frequency, and bandwidth of each.

TRANSFER FUNCTION FOR BANDPASS FILTERS

$$H(s) = \frac{\beta s}{s^2 + \beta s + \omega_o^2}. \qquad (14.18)$$

Any circuit with the transfer function in Eq. 14.18 acts as a bandpass filter with a center frequency ω_o and a bandwidth β.

In Example 14.7, we saw that the transfer function can also be written in the form

$$H(s) = \frac{K\beta s}{s^2 + \beta s + \omega_o^2},$$

where the values for K and β depend on whether the series resistance of the voltage source is zero or nonzero.

Relating the Frequency Domain to the Time Domain

It should come as no surprise that the parameters characterizing the frequency response of *RLC* bandpass filters and the parameters characterizing the time response of *RLC* circuits are related. Consider the series *RLC* circuit in Fig. 14.19(a). In Chapter 8 we discovered that the natural response of this circuit is characterized by the neper frequency (α) and the resonant frequency (ω_o). These parameters were expressed in terms of the circuit components in Eqs. 8.29 and 8.30, which are repeated here for convenience:

$$\alpha = \frac{R}{2L},$$

$$\omega_o = \sqrt{\frac{1}{LC}}.$$

We see that the same parameter ω_o is used to characterize both the time response and the frequency response. That's why the center frequency is also called the resonant frequency. The bandwidth and the neper frequency are related by the equation

$$\beta = 2\alpha. \qquad (14.19)$$

Recall that the natural response of a series *RLC* circuit may be underdamped, overdamped, or critically damped. The transition from overdamped to underdamped occurs when $\omega_o^2 = \alpha^2$. Consider the relationship between α and β from Eq. 14.19 and the definition of the quality factor Q. The transition from an overdamped to an underdamped response occurs when $Q = 1/2$. Thus, a circuit whose frequency response contains a sharp peak at ω_o, indicating a high Q and a narrow bandwidth, will have an underdamped natural response. Conversely, a circuit whose frequency response has a broad bandwidth and a low Q will have an overdamped natural response.

◨ ASSESSMENT PROBLEMS

Objective 3—Know the *RLC* circuit configurations that act as bandpass filters

14.6 Using the circuit in Fig. 14.19(a), compute the values of R and L to give a bandpass filter with a center frequency of 12 kHz and a quality factor of 6. Use a 0.1 μF capacitor.

Answer: $L = 1.76$ mH, $R = 22.10\ \Omega$.

14.7 Using the circuit in Fig. 14.22, compute the values of L and C to give a bandpass filter with a center frequency of 2 kHz and a bandwidth of 500 Hz. Use a 250 Ω resistor.

Answer: $L = 4.97$ mH, $C = 1.27\ \mu$F.

14.8 Recalculate the component values for the circuit in Example 14.6(d) so that the frequency response of the resulting circuit is unchanged using a 0.2 μF capacitor.

Answer: $L = 5.07$ mH, $R = 3.98$ kΩ.

14.9 Recalculate the component values for the circuit in Example 14.6(d) so that the quality factor of the resulting circuit is unchanged but the center frequency has been moved to 2 kHz. Use a 0.2 μF capacitor.

Answer: $R = 9.95$ kΩ, $L = 31.66$ mH.

SELF-CHECK: Also try Chapter Problems 14.26 and 14.30.

14.5 Bandreject Filters

We turn now to the last of the four filter categories—the bandreject filter. This filter passes source voltages outside the band between the two cutoff frequencies to the output (the passband) and attenuates source voltages before they reach the output at frequencies between the two cutoff frequencies (the stopband). Bandpass filters and bandreject filters thus perform complementary functions in the frequency domain.

Bandreject filters and bandpass filters have the same characteristic parameters: the two cutoff frequencies, the center frequency, the bandwidth, and the quality factor. Again, only two of these five parameters can be specified independently.

We examine two circuits that function as bandreject filters and then derive equations relating the circuit component values to the characteristic parameters for each circuit.

The Series *RLC* Circuit—Qualitative Analysis

Figure 14.28(a) shows a series *RLC* circuit. Although the circuit components and connections are identical to those in the series *RLC* bandpass filter in Fig. 14.19(a), the output voltage is now defined across the inductor-capacitor pair. As we saw in the case of low- and high-pass filters, the same circuit may perform two different filtering functions, depending on the definition of the output voltage.

We have already noted that at $\omega = 0$, the inductor behaves like a short circuit and the capacitor behaves like an open circuit, as shown in Fig, 14.28(b). At $\omega = \infty$, these roles switch, as shown in Fig. 14.28(c). In both equivalent circuits, the output voltage is defined across an open circuit, so the output and input voltages have the same magnitude. This series *RLC* bandreject filter circuit then has two passbands—one below the lower cutoff frequency and the other above the upper cutoff frequency.

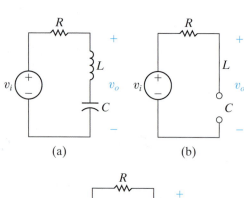

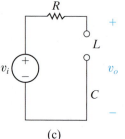

Figure 14.28 ▲ (a) A series *RLC* bandreject filter. (b) The equivalent circuit for $\omega = 0$. (c) The equivalent circuit for $\omega = \infty$.

Between these two passbands, both the inductor and the capacitor have finite impedances of opposite signs. As the frequency is increased from zero, the inductor impedance increases and the capacitor impedance decreases. Therefore, the phase shift between the input and the output approaches $-90°$ as ωL approaches $1/\omega C$. As soon as ωL exceeds $1/\omega C$, the phase shift jumps to $+90°$ and then approaches zero as ω continues to increase.

At some frequency between the two passbands, the impedances of the inductor and capacitor are equal but have opposite signs, so their sum is zero. Thus, at this frequency, the series combination of the inductor and capacitor acts like a short circuit, and the output voltage magnitude is zero. This is the center frequency of the series RLC bandreject filter.

Figure 14.29 presents a sketch of the frequency response for the series RLC bandreject filter from Fig. 14.28(a). Note that the magnitude plot is overlaid with that of the ideal bandreject filter from Fig. 14.3(b). Our qualitative analysis has confirmed the shape of the magnitude and phase angle plots. We now turn to a quantitative analysis of the circuit to confirm this frequency response and to compute values for the parameters that characterize this response.

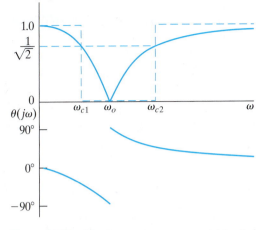

Figure 14.29 ▲ The frequency response plot for the series RLC bandreject filter circuit in Fig. 14.28(a).

The Series *RLC* Circuit—Quantitative Analysis

After transforming to the s domain, as shown in Fig. 14.30, we use voltage division to find the transfer function equation:

$$H(s) = \frac{sL + \dfrac{1}{sC}}{R + sL + \dfrac{1}{sC}} = \frac{s^2 + \dfrac{1}{LC}}{s^2 + \dfrac{R}{L}s + \dfrac{1}{LC}}.$$

Substituting $j\omega$ for s in $H(s)$, we generate equations for the transfer function magnitude and the phase angle:

$$|H(j\omega)| = \frac{\left|\dfrac{1}{LC} - \omega^2\right|}{\sqrt{\left(\dfrac{1}{LC} - \omega^2\right)^2 + \left(\dfrac{\omega R}{L}\right)^2}},$$

$$\theta(j\omega) = -\tan^{-1}\left(\frac{\dfrac{\omega R}{L}}{\dfrac{1}{LC} - \omega^2}\right).$$

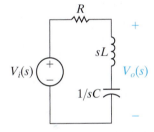

Figure 14.30 ▲ The s-domain equivalent of the circuit in Fig. 14.28(a).

Note that the equations for the transfer function magnitude and phase angle confirm the frequency response shape pictured in Fig. 14.29, which we developed from the qualitative analysis.

We use the circuit in Fig. 14.30 to calculate the center frequency. For the bandreject filter, the center frequency is still defined as the frequency for which the sum of the impedances of the capacitor and inductor is zero. In the bandpass filter, the magnitude at the center frequency was a maximum, but in the bandreject filter, this magnitude is a minimum. This is because in the bandreject filter, the center frequency is in the stopband, not in the passband. Because the sum of the capacitor and inductor impedances is zero at the center frequency,

$$\omega_o = \sqrt{\frac{1}{LC}}.$$

Substituting $\sqrt{1/LC}$ for ω_o in the equation for the transfer function magnitude shows that $|H(j\omega_o)| = 0$.

The cutoff frequencies, the bandwidth, and the quality factor are defined and calculated for the bandreject filter and the bandpass filter in exactly the same way. Note that for the bandreject filter, $H_{max} = |H(j0)| = |H(j\infty)|$, and for the series RLC bandreject filter in Fig. 14.28(a), $H_{max} = 1$. Thus,

$$\omega_{c1} = -\frac{R}{2L} + \sqrt{\left(\frac{R}{2L}\right)^2 + \frac{1}{LC}},$$

$$\omega_{c2} = \frac{R}{2L} + \sqrt{\left(\frac{R}{2L}\right)^2 + \frac{1}{LC}}.$$

These equations are the same as Eqs. 14.14 and 14.15.

Use the cutoff frequencies to generate an expression for the bandwidth, β:

$$\beta = R/L.$$

This equation is the same as Eq. 14.16.

Finally, the center frequency and the bandwidth produce an equation for the quality factor, Q:

$$Q = \sqrt{\frac{L}{R^2 C}}.$$

This equation and Eq. 14.17 are the same.

Again, we can represent the expressions for the two cutoff frequencies in terms of the bandwidth and center frequency, as we did for the bandpass filter:

$$\omega_{c1} = -\frac{\beta}{2} + \sqrt{\left(\frac{\beta}{2}\right)^2 + \omega_o^2},$$

$$\omega_{c2} = \frac{\beta}{2} + \sqrt{\left(\frac{\beta}{2}\right)^2 + \omega_o^2}.$$

Alternative forms for these equations express the cutoff frequencies in terms of the quality factor and the center frequency:

$$\omega_{c1} = \omega_o \cdot \left[-\frac{1}{2Q} + \sqrt{1 + \left(\frac{1}{2Q}\right)^2} \right],$$

$$\omega_{c2} = \omega_o \cdot \left[\frac{1}{2Q} + \sqrt{1 + \left(\frac{1}{2Q}\right)^2} \right].$$

Example 14.8 presents the design of a series RLC bandreject filter.

| EXAMPLE 14.8 | Designing a Series *RLC* Bandreject Filter |

Using the series RLC circuit in Fig. 14.28(a), compute the component values that yield a bandreject filter with a bandwidth of 250 Hz and a center frequency of 750 Hz. Use a 100 nF capacitor. Compute values for R, L, ω_{c1}, ω_{c2}, and Q.

Solution

We begin by using the definition of quality factor to compute its value for this filter:

$$Q = \omega_o/\beta = 3.$$

Use Eq. 14.13 to compute L, remembering to convert ω_o to radians per second:

$$L = \frac{1}{\omega_o^2 C} = \frac{1}{[2\pi(750)]^2(100 \times 10^{-9})} = 450 \text{ mH}.$$

Use Eq. 14.16 to calculate R:

$$R = \beta L = 2\pi(250)(450 \times 10^{-3}) = 707 \ \Omega.$$

The values for the center frequency and bandwidth can be used to compute the two cutoff frequencies:

$$\omega_{c1} = -\frac{\beta}{2} + \sqrt{\left(\frac{\beta}{2}\right)^2 + \omega_o^2} = 3992.0 \text{ rad/s,}$$

$$\omega_{c2} = \frac{\beta}{2} + \sqrt{\left(\frac{\beta}{2}\right)^2 + \omega_o^2} = 5562.8 \text{ rad/s.}$$

The cutoff frequencies are at 635.3 Hz and 885.3 Hz. Their difference is $885.3 - 635.3 = 250$ Hz, confirming the specified bandwidth. The geometric mean is $\sqrt{(635.3)(885.3)} = 750$ Hz, confirming the specified center frequency.

As you might suspect by now, another configuration that produces a bandreject filter is a parallel RLC circuit. The analysis details of the parallel RLC circuit are left to Problem 14.40, and the results are summarized in Fig. 14.31, along with the series RLC bandreject filter. As we did for other categories of filters, we can state a general form for the transfer functions of bandreject filters, replacing the constant terms with β and ω_o:

TRANSFER FUNCTION FOR BANDREJECT FILTERS

$$H(s) = \frac{s^2 + \omega_o^2}{s^2 + \beta s + \omega_o^2}. \tag{14.20}$$

Equation 14.20 is useful in filter design because any circuit with a transfer function in this form can be used as a bandreject filter.

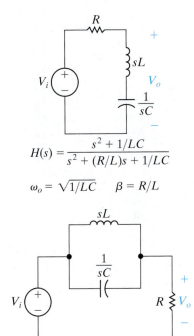

$$H(s) = \frac{s^2 + 1/LC}{s^2 + (R/L)s + 1/LC}$$

$$\omega_o = \sqrt{1/LC} \qquad \beta = R/L$$

$$H(s) = \frac{s^2 + 1/LC}{s^2 + s/RC + 1/LC}$$

$$\omega_o = \sqrt{1/LC} \qquad \beta = 1/RC$$

Figure 14.31 ▲ Two RLC bandreject filters, together with equations for the transfer function, center frequency, and bandwidth of each.

◼ ASSESSMENT PROBLEMS

Objective 4—Know the *RLC* circuit configurations that act as bandreject filters

14.10 Design the component values for the series RLC bandreject filter shown in Fig. 14.28(a) so that the center frequency is 4 kHz and the quality factor is 5. Use a 500 nF capacitor.

Answer: $L = 3.17$ mH,
$\quad\quad\quad R = 15.92\ \Omega.$

14.11 Recompute the component values for Assessment Problem 14.10 to achieve a bandreject filter with a center frequency of 20 kHz. The filter has a 100 Ω resistor. The quality factor remains at 5.

Answer: $L = 3.98$ mH,
$\quad\quad\quad C = 15.92$ nF.

SELF-CHECK: Also try Chapter Problems 14.41 and 14.42.

■ Practical Perspective

Pushbutton Telephone Circuits

In the Practical Perspective at the start of this chapter, we described the dual-tone-multiple-frequency (DTMF) system used to signal that a button has been pushed on a pushbutton telephone. A key element of the DTMF system is the DTMF receiver—a circuit that decodes the tones produced by pushing a button and determines which button was pushed.

In order to design a DTMF receiver, we need a better understanding of the DTMF system. As you can see from Fig. 14.32, the buttons on the telephone are organized into rows and columns. The pair of tones generated by pushing a button depends on the button's row and column. The button's row determines its low-frequency tone, and the button's column determines its high-frequency tone.[1] For example, pressing the "6" button produces sinusoidal tones with the frequencies 770 Hz and 1477 Hz.

At the telephone switching facility, bandpass filters in the DTMF receiver first detect whether tones from both the low-frequency and high-frequency groups are simultaneously present. This test rejects many extraneous audio signals that are not DTMF. If tones are present in both bands, other filters are used to select among the possible tones in each band so that the frequencies can be decoded, identifying a unique button. Additional tests are performed to prevent false button detection. For example, only one tone per frequency band is allowed; the high- and low-band frequencies must start and stop within a few milliseconds of one another to be considered valid; and the high- and low-band signal amplitudes must be sufficiently close to each other.

You may wonder why bandpass filters are used instead of a high-pass filter for the high-frequency group of DTMF tones and a low-pass filter for the low-frequency group of DTMF tones. The reason is that the telephone system uses frequencies outside of the $300 - 3\,\text{kHz}$ band for other signaling purposes, such as ringing the phone's bell. Bandpass filters prevent the DTMF receiver from erroneously detecting these other signals.

SELF-CHECK: Assess your understanding of this Practical Perspective by trying Chapter Problems 14.51–14.53.

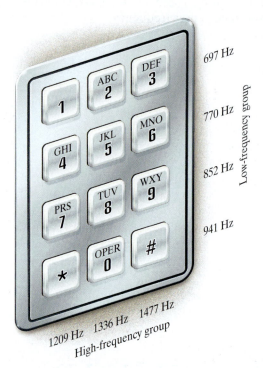

Figure 14.32 ▲ Tones generated by the rows and columns of telephone pushbuttons.

■ Summary

- A **frequency selective circuit**, or **filter**, enables signals at certain frequencies to reach the output, and it attenuates signals at other frequencies to prevent them from reaching the output. The **passband** contains the frequencies of those signals that are passed; the **stopband** contains the frequencies of those signals that are attenuated. (See page 538.)

- The **cutoff frequency**, ω_c, separates frequencies in the stopband from frequencies in the passband. At the cutoff frequency, the magnitude of the transfer function equals $(1/\sqrt{2})H_{\text{max}}$. (See page 540.)

- A **low-pass filter** passes voltages at frequencies below ω_c and attenuates frequencies above ω_c. Any circuit with the transfer function

$$H(s) = \frac{\omega_c}{s + \omega_c}$$

functions as a low-pass filter. (See page 545.)

- A **high-pass filter** passes voltages at frequencies above ω_c and attenuates voltages at frequencies below ω_c. Any circuit with the transfer function

---

[1]A fourth high-frequency tone is reserved at 1633 Hz. This tone is used infrequently and is not produced by a standard 12-button telephone.

$$H(s) = \frac{s}{s + \omega_c}$$

functions as a high-pass filter. (See page 549.)

- Bandpass filters and bandreject filters each have two cutoff frequencies, ω_{c1} and ω_{c2}. These filters are further characterized by their **center frequency** (ω_o), **bandwidth** (β), and **quality factor** (Q). These quantities are defined as

$$\omega_o = \sqrt{\omega_{c1} \cdot \omega_{c2}},$$

$$\beta = \omega_{c2} - \omega_{c1},$$

$$Q = \omega_o / \beta.$$

(See pages 553 and 554.)

- A **bandpass filter** passes voltages at frequencies within the passband, which is between ω_{c1} and ω_{c2}. It attenuates frequencies outside of the passband. Any circuit with the transfer function

$$H(s) = \frac{\beta s}{s^2 + \beta s + \omega_o^2}$$

functions as a bandpass filter. (See page 559.)

- A **bandreject filter** attenuates voltages at frequencies within the stopband, which is between ω_{c1} and ω_{c2}. It passes frequencies outside of the stopband. Any circuit with the transfer function

$$H(s) = \frac{s^2 + \omega_o^2}{s^2 + \beta s + \omega_o^2}$$

functions as a bandreject filter. (See page 563.)

- Adding a load to the output of a passive filter changes its filtering properties by altering the location and magnitude of the passband. Replacing an ideal voltage source with one whose source resistance is nonzero also changes the filtering properties of the rest of the circuit, again by altering the location and magnitude of the passband. (See page 548 and 557.)

Problems

Section 14.2

14.1 a) Find the cutoff frequency in hertz for the RL filter shown in Fig. P14.1.

b) Calculate $H(j\omega)$ at ω_c, $0.2\omega_c$, and $5\omega_c$.

c) If $v_i = 10 \cos \omega t$ V, write the steady-state expression for v_o when $\omega = \omega_c$, $\omega = 0.2\omega_c$, and $\omega = 5\omega_c$.

Figure P14.1

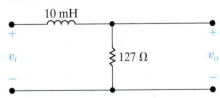

14.2 Consider the low-pass filter in Fig. P14.2, which has a load resistor R_L.

a) What is the transfer function of the unloaded filter?

b) What is the transfer function of the loaded filter?

c) Compare the transfer function of the unloaded filter in part (a) and the transfer function of the loaded filter in part (b). Are the cutoff frequencies different? Are the passband gains different?

d) If $R = 330 \, \Omega$ and $L = 10$ mH, what is the cutoff frequency of the unloaded filter in rad/s?

e) What is the smallest value of load resistance that can be used with the filter components in part (d) so that the cutoff frequency of the resulting filter is no more than 5% different from the unloaded filter?

Figure P14.2

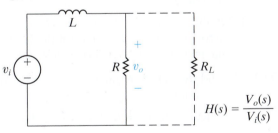

14.3 Use a 5 mH inductor to design a low-pass, RL, passive filter with a cutoff frequency of 1 kHz.

DESIGN
PROBLEM

PSPICE

MULTISIM

a) Specify the value of the resistor.

b) A load having a resistance of 270 Ω is connected across the output terminals of the filter. What is the corner, or cutoff, frequency of the loaded filter in hertz?

c) If you must use a single resistor from Appendix H for part (a), what resistor should you use? What is the resulting cutoff frequency of the filter?

14.4 A resistor, denoted as R_l, is added in series with the inductor in the circuit in Fig. 14.4(a). The new low-pass filter circuit is shown in Fig. P14.4.

a) Derive the expression for $H(s)$ where $H(s) = V_o/V_i$.

b) At what frequency will the magnitude of $H(j\omega)$ be maximum?

c) What is the maximum value of the magnitude of $H(j\omega)$?

d) At what frequency will the magnitude of $H(j\omega)$ equal its maximum value divided by $\sqrt{2}$?

e) Assume a resistance of $75\,\Omega$ is added in series with the 10 mH inductor in the circuit in Fig. P14.1. Find $\omega_c, H(j0), H(j\omega_c), H(j0.2\omega_c)$, and $H(j5\omega_c)$.

Figure P14.4

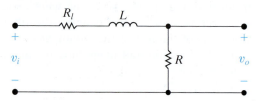

14.5 Use a $250\,\Omega$ resistor to design a low-pass passive filter with a cutoff frequency of 8 krad/s.

a) Specify the value of the filter's inductor.

b) A load resistor of $1\,k\Omega$ is connected across the output terminals of the filter. What is the cutoff frequency of the loaded filter, in rad/s?

14.6 Consider the low-pass filter designed in Problem 14.5.

a) Assume the cutoff frequency cannot decrease by more than 10% from the specified value, 8 krad/s. What is the smallest value of load resistance that can be connected across the output terminals of the filter?

b) If the resistor found in part (a) is connected across the output terminals of the filter, what is the magnitude of $H(j\omega)$ when $\omega = 0$?

14.7 a) Find the cutoff frequency (in hertz) of the low-pass filter shown in Fig. P14.7.

b) Calculate $H(j\omega)$ at $\omega_c, 0.1\omega_c$, and $10\omega_c$.

c) If $v_i = 200 \cos \omega t$ mV, write the steady-state expression for v_o when $\omega = \omega_c, 0.1\omega_c$, and $10\omega_c$.

Figure P14.7

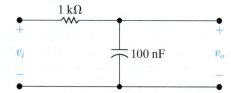

14.8 A resistor denoted as R_L is connected in parallel with the capacitor in the circuit in Fig. 14.7. The loaded low-pass filter circuit is shown in Fig. P14.8.

a) Derive the expression for the voltage transfer function V_o/V_i.

b) At what frequency will the magnitude of $H(j\omega)$ be maximum?

c) What is the maximum value of the magnitude of $H(j\omega)$?

d) At what frequency will the magnitude of $H(j\omega)$ equal its maximum value divided by $\sqrt{2}$?

e) Assume a resistance of $10\,k\Omega$ is added in parallel with the 100 nF capacitor in the circuit in Fig. P14.7. Find $\omega_c, H(j0), H(j\omega_c), H(j0.1\omega_c)$, and $H(j10\omega_c)$.

Figure P14.8

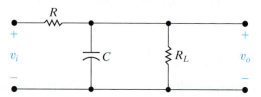

14.9 Use a 500 nF capacitor to design a low-pass passive filter with a cutoff frequency of 50 krad/s.

a) Specify the cutoff frequency in hertz.

b) Specify the value of the filter resistor.

c) Assume the cutoff frequency cannot increase by more than 5%. What is the smallest value of load resistance that can be connected across the output terminals of the filter?

d) If the resistor found in (c) is connected across the output terminals, what is the magnitude of $H(j\omega)$ when $\omega = 0$?

14.10 Design a passive RC low pass filter (see Fig. 14.7) with a cutoff frequency of 100 Hz using a $4.7\,\mu F$ capacitor.

a) What is the cutoff frequency in rad/s?

b) What is the value of the resistor?

c) Draw your circuit, labeling the component values and output voltage.

d) What is the transfer function of the filter in part (c)?

e) If the filter in part (c) is loaded with a resistor whose value is the same as the resistor part (b), what is the transfer function of this loaded filter?

f) What is the cutoff frequency of the loaded filter from part (e)?

g) What is the gain in the pass band of the loaded filter from part (e)?

Section 14.3

14.11 Consider the circuit shown in Fig. P14.11.

a) What is the transfer function, $H(s) = V_o(s)/V_i(s)$, of this filter?

b) What is the cutoff frequency of this filter?

c) What is the magnitude of the filter's transfer function at $s = j\omega_c$?

Figure P14.11

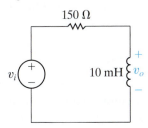

150 Ω

v_i 10 mH v_o

14.12 Suppose a 150 Ω load resistor is attached to the filter in Fig. P14.11.

a) What is the transfer function, $H(s) = V_o(s)/V_i(s)$, of this filter?

b) What is the cutoff frequency of this filter?

c) How does the cutoff frequency of the loaded filter compare with the cutoff frequency of the unloaded filter in Fig. P14.11?

d) What else is different for these two filters?

14.13 Using a 5 mH inductor, design a high-pass, RL, passive filter with a cutoff frequency of 25 krad/s.

a) Specify the value of the resistance.

b) Assume the filter is connected to a pure resistive load. The cutoff frequency is not to drop below 24 krad/s. What is the smallest load resistor that can be connected across the output terminals of the filter?

14.14 Design a passive RC high pass filter (see Fig. 14.10[a]) with a cutoff frequency of 500 Hz using a 220 pF capacitor.

a) What is the cutoff frequency in rad/s?

b) What is the value of the resistor?

c) Draw your circuit, labeling the component values and output voltage.

d) What is the transfer function of the filter in part (c)?

e) If the filter in part (c) is loaded with a resistor whose value is the same as the resistor in (b), what is the transfer function of this loaded filter?

f) What is the cutoff frequency of the loaded filter from part (e)?

g) What is the gain in the pass band of the loaded filter from part (e)?

14.15 a) Find the cutoff frequency (in hertz) for the high-pass filter shown in Fig. P14.15.

b) Find $H(j\omega)$ at $\omega_c, 0.2\omega_c$, and $5\omega_c$.

c) If $v_1 = 0.5 \cos \omega t$ V, write the steady-state expression for v_o when $\omega = \omega_c$, $\omega = 0.2\omega_c$, and $\omega = 5\omega_c$.

Figure P14.15

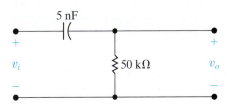

5 nF

v_i 50 kΩ v_o

14.16 A resistor, R_c, is connected in series with the capacitor in the circuit in Fig. 14.10(a). The new high-pass filter circuit is shown in Fig. P14.16.

a) Derive the expression for $H(s)$ where $H(s) = V_o/V_i$.

b) At what frequency will the magnitude of $H(j\omega)$ be maximum?

c) What is the maximum value of the magnitude of $H(j\omega)$?

d) At what frequency will the magnitude of $H(j\omega)$ equal its maximum value divided by $\sqrt{2}$?

e) Assume a resistance of 12.5 kΩ is connected in series with the 5 nF capacitor in the circuit in Fig. P14.15. Calculate $\omega_c, H(j\omega_c), H(j0.2\omega_c)$, and $H(j5\omega_c)$.

Figure P14.16

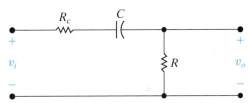

R_c C

v_i R v_o

14.17 Using a 100 nF capacitor, design a high-pass passive filter with a cutoff frequency of 300 Hz.

a) Specify the value of R in kilohms.

b) A 47 kΩ resistor is connected across the output terminals of the filter. What is the cutoff frequency, in hertz, of the loaded filter?

Section 14.4

14.18 a) Using the relationship between the bandwidth and the cutoff frequencies and the relationship between the center frequency and the cutoff frequencies, show that

$$\omega_{c1} = -\frac{\beta}{2} + \sqrt{\left(\frac{\beta}{2}\right)^2 + \omega_o^2}; \quad \omega_{c2} = \frac{\beta}{2} + \sqrt{\left(\frac{\beta}{2}\right)^2 + \omega_o^2}.$$

b) Using the expressions for the two cutoff frequencies in terms of the center frequency and the bandwidth given in part (a), show that

$$\omega_{c1} = \omega_o \cdot \left[-\frac{1}{2Q} + \sqrt{1 + \left(\frac{1}{2Q}\right)^2} \right],$$

$$\omega_{c2} = \omega_o \cdot \left[\frac{1}{2Q} + \sqrt{1 + \left(\frac{1}{2Q}\right)^2} \right].$$

14.19 A bandpass filter has a center, or resonant, frequency of 50 krad/s and a quality factor of 4. Find the bandwidth, the upper cutoff frequency, and the lower cutoff frequency. Express all answers in kilohertz.

14.20 Calculate the center frequency, the bandwidth, and the quality factor of a bandpass filter that has an upper cutoff frequency of 121 krad/s and a lower cutoff frequency of 100 krad/s.

14.21 Design a series RLC bandpass filter (see Fig. 14.19[a]) with a quality of 2 and a center frequency of 8 kHz, using a 5 nF capacitor.

  a) Draw your circuit, labeling the component values and output voltage.

  b) For the filter in part (a), calculate the bandwidth and the values of the two cutoff frequencies.

14.22 The input to the series RLC bandpass filter designed in Problem 14.21 is $20 \cos \omega t$ V. Find the voltage drop across the resistor when (a) $\omega = \omega_o$; (b) $\omega = \omega_{c1}$; (c) $\omega = \omega_{c2}$; (d) $\omega = 0.1\omega_o$; (e) $\omega = 10\omega_o$.

14.23 The input to the series RLC bandpass filter designed in Problem 14.21 is $20 \cos \omega t$ V. Find the voltage drop across the series combination of the inductor and capacitor when (a) $\omega = \omega_o$; (b) $\omega = \omega_{c1}$; (c) $\omega = \omega_{c2}$; (d) $\omega = 0.1\omega_o$; (e) $\omega = 10\omega_o$.

14.24 For the bandpass filter shown in Fig. P14.24, calculate the following: (a) f_o; (b) Q; (c) f_{c1}; (d) f_{c2}; and (e) β.

Figure P14.24

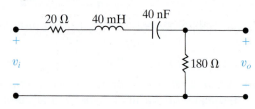

14.25 The input voltage in the circuit in Fig. P14.24 is $0.5 \cos \omega t$ V. Calculate the output voltage when (a) $\omega = \omega_o$; (b) $\omega = \omega_{c1}$; and (c) $\omega = \omega_{c2}$.

14.26 Using a 50 nF capacitor in the bandpass circuit shown in Fig. 14.22, design a filter with a quality factor of 5 and a center frequency of 20 krad/s.

  a) Specify the numerical values of R and L.

  b) Calculate the upper and lower cutoff frequencies in kilohertz.

  c) Calculate the bandwidth in hertz.

14.27 Design a series RLC bandpass filter using only three components from Appendix H that comes closest to meeting the filter specifications in Problem 14.26.

  a) Draw your filter, labeling all component values and the input and output voltages.

  b) Calculate the percent error in this new filter's center frequency and quality factor when compared to the values specified in Problem 14.26.

14.28 Use a 5 nF capacitor to design a series RLC bandpass filter, as shown at the top of Fig. 14.27. The center frequency of the filter is 8 kHz, and the quality factor is 2.

  a) Specify the values of R and L.

  b) What is the lower cutoff frequency in kilohertz?

  c) What is the upper cutoff frequency in kilohertz?

  d) What is the bandwidth of the filter in kilohertz?

14.29 Design a series RLC bandpass filter using only three components from Appendix H that comes closest to meeting the filter specifications in Problem 14.28.

  a) Draw your filter, labeling all component values and the input and output voltages.

  b) Calculate the percent error in this new filter's center frequency and quality factor when compared to the values specified in Problem 14.28.

14.30 For the bandpass filter shown in Fig. P14.30, find (a) ω_o, (b) f_o, (c) Q, (d) ω_{c1}, (e) f_{c1}, (f) ω_{c2}, (g) f_{c2}, and (h) β.

Figure P14.30

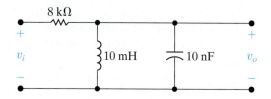

14.31 Consider the circuit shown in Fig. P14.31.

  a) Find ω_o.

  b) Find β.

  c) Find Q.

  d) Find the steady-state expression for v_o when $v_i = 250 \cos \omega_o t$ mV.

  e) Show that if R_L is expressed in kilohms the Q of the circuit in Fig. P14.31 is

$$Q = \frac{20}{1 + 100/R_L}.$$

  f) Plot Q versus R_L for $20\,\text{k}\Omega \leq R_L \leq 2\,\text{M}\Omega$.

Figure P14.31

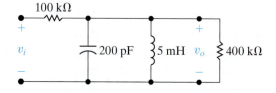

14.32 A block diagram of a system consisting of a sinusoidal voltage source, a series RLC bandpass filter, and a load is shown in Fig. P14.32. The internal impedance of the sinusoidal source is $80 + j0\,\Omega$, and the impedance of the load is $480 + j0\,\Omega$.

The RLC series bandpass filter has a 20 nF capacitor, a center frequency of 50 krad/s, and a quality factor of 6.25.

a) Draw a circuit diagram of the system.

b) Specify the numerical values of L and R for the filter section of the system.

c) What is the quality factor of the interconnected system?

d) What is the bandwidth (in hertz) of the interconnected system?

Figure P14.32

14.33 The purpose of this problem is to investigate how a resistive load connected across the output terminals of the bandpass filter shown in Fig. 14.19 affects the quality factor and hence the bandwidth of the filtering system. The loaded filter circuit is shown in Fig. P14.33.

a) Calculate the transfer function V_o/V_i for the circuit shown in Fig. P14.33.

b) What is the expression for the bandwidth of the system?

c) What is the expression for the loaded bandwidth (β_L) as a function of the unloaded bandwidth (β_U)?

d) What is the expression for the quality factor of the system?

e) What is the expression for the loaded quality factor (Q_L) as a function of the unloaded quality factor (Q_U)?

f) What are the expressions for the cutoff frequencies ω_{c1} and ω_{c2}?

Figure P14.33

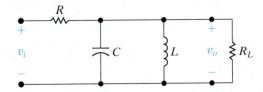

14.34 The parameters in the circuit in Fig. P14.33 are $R = 2.4\,k\Omega$, $C = 50$ pF, and $L = 2\,\mu H$. The quality factor of the circuit is not to drop below 7.5. What is the smallest permissible value of the load resistor R_L?

Section 14.5

14.35 For the bandreject filter in Fig. P14.35, calculate (a) ω_o; (b) f_o; (c) Q; (d) β in hertz; (e) ω_{c1}; (f) f_{c1}; (g) ω_{c2}; and (h) f_{c2}.

Figure P14.35

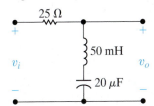

14.36 For the bandreject filter in Fig. P14.35,

a) Find $H(j\omega)$ at $\omega_o, \omega_{c1}, \omega_{c2}, 0.125\omega_o$, and $8\omega_o$.

b) If $v_i = 5\cos\omega t$ V, write the steady-state expression for v_o when $\omega = \omega_o$, $\omega = \omega_{c1}$, $\omega = \omega_{c2}$, $\omega = 0.125\omega_o$, and $\omega = 8\omega_o$.

14.37 Design an RLC bandreject filter (see Fig. 14.28[a]) with a quality of 5 and a center frequency of 20 krad/s, using a 50 nF capacitor.

a) Draw your circuit, labeling the component values and output voltage.

b) For the filter in part (a), calculate the bandwidth and the values of the two cutoff frequencies.

14.38 The input to the RLC bandreject filter designed in Problem 14.37 is $50\cos\omega t$ V. Find the voltage drop across the series combination of the inductor and capacitor when (a) $\omega = \omega_o$; (b) $\omega = \omega_{c1}$; (c) $\omega = \omega_{c2}$; (d) $\omega = 0.1\omega_o$; (e) $\omega = 10\omega_o$.

14.39 The input to the RLC bandreject filter designed in Problem 14.37 is $50\cos\omega t$ V. Find the voltage drop across the resistor when (a) $\omega = \omega_o$; (b) $\omega = \omega_{c1}$; (c) $\omega = \omega_{c2}$; (d) $\omega = 0.1\omega_o$; (e) $\omega = 10\omega_o$.

14.40 a) Show (via a qualitative analysis) that the circuit in Fig. P14.40 is a bandreject filter.

b) Support the qualitative analysis of (a) by finding the voltage transfer function of the filter.

c) Derive the expression for the center frequency of the filter.

d) Derive the expressions for the cutoff frequencies ω_{c1} and ω_{c2}.

e) What is the expression for the bandwidth of the filter?

f) What is the expression for the quality factor of the circuit?

Figure P14.40

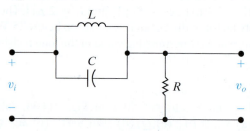

14.41 For the bandreject filter in Fig. P14.41, calculate
PSPICE (a) ω_o; (b) f_o; (c) Q; (d) ω_{c1}; (e) f_{c1}; (f) ω_{c2}; (g) f_{c2}; and
MULTISIM (h) β in kilohertz.

Figure P14.41

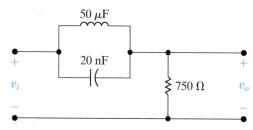

14.42 Use a 500 nF capacitor to design a bandreject filter,
DESIGN as shown in Fig. P14.42. The filter has a center fre-
PROBLEM quency of 4 kHz and a quality factor of 5.

PSPICE
MULTISIM a) Specify the numerical values of R and L.

b) Calculate the upper and lower corner, or cutoff, frequencies in kilohertz.

c) Calculate the filter bandwidth in kilohertz.

Figure P14.42

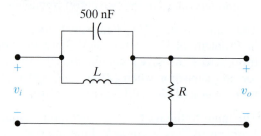

14.43 Assume the bandreject filter in Problem 14.42 is
PSPICE loaded with a $1\,k\Omega$ resistor.
MULTISIM

a) What is the quality factor of the loaded circuit?

b) What is the bandwidth (in kilohertz) of the loaded circuit?

c) What is the upper cutoff frequency in kilohertz?

d) What is the lower cutoff frequency in kilohertz?

14.44 Design a parallel RLC bandreject filter using only three components from Appendix H that comes closest to meeting the filter specifications in Problem 14.42.

a) Draw your filter, labeling all component values and the input and output voltages.

b) Calculate the percent error in this new filter's center frequency and quality factor when compared to the values specified in Problem 14.42.

14.45 The purpose of this problem is to investigate how a resistive load connected across the output terminals of the bandreject filter shown in Fig. 14.28(a) affects the behavior of the filter. The loaded filter circuit is shown in Fig. P14.45.

a) Find the voltage transfer function V_o/V_i.

b) What is the expression for the center frequency?

c) What is the expression for the bandwidth?

d) What is the expression for the quality factor?

e) Evaluate $H(j\omega_o)$.

f) Evaluate $H(j0)$.

g) Evaluate $H(j\infty)$.

h) What are the expressions for the corner frequencies ω_{c1} and ω_{c2}?

Figure P14.45

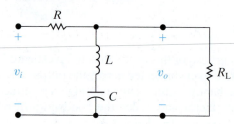

14.46 The parameters in the circuit in Fig. P14.45 are
PSPICE $R = 30\,\Omega$, $L = 1\,\mu H$, $C = 4\,pF$, and $R_L = 150\,\Omega$.
MULTISIM

a) Find ω_o, β (in kilohertz), and Q.

b) Find $H(j0)$ and $H(j\infty)$.

c) Find f_{c2} and f_{c1}.

d) Show that if R_L is expressed in ohms the Q of the circuit is

$$Q = \frac{50}{3}[1 + (30/R_L)].$$

e) Plot Q versus R_L for $10\,\Omega \le R_L \le 300\,\Omega$.

14.47 A $500\,\Omega$ load is added to the bandreject filter shown
PSPICE in Fig. P14.42. The center frequency of the filter is
MULTISIM $25\,krad/s$, and the capacitor is 25 nF. At very low and very high frequencies, the amplitude of the sinusoidal output voltage should be at least 90% of the amplitude of the sinusoidal input voltage.

a) Specify the numerical values of R and L.

b) What is the quality factor of the circuit?

Sections 14.1–14.5

14.48 Consider the following voltage transfer function:

$$H(s) = \frac{V_o}{V_i}$$

$$= \frac{10^{10}}{s^2 + 50,000 + 10^{10}}.$$

a) At what frequencies (in radians per second) is the magnitude of the transfer function equal to unity?

b) At what frequency is the magnitude of the transfer function maximum?

c) What is the maximum value of the transfer function magnitude?

14.49 Consider the series RLC circuit shown in Fig. P14.49. When the output is the voltage across the resistor, we know this circuit is a bandpass filter. When the output is the voltage across the series combination of the inductor and capacitor, we know this circuit is a bandreject filter. This problem investigates the behavior of this circuit when the output is across the inductor.

a) Find the transfer function, $H(s) = V_o(s)/V_i(s)$ when $V_o(s)$ is the voltage across the inductor.

b) Find the magnitude of the transfer function in part (a) for very low frequencies.

c) Find the magnitude of the transfer function in part (a) for very high frequencies.

d) Based on your answers in parts (b) and (c), what type of filter is this?

e) Suppose $R = 600\,\Omega$, $L = 400\,\text{mH}$, $C = 2.5\,\mu\text{F}$. Calculate the cutoff frequency of this filter, that is, the frequency at which the magnitude of the transfer function is $1/\sqrt{2}$.

Figure P14.49

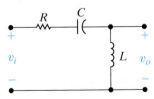

14.50 Repeat Problem 14.49 for the circuit shown in Fig. P14.50. Note that the output voltage is now the voltage across the capacitor.

Figure P14.50

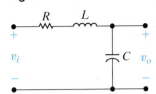

14.51 Design a series RLC bandpass filter (see Fig. 14.27)
PRACTICAL
PERSPECTIVE
for detecting the low-frequency tone generated by pushing a telephone button as shown in Fig. 14.32.

DESIGN
PROBLEM
a) Calculate the values of L and C that place the cutoff frequencies at the edges of the DTMF low-frequency band. Note that the resistance in standard telephone circuits is always $R = 600\,\Omega$.

b) What is the output amplitude of this circuit at each of the low-band frequencies, relative to the peak amplitude of the bandpass filter?

c) What is the output amplitude of this circuit at the lowest of the high-band frequencies?

14.52 Design a DTMF high-band bandpass filter similar to
PRACTICAL
PERSPECTIVE
the low-band filter design in Problem 14.51. Be sure to include the fourth high-frequency tone, 1633 Hz,
DESIGN
PROBLEM
in your design. What is the response amplitude of your filter to the highest of the low-frequency DTMF tones?

14.53 The 20 Hz signal that rings a telephone's bell has
PRACTICAL
PERSPECTIVE
to have a very large amplitude to produce a loud enough bell signal. How much larger can the ringing
DESIGN
PROBLEM
signal amplitude be, relative to the low-band DTMF signal, so that the response of the filter in Problem 14.51 is no more than half as large as the largest of the DTMF tones?

CHAPTER

15

CHAPTER CONTENTS

15.1 **First-Order Low-Pass and High-Pass Filters** *p. 574*

15.2 **Scaling** *p. 577*

15.3 **Op Amp Bandpass and Bandreject Filters** *p. 580*

15.4 **Higher-Order Op Amp Filters** *p. 587*

15.5 **Narrowband Bandpass and Bandreject Filters** *p. 600*

CHAPTER OBJECTIVES

1 Know the op amp circuits that behave as first-order low-pass and high-pass filters and be able to calculate component values for these circuits to meet specifications of cutoff frequency and passband gain.

2 Be able to design filter circuits starting with a prototype circuit and use scaling to achieve desired frequency response characteristics and component values.

3 Understand how to use cascaded first- and second-order Butterworth filters to implement low-pass, high-pass, bandpass, and bandreject filters of any order.

4 Be able to use the design equations to calculate component values for prototype narrowband bandpass, and narrowband bandreject filters to meet desired filter specifications.

Active Filter Circuits

Active filter circuits, which employ op amps, can be configured as low-pass filters, high-pass filters, bandpass filters and bandreject filters, just like the passive circuits we analyzed in Chapter 14. Active filter circuits have several advantages over their passive counterparts.

- Active circuits can produce bandpass and bandreject filters without using inductors. This is advantageous because inductors are usually large, heavy, and costly, and they can introduce electromagnetic field effects that compromise the desired frequency response characteristics.

- Active filters provide control over passband amplification that is not available in passive filter circuits. Examine the transfer functions of all the filter circuits from Chapter 14 and you will notice that the maximum magnitude does not exceed 1. This is not surprising, since most of the transfer functions in Chapter 14 were derived using voltage division. Active filters permit us to specify both the filtering characteristics and a passband gain, an advantage over passive filters.

- Active filters can have resistive loads at their outputs whose presence *does not* alter the filter characteristics due to the properties of ideal op amps. This is another advantage of active filters, as both the cutoff frequency and the passband magnitude of passive filters can change when a resistive load is added at the output.

Thus, we can implement filter designs using active circuits when physical size, passband amplification, and load variation are important parameters in the design specifications.

In this chapter, we analyze a few of the many filter circuits that use op amps. You will see how these op amp circuits overcome the disadvantages of passive filter circuits. We will also combine basic op amp filter circuits to achieve specific frequency responses and to attain a more nearly ideal filter response. Note that throughout this chapter we assume that every op amp behaves as an ideal op amp.

■ Practical Perspective

Bass Volume Control

The circuits we analyze in this chapter are frequency selective, which means that the circuit's behavior depends on its sinusoidal input frequency. These filter circuits play important roles in many audioelectronic applications.

Audioelectronic systems such as radios, CD players, and home stereo systems often provide separate volume controls labeled "treble" and "bass." These controls allow you to set the volume of high-frequency audio signals ("treble") and the volume of low-frequency audio signals ("bass") independently. Adjusting the amplification (boost) or attenuation (cut) in these two frequency bands allows you to customize the sound with more precision than you get using a single volume control.

The Practical Perspective example at the end of this chapter presents a bass volume control circuit composed of a single op amp together with resistors and capacitors. An adjustable resistor controls the amplification in the bass frequency range.

Be Good/Shutterstock

Juraj Kovac/Shutterstock

Arnut09Job/Shutterstock

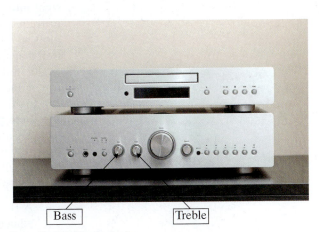

Peter Gudella/Shutterstock

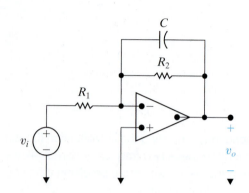

Figure 15.1 ▲ A first-order low-pass filter.

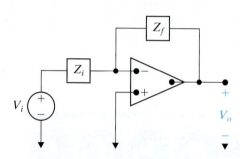

Figure 15.2 ▲ A general op amp circuit.

15.1 First-Order Low-Pass and High-Pass Filters

Consider the circuit in Fig. 15.1. When the frequency of the source is varied, only the impedance of the capacitor is affected. At very low frequencies, the capacitor acts like an open circuit, and the op amp circuit acts like an inverting amplifier with a gain of $-R_2/R_1$. At very high frequencies, the capacitor acts like a short circuit that connects the op amp's output to ground. The op amp circuit in Fig. 15.1 thus functions as a low-pass filter with a passband gain of $-R_2/R_1$.

To confirm this qualitative assessment, we compute the transfer function $H(s) = V_o(s)/V_i(s)$. Note that the circuit in Fig. 15.1 has the general form of the circuit shown in Fig. 15.2, where the impedance in the input path (Z_i) is the resistor R_1, and the impedance in the feedback path (Z_f) is the parallel combination of the resistor R_2 and the capacitor impedance $1/sC$.

The circuit in Fig. 15.2 has the same configuration as the inverting amplifier circuit from Chapter 5, so its transfer function is $-Z_f/Z_i$. Therefore, the transfer function for the circuit in Fig. 15.1 is

$$H(s) = \frac{-Z_f}{Z_i} = \frac{-R_2 \left\| \left(\dfrac{1}{sC} \right)\right.}{R_1} = -K\frac{\omega_c}{s + \omega_c}, \tag{15.1}$$

where

$$K = \frac{R_2}{R_1}, \tag{15.2}$$

and

$$\omega_c = \frac{1}{R_2C}. \tag{15.3}$$

Note that Eq. 15.1 has the same form as the general equation for low-pass filters given in Eq. 14.6, with an important difference: the gain in the passband, K, is set by the ratio R_2/R_1. The op amp low-pass filter thus permits the passband gain and the cutoff frequency to be specified independently.

A Note About Frequency Response Plots

Frequency response plots, introduced in Chapter 14, provide valuable insight into the way a filter circuit functions. Thus, we make extensive use of frequency response plots in this chapter, too. The frequency response plots in Chapter 14 have two components—a plot of the transfer function magnitude versus frequency and a plot of the transfer function phase angle, in degrees, versus frequency. The two plots are usually stacked on top of one another so that they can share the same frequency axis.

In this chapter, we use a special type of frequency response plot called the **Bode plot**. Bode plots are discussed in detail in Appendix E, which includes information about how to construct these plots by hand. You will probably use a computer to construct Bode plots, so here we summarize the special features of these plots. Bode plots differ from the frequency response plots in Chapter 14 in two important ways.

- A Bode plot uses a logarithmic axis for the frequency values instead of a linear axis. This permits us to plot a wider range of frequencies of interest. Normally, we plot three or four decades of frequencies, say from 10^2 rad/s to 10^6 rad/s, or 1 kHz to 1 MHz, choosing the frequency range where the transfer function characteristics are changing. If we plot both the magnitude and phase angle plots, they share the frequency axis.

• The Bode magnitude is plotted in decibels (dB) versus the log of the frequency, instead of plotting the absolute magnitude versus linear frequency. The decibel is discussed in Appendix D. Briefly, if the magnitude of the transfer function is $|H(j\omega)|$, its value in dB is given by

$$A_{dB} = 20\log_{10}|H(j\omega)|.$$

Note that although $|H(j\omega)|$ is an unsigned quantity, A_{dB} is a signed quantity. When $A_{dB} = 0$, the transfer function magnitude is 1, since $20\log_{10}(1) = 0$. When $A_{dB} < 0$, the transfer function magnitude is between 0 and 1, and when $A_{dB} > 0$, the transfer function magnitude is greater than 1. Finally, note that

$$20\log_{10}|1/\sqrt{2}| = -3 \text{ dB}.$$

Recall that at a filter's cutoff frequency the transfer function's magnitude has been reduced from its maximum value by $1/\sqrt{2}$. Translating this definition to magnitude in dB, we find that a transfer function's magnitude at its cutoff frequency has been reduced from its maximum magnitude by 3 dB. For example, if the magnitude of a low-pass filter in its passband is 26 dB, the magnitude used to find the cutoff frequency is $26 - 3 = 23$ dB.

Example 15.1 illustrates the design of a first-order low-pass filter to meet desired specifications of passband gain and cutoff frequency; it also illustrates a Bode magnitude plot of the filter's transfer function.

EXAMPLE 15.1	**Designing a Low-Pass Op Amp Filter**

Using the circuit shown in Fig. 15.1, calculate values for C and R_2 that, together with $R_1 = 1 \ \Omega$, produce a low-pass filter having a gain of 1 in the passband and a cutoff frequency of 1 rad/s. Find the transfer function for this filter and use it to sketch a Bode magnitude plot of the filter's frequency response.

Solution

Equation 15.2 gives the passband gain in terms of R_1 and R_2, so we use it to calculate the required value of R_2:

$$R_2 = KR_1 = (1)(1) = 1 \ \Omega.$$

Now use Eq. 15.3 to calculate C and satisfy the cutoff frequency specification:

$$C = \frac{1}{R_2\omega_c} = \frac{1}{(1)(1)} = 1 \text{ F}.$$

The transfer function for the low-pass filter is given by Eq. 15.1:

$$H(s) = -K\frac{\omega_c}{s + \omega_c} = \frac{-1}{s + 1}.$$

The Bode plot of $|H(j\omega)|$ is shown in Fig. 15.3. We have just designed a **prototype** low-pass op amp filter. It uses a resistor value of 1 Ω and a capacitor value of 1 F, and it has a cutoff frequency of 1 rad/s.

As we shall see in the next section, prototype filters provide a useful starting point for the design of filters that use more realistic component values to achieve a desired frequency response.

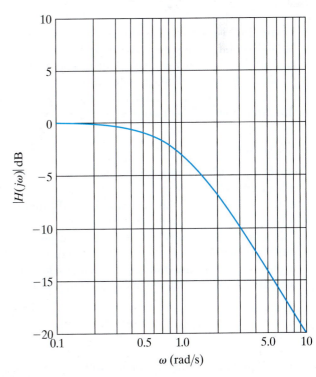

Figure 15.3 ▲ The Bode magnitude plot of the low-pass filter from Example 15.1.

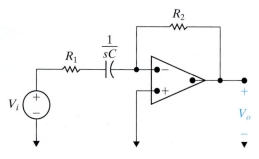

Figure 15.4 ▲ A first-order high-pass filter.

You may have recognized the circuit in Fig. 15.1 as the integrating amplifier circuit introduced in Chapter 7. They are indeed the same circuit, so integrating in the time domain corresponds to low-pass filtering in the frequency domain. This relationship between integration and low-pass filtering is also evident from the operational Laplace transform for integration derived in Chapter 12.

The circuit in Fig. 15.4 is an active first-order high-pass filter. This circuit also has the general form of the circuit in Fig. 15.2, only now the impedance in the input path is the series combination of R_1 and $1/sC$, and the impedance in the feedback path is the resistor R_2. The transfer function for the circuit in Fig 15.4 is thus

$$H(s) = \frac{-Z_f}{Z_i} = \frac{-R_2}{R_1 + \dfrac{1}{sC}} = -K\frac{s}{s + \omega_c}, \tag{15.4}$$

where

$$K = \frac{R_2}{R_1}, \tag{15.5}$$

and

$$\omega_c = \frac{1}{R_1 C}. \tag{15.6}$$

The form of the transfer function given in Eq. 15.4 is the same as that in Eq. 14.10, the equation for passive high-pass filters. Again, the active filter can have a passband gain greater than 1.

Example 15.2 designs an active high-pass filter that must meet frequency response specifications from a Bode plot.

EXAMPLE 15.2   Designing a High-Pass Op Amp Filter

Figure 15.5 shows the Bode magnitude plot of a high-pass filter. Using the active high-pass filter circuit in Fig. 15.4, calculate values of R_1 and R_2 that produce the desired magnitude response. Use a 0.1 μF capacitor. If a 10 kΩ load resistor is added to this filter, how will the magnitude response change?

Solution

Begin by writing a transfer function that has the magnitude plot shown in Fig. 15.5. To do this, note that the gain in the passband is 20 dB; therefore, $K = 10$. Also note that the 3 dB point is 500 rad/s, which must be the filter's cutoff frequency. Equation 15.4 is the transfer function for a high-pass filter, so the transfer function that has the magnitude response shown in Fig. 15.5 is given by

$$H(s) = \frac{-10s}{s + 500}.$$

Next, equate this transfer function with Eq. 15.4:

$$H(s) = \frac{-10s}{s + 500} = \frac{-(R_2/R_1)s}{s + (1/R_1C)}.$$

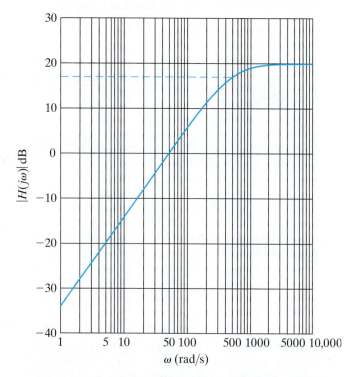

Figure 15.5 ▲ The Bode magnitude plot of the high-pass filter for Example 15.2.

Equating the numerators and denominators and then simplifying, we get two equations:

$$10 = \frac{R_2}{R_1}, \quad 500 = \frac{1}{R_1 C}.$$

Using the specified value of C (0.1 μF), we find

$$R_1 = 20\ \text{k}\Omega, \quad R_2 = 200\ \text{k}\Omega.$$

The circuit is shown in Fig. 15.6.

Because we have made the assumption that the op amp is ideal, adding any load resistor, regardless of its resistance, has no effect on the filter circuit. Thus, the magnitude response of a high-pass filter with a load resistor is the same as that of a high-pass filter with no load resistor, which is depicted in Fig. 15.5.

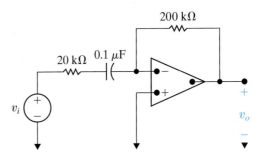

Figure 15.6 ▲ The high-pass filter for Example 15.2.

ASSESSMENT PROBLEMS

Objective 1—Know the op amp circuits that behave as first-order low-pass and high-pass filters and be able to calculate their component values

15.1 Compute the values for R_2 and C that yield a high-pass filter with a passband gain of 1 and a cutoff frequency of 1 rad/s if R_1 is 1 Ω. (*Note:* This is the prototype high-pass filter.)

Answer: $R_2 = 1\ \Omega$, $C = 1$ F.

15.2 Compute the resistor values needed for the low-pass filter circuit in Fig. 15.1 to produce the transfer function

$$H(s) = \frac{-20,000}{s + 5000}.$$

Use a 5 μF capacitor.

Answer: $R_1 = 10\ \Omega$, $R_2 = 40\ \Omega$.

SELF-CHECK: Also try Chapter Problems 15.1 and 15.8.

15.2 Scaling

When designing and analyzing filter circuits, it is convenient to use component values like 1 Ω, 1 H, and 1 F because any computations required are simple. Unfortunately, these values are unrealistic for specifying practical components, and they result in filters with undesirable characteristics, like a cutoff frequency of 1 rad/s. But we can use **scaling** to transform the convenient component values into realistic values and transform undesirable filter characteristics into desirable ones. There are two types of scaling: magnitude and frequency.

Magnitude Scaling

We scale a circuit in magnitude by multiplying the impedance at a given frequency by the scale factor k_m. Thus, we multiply all resistors and inductors by k_m and all capacitors by $1/k_m$. If we let unprimed variables represent the initial values of the parameters, and we let primed variables represent the scaled values of the variables, we have

$$R' = k_m R, \quad L' = k_m L, \quad \text{and} \quad C' = C/k_m.$$

Note that k_m is by definition a positive real number that can be either less than or greater than 1.

Frequency Scaling

We scale a circuit in frequency by changing the circuit's component values so that at the new frequency, the impedance of each element is the same as it was at the original frequency. Because resistive impedance is independent of frequency, resistors are unaffected by frequency scaling. If we let k_f denote the frequency scale factor, inductor and capacitor values are both multiplied by $1/k_f$. Thus, for frequency scaling,

$$R' = R, \quad L' = L/k_f, \quad \text{and} \quad C' = C/k_f.$$

The frequency scale factor k_f is also a positive real number that can be less than or greater than 1.

A circuit can be scaled simultaneously in both magnitude and frequency. The scaled values (primed) in terms of the original values (unprimed) are

COMPONENT SCALE FACTORS

$$R' = k_m R,$$

$$L' = \frac{k_m}{k_f} L, \tag{15.7}$$

$$C' = \frac{1}{k_m k_f} C.$$

The Use of Scaling in the Design of Filters

When designing filters, follow these steps to use scaling:

- Select the cutoff frequency, ω_c, to be 1 rad/s (if you are designing low- or high-pass filters), or select the center frequency, ω_o, to be 1 rad/s (if you are designing bandpass or bandreject filters).
- Select a 1 F capacitor and calculate the values of the resistors needed to give the desired passband gain (if you are designing an active filter) and the 1 rad/s cutoff or center frequency.
- Use scaling to compute more realistic component values that give the desired cutoff or center frequency.

Example 15.3 illustrates the scaling process in general, and Example 15.4 illustrates the use of scaling in the design of a low-pass filter.

EXAMPLE 15.3 | **Scaling a Series *RLC* Filter**

The passive series RLC filter shown in Fig. 15.7 has a center frequency of $\sqrt{1/LC} = 1$ rad/s, a bandwidth of $R/L = 1$ rad/s, and thus a quality factor of 1. Use scaling to compute new values of R and L that yield a circuit with the same quality factor but with a center frequency of 500 Hz. Use a 2 μF capacitor.

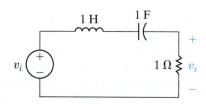

Figure 15.7 ▲ The series RLC circuit for Example 15.3.

Solution

Compute the frequency scale factor that will shift the center frequency from 1 rad/s to 500 Hz. Remember, the unprimed variables represent values before scaling, whereas the primed variables represent values after scaling.

$$k_f = \frac{\omega'_o}{\omega_o} = \frac{2\pi(500)}{1} = 3141.59.$$

Now, use Eq. 15.7 to compute the magnitude scale factor that, together with the frequency scale factor, will yield a capacitor value of 2 μF:

$$k_m = \frac{1}{k_f}\frac{C}{C'} = \frac{1}{(3141.59)(2 \times 10^{-6})} = 159.155.$$

Use Eq. 15.7 again to compute the magnitude- and frequency-scaled values of R and L:

$$R' = k_m R = 159.155 \ \Omega,$$

$$L' = \frac{k_m}{k_f} L = 50.66 \text{ mH}.$$

With these component values, the center frequency of the series RLC circuit is

$$\sqrt{1/LC} = 3141.61 \text{ rad/s or } 500 \text{ Hz},$$

and the bandwidth is

$$R/L = 3141.61 \text{ rad/s or } 500 \text{ Hz};$$

thus, the quality factor is still 1.

EXAMPLE 15.4 Scaling a Prototype Low-Pass Op Amp Filter

Use the prototype low-pass op amp filter from Example 15.1, along with magnitude and frequency scaling, to compute the resistor values for a low-pass filter with a gain of 5, a cutoff frequency of 1000 Hz, and a feedback capacitor of 10 nF. Construct a Bode plot of the resulting transfer function's magnitude.

Solution

To begin, use frequency scaling to place the cutoff frequency at 1000 Hz:

$$k_f = \omega'_c/\omega_c = 2\pi(1000)/1 = 6283.185,$$

where the primed variable has the new value and the unprimed variable has the old value of the cutoff frequency. Then compute the magnitude scale factor that, together with $k_f = 6283.185$, will scale the capacitor to 10 nF:

$$k_m = \frac{1}{k_f}\frac{C}{C'} = \frac{1}{(6283.185)(10^{-8})} = 15,915.5.$$

Since resistors are scaled only by using magnitude scaling,

$$R'_1 = R'_2 = k_m R = (15,915.5)(1) = 15,915.5 \ \Omega.$$

Finally, we need to meet the passband gain specification. We can adjust the values of either R'_1 or R'_2 because $K = R'_2/R'_1$. If we adjust R'_2, we will change the cutoff frequency because $\omega'_c = 1/R'_2 C'$. Therefore, we can adjust the value of R'_1 to alter only the passband gain:

$$R'_1 = R'_2/K = (15,915.5)/(5) = 3183.1 \ \Omega.$$

The final component values are

$$R'_1 = 3183.1 \ \Omega, R'_2 = 15,915.5 \ \Omega, C' = 10 \text{ nF}.$$

The transfer function of the filter is given by

$$H(s) = \frac{-31,415.93}{s + 6283.185}.$$

The Bode plot of this transfer function's magnitude is shown in Fig. 15.8.

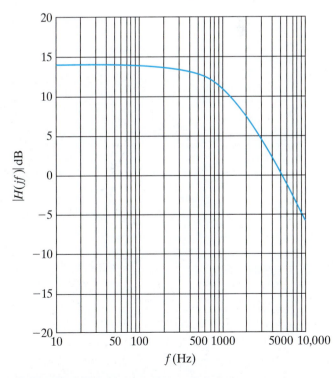

Figure 15.8 ▲ The Bode magnitude plot of the low-pass filter from Example 15.4.

ASSESSMENT PROBLEM

Objective 2—Be able to design filter circuits starting with a prototype and use scaling to achieve desired frequency response and component values

15.3   What magnitude and frequency scale factors will transform the prototype high-pass filter into a high-pass filter with a 0.5 μF capacitor and a cutoff frequency of 10 kHz?

Answer: $k_f = 62{,}831.85$, $k_m = 31.831$.

SELF-CHECK: Also try Chapter Problems 15.15 and 15.16.

15.3 Op Amp Bandpass and Bandreject Filters

We now analyze and design op amp circuits that act as bandpass and band-reject filters. Our initial approach is motivated by the Bode plot construction shown in Fig. 15.9. We can see from the plot that the bandpass filter consists of three separate elements:

- A unity-gain low-pass filter whose cutoff frequency is ω_{c2}, the larger of the two cutoff frequencies.
- A unity-gain high-pass filter whose cutoff frequency is ω_{c1}, the smaller of the two cutoff frequencies.
- A gain component to provide the desired passband gain.

These three subcircuits are cascaded in series. The subcircuit transfer functions are multiplied to form the cascade transfer function, and the subcircuit magnitude plots are added to create the cascade magnitude plot.

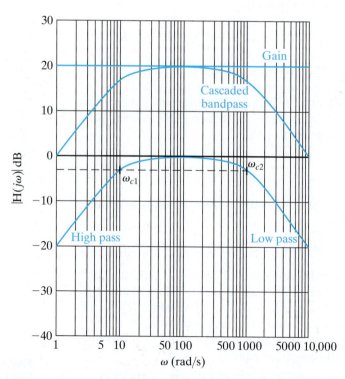

Figure 15.9 ▲ Constructing the Bode magnitude plot of a bandpass filter.

This method of constructing an active bandpass filter and its corresponding magnitude response assumes that the lower cutoff frequency (ω_{c1}) is significantly smaller than the upper cutoff frequency (ω_{c2}). The resulting filter is called a **broadband** bandpass filter because the band of frequencies passed is wide. The formal definition of a broadband filter requires that the two cutoff frequencies satisfy the equation

$$\frac{\omega_{c2}}{\omega_{c1}} \geq 2.$$

As illustrated by the Bode plot construction in Fig. 15.9, we require that the magnitude of the high-pass filter be unity at the cutoff frequency of the low-pass filter and that the magnitude of the low-pass filter be unity at the cutoff frequency of the high-pass filter. Then the bandpass filter will have the cutoff frequencies specified by the low-pass and high-pass filters. We need to determine the relationship between ω_{c1} and ω_{c2} that will satisfy the requirements illustrated in Fig. 15.9.

We can construct a circuit consisting of three subcircuits by cascading a low-pass op amp filter, a high-pass op amp filter, and an inverting amplifier (see Section 5.3), as shown in Fig. 15.10(a). Figure 15.10(a) is called a **block diagram**. Each block represents a component or subcircuit, and the output of one block is the input to the next, in the direction indicated. We want to establish the relationship between ω_{c1} and ω_{c2} that will permit each subcircuit to be designed independently, without concern for the other subcircuits in the cascade. This reduces the bandpass filter design to the design of a unity-gain first-order low-pass filter, a unity-gain first-order high-pass filter, and an inverting amplifier, each of which is a simple circuit.

The transfer function of the cascaded bandpass filter is the product of the transfer functions of the three cascaded subcircuits:

$$H(s) = \frac{V_o}{V_i}$$

$$= \left(\frac{-\omega_{c2}}{s + \omega_{c2}} \right) \left(\frac{-s}{s + \omega_{c1}} \right) \left(\frac{-R_f}{R_i} \right)$$

$$= \frac{-K\omega_{c2}s}{(s + \omega_{c1})(s + \omega_{c2})}$$

$$= \frac{-K\omega_{c2}s}{s^2 + (\omega_{c1} + \omega_{c2})s + \omega_{c1}\omega_{c2}}. \tag{15.8}$$

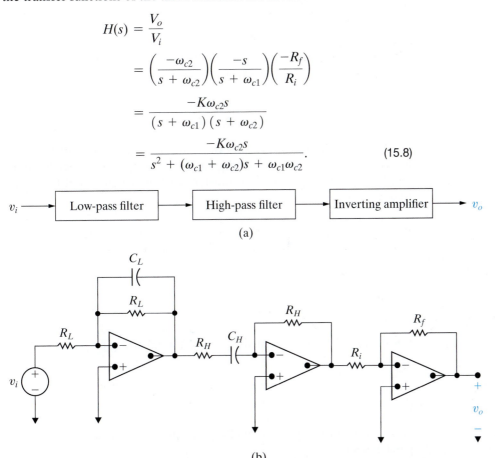

(a)

(b)

Figure 15.10 ▲ A cascaded op amp bandpass filter. (a) The block diagram. (b) The circuit.

Equation 15.8 is not in the standard form for the transfer function of a bandpass filter discussed in Chapter 14, namely,

$$H_{BP} = \frac{\beta s}{s^2 + \beta s + \omega_o^2}.$$

In order to convert Eq. 15.8 into the form of the standard transfer function for a bandpass filter, we require that

$$\omega_{c2} \gg \omega_{c1}.$$

When the upper cutoff frequency is much larger than the lower cutoff frequency,

$$(\omega_{c1} + \omega_{c2}) \approx \omega_{c2},$$

and the transfer function for the cascaded bandpass filter in Eq. 15.8 becomes

$$H(s) = \frac{-K\omega_{c2}s}{s^2 + \omega_{c2}s + \omega_{c1}\omega_{c2}}.$$

Thus, if the bandpass filter specifications include an upper cutoff frequency that is much larger than the lower cutoff frequency, we can design each subcircuit of the cascaded circuit independently. We compute the values of R_L and C_L in the low-pass filter to give us the desired upper cutoff frequency, ω_{c2}:

$$\omega_{c2} = \frac{1}{R_L C_L}.$$

We compute the values of R_H and C_H in the high-pass filter to give us the desired lower cutoff frequency, ω_{c1}:

$$\omega_{c1} = \frac{1}{R_H C_H}.$$

Now we compute the values of R_i and R_f in the inverting amplifier to provide the desired passband gain. To do this, we consider the magnitude of the bandpass filter's transfer function, evaluated at the center frequency, ω_o:

$$|H(j\omega_o)| = \left| \frac{-K\omega_{c2}(j\omega_o)}{(j\omega_o)^2 + \omega_{c2}(j\omega_o) + \omega_{c1}\omega_{c2}} \right|$$

$$= \frac{K\omega_{c2}}{\omega_{c2}}$$

$$= K.$$

Recall from Chapter 5 that the gain of the inverting amplifier is R_f/R_i. Therefore,

$$|H(j\omega_o)| = \frac{R_f}{R_i}.$$

Any choice of resistors that satisfies this equation will produce the desired passband gain.

Example 15.5 illustrates the design process for the cascaded bandpass filter.

EXAMPLE 15.5 | Designing a Broadband Bandpass Op Amp Filter

Design a bandpass filter for a graphic equalizer to provide an amplification of 2 within the band of frequencies between 100 and 10,000 Hz. Use 0.2 μF capacitors.

Solution

We can design each subcircuit in the cascade and meet the specified cutoff frequency values only if the upper cutoff frequency is much larger than the lower cutoff frequency. In this case, $\omega_{c2} = 100\omega_{c1}$, so we can say that $\omega_{c2} \gg \omega_{c1}$. Begin with the low-pass filter, whose cutoff frequency is ω_{c2}. From Eq. 15.3,

$$\omega_{c2} = \frac{1}{R_L C_L} = 2\pi(10000),$$

$$R_L = \frac{1}{[2\pi(10000)](0.2 \times 10^{-6})}$$

$$\approx 80 \ \Omega.$$

Next, we turn to the high-pass filter, whose cutoff frequency is ω_{c1}. From Eq. 15.6,

$$\omega_{c1} = \frac{1}{R_H C_H} = 2\pi(100),$$

$$R_H = \frac{1}{[2\pi(100)](0.2 \times 10^{-6})}$$

$$\approx 7958 \ \Omega.$$

Finally, we need the gain stage. Two resistors are required, so one of the resistors can be selected arbitrarily. Let's select a 1 kΩ resistor for R_i. Then,

$$R_f = 2(1000)$$

$$= 2000 \ \Omega = 2 \ \text{k}\Omega.$$

The resulting circuit is shown in Fig. 15.11. We leave it to you to show that the magnitude of this circuit's transfer function is reduced by $1/\sqrt{2}$ from its maximum value at both cutoff frequencies, verifying the validity of the assumption $\omega_{c2} \gg \omega_{c1}$.

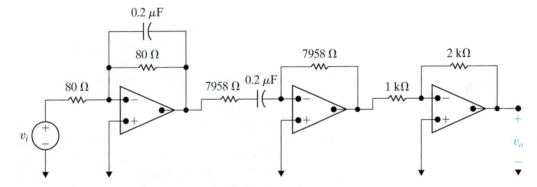

Figure 15.11 ▲ The cascaded op amp bandpass filter designed in Example 15.5.

We can use a subcircuit approach when designing op amp bandreject filters, too, as illustrated in Fig. 15.12. Like the bandpass filter, the bandreject filter consists of three separate elements.

- A unity-gain low-pass filter with a cutoff frequency of ω_{c1}, which is the smaller of the two cutoff frequencies.
- A unity-gain high-pass filter with a cutoff frequency of ω_{c2}, which is the larger of the two cutoff frequencies.
- A summing amplifier that provides the desired gain in the passbands.

There are important differences between the three subcircuits that comprise the bandpass filter and those that comprise the bandreject filter. The cutoff frequencies of the low-pass and high-pass filters are obviously different—in the bandpass filter, the low-pass filter subcircuit has a cutoff frequency of ω_{c2}, and in the bandreject filter, the low-pass filter subcircuit has a cutoff frequency of ω_{c1}. The cutoff frequencies for the high-pass filter subcircuit are also reversed. The most important difference is that the

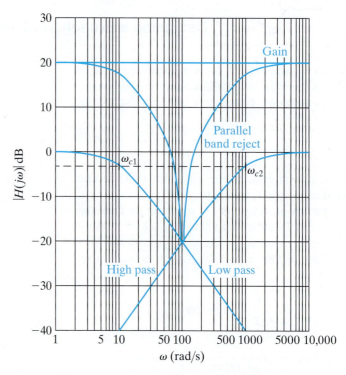

Figure 15.12 ▲ Constructing the Bode magnitude plot of a bandreject filter.

three subcircuits in the bandreject filter cannot be cascaded in series because they do not combine additively on the Bode plot. Instead, the low-pass and high-pass filters act in parallel, and a summing amplifier combines their outputs and provides the passband gain. Figure 15.13 shows this design both in block diagram form and as a circuit.

Again, it is assumed that the two cutoff frequencies for the band-reject filter are widely separated so that the resulting design is a broad-band bandreject filter, and $\omega_{c2} \gg \omega_{c1}$. Then each subcircuit in the parallel design can be created independently, and the cutoff frequency specifications will be satisfied. The transfer function of the resulting circuit is the sum of the low-pass and high-pass filter transfer functions. From Fig. 15.13(b),

$$
\begin{aligned}
H(s) &= \left(-\frac{R_f}{R_i} \right)\left[\frac{-\omega_{c1}}{s + \omega_{c1}} + \frac{-s}{s + \omega_{c2}} \right] \\
&= \frac{R_f}{R_i}\left(\frac{\omega_{c1}(s + \omega_{c2}) + s(s + \omega_{c1})}{(s + \omega_{c1})(s + \omega_{c2})} \right) \\
&= \frac{R_f}{R_i}\left(\frac{s^2 + 2\omega_{c1}s + \omega_{c1}\omega_{c2}}{(s + \omega_{c1})(s + \omega_{c2})} \right).
\end{aligned}
\tag{15.9}
$$

As we saw in the cascaded bandpass filter design, the two cutoff frequencies for the transfer function in Eq. 15.9 are ω_{c1} and ω_{c2} only if $\omega_{c2} \gg \omega_{c1}$. Then the cutoff frequencies are given by the equations

$$\omega_{c1} = \frac{1}{R_L C_L},$$

$$\omega_{c2} = \frac{1}{R_H C_H}.$$

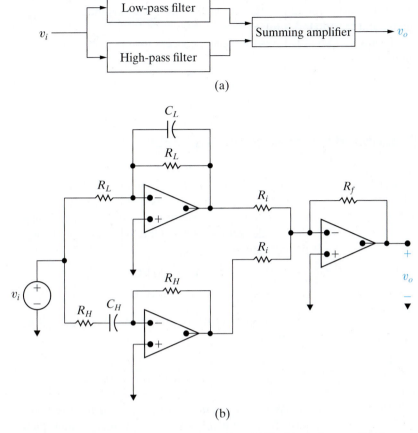

Figure 15.13 ▲ A parallel op amp bandreject filter. (a) The block diagram. (b) The circuit.

In the two passbands (as $s \rightarrow 0$ and $s \rightarrow \infty$), the gain of the transfer function is R_f/R_i. Therefore,

$$K = \frac{R_f}{R_i}.$$

As with the design of the cascaded bandpass filter, we have six unknowns and three equations. Typically, we choose a commercially available capacitor value for C_L and C_H. Then we use the equations for the cutoff frequencies of the low-pass and high-pass filters, calculating R_L and R_H to meet the specified cutoff frequencies. Finally, we choose a value for either R_f or R_i and then use the equation for K to compute the other resistance to meet the passband gain specification.

The magnitude of the transfer function in Eq. 15.9 at the center frequency, $\omega_o = \sqrt{\omega_{c1}\omega_{c2}}$, is

$$|H(j\omega_o)| = \left| \frac{R_f}{R_i} \left(\frac{(j\omega_o)^2 + 2\omega_{c1}(j\omega_o) + \omega_{c1}\omega_{c2}}{(j\omega_o)^2 + (\omega_{c1} + \omega_{c2})(j\omega_o) + \omega_{c1}\omega_{c2}} \right) \right|$$

$$= \frac{R_f}{R_i} \frac{2\omega_{c1}}{(\omega_{c1} + \omega_{c2})}$$

$$\approx \frac{R_f}{R_i} \frac{2\omega_{c1}}{\omega_{c2}}.$$

If $\omega_{c2} \gg \omega_{c1}$, then $|H(j\omega_o)| \ll 2R_f/R_i$ (because $\omega_{c1}/\omega_{c2} \ll 1$), so the magnitude at the center frequency is much smaller than the passband magnitude. Thus, the bandreject filter successfully rejects frequencies near the center frequency but only if the specifications meet the requirements of a broadband filter.

Example 15.6 illustrates the design process for the parallel bandreject filter.

EXAMPLE 15.6 | **Designing a Broadband Bandreject Op Amp Filter**

Design a circuit based on the parallel bandreject op amp filter in Fig. 15.13(b). The Bode magnitude plot of the filter is shown in Fig. 15.14. Use 0.5 μF capacitors in your design.

Solution

From the Bode magnitude plot in Fig. 15.14, we see that the bandreject filter's cutoff frequencies are 100 rad/s and 2000 rad/s and its passband gain is 3. Thus, $\omega_{c2} = 20\omega_{c1}$, so we make the assumption that $\omega_{c2} \gg \omega_{c1}$. Let's begin with the prototype low-pass filter from Example 15.1 and use scaling to meet the specifications for cutoff frequency and capacitor value. The frequency scale factor k_f is 100, which shifts the cutoff frequency from

1 rad/s to 100 rad/s. The magnitude scale factor k_m is 20,000, which, together with the frequency scale factor, scales the capacitor from 1 F to 0.5 μF. Using these scale factors results in the following scaled component values:

$$R_L = 20 \text{ k}\Omega,$$

$$C_L = 0.5 \mu\text{F}.$$

The resulting cutoff frequency of the low-pass filter component is

$$\omega_{c1} = \frac{1}{R_L C_L} = \frac{1}{(20 \times 10^3)(0.5 \times 10^{-6})} = 100 \text{ rad/s}.$$

We use the same approach to design the high-pass filter, starting with the prototype high-pass op amp filter. Here, the frequency scale factor is $k_f = 2000$, shifting the cutoff frequency from 1 rad/s to 2000 rad/s. The magnitude scale factor is $k_m = 1000$, which, together with the frequency scale factor, scales the capacitor from 1 F to 0.5 μF. The scaled component values are

$$R_H = 1 \text{ k}\Omega,$$

$$C_H = 0.5 \mu\text{F}.$$

Finally, because the cutoff frequencies are widely separated, we can use the ratio R_f/R_i to create the passband gain of 3. Let's choose $R_i = 1 \text{ k}\Omega$, as we are already using that resistance for R_H. Then $R_f = 3 \text{ k}\Omega$, and $K = R_f/R_i = 3000/1000 = 3$. The resulting active broadband bandreject filter circuit is shown in Fig. 15.15.

Now let's check our assumption that $\omega_{c2} \gg \omega_{c1}$ by calculating the actual gain at the specified cutoff frequencies. We do this by making the substitutions $s = j100$ and $s = j2000$ into the transfer function for the parallel bandreject filter, Eq. 15.9, and calculating the resulting magnitude. You should verify that the magnitude at the specified cutoff frequencies is greater than the magnitude of $3/\sqrt{2} = 2.12$ that we expect. Therefore, our rejecting band is more narrow than specified in the problem statement.

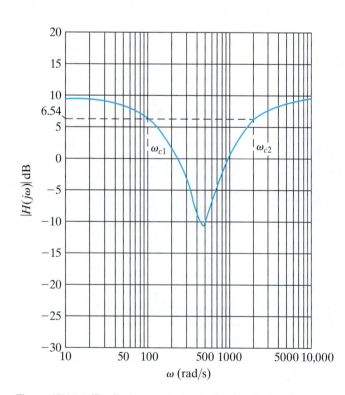

Figure 15.14 ▲ The Bode magnitude plot for the circuit to be designed in Example 15.6.

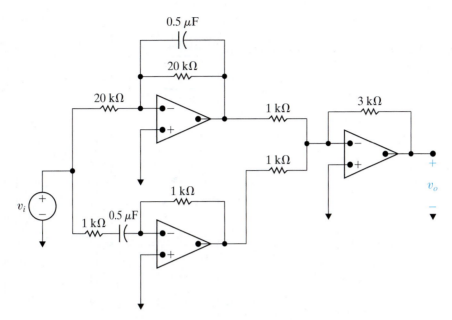

Figure 15.15 ▲ The resulting bandreject filter circuit designed in Example 15.6.

SELF-CHECK: Assess your understanding of this material by trying Chapter Problems 15.30 and 15.31.

15.4 Higher-Order Op Amp Filters

None of the filter circuits we have examined so far, whether passive or active, are ideal. Remember from Chapter 14 that an ideal filter has a discontinuity at the cutoff frequency, which sharply divides the passband and the stopband. Although we cannot hope to construct a linear circuit with a discontinuous frequency response, we can construct circuits with a sharper, yet still continuous, transition at the cutoff frequency.

Cascading Identical Filters

How can we obtain a sharper transition between the passband and the stopband? One approach is suggested by the Bode magnitude plots in Fig. 15.16. This figure shows the Bode magnitude plots of a cascade of identical prototype low-pass filters and includes plots of a single filter, two in cascade, three in cascade, and four in cascade. You can see that the transition from the passband to the stopband becomes sharper as more filters are added to the cascade. The rules for constructing Bode plots (from Appendix E) tell us that with one filter, the transition occurs with an asymptotic slope of 20 decibels per decade (dB/dec). Because circuits in cascade are additive on a Bode magnitude plot, a cascade with two filters has a transition with an asymptotic slope of $20 + 20 = 40$ dB/dec; for three filters, the asymptotic slope is 60 dB/dec, and for four filters, it is 80 dB/dec, as seen in Fig. 15.16.

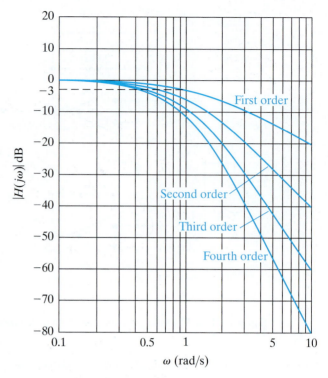

Figure 15.16 ▲ The Bode magnitude plot of a cascade of identical prototype first-order filters.

In general, an n-element cascade of identical low-pass filters will transition from the passband to the stopband with a slope of $20n$ dB/dec. Figure 15.17 shows both the block diagram and the circuit diagram for such a cascade. We compute the transfer function for a cascade of n prototype low-pass filters by multiplying the individual transfer functions:

$$H(s) = \left(\frac{-1}{s+1}\right)\left(\frac{-1}{s+1}\right)\cdots\left(\frac{-1}{s+1}\right)$$

$$= \frac{(-1)^n}{(s+1)^n}. \tag{15.10}$$

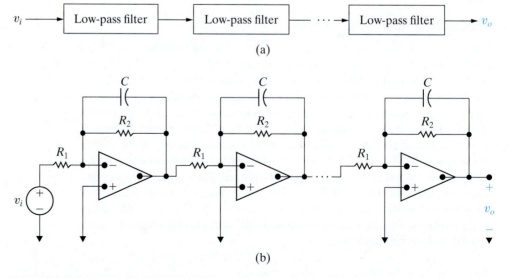

Figure 15.17 ▲ A cascade of identical unity-gain low-pass filters. (a) The block diagram. (b) The circuit.

A cascade of first-order low-pass filters yields a higher-order filter. From Eq. 15.10, a cascade of n first-order filters produces an nth-order filter because the transfer function has n poles, and the filter has a final slope of $20n$ dB/dec in the transition band.

There is an important issue yet to be resolved, as you will see if you look closely at Fig. 15.16. As the order of the low-pass filter is increased by adding prototype low-pass filters to the cascade, the cutoff frequency changes. For example, in a cascade of two first-order low-pass filters, the magnitude of the resulting second-order filter at 1 rad/s, the cutoff frequency of the single prototype low-pass filer, is -6 dB. The cutoff frequency of the second-order filter is not 1 rad/s. In fact, the cutoff frequency is less than 1 rad/s.

Therefore, we need to find the cutoff frequency of the nth-order filter formed in the cascade of n first-order filters. Then we can use frequency scaling to calculate component values that move the cutoff frequency to its specified location. The cutoff frequency, ω_{cn}, satisfies the equation $|H(j\omega)| = 1/\sqrt{2}$:

$$H(s) = \frac{(-1)^n}{(s+1)^n},$$

$$|H(j\omega_{cn})| = \left| \frac{1}{(j\omega_{cn}+1)^n} \right| = \frac{1}{\sqrt{2}},$$

$$\frac{1}{(\sqrt{\omega_{cn}^2+1})^n} = \frac{1}{\sqrt{2}},$$

$$\frac{1}{\omega_{cn}^2+1} = \left(\frac{1}{\sqrt{2}} \right)^{2/n},$$

$$\sqrt[n]{2} = \omega_{cn}^2 + 1,$$

$$\omega_{cn} = \sqrt{\sqrt[n]{2} - 1}. \tag{15.11}$$

Let's use Eq. 15.11 to find the cutoff frequency of a fourth-order unity-gain low-pass filter constructed from a cascade of four prototype low-pass filters:

$$\omega_{c4} = \sqrt{\sqrt[4]{2} - 1} = 0.435 \text{ rad/s.}$$

Thus, we can design a fourth-order low-pass filter with the cutoff frequency ω_c by starting with a fourth-order cascade consisting of prototype low-pass filters and then scaling the components by $k_f = \omega_c/0.435$.

We can build a higher-order low-pass filter with a nonunity gain by adding an inverting amplifier circuit to the cascade. Example 15.7 illustrates the design of a fourth-order low-pass filter with nonunity gain.

EXAMPLE 15.7 **Designing a Fourth-Order Low-Pass Active Filter**

Design a fourth-order low-pass filter with a cutoff frequency of 500 Hz and a passband gain of 10. Use 1 μF capacitors. Sketch the Bode magnitude plot for this filter.

Solution

Our design cascades four prototype low-pass filters. We have already used Eq. 15.11 to calculate the cutoff frequency for the resulting fourth-order

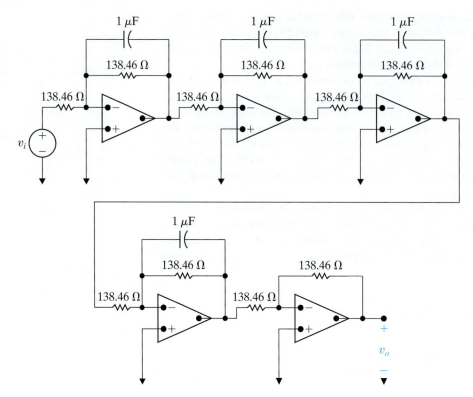

Figure 15.18 ▲ The cascade circuit for the fourth-order low-pass filter designed in Example 15.7.

low-pass filter as 0.435 rad/s. A frequency scale factor of $k_f = 7222.39$ will scale the component values to give a 500 Hz cutoff frequency. A magnitude scale factor of $k_m = 138.46$, together with the frequency scale factor, scales the capacitor value from 1 F to 1 μF. The scaled component values are thus

$$R = 138.46 \ \Omega; \quad C = 1 \ \mu F.$$

Add an inverting amplifier stage with a gain of $R_f/R_i = 10$ to satisfy the passband gain specification. As usual, we can arbitrarily select one of the two resistor values. Because we are already using 138.46 Ω resistors, let $R_i = 138.46 \ \Omega$; then,

$$R_f = 10R_i = 1384.6 \ \Omega.$$

The circuit for this cascaded fourth-order low-pass filter is shown in Fig. 15.18. It has the transfer function

$$H(s) = -10\left[\frac{7222.39}{s + 7222.39}\right]^4.$$

The Bode magnitude plot for this transfer function is sketched in Fig. 15.19.

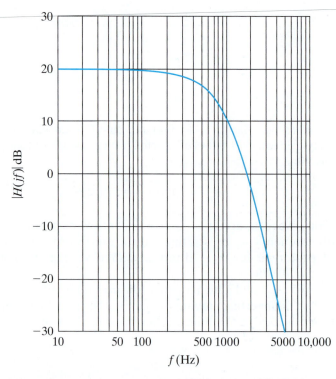

Figure 15.19 ▲ The Bode magnitude plot for the fourth-order low-pass filter designed in Example 15.7.

By cascading identical low-pass filters, we can increase the asymptotic slope in the transition between passband and stopband, but our approach has a serious shortcoming: The gain of the filter is not constant between zero and the cutoff frequency ω_c. Remember that in an ideal low-pass filter, the passband magnitude is 1 for all frequencies below the cutoff frequency. But in Fig. 15.16, we see that the passband magnitude is less than 1 (0 dB), even for frequencies much less than the cutoff frequency.

We can understand why the passband magnitude is not ideal by looking at the magnitude of the transfer function for a unity-gain low-pass nth-order cascade. Because

$$H(s) = \frac{\omega_{cn}^n}{(s + \omega_{cn})^n},$$

the magnitude is

$$|H(j\omega)| = \frac{\omega_{cn}^n}{(\sqrt{\omega^2 + \omega_{cn}^2})^n}$$

$$= \frac{1}{(\sqrt{(\omega/\omega_{cn})^2 + 1})^n}.$$

As we can see from this expression, when $\omega \ll \omega_{cn}$, the denominator is approximately 1, and the magnitude of the transfer function is also nearly 1. But as $\omega \to \omega_{cn}$, the denominator becomes larger than 1, so the magnitude becomes smaller than 1. Because the cascade of low-pass filters results in this nonideal behavior in the passband, other approaches are used when designing higher-order filters. One such approach is examined next.

Butterworth Filters

A unity-gain **Butterworth low-pass filter** has a transfer function whose magnitude is

$$|H(j\omega)| = \frac{1}{\sqrt{1 + (\omega/\omega_c)^{2n}}}, \tag{15.12}$$

where n is an integer that denotes the order of the filter.[1]

When studying Eq. 15.12, note the following:

- The cutoff frequency is ω_c rad/s for all values of n.
- If n is large enough, the denominator is always close to unity when $\omega < \omega_c$.
- In the expression for $|H(j\omega)|$, the exponent of ω/ω_c is always even.

This last observation is important because only even exponents exist in circuits with resistors, inductors, capacitors and op amps. (See Problem 15.33.)

Given an equation for the magnitude of the transfer function, how do we find $H(s)$? We can simplify the derivation for $H(s)$ by using a prototype filter, so we set ω_c equal to 1 rad/s in Eq. 15.12. As before, we will use scaling to transform the prototype filter to a filter that meets the given specifications.

To find $H(s)$, first note that if N is a complex quantity, then $|N|^2 = NN^*$, where N^* is the conjugate of N. It follows that

$$|H(j\omega)|^2 = H(j\omega) H(-j\omega).$$

But because $s = j\omega$, we can write

$$|H(j\omega)|^2 = H(s) H(-s).$$

---

[1]This filter was developed by the British engineer S. Butterworth and reported in *Wireless Engineering 7* (1930): 536–541.

Now observe that $s^2 = -\omega^2$. Thus,

$$|H(j\omega)|^2 = \frac{1}{1 + \omega^{2n}}$$

$$= \frac{1}{1 + (\omega^2)^n}$$

$$= \frac{1}{1 + (-s^2)^n}$$

$$= \frac{1}{1 + (-1)^n s^{2n}},$$

or

$$H(s)\,H(-s) = \frac{1}{1 + (-1)^n s^{2n}}.$$

Follow these steps to find $H(s)$ for a given value of n:

1. Find the roots of the polynomial

$$1 + (-1)^n s^{2n} = 0.$$

2. Assign the left-half plane roots to $H(s)$ and the right-half plane roots to $H(-s)$.

3. Combine terms in the denominator of $H(s)$ to form first- and second-order factors.

Example 15.8 illustrates this process.

EXAMPLE 15.8   Calculating Butterworth Transfer Functions

Find the Butterworth transfer functions for $n = 2$ and $n = 3$.

Solution

For $n = 2$, we find the roots of the polynomial

$$1 + (-1)^2 s^4 = 0.$$

Rearranging terms, we find

$$s^4 = -1 = 1\underline{/180°}.$$

Therefore, the four roots are

$$s_1 = 1\underline{/45°} = 1/\sqrt{2} + j/\sqrt{2},$$

$$s_2 = 1\underline{/135°} = -1/\sqrt{2} + j/\sqrt{2},$$

$$s_3 = 1\underline{/225°} = -1/\sqrt{2} + -j/\sqrt{2},$$

$$s_4 = 1\underline{/315°} = 1/\sqrt{2} + -j/\sqrt{2}.$$

Roots s_2 and s_3 are in the left-half plane. Thus,

$$H(s) = \frac{1}{(s + 1/\sqrt{2} - j/\sqrt{2})(s + 1/\sqrt{2} + j/\sqrt{2})}$$

$$= \frac{1}{(s^2 + \sqrt{2}s + 1)}.$$

For $n = 3$, we find the roots of the polynomial

$$1 + (-1)^3 s^6 = 0.$$

Rearranging terms,

$$s^6 = 1\underline{/0°} = 1\underline{/360°}.$$

Therefore, the six roots are

$$s_1 = 1\underline{/0°} = 1,$$

$$s_2 = 1\underline{/60°} = 1/2 + j\sqrt{3}/2,$$

$$s_3 = 1\underline{/120°} = -1/2 + j\sqrt{3}/2,$$

$$s_4 = 1\underline{/180°} = -1 + j0,$$

$$s_5 = 1\underline{/240°} = -1/2 + -j\sqrt{3}/2,$$

$$s_6 = 1\underline{/300°} = 1/2 + -j\sqrt{3}/2.$$

Roots s_3, s_4, and s_5 are in the left-half plane. Thus,

$$H(s) = \frac{1}{(s + 1)(s + 1/2 - j\sqrt{3}/2)(s + 1/2 + j\sqrt{3}/2)}$$

$$= \frac{1}{(s + 1)(s^2 + s + 1)}.$$

Note that the roots of the Butterworth polynomial are always equally spaced around the unit circle in the s plane. To assist in the design of Butterworth filters, Table 15.1 lists the Butterworth polynomials up to $n = 8$.

TABLE 15.1 Normalized (so that $\omega_c = 1$ rad/s) Butterworth Polynomials up to the Eighth Order

n	nth-Order Butterworth Polynomial
1	$(s + 1)$
2	$(s^2 + \sqrt{2}s + 1)$
3	$(s + 1)(s^2 + s + 1)$
4	$(s^2 + 0.765s + 1)(s^2 + 1.848s + 1)$
5	$(s + 1)(s^2 + 0.618s + 1)(s^2 + 1.618s + 1)$
6	$(s^2 + 0.518s + 1)(s^2 + \sqrt{2} + 1)(s^2 + 1.932s + 1)$
7	$(s + 1)(s^2 + 0.445s + 1)(s^2 + 1.247s + 1)(s^2 + 1.802s + 1)$
8	$(s^2 + 0.390s + 1)(s^2 + 1.111s + 1)(s^2 + 1.6663s + 1)(s^2 + 1.962s + 1)$

Butterworth Filter Circuits

Now that we know how to specify the transfer function for a Butterworth filter circuit (either by calculating the poles of the transfer function directly or by using Table 15.1), we need to design a circuit with such a transfer function. Notice the form of the Butterworth polynomials in Table 15.1. They are the product of first- and second-order factors; therefore, we can construct a circuit whose transfer function has a Butterworth polynomial in its denominator by cascading active filter circuits, each of which provides one of the needed factors. Figure 15.20 presents a block diagram of a cascade whose transfer function has a fifth-order Butterworth polynomial in its denominator.

All odd-order Butterworth polynomials include the factor $(s + 1)$, so all odd-order Butterworth filter circuits must have a subcircuit with the transfer function $H(s) = 1/(s + 1)$. This is the transfer function of the prototype low-pass active filter from Fig. 15.1. So what remains is to find a circuit whose transfer function has the form $H(s) = 1/(s^2 + b_1s + 1)$.

Such a circuit is shown in Fig. 15.21. To analyze this circuit, write the s-domain KCL equations at the noninverting terminal of the op amp and at the node labeled V_a:

$$\frac{V_a - V_i}{R} + (V_a - V_o)sC_1 + \frac{V_a - V_o}{R} = 0,$$

$$V_o sC_2 + \frac{V_o - V_a}{R} = 0.$$

Simplifying the two KCL equations yields

$$(2 + RC_1s)V_a - (1 + RC_1s)V_o = V_i,$$

$$-V_a + (1 + RC_2s)V_o = 0.$$

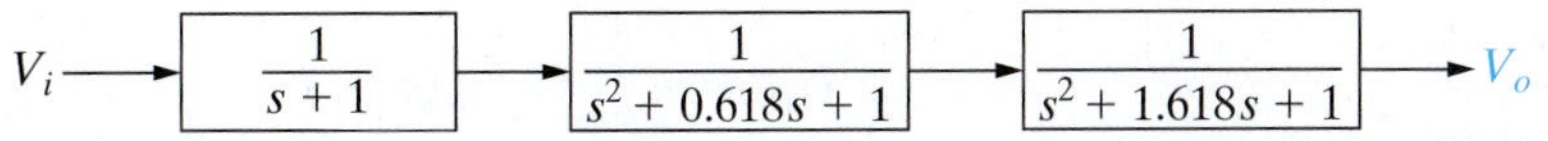

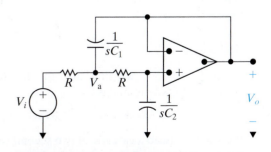

Figure 15.21 ▲ A circuit that provides the second-order transfer function for the Butterworth filter cascade.

$$V_i \longrightarrow \boxed{\frac{1}{s + 1}} \longrightarrow \boxed{\frac{1}{s^2 + 0.618s + 1}} \longrightarrow \boxed{\frac{1}{s^2 + 1.618s + 1}} \longrightarrow V_o$$

Figure 15.20 ▲ A cascade of first- and second-order circuits with the indicated transfer functions yielding a fifth-order low-pass Butterworth filter with $\omega_c = 1$ rad/s.

Use back-substitution to eliminate V_a:

$$(2 + RC_1s)(1 + RC_2s)V_o - (1 + RC_1s)V_o = V_i.$$

Then, rearrange this equation to write the transfer function for the circuit in Fig. 15.21:

$$H(s) = \frac{V_o}{V_i} = \frac{\dfrac{1}{R^2C_1C_2}}{s^2 + \dfrac{2}{RC_1}s + \dfrac{1}{R^2C_1C_2}}.$$

Finally, set $R = 1\ \Omega$; then

$$H(s) = \frac{\dfrac{1}{C_1C_2}}{s^2 + \dfrac{2}{C_1}s + \dfrac{1}{C_1C_2}}. \tag{15.13}$$

Equation 15.13 has the form required for the second-order circuit in the Butterworth cascade. To get a transfer function of the form

$$H(s) = \frac{1}{s^2 + b_1s + 1},$$

we use the circuit in Fig. 15.21 and choose capacitor values so that

$$b_1 = \frac{2}{C_1} \quad \text{and} \quad 1 = \frac{1}{C_1C_2}. \tag{15.14}$$

We have thus outlined the procedure for designing an nth-order Butterworth low-pass filter circuit with a cutoff frequency of $\omega_c = 1$ rad/s and a gain of 1 in the passband. We can use frequency scaling to calculate revised capacitor values that yield any specified cutoff frequency, and we can use magnitude scaling to provide more realistic or practical component values in our design. Cascading an inverting amplifier circuit provides a passband gain other than 1.

Example 15.9 illustrates this design process.

EXAMPLE 15.9 | **Designing a Fourth-Order Low-Pass Butterworth Filter**

Design a fourth-order Butterworth low-pass filter with a cutoff frequency of 500 Hz and a passband gain of 10. Use as many $1\ \text{k}\Omega$ resistors as possible. Compare the Bode magnitude plot for this Butterworth filter with that of the identical cascade filter in Example 15.7.

Solution

From Table 15.1, we find that the fourth-order Butterworth polynomial is

$$(s^2 + 0.765s + 1)(s^2 + 1.848s + 1).$$

We will thus need a cascade of two second-order filters to get the fourth-order transfer function and an inverting amplifier circuit to get a passband gain of 10. The circuit is shown in Fig. 15.22.

The first stage of the cascade is a circuit whose transfer function has the polynomial $(s^2 + 0.765s + 1)$ in its denominator. From Eq. 15.14,

$$C_{1a} = \frac{2}{0.765} = 2.61\ \text{F},$$

$$C_{2a} = \frac{1}{2.61} = 0.38\ \text{F}.$$

The second stage of the cascade is a circuit whose the transfer function has the polynomial $(s^2 + 1.848s + 1)$ in its denominator. From Eq. 15.14,

$$C_{1b} = \frac{2}{1.848} = 1.08\ \text{F},$$

$$C_{2b} = \frac{1}{1.08} = 0.924\ \text{F}.$$

These values for C_{1a}, C_{2a}, C_{1b}, and C_{2b} yield a fourth-order Butterworth filter with a cutoff frequency of 1 rad/s. A frequency scale factor of $k_f = 3141.6$ will move the cutoff frequency to 500 Hz. A magnitude scale factor of $k_m = 1000$ allows us to use 1 kΩ resistors in place of 1 Ω resistors. The resulting scaled component values are

$$R = 1\ \text{k}\Omega,$$

$$C_{1a} = 831\ \text{nF},$$

$$C_{2a} = 121\ \text{nF},$$

$$C_{1b} = 344\ \text{nF},$$

$$C_{2b} = 294\ \text{nF}.$$

Finally, we need to specify the resistor values in the inverting amplifier stage to yield a passband gain of 10. Let $R_1 = 1\ \text{k}\Omega$; then

$$R_f = 10R_1 = 10\ \text{k}\Omega.$$

Figure 15.23 compares the magnitude responses of the fourth-order identical cascade filter from Example 15.7 and the Butterworth filter we just designed. Note that both filters provide a passband gain of 10 (20 dB) and a cutoff frequency of 500 Hz, but the Butterworth filter is closer to an ideal low-pass filter due to its flatter passband and steeper rolloff at the cutoff frequency. Thus, the Butterworth design is preferred to the identical cascade design.

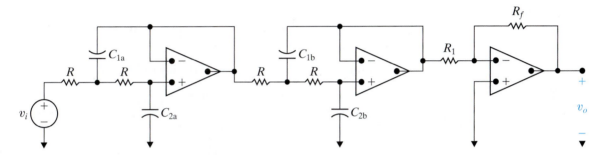

Figure 15.22 ▲ A fourth-order Butterworth filter with nonunity gain.

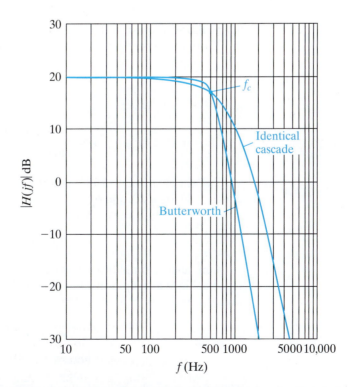

Figure 15.23 ▲ A comparison of the magnitude responses for a fourth-order low-pass filter using the identical cascade and Butterworth designs.

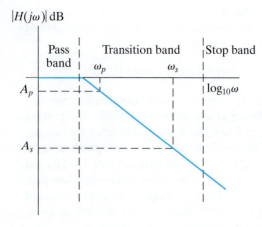

$|H(j\omega)|$ dB

Pass band

Transition band

Stop band

ω_p

ω_s

$\log_{10}\omega$

A_p

A_s

Figure 15.24 ▲ Defining the transition region for a low-pass filter.

The Order of a Butterworth Filter

You have probably noticed that the higher the order of the Butterworth filter, the closer the transfer function magnitude is to that of an ideal low-pass filter. In other words, as n increases, the magnitude stays close to unity in the passband, the transition band narrows, and the magnitude stays close to zero in the stopband. At the same time, as the order increases, the number of circuit components increases. Thus, when designing a filter, determining the smallest value of n that will meet the filtering specifications is an important first step.

The filtering specifications for a low-pass filter usually define the abruptness of the transition region, as shown in Fig. 15.24. Once A_p, ω_p, A_s, and ω_s are specified, the order of the Butterworth filter can be determined.

For the Butterworth filter,

$$A_p = 20 \log_{10} \frac{1}{\sqrt{1 + \omega_p^{2n}}} = -10 \log_{10}(1 + \omega_p^{2n}),$$

$$A_s = 20 \log_{10} \frac{1}{\sqrt{1 + \omega_s^{2n}}} = -10 \log_{10}(1 + \omega_s^{2n}).$$

It follows from the definition of the logarithm that

$$10^{-0.1A_p} = 1 + \omega_p^{2n},$$

$$10^{-0.1A_s} = 1 + \omega_s^{2n}.$$

Now, solve for ω_p^n and ω_s^n and find the ratio $(\omega_s/\omega_p)^n$. We get

$$\left(\frac{\omega_s}{\omega_p}\right)^n = \frac{\sqrt{10^{-0.1A_s} - 1}}{\sqrt{10^{-0.1A_p} - 1}} = \frac{\sigma_s}{\sigma_p},$$

where the symbols σ_s and σ_p have been introduced for convenience. From the expression for σ_s/σ_p we can write

$$n \log_{10}(\omega_s/\omega_p) = \log_{10}(\sigma_s/\sigma_p),$$

or

$$n = \frac{\log_{10}(\sigma_s/\sigma_p)}{\log_{10}(\omega_s/\omega_p)}. \tag{15.15}$$

We can simplify Eq. 15.15 if ω_p is the cutoff frequency because then A_p equals $-20 \log_{10}\sqrt{2}$, and $\sigma_p = 1$. Hence

$$n = \frac{\log_{10}\sigma_s}{\log_{10}(\omega_s/\omega_p)}.$$

One further simplification is possible. We are using a Butterworth filter to achieve a steep transition region. Therefore, the filtering specification will make $10^{-0.1A_s} \gg 1$. Thus

$$\sigma_s \approx 10^{-0.05A_s},$$

$$\log_{10}\sigma_s \approx -0.05A_s.$$

Therefore, a good approximation for the calculation of n is

$$n = \frac{-0.05A_s}{\log_{10}(\omega_s/\omega_p)}. \tag{15.16}$$

Note that $\omega_s/\omega_p = f_s/f_p$, so we can work with frequencies specified in either radians per second or hertz to calculate n.

The order of the filter must be an integer; hence, in using either Eq. 15.15 or Eq. 15.16, we select the nearest integer value greater than the result given by the equation. Examples 15.10 and 15.11 illustrate the use of Eqs. 15.15 and 15.16.

EXAMPLE 15.10 | **Determining the Order of a Butterworth Filter**

a) Determine the order of a Butterworth filter that has a cutoff frequency of 1000 Hz and a gain of no more than -50 dB at 6000 Hz.

b) What is the actual gain in dB at 6000 Hz?

Solution

a) Because the cutoff frequency is specified, we know $\sigma_p = 1$. We also note from the specification that $10^{-0.1(-50)}$ is much greater than 1. Hence, we can use Eq. 15.16 with confidence:

$$n = \frac{(-0.05)(-50)}{\log_{10}(6000/1000)} = 3.21.$$

Therefore, we need a fourth-order Butterworth filter.

b) We can use Eq. 15.12 to calculate the actual gain at 6000 Hz. The gain in decibels will be

$$K = 20\log_{10}\left(\frac{1}{\sqrt{1+6^8}}\right) = -62.25 \text{ dB.}$$

EXAMPLE 15.11 | **An Alternate Approach to Determining the Order of a Butterworth Filter**

a) Determine the order of a Butterworth filter whose magnitude is 10 dB less than the passband magnitude at 500 Hz and at least 60 dB less than the passband magnitude at 5000 Hz.

b) Determine the cutoff frequency of the filter (in hertz).

c) What is the actual gain of the filter (in decibels) at 5000 Hz?

Solution

a) Because the cutoff frequency is not given, we use Eq. 15.15 to determine the order of the filter:

$$\sigma_p = \sqrt{10^{-0.1(-10)} - 1} = 3,$$

$$\sigma_s = \sqrt{10^{-0.1(-60)} - 1} \approx 1000,$$

$$\omega_s/\omega_p = f_s/f_p = 5000/500 = 10,$$

$$n = \frac{\log_{10}(1000/3)}{\log_{10}(10)} = 2.52.$$

Therefore, we need a third-order Butterworth filter to meet the specifications.

b) Knowing that the gain at 500 Hz is -10 dB, we can determine the cutoff frequency. From Eq. 15.12 we can write

$$-10\log_{10}\left[1 + (\omega/\omega_c)^6\right] = -10,$$

where $\omega = 1000\pi$ rad/s. Therefore

$$1 + (\omega/\omega_c)^6 = 10,$$

and

$$\omega_c = \frac{\omega}{\sqrt[6]{9}}$$

$$= 2178.26 \text{ rad/s.}$$

It follows that

$$f_c = 346.68 \text{ Hz.}$$

c) The actual gain of the filter at 5000 Hz is

$$K = -10\log_{10}\left[1 + (5000/346.68)^6\right]$$

$$= -69.54 \text{ dB.}$$

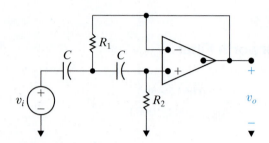

Figure 15.25 ▲ A second-order Butterworth high-pass filter circuit.

Butterworth High-Pass, Bandpass, and Bandreject Filters

An nth-order Butterworth high-pass filter has a transfer function with the nth-order Butterworth polynomial in the denominator, just like the nth-order Butterworth low-pass filter. But in the high-pass filter, the numerator of the transfer function is s^n, whereas in the low-pass filter, the numerator is 1. Again, we use a cascade approach in designing the Butterworth high-pass filter. The first-order factor is achieved by including a prototype high-pass filter (Fig. 15.4, with $R_1 = R_2 = 1\ \Omega$, and $C = 1$ F) in the cascade.

To produce the second-order factors in the Butterworth polynomial, we need a circuit with a transfer function of the form

$$H(s) = \frac{s^2}{s^2 + b_1 s + 1}.$$

Such a circuit is shown in Fig. 15.25.

This circuit has the transfer function

$$H(s) = \frac{V_o}{V_i} = \frac{s^2}{s^2 + \dfrac{2}{R_2 C} s + \dfrac{1}{R_1 R_2 C^2}}.$$

Setting $C = 1$ F yields

$$H(s) = \frac{s^2}{s^2 + \dfrac{2}{R_2} s + \dfrac{1}{R_1 R_2}}. \tag{15.17}$$

Thus, we can realize any second-order factor in a Butterworth polynomial of the form $(s^2 + b_1 s + 1)$ by including in the cascade the second-order circuit in Fig. 15.25 with resistor values that satisfy Eq. 15.18:

$$b_1 = \frac{2}{R_2} \quad \text{and} \quad 1 = \frac{1}{R_1 R_2}. \tag{15.18}$$

At this point, we pause to make a couple of observations about the circuits in Figs. 15.21 and 15.25 and their prototype transfer functions $1/(s^2 + b_1 s + 1)$ and $s^2/(s^2 + b_1 s + 1)$.

- The high-pass circuit in Fig. 15.25 was obtained from the low-pass circuit in Fig. 15.21 by interchanging resistors and capacitors.
- The prototype transfer function of a high-pass filter can be obtained from that of a low-pass filter by replacing s in the low-pass expression with $1/s$ (see Problem 15.46).

These observations are important because they are true in general.

We can use frequency and magnitude scaling to design a Butterworth high-pass filter with practical component values and a cutoff frequency other than 1 rad/s. Adding an inverting amplifier to the cascade will accommodate designs with nonunity passband gains. The problems at the end of the chapter include several Butterworth high-pass filter designs.

Now that we can design both nth-order low-pass and high-pass Butterworth filters with arbitrary cutoff frequencies and passband gains, we can combine these filters in cascade (as we did in Section 15.3) to produce nth-order Butterworth bandpass filters. We can combine these filters in parallel with a summing amplifier (again, as we did in Section 15.3) to produce nth-order Butterworth bandreject filters. Example 15.12 illustrates the design of a Butterworth bandpass filter.

EXAMPLE 15.12 | Designing a Butterworth Bandpass Filter

Design the bandpass filter from Example 15.5 using a cascade of a fourth-order low-pass Butterworth filter, a fourth-order high-pass Butterworth filter, and an inverting amplifier. The filter should provide an amplification of 2 within the band of frequencies between 100 and 10,000 Hz and use as many 2 kΩ resistors and 0.2 μF capacitors as possible.

Solution

The circuit is shown in Fig. 15.26. The first two stages in the cascade form the fourth-order low-pass Butterworth filter, which should have a cutoff frequency of 10,000 Hz. We have already designed the prototype fourth-order low-pass Butterworth filter in Example 15.9, so $R = 1\ \Omega$ and

$$C_{1a} = 2.61\ \text{F}, \quad C_{2a} = 0.38\ \text{F},$$

$$C_{1b} = 1.08\ \text{F}, \quad C_{2b} = 0.924\ \text{F}.$$

This prototype low-pass filter has a cutoff frequency of 1 rad/s, so to move the cutoff frequency to 10,000 Hz we need a frequency scale factor $k_f = 20{,}000\pi$. To use 2 kΩ resistors, we need to scale the 1 Ω resistors using a magnitude scale factor $k_m = 2000$. The scaled capacitor values are

$$C_{1a} = 20.77\ \text{nF}, \quad C_{2a} = 3.02\ \text{nF},$$

$$C_{1b} = 8.59\ \text{nF}, \quad C_{2b} = 7.35\ \text{nF}.$$

The next two stages in the cascade form the fourth-order high-pass Butterworth filter, which

should have a cutoff frequency of 100 Hz. To design the prototype filter, let $C = 1$ F. The first stage of the prototype high-pass filter must have a transfer function whose denominator is the Butterworth polynomial $(s^2 + 0.765\,s + 1)$. From Eq. 15.18,

$$R_{2a} = \frac{2}{0.765} = 2.61\ \Omega, \quad R_{1a} = \frac{1}{R_{2a}} = \frac{1}{2.61} = 0.38\ \Omega.$$

The second stage of the prototype high-pass filter must have a transfer function whose denominator is the Butterworth polynomial $(s^2 + 1.848\,s + 1)$. From Eq. 15.18,

$$R_{2b} = \frac{2}{1.848} = 1.08\ \Omega, \quad R_{1b} = \frac{1}{R_{2b}} = \frac{1}{1.08} = 0.924\ \Omega.$$

This prototype high-pass filter has a cutoff frequency of 1 rad/s, so to move the cutoff frequency to 100 Hz, we need a frequency scale factor $k_f = 200\pi$. To use 0.2 μF capacitors, we need a magnitude scale factor $k_m = 7957.747$. The scaled resistor values are

$$R_{1a} = 3023.94\ \Omega, \quad R_{2a} = 20{,}769.72\ \Omega,$$

$$R_{1b} = 7352.96\ \Omega, \quad R_{2b} = 8594.37\ \Omega.$$

The final stage of the cascade in Fig. 15.26 is the inverting amplifier, which must have a gain of 2. Let's use $R = 2$ kΩ, so $R_f = 2R = 4$ kΩ.

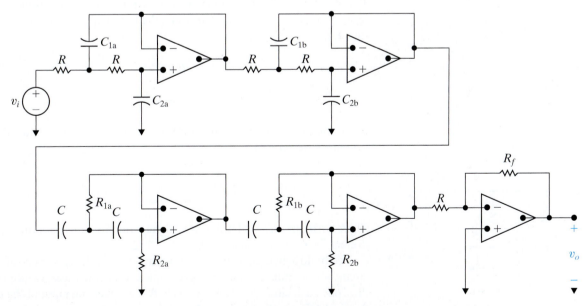

Figure 15.26 ▲ The fourth-order Butterworth bandpass filter for Example 15.12.

ASSESSMENT PROBLEM

Objective 3—Understand how to use cascaded first- and second-order Butterworth filters

15.4 For the circuit in Fig. 15.25, find values of R_1 and R_2 that yield a second-order prototype Butterworth high-pass filter.

Answer: $R_1 = 0.707 \ \Omega$, $R_2 = 1.41 \ \Omega$.

SELF-CHECK: Also try Chapter Problems 15.34, 15.36, and 15.40.

15.5 Narrowband Bandpass and Bandreject Filters

We can only implement broadband or low-Q bandpass and bandreject filters when using the cascade and parallel subcircuit designs from the previous sections. This limitation is due principally to the form of the transfer functions for cascaded bandpass and parallel bandreject filters—they only have discrete real poles. The largest quality factor we can achieve with discrete real poles arises when the cutoff frequencies, and thus the pole locations, are the same. As an example, consider the transfer function for a bandpass filter created from a cascade of a first-order low-pass filter and a first-order high-pass filter, where both filters have the same cutoff frequency:

$$H(s) = \left(\frac{-\omega_c}{s + \omega_c} \right)\left(\frac{-s}{s + \omega_c} \right)$$

$$= \frac{s\omega_c}{s^2 + 2\omega_c s + \omega_c^2}$$

$$= \frac{0.5\beta s}{s^2 + \beta s + \omega_c^2}.$$

This equation is in the standard form of the transfer function of a bandpass filter, and thus we can determine the bandwidth and center frequency directly:

$$\beta = 2\omega_c,$$

$$\omega_o^2 = \omega_c^2.$$

From the equations for bandwidth and center frequency and the definition of Q, we see that

$$Q = \frac{\omega_o}{\beta} = \frac{\omega_c}{2\omega_c} = \frac{1}{2}.$$

Thus, with discrete real poles, the highest quality bandpass filter (or bandreject filter) we can achieve has $Q = 1/2$.

To build active filters with high-quality factor values, we need an op amp circuit whose transfer function has complex conjugate poles. Figure 15.27 depicts one such circuit for us to analyze. At the inverting input of the op amp, write a KCL equation to get

$$\frac{V_a}{1/sC} = \frac{-V_o}{R_3}.$$

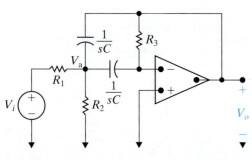

Figure 15.27 ▲ An active high-Q bandpass filter.

Solving for V_a,

$$V_a = \frac{-V_o}{sR_3C}.$$

At the node labeled V_a, write a KCL equation to get

$$\frac{V_i - V_a}{R_1} = \frac{V_a - V_o}{1/sC} + \frac{V_a}{1/sC} + \frac{V_a}{R_2}.$$

Solving for V_i,

$$V_i = (1 + 2sR_1C + R_1/R_2)V_a - sR_1CV_o.$$

Substituting the equation for V_a into the equation for V_i and then rearranging, we get an expression for the transfer function V_o/V_i:

$$H(s) = \frac{\dfrac{-s}{R_1C}}{s^2 + \dfrac{2}{R_3C}s + \dfrac{1}{R_{eq}R_3C^2}}, \tag{15.19}$$

where

$$R_{eq} = R_1 \| R_2 = \frac{R_1R_2}{R_1 + R_2}.$$

Equation 15.19 is in the standard form for a bandpass filter transfer function; that is,

$$H(s) = \frac{-K\beta s}{s^2 + \beta s + \omega_o^2},$$

so we can equate terms and solve for the values of the resistors that will achieve a specified center frequency (ω_o), quality factor (Q), and passband gain (K):

$$\beta = \frac{2}{R_3C};$$

$$K\beta = \frac{1}{R_1C};$$

$$\omega_o^2 = \frac{1}{R_{eq}R_3C^2}.$$

Let's define the prototype version of the circuit in Fig. 15.27 as a bandpass filter with $\omega_o = 1$ rad/s and $C = 1$ F. Then the expressions for R_1, R_2, and R_3 can be given in terms of the desired quality factor and passband gain. You should verify (in Problem 15.57) that for the prototype circuit, the expressions for R_1, R_2, and R_3 are

$$R_1 = Q/K,$$

$$R_2 = Q/(2Q^2 - K), \tag{15.20}$$

$$R_3 = 2Q.$$

Use scaling to achieve the desired center frequency and to specify practical values for the circuit components. This design process is illustrated in Example 15.13.

EXAMPLE 15.13   Designing a High-Q Bandpass Filter

Design a bandpass filter, using the circuit in Fig. 15.27, that has a center frequency of 3000 Hz, a quality factor of 10, and a passband gain of 2. Use 0.01 μF capacitors in your design. Compute the transfer function of your circuit, and sketch a Bode plot of its magnitude response.

Solution

Since $Q = 10$ and $K = 2$, the values for R_1, R_2, and R_3 in the prototype circuit, from Eq. 15.20, are

$$R_1 = \frac{Q}{K} = \frac{10}{2} = 5 \, \Omega,$$

$$R_2 = \frac{Q}{2Q^2 - K} = \frac{10}{200 - 2} = \frac{10}{198} \, \Omega,$$

$$R_3 = 2Q = 2(10) = 20 \, \Omega.$$

The scaling factors are $k_f = 6000\pi$, to move the center frequency from 1 rad/s to 3000 Hz, and $k_m = 10^8/k_f$, so we can use 0.01 μF capacitors. After scaling,

$$R_1 = 26.5 \, \text{k}\Omega,$$

$$R_2 = 268.0 \, \Omega,$$

$$R_3 = 106.1 \, \text{k}\Omega.$$

The circuit is shown in Fig. 15.28.

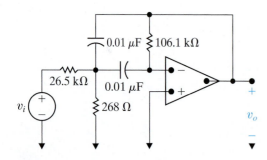

Figure 15.28 ▲ The high-Q bandpass filter designed in Example 15.13.

Substituting the values of resistance and capacitance in Eq. 15.19 gives the transfer function for this circuit:

$$H(s) = \frac{-3770s}{s^2 + 1885.0s + 355 \times 10^6}.$$

It is easy to see that this transfer function meets the specification of the bandpass filter defined in the example. A Bode plot of its magnitude response is sketched in Fig. 15.29.

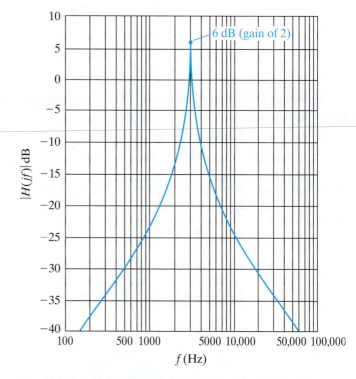

Figure 15.29 ▲ The Bode magnitude plot for the high-Q bandpass filter designed in Example 15.13.

The parallel implementation of a bandreject filter that combines low-pass and high-pass filter components with a summing amplifier has the same low-Q restriction as the cascaded bandpass filter. The circuit in Fig. 15.30 is an active high-Q bandreject filter known as the **twin-T notch filter** because of the two T-shaped parts of the circuit at the nodes labeled a and b.

We begin analyzing this circuit by writing a KCL equation at the node labeled V_a:

$$(V_a - V_i)sC + (V_a - V_o)sC + \frac{2(V_a - \sigma V_o)}{R} = 0$$

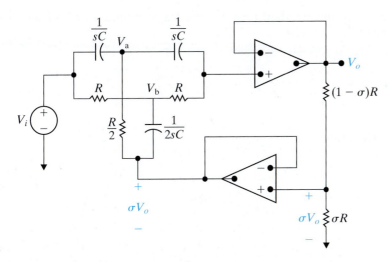

Figure 15.30 ▲ A high-Q active bandreject filter.

or

$$V_a[2sCR + 2] - V_o[sCR + 2\sigma] = sCRV_i.$$

Writing a KCL equation at the node labeled V_b yields

$$\frac{V_b - V_i}{R} + \frac{V_b - V_o}{R} + (V_b - \sigma V_o)2sC = 0$$

or

$$V_b[2 + 2RCs] - V_o[1 + 2\sigma RCs] = V_i.$$

Writing a KCL equation at the noninverting input terminal of the top op amp gives

$$(V_o - V_a)sC + \frac{V_o - V_b}{R} = 0$$

or

$$-sRCV_a - V_b + (sRC + 1)V_o = 0.$$

From the three KCL equations, we can use Cramer's rule to solve for V_o:

$$V_o = \frac{\begin{vmatrix} 2(RCs + 1) & 0 & sCRV_i \\ 0 & 2(RCs + 1) & V_i \\ -RCs & -1 & 0 \end{vmatrix}}{\begin{vmatrix} 2(RCs + 1) & 0 & -(RCs + 2\sigma) \\ 0 & 2(RCs + 1) & -(2\sigma RCs + 1) \\ -RCs & -1 & RCs + 1 \end{vmatrix}}$$

$$= \frac{(R^2C^2s^2 + 1)V_i}{R^2C^2s^2 + 4RC(1 - \sigma)s + 1}.$$

Rearranging this equation for V_o, we can solve for the transfer function:

$$H(s) = \frac{V_o}{V_i} = \frac{\left(s^2 + \dfrac{1}{R^2C^2}\right)}{\left[s^2 + \dfrac{4(1 - \sigma)}{RC}s + \dfrac{1}{R^2C^2}\right]}, \qquad (15.21)$$

which is in the standard form for the transfer function of a bandreject filter:

$$H(s) = \frac{s^2 + \omega_0^2}{s^2 + \beta s + \omega_0^2}.$$

Therefore,

$$\omega_o^2 = \frac{1}{R^2 C^2},$$

$$\beta = \frac{4(1 - \sigma)}{RC}.$$

In this circuit, we have three parameters (R, C, and σ) and two design constraints (ω_o and β). Thus, one parameter is chosen arbitrarily; it is usually the capacitor value because it normally has the fewest commercially available options. Once C is chosen,

$$R = \frac{1}{\omega_o C},$$

$$\sigma = 1 - \frac{\beta}{4\omega_o} = 1 - \frac{1}{4Q}.$$

(15.22)

Example 15.14 illustrates the design of a high-Q active bandreject filter.

EXAMPLE 15.14 | **Designing a High-Q Bandreject Filter**

Design a high-Q active bandreject filter (based on the circuit in Fig. 15.30) with a center frequency of 5000 rad/s and a bandwidth of 1000 rad/s. Use 1 μF capacitors in your design.

100 Ω ($R/2$), 190 Ω (σR), and 10 Ω [$(1 - \sigma)R$]. The final design is depicted in Fig. 15.31, and the Bode magnitude plot is shown in Fig. 15.32.

Solution

In the bandreject prototype filter, $\omega_o = 1$ rad/s, $R = 1\,\Omega$, and $C = 1$ F. Once ω_o and Q are determined from the filter specifications, C can be chosen arbitrarily, and R and σ can be found from Eqs. 15.22. From the specifications, $\omega_o = 5000$ rad/s and $Q = 5$. Using Eqs. 15.22, we see that

$$R = \frac{1}{\omega_o C} = \frac{1}{(5000)(10^{-6})} = 200\,\Omega,$$

$$\sigma = 1 - \frac{1}{4Q} = 1 - \frac{1}{4(5)} = 0.95.$$

Therefore, we need resistors with the values 200 Ω (R),

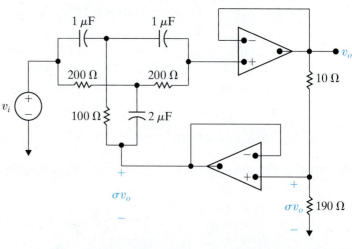

Figure 15.31 ▲ The high-Q active bandreject filter designed in Example 15.14.

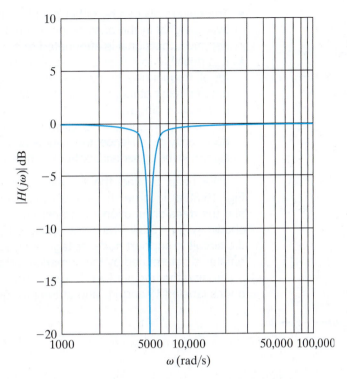

Figure 15.32 ▲ The Bode magnitude plot for the high-Q active bandreject filter designed in Example 15.14.

ASSESSMENT PROBLEMS

Objective 4—Be able to use design equations to calculate component values for prototype narrowband bandpass, and bandreject filters

15.5 Design an active bandpass filter with $Q = 8$, $K = 5$, and $\omega_o = 1000$ rad/s. Use 1 μF capacitors, and specify the values of all resistors.

Answer: $R_1 = 1.6$ kΩ, $R_2 = 65.04$ Ω, $R_3 = 16$ kΩ.

15.6 Design an active unity-gain bandreject filter with $\omega_o = 1000$ rad/s and $Q = 4$. Use 2 μF capacitors in your design, and specify the values of R and σ.

Answer: $R = 500$ Ω, $\sigma = 0.9375$.

SELF-CHECK: Also try Chapter Problems 15.58 and 15.60.

■ **Practical Perspective**

Bass Volume Control

We now look at an op amp circuit that can be used to control audio signal amplification in the bass range. Signals in the audio range have frequencies from 20 Hz to 20 kHz. The bass range includes frequencies up to 300 Hz. The volume control circuit and its frequency response are shown in Fig. 15.33. We can select a desired frequency response curve from the family of curves in Fig. 15.33(b) by adjusting the potentiometer in Fig. 15.33(a).

Study the frequency response curves in Fig. 15.33(b) and note the following.

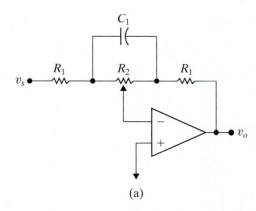

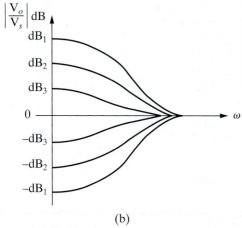

Figure 15.33 ▲ (a) Bass volume control circuit;
(b) Bass volume control circuit frequency response.

- The gain in dB can be either positive or negative. If the gain is positive, a signal in the bass range is amplified or boosted. If the gain is negative, the signal is attenuated or cut.
- It is possible to select a response curve that has unity gain (zero dB) for all frequencies in the bass range. As we shall see, if the potentiometer is set at its midpoint, the circuit will have no effect on signals in the bass range.
- As the frequency increases, all frequency response curves approach zero dB or unity gain. Hence, the volume control circuit will have no effect on signals in the upper end, or treble range, of the audio frequencies.

We need to find the transfer function, V_o/V_s, for the circuit in Fig. 15.33(a) in order to analyze the circuit's frequency response. To find the transfer function, we transform the circuit into the s domain as shown in Fig. 15.34. The node voltages V_a and V_b have been labeled in the circuit to support node-voltage analysis. The position of the potentiometer is determined by the numerical value of α, as noted in Fig. 15.34.

To find the transfer function, we write the three KCL equations at the nodes labeled V_a and V_b, and at the inverting input of the op amp:

$$\frac{V_a}{(1-\alpha)R_2} + \frac{V_a - V_s}{R_1} + (V_a - V_b)s\,C_1 = 0;$$

$$\frac{V_b}{\alpha R_2} + (V_b - V_a)s\,C_1 + \frac{V_b - V_o}{R_1} = 0;$$

$$\frac{V_a}{(1-\alpha)R_2} + \frac{V_b}{\alpha R_2} = 0.$$

Solve the three node-voltage equations to find V_o as a function of V_s and hence the transfer function $H(s)$:

$$H(s) = \frac{V_o}{V_s} = \frac{-(R_1 + \alpha R_2 + R_1 R_2 C_1 s)}{R_1 + (1-\alpha)R_2 + R_1 R_2 C_1 s}.$$

It follows directly that

$$H(j\omega) = \frac{-(R_1 + \alpha R_2 + j\omega R_1 R_2 C_1)}{[R_1 + (1-\alpha)R_2 + j\omega R_1 R_2 C_1]}.$$

Now let's verify that this transfer function generates the family of frequency response curves depicted in Fig. 15.33(b). First note that when $\alpha = 0.5$ the magnitude of $H(j\omega)$ is unity for all frequencies, that is,

$$|H(j\omega)| = \frac{|R_1 + 0.5R_2 + j\omega R_1 R_2 C_1|}{|R_1 + 0.5R_2 + j\omega R_1 R_2 C_1|} = 1.$$

When $\omega = 0$, we have

$$|H(j0)| = \frac{R_1 + \alpha R_2}{R_1 + (1-\alpha)R_2}.$$

Observe that $|H(j0)|$ at $\alpha = 1$ is the reciprocal of $|H(j0)|$ at $\alpha = 0$; that is

$$|H(j0)|_{\alpha=1} = \frac{R_1 + R_2}{R_1} = \frac{1}{|H(j0)|_{\alpha=0}}.$$

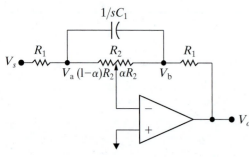

Figure 15.34 ▲ The s-domain circuit for the bass volume control. Note that α determines the potentiometer setting, so $0 \leq \alpha \leq 1$.

In fact, this reciprocal relationship holds for all frequencies, not just $\omega = 0$. For example, $\alpha = 0.4$ and $\alpha = 0.6$ are symmetric about $\alpha = 0.5$ and

$$H(j\omega)_{\alpha=0.4} = \frac{-(R_1 + 0.4R_2) + j\omega R_1 R_2 C_1}{(R_1 + 0.6R_2) + j\omega R_1 R_2 C_1}$$

while

$$H(j\omega)_{\alpha=0.6} = \frac{-(R_1 + 0.6R_2) + j\omega R_1 R_2 C_1}{(R_1 + 0.4R_2) + j\omega R_1 R_2 C_1}.$$

Hence

$$H(j\omega)_{\alpha=0.4} = \frac{1}{H(j\omega)_{\alpha=0.6}}.$$

The volume control circuit can either amplify or attenuate its input signal, depending on the value of α.

The numerical values of R_1, R_2, and C_1 are based on two design specifications. The first specification is the passband amplification or attenuation in the bass range (as $\omega \to 0$). The second specification is the frequency at which this passband amplification or attenuation is changed by 3 dB. The component values that satisfy the specifications are calculated with α equal to either 1 or 0.

As we have already observed, the maximum gain will be $(R_1 + R_2)/R_1$, and the maximum attenuation will be $R_1/(R_1 + R_2)$. If we assume $(R_1 + R_2)/R_1 \gg 1$, then the amplification (or attenuation) will differ by 3 dB from its maximum value when $\omega = 1/R_2 C_1$. This can be seen by noting that

$$\left| H\left(j\frac{1}{R_2 C_1}\right) \right|_{\alpha=1} = \frac{|R_1 + R_2 + jR_1|}{|R_1 + jR_1|}$$

$$= \frac{\left| \dfrac{R_1 + R_2}{R_1} \right|}{|1 + j1|} \approx \frac{1}{\sqrt{2}}\left(\frac{R_1 + R_2}{R_1}\right)$$

and $\left| H\left(j\frac{1}{R_2 C_1}\right) \right|_{\alpha=0} = \frac{|R_1 + jR_1|}{|R_1 + R_2 + jR_1|}$

$$= \frac{|1 + j1|}{\left| \left| \dfrac{R_1 + R_2}{R_1} + j1 \right| \right|} \approx \sqrt{2}\left(\frac{R_1}{R_1 + R_2}\right).$$

SELF-CHECK: Assess your understanding of this Practical Perspective by trying Chapter Problems 15.61 and 15.62.

■ Summary

- Active filters consist of op amps, resistors, and capacitors. They can be configured as low-pass, high-pass, bandpass, and bandreject filters. They overcome many of the disadvantages associated with passive filters. (See page 572.)

- A **prototype low-pass filter** has component values of $R_1 = R_2 = 1\ \Omega$ and $C = 1\ F$, and it produces a unity passband gain and a cutoff frequency of 1 rad/s. The **prototype high-pass filter** has the same component values and also produces a unity passband gain and a cutoff frequency of 1 rad/s. (See pages 575 and 576.)

- **Magnitude scaling** can be used to alter component values without changing the frequency response of a circuit. For a magnitude scale factor of k_m, the scaled (primed) values of resistance, capacitance, and inductance are

$$R' = k_m R, \quad L' = k_m L, \quad \text{and} \quad C' = C/k_m.$$

(See page 577.)

- **Frequency scaling** can be used to shift the frequency response of a circuit to another frequency region without changing the overall shape of the frequency response. For a frequency scale factor of k_f, the scaled (primed) values of resistance, capacitance, and inductance are

$$R' = R, \quad L' = L/k_f, \quad \text{and} \quad C' = C/k_f.$$

(See page 578.)

- Components can be scaled in both magnitude and frequency, with the scaled (primed) component values given by

$$R' = k_m R, \quad L' = (k_m/k_f)L, \quad \text{and} \quad C' = C/(k_m k_f).$$

(See page 578.)

- The design of active low-pass and high-pass filters can begin with a prototype filter circuit. Scaling can then be applied to shift the frequency response to the desired cutoff frequency, using component values that are commercially available. (See page 578.)

- An active broadband bandpass filter can be constructed using a cascade of a low-pass filter with the bandpass filter's upper cutoff frequency, a high-pass filter with the bandpass filter's lower cutoff frequency, and (optionally) an inverting amplifier gain stage to achieve nonunity gain in the passband. Bandpass filters implemented in this fashion must be broadband filters ($\omega_{c2} \gg \omega_{c1}$), so that each op amp circuit in the cascade can be specified independently. (See page 579.)

- An active broadband bandreject filter can be constructed using a parallel combination of a low-pass filter with the bandreject filter's lower cutoff frequency and a high-pass filter with the bandreject filter's upper cutoff frequency. The outputs are then fed into a summing amplifier, which can produce nonunity gain in the passband. Bandreject filters implemented in this way must be broadband filters ($\omega_{c2} \gg \omega_{c1}$), so that the low-pass and high-pass filter circuits can be designed independently. (See page 585.)

- Higher-order active filters have multiple poles in their transfer functions, resulting in a sharper transition from the passband to the stopband and thus a more nearly ideal frequency response. (See page 588.)

- The transfer function of an nth-order Butterworth low-pass filter with a cutoff frequency of 1 rad/s can be determined from the equation

$$H(s)H(-s) = \frac{1}{1 + (-1)^n s^{2n}}$$

by

- finding the roots of the denominator polynomial

- assigning the left-half plane roots to $H(s)$

- writing the denominator of $H(s)$ as a product of first- and second-order factors

(See pages 591–592.)

- The fundamental problem in the design of a Butterworth filter is to determine the order of the filter. The filter specification usually defines the sharpness of the transition band in terms of the quantities A_p, ω_p, A_s, and ω_s. From these quantities, we calculate the smallest integer larger than the solution to either Eq. 15.15 or Eq. 15.16. (See page 596.)

- A cascade of second-order low-pass op amp filters (Fig. 15.21) with 1 Ω resistors and capacitor values chosen to produce each factor in the Butterworth polynomial will produce an even-order Butterworth low-pass filter. Adding a prototype low-pass op amp filter will produce an odd-order Butterworth low-pass filter. (See page 593.)

- A cascade of second-order high-pass op amp filters (Fig. 15.25) with 1 F capacitors and resistor values chosen to produce each factor in the Butterworth polynomial will produce an even-order Butterworth high-pass filter. Adding a prototype high-pass op amp filter will produce an odd-order Butterworth high-pass filter. (See page 598.)

- For both high- and low-pass Butterworth filters, frequency and magnitude scaling can be used to shift the cutoff frequency from 1 rad/s and to include realistic component values in the design. Cascading an inverting amplifier will produce a nonunity passband gain. (See page 598.)

- Butterworth low-pass and high-pass filters can be cascaded to produce Butterworth bandpass filters of any order n. Butterworth low-pass and high-pass filters can be combined in parallel with a summing amplifier to produce a Butterworth bandreject filter of any order n. (See page 599.)

- If a high-Q, or narrowband, bandpass, or bandreject filter is needed, the cascade or parallel combination will not work. Instead, the circuits shown in Figs. 15.27 and 15.30 are used with the appropriate design equations. Typically, capacitor values are chosen from those commercially available, and the design equations are used to specify the resistor values. (See pages 600 and 603.)

Problems

Section 15.1

15.1 a) Using the circuit in Fig. 15.1, design a low-pass filter with a passband gain of 5 dB and a cutoff frequency of 10 kHz. Assume a 50 nF capacitor is available.

DESIGN PROBLEM

    b) Draw the circuit diagram and label all components.

15.2 a) Using only three components from Appendix H, design a low-pass filter with a cutoff frequency and passband gain as close as possible to the specifications in Problem 15.1(a). Draw the circuit diagram and label all component values.

    b) Calculate the percent error in this new filter's cutoff frequency and passband gain when compared to the values that are specified in Problem 15.1(a).

15.3 Design an op amp-based low-pass filter with a cutoff frequency of 500 Hz and a passband gain of 8 using a 5 μF capacitor.

    a) Draw your circuit, labeling the component values and output voltage.

    b) If the value of the feedback resistor in the filter is changed but the value of the resistor in the forward path is unchanged, what characteristic of the filter is changed?

15.4 The low-pass filter designed in Problem 15.3 has an input of 2 cos ωt V.

    a) Suppose the power supplies are $\pm V_{cc}$. What is the smallest value of V_{cc} that will still cause the op amp to operate in its linear region?

    b) Find the output voltage when $\omega = \omega_c$.

    c) Find the output voltage when $\omega = 0.1\omega_c$.

    d) Find the output voltage when $\omega = 10\omega_c$.

15.5 Find the transfer function V_o/V_i for the circuit shown in Fig. P15.5 if Z_f is the equivalent impedance of the feedback circuit, Z_i is the equivalent impedance of the input circuit, and the operational amplifier is ideal.

Figure P15.5

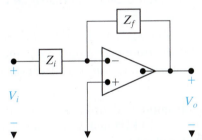

15.6 a) Use the results of Problem 15.5 to find the transfer function of the circuit shown in Fig. P15.6.

    b) What is the gain of the circuit as $\omega \to 0$?

    c) What is the gain of the circuit as $\omega \to \infty$?

    d) Do your answers to (b) and (c) make sense in terms of known circuit behavior?

Figure P15.6

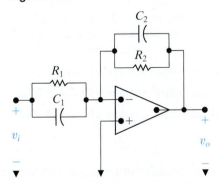

15.7 Repeat Problem 15.6, using the circuit shown in Fig. P15.7.

Figure P15.7

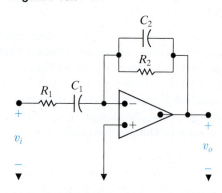

15.8 a) Use the circuit in Fig. 15.4 to design a high-pass
filter with a cutoff frequency of 2.5 kHz and a
passband gain of 10 dB. Use a 50 nF capacitor in
the design.

b) Draw the circuit diagram of the filter and label
all the components.

15.9 Using only three components from Appendix H,
design a high-pass filter with a cutoff frequency and
passband gain as close as possible to the specifications in Problem 15.8.

a) Draw the circuit diagram and label all component values.

b) Calculate the percent error in this new filter's cutoff frequency and passband gain
when compared to the values specified in
Problem 15.8(a).

15.10 Design an op amp-based high-pass filter with a cutoff frequency of 10 kHz and a passband gain of 4
using a 75 nF capacitor.

a) Draw your circuit, labeling the component values and the output voltage.

b) If the value of the feedback resistor in the filter
is changed but the value of the resistor in the
forward path is unchanged, what characteristic
of the filter is changed?

15.11 The input to the high-pass filter designed in
Problem 15.10 is $4 \cos \omega t$ V.

a) Suppose the power supplies are $\pm V_{cc}$. What is
the smallest value of V_{cc} that will still cause the
op amp to operate in its linear region?

b) Find the output voltage when $\omega = \omega_c$.

c) Find the output voltage when $\omega = 0.1\omega_c$.

d) Find the output voltage when $\omega = 10\omega_c$.

Section 15.2

15.12 The voltage transfer function for either high-pass
prototype filter shown in Fig. P15.12 is

$$H(s) = \frac{s}{s + 1}.$$

Show that if either circuit is scaled in both magnitude and frequency, the scaled transfer function is

$$H'(s) = \frac{(s/k_f)}{(s/k_f) + 1}.$$

Figure P15.12

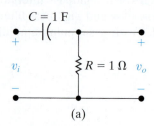

(a)

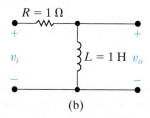

(b)

15.13 The voltage transfer function of either low-pass
prototype filter shown in Fig. P15.13 is

$$H(s) = \frac{1}{s + 1}.$$

Show that if either circuit is scaled in both magnitude and frequency, the scaled transfer function is

$$H'(s) = \frac{1}{(s/k_f) + 1}.$$

Figure P15.13

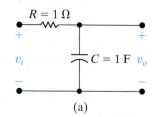

(a)

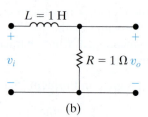

(b)

15.14 The voltage transfer function of the prototype
bandpass filter shown in Fig. P15.14 is

$$H(s) = \frac{\left(\dfrac{1}{Q}\right)s}{s^2 + \left(\dfrac{1}{Q}\right)s + 1}.$$

Show that if the circuit is scaled in both magnitude and frequency, the scaled transfer function is

$$H'(s) = \frac{\left(\dfrac{1}{Q}\right)\left(\dfrac{s}{k_f}\right)}{\left(\dfrac{s}{k_f}\right)^2 + \left(\dfrac{1}{Q}\right)\left(\dfrac{s}{k_f}\right) + 1}.$$

Figure P15.14

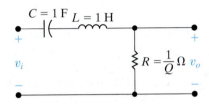

15.15 a) Specify the component values for the prototype passive bandpass filter described in Problem 15.14 if the quality factor of the filter is 10.

b) Specify the component values for the scaled bandpass filter described in Problem 15.14 if the quality factor is 10; the center, or resonant, frequency is 75 krad/s; and the impedance at resonance is 2.5 kΩ.

c) Draw a circuit diagram of the scaled filter and label all the components.

15.16 An alternative to the prototype bandpass filter illustrated in Fig. P15.14 is to make $\omega_o = 1$ rad/s, $R = 1\ \Omega$, and $L = Q$ henrys.

a) What is the value of C in the prototype filter circuit?

b) What is the transfer function of the prototype filter?

c) Use the alternative prototype circuit just described to design a passive bandpass filter that has a quality factor of 12, a center frequency of 40 krad/s, and an impedance of 50 kΩ at resonance.

d) Draw a diagram of the scaled filter and label all the components.

e) Use the results obtained in Problem 15.14 to write the transfer function of the scaled circuit.

15.17 The passive bandpass filter illustrated in Fig. 14.22 has two prototype circuits. In the first prototype circuit, $\omega_o = 1$ rad/s, $C = 1$ F, $L = 1$ H, and $R = Q$ ohms. In the second prototype circuit, $\omega_o = 1$ rad/s, $R = 1\ \Omega$, $C = Q$ farads, and $L = (1/Q)$ henrys.

DESIGN PROBLEM

a) Use one of these prototype circuits (your choice) to design a passive bandpass filter that has a quality factor of 15 and a center frequency of 100 krad/s. The resistor R is 10 kΩ.

b) Draw a circuit diagram of the scaled filter and label all components.

15.18 The passive bandreject filter illustrated in Fig. 14.28(a) has the two prototype circuits shown in Fig. P15.18.

a) Show that for both circuits, the transfer function is

$$H(s) = \frac{s^2 + 1}{s^2 + \left(\dfrac{1}{Q}\right)s + 1}.$$

b) Write the transfer function for a bandreject filter that has a center frequency of 10 krad/s and a quality factor of 8.

Figure P15.18

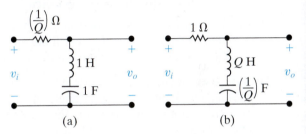

(a)   (b)

15.19 The transfer function for the bandreject filter shown in Fig. 14.28(a) is

$$H(s) = \frac{s^2 + \left(\dfrac{1}{LC}\right)}{s^2 + \left(\dfrac{R}{L}\right)s + \left(\dfrac{1}{LC}\right)}.$$

Show that if the circuit is scaled in both magnitude and frequency, the transfer function of the scaled circuit is equal to the transfer function of the unscaled circuit with s replaced by (s/k_f), where k_f is the frequency scale factor.

15.20 Show that the observation made in Problem 15.19 with respect to the transfer function for the circuit in Fig. 14.28(a) also applies to the bandreject filter circuit (lower one) in Fig. 14.31.

15.21 The two prototype versions of the passive bandreject filter shown in Fig. 14.31 (lower circuit) are shown in Fig. P15.21(a) and (b).

Show that the transfer function for either version is

$$H(s) = \frac{s^2 + 1}{s^2 + \left(\dfrac{1}{Q}\right)s + 1}.$$

Figure P15.21

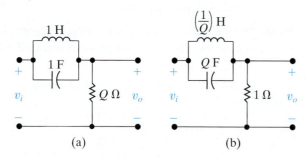

(a)   (b)

15.22 The circuit in Fig. P9.27 is scaled so that the 100 Ω resistor is replaced by 500 Ω resistor and the 160 μH inductor is replaced by a 2 mH inductor.

  a) What is the scaled value of the capacitor?

  b) Find the frequency for which the impedance Z_{ab} is purely resistive for the scaled circuit.

  c) How is the frequency you found in part (b) related to the frequency for which the impedance Z_{ab} is purely resistive in the un-scaled circuit?

15.23 Scale the inductor and capacitor in Fig. P9.66 so that the magnitude and the phase angle of the output current do not change when the input frequency is changed from 20 krad/s to 500 rad/sec.

  a) What are the scaled values of the inductor and capacitor?

  b) What is the steady-state value of the output current, i_o, when the input current is 400 cos 500t mA?

15.24 Scale the bandpass filter in Problem 14.30 so that the center frequency is 200 kHz and the quality factor is still 8, using a 2.5 nF capacitor. Determine the values of the resistor and the inductor, and the two cutoff frequencies of the scaled filter.

15.25 Scale the bandreject filter in Problem 14.41 to get a center frequency of 50 krad/s, using a 200 μH inductor. Determine the values of the resistor, the capacitor, and the bandwidth of the scaled filter.

15.26 a) Show that if the low-pass filter circuit illustrated in Fig. 15.1 is scaled in both magnitude and frequency, the transfer function of the scaled circuit is the same as Eq. 15.1 with s replaced by s/k_f, where k_f is the frequency scale factor.

  b) In the prototype version of the low-pass filter circuit in Fig. 15.1, $\omega_c = 1$ rad/s, $C = 1$ F, $R_2 = 1$ Ω, and $R_1 = 1/K$ ohms. What is the transfer function of the prototype circuit?

  c) Using the result obtained in (a), derive the transfer function of the scaled filter.

15.27 a) Show that if the high-pass filter illustrated in Fig. 15.4 is scaled in both magnitude and frequency, the transfer function is the same as Eq. 15.4 with

s replaced by s/k_f, where k_f is the frequency scale factor.

  b) In the prototype version of the high-pass filter circuit in Fig. 15.4, $\omega_c = 1$ rad/s, $R_1 = 1$ Ω, $C = 1$ F, and $R_2 = K$ ohms. What is the transfer function of the prototype circuit?

  c) Using the result in (a), derive the transfer function of the scaled filter.

Section 15.3

15.28 a) Using 100 nF capacitors, design an active broadband first-order bandpass filter that has a lower cutoff frequency of 1000 Hz, an upper cutoff frequency of 5000 Hz, and a passband gain of 0 dB. Use prototype versions of the low-pass and high-pass filters in the design process (see Problems 15.26 and 15.27).

  b) Write the transfer function for the scaled filter.

  c) Use the transfer function derived in part (b) to find $H(j\omega_o)$, where ω_o is the center frequency of the filter.

  d) What is the passband gain (in decibels) of the filter at ω_o?

  e) Using a computer program of your choice, make a Bode magnitude plot of the filter.

15.29 a) Using 10 nF capacitors, design an active broadband first-order bandreject filter with a lower cutoff frequency of 400 Hz, an upper cutoff frequency of 4000 Hz, and a passband gain of 0 dB. Use the prototype filter circuits introduced in Problems 15.26 and 15.27 in the design process.

  b) Draw the circuit diagram of the filter and label all of the components.

  c) What is the transfer function of the scaled filter?

  d) Evaluate the transfer function derived in (c) at the center frequency of the filter.

  e) What is the gain (in decibels) at the center frequency?

  f) Using a computer program of your choice, make a Bode magnitude plot of the filter transfer function.

15.30 Design a bandpass filter, using a cascade connection, to give a center frequency of 500 Hz, a bandwidth of 4 kHz, and a passband gain of 4. Use 250 nF capacitors. Specify f_{c1}, f_{c2}, R_L, and R_H.

15.31 Design a unity-gain parallel bandreject filter with a center frequency of 250 rad/s and a bandwidth of 2000 rad/s. Use 1 μF capacitors, and specify all resistor values.

15.32 Show that the circuit in Fig. P15.32 behaves as a bandpass filter. (*Hint*—find the transfer function for this circuit and show that it has the same form as the transfer function for a bandpass filter. Use the result from Problem 15.1.)

a) Find the center frequency, bandwidth and gain for this bandpass filter.

b) Find the cutoff frequencies and the quality factor for this bandpass filter.

Figure P15.32

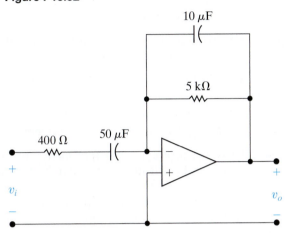

15.33 For circuits consisting of resistors, capacitors, inductors, and op amps, $|H(j\omega)|^2$ involves only even powers of ω. To illustrate this, compute $|H(j\omega)|^2$ for the three circuits in Fig. P15.33 when

$$H(s) = \frac{V_o}{V_i}.$$

Figure P15.33

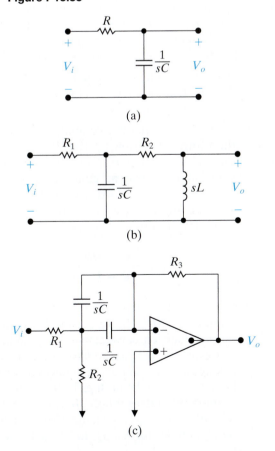

(a)

(b)

(c)

Section 15.4

15.34 a) Determine the order of a low-pass Butterworth filter that has a cutoff frequency of 800 Hz and a gain of no more than -25 dB at 2500 Hz.

b) What is the actual gain, in decibels, at 2500 Hz?

15.35 The circuit in Fig. 15.21 has the transfer function

$$H(s) = \frac{\dfrac{1}{R^2 C_1 C_2}}{s^2 + \dfrac{2}{RC_1} s + \dfrac{1}{R^2 C_1 C_2}}.$$

Show that if the circuit in Fig. 15.21 is scaled in both magnitude and frequency, the transfer function of the scaled circuit is

$$H'(s) = \frac{\dfrac{1}{R^2 C_1 C_2}}{\left(\dfrac{s}{k_f}\right)^2 + \dfrac{2}{RC_1}\left(\dfrac{s}{k_f}\right) + \dfrac{1}{R^2 C_1 C_2}}.$$

15.36 a) Write the transfer function for the prototype Butterworth filter using the filter order calculated in Problem 15.34(a).

b) Use the filter specifications in Problem 15.34(a) to calculate the frequency scale factor, then use it to scale the transfer function in Problem 15.36(a). (*Hint*: see Problem 15.35.)

c) Check the expression derived in part (b) by using it to calculate the gain (in decibels) at 2500 Hz. Compare your result with that found in Problem 15.34(b).

15.37 a) Using 2.7 kΩ resistors and ideal op amps, design a circuit that will implement the low-pass Butterworth filter specified in Problem 15.34. The gain in the passband is one.

DESIGN
PROBLEM

b) Construct the circuit diagram and label all component values.

15.38 The purpose of this problem is to illustrate the advantage of an nth-order low-pass Butterworth filter over the cascade of n identical low-pass sections by calculating the slope (in decibels per decade) of each magnitude plot at the corner frequency ω_c. To facilitate the calculation, let y represent the magnitude of the plot (in decibels), and let $x = \log_{10}\omega$. Then calculate dy/dx at ω_c for each plot.

a) Show that at the corner frequency ($\omega_c = 1$ rad/s) of an nth-order low-pass prototype Butterworth filter,

$$\frac{dy}{dx} = -10n \text{ dB/dec}.$$

b) Show that for a cascade of n identical low-pass prototype sections, the slope at ω_c is

$$\frac{dy}{dx} = \frac{-20n(2^{1/n} - 1)}{2^{1/n}} \text{ dB/dec.}$$

c) Compute dy/dx for each type of filter for $n = 1, 2, 3, 4,$ and ∞.

d) Discuss the significance of the results obtained in part (c).

15.39 Verify the entries in Table 15.1 for $n = 5$ and $n = 6$.

15.40 The circuit in Fig. 15.25 has the transfer function

$$H(s) = \frac{s^2}{s^2 + \dfrac{2}{R_2 C}s + \dfrac{1}{R_1 R_2 C^2}}.$$

Show that if the circuit is scaled in both magnitude and frequency, the transfer function of the scaled circuit is

$$H'(s) = \frac{\left(\dfrac{s}{k_f}\right)^2}{\left(\dfrac{s}{k_f}\right)^2 + \dfrac{2}{R_2 C}\left(\dfrac{s}{k_f}\right) + \dfrac{1}{R_1 R_2 C^2}}.$$

Hence the transfer function of a scaled circuit is obtained from the transfer function of an unscaled circuit by simply replacing s in the unscaled transfer function by s/k_f, where k_f is the frequency scaling factor.

15.41 a) Using 1 kΩ resistors and ideal op amps, design a low-pass unity-gain Butterworth filter that has a cutoff frequency of 8 kHz and is down at least 48 dB at 32 kHz.

DESIGN PROBLEM

b) Draw a circuit diagram of the filter and label all the components.

15.42 a) Using 10 nF capacitors and ideal op amps, design a high-pass unity-gain Butterworth filter with a cutoff frequency of 2 kHz and a gain of at least −48 dB at 500 Hz.

DESIGN PROBLEM

b) Draw a circuit diagram of the filter and label all component values.

15.43 The low-pass filter designed in Problem 15.41 is cascaded with the high-pass filter designed in Problem 15.42.

a) Describe the type of filter formed by this interconnection.

b) Specify the cutoff frequencies, the midfrequency, and the quality factor of the filter.

c) Use the results of Problems 15.35 and 15.40 to derive the scaled transfer function of the filter.

d) Check the derivation of (c) by using it to calculate $H(j\omega_o)$, where ω_o is the center frequency of the filter.

15.44 a) Design a broadband Butterworth bandpass filter with a lower cutoff frequency of 500 Hz and an upper cutoff frequency of 4500 Hz. The passband gain of the filter is 20 dB. The gain should be down at least 20 dB at 200 Hz and 11.25 kHz. Use 15 nF capacitors in the high-pass circuit and 10 kΩ resistors in the low-pass circuit.

DESIGN PROBLEM

b) Draw a circuit diagram of the filter and label all the components.

15.45 a) Derive the transfer function for the filter designed in Problem 15.44.

b) Using the expression derived in (a), find the gain (in decibels) at 200 Hz and 1500 Hz.

c) Do the values obtained in part (b) satisfy the filtering specifications given in Problem 15.44?

15.46 Derive the prototype transfer function for a sixth-order high-pass Butterworth filter by first writing the transfer function for a sixth-order prototype low-pass Butterworth filter and then replacing s by $1/s$ in the low-pass expression.

15.47 The sixth-order Butterworth filter in Problem 15.46 is used in a system where the cutoff frequency is 25 krad/s.

a) What is the scaled transfer function for the filter?

b) Test your expression by finding the gain (in decibels) at the cutoff frequency.

15.48 The purpose of this problem is to guide you through the analysis necessary to establish a design procedure for determining the circuit components in a filter circuit. The circuit to be analyzed is shown in Fig. P15.48.

DESIGN PROBLEM

a) Analyze the circuit qualitatively and convince yourself that the circuit is a low-pass filter with a passband gain of R_2/R_1.

b) Support your qualitative analysis by deriving the transfer function V_o/V_i. (*Hint:* In deriving the transfer function, represent the resistors with their equivalent conductances, that is, $G_1 = 1/R_1$, and so forth.) To make the transfer function useful in terms of the entries in Table 15.1, put it in the form

$$H(s) = \frac{-Kb_o}{s^2 + b_1 s + b_o}.$$

c) Now observe that we have five circuit components—R_1, R_2, R_3, C_1, and C_2—and three transfer function constraints—K, b_1, and b_o. At first glance, it appears we have two free choices among the five components. However, when we investigate the relationships between the circuit components and the transfer function constraints, we see that if C_2 is chosen, there is an upper limit on C_1 in order for $R_2(G_2)$ to be realizable. With this in mind, show that if $C_2 = 1$ F, the three conductances are given by the expressions

$$G_1 = KG_2;$$

$$G_3 = \left(\frac{b_o}{G_2}\right)C_1;$$

$$G_2 = \frac{b_1 \pm \sqrt{b_1^2 - 4b_o(1 + K)C_1}}{2(1 + K)}.$$

For G_2 to be realizable,

$$C_1 \leq \frac{b_1^2}{4b_o(1 + K)}.$$

d) Based on the results obtained in (c), outline the design procedure for selecting the circuit components once K, b_o, and b_1 are known.

Figure P15.48

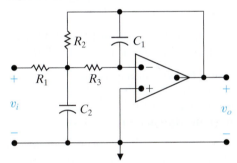

15.49 Assume the circuit analyzed in Problem 15.48 is part of a third-order low-pass Butterworth filter having a passband gain of 4. (*Hint:* implement the gain of 4 in the second-order section of the filter.)

DESIGN
PROBLEM

a) If $C_2 = 1$ F in the prototype second-order section, what is the upper limit on C_1?

b) If the limiting value of C_1 is chosen, what are the prototype values of R_1, R_2, and R_3?

c) If the corner frequency of the filter is 2.5 kHz and C_2 is chosen to be 10 nF, calculate the scaled values of C_1, R_1, R_2, and R_3.

d) Specify the scaled values of the resistors and the capacitor in the first-order section of the filter.

e) Construct a circuit diagram of the filter and label all the component values on the diagram.

15.50 Interchange the Rs and Cs in the circuit in Fig. P15.48; that is, replace R_1 with C_1, R_2 with C_2, R_3 with C_3, C_1 with R_1, and C_2 with R_2.

DESIGN
PROBLEM

a) Describe the type of filter implemented as a result of the interchange.

b) Confirm the filter type described in (a) by deriving the transfer function V_o/V_i. Write the transfer function in a form that makes it compatible with Table 15.1.

c) Set $C_2 = C_3 = 1$ F and derive the expressions for C_1, R_1, and R_2 in terms of K, b_1, and b_o. (See Problem 15.48 for the definition of b_1 and b_o.)

d) Assume the filter described in (a) is used in a third-order Butterworth filter that has a passband gain of 8. With $C_2 = C_3 = 1$ F, calculate the prototype values of C_1, R_1, and R_2 in the second-order section of the filter.

15.51 a) Use the circuits analyzed in Problems 15.48 and 15.50 to implement a broadband bandreject filter having a passband gain of 0 dB, a lower corner frequency of 400 Hz, an upper corner frequency of 6400 Hz, and an attenuation of at least 30 dB at both 1000 Hz and 2560 Hz. Use 10 nF capacitors whenever possible.

DESIGN
PROBLEM

b) Draw a circuit diagram of the filter and label all the components.

15.52 a) Derive the transfer function for the bandreject filter described in Problem 15.51.

b) Use the transfer function derived in part (a) to find the attenuation (in decibels) at the center frequency of the filter.

15.53 The purpose of this problem is to develop the design equations for the circuit in Fig. P15.53. (See Problem 15.48 for suggestions on the development of design equations.)

DESIGN
PROBLEM

a) Based on a qualitative analysis, describe the type of filter implemented by the circuit.

b) Verify the conclusion reached in (a) by deriving the transfer function V_o/V_i. Write the transfer function in a form that makes it compatible with the entries in Table 15.1.

c) How many free choices are there in the selection of the circuit components?

d) Derive the expressions for the conductances $G_1 = 1/R_1$ and $G_1 = 1/R_2$ in terms of C_1, C_2, and the coefficients b_o and b_1. (See Problem 15.48 for the definition of b_o and b_1.)

e) Are there any restrictions on C_1 or C_2?

f) Assume the circuit in Fig. P15.53 is used to design a fourth-order low-pass unity-gain Butterworth filter. Specify the prototype values of R_1 and R_2 in each second-order section if 1 F capacitors are used in the prototype circuit.

Figure P15.53

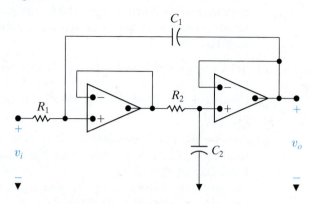

15.54 The fourth-order low-pass unity-gain Butterworth
DESIGN filter in Problem 15.53 is used in a system where
PROBLEM the cutoff frequency is 3 kHz. The filter has 4.7 nF
capacitors.

    a) Specify the numerical values of R_1 and R_2 in
each section of the filter.

    b) Draw a circuit diagram of the filter and label all
the components.

15.55 Interchange the Rs and Cs in the circuit in
DESIGN Fig. P15.53, that is, replace R_1 with C_1, R_2 with C_2,
PROBLEM and vice versa.

    a) Analyze the circuit qualitatively and predict the
type of filter implemented by the circuit.

    b) Verify the conclusion reached in (a) by deriving
the transfer function V_o/V_i. Write the transfer
function in a form that makes it compatible with
the entries in Table 15.1.

    c) How many free choices are there in the selection
of the circuit components?

    d) Find R_1 and R_2 as functions of b_o, b_1, C_1, and C_2.

    e) Are there any restrictions on C_1 and C_2?

    f) Assume the circuit is used in a third-order
Butterworth filter of the type found in (a).
Specify the prototype values of R_1 and R_2
in the second-order section of the filter if
$C_1 = C_2 = 1$ F.

15.56 a) The circuit in Problem 15.55 is used in a third-
DESIGN order high-pass unity-gain Butterworth filter
PROBLEM that has a cutoff frequency of 5 kHz. Specify the
values of R_1 and R_2 if 75 nF capacitors are available to construct the filter.

    b) Specify the values of resistance and capacitance
in the first-order section of the filter.

    c) Draw the circuit diagram and label all the com-
ponents.

    d) Give the numerical expression for the scaled
transfer function of the filter.

    e) Use the scaled transfer function derived in (d) to
find the gain in dB at the cutoff frequency.

Section 15.5

15.57 Show that if $\omega_o = 1$ rad/s and $C = 1$ F in the circuit
in Fig. 15.27, the prototype values of R_1, R_2, and R_3 are

$$R_1 = \frac{Q}{K},$$

$$R_2 = \frac{Q}{2Q^2 - K},$$

$$R_3 = 2Q.$$

15.58 a) Use 20 nF capacitors in the circuit in Fig. 15.27
DESIGN to design a bandpass filter with a quality factor
PROBLEM of 16, a center frequency of 6.4 kHz, and a pass-
band gain of 20 dB.

    b) Draw the circuit diagram of the filter and label
all the components.

15.59 a) Show that the transfer function for a prototype
narrow band bandreject filter is

$$H(s) = \frac{s^2 + 1}{s^2 + (1/Q)s + 1}.$$

    b) Use the result found in (a) to find the transfer
function of the filter designed in Example 15.14.

15.60 a) Using the circuit shown in Fig. 15.30, design a
DESIGN narrow-band bandreject filter having a center
PROBLEM frequency of 1 kHz and a quality factor of 20.
Base the design on $C = 15$ nF.

    b) Draw the circuit diagram of the filter and label
all component values on the diagram.

    c) What is the scaled transfer function of the filter?

Sections 15.1–15.5

15.61 Using the circuit in Fig. 15.33(a) design a volume
PRACTICAL control circuit to give a maximum gain of 20 dB
PERSPECTIVE and a gain of 17 dB at a frequency of 40 Hz. Use a
DESIGN 11.1 kΩ resistor and a 100 kΩ potentiometer. Test
PROBLEM your design by calculating the maximum gain at
$\omega = 0$ and the gain at $\omega = 1/R_2C_1$ using the selected
values of R_1, R_2, and C_1.

15.62 Use the circuit in Fig. 15.33(a) to design a bass vol-
PRACTICAL ume control circuit that has a maximum gain of
PERSPECTIVE 13.98 dB that drops off 3 dB at 50 Hz.
DESIGN
PROBLEM

15.63 Plot the maximum gain in decibels versus α when
PRACTICAL $\omega = 0$ for the circuit designed in Problem 15.61. Let
PERSPECTIVE α vary from 0 to 1 in increments of 0.1.

15.64 a) Show that the circuits in Fig. P15.64(a) and (b)
PRACTICAL are equivalent.
PERSPECTIVE

    b) Show that the points labeled x and y in
Fig. P15.64(b) are always at the same potential.

    c) Using the information in (a) and (b), show that
the circuit in Fig. 15.34 can be drawn as shown in
Fig. P15.64(c).

d) Show that the circuit in Fig. P15.64(c) is in the form of the circuit in Fig. 15.2, where

$$Z_i = \frac{R_1 + (1 - \alpha)R_2 + R_1 R_2 C_1 s}{1 + R_2 C_1 s},$$

$$Z_f = \frac{R_1 + \alpha R_2 + R_1 R_2 C_1 s}{1 + R_2 C_1 s}.$$

Figure P15.64

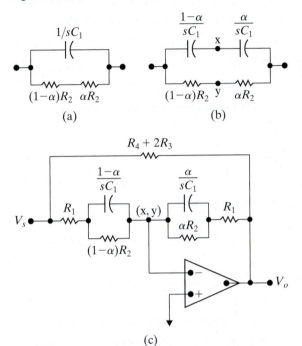

(a)   (b)

(c)

15.65 An engineering project manager has received a proposal from a subordinate who claims the circuit shown in Fig. P15.65 could be used as a treble volume control circuit if $R_4 \gg R_1 + R_3 + 2R_2$. The subordinate further claims that the voltage transfer function for the circuit is

$$H(s) = \frac{V_o}{V_s}$$

$$= \frac{-\{(2R_3 + R_4) + [(1 - \beta)R_4 + R_o](\beta R_4 + R_3)C_2 s\}}{\{(2R_3 + R_4) + [(1 - \beta)R_4 + R_3](\beta R_4 + R_o)C_2 s\}}$$

PRACTICAL PERSPECTIVE

where $R_o = R_1 + R_3 + 2R_2$. Fortunately the project engineer has an electrical engineering undergraduate student as an intern and therefore asks the student to check the subordinate's claim.

The student is asked to check the behavior of the transfer function as $\omega \to 0$; as $\omega \to \infty$; and the behavior when $\omega = \infty$ and β varies between 0 and 1. Based on your testing of the transfer function do you think the circuit could be used as a treble volume control? Explain.

Figure P15.65

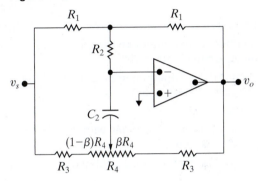

15.66 In the circuit of Fig. P15.65 the component values are $R_1 = R_2 = 20 \text{ k}\Omega$, $R_3 = 5.9 \text{ k}\Omega$, $R_4 = 500 \text{ k}\Omega$, and $C_2 = 2.7 \text{ nF}$.

PRACTICAL PERSPECTIVE

a) Calculate the maximum boost in decibels.

b) Calculate the maximum cut in decibels.

c) Is R_4 significantly greater than R_o?

d) When $\beta = 1$, what is the boost in decibels when $\omega = 1/R_3 C_2$?

e) When $\beta = 0$, what is the cut in decibels when $\omega = 1/R_3 C_2$?

f) Based on the results obtained in (d) and (e), what is the significance of the frequency $1/R_3 C_2$ when $R_4 \gg R_0$?

15.67 Using the component values given in Problem 15.66, plot the maximum gain in decibels versus β when ω is infinite. Let β vary from 0 to 1 in increments of 0.1.

PRACTICAL PERSPECTIVE

Fourier Series

CHAPTER CONTENTS

16.1 **Fourier Series Analysis: An Overview** *p. 621*

16.2 **The Fourier Coefficients** *p. 622*

16.3 **The Effect of Symmetry on the Fourier Coefficients** *p. 625*

16.4 **An Alternative Trigonometric Form of the Fourier Series** *p. 631*

16.5 **An Application** *p. 633*

16.6 **Average-Power Calculations with Periodic Functions** *p. 639*

16.7 **The rms Value of a Periodic Function** *p. 641*

16.8 **The Exponential Form of the Fourier Series** *p. 642*

16.9 **Amplitude and Phase Spectra** *p. 645*

CHAPTER OBJECTIVES

1 Be able to calculate the trigonometric form of the Fourier coefficients for a periodic waveform using the definition of the coefficients and the simplifications possible if the waveform exhibits one or more types of symmetry.

2 Know how to analyze a circuit's response to a periodic waveform using Fourier coefficients and superposition.

3 Be able to estimate the average power delivered to a resistor using a small number of Fourier coefficients.

4 Be able to calculate the exponential form of the Fourier coefficients for a periodic waveform and use them to generate magnitude and phase spectrum plots for that waveform.

In this chapter, we find the steady-state response of circuits to periodic, nonsinusoidal, inputs. A **periodic function** repeats itself every T seconds, so T is the period of the function. For example, the function plotted in Fig. 16.1 on page 620 is a periodic waveform that is not a sinusoid.

A periodic function satisfies the relationship

$$f(t) = f(t \pm nT), \tag{16.1}$$

where n is an integer $(1, 2, 3, \dots)$ and T is the period. The function shown in Fig. 16.1 is periodic because

$$f(t_0) = f(t_0 - T) = f(t_0 + T) = f(t_0 + 2T) = \cdots$$

for any arbitrarily chosen value of t_0. Note that T is the smallest time interval that a periodic function may be shifted (either left or right) to produce a function that is identical to itself.

Why are we interested in the response of circuits to inputs that are periodic but not sinusoids? One reason is that many electrical sources of practical value generate such waveforms. Here are a few examples.

- A nonfiltered electronic rectifier driven from a sinusoidal source produces a rectified sine wave that is not sinusoidal but is periodic. Figures 16.2(a) and (b) on page 620 show the waveforms of the full-wave and half-wave sinusoidal rectifiers, respectively.
- The sweep generator used to control the electron beam of a cathode-ray oscilloscope produces a periodic triangular wave like the one shown in Fig. 16.3 on page 620.
- Function generators, which are used to test equipment in a laboratory, are designed to produce nonsinusoidal periodic waveforms, including square waves, triangular waves, and rectangular waves. Figure 16.4 on page 620 illustrates typical waveforms output by a function generator.
- A power generator is designed to produce a sinusoidal waveform but cannot produce a pure sine wave. Instead, it

■ Practical Perspective

Active High-Q Filters

An important characteristic of bandpass and bandreject filters is the quality factor, Q, as we discovered in Chapters 14 and 15. The quality factor provides a measure of how selective the filter is at and near its center frequency. For example, an active bandpass filter with a large value of Q will amplify signals at or near its center frequency and will attenuate signals at all other frequencies. In contrast, a bandreject filter with a small value of Q will not effectively distinguish between signals at the center frequency and signals at frequencies quite different from the center frequency.

We can test the quality factor of a bandpass or bandreject filter using a periodic signal. For example, to test the quality factor of a bandpass filter, input a square wave whose frequency is the same as the center frequency of

the bandpass filter and analyze the output. In this chapter, we learn that any periodic signal can be represented as a sum of sinusoids. The frequencies of the sinusoids include the frequency of the periodic signal and integer multiples of that frequency. So the input square wave, whose frequency is ω, consists of a sum of sinusoids at the frequencies ω, 2ω, 3ω, and so on. If the bandpass filter has a high-quality factor, its output will be nearly sinusoidal because it filtered out all of the sinusoids that make up the square wave except the one at ω. If the filter has a low-quality factor, its output will still look like a square wave because the filter passed many of the sinusoids that make up the input square wave to the output. We present an example at the end of this chapter.

The sum of sinusoids whose frequencies are integer multiples of the fundamental frequency

The sum is a square wave at the fundamental frequency

The center frequency of the filter is the fundamental frequency

High-Q Bandpass Filter

The filter extracts the sinusoid at the fundamental frequency from the square wave

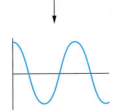

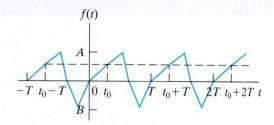

Figure 16.1 ▲ A periodic waveform.

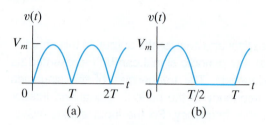

Figure 16.2 ▲ Output waveforms of a nonfiltered sinusoidal rectifier. (a) Full-wave rectification. (b) Half-wave rectification.

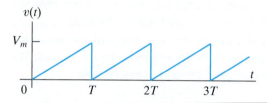

Figure 16.3 ▲ The triangular waveform of a cathode-ray oscilloscope sweep generator.

produces a distorted sinusoidal wave that is still periodic. We can use the analysis techniques in this chapter to determine the consequences of a circuit with a slightly distorted sinusoidal voltage.

• Any nonlinearity in an otherwise linear circuit creates a nonsinusoidal periodic function. The rectifier circuit mentioned earlier is one example of this phenomenon. Magnetic saturation, which occurs in both machines and transformers, is another example of a nonlinearity that generates a nonsinusoidal periodic function. An electronic clipping circuit, which uses transistor saturation, is yet another example.

Nonsinusoidal periodic functions are also important when analyzing nonelectrical systems. Problems involving mechanical vibration, fluid flow, and heat flow all make use of periodic functions. In fact, the study and analysis of heat flow in a metal rod led the French mathematician Jean Baptiste Joseph Fourier (1768–1830) to the trigonometric series representation of a periodic function. This series bears his name and is the starting point for finding the steady-state response of a circuit to a periodic input.

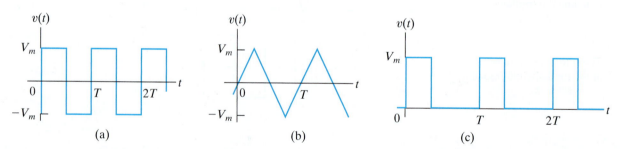

Figure 16.4 ▲ Waveforms produced by function generators used in laboratory testing. (a) Square wave. (b) Triangular wave. (c) Rectangular pulse.

16.1 Fourier Series Analysis: An Overview

A periodic function can be represented by an infinite sum of sine and cosine functions that are harmonically related, as Fourier discovered while investigating heat-flow problems. Specifically, the frequency of every trigonometric term in the infinite series is an integer multiple, or harmonic, of the fundamental frequency, ω_0, of the periodic function. Thus, if $f(t)$ is periodic, Fourier showed that it can be expressed as

FOURIER SERIES REPRESENTATION OF A PERIODIC FUNCTION

$$f(t) = a_v + \sum_{n=1}^{\infty} a_n \cos n\omega_0 t + b_n \sin n\omega_0 t, \qquad (16.2)$$

where n is the integer sequence $1, 2, 3, \ldots$.

In Eq. 16.2, a_v, a_n, and b_n are known as the **Fourier coefficients** and are calculated from $f(t)$. We discuss these calculations in Section 16.2. The term ω_0 (which equals $2\pi/T$) represents the **fundamental frequency** of the periodic function $f(t)$. The integer multiples of ω_0—that is, $2\omega_0, 3\omega_0, 4\omega_0$, and so on—are known as the **harmonic frequencies** of $f(t)$. Thus, $2\omega_0$ is the second harmonic, $3\omega_0$ is the third harmonic, and $n\omega_0$ is the nth harmonic of $f(t)$.

Before learning how to find a circuit's response to a periodic input using a Fourier series representation of that input, we first look at the process in general terms. We can express all the periodic functions of interest in circuit analysis using a Fourier series. But not every periodic function has a Fourier series representation. The conditions on a periodic function $f(t)$ that ensure it can be expressed as a convergent Fourier series (known as **Dirichlet's conditions**) are as follows.

1. $f(t)$ is single-valued.

2. $f(t)$ has a finite number of discontinuities in the periodic interval.

3. $f(t)$ has a finite number of maxima and minima in the periodic interval.

4. The integral

$$\int_{t_0}^{t_0+T} |f(t)| \, dt$$

     exists.

All periodic functions encountered in circuit analysis satisfy Dirichlet's conditions. These are **sufficient**, not **necessary** conditions. Thus, if $f(t)$ meets these requirements, we know that we can express it as a Fourier series. However, if $f(t)$ does not meet these requirements, we still may be able to express it as a Fourier series. The necessary conditions on $f(t)$ are not known.

After we have determined $f(t)$ and calculated the Fourier coefficients (a_v, a_n, and b_n), we replace the periodic source with a dc source (a_v) and sinusoidal sources, all connected in series. Each source represents a term in the Fourier series representation of the periodic input. Then, we use superposition to calculate the response to each source. The sum of the individual responses gives us the total steady-state response. The steady-state response to each sinusoidal source is most easily found using the phasor methods of Chapter 9.

The procedure just outlined involves no new circuit analysis techniques. It produces the Fourier series representation of the steady-state response; consequently, the functional form of the response is unknown. Furthermore, the output waveform is expressed as an infinite sum, so we can only estimate its shape by adding a sufficient number of its terms together. While the Fourier series method for finding the steady-state response of a circuit to a periodic input does have some drawbacks, it introduces a way of thinking about a problem that is as important as getting quantitative results. In fact, the conceptual picture is even more important, in some respects, than the quantitative one.

16.2 The Fourier Coefficients

Using the definition of a periodic function over its fundamental period, we determine the Fourier coefficients from the relationships

FOURIER COEFFICIENTS

$$a_v = \frac{1}{T} \int_{t_0}^{t_0 + T} f(t)\, dt, \tag{16.3}$$

$$a_k = \frac{2}{T} \int_{t_0}^{t_0 + T} f(t) \cos k\omega_0 t\, dt, \tag{16.4}$$

$$b_k = \frac{2}{T} \int_{t_0}^{t_0 + T} f(t) \sin k\omega_0 t\, dt. \tag{16.5}$$

In Eqs. 16.4 and 16.5, the subscript k indicates the kth coefficient in the integer sequence 1, 2, 3, ... Note that a_v is the average value of $f(t)$, a_k is twice the average value of $f(t) \cos k\omega_0 t$, and b_k is twice the average value of $f(t) \sin k\omega_0 t$.

We derive Eqs. 16.3–16.5 from Eq. 16.2 by recalling the following integral relationships, which hold when m and n are integers:

$$\int_{t_0}^{t_0 + T} \sin m\omega_0 t\, dt = 0, \quad \text{for all } m,$$

$$\int_{t_0}^{t_0 + T} \cos m\omega_0 t\, dt = 0, \quad \text{for all } m,$$

$$\int_{t_0}^{t_0 + T} \cos m\omega_0 t \sin n\omega_0 t\, dt = 0, \quad \text{for all } m \text{ and } n,$$

$$\int_{t_0}^{t_0 + T} \sin m\omega_0 t \sin n\omega_0 t\, dt = 0, \quad \text{for all } m \neq n,$$

$$= \frac{T}{2}, \quad \text{for } m = n,$$

$$\int_{t_0}^{t_0 + T} \cos m\omega_0 t \cos n\omega_0 t\, dt = 0, \quad \text{for all } m \neq n,$$

$$= \frac{T}{2}, \quad \text{for } m = n.$$

We leave you to verify these equations in Problem 16.5.

To derive Eq. 16.3, integrate both sides of Eq. 16.2 over one period:

$$\int_{t_0}^{t_0+T} f(t) \, dt = \int_{t_0}^{t_0+T} \left(a_v + \sum_{n=1}^{\infty} a_n \cos n\omega_0 t + b_n \sin n\omega_0 t \right) dt$$

$$= \int_{t_0}^{t_0+T} a_v \, dt + \sum_{n=1}^{\infty} \int_{t_0}^{t_0+T} (a_n \cos n\omega_0 t + b_n \sin n\omega_0 t) \, dt$$

$$= a_v T + 0.$$

Solving for a_v gives us Eq. 16.3.

To derive the expression for the kth value of a_n, we first multiply Eq. 16.2 by $\cos k\omega_0 t$ and then integrate both sides over one period of $f(t)$:

$$\int_{t_0}^{t_0+T} f(t) \cos k\omega_0 t \, dt = \int_{t_0}^{t_0+T} a_v \cos k\omega_0 t \, dt$$

$$+ \sum_{n=1}^{\infty} \int_{t_0}^{t_0+T} (a_n \cos n\omega_0 t \cos k\omega_0 t + b_n \sin n\omega_0 t \cos k\omega_0 t) \, dt$$

$$= 0 + a_k \left(\frac{T}{2} \right) + 0.$$

Solving for a_k yields the expression in Eq. 16.4.

We obtain the expression for the kth value of b_n by first multiplying both sides of Eq. 16.2 by $\sin k\omega_0 t$ and then integrating each side over one period of $f(t)$. You should complete the derivation of Eq. 16.5 in Problem 16.6. In Example 16.1 we use Eqs. 16.3–16.5 to find the Fourier coefficients for a specific periodic function.

EXAMPLE 16.1   **Finding the Fourier Series of a Triangular Waveform**

Find the Fourier series for the periodic voltage shown in Fig. 16.5.

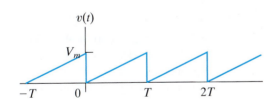

Figure 16.5 ▲ The periodic voltage for Example 16.1.

Solution

To find a_v, a_k, and b_k using Eqs. 16.3–16.5, we must choose the starting time t_0. For the periodic voltage of Fig. 16.5, the best choice for t_0 is zero. Any other choice makes the required integrations more cumbersome. The expression for $v(t)$ between 0 and T is

$$v(t) = \left(\frac{V_m}{T} \right) t.$$

The equation for a_v is

$$a_v = \frac{1}{T} \int_0^T \left(\frac{V_m}{T} \right) t \, dt = \frac{1}{2} V_m.$$

This is clearly the average value of the waveform in Fig. 16.5.

The equation for the kth value of a_n is

$$a_k = \frac{2}{T} \int_0^T \left(\frac{V_m}{T} \right) t \cos k\omega_0 t \, dt$$

$$= \frac{2V_m}{T^2} \left(\frac{1}{k^2 \omega_0^2} \cos k\omega_0 t + \frac{t}{k\omega_0} \sin k\omega_0 t \right) \Bigg|_0^T$$

$$= \frac{2V_m}{T^2} \left[\frac{1}{k^2 \omega_0^2} (\cos 2\pi k - 1) \right] = 0 \quad \text{for all } k.$$

Notice that we used the relationship between T and ω_0, $T = 2\pi/\omega_0$, when evaluating the integral at its upper limit.

The equation for the kth value of b_n is

$$b_k = \frac{2}{T} \int_0^T \left(\frac{V_m}{T}\right) t \sin k\omega_0 t \, dt$$

$$= \frac{2V_m}{T^2} \left(\frac{1}{k^2\omega_0^2} \sin k\omega_0 t - \frac{t}{k\omega_0} \cos k\omega_0 t\right)\Big|_0^T$$

$$= \frac{2V_m}{T^2} \left(0 - \frac{T}{k\omega_0} \cos 2\pi k\right)$$

$$= \frac{-V_m}{\pi k}.$$

The Fourier series for $v(t)$ is

$$v(t) = \frac{V_m}{2} - \frac{V_m}{\pi} \sum_{n=1}^{\infty} \frac{1}{n} \sin n\omega_0 t$$

$$= \frac{V_m}{2} - \frac{V_m}{\pi} \sin \omega_0 t - \frac{V_m}{2\pi} \sin 2\omega_0 t - \frac{V_m}{3\pi} \sin 3\omega_0 t - \cdots .$$

ASSESSMENT PROBLEMS

Objective 1 — Be able to calculate the trigonometric form of the Fourier coefficients for a periodic waveform

16.1 Derive the expressions for a_v, a_k, and b_k for the periodic voltage function shown if $V_m = 9\pi$ V.

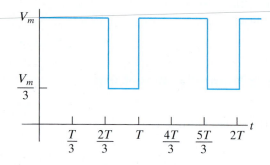

Answer: $a_v = 21.99$ V,

$$a_k = \frac{6}{k} \sin \frac{4k\pi}{3} \text{ V},$$

$$b_k = \frac{6}{k}\left(1 - \cos \frac{4k\pi}{3}\right)\text{V}.$$

SELF-CHECK: Also try Chapter Problems 16.1–16.3.

16.2 Refer to Assessment Problem 16.1.
a) What is the average value of the periodic voltage?
b) Compute the numerical values of $a_1 - a_5$ and $b_1 - b_5$.
c) If $T = 125.66$ ms, what is the fundamental frequency in radians per second?
d) What is the frequency of the third harmonic in hertz?
e) Write the Fourier series up to and including the fifth harmonic.

Answer: (a) 21.99 V;
(b) −5.2 V, 2.6 V, 0 V, −1.3, and 1.04 V;
9 V, 4.5 V, 0 V, 2.25 V, and 1.8 V;
(c) 50 rad/s;
(d) 23.87 Hz;
(e) $v(t) = 21.99 - 5.2 \cos 50t + 9 \sin 50t$
$+ 2.6 \cos 100t + 4.5 \sin 100t$
$- 1.3 \cos 200t + 2.25 \sin 200t$
$+ 1.04 \cos 250t + 1.8 \sin 250t$ V.

Finding the Fourier coefficients, in general, is tedious. Therefore, anything that simplifies the task is beneficial. Fortunately, when a periodic function possesses certain types of symmetry, we can find its Fourier coefficients with fewer computations. In Section 16.3, we discuss how symmetry affects the coefficients in a Fourier series.

16.3 The Effect of Symmetry on the Fourier Coefficients

Four types of symmetry make the task of evaluating the Fourier coefficients easier:

- even-function symmetry,
- odd-function symmetry,
- half-wave symmetry,
- quarter-wave symmetry.

The effect of each type of symmetry on the Fourier coefficients is discussed in the following sections.

Even-Function Symmetry

A function is defined as *even* if

EVEN FUNCTION

$$f(t) = f(-t). \tag{16.6}$$

Figure 16.6 illustrates an even periodic function. Functions that satisfy Eq. 16.6 are said to be even because polynomial functions with only even exponents possess this characteristic.

For even periodic functions, the equations for the Fourier coefficients reduce to

$$a_v = \frac{2}{T} \int_0^{T/2} f(t)\, dt, \tag{16.7}$$

$$a_k = \frac{4}{T} \int_0^{T/2} f(t) \cos k\omega_0 t\, dt, \tag{16.8}$$

$$b_k = 0, \quad \text{for all } k. \tag{16.9}$$

Note that all the b coefficients are zero if the periodic function is even. This means that the Fourier series representation of an even periodic function consists only of the constant term and cosine terms—there are no sine terms. This is not surprising because the cosine function is even, but the sine function is not.

The derivations of Eqs. 16.7–16.9 follow directly from Eqs. 16.3–16.5. In each derivation, we select $t_0 = -T/2$ and then break the interval of integration into the range from $-T/2$ to 0 and 0 to $T/2$. For example,

$$a_v = \frac{1}{T} \int_{-T/2}^{T/2} f(t)\, dt$$

$$= \frac{1}{T} \int_{-T/2}^{0} f(t)\, dt + \frac{1}{T} \int_{0}^{T/2} f(t)\, dt.$$

Now we change the variable of integration in the first integral on the right-hand side of the equation for a_v. We let $t = -x$ and note that $f(t) = f(-x) = f(x)$ because the function is even. Note that $x = T/2$ when $t = -T/2$ and $dt = -dx$. Then

$$\int_{-T/2}^{0} f(t)\, dt = \int_{T/2}^{0} f(x)\,(-dx) = \int_{0}^{T/2} f(x)\, dx,$$

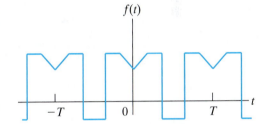

$f(t)$

Figure 16.6 ▲ An even periodic function, $f(t) = f(-t)$.

which shows that the integral from $-T/2$ to 0 is identical to the integral from 0 to $T/2$. This completes the derivation of Eq. 16.7.

The derivation of Eq. 16.8 proceeds along similar lines. Here,

$$a_k = \frac{2}{T} \int_{-T/2}^{0} f(t) \cos k\omega_0 t \, dt + \frac{2}{T} \int_{0}^{T/2} f(t) \cos k\omega_0 t \, dt,$$

but

$$\int_{-T/2}^{0} f(t) \cos k\omega_0 t \, dt = \int_{T/2}^{0} f(x) \cos(-k\omega_0 x)(-dx)$$

$$= \int_{0}^{T/2} f(x) \cos k\omega_0 x \, dx.$$

As before, the integral from $-T/2$ to 0 is the same as the integral from 0 to $T/2$. Replacing the first integral in the equation for a_k with a copy of the second integral and simplifying yields Eq. 16.8.

All the b coefficients are zero when $f(t)$ is an even periodic function because the integral from $-T/2$ to 0 is the negative of the integral from 0 to $T/2$; that is,

$$\int_{-T/2}^{0} f(t) \sin k\omega_0 t \, dt = \int_{T/2}^{0} f(x) \sin(-k\omega_0 x)(-dx)$$

$$= -\int_{0}^{T/2} f(x) \sin k\omega_0 x \, dx.$$

When we use Eqs. 16.7 and 16.8 to find the Fourier coefficients, the interval of integration must be between 0 and $T/2$.

Odd-Function Symmetry

A function is defined as odd if

ODD FUNCTION

$$f(t) = -f(-t). \tag{16.10}$$

Functions that satisfy Eq. 16.10 are said to be odd because polynomial functions with only odd exponents have this characteristic. Figure 16.7 shows an odd periodic function.

The expressions for the Fourier coefficients are

$$a_v = 0; \tag{16.11}$$

$$a_k = 0, \quad \text{for all } k; \tag{16.12}$$

$$b_k = \frac{4}{T} \int_{0}^{T/2} f(t) \sin k\omega_0 t \, dt. \tag{16.13}$$

Note that all the a coefficients are zero if the periodic function is odd. This means that the Fourier series representation of an odd periodic function consists only of sine terms; there are no cosine terms. This is not surprising because the sine function is odd, but the cosine function is even. It also

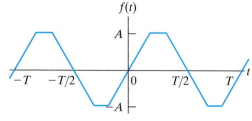

Figure 16.7 ▲ An odd periodic function $f(t) = -f(-t)$.

means that the average value of an odd function is always zero because the constant term, a_v, is always zero.

We use the same process to derive Eqs. 16.11–16.13 that we used to derive Eqs. 16.7–16.9. We leave the derivations to you in Problem 16.7.

If a periodic function has even or odd symmetry, the symmetry can be destroyed by shifting the function along the time axis. This also means that if a periodic function lacks even or odd symmetry, it may be possible to shift the function along the time axis to create this symmetry. For example, the triangular function shown in Fig. 16.8(a) is neither even nor odd. However, we can make the function even by shifting it left, as shown in Fig. 16.8(b), or odd by shifting it right, as shown in Fig. 16.8(c).

Half-Wave Symmetry

A periodic function possesses half-wave symmetry if it satisfies the constraint

HALF-WAVE SYMMETRY

$$f(t) = -f(t - T/2). \qquad (16.14)$$

Equation 16.14 states that a periodic function has half-wave symmetry if, after it is shifted one-half period and inverted, it is identical to the original function. For example, the functions shown in Figs. 16.7 and 16.8 have half-wave symmetry, whereas those in Figs. 16.5 and 16.6 do not. Note that half-wave symmetry is not determined by where $t = 0$, as seen in Fig. 16.8.

If a periodic function has half-wave symmetry, both a_k and b_k are zero for even values of k. Moreover, a_v also is zero because the average value of a function with half-wave symmetry is zero. The expressions for the Fourier coefficients are

$$a_v = 0; \qquad (16.15)$$

$$a_k = 0, \qquad \text{for } k \text{ even};$$

$$\qquad\qquad\qquad\qquad\qquad (16.16)$$

$$a_k = \frac{4}{T}\int_0^{T/2} f(t) \cos k\omega_0 t \, dt, \quad \text{for } k \text{ odd};$$

$$b_k = 0, \qquad \text{for } k \text{ even};$$

$$\qquad\qquad\qquad\qquad\qquad (16.17)$$

$$b_k = \frac{4}{T}\int_0^{T/2} f(t) \sin k\omega_0 t \, dt, \quad \text{for } k \text{ odd}.$$

We derive Eqs. 16.15–16.17 by starting with Eqs. 16.3–16.5 and choosing the interval of integration as $-T/2$ to $T/2$. We then divide this range into the intervals $-T/2$ to 0 and 0 to $T/2$. For example, the derivation for a_k is

$$a_k = \frac{2}{T}\int_{t_0}^{t_0+T} f(t) \cos k\omega_0 t \, dt$$

$$= \frac{2}{T}\int_{-T/2}^{T/2} f(t) \cos k\omega_0 t \, dt$$

$$= \frac{2}{T}\int_{-T/2}^{0} f(t) \cos k\omega_0 t \, dt$$

$$+ \frac{2}{T}\int_{0}^{T/2} f(t) \cos k\omega_0 t \, dt.$$

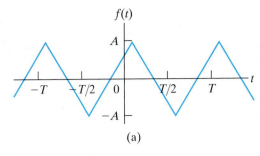

(a)

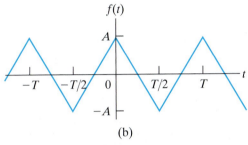

(b)

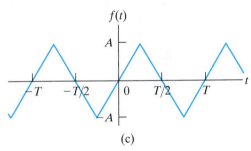

(c)

Figure 16.8 ▲ How the choice of where $t = 0$ can make a periodic function even, odd, or neither. (a) A periodic triangular wave that is neither even nor odd. (b) The triangular wave of (a) made even by shifting the function along the t axis. (c) The triangular wave of (a) made odd by shifting the function along the t axis.

Now we change a variable in the first integral on the right-hand side of the equation for a_k. Specifically, we let

$$t = x - T/2.$$

Then

$$x = T/2, \qquad \text{when } t = 0;$$

$$x = 0, \qquad \text{when } t = -T/2;$$

$$dt = dx.$$

We rewrite the first integral as

$$\int_{-T/2}^{0} f(t) \cos k\omega_0 t \, dt = \int_{0}^{T/2} f(x - T/2) \cos k\omega_0 (x - T/2) \, dx.$$

Note that

$$\cos k\omega_0 (x - T/2) = \cos (k\omega_0 x - k\pi) = \cos k\pi \cos k\omega_0 x$$

and that, by hypothesis,

$$f(x - T/2) = -f(x).$$

Therefore,

$$\int_{-T/2}^{0} f(t) \cos k\omega_0 t \, dt = \int_{0}^{T/2} [-f(x)] \cos k\pi \cos k\omega_0 x \, dx.$$

Incorporating this integral into the equation for a_k gives

$$a_k = \frac{2}{T} (1 - \cos k\pi) \int_{0}^{T/2} f(t) \cos k\omega_0 t \, dt.$$

But $\cos k\pi$ is 1 when k is even and -1 when k is odd. Therefore, we get the expressions for a_k given in Eqs. 16.16.

We leave it to you to verify that this same process can be used to derive Eqs. 16.17 (see Problem 16.8).

We summarize our observations by noting that the Fourier series representation of a periodic function with half-wave symmetry has zero average value and contains only odd harmonics.

Quarter-Wave Symmetry

The term **quarter-wave symmetry** describes a periodic function that has half-wave symmetry and, in addition, symmetry about the midpoint of the positive and negative half-cycles. The function illustrated in Fig. 16.9(a)

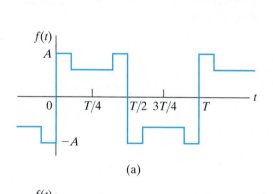

(a)

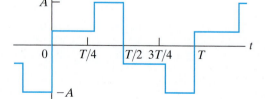

(b)

Figure 16.9 ▲ (a) A function that has quarter-wave symmetry. (b) A function that does not have quarterwave symmetry.

has quarter-wave symmetry about the midpoint of the positive and negative half-cycles. The function in Fig. 16.9(b) does not have quarter-wave symmetry, although it does have half-wave symmetry.

A periodic function that has quarter-wave symmetry can always be made either even or odd by the proper choice of the point where $t = 0$. For example, the function shown in Fig. 16.9(a) is odd and can be made even by shifting the function $T/4$ units either right or left along the t axis. However, the function in Fig. 16.9(b) can never be made either even or odd.

To take advantage of quarter-wave symmetry when calculating the Fourier coefficients, you must choose the point where $t = 0$ to make the function either even or odd. If the function is made even, then

$a_v = 0$, because of half-wave symmetry;

$a_k = 0$, for k even, because of half-wave symmetry;

(16.18)

$$a_k = \frac{8}{T} \int_0^{T/4} f(t) \cos k\omega_0 t \, dt, \quad \text{for } k \text{ odd};$$

$b_k = 0$, for all k, because the function is even.

Equations 16.18 result from the function's quarter-wave symmetry in addition to its being even. Remember that if a function has quarter-wave symmetry, it also has half-wave symmetry, so we can eliminate a_v and a_k for k even. Comparing the expression for a_k, k odd, in Eqs. 16.18 with Eqs. 16.16, shows that when an even function also has quarter-wave symmetry, we can shorten the range of integration from 0 to $T/2$ to 0 to $T/4$. We leave the derivation of Eqs. 16.18 to you in Problem 16.9.

If the quarter-wave symmetric function is made odd,

$a_v = 0$, because the function is odd;

$a_k = 0$, for all k, because the function is odd;

(16.19)

$b_k = 0$, for k even, because of half-wave symmetry;

$$b_k = \frac{8}{T} \int_0^{T/4} f(t) \sin k\omega_0 t \, dt, \quad \text{for } k \text{ odd}.$$

Equations 16.19 are a direct consequence of the function being both odd and quarter-wave symmetric. Again, quarter-wave symmetry allows us to shorten the interval of integration from 0 to $T/2$ to 0 to $T/4$. We leave the derivation of Eqs. 16.19 to you in Problem 16.10.

Example 16.2 shows how symmetry simplifies the task of finding the Fourier coefficients.

EXAMPLE 16.2 · Finding the Fourier Series of a Periodic Function with Symmetry

Find the Fourier series representation for the current waveform shown in Fig. 16.10.

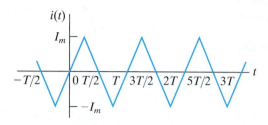

Figure 16.10 ▲ The periodic waveform for Example 16.2.

Solution

We begin by looking for symmetry in the waveform. We find that the function is odd and has half-wave and quarter-wave symmetry. Because the function is odd, all the a coefficients are zero; that is, $a_v = 0$ and $a_k = 0$ for all k. Because the function has half-wave symmetry, $b_k = 0$ for even values of k. Because the function has quarter-wave symmetry, the expression for b_k for odd values of k is

$$b_k = \frac{8}{T}\int_0^{T/4} i(t)\sin k\omega_0 t\, dt.$$

In the interval $0 \le t \le T/4$, the expression for $i(t)$ is

$$i(t) = \frac{4I_m}{T}t.$$

Thus

$$b_k = \frac{8}{T}\int_0^{T/4}\frac{4I_m}{T}t\sin k\omega_0 t\, dt$$

$$= \frac{32I_m}{T^2}\left(\frac{\sin k\omega_0 t}{k^2\omega_0^2} - \frac{t\cos k\omega_0 t}{k\omega_0}\bigg|_0^{T/4}\right)$$

$$= \frac{8I_m}{\pi^2 k^2}\sin\frac{k\pi}{2} \quad (k \text{ is odd}).$$

The Fourier series representation of $i(t)$ is

$$i(t) = \frac{8I_m}{\pi^2}\sum_{n=1,3,5,\ldots}^{\infty}\frac{1}{n^2}\sin\frac{n\pi}{2}\sin n\omega_0 t$$

$$= \frac{8I_m}{\pi^2}\left(\sin\omega_0 t - \frac{1}{9}\sin 3\omega_0 t\right.$$

$$\left. + \frac{1}{25}\sin 5\omega_0 t - \frac{1}{49}\sin 7\omega_0 t + \cdots\right).$$

◻ ASSESSMENT PROBLEM

Objective 1—Be able to calculate the trigonometric form of the Fourier coefficients for a periodic waveform

16.3 Derive the Fourier series for the periodic voltage shown.

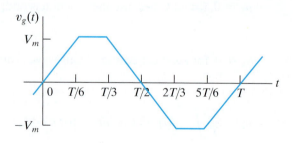

Answer: $v_g(t) = \dfrac{12V_m}{\pi^2}\displaystyle\sum_{n=1,3,5,\ldots}^{\infty}\dfrac{\sin(n\pi/3)}{n^2}\sin n\omega_0 t.$

SELF-CHECK: Also try Chapter Problems 16.11 and 16.12.

16.4 An Alternative Trigonometric Form of the Fourier Series

When analyzing a circuit with a periodic input voltage, we could replace the input voltage with a collection of series-connected voltage sources, where each source corresponds to a term in the Fourier series representation of the periodic voltage. This would mean that for every harmonic frequency, there is a source for the sine term and a source for the cosine term at that frequency.

Instead, to simplify the circuit and our analysis of it, we combine the sine and cosine terms at the same harmonic frequency into a single term. Then, we can transform the circuit into the phasor domain for each harmonic frequency, where the combined source is represented as a single phasor quantity. Thus, we write the Fourier series in Eq. 16.2 as

$$f(t) = a_v + \sum_{n=1}^{\infty} A_n \cos(n\omega_0 t - \theta_n), \tag{16.20}$$

where A_n and θ_n are defined as

$$a_n - jb_n = \sqrt{a_n^2 + b_n^2} \underline{/-\theta_n} = A_n \underline{/-\theta_n}. \tag{16.21}$$

We derive Eqs. 16.20 and 16.21 using the phasor method to add the cosine and sine terms in Eq. 16.2. We begin by expressing the sine functions as cosine functions; that is, we rewrite Eq. 16.2 as

$$f(t) = a_v + \sum_{n=1}^{\infty} a_n \cos n\omega_0 t + b_n \cos(n\omega_0 t - 90°).$$

Adding the terms under the summation sign by using phasors gives

$$\mathcal{P}\{a_n \cos n\omega_0 t\} = a_n \underline{/0°}$$

and

$$\mathcal{P}\{b_n \cos(n\omega_0 t - 90°)\} = b_n \underline{/-90°} = -jb_n.$$

Then

$$\mathcal{P}\{a_n \cos(n\omega_0 t + b_n \cos(n\omega_0 t - 90°)\} = a_n - jb_n$$
$$= \sqrt{a_n^2 + b_n^2} \underline{/-\theta_n}$$
$$= A_n \underline{/-\theta_n}.$$

The right-hand sides of this expression correspond to Eq. 16.21. When we inverse phasor-transform both sides of this expression, we get

$$a_n \cos n\omega_0 t + b_n \cos(n\omega_0 t - 90°) = \mathcal{P}^{-1}\{A_n \underline{/-\theta_n}\}$$
$$= A_n \cos(n\omega_0 t - \theta_n).$$

After substituting the right-hand side for the argument of the summation in the expression for $f(t)$, we get Eq. 16.20. If the periodic function is either even or odd, A_n reduces to either a_n (even) or b_n (odd), and θ_n is either 0° (even) or 90° (odd).

Example 16.3 calculates the alternative form of the Fourier series for a specific periodic function.

<table>
<tr><td colspan="2">

EXAMPLE 16.3 | **Calculating Forms of the Trigonometric Fourier Series for Periodic Voltage**

</td></tr>
</table>

a) Derive the expressions for a_k and b_k for the periodic function shown in Fig. 16.11.

b) Write the first four terms of the Fourier series representation of $v(t)$ using the format of Eq. 16.20.

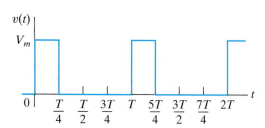

Figure 16.11 ▲ The periodic function for Example 16.3.

Solution

a) The voltage $v(t)$ is neither even nor odd, nor does it have half-wave symmetry. Therefore, we use Eqs. 16.4 and 16.5 to find a_k and b_k. Choosing t_0 as zero, we obtain

$$a_k = \frac{2}{T}\left[\int_0^{T/4} V_m \cos k\omega_0 t \, dt + \int_{T/4}^{T} (0) \cos k\omega_0 t \, dt\right]$$

$$= \frac{2V_m}{T} \frac{\sin k\omega_0 t}{k\omega_0}\Big|_0^{T/4} = \frac{V_m}{k\pi} \sin \frac{k\pi}{2}$$

and

$$b_k = \frac{2}{T} \int_0^{T/4} V_m \sin k\omega_0 t \, dt$$

$$= \frac{2V_m}{T}\left(\frac{-\cos k\omega_0 t}{k\omega_0}\Big|_0^{T/4}\right)$$

$$= \frac{V_m}{k\pi}\left(1 - \cos\frac{k\pi}{2}\right).$$

b) The average value of $v(t)$ is

$$a_v = \frac{V_m(T/4)}{T} = \frac{V_m}{4}.$$

The values of $a_k - jb_k$ for $k = 1, 2,$ and 3 are

$$a_1 - jb_1 = \frac{V_m}{\pi} - j\frac{V_m}{\pi} = \frac{\sqrt{2}V_m}{\pi}\underline{/-45°},$$

$$a_2 - jb_2 = 0 - j\frac{V_m}{\pi} = \frac{V_m}{\pi}\underline{/-90°},$$

$$a_3 - jb_3 = \frac{-V_m}{3\pi} - j\frac{V_m}{3\pi} = \frac{\sqrt{2}V_m}{3\pi}\underline{/-135°}.$$

Thus, the first four terms in the Fourier series representation of $v(t)$ are

$$v(t) = \frac{V_m}{4} + \frac{\sqrt{2}V_m}{\pi}\cos(\omega_0 t - 45°)$$

$$+ \frac{V_m}{\pi}\cos(2\omega_0 t - 90°)$$

$$+ \frac{\sqrt{2}V_m}{3\pi}\cos(3\omega_0 t - 135°) + \cdots.$$

⬛ ASSESSMENT PROBLEM

Objective 1—Be able to calculate the trigonometric form of the Fourier coefficients for a periodic waveform

16.4 a) Compute $A_1 - A_5$ and $\theta_1 - \theta_5$ for the periodic function shown if $V_m = 9\pi$ V.

b) Using the format of Eq. 16.20, write the Fourier series for $v(t)$ up to and including the fifth harmonic assuming $T = 125.66$ ms.

Answer: (a) 10.4, 5.2, 0, 2.6, 2.1 V, and −120°, −60°, not defined, −120°, −60°;

(b) $v(t) = 21.99 + 10.4\cos(50t - 120°)$
$+ 5.2\cos(100t - 60°)$
$+ 2.6\cos(200t - 120°)$
$+ 2.1\cos(250t - 60°)$ V.

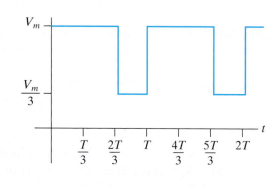

SELF-CHECK: Also try Chapter Problem 16.22.

16.5 An Application

Let's use a Fourier series representation of a periodic input voltage to find the steady-state output voltage of a linear circuit, which is the *RC* circuit shown in Fig. 16.12(a). The circuit's input voltage is the periodic square wave shown in Fig. 16.12(b). The output voltage is defined across the capacitor.

First, we represent the periodic voltage with its Fourier series. The waveform in Fig. 16.12(b) has odd, half-wave, and quarter-wave symmetry, so the only nonzero Fourier coefficients are the b_k coefficients for odd values of k:

$$b_k = \frac{8}{T} \int_0^{T/4} V_m \sin k\omega_0 t \, dt$$

$$= \frac{4V_m}{\pi k} \quad (k \text{ is odd}).$$

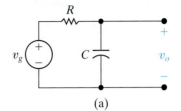

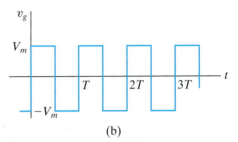

(b)

Figure 16.12 ▲ An *RC* circuit excited by a periodic voltage. (a) The *RC* circuit. (b) The square-wave voltage.

Then the Fourier series representation of v_g is

$$v_g = \frac{4V_m}{\pi} \sum_{n=1,3,5,\dots}^{\infty} \frac{1}{n} \sin n\omega_0 t.$$

Writing the series in expanded form, we have

$$v_g = \frac{4V_m}{\pi} \sin \omega_0 t + \frac{4V_m}{3\pi} \sin 3\omega_0 t$$

$$+ \frac{4V_m}{5\pi} \sin 5\omega_0 t + \frac{4V_m}{7\pi} \sin 7\omega_0 t + \cdots .$$

The voltage source, v_g, in Fig. 16.12(a) can be replaced by infinitely many series-connected sinusoidal sources. Each source is a sine function whose frequency is an odd multiple of the square wave's frequency. We use the principle of superposition to find the contribution of each source to the output voltage.

For each sinusoidal source, the phasor-domain expression for the output voltage is

$$\mathbf{V}_o = \frac{\mathbf{V}_g}{1 + j\omega RC}.$$

Since all of the voltage sources are expressed as sine functions, we interpret a phasor using the sine instead of the cosine. In other words, when we go from the phasor domain back to the time domain, we simply write the time-domain expressions as $\sin(\omega t + \theta)$ instead of $\cos(\omega t + \theta)$.

The phasor output voltage due to the sinusoidal source at the fundamental frequency is

$$\mathbf{V}_{o1} = \frac{(4V_m/\pi)\underline{/0°}}{1 + j\omega_0 RC}.$$

Writing $\mathbf{V}_{o1}$ in polar form gives

$$\mathbf{V}_{o1} = \frac{(4V_m)}{\pi\sqrt{1 + \omega_0^2 R^2 C^2}} \underline{/-\beta_1},$$

where

$$\beta_1 = \tan^{-1} \omega_0 RC.$$

Inverse-phasor-transform $\mathbf{V}_{o1}$ to get the time-domain expression for the fundamental frequency component of v_o:

$$v_{o1} = \frac{4V_m}{\pi\sqrt{1 + \omega_0^2 R^2 C^2}} \sin(\omega_0 t - \beta_1).$$

We derive the third-harmonic component of the output voltage in a similar manner. The third-harmonic phasor output voltage is

$$\mathbf{V}_{o3} = \frac{(4V_m/3\pi)\underline{/0^\circ}}{1 + j3\omega_0 RC}$$

$$= \frac{4V_m}{3\pi\sqrt{1 + 9\omega_0^2 R^2 C^2}}\underline{/-\beta_3},$$

where

$$\beta_3 = \tan^{-1} 3\omega_0 RC.$$

The time-domain expression for the third-harmonic output voltage is

$$v_{o3} = \frac{4V_m}{3\pi\sqrt{1 + 9\omega_0^2 R^2 C^2}} \sin(3\omega_0 t - \beta_3).$$

Hence, the expression for the kth-harmonic component of the output voltage is

$$v_{ok} = \frac{4V_m}{k\pi\sqrt{1 + k^2\omega_0^2 R^2 C^2}} \sin(k\omega_0 t - \beta_k) \quad (k \text{ is odd}),$$

where

$$\beta_k = \tan^{-1} k\omega_0 RC \quad (k \text{ is odd}).$$

We now write down the Fourier series representation of the output voltage:

$$v_o(t) = \frac{4V_m}{\pi} \sum_{n=1,3,5,\ldots}^{\infty} \frac{\sin(n\omega_0 t - \beta_n)}{n\sqrt{1 + (n\omega_0 RC)^2}}. \tag{16.22}$$

We derived an analytic expression for the steady-state output, but the shape of $v_o(t)$, and its functional form, are not apparent from the equation. As we mentioned earlier, this is a disadvantage of the Fourier series approach.

The Fourier series representation of $v_o(t)$ in Eq. 16.22 does provide some information about the steady-state output voltage, if we focus on the frequency response of the circuit. For example, if C is large, $1/n\omega_0 C$ is small for the higher-order harmonics. Thus, the capacitor short circuits the high-frequency components of the input waveform, and the higher-order harmonics in the Fourier series representation of $v_o(t)$ are negligible compared to the lower-order harmonics. For large C, we can approximate $v_o(t)$ as

$$v_o \approx \frac{4V_m}{\pi\omega_0 RC} \sum_{n=1,3,5,\ldots}^{\infty} \frac{1}{n^2} \sin(n\omega_0 t - 90^\circ)$$

$$\approx \frac{-4V_m}{\pi\omega_0 RC} \sum_{n=1,3,5,\ldots}^{\infty} \frac{1}{n^2} \cos n\omega_0 t. \tag{16.23}$$

Equation 16.23 shows that the amplitude of the harmonic in the output voltage is decreasing by $1/n^2$, while the amplitude of the harmonic in the input voltage decreases by $1/n$. If C is so large that only the fundamental component is significant, then to a first approximation

$$v_o(t) \approx \frac{-4V_m}{\pi\omega_0 RC} \cos \omega_0 t,$$

and Fourier analysis tells us that the square-wave input is deformed into a sinusoidal output.

Now let's see what happens as $C \to 0$. The circuit shows that v_o and v_g are the same when $C = 0$ because the capacitive branch looks like an open circuit at all frequencies. Equation 16.22 predicts the same result because, as $C \to 0$,

$$v_o = \frac{4V_m}{\pi} \sum_{n=1,3,5,\ldots}^{\infty} \frac{1}{n} \sin n\omega_0 t.$$

Therefore $v_o \to v_g$ as $C \to 0$.

Thus, Fourier series representation for $v_o(t)$ predicts that the output will be a highly distorted replica of the input waveform if C is large, and a reasonable replica if C is small. In Chapter 13, we looked at the distortion between the input and output in terms of how much memory the system weighting function had. In the frequency domain, we look at the distortion between the steady-state input and output in terms of how the amplitude and phase of each harmonic component are altered as it is transmitted through the circuit. When the circuit significantly alters the amplitude and phase relationships among the harmonics at the output relative to that at the input, the output is a distorted version of the input. Thus, in the frequency domain, we speak of amplitude distortion and phase distortion.

For the RC circuit in Fig. 16.12(a), amplitude distortion is present because the amplitudes of the input harmonics decrease as $1/n$, whereas the amplitudes of the output harmonics decrease as

$$\frac{1}{n} \frac{1}{\sqrt{1 + (n\omega_0 RC)^2}}.$$

This circuit also exhibits phase distortion because the phase angle of each input harmonic is zero, whereas that of the nth harmonic in the output signal is $-\tan^{-1} n\omega_0 RC$.

An Application of the Direct Approach to the Steady-State Response

For the simple RC circuit shown in Fig. 16.12(a), we can find the steady-state response without using the Fourier series representation of the periodic input voltage. Doing this extra analysis here adds to our understanding of the Fourier series approach.

To find the steady-state expression for v_o, we reason as follows. The square-wave input voltage alternates between charging the capacitor toward $+V_m$ and $-V_m$. After the circuit reaches steady-state operation, this alternate charging becomes periodic. From our analysis of the first-order RC circuit in Chapter 7, we know that an abrupt change in the input voltage results in an output voltage that is exponential. Thus, the steady-state waveform of the capacitor voltage for the RC circuit shown in Fig. 16.12(a) looks like the waveform shown in Fig. 16.13.

The analytic expressions for $v_o(t)$ in the time intervals $0 \le t \le T/2$ and $T/2 \le t \le T$ are

$$v_o = V_m + (V_1 - V_m)e^{-t/RC}, \qquad 0 \le t \le T/2;$$
$$v_o = -V_m + (V_2 + V_m)e^{-[t-(T/2)]/RC}, \quad T/2 \le t \le T.$$

We derive these equations by using the methods of Chapter 7, as summarized by Eq. 7.23. The value of V_2 is the value of $v_o(t)$ at the end of the interval $0 \le t \le T/2$:

$$v_o(T/2) = V_2 = V_m + (V_1 - V_m)e^{-T/2RC},$$

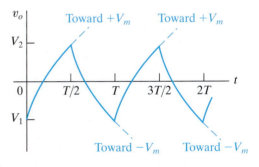

Figure 16.13 ▲ The steady-state waveform of v_o for the circuit in Fig. 16.12(a).

and the value of V_1 is the value of $v_o(t)$ at the end of the interval $T/2 \leq t \leq T$:

$$v_o(T) = V_1 = -V_m + (V_2 + V_m)e^{-T/2RC}.$$

Solving these equations for V_1 and V_2 yields

$$V_2 = -V_1 = \frac{V_m(1 - e^{-T/2RC})}{1 + e^{-T/2RC}}.$$

Substituting the expressions for V_1 and V_2 into the equations for $v_o(t)$ gives

$$v_o = V_m - \frac{2V_m}{1 + e^{-T/2RC}} e^{-t/RC}, \quad 0 \leq t \leq T/2, \tag{16.24}$$

and

$$v_o = -V_m + \frac{2V_m}{1 + e^{-T/2RC}} e^{-[t-(T/2)]/RC}, \quad T/2 \leq t \leq T. \tag{16.25}$$

Equations 16.24 and 16.25 indicate that $v_o(t)$ has half-wave symmetry, so the average value of v_o is zero. This result agrees with the Fourier series solution for the steady-state response—namely, that because the input voltage has no dc component, the output voltage cannot have a dc component. Equations 16.24 and 16.25 also show the effect of changing the size of the capacitor. If C is small, the exponential functions quickly vanish, $v_o = V_m$ between 0 and $T/2$, and $v_o = -V_m$ between $T/2$ and T. In other words, $v_o \rightarrow v_g$ as $C \rightarrow 0$. If C is large, the output waveform becomes triangular in shape, as Fig. 16.14 shows. Note that for large C, we may approximate the exponential terms $e^{-t/RC}$ and $e^{-[t-(T/2)]/RC}$ by the linear terms $1 - (t/RC)$ and $1 - \{[t - (T/2)]/RC\}$, respectively. Equation 16.23 gives the Fourier series of this triangular waveform.

Figure 16.14 summarizes the results. The dashed line in Fig. 16.14 is the input voltage, the solid colored line depicts the output voltage when C is small, and the solid black line depicts the output voltage when C is large.

Finally, we verify that the steady-state response of Eqs. 16.24 and 16.25 is equivalent to the Fourier series solution in Eq. 16.22. To do so, we derive the Fourier series representation of the periodic function described by Eqs. 16.24 and 16.25. We have already noted that the periodic voltage response has half-wave symmetry. Therefore, the Fourier series contains only odd harmonics. For k odd,

$$a_k = \frac{4}{T} \int_0^{T/2} \left(V_m - \frac{2V_m e^{-t/RC}}{1 + e^{-T/2RC}} \right) \cos k\omega_0 t \, dt$$

$$= \frac{-8RCV_m}{T[1 + (k\omega_0 RC)^2]} \quad (k \text{ is odd}), \tag{16.26}$$

$$b_k = \frac{4}{T} \int_0^{T/2} \left(V_m - \frac{2V_m e^{-t/RC}}{1 + e^{-T/2RC}} \right) \sin k\omega_0 t \, dt$$

$$= \frac{4V_m}{k\pi} - \frac{8k\omega_0 V_m R^2 C^2}{T[1 + (k\omega_0 RC)^2]} \quad (k \text{ is odd}). \tag{16.27}$$

To show that Eqs. 16.26 and 16.27 are consistent with Eq. 16.22, we must prove that

$$\sqrt{a_k^2 + b_k^2} = \frac{4V_m}{k\pi} \frac{1}{\sqrt{1 + (k\omega_0 RC)^2}}, \tag{16.28}$$

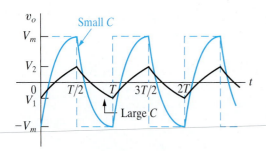

Figure 16.14 ▲ The effect of capacitor size on the steady-state response.

and that

$$\frac{a_k}{b_k} = -k\omega_0 RC. \qquad (16.29)$$

You can verify Eqs. 16.26–16.29 in Problems 16.23 and 16.24. Equations 16.28 and 16.29 are used with Eqs. 16.20 and 16.21 to derive the Fourier series expression in Eq. 16.22; we leave the details to you in Problem 16.25.

With this illustrative circuit, we showed how to use the Fourier series in conjunction with the principle of superposition to find the steady-state response to a periodic input voltage. Again, the principal shortcoming of the Fourier series approach is the difficulty of ascertaining the output waveform. However, by considering a circuit's frequency response, we can deduce a reasonable approximation of the steady-state response, using a finite number of terms in the Fourier series representation. (See Problems 16.29 and 16.30.) Example 16.4 provides another example of the Fourier series circuit analysis method.

| EXAMPLE 16.4 | **Finding the Response of an *RLC* Circuit to a Square-Wave Voltage** |

The square wave in Fig. 16.12(b) has a magnitude of 4π V and a frequency of 100 rad/s. This periodic voltage is input to the circuit in Fig. 16.15.

a) Find the first five nonzero terms in the Fourier series representation of the square-wave input voltage.

b) Find the output voltage, v_o, for the circuit in Fig. 16.15, using the five terms calculated in part (a).

c) Use the frequency response of the circuit in Fig. 16.15 to explain the output voltage terms you calculated in part (b).

Solution

a) The Fourier series representation of v_g is

$$v_g = \frac{4V_m}{\pi} \sum_{n=1,3,5,\ldots}^{\infty} \frac{1}{n} \sin n\omega_o t$$

$$= 16 \sin 100t + 5.33 \sin 300t + 3.2 \sin 500t$$

$$+ 2.29 \sin 700t + 1.78 \sin 900t + \ldots \text{ V.}$$

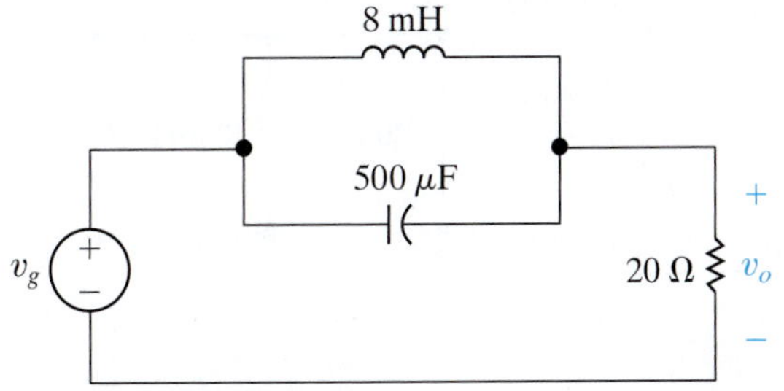

Figure 16.15 ▲ The circuit for Example 16.4.

b) The circuit in Fig. 16.15 is a parallel *RLC* bandreject filter. From Fig. 14.31, the transfer function is

$$H(s) = \frac{s^2 + 1/LC}{s^2 + s/RC + 1/LC}$$

$$= \frac{s^2 + 25 \times 10^4}{s^2 + 100s + 25 \times 10^4}.$$

To find the terms in the Fourier series representation of v_o, evaluate $H(s)$ for $s = j\omega$, where ω is the frequency of the term in v_g. Then, multiply by the phasor representation of that term in v_g, and inverse-phasor-transform the result to find the corresponding term in v_o. Note that we are using the sine function, instead of the cosine function for the phasor transform and its inverse. For $\omega = 100$ rad/s:

$$H(j100) = \frac{(j100)^2 + 25 \times 10^4}{(j100)^2 + 100(j100) + 25 \times 10^4}$$

$$= 0.999\underline{/-2.4°},$$

$$V_o(j100) = (0.998\underline{/-2.4°})(16) = 15.99\underline{/-2.4°}.$$

For $\omega = 300$ rad/s:

$$H(j300) = \frac{(j300)^2 + 25 \times 10^4}{(j300)^2 + 100(j300) + 25 \times 10^4}$$

$$= 0.983\underline{/-10.6°},$$

$$V_o(j300) = (0.983\underline{/-10.6°})(5.33) = 5.24\underline{/-10.6°}.$$

For $\omega = 500$ rad/s:

$$H(j500) = \frac{(j500)^2 + 25 \times 10^4}{(j500)^2 + 100(j500) + 25 \times 10^4} = 0,$$

$$V_o(j500) = (0)(3.2) = 0.$$

For $\omega = 700$ rad/s:

$$H(j700) = \frac{(j700)^2 + 25 \times 10^4}{(j700)^2 + 100(j700) + 25 \times 10^4}$$

$$= 0.96\underline{/16.26°},$$

$$V_o(j700) = (0.96\underline{/16.26°})(2.29) = 2.2\underline{/16.26°}.$$

For $\omega = 900$ rad/s:

$$H(j900) = \frac{(j900)^2 + 25 \times 10^4}{(j900)^2 + 100(j900) + 25 \times 10^4}$$

$$= 0.99\underline{/9.13°},$$

$$V_o(j900) = (0.99\underline{/9.13°})(1.78) = 1.76\underline{/9.13°}.$$

Therefore, the first five terms in the Fourier series representation of the steady-state output voltage are

$$v_o = 15.99 \sin(100t - 2.4°) + 5.24 \sin(300t - 10.6°)$$

$$+ 0 \sin 500t + 2.2 \sin(700t + 16.26°)$$

$$+ 1.76 \sin(900t + 9.13°) + \dots \text{V}.$$

c) As noted in part (a), the circuit in Fig. 16.15 is a bandreject filter. It has a center frequency of $\sqrt{1/LC} = 500$ rad/s, a bandwidth of $1/RC = 100$ rad/s, and a quality of 5, so it is a selective filter. Its cutoff frequencies are 452.494 rad/s and 552.494 rad/s. The magnitude of v_o at 500 rad/s, the center frequency of the filter, is zero, exactly what we should expect from a bandreject filter. The magnitudes of v_o at the frequencies outside the passband defined by the cutoff frequencies are very close to the magnitudes of v_g at those frequencies. This is also the behavior we expect from a selective bandreject filter.

◨ ASSESSMENT PROBLEMS

Objective 2—Know how to analyze a circuit's response to a periodic waveform

16.5 The periodic triangular-wave voltage seen on the left is applied to the circuit shown on the right. Derive the first three nonzero terms in the Fourier series that represent the steady-state voltage v_o if $V_m = 281.25\pi^2$ mV and the period of the input voltage is 200π ms.

Answer: $2238.83 \cos(10t - 5.71°)$
$+ 239.46 \cos(30t - 16.70°)$
$+ 80.50 \cos(50t - 26.57°) + \dots$ mV.

16.6 The periodic square-wave shown on the top is applied to the circuit shown on the bottom.
a) Derive the first four nonzero terms in the Fourier series that represents the steady-state voltage v_o if $V_m = 210\pi$ V and the period of the input voltage is 0.2π ms.
b) Which harmonic dominates the output voltage? Explain why.

Answer: (a) $17.5 \cos(10,000t + 88.81°)$
$+ 26.14 \cos(30,000t - 95.36°)$
$+ 168 \cos(50,000t)$
$+ 17.32 \cos(70,000t + 98.30°) + \dots$ V;
(b) The fifth harmonic, at 50,000 rad/s, because the circuit is a bandpass filter with a center frequency of 50,000 rad/s and a quality factor of 10.

SELF-CHECK: Also try Chapter Problems 16.28 and 16.29.

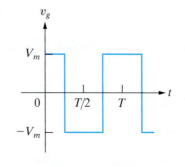

16.6 Average-Power Calculations with Periodic Functions

Given the Fourier series representation of the voltage and current at a pair of terminals in a linear lumped-parameter circuit, we can express the average power at the terminals as a function of the harmonic voltages and currents. Using the trigonometric form of the Fourier series from Eq. 16.20, we find that the periodic voltage and current at the circuit's terminals are

$$v = V_{dc} + \sum_{n=1}^{\infty} V_n \cos(n\omega_0 t - \theta_{vn}),$$

$$i = I_{dc} + \sum_{n=1}^{\infty} I_n \cos(n\omega_0 t - \theta_{in}).$$

The notation used in the equations for voltage and current is defined as follows:

V_{dc} = the amplitude of the dc voltage component,

V_n = the amplitude of the nth-harmonic voltage,

θ_{vn} = the phase angle of the nth-harmonic voltage,

I_{dc} = the amplitude of the dc current component,

I_n = the amplitude of the nth-harmonic current,

θ_{in} = the phase angle of the nth-harmonic current.

We assume that the current is in the direction of the voltage drop across the terminals, so, using the passive sign convention, the instantaneous power at the terminals is vi. The average power is

$$P = \frac{1}{T} \int_{t_0}^{t_0+T} p \, dt = \frac{1}{T} \int_{t_0}^{t_0+T} vi \, dt.$$

To find the expression for the average power, we substitute the expressions for voltage and current into the equation for average power and integrate. At first glance, this appears to be a formidable task because the product vi requires multiplying two infinite series. However, the only terms to survive integration are the products of voltage and current at the same frequency. (See Problem 16.5.) Therefore, the equation for average power reduces to

$$P = \frac{1}{T} V_{dc} I_{dc} t \Big|_{t_0}^{t_0+T}$$

$$+ \sum_{n=1}^{\infty} \frac{1}{T} \int_{t_0}^{t_0+T} V_n I_n \cos(n\omega_0 t - \theta_{vn}) \cos(n\omega_0 t - \theta_{in}) \, dt.$$

Now, using the trigonometric identity

$$\cos \alpha \cos \beta = \frac{1}{2} \cos(\alpha - \beta) + \frac{1}{2} \cos(\alpha + \beta).$$

we simplify the expression for P to

$$P = V_{dc} I_{dc}$$

$$+ \frac{1}{T} \sum_{n=1}^{\infty} \frac{V_n I_n}{2} \int_{t_0}^{t_0+T} [\cos(\theta_{vn} - \theta_{in}) + \cos(2n\omega_0 t - \theta_{vn} - \theta_{in})] dt.$$

The second term under the integral sign integrates to zero, so

$$P = V_{dc}I_{dc} + \sum_{n=1}^{\infty} \frac{V_n I_n}{2} \cos(\theta_{vn} - \theta_{in}). \qquad (16.30)$$

Equation 16.30 states that the total average power delivered by a periodic signal is the sum of the average powers associated with each harmonic voltage and current. Currents and voltages of different frequencies do not interact to produce average power. Example 16.5 computes the average power delivered by a periodic voltage.

EXAMPLE 16.5 | **Calculating Average Power for a Circuit with a Periodic Voltage Source**

Assume that the periodic square-wave voltage in Example 16.3 is applied across the terminals of a 15 Ω resistor. The value of V_m is 60 V, and the value of T is 5 ms.

a) Write the first five nonzero terms of the Fourier series representation of $v(t)$. Use the trigonometric form given in Eq. 16.20.

b) Calculate the average power associated with each term in (a).

c) Calculate the total average power delivered to the 15 Ω resistor.

d) What percentage of the total power is delivered by the first five terms of the Fourier series?

Solution

a) The dc component of $v(t)$ is

$$a_v = \frac{(60)(T/4)}{T} = 15 \text{ V.}$$

From Example 16.3 we have

$$A_1 = \sqrt{2}\,60/\pi = 27.01 \text{ V,}$$

$$\theta_1 = 45°,$$

$$A_2 = 60/\pi = 19.10 \text{ V,}$$

$$\theta_2 = 90°,$$

$$A_3 = 20\sqrt{2}/\pi = 9.00 \text{ V,}$$

$$\theta_3 = 135°,$$

$$A_4 = 0,$$

$$\theta_4 = 0°,$$

$$A_5 = 5.40 \text{ V,}$$

$$\theta_5 = 45°,$$

$$\omega_0 = \frac{2\pi}{T} = \frac{2\pi}{0.005} = 400\pi \text{ rad/s.}$$

Thus, using the first five nonzero terms of the Fourier series,

$$v(t) = 15 + 27.01 \cos(400\pi t - 45°)$$
$$+ 19.10 \cos(800\pi t - 90°)$$
$$+ 9.00 \cos(1200\pi t - 135°)$$
$$+ 5.40 \cos(2000\pi t - 45°) + \cdots \text{ V.}$$

b) The voltage is applied to the terminals of a resistor, so we can find the power associated with each term as follows:

$$P_{dc} = \frac{15^2}{15} = 15 \text{ W,}$$

$$P_1 = \frac{1}{2}\frac{27.01^2}{15} = 24.32 \text{ W,}$$

$$P_2 = \frac{1}{2}\frac{19.10^2}{15} = 12.16 \text{ W,}$$

$$P_3 = \frac{1}{2}\frac{9^2}{15} = 2.70 \text{ W,}$$

$$P_5 = \frac{1}{2}\frac{5.4^2}{15} = 0.97 \text{ W.}$$

c) To obtain the total average power delivered to the 15 Ω resistor, we first calculate the rms value of $v(t)$:

$$V_{rms} = \sqrt{\frac{(60)^2(T/4)}{T}} = \sqrt{900} = 30 \text{ V.}$$

The total average power delivered to the 15 Ω resistor is

$$P_T = \frac{30^2}{15} = 60 \text{ W.}$$

d) The total power delivered by the first five nonzero terms is

$$P = P_{dc} + P_1 + P_2 + P_3 + P_5 = 55.15 \text{ W.}$$

This is $(55.15/60)(100)$, or 91.92% of the total.

ASSESSMENT PROBLEM

Objective 3—Be able to estimate the average power delivered to a resistor using a small number of Fourier coefficients

16.7 The periodic voltage function in Assessment Problem 16.3 is applied to the circuit shown. If $12V_m = 296.09$ V and $T = 2094.4$ ms, estimate the average power delivered to the 2 Ω resistor.

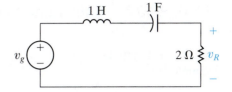

Answer: 60.75 W.

SELF-CHECK: Also try Chapter Problems 16.34 and 16.35.

16.7 The rms Value of a Periodic Function

The rms value of a periodic function can be expressed in terms of the Fourier coefficients; by definition,

$$F_{\text{rms}} = \sqrt{\frac{1}{T} \int_{t_0}^{t_0+T} f(t)^2 \, dt}.$$

Replacing $f(t)$ with its Fourier series representation gives

$$F_{\text{rms}} = \sqrt{\frac{1}{T} \int_{t_0}^{t_0+T} \left[a_v + \sum_{n=1}^{\infty} A_n \cos(n\omega_0 t - \theta_n) \right]^2 \, dt}.$$

The integral of the squared time function simplifies because the only terms whose integral over one period are not zero are the square of the dc term and the square of the terms at each frequency. Therefore,

$$F_{\text{rms}} = \sqrt{\frac{1}{T}\left(a_v^2 T + \sum_{n=1}^{\infty} \frac{T}{2} A_n^2 \right)}$$

$$= \sqrt{a_v^2 + \sum_{n=1}^{\infty} \frac{A_n^2}{2}}$$

$$= \sqrt{a_v^2 + \sum_{n=1}^{\infty} \left(\frac{A_n}{\sqrt{2}} \right)^2}. \qquad (16.31)$$

We can use Eq. 16.31 to find the rms value of a periodic function by adding the square of the rms value of each harmonic to the square of the dc value and taking the square root of that sum. For example, let's assume that a periodic voltage is represented by the finite series

$$v = 10 + 30 \cos(\omega_0 t - \theta_1) + 20 \cos(2\omega_0 t - \theta_2)$$
$$+ 5 \cos(3\omega_0 t - \theta_3) + 2 \cos(5\omega_0 t - \theta_5) \, \text{V}.$$

The rms value of this voltage is

$$V_{\text{rms}} = \sqrt{10^2 + (30/\sqrt{2})^2 + (20/\sqrt{2})^2 + (5/\sqrt{2})^2 + (2/\sqrt{2})^2}$$
$$= \sqrt{764.5} = 27.65 \, \text{V}.$$

Because the number of terms in a Fourier series representation of a periodic function is usually infinite, Eq. 16.31 yields an estimate of the actual rms value. We illustrate this result in Example 16.6.

EXAMPLE 16.6 | **Estimating the rms Value of a Periodic Function**

Use Eq. 16.31 to estimate the rms value of the voltage in Example 16.5.

Solution

From Example 16.5,

$V_{dc} = 15$ V,

$V_1 = 27.01/\sqrt{2}$ V, the rms value of the fundamental,

$V_2 = 19.10/\sqrt{2}$ V, the rms value of the second harmonic,

$V_3 = 9.00/\sqrt{2}$ V, the rms value of the third harmonic,

$V_5 = 5.40/\sqrt{2}$ V, the rms value of the fifth harmonic.

Therefore,

$$V_{rms} = \sqrt{15^2 + \left(\frac{27.01}{\sqrt{2}}\right)^2 + \left(\frac{19.10}{\sqrt{2}}\right)^2 + \left(\frac{9.00}{\sqrt{2}}\right)^2 + \left(\frac{5.40}{\sqrt{2}}\right)^2}$$

$$= 28.76 \text{ V}.$$

From Example 16.5, the actual rms value is 30 V. We can get closer to this value by including more and more harmonics in Eq. 16.31. For example, if we include the harmonics through $k = 9$, the estimated rms value is 29.32 V.

SELF-CHECK: Assess your understanding of this material by trying Chapter Problems 16.37 and 16.39.

16.8 The Exponential Form of the Fourier Series

The exponential form of the Fourier series is a concise representation of the series, given by

$$f(t) = \sum_{n=-\infty}^{\infty} C_n e^{jn\omega_0 t}, \tag{16.32}$$

where

$$C_n = \frac{1}{T} \int_{t_0}^{t_0 + T} f(t) e^{-jn\omega_0 t} \, dt. \tag{16.33}$$

To derive Eqs. 16.32 and 16.33, start with Eq. 16.2 and replace the cosine and sine functions with their exponential equivalents:

$$\cos n\omega_0 t = \frac{e^{jn\omega_0 t} + e^{-jn\omega_0 t}}{2},$$

$$\sin n\omega_0 t = \frac{e^{jn\omega_0 t} - e^{-jn\omega_0 t}}{2j}.$$

Substituting the exponential equivalents into Eq. 16.2 gives

$$f(t) = a_v + \sum_{n=1}^{\infty} \frac{a_n}{2}(e^{jn\omega_0 t} + e^{-jn\omega_0 t}) + \frac{b_n}{2j}(e^{jn\omega_0 t} - e^{-jn\omega_0 t})$$

$$= a_v + \sum_{n=1}^{\infty} \left(\frac{a_n - jb_n}{2}\right)e^{jn\omega_0 t} + \left(\frac{a_n + jb_n}{2}\right)e^{-jn\omega_0 t}. \tag{16.34}$$

Now we define C_n as

$$C_n = \frac{1}{2}(a_n - jb_n) = \frac{A_n}{2}\angle{-\theta_n}, \quad n = 1, 2, 3, \cdots .$$

From the definition of C_n,

$$C_n = \frac{1}{2}\left[\frac{2}{T}\int_{t_0}^{t_0+T} f(t)\cos n\omega_0 t\, dt - j\frac{2}{T}\int_{t_0}^{t_0+T} f(t)\sin n\omega_0 t\, dt\right]$$

$$= \frac{1}{T}\int_{t_0}^{t_0+T} f(t)\left(\cos n\omega_0 t - j\sin n\omega_0 t\right) dt$$

$$= \frac{1}{T}\int_{t_0}^{t_0+T} f(t)e^{-jn\omega_0 t}\, dt,$$

which completes the derivation of Eq. 16.33. To complete the derivation of Eq. 16.32, we first observe from Eq. 16.33 that

$$C_0 = \frac{1}{T}\int_{t_0}^{t_0+T} f(t)dt = a_v.$$

Next we note that

$$C_{-n} = \frac{1}{T}\int_{t_0}^{t_0+T} f(t)e^{jn\omega_0 t}dt = C_n^* = \frac{1}{2}(a_n + jb_n).$$

Substituting the expressions for C_0, C_n, and C_n^* into Eq. 16.34 yields

$$f(t) = C_0 + \sum_{n=1}^{\infty}(C_n e^{jn\omega_0 t} + C_n^* e^{-jn\omega_0 t})$$

$$= \sum_{n=0}^{\infty} C_n e^{jn\omega_0 t} + \sum_{n=1}^{\infty} C_n^* e^{-jn\omega_0 t}.$$

Note that the second summation on the right-hand side is equivalent to summing $C_n e^{jn\omega_0 t}$ from -1 to $-\infty$; that is,

$$\sum_{n=1}^{\infty} C_n^* e^{-jn\omega_0 t} = \sum_{n=-1}^{-\infty} C_n e^{jn\omega_0 t}.$$

Because the summation from -1 to $-\infty$ is the same as the summation from $-\infty$ to -1, we can rewrite $f(t)$ as

$$f(t) = \sum_{n=0}^{\infty} C_n e^{jn\omega_0 t} + \sum_{-\infty}^{-1} C_n e^{jn\omega_0 t}$$

$$= \sum_{-\infty}^{\infty} C_n e^{jn\omega_0 t},$$

which completes the derivation of Eq. 16.32.

We can also find the rms value of a periodic function using the complex Fourier coefficients. From Eqs. 16.21 and 16.31,

$$F_{\text{rms}} = \sqrt{a_v^2 + \sum_{n=1}^{\infty}\frac{a_n^2 + b_n^2}{2}},$$

$$|C_n| = \frac{\sqrt{a_n^2 + b_n^2}}{2},$$

$$C_0^2 = a_v^2.$$

Therefore:

$$F_{\text{rms}} = \sqrt{C_0^2 + 2\sum_{n=1}^{\infty}|C_n|^2}. \tag{16.35}$$

Example 16.7 finds the exponential Fourier series representation of a periodic function.

EXAMPLE 16.7 | Finding the Exponential Form of the Fourier Series

Find the exponential Fourier series for the periodic voltage shown in Fig. 16.16.

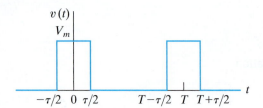

Figure 16.16 ▲ The periodic voltage for Example 16.7.

Solution

Using $-\tau/2$ as the starting point for the integration, we have, from Eq. 16.33,

$$C_n = \frac{1}{T} \int_{-\tau/2}^{\tau/2} V_m e^{-jn\omega_0 t} \, dt$$

$$= \frac{V_m}{T} \left(\frac{e^{-jn\omega_0 t}}{-jn\omega_0} \right)\Bigg|_{-\tau/2}^{\tau/2}$$

$$= \frac{jV_m}{n\omega_0 T} \left(e^{-jn\omega_0 \tau/2} - e^{jn\omega_0 \tau/2} \right)$$

$$= \frac{2V_m}{n\omega_0 T} \sin n\omega_0 \tau/2.$$

Since $v(t)$ has even symmetry, $b_n = 0$ for all n, and we expect C_n to be real. Moreover, the amplitude of C_n follows a $(\sin x)/x$ distribution, as indicated when we rewrite

$$C_n = \frac{V_m \tau}{T} \frac{\sin (n\omega_0 \tau/2)}{n\omega_0 \tau/2}.$$

We say more about this subject in Section 16.9. The exponential Fourier series representation of $v(t)$ is

$$v(t) = \sum_{n=-\infty}^{\infty} \left(\frac{V_m \tau}{T} \right) \frac{\sin (n\omega_0 \tau/2)}{n\omega_0 \tau/2} e^{jn\omega_0 t}$$

$$= \left(\frac{V_m \tau}{T} \right) \sum_{n=-\infty}^{\infty} \frac{\sin (n\omega_0 \tau/2)}{n\omega_0 \tau/2} e^{jn\omega_0 t}.$$

ASSESSMENT PROBLEMS

Objective 4—Be able to calculate the exponential form of the Fourier coefficients for a periodic waveform

16.8 Derive the expression for the Fourier coefficients C_n for the periodic function shown. *Hint:* Take advantage of symmetry by using the fact that $C_n = (a_n - jb_n)/2$.

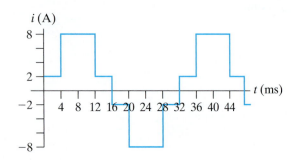

Answer: $C_n = -j\frac{4}{\pi n}\left(1 + 3\cos\frac{n\pi}{4}\right)$, n odd

SELF-CHECK: Also try Chapter Problems 16.44 and 16.45.

16.9 a) Calculate the rms value of the periodic current in Assessment Problem 16.8.
   b) Using C_1–C_{11}, estimate the rms value.
   c) What is the percentage of error in the value obtained in (b), based on the true value found in (a)?
   d) For this periodic function, could fewer terms be used to estimate the rms value and still insure the error is less than 1%?

Answer: (a) $\sqrt{34}$ A;

   (b) 5.777 A;

   (c) −0.93 %;

   (d) yes; if C_1–C_9 are used, the error is −0.98 %.

16.9 Amplitude and Phase Spectra

As we have noted, we generally cannot visualize what a periodic function looks like in the time domain from its Fourier series representation. Nevertheless, we recognize that the Fourier coefficients characterize the periodic function completely.

We can describe a periodic function by plotting the amplitude and phase angle of each term in its Fourier series versus frequency. The plot of each term's amplitude versus the frequency is called the **amplitude spectrum** of $f(t)$, and the plot of each term's phase angle versus the frequency is called the **phase spectrum** of $f(t)$. Because the amplitude and phase angle are defined at discrete frequency values (that is, at $\omega_0, 2\omega_0, 3\omega_0, \dots$), these plots are also called **line spectra**.

Amplitude and phase spectra plots are based on either Eq. 16.20 (A_n and $-\theta_n$) or Eq. 16.32 (C_n). We focus on Eq. 16.32 and leave the plots based on Eq. 16.20 to the problems. For example, consider the periodic voltage in Fig. 16.16. From Example 16.6,

$$C_n = \frac{V_m\tau}{T}\frac{\sin(n\omega_0\tau/2)}{n\omega_0\tau/2},$$

From the expression for C_n, we see that the amplitude spectrum is bounded by the envelope of the $|(\sin x)/x|$ function. Figure 16.17 provides the plot of $|(\sin x)/x|$ versus x, where x is in radians. It shows that the function's value is zero whenever x is an integral multiple of π. Replacing ω_0 with $2\pi/T$ in the argument of the sine function in the expression for C_n,

$$n\omega_0\left(\frac{\tau}{2}\right) = \frac{n\pi\tau}{T} = \frac{n\pi}{T/\tau}.$$

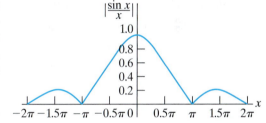

Figure 16.17 ▲ The plot of $(\sin x)/x$ versus x.

We can therefore deduce that the amplitude spectrum goes through zero whenever $n\tau/T$ is an integer. As the reciprocal of τ/T becomes an increasingly larger integer, the number of harmonics between every π radians increases. If $n\pi/T$ is not an integer, the amplitude spectrum still follows the $|(\sin x)/x|$ envelope. However, the envelope is not zero at an integral multiple of ω_0. Because C_n is real for all n, the phase angle associated with C_n is either zero or 180°, depending on its algebraic sign.

Now, what happens to the amplitude and phase spectra if $f(t)$ is shifted along the time axis? To find out, we shift the periodic voltage in Example 16.6 t_0 units to the right. By hypothesis,

$$v(t) = \sum_{n=-\infty}^{\infty} C_n e^{jn\omega_0 t}.$$

Therefore

$$v(t - t_0) = \sum_{n=-\infty}^{\infty} C_n e^{jn\omega_0(t-t_0)} = \sum_{n=-\infty}^{\infty} C_n e^{-jn\omega_0 t_0} e^{jn\omega_0 t},$$

which indicates that shifting the origin has no effect on the amplitude spectrum, because

$$|C_n| = |C_n e^{-jn\omega_0 t_0}|.$$

However, the phase spectrum has changed to $-(\theta_n + n\omega_0 t_0)$ rads.

Example 16.8 plots the amplitude and phase spectra for a specific instance of the periodic voltage in Example 16.7.

EXAMPLE 16.8 **Plotting the Amplitude and Phase Spectra for a Periodic Voltage**

Suppose that for the periodic voltage in Fig. 16.16, $V_m = 5$ V, $T = 10$ ms, and $\tau = T/5$.

a) Plot the amplitude and phase spectra versus frequency (in Hz) for $-10 \le n \le +10$, using the expressions for C_n given by Eq. 16.32.

b) Repeat part (a) if the periodic voltage in Fig. 16.16 is shifted $\tau/2$ units to the right.

Solution

a) Substituting the values for V_m and τ into the expression for C_n from Example 16.7 and simplifying, we get

$$C_n = \frac{\sin(n\pi/5)}{n\pi/5}.$$

The values of C_n for n between -10 and $+10$ are tabulated at right. Since C_n is a real number for all values of n, its magnitude is the absolute value

of C_n and its phase angle is $0°$ if C_n is positive and $180°$ if C_n is negative. The amplitude and phase spectra are plotted in Fig. 16.18.

n	C_n	n	C_n
-10	0.000	1	0.935
-9	-0.104	2	0.757
-8	-0.189	3	0.505
-7	-0.216	4	0.234
-6	-0.156	5	0.000
-5	0.000	6	-0.156
-4	0.234	7	-0.216
-3	0.505	8	-0.189
-2	0.757	9	-0.104
-1	0.935	10	0.000
0	1.000		

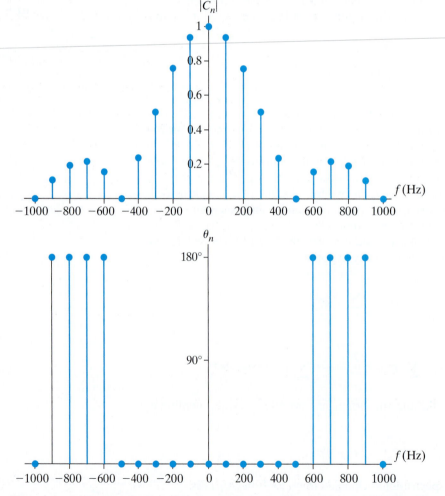

Figure 16.18 ▲ The amplitude and phase spectra for the periodic waveform in Fig. 16.16, with $V_m = 5$ V, $T = 10$ ms, and $\tau = T/5$.

b) When the periodic voltage in Fig. 16.16 is shifted to the right by $\tau/2$, we know that the amplitude spectrum is unchanged. Since $\tau = T/5$, the new phase angle θ'_n is

$$\theta'_n = -(\theta_n + n\pi/5).$$

The plot of the amplitude spectra for this shifted periodic voltage is the same as the one shown in Fig. 16.18, while the plot of the phase spectra for the shifted period voltage is shown in Fig. 16.19, for $-10 \le n \le +10$.

The periodic waveform in Fig. 16.16 is important because it provides an excellent way to illustrate the transition from the Fourier series representation of a periodic function to the Fourier transform representation of a nonperiodic function. We discuss the Fourier transform in Chapter 17.

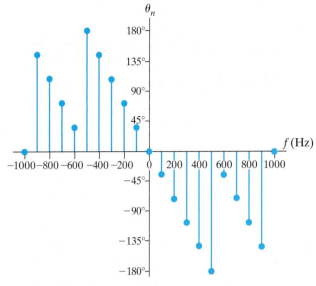

Figure 16.19 ▲ The phase spectra for the periodic waveform in Fig. 16.16, shifted to the right by $\tau/2$, with $V_m = 5$ V, $T = 10$ ms, and $\tau = T/5$.

ASSESSMENT PROBLEM

Objective 4—Be able to calculate the exponential form of the Fourier coefficients for a periodic waveform

16.10 The function in Assessment Problem 16.8 is shifted along the time axis 8 ms to the right. Write the exponential Fourier series for the periodic current.

Answer:

$$i(t) = \frac{4}{\pi} \sum_{n=-\infty \,(\text{odd})}^{\infty} \frac{1}{n}\left(1 + 3\cos\frac{n\pi}{4}\right)e^{-(j\pi/2)(n+1)}e^{jn\omega_0 t}\ \text{A}.$$

SELF-CHECK: Also try Chapter Problems 16.49 and 16.50.

■ Practical Perspective

Active High-Q Filters

Consider the narrowband op amp bandpass filter shown in Fig. 16.20(a). The square-wave voltage shown in Fig. 16.20(b) is the input to the filter. We know that the square wave is composed of an infinite sum of sinusoids, one sinusoid at the same frequency as the square wave and all of the remaining sinusoids at integer multiples of that frequency. What effect will the filter have on this input sum of sinusoids?

The Fourier series representation of the square wave in Fig. 16.20(b) is given by

$$v_g(t) = \frac{4A}{\pi} \sum_{n=1,3,5,\dots}^{\infty} \frac{1}{n}\sin\frac{n\pi}{2}\cos n\omega_0 t$$

where $A = 15.65\pi$ V. The first three terms of this Fourier series are given by

$$v_g(t) = 62.6\cos\omega_0 t - 20.87\cos 3\omega_0 t + 12.52\cos 5\omega_0 t - \dots \text{V}.$$

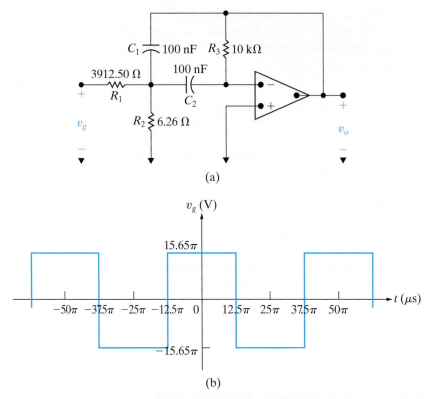

(a)

(b)

Figure 16.20 ▲ (a) narrowband bandpass filter; (b) square wave input.

The period of the square wave is 50π μs, so the frequency of the square wave is 40,000 rad/s.

The transfer function for the bandpass filter in Fig. 16.20(a) is

$$H(s) = \frac{K\beta s}{s^2 + \beta s + \omega_0^2}$$

where $K = 400/313$, $\beta = 2000$ rad/s, $\omega_0 = 40,000$ rad/s. This filter has a quality factor of $40,000/2000 = 20$. Note that the center frequency of the bandpass filter equals the frequency of the input square wave.

Multiply each term of the Fourier series representation of the square wave, represented as a phasor, by the transfer function $H(s)$ evaluated at the frequency of the term in the Fourier series to get the representation of the output voltage of the filter as a Fourier series:

$$v_o(t) = -80 \cos \omega_0 t - 0.5 \cos 3\omega_0 t + 0.17 \cos 5\omega_0 t + \cdots \text{ V}.$$

Notice the selective nature of the bandpass filter, which effectively amplifies the fundamental frequency component of the input square wave and attenuates all of the harmonic components.

Now make the following changes to the bandpass filter of Fig. 16.20(a)—let $R_1 = 391.25$ Ω, $R_2 = 74.4$ Ω, $R_3 = 1$ kΩ, and $C_1 = C_2 = 0.1$ μF. The transfer function for the filter, $H(s)$, has the same form given above, but now $K = 400/313$, $\beta = 20,000$ rad/s, $\omega_0 = 40,000$ rad/s. The passband gain and center frequency are unchanged, but the bandwidth has increased by a factor of 10. This makes the quality factor 2, and the resulting bandpass filter is less selective than the original filter. We can see this by looking at the output voltage of the filter as a Fourier series:

$$v_o(t) = -80 \cos \omega_0 t - 5 \cos 3\omega_0 t + 1.63 \cos 5\omega_0 t + \cdots \text{ V}.$$

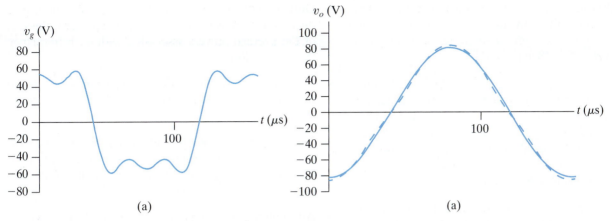

Figure 16.21 ▲ (a) The first three terms of the Fourier series of the square wave in Fig. 16.20(b); (b) the first three terms of the Fourier series of the output from the bandpass filter in Fig. 16.20(a), where $Q = 20$ (solid line); the first three terms of the Fourier series of the output from the bandpass filter in Fig. 16.20(a) with component values changed to give $Q = 2$ (dashed line).

At the fundamental frequency, the output has the same amplification factor, but the higher harmonic components have not been attenuated as significantly as they were when the filter with $Q = 20$ was used. Figure 16.21 plots the first three terms of the Fourier series representations of the input square wave and the resulting output waveforms for the two bandpass filters. Note the nearly perfect replication of a sinusoid in the solid line plot of Fig. 16.21(b) and the distortion that results when using a less selective filter in the dashed line plot of Fig. 16.21(b).

SELF-CHECK: Assess your understanding of the Practical Perspective by solving Chapter Problems 16.56 and 16.57.

■ Summary

- A **periodic function** is a function that repeats itself every T seconds. A period is the smallest time interval (T) that a periodic function can be shifted to produce a function identical to itself. (See page 618.)

- The Fourier series is an infinite series used to represent a periodic function. The series consists of a constant term and infinitely many harmonically related cosine and sine terms. (See page 621.)

- The **fundamental frequency** is the frequency determined by the fundamental period ($f_0 = 1/T$ or $\omega_0 = 2\pi f_0$). (See page 621.)

- The **harmonic frequency** is an integer multiple of the fundamental frequency. (See page 621.)

- The **Fourier coefficients** are the constant term and the coefficient of each cosine and sine term in the series. (See Eqs. 16.3–16.5.) (See page 622.)

- Five types of symmetry are used to simplify the computation of the Fourier coefficients:

  - *even,* in which all sine terms in the series are zero;

  - *odd,* in which all cosine terms and the constant term are zero;

  - *half-wave,* in which all even harmonics are zero;

  - *quarter-wave, half-wave, even,* in which the series contains only odd harmonic cosine terms;

  - *quarter-wave, half-wave, odd,* in which the series contains only odd harmonic sine terms.

  (See page 625.)

- In the alternative form of the Fourier series, each harmonic represented by the sum of a cosine and sine term is combined into a single term of the form $A_n \cos(n\omega_0 t - \theta_n)$. (See page 631.)

- For steady-state response, the Fourier series of the output signal is determined by first finding the output for each component of the input signal. The individual responses are added (superimposed) to form the Fourier series of the output signal. The response to the individual terms in the input series is found by either frequency domain or s-domain analysis. (See page 633.)

- The waveform of the output signal is difficult to obtain without the aid of a computer. Sometimes the frequency response (or filtering) characteristics of the circuit can be used to ascertain how closely the output waveform matches the input waveform. (See page 635.)

- Only harmonics of the same frequency interact to produce average power. The total average power is the sum of the average powers associated with each frequency. (See page 640.)

- The rms value of a periodic function can be estimated from the Fourier coefficients. (See Eqs. 16.31 and 16.35.) (See page 641.)

- The Fourier series of a periodic signal may also be written in exponential form by using Euler's identity to replace the cosine and sine terms with their exponential equivalents. (See page 642.)

- An amplitude spectrum plots the amplitudes of a function's Fourier series representation versus discrete frequencies. A phase spectrum plots the phase angles of a function's Fourier series representation versus discrete frequencies. These plots help to visualize the transformation of a circuit's input signal to its output signal. (See page 645.)

Problems

Sections 16.1–16.2

16.1 For each of the periodic functions in Fig. P16.1, specify

  a) ω_o in radians per second;

  b) f_o in hertz;

  c) the value of a_v;

  d) the equations for a_k and b_k;

  e) $v(t)$ as a Fourier series.

Figure P16.1

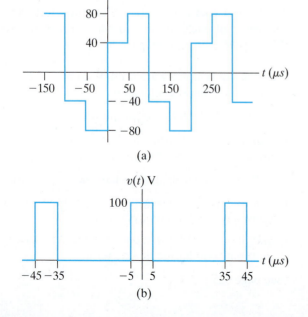

(a)

(b)

16.2 Derive the Fourier series for the periodic voltage shown in Fig. P16.2, given that

$$v(t) = 100 \sin \frac{2\pi}{T} t \text{ V}, \quad 0 \le t \le \frac{T}{2};$$

$$v(t) = 60 \sin \frac{2\pi}{T} (t - T/2) \text{ V}, \quad \frac{T}{2} \le t \le T.$$

Figure P16.2

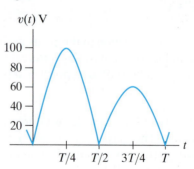

16.3 Find the Fourier series expressions for the periodic voltage functions shown in Fig. P16.3. Note that Fig. P16.3(a) illustrates the square wave; Fig. P16.3(b) illustrates the full-wave rectified sine wave, where $v(t) = V_m \sin(\pi/T)t, 0 \le t \le T$; and Fig. P16.3(c) illustrates the half-wave rectified sine wave, where $v(t) = V_m \sin(2\pi/T)t, 0 \le t \le T/2$.

Figure P16.3

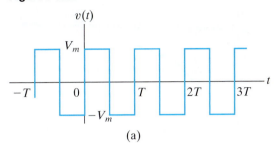

(a)

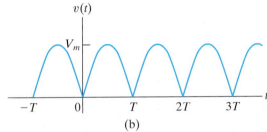

(b)

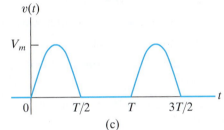

(c)

16.4 Derive the expressions for a_v, a_k, and b_k for the periodic voltage shown in Fig. P16.4 if $V_m = 100\pi$ V.

Figure P16.4

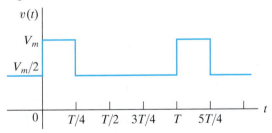

16.5 a) Verify the following equations:

$$\int_{t_0}^{t_0+T} \sin m\omega_0 t \, dt = 0, \quad \text{for all } m,$$

$$\int_{t_0}^{t_0+T} \cos m\omega_0 t \, dt = 0, \quad \text{for all } m.$$

b) Verify the following equation:

$$\int_{t_0}^{t_0+T} \cos m\omega_0 t \sin n\omega_0 t \, dt = 0, \quad \text{for all } m \text{ and } n.$$

Hint: Use the trigonometric identity

$$\cos \alpha \sin \beta = \frac{1}{2} \sin(\alpha + \beta) - \frac{1}{2} \sin(\alpha - \beta).$$

c) Verify the following equation:

$$\int_{t_0}^{t_0+T} \sin m\omega_0 t \sin n\omega_0 t \, dt = 0, \quad \text{for all } m \neq n,$$

$$= \frac{T}{2}, \quad \text{for } m = n.$$

Hint: Use the trigonometric identity

$$\sin \alpha \sin \beta = \frac{1}{2} \cos(\alpha - \beta) - \frac{1}{2} \cos(\alpha + \beta).$$

d) Verify the following equation:

$$\int_{t_0}^{t_0+T} \cos m\omega_0 t \cos n\omega_0 t \, dt = 0, \quad \text{for all } m \neq n,$$

$$= \frac{T}{2}, \quad \text{for } m = n.$$

Hint: Use the trigonometric identity

$$\cos \alpha \cos \beta = \frac{1}{2} \cos(\alpha - \beta) + \frac{1}{2} \cos(\alpha + \beta).$$

16.6 Derive Eq. 16.5.

Section 16.3

16.7 Derive the expressions for the Fourier coefficients of an odd periodic function. *Hint:* Use the same technique as used in the text in deriving Eqs. 16.7–16.9.

16.8 Show that if $f(t) = -f(t - T/2)$, the Fourier coefficients b_k are given by the expressions

$$b_k = 0 \qquad \text{for } k \text{ even;}$$

$$b_k = \frac{4}{T} \int_0^{T/2} f(t) \sin k\omega_o t \, dt, \quad \text{for } k \text{ odd.}$$

Hint: Use the same technique as used in the text to derive Eqs, 16.16.

16.9 Derive Eqs. 16.18. *Hint:* Start with Eq. 16.16 and divide the interval of integration into 0 to $T/4$ and $T/4$ to $T/2$. Note that because of evenness and quarter-wave symmetry, $f(t) = -f(T/2 - t)$ in the interval $T/4 \leq t \leq T/2$. Let $x = T/2 - t$ in the second interval and integrate between 0 and $T/4$.

16.10 Derive Eqs. 16.19. Follow the hint given in Problem 16.9 except that because of oddness and quarter-wave symmetry, $f(t) = f(T/2 - t)$ in the interval $T/4 \leq t \leq T/2$.

16.11 It is given that $v(t) = 20 \sin \pi|t|$ V over the interval $-1 \leq t \leq 1$ s. The function then repeats itself.

a) What is the fundamental frequency in radians per second?

b) Is the function even?

c) Is the function odd?

d) Does the function have half-wave symmetry?

16.12 One period of a periodic function is described by the following equations:

$i(t) = 5t$ A, -2 ms $\leq t \leq 2$ ms;

$i(t) = 10$ mA, 2 ms $\leq t \leq 6$ ms;

$i(t) = 0.04 - 5t$ A, 6 ms $\leq t \leq 10$ ms;

$i(t) = -10$ mA, 10 ms $\leq t \leq 14$ ms.

a) What is the fundamental frequency in hertz?

b) Is the function even?

c) Is the function odd?

d) Does the function have half-wave symmetry?

e) Does the function have quarter-wave symmetry?

f) Give the numerical expressions for a_v, a_k, and b_k.

16.13 Find the Fourier series of each periodic function shown in Fig. P16.13.

Figure P16.13

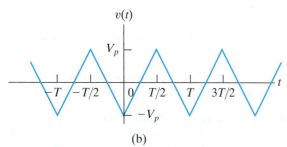

(a)

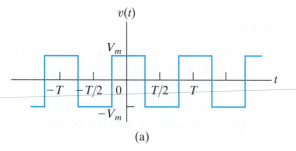

(b)

16.14 The periodic function shown in Fig. P16.14 is even and has both half-wave and quarter-wave symmetry.

a) Sketch one full cycle of the function over the interval $-T/4 \leq t \leq 3T/4$.

b) Derive the expression for the Fourier coefficients a_v, a_k, and b_k.

c) Write the first three nonzero terms in the Fourier expansion of $f(t)$.

d) Use the first three nonzero terms to estimate $f(T/4)$.

Figure P16.14

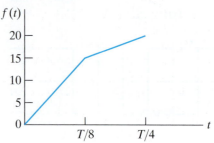

16.15 It is given that $f(t) = 4t^2$ over the interval -2 s $< t < 2$ s.

a) Construct a periodic function that satisfies this $f(t)$ between -2 s and $+2$ s, has a period of 8 s, and has half-wave symmetry.

b) Is the function even or odd?

c) Does the function have quarter-wave symmetry?

d) Derive the Fourier series for $f(t)$.

e) Write the Fourier series for $f(t)$ if $f(t)$ is shifted 2 s to the right.

16.16 Repeat Problem 16.15 given that $f(t) = t^3$ over the interval -2 s $< t < 2$ s.

16.17 a) Derive the Fourier series for the periodic current shown in Fig. P16.17.

b) Repeat (a) if the vertical reference axis is shifted $T/2$ units to the left.

Figure P16.17

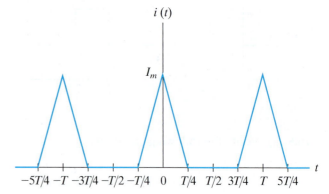

16.18 It is sometimes possible to use symmetry to find the Fourier coefficients, even though the original function is not symmetrical! With this thought in mind, consider the function in Fig P16.4. Observe that $v(t)$ can be divided into the two functions illustrated in Fig. P16.18(a) and (b). Furthermore, we can make

$v_2(t)$ an even function by shifting it $T/8$ units to the left. This is illustrated in Fig. P16.18(c). At this point we note that $v(t) = v_1(t) + v_2(t)$ and that the Fourier series of $v_1(t)$ is a single-term series consisting of $V_m/2$. To find the Fourier series of $v_2(t)$, we first find the Fourier series of $v_2(t + T/8)$ and then shift this series $T/8$ units to the right. Use the technique just outlined to verify the Fourier coefficients found in Problem 16.4.

Figure P16.18

$v_1(t)$

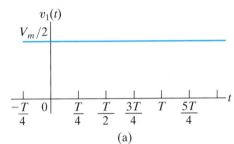

(a)

$v_2(t)$

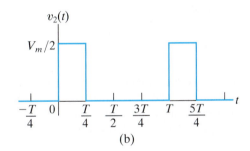

(b)

$v_2(t + T/8)$

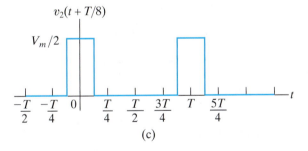

(c)

Section 16.4

16.19 Derive the Fourier series for the periodic function shown in Fig. P16.2, using the form of Eq. 16.20.

16.20 Derive the Fourier series for the periodic function described in Problem 16.12, using the form of Eq. 16.20.

16.21 Derive the Fourier series for the periodic function constructed in Problem 16.15, using the form of Eq. 16.20.

16.22 a) Derive the Fourier series for the periodic function shown in Fig. P16.22 when $I_m = 5\pi^2$ A. Write the series in the form of Eq. 16.20.

b) Use the first five nonzero terms to estimate $i(T/4)$.

Figure P16.22

$i(t)$

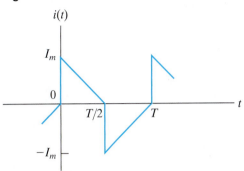

Section 16.5

16.23 Derive Eqs. 16.26 and 16.27.

16.24 a) Derive Eq. 16.28.
   Hint: Note that $b_k = 4V_m/\pi k + k\omega_o RCa_k$. Use this expression for b_k to find $a_k^2 + b_k^2$ in terms of a_k. Then use the expression for a_k to derive Eq. 16.28.

b) Derive Eq. 16.29.

16.25 Show that when we combine Eqs. 16.28 and 16.29 with Eqs. 16.20 and 16.21, the result is Eq. 16.22. *Hint:* Note from the definition of β_k that

$$\frac{a_k}{b_k} = -\tan \beta_k,$$

and from the definition of θ_k that

$$\tan \theta_k = -\cot \beta_k.$$

Now use the trigonometric identity

$$\tan x = \cot (90 - x)$$

to show that $\theta_k = 90 + \beta_k$.

16.26 a) Show that for large values of C, Eq. 16.24 can be approximated by the expression

$$v_o(t) \approx \frac{-V_m T}{4RC} + \frac{V_m}{RC} t.$$

Note that this expression is the equation of the triangular wave for $0 \le t \le T/2$. *Hints:* (1) Let $e^{-t/RC} \approx 1 - (t/RC)$ and $e^{-T/2RC} \approx 1 - (T/2RC)$; (2) put the resulting expression over the common denominator $2 - (T/2RC)$; (3) simplify the numerator; and (4) for large C, assume that $T/2RC$ is much less than 2.

b) Substitute the peak value of the triangular wave into the solution for Problem 16.13 (see Fig. P16.13(b)) and show that the result is Eq. 16.23.

16.27 The square-wave voltage shown in Fig. P16.27(a) is applied to the circuit shown in Fig. P16.27(b).

a) Find the Fourier series representation of the steady-state current i.

b) Find the steady-state expression for i by straightforward circuit analysis.

Figure P16.27

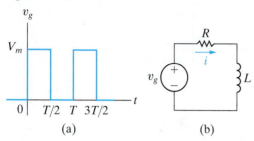

(a)　　　　　(b)

16.28 The periodic square-wave voltage seen in Fig. P16.28(a) is applied to the circuit shown in Fig. P16.28(b). Derive the first three nonzero terms in the Fourier series that represents the steady-state voltage v_o if $V_m = 15\pi$ V and the period of the input voltage is 4π ms.

Figure P16.28

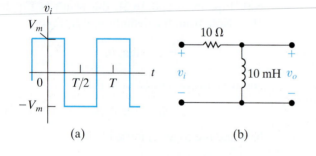

(a)　　　　　(b)

16.29 The periodic square-wave voltage shown in Fig P16.13(a) with $V_m = 10.5\pi$ V and $T = \pi$ ms is applied to the circuit shown in Fig. P16.29.

a) Derive the first four nonzero terms in the Fourier series that represents the steady-state voltage v_o.

b) Which frequency component in the input voltage is eliminated from the output voltage? Explain why.

Figure P16.29

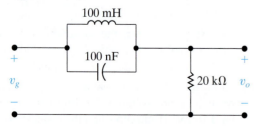

16.30 The full-wave rectified sine-wave voltage shown in Fig. P16.30(a) is applied to the circuit shown in Fig. P16.30(b).

a) Find the first five nonzero terms in the Fourier series representation of i_o.

b) Does your solution for i_o make sense? Explain.

Figure P16.30

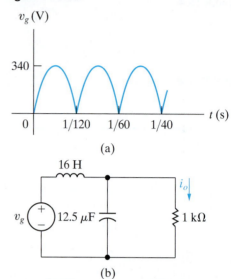

(a)

(b)

16.31 The periodic current described below is used to energize the circuit shown in Fig. P16.31. Write the time-domain expression for the third-harmonic component in the expression for v_o.

$$i_g = 500t \text{ A}, \qquad -2 \text{ ms} \le t \le 2 \text{ ms};$$
$$= 1 \text{ A}, \qquad 2 \text{ ms} \le t \le 8 \text{ ms};$$
$$= 5 - 500t \text{ A}, \quad 8 \text{ ms} \le t \le 12 \text{ ms};$$
$$= -1 \text{ A}, \qquad 12 \text{ ms} \le t \le 18 \text{ ms}.$$

Figure P16.31

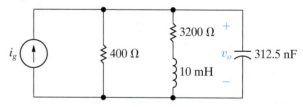

16.32 A periodic voltage having a period of 10π μs is given by the following Fourier series:

$$v_g = 150 \sum_{n=1,3,5,\dots}^{\infty} \frac{1}{n} \sin \frac{n\pi}{2} \cos n\omega_o t \text{ V}.$$

This periodic voltage is applied to the circuit shown in Fig. P16.32. Find the amplitude and phase angle of the components of v_o that have frequencies of 3 and 5 Mrad/s.

Figure P16.32

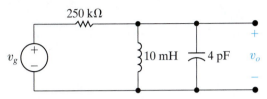

Section 16.6

16.33 The periodic current shown in Fig. P16.33 is applied to a 1 kΩ resistor.

a) Use the first three nonzero terms in the Fourier series representation of $i(t)$ to estimate the average power dissipated in the 1 kΩ resistor.

b) Calculate the exact value of the average power dissipated in the 1 kΩ resistor.

c) What is the percentage of error in the estimated value of the average power?

Figure P16.33

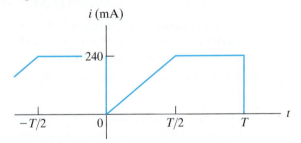

16.34 The periodic voltage across a 10 Ω resistor is shown in Fig. P16.34.

a) Use the first three nonzero terms in the Fourier series representation of $v(t)$ to estimate the average power dissipated in the 10 Ω resistor.

b) Calculate the exact value of the average power dissipated in the 10 Ω resistor.

c) What is the percentage error in the estimated value of the average power dissipated?

Figure P16.34

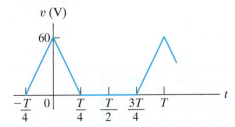

16.35 The triangular-wave voltage source, shown in Fig. P16.35(b), is applied to the circuit in Fig. P16.35(a). Estimate the average power delivered to the $50\sqrt{2}$ Ω resistor when the circuit is in steady-state operation.

Figure P16.35

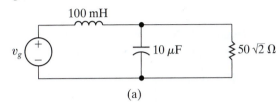

(a)

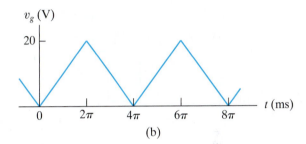

(b)

Section 16.7

16.36 The voltage and current at the terminals of a network are

$$v = 15 + 400 \cos 500t + 100 \sin 1500t \text{ V},$$

$$i = 2 + 5 \sin (500t + 60°) + 3 \cos(1500t - 15°) \text{ A}.$$

The current is in the direction of the voltage drop across the terminals.

a) What is the average power at the terminals?

b) What is the rms value of the voltage?

c) What is the rms value of the current?

16.37 a) Find the rms value of the voltage shown in Fig. P16.37 for $V_m = 100$ V. Note that the Fourier series for this periodic voltage was found in Assessment Problem 16.3.

b) Estimate the rms value of the voltage, using the first three nonzero terms in the Fourier series representation of $v_g(t)$.

Figure P16.37

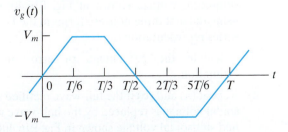

16.38 a) Use the first four nonzero terms in the Fourier series approximation of the periodic voltage shown in Fig. P16.38 to estimate its rms value.

b) Calculate the true rms value of the voltage.

c) Calculate the percentage of error in the estimated value.

Figure P16.38

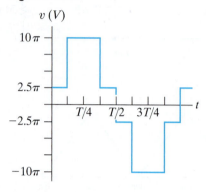

16.39 a) Estimate the rms value of the periodic square-wave voltage shown in Fig. P16.39(a) by using the first five nonzero terms in the Fourier series representation of $v(t)$.

b) Calculate the percentage of error in the estimation.

c) Repeat parts (a) and (b) if the periodic square-wave voltage is replaced by the periodic triangular voltage shown in Fig. P16.39(b).

Figure P16.39

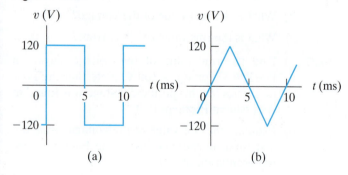

(a)   (b)

16.40 a) Estimate the rms value of the full-wave rectified sinusoidal voltage shown in Fig. P16.40(a) by using the first three nonzero terms in the Fourier series representation of $v(t)$.

b) Calculate the percentage of error in the estimation.

c) Repeat (a) and (b) if the full-wave rectified sinusoidal voltage is replaced by the half-wave rectified sinusoidal voltage shown in Fig. P16.40(b).

Figure P16.40

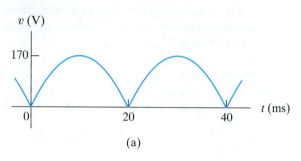

(a)

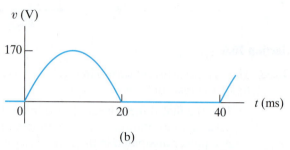

(b)

16.41 Assume the periodic function described in Problem 16.14 is a voltage v with a peak amplitude of 20 V.

a) Find the rms value of the voltage.

b) If this voltage is applied to a 15 Ω resistor, what is the average power dissipated in the resistor?

c) If v is approximated by using just the fundamental frequency term of its Fourier series, what is the average power delivered to the 15 Ω resistor?

d) What is the percentage of error in the estimation of the power dissipated?

16.42 a) Derive the expressions for the Fourier coefficients for the periodic current shown in Fig. P16.42.

b) Write the first four nonzero terms of the series using the alternative trigonometric form given by Eq. 16.20.

c) Use the first four nonzero terms of the expression derived in (b) to estimate the rms value of i_g.

d) Find the exact rms value of i_g.

e) Calculate the percentage of error in the estimated rms value.

Figure P16.42

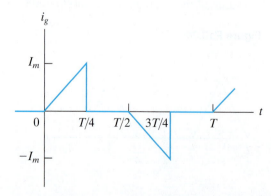

16.43 The rms value of any periodic triangular wave having the form depicted in Fig. P16.43(a) is independent of t_a and t_b. Note that for the function to be single valued, $t_a \le t_b$. The rms value is equal to $V_p/\sqrt{3}$. Verify this observation by finding the rms value of the three waveforms depicted in Fig. P16.43(b)–(d).

Figure P16.43

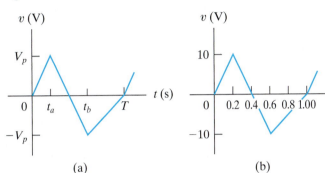

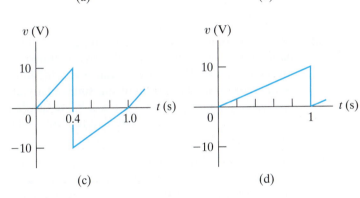

(a)　(b)

(c)　(d)

Section 16.8

16.44 Derive the expression for the complex Fourier coefficients for the periodic voltage shown in Fig. P16.44.

Figure P16.44

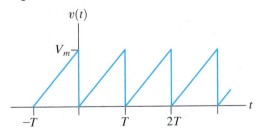

16.45 a) The periodic voltage in Problem 16.44 is applied to a 10 Ω resistor. If $V_m = 120$ V what is the average power delivered to the resistor?

b) Assume $v(t)$ is approximated by a truncated exponential form of the Fourier series consisting of the first seven nonzero terms, that is, $n = 0, 1, 2, 3, 4, 5, 6$, and 7. What is the rms value of the voltage, using this approximation?

c) If the approximation in part (b) is used to represent v, what is the percentage of error in the calculated power?

16.46 Use the exponential form of the Fourier series to write an expression for the voltage shown in Fig. P16.46.

Figure P16.46

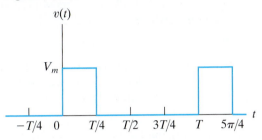

16.47 The periodic voltage source in the circuit shown in Fig. P16.47(a) has the waveform shown in Fig. P16.47(b).

a) Derive the expression for C_n.

b) Find the values of the complex coefficients $C_o, C_{-1}, C_1, C_{-2}, C_2, C_{-3}, C_3, C_{-4}$, and C_4 for the input voltage v_g if $V_m = 54$ V and $T = 10\pi \, \mu s$.

c) Repeat (b) for v_o.

d) Use the complex coefficients found in (c) to estimate the average power delivered to the 250 kΩ resistor.

Figure P16.47

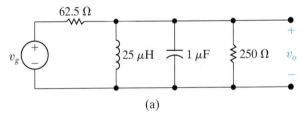

(a)

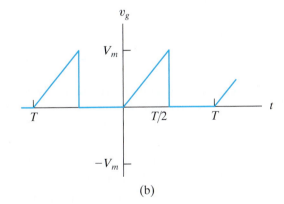

(b)

16.48 a) Find the rms value of the periodic voltage in Fig. P16.47(b).

b) Estimate the rms value of v_g using the complex coefficients derived in Problem 16.47(b).

c) What is the percentage of error in the estimated rms value of v_g?

Section 16.9

16.49 a) Make an amplitude and phase plot, based on Eq. 16.20, for the periodic voltage in Example 16.3. Assume V_m is 40 V. Plot both amplitude and phase versus $n\omega_o$, where $n = 0, 1, 2, 3, \ldots$

b) Repeat (a), but base the plots on Eq. 16.32.

16.50 a) Make amplitude and phase plots for the periodic voltage in Problem 16.33, based on Eq. 16.20. Plot both amplitude and phase versus $n\omega_o$ where $n = 0, 1, 2, \ldots$

b) Repeat (a), but base the plots on Eq. 16.32.

16.51 A periodic voltage is represented by a truncated Fourier series. The amplitude and phase spectra are shown in Fig. P16.51(a) and (b), respectively.

a) Write an expression for the periodic voltage using the form given by Eq. 16.20.

b) Is the voltage an even or odd function of t?

c) Does the voltage have half-wave symmetry?

d) Does the voltage have quarter-wave symmetry?

Figure P16.51

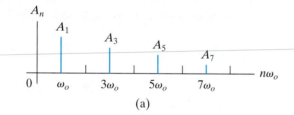

(a)

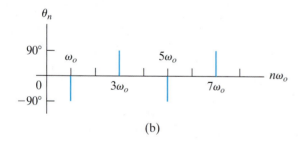

(b)

16.52 A periodic function is represented by a Fourier series that has a finite number of terms. The amplitude and phase spectra are shown in Fig. P16.52(a) and (b), respectively.

a) Write the expression for the periodic current using the form given by Eq. 16.20.

b) Is the current an even or odd function of t?

c) Does the current have half-wave symmetry?

d) Calculate the rms value of the current in milliamperes.

e) Write the exponential form of the Fourier series.

f) Make the amplitude and phase spectra plots on the basis of the exponential series.

Figure P16.52

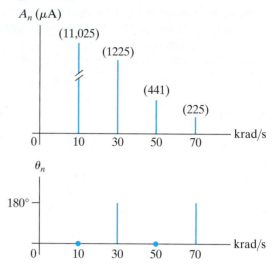

Sections 16.1–16.9

16.53 The input signal to a unity-gain third-order low-pass Butterworth filter is a half-wave rectified sinusoidal voltage. The corner frequency of the filter is 100 rad/s. The amplitude of the sinusoidal voltage is 54π V and its period is 5π ms. Write the first three terms of the Fourier series that represents the steady-state output voltage of the filter.

16.54 The input signal to a unity-gain second-order low-pass Butterworth filter is the periodic triangular-wave voltage shown in Fig P16.54. The corner frequency of the filter is 2 krad/s. Write the first three terms of the Fourier series that represents the steady-state output voltage of the filter.

Figure P16.54

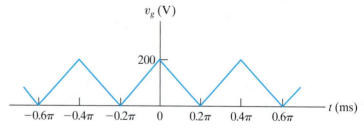

16.55 The input signal to a unity-gain second-order high-pass Butterworth filter is a full-wave rectified sine wave with an amplitude of 2.5π V and a fundamental frequency of 5000 rad/s. The corner frequency of the filter is 1 krad/s. Write the first two terms in the Fourier series that represents the steady-state output voltage of the filter.

16.56 The transfer function (V_o/V_g) for the narrowband bandpass filter circuit in Fig. P16.56(a) is

$$H(s) = \frac{-K_o \beta s}{s^2 + \beta s + \omega_o^2}.$$

a) Find K_o, β, and ω_o^2 as functions of the circuit parameters R_1, R_2, R_3, C_1, and C_2.

b) Write the first three terms in the Fourier series that represents v_o if v_g is the periodic voltage in Fig. P16.56(b).

c) Predict the value of the quality factor for the filter by examining the result in part (b).

d) Calculate the quality factor for the filter using β and ω_o and compare the value to your prediction in part (c).

16.57 a) Find the values for K, β, and ω_o^2 for the bandpass filter shown in Fig. P16.57(a).

b) Find the first three terms in the Fourier series for v_o in Fig. P16.57(a) if the input to the filter is the waveform shown in Fig. P16.57(b).

Figure P16.56

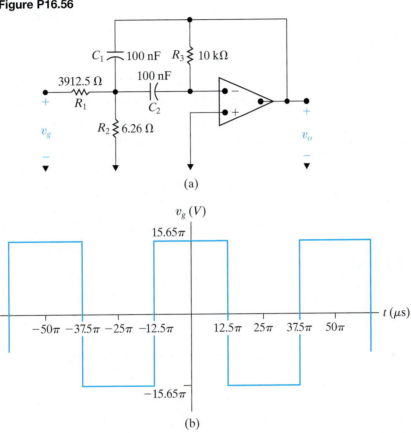

(a)

(b)

Figure P16.57

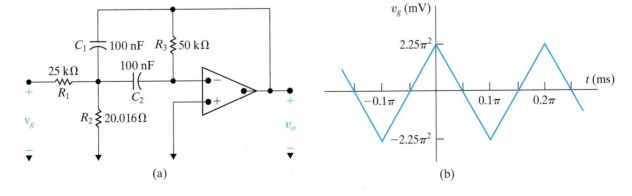

(a)

(b)

CHAPTER

17

The Fourier Transform

CHAPTER CONTENTS

17.1 **The Derivation of the Fourier Transform** *p. 662*

17.2 **The Convergence of the Fourier Integral** *p. 664*

17.3 **Using Laplace Transforms to Find Fourier Transforms** *p. 666*

17.4 **Fourier Transforms in the Limit** *p. 668*

17.5 **Some Mathematical Properties** *p. 671*

17.6 **Operational Transforms** *p. 672*

17.7 **Circuit Applications** *p. 677*

17.8 **Parseval's Theorem** *p. 679*

CHAPTER OBJECTIVES

1 Be able to calculate the Fourier transform of a function using any or all of the following:

- The definition of the Fourier transform;
- Laplace transforms;
- Mathematical properties of the Fourier transform;
- Operational transforms.

2 Know how to use the Fourier transform to find the response of a circuit.

3 Understand Parseval's theorem and be able to use it to answer questions about the energy contained within specific frequency bands.

In Chapter 16, we represented periodic functions with a Fourier series. This series describes the periodic function using the frequency-domain attributes of amplitude and phase angle. The **Fourier transform** extends this frequency-domain representation to functions that are not periodic. But we have already transformed aperiodic functions from the time domain to the frequency domain, using the Laplace transform. You may wonder, then, why yet another type of transformation is necessary.

Strictly speaking, the Fourier transform is not a new transform. It is a special case of the bilateral Laplace transform, with the real part of the complex frequency set to zero. However, in terms of a physical interpretation, the Fourier transform is better viewed as a limiting case of a Fourier series. We present this point of view in Section 17.1, where we derive the Fourier transform equations.

The Fourier transform is more useful than the Laplace transform in certain communications theory and signal-processing applications. Although we cannot pursue the Fourier transform in depth, its introduction here seems appropriate while the ideas underlying the Laplace transform and the Fourier series are still fresh in your mind.

■ Practical Perspective

Filtering Digital Signals

It is common to use telephone lines to communicate information from one computer to another. As you may know, computers represent all information as collections of 1's and 0's. The value 1 is represented as a voltage, usually 5 V, and 0 is represented as 0 V, as shown below.

The telephone line has a frequency response characteristic that is similar to a low-pass filter. We can use Fourier transforms to understand the effect of transmitting a digital value using a telephone line that behaves like a filter.

0111010010

Squareplum/Shutterstock

0111010010

17.1 The Derivation of the Fourier Transform

We derive the Fourier transform, viewed as a limiting case of a Fourier series and using the exponential form of the series:

$$f(t) = \sum_{n=-\infty}^{\infty} C_n e^{jn\omega_0 t}, \tag{17.1}$$

where

$$C_n = \frac{1}{T} \int_{-T/2}^{T/2} f(t) e^{-jn\omega_0 t} \, dt. \tag{17.2}$$

In Eq. 17.2, we elected to start the integration at $t_0 = -T/2$.

If the period T increases without limit, $f(t)$ transitions from a periodic to an aperiodic function. In other words, if T is infinite, $f(t)$ never repeats itself and hence is aperiodic. As T increases, the separation between adjacent harmonic frequencies becomes smaller and smaller. In particular,

$$\Delta\omega = (n+1)\omega_0 - n\omega_0 = \omega_0 = \frac{2\pi}{T},$$

and as T gets larger and larger, the incremental separation $\Delta\omega$ approaches a differential separation $d\omega$. Therefore,

$$\frac{1}{T} \rightarrow \frac{d\omega}{2\pi} \quad \text{as} \quad T \rightarrow \infty.$$

As the period increases, the frequency is no longer a discrete variable but is instead a continuous variable, or

$$n\omega_0 \rightarrow \omega \quad \text{as} \quad T \rightarrow \infty.$$

We can also see from Eq. 17.2 that, as the period increases, the Fourier coefficients C_n get smaller. In the limit, $C_n \rightarrow 0$ as $T \rightarrow \infty$. This result makes sense because we expect the Fourier coefficients to vanish when the function is no longer periodic. Note, however, the limiting value of the product $C_n T$; that is,

$$C_n T \rightarrow \int_{-\infty}^{\infty} f(t) e^{-j\omega t} \, dt \quad \text{as} \quad T \rightarrow \infty.$$

In writing the limiting value of $C_n T$, we replaced $n\omega_0$ with ω. The integral represents the **Fourier transform** of $f(t)$ and is denoted

FOURIER TRANSFORM

$$F(\omega) = \mathscr{F}\{f(t)\} = \int_{-\infty}^{\infty} f(t) e^{-j\omega t} \, dt. \tag{17.3}$$

We obtain an explicit expression for the inverse Fourier transform by investigating the limiting form of Eq. 17.1 as $T \to \infty$. We begin by multiplying and dividing Eq. 17.1 by T:

$$f(t) = \sum_{n=-\infty}^{\infty} (C_n T) e^{jn\omega_0 t} \left(\frac{1}{T}\right).$$

As $T \to \infty$, the summation approaches integration, $C_n T \to F(\omega)$, $n\omega_0 \to \omega$, and $1/T \to d\omega/2\pi$. Thus, in the limit, $f(t)$ becomes

INVERSE FOURIER TRANSFORM

$$f(t) = \frac{1}{2\pi} \int_{-\infty}^{\infty} F(\omega) e^{j\omega t} \, d\omega. \qquad (17.4)$$

Equations 17.3 and 17.4 define the Fourier transform. Equation 17.3 transforms the time-domain expression $f(t)$ into its corresponding frequency-domain expression $F(\omega)$. Equation 17.4 defines the inverse operation, transforming $F(\omega)$ into $f(t)$.

Let's now derive the Fourier transform of the pulse shown in Fig. 17.1. Note that this pulse corresponds to the periodic voltage in Example 16.7 if we let $T \to \infty$. The Fourier transform of $v(t)$ comes directly from Eq. 17.3:

$$V(\omega) = \int_{-\tau/2}^{\tau/2} V_m e^{-j\omega t} \, dt$$

$$= V_m \frac{e^{-j\omega t}}{(-j\omega)} \Bigg|_{-\tau/2}^{\tau/2}$$

$$= \frac{V_m}{-j\omega} \left(-2j \sin \frac{\omega \tau}{2} \right),$$

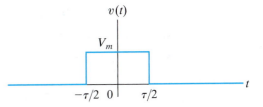

Figure 17.1 ▲ A voltage pulse.

which can be written in the form $(\sin x)/x$ by multiplying the numerator and denominator by $\tau/2$. Then,

$$V(\omega) = V_m \tau \frac{\sin \omega \tau/2}{\omega \tau/2}. \qquad (17.5)$$

For the periodic voltage pulses in Example 16.7, the expression for the Fourier coefficients is

$$C_n = \frac{V_m \tau}{T} \frac{\sin n\omega_0 \tau/2}{n\omega_0 \tau/2}. \qquad (17.6)$$

Compare Eqs. 17.5 and 17.6. Note that as the time-domain function goes from periodic to aperiodic, the amplitude spectrum goes from a discrete line spectrum to a continuous spectrum. Furthermore, the envelope of the line spectrum has the same shape as the continuous spectrum. Thus, as T increases, the lines in the spectrum get closer together and their amplitudes get smaller, but their envelope doesn't change shape. The physical interpretation

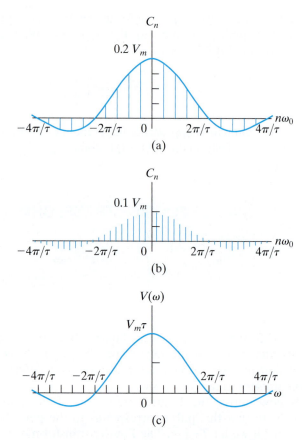

Figure 17.2 ▲ Transition of the amplitude spectrum as $f(t)$ goes from periodic to aperiodic. (a) C_n versus $n\omega_0$, $T/\tau = 5$; (b) C_n versus $n\omega_0$, $T/\tau = 10$; (c) $V(\omega)$ versus ω.

of the Fourier transform $V(\omega)$ is therefore a measure of the frequency content of $v(t)$. Figure 17.2 illustrates these observations. The amplitude spectrum plot is based on the assumption that τ is constant and T is increasing.

17.2 The Convergence of the Fourier Integral

A function of time $f(t)$ has a Fourier transform if the integral in Eq. 17.3 converges. If $f(t)$ is a well-behaved function that differs from zero over a finite interval of time, convergence is no problem. *Well-behaved* implies that $f(t)$ is single valued and encloses a finite area over the range of integration. In practical terms, all pulses of finite duration that interest us are well-behaved functions. The evaluation of the Fourier transform of the rectangular pulse discussed in Section 17.1 illustrates this point.

If $f(t)$ is different from zero over an infinite interval, the convergence of the Fourier integral depends on the behavior of $f(t)$ as $t \rightarrow \infty$. A single-valued function that is nonzero over an infinite interval has a Fourier transform if the integral

$$\int_{-\infty}^{\infty} |f(t)|\, dt$$

exists and if any discontinuities in $f(t)$ are finite. An example is the decaying exponential function illustrated in Fig. 17.3. The Fourier transform of $f(t)$ is

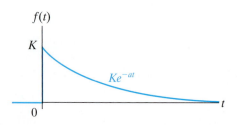

Figure 17.3 ▲ The decaying exponential function $Ke^{-at}u(t)$.

$$F(\omega) = \int_{-\infty}^{\infty} f(t)e^{-j\omega t}\,dt = \int_{0}^{\infty} Ke^{-at}e^{-j\omega t}\,dt$$

$$= \frac{Ke^{-(a+j\omega)t}}{-(a+j\omega)}\bigg|_{0}^{\infty} = \frac{K}{-(a+j\omega)}\,(0-1)$$

$$= \frac{K}{a+j\omega},\quad a>0.$$

A third important group of functions has great practical interest, but these functions do not, in a strict sense, have Fourier transforms. For example, the integral in Eq. 17.3 doesn't converge if $f(t)$ is a constant. The same can be said if $f(t)$ is a sinusoidal function, $\cos \omega_0 t$, or a step function, $Ku(t)$. These functions are very important in circuit analysis, but, to include them in Fourier analysis, we must resort to the following mathematical manipulations.

- Create a function in the time domain that has a Fourier transform and at the same time can be made arbitrarily close to the function of interest. Call the approximating function $f'(t)$.
- Find the Fourier transform of the approximating function, $F'(\omega)$, and then evaluate the limiting value of $F'(\omega)$ as $f'(t)$ approaches $f(t)$.
- Define the limiting value of $F'(\omega)$ as the Fourier transform of $f(t)$.

Example 17.1 illustrates this three-step process.

EXAMPLE 17.1 | **Finding the Fourier Transform of a Constant**

Find the Fourier transform of a constant function, $f(t) = A$.

Solution

Approximate a constant with the exponential function

$$f'(t) = Ae^{-\epsilon|t|},\quad \epsilon > 0.$$

As $\epsilon \to 0$, $f'(t) \to A$. Figure 17.4 shows the approximation graphically.

The Fourier transform of $f'(t)$ is

$$F'(\omega) = \int_{-\infty}^{0} Ae^{\epsilon t}e^{-j\omega t}\,dt + \int_{0}^{\infty} Ae^{-\epsilon t}e^{-j\omega t}\,dt.$$

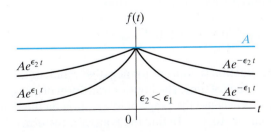

Figure 17.4 ▲ The approximation of a constant with an exponential function.

Evaluating the integrals gives

$$F'(\omega) = \frac{A}{\epsilon - j\omega} + \frac{A}{\epsilon + j\omega} = \frac{2\epsilon A}{\epsilon^2 + \omega^2}.$$

This function generates an impulse function at $\omega = 0$ as $\epsilon \to 0$. You can verify this result by showing that

- $F'(\omega)$ approaches infinity at $\omega = 0$ as $\epsilon \to 0$.
- The width of $F'(\omega)$ approaches zero as $\epsilon \to 0$.
- The area under $F'(\omega)$ is independent of ϵ.

The area under $F'(\omega)$ is the strength of the impulse and is

$$\int_{-\infty}^{\infty} \frac{2\epsilon A}{\epsilon^2 + \omega^2}\,d\omega = 4\epsilon A \int_{0}^{\infty} \frac{d\omega}{\epsilon^2 + \omega^2} = 2\pi A.$$

In the limit, $f'(t)$ approaches a constant A, and $F'(\omega)$ approaches an impulse function $2\pi A\delta(\omega)$. Therefore, the Fourier transform of a constant A is defined as $2\pi A\delta(\omega)$, or

$$\mathscr{F}\{A\} = 2\pi A\delta(\omega).$$

ASSESSMENT PROBLEMS

Objective 1—Be able to calculate the Fourier transform of a function

17.1 Use the defining integral to find the Fourier transform of the following functions:

a) $f(t) = -A, \quad -\tau/2 \le t < 0;$

$f(t) = A, \quad 0 < t \le \tau/2;$

$f(t) = 0 \quad$ elsewhere.

b) $f(t) = 0, \quad t < 0;$

$f(t) = te^{-at}, \quad t \ge 0, a > 0.$

Answer: (a) $-j\left(\dfrac{2A}{\omega}\right)\left(1 - \cos\dfrac{\omega\tau}{2}\right);$

(b) $\dfrac{1}{(a + j\omega)^2}.$

17.2 The Fourier transform of $f(t)$ is given by

$F(\omega) = 0, \quad -\infty \le \omega < -3 \text{ s};$

$F(\omega) = 4, \quad -3 \text{ s} < \omega < -2 \text{ s};$

$F(\omega) = 1, \quad -2 \text{ s} < \omega < 2 \text{ s};$

$F(\omega) = 4, \quad 2 \text{ s} < \omega < 3 \text{ s};$

$F(\omega) = 0, \quad 3 \text{ s} < \omega \le \infty.$

Find $f(t)$.

Answer: $f(t) = \dfrac{1}{\pi t}(4 \sin 3t - 3 \sin 2t).$

SELF-CHECK: Also try Chapter Problems 17.1 and 17.2.

In Section 17.4, we say more about Fourier transforms defined through a limit process. Before doing so, in Section 17.3, we use the Laplace transform to find the Fourier transform of functions for which the Fourier integral converges.

17.3 Using Laplace Transforms to Find Fourier Transforms

We can use a table of unilateral, or one-sided, Laplace transform pairs to find the Fourier transform for functions whose Fourier integral converges. The Fourier integral converges when the poles of $F(s)$ lie in the left half of the s plane. Note that if $F(s)$ has poles in the right half of the s plane or along the imaginary axis, $f(t)$ does not satisfy the constraint that $\int_{-\infty}^{\infty}|f(t)|\,dt$ exists.

The following rules apply when using Laplace transforms to find the Fourier transforms of such functions.

1. If $f(t)$ is zero for $t \le 0^-$, replace s by $j\omega$ in the Laplace transform of $f(t)$ to get the Fourier transform of $f(t)$. Thus

$$\mathscr{F}\{f(t)\} = \mathscr{L}\{f(t)\}_{s=j\omega}. \tag{17.7}$$

2. A negative-time function is nonzero for negative values of time and zero for positive values of time. The Fourier transform of a negative-time function exists because the range of integration on the Fourier integral goes from $-\infty$ to $+\infty$. To find the Fourier transform of such a negative-time function, reflect the function over to the positive time domain and find its one-sided Laplace transform. Replace s with $-j\omega$ in the Laplace transform to get the Fourier transform of the original time function. Therefore,

$$\mathcal{F}\{f(t)\} = \mathcal{L}\{f(-t)\}_{s=-j\omega}. \qquad (17.8)$$

3. Functions that are nonzero over all time can be resolved into positive- and negative-time functions. We use Eqs. 17.7 and 17.8 to find the Fourier transform of the positive- and negative-time functions, respectively. The Fourier transform of the original function is the sum of the two transforms. Thus, if we let

$$f^+(t) = f(t) \quad (\text{for } t > 0),$$
$$f^-(t) = f(t) \quad (\text{for } t < 0),$$

then

$$f(t) = f^+(t) + f^-(t)$$

and

$$\mathcal{F}\{f(t)\} = \mathcal{F}\{f^+(t)\} + \mathcal{F}\{f^-(t)\}$$
$$= \mathcal{L}\{f^+(t)\}_{s=j\omega} + \mathcal{L}\{f^-(-t)\}_{s=-j\omega}. \qquad (17.9)$$

If $f(t)$ is even, Eq. 17.9 reduces to

$$\mathcal{F}\{f(t)\} = \mathcal{L}\{f(t)\}_{s=j\omega} + \mathcal{L}\{f(t)\}_{s=-j\omega}.$$

If $f(t)$ is odd, then Eq. 17.9 becomes

$$\mathcal{F}\{f(t)\} = \mathcal{L}\{f(t)\}_{s=j\omega} - \mathcal{L}\{f(t)\}_{s=-j\omega}.$$

Example 17.2 uses the Laplace transform to find the Fourier transform.

EXAMPLE 17.2 | **Finding the Fourier Transform from the Laplace Transform**

Use the Laplace transform to find the Fourier transform for each of the following functions:

a) $f(t) = 0$, $t \le 0^-$;
   $f(t) = e^{-at}\cos\omega_0 t$, $t \ge 0^+$.

b) $f(t) = 0$, $t \ge 0^+$;
   $f(t) = e^{at}\cos\omega_0 t$, $t \le 0^-$.

c) $f(t) = e^{-a|t|}$.

Solution

a) Using Rule 1, find the Laplace transform of $f(t)$ and substitute $j\omega$ for s to get $F(\omega)$. Therefore,

$$\mathcal{F}\{f(t)\} = \left. \frac{s+a}{(s+a)^2 + \omega_0^2} \right|_{s=j\omega} = \frac{j\omega + a}{(j\omega + a)^2 + \omega_0^2}.$$

b) This is a negative-time function. Using Rule 2, reflect the function over to the positive-time domain to get

$$f(-t) = 0, \qquad (\text{for } t \leq 0^-);$$

$$f(-t) = e^{-at} \cos \omega_0 t, \quad (\text{for } t \geq 0^+).$$

Both $f(t)$ and its mirror image are plotted in Fig. 17.5. The Fourier transform of $f(t)$ is

$$\mathcal{F}\{f(t)\} = \mathcal{L}\{f(-t)\}_{s=-j\omega} = \left.\frac{s+a}{(s+a)^2 + \omega_0^2}\right|_{s=-j\omega}$$

$$= \frac{-j\omega + a}{(-j\omega + a)^2 + \omega_0^2}.$$

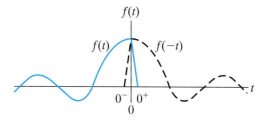

Figure 17.5 ▲ The reflection of a negative-time function over to the positive-time domain.

c) This function is defined for all positive and negative time. Using Rule 3, we find that the positive- and negative-time functions are

$$f^+(t) = e^{-at} \quad \text{and} \quad f^-(t) = e^{at}.$$

Find the Laplace transform of $f^+(t)$ and $f^-(-t)$:

$$\mathcal{L}\{f^+(t)\} = \frac{1}{s+a},$$

$$\mathcal{L}\{f^-(-t)\} = \frac{1}{s+a}.$$

Therefore, from Eq. 17.9,

$$\mathcal{F}\{e^{-a|t|}\} = \left.\frac{1}{s+a}\right|_{s=j\omega} + \left.\frac{1}{s+a}\right|_{s=-j\omega}$$

$$= \frac{1}{j\omega + a} + \frac{1}{-j\omega + a}$$

$$= \frac{2a}{\omega^2 + a^2}.$$

◧ ASSESSMENT PROBLEM

Objective 1 — Be able to calculate the Fourier transform of a function

17.3 Find the Fourier transform of each function. In each case, a is a positive real constant.

a) $f(t) = 0, \qquad t < 0;$
$f(t) = e^{-at} \sin \omega_0 t, \quad t \geq 0.$

b) $f(t) = 0, \qquad t > 0;$
$f(t) = -te^{at}, \qquad t \leq 0.$

c) $f(t) = te^{-at}, \qquad t \geq 0;$
$f(t) = te^{at}, \qquad t \leq 0.$

Answer: (a) $\dfrac{\omega_0}{(a+j\omega)^2 + \omega_0^2}$;

(b) $\dfrac{1}{(a-j\omega)^2}$;

(c) $\dfrac{-j4a\omega}{(a^2 + \omega^2)^2}$.

SELF-CHECK: Also try Chapter Problem 17.5.

17.4 Fourier Transforms in the Limit

As we pointed out in Section 17.2, the Fourier transforms of several practical functions must be defined by a limit process. We now return to these types of functions and develop their transforms.

The Fourier Transform of a Signum Function

We showed that the Fourier transform of a constant A is $2\pi A\delta(\omega)$ in Example 17.1. Another function of interest is the signum function, defined as $+1$ for $t > 0$ and -1 for $t < 0$. The signum function is denoted $\text{sgn}(t)$ and can be expressed in terms of unit step functions as

$$\text{sgn}(t) = u(t) - u(-t). \qquad (17.10)$$

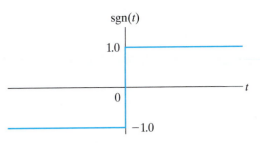

Figure 17.6 ▲ The signum function.

Figure 17.6 shows the function graphically.

To find the signum function's Fourier transform, create a function that approaches the signum function in the limit:

$$\text{sgn}(t) = \lim_{\epsilon \to 0}[e^{-\epsilon t}u(t) - e^{\epsilon t}u(-t)], \quad \epsilon > 0.$$

The function inside the brackets, plotted in Fig. 17.7, has a Fourier transform because the Fourier integral converges. Since $f(t)$ is an odd function, its Fourier transform is:

$$\mathcal{F}\{f(t)\} = \left.\frac{1}{s + \epsilon}\right|_{s=j\omega} - \left.\frac{1}{s + \epsilon}\right|_{s=-j\omega}$$

$$= \frac{1}{j\omega + \epsilon} - \frac{1}{-j\omega + \epsilon}$$

$$= \frac{-2j\omega}{\omega^2 + \epsilon^2}.$$

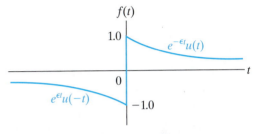

Figure 17.7 ▲ A function that approaches sgn (t) as ϵ approaches zero.

As $\epsilon \to 0$, $f(t) \to \text{sgn}(t)$, and $\mathcal{F}\{f(t)\} \to 2/j\omega$. Therefore,

$$\mathcal{F}\{\text{sgn}(t)\} = \frac{2}{j\omega}.$$

The Fourier Transform of a Unit Step Function

To find the Fourier transform of a unit step function, we use the Fourier transforms of a constant and the signum function. Note that the unit step function can be expressed as

$$u(t) = \frac{1}{2} + \frac{1}{2}\text{sgn}(t).$$

Thus,

$$\mathcal{F}\{u(t)\} = \mathcal{F}\left\{\frac{1}{2}\right\} + \mathcal{F}\left\{\frac{1}{2}\text{sgn}(t)\right\}$$

$$= \pi\delta(\omega) + \frac{1}{j\omega}.$$

The Fourier Transform of a Cosine Function

To find the Fourier transform of $\cos \omega_0 t$, we return to the inverse-transform integral of Eq. 17.4 and observe that if

$$F(\omega) = 2\pi\delta(\omega - \omega_0),$$

then

$$f(t) = \frac{1}{2\pi} \int_{-\infty}^{\infty} [2\pi\delta(\omega - \omega_0)]e^{j\omega t} \, d\omega.$$

Using the sifting property of the impulse function, we simplify $f(t)$ to

$$f(t) = e^{j\omega_0 t}.$$

Therefore,

$$\mathcal{F}\{e^{j\omega_0 t}\} = 2\pi\delta(\omega - \omega_0).$$

We now use the Fourier transform of $e^{j\omega_0 t}$ to find the Fourier transform of $\cos \omega_0 t$ because

$$\cos \omega_0 t = \frac{e^{j\omega_0 t} + e^{-j\omega_0 t}}{2}.$$

Thus,

$$\mathcal{F}\{\cos \omega_0 t\} = \frac{1}{2}\left(\mathcal{F}\{e^{j\omega_0 t}\} + \mathcal{F}\{e^{-j\omega_0 t}\}\right)$$

$$= \frac{1}{2}[2\pi\delta(\omega - \omega_0) + 2\pi\delta(\omega + \omega_0)]$$

$$= \pi\delta(\omega - \omega_0) + \pi\delta(\omega + \omega_0).$$

The Fourier transform of $\sin \omega_0 t$ involves similar manipulation, which we leave for Problem 17.4. Table 17.1 presents a summary of the transform pairs of the important elementary functions.

We now turn to the properties of the Fourier transform that help us describe aperiodic time-domain behavior in terms of frequency-domain behavior.

TABLE 17.1	Fourier Transforms of Elementary Functions	
Type	$f(t)$	$F(\omega)$
impulse	$\delta(t)$	1
constant	A	$2\pi A\delta(\omega)$
signum	$\text{sgn}(t)$	$2/j\omega$
step	$u(t)$	$\pi\delta(\omega) + 1/j\omega$
positive-time exponential	$e^{-at}u(t)$	$1/(a + j\omega), a > 0$
negative-time exponential	$e^{at}u(-t)$	$1/(a - j\omega), a > 0$
positive- and negative-time exponential	$e^{-a|t|}$	$2a/(a^2 + \omega^2), a > 0$
complex exponential	$e^{j\omega_0 t}$	$2\pi\delta(\omega - \omega_0)$
cosine	$\cos \omega_0 t$	$\pi[\delta(\omega + \omega_0) + \delta(\omega - \omega_0)]$
sine	$\sin \omega_0 t$	$j\pi[\delta(\omega + \omega_0) - \delta(\omega - \omega_0)]$

17.5 Some Mathematical Properties

We begin by noting that $F(\omega)$ is a complex quantity and can be expressed in either rectangular or polar form. Thus, from the defining integral,

$$F(\omega) = \int_{-\infty}^{\infty} f(t)e^{-j\omega t}dt$$

$$= \int_{-\infty}^{\infty} f(t)(\cos \omega t - j\sin \omega t)\, dt$$

$$= \int_{-\infty}^{\infty} f(t)\cos \omega t\, dt - j\int_{-\infty}^{\infty} f(t)\sin \omega t\, dt.$$

Now we let

$$A(\omega) = \int_{-\infty}^{\infty} f(t)\cos \omega t\, dt,$$

$$B(\omega) = -\int_{-\infty}^{\infty} f(t)\sin \omega t\, dt.$$

Thus, using the definitions for $A(\omega)$ and $B(\omega)$, we get

$$F(\omega) = A(\omega) + jB(\omega) = |F(\omega)|e^{j\theta(\omega)}.$$

We can make the following observations about $F(\omega)$:

- The real part of $F(\omega)$—that is, $A(\omega)$—is an even function of ω, so $A(\omega) = A(-\omega)$.
- The imaginary part of $F(\omega)$—that is, $B(\omega)$—is an odd function of ω, so $B(\omega) = -B(-\omega)$.
- The magnitude of $F(\omega)$—that is, $\sqrt{A^2(\omega) + B^2(\omega)}$—is an even function of ω.
- The phase angle of $F(\omega)$—that is, $\theta(\omega) = \tan^{-1}B(\omega)/A(\omega)$—is an odd function of ω.
- Replacing ω by $-\omega$ generates the conjugate of $F(\omega)$, so $F(-\omega) = F^*(\omega)$.

Hence, if $f(t)$ is an even function, $F(\omega)$ is real, and if $f(t)$ is an odd function, $F(\omega)$ is imaginary. If $f(t)$ is even,

$$A(\omega) = 2\int_0^{\infty} f(t)\cos \omega t\, dt \qquad (17.11)$$

and

$$B(\omega) = 0.$$

If $f(t)$ is an odd function,

$$A(\omega) = 0$$

and

$$B(\omega) = -2 \int_0^\infty f(t) \sin \omega t \, dt.$$

We leave the derivations for you as Problems 17.10 and 17.11.

If $f(t)$ is an even function, its Fourier transform is an even function, and if $f(t)$ is an odd function, its Fourier transform is an odd function. Moreover, if $f(t)$ is an even function, from the inverse Fourier integral,

$$f(t) = \frac{1}{2\pi} \int_{-\infty}^\infty F(\omega) e^{j\omega t} d\omega = \frac{1}{2\pi} \int_{-\infty}^\infty A(\omega) e^{j\omega t} d\omega$$

$$= \frac{1}{2\pi} \int_{-\infty}^\infty A(\omega)(\cos \omega t + j \sin \omega t) d\omega$$

$$= \frac{1}{2\pi} \int_{-\infty}^\infty A(\omega) \cos \omega t \, d\omega + 0$$

$$= \frac{2}{2\pi} \int_0^\infty A(\omega) \cos \omega t \, d\omega. \tag{17.12}$$

Now compare Eq. 17.12 with Eq. 17.11. Note that, except for a factor of $1/2\pi$, these two equations have the same form. Thus, the waveforms of $A(\omega)$ and $f(t)$ become interchangeable if $f(t)$ is an even function.

For example, we have already observed that a rectangular pulse in the time domain produces a frequency spectrum of the form $(\sin \omega)/\omega$. Specifically, Eq. 17.5 expresses the Fourier transform of the voltage pulse shown in Fig. 17.1. Hence, a rectangular pulse in the frequency domain must be generated by a time-domain function of the form $(\sin t)/t$. We can illustrate this transform by finding the time-domain function $f(t)$ corresponding to the frequency spectrum shown in Fig. 17.8. From Eq. 17.12,

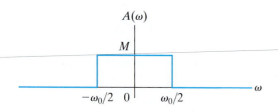

Figure 17.8 ▲ A rectangular frequency spectrum.

$$f(t) = \frac{2}{2\pi} \int_0^{\omega_0/2} M \cos \omega t \, d\omega = \frac{2M}{2\pi} \left(\frac{\sin \omega t}{t} \right) \Big|_0^{\omega_0/2}$$

$$= \frac{1}{2\pi} \left(M \frac{\sin \omega_0 t/2}{t/2} \right)$$

$$= \frac{1}{2\pi} \left(M\omega_0 \frac{\sin \omega_0 t/2}{\omega_0 t/2} \right).$$

We say more about the frequency spectrum of a rectangular pulse in the time domain versus the rectangular frequency spectrum of $(\sin t)/t$ after we introduce Parseval's theorem.

17.6 Operational Transforms

Fourier transforms, like Laplace transforms, can be classified as functional and operational. So far, we have concentrated on the functional transforms. We now discuss some of the important operational Fourier transforms, which are similar to the operational Laplace transforms in Chapter 12. Hence, we leave their proofs to you as Problems 17.12–17.19.

Multiplication by a Constant

From the defining integral, if

$$\mathscr{F}\{f(t)\} = F(\omega),$$

then

$$\mathscr{F}\{Kf(t)\} = KF(\omega).$$

Thus, multiplication of $f(t)$ by a constant corresponds to multiplying $F(\omega)$ by that same constant.

Addition (Subtraction)

Addition (subtraction) in the time domain translates into addition (subtraction) in the frequency domain. Thus, if

$$\mathscr{F}\{f_1(t)\} = F_1(\omega),$$
$$\mathscr{F}\{f_2(t)\} = F_2(\omega),$$
$$\mathscr{F}\{f_3(t)\} = F_3(\omega),$$

then

$$\mathscr{F}\{f_1(t) - f_2(t) + f_3(t)\} = F_1(\omega) - F_2(\omega) + F_3(\omega),$$

which is derived by substituting the algebraic sum of time-domain functions into the defining integral.

Differentiation

The Fourier transform of the first derivative of $f(t)$ is

$$\mathscr{F}\left\{\frac{df(t)}{dt}\right\} = j\omega F(\omega).$$

The nth derivative of $f(t)$ is

$$\mathscr{F}\left\{\frac{d^n f(t)}{dt^n}\right\} = (j\omega)^n F(\omega).$$

These equations are valid if $f(t)$ is zero at $\pm\infty$.

Integration

If

$$g(t) = \int_{-\infty}^{t} f(x)\, dx,$$

then

$$\mathscr{F}\{g(t)\} = \frac{F(\omega)}{j\omega}.$$

This equation is valid if

$$\int_{-\infty}^{\infty} f(x)\, dx = 0.$$

Scale Change

Dimensionally, time and frequency are reciprocals. Therefore, when time is expanded, frequency is compressed (and vice versa), as reflected in the functional transform

$$\mathcal{F}\{f(at)\} = \frac{1}{a}F\left(\frac{\omega}{a}\right), \quad a > 0.$$

Note that when $0 < a < 1.0$, time is expanded, whereas when $a > 1.0$, time is compressed.

Translation in the Time Domain

Translating a function in the time domain alters the phase spectrum and leaves the amplitude spectrum unchanged. Thus

$$\mathcal{F}\{f(t - a)\} = e^{-j\omega a}F(\omega).$$

If a is positive in this equation, the time function is delayed (shifted to the right on the time axis), and if a is negative, the time function is advanced (shifted to the left on the time axis).

Translation in the Frequency Domain

Translation in the frequency domain corresponds to multiplication by the complex exponential in the time domain:

$$\mathcal{F}\{e^{j\omega_0 t}f(t)\} = F(\omega - \omega_0).$$

Modulation

Amplitude modulation varies the amplitude of a sinusoidal carrier. If the carrier signal is $f(t)$, the modulated carrier is $f(t) \cos \omega_0 t$. The amplitude spectrum of the modulated carrier is one-half the amplitude spectrum of $f(t)$ and is centered at $\pm\omega_0$, so,

$$\mathcal{F}\{f(t) \cos \omega_0 t\} = \frac{1}{2}F(\omega - \omega_0) + \frac{1}{2}F(\omega + \omega_0).$$

Convolution in the Time Domain

Convolution in the time domain corresponds to multiplication in the frequency domain. In other words,

$$y(t) = \int_{-\infty}^{\infty} x(\lambda)h(t - \lambda)\, d\lambda$$

becomes

$$\mathcal{F}\{y(t)\} = Y(\omega) = X(\omega)H(\omega). \tag{17.13}$$

Equation 17.13 is important in Fourier transform applications because it states that the output's transform $Y(\omega)$ is the product of the input's transform $X(\omega)$ and the system function $H(\omega)$. We say more about this relationship in Section 17.7.

Convolution in the Frequency Domain

Convolution in the frequency domain corresponds to finding the Fourier transform of the product of two time functions. Thus, if

$$f(t) = f_1(t)f_2(t),$$

then

$$F(\omega) = \frac{1}{2\pi} \int_{-\infty}^{\infty} F_1(u)F_2(\omega - u)\, du.$$

Table 17.2 summarizes these 10 operational transforms and another operational transform that we introduce in Example 17.3.

TABLE 17.2 **Operational Transforms**

$f(t)$	$F(\omega)$
$Kf(t)$	$KF(\omega)$
$f_1(t) - f_2(t) + f_3(t)$	$F_1(\omega) - F_2(\omega) + F_3(\omega)$
$d^n f(t)/dt^n$	$(j\omega)^n F(\omega)$
$\displaystyle\int_{-\infty}^{t} f(x)\, dx$	$F(\omega)/j\omega$
$f(at)$	$\dfrac{1}{a} F\!\left(\dfrac{\omega}{a}\right), \quad a > 0$
$f(t - a)$	$e^{-j\omega a} F(\omega)$
$e^{j\omega_0 t} f(t)$	$F(\omega - \omega_0)$
$f(t) \cos \omega_0 t$	$\dfrac{1}{2} F(\omega - \omega_0) + \dfrac{1}{2} F(\omega + \omega_0)$
$\displaystyle\int_{-\infty}^{\infty} x(\lambda)h(t - \lambda)\, d\lambda$	$X(\omega)H(\omega)$
$f_1(t)f_2(t)$	$\dfrac{1}{2\pi} \displaystyle\int_{-\infty}^{\infty} F_1(u)F_2(\omega - u)\, du$
$t^n f(t)$	$(j)^n \dfrac{d^n F(\omega)}{d\omega^n}$

EXAMPLE 17.3 **Deriving an Operational Fourier Transform**

Find the Fourier transform for the function $f(t) = t^n f(t)$, assuming that $F(\omega) = \mathscr{F}\{f(t)\}$ is known.

Solution

From the definition of the Fourier transform in Eq. 17.3,

$$F(\omega) = \int_{-\infty}^{\infty} f(t)e^{-j\omega t}\, dt.$$

Take the derivative of both sides with respect to the frequency ω and simplify:

$$\frac{dF(\omega)}{d\omega} = \int_{-\infty}^{\infty} \frac{d}{d\omega}\left[f(t)e^{-j\omega t}\, dt \right]$$

$$= -j \int_{-\infty}^{\infty} tf(t)e^{-j\omega t}\, dt = -j\mathscr{F}\{tf(t)\}.$$

Therefore,

$$\mathcal{F}\{tf(t)\} = j\frac{dF(\omega)}{d\omega}.$$

Repeat this process, taking the second derivative of both sides of Eq. 17.3 with respect to ω:

$$\frac{d^2F(\omega)}{d\omega^2} = \int_{-\infty}^{\infty}\frac{d^2}{d\omega^2}\left[f(t)e^{-j\omega t}dt\right]$$

$$= \int_{-\infty}^{\infty}(-jt)(-jt)f(t)e^{-j\omega t}dt$$

$$= (-j)^2\mathcal{F}\{t^2f(t)\}.$$

Therefore,

$$\mathcal{F}\{t^2f(t)\} = (j)^2\frac{d^2F(\omega)}{d\omega^2}.$$

Now take the nth derivative of both sides of Eq. 17.3 with respect to ω:

$$\frac{d^nF(\omega)}{d\omega^n} = \int_{-\infty}^{\infty}\frac{d^n}{d\omega^n}\left[f(t)e^{-j\omega t}dt\right]$$

$$= \int_{-\infty}^{\infty}(-jt)^nf(t)e^{-j\omega t}dt$$

$$= (-j)^n\mathcal{F}\{t^nf(t)\}.$$

Replacing $(-j)^n$ with $1/j^n$, we get

$$\mathcal{F}\{t^nf(t)\} = (j)^n\frac{d^nF(\omega)}{d\omega^n}.$$

Therefore, the effect of multiplying a function by t^n in the time domain corresponds to finding the nth derivative of the function's Fourier transform with respect to ω.

ASSESSMENT PROBLEMS

Objective 1—Be able to calculate the Fourier transform of a function

17.4 Suppose $f(t)$ is defined as follows:

$$f(t) = \frac{2A}{\tau}t + A, \qquad -\frac{\tau}{2} \le t \le 0;$$

$$f(t) = -\frac{2A}{\tau}t + A, \qquad 0 \le t \le \frac{\tau}{2},$$

$$f(t) = 0, \qquad\qquad \text{elsewhere.}$$

a) Find the second derivative of $f(t)$.

b) Find the Fourier transform of the second derivative.

c) Use the result obtained in (b) to find the Fourier transform of the function in (a). (*Hint:* Use the operational transform of differentiation.)

Answer: (a) $\dfrac{2A}{\tau}\delta\left(t + \dfrac{\tau}{2}\right) - \dfrac{4A}{\tau}\delta(t)$

$\qquad\qquad + \dfrac{2A}{\tau}\delta\left(t - \dfrac{\tau}{2}\right);$

(b) $\dfrac{4A}{\tau}\left(\cos\dfrac{\omega\tau}{2} - 1\right);$

(c) $\dfrac{4A}{\omega^2\tau}\left(1 - \cos\dfrac{\omega\tau}{2}\right).$

17.5 The rectangular pulse shown can be expressed as the difference between two step voltages; that is,

$$v(t) = V_m u\left(t + \frac{\tau}{2}\right) - V_m u\left(t - \frac{\tau}{2}\right)\text{ V.}$$

Use the operational transform for translation in the time domain to find the Fourier transform of $v(t)$.

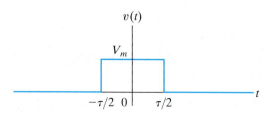

Answer: $V(\omega) = V_m\tau\dfrac{\sin(\omega\tau/2)}{(\omega\tau/2)}.$

SELF-CHECK: Also try Chapter Problem 17.19.

17.7 Circuit Applications

The Laplace transform is used more often than the Fourier transform when finding a circuit's response to a nonperiodic input. This is because the Laplace transform integral converges for a wider range of input functions and it accommodates initial conditions. We can certainly use the Fourier transform to find a circuit's response to an input using Eq. 17.13, which relates the transform of the output $Y(\omega)$ to the transform of the input $X(\omega)$ and the transfer function $H(\omega)$ of the circuit. Note that $H(\omega)$ is the familiar $H(s)$, with s replaced by $j\omega$.

Example 17.4 uses the Fourier transform to find the response of a circuit.

EXAMPLE 17.4 **Using the Fourier Transform to Find the Transient Response**

Use the Fourier transform to find $i_o(t)$ in the circuit shown in Fig. 17.9. The current source $i_g(t)$ is the signum function 20 sgn(t) A.

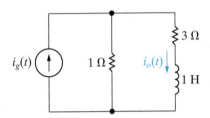

Figure 17.9 ▲ The circuit for Example 17.4.

Solution

The Fourier transform of the input is

$$I_g(\omega) = \mathcal{F}\{20 \operatorname{sgn}(t)\}$$

$$= 20\left(\frac{2}{j\omega}\right)$$

$$= \frac{40}{j\omega}.$$

The transfer function of the circuit is the ratio of I_o to I_g, which we can find using current division in the s domain:

$$I_o = \frac{1\|(3 + s)}{3 + s} I_g$$

$$= \frac{1}{4 + s} I_g.$$

Therefore,

$$H(s) = \frac{1}{4 + s}$$

and

$$H(\omega) = \frac{1}{4 + j\omega}.$$

The Fourier transform of $i_o(t)$ is

$$I_o(\omega) = I_g(\omega)H(\omega)$$

$$= \frac{40}{j\omega(4 + j\omega)}.$$

Expanding $I_o(\omega)$ into a sum of partial fractions yields

$$I_o(\omega) = \frac{K_1}{j\omega} + \frac{K_2}{4 + j\omega}.$$

Evaluating K_1 and K_2 gives

$$K_1 = \frac{40}{4} = 10,$$

$$K_2 = \frac{40}{-4} = -10.$$

Therefore

$$I_o(\omega) = \frac{10}{j\omega} - \frac{10}{4 + j\omega}.$$

Find the output current in the time domain using the functional and operational Fourier transform tables, Tables 17.1 and 17.2. The result is

$$i_o(t) = \mathcal{F}^{-1}[I_o(\omega)]$$

$$= 5 \operatorname{sgn}(t) - 10e^{-4t}u(t).$$

Figure 17.10 shows the response. Does the solution make sense in terms of known circuit behavior? The answer is yes, based on the following analysis. The current source delivers −20 A to the circuit between −∞ and 0, and because the current is constant, the inductor behaves like a short circuit over this time interval. Therefore, the resistance in each branch determines how the −20 A divides between the two branches. One-fourth of

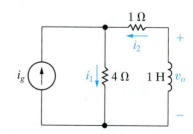

Figure 17.10 ▲ The plot of $i_o(t)$ versus t.

the -20 A appears in the i_o branch, so i_o is -5 A for $t < 0$. When the current source jumps from -20 A to $+20$ A at $t = 0$, i_o approaches its final value of $+5$ A exponentially. The equivalent resistance with respect to the inductor's terminals is $4\,\Omega$, so the time constant of the exponential rise for $t > 0$ is $\frac{1}{4}$ s.

An important characteristic of the Fourier transform is that it gives us the steady-state response to a sinusoidal input. There is no transient component in the response because the Fourier transform of $\cos \omega_0 t$ assumes that the function exists over all time. Example 17.5 illustrates this feature.

EXAMPLE 17.5 Using the Fourier Transform to Find the Sinusoidal Steady-State Response

Suppose the current source in the circuit in Example 17.4 (Fig. 17.9) changes to a sinusoid, given by

$$i_g(t) = 50\cos 3t \text{ A}.$$

Use the Fourier transform method to find $i_o(t)$.

Solution

The transform of the input current is

$$I_g(\omega) = 50\pi[\delta(\omega - 3) + \delta(\omega + 3)].$$

As before, the transfer function of the circuit is

$$H(\omega) = \frac{1}{4 + j\omega}.$$

The transform of the current response then is

$$I_o(\omega) = 50\pi \frac{\delta(\omega - 3) + \delta(\omega + 3)}{4 + j\omega}.$$

To find the inverse transform of $I_o(\omega)$ we begin with the inverse Fourier transform intergral in Eq. 17.4, and then use the sifting property of the impulse function:

$$i_o(t) = \mathcal{F}^{-1}\{I_o(\omega)\}$$

$$= \frac{50\pi}{2\pi} \int_{-\infty}^{\infty} \left[\frac{\delta(\omega - 3) + \delta(\omega + 3)}{4 + j\omega}\right] e^{j\omega t}\, d\omega$$

$$= 25\left(\frac{e^{j3t}}{4 + j3} + \frac{e^{-j3t}}{4 - j3}\right)$$

$$= 25\left(\frac{e^{j3t}e^{-j36.87°}}{5} + \frac{e^{-j3t}e^{j36.87°}}{5}\right)$$

$$= 5[2\cos(3t - 36.87°)]$$

$$= 10\cos(3t - 36.87°) \text{ A}.$$

You should verify that the solution for $i_o(t)$ is identical to that obtained by phasor analysis.

ASSESSMENT PROBLEMS

Objective 2—Know how to use the Fourier transform to find the response of a circuit

17.6 The current source in the circuit shown delivers a current of $10\text{sgn}(t)$ A. The output is the voltage across the 1 H inductor. Compute
(a) $I_g(\omega)$; (b) $H(j\omega)$; (c) $V_o(\omega)$; (d) $v_o(t)$;
(e) $i_1(0^-)$; (f) $i_1(0^+)$; (g) $i_2(0^-)$; (h) $i_2(0^+)$;
(i) $v_o(0^-)$; and (j) $v_o(0^+)$.

Answer: (a) $20/j\omega$;
(b) $4j\omega/(5 + j\omega)$;
(c) $80/(5 + j\omega)$;
(d) $80e^{-5t}u(t)$ V;
(e) -2 A;
(f) 18 A;
(g) 8 A;
(h) 8 A;
(i) 0 V;
(j) 80 V.

17.7 The voltage source in the circuit shown is generating the voltage

$$v_g = e^t u(-t) + u(t) \text{ V.}$$

a) Use the Fourier transform method to find v_a.

SELF-CHECK: Also try Chapter Problems 17.20, 17.28, and 17.30.

b) Compute $v_a(0^-)$, $v_a(0^+)$, and $v_a(\infty)$.

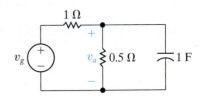

Answer: (a) $\dfrac{1}{4} e^t u(-t) - \dfrac{1}{12} e^{-3t}u(t) + \dfrac{1}{6}$

$$+ \frac{1}{6} \operatorname{sgn}(t) \text{ V;}$$

(b) $\dfrac{1}{4}$ V, $\dfrac{1}{4}$ V, $\dfrac{1}{3}$ V.

17.8 Parseval's Theorem

Parseval's theorem relates the energy of a time-domain function having finite energy to the Fourier transform of that function. Imagine that the time-domain function $f(t)$ is either the voltage across or the current in a 1 Ω resistor. The energy of this function then is

$$W_{1\Omega} = \int_{-\infty}^{\infty} f^2(t)\, dt.$$

Parseval's theorem says that we can also calculate this energy using a frequency-domain integral:

$$W_{1\Omega} = \int_{-\infty}^{\infty} f^2(t)\, dt = \frac{1}{2\pi} \int_{-\infty}^{\infty} |F(\omega)|^2\, d\omega. \qquad (17.14)$$

Therefore, we can calculate the 1 Ω energy associated with $f(t)$ either by integrating the square of $f(t)$ over all time or by integrating the square of the magnitude of the Fourier transform of $f(t)$ over all frequencies and multiplying by $1/2\pi$. Parseval's theorem is valid if both integrals exist.

The average power of time-domain signals with finite energy is zero when averaged over all time. Therefore, when comparing such signals, we use the energy content of the signals instead of their average power. Using a 1 Ω resistor when making the energy calculation is convenient and lets us compare the energy content of voltages and currents.

We begin deriving Eq. 17.14 by rewriting the left-hand side as $f(t)$ times itself and then expressing one $f(t)$ in terms of the inversion integral:

$$\int_{-\infty}^{\infty} f^2(t)\, dt = \int_{-\infty}^{\infty} f(t)\, f(t)\, dt$$

$$= \int_{-\infty}^{\infty} f(t) \left[\frac{1}{2\pi} \int_{-\infty}^{\infty} F(\omega) e^{j\omega t}\, d\omega \right] dt.$$

We move $f(t)$ inside the interior integral because the integration is with respect to ω, and then we factor the constant $1/2\pi$ outside both integrals. Then,

$$\int_{-\infty}^{\infty} f^2(t)\, dt = \frac{1}{2\pi} \int_{-\infty}^{\infty} \left[\int_{-\infty}^{\infty} F(\omega) f(t) e^{j\omega t}\, d\omega \right] dt.$$

We reverse the order of integration and then factor $F(\omega)$ out of the integral with respect to t. Thus

$$\int_{-\infty}^{\infty} f^2(t)\, dt = \frac{1}{2\pi} \int_{-\infty}^{\infty} F(\omega) \left[\int_{-\infty}^{\infty} f(t) e^{j\omega t}\, dt \right] d\omega.$$

The bracketed integral is $F(-\omega)$, so

$$\int_{-\infty}^{\infty} f^2(t)\, dt = \frac{1}{2\pi} \int_{-\infty}^{\infty} F(\omega) F(-\omega)\, d\omega.$$

In Section 17.5, we noted that $F(-\omega) = F^*(\omega)$. Thus, the product $F(\omega) F(-\omega)$ is the magnitude of $F(\omega)$ squared, completing the derivation of Eq. 17.14. We also noted that $|F(\omega)|$ is an even function of ω. Therefore, we can also write Eq. 17.14 as

$$\int_{-\infty}^{\infty} f^2(t)\, dt = \frac{1}{\pi} \int_{0}^{\infty} |F(\omega)|^2\, d\omega. \tag{17.15}$$

Demonstrating Parseval's Theorem

To demonstrate Eq. 17.15, suppose

$$f(t) = e^{-a|t|}.$$

Substituting this $f(t)$ into the left-hand side of Eq. 17.15 and evaluating the left-hand side, we get

$$\int_{-\infty}^{\infty} e^{-2a|t|}\, dt = \int_{-\infty}^{0} e^{2at}\, dt + \int_{0}^{\infty} e^{-2at}\, dt$$

$$= \frac{e^{2at}}{2a}\bigg|_{-\infty}^{0} + \frac{e^{-2at}}{-2a}\bigg|_{0}^{\infty}$$

$$= \frac{1}{2a} + \frac{1}{2a} = \frac{1}{a}.$$

The Fourier transform of $f(t)$ is

$$F(\omega) = \frac{2a}{a^2 + \omega^2},$$

and therefore the right-hand side of Eq. 17.15 evaluates to

$$\frac{1}{\pi} \int_{0}^{\infty} \frac{4a^2}{(a^2 + \omega^2)^2}\, d\omega = \frac{4a^2}{\pi} \frac{1}{2a^2} \left(\frac{\omega}{\omega^2 + a^2} + \frac{1}{a} \tan^{-1} \frac{\omega}{a} \right)\bigg|_{0}^{\infty}$$

$$= \frac{2}{\pi} \left(0 + \frac{\pi}{2a} - 0 - 0 \right)$$

$$= \frac{1}{a}.$$

Note that both the left-hand side and the right-hand side of Eq. 17.15 evaluate to $1/a$.

Interpreting Parseval's Theorem

A physical interpretation of Parseval's theorem tells us that the magnitude of the Fourier transform squared, $|F(\omega)|^2$, is an energy density (in joules per hertz). To see this, change the variable of integration on the right-hand side of Eq. 17.15, using $\omega = 2\pi f$. The result is

$$\frac{1}{\pi} \int_0^\infty |F(2\pi f)|^2 2\pi\, df = 2 \int_0^\infty |F(2\pi f)|^2\, df,$$

where $|F(2\pi f)|^2 df$ is the energy in an infinitesimal band of frequencies (df), and the total $1\,\Omega$ energy associated with $f(t)$ is the sum (integral) of $|F(2\pi f)|^2 df$ over all frequencies.

We can also calculate the portion of the total energy for a specified frequency band. For example, the 1Ω energy in the frequency band from ω_1 to ω_2 is

$$W_{1\Omega} = \frac{1}{\pi} \int_{\omega_1}^{\omega_2} |F(\omega)|^2\, d\omega. \tag{17.16}$$

Note that if we write the frequency-domain integral using

$$\frac{1}{2\pi} \int_{-\infty}^\infty |F(\omega)|^2\, d\omega$$

instead of

$$\frac{1}{\pi} \int_0^\infty |F(\omega)|^2\, d\omega,$$

we can rewrite Eq. 17.16 as

$$W_{1\Omega} = \frac{1}{2\pi} \int_{-\omega_2}^{-\omega_1} |F(\omega)|^2\, d\omega + \frac{1}{2\pi} \int_{\omega_2}^{\omega_1} |F(\omega)|^2\, d\omega. \tag{17.17}$$

Figure 17.11 shows a graphical interpretation of Eq. 17.17.
Examples 17.6–17.9 illustrate calculations using Parseval's theorem.

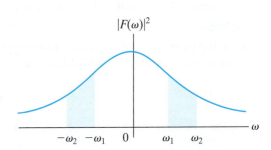

Figure 17.11 ▲ The graphic interpretation of Eq. 17.17.

EXAMPLE 17.6 | Applying Parseval's Theorem

The current in a 40 Ω resistor is

$$i = 20e^{-2t}u(t) \text{ A}.$$

What percentage of the total energy dissipated in the resistor is associated with the frequency band $0 \le \omega \le 2\sqrt{3} \text{ rad/s}$?

Solution

The total energy dissipated in the 40 Ω resistor is

$$W_{40\Omega} = 40 \int_0^\infty 400e^{-4t}\, dt$$

$$= 16{,}000 \left. \frac{e^{-4t}}{-4} \right|_0^\infty = 4000 \text{ J}.$$

We can check this total energy calculation with Parseval's theorem:

$$F(\omega) = \frac{20}{2 + j\omega}.$$

Therefore

$$|F(\omega)| = \frac{20}{\sqrt{4 + \omega^2}}$$

and

$$W_{40\Omega} = \frac{40}{\pi} \int_0^\infty \frac{400}{4 + \omega^2} \, d\omega$$

$$= \frac{16,000}{\pi} \left(\frac{1}{2} \tan^{-1} \frac{\omega}{2} \Big|_0^\infty \right)$$

$$= \frac{8000}{\pi} \left(\frac{\pi}{2} \right) = 4000 \text{ J}.$$

The energy associated with the frequency band $0 \le \omega \le 2\sqrt{3}$ rad/s is

$$W_{40\Omega} = \frac{40}{\pi} \int_0^{2\sqrt{3}} \frac{400}{4 + \omega^2} \, d\omega$$

$$= \frac{16,000}{\pi} \left(\frac{1}{2} \tan^{-1} \frac{\omega}{2} \Big|_0^{2\sqrt{3}} \right)$$

$$= \frac{8000}{\pi} \left(\frac{\pi}{3} \right) = \frac{8000}{3} \text{ J}.$$

Hence, the percentage of the total energy associated with this range of frequencies is

$$\eta = \frac{8000/3}{4000} \times 100 = 66.67\%.$$

EXAMPLE 17.7 Applying Parseval's Theorem to an Ideal Bandpass Filter

The input voltage for an ideal bandpass filter is

$$v(t) = 120e^{-24t}u(t) \text{ V}.$$

The filter passes all frequencies that lie between 24 and 48 rad/s, without attenuation, and completely rejects all frequencies outside this passband.

a) Sketch $|V(\omega)|^2$ for the filter input voltage.
b) Sketch $|V_o(\omega)|^2$ for the filter output voltage.
c) What percentage of the total 1 Ω energy content of the input voltage is available at the output?

Solution

a) The Fourier transform of the filter input voltage is

$$V(\omega) = \frac{120}{24 + j\omega}.$$

Therefore

$$|V(\omega)|^2 = \frac{14,400}{576 + \omega^2}.$$

Figure 17.12 shows the sketch of $|V(\omega)|^2$ versus ω.

b) The ideal bandpass filter rejects all frequencies outside the passband. The plot of $|V_o(\omega)|^2$ versus ω looks just like the plot in Fig. 17.12 between -48 and -24 rad/s and between 24 and 48 rad/s, and is zero for all other frequencies, as shown in Fig. 17.13.

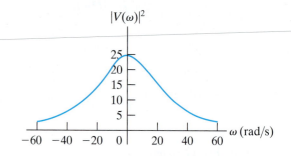

Figure 17.12 ▲ $|V(\omega)|^2$ versus ω for Example 17.7.

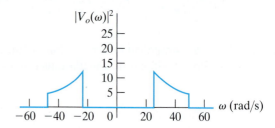

Figure 17.13 ▲ $|V_o(\omega)|^2$ versus ω for Example 17.7.

c) The total 1 Ω energy for the input voltage is

$$W_i = \frac{1}{\pi} \int_0^\infty \frac{14,400}{576 + \omega^2} \, d\omega = \frac{14,400}{\pi} \left(\frac{1}{24} \tan^{-1} \frac{\omega}{24} \Big|_0^\infty \right)$$

$$= \frac{600}{\pi} \frac{\pi}{2} = 300 \text{ J}.$$

The total 1 Ω energy for the filter's output is

$$W_o = \frac{1}{\pi} \int_{24}^{48} \frac{14{,}400}{576 + \omega^2} \, d\omega = \frac{600}{\pi} \tan^{-1} \frac{\omega}{24} \Big|_{24}^{48}$$

$$= \frac{600}{\pi} \left(\tan^{-1} 2 - \tan^{-1} 1 \right) = \frac{600}{\pi} \left(\frac{\pi}{2.84} - \frac{\pi}{4} \right)$$

$$= 61.45 \text{ J}.$$

The percentage of the input energy available at the output is

$$\eta = \frac{61.45}{300} \times 100 = 20.48\%.$$

EXAMPLE 17.8   Applying Parseval's Theorem to a Low-Pass Filter

We can use Parseval's theorem to calculate the energy available at a filter's output even if the time-domain expression for $v_o(t)$ is unknown. Suppose the input voltage to the low-pass RC filter circuit shown in Fig. 17.14 is

$$v_i(t) = 15e^{-5t} u(t) \text{ V}.$$

a) What percentage of the input signal's 1 Ω energy is available in the output signal?

b) What percentage of the output energy is associated with the frequency range $0 \le \omega \le 10 \text{ rad/s}$?

Solution

a) The 1 Ω energy in the input signal is

$$W_i = \int_0^{\infty} (15e^{-5t})^2 \, dt = 225 \frac{e^{-10t}}{-10} \Big|_0^{\infty} = 22.5 \text{ J}.$$

The Fourier transform of the output voltage is

$$V_o(\omega) = V_i(\omega) H(\omega),$$

where

$$V_i(\omega) = \frac{15}{5 + j\omega},$$

$$H(\omega) = \frac{1/RC}{1/RC + j\omega} = \frac{10}{10 + j\omega}.$$

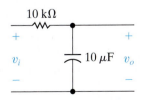

Figure 17.14 ▲ The low-pass RC filter for Example 17.8.

Hence

$$V_o(\omega) = \frac{150}{(5 + j\omega)(10 + j\omega)},$$

$$|V_o(\omega)|^2 = \frac{22{,}500}{(25 + \omega^2)(100 + \omega^2)}.$$

The 1 Ω energy available in the filter's output is

$$W_o = \frac{1}{\pi} \int_0^{\infty} \frac{22{,}500}{(25 + \omega^2)(100 + \omega^2)} \, d\omega.$$

We can evaluate the integral using a sum of partial fractions:

$$\frac{22{,}500}{(25 + \omega^2)(100 + \omega^2)} = \frac{300}{25 + \omega^2} - \frac{300}{100 + \omega^2}.$$

Then

$$W_o = \frac{300}{\pi} \left\{ \int_0^{\infty} \frac{d\omega}{25 + \omega^2} - \int_0^{\infty} \frac{d\omega}{100 + \omega^2} \right\}$$

$$= \frac{300}{\pi} \left[\frac{1}{5} \left(\frac{\pi}{2} \right) - \frac{1}{10} \left(\frac{\pi}{2} \right) \right] = 15 \text{ J}.$$

The energy available in the output signal therefore is 66.67% of the energy available in the input signal; that is,

$$\eta = \frac{15}{22.5} (100) = 66.67\%.$$

b) The output energy associated with the frequency range $0 \leq \omega \leq 10$ rad/s is

$$W'_o = \frac{300}{\pi} \left\{ \int_0^{10} \frac{d\omega}{25 + \omega^2} - \int_0^{10} \frac{d\omega}{100 + \omega^2} \right\}$$

$$= \frac{300}{\pi} \left(\frac{1}{5} \tan^{-1} \frac{10}{5} - \frac{1}{10} \tan^{-1} \frac{10}{10} \right) = \frac{30}{\pi} \left(\frac{2\pi}{2.84} - \frac{\pi}{4} \right)$$

$$= 13.64 \text{ J}.$$

The total $1\ \Omega$ energy in the output signal is 15 J, so the percentage associated with the frequency range 0 to 10 rad/s is 90.97%.

EXAMPLE 17.9 | **Calculating Energy Contained in a Rectangular Voltage Pulse**

A voltage pulse, $v(t)$, is shown in Fig. 17.15(a). Use Parseval's theorem to calculate the fraction of the total energy associated with $v(t)$ that lies in the frequency range $0 \leq \omega \leq 2\pi/\tau$. Recall from Section 17.1 that we found the Fourier transform of the voltage pulse to be

$$V(\omega) = V_m\tau \frac{\sin \omega\tau/2}{\omega\tau/2}.$$

The Fourier transform of the voltage pulse is plotted in Fig. 17.15(b).

Solution

To begin, substitute the Fourier transform of the voltage pulse into Eq. 17.16:

$$W = \frac{1}{\pi} \int_0^{2\pi/\tau} V_m^2\tau^2 \frac{\sin^2 \omega\tau/2}{(\omega\tau/2)^2} \, d\omega.$$

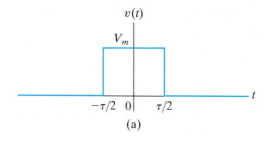

V_m

$-\tau/2 \quad 0 \quad \tau/2$

(a)

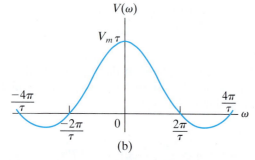

$V(\omega)$

$V_m\tau$

$\dfrac{-4\pi}{\tau}$ $\dfrac{-2\pi}{\tau}$ 0 $\dfrac{2\pi}{\tau}$ $\dfrac{4\pi}{\tau}$ ω

(b)

Figure 17.15 ▲ The rectangular voltage pulse and its Fourier transform. (a) The rectangular voltage pulse. (b) The Fourier transform of $v(t)$.

To evaluate this integral, let $x = \omega\tau/2$, so $dx = (\tau/2)d\omega$. Note that when $\omega = 2\pi/\tau$, $x = \pi$. Using these substitutions, the energy equation becomes

$$W = \frac{2V_m^2\tau}{\pi} \int_0^{\pi} \frac{\sin^2 x}{x^2} \, dx.$$

Now we can integrate by parts. Let $u = \sin^2 x$ and $dv = dx/x^2$, so $du = 2\sin x \cos x\, dx = \sin 2x\, dx$ and $v = -1/x$. Therefore,

$$\int_0^{\pi} \frac{\sin^2 x}{x^2} \, dx = -\frac{\sin^2 x}{x} \bigg|_0^{\pi} - \int_0^{\pi} -\frac{1}{x} \sin 2x \, dx$$

$$= 0 + \int_0^{\pi} \frac{\sin 2x}{x} \, dx$$

and

$$W = \frac{4V_m^2\tau}{\pi} \int_0^{\pi} \frac{\sin 2x}{2x} \, dx.$$

To evaluate this integral, make another substitution of variables to get the form $\sin y/y$. Let $y = 2x$ so $dy = 2\, dx$, and $y = 2\pi$ when $x = \pi$. The resulting equation is

$$W = \frac{2V_m^2\tau}{\pi} \int_0^{2\pi} \frac{\sin y}{y} \, dy.$$

The integral's value can be found using an on-line calculator.[1] Its value is 1.41815, so

$$W = \frac{2V_m^2\tau}{\pi} (1.41815).$$

To find the total $1\ \Omega$ energy associated with $v(t)$, use either the time-domain integral or the frequency-domain integral in Eq. 17.14. The total energy is

$$W_t = V_m^2\tau.$$

---

[1] http://www.wolframalpha.com/input/?i=integrate+sin+x+%2F+x+ from+0+to+2*pi

The fraction of the total energy associated with the band of frequencies between 0 and $2\pi/\tau$ is

$$\eta = \frac{W}{W_t}$$

$$= \frac{2V_m^2\tau(1.41815)}{\pi(V_m^2\tau)}$$

$$= 0.9028.$$

Therefore, approximately 90% of the energy associated with $v(t)$ is contained in the dominant portion of the amplitude spectrum.

Note that the plots in Fig. 17.15 show that as the width of the voltage pulse (τ) becomes smaller, the dominant portion of the amplitude spectrum (that is, the spectrum from $-2\pi/\tau$ to $2\pi/\tau$) spreads out over a wider range of frequencies. This result agrees with our earlier comments about the operational transform involving a scale change—when time is compressed, frequency is expanded and vice versa. To transmit a single rectangular pulse with reasonable fidelity, the bandwidth of the system must be at least wide enough to accommodate the dominant portion of the amplitude spectrum. Thus, the cutoff frequency should be at least $2\pi/\tau$ rad/s, or $1/\tau$ Hz.

ASSESSMENT PROBLEMS

Objective 3—Understand Parseval's theorem and be able to use it

17.8 The voltage across a 50 Ω resistor is

$$v = 4te^{-t}u(t) \text{ V}.$$

What percentage of the total energy dissipated in the resistor can be associated with the frequency band $0 \le \omega \le \sqrt{3}$ rad/s?

Answer: 94.23%.

17.9 Assume that the magnitude of the Fourier transform of $v(t)$ is as shown. This voltage is applied to a 6 kΩ resistor. Calculate the total energy delivered to the resistor.

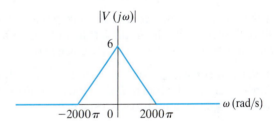

Answer: 4 J.

SELF-CHECK: Also try Chapter Problem 17.40.

■ **Practical Perspective**

Filtering Digital Signals

To understand the effect of transmitting a digital signal on a telephone line, consider a pulse that represents a digital value of 1, as shown in Fig. 17.15(a), with $V_m = 5$ V and $\tau = 1$ μs. The Fourier transform of this pulse is shown in Fig. 17.15(b), where the amplitude $V_m\tau = 5$ μV and the first positive zero-crossing on the frequency axis is $2\pi/\tau = 6.28$ Mrad/s $= 1$ MHz.

Note that the digital pulse representing the value 1 is ideally a sum of an infinite number of frequency components. But the telephone line cannot transmit all of these frequency components. Typically, the line has a bandwidth of 10 MHz, meaning that it is capable of transmitting only those frequency components below 10 MHz. This causes the original pulse to be distorted once it is received by the computer on the other end of the telephone line, as seen in Fig. 17.16.

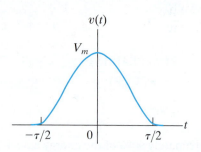

Figure 17.16 ▲ The effect of sending a square voltage pulse through a bandwidth-limited filter, causing distortion of the resulting output signal in the time domain.

Summary

- The **Fourier transform** gives a frequency-domain description of an aperiodic time-domain function. Depending on the nature of the time-domain signal, one of three approaches to finding its Fourier transform may be used:

  - If the time-domain signal is a well-behaved pulse of finite duration, the integral that defines the Fourier transform is used. (See page 662.)

  - If the one-sided Laplace transform of $f(t)$ exists and all the poles of $F(s)$ lie in the left half of the s plane, $F(s)$ may be used to find $F(\omega)$. (See page 666.)

  - If $f(t)$ is a constant, a signum function, a step function, or a sinusoidal function, the Fourier transform is found by using a limit process. (See page 668.)

- Functional and operational Fourier transforms that are useful in circuit analysis are tabulated in Tables 17.1 and 17.2. (See pages 670 and 675.)

- The Fourier transform accommodates both negative-time and positive-time functions and therefore is suited to problems whose signals start at $t = -\infty$. In contrast, the unilateral Laplace transform is suited to problems with initial conditions and signals that exist for $t > 0$. (See page 677.)

- The Fourier transform of a response signal $y(t)$ is

  $$Y(\omega) = X(\omega)H(\omega),$$

  where $X(\omega)$ is the Fourier transform of the input signal $x(t)$ and $H(\omega)$ is the transfer function $H(s)$ evaluated at $s = j\omega$. (See page 677.)

- The magnitude of the Fourier transform squared is a measure of the energy density (joules per hertz) in the frequency domain (Parseval's theorem). Thus, the Fourier transform permits us to associate a fraction of the total energy contained in $f(t)$ with a specified band of frequencies. (See page 681.)

Problems

Sections 17.1–17.2

17.1 a) Find the Fourier transform of the function shown in Fig. P17.1.

  b) Find $F(\omega)$ when $\omega = 0$.

  c) Sketch $|F(\omega)|$ versus ω when $A = 10$ and $\tau = 0.1$. *Hint:* Remember that $|F(\omega)|$ is an even function of ω.

Figure P17.1

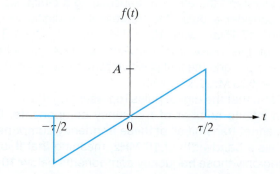

17.2 The Fourier transform of $f(t)$ is shown in Fig. P17.2.

a) Find $f(t)$.

b) Evaluate $f(0)$.

c) Sketch $f(t)$ for -150 s $\le t \le 150$ s when $A = 5\pi$ and $\omega_0 = 100$ rad/s. *Hint:* Remember that $f(t)$ is even.

Figure P17.2

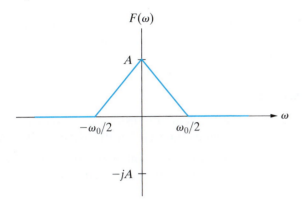

17.3 Use the defining integral to find the Fourier transform of the following functions:

a) $f(t) = A \sin \dfrac{\pi}{2} t, \quad -2 \le t \le 2;$

$\quad f(t) = 0, \qquad\qquad$ elsewhere.

b) $f(t) = \dfrac{2A}{\tau} t + A, \quad -\dfrac{\tau}{2} \le t \le 0;$

$\quad f(t) = -\dfrac{2A}{\tau} t + A, \quad 0 \le t \le \dfrac{\tau}{2};$

$\quad f(t) = 0, \qquad\qquad$ elsewhere.

Sections 17.3–17.5

17.4 Derive $\mathcal{F}\{\sin \omega_0 t\}$.

17.5 Find the Fourier transform of each of the following functions. In all of the functions, a is a positive real constant and $-\infty \le t \le \infty$.

a) $f(t) = |t| e^{-a|t|};$

b) $f(t) = t^3 e^{-a|t|};$

c) $f(t) = e^{-a|t|} \cos \omega_0 t;$

d) $f(t) = e^{-a|t|} \sin \omega_0 t;$

e) $f(t) = \delta(t - t_0).$

17.6 If $f(t)$ is a real function of t, show that the inversion integral reduces to

$$f(t) = \frac{1}{2\pi} \int_{-\infty}^{\infty} [A(\omega) \cos \omega t - B(\omega) \sin \omega t]\, d\omega.$$

17.7 If $f(t)$ is a real, odd function of t, show that the inversion integral reduces to

$$f(t) = -\frac{1}{2\pi} \int_{-\infty}^{\infty} B(\omega) \sin \omega t\, d\omega.$$

17.8 Use the inversion integral (Eq. 17.4) to show that $\mathcal{F}^{-1}\{2/j\omega\} = \mathrm{sgn}(t)$. *Hint:* Use Problem 17.7.

17.9 Find $\mathcal{F}\{\cos \omega_0 t\}$ by using the approximating function

$$f(t) = e^{-\epsilon|t|} \cos \omega_0 t,$$

where ϵ is a positive real constant.

17.10 Show that if $f(t)$ is an odd function,

$$A(\omega) = 0,$$

$$B(\omega) = -2 \int_{0}^{\infty} f(t) \sin \omega t\, dt.$$

17.11 Show that if $f(t)$ is an even function,

$$A(\omega) = 2 \int_{0}^{\infty} f(t) \cos \omega t\, dt,$$

$$B(\omega) = 0.$$

Section 17.6

17.12 a) Show that $\mathcal{F}\{df(t)/dt\} = j\omega F(\omega)$, where $F(\omega) = \mathcal{F}\{f(t)\}$. *Hint:* Use the defining integral and integrate by parts.

b) What is the restriction on $f(t)$ if the result given in (a) is valid?

c) Show that $\mathcal{F}\{d^n f(t)/dt^n\} = (j\omega)^n F(\omega)$, where $F(\omega) = \mathcal{F}\{f(t)\}$.

17.13 a) Show that

$$\mathcal{F}\left\{ \int_{-\infty}^{t} f(x)dx \right\} = \frac{F(\omega)}{j\omega},$$

where $F(\omega) = \mathcal{F}\{f(x)\}$. *Hint:* Use the defining integral and integrate by parts.

b) What is the restriction on $f(x)$ if the result given in (a) is valid?

c) If $f(x) = e^{-ax} u(x)$, can the operational transform in (a) be used? Explain.

17.14 a) Show that

$$\mathcal{F}\{f(at)\} = \frac{1}{a}F\left(\frac{\omega}{a}\right), \quad a > 0.$$

b) Given that $f(at) = e^{-a|t|}$ for $a > 0$, sketch $F(\omega) = \mathcal{F}\{f(at)\}$ for $a = 0.5, 1.0,$ and 2.0. Do your sketches reflect the observation that *compression* in the time domain corresponds to *stretching* in the frequency domain?

17.15 Derive each of the following operational transforms:

a) $\mathcal{F}\{f(t - a)\} = e^{-j\omega a}F(\omega);$

b) $\mathcal{F}\{e^{j\omega_0 t}f(t)\} = F(\omega - \omega_0);$

c) $\mathcal{F}\{f(t)\cos\omega_0 t\} = \frac{1}{2}F(\omega - \omega_0) + \frac{1}{2}F(\omega + \omega_0).$

17.16 Given

$$y(t) = \int_{-\infty}^{\infty} x(\lambda)h(t - \lambda)\,d\lambda,$$

show that $Y(\omega) = \mathcal{F}\{y(t)\} = X(\omega)H(\omega)$, where $X(\omega) = \mathcal{F}\{x(t)\}$ and $H(\omega) = \mathcal{F}\{h(t)\}$. *Hint:* Use the defining integral to write

$$\mathcal{F}\{y(t)\} = \int_{-\infty}^{\infty}\left[\int_{-\infty}^{\infty} x(\lambda)h(t - \lambda)\,d\lambda\right]e^{-j\omega t}\,dt.$$

Next, reverse the order of integration and then make a change in the variable of integration; that is, let $u = t - \lambda$.

17.17 Given $f(t) = f_1(t)f_2(t)$, show that

$$F(\omega) = (1/2\pi)\int_{-\infty}^{\infty} F_1(u)F_2(\omega - u)\,du.$$

Hint: First, use the defining integral to express $F(\omega)$ as

$$F(\omega) = \int_{-\infty}^{\infty} f_1(t)f_2(t)e^{-j\omega t}\,dt.$$

Second, use the inversion integral to write

$$f_1(t) = \frac{1}{2\pi}\int_{-\infty}^{\infty} F_1(u)e^{j\omega t}\,du.$$

Third, substitute the expression for $f_1(t)$ into the defining integral and then interchange the order of integration.

17.18 Use the Fourier transform of $t^n f(t)$ to find each of the following Fourier transforms (assuming $a > 0$):

$$\mathcal{F}\{te^{-at}u(t)\},$$

$$\mathcal{F}\{|t|e^{-a|t|}\},$$

$$\mathcal{F}\{te^{-a|t|}\}.$$

17.19 Suppose that $f(t) = f_1(t)f_2(t)$, where

$$f_1(t) = \cos\omega_0 t,$$

$$f_2(t) = 1, \quad -\tau/2 < t < \tau/2;$$

$$f_2(t) = 0, \quad \text{elsewhere.}$$

a) Use convolution in the frequency domain to find $F(\omega)$.

b) What happens to $F(\omega)$ as the width of $f_2(t)$ increases so that $f(t)$ includes more and more cycles of $f_1(t)$?

Section 17.7

17.20 a) Use the Fourier transform method to find $i_o(t)$ in the circuit shown in Fig. P17.20 if $v_g = 36\,\text{sgn}(t)$ V.

PSPICE
MULTISIM

b) Does your solution make sense in terms of known circuit behavior? Explain.

Figure P17.20

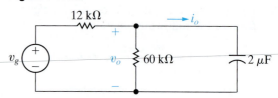

17.21 Repeat Problem 17.20 except replace $i_o(t)$ with $v_o(t)$.

PSPICE
MULTISIM

17.22 a) Use the Fourier transform method to find $v_o(t)$ in the circuit shown in Fig. P17.22. The initial value of $v_o(t)$ is zero, and the source voltage is $100u(t)$ V.

PSPICE
MULTISIM

b) Sketch $v_o(t)$ versus t.

Figure P17.22

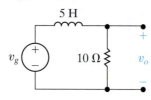

17.23 Repeat Problem 17.22 if the input voltage (v_g) is changed to $100\,\text{sgn}(t)$.

PSPICE
MULTISIM

17.24 a) Use the Fourier transform to find i_o in the circuit in Fig. P17.24 if $i_g = 0.2\,\text{sgn}(t)$ mA.

PSPICE
MULTISIM

b) Does your solution make sense in terms of known circuit behavior? Explain.

Figure P17.24

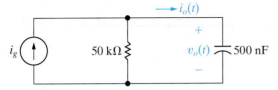

17.25 Repeat Problem 17.24 except replace i_o with v_o.

PSPICE
MULTISIM

17.26 The voltage source in the circuit in Fig. P17.26 is
PSPICE given by the expression
MULTISIM

$$v_g = 3 \, \text{sgn}(t) \, \text{V}.$$

a) Find $v_o(t)$.

b) What is the value of $v_o(0^-)$?

c) What is the value of $v_o(0^+)$?

d) Use the Laplace transform method to find $v_o(t)$ for $t > 0^+$.

e) Does the solution obtained in (d) agree with $v_o(t)$ for $t > 0^+$ from (a)?

Figure P17.26

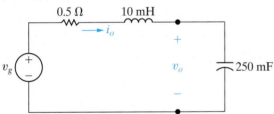

17.27 Repeat Problem 17.26 except replace $v_o(t)$ with $i_o(t)$.
PSPICE
MULTISIM

17.28 a) Use the Fourier transform to find v_o in the cir-
PSPICE   cuit in Fig. P17.28 if i_g equals $3e^{-5|t|}$ A.
MULTISIM

b) Find $v_o(0^-)$.

c) Find $v_o(0^+)$.

d) Use the Laplace transform method to find v_o for $t \geq 0$.

e) Does the solution obtained in (d) agree with v_o for $t > 0^+$ from (a)?

Figure P17.28

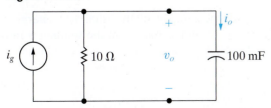

17.29 a) Use the Fourier transform to find i_o in the circuit
PSPICE   in Fig. P17.28 if i_g equals $3e^{-5|t|}$ A.
MULTISIM

b) Find $i_o(0^-)$.

c) Find $i_o(0^+)$.

d) Use the Laplace transform method to find i_o for $t \geq 0$.

e) Does the solution obtained in (d) agree with i_o for $t > 0^+$ from (a)?

17.30 Use the Fourier transform method to find i_o in the
PSPICE circuit in Fig. P17.30 if $v_g = 300 \cos 5000t$ V.
MULTISIM

Figure P17.30

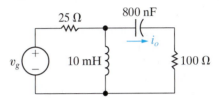

17.31 a) Use the Fourier transform method to find i_o in
PSPICE   the circuit in Fig. P17.31 if $v_g = 125 \cos 40,000t$ V.
MULTISIM

b) Check the answer obtained in (a) by finding the steady-state expression for i_o using phasor domain analysis.

Figure P17.31

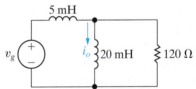

17.32 a) Use the Fourier transform method to find v_o
PSPICE   in the circuit shown in Fig. P17.32. The voltage
MULTISIM   source generates the voltage

$$v_g = 45e^{-500|t|} \, \text{V}.$$

b) Calculate $v_o(0^-)$, $v_o(0^+)$, and $v_o(\infty)$.

c) Find $i_L(0^-)$; $i_L(0^+)$; $v_C(0^-)$; and $v_C(0^+)$.

d) Do the results in part (b) make sense in terms of known circuit behavior? Explain.

Figure P17.32

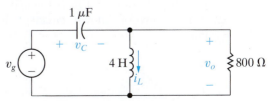

17.33 The voltage source in the circuit in Fig. P17.33 is generating the signal

PSPICE
MULTISIM

$$v_g = 5\,\text{sgn}(t) - 5 + 30e^{-5t}u(t)\ \text{V}.$$

a) Find $v_o(0^-)$ and $v_o(0^+)$.
b) Find $i_o(0^-)$ and $i_o(0^+)$.
c) Find v_o.

Figure P17.33

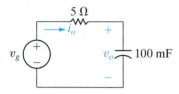

17.34 a) Use the Fourier transform method to find v_o in the circuit in Fig. P17.34 when

PSPICE
MULTISIM

$$v_g = 36e^{4t}u(-t) - 36e^{-4t}u(t)\ \text{V}.$$

b) Find $v_o(0^-)$.
c) Find $v_o(0^+)$.

Figure P17.34

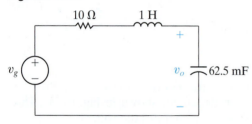

17.35 a) Use the Fourier transform method to find v_o in the circuit in Fig. P17.35 when

PSPICE
MULTISIM

$$i_g = 18e^{10t}u(-t) - 18e^{-10t}u(t)\ \text{A}.$$

b) Find $v_o(0^-)$.
c) Find $v_o(0^+)$.
d) Do the answers obtained in (b) and (c) make sense in terms of known circuit behavior? Explain.

Figure P17.35

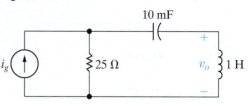

17.36 When the input voltage to the system shown in Fig. P17.36 is $15u(t)$ V, the output voltage is

$$v_o = [10 + 30e^{-20t} - 40e^{-30t}]u(t)\ \text{V}.$$

What is the output voltage if $v_i = 15\,\text{sgn}(t)$ V?

Figure P17.36

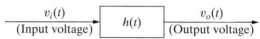

Section 17.8

17.37 It is given that $F(\omega) = e^{\omega}u(-\omega) + e^{-\omega}u(\omega)$.
a) Find $f(t)$.
b) Find the 1 Ω energy associated with $f(t)$ via time-domain integration.
c) Repeat (b) using frequency-domain integration.
d) Find the value of ω_1 if $f(t)$ has 90% of the energy in the frequency band $0 \le |\omega| \le \omega_1$.

17.38 The circuit shown in Fig. P17.38 is driven by the current

$$i_g = 12e^{-10t}u(t)\ \text{A}.$$

What percentage of the total 1 Ω energy content in the output current i_o lies in the frequency range $0 \le |\omega| \le 100\ \text{rad/s}$?

Figure P17.38

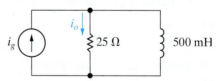

17.39 The input current signal in the circuit seen in Fig. P17.39 is

$$i_g = 30e^{-2t}u(t)\ \mu\text{A}, \quad t \ge 0^+.$$

What percentage of the total 1 Ω energy content in the output signal lies in the frequency range 0 to 4 rad/s?

Figure P17.39

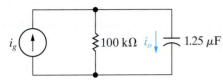

17.40 The input voltage in the circuit in Fig. P17.40 is $v_g = 30e^{-|t|}$ V.

a) Find $v_o(t)$.

b) Sketch $|V_g(\omega)|$ for $-5 \le \omega \le 5$ rad/s.

c) Sketch $|V_o(\omega)|$ for $-5 \le \omega \le 5$ rad/s.

d) Calculate the 1 Ω energy content of v_g.

e) Calculate the 1 Ω energy content of v_o.

f) What percentage of the 1 Ω energy content in v_g lies in the frequency range $0 \le |\omega| \le 2$ rad/s?

g) Repeat (f) for v_o.

Figure P17.40

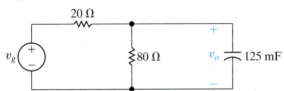

17.41 The amplitude spectrum of the input voltage to the high-pass RC filter in Fig. P17.41 is

$$V_i(\omega) = \frac{200}{|\omega|}, \quad 100 \text{ rad/s} \le |\omega| \le 200 \text{ rad/s};$$

$$V_i(\omega) = 0, \quad \text{elsewhere.}$$

a) Sketch $|V_i(\omega)|^2$ for $-300 \le \omega \le 300$ rad/s.

b) Sketch $|V_o(\omega)|^2$ for $-300 \le \omega \le 300$ rad/s.

c) Calculate the 1 Ω energy in the signal at the input of the filter.

d) Calculate the 1 Ω energy in the signal at the output of the filter.

Figure P17.41

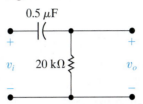

17.42 The input voltage to the high-pass RC filter circuit in Fig. P17.42 is

$$v_i(t) = Ae^{-at}u(t).$$

Let α denote the corner frequency of the filter, that is, $\alpha = 1/RC$.

a) What percentage of the energy in the signal at the output of the filter is associated with the frequency band $0 \le |\omega| \le \alpha$ if $\alpha = a$?

b) Repeat (a), given that $\alpha = \sqrt{3}a$.

c) Repeat (a), given that $\alpha = a/\sqrt{3}$.

Figure P17.42

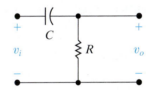

CHAPTER
18

Two-Port Circuits

 CHAPTER CONTENTS

18.1 **The Terminal Equations** *p. 694*

18.2 **The Two-Port Parameters** *p. 695*

18.3 **Analysis of the Terminated Two-Port Circuit** *p. 703*

18.4 **Interconnected Two-Port Circuits** *p. 708*

CHAPTER OBJECTIVES

1 Be able to calculate any set of two-port parameters with any of the following methods:

- Circuit analysis;
- Measurements made on a circuit;
- Converting from another set of two-port parameters using Table 18.1.

2 Be able to analyze a terminated two-port circuit to find currents, voltages, impedances, and ratios of interest using Table 18.2.

3 Know how to analyze a cascade interconnection of two-port circuits.

We have frequently focused on the behavior of a circuit at a specified pair of terminals. We introduced the Thévenin and Norton equivalent circuits to simplify circuit analysis relative to one pair of terminals. But in some electrical systems, a signal is fed into one pair of terminals, processed by the system, and extracted at a second pair of terminals. We can simplify the analysis of such systems using two pairs of terminals, each representing the points, or **ports**, where signals are either input or output.

In this chapter, we present circuits that have one input and one output port. Figure 18.1 on page 694 illustrates the basic two-port building block. We make several assumptions when using the two-port model in Fig. 18.1 to represent a circuit:

- There is no energy stored within the circuit.
- There are no independent sources within the circuit; dependent sources, however, are permitted.
- The current into a given port equals the current out of that port; that is, $i_1 = i'_1$ and $i_2 = i'_2$.
- All external connections must be made to either the input port or the output port; no connections can be made between ports, that is, between terminals a and c, a and d, b and c, or b and d.

These assumptions limit the types of circuits that can be represented by a two-port model.

When we represent a circuit using a two-port model, we are only interested in the circuit's terminal variables (i_1, v_1, i_2, and v_2). We cannot use the two-port model to find the currents and voltages inside the circuit. We have already used such terminal behavior when analyzing operational amplifier circuits. In this chapter, we formalize that approach by introducing the two-port parameters.

■ Practical Perspective

Characterizing an Unknown Circuit

Up to this point, whenever we wanted to create a model of a circuit, we needed to know what types of components make up the circuit, the values of those components, and the interconnections among those components. But what if we want to model a circuit that is inside a "black box," where the components, their values, and their interconnections are hidden?

In this chapter, we will discover that by making two simple measurements on such a black box, we can create a model consisting of just four values—the two-port parameter model for the circuit. We can then use the two-port parameter model to predict the behavior of the circuit once we have attached a power source to one of its ports and a load to the other port.

In this example, suppose we have found a circuit, enclosed in a casing, with two wires extending from each side, as shown below. The casing is labeled "amplifier," and we want to determine whether or not it would be safe to use this amplifier to connect a cd player modeled as a 2 V source to a speaker modeled as a 32 Ω resistor with a power rating of 100 W.

HSNphotography/Shutterstock

Ensuper/Shutterstock

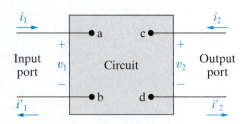

Figure 18.1 ▲ The two-port building block.

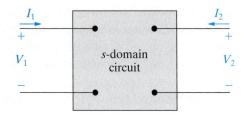

Figure 18.2 ▲ The s-domain two-port basic building block.

18.1 The Terminal Equations

When we consider a circuit to be a two-port network, we want to relate the current and voltage at one port to the current and voltage at the other port. Figure 18.1 shows the reference polarities of the terminal voltages and the reference directions of the terminal currents. The references at each port are symmetric with respect to each other; that is, at each port the current is directed into the upper terminal, and at each port the voltage rises from the lower to the upper terminal.

The most general two-port network describes the circuit in the s domain. For purely resistive circuits, we can analyze the time-domain circuit to find its two-port network description. Two-port networks used to find sinusoidal steady-state responses can be constructed by finding the appropriate s-domain expressions and then replacing s with $j\omega$, or by using phasor techniques directly. Here, we write all equations in the s domain; resistive networks and sinusoidal steady-state solutions become special cases. Figure 18.2 shows the basic building block in terms of the s-domain variables I_1, V_1, I_2, and V_2.

Only two of the four terminal variables are independent. Thus, for any circuit, once we specify two of the variables, we can find the two remaining unknowns. For example, knowing V_1 and V_2 and the circuit within the box, we can find I_1 and I_2. The two-port network description consists of two simultaneous equations. However, there are six different ways in which to combine the four variables:

$$\begin{aligned} V_1 &= z_{11}I_1 + z_{12}I_2, \\ V_2 &= z_{21}I_1 + z_{22}I_2; \end{aligned} \tag{18.1}$$

$$\begin{aligned} I_1 &= y_{11}V_1 + y_{12}V_2, \\ I_2 &= y_{21}V_1 + y_{22}V_2; \end{aligned} \tag{18.2}$$

$$\begin{aligned} V_1 &= a_{11}V_2 - a_{12}I_2, \\ I_1 &= a_{21}V_2 - a_{22}I_2; \end{aligned} \tag{18.3}$$

$$\begin{aligned} V_2 &= b_{11}V_1 - b_{12}I_1, \\ I_2 &= b_{21}V_1 - b_{22}I_1; \end{aligned} \tag{18.4}$$

$$\begin{aligned} V_1 &= h_{11}I_1 + h_{12}V_2, \\ I_2 &= h_{21}I_1 + h_{22}V_2; \end{aligned} \tag{18.5}$$

$$\begin{aligned} I_1 &= g_{11}V_1 + g_{12}I_2, \\ V_2 &= g_{21}V_1 + g_{22}I_2. \end{aligned} \tag{18.6}$$

These six sets of equations may also be considered as three pairs of mutually inverse relations. The first set, Eqs. 18.1, gives the input and output voltages as functions of the input and output currents. The second set, Eqs. 18.2, gives the inverse relationship, that is, the input and output currents as functions of the input and output voltages. Equations 18.3 and 18.4 are inverse relations, as are Eqs. 18.5 and 18.6.

The coefficients of the current and/or voltage variables on the right-hand side of Eqs. 18.1–18.6 are called the **parameters** of the two-port circuit. Thus, when using Eqs. 18.1, we refer to the z parameters of the circuit. Similarly, we refer to the y parameters, the a parameters, the b parameters, the h parameters, and the g parameters of the network.

18.2 The Two-Port Parameters

We can determine the parameters for any circuit by computation or measurement. The computations or measurements we need come from the parameter equations. For example, suppose we want to find the z parameters for a circuit. From Eqs. 18.1,

$$z_{11} = \left.\frac{V_1}{I_1}\right|_{I_2=0} \Omega, \tag{18.7}$$

$$z_{12} = \left.\frac{V_1}{I_2}\right|_{I_1=0} \Omega, \tag{18.8}$$

$$z_{21} = \left.\frac{V_2}{I_1}\right|_{I_2=0} \Omega, \tag{18.9}$$

$$z_{22} = \left.\frac{V_2}{I_2}\right|_{I_1=0} \Omega. \tag{18.10}$$

Equations 18.7–18.10 reveal that the four z parameters can be described as follows:

- z_{11} is the impedance seen looking into port 1 when port 2 is open.
- z_{12} is a transfer impedance. It is the ratio of the port 1 voltage to the port 2 current when port 1 is open.
- z_{21} is a transfer impedance. It is the ratio of the port 2 voltage to the port 1 current when port 2 is open.
- z_{22} is the impedance seen looking into port 2 when port 1 is open.

Therefore, the impedance parameters may be either calculated or measured by first opening port 2 and determining the ratios V_1/I_1 and V_2/I_1, and then opening port 1 and determining the ratios V_1/I_2 and V_2/I_2.

Equations 18.7–18.10 show why the parameters in Eqs. 18.1 are called the z parameters. Each parameter is the ratio of a voltage to a current and therefore is an impedance with the dimension of ohms.

We use the same process to determine the remaining two-port parameters, which are either calculated or measured. Finding a two-port parameter requires that a port be either opened or shorted. The two-port parameters are either impedances, admittances, or dimensionless ratios. The dimensionless ratios are either ratios of two voltages or two currents. Equations 18.11–18.15 summarize these observations.

$$y_{11} = \left.\frac{I_1}{V_1}\right|_{V_2=0} S, \qquad y_{12} = \left.\frac{I_1}{V_2}\right|_{V_1=0} S,$$
$$ \tag{18.11}$$
$$y_{21} = \left.\frac{I_2}{V_1}\right|_{V_2=0} S, \qquad y_{22} = \left.\frac{I_2}{V_2}\right|_{V_1=0} S.$$

$$a_{11} = \left.\frac{V_1}{V_2}\right|_{I_2=0}, \qquad a_{12} = -\left.\frac{V_1}{I_2}\right|_{V_2=0} \Omega,$$
$$ \tag{18.12}$$
$$a_{21} = \left.\frac{I_1}{V_2}\right|_{I_2=0} S, \qquad a_{22} = -\left.\frac{I_1}{I_2}\right|_{V_2=0}.$$

$$b_{11} = \left.\frac{V_2}{V_1}\right|_{I_1=0}, \qquad b_{12} = -\left.\frac{V_2}{I_1}\right|_{V_1=0} \Omega,$$
$$ \tag{18.13}$$
$$b_{21} = \left.\frac{I_2}{V_1}\right|_{I_1=0} S, \qquad b_{22} = -\left.\frac{I_2}{I_1}\right|_{V_1=0}.$$

$$h_{11} = \left.\frac{V_1}{I_1}\right|_{V_2=0} \Omega, \qquad h_{12} = \left.\frac{V_1}{V_2}\right|_{I_1=0},$$

$$h_{21} = \left.\frac{I_2}{I_1}\right|_{V_2=0}, \qquad h_{22} = \left.\frac{I_2}{V_2}\right|_{I_1=0} S. \qquad (18.14)$$

$$g_{11} = \left.\frac{I_1}{V_1}\right|_{I_2=0} S, \qquad g_{12} = \left.\frac{I_1}{I_2}\right|_{V_1=0},$$

$$g_{21} = \left.\frac{V_2}{V_1}\right|_{I_2=0}, \qquad g_{22} = \left.\frac{V_2}{I_2}\right|_{V_1=0} \Omega. \qquad (18.15)$$

We also give descriptive names to the reciprocal sets of two-port parameters. The impedance and admittance parameters are grouped into the **immittance** parameters. An **immittance** is either an impedance or an admittance. The a and b parameters are called the **transmission** parameters because they describe the voltage and current at one end of the two-port network in terms of the voltage and current at the other end. The immittance and transmission parameters are the natural choices for relating the two-port variables because they relate either voltage to current variables or input to output variables. In contrast, the h and g parameters relate an input voltage and output current to an output voltage and input current. Therefore, the h and g parameters are called **hybrid** parameters.

Example 18.1 calculates the z parameters for a resistive circuit by analyzing the circuit. Example 18.2 illustrates how a set of measurements made at the terminals of a two-port circuit can be used to calculate the a parameters.

EXAMPLE 18.1   Finding the *z* Parameters of a Two-Port Circuit

Find the z parameters for the circuit shown in Fig. 18.3.

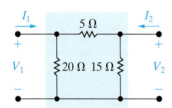

Figure 18.3 ▲ The circuit for Example 18.1.

Solution

The circuit is purely resistive, so the s-domain circuit is also purely resistive. There are many ways to find the two-port parameter values. Here, we note that with port 2 open, that is, $I_2 = 0$, we can apply a 1 A current source at port 1 and use circuit analysis to find V_1 and V_2. With those two voltages, we can find z_{11} and z_{21}. The circuit is shown in Fig. 18.4.

To find V_1 for the circuit in Fig. 18.4, find the equivalent resistance seen by the 1 A current source:

$$Z_{\text{eq}} = 20 \| (5 + 15) = 10 \ \Omega.$$

Then,

$$V_1 = Z_{\text{eq}}(1) = 10 \text{ V}$$

and

$$z_{11} = \left.\frac{V_1}{I_1}\right|_{I_2=0} = \frac{10}{1} = 10 \ \Omega.$$

Now use voltage division to find V_2 for the circuit in Fig. 18.4, noting that the voltage across the series combination of the 5 Ω and 15 Ω resistors is $V_1 = 10$ V:

$$V_2 = \frac{15}{15 + 5} V_1 = 7.5 \text{ V},$$

so

$$z_{21} = \left.\frac{V_2}{I_1}\right|_{I_2=0} = \frac{7.5}{1} = 7.5 \ \Omega.$$

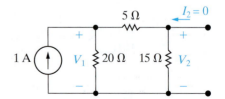

Figure 18.4 ▲ The circuit used to find z_{11} and z_{21} for the circuit in Fig. 18.3.

Next we leave port 1 open, that is, $I_1 = 0$, apply a 1 A current source at port 2, and use circuit analysis to find V_1 and V_2. With those two voltages, we can find z_{22} and z_{12}. The circuit is shown in Fig. 18.5. To find V_2 for the circuit in Fig. 18.5, find the equivalent resistance seen by the 1 A current source:

$$Z_{eq} = 15 \| (5 + 20) = 9.375 \ \Omega.$$

Then,

$$V_2 = Z_{eq}(1) = 9.375 \ V$$

and

$$z_{22} = \frac{V_2}{I_2}\bigg|_{I_1=0} = \frac{9.375}{1} = 9.375 \ \Omega.$$

Now use voltage division to find V_1 for the circuit in Fig. 18.5, noting that the voltage across the

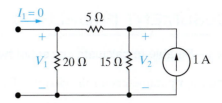

Figure 18.5 ▲ The circuit used to find z_{12} and z_{22} for the circuit in Fig. 18.3.

series combination of the 5 Ω and 20 Ω resistors is $V_2 = 9.375$ V:

$$V_1 = \frac{20}{20 + 5} V_2 = 7.5 \ V,$$

so

$$z_{12} = \frac{V_1}{I_2}\bigg|_{I_1=0} = \frac{7.5}{1} = 7.5 \ \Omega.$$

EXAMPLE 18.2   Finding the *a* Parameters from Measurements

A two-port circuit is operating in the sinusoidal steady state. With port 2 open, a voltage of $150 \cos 4000t$ V is applied to port 1. Two measurements are made: the current into port 1 is $25 \cos(4000t - 45°)$ A, and the voltage across port 2 is $100 \cos(4000t + 15°)$ V. Then, port 2 is short-circuited, and a voltage of $30 \cos 4000t$ V is applied to port 1. Two more measurements are made: the current into port 1 is $1.5 \cos(4000t + 30°)$ A, and the current into port 2 is $0.25 \cos(4000t + 150°)$ A. Find the *a* parameters that describe the sinusoidal steady-state behavior of the circuit.

Solution

Phasor-transforming the first set of measurements gives

$$\mathbf{V}_1 = 150 \underline{/0°} \ V, \quad \mathbf{I}_1 = 25 \underline{/-45°} \ A,$$

$$\mathbf{V}_2 = 100 \underline{/15°} \ V, \quad \mathbf{I}_2 = 0 \ A.$$

From Eqs. 18.12,

$$a_{11} = \frac{\mathbf{V}_1}{\mathbf{V}_2}\bigg|_{I_2=0} = \frac{150\underline{/0°}}{100\underline{/15°}} = 1.5 \underline{/-15°},$$

$$a_{21} = \frac{\mathbf{I}_1}{\mathbf{V}_2}\bigg|_{I_2=0} = \frac{25\underline{/-45°}}{100\underline{/15°}} = 0.25 \underline{/-60°} \ S.$$

Phasor-transforming the second set of measurements gives

$$\mathbf{V}_1 = 30\underline{/0°} \ V, \quad \mathbf{I}_1 = 1.5 \underline{/30°} \ A,$$

$$\mathbf{V}_2 = 0 \ V, \quad \mathbf{I}_2 = 0.25 \underline{/150°} \ A.$$

Therefore

$$a_{12} = -\frac{\mathbf{V}_1}{\mathbf{I}_2}\bigg|_{V_2=0} = \frac{-30\underline{/0°}}{0.25\underline{/150°}} = 120 \underline{/30°} \ \Omega,$$

$$a_{21} = -\frac{\mathbf{I}_1}{\mathbf{I}_2}\bigg|_{V_2=0} = \frac{-1.5\underline{/30°}}{0.25\underline{/150°}} = 6 \underline{/60°}.$$

ASSESSMENT PROBLEMS

Objective 1—Be able to calculate any set of two-port parameters

18.1 Find the y parameters for the circuit in Fig. 18.3.

Answer: $y_{11} = 0.25$ S,

$y_{12} = y_{21} = -0.2$ S,

$y_{22} = \dfrac{4}{15}$ S.

18.2 Find the g and h parameters for the circuit in Fig. 18.3.

Answer: $g_{11} = 0.1$ S; $g_{12} = -0.75$; $g_{21} = 0.75$; $g_{22} = 3.75$ Ω; $h_{11} = 4$ Ω; $h_{12} = 0.8$; $h_{21} = -0.8$; $h_{22} = 0.1067$ S.

18.3 The following measurements were made on a two-port resistive circuit. With 50 mV applied to port 1 and port 2 open, the current into port 1 is 5 μA, and the voltage across port 2 is 200 mV. With port 1 short-circuited and 10 mV applied to port 2, the current into port 1 is 2 μA, and the current into port 2 is 0.5 μA. Find the g parameters of the network.

Answer: $g_{11} = 0.1$ mS;

$g_{12} = 4$;

$g_{21} = 4$;

$g_{22} = 20$ kΩ.

SELF-CHECK: Also try Chapter Problems 18.2, 18.4, and 18.10.

Relationships Among the Two-Port Parameters

All six sets of two-port parameter equations relate to the same variables. Therefore, every set of two-port parameters must be related to every other set of these parameters. In other words, if we know one set of two-port parameters, we can derive all the other sets from the known set. The equations used to calculate any single two-port parameter from any set of two-port parameters are given in Table 18.1.

We do not derive all the relationships listed in Table 18.1 because of the amount of algebra involved. To illustrate the general process, we derive the relationships between the z and y parameters and between the z and a parameters. To find the z parameters as functions of the y parameters, begin by solving Eqs. 18.2 for V_1 and V_2. We then compare the coefficients of I_1 and I_2 in the resulting expressions to the coefficients of I_1 and I_2 in Eqs. 18.1. From Eqs. 18.2,

$$V_1 = \frac{\begin{vmatrix} I_1 & y_{12} \\ I_2 & y_{22} \end{vmatrix}}{\begin{vmatrix} y_{11} & y_{12} \\ y_{21} & y_{22} \end{vmatrix}} = \frac{y_{22}}{\Delta y} I_1 - \frac{y_{12}}{\Delta y} I_2,$$

$$V_2 = \frac{\begin{vmatrix} y_{11} & I_1 \\ y_{21} & I_2 \end{vmatrix}}{\Delta y} = -\frac{y_{21}}{\Delta y} I_1 + \frac{y_{11}}{\Delta y} I_2.$$

Comparing these expressions for V_1 and V_2 with Eqs. 18.1 shows

$$z_{11} = \frac{y_{22}}{\Delta y},$$

$$z_{12} = -\frac{y_{12}}{\Delta y},$$

$$z_{21} = -\frac{y_{21}}{\Delta y},$$

$$z_{22} = \frac{y_{11}}{\Delta y}.$$

TABLE 18.1 Parameter Conversion Table

$$z_{11} = \frac{y_{22}}{\Delta y} = \frac{a_{11}}{a_{21}} = \frac{b_{22}}{b_{21}} = \frac{\Delta h}{h_{22}} = \frac{1}{g_{11}}$$

$$z_{12} = -\frac{y_{12}}{\Delta y} = \frac{\Delta a}{a_{21}} = \frac{1}{b_{21}} = \frac{h_{12}}{h_{22}} = -\frac{g_{12}}{g_{11}}$$

$$z_{21} = \frac{-y_{21}}{\Delta y} = \frac{1}{a_{21}} = \frac{\Delta b}{b_{21}} = -\frac{h_{21}}{h_{22}} = \frac{g_{21}}{g_{11}}$$

$$z_{22} = \frac{y_{11}}{\Delta y} = \frac{a_{22}}{a_{21}} = \frac{b_{11}}{b_{21}} = \frac{1}{h_{22}} = \frac{\Delta g}{g_{11}}$$

$$y_{11} = \frac{z_{22}}{\Delta z} = \frac{a_{22}}{a_{12}} = \frac{b_{11}}{b_{12}} = \frac{1}{h_{11}} = \frac{\Delta g}{g_{22}}$$

$$y_{12} = -\frac{z_{12}}{\Delta z} = -\frac{\Delta a}{a_{12}} = -\frac{1}{b_{12}} = -\frac{h_{12}}{h_{11}} = \frac{g_{12}}{g_{22}}$$

$$y_{21} = -\frac{z_{21}}{\Delta z} = -\frac{1}{a_{12}} = -\frac{\Delta b}{b_{12}} = \frac{h_{21}}{h_{11}} = -\frac{g_{21}}{g_{22}}$$

$$y_{22} = \frac{z_{11}}{\Delta z} = \frac{a_{11}}{a_{12}} = \frac{b_{22}}{b_{12}} = \frac{\Delta h}{h_{11}} = \frac{1}{g_{22}}$$

$$a_{11} = \frac{z_{11}}{z_{21}} = -\frac{y_{22}}{y_{21}} = \frac{b_{22}}{\Delta b} = -\frac{\Delta h}{h_{21}} = \frac{1}{g_{21}}$$

$$a_{12} = \frac{\Delta z}{z_{21}} = -\frac{1}{y_{21}} = \frac{b_{12}}{\Delta b} = -\frac{h_{11}}{h_{21}} = \frac{g_{22}}{g_{21}}$$

$$a_{21} = \frac{1}{z_{21}} = -\frac{\Delta y}{y_{21}} = \frac{b_{21}}{\Delta b} = -\frac{h_{22}}{h_{21}} = \frac{g_{11}}{g_{21}}$$

$$a_{22} = \frac{z_{22}}{z_{21}} = -\frac{y_{11}}{y_{21}} = \frac{b_{11}}{\Delta b} = -\frac{1}{h_{21}} = \frac{\Delta g}{g_{21}}$$

$$b_{11} = \frac{z_{22}}{z_{12}} = -\frac{y_{11}}{y_{12}} = \frac{a_{22}}{\Delta a} = \frac{1}{h_{12}} = -\frac{\Delta g}{g_{12}}$$

$$b_{12} = \frac{\Delta z}{z_{12}} = -\frac{1}{y_{12}} = \frac{a_{12}}{\Delta a} = \frac{h_{11}}{h_{12}} = -\frac{g_{22}}{g_{12}}$$

$$b_{21} = \frac{1}{z_{12}} = -\frac{\Delta y}{y_{12}} = \frac{a_{21}}{\Delta a} = \frac{h_{22}}{h_{12}} = -\frac{g_{11}}{g_{12}}$$

$$b_{22} = \frac{z_{11}}{z_{12}} = \frac{y_{22}}{y_{12}} = \frac{a_{11}}{\Delta a} = \frac{\Delta h}{h_{12}} = -\frac{1}{g_{12}}$$

$$h_{11} = \frac{\Delta z}{z_{22}} = \frac{1}{y_{11}} = \frac{a_{12}}{a_{22}} = \frac{b_{12}}{b_{11}} = \frac{g_{22}}{\Delta g}$$

$$h_{12} = \frac{z_{12}}{z_{22}} = -\frac{y_{12}}{y_{11}} = \frac{\Delta a}{a_{22}} = \frac{1}{b_{11}} = -\frac{g_{12}}{\Delta g}$$

$$h_{21} = -\frac{z_{21}}{z_{22}} = \frac{y_{21}}{y_{11}} = -\frac{1}{a_{22}} = -\frac{\Delta b}{b_{11}} = -\frac{g_{21}}{\Delta g}$$

$$h_{22} = \frac{1}{z_{22}} = \frac{\Delta y}{y_{11}} = \frac{a_{21}}{a_{22}} = \frac{b_{21}}{b_{11}} = \frac{g_{11}}{\Delta g}$$

$$g_{11} = \frac{1}{z_{11}} = \frac{\Delta y}{y_{22}} = \frac{a_{21}}{a_{11}} = \frac{b_{21}}{b_{22}} = \frac{h_{22}}{\Delta h}$$

$$g_{12} = -\frac{z_{12}}{z_{11}} = \frac{y_{12}}{y_{22}} = -\frac{\Delta a}{a_{11}} = -\frac{1}{b_{22}} = -\frac{h_{12}}{\Delta h}$$

$$g_{21} = \frac{z_{21}}{z_{11}} = -\frac{y_{21}}{y_{22}} = \frac{1}{a_{11}} = \frac{\Delta b}{b_{22}} = -\frac{h_{21}}{\Delta h}$$

$$g_{22} = \frac{\Delta z}{z_{11}} = \frac{1}{y_{22}} = \frac{a_{12}}{a_{11}} = \frac{b_{12}}{b_{22}} = \frac{h_{11}}{\Delta h}$$

$$\Delta z = z_{11}z_{22} - z_{12}z_{21}$$

$$\Delta y = y_{11}y_{22} - y_{12}y_{21}$$

$$\Delta a = a_{11}a_{22} - a_{12}a_{21}$$

$$\Delta b = b_{11}b_{22} - b_{12}b_{21}$$

$$\Delta h = h_{11}h_{22} - h_{12}h_{21}$$

$$\Delta g = g_{11}g_{22} - g_{12}g_{21}$$

To find the z parameters as functions of the a parameters, we rearrange Eqs. 18.3 in the form of Eqs. 18.1 and then compare coefficients. From the second equation in Eqs. 18.3,

$$V_2 = \frac{1}{a_{21}} I_1 + \frac{a_{22}}{a_{21}} I_2.$$

Therefore, substituting this expression for V_2 into the first equation of Eqs. 18.3 yields

$$V_1 = \frac{a_{11}}{a_{21}} I_1 + \left(\frac{a_{11}a_{22}}{a_{21}} - a_{12} \right) I_2.$$

Comparing these expressions for V_1 and V_2 with Eqs. 18.1 gives

$$z_{11} = \frac{a_{11}}{a_{21}},$$

$$z_{12} = \frac{\Delta a}{a_{21}}.$$

$$z_{21} = \frac{1}{a_{21}},$$

$$z_{22} = \frac{a_{22}}{a_{21}}.$$

Example 18.3 illustrates the usefulness of the parameter conversion table.

EXAMPLE 18.3   Finding *h* Parameters from Measurements and Table 18.1

Two sets of measurements are made on a two-port resistive circuit. The first set is made with port 2 open, and the second set is made with port 2 short-circuited. The results are as follows:

Port 2 Open	Port 2 Short-Circuited
$V_1 = 10$ mV	$V_1 = 24$ mV
$I_1 = 10$ μA	$I_1 = 20$ μA
$V_2 = -40$ V	$I_2 = 1$ mA

Find the *h* parameters of the circuit.

Solution

We can find h_{11} and h_{21} directly from the short-circuit test:

$$h_{11} = \frac{V_1}{I_1}\bigg|_{V_2=0}$$

$$= \frac{24 \times 10^{-3}}{20 \times 10^{-6}} = 1.2 \text{ k}\Omega,$$

$$h_{21} = \frac{I_2}{I_1}\bigg|_{V_2=0}$$

$$= \frac{10^{-3}}{20 \times 10^{-6}} = 50.$$

The parameters h_{12} and h_{22} cannot be obtained directly from the open-circuit test. However, a check of Eqs. 18.7–18.15 indicates that the four *a* parameters can be derived from the test data. Therefore, h_{12} and h_{22} can be obtained through the conversion table. Specifically,

$$h_{12} = \frac{\Delta a}{a_{22}},$$

$$h_{22} = \frac{a_{21}}{a_{22}}.$$

The *a* parameters are

$$a_{11} = \frac{V_1}{V_2}\bigg|_{I_2=0} = \frac{10 \times 10^{-3}}{-40} = -0.25 \times 10^{-3},$$

$$a_{21} = \frac{I_1}{V_2}\bigg|_{I_2=0} = \frac{10 \times 10^{-6}}{-40} = -0.25 \times 10^{-6}\text{ S},$$

$$a_{12} = -\frac{V_1}{I_2}\bigg|_{V_2=0} = -\frac{24 \times 10^{-3}}{10^{-3}} = -24 \text{ }\Omega,$$

$$a_{22} = -\frac{I_1}{I_2}\bigg|_{V_2=0} = -\frac{20 \times 10^{-6}}{10^{-3}} = -20 \times 10^{-3}.$$

The numerical value of Δa is

$$\Delta a = a_{11}a_{22} - a_{12}a_{21}$$

$$= 5 \times 10^{-6} - 6 \times 10^{-6} = -10^{-6}.$$

Thus

$$h_{12} = \frac{\Delta a}{a_{22}}$$

$$= \frac{-10^{-6}}{-20 \times 10^{-3}} = 5 \times 10^{-5},$$

$$h_{22} = \frac{a_{21}}{a_{22}}$$

$$= \frac{-0.25 \times 10^{-6}}{-20 \times 10^{-3}} = 12.5 \text{ }\mu\text{S}.$$

ASSESSMENT PROBLEM

Objective 1—Be able to calculate any set of two-port parameters

18.4 The following measurements were made on a two-port resistive circuit: With port 1 open, $V_2 = 15$ V, $V_1 = 10$ V, and $I_2 = 30$ A; with port 1 short-circuited, $V_2 = 10$ V, $I_2 = 4$ A, and $I_1 = -5$ A. Calculate the *z* parameters.

Answer: $z_{11} = (4/15)$ Ω;

$z_{12} = (1/3)$ Ω;

$z_{21} = -1.6$ Ω;

$z_{22} = 0.5$ Ω.

SELF-CHECK: Also try Chapter Problem 18.13.

Reciprocal Two-Port Circuits

A two-port circuit is **reciprocal** if interchanging an ideal voltage source at one port with an ideal ammeter at the other port produces the same ammeter reading. A two-port circuit is also reciprocal if interchanging an ideal current source at one port with an ideal voltmeter at the other port produces the same voltmeter reading. Relationships exist among reciprocal two-port parameters, as shown in Table 18.2. For a reciprocal two-port circuit, only three calculations or measurements are needed to determine a set of parameters.

A reciprocal two-port circuit is **symmetric** if its ports can be interchanged without altering terminal current and voltage values. Figure 18.6 shows four examples of symmetric two-port circuits. In these circuits, additional relationships exist among the two-port parameters, as shown in Table 18.2. If a circuit is both reciprocal and symmetric, then only two calculations or measurements are necessary to determine all the two-port parameters. Example 18.4 determines whether a circuit is reciprocal and symmetric.

TABLE 18.2	Two-Port Parameter Relationships for Reciprocal Circuits

Reciprocal Circuits

$z_{12} = z_{21}$

$y_{12} = y_{21}$

$a_{11}a_{22} - a_{12}a_{21} = \Delta a = 1$

$b_{11}b_{22} - b_{12}b_{21} = \Delta b = 1$

$h_{12} = -h_{21}$

$g_{12} = -g_{21}$

Reciprocal and Symmetric Circuits

$z_{11} = z_{22}$

$y_{11} = y_{22}$

$a_{11} = a_{22}$

$b_{11} = b_{22}$

$h_{11}h_{22} - h_{12}h_{21} = \Delta h = 1$

$g_{11}g_{22} - g_{12}g_{21} = \Delta g = 1$

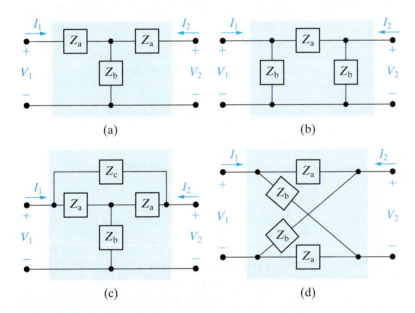

Figure 18.6 ▲ Four examples of symmetric two-port circuits. (a) A symmetric tee. (b) A symmetric pi. (c) A symmetric bridged tee. (d) A symmetric lattice.

EXAMPLE 18.4   **Determining Whether a Circuit Is Reciprocal and Symmetric**

Consider the resistive circuit shown in Fig. 18.7.

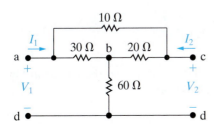

Figure 18.7 ▲ The circuit for Example 18.4.

a) Attach a 15 V source to the terminals a and d and calculate the reading on an ammeter attached between terminals c and d. Then attach a 15 V source to the terminals c and d and calculate the reading on an ammeter attached between terminals a and d. Is the circuit reciprocal?

b) Find the *y* parameters for the circuit and determine whether or not it is symmetric.

Solution

a) A 15 V source is attached between nodes a and d, and an ammeter is attached between terminals c and d. The resulting circuit is shown in Fig. 18.8. To find the ammeter current, begin by writing a KCL equation at node b to determine the voltage between nodes b and d, V_{bd}. Remember that an ideal ammeter behaves like a short circuit, so

$$\frac{V_{bd}}{60} + \frac{V_{bd} - 15}{30} + \frac{V_{bd}}{20} = 0,$$

and $V_{bd} = 5$ V. Therefore, the current I in the ammeter is the sum of the current in the 10 Ω resistor and the current in the 20 Ω resistor:

$$I = \frac{5}{20} + \frac{15}{10} = 1.75 \text{ A}.$$

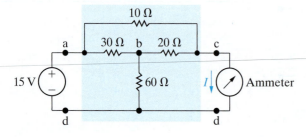

Figure 18.8 ▲ The circuit in Fig. 18.7, with a voltage source and an ammeter attached.

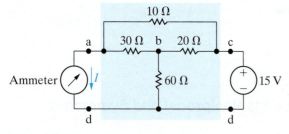

Figure 18.9 ▲ The circuit shown in Fig. 18.7, with the voltage source and ammeter interchanged.

Now interchange the voltage source and ammeter as shown in Fig. 18.9. Again, we write a KCL equation at node b to determine the voltage V_{bd},

remembering that the ideal ammeter behaves like a short circuit. Then,

$$\frac{V_{bd}}{60} + \frac{V_{bd}}{30} + \frac{V_{bd} - 15}{20} = 0,$$

so $V_{bd} = 7.5$ V. The ammeter current I is the sum of the current in the 30 Ω resistor and the current in the 20 Ω resistor:

$$I = \frac{7.5}{30} + \frac{15}{10} = 1.75 \text{ A}.$$

Since both ammeter readings are the same, the circuit in Fig. 18.7 is reciprocal.

b) We can use the analysis results for the circuit in Fig. 18.8 to find two of the y parameters. Attaching the ammeter between nodes c and d is the same as setting $V_2 = 0$ in Fig. 18.7. Also, I_2 in Fig. 18.7 equals $-I$ in Fig. 18.8, and I_1 in Fig. 18.7 is the sum of the currents in the 30 Ω and 20 Ω resistors in Fig. 18.8. Therefore,

$$I_2 = -I = -1.75 \text{ A},$$

$$I_1 = \frac{15 - 5}{30} + \frac{15}{10} = \frac{55}{30} \text{ A},$$

so

$$y_{11} = \frac{I_1}{V_1}\bigg|_{V_2=0} = \frac{55/30}{15} = \frac{11}{99} \text{ S},$$

$$y_{21} = \frac{I_2}{V_1}\bigg|_{V_2=0} = \frac{-1.75}{15} = -\frac{7}{60} \text{ S}.$$

We have already shown that the circuit in Fig. 18.7 is reciprocal, so $y_{12} = y_{21} = -7/60$ S.

We can use the analysis results for the circuit in Fig. 18.9 to find y_{22}, because attaching the ammeter between nodes a and d is the same as setting $V_1 = 0$ in Fig. 18.7. Also, I_1 in Fig. 18.7 is the sum of the currents in the 10 Ω and 20 Ω resistors in Fig. 18.9. Therefore,

$$I_1 = \frac{15 - 7.5}{20} + \frac{15}{10} = 1.875 \text{ A}$$

so

$$y_{22} = \frac{I_2}{V_2}\bigg|_{V_1=0} = \frac{1.875}{15} = 0.125 \text{ S}.$$

Note that $y_{11} \neq y_{22}$, so the circuit in Fig. 18.7 is not symmetric.

ASSESSMENT PROBLEM

Objective 1—Be able to calculate any set of two-port parameters

18.5 The following measurements were made on a resistive two-port network that is symmetric and reciprocal: With port 2 open, $V_1 = 95$ V and $I_1 = 5$ A; with a short circuit across port 2,

$V_1 = 11.52$ V and $I_2 = -2.72$ A. Calculate the z parameters of the two-port network.

Answer: $z_{11} = z_{22} = 19\ \Omega,\ z_{12} = z_{21} = 17\ \Omega.$

SELF-CHECK: Also try Chapter Problem 18.14.

18.3 Analysis of the Terminated Two-Port Circuit

A terminated two-port circuit usually has a source attached at port 1 and a load attached at port 2, as shown in the *s*-domain circuit of Fig. 18.10. In this circuit, Z_g is the internal source impedance, V_g the internal source voltage, and Z_L is the load impedance. To analyze this circuit, we find the terminal currents and voltages as functions of the two-port parameters, V_g, Z_g, and Z_L.

Six characteristics of the terminated two-port circuit define its terminal behavior:

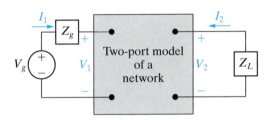

Figure 18.10 ▲ A terminated two-port model.

- the input impedance $Z_{\text{in}} = V_1/I_1$, or the admittance $Y_{\text{in}} = I_1/V_1$;
- the output current I_2;
- the Thévenin voltage and impedance (V_{Th}, Z_{Th}) with respect to port 2;
- the current gain I_2/I_1;
- the voltage gain V_2/V_1;
- the voltage gain V_2/V_g.

The Six Characteristics in Terms of the *z* Parameters

To illustrate how these six characteristics are derived, we find their expressions using the z parameters of the two-port portion of the circuit. Table 18.3 summarizes the expressions involving the y, a, b, h, and g parameters.

We find the expression for each characteristic using one set of two-port equations, along with the two constraint equations imposed by the source applied at port 1 and the load applied at port 2. Using the z parameters, the circuit in Fig. 18.10 is described by the following four equations:

$$V_1 = z_{11}I_1 + z_{12}I_2, \tag{18.16}$$

$$V_2 = z_{21}I_1 + z_{22}I_2, \tag{18.17}$$

$$V_1 = V_g - I_1Z_g, \tag{18.18}$$

$$V_2 = -I_2Z_L. \tag{18.19}$$

TABLE 18.3 Terminated Two-Port Equations

z Parameters

$$Z_{in} = z_{11} - \frac{z_{12}z_{21}}{z_{22} + Z_L}$$

$$I_2 = \frac{-z_{21}V_g}{(z_{11} + Z_g)(z_{22} + Z_L) - z_{12}z_{21}}$$

$$V_{Th} = \frac{z_{21}}{z_{11} + Z_g}V_g$$

$$Z_{Th} = z_{22} - \frac{z_{12}z_{21}}{z_{11} + Z_g}$$

$$\frac{I_2}{I_1} = \frac{-z_{21}}{z_{22} + Z_L}$$

$$\frac{V_2}{V_1} = \frac{z_{21}Z_L}{z_{11}Z_L + \Delta z}$$

$$\frac{V_2}{V_g} = \frac{z_{21}Z_L}{(z_{11} + Z_g)(z_{22} + Z_L) - z_{12}z_{21}}$$

y Parameters

$$Y_{in} = y_{11} - \frac{y_{12}y_{21}Z_L}{1 + y_{22}Z_L}$$

$$I_2 = \frac{y_{21}V_g}{1 + y_{22}Z_L + y_{11}Z_g + \Delta y Z_g Z_L}$$

$$V_{Th} = \frac{-y_{21}V_g}{y_{22} + \Delta y Z_g}$$

$$Z_{Th} = \frac{1 + y_{11}Z_g}{y_{22} + \Delta y Z_g}$$

$$\frac{I_2}{I_1} = \frac{y_{21}}{y_{11} + \Delta y Z_L}$$

$$\frac{V_2}{V_1} = \frac{-y_{21}Z_L}{1 + y_{22}Z_L}$$

$$\frac{V_2}{V_g} = \frac{y_{21}Z_L}{y_{12}y_{21}Z_gZ_L - (1 + y_{11}Z_g)(1 + y_{22}Z_L)}$$

a Parameters

$$Z_{in} = \frac{a_{11}Z_L + a_{12}}{a_{21}Z_L + a_{22}}$$

$$I_2 = \frac{-V_g}{a_{11}Z_L + a_{12} + a_{21}Z_g Z_L + a_{22}Z_g}$$

$$V_{Th} = \frac{V_g}{a_{11} + a_{21}Z_g}$$

$$Z_{Th} = \frac{a_{12} + a_{22}Z_g}{a_{11} + a_{21}Z_g}$$

$$\frac{I_2}{I_1} = \frac{-1}{a_{21}Z_L + a_{22}}$$

$$\frac{V_2}{V_1} = \frac{Z_L}{a_{11}Z_L + a_{12}}$$

$$\frac{V_2}{V_g} = \frac{Z_L}{(a_{11} + a_{21}Z_g)Z_L + a_{12} + a_{22}Z_g}$$

b Parameters

$$Z_{in} = \frac{b_{22}Z_L + b_{12}}{b_{21}Z_L + b_{11}}$$

$$I_2 = \frac{-V_g\Delta b}{b_{11}Z_g + b_{21}Z_g Z_L + b_{22}Z_L + b_{12}}$$

$$V_{Th} = \frac{V_g\Delta b}{b_{22} + b_{21}Z_g}$$

$$Z_{Th} = \frac{b_{11}Z_g + b_{12}}{b_{21}Z_g + b_{22}}$$

$$\frac{I_2}{I_1} = \frac{-\Delta b}{b_{11} + b_{21}Z_L}$$

$$\frac{V_2}{V_1} = \frac{\Delta b Z_L}{b_{12} + b_{22}Z_L}$$

$$\frac{V_2}{V_g} = \frac{\Delta b Z_L}{b_{12} + b_{11}Z_g + b_{22}Z_L + b_{21}Z_g Z_L}$$

h Parameters

$$Z_{in} = h_{11} - \frac{h_{12}h_{21}Z_L}{1 + h_{22}Z_L}$$

$$I_2 = \frac{h_{21}V_g}{(1 + h_{22}Z_L)(h_{11} + Z_g) - h_{12}h_{21}Z_L}$$

$$V_{Th} = \frac{-h_{21}V_g}{h_{22}Z_g + \Delta h}$$

$$Z_{Th} = \frac{Z_g + h_{11}}{h_{22}Z_g + \Delta h}$$

$$\frac{I_2}{I_1} = \frac{h_{21}}{1 + h_{22}Z_L}$$

$$\frac{V_2}{V_1} = \frac{-h_{21}Z_L}{\Delta h Z_L + h_{11}}$$

$$\frac{V_2}{V_g} = \frac{-h_{21}Z_L}{(h_{11} + Z_g)(1 + h_{22}Z_L) - h_{12}h_{21}Z_L}$$

g Parameters

$$Y_{in} = g_{11} - \frac{g_{12}g_{21}}{g_{22} + Z_L}$$

$$I_2 = \frac{-g_{21}V_g}{(1 + g_{11}Z_g)(g_{22} + Z_L) - g_{12}g_{21}Z_g}$$

$$V_{Th} = \frac{g_{21}V_g}{1 + g_{11}Z_g}$$

$$Z_{Th} = g_{22} - \frac{g_{12}g_{21}Z_g}{1 + g_{11}Z_g}$$

$$\frac{I_2}{I_1} = \frac{-g_{21}}{g_{11}Z_L + \Delta g}$$

$$\frac{V_2}{V_1} = \frac{g_{21}Z_L}{g_{22} + Z_L}$$

$$\frac{V_2}{V_g} = \frac{g_{21}Z_L}{(1 + g_{11}Z_g)(g_{22} + Z_L) - g_{12}g_{21}Z_g}$$

Let's find the impedance seen looking into port 1, that is, $Z_{\text{in}} = V_1/I_1$. In Eq. 18.17, replace V_2 with $-I_2 Z_L$ and solve the resulting expression for I_2:

$$I_2 = \frac{-z_{21}I_1}{Z_L + z_{22}}. \qquad (18.20)$$

Then substitute this equation into Eq. 18.16 and solve for Z_{in}:

$$Z_{\text{in}} = z_{11} - \frac{z_{12}z_{21}}{z_{22} + Z_L}.$$

To find the output current I_2, we first solve Eq. 18.16 for I_1 after replacing V_1 with the right-hand side of Eq. 18.18. The result is

$$I_1 = \frac{V_g - z_{12}I_2}{z_{11} + Z_g}. \qquad (18.21)$$

We now substitute Eq. 18.21 into Eq. 18.20 and solve the resulting equation for I_2:

$$I_2 = \frac{-z_{21}V_g}{(z_{11} + Z_g)(z_{22} + Z_L) - z_{12}z_{21}}.$$

The Thévenin voltage with respect to port 2 equals V_2 when $I_2 = 0$. With $I_2 = 0$, Eq. 18.17 reduces to

$$V_2\big|_{I_2=0} = z_{21}I_1. \qquad (18.22)$$

But when $I_2 = 0$, Eq. 18.21 becomes $I_1 = V_g/(z_{11} + Z_g)$. Substituting this expression for I_1 into Eq. 18.22 yields the open-circuit value of V_2:

$$V_2\big|_{I_2=0} = V_{\text{Th}} = \frac{z_{21}}{Z_g + z_{11}}V_g.$$

The Thévenin, or output, impedance is the ratio V_2/I_2 when V_g is replaced by a short circuit. When V_g is zero, Eq. 18.18 reduces to

$$V_1 = -I_1 Z_g. \qquad (18.23)$$

Substituting Eq. 18.23 into Eq. 18.16 and solving for I_1 gives

$$I_1 = \frac{-z_{12}I_2}{z_{11} + Z_g}.$$

Now use this expression to replace I_1 in Eq. 18.17, resulting in

$$\frac{V_2}{I_2}\bigg|_{V_g=0} = Z_{\text{Th}} = z_{22} - \frac{z_{12}z_{21}}{z_{11} + Z_g}.$$

The current gain I_2/I_1 comes directly from Eq. 18.20:

$$\frac{I_2}{I_1} = \frac{-z_{21}}{Z_L + z_{22}}.$$

To derive the expression for the voltage gain V_2/V_1, we start by replacing I_2 in Eq. 18.17 with its value from Eq. 18.19; thus

$$V_2 = z_{21}I_1 + z_{22}\left(\frac{-V_2}{Z_L}\right). \tag{18.24}$$

Next, solve Eq. 18.16 for I_1 as a function of V_1 and V_2:

$$z_{11}I_1 = V_1 - z_{12}\left(\frac{-V_2}{Z_L}\right)$$

or

$$I_1 = \frac{V_1}{z_{11}} + \frac{z_{12}V_2}{z_{11}Z_L}. \tag{18.25}$$

Finally, replace I_1 in Eq. 18.24 with Eq. 18.25 and solve the resulting expression for V_2/V_1:

$$\frac{V_2}{V_1} = \frac{z_{21}Z_L}{z_{11}Z_L + z_{11}z_{22} - z_{12}z_{21}}$$

$$= \frac{z_{21}Z_L}{z_{11}Z_L + \Delta z}.$$

To derive the voltage ratio V_2/V_g, we first combine Eqs. 18.16, 18.18, and 18.19 to find I_1 as a function of V_2 and V_g:

$$I_1 = \frac{z_{12}V_2}{Z_L(z_{11} + Z_g)} + \frac{V_g}{z_{11} + Z_g}. \tag{18.26}$$

Then, starting with Eq. 18.17, use Eq. 18.26 to substitute for I_1 and Eq. 18.19 to substitute for I_2. The result is an expression involving only V_2 and V_g:

$$V_2 = \frac{z_{21}z_{12}V_2}{Z_L(z_{11} + Z_g)} + \frac{z_{21}V_g}{z_{11} + Z_g} - \frac{z_{22}}{Z_L}V_2,$$

which we can rearrange to get the desired voltage ratio:

$$\frac{V_2}{V_g} = \frac{z_{21}Z_L}{(z_{11} + Z_g)(z_{22} + Z_L) - z_{12}z_{21}}.$$

Example 18.5 illustrates the usefulness of the relationships listed in Table 18.3.

EXAMPLE 18.5 | Analyzing a Terminated Two-Port Circuit

We describe the two-port circuit shown in Fig. 18.11 using its b parameters, whose values are

$$b_{11} = -20, \qquad b_{12} = -3000 \ \Omega,$$

$$b_{21} = -2 \ \text{mS}, \quad b_{22} = -0.2.$$

a) Find the phasor voltage $\mathbf{V}_2$.
b) Find the average power delivered to the 5 kΩ load.
c) Find the average power delivered to the input port.
d) Find the load impedance for maximum average power transfer.
e) Find the maximum average power delivered to the load in (d).

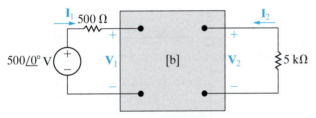

Figure 18.11 ▲ The circuit for Example 18.5.

Solution

a) To find $\mathbf{V}_2$, we have two choices from the entries in Table 18.3. We could find $\mathbf{I}_2$ and then find $\mathbf{V}_2$ from the relationship $\mathbf{V}_2 = -\mathbf{I}_2 Z_L$, or we could find the voltage gain $\mathbf{V}_2/\mathbf{V}_g$ and calculate $\mathbf{V}_2$ from the gain. Let's use the latter approach. For the b-parameter values given, we have

$$\Delta b = b_{11}b_{22} - b_{12}b_{21}$$

$$= (-20)(-0.2) - (-3000)(-0.002) = -2.$$

From Table 18.3,

$$\frac{\mathbf{V}_2}{\mathbf{V}_g} = \frac{\Delta b Z_L}{b_{12} + b_{11}Z_g + b_{22}Z_L + b_{21}Z_g Z_L}$$

$$= \frac{(-2)(5000)}{-3000 + (-20)500 + (-0.2)5000 + [-0.002(500)(5000)]}$$

$$= \frac{10}{19}.$$

Then,

$$\mathbf{V}_2 = \left(\frac{10}{19}\right)500 = 263.16\underline{/0°} \ \text{V}.$$

b) The average power delivered to the 5000 Ω load is

$$P_2 = \frac{|\mathbf{V}_2|^2}{2(5000)} = \frac{263.16^2}{2(5000)} = 6.93 \ \text{W}.$$

c) To find the average power delivered to the input port, we first find the input impedance Z_{in}. From Table 18.3,

$$Z_{\text{in}} = \frac{b_{22}Z_L + b_{12}}{b_{21}Z_L + b_{11}}$$

$$= \frac{(-0.2)(5000) - 3000}{-2 \times 10^{-3}(5000) - 20}$$

$$= \frac{400}{3} = 133.33 \ \Omega.$$

Now $\mathbf{I}_1$ follows directly:

$$\mathbf{I}_1 = \frac{\mathbf{V}_g}{Z_g + Z_{\text{in}}} = \frac{500}{500 + 133.33} = 789.47 \ \text{mA}.$$

The average power delivered to the input port is

$$P_1 = \frac{|\mathbf{I}_1|^2 Z_{\text{in}}}{2} = \frac{(0.78947)^2(133.33)}{2} = 41.55 \ \text{W}.$$

d) The load impedance for maximum power transfer equals the conjugate of the Thévenin impedance seen looking into port 2. From Table 18.3,

$$Z_{\text{Th}} = \frac{b_{11}Z_g + b_{12}}{b_{21}Z_g + b_{22}}$$

$$= \frac{(-20)(500) - 3000}{(-2 \times 10^{-3})(500) - 0.2}$$

$$= \frac{13,000}{1.2} = 10,833.33 \ \Omega.$$

Therefore $Z_L = Z_{\text{Th}}^* = 10,833.33 \ \Omega.$

e) To find the maximum average power delivered to Z_L, we first find $\mathbf{V}_2$ from the voltage-gain expression $\mathbf{V}_2/\mathbf{V}_g$. When Z_L is 10,833.33 Ω, this gain is

and

$$P_L(\text{maximum}) = \frac{1}{2}\frac{416.67^2}{10,833.33}$$
$$= 8.01 \text{ W}.$$

$$\frac{\mathbf{V}_2}{\mathbf{V}_g} = \frac{\Delta b Z_L}{b_{12} + b_{11}Z_g + b_{22}Z_L + b_{21}Z_g Z_L}$$

$$= \frac{(-2)(10{,}833.33)}{(-3000) + (-20)(500) + (-0.2)(10{,}833.33) + (-0.002)(500)(10{,}833.33)}$$

$$= 0.8333.$$

Thus

$$\mathbf{V}_2 = (0.8333)(500) = 416.67 \text{ V},$$

ASSESSMENT PROBLEM

Objective 2—Be able to analyze a terminated two-port circuit to find currents, voltages, and ratios of interest

18.6 The a parameters of the two-port network shown are $a_{11} = 5 \times 10^{-4}$, $a_{12} = 10\ \Omega$, $a_{21} = 10^{-6}$ S, and $a_{22} = -3 \times 10^{-2}$. The network is driven by a sinusoidal voltage source having a maximum amplitude of 50 mV and an internal impedance of $100 + j0\ \Omega$. It is terminated in a resistive load of 5 kΩ.
  a) Calculate the average power delivered to the load resistor.
  b) Calculate the load resistance for maximum average power.
  c) Calculate the maximum average power delivered to the resistor in (b).

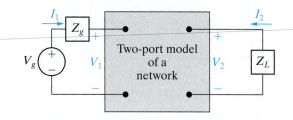

Answer: (a) 62.5 mW;
(b) 70/6 kΩ;
(c) 74.4 mW.

SELF-CHECK: Also try Chapter Problems 18.29, 18.31, and 18.33.

18.4 Interconnected Two-Port Circuits

Synthesizing a large, complex system is usually made easier by first designing subsections of the system and then interconnecting the subsections. If the subsections are modeled by two-port circuits, we will need to analyze interconnected two-port circuits to complete the system design.

Two-port subsystem circuits can be interconnected in five ways: (1) in cascade, (2) in series, (3) in parallel, (4) in series-parallel, and (5) in parallel-series. Figure 18.12 depicts these five basic interconnections.

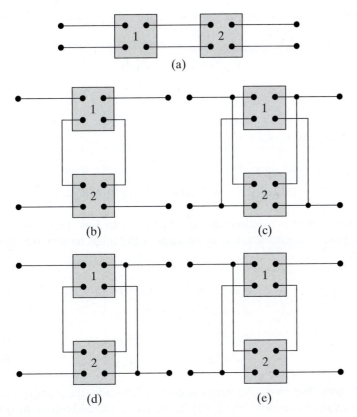

Figure 18.12 ▲ The five basic interconnections of two-port circuits. (a) Cascade. (b) Series. (c) Parallel. (d) Series-parallel. (e) Parallel-series.

We analyze and illustrate only the cascade connection in this section. The cascade connection is important because it occurs frequently in the modeling of large systems. The a parameters are best suited for describing the cascade connection. We analyze the cascade connection using the circuit shown in Fig. 18.13, where a single prime denotes a parameters in the first circuit and a double prime denotes a parameters in the second circuit. The output voltage and current of the first circuit are labeled V'_2 and I'_2, and the input voltage and current of the second circuit are labeled V'_1 and I'_1. The problem is to derive the a-parameter equations that relate V_2 and I_2 to V_1 and I_1. In other words, we want to construct the equations

$$V_1 = a_{11}V_2 - a_{12}I_2,$$
$$I_1 = a_{21}V_2 - a_{22}I_2, \tag{18.27}$$

where the a parameters are given explicitly in terms of the a parameters of the individual circuits.

We begin the derivation by noting from Fig. 18.13 that

$$V_1 = a'_{11}V'_2 - a'_{12}I'_2,$$
$$I_1 = a'_{21}V'_2 - a'_{22}I'_2. \tag{18.28}$$

The interconnection means that $V'_2 = V'_1$ and $I'_2 = -I'_1$. Substituting these constraints into Eqs. 18.28 yields

$$V_1 = a'_{11}V'_1 + a'_{12}I'_1,$$
$$I_1 = a'_{21}V'_1 + a'_{22}I'_1. \tag{18.29}$$

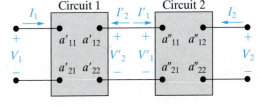

Figure 18.13 ▲ A cascade connection.

The voltage V'_1 and the current I'_1 are related to V_2 and I_2 through the a parameters of the second circuit:

$$V'_1 = a''_{11}V_2 - a''_{12}I_2,$$
$$I'_1 = a''_{21}V_2 - a''_{22}I_2. \tag{18.30}$$

We substitute Eqs. 18.30 into Eqs. 18.29 to generate the relationships between V_1, I_1 and V_2, I_2:

$$V_1 = (a'_{11}a''_{11} + a'_{12}a''_{21})V_2 - (a'_{11}a''_{12} + a'_{12}a''_{22})I_2,$$
$$I_1 = (a'_{21}a''_{11} + a'_{22}a''_{21})V_2 - (a'_{21}a''_{12} + a'_{22}a''_{22})I_2.$$

By comparing these expressions for V_1 and I_1 to Eqs. 18.27, we get the desired expressions for the a parameters of the interconnected networks, namely,

$$a_{11} = a'_{11}a''_{11} + a'_{12}a''_{21}, \tag{18.31}$$

$$a_{12} = a'_{11}a''_{12} + a'_{12}a''_{22}, \tag{18.32}$$

$$a_{21} = a'_{21}a''_{11} + a'_{22}a''_{21}, \tag{18.33}$$

$$a_{22} = a'_{21}a''_{12} + a'_{22}a''_{22}. \tag{18.34}$$

If more than two units are connected in cascade, the a parameters of the equivalent two-port circuit can be found by successively reducing the original set of two-port circuits one pair at a time.

Example 18.6 illustrates how to use Eqs. 18.31–18.34 to analyze a cascade connection with two amplifier circuits.

EXAMPLE 18.6   Analyzing Cascaded Two-Port Circuits

Two identical amplifiers are connected in cascade, as shown in Fig. 18.14. Each amplifier is described using its h parameters. The values are $h_{11} = 1000\ \Omega$, $h_{12} = 0.0015$, $h_{21} = 100$, and $h_{22} = 100\ \mu S$. Find the voltage gain V_2/V_g.

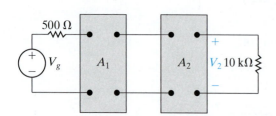

Figure 18.14 ▲ The circuit for Example 18.6.

Solution

The first step in finding V_2/V_g is to convert from h parameters to a parameters. The amplifiers are identical, so one set of a parameters describes the amplifiers:

$$a'_{11} = \frac{-\Delta h}{h_{21}} = \frac{+0.05}{100} = 5 \times 10^{-4},$$

$$a'_{12} = \frac{-h_{11}}{h_{21}} = \frac{-1000}{100} = -10\ \Omega,$$

$$a'_{21} = \frac{-h_{22}}{h_{21}} = \frac{-100 \times 10^{-6}}{100} = -10^{-6}\ S,$$

$$a'_{22} = \frac{-1}{h_{21}} = \frac{-1}{100} = -10^{-2}.$$

Next we use Eqs. 18.31–18.34 to compute the a parameters of the cascaded amplifiers:

$$a_{11} = a'_{11}a'_{11} + a'_{12}a'_{21}$$
$$= 25 \times 10^{-8} + (-10)(-10^{-6})$$
$$= 10.25 \times 10^{-6},$$

$$a_{12} = a'_{11}a'_{12} + a'_{12}a'_{22}$$

$$= (5 \times 10^{-4})(-10) + (-10)(-10^{-2})$$

$$= 0.095 \ \Omega,$$

$$a_{21} = a'_{21}a'_{11} + a'_{22}a'_{21}$$

$$= (-10^{-6})(5 \times 10^{-4}) + (-0.01)(-10^{-6})$$

$$= 9.5 \times 10^{-9} \ \text{S},$$

$$a_{22} = a'_{21}a'_{12} + a'_{22}a'_{22}$$

$$= (-10^{-6})(-10) + (-10^{-2})^2$$

$$= 1.1 \times 10^{-4}.$$

From Table 18.3,

$$\frac{V_2}{V_g} = \frac{Z_L}{(a_{11} + a_{21}Z_g)Z_L + a_{12} + a_{22}Z_g}$$

$$= \frac{10^4}{[10.25 \times 10^{-6} + 9.5 \times 10^{-9}(500)]10^4 + 0.095 + 1.1 \times 10^{-4}(500)}$$

$$= \frac{10^4}{0.15 + 0.095 + 0.055}$$

$$= \frac{10^5}{3}$$

$$= 33{,}333.33.$$

Thus, an input signal of $150 \ \mu\text{V}$ is amplified to an output signal of 5 V.

ASSESSMENT PROBLEM

Objective 3—Know how to analyze a cascade interconnection of two-port circuits

18.7 Each element in the symmetric bridged-tee circuit shown is a 15 Ω resistor. Two of these bridged tees are connected in cascade between a dc voltage source and a resistive load. The dc voltage source has a no load voltage of 100 V and an internal resistance of 8 Ω. The load resistor is adjusted until maximum power is delivered to the load. Calculate (a) the load resistance, (b) the load voltage, and (c) the load power.

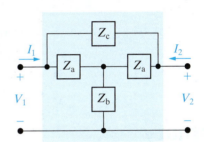

Answer: (a) 14.44 Ω;
(b) 16 V;
(c) 17.73 W.

SELF-CHECK: Also try Chapter Problem 18.38.

Practical Perspective

Characterizing an Unknown Circuit

We make the following measurements to find the h parameters for our "black box" amplifier:

- With Port 1 open, apply 50 V at Port 2. Measure the voltage at Port 1 and the current at Port 2:

$$V_1 = 50 \ \text{mV}; \quad I_2 = 2.5 \ \text{A}.$$

- With Port 2 short-circuited, apply 2.5 mA at Port 1. Measure the voltage at Port 1 and the current at Port 2:

$$V_1 = 1.25 \ \text{V}; \quad I_2 = 3.75 \ \text{A}.$$

Calculate the *h* parameters according to Eqs. 18.14:

$$h_{11} = \left|\frac{V_1}{I_1}\right|_{V_2=0} = \frac{1.25}{0.0025} = 500 \ \Omega; \quad h_{12} = \left|\frac{V_1}{V_2}\right|_{I_1=0} = \frac{0.05}{50} = 10^{-3};$$

$$h_{21} = \left|\frac{I_2}{I_1}\right|_{V_2=0} = \frac{3.75}{0.0025} = 1500; \quad h_{22} = \left|\frac{I_2}{V_2}\right|_{I_1=0} = \frac{2.5}{50} = 50 \text{ mS}.$$

Now we use the terminated two-port equations to determine whether or not it is safe to attach a 2 V(rms) source with a 100 Ω internal impedance to Port 1 and use this source together with the amplifier to drive a speaker modeled as a 32 Ω resistance with a power rating of 100 W. Begin by finding the value of I_2 from Table 18.3:

$$I_2 = \frac{h_{21}V_g}{(1 + h_{22}Z_L)(h_{11} + Z_g) - h_{12}h_{21}Z_L}$$

$$= \frac{1500(2)}{[1 + (0.05)(32)][500 + 100] - (1500)(10^{-3})(32)}$$

$$= 1.98 \text{ A (rms)}.$$

Then, calculate the power delivered to the 32 Ω speaker:

$$P = RI_2^2 = (32)(1.98)^2 = 126 \text{ W}.$$

The amplifier would thus deliver 126 W to the speaker, which is rated at 100 W, so it would be better to use a different amplifier or buy a more powerful speaker.

SELF-CHECK: Also try Chapter Problem 18.46

Summary

- The **two-port model** is used to describe the performance of a circuit in terms of the voltage and current at its input and output ports. (See page 694.)

- The model is limited to circuits in which

  - no independent sources are inside the circuit between the ports;

  - no energy is stored inside the circuit between the ports;

  - the current into the port is equal to the current out of the port; and

  - no external connections exist between the input and output ports.

  (See page 692.)

- Two of the four terminal variables (V_1, I_1, V_2, I_2) are independent; therefore, only two simultaneous equations involving the four variables are needed to describe the circuit. (See page 694.)

- The six possible sets of simultaneous equations involving the four terminal variables are called the *z*-, *y*-, *a*-, *b*-, *h*-,

and *g*-parameter equations. See Eqs. 18.1–18.6. (See page 694.)

- The equations are written in the *s* domain. The dc values of the parameters are obtained by setting $s = 0$, and the sinusoidal steady-state values are obtained by setting $s = j\omega$. (See page 694.)

- Any set of two-port parameters can be calculated or measured by invoking appropriate short-circuit and open-circuit conditions at the input and output ports. See Eqs. 18.7–18.15. (See pages 695 and 696.)

- The relationships among the six sets of two-port parameters are given in Table 18.1. (See page 699.)

- A two-port circuit is **reciprocal** if interchanging an ideal voltage source at one port with an ideal ammeter at the other port produces the same ammeter reading. The effect of reciprocity on the two-port parameters is given in Table 18.2. (See page 701.)

- A reciprocal two-port circuit is **symmetric** if its ports can be interchanged without disturbing the values of the terminal currents and voltages. The added effect of symmetry on the two-port parameters is given in Table 18.2. (See page 701.)

- The performance of a two-port circuit connected to a Thévenin equivalent source and a load is summarized by the relationships given in Table 18.3. (See page 704.)

- Large networks can be divided into subnetworks by means of interconnected two-port models. The cascade connection was used in this chapter to illustrate the analysis of interconnected two-port circuits. (See page 709.)

Problems

Sections 18.1–18.2

18.1 Find the *h* and *g* parameters for the circuit in Example 18.1.

18.2 Find the *z* parameters for the circuit in Fig. P18.2.

Figure P18.2

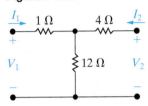

18.3 Use the results obtained in Problem 18.2 to calculate the *y* parameters for the circuit in Fig. P18.2.

18.4 Find the *b* parameters for the circuit shown in Fig. P18.4.

Figure P18.4

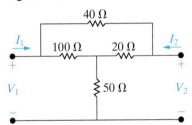

18.5 Find the *y* parameters for the circuit shown in Fig. P18.5.

Figure P18.5

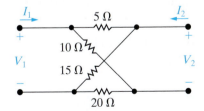

18.6 Find the *g* parameters for the circuit in Fig. P18.6.

Figure P18.6

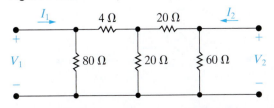

18.7 Select the values of R_1, R_2, and R_3 in the circuit in Fig. P18.7 so that $h_{11} = 4$ Ω, $h_{12} = 0.8$, $h_{21} = -0.8$, and $h_{22} = 0.14$ S.

Figure P18.7

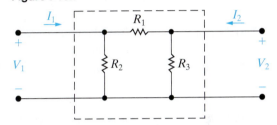

18.8 The operational amplifier in the circuit shown in Fig. P18.8 is ideal. Find the *h* parameters of the circuit.

Figure P18.8

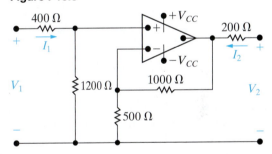

18.9 Find the *g* parameters for the operational amplifier circuit shown in Fig. P18.9.

Figure P18.9

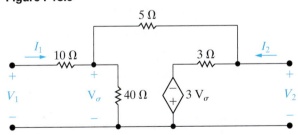

18.10 Find the *a* parameters for the circuit in Fig. P18.10.

Figure P18.10

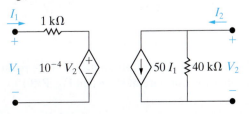

18.11 Use the results obtained in Problem 18.10 to calculate the g parameters of the circuit in Fig. P18.10.

18.12 Find the h parameters of the two-port circuit shown in Fig. P18.12.

Figure P18.12

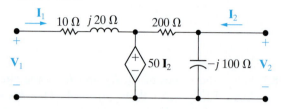

18.13 The following direct-current measurements were made on the two-port network shown in Fig. P18.13.

Port 2 Open	Port 2 Short-Circuited
$V_1 = 20$ mV	$I_1 = 200\ \mu$A
$V_2 = -5$ V	$I_2 = 50\ \mu$A
$I_1 = 0.25\ \mu$A	$V_1 = 10$ V

Calculate the g parameters for the network.

Figure P18.13

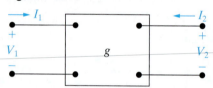

18.14 a) Use the measurements given in Problem 18.13 to find the y parameters for the network.

b) Check your calculations by finding the y parameters directly from the g parameters found in Problem 18.13.

18.15 Derive the expressions for the h parameters as functions of the g parameters.

18.16 Derive the expressions for the b parameters as functions of the h parameters.

18.17 Derive the expressions for the g parameters as functions of the z parameters.

18.18 Find the s-domain expressions for the a parameters of the two-port circuit shown in Fig. P18.18.

Figure P18.18

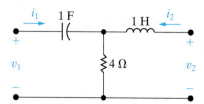

18.19 Find the s-domain expressions for the z parameters of the two-port circuit shown in Fig. P18.19.

Figure P18.19

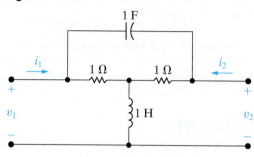

18.20 a) Use the defining equations to find the s-domain expressions for the h parameters for the circuit in Fig. P18.20.

b) Show that the results obtained in (a) agree with the h-parameter relationships for a reciprocal symmetric network.

Figure P18.20

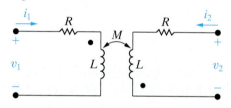

18.21 Find the frequency-domain values of the a parameters for the two-port circuit shown in Fig. P18.21.

Figure P18.21

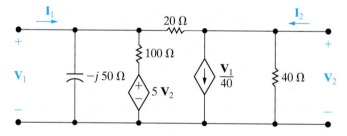

18.22 Find the h parameters for the two-port circuit shown in Fig. P18.21.

18.23 Is the two-port circuit shown in Fig. P18.23 symmetric? Justify your answer.

Figure P18.23

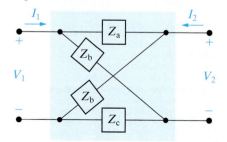

Section 18.3

18.24 Derive the expression for the input impedance, $Z_{in} = V_1/I_1$ for the circuit in Fig. 18.10 in terms of the b parameters.

18.25 Derive the expression for the current gain I_2/I_1 of the circuit in Fig. 18.10 in terms of the g parameters.

18.26 Derive the expression for the voltage gain V_2/V_1 of the circuit in Fig. 18.10 in terms of the y parameters.

18.27 Find the Thévenin equivalent circuit with respect to port 2 of the circuit in Fig. 18.10 in terms of the z parameters.

18.28 Derive the expression for the voltage gain V_2/V_g of the circuit in Fig. 18.10 in terms of the h parameters.

18.29 The following dc measurements were made on the resistive network shown in Fig. P18.29.

Measurement 1	**Measurement 2**
$V_1 = 4$ V	$V_1 = 20$ mV
$I_1 = 5$ mA	$I_1 = 20\ \mu$A
$V_2 = 0$	$V_2 = 40$ V
$I_2 = -200$ mA	$I_2 = 0$

A variable resistor R_o is connected across port 2 and adjusted for maximum power transfer to R_o. Find the maximum power.

Figure P18.29

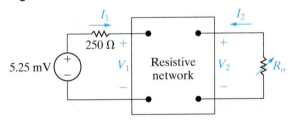

18.30 The b parameters for the two-port circuit in Fig. P18.30 are

$$b_{11} = 1 + j\frac{1}{3}; \quad b_{12} = -1 + j4\ \Omega;$$

$$b_{21} = \frac{1}{3}\ \text{S}; \quad b_{22} = 1 + j1.$$

The load impedance Z_L is adjusted for maximum average power transfer to Z_L. The ideal voltage source is generating a sinusoidal voltage of

$$v_g = 90 \cos 8000t\ \text{V}.$$

a) Find the rms value of V_2.

b) Find the average power delivered to Z_L.

c) What percentage of the average power developed by the ideal voltage source is delivered by Z_L?

Figure P18.30

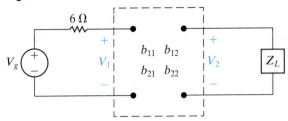

18.31 The h parameters for the two-port amplifier circuit in Fig. P18.31 are

$$h_{11} = 500\ \Omega; \quad h_{12} = 10^{-3};$$

$$h_{21} = 50; \quad h_{22} = 50\ \mu\text{S}.$$

The internal impedance of the source is $1500 + j0\ \Omega$, and the load impedance is $10,000 + j0\ \Omega$. The ideal voltage source is generating a voltage

$$v_g = 250 \cos 40,000t\ \text{mV}.$$

a) Find the rms value of V_2.

b) Find the average power delivered to Z_L.

c) Find the average power developed by the ideal voltage source.

Figure P18.31

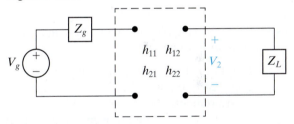

18.32 For the terminated two-port amplifier circuit in Fig. P18.31, find

a) the value of Z_L for maximum average power transfer to Z_L.

b) the maximum average power delivered to Z_L.

c) the average power developed by the ideal voltage source when maximum power is delivered to Z_L.

18.33 The b parameters of the amplifier in the circuit shown in Fig. P18.33 are

$$b_{11} = 25; \quad b_{12} = 1\ \text{k}\Omega;$$

$$b_{21} = -1.25\ \text{S}; \quad b_{22} = -40.$$

Find the ratio of the output power to that supplied by the ideal voltage source.

Figure P18.33

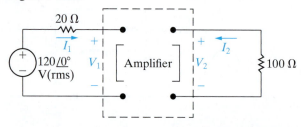

18.34 The linear transformer in the circuit shown in Fig. P18.34 has a coefficient of coupling of 0.75. The transformer is driven by a sinusoidal voltage source whose internal voltage is $v_g = 260 \cos 4000t$ V. The internal impedance of the source is $25 + j0 \ \Omega$.

a) Find the frequency-domain a parameters of the linear transformer.

b) Use the a parameters to derive the Thévenin equivalent circuit with respect to the terminals of the load.

c) Derive the steady-state time-domain expression for v_2.

Figure P18.34

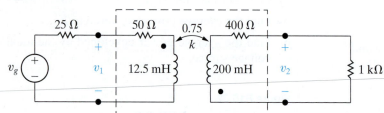

18.35 The following measurements were made on a resistive two-port network:

Condition 1 – create a short circuit at port 2 and apply 20 V to port 1:
Measurements: $I_1 = 1$ A; $I_2 = -1$ A.
Condition 2 – create an open circuit at port 1 and apply 80 V to port 2:
Measurements: $V_1 = 400$ V; $I_2 = 3$ A.

Find the maximum power that this two-port circuit can deliver to a resistive load at port 2 when port 1 is driven by a 4 A dc current source with an internal resistance of 60 Ω.

18.36 a) Find the z parameters for the two-port network in Fig. P18.36.

b) Find v_2 for $t > 0$ when $v_g = 50u(t)$ V.

Figure P18.36

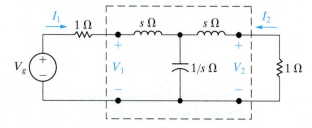

18.37 a) Find the s-domain expressions for the h parameters of the circuit in Fig. P18.37.

b) Port 2 in Fig. P18.37 is terminated in a resistance of 400 Ω, and port 1 is driven by a step voltage source $v_1(t) = 30u(t)$ V. Find $v_2(t)$ for $t > 0$ if $C = 200$ nF and $L = 200$ mH.

Figure P18.37

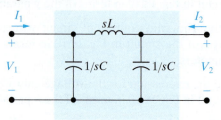

Section 18.4

18.38 The g and h parameters for the resistive two-ports in Fig. P18.38 are given by

$$g_{11} = \frac{3}{35} \text{ S}; \quad h_{11} = 5 \text{ k}\Omega;$$

$$g_{12} = \frac{20}{7}; \quad h_{12} = -0.2;$$

$$g_{21} = \frac{800}{7}; \quad h_{21} = -4;$$

$$g_{22} = \frac{50}{7} \text{ k}\Omega; \quad h_{22} = 200 \ \mu\text{S};$$

Calculate v_o if $v_g = 30$ V dc.

Figure P18.38

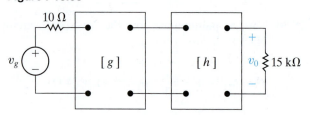

18.39 The h parameters of the first two-port circuit in Fig. P18.39(a) are

$$h_{11} = 1000 \ \Omega; \quad h_{12} = 5 \times 10^{-4};$$

$$h_{21} = 40; \quad h_{22} = 25 \ \mu\text{S}.$$

The circuit in the second two-port circuit is shown in Fig. P18.39(b), where $R = 72$ kΩ. Find v_o if $v_g = 9$ mV dc.

Figure P18.39

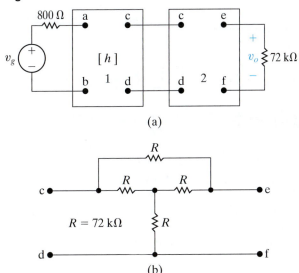

(a)

(b)

18.40 The networks A and B in the circuit in Fig. P18.40 are reciprocal and symmetric. For network A, it is known that $a'_{11} = 2$ and $a'_{12} = 1\ \Omega$.

a) Find the a parameters of network B.

b) Find $\mathbf{V}_2$ when $\mathbf{V}_g = 75\underline{/0°}$ V, $Z_g = 1\underline{/0°}\ \Omega$, and $Z_L = 14\underline{/0°}\ \Omega$.

Figure P18.40

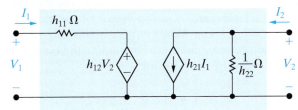

Sections 18.1–18.4

18.41 a) Show that the circuit in Fig. P18.41 is an equivalent circuit satisfied by the h-parameter equations.

b) Use the h-parameter equivalent circuit of (a) to find the voltage gain V_2/V_g in the circuit in Fig. 18.14, using the h-parameter values given in Example 18.6.

Figure P18.41

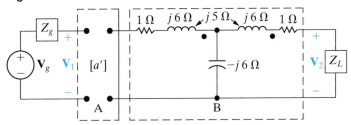

18.42 a) Show that the circuit in Fig. P18.42 is an equivalent circuit satisfied by the z-parameter equations.

b) Assume that the equivalent circuit in Fig. P18.42 is driven by a voltage source having an internal

impedance of Z_g ohms. Calculate the Thévenin equivalent circuit with respect to port 2. Check your results against the appropriate entries in Table 18.3.

Figure P18.42

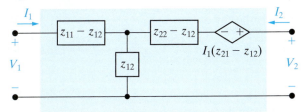

18.43 a) Show that the circuit in Fig. P18.43 is also an equivalent circuit satisfied by the z-parameter equations.

b) Assume that the equivalent circuit in Fig. P18.43 is terminated in an impedance of Z_L ohms at port 2. Find the input impedance V_1/I_1. Check your results against the appropriate entry in Table 18.3.

Figure P18.43

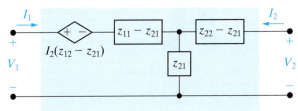

18.44 a) Derive two equivalent circuits that are satisfied by the y-parameter equations. *Hint:* Start with Eqs. 18.2. Add and subtract $y_{21}V_2$ to the first equation of the set. Construct a circuit that satisfies the resulting set of equations, by thinking in terms of node voltages. Derive an alternative equivalent circuit by first altering the second equation in Eq. 18.2.

b) Assume that port 1 is driven by a voltage source having an internal impedance Z_g, and port 2 is loaded with an impedance Z_L. Find the current gain I_2/I_1. Check your results against the appropriate entry in Table 18.3.

18.45 a) Derive the equivalent circuit satisfied by the g-parameter equations.

b) Use the g-parameter equivalent circuit derived in part (a) and the h-parameter equivalent circuit derived in Problem 18.41 to solve for the output voltage v_o in Problem 18.38.

18.46 a) What conditions and measurements will allow you to calculate the b parameters for the "black box" amplifier described in the Practical Perspective?

PRACTICAL
PERSPECTIVE

b) What measurements will be made if the resulting b parameters are equivalent to the h parameters calculated in the Practical Perspective?

18.47 Repeat Problem 18.46 for the y parameters.

PRACTICAL
PERSPECTIVE

APPENDIX

A

The Solution of Linear Simultaneous Equations

Circuit analysis frequently requires us to solve a set of linear simultaneous equations. We present several different solution methods here—some that employ engineering calculators or personal computers, and others that require just a pencil and paper. Most of the methods begin by placing the system of equations in matrix form. You should review matrices and matrix arithmetic, topics found in most intermediate-level algebra texts.

A.1 Preliminary Steps

To solve a set of simultaneous equations, we begin by organizing the equations into a standard form. To do this, collect all terms containing an unknown variable on the left-hand side of each equation and place all constants on the right-hand side. Then, arrange the equations in a vertical stack such that each variable occupies the same horizontal position in every equation. For example, in Eqs. A.1, the variables i_1, i_2, and i_3 occupy the first, second, and third position, respectively, on the left-hand side of each equation:

$$21i_1 - 9i_2 - 12i_3 = -33,$$

$$-3i_1 + 6i_2 - 2i_3 = 3, \tag{A.1}$$

$$-8i_1 - 4i_2 + 22i_3 = 50.$$

Once the equations are in this standard form, you can write the equations using matrix notation as

$$\begin{bmatrix} 21 & -9 & -12 \\ -3 & 6 & -2 \\ -8 & -4 & 22 \end{bmatrix} \begin{bmatrix} i_1 \\ i_2 \\ i_3 \end{bmatrix} = \begin{bmatrix} -33 \\ 3 \\ 50 \end{bmatrix}.$$

We can abbreviate the set of equations in matrix form as $AX = B$, where A is the matrix of coefficients that multiply the variables, X is the vector of the variables, and B is the vector of constants from the right-hand side of the equations. When written in this abbreviated form, we can solve for the vector of unknowns by finding the inverse of the A matrix and multiplying it by the B vector:

$$X = A^{-1}B.$$

If any equation is missing one or more variables, the missing variables can be inserted by making their coefficients zero. Thus, Eqs. A.2 are written in standard form as shown by Eqs. A.3:

$$2v_1 - v_2 = 4,$$

$$4v_2 + 3v_3 = 16, \tag{A.2}$$

$$7v_1 + 2v_3 = 5;$$

$$2v_1 - v_2 + 0v_3 = 4,$$

$$0v_1 + 4v_2 + 3v_3 = 16, \tag{A.3}$$

$$7v_1 + 0v_2 + 2v_3 = 5.$$

Equation A.3 is written using matrix form as

$$\begin{bmatrix} 3 & -1 & 0 \\ 0 & 4 & 3 \\ 7 & 0 & 2 \end{bmatrix} \begin{bmatrix} v_1 \\ v_2 \\ v_3 \end{bmatrix} = \begin{bmatrix} 4 \\ 16 \\ 5 \end{bmatrix}.$$

A.2 Calculator and Computer Methods

Most calculators recommended for engineering students can solve a set of simultaneous algebraic equations. Because there are many different calculators, it is impractical to provide directions or an example here. Instead, we provide the general steps for using your calculator to solve equations; you should refer to the manual for your specific calculator or search for instructions on the web.

1. Input the problem dimension by specifying the number of unknowns.

2. Create the A matrix, which is the matrix of coefficients that multiply the unknowns on the left-hand sides of the equations.

3. Create the B array, which is the array of constants on the right-hand sides of the equations.

4. Use the "solve" function (which has different names for different calculators) for the matrix A and the array B, to calculate the array X that contains the values of the unknowns.

Most calculators can solve simultaneous equations that have real number coefficients and complex number coefficients. Some can even solve simultaneous equations whose coefficients include symbols.

There are many different computer programs that can solve a set of algebraic equations. We present two examples: Excel, the spreadsheet application, and MATLAB, the matrix-based programming language. You should explore the options available and pick the software that works best for you.

Using Excel

Figure A.1 uses Excel to solve the simultaneous equations given in Eqs. A.1. Note that Excel can only solve simultaneous equations whose coefficients are real numbers. To begin, enter the A matrix in a square collection of cells and enter the B vector in a column of cells. You can

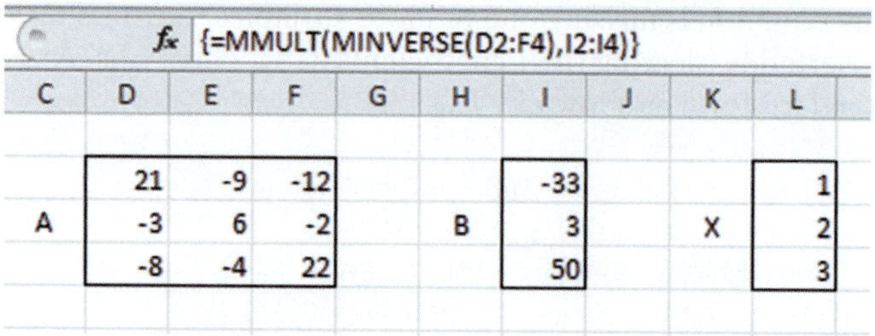

Figure A.1 ▲ Using Excel to solve the simultaneous equations in Eqs. A.1.

label the matrix and vector, as shown in the figure, but this is not required. Then highlight a column of cells for the vector X, which will contain the values of the unknowns. The number of cells in this column must equal the number of unknowns. Type the following function in the function box:

$$= \text{MMULT}(\text{MINVERSE}(\text{start\_cell, end\_cell}), b_1:b_n)$$

and simultaneously press the Ctrl-Shift-Enter keys. This will enter the values for the X vector into the highlighted cells and also will surround the function with braces, as shown in Fig. A.1. The function MINVERSE(start_cell, end_cell) calculates the inverse of a matrix. The matrix values occupy square collection of cells whose upper left cell is start_cell and whose lower right cell is end_cell. The MMULT(R, S) multiplies two matrices, R and S, supplied as arguments.

Using MATLAB

There are several ways to solve a system of simultaneous equations using MATLAB. Figure A.2 illustrates one method to solve Eqs. A.1. It begins

```
>> syms  i1 i2 i3
>> eq1=21*i1 - 9*i2 - 12*i3 == -33;
>> eq2=-3*i1 + 6*i2 - 2*i3 ==3;
>> eq3=-8*i1 - 4*i2 + 22*i3 == 50;
>> [A,B] = equationsToMatrix([eq1, eq2, eq3], [i1, i2, i3])


A =


[ 21, -9, -12]
[ -3,  6,  -2]
[ -8, -4, -22]


B =


 -33
   3
  50


>> X = linsolve(A,B)

X =


1
2
3
```

Figure A.2 ▲ Using MATLAB to solve the simultaneous equations in Eqs. A.1.

by defining the three unknown variables and the three simultaneous equations. Then, the function `equationsToMatrix(eq, var)` constructs the A matrix and the B vector from the equations supplied in the function's first argument, using the variables specified in the function's second argument. Finally, the function `linsolve(A, B)` calculates the values of the unknowns by inverting the matrix supplied in the function's first argument and then multiplying by the vector supplied in the function's second argument. Note that, unlike Excel, the simultaneous equations specified in MATLAB can have complex numbers as coefficients on the left-hand side and constants on the right-hand side.

A.3 Paper-and-Pencil Methods

We present two methods that do not require an engineering calculator or computer software: back-substitution and Cramer's method. Both methods are easy to use when solving two or three simultaneous equations. If you have four or more simultaneous equations, you should solve them with your calculator or computer because the paper-and-pencil methods are quite complicated. Both back-substitution and Cramer's method work for equations with real numbers, with complex numbers, or even with symbols as coefficients and constants.

Back-Substitution

The back-substitution method picks one equation and solves it for one unknown in terms of the remaining unknowns. The solution is used to eliminate that unknown in the remaining equations. This process is repeated until only one equation and one unknown remain. To illustrate, we will solve Eqs. A.2 using back-substitution. Begin by solving the third equation for v_3, to get

$$v_3 = 2.5 - 3.5v_1.$$

Now eliminate v_3 in the remaining equations:

$$2v_1 - v_2 = 4,$$

$$4v_2 + 3(2.5 - 3.5v_1) = 16.$$

Next, solve the first of the two remaining equations for v_2, to get

$$v_2 = 2v_1 - 4.$$

Eliminate v_2 in the other equation:

$$4(2v_1 - 4) + 3(2.5 - 3.5v_1) = 16.$$

Simplify and solve for v_1:

$$-2.5v_1 = 24.5 \quad \text{so} \quad v_1 = \frac{24.5}{-2.5} = -9.8 \text{ V}.$$

Finally, use this value for v_1 to find the remaining unknowns:

$$v_2 = 2v_1 - 4 = 2(-9.8) - 4 = -23.6 \text{ V},$$

$$v_3 = 2.5 - 3.5v_1 = 2.5 - 3.5(-9.8) = 36.8 \text{ V}.$$

There is another way to solve a set of simultaneous equations with the back-substitution method. Begin by picking any two equations, and multiply one or both equations by a different constant such that when the resulting equations are added together, one of the unknowns is eliminated. For example, consider the first two equations in Eqs. A.3. If we multiply the first equation by the constant 4 and add the resulting equation to the second equation, we eliminate v_2:

$$
\begin{aligned}
8v_1 - 4v_2 + 0v_3 &= 16 \\
+\ 0v_1 + 4v_2 + 3v_3 &= 16 \\
\hline
8v_1 + 0v_2 + 3v_3 &= 32.
\end{aligned}
$$

Now multiply this new equation by 2, and multiply the third equation in Eqs. A.3 by −3. Then add the two equations together to eliminate v_3:

$$
\begin{aligned}
16v_1 + 6v_3 &= 64 \\
+\ {-21}v_1 + {-6}v_3 &= -15 \\
\hline
-5v_1 + 0v_3 &= 49.
\end{aligned}
$$

Therefore,

$$
v_1 = \frac{49}{-5} = -9.8 \text{ V}.
$$

We can substitute the value of v_1 back into the first equation in Eqs. A.3 to find v_2 and then substitute the value of v_2 back into the second equation in Eqs. A.3 to find v_3. You should complete these final steps to verify the values.

Cramer's Method

We can also use **Cramer's method** to solve a set of simultaneous equations. The value of each unknown variable is the ratio of two determinants. If we let N, with an appropriate subscript, represent the numerator determinant and Δ represent the denominator determinant, then the kth unknown x_k is

$$
x_k = \frac{N_k}{\Delta}. \tag{A.4}
$$

The denominator determinant Δ is the same for every unknown variable and is called the **characteristic determinant** of the set of equations. The numerator determinant N_k varies with each unknown.

The characteristic determinant is the determinant of the A matrix. For example, the characteristic determinant of Eqs. A.3 is

$$
\Delta = \begin{vmatrix} 2 & -1 & 0 \\ 0 & 4 & 3 \\ 7 & 0 & 2 \end{vmatrix}.
$$

To find the determinant, rewrite the first two columns to the right of the determinant to get

$$
\Delta = \begin{vmatrix} 2 & -1 & 0 \\ 0 & 4 & 3 \\ 7 & 0 & 2 \end{vmatrix} \begin{matrix} 2 & -1 \\ 0 & 4 \\ 7 & 0 \end{matrix}.
$$

There are now five columns. Sum the products of the left-to-right diagonals for the first three columns; then subtract the sum of products of the right-to-left diagonals for the last three columns:

$$
\Delta = (2 \cdot 4 \cdot 2) + (-1 \cdot 3 \cdot 7) + (0 \cdot 0 \cdot 0) - (0 \cdot 4 \cdot 7) - (2 \cdot 3 \cdot 0) - (-1 \cdot 0 \cdot 2)
$$

$$
= 16 - 21 + 0 - 0 - 0 - 0 = -5.
$$

Note that this shortcut method for finding a determinant works only for square matrices of dimension 3. To find the determinant for a matrix whose dimension is larger than 3, consult a reference on determinants.

To construct the numerator determinant N_k, replace the kth column in the characteristic determinant with the values in the B vector. For example, the numerator determinants for evaluating v_1, v_2, and v_3 in Eqs. A.3 are

$$N_1 = \begin{vmatrix} 4 & -1 & 0 \\ 16 & 4 & 3 \\ 5 & 0 & 2 \end{vmatrix}$$

$$= (4 \cdot 4 \cdot 2) + (-1 \cdot 3 \cdot 5) + (0 \cdot 16 \cdot 0) - (0 \cdot 4 \cdot 5) - (4 \cdot 3 \cdot 0) - (-1 \cdot 16 \cdot 2)$$

$$= 49,$$

$$N_2 = \begin{vmatrix} 2 & 4 & 0 \\ 0 & 16 & 3 \\ 7 & 5 & 2 \end{vmatrix}$$

$$= (2 \cdot 16 \cdot 2) + (4 \cdot 3 \cdot 7) + (0 \cdot 0 \cdot 5) - (0 \cdot 16 \cdot 7) - (2 \cdot 3 \cdot 5) - (4 \cdot 0 \cdot 2)$$

$$= 118,$$

and

$$N_3 = \begin{vmatrix} 2 & -1 & 4 \\ 0 & 4 & 16 \\ 7 & 0 & 5 \end{vmatrix}$$

$$= (2 \cdot 4 \cdot 5) + (-1 \cdot 16 \cdot 7) + (4 \cdot 0 \cdot 0) - (4 \cdot 4 \cdot 7) - (2 \cdot 16 \cdot 0) - (-1 \cdot 0 \cdot 5)$$

$$= -184.$$

Using Cramer's method in Eq. A.4, we can solve for v_1, v_2, and v_3:

$$v_1 = \frac{N_1}{\Delta} = \frac{49}{-5} = -9.8 \text{ V},$$

$$v_2 = \frac{N_2}{\Delta} = \frac{118}{-5} = -23.6 \text{ V},$$

and

$$v_3 = \frac{N_3}{\Delta} = \frac{-184}{-5} = 36.8 \text{ V}.$$

A.4 Applications

The following examples demonstrate the various techniques for solving a system of simultaneous equations generated from circuit analysis.

EXAMPLE A.1

Use Cramer's method to solve for the node voltages v_1 and v_2 in Eqs. 4.1 and 4.2.

Solution

The first step is to rewrite Eqs. 4.1 and 4.2 in standard form. Collecting the coefficients of v_1 and v_2 on the left-hand side and moving the constant terms to the right-hand side of the equations gives us

$$1.7v_1 - 0.5v_2 = 10,$$

$$-0.5v_1 + 0.6v_2 = 2.$$

Rewriting this set of equations in $AX = B$ format gives us

$$\begin{bmatrix} 1.7 & -0.5 \\ -0.5 & 0.6 \end{bmatrix} \begin{bmatrix} v_1 \\ v_2 \end{bmatrix} = \begin{bmatrix} 10 \\ 2 \end{bmatrix}.$$

Using Cramer's method (Eq. A.4), we can write expressions for the unknown voltages:

$$v_1 = \frac{\begin{vmatrix} 10 & -0.5 \\ 2 & 0.6 \end{vmatrix}}{\begin{vmatrix} 1.7 & -0.5 \\ -0.5 & 0.6 \end{vmatrix}},$$

$$v_2 = \frac{\begin{vmatrix} 1.7 & 10 \\ -0.5 & 2 \end{vmatrix}}{\begin{vmatrix} 1.7 & -0.5 \\ -0.5 & 0.6 \end{vmatrix}}.$$

The shortcut for calculating the determinant for a matrix of dimension 3 does not work for matrices of dimension 2. Instead, there is a different shortcut for these matrices. Starting at the top of the first column, find the product along the left-to-right diagonal. Then subtract the product along the right-to-left diagonal that starts at the top of the second column. Thus,

$$v_1 = \frac{\begin{vmatrix} 10 & -0.5 \\ 2 & 0.6 \end{vmatrix}}{\begin{vmatrix} 1.7 & -0.5 \\ -0.5 & 0.6 \end{vmatrix}} = \frac{(10)(0.6) - (-0.5)(2)}{(1.7)(0.6) - (-0.5)(-0.5)}$$

$$= \frac{6 + 1}{1.02 - 0.25} = 9.09 \text{ V},$$

$$v_2 = \frac{\begin{vmatrix} 1.7 & 10 \\ -0.5 & 2 \end{vmatrix}}{\begin{vmatrix} 1.7 & -0.5 \\ -0.5 & 0.6 \end{vmatrix}} = \frac{(1.7)(2) - (10)(-0.5)}{(1.7)(0.6) - (-0.5)(-0.5)}$$

$$= \frac{3.4 + 5}{1.02 - 0.25} = 10.91 \text{ V}.$$

EXAMPLE A.2

Use Excel to find the three mesh currents in the circuit in Fig. 4.24.

Solution

The equations that describe the circuit in Fig. 4.24 were derived in Example 4.7. There are three KVL equations:

$$5(i_1 - i_2) + 20(i_1 - i_3) - 50 = 0,$$

$$5(i_2 - i_1) + 1i_2 + 4(i_2 - i_3) = 0,$$

$$20(i_3 - i_1) + 4(i_3 - i_2) + 15i_\phi = 0.$$

There is also a dependent source constraint equation:

$$i_\phi = i_1 - i_3.$$

Putting these four equations in standard form, we get

$$\begin{aligned} 25i_1 &- 5i_2 - 20i_3 + 0i_\phi = 50, \\ -5i_1 &+ 10i_2 - 4i_3 + 0i_\phi = 0, \\ -20i_1 &- 4i_2 + 24i_3 + 15i_\phi = 0, \\ i_1 &- 0i_2 - i_3 - i_\phi = 0. \end{aligned}$$

Rewriting this set of equations in $AX = B$ format gives us

$$\begin{bmatrix} 25 & -5 & -20 & 0 \\ -5 & 10 & -4 & 0 \\ -20 & -4 & 24 & 15 \\ 1 & 0 & -1 & -1 \end{bmatrix} \begin{bmatrix} i_1 \\ i_2 \\ i_3 \\ i_\phi \end{bmatrix} = \begin{bmatrix} 50 \\ 0 \\ 0 \\ 0 \end{bmatrix}.$$

The Excel solution is shown in Fig. A.3. The mesh currents are $i_1 = 29.6$ A, $i_2 = 26$ A, and $i_3 = 28$ A, and the controlling current for the dependent source is $i_\phi = 1.6$ A.

fx {=MMULT(MINVERSE(D2:G5),J2:J5)}

C	D	E	F	G	H	I	J	K	L	M
	25	-5	-20	0			50			29.6
A	-5	10	-4	0		B	0		X	26
	-20	-4	24	15			0			28
	1	0	-1	-1			0			1.6

Figure A.3 ▲ Using Excel to solve the simultaneous equations in Example A.2.

EXAMPLE A.3

Use MATLAB to find the phasor mesh currents $\mathbf{I}_1$ and $\mathbf{I}_2$ in the circuit in Fig. 9.40.

Solution

Summing the voltages around mesh 1 generates the equation

$$(1 + j2)\mathbf{I}_1 + (12 - j16)(\mathbf{I}_1 - \mathbf{I}_2) = 150\underline{/0^\circ}.$$

Summing the voltages around mesh 2 produces the equation

$$(12 - j16)(\mathbf{I}_2 - \mathbf{I}_1) + (1 + j3)\mathbf{I}_2 + 39\mathbf{I}_x = 0.$$

The current controlling the dependent voltage source is

$$\mathbf{I}_x = (\mathbf{I}_1 - \mathbf{I}_2).$$

Converting these three equations into standard form, we get

$$\begin{aligned}(13 + j14)\mathbf{I}_1 + (-12 + j16)\mathbf{I}_2 + \ 0\mathbf{I}_x &= 150,\\ (-12 + j16)\mathbf{I}_1 + \ (13 + j13)\mathbf{I}_2 + 39\mathbf{I}_x &= 0,\\ \mathbf{I}_1 - \qquad \mathbf{I}_2 - \mathbf{I}_x &= 0.\end{aligned}$$

The MATLAB commands used to solve this set of simultaneous equations are shown in Fig. A.4. Note the

```
>> syms  i1 i2 ix
>> eq1 = complex(13,-14)*i1 + complex(-12,16)*i2 == 150;
>> eq2 = complex(-12,16)*i1 + complex(13,-13)*i2 + 39*ix == 0;
>> eq3 = i1 - i2 - ix ==0,
>> [A,B] = equationsToMatrix([eq1, eq2, eq3], [i1, i2, ix])

A =

[   13 - 14i, - 12 + 16i,  0]
[  -12 + 16i,   13 - 13i, 39]
[          1,         -1, -1]


B =

 150
   0
   0


>> X = linsolve(A,B)

X =

 - 26 - 52i
 - 24 - 58i
 - 2 +   6i
```

Figure A.4 ▲ Using MATLAB to solve the simultaneous equations in Example A.3.

use of the `complex(a,b)` function to construct the complex coefficients, where a is the real part of the coefficient and b is the imaginary part of the coefficient. MATLAB gives us the solution

$$\mathbf{I}_1 = (-26 - j52) = 58.14\underline{/-116.57°} \text{ A},$$

$$\mathbf{I}_2 = (-24 - j58) = 62.77\underline{/-122.48°} \text{ A}.$$

In the first three examples, the matrix elements have been numbers—real numbers in Examples A.1 and A.2, and complex numbers in Example A.3. It is also possible for the elements to be symbols. Example A.4 illustrates the use of back substitution in a circuit problem where the elements in the coefficient matrix are symbols.

EXAMPLE A.4

Use back-substitution to derive expressions for the node voltages V_1 and V_2 in the circuit in Fig. A.5.

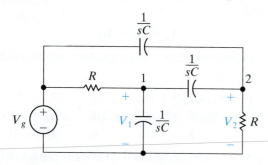

Figure A.5 ▲ The circuit for Example A.4.

Solution

Summing the currents leaving nodes 1 and 2 generates the following set of equations:

$$\frac{V_1 - V_g}{R} + V_1 sC + (V_1 - V_2)sC = 0,$$

$$\frac{V_2}{R} + (V_2 - V_1)sC + (V_2 - V_g)sC = 0.$$

Let $G = 1/R$ and collect the coefficients of V_1 and V_2 to get

$$(G + 2sC)V_1 - sCV_2 = GV_g,$$

$$-sCV_1 + (G + 2sC)V_2 = sCV_g.$$

Solve the first equation for V_2 to get

$$V_2 = \frac{(G + 2sC)}{sC}V_1 - \frac{G}{sC}V_g.$$

Substitute this expression into the second equation to eliminate V_2:

$$-sCV_1 + (G + 2sC)\left[\frac{(G + 2sC)}{sC}V_1 - \frac{G}{sC}V_g\right] = sCV_g.$$

Rearrange and simplify this equation to find the expression for V_1:

$$\frac{(G + 2sC)^2 - (sC)^2}{sC}V_1 = \frac{G(G + 2sC) + (sC)^2}{sC}V_g,$$

so

$$V_1 = \frac{G(G + 2sC) + (sC)^2}{(G + 2sC)^2 - (sC)^2}V_g$$

$$= \frac{(G^2 + 2sCG + s^2C^2)}{(G^2 + 4sCG + 3s^2C^2)}V_g.$$

Finally, substitute this expression for V_1 into the equation for V_2; rearrange and simplify to find the expression for V_2:

$$V_2 = \frac{(G + 2sC)}{sC}\left[\frac{(G^2 + 2sCG + s^2C^2)}{(G^2 + 4sCG + 3s^2C^2)}V_g\right] - \frac{G}{sC}V_g,$$

so

$$V_2 = \left[\frac{(G+2sC)(G^2+2sCG+s^2C^2)-G(G^2+4sCG+3s^2C^2)}{sC(G^2+4sCG+3s^2C^2)}\right]V_g$$

$$= \frac{2sC(G + sC)}{(G^2 + 4sCG + 3s^2C^2)}V_g.$$

B

Complex Numbers

Complex numbers allow us to find the square root of negative numbers. For example, consider the equation $x^2 + 8x + 41 = 0$. From the quadratic formula, we know that the two values of x that satisfy this equation are

$$x_{1,2} = \frac{-8 \pm \sqrt{8^2 - 4(41)}}{2} = -4 \pm \sqrt{-25}.$$

Therefore, this equation has no solution in a number system that excludes complex numbers. Complex numbers, and the ability to manipulate them algebraically, are extremely useful in circuit analysis.

B.1 Notation

There are two ways to designate a complex number: with the cartesian, or rectangular, form or with the polar, or trigonometric, form. In **rectangular form**, a complex number is written as a sum of a real component and an imaginary component; hence

$$n = a + jb,$$

where a is the real component, b is the imaginary component, and j is, by definition, $\sqrt{-1}$.[1]

In **polar form**, a complex number is written in terms of its magnitude (or modulus) and angle (or argument); hence

$$n = ce^{j\theta}$$

where c is the magnitude, θ is the angle, e is the base of the natural logarithm, and, as before, $j = \sqrt{-1}$. The symbol $\underline{/\theta^\circ}$ is frequently used in place of $e^{j\theta}$, so the polar form can also be written as

$$n = c\underline{/\theta}.$$

When we write a complex number in polar form, we typically use the angle notation. But when we perform mathematical operations using complex numbers in polar form, we use the exponential notation because the rules for manipulating an exponential quantity are well known. For example, because $(y^x)^n = y^{xn}$, then $(e^{j\theta})^n = e^{jn\theta}$; because $y^{-x} = 1/y^x$, then $e^{-j\theta} = 1/e^{j\theta}$; and so forth.

---

[1] You may be more familiar with the notation $i = \sqrt{-1}$. In electrical engineering, i is used as the symbol for current, and hence in electrical engineering literature, j is used to denote $\sqrt{-1}$.

Since there are two ways of expressing the same complex number, we need to relate one form to the other. The transition from polar to rectangular form makes use of Euler's identity:

$$e^{\pm j\theta} = \cos\theta \pm j\sin\theta.$$

A complex number in polar form can be put in rectangular form by writing

$$ce^{j\theta} = c(\cos\theta + jc\sin\theta)$$
$$= c(\cos\theta + jc\sin\theta)$$
$$= a + jb.$$

The transition from rectangular to polar form uses right triangle geometry, namely,

$$a + jb = \left(\sqrt{a^2 + b^2}\right)e^{j\theta}$$
$$= ce^{j\theta},$$

where

$$\tan\theta = b/a.$$

The expression for $\tan\theta$ does not specify the quadrant where the angle θ is located. We can determine the location of θ using a graphical representation of the complex number.

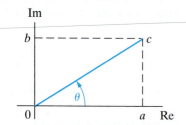

Figure B.1 ▲ The graphical representation of $a + jb$ when a and b are both positive.

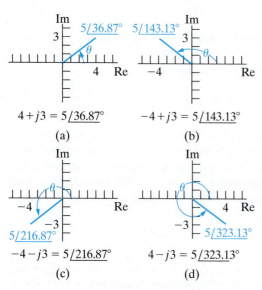

$4 + j3 = 5\underline{/36.87°}$

(a)

$-4 + j3 = 5\underline{/143.13°}$

(b)

$-4 - j3 = 5\underline{/216.87°}$

(c)

$4 - j3 = 5\underline{/323.13°}$

(d)

Figure B.2 ▲ The graphical representation of four complex numbers.

B.2 The Graphical Representation of a Complex Number

A complex number is represented graphically on a complex-number plane, where the horizontal axis represents the real component and the vertical axis represents the imaginary component. The angle of the complex number is measured counterclockwise from the positive real axis. The plot of the complex number $n = a + jb = c\underline{/\theta}$, assuming that a and b are both positive, is shown in Fig. B.1.

This plot makes very clear the relationship between the rectangular and polar forms. Any point in the complex-number plane is uniquely defined by giving either its distance from each axis (that is, a and b) or its radial distance from the origin (c) and the angle with respect to the positive real axis, θ.

It follows from Fig. B.1 that θ is in the first quadrant when a and b are both positive, in the second quadrant when a is negative and b is positive, in the third quadrant when a and b are both negative, and in the fourth quadrant when a is positive and b is negative. These observations are illustrated in Fig. B.2, where we have plotted $4 + j3$, $-4 + j3$, $-4 - j3$, and $4 - j3$.

Note that we can also specify θ as a clockwise angle from the positive real axis. Thus, in Fig. B.2(c), we could also designate $-4 - j3$ as $5\underline{/-143.13°}$. In Fig. B.2(d), we observe that $5\underline{/323.13°} = 5\underline{/-36.87°}$. It is customary to express θ in terms of negative values when θ lies in the third or fourth quadrant, so that $-180° \leq \theta \leq 180°$.

The **conjugate of a complex number** is formed by reversing the sign of its imaginary component. Thus, the conjugate of $a + jb$ is $a - jb$, and the conjugate of $-a + jb$ is $-a - jb$. When we write a complex number in polar form, we construct its conjugate by reversing the sign of the angle θ. Therefore, the conjugate of $c\underline{/\theta}$ is $c\underline{/-\theta}$. The conjugate of a complex number is designated with an asterisk, so $n*$ is understood to be the conjugate of n. Figure B.3 shows two complex numbers and their conjugates plotted on the complex-number plane. Note that conjugation reflects a complex number about the real axis.

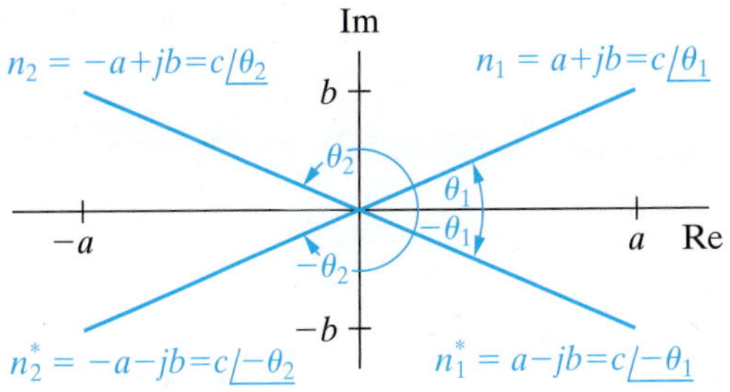

Figure B.3 ▲ The complex numbers n_1 and n_2 and their conjugates $n*_1$ and $n*_2$.

B.3 Arithmetic Operations

Addition (Subtraction)

When adding or subtracting complex numbers, we express the numbers in rectangular form. Addition involves adding the real parts of the complex numbers to form the real part of the sum and adding the imaginary parts to form the imaginary part of the sum. Thus, if we are given

$$n_1 = 8 + j16$$

and

$$n_2 = 12 - j3,$$

then

$$n_1 + n_2 = (8 + 12) + j(16 - 3) = 20 + j13.$$

Subtraction follows the same rule. Thus

$$n_2 - n_1 = (12 - 8) + j(-3 - 16) = 4 - j19.$$

If the numbers to be added or subtracted are given in polar form, they are first converted to rectangular form. For example, if

$$n_1 = 10\underline{/53.13°}$$

and

$$n_2 = 5\underline{/-135°},$$

then

$$n_1 + n_2 = 6 + j8 - 3.535 - j3.535$$

$$= (6 - 3.535) + j(8 - 3.535)$$

$$= 2.465 + j4.465 = 5.10\underline{/61.10°},$$

and

$$n_1 - n_2 = 6 + j8 - (-3.535 - j3.535)$$

$$= 9.535 + j11.535$$

$$= 14.966\underline{/50.42°}.$$

Multiplication (Division)

When multiplying or dividing complex numbers, the numbers can be written in either rectangular or polar form. In most cases, polar form is more convenient. As an example, let's find the product $n_1 n_2$ when $n_1 = 8 + j10$ and $n_2 = 5 - j4$. Using rectangular form, we have

$$n_1 n_2 = (8 + j10)(5 - j4) = 40 - j32 + j50 + 40$$

$$= 80 + j18$$

$$= 82 \underline{/12.68°}.$$

If we use polar form, we find the product of two complex numbers by multiplying their magnitudes and adding their angles. For example,

$$n_1 n_2 = (12.81 \underline{/51.34°})(6.40 \underline{/-38.66°})$$

$$= 82 \underline{/12.68°}$$

$$= 80 + j18.$$

The first step in dividing two complex numbers in rectangular form is to multiply the numerator and denominator by the conjugate of the denominator. This makes the denominator a real number. We then divide the real number into the new numerator. As an example, let's calculate n_1/n_2, where $n_1 = 6 + j3$ and $n_2 = 3 - j1$. We have

$$\frac{n_1}{n_2} = \frac{6 + j3}{3 - j1} = \frac{(6 + j3)(3 + j1)}{(3 - j1)(3 + j1)}$$

$$= \frac{18 + j6 + j9 - 3}{9 + 1}$$

$$= \frac{15 + j15}{10} = 1.5 + j1.5$$

$$= 2.12 \underline{/45°}.$$

In polar form, we calculate n_1/n_2 by dividing the magnitudes and subtracting the angles. For example,

$$\frac{n_1}{n_2} = \frac{6.71 \underline{/26.57°}}{3.16 \underline{/-18.43°}} = 2.12 \underline{/45°}$$

$$= 1.5 + j1.5.$$

B.4 Useful Identities

In working with complex numbers and quantities, the following identities are very useful:

$$\pm j^2 = \mp 1,$$

$$(-j)(j) = 1,$$

$$j = \frac{1}{-j},$$

$$e^{\pm j\pi} = -1,$$

$$e^{\pm j\pi/2} = \pm j.$$

Given that $n = a + jb = c\underline{/\theta}$, it follows that

$$nn^* = a^2 + b^2 = c^2,$$

$$n + n^* = 2a,$$

$$n - n^* = j2b,$$

$$n/n^* = 1\underline{/2\theta}.$$

B.5 The Integer Power of a Complex Number

To raise a complex number to an integer power k, begin by expressing the complex number in polar form. Then, to find the kth power of a complex number, raise its magnitude to the kth power and multiply its angle by k. Thus

$$n^k = (a + jb)^k$$

$$= (ce^{j\theta})^k = c^k e^{jk\theta}$$

$$= c^k(\cos k\theta + j \sin k\theta).$$

For example,

$$(2e^{j12°})^5 = 2^5 e^{j60°} = 32e^{j60°}$$

$$= 16 + j27.71,$$

and

$$(3 + j4)^4 = (5e^{j53.13°})^4 = 5^4 e^{j212.52°}$$

$$= 625e^{j212.52°}$$

$$= -527 - j336.$$

B.6 The Roots of a Complex Number

To find the kth root of a complex number, we solve the equation

$$x^k - ce^{j\theta} = 0,$$

which is an equation of the kth degree and therefore has k roots.

To find the k roots $(x_1, x_2, ..., x_k)$, we first note that

$$ce^{j\theta} = ce^{j(\theta + 2\pi)} = ce^{j(\theta + 4\pi)} = \cdots.$$

It follows that

$$x_1 = (ce^{j\theta})^{1/k} = c^{1/k}e^{j\theta/k},$$

$$x_2 = [ce^{j(\theta + 2\pi)}]^{1/k} = c^{1/k}e^{j(\theta + 2\pi)/k},$$

$$x_3 = [ce^{j(\theta + 4\pi)}]^{1/k} = c^{1/k}e^{j(\theta + 4\pi)/k},$$

$$\vdots$$

We continue this process until the roots start repeating. This will happen when the multiple of π is equal to $2k$. For example, let's find the four roots of $81e^{j60°}$. We have

$$x_1 = 81^{1/4}e^{j60/4} = 3e^{j15°},$$

$$x_2 = 81^{1/4}e^{j(60 + 360)/4} = 3e^{j105°},$$

$$x_3 = 81^{1/4}e^{j(60 + 720)/4} = 3e^{j195°},$$

$$x_4 = 81^{1/4}e^{j(60 + 1080)/4} = 3e^{j285°},$$

$$x_5 = 81^{1/4}e^{j(60 + 1440)/4} = 3e^{j375°} = 3e^{j15°}.$$

Here, x_5 is the same as x_1, so the roots have started to repeat. Therefore, we know the four roots of $81e^{j60°}$ are the values given by x_1, x_2, x_3, and x_4.

Note that the roots of a complex number lie on a circle in the complex-number plane. The radius of the circle is $c^{1/k}$. The roots are uniformly distributed around the circle, and the angle between adjacent roots is $2\pi/k$ radians or $360/k$ degrees. The four roots of $81e^{j60°}$ are plotted in Fig. B.4.

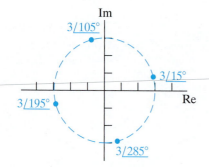

Figure B.4 ▲ The four roots of $81e^{j60°}$.

More on Magnetically Coupled Coils and Ideal Transformers

C.1 Equivalent Circuits for Magnetically Coupled Coils

It is sometimes convenient to model magnetically coupled coils with an equivalent circuit that does not include magnetic coupling. Consider the two magnetically coupled coils shown in Fig. C.1. The resistances R_1 and R_2 represent the winding resistance of each coil. The goal is to replace the magnetically coupled coils inside the shaded area with a set of inductors that are not magnetically coupled. Before deriving the equivalent circuits, we must point out an important restriction: The voltage between terminals b and d must be zero. In other words, we must be able to short together terminals b and d without disturbing the voltages and currents in the original circuit. This restriction is imposed because, while the equivalent circuits we develop have four terminals, two of those four terminals are shorted together. Thus, the same requirement is placed on the original circuits.

We begin by writing the two equations that relate the terminal voltages v_1 and v_2 to the terminal currents i_1 and i_2. For the given references and polarity dots,

$$v_1 = L_1 \frac{di_1}{dt} + M \frac{di_2}{dt} \tag{C.1}$$

and

$$v_2 = M \frac{di_1}{dt} + L_2 \frac{di_2}{dt}. \tag{C.2}$$

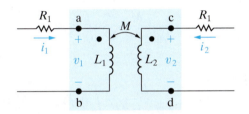

Figure C.1 ▲ The circuit used to develop an equivalent circuit for magnetically coupled coils.

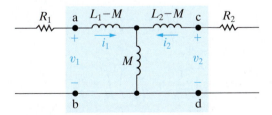

Figure C.2 ▲ The T-equivalent circuit for the magnetically coupled coils of Fig. C.1.

The T-Equivalent Circuit

How can we arrange uncoupled inductors in a circuit that can be described by Eqs. C.1 and C.2? If we regard Eqs. C.1 and C.2 as mesh-current equations with i_1 and i_2 as the mesh currents, we need one mesh with a total inductance of L_1 and a second mesh with a total inductance of L_2. Also, the two meshes must have a common inductance of M. The T-equivalent circuit shown in Fig. C.2 satisfies these requirements.

You should verify that the equations relating v_1 and v_2 to i_1 and i_2 reduce to Eqs. C.1 and C.2. Note the absence of magnetic coupling between the inductors and the zero voltage between b and d.

The π-Equivalent Circuit

We can derive a π-equivalent circuit for the magnetically coupled coils shown in Fig. C.1 by solving Eqs. C.1 and C.2 for the derivatives di_1/dt and di_2/dt. We treat the resulting expressions as a pair of node-voltage equations. Find di_1/dt and di_2/dt using Cramer's method:

$$\frac{di_1}{dt} = \frac{\begin{vmatrix} v_1 & M \\ v_2 & L_2 \end{vmatrix}}{\begin{vmatrix} L_1 & M \\ M & L_2 \end{vmatrix}} = \frac{L_2}{L_1 L_2 - M^2} v_1 - \frac{M}{L_1 L_2 - M^2} v_2; \tag{C.3}$$

$$\frac{di_2}{dt} = \frac{\begin{vmatrix} L_1 & v_1 \\ M & v_2 \end{vmatrix}}{L_1 L_2 - M^2} = \frac{-M}{L_1 L_2 - M^2} v_1 + \frac{L_1}{L_1 L_2 - M^2} v_2. \tag{C.4}$$

Now we solve for i_1 and i_2 by multiplying both sides of Eqs. C.3 and C.4 by dt and then integrating:

$$i_1 = i_1(0) + \frac{L_2}{L_1 L_2 - M^2} \int_0^t v_1 \, d\tau - \frac{M}{L_1 L_2 - M^2} \int_0^t v_2 \, d\tau \tag{C.5}$$

and

$$i_2 = i_2(0) - \frac{M}{L_1 L_2 - M^2} \int_0^t v_1 \, d\tau + \frac{L_1}{L_1 L_2 - M^2} \int_0^t v_2 \, d\tau. \tag{C.6}$$

If we regard v_1 and v_2 as node voltages, Eqs. C.5 and C.6 describe a circuit like the one shown in Fig. C.3.

To find L_A, L_B, and L_C as functions of L_1, L_2, and M, write the equations for i_1 and i_2 in Fig. C.3 and compare them with Eqs. C.5 and C.6. Thus

$$i_1 = i_1(0) + \frac{1}{L_A} \int_0^t v_1 \, d\tau + \frac{1}{L_B} \int_0^t (v_1 - v_2) \, d\tau$$

$$= i_1(0) + \left(\frac{1}{L_A} + \frac{1}{L_B} \right) \int_0^t v_1 d\tau - \frac{1}{L_B} \int_0^t v_2 d\tau$$

and

$$i_2 = i_2(0) + \frac{1}{L_C} \int_0^t v_2 d\tau + \frac{1}{L_B} \int_0^t (v_1 - v_2) \, d\tau$$

$$= i_2(0) + \frac{1}{L_B} \int_0^t v_1 d\tau + \left(\frac{1}{L_B} + \frac{1}{L_C} \right) \int_0^t v_2 d\tau.$$

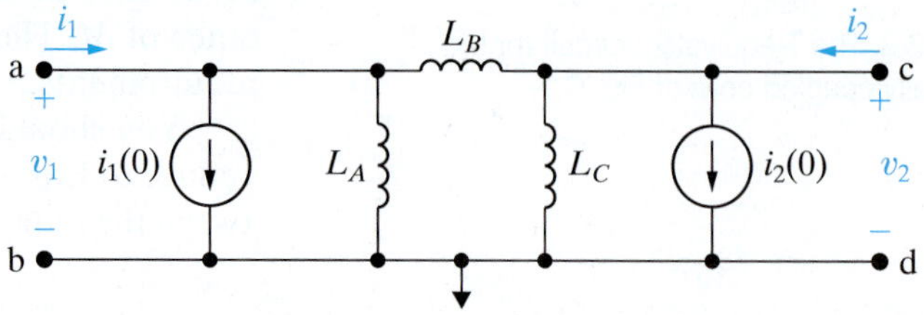

Figure C.3 ▲ The circuit used to derive the -equivalent circuit for magnetically coupled coils.

Then

$$\frac{1}{L_B} = \frac{M}{L_1L_2 - M^2},$$

$$\frac{1}{L_A} = \frac{L_2 - M}{L_1L_2 - M^2},$$

$$\frac{1}{L_C} = \frac{L_1 - M}{L_1L_2 - M^2}.$$

Incorporate the expressions for L_A, L_B, and L_C into the circuit shown in Fig. C.3 to get the π-equivalent circuit for the magnetically coupled coils shown in Fig. C.1. The result is shown in Fig. C.4.

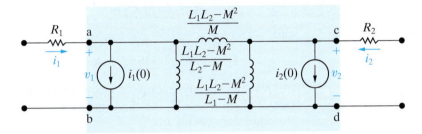

Figure C.4 ▲ The π-equivalent circuit for the magnetically coupled coils of Fig. C.1.

Note that the initial values of i_1 and i_2 are explicit in the π-equivalent circuit but implicit in the T-equivalent circuit. If we want to find the sinusoidal steady-state behavior of circuits containing mutual inductance, we can assume that the initial values of i_1 and i_2 are zero. We can thus eliminate the current sources in the π-equivalent circuit, and the circuit shown in Fig. C.4 simplifies to the one shown in Fig. C.5.

The mutual inductance carries its own algebraic sign in the T- and π-equivalent circuits. In other words, if the magnetic polarity of the coupled coils is reversed from that given in Fig. C.1, the algebraic sign of M reverses. A reversal in magnetic polarity requires moving one polarity dot without changing the reference polarities of the terminal currents and voltages.

Example C.1 illustrates the application of the T-equivalent circuit.

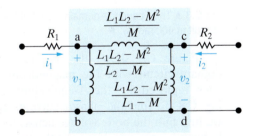

Figure C.5 ▲ The π-equivalent circuit used for sinusoidal steady-state analysis.

EXAMPLE C.1

a) Replace the magnetically coupled coils shown in Fig. C.6 with a T-equivalent circuit. Then find the phasor currents I_1 and I_2. The source frequency is 400 rad/s.

b) Repeat (a), but with the polarity dot on the secondary winding moved to the lower terminal.

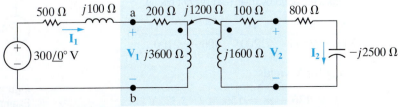

Figure C.6 ▲ The frequency-domain equivalent circuit for Example C.1.

Solution

a) For the polarity dots shown in Fig. C.6, M has a value of $+3$ H in the T-equivalent circuit. Therefore, the three inductances in the equivalent circuit are

$$L_1 - M = 9 - 3 = 6\text{ H};$$

$$L_2 - M = 4 - 3 = 1\text{ H};$$

$$M = 3\text{ H}.$$

Figure C.7 shows the T-equivalent circuit, and Fig. C.8 shows the frequency-domain equivalent circuit at a frequency of 400 rad/s.

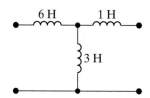

Figure C.7 ▲ The T-equivalent circuit for the magnetically coupled coils in Example C.1.

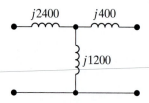

Figure C.8 ▲ The frequency-domain model of the equivalent circuit at 400 rad/s.

Figure C.9 shows the original frequency-domain circuit with the magnetically coupled coils replaced by the T-equivalent circuit in Fig. C.8. To find the phasor currents $\mathbf{I}_1$ and $\mathbf{I}_2$, we first find the node voltage across the 1200 Ω inductive reactance. If we use the lower node as the reference, the single KCL equation is

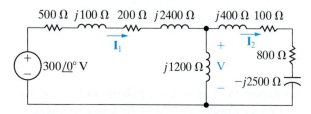

Figure C.9 ▲ The circuit of Fig. C.6, with the magnetically coupled coils replaced by their T-equivalent circuit.

$$\frac{\mathbf{V} - 300}{700 + j2500} + \frac{\mathbf{V}}{j1200} + \frac{\mathbf{V}}{900 - j2100} = 0.$$

Solving for $\mathbf{V}$ yields

$$\mathbf{V} = 136 - j8 = 136.24 \,\underline{/-3.37°}\text{ V}.$$

Then

$$\mathbf{I}_1 = \frac{300 - (136 - j8)}{700 + j2500} = 63.25 \,\underline{/-71.57°}\text{ mA}$$

and

$$\mathbf{I}_2 = \frac{136 - j8}{900 - j2100} = 59.63 \,\underline{/63.43°}\text{ mA}.$$

b) When the polarity dot is moved to the lower terminal of the secondary coil, M has a value of -3 H in the T-equivalent circuit. Before analyzing the new T-equivalent circuit, we note that reversing the algebraic sign of M has no effect on the solution for $\mathbf{I}_1$ and shifts $\mathbf{I}_2$ by 180°. Therefore, we anticipate that

$$\mathbf{I}_1 = 63.25 \,\underline{/-71.57°}\text{ mA}$$

and

$$\mathbf{I}_2 = 59.63 \,\underline{/-116.57°}\text{ mA}.$$

Now let's analyze the new T-equivalent circuit. With $M = -3$ H, the three inductances in the equivalent circuit are

$$L_1 - M = 9 - (-3) = 12\text{ H};$$

$$L_2 - M = 4 - (-3) = 7\text{ H};$$

$$M = -3\text{ H}.$$

At an operating frequency of 400 rad/s, the frequency-domain equivalent circuit requires two inductors and a capacitor, as shown in Fig. C.10.

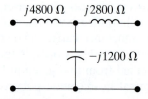

Figure C.10 ▲ The frequency-domain equivalent circuit for $M = -3$ H and $\omega = 400$ rad/s.

The resulting frequency-domain circuit for the original system appears in Fig. C.11. As before, we first find the voltage across the center branch, which contains a capacitive reactance of $-j1200$ Ω. If we use the lower node as the reference, the KCL equation is

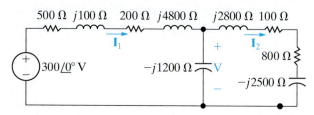

Figure C.11 ▲ The frequency-domain equivalent circuit for Example C.1(b).

$$\frac{V - 300}{700 + j4900} + \frac{V}{-j1200} + \frac{V}{900 + j300} = 0.$$

Solving for V gives

$$V = -8 - j56$$

$$= 56.57 \underline{/-98.13°} \, V.$$

Then

$$I_1 = \frac{300 - (-8 - j56)}{700 + j4900}$$

$$= 63.25 \underline{/-71.57°} \, mA.$$

and

$$I_2 = \frac{-8 - j56}{900 + j300}$$

$$= 59.63 \underline{/-116.57°} \, mA.$$

C.2 The Need for Ideal Transformers in the Equivalent Circuits

The inductors in the T- and π-equivalent circuits of magnetically coupled coils can have negative values. For example, if $L_1 = 3$ mH, $L_2 = 12$ mH, and $M = 5$ mH, the T-equivalent circuit requires an inductor of -2 mH, and the π-equivalent circuit requires an inductor of -5.5 mH. These negative inductance values do not create a problem if you use the equivalent circuits in computations. However, if you want to build the equivalent circuits using circuit components, the negative reactance can only be achieved using capacitors. But whenever the frequency of the sinusoidal source changes, you must change the capacitor used to generate the negative reactance. For example, at a frequency of 50 krad/s, a -2 mH inductor has an impedance of $-j100 \, \Omega$. This impedance can be modeled with a 0.2 μF capacitor. If the frequency changes to 25 krad/s, the -2 mH inductive impedance changes to $-j50 \, \Omega$ and we need a 0.8 μF capacitor. If the frequency is varied continuously, using a capacitor to simulate negative inductance is impractical.

Instead of using capacitors to create negative reactance, you can include an ideal transformer in the equivalent circuit. This doesn't completely solve the problem because ideal transformers can only be approximated. However, in some situations the approximation is satisfactory, so knowing how to use an ideal transformer in the T- and π-equivalent circuits of magnetically coupled coils is an important tool.

An ideal transformer can be used in two different ways in either the T-equivalent or the π-equivalent circuit. Figure C.12 shows the two arrangements for each type of equivalent circuit. To verify any of the equivalent circuits in Fig. C.12, we must show that the equations relating v_1 and v_2 to di_1/dt and di_2/dt are identical to Eqs. C.1 and C.2. To demonstrate, we validate the circuit shown in Fig. C.12(a); we leave it to you to verify the circuits in Figs. C.12(b), (c), and (d).

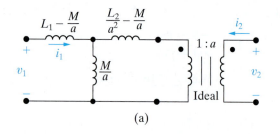

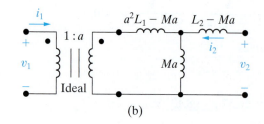

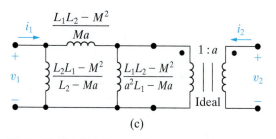

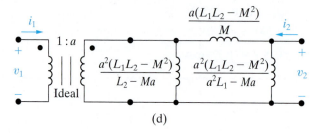

Figure C.12 ▲ The four ways of using an ideal transformer in the T- and π-equivalent circuit for magnetically coupled coils.

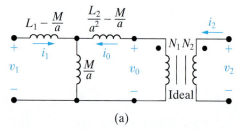

(a)

Figure C.13 ▲ The circuit of Fig. C.12(a) with i_0 and v_0 defined.

We redrew the circuit shown in Fig. C.12(a) as Fig. C.13, adding the variables i_0 and v_0 to aid the discussion. From this circuit,

$$v_1 = \left(L_1 - \frac{M}{a}\right)\frac{di_1}{dt} + \frac{M}{a}\frac{d}{dt}(i_1 + i_0)$$

and

$$v_0 = \left(\frac{L_2}{a^2} - \frac{M}{a}\right)\frac{di_0}{dt} + \frac{M}{a}\frac{d}{dt}(i_0 + i_1).$$

The ideal transformer imposes constraints on v_0 and i_0:

$$v_0 = \frac{v_2}{a};$$

$$i_0 = ai_2.$$

Substituting the constraint equations into the expressions for v_1 and v_0 from the circuit gives

$$v_1 = L_1\frac{di_1}{dt} + \frac{M}{a}\frac{d}{dt}(ai_2)$$

and

$$\frac{v_2}{a} = \frac{L_2}{a^2}\frac{d}{dt}(ai_2) + \frac{M}{a}\frac{di_1}{dt}.$$

Simplifying, we get,

$$v_1 = L_1\frac{di_1}{dt} + M\frac{di_2}{dt}$$

and

$$v_2 = M\frac{di_1}{dt} + L_2\frac{di_2}{dt}.$$

These expressions for v_1 and v_2 are identical to Eqs. C.1 and C.2. Thus, the circuit shown in Fig. C.13 is equivalent to the magnetically coupled coils shown inside the box in Fig. C.1 because the terminal behavior of these two circuits is the same.

When we showed that the circuit in Fig. C.13 is equivalent to the magnetically coupled coils in Fig. C.1, we placed no restrictions on the turns ratio a. Therefore, an infinite number of equivalent circuits are possible. However, we can always find a turns ratio that makes all of the inductances positive. Two values of a are of particular interest:

$$a = \frac{M}{L_1}, \tag{C.7}$$

and

$$a = \frac{L_2}{M}. \tag{C.8}$$

The value of a given by Eq. C.7 eliminates the inductances $L_1 - M/a$ and $a^2L_1 - aM$ from the T-equivalent circuits and the inductances $(L_1L_2 - M^2)/(a^2L_1 - aM)$ and $a^2(L_1L_2 - M^2)/(a^2L_1 - aM)$ from the π-equivalent circuits. The value of a given by Eq. C.8 eliminates the inductances $(L_2/a^2) - (M/a)$ and $L_2 - aM$ from the T-equivalent circuits and the inductances $(L_1L_2 - M^2)/(L_2 - aM)$ and $a^2(L_1L_2 - M^2)/(L_2 - aM)$ from the π-equivalent circuits.

Also note that when $a = M/L_1$, the circuits in Figs. C.12(a) and (c) are identical, and when $a = L_2/M$, the circuits in Figs. C.12(b) and (d) are identical. Figures C.14 and C.15 summarize these observations. We can use the relationship $M = k\sqrt{L_1L_2}$ to derive the inductor values in Figs. C.14 and C.15 as functions of the self-inductances L_1 and L_2, and the coupling coefficient k. Then, the values of a given by Eqs. C.7 and C.8 will reduce the number of inductances needed in the equivalent circuit and guarantee that all the inductances will be positive.

The values of a given by Eqs. C.7 and C.8 can be determined experimentally from the magnetically coupled coils. To find the ratio M/L_1, drive the coil with N_1 turns using a sinusoidal voltage source and leave the N_2 coil open. Use a source frequency that guarantees $\omega L_1 \gg R_1$. Figure C.16 shows the resulting circuit.

Because the N_2 coil is open,

$$\mathbf{V}_2 = j\omega M\mathbf{I}_1.$$

Since $\omega L_1 \gg R_1$, the current I_1 is

$$\mathbf{I}_1 = \frac{\mathbf{V}_1}{j\omega L_1}.$$

Substituting the expression for $\mathbf{I}_1$ into the equation for $\mathbf{V}_2$ and rearranging yields

$$\left(\frac{\mathbf{V}_2}{\mathbf{V}_1}\right)_{I_2=0} = \frac{M}{L_1}.$$

Thus, the ratio M/L_1 equals the ratio of the terminal voltages when $\mathbf{I}_2 = 0$.

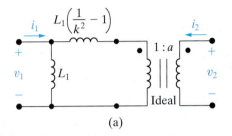

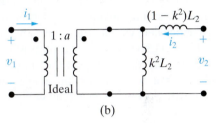

Figure C.14 ▲ Two equivalent circuits when $a = M/L_1$.

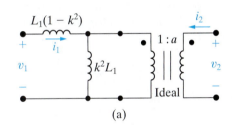

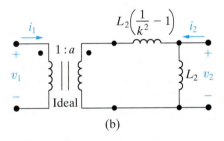

Figure C.15 ▲ Two equivalent circuits when $a = L_2/M$.

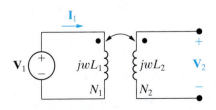

Figure C.16 ▲ Experimental determination of the ratio M/L_1.

To experimentally determine the ratio L_2/M, reverse the procedure; that is, coil 2 is energized and coil 1 is left open. Then

$$\frac{L_2}{M} = \left(\frac{\mathbf{V}_2}{\mathbf{V}_1}\right)_{I_1=0}.$$

Example C.2 illustrates how to replace magnetically coupled coils with an equivalent circuit that includes an ideal transformer and positive inductor values.

EXAMPLE C.2

Consider the circuit we analyzed in part (b) of Example C.1, which replaced the magnetically coupled coils in Fig. C.6 (with the polarity dot on the secondary winding moved to the bottom terminal) with a T-equivalent circuit. The equivalent circuit is shown in Fig. C.11 and must use a capacitor to represent the negative reactance. Repeat the analysis in Example C.1(b). but now replace the magnetically coupled coils with the equivalent circuit in Fig. C.14(a).

Solution

From Example C.1(b) we know that $L_1 = 9\,\text{H}$, $L_2 = 4\,\text{H}$, and $M = -3\,\text{H}$. We need to calculate the coupling coefficient k, the turns ratio a, and the inductor values for the equivalent circuit in Fig. C.14(a):

$$k = \frac{M}{\sqrt{L_1 L_2}} = \frac{3}{\sqrt{(9)(4)}} = 0.5,$$

$$a = \frac{M}{L_1} = \frac{-3}{9} = -\frac{1}{3},$$

$$L_1\left(\frac{1}{k^2} - 1\right) = 9\left(\frac{1}{0.5^2} - 1\right) = 27\,\text{H}.$$

Note that in calculating the coupling coefficient, we ignore the sign of the mutual inductance because this sign represents the location of the polarity marks and has no effect on the amount of coupling between the coils.

Using the values we calculated, we can replace the magnetically coupled coils in Fig. C.6 with the

equivalent circuit in Fig. C.14(a). Remember that the source frequency is 400 rad/sec. The result is the circuit in Fig. C.17. We have defined the currents and voltages for the ideal transformer, so we can write the equations for this circuit. There are three meshes, so we will write three KVL equations using the mesh currents $\mathbf{I}_1$, $\mathbf{I}_a$, and $\mathbf{I}_b$. Then we use the dot convention for ideal transformers to write the equations relating the phasor currents $\mathbf{I}_a$ and $\mathbf{I}_b$, and the phasor voltages $\mathbf{V}_a$ and $\mathbf{V}_b$:

$$(700 + j3700)\mathbf{I}_1 - j3600\mathbf{I}_a - 300 = 0,$$

$$-j3600\mathbf{I}_1 - j14{,}400\mathbf{I}_a + \mathbf{V}_a = 0,$$

$$(900 - j2500)\mathbf{I}_b + \mathbf{V}_b = 0,$$

$$\mathbf{V}_a = \frac{\mathbf{V}_b}{-(1/3)},$$

$$\mathbf{I}_a = \frac{1}{3}\mathbf{I}_b.$$

Solving for $\mathbf{I}_1$ and $\mathbf{I}_b$,

$$\mathbf{I}_1 = 0.02 - j0.06 = 63.25\underline{/-71.57°}\,\text{mA},$$

$$\mathbf{I}_b = 0.0267 + j0.0533 = 59.63\underline{/63.43°}\,\text{mA}.$$

Because $\mathbf{I}_2 = -\mathbf{I}_b$,

$$\mathbf{I}_2 = -0.0267 - j0.0533 = 59.63\underline{/-116.57°}\,\text{mA}.$$

The values for $\mathbf{I}_1$ and $\mathbf{I}_2$ match those found in part (b) of Example C.1.

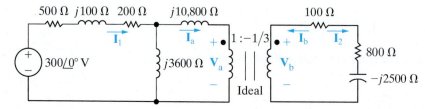

Figure C.17 ▲ The circuit of Fig. C.6, with the magnetically coupled coils replaced by the equivalent circuit in Fig. C.14(a)

APPENDIX
D

The Decibel

The decibel was introduced by telephone engineers to characterize the power loss across the cascaded circuits used to transmit telephone signals. Figure D.1 defines the problem.

There, p_i is the power input to the system, p_1 is the power output of circuit A, p_2 is the power output of circuit B, and p_o is the power output of the system. The power gain of each circuit is the ratio of the power out to the power in. Thus

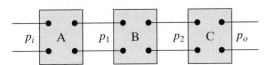

Figure D.1 ▲ Three cascaded circuits.

$$\sigma_A = \frac{p_1}{p_i}, \quad \sigma_B = \frac{p_2}{p_1}, \quad \text{and} \quad \sigma_C = \frac{p_o}{p_2}.$$

The overall power gain of the system is the product of the individual gains, or

$$\frac{p_o}{p_i} = \frac{p_1}{p_i}\frac{p_2}{p_1}\frac{p_o}{p_2} = \sigma_A\sigma_B\sigma_C.$$

We can replace the multiplication of power ratios with addition if we use the logarithm; that is,

$$\log_{10}\frac{p_o}{p_i} = \log_{10}\sigma_A + \log_{10}\sigma_B + \log_{10}\sigma_C.$$

The log of a power ratio was named the **bel**, in honor of Alexander Graham Bell. Thus, we calculate the overall power gain, in bels, by summing the power gains, also in bels, of each segment of the transmission system. In practice, the bel is an inconveniently large quantity. One-tenth of a bel is a more useful measure of power gain—hence the **decibel**. The number of decibels equals 10 times the number of bels, so

$$\text{Number of decibels} = 10\log_{10}\frac{p_o}{p_i}.$$

When we use the decibel as a measure of power ratios, in some situations the resistance seen looking into the circuit equals the resistance loading the circuit, as illustrated in Fig. D.2. When the input resistance equals the load resistance, we can convert the power ratio to either a voltage ratio or a current ratio:

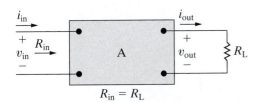

Figure D.2 ▲ A circuit in which the input resistance equals the load resistance.

$$\frac{p_o}{p_i} = \frac{v_{out}^2/R_L}{v_{in}^2/R_{in}} = \left(\frac{v_{out}}{v_{in}}\right)^2$$

or

$$\frac{p_o}{p_i} = \frac{i_{out}^2 R_L}{i_{in}^2/R_{in}} = \left(\frac{i_{out}}{i_{in}}\right)^2.$$

These equations show that the number of decibels becomes

$$\text{Number of decibels} = 20 \log_{10} \frac{v_{\text{out}}}{v_{\text{in}}} \qquad \text{(D.1)}$$

$$= 20 \log_{10} \frac{i_{\text{out}}}{i_{\text{in}}}.$$

The definition of the decibel used in Bode diagrams (see Appendix E) is borrowed from the results expressed by Eq. D.1, since these results apply to any transfer function involving a voltage ratio, a current ratio, a voltage-to-current ratio, or a current-to-voltage ratio. You should keep the original definition of the decibel firmly in mind because it is of fundamental importance in many engineering applications.

When you are working with transfer function amplitudes expressed in decibels, having a table that translates the decibel value to the actual value of the output/input ratio is helpful. Table D.1 gives some useful pairs. The ratio corresponding to a negative decibel value is the reciprocal of the positive ratio. For example, -3 dB corresponds to an output/input ratio of $1/1.41$, or 0.707. Interestingly, -3 dB corresponds to the half-power frequencies of the filter circuits discussed in Chapters 14 and 15.

TABLE D.1	Some dB-Ratio Pairs		
dB	**Ratio**	**dB**	**Ratio**
0	1.00	30	31.62
3	1.41	40	100.00
6	2.00	60	10^3
10	3.16	80	10^4
15	5.62	100	10^5
20	10.00	120	10^6

The decibel is also used as a unit of power when it expresses the ratio of a known power to a reference power. Usually, the reference power is 1 mW, and the power unit is written dBm, which stands for "decibels relative to one milliwatt." For example, a power of 20 mW corresponds to ± 13 dBm.

AC voltmeters commonly provide dBm readings that assume not only a 1 mW reference power but also a 600 Ω reference resistance (a value commonly used in telephone systems). Since a power of 1 mW in 600 Ω corresponds to 0.7746 V (rms), that voltage is read as 0 dBm on the meter. For analog meters, there usually is exactly a 10 dB difference between adjacent ranges. Although the scales may be marked 0.1, 0.3, 1, 3, 10, and so on, in fact 3.16 V on the 3 V scale lines up with 1 V on the 1 V scale.

Some voltmeters provide a switch to choose a reference resistance (50, 135, 600, or 900 Ω) or to select dBm or dBV (decibels relative to one volt).

Bode Diagrams

The frequency response plot is a very important tool for analyzing a circuit's behavior. Up to this point, we have shown qualitative sketches of the frequency response without discussing how to create these plots. The best way to generate and plot the amplitude and phase data is to use a computer; we can rely on it to give us accurate numerical plots of $|H(j\omega)|$ and $\theta(j\omega)$ versus ω. But we can create preliminary sketches of the frequency response using Bode diagrams.

A Bode diagram is a graphical technique that approximates a system's frequency response. These diagrams are named to recognize the pioneering work of H. W. Bode.[1] They are most useful for systems whose transfer function poles and zeros are reasonably well separated.

A Bode diagram consists of two separate plots: One shows how the transfer function amplitude varies with frequency, and the other shows how the transfer function phase angle varies with frequency. The Bode diagram plots are constructed using semilog graph paper to accommodate a wide range of frequencies. In both the amplitude and phase plots, the frequency is plotted on the horizontal log scale. The amplitude and phase angle are plotted on the linear vertical scale.

E.1 Real, First-Order Poles and Zeros

First, we consider a transfer function, $H(s)$, with poles and zeros that are real and distinct. We introduce the procedure for constructing a Bode diagram using

$$H(s) = \frac{K(s + z_1)}{s(s + p_1)},$$

from which

$$H(j\omega) = \frac{K(j\omega + z_1)}{j\omega(j\omega + p_1)}.$$

To begin, put the expression for $H(j\omega)$ in a **standard form**, which we derive by dividing out the poles and zeros:

$$H(j\omega) = \frac{Kz_1(1 + j\omega/z_1)}{p_1(j\omega)(1 + j\omega/p_1)}.$$

Next we let K_o equal Kz_1/p_1, and express $H(j\omega)$ in polar form:

$$H(j\omega) = \frac{K_o|1 + j\omega/z_1|\underline{/\psi_1}}{(|\omega|\underline{/90°})(|1 + j\omega/p_1|\underline{/\beta_1})}.$$

---

[1]See H. W. Bode, *Network Analysis and Feedback Design* (New York: Van Nostrand, 1945).

$$= \frac{K_o|1 + j\omega/z_1|}{|\omega||1 + j\omega/p_1|} \underline{/(\psi_1 - 90° - \beta_1)} .$$

Therefore,

$$|H(j\omega)| = \frac{K_o|1 + j\omega/z_1|}{\omega|1 + j\omega/p_1|}, \tag{E.1}$$

$$\theta(j\omega) = \psi_1 - 90° - \beta_1. \tag{E.2}$$

By definition, the phase angles ψ_1 and β_1 are

$$\psi_1 = \tan^{-1}\omega/z_1;$$

$$\beta_1 = \tan^{-1}\omega/p_1.$$

The Bode diagram includes the plots of Eq. E.1 (amplitude) and Eq. E.2 (phase) as functions of ω.

E.2 Straight-Line Amplitude Plots

Notice from Eq. E.1 that the amplitude plot requires us to multiply and divide values associated with the poles and zeros of $H(s)$. We can transform the multiplication and division into addition and subtraction if we express $|H(j\omega)|$ as a logarithmic value: the decibel (dB).[2] The amplitude of $H(j\omega)$ in decibels is

$$A_{dB} = 20 \log_{10}|H(j\omega)|.$$

Expressing Eq. E.1 in terms of decibels gives

$$A_{dB} = 20 \log_{10}\frac{K_o|1 + j\omega/z_1|}{\omega|1 + j\omega/p_1|} \tag{E.3}$$

$$= 20 \log_{10}K_o + 20 \log_{10}|1 + j\omega/z_1|$$

$$- 20 \log_{10}\omega - 20 \log_{10}|1 + j\omega/p_1|.$$

To construct a plot of Eq. E.3 versus frequency ω, we plot each term in the equation separately and then combine the separate plots graphically. The terms in Eq. E.3 are easy to plot because they can be approximated by straight lines.

The plot of $20 \log_{10}K_o$ is a horizontal straight line because K_o is not a function of frequency. The value of this term is positive for $K_o > 1$, zero for $K_o = 1$, and negative for $K_o < 1$.

Two straight lines approximate the plot of $20 \log_{10}|1 + j\omega/z_1|$. For small values of ω, the magnitude $|1 + j\omega/z_1|$ is approximately 1, and therefore

$$20 \log_{10}|1 + j\omega/z_1| \rightarrow 0 \quad \text{as } \omega \rightarrow 0.$$

---

[2]See Appendix D for more information regarding the decibel.

For large values of ω, the magnitude $|1 + j\omega/z_1|$ is approximately ω/z_1, and therefore

$$20 \log_{10}|1 + j\omega/z_1| \rightarrow 20 \log_{10}(\omega/z_1) \text{ as } \omega \rightarrow \infty.$$

On a log frequency scale, $20 \log_{10}(\omega/z_1)$ is a straight line with a slope of 20 dB/decade (a decade is a 10-to-1 change in frequency). This straight line intersects the 0 dB axis at $\omega = z_1$. This value of ω is called the **corner frequency**. Thus, two straight lines can approximate the amplitude plot of a first-order zero, as shown in Fig. E.1.

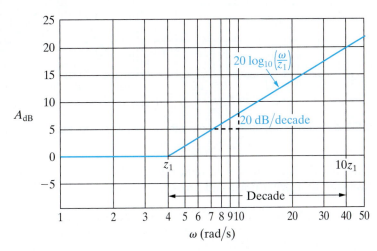

Figure E.1 ▲ A straight-line approximation of the amplitude plot of a first-order zero.

The plot of $-20 \log_{10}\omega$ is a straight line having a slope of -20 dB/decade that intersects the 0 dB axis at $\omega = 1$.

Two straight lines approximate the plot of $-20 \log_{10}|1 + j\omega/p_1|$ and intersect on the 0 dB axis at $\omega = p_1$. For large values of ω, the straight line $20 \log_{10}(\omega/p_1)$ has a slope of -20 dB/decade. Figure E.2 shows the straight-line approximation of the amplitude plot for a first-order pole.

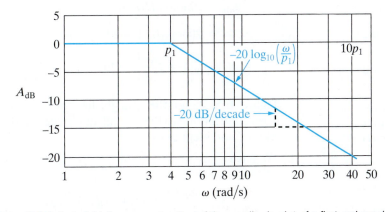

Figure E.2 ▲ A straight-line approximation of the amplitude plot of a first-order pole.

Figure E.3 shows a plot of Eq. E.3 for $K_o = \sqrt{10}$, $z_1 = 0.1$ rad/s, and $p_1 = 5$ rad/s. Each term in Eq. E.3 is also plotted in Fig. E.3, so you can verify that the individual terms sum to create the plot of the transfer function's amplitude, labeled $20 \log_{10}|H(j\omega)|$.

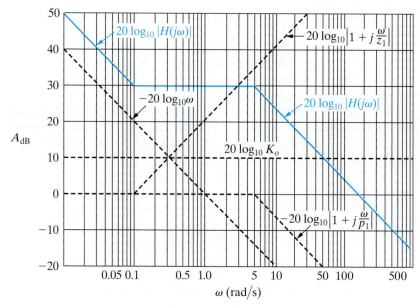

Figure E.3 ▲ A straight-line approximation of the amplitude plot for Eq. E.3.

Example E.1 constructs a straight-line amplitude plot for a transfer function with first-order poles and zeros.

EXAMPLE E.1

For the circuit in Fig. E.4:

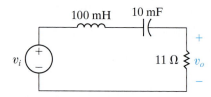

Figure E.4 ▲ The circuit for Example E.1.

a) Find the transfer function, $H(s) = V_o(s)/V_i(s)$.

b) Construct a straight-line approximation of the Bode amplitude plot.

c) Calculate $20 \log_{10}|H(j\omega)|$ at $\omega = 50$ rad/s and $\omega = 1000$ rad/s. Identify these two values on the plot you constructed in (b).

d) If $v_i(t) = 5 \cos (500t + 15°)$ V, use the Bode plot you constructed to predict the amplitude of $v_o(t)$ in the steady state.

Solution

a) Transform the circuit in Fig. E.4 into the s-domain and then use voltage division to get

$$H(s) = \frac{(R/L)s}{s^2 + (R/L)s + \dfrac{1}{LC}}.$$

Substituting the numerical values from the circuit, we get

$$H(s) = \frac{110s}{s^2 + 110s + 1000} = \frac{110s}{(s + 10)(s + 100)}.$$

b) Write $H(j\omega)$ in standard form:

$$H(j\omega) = \frac{0.11j\omega}{[1 + j(\omega/10)][1 + j(\omega/100)]}.$$

The expression for the amplitude of $H(j\omega)$ in decibels is

$$A_{dB} = 20 \log_{10}|H(j\omega)|$$

$$= 20 \log_{10}0.11 + 20 \log_{10}|j\omega|$$

$$- 20 \log_{10}\left|1 + j\frac{\omega}{10}\right| - 20 \log_{10}\left|1 + j\frac{\omega}{100}\right|.$$

Figure E.5 shows the straight-line plot. Each term contributing to the overall amplitude is also plotted and identified.

c) We have

$$H(j50) = \frac{0.11(j50)}{(1 + j5)(1 + j0.5)}$$

$$= 0.9648\underline{/-15.25°},$$

$$20 \log_{10}|H(j50)| = 20 \log_{10}0.9648$$

$$= -0.311 \text{ dB};$$

$$H(j1000) = \frac{0.11(j1000)}{(1 + j100)(1 + j10)}$$

$$= 0.1094\underline{/-83.72°};$$

$$20 \log_{10} 0.1094 = -19.22 \text{ dB}.$$

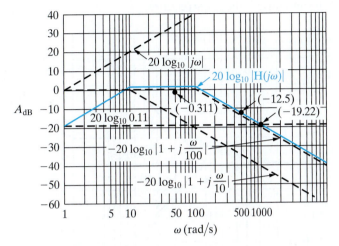

Figure E.5 ▲ The straight-line amplitude plot for the transfer function of the circuit in Fig. E.4.

These two magnitude values are plotted in Fig. E.5.

d) The frequency of v_i is 500 rad/sec. As we can see from the Bode plot in Fig. E.5, the value of A_{dB} at $\omega = 500$ rad/s is approximately -12.5 dB. Therefore,

$$|H(j500)| = 10^{(-12.5/20)} = 0.24$$

and

$$V_o = |H(j500)|V_i = (0.24)(5) = 1.19 \text{ V}.$$

To calculate the actual value of $|H(j\omega)|$, substitute $\omega = 500$ into the equation for $|H(j\omega)|$:

$$H(j500) = \frac{0.11(j500)}{(1 + j50)(1 + j5)} = 0.22\underline{/-77.54°}.$$

Thus, the actual output voltage magnitude for the specified signal source at a frequency of 500 rad/s is

$$V_o = |H(j500)|V_i = (0.22)(5) = 1.1 \text{ V}.$$

E.3 More Accurate Amplitude Plots

We can make the straight-line plots for first-order poles and zeros more accurate by correcting the amplitude values at the corner frequency, one-half the corner frequency and twice the corner frequency. At the corner frequency, the actual value in decibels is

$$A_{dB_c} = \pm 20 \log_{10}|1 + j1|$$

$$= \pm 20 \log_{10}\sqrt{2}$$

$$\approx \pm 3 \text{ dB}.$$

The actual value at one-half the corner frequency is

$$A_{dB_{c/2}} = \pm 20 \log_{10}\left|1 + j\frac{1}{2}\right|$$

$$= \pm 20 \log_{10}\sqrt{\frac{5}{4}}$$

$$\approx \pm 1 \text{ dB}.$$

At twice the corner frequency, the actual value in decibels is

$$A_{dB_{2c}} = \pm 20 \log_{10}|1 + j2|$$

$$= \pm 20 \log_{10}\sqrt{5}$$

$$\approx \pm 7 \text{ dB}.$$

In these three equations, the plus sign applies to a first-order zero, and the minus sign applies to a first-order pole. The straight-line approximation of the amplitude plot is 0 dB at both the corner frequency and at one-half the corner frequency, and is ± 6 dB at twice the corner frequency. Hence, the corrections are ± 3 dB at the corner frequency and ± 1 dB at both one-half the corner frequency and twice the corner frequency. Figure E.6 summarizes these corrections.

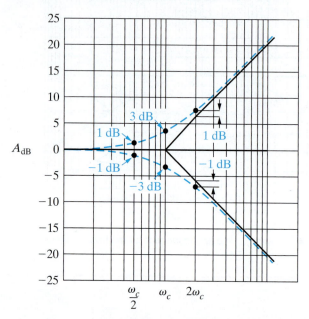

Figure E.6 ▲ Corrected amplitude plots for a first-order zero and pole.

If the poles and zeros of $H(s)$ are widely separated, you can include these corrections into the overall amplitude plot and produce a reasonably accurate curve. However, if the poles and zeros are close together, the overlapping corrections are difficult to evaluate, so estimating the amplitude characteristic using the uncorrected straight-line plot is a better choice. Use a computer to create an accurate amplitude plot in the frequency range of interest.

E.4 Straight-Line Phase Angle Plots

We can also use straight-line approximations to plot the transfer function phase angle versus frequency. Using the transfer function phase angle in Eq. E.2, we know that the phase angle associated with the constant K_o is zero, and the phase angle associated with a first-order zero or pole at the origin is a constant $\pm 90°$. For a first-order zero or pole not at the origin, the straight-line approximations are as follows:

- For frequencies less than one-tenth the corner frequency, the phase angle is approximated by zero.
- For frequencies greater than 10 times the corner frequency, the phase angle is approximated by $\pm 90°$.
- Between one-tenth the corner frequency and 10 times the corner frequency, the phase angle plot is a straight line with the value 0° at one-tenth the corner frequency, $\pm 45°$ at the corner frequency, and $\pm 90°$ at 10 times the corner frequency.

Throughout, the plus sign applies to the first-order zero and the minus sign to the first-order pole. Figure E.7 depicts the straight-line approximation for a first-order zero and pole. The dashed curves show the exact plot of the phase angle versus frequency. Note how closely the straight-line plot approximates the actual phase angle plot. The maximum deviation between the straight-line plot and the actual plot is approximately 6°.

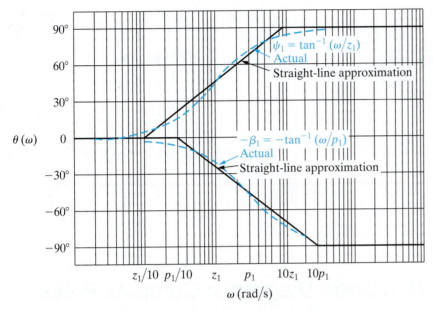

Figure E.7 ▲ Phase angle plots for a first-order zero and pole.

We construct a phase angle plot using a straight-line approximation in Example E.2.

EXAMPLE E.2

a) Make a straight-line phase angle plot for the transfer function in Example E.1.

b) Compute the phase angle $\theta(j\omega)$ at $\omega = 50, 500,$ and 1000 rad/s. Add these phase angle values to the plot you constructed in (a).

c) Using the results from Example E.1(e) and Example E.2(b), compute the steady-state output voltage if the source voltage is given by $v_i(t) = 10\cos(500t - 25°)$ V.

Solution

a) From Example E.1,

$$H(j\omega) = \frac{0.11(j\omega)}{[1 + j(\omega/10)][1 + j(\omega/100)]}$$

$$= \frac{0.11|j\omega|}{|1 + j(\omega/10)||1 + j(\omega/100)|}\underline{/(\psi_1 - \beta_1 - \beta_2)}.$$

Therefore,

$$\theta(j\omega) = \psi_1 - \beta_1 - \beta_2,$$

where

$$\psi_1 = 90°,$$

$$\beta_1 = \tan^{-1}(\omega/10), \quad \text{and}$$

$$\beta_2 = \tan^{-1}(\omega/100).$$

We construct the straight-line approximation of $\theta(j\omega)$ by adding $\psi_1 = 90°$ and the straight-line approximations for $-\beta_1$ and $-\beta_2$, all of which are depicted in Fig. E.8.

b) From the transfer function we have

$$H(j50) = 0.96\underline{/-15.25°},$$

$$H(j500) = 0.22\underline{/-77.54°},$$

$$H(j1000) = 0.11\underline{/-83.72°}.$$

Thus,

$$\theta(j50) = -15.25°,$$

$$\theta(j500) = -77.54°,$$

and

$$\theta(j1000) = -83.72°.$$

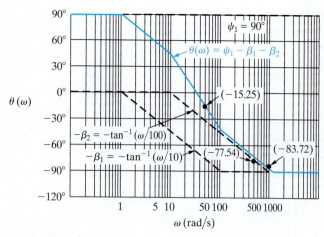

Figure E.8 ▲ A straight-line approximation of $\theta(\omega)$ for Example E.2.

These three phase angles are identified in Fig. E.8.

c) The frequency of the input voltage is 500 rad/sec. Therefore,

$$V_o = |H(j500)|V_i$$

$$= (0.22)(10)$$

$$= 2.2 \text{ V,}$$

and

$$\theta_o = \theta(j500) + \theta_i$$

$$= -77.54° - 25°$$

$$= -102.54°.$$

Thus,

$$v_o(t) = 2.2 \cos(500t - 102.54°) \text{ V.}$$

E.5 Bode Diagrams: Complex Poles and Zeros

We now consider how to construct a Bode diagram when the transfer function has complex poles and zeros. Let's focus on constructing the amplitude and phase angle plots for a transfer function with a pair of complex poles. Once you understand the rules for handling complex poles, you can apply these rules, with minor adjustments, to create plots for a pair of complex zeros.

The complex poles and zeros of $H(s)$ always appear in conjugate pairs. When $H(s)$ has complex poles, the first step in constructing a Bode diagram for $H(s)$ is to combine the conjugate pair into a single quadratic term. Thus, for

$$H(s) = \frac{K}{(s + \alpha - j\beta)(s + \alpha + j\beta)},$$

we first rewrite the product $(s + \alpha - j\beta)(s + \alpha + j\beta)$ as

$$(s + \alpha)^2 + \beta^2 = s^2 + 2\alpha s + \alpha^2 + \beta^2.$$

When making Bode diagrams, we write the quadratic term in a more convenient form:

$$s^2 + 2\alpha s + \alpha^2 + \beta^2 = s^2 + 2\zeta\omega_n s + \omega_n^2.$$

Comparing the two forms shows that

$$\omega_n^2 = \alpha^2 + \beta^2$$

and

$$\zeta\omega_n = \alpha.$$

The term ω_n is the corner frequency of the quadratic term, and ζ is the damping coefficient of the quadratic term. If $\zeta < 1$, the roots of the quadratic term are complex, and if $\zeta \geq 1$, the roots are real. Assuming that $\zeta < 1$, we rewrite the transfer function as

$$H(s) = \frac{K}{s^2 + 2\zeta\omega_n s + \omega_n^2}.$$

We then write the transfer function in standard form by dividing through by ω_n^2, so

$$H(s) = \frac{K}{\omega_n^2} \frac{1}{1 + (s/\omega_n)^2 + 2\zeta(s/\omega_n)},$$

from which

$$H(j\omega) = \frac{K_o}{1 - (\omega^2/\omega_n^2) + j(2\zeta\omega/\omega_n)},$$

where

$$K_o = \frac{K}{\omega_n^2}.$$

We replace the ratio ω/ω_n by a new variable, u. Then

$$H(j\omega) = \frac{K_o}{1 - u^2 + j2\zeta u}.$$

Now we write $H(j\omega)$ in polar form:

$$H(j\omega) = \frac{K_o}{|(1 - u^2) + j2\zeta u|\underline{/\beta_1}},$$

from which

$$A_{\text{dB}} = 20 \log_{10}|H(j\omega)|$$

$$= 20 \log_{10}K_o - 20 \log_{10}|(1 - u^2) + j2\zeta u|,$$

and

$$\theta(j\omega) = -\beta_1 = -\tan^{-1}\frac{2\zeta u}{1 - u^2}.$$

E.6 Straight-Line Amplitude Plots for Complex Poles

The expression $-20 \log_{10}|1 - u^2 + j2\zeta u|$ represents the quadratic term's contribution to the amplitude of $H(j\omega)$. Because $u = \omega/\omega_n$, $u \to 0$ as $\omega \to 0$, and $u \to \infty$ as $\omega \to \infty$. To see how the term behaves as ω ranges from 0 to ∞, we note that

$$-20 \log_{10}|(1 - u^2) + j2\zeta u| = -20 \log_{10}\sqrt{(1 - u^2)^2 + 4\zeta^2 u^2}$$

$$= -10 \log_{10}[u^4 + 2u^2(2\zeta^2 - 1) + 1]. \tag{E.4}$$

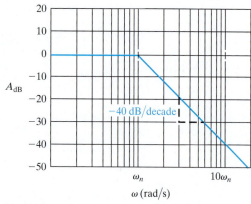

Figure E.9 ▲ The amplitude plot for a pair of complex poles.

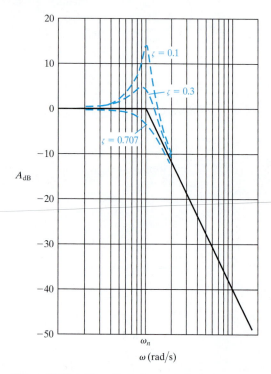

Figure E.10 ▲ The effect of ζ on the amplitude plot.

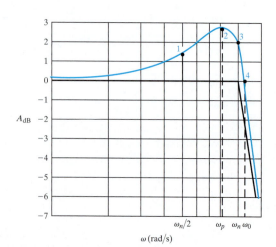

Figure E.11 ▲ Four points on the corrected amplitude plot for a pair of complex poles.

Then, as $u \to 0$,

$$-10 \log_{10}[u^4 + 2u^2(2\zeta^2 - 1) + 1] \to 0,$$

and as $u \to \infty$,

$$-10 \log_{10}[u^4 + 2u^2(2\zeta^2 - 1) + 1] \to -40 \log_{10}u.$$

From these limiting expressions, we see that the approximate amplitude plot consists of two straight lines. For $\omega < \omega_n$, the straight line lies along the 0 dB axis, and for $\omega > \omega_n$, the straight line has a slope of -40 dB/decade. These two straight lines join on the 0 dB axis at $u = 1$ or $\omega = \omega_n$. Figure E.9 shows the straight-line approximation for a quadratic term with $\zeta < 1$.

E.7 Correcting Straight-Line Amplitude Plots for Complex Poles

Corrections to the straight-line amplitude plot for a pair of complex poles depend on the damping coefficient ζ. Figure E.10 shows the effect of ζ on the amplitude plot. Note that when ζ is very small, a large peak in the amplitude occurs in the neighborhood of the corner frequency $\omega_n(u = 1)$. When $\zeta \geq 1/\sqrt{2}$, the corrected amplitude plot lies entirely below the straight-line approximation. The straight-line amplitude plot can be corrected by locating four points on the actual curve. These four points correspond to (1) one-half the corner frequency, (2) the frequency at which the amplitude reaches its peak value, (3) the corner frequency, and (4) the frequency at which the amplitude is zero.

At one-half the corner frequency (point 1), the actual amplitude is

$$A_{dB}(\omega_n/2) = -10 \log_{10}(\zeta^2 + 0.5625). \tag{E.5}$$

The amplitude peaks (point 2) at a frequency of

$$\omega_p = \omega_n\sqrt{1 - 2\zeta^2}, \tag{E.6}$$

and it has a peak amplitude of

$$A_{dB}(\omega_p) = -10 \log_{10}[4\zeta^2(1 - \zeta^2)]. \tag{E.7}$$

At the corner frequency (point 3), the actual amplitude is

$$A_{dB}(\omega_n) = -20 \log_{10}2\zeta. \tag{E.8}$$

The corrected amplitude plot crosses the 0 dB axis (point 4) at

$$\omega_o = \omega_n\sqrt{2(1 - 2\zeta^2)} = \sqrt{2}\omega_p. \tag{E.9}$$

Figure E.11 shows these four points.

Evaluating Eq. E.4 at $u = 0.5$ and $u = 1.0$, respectively, yields Eqs. E.5 and E.8. Equation E.9 results from finding the value of u that makes $u^4 + 2u^2(2\zeta^2 - 1) + 1 = 1$. To derive Eq. E.6, differentiate Eq. E.4 with respect to u and find the value of u where the derivative is zero. To derive E.7, find the value of u for the frequency in Eq. E.6, then evaluate Eq. E.4 at that value of u.

Example E.3 illustrates the amplitude plot for a transfer function with a pair of complex poles.

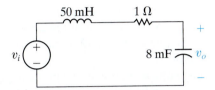

EXAMPLE E.3

Compute the transfer function for the circuit shown in Fig. E.12.

Figure E.12 ▲ The circuit for Example E.3.

a) Find the value of the corner frequency in radians per second.

b) Find the value of K_o.

c) Find the value of the damping coefficient.

d) Make a straight-line amplitude plot for frequencies from 10 to 500 rad/s.

e) Calculate the actual amplitude in decibels at $\omega_n/2$, ω_p, ω_n, and ω_o. Use these values to correct the plot in (d).

f) Using the corrected amplitude plot, describe the type of filter represented by the circuit in Fig. E.12 and estimate its cutoff frequency, ω_c.

Solution

Transform the circuit in Fig. E.12 to the s-domain and then use voltage division to get

$$H(s) = \frac{\dfrac{1}{LC}}{s^2 + \left(\dfrac{R}{L}\right)s + \dfrac{1}{LC}}.$$

Substituting the component values,

$$H(s) = \frac{2500}{s^2 + 20s + 2500}.$$

a) From the expression for $H(s)$, $\omega_n^2 = 2500$; therefore, $\omega_n = 50$ rad/s.

b) By definition, K_o is $2500/\omega_n^2$, or 1.

c) The coefficient of s in the denominator equals $2\zeta\omega_n$; therefore

$$\zeta = \frac{20}{2\omega_n} = 0.20.$$

d) See Fig. E.13.

e) The actual amplitudes are

$$A_{dB}(\omega_n/2) = -10 \log_{10}(0.6025) = 2.2 \text{ dB},$$

$$\omega_p = 50\sqrt{0.92} = 47.96 \text{ rad/s},$$

$$A_{dB}(\omega_p) = -10 \log_{10}(0.16)(0.96) = 8.14 \text{ dB},$$

$$A_{dB}(\omega_n) = -20 \log_{10}(0.4) = 7.96 \text{ dB},$$

$$\omega_o = \sqrt{2}\omega_p = 67.82 \text{ rad/s},$$

$$A_{dB}(\omega_o) = 0 \text{ dB}.$$

Figure E.13 shows these four points, identified as follows:

- Point 1 has the coordinates (25 rad/s, 2.2 dB),
- Point 2 has the coordinates (47.96 rad/s, 8.14 dB),
- Point 3 has the coordinates (50 rad/s, 7.96 dB), and
- Point 4 has the coordinates (67.82 rad/s, 0 dB).

Figure E.13 also shows the corrected plot, which is the dashed line that passes through these four points.

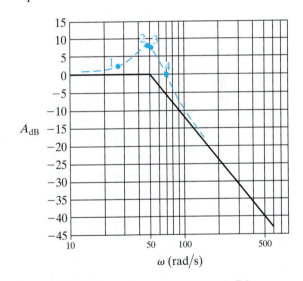

Figure E.13 ▲ The amplitude plot for Example E.3.

f) The amplitude plot in Fig. E.13 identifies the circuit as a low-pass filter. At the cutoff frequency, the magnitude of the transfer function, $|H(j\omega_c)|$, is 3 dB less than the maximum magnitude. From the corrected plot, the cutoff frequency appears to be about 55 rad/s, almost the same as that predicted by the straight-line Bode diagram.

E.8 Phase Angle Plots for Complex Poles

The phase angle plot for a pair of complex poles is a plot of $\theta(j\omega) = -\tan^{-1}[2\zeta u/(1 - u^2)]$. The phase angle is zero at zero frequency and is $-90°$ at the corner frequency. It approaches $-180°$ as ω (and, therefore, u) becomes large. As in the case of the amplitude plot, ζ determines the exact shape of the phase angle plot. Figure E.14 shows the effect of ζ on the phase angle plot.

We can make a straight-line approximation of the phase angle plot for a pair of complex poles. We do so by drawing three line segments:

- For $\omega \leq (4.81^{-\zeta})\omega_n$, draw a horizontal line at $0°$;
- For $\omega \geq (4.81^{\zeta})\omega_n$, draw a horizontal line at $-180°$;
- For $(4.81^{-\zeta})\omega_n \leq \omega \leq (4.81^{\zeta})\omega_n$, draw a straight line connecting the $0°$ phase angle at $(4.81^{-\zeta})\omega_n$ to the $-180°$ phase angle at $(4.81^{\zeta})\omega_n$. This line passes through $-90°$ at ω_n and has a slope of $-132/\zeta$ degrees/decade ($-2.3/\zeta$ rad/decade).

Figure E.15 depicts the straight-line approximation for $\zeta = 0.3$ and shows the actual phase angle plot. We see that the straight line is a good approximation of the actual curve near the corner frequency, but the error is quite large near the two points where the straight lines intersect. In Example E.4, we summarize our discussion of Bode diagrams.

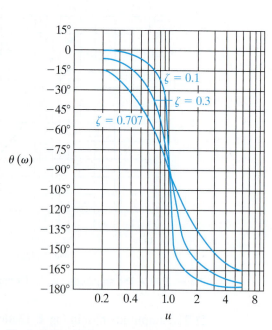

Figure E.14 ▲ The effect of ζ on the phase angle plot.

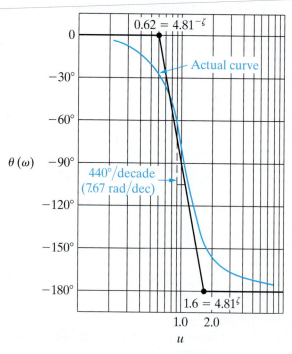

Figure E.15 ▲ A straight-line approximation of the phase angle for a pair of complex poles.

EXAMPLE E.4

a) Compute the transfer function, $H(s)$, for the circuit shown in Fig. E.16.

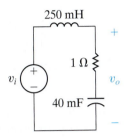

Figure E.16 ▲ The circuit for Example E.4.

b) Make a straight-line amplitude plot of $20 \log_{10}|H(j\omega)|$.

c) Use the straight-line amplitude plot to determine the type of filter represented by this circuit and then estimate its cutoff frequency.

d) Find the actual cutoff frequency.

e) Make a straight-line phase angle plot of $H(j\omega)$.

f) Using the plot in (e), find the phase angle at the estimated cutoff frequency found in (c).

g) Find the phase angle at the actual cutoff frequency found in (d).

Solution

a) Transform the circuit in Fig. E.16 to the s-domain and then use voltage division to get

$$H(s) = \frac{\dfrac{R}{L}s + \dfrac{1}{LC}}{s^2 + \dfrac{R}{L}s + \dfrac{1}{LC}}.$$

Substituting the component values from the circuit gives

$$H(s) = \frac{4(s + 25)}{s^2 + 4s + 100}.$$

b) The first step in making Bode diagrams is to put $H(j\omega)$ in standard form. Because $H(s)$ contains a quadratic factor, we first check the value of ζ. We find that $\zeta = 0.2$ and $\omega_n = 10$, so

$$H(s) = \frac{s/25 + 1}{1 + (s/10)^2 + 0.4(s/10)},$$

from which

$$H(j\omega) = \frac{|1 + j\omega/25|\ \underline{/\psi_1}}{|1 - (\omega/10)^2 + j0.4\omega/10|\ \underline{/\beta_1}}.$$

Note that for the quadratic term, $u = \omega/10$. The amplitude of $H(j\omega)$ in decibels is

$$A_{\mathrm{dB}} = 20 \log_{10}|1 + j\omega/25|$$
$$- 20 \log_{10}\left[\left|1 - \left(\frac{\omega}{10}\right)^2 + j0.4\left(\frac{\omega}{10}\right)\right|\right],$$

and the phase angle is

$$\theta(j\omega) = \psi_1 - \beta_1,$$

where

$$\psi_1 = \tan^{-1}(\omega/25),$$

$$\beta_1 = \tan^{-1}\frac{0.4(\omega/10)}{1 - (\omega/10)^2}.$$

Figure E.17 shows the amplitude plot and includes the plots of the two terms that make up A_{dB}.

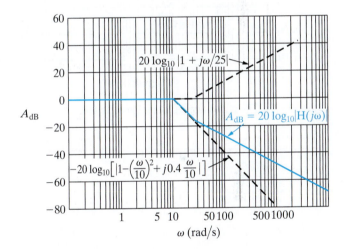

Figure E.17 ▲ The amplitude plot for Example E.4.

c) The straight-line amplitude plot in Fig. E.17 indicates that the circuit is a low-pass filter. At the cutoff frequency, the amplitude of $H(j\omega)$ is 3 dB less than the amplitude in the passband. From the plot, we predict that the cutoff frequency is approximately 13 rad/s.

d) To solve for the actual cutoff frequency, replace s with $j\omega$ in $H(s)$, compute the expression for $|H(j\omega)|$, set $|H(j\omega_c)| = (1/\sqrt{2})H_{\max} = 1/\sqrt{2}$, and solve for ω_c.

First,

$$H(j\omega) = \frac{4(j\omega) + 100}{(j\omega)^2 + 4(j\omega) + 100}.$$

Then,

$$|H(j\omega_c)| = \frac{\sqrt{(4\omega_c)^2 + 100^2}}{\sqrt{(100 - \omega_c^2)^2 + (4\omega_c)^2}} = \frac{1}{\sqrt{2}}.$$

Solving for ω_c gives us

$$\omega_c = 16 \text{ rad/s}.$$

e) Figure E.18 shows the phase angle plot, as well as plots of the two angles ψ_1 and $-\beta_1$. Note that the straight-line segment of $\theta(\omega)$ between 1.0 and 2.5 rad/s does not have the same slope as the segment between 2.5 and 100 rad/s.

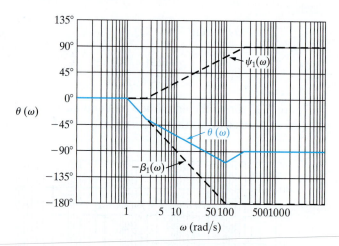

Figure E.18 ▲ The phase angle plot for Example E.4.

f) From the phase angle plot in Fig. E.18, we estimate the phase angle at the estimated cutoff frequency of 13 rad/s to be $-65°$.

g) We can compute the exact phase angle at the actual cutoff frequency by substituting $s = j16$ into the transfer function $H(s)$:

$$H(j16) = \frac{4(j16 + 25)}{(j16)^2 + 4(j16) + 100}.$$

Computing the phase angle, we see

$$\theta(j\omega_c) = \theta(j16) = -125.0°.$$

Note the large error in the predicted angle. In general, straight-line phase angle plots do not give satisfactory results in the frequency band where the phase angle is changing. The straight-line phase angle plot is useful only in predicting the general behavior of the phase angle, not in estimating actual phase angle values at particular frequencies.

An Abbreviated Table of Trigonometric Identities

1. $\sin(\alpha \pm \beta) = \sin\alpha \cos\beta \pm \cos\alpha \sin\beta$

2. $\cos(\alpha \pm \beta) = \cos\alpha \cos\beta \mp \sin\alpha \sin\beta$

3. $\sin\alpha + \sin\beta = 2 \sin\dfrac{\alpha + \beta}{2} \cos\dfrac{\alpha - \beta}{2}$

4. $\sin\alpha - \sin\beta = 2 \cos\left(\dfrac{\alpha + \beta}{2}\right) \sin\left(\dfrac{\alpha - \beta}{2}\right)$

5. $\cos\alpha + \cos\beta = 2 \cos\left(\dfrac{\alpha + \beta}{2}\right) \cos\left(\dfrac{\alpha - \beta}{2}\right)$

6. $\cos\alpha - \cos\beta = -2 \sin\left(\dfrac{\alpha + \beta}{2}\right) \sin\left(\dfrac{\alpha - \beta}{2}\right)$

7. $2 \sin\alpha \sin\beta = \cos(\alpha - \beta) - \cos(\alpha + \beta)$

8. $2 \cos\alpha \cos\beta = \cos(\alpha - \beta) + \cos(\alpha + \beta)$

9. $2 \sin\alpha \cos\beta = \sin(\alpha + \beta) + \sin(\alpha - \beta)$

10. $\sin 2\alpha = 2 \sin\alpha \cos\alpha$

11. $\cos 2\alpha = 2 \cos^2\alpha - 1 = 1 - 2 \sin^2\alpha$

12. $\cos^2\alpha = \dfrac{1}{2} + \dfrac{1}{2} \cos 2\alpha$

13. $\sin^2\alpha = \dfrac{1}{2} - \dfrac{1}{2} \cos 2\alpha$

14. $\tan(\alpha \pm \beta) = \dfrac{\tan\alpha \pm \tan\beta}{1 \mp \tan\alpha \tan\beta}$

15. $\tan 2\alpha = \dfrac{2 \tan\alpha}{1 - \tan^2\alpha}$

An Abbreviated Table of Integrals

1. $\displaystyle\int xe^{ax}dx = \frac{e^{ax}}{a^2}(ax - 1)$

2. $\displaystyle\int x^2 e^{ax}\,dx = \frac{e^{ax}}{a^3}(a^2x^2 - 2ax + 2)$

3. $\displaystyle\int x\,\sin ax\,dx = \frac{1}{a^2}\sin ax - \frac{x}{a}\cos ax$

4. $\displaystyle\int x\,\cos ax\,dx = \frac{1}{a^2}\cos ax + \frac{x}{a}\sin ax$

5. $\displaystyle\int e^{ax}\sin bx\,dx = \frac{e^{ax}}{a^2 + b^2}(a\,\sin bx - b\,\cos bx)$

6. $\displaystyle\int e^{ax}\cos bx\,dx = \frac{e^{ax}}{a^2 + b^2}(a\,\cos bx + b\,\sin bx)$

7. $\displaystyle\int \frac{dx}{x^2 + a^2} = \frac{1}{a}\tan^{-1}\frac{x}{a}$

8. $\displaystyle\int \frac{dx}{(x^2 + a^2)^2} = \frac{1}{2a^2}\left(\frac{x}{x^2 + a^2} + \frac{1}{a}\tan^{-1}\frac{x}{a}\right)$

9. $\displaystyle\int \sin ax\,\sin bx\,dx = \frac{\sin(a - b)x}{2(a - b)} - \frac{\sin(a + b)x}{2(a + b)}, \quad a^2 \neq b^2$

10. $\displaystyle\int \cos ax\,\cos bx\,dx = \frac{\sin(a - b)x}{2(a - b)} + \frac{\sin(a + b)x}{2(a + b)}, \quad a^2 \neq b$

11. $\displaystyle\int \sin ax \cos bx\, dx = -\frac{\cos(a-b)x}{2(a-b)} - \frac{\cos(a+b)x}{2(a+b)}, \quad a^2 \neq b^2$

12. $\displaystyle\int \sin^2 ax\, dx = \frac{x}{2} - \frac{\sin 2ax}{4a}$

13. $\displaystyle\int \cos^2 ax\, dx = \frac{x}{2} + \frac{\sin 2ax}{4a}$

14. $\displaystyle\int_0^\infty \frac{a\, dx}{a^2 + x^2} = \begin{cases} \frac{\pi}{2}, a > 0; \\ 0, a = 0; \\ \frac{-\pi}{2}, a < 0 \end{cases}$

15. $\displaystyle\int_0^\infty \frac{\sin ax}{x}\, dx = \begin{cases} \frac{\pi}{2}, a > 0; \\ \frac{-\pi}{2}, a < 0 \end{cases}$

16. $\displaystyle\int x^2 \sin ax\, dx = \frac{2x}{a^2} \sin ax - \frac{a^2 x^2 - 2}{a^3} \cos ax$

17. $\displaystyle\int x^2 \cos ax\, dx = \frac{2x}{a^2} \cos ax + \frac{a^2 x^2 - 2}{a^3} \sin ax$

18. $\displaystyle\int e^{ax} \sin^2 bx\, dx = \frac{e^{ax}}{a^2 + 4b^2}\left[(a \sin bx - 2b \cos bx) \sin bx + \frac{2b^2}{a}\right]$

19. $\displaystyle\int e^{ax} \cos^2 bx\, dx = \frac{e^{ax}}{a^2 + 4b^2}\left[(a \cos bx + 2b \sin bx) \cos bx + \frac{2b^2}{a}\right]$

Common Standard Component Values

Resistors (5% tolerance)[Ω]					
10	100	1.0 k	10 k	100 k	1.0 M
	120	1.2 k	12 k	120 k	
15	150	1.5 k	15 k	150 k	1.5 M
	180	1.8 k	18 k	180 k	
22	220	2.2 k	22 k	220 k	2.2 M
	270	2.7 k	27 k	270 k	
33	330	3.3 k	33 k	330 k	3.3 M
	390	3.9 k	39 k	390 k	
47	470	4.7 k	47 k	470 k	4.7 M
	560	5.6 k	56 k	560 k	
68	680	6.8 k	68 k	680 k	6.8 M

Capacitors		
10 pF	22 pF	47 pF
100 pF	220 pF	470 pF
0.001 μF	0.0022 μF	0.0047 μF
0.01 μF	0.022 μF	0.047 μF
0.1 μF	0.22 μF	0.47 μF
1 μF	2.2 μF	4.7 μF
10 μF	22 μF	47 μF
100 μF	220 μF	470 μF

Inductors	
Value	**Current Rating**
10 μH	3 A
100 μH	0.91 A
1 mH	0.15 A
10 mH	0.04 A

Answers to Selected Problems

Chapter 1

1.3 a) 3.456 s

    b) 175,000 cell lengths/week

1.6 104.4 gigawatt-hours

1.15 a) -100 W; power is being delivered by the box

    b) Leaving

    c) Gain

1.18 a) 937.5 mW

    b) 1.875 mJ

1.25 a) 5.39 W

    b) 9 J

1.34 $\sum P_{del} = \sum P_{abs} = 2280$ W

Chapter 2

2.7 a) 20 V

    b) 8 W (absorbed)

2.11 a) -16 mA

    b) 640 mW

    c) 16 mA; 640 mW

2.16 100 Ω resistor

2.17 a) 1.2 A, 0.3 A

    b) 120 V

    c) $\sum P_{dev} = \sum P_{diss} = 180$ W

2.29 a) 10 A in parallel with 5 Ω

    b) 80 W

2.33 15 V, 1.4167 W

2.42 1800 W, which is $1/2$ the power for the circuit in Fig. 2.41

Chapter 3

3.2 a) 21.2 Ω, 10 kΩ, 1600 Ω, 120 Ω

    b) 4.72 W, 90 mW, 25 μW, 108 mW

3.13 a) 15 V

    b) 1.8 W, 450 mW

    c) 3600 Ω, 900 Ω

3.14 a) 1200 Ω, 300 Ω

    b) 1 W

3.27 a) 150 mA

    b) 5.4 V

    c) 3.6 V

    d) 1 V

3.33 7.5 A

3.37 a) 49,980 Ω

    b) 4980 Ω

    c) 230 Ω

    d) 5 Ω

3.51 a) 900 Ω

    b) 25 mA

    c) 800 Ω, 180 mW

    d) 900 Ω, 90 mW

3.60 2.4 A, 72.576 W

3.63 a) 80 Ω

    b) 279 W

3.73 a) 0.2, 0.75

    b) 384, 200

Chapter 4

4.2 a) 8

    b) 3

    c) 4

    d) Top-most and left-most meshes cannot be used, as they both contain current sources

4.5 a) $i_4 + i_6 - i_5 - I = 0$

    b) derivation

4.10 120 V, 96 V

4.12 a) 8 A, 2 A, 6 A, 2 A, 4 A

    b) 360 W

4.17 750 W

4.27 a) $-37.5\,\text{V}, 75\,\text{W}$

b) $-37.5\,\text{V}, 75\,\text{W}$

c) Part (b), fewer equations

4.36 a) $0.1\,\text{A}, 0.3\,\text{A}, 0.2\,\text{A}$

b) $0.38\,\text{A}, 0.02\,\text{A}, -0.36\,\text{A}$

4.41 $2700\,\text{W}$

4.46 a) $2\,\text{mA}$

b) $304\,\text{mW}$

c) $0.9\,\text{mW}$

4.50 $740\,\text{W}$

4.56 a) Mesh current method

b) $4\,\text{mW}$

c) No

d) $200\,\text{mW}$

4.62 a) $-0.85\,\text{A}$

b) $-0.85\,\text{A}$

4.68 $8\,\text{mA}$ down in parallel with $10\,\text{k}\Omega$

4.72 a) $51.3\,\text{V}$

b) -5%

4.79 $150\,\Omega$

4.87 $2.5\,\Omega$ and $22.5\,\Omega$

4.92 a) $50\,\text{V}$

b) $250\,\text{W}$

4.105 $39.583\,\text{V}, 102.5\,\text{V}$

Chapter 5

5.3 $-1\,\text{mA}$

5.10 a) $0 \leq \sigma \leq 0.40$

b) $556.25\,\mu\text{A}$

5.14 $0 \leq R_f \leq 75\,\text{k}\Omega$

5.20 a) $7.56\,\text{V}$

b) $-3.97\,\text{V} \leq v_g \leq 3.97\,\text{V}$

c) $35\,\text{k}\Omega$

5.27 a) $-1.53\,\text{mV}$

b) $33.5\,\text{k}\Omega$

c) $80\,\text{k}\Omega$

5.31 a) $16\,\text{V}$

b) $-4.2\,\text{V} \leq v_b \leq 3.8\,\text{V}$

5.33 $19.93\,\text{k}\Omega \leq R_x \leq 20.07\,\text{k}\Omega$

5.44 a) -19.9844

b) $736.1\,\mu\text{V}$

c) $5003.68\,\Omega$

d) $-20, 0, 5000\,\Omega$

5.49 a) $2\,\text{k}\Omega$

b) $12\,\text{m}\Omega$

Chapter 6

6.3 a) $i = 0, t \leq 0$

$i = 50t\,\text{A}, 0 \leq t \leq 5\,\text{ms}$

$i = (5 - 50t)\,\text{A}, 5\,\text{ms} \leq t \leq 10\,\text{ms}$

$i = 0, t \geq 10\,\text{ms}$

b) $v = 0, t \leq 0$

$v = 1\,\text{V}, 0 \leq t \leq 5\,\text{ms}$

$v = -1\,\text{V}, 5\,\text{ms} \leq t \leq 10\,\text{ms}$

$v = 0, t \geq 10\,\text{ms}$

$p = 0, t \leq 0$

$p = 50t\,\text{W}, 0 \leq t \leq 5\,\text{ms}$

$p = (50t - 0.5)\,\text{W}, 5\,\text{ms} \leq t \leq 10\,\text{ms}$

$p = 0, t \geq 10\,\text{ms}$

$w = 0, t \leq 0$

$w = 25t^2\,\text{J}, 0 \leq t \leq 5\,\text{ms}$

$w = (25t^2 - 0.5t + 0.0025)\,\text{J}, 5\,\text{ms} \leq t \leq 10\,\text{ms}$

$w = 0, t \geq 10\,\text{ms}$

6.21 a) $(8 \times 10^5 t + 12)\,\text{V}$

b) $(-4 \times 10^5 t + 18)\,\text{V}$

c) $(12 \times 10^5 t - 14)\,\text{V}$

d) $(8 \times 10^5 t - 4)\,\text{V}$

e) $24\,\text{V}$

f)

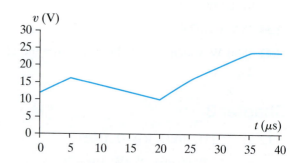

6.30 $5\,\text{nF}$ with an initial voltage drop of $15\,\text{V}$; $10\,\mu\text{F}$ with an initial voltage drop of $25\,\text{V}$

6.31 a) $10e^{-t}\,\text{V}, t \geq 0$

b) $(6.67e^{-t} - 2.67)\,\text{V}, t \geq 0$

c) $(3.33e^{-t} + 2.67)$ V, $t \geq 0$

d) $100 \, \mu$J

e) $132 \, \mu$J

f) $32 \, \mu$J

g) $100 + 32 = 132 \, \mu$J

6.41 a) $0.2\dfrac{di_2}{dt} + 10i_2 = -0.5\dfrac{di_g}{dt}$

b) $0.2\dfrac{di_2}{dt} + 10i_2 = 5e^{-10t}$ and $-0.5\dfrac{di_g}{dt} = 5e^{-10t}$

c) $(-53.125e^{-10t} + 6.25e^{-50t})$ V, $t \geq 0$

d) -46.875 V

6.48 a) 0.95

b) 24 mH

c) 2.5

6.49 0.8 nWb/A, 1.2 nWb/A

6.53 a) $(2.1, 4.3)$; $(3.2, 2.5)$; $(2.1, 2.5)$; $(3.2, 4.3)$

b) Zoom in

c) Zoom out

Chapter 7

7.2 a) 0.5 A

b) 2 ms

c) $0.5e^{-500t}$ A, $t \geq 0$; $-80e^{-500t}$ V, $t \geq 0^+$; $-35e^{-500t}$ V, $t \geq 0^+$

d) 35.6%

7.4 a) 0 A, 100 mA, 0 V

b) 400 mA, 100 mA, -20 V

c) 500 mA, 0 A, 0 V

d) $0.1e^{-4000t}$ A

e) $(0.5 - 0.1e^{-4000t})$ A

f) $-20e^{-4000t}$ V

7.18 a) $-10e^{-5000t}$ A, $t \geq 0$

b) 80 mJ

c) 1.498τ

7.22 a) $59.4e^{-1000t}$ V, $t \geq 0$

b) $9.9e^{-1000t}$ mA, $t \geq 0^+$

7.27 a) 4 kΩ

b) 10 μF

c) 40 ms

d) 11.52 mJ

e) 10.95 mJ

7.38 a) $(-0.8 + 2.4e^{-4000t})$ A, $t \geq 0$; $(41.6 + 19.2e^{-4000t})$ V, $t \geq 0^+$

b) -48 V, 60.8 V

7.46 17.33 ms

7.55 a) 90 V

b) -60 V

c) 1000 μs

d) 916.3 μs

7.59 $(-100 + 130e^{-200t})$ V, $t \geq 0$

7.63 3.67 ms

7.69 a) $(6 - 6e^{-10t})$ A, $t \geq 0$

b) $100e^{-10t}$ V, $t \geq 0^+$

c) $(7 - 7e^{-10t})$ A, $t \geq 0$

d) $(-1 + e^{-10t})$ A, $t \geq 0$

e) Yes

7.73 -5.013 V

7.78 10 V, $0 \leq t \leq 50$ ms; $10e^{-1000(t - 0.05)}$ V, $t \geq 50$ ms

7.87 83.09 ms

7.94 a) $\dfrac{1}{RC}\displaystyle\int_0^t (v_b - v_a)\,dy$

b) Output is the integral of the difference between v_b and v_a, scaled by a factor of $1/RC$

c) 120 ms

7.105 73 beats per minute

Chapter 8

8.3 $(75{,}000t + 15)e^{-10{,}000t}$ V, $t \geq 0$

8.9 a) 1 kΩ, 1 μF, 6000 V/s, 8 V

b) $(-3000t + 2)e^{-500t}$ mA, $t \geq 0^+$

8.14 $(5e^{-2000t} + 10e^{-8000t})$ V, $t \geq 0$

8.16 $(15e^{-2500t}\cos 3122.5t + 7.21e^{-2500t}\sin 3122.5t)$ V, $t \geq 0$

8.33 $(2 + 0.33e^{-400t} - 1.33e^{-1600t})$ A, $t \geq 0$

8.41 8 kΩ, 2 H, 7.5 mA, 0

8.51 $(20 - 10{,}000te^{-500t} - 20e^{-500t})$ V, $t \geq 0$

8.56 a) $75e^{-37{,}500t}\sin 50{,}000t$ V

b) 18.55 μs

c) 29.93 V

d) $601.08e^{-3750t}\sin 62{,}387.4t$ V, 24.22 μs, 547.92 V

8.63 a) $0 \le t \le 0.5^-$ s: $v_{o1} = -1.6t$ V, $v_o = 10t^2$ V

0.5^+ s $\le t \le t_{sat}$: $v_{o1} = (0.8t - 12)$ V,

$v_o = (-5t^2 + 15t - 3.75)$ V

b) 3.5 s

8.64 $0 \le t \le 0.5$ s: $v_{o1} = (-0.8e^{-t} + 0.8e^{-2t})$ V,

$v_o = (10 - 20e^{-t} + 10e^{-2t})$ V

$t \ge 0.5$ s: $v_{o1} = (0.4 - 0.91e^{-2(t-0.5)})$ V,

$v_o = (-5 + 19.42e^{-(t-0.5)} - 12.87e^{-2(t-0.5)})$ V

8.67 a) 6.33 pF

b) $5.03 \sin(4\pi \times 10^9 t)$ V, $t \ge 0$

Chapter 9

9.1 a) 40 V

b) 50 Hz

c) 314.159 rad/s

d) 1.05 rad

e) 60°

f) 20 ms

g) 6.67 ms

h) $40 \cos 100\pi t$ V

i) 8.33 ms

9.10 a) $[-195.72e^{-1066.67t} + 200 \cos(800t$

$- 11.87°)]$ mA, $t \ge 0$

b) transient: $-195.72e^{-1066.67t}$ mA;

steady-state: $200 \cos(800t - 11.87°)$ mA

c) 28.39 mA

d) 200 mA, 800 rad/s, $-11.87°$

e) Current lags voltage by 36.87°

9.11 a) $111.8 \cos(500t - 3.43°)$

b) $102.99 \cos(377t + 40.29°)$

c) $161.57 \cos(100t - 29.96°)$

d) 0

9.13 a) 400 Hz

b) $-90°$

c) 5 Ω

d) 1.99 mH

e) $j5$ Ω

9.15 a)

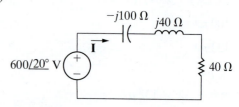

b) $8.32\underline{/76.31°}$ A

c) $8.32 \cos(8000t + 76.31°)$ A

9.17 a) $160 + j120$ mS

b) 160 mS

c) 120 mS

d) 10 A

9.28 $42.43 \sin(50{,}000t + 45°)$ V

9.36 4000 rad/s

9.42 2/3 Ω

9.48 $(6 + j4)$ A in parallel with $(-20 + j20)$ Ω

9.55 $188.43\underline{/-42.88°}$ V

9.64 $178.89 \cos(5000t + 153.43°)$ V

9.77 a) 0.3536

b) 2 A

9.79 $1250\underline{/-45°}$ Ω

9.83 a) $(247 + j7.25)$ V

b) $-j32$ Ω, $(241 + j8)$ V

c) -26.90 Ω

9.87 a) 0 A

b) $0.436\underline{/0°}$ A

c) Yes

Chapter 10

10.1 a) 409.58 W(abs), 286.79 var(abs)

b) 103.53 W(abs), -386.37 var(del)

c) -1000 W(del), -1732.05 var (del)

d) -250 W(del), 433.01 var(abs)

10.2 a) No [40.1 A(rms)]

b) Yes [66.2 A(rms)]

10.8 5 mW

10.14 a) 3 V(rms)

b) 4 W

10.17 a) $227.81\underline{/2.52°}\,\text{V(rms)}; 24.54\underline{/24.05°}\,\text{V(rms)}$

b)

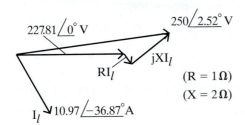

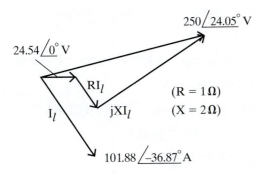

10.27 a) $0.96\,(\text{lag}), 0.28; 0.6\,(\text{lead}), -0.8; 0.8\,(\text{lead}), -0.6$

    b) $0.74\,(\text{lead}), -0.67$

10.43 a) $(20 + j20)\,\Omega$

    b) 45 W

    c) $R = 22\,\Omega$ and $L = 1$ mH gives 42.5 W

10.49 a) 360 mW

    b) $4000\,\Omega, 0.1\,\mu\text{F}$

    c) 443.1 mW

    d) 450 mW

    e) $4000\,\Omega, 66.67$ nF

    f) Yes

10.52 a) $-21\underline{/0°}\,\text{V(rms)}$

    b) 63 W

    c) 9.72%

10.60 a) 125

    b) 26.28125 W

10.67 a) 15.63 kWh

    b) 11.72 kWh

    c) 9.16 kWh

    d) 6.64 kWh

Chapter 11

11.1 a) acb

    b) abc

11.11 a) 10 A(rms)

    b) 669.3 V(rms)

11.12 a) $5\underline{/-36.87°}\,\text{A(rms)}, 5\underline{/83.13°}\,\text{A(rms)},$
$5\underline{/-156.87°}\,\text{A(rms)}$

    b) $216.51\underline{/-30°}\,\text{V(rms)}, 216.51\underline{/90°}\,\text{V(rms)},$
$216.51\underline{/-150°}\,\text{V(rms)}$

    c) $122.23\underline{/-1.36°}\,\text{V(rms)}, 122.23\underline{/118.64°}\,\text{V(rms)},$
$122.23\underline{/-121.36°}\,\text{V(rms)}$

    d) $211.72\underline{/-31.36°}\,\text{V(rms)}, 211.72\underline{/88.64°}\,\text{V(rms)},$
$211.72\underline{/-151.36°}\,\text{V(rms)}$

11.15 $21.64\underline{/121.34°}\,\text{A(rms)}$

11.16 $159.5\underline{/29.34°}\,\text{V(rms)}$

11.25 $6120\underline{/36.61°}\,\text{VA}$

11.27 a) $1833.46\underline{/22°}\,\text{VA}$

    b) 519.62 V(rms)

11.35 6990.62 V(rms)

11.45 a) proof

    b) 2592 var, -2592 var, 3741.23 var, -4172.80 var

11.54 a) $16.71\,\mu\text{F}$

    b) $50.14\,\mu\text{F}$

Chapter 12

12.1 a) $(t + 10)u(t + 10) - 2tu(t) + (t - 10)u(t - 10)$

    b) $-8(t + 3)u(t + 3) + 8(t + 2)u(t + 2)$
$+ 8(t + 1)u(t + 1) - 8(t - 1)u(t - 1)$
$- 8(t - 2)u(t - 2) + 8(t - 3)u(t - 3)$

12.10 2/9

12.20 a) $\dfrac{1}{(s + a)^2}$

    b) $\dfrac{\omega}{s^2 + \omega^2}$

    c) $\dfrac{\omega \cos\theta + s \sin\theta}{s^2 + \omega^2}$

    d) $\dfrac{1}{s^2}$

    e) $\dfrac{\sinh\theta + s \cosh\theta}{(s^2 - 1)}$

12.22 a) $\dfrac{\omega s}{s^2 + \omega^2}$

    b) $\dfrac{-\omega^2}{s^2 + \omega^2}$

    c) 2

    d) Check

12.26 $\dfrac{1.92s^2}{(s^2 + 1.6s + 1)(s^2 + 1)}$

12.40 a) $[e^{-t} + 5e^{-2t} + 2e^{-4t}]u(t)$

b) $[10 + 5e^{-2t} - 8e^{-3t} + e^{-5t}]u(t)$

12.41 c) $[10e^{-6t} + 5.66e^{-2t}\cos(4t + 45°)]u(t)$

d) $[9.25e^{-5t}\cos(3t - 40.05°)$
$+ 7.21e^{-4t}\cos(2t + 168.93°)]u(t)$

12.42 b) $[60 - 40te^{-2t} - 60e^{-2t}]u(t)$

c) $[12te^{-t} + 0.6e^{-t}$
$+ 3.35e^{-3t}\cos(4t + 100.305°)]u(t)$

12.43 b) $[10te^{-t}\cos(t - 90°) + 10e^{-t}\cos(t - 90°)]u(t)$

d) $5\delta'(t) - 15\delta(t) + [10e^{-2t} - 4e^{-5t}]u(t)$

12.46 zero at -3 rad/s, poles at 0 and two at -2 rad/s;
zero at -5 rad/s, poles at $(-3 + j4)$ rad/s,
$(-3 - j4)$ rad/s and two at -1 rad/s

12.52 a) $8, 0$

b) $8, 10$

c) $22, 0$

d) $250, 490$

12.57 0.947

Chapter 13

13.4 a) $\dfrac{0.025(s^2 + 16 \times 10^4 s + 10^{10})}{s}$

b) zeros at $(-80{,}000 + j60{,}000)$ rad/s and
$(-80{,}000 - j60{,}000)$ rad/s, pole at 0

13.5 a) $\dfrac{4 \times 10^6 s}{s^2 + 2000s + 64 \times 10^4}$

b) zero at 0, poles at -400 rad/s and -1600 rad/s

13.9 a)

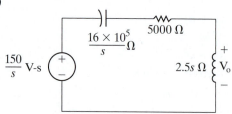

b) $\dfrac{-150}{(s + 400)(s + 1600)}$

c) $[50e^{-400t} - 200e^{-1600t}]u(t)$ V

13.15 a)

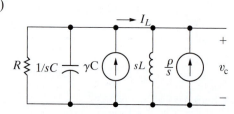

where $R = 20\,\Omega$, $C = 6.25\,\mu\text{F}$, $\gamma = -48\,\text{V}$,
$L = 6.4\,\text{mH}, \rho = -2.4\,\text{A}$

b) $\dfrac{-48(s + 8000)}{s^2 + 8000s + 25 \times 10^6}$

c) $\dfrac{2.4(s + 4875)}{s^2 + 8000s + 25 \times 10^6}$

d) $[80e^{-4000t}\cos(3000t + 126.87°)]u(t)$ V

e) $[2.5e^{-4000t}\cos(3000t - 16.26°)]u(t)$ A

13.20 a) $[1 - 1e^{-250t}\cos 250t]u(t)$ A

b) $[141.42e^{-250t}\cos(250t - 45°)]u(t)$ V

c) Yes

13.23 $[17.68\cos(50t - 135°) + 18.02e^{-8.33t}\cos(39.965t + 46.194°)]u(t)$ mA

13.37 $[63.25e^{-150t}\cos(50t + 71.57°)]u(t)$ mA

13.39 $[96e^{-5t} - 96e^{-20t}]u(t)$ V

13.44 a) $\dfrac{480(s + 2.5)}{s(s + 4)(s + 6)}$

b) $[50 + 90e^{-4t} - 140e^{-6t}]u(t)$ V

13.51 a) $\dfrac{200}{(s + 200)}$; no zeros, pole at -200 rad/s

b) $\dfrac{s}{(s + 200)}$; zero at 0, pole at -200 rad/s

c) $\dfrac{s}{(s + 8000)}$; zero at 0, pole at -8000 rad/s

d) $\dfrac{8000}{(s + 8000)}$; no zeros, pole at -8000 rad/s

e) $\dfrac{100}{(s + 500)}$; no zeros, pole at -500 rad/s

13.56 a) $\dfrac{-10^4(s + 5000)}{s(s + 25{,}000)}$

b) zero at -5000 rad/s, poles at 0
and $-25{,}000$ rad/s

c) $[-2000t - 0.32 + 0.32e^{-25{,}000t}]u(t)$ V

13.62 e^{-t} V, $0 \le t \le 1$ s; $(1 - e)e^{-t}$ V, 1 s $\le t \le \infty$

13.79 $16.97\cos(3t + 8.13°)$ V

13.91 a) 0.8 A

b) 0.6 A

c) 0.2 A

d) -0.6 A

e) $0.6e^{-2 \times 10^6 t}u(t)$ A

f) $-0.6e^{-2 \times 10^6 t}u(t)$ A

g) $[-1.6 \times 10^{-3}\delta(t) - 7200e^{-2 \times 10^6 t}u(t)]$ V

13.94 a) $0\,\text{A}, 25\sqrt{2}\,\text{A}$

b) $\dfrac{1449\pi(122.92\sqrt{2}s - 3000\pi\sqrt{2})}{(s + 1475\pi)(s^2 + 14{,}400\pi^2)}$

$+ \dfrac{300\sqrt{2}}{(s + 1475\pi)}, [178.82\sqrt{2}e^{-1475\pi t}$

$+ 122.06\sqrt{2}\cos(120\pi t + 6.85°)]\,\text{V},$

$300\sqrt{2}\,\text{V}$

c) $122.06\sqrt{2}\cos(120\pi t + 6.85°)\,\text{V}$

d)

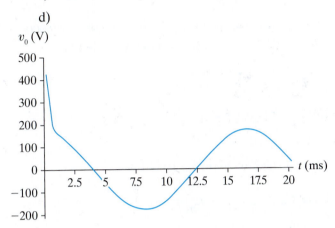

Chapter 14

14.1 a) 2021.27 Hz

b) $0.7071\underline{/-45°}, 0.981\underline{/-11.31°},$
$0.196\underline{/-78.69°}$

c) $7.07\cos(12{,}700t - 45°)\,\text{V},$
$9.81\cos(2540t - 11.31°)\,\text{V},$
$1.96\cos(63{,}500t - 78.69°)\,\text{V}$

14.3 a) $31.42\,\Omega$

b) 895.77 Hz

c) With $33\,\Omega$ resistor, 1050.4 Hz

14.13 a) $125\,\Omega$

b) $3\,\text{k}\Omega$

14.17 a) $5.305\,\text{k}\Omega$

b) 333.86 Hz

14.26 a) $5\,\text{k}\Omega, 50\,\text{mH}$

b) 3.52 kHz, 2.88 kHz

c) 636.62 Hz

14.34 $4\,\text{k}\Omega$

14.42 a) $397.89\,\Omega, 3.17\,\text{mH}$

b) 4.42 kHz, 3.62 kHz

c) 800 Hz

14.51 a) $0.39\,\text{H}, 0.1\,\mu\text{F}$

b) $|V_{697\,\text{Hz}}| = |V_{941\,\text{Hz}}| = 0.707|V_{\text{peak}}|;$
$|V_{770\,\text{Hz}}| = |V_{852\,\text{Hz}}| = 0.948|V_{\text{peak}}|$

c) $0.344|V_{\text{peak}}|$

Chapter 15

15.1 a) $179\,\Omega, 318.31\,\Omega$

b)

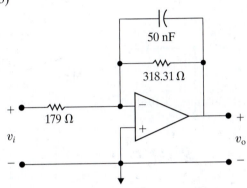

15.8 a) $1273.24\,\Omega, 4026.337\,\Omega$

b)

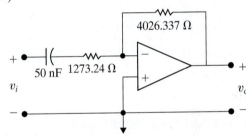

15.15 a) $1\,\text{H}, 0.1\,\Omega, 1\,\text{F}$

b) $333.33\,\text{mH}, 2.5\,\text{k}\Omega, 533.33\,\text{pF}$

c)

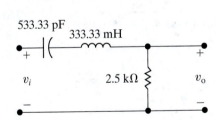

15.23 a) $2.4\,\text{H}, 5\,\mu\text{F}$

b) $0.57\cos(500t + 15.14°)\,\text{A}$

15.30 61.553 Hz, 4061.553 Hz, $156.7\,\Omega, 10.34\,\text{k}\Omega$

15.31 $R_L = 32.49\,\text{k}\Omega, R_H = 492.4\,\Omega, R_i = R_f = 1\,\text{k}\Omega$

15.34 a) 3

b) $-29.7\,\text{dB}$

15.58 a) $R_1 = 1.99\,\text{k}\Omega, R_2 = 39.63\,\text{k}\Omega, R_3 = 39.79\,\text{k}\Omega$

b)

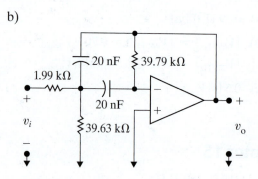

15.60 a) See component values in part (b)

b)

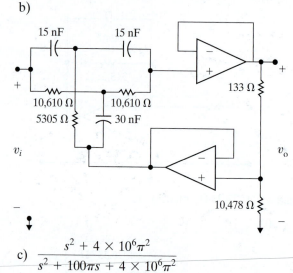

c) $\dfrac{s^2 + 4 \times 10^6 \pi^2}{s^2 + 100\pi s + 4 \times 10^6 \pi^2}$

15.62 $R_1 = 100\,\text{k}\Omega, R_2 = 400\,\text{k}\Omega, C_1 = 7.96\,\text{nF}$

Chapter 16

16.1 a) 31,415.93 rad/s, 3978.87 rad/s

b) 5 kHz, 25 kHz

c) 0, 25 V

d) $a_{va} = 0$; for k even, $a_{ka} = 0, b_{ka} = 0$;

for k odd, $a_{ka} = \dfrac{-80}{\pi k}\sin\dfrac{\pi k}{2}, b_{ka} = \dfrac{240}{\pi k}$;

$a_{vb} = 25; b_{kb} = 0$ for all k;

$a_{kb} = \dfrac{200}{\pi k}\sin\dfrac{\pi k}{4}$ for all k

e) $\dfrac{80}{\pi}\displaystyle\sum_{n=1,3,5,\ldots}^{\infty}\left(-\dfrac{1}{n}\sin\dfrac{n\pi}{2}\cos n\omega_0 t + \dfrac{3}{n}\sin n\omega_0 t\right)$ V;

$\left[25 + \dfrac{200}{\pi}\displaystyle\sum_{n=1}^{\infty}\left(\dfrac{1}{n}\sin\dfrac{n\pi}{4}\cos n\omega_0 t\right)\right]$ V

16.2 $\left[\dfrac{160}{\pi} + 20\sin\omega_0 t - \dfrac{320}{\pi}\displaystyle\sum_{n=2,4,6,\ldots}^{\infty}\dfrac{\cos n\omega_0 t}{n^2 - 1}\right]$ V

16.11 a) 2π rad/s

b) yes

c) no

d) no

16.22 a) $10\displaystyle\sum_{n=1,3,5,\ldots}^{\infty}\dfrac{\sqrt{(n\pi)^2 + 4}}{n^2}\cos(n\omega_0 t - \theta_n)$ A,

$\theta_n = \tan^{-1}\dfrac{n\pi}{2}$

b) 26.23 A

16.29 a) $[41.998\cos(2000t - 0.60°) + 13.985\cos(6000t - 177.32°) + 5.984\cos(14{,}000t + 184.17°)]$ V

b) Fifth harmonic

16.35 1.85 W

16.39 a) 117.55 V(rms)

b) -2.04%

c) 69.2765 V(rms), -0.0081%

16.44 $C_0 = \dfrac{V_m}{2}, C_n = j\dfrac{V_m}{2n\pi}$

for $n = \pm 1, \pm 2, \pm 3, \ldots$

16.50 a)

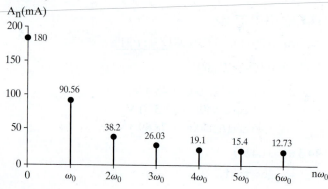

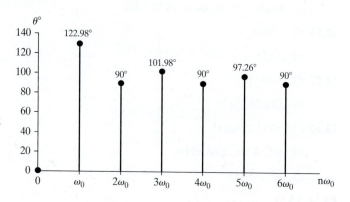

b)

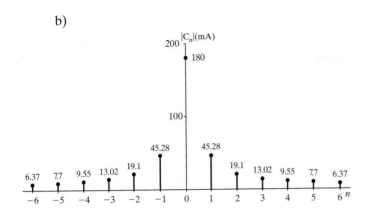

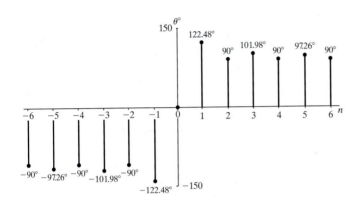

16.57 a) $\dfrac{400}{313}$, 2000 rad/s, 16×10^8 rad^2/sec^2

b) $[-80 \cos \omega_0 t - 0.50 \cos(3\omega_0 t + 91.07°)$
$+ 0.17 \cos(5\omega_0 t + 90.60°)]$ V

Chapter 17

17.1 a) $j\dfrac{2A}{\tau}\left[\dfrac{\omega\tau\cos(\omega\tau/2) - 2\sin(\omega\tau/2)}{\omega^2}\right]$

b) 0

c)

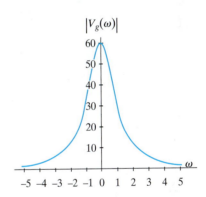

17.5 a) $\dfrac{2(a^2 - \omega^2)}{(a^2 + \omega^2)^2}$

b) $-j48a\omega\dfrac{(a^2 - \omega^2)}{(a^2 + \omega^2)^4}$

c) $\dfrac{a}{a^2 + (\omega - \omega_0)^2} + \dfrac{a}{a^2 + (\omega + \omega_0)^2}$

d) $\dfrac{-ja}{a^2 + (\omega - \omega_0)^2} + \dfrac{ja}{a^2 + (\omega + \omega_0)^2}$

e) $e^{-j\omega t_0}$

17.19 a) $\dfrac{\tau}{2} \cdot \dfrac{\sin[(\omega + \omega_0)(\tau/2)]}{(\omega + \omega_0)(\tau/2)}$
$+ \dfrac{\tau}{2} \cdot \dfrac{\sin[(\omega - \omega_0)(\tau/2)]}{(\omega - \omega_0)(\tau/2)}$

b) $F(\omega) \to \pi[\delta(\omega - \omega_0) + \delta(\omega + \omega_0)]$

17.28 a) $[12.5e^{-t} - 7.5e^{-5t}]u(t) + 5e^{5t}u(-t)$ V

b) 5 V

c) 5 V

d) $[12.5e^{-t} - 7.5e^{-5t}]u(t)$ V

e) Yes

17.30 $1\cos(5000t + 90°)$ A

17.40 a) $[(-24e^{-t} + 32e^{-5t/2})u(t) + 8e^t u(-t)]$ V

b)

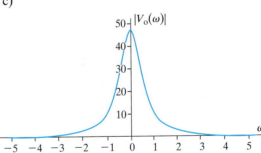

c)

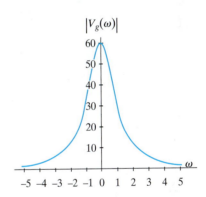

d) 900 J

e) 320 J

f) 95.95%

g) 99.75%

Chapter 18

18.2 $z_{11} = 13\ \Omega; z_{12} = 12\ \Omega; z_{21} = 12\ \Omega; z_{22} = 16\ \Omega$

18.10 $a_{11} = -4 \times 10^{-4}; a_{12} = -20\ \Omega;$

$\quad a_{21} = -0.5\ \mu S; a_{22} = -0.02$

18.13 $g_{11} = 12.5\ \mu S; g_{12} = 1.5;$

$\quad g_{21} = -250; g_{22} = 50\ M\Omega$

18.14 a) $y_{11} = 20\ \mu S; y_{12} = 30\ nS;$

$\quad\quad y_{21} = 5\ \mu S; y_{22} = 20\ nS$

b) $y_{11} = 20\ \mu S; y_{12} = 30\ nS;$

$\quad\quad y_{21} = 5\ \mu S; y_{22} = 20\ nS$

18.21 $a_{11} = 3; a_{12} = 40\ \Omega; a_{21} = 80 + j60\ mS;$

$\quad a_{22} = 2.4 + j0.8$

18.31 a) 35.36 V(rms)

b) 125 mW

c) 18.75 μW

18.33 12.5

18.38 3750 V

Index

A

a-, b-, and c-phase voltage references, 414
ac circuits, balancing power in, 391–392
Active circuit elements, 30
Active filter circuits, 572–617
  advantages of, 572, 608
  bandpass filters, 580–583, 598–599, 600–602, 608–609
  bandreject filters, 583–587, 598, 602–605, 608–609
  bass volume control, 573, 605–607
  block diagram for, 581
  Bode plots for, 574–575, 580, 583–584, 587–588, 602, 605
  broadband (low-Q) filters, 581–587, 608
  Butterworth filters, 591–600, 608–609
  cascading filters, 581–583, 587–591, 608–609
  first-order, 574–577, 608
  fourth-order, 589–590, 594–595
  higher-order, 587–600, 608
  high-pass filters, 576–577, 598, 608
  low-pass filters, 574–575, 579, 587–591, 593–595, 608
  narrowband (high-Q) filters, 600–605, 609
  op amp filters, 575–577, 579, 580–600, 608
  prototypes, 575–577, 608
  scaling, 577–580, 608
Active high-Q filter response, 619, 647–649
Addition/subtraction operations, 454, 673, 729
Admittance (Y), 336, 361
Ammeter, 70, 71–72, 80
Amplifier circuit analysis, 47, 102–103, 111, 125–126, 252–254, 256, 303–308, 310. *See also* Operational amplifiers
  cascading connections, 303–308, 310
  integrating-amplifier, 252–254, 256, 303–308, 310
  Kirchhoff's laws for, 47
  mesh-current method, 111
  node-voltage method, 102–103
  Ohm's law for, 47

resistor-inductor-capacitor (RLC) circuits for, 303–308, 310
Thévenin equivalent circuits for, 125–126
Amplitude plots, 744–748, 751–753
  accuracy of, 747–748
  complex poles, 751–753
  corrections to, 752–753
  straight-line, 744–748, 751–753
Amplitude spectrum, 645–647, 650, 663–664
  Fourier series and, 645–647, 650
  Fourier transform transition of, 663–664
Amplitude, 320
Analog meters, 71–72, 80
Analysis methods, 97, 99, 102, 105, 107, 110, 154, 224, 230, 234, 238, 242, 280, 283, 287, 292, 299, 487
  analyzing a circuit with an ideal op amp, 154
  basics version of the mesh-current method, 105
  basic version of the node-voltage method, 97
  complete form of the mesh-current method, 110
  complete form of the node-voltage method ,102
  finding the RC natural response, 230
  finding the RC step response, 238
  finding the RL and RC natural and step response, 242
  finding the RL natural response, 224
  finding the RL step response, 234
  Laplace-transform circuit analysis method, 487
  modified step 3 for the mesh-current method, 107
  modified step 3 for the node-voltage method, 99
  natural response of an overdamped or underdamped parallel RLC circuit, 283
  natural response of an overdamped parallel RLC circuit, 280
  natural response of parallel RLC circuits, 287
  natural response of series RLC circuits, 299

step response of parallel RLC circuits, 292
Angular frequency (ω), 320
Appliance power ratings, 381–382
Artificial pacemaker design, 221, 255
Attenuation, 536
Average power (P), 377–384, 389, 394, 401, 425–426, 430–433, 435, 639–641, 650
  absorbed, 394
  appliance power ratings, 381–382
  balanced three-phase circuits, 425–426, 430–433, 435
  calculations for, 377–382, 389, 401, 425–426, 639–641
  Fourier series applications, 639–641, 650
  maximum power transfer (P_{max}), 394
  measurement of, 430–433, 435
  periodic functions ($f(t)$) and, 639–641, 650
  power factor (pf) for, 379, 401
  root-mean-square (rms) value for, 382–384
  sinusoidal steady-state circuits, 377–382, 389, 394, 401
  wye (Y) loads, 425–426

B

Back-substitution method, 721–722, 726
Balanced three-phase circuits, 412, 416–417, 435. *See also* Three-phase circuits
Bandpass filters, 538–539, 550–560, 565, 580–583, 598–599, 600–602, 608–609, 682–683
  active, 580–583, 598–599, 600–602, 608
  bandwidth (β) for, 550, 553–554, 556, 565
  block diagram for, 581
  Bode plot for, 580, 601
  broadband, 581–583, 608
  Butterworth, 598–599, 608
  cascaded, 581–583
  center frequency (ω_o) for, 550, 552–554, 565

Bandpass filters (*Continued*)
  cutoff frequency (ω_c) for, 552–553, 556
  design of, 555–557
  frequency and time domain relationships, 559–560
  frequency of, 538–539
  frequency response plot, 539
  narrowband, 600–602, 609
  op amp filters, 580–583, 598–599, 608
  parallel *RLC* circuit, 555–557
  parameters of, 550
  Parseval's theorem for, 682–683
  qualitative analysis, 551
  quality factor (*Q*) for, 550, 554, 556
  quantitative analysis, 551–554
  resonant frequency (ω_o) for, 550
  series *RLC* circuit, 551–555, 557–559
  transfer function (*H*(*s*)) for, 559
Bandreject filter, 538–539, 560–563, 565, 583–587, 598, 602–605, 608–609
  active, 583–587, 598, 602–605, 608–609
  Bode plots for, 584, 605
  broadband, 583–587, 608
  Butterworth, 598
  center frequency (ω_o) for, 562
  cutoff frequency (ω_c) for, 562
  design of, 562–563
  frequency of, 538–539
  frequency response plot, 539
  narrowband, 602–605, 609
  op amp filters, 583–587, 608
  parallel *RLC* circuit, 563
  qualitative analysis, 560–561
  quantitative analysis, 561–562
  series *RLC* circuit, 560–563
  transfer function (*H*(*s*)) for, 561, 563
Bandwidth (β), 550, 553–554, 556, 562, 565
Bass volume control circuit analysis, 573, 605–607
Bilateral configurations, 115–116
Black box concept, 60
Block diagrams, 581
Bode plots (diagrams), 574–576, 580, 583–584, 587–588, 602, 605, 743–756
  active filter analysis using, 574–576, 580, 583–584, 588
  amplitude plots, 744–748, 751–753
  bandpass filters, 580, 602
  bandreject filters, 583–584, 605
  broadband (low-*Q*) filters, 581, 583–584
  cascading filters, 587–588
  complex poles, 750–756
  complex zeros, 750–751
  corner frequency for, 747–748

high-pass filters, 576
low-pass filters, 574–576, 588
magnitude plots, 587–588
narrowband (high-*Q*) filters, 602, 605
phase angle plots, 748–750, 754–756
real, first-order poles and zeros, 743–744
straight-line plots, 744–756
Branches, 94–96
Broadband (low-*Q*) filters, 581–587, 608
  bandpass, 581–583, 608
  bandreject, 583–587, 608
  Bode plots for, 581, 583–584
Butterworth filters, 591–600, 608–609
  bandpass, 598–599, 608
  bandreject, 598, 608
  cascading, 591–600, 608–609
  circuit analysis, 591, 593–595
  high-pass, 598, 608
  low-pass, 591, 593–595
  order of, 596–597
  transfer functions (*H*(*s*)) for, 592–593
  transition region, 596–597

C

Capacitance (*C*), 183, 189, 195–198, 209–210, 211
  equivalent, 196–197
  series–parallel combinations, 195–198
  touch screens values of, 183, 209–210
Capacitive circuits, power for, 379
Capacitors, 182, 189–193, 195–196, 198–199, 211, 329–330, 361, 485–486, 514–515, 521, 760
  circuit analysis of, 189–193, 514–515
  circuit component values, 760
  circuit parameter of, 189
  current to voltage (*i*–*v*) relationships, 190
  displacement current, 189–190
  duality (symmetry) of, 198–199, 211
  energy in, 182, 191, 211
  equivalent circuits for, 196–197, 485–486, 521
  impulse function ($K\delta(t)$) for, 514–515
  in series, 195
  Laplace transform method for, 485–486, 514–515, 521
  multiple, 196–197
  parallel, 195–196
  passive behavior of, 182, 211
  phasor relationships, 329–330, 361
  power in, 182, 190–191, 211
  s domain representation, 485–486, 521
  symbols for, 189

voltage to current (*v*–*i*) relationships, 190, 329–330, 361
Cascading connections, 303–308, 310, 581–583, 587–600, 608–609, 708–711
  active filters, 581–583, 587–600, 608–609
  bandpass filters, 581–583
  Bode magnitude plot for, 588
  Butterworth filters, 591–600, 608–609
  fifth-order, 593
  fourth-order filters, 589–590
  high-pass filters, 598, 608
  identical first-order filters, 587–591
  integrating-amplifiers, 303–308, 310
  low-pass filters, 587–591, 593–595, 608
  op amp filters, 587–591
  two-port circuits, 708–711
Center frequency (ω_o), 550, 552–554, 562, 565
Characteristic determinant, 722–723
Characteristic equations, 275–276, 293, 297, 299–301, 309
  parallel *RLC* circuits, 275–276, 293, 309
  series *RLC* circuits, 297, 299–301
Circuit analysis, 11–15, 40–48, 68–70, 75–80, 92–149, 150–181, 220, 224, 230, 234, 238, 241–242, 246–254, 256, 273, 308–309, 318–373, 482–535, 536–571, 573, 605–607, 677–679, 681–685
  amplifiers, 47, 102–103, 111, 125–126
  bass volume control, 573, 605–607
  circuit design and, 11–12
  circuit models for, 13
  clock for computer timing, 273, 308–309
  component models for, 13
  current division, 69–70, 79–80, 336
  delta-to-wye (Δ-to-Y) transformations, 75–78, 80, 338–340
  dependent sources and, 45–48, 98–100, 107–108, 122
  Fourier transforms for, 677–679, 681–685
  frequency-selective circuits, 536–571
  household distribution, 319, 359–360
  ideal basic circuit element, 14
  ideal circuit components, 11–12
  integrating-amplifiers, 252–254, 256
  Kirchhoff's laws for, 40–44, 487
  Laplace transform in, 482–535
  Laplace transform method for, 487–488, 521

maximum power transfer, 92, 126–128, 135

mesh-current method, 92, 104–115, 134–135, 354–347

natural response method, 220, 224, 230, 241–242, 256

node-voltage method, 92, 96–104, 112–115, 134–135, 344–345

nonplanar circuits for, 94

Norton equivalent circuits for, 92, 120–123, 135

Ohm's law for, 42–43, 486

operational amplifiers (op amps), 150–181

passive sign convention for, 14–15

physical prototypes, 12

planar circuits for, 94

resistive circuits, 68–70, 79–80, 93, 115–118, 131–134

s domain, 486–488, 521

sensitivity analysis, 93, 131–134

sequential switching and, 246–250, 256

simultaneous equations for, 94–96

sinusoidal steady-state, 318–373

source transformation, 92, 115–118, 121, 134, 340–343

step response method, 220, 234, 238, 241–242, 256

superposition, 92, 129–131, 135

surge suppressors and, 483, 520

terminal behavior and, 118–126, 152–156

terminals, 118–126

Thévenin equivalent circuits for, 92, 118–126, 135

transformers, 347–356, 361

voltage division, 68–70, 79, 334

Circuit components, 11–13, 37, 324, 578, 760

electrical behavior of, 37

ideal, 11–12

models, 13, 37

scale factors, 578

steady-state current, 324

transient current, 324

values, 760

Circuit elements, 26–57, 194–209, 211, 484–486, 521

active, 30

capacitance (C), 195–198, 211, 485–486, 521

current sources, 28–31, 50

dependent sources, 28, 31, 45–48, 50

electrical radiator examples of, 27, 48–49

electrical sources, 28

equivalent circuits in s domain, 485–486, 521

in series, 41, 50

inductance (L), 194–209, 211, 484–485, 521

Kirchhoff's laws for, 39–44, 45–47, 50

Laplace transform method and, 484–487

loops (closed path), 40, 50

model construction, 36–38

nodes, 39, 50

Ohm's law for, 32–33, 42–43, 45–47, 50, 484

passive, 30

resistance (R), 32–35, 484, 521

s domain representation, 484–486, 521

symbols for, 28–29

voltage sources, 28–31, 50

Circuit models, 13, 27, 32–38, 47–49, 154–155, 167–170, 172–173, 692, 712

advantage of, 13

amplifier, 47

approximation for, 38

construction of, 36–38, 43–44

electrical behavior of components, 37

electrical effects from, 37

electrical radiator, 27, 48–49

flashlight (electrical system), 36–38, 39–41

operational amplifiers (op amps), 154–155, 167–170, 172–173

resistors, 32–35

terminal measurements for, 38, 43–44

two-port, 692, 712

Circuit theory, 7–8

Circuits, 2–25, 32–35, 58–91, 445, 474–476, 483, 520, 537, 564

analysis of, 11–12

current (i), 12–13, 15–17, 33

current-divider, 67, 79

delta-to-wye (Δ-to-Y) equivalent, 75–78, 80

electrical charge, 12–13

electrical effects, 7

electrical engineering and, 2–9

energy and, 15–17

frequency (f) and, 7

ideal basic element, 14–15

International System of Units (SI) for, 9–11

lumped-parameter systems, 7–8

magnetic coupling, 7

net charge, 7

open, 37

passive sign convention for, 14–17

power, 3, 15–20, 33–34

pushbutton telephone, 537, 564

resistive, 32–35, 58–91

short, 37

surge suppressors for, 483, 520

transient effects on, 445, 474–476

voltage (v), 12–13, 15–17, 34

voltage-divider, 64–66, 79

Wheatstone bridge, 73–75, 80

Clock for computer timing, circuit analysis of, 273, 308–309

Closed path (loop), 40, 50

Coefficient of coupling, 207–208, 211

Common mode input, 164

Common mode rejection ratio (CMRR), 165–167, 173

Communication systems, 4–5

Complex numbers, 727–732

addition/subtraction of, 729

arithmetic operations, 729–730

graphical representation of, 728–729

identities for, 730–731

integer power of, 731

multiplication/division of, 729

notation for, 727–728

polar form, 727–728

rectangular form, 727–728

roots of, 731–732

Complex power, 384–393, 401, 426–427

apparent power (magnitude), 385, 401

balanced three-phase circuits, 426–427

balancing power in ac circuits, 391–392

calculations for, 384–393, 401

defined, 384

delta (Δ) loads, 426–427

parallel loads and, 390–391

phasors for, 387–388

power calculations for, 384–393, 401, 426–427

power triangle relationship, 385

units for, 384–385, 401

wye (Y) loads, 426

Computer systems, 5

Condition of equivalence, 116

Conductance (G), 33, 336

Continuous functions, 446–447

Control systems, 5

Convolution, 505–511, 521, 674–675

Fourier operational transforms, 674–675

frequency domain, 675

integral, 505–511, 521, 675

memory, concept of using, 510–511

output signal from, 509–510

time domain, 674

Convolution (*Continued*)
  transfer function ($H(s)$) in,
    505–511, 521
  weighting function and, 510
Cosine functions, 631–632, 650,
    669–670
Cramer's method, 722–724
Critically damped response, 286–289,
    291–293, 297–299, 300–301, 309
  natural response, 286–289, 297–300,
    309
  parallel *RLC* circuits, 286–289,
    291–293, 309
  series *RLC* circuits, 297, 297–299, 301
  step response, 291–293, 300–301
Current (*i*), 12–17, 33, 39–40, 41, 50, 58,
    70–73, 80, 152–156, 172, 184–186,
    189–193, 222–224, 230, 289–290,
    318, 320–321, 324, 327–333,
    353–355, 417, 418–419, 422–423,
    435, 639–641
  balanced three-phase circuits, 417,
    418–419, 422–423, 435
  capacitors, 189–193
  defined, 13
  displacement, 189–190
  dot convention for, 354–355
  electric charge and, 12–14
  Fourier series applications, 639–641,
  frequency domain and, 327–333,
    353–355
  ideal transformer ratios, 353–355
  inductor relationships, 184–186
  initial inductor (I_0), 223
  input constraint, 154
  Kirchhoff's current law (KCL),
    39–40, 41, 50, 332–333
  line, 417, 418–419, 422–423, 435
  measurement of, 58, 70–73, 80
  natural response and, 222–224, 230
  Ohm's law for, 33, 50
  op-amp terminals, 152–156, 172
  periodic, 639–641
  phase, 418–419, 422–423, 435
  polarity of, 354–355
  polarity reference, 14
  power and energy relationship to,
    15–17
  resistor power in terms of, 33
  resistor-capacitor (*RC*) circuit
    expression, 230
  resistor-inductor (*RL*) circuit
    expression, 222–224
  sinusoidal source, 320–321
  steady-state analysis and behavior
    of, 318, 324, 327–333, 353–355
  steady-state component, 324
  step response inductor, 289–290

  transient component, 324
Current coil, 430, 435
Current-divider circuit, 67, 79
Current division, 69–70, 79–80, 336
  frequency domain, 336
  resistive circuit analysis, 69–70, 79–80
Current sources, 28–31, 50
Current to voltage (*i–v*) relationships,
    185–186, 190
Cutoff frequency (ω_c), 540–542, 544,
    552–553, 556, 562, 564
  bandpass filters, 552–553, 556
  bandreject filter, 562
  bandwidth (β) relationship to, 553
  center frequency (ω_o) relationship
    to, 553
  defined, 540–542, 564
  half-power frequency, 541
  low-pass filters, 540–542, 544
  RL circuit filters, 544
  RLC circuit filters, 553, 556

D

d'Arsonval meter movement, 71
Damped radian frequency (ω_d), 282,
    293, 299, 301
Decaying exponential function, 452
Decibels (dB), 741–742
Delta (Δ) interconnection, 76
Delta (Δ) loads, 426–427
Delta-to-wye (Δ-to-Y) transforma-
    tions, 75–78, 80, 338–340
  equivalent circuits, 75–78, 80
  frequency domain, 338–340
Dependent sources, 28, 31, 45–48,
    98–100, 107–108, 122, 130–131
  analysis of circuits with, 45–48,
    98–100, 107–108, 130–131
  circuit elements as, 28, 50
  interconnections of, 31
  mesh-current method for, 107–108
  node-voltage method for, 98–100
  Ohm's law for, 45–47
  superposition for, 130–131
  Thévenin equivalent circuits of, 122
Derived units, 10
Difference-amplifier circuit,
    162–167, 173
  common mode input, 164
  common mode rejection ratio
    (CMRR) for, 165–167, 173
  differential mode input, 164
  ideal op-amp model for, 162–167, 173
  negative feedback in, 162
Differential mode input, 164
Differentiation, operational trans-
    forms for, 454–455, 673

Digital meters, 73, 80
Digital signal filtering, 661, 685–686
Direct approach for Fourier series,
    635–637
Dirichlet's conditions, 621
Discontinuities of circuits, 447–450,
    476. *See also* Impulse function;
    Step function
Discontinuous functions, 446–447
Displacement current, 189–190
Domain translation, Laplace trans-
    form for, 456–457
Dot convention, 199–201, 211, 354–355
  ideal transformers, 354–355
  mutual inductance, 199–201, 211
  polarity and, 199–201, 211, 354–355
  procedure for determining, 200–201
Duality, 104, 198–199, 211. *See also*
    Symmetry
Dual-tone multiple-frequency
    (DTMF) design, 537
Dynamo, 28

E

Effective value, 383
Efficiency, power system optimization
    for, 126
Electric power transmission and
    distribution, 413, 433–435
Electrical charge, 12–13
Electrical engineering, 2–9
  balancing power, 3
  circuit theory for, 7–8
  communication systems, 4–5
  computer systems, 5
  control systems, 5
  interaction of systems, 6
  power systems, 5
  problem-solving strategy, 8–9
  profession of, 2
  signal-processing systems, 5
Electrical radiator circuits, 27, 48–49
Electrical sources, 28. *See also* Sources
Electrodynamic wattmeter, 430–433,
    435
Energy, 15–17, 182, 187–189, 191,
    207–209, 211, 226, 230, 374,
    679–685, 686
  capacitors and, 182, 191, 211
  inductors and, 182, 187–189, 211
  mutual inductance and, 207–209, 211
  natural response and, 226, 230
  power and, 15–17
  Parseval's theorem for, 679–685, 686
  power calculations for delivery of, 374
  resistor-capacitor (*RC*) circuit
    expression, 230

resistor-inductor (RL) circuit expression, 226
storage in magnetically coupled coils, 207–209, 211
Equivalent capacitance (C_{eq}), 197
Equivalent circuits, 61–62, 194–197, 418–421, 435, 484–486, 521, 733–740
capacitance (C_{eq}), 196–197, 485–486, 521
ideal transformers in, 737–740
inductance (L_{eq}), 194–195, 196, 484–485, 521
magnetically coupled coils and, 733–737
π-equivalent circuit, 734–735
resistance (R_{eq}), 61–62, 484, 521
s domain representation, 484–486, 521
single-phase, 418–421, 435
T-equivalent circuit, 733
Equivalent inductance (L_{eq}), 196
Essential branches and nodes, 94–96
Even periodic function, 625–626
Exponential form of Fourier series, 642–644, 650

F

Faraday's law, 203–204
Feedback, *see* Negative feedback
Feedback resistors, 305–308, 310
Filters, 536, 538–539. *See also* Active filter circuits; Frequency; Frequency-selective circuits
Final-value theorem, 472–474, 477
First-order active filters, 574–577, 608
Bode plots for, 574–576
frequency response plots for, 574–575
high-pass filters, 576–577, 608
low-pass filters, 574–575, 608
op amp filter design, 575, 576–577
prototypes, 575–577, 608
First-order circuits, 220, 222, 256. *See also* Resistor-capacitor (RC) circuits; Resistor-inductor (RL) circuits
Flashlight (electrical system) circuit model, 36–38, 39–41
Fourier coefficients, 621–630, 649
even-function symmetry, 625–626
Fourier series of periodic function found with, 630
Fourier series of triangular waveform using, 623–624
half-wave symmetry, 627–628

odd-function symmetry, 626–627
periodic functions ($f(t)$) for, 621–630
quarter-wave symmetry, 628–629
symmetry effects on, 625–630, 649
trigonometric form of, 622–624
Fourier series, 618–659
active high-Q filter response, 619, 647–649
amplitude spectrum, 645–647, 650
average-power calculations, 639–641, 650
direct approach, 635–637
exponential form of, 642–644, 650
Fourier coefficients, 621–630, 649
fundamental frequency (ω_0), 621, 649
harmonic frequency, 621, 649
periodic functions ($f(t)$), 618, 620–622, 639–642, 649
periodic response and, 618, 620
periodic voltage applications, 631–637, 639–641, 645–647
phase spectrum, 645–647, 650
phasor domain circuit transformation, 631–632, 650
quality factor (Q), 619
RC circuit application, 633–637
root-mean-square (rms) value, 641–642, 650
sine and cosine terms for, 631–632, 650
steady-state response and, 621–622, 633–638, 650
symmetry functions, 625–630, 649
waveforms, 618, 620, 623–629, 633, 635–636, 650
Fourier transform, 660–691
amplitude spectrum, 663–664
circuit applications, 677–679, 681–685
convergence of integral, 664–666
derivation of, 662–664
digital signal filtering, 661, 685–686
frequency-domain ($F(\omega)$), 662–663, 671–672
frequency-domain representation, 660, 686
integrals used in, 662, 664–666, 671–672
inverse, 663
Laplace transforms for, 666–668
limiting values, 662–663, 665–670
mathematical properties of, 671–672
operational transforms, 672–676
Parseval's theorem for, 679–685, 686
periodic to aperiodic transition, 660, 663, 686
periodic voltage pulse from, 663

steady-state response from, 678
time-domain ($f(t)$), 662–663, 671–672
transient response from, 677–678
Fourth-order filters, 589–590, 594–595
Frequency (ω), 324, 361, 538–542, 544, 550, 552–554, 564–565, 621, 649
bandpass filter, 538–539
bandreject filter, 538–539
center (ω_o), 550, 552–554, 565
cutoff (ω_c), 540–542, 544, 552–553, 562, 564
fundamental (ω_0), 621, 649
half-power, 541
harmonic, 621, 649
high-pass filter, 538
infinite, 544
low-pass filter, 538–542, 544
passband, 538, 564
passive filters, 539, 549, 559, 565
resonant (ω_o), 550
steady-state response, 324, 361
stopband, 538, 564
zero, 544
Frequency domain, 327–357, 361, 446, 457, 460–474, 476–477, 545, 559–560, 660, 662–663, 671–672, 674–675, 686
bandpass filters, 559–560
convolution in, 675
current division in, 336
delta-to-wye (Δ-to-Y) transformations, 338–340
final-value theorem for, 472–474, 477
Fourier transform ($F(\omega)$), 660, 662–663, 671–672, 674–675, 686
frequency-selective circuit analysis, 545, 559–560
impedance (Z) and, 330–331, 333–338, 340–343, 361
initial-value theorem for, 472–474, 477
inverse Laplace transforms for, 460–470, 476
Kirchhoff's laws in, 332–333
Laplace transform ($F(s)$), 446, 457, 460–474, 476–477
low-pass filters, 545
node-voltage method in, 344–345
Norton equivalent circuits for, 340–341
operational Laplace transform for, 457
operational transforms, 674–675
parallel impedance combination, 335, 337
partial fraction expansion, 461–470
passive circuit elements in, 327–331
phasor transforms and, 325, 327–331

Frequency domain (*Continued*)
poles of $F(s)$, 470–472, 476
rational functions ($F(s)$) and,
460–470
series impedance combination,
334–335, 337
source transformations for,
340–343
steady-state circuit analysis in,
327–357, 361
Thévenin equivalent circuits for,
340–341, 343
time domain (t) relationships, 446,
472–474, 477, 545
translation in, 457, 674
voltage division in, 334
voltage to current (v–i) relationships in, 330–331
zeros of $F(s)$, 470–472, 476
Frequency of lumped-parameter
systems, 7
Frequency response, 536
Frequency response plots, 538–539,
574–576, 580, 584
bandpass filter, 539
bandreject filter, 539
Bode plots, 574–576, 580, 584
first-order active filters, 574–576
high-pass filter, 538, 576
ideal, 538–539
low-pass filter, 538, 575
magnitude plot, 538
phase angle plot, 538
Frequency scaling, 578, 608
Frequency-selective circuits, 536–571
attenuation of, 536
bandpass filters, 550–560, 565
bandreject filters, 560–563, 565
center frequency (ω_o) for, 550,
552–554, 565
cutoff frequency (ω_c) for, 540–542,
544, 552–553, 556, 562, 564
filter categories, 538–539
filters as, 536, 564
frequency and time domain relationships, 545, 559–560
high-pass filters, 545–550, 564–565
low-pass filters, 539–545, 564
parallel *RLC*, 555–557, 563
pushbutton telephone circuits,
537, 564
qualitative analysis, 539–541, 544,
546, 551, 560–561
quantitative analysis, 542–543,
546–547, 551–554, 561–562
series *RC*, 543–544, 546–547
series *RL*, 539–540, 542–543,
547–548

series *RLC*, 551–555, 557–559,
560–563
transfer function ($H(s)$) for, 545,
549, 559, 561, 563
Functional Laplace transforms (t), 447,
452–453, 476
Fundamental frequency (ω_0), 621, 649

G

Gain, 153
Galvanometer, 73
Generator, 28
Graphical representation of complex
numbers, 728–729

H

Half-power frequency, 541
Half-wave periodic function, 627–628
Harmonic frequency, 621, 649
High-pass filters, 538, 545–550,
564–565, 576–577, 598, 608
active, 576–577, 598, 608
Bode plot for, 576
Butterworth, 598, 608
cascading, 598, 608
design of, 547–548, 576–577
first-order, 576–577, 608
frequency of, 538
frequency response plots for,
538, 576
frequency-selective circuit analysis,
538, 545–550, 564–565
loading, 548
op amp filter design, 576–577
prototypes, 576–577, 608
qualitative analysis, 546
quantitative analysis, 546–547
second-order, 598, 608
series *RC* circuits, 546–547
series *RL* circuits, 547–548
transfer function ($H(s)$) for, 549, 565
Household appliance ratings, 381–382
Household distribution circuit, 319,
359–360
Hybrid parameters, 696, 697

I

Ideal basic circuit element, 14–15
Ideal transformers, 351–356, 361,
397–398, 737–740
current (i) ratios, 353–355
dot convention for, 354–355
equivalent circuits with, 737–740
frequency domain analysis, 351–356
impedance matching using, 356

limiting values of, 351–353
maximum power transfer in, 397–398
polarity of voltage and current,
354–355
steady-state analysis of, 351–356, 361
symbol for, 354
voltage (v) ratios, 353–355
Ideal versus realistic op-amp models,
173, 14–15
Identities for complex numbers,
730–731
Immitance, 696
Impedance (Z), 330–331, 333–338,
340–343, 348–349, 361, 393–397, 422
admittance (Y) and, 336, 361
balanced three-phase circuits, 422
conductance (G) and, 336
current division and, 336
defined, 330
frequency domain simplifications,
330–331, 333–338, 340–343, 361
linear transformer circuit analysis
using, 348–349, 361
maximum power transfer conditions
and restrictions from, 393–397
parallel combination, 335, 337
passive circuit elements, 330–331
phasors and, 330–331
reactance and, 331
reflected (Z_r), 348–349, 361
self-, 348
series combination, 334–335, 337
source transformations using, 340–343
steady-state analysis using, 330–331,
333–338, 361, 348–349
susceptance (B) and, 336
voltage division and, 334
voltage to current (v–i) relationships and, 330–331
wye- and delta-connected relationships, 422
Impedance matching, 356
Improper rational functions, 460,
469–470
Impulse function ($K\delta(t)$), 449–451,
476, 514–520, 521
capacitor circuit analysis, 514–515
circuit analysis using, 514–520, 521
derivatives of, 451
discontinuities of circuits and,
449–450
impulsive sources, 517–520
inductor circuit analysis, 515–517
Laplace transform method and,
514–520, 521
Laplace transform of, 450–451
sifting property, 450–451
strength (K) of, 449, 476

switching operations, 514–517
unit impulse function ($\delta(t)$), 449, 476
variable parameter function, 449–450
voltage drop and, 516–517
Independent sources, 28, 31, 50
Induced voltage, 199–201, 203–204, 211
Inductance (L), 184, 194–209, 211, 347–348, 351–352
  circuit parameter of, 184
  equivalent, 194–195, 196
  mutual, 182, 199–209, 211
  series–parallel combinations, 194–198
  self-, 199–200, 203–204, 206–207, 211, 347–348, 351–352
  steady-state transformer analysis and, 347–348, 351–352
Inductive circuits, power for, 378–379
Inductor current, 289–290
Inductors, 182, 184–189, 194–195, 198–199, 211, 328–329, 361, 484–485, 515–517, 521, 760
  circuit analysis of, 184–189, 515–517
  circuit component values, 760
  current to voltage (i–v) relationships, 185–186
  duality (symmetry) of, 198–199, 211
  energy in, 182, 187–189, 211
  equivalent circuits for, 194–195, 196, 484–485, 521
  impulse function ($K\delta(t)$) for, 515–517
  in series, 194
  Laplace transform method for, 484–485, 515–517, 521
  magnetic field and, 184
  multiple, 194–195, 196
  parallel, 194–195
  passive behavior of, 182, 211
  phasor relationships, 328–329, 361
  power in, 182, 186–187, 211
  s domain representation, 484–485, 521
  symbols for, 184
  voltage to current (v–i) relationships, 184–185, 328–329, 361
Infinite frequency, 544
Initial-value theorem, 472–474, 477
Input constraints, 153–154, 172
In-series circuit elements, 41, 50. *See also* Series-connected circuits
Instantaneous power, 376–377, 378, 401, 427–428
  balanced three-phase circuits, 427–428
  calculations for, 376–377, 378, 401, 427–428

sinusoidal steady-state circuits, 376–377, 378, 401
  time-invariant, 427–428
Instantaneous real power, 378
Integer power of complex numbers, 731
Integrals, 446, 455–456, 505–511, 521, 662–663, 664–666, 671–672, 758–759
  convergence of, 664–666
  convolution, 505–511, 521
  Fourier transforms, 662, 664–666, 671–672
  frequency-domain function ($F(\omega)$) and, 671–672
  Laplace transforms, 446, 455–456
  table of, 758–759
  time-domain function ($F(t)$) and, 671–672
  transfer function ($H(s)$) use of, 505–511, 521
Integrating-amplifier, 252–254, 256, 303–308, 310
  cascading connections, 303–308, 310
  feedback resistors and, 305–308, 310
  first-order circuit analysis, 252–254, 256
  second-order circuit analysis, 303–308, 310
  step response of, 305
Integration, operational transforms for, 673
Interconnections, 30–31, 36–37, 58, 60–64, 75–78, 79–80
  circuit model creation for, 36–37
  delta (Δ), 76
  delta-to-wye (Δ-to-Y) equivalent circuits, 75–78, 80
  dependent sources, 31
  independent sources, 31
  parallel-connected resistors, 58, 61–64, 79
  pi (π), 76
  pi-to-tee (π-to-T) equivalent circuits, 75–78, 80
  resistive circuits, 58, 60–64, 75–78, 79–80
  series-connected resistors, 58, 60, 79
  series–parallel simplification, 62–63
  tee (T), 76
  testing ideal sources, 30–31
  wye (Y), 76
International System of Units (SI), 9–11
Inverse Fourier transform, 663
Inverse Laplace transforms, 460–470, 476
  distinct complex roots of $D(s)$, 463–465
  distinct roots of $D(s)$, 461–462

improper rational functions, 460, 469–470
partial fraction expansion, 461–470
proper rational functions, 461–469
rational functions ($F(s)$) and, 460–470
repeated complex roots of $D(s)$, 468–468
repeated real roots of $D(s)$, 466–467
s-domain and, 460–470, 476
transform pairs for, 469
Inverse phasor transform, 326
Inverting-amplifier circuit, 156–158, 168, 172–173
  ideal op-amp model for, 156–158, 172
  negative feedback in, 157
  realistic op-amp model for, 168, 173

K

Kirchhoff's laws, 39–44, 45–47, 50, 60–61, 94–95, 332–333, 487
  amplifier circuit analysis using, 47
  circuit analysis using, 40–47, 50
  current law (KCL), 39–40, 41, 50, 60–61, 332–333
  dependent sources and, 45–47
  frequency domain and, 332–333
  loop (closed path) for, 40
  nodes for, 39
  Ohm's law and, 42–43
  parallel-connected circuits and, 61
  s domain use of, 487
  series-connected (in-series) circuits and, 60
  simultaneous equations from, 94–95
  steady-state analysis using, 332–333
  unknown voltage found from, 46
  voltage law (KVL), 40, 42, 50, 60, 332–333

L

Lagging/leading power factors, 379, 401
Laplace transform method, 482–535
  circuit analysis in s domain, 486–488
  circuit analysis using, 482–535
  circuit elements in s domain, 484–486
  convolution integral and, 505–511, 521
  equivalent circuits for time and frequency domains, 485–486
  impulse function and, 514–520, 521
  impulsive sources and, 517–520
  Kirchhoff's laws in s domain, 487

Laplace transform method (*Continued*)
  multiple mesh circuit analysis, 493–494
  mutual inductance circuit analysis, 497–498
  natural response using, 489
  Ohm's law in *s* domain, 484, 486
  partial fraction expansions for, 502–505
  procedure for, 487–488, 521
  RC circuit analysis, 489
  RLC circuit analysis, 489–490
  s-domain applications, 495–496, 499–500
  s-domain equivalent circuits, 485–486
  sinusoidal source circuit analysis, 491–492
  sinusoidal steady-state response and, 511–513, 521
  step response using, 489–490
  superposition applications, 499–500
  surge suppressor analysis, 483, 520
  Thévenin equivalent circuit from, 495–496
  time-invariant circuits, 504–505, 521
  transfer function (*H(s)*) and, 500–513, 521
  voltage to current (*v-i*) equations for, 484–485, 521
Laplace transform, 444–481, 666–668
  applications of, 458–460
  continuous and discontinuous functions, 446–447
  defined, 446
  final-value theorem for, 472–474, 477
  Fourier transforms from, 666–668
  frequency domain (*F(s)*), 446, 452–458, 460–474, 476–477
  functional transforms, 447, 452–453, 476
  impulse function (*Kδ(t)*), 449–451, 476
  initial-value theorem for, 472–474, 477
  integral of, 446, 455–456
  inverse transforms, 460–470, 476
  lumped-parameter circuits and, 458–460, 476
  operational transforms, 447, 453–458, 476
  partial fraction expansion for, 460–470
  poles of *F(s)*, 470–472, 476
  problem-solving uses, 444
  s-domain, 446, 452–458, 460–474, 476–477
  step function (*Ku(t)*), 447–448, 476

time domain (*f(t)*), 446–458, 472–474, 476–477
  transform pairs, 452–453, 469
  transient effects on circuits, 445, 474–476
  unilateral (one-sided) behavior of, 446–447
  unit impulse function (*δ(t)*), 449, 476
  unit step function (*u(t)*), 447, 476
  zeros of *F(s)*, 470–472, 476
Limiting values, 662–663, 665–670
  cosine functions, 669–670
  elementary functions, 670
  Fourier transforms derived using, 662–663
  Fourier transforms of, 665–670
  signum functions, 669
  unit step function, 669
Line current, 417, 418–419, 422–423, 435
Line spectra, 645
Line voltage, 417, 418–419, 435
Linear simultaneous equations, 718–726. *See also* Simultaneous equations
Linear transformer circuits, 347–351, 361
  frequency domain analysis of, 347–351
  reflected impedance (*Z_r*), 348–349, 361
  self-impedance, 348
  steady-state analysis, 347–351, 361
  winding (primary and secondary), 347
Loads, 127–128, 390–391, 393–397, 425–430
  balanced three-phase circuits, 425–430
  delta (Δ), 426–427
  impedance (*Z*) conditions and restrictions, 393–397
  maximum power transfer, 127–128, 393–397
  parallel, 390–391
  power calculations for, 390–391, 393–397, 425–430
  resistive, 127–128
  unspecified, 429–430
  wye (Y), 425–426
Loop (closed path), 40, 50, 94
Low-pass filters, 538, 539–545, 564, 574–575, 587–591, 593–595, 608, 683–684
  active, 574–575, 587–591, 593–595, 608
  Bode plots for, 574–576, 588
  Butterworth, 593–595, 608
  cascading connections, 587–591, 593–595, 608

cutoff frequency (*ω_c*) for, 540–542, 544
  design of, 543, 544, 575
  first-order active, 574–575, 608
  fourth-order active, 589–590, 594–595
  frequency and time domain relationships, 545
  frequency of, 538
  frequency response plots for, 538, 574–576
  frequency-selective circuit analysis, 539–545, 564
  half-power frequency, 541
  infinite frequency, 544
  op amp filter design, 575
  Parseval's theorem for, 683–684
  prototype, 575, 608
  qualitative analysis, 539–541, 544
  quantitative analysis, 542–543
  series *RC* circuits, 543–544
  series *RL* circuits, 539–540, 542–543
  transfer function (*H(s)*) for, 545, 564
  zero frequency, 544
Lumped-parameter circuits, 7–8, 458–460, 476
  frequency (*f*) of, 7
  Laplace transform for, 458–460, 476
  systems of, 7–8

M

Magnetic coupling, 7
Magnetic fields, inductors and, 184
Magnetically coupled coils, 207–209, 211, 245–246, 733–737
  energy storage in, 207–209, 211
  equivalent circuits for, 733–737
  mutual inductance and, 207–209, 211
  π-equivalent circuit, 734–735
  step response of circuit with, 245–246
  T-equivalent circuit, 733
Magnitude plot, 538
Magnitude scaling, 577–578, 608
Maximum power transfer
Maximum power transfer (*P_max*), 92, 126–128, 135, 393–399, 401
  average power absorbed, 394
  circuit analysis for, 92, 126–128, 135
  ideal transformer analysis, 397–398
  impedance (*Z*) conditions and restrictions, 393–397
  power calculations for, 393–399, 401
  resistive loads, 127–128

sinusoidal steady-state analysis of, 393–399, 401
system optimization and, 126
with load restrictions, 396
without load restrictions, 395
Measurement, 9–11, 58, 70–73, 80, 430–433, 435, 697, 700, 741–742
  ammeter for, 70, 71–72, 80
  analog meters for, 71–72, 80
  current, 58, 70–73, 80
  d'Arsonval meter movement, 71
  decibels (dB), 741–742
  digital meters for, 73, 80
  electrodynamic wattmeter for, 430–433, 435
  International System of Units (SI), 9–11
  power, 430–433, 741–742
  resistance, 73–75, 80
  three-phase circuits, 430–433,
  two-port circuit parameters from, 697, 700
  unit prefixes, 10–11
  voltage, 58, 70–73, 80
  voltmeter for, 70, 72, 80
  Wheatstone bridge, 73–75, 80
Memory, concept of using convolution integral, 510–511
Mesh, 94
Mesh circuit analysis, Laplace transform method for, 493–494
Mesh current, 104–105
Mesh-current method, 92, 104–115, 134–135, 199–202, 345–347
  amplifier circuit analysis, 111
  circuit analysis process, 92, 104–115, 134–135
  dependent sources and, 107–108
  duality of, 104
  frequency-domain circuit analysis, 345–347
  mutual inductance and, 199–202
  node-voltage method compared to, 112–115
  special cases for, 108–112
  steady-state circuit analysis, 345–347
  supermesh and, 109–110
Modulation, 674
Motor, 28
Multiplication operations, 453, 673, 730
Mutual inductance, 182, 199–209, 211, 497–498
  coefficient of coupling for, 207–208, 211
  concept of, 204–206
  dot convention, 199–201, 211

energy storage in magnetically coupled coils, 207–209, 211
Laplace transform method for, 497–498
mesh-current method for, 199–202
polarity of induced voltages, 199–201, 204, 211
procedure for determining dot markings, 200–201
s domain circuit, 497–498
self-inductance and, 199–200, 203–204, 206–207, 211

N

Narrowband (high-*Q*) filters, 600–605, 609
  bandpass, 600–602, 608
  bandreject, 602–605, 609
  Bode plots for, 602, 605
Natural response, 220, 222–232, 241–246, 256, 274–289, 296–303, 308–310, 489
  characteristic equation for, 275–276, 297, 299, 309
  circuit phase analysis using, 220
  clock analysis for computer timing, 273, 308–309
  critically damped response, 286–289, 297–300, 309
  current (*i*) expression for, 222–224, 230
  damped radian frequency (ω_d), 282, 293, 299–300
  energy (*w*) expression for, 226, 230
  general solution for, 241–246, 256
  initial inductor current (I_0), 223
  Laplace transform method for, 489
  method for, 224, 230, 241–242, 256, 280, 283, 298–299
  Neper frequency (α), 276, 297, 299, 309
  overdamped response, 279–282, 297–300, 309
  parallel *RLC* circuits, 272, 274–296, 309
  power (*p*) expression for, 226, 230
  resistor-capacitor (*RC*) circuits, 220, 228–232, 241–246, 256, 489
  resistor-inductor (*RL*) circuits, 220, 222–228, 241–246, 256
  resistor-inductor-capacitor (*RLC*) circuits, 274–289, 296–300, 302, 308–310
  resonant radian frequency (ω_0), 276, 293, 297, 299
  series *RLC* circuits, 272, 296–300, 302, 310

steady-state response for, 227
symbols for, 274
time constant (τ), 223–224, 226–227, 229, 256
transient response of, 227
underdamped response, 282–286, 297–300, 309
voltage (*v*) expressions for, 226, 229, 274
Negative feedback, 153–155, 157, 158, 160, 162, 172
  difference-amplifier circuit, 162
  input voltage constraints and, 153–154
  inverting-amplifier circuit, 157
  op-amp circuit analysis and, 153–155
  summing-amplifier circuit, 158
  voltage constraint and, 153–154, 172
Neper frequency (α), 276, 293, 297, 299, 301, 309
  parallel *RLC* circuits, 276, 293, 309
  series *RLC* circuits, 297, 299, 301
Net charge, 7
Neutral terminal, 415
Node voltage, 97
Node-voltage method, 92, 96–104, 112–115, 134–135, 344–345
  amplifier circuit analysis, 102–103
  circuit analysis process, 96–98, 101–102, 134–135
  dependent sources and, 98–100
  duality of, 104
  frequency-domain circuit analysis, 344–345
  mesh-current method compared to, 112–115
  special cases for, 100–104
  steady-state circuit analysis, 344–345
  supernodes and, 101–102
Nodes, 39, 50, 94–96
Noninverting-amplifier circuit, 159–162, 168–170, 172–173
  ideal op-amp model for, 159–162, 172
  negative feedback in, 160
  realistic op-amp model for, 168–170, 173
Nonplanar circuits, 94
Norton equivalent circuits, 92, 120–123, 135, 340–341
  frequency domain simplification, 340–341
  impedance (*Z*) in, 340–341
  source transformation, 92, 120–123, 135, 340–341
  terminal circuit simplification using, 92, 120–123, 135

O

Odd periodic function, 626–627
Ohm's law, 32–33, 42–43, 45–47, 50, 61, 484, 486
 amplifier circuit analysis using, 47
 circuit analysis using, 42–43, 45–47, 50
 dependent sources and, 45–47
 electrical resistance and, 32–33, 50
 Kirchhoff's law and, 42–43
 parallel-connected circuits and, 61
 s domain use of, 484, 486
Op amp filters, 575–577, 579, 580–600, 608
 bandpass, 580–583, 598–599, 608
 bandreject, 583–587, 608
 Butterworth, 591–600
 cascading, 587–591
 design of, 575, 576–577
 first-order active, 575, 576–577, 587–591
 higher-order active, 587–600
 high-pass, 576–577
 low-pass, 575, 579
 scaling, 579
Open circuit, 37
Open-loop operation, 157
Operational amplifiers (op amps), 150–181. *See also* Op amp filters
 common mode rejection ratio (CMRR) for, 165–167, 173
 current (i), 152–156, 172
 difference-amplifier circuit, 162–167, 173
 gain, 153
 ideal model, analysis of, 154–155, 172–173
 input constraints, 153–154, 172
 inverting-amplifier circuit, 156–158, 168, 172
 negative feedback in, 153–154, 157, 158, 160, 162, 172
 noninverting-amplifier circuit, 159–162, 168–170, 172
 open-loop operation, 157
 realistic models, analysis of, 167–170, 173
 strain gage analysis, 151, 171–172
 summing-amplifier circuit, 158–159, 172
 symbols for, 152
 terminals, 152–156, 172
 transducers, 151, 171–172
 voltage (v), 152–156, 172
Operational transforms, 447, 453–458, 476, 672–676
 addition and subtraction, 454, 673

convolution in frequency domain, 675
convolution in time domain, 674
defined, 447, 476
differentiation, 454–455, 673
Fourier, 672–676
integration, 455–456, 673
Laplace, 447, 453–458, 476
modulation, 674
multiplication by a constant, 453, 673
scale changing, 457, 674
translation in frequency domain, 457, 674
translation in time domain, 456–457, 674
Overdamped response, 279–282, 291–293, 297–299, 300–301, 309
 natural response, 279–282, 297–300, 309
 parallel *RLC* circuits, 279–282, 291–293, 309
 series *RLC* circuits, 297–299, 301
 step response, 291–293, 300–301

P

Parallel-connected circuits, 58, 61–64, 79, 194–196, 335, 337, 390–391, 708–709. *See also* Parallel *RLC* circuits
 capacitors, 195–196
 circuit elements, 61
 combining, 61–62
 frequency domain, 335, 337
 impedance (Z) combined in, 335, 337
 inductors, 194–195
 Kirchhoff's current law for, 61
 Ohm's law for, 61
 power calculations for, 390–391
 resistors, 58, 61–64, 79
 series–parallel simplification, 62–63
 two-port, 708–709
Parallel *RLC* circuits, 272, 274–296, 309, 555–557, 563
 bandpass filters, 555–557
 bandreject filters, 563
 bandwidth (β), 556
 characteristic equation for, 275–276, 293, 309
 critically damped voltage response, 286–289, 291–293, 309
 cutoff frequency (ω_c), 556
 damped radian frequency (ω_d), 282, 293
 frequency-selective circuits, 555–557, 563

natural response of, 274–289, 309
Neper frequency (α), 276, 293, 309
overdamped response, 279–282, 291–293, 309
parameters of, 276–277
quality factor (Q), 556
resonant radian frequency (ω_0), 276, 293
second-order differential equations for, 274–277
step response of, 289–296, 309
symbols for, 274
underdamped response, 282–286, 291–293, 309
Parasitic resistance, 37
Parseval's theorem, 679–685, 686
 bandpass filter application, 662–663
 energy calculations using, 679–685, 686
 Fourier transform time-domain functions, 679–685, 686
 graphic interpretation of, 681
 low-pass filter application, 683–684
 rectangular voltage pulse application, 684–685
Partial fraction expansion, 461–470, 502–505
 distinct complex roots of $D(s)$, 463–465
 distinct roots of $D(s)$, 461–462
 improper rational functions, 460, 469–470
 inverse Laplace transforms and, 461–470
 proper rational functions, 461–469
 repeated complex roots of $D(s)$, 468–468
 repeated real roots of $D(s)$, 466–467
 s domain use of, 461–470, 502–505
 transfer function ($H(s)$) in, 502–505
 transform pairs for, 469
Passband frequency, 538, 564
Passive circuit elements, 30, 182, 211, 327–331, 361
 capacitors, 182, 211, 329–330, 361
 defined, 30
 frequency domain, 327–331
 impedance (Z) and, 330–331
 inductors, 182, 211
 phasor transforms and, 325, 327–331
 reactance and, 331
 resistors, 327–328, 361
 voltage to current (v–i) relationships in, 330–331
Passive filters, 539, 549, 559, 565

Passive sign convention, 14–17
Period of time (*T*), 320
Periodic current, average-power calculations and, 639–641
Periodic functions (*f*(*t*)), 618, 620–630, 639–642, 649–650
 average power calculations with, 639–641, 650
 defined, 618, 649
 Dirichlet's conditions, 621
 even, 625–626
 Fourier coefficients and, 621–623, 649
 Fourier series of found with symmetry, 630
 Fourier series representation, 621–622, 639–641
 fundamental frequency (ω_0), 621, 649
 half, 627–628
 harmonic frequency, 621, 649
 odd, 626–627
 periodic voltage and, 639–641
 quarter, 628–629
 root-mean-square (rms) value of, 641–642, 650
 steady-state response from, 621–622
 symmetry effects, 625–630, 649
 waveforms, 618, 620–629
Periodic response, 618, 620
Periodic to aperiodic transition, 660, 663, 686
Periodic voltage, 631–637, 639–641, 645–647, 663
 amplitude spectra for, 645–647
 average-power calculations expressed from, 639–641
 Fourier series applications, 631–637, 639–641, 645–647
 inverse Fourier transform and, 663
 phase spectra for, 645–647
 phasor domain circuit transformation and, 631–632
 sine and cosine terms for, 631–632
 steady-state response and, 633–637
 waveforms, 633, 635–636
Periodic waveforms, 618, 620
Phase angle (ϕ) of, 320
Phase angle plots, 538, 748–750, 754–756
 complex poles, 754–756
 frequency response and, 538
 straight-line, 748–750, 754–756
Phase current, 418–419, 422–423, 435
Phase sequences for three-phase circuits, 414, 435
Phase spectrum, 645–647, 650
Phase voltage, 418–419, 435
Phase windings, 415
Phasor diagrams, 357–359, 414, 419

Phasor transform, 325–333
 frequency domain and, 325, 327–331
 inverse, 326
 phasor representation as, 325
 voltage to current (*v*–*i*) relationships, 327–331, 361
Phasors, 324–331, 387–388, 631–632, 650. *See also* Phasor transforms
 capacitor voltage to current (*v*–*i*) relationships, 329–330, 361
 complex power calculations using, 387–388
 concept of, 324–325
 Fourier series transformation to phasor domain, 631–632, 650
 impedance (*Z*) and, 330–331
 inductor voltage to current (*v*–*i*) relationships, 328–329, 361
 reactance and, 331
 representation, 325
 resistor voltage to current (*v*–*i*) relationships, 327–328, 361
 sinusoidal functions and, 324–327
 steady-state analysis using, 324–330
Pi (π-equivalent circuit, 734–735
Pi (π) interconnection, 76
Pi-to-tee (π-to-T) equivalent circuits, 75–78, 80
Planar circuits, 94
Polar form of complex numbers, 727–728
Polarity, 14, 16–17, 199–201, 204, 211, 354–355
 arrows for reference of, 14, 16
 coil current and voltage, 354–355
 dot convention for, 199–201, 211, 354–355
 ideal transformers, 354–355
 induced voltages, 199–201, 204, 211
 mutual inductance, 199–201, 211
 power reference, 16–17
 self-inductance, 204
 voltage and current references, 14
Poles, 470–472, 476, 502, 743–744, 750–756
 amplitude plots, 750–753
 complex, 750–756
 frequency domain (*F*(*s*)), 470–472, 476
 phase angle plots, 754–756
 real, first-order, 743–744
 transfer functions (*H*(*s*)), 502
Ports, 692
Potential coil, 430, 435
Power, 3, 15–20, 33–34, 126–128, 135, 182, 186–187, 190–191, 211, 226, 230, 378–379, 384–385, 391–392, 401, 413, 427–428, 433–435

ac circuits, 391–392
 algebraic sign of, 16–17
 balance of in circuits, 3, 19–20
 balanced three-phase circuits, 413, 433–435
 capacitive circuits, 379
 capacitors and, 182, 190–191, 211
 current and voltage relationship to, 15–17
 defined, 16
 electric transmission and distribution, 413, 433–435
 energy and, 15–17
 inductive circuits, 378–379
 inductors and, 182, 186–187, 211
 maximum power transfer, 126–128, 135
 natural response and, 226, 230
 passive sign convention for, 16–17
 polarity reference, 16–17
 resistive circuits, 378
 resistive load transfer, 127–128
 resistor-capacitor (*RC*) circuit expression, 230
 resistor-inductor (*RL*) circuit expression, 226
 resistors, 33–34
 time-invariant, 427–428
 units for, 379, 384–385, 401
Power calculations, 374–411, 425–430, 435
 apparent power, 385, 401
 appliance ratings for, 381–382
 average power (*P*), 377–384, 389, 394, 401, 425–426
 balanced three-phase circuits, 425–430, 435
 balancing power in ac circuits, 391–392
 complex power, 384–393, 401, 426–427
 delta (Δ) loads, 426–427
 energy delivery and, 374
 instantaneous power, 376–377, 378, 401, 427–428
 lagging/leading factors for, 379, 401
 maximum power transfer (P_{max}), 393–399, 401
 parallel loads and, 390–391
 phasors for, 387–388
 power factor (pf) for, 379, 401
 reactive factor (rf) for, 379, 401
 reactive power (*Q*), 377–382, 389, 401, 426–427
 root-mean-square (rms) value for, 382–384
 sinusoidal steady-state analysis, 374–411

Power calculations (*Continued*)
  standby (vampire) power, 375, 399–400
  unspecified loads, 429–430
  wye (Y) loads, 425–426
  wye-delta (Y-Δ) circuits, 428–429
  wye-wye (Y-Y) circuits, 428
Power consumption, 399
Power equation, 16
Power factor (pf), 379, 401
Power measurement, 430–433, 435, 741–742
  balanced three-phase circuits, 430–433, 435
  bels, 741
  decibels (dB), 741–742
  electrodynamic wattmeter for, 430–433, 435
  current coil, 430, 435
  potential coil, 430, 435
  power gain, 741–742
  two-wattmeter method, 431–432, 435
  wattmeter reading calculations, 432–433
Power systems, 5, 319, 359–360
Power triangle, 385
Problem-solving strategy, 8–9
Proper rational functions, 460–469.
    See also Partial fraction expansion
Prototypes, 12, 575–577, 608
Pushbutton telephone circuits, 537, 564

Q

Qualitative analysis, 539–541, 544, 546, 551, 560–561
  bandpass filters, 551
  bandreject filters, 560–561
  high-pass filters, 546
  low-pass filters, 539–541, 544
Quality factor (Q), 550, 554, 562, 556, 619
Quantitative analysis, 542–543, 546–547, 551–554, 561–562
  bandpass filters, 551–554
  bandreject filters, 561–562
  high-pass filters, 546–547
  low-pass filters, 542–543
Quarter-wave periodic function, 628–629

R

Rational functions ($F(s)$), 460–470
  improper, 460, 469–470
  inverse Laplace transforms and, 469–470
  proper, 461–469

transform pairs for, 469
Reactance, 331
Reactive factor (rf), 379, 401
Reactive power (Q), 377–382, 389, 401, 426–427
  balanced three-phase circuits, 426–427
  calculations for, 377–382, 389, 401, 426–427
  delta (Δ) loads, 426–427
  sinusoidal steady-state analysis, 377–382, 389
  wye (Y) loads, 426
Reciprocal two-port circuits, 701–702, 712
Rectangular form of complex numbers, 727–728
Rectangular waveforms, 620
Reflected impedance (Z_r), 348–349, 361
Resistance (R), 32–35, 37, 50, 61–62, 73–75, 80
  conductance (G) and, 33
  equivalent (R_{eq}), 61–62
  measurement of, 73–75, 80
  Ohm's law and, 32–33, 50
  parasitic, 37
  resistors as models of, 32–35
  Wheatstone bridge circuit for, 73–75, 80
Resistive circuits, 58–91, 115–118, 121, 134, 378
  analysis of, 68–70, 79–80, 115–118, 121, 134
  current-divider circuit, 67, 79
  current division, 69–70, 79–80
  delta-to-wye (Δ-to-Y) equivalent circuits, 75–78, 80
  interconnections, 58, 60–64, 75–78
  measurement of voltage and current, 58, 70–73, 80
  parallel connections, 58, 61–64, 79
  pi-to-tee (π-to-T) equivalent circuits, 75–78, 80
  power for, 378
  resistor value measurements, 73–75, 80
  series connections, 58, 60, 79
  series–parallel simplification, 62–63
  source transformation, 115–118, 121, 134
  touch screens, 59, 78–80
  voltage-divider circuit, 64–66, 79
  voltage division, 68–70, 79
  Wheatstone bridge, 73–75, 80
Resistive loads, 127–128
Resistor-capacitor (RC) circuits, 220–222, 228–232, 238–246,

249–255, 489, 543–544, 546–547, 633–637
  analysis phases for, 220
  artificial pacemaker design, 221, 255
  current (i) expression, 230
  cutoff frequency (ω_c), 544
  energy (w) expression, 230
  first-order circuits as, 220, 222, 256
  Fourier series application, 633–637
  frequency-selective analysis of, 543–544, 546–547
  general solution for, 241–246, 256
  high-pass filters, 546–547
  integrating-amplifier circuit analysis, 252–254, 256
  Laplace transform method for, 489
  low-pass filters, 543–544
  natural response of, 220, 228–232, 241–246, 256, 489
  periodic voltage in, 633–637
  power (p) expression, 230
  sequential switching, 246, 249–250, 256
  steady state response, 633–637
  step response of, 220, 238–246, 256
  time constant (τ), 229, 256
  unbounded response, 250–251, 256
  voltage (v) expression, 229
Resistor-inductor (RL) circuits, 220–228, 233–237, 241–251, 256, 539–540, 542–543, 547–548
  analysis phases for, 220
  current (i) expression, 222–224
  cutoff frequency (ω_c), 542
  energy (w) expression, 226
  first-order circuits as, 220, 222, 256
  frequency-selective analysis of, 539–540, 542–543, 547–548
  general solution for, 241–246, 256
  high-pass filters, 547–548
  low-pass filters, 539–540, 542–543
  natural response of, 220, 222–228, 241–246, 256
  power (p) expression, 226
  qualitative analysis, 546
  quantitative analysis, 546–547
  sequential switching, 246–249, 256
  step response of, 220, 233–237, 241–246, 256
  time constant (τ), 223–224, 226–227, 256
  unbounded response, 250–251, 256
  voltage (v) expression, 226
Resistor-inductor-capacitor (RLC) circuits, 272–317, 489–490, 551–559, 560–563, 637–638
  bandwidth (β), 553–554, 556

center frequency (ω_o), 552–554

characteristic equations for, 275–276, 293, 297, 299, 301, 309

clock for computer timing, 273, 308–309

critically damped voltage response, 286–289, 291–293, 298–299, 300–301, 309

cutoff frequency (ω_c), 553, 556

damped radian frequency (ω_d), 282, 293, 301

direct approach for, 291–292, 637–638

Fourier series approach for, 637–638

frequency-selective circuit analysis, 551–559, 560–563

indirect approach for, 290–291

inductor current for, 289–290

integrating amplifiers in cascade, 303–308, 310

Laplace transform method for, 489–490

natural response of, 274–289, 296–300, 302, 308–310

Neper frequency (α), 276, 293, 297, 301, 309

overdamped response, 279–282, 291–293, 298, 300–301, 309

parallel, 272, 274–296, 309, 555–557, 563

quality factor (Q) for, 554, 556

resonant radian frequency (ω_0), 276, 293, 297, 301

second-order differential equations for, 274–277

series-connected, 272, 274, 296–303, 310, 551–555, 557–559, 560–563

square-wave voltage and, 637–638

steady-state response of, 637–638

step response of, 289–296, 300–303, 305, 309–310, 489–490

symbols for, 274, 289

timing signals, 273

underdamped voltage response, 282–286, 291–293, 298–299, 300–301, 309

voltage expressions for, 274

Resistors, 32–35, 50, 61–64, 93, 131–134, 327–328, 361, 484, 521, 760

circuit component values, 760

conductance (G) and, 33

equivalent circuits for, 61–62, 484, 521

multiple, 61–62

Ohm's law for, 32–33, 50

phasor relationships, 327–328, 361

power in terms of current, 33

power in terms of voltage, 34

resistance (R) models, 32–35

s domain representation, 484, 521

sensitivity analysis of, 93, 131–134

series–parallel simplification, 62–63

signals in phase, 328

voltage to current (v–i) relationships, 327–328, 361

Resonant frequency (ω_o), 550

Resonant radian frequency (ω_0), 276, 293, 297, 299, 301

parallel RLC circuits, 276, 293

series RLC circuits, 297, 299, 301

Response, 220, 222–246, 241–246, 250–251, 256, 272–317, 323–324, 361, 504–513, 521, 677–678

critically damped, 286–289, 291–293, 297–299, 300–301, 309

damped radian frequency (ω_d), 282, 293, 299, 301

Fourier transforms for, 677–678

general solution for, 241–246, 256

memory, concept of, 510–511

natural, 220, 222–232, 241–246, 256, 274–289, 296–300, 302, 308–310

overdamped, 279–282, 291–293, 297–299, 300–301, 309

resistor-capacitor (RC) circuits, 220, 228–232, 238–246, 250–251, 256

resistor-inductor (RL) circuits, 220, 222–228, 233–237, 241–251, 256

resistor-inductor-capacitor (RLC) circuits, 272–317

sinusoidal, 323–324, 361, 504–513, 521

steady-state analysis of, 318, 323–324, 361

steady-state, 227, 318, 323–324, 361, 504–513, 521, 678

step, 220, 233–246, 256, 289–296, 300–303, 309–310

transfer function ($H(s)$) and, 504–513, 521

transient, 227, 677–678

unbounded, 250–251, 256

underdamped, 282–286, 291–293, 297–299, 300–301, 309

unit impulse ($h(t)$), 504–511

weighting function for, 510

Root-mean-square (rms) value, 321, 382–384, 641–642, 650

effective value as, 383

periodic functions ($f(t)$), 641–642, 650

power calculations using, 382–384

sinusoidal sources and, 321

Roots of complex numbers, 731–732

S

s domain, 446, 452–458, 460–474, 476–477, 482–535. *See also* Frequency domain

circuit analysis in, 486–488

circuit elements in, 484–486

final-value theorem for, 472–474, 477

initial-value theorem for, 472–474, 477

inverse Laplace transforms for, 460–470, 476

Kirchhoff's laws in, 487

Laplace transform method applications, 482–535

Laplace transform ($F(s)$) of, 460–474, 476–477

mutual inductance circuit in, 497–498

Ohm's law in, 484, 486

operational transforms for, 447, 453–458, 476

partial fraction expansion, 461–470, 502–505

poles of $F(s)$, 470–472, 476, 502

rational functions ($F(s)$) and, 460–470

superposition applications in, 499–500

Thévenin equivalent circuit in, 495–496

time domain (t) relationships, 446, 453–458, 472–474, 477

transfer function ($H(s)$), 500–513, 521

transform pairs, 452–453, 469

zeros of $F(s)$, 470–472, 476, 502

Scale change, 457, 674

Scaling, 577–580, 608

circuit component scale factors, 578

filter design using, 578

frequency, 578, 608

low-pass op-amp filter, 579

magnitude, 577–578, 608

series RLC filter, 578–579

Second-order circuits, 274. *See also* Resistor-inductor-capacitor (RLC) circuits

Second-order filters, 598

Self-impedance, 348

Self-inductance, 199–200, 203–204, 206–207, 211, 347–348, 351–352

Faraday's law for, 203–204

mutual inductance and, 199–200, 206–207, 211

polarity of induced voltages, 199–201, 204, 211

Self-inductance (*Continued*)
  steady-state transformer analysis and, 347–348, 351–352
  voltage drop, 199
Sensitivity analysis of resistors, 93, 131–134
Sequential switching, 246–250, 256
  circuit analysis and, 246–250, 256
  defined, 246
  resistor-capacitor (*RC*) circuits with, 246, 249–250, 256
  resistor-inductor (*RL*) circuits with, 246–249, 256
Series-connected (in-series) circuits, 41, 50, 58, 60, 79, 194–195, 334–335, 337, 539–540, 542–544, 546–548, 708–709. *See also* Series *RLC* circuits
  black box concept, 60
  capacitors, 195
  circuit elements, 41, 50
  combining, 60
  frequency domain, 334–335, 337
  frequency-selective circuits, 539–540, 542–544, 546–548
  high-pass filters, 546–548
  impedances (*Z*) combined in, 334–335, 337
  inductors, 194
  Kirchhoff's laws for, 60
  low-pass filters, 539–540, 542–544
  resistors, 58, 60, 79
  two-port circuits, 708–709
Series–parallel connections, 62–63, 708–709
  simplification of, 62–63
  two-port circuits, 708–709
Series *RLC* circuits, 272, 274, 296–303, 310, 551–555, 557–563, 578–579
  bandpass filters, 551–555, 557–559
  bandreject filter, 560–563
  bandwidth (β), 553, 562
  center frequency (ω_o), 552, 562
  characteristic equation for, 297, 299–301
  critically damped response, 297, 297–299, 301
  cutoff frequency (ω_c), 552–553, 562
  frequency-selective circuits, 551–555, 557–563
  natural response of, 272, 274, 296–303, 310
  Neper frequency (α), 297, 299, 301
  overdamped response, 297–299, 301
  quality factor (*Q*), 554, 562
  resonant radian frequency (ω_0), 297, 299, 301
  scaling, 578–579

step response of, 300–303, 310
symbols for, 274
underdamped response, 297–299, 301
Short circuit, 37
Sifting property, 450–451
Signal-processing systems, 5
Signals in phase, 328
Signum functions, 669
Simplification techniques, 62–63, 75–78, 115–118, 121
  delta-to-wye (Δ-to-Y) transformation, 75–78
  series–parallel simplification, 62–63
  source transformation, 115–118, 121
Simultaneous equations, 94–96, 718–726
  applications of, 723–726
  back-substitution method for, 721–722, 726
  calculator and computer methods for, 719–721, 724–726
  characteristic determinant of, 722–723
  circuit analysis using, 94–96
  Cramer's method for, 722–724
  essential nodes and branches for, 95–96
  Kirchhoff's laws for, 94–95
  linear, 718–726
  number of, 94–96
  solution of, 718–726
Sine functions, 631–632, 650
Single-phase equivalent circuits, 418–421, 435
Sinusoidal circuits, 318–373, 374–411
  power calculations, 374–411
  steady-state analysis, 318–373
Sinusoidal function, 452
Sinusoidal rectifiers, 618, 620
Sinusoidal response, 323–324, 361, 511–513, 521
  frequency (ω) of, 324, 361
  steady-state analysis of, 323–324, 361
  steady-state current component, 324
  steady-state solution characteristics, 324
  transient current component, 324
  transfer function (*H(s)*) and, 511–513, 521
Sinusoidal sources, 318–323, 361, 491–492
  amplitude of, 320
  angular frequency (ω), 320
  current behavior and, 318
  current (*i*), 320–321
  Laplace transform method for, 491–492

period of time (*T*), 320
phase angle (ϕ) of, 320
root-mean-square (rms) value, 321–323
steady-state analysis and, 318–323, 361
steady-state response from, 319, 323–324
voltage (*v*), 320–322
Source transformation, 92, 115–118, 121, 134, 340–343
  bilateral configurations, 115–116
  condition of equivalence for, 116
  defined, 115
  frequency-domain circuit simplification, 340–343
  impedance (*Z*) for, 340–343
  Norton equivalent circuits from, 121, 340–341
  resistive circuit simplification, 92, 115–118, 121, 134
  steady-state circuit analysis, 340–343
  Thévenin equivalent circuits from, 121, 340–341, 343
Sources, 28–31, 45–48, 318–323, 361, 491–492, 517–520
  active circuit elements, 30
  current, 28–31, 50
  dependent, 28, 31, 45–48, 50
  direct current (dc), 30
  electrical, 28
  impulsive, 517–520
  ideal, 28–31, 50
  independent, 28, 31, 50
  interconnections of, 30–31
  Laplace transform method for, 517–520
  passive circuit elements, 30
  sinusoidal, 318–323, 361, 491–492
  symbols for, 28–29
  voltage, 28–31, 50
Square-wave voltage, 633, 637–638
Square waveforms, 620, 633
Standby (vampire) power analysis, 375, 399–400
Steady-state analysis, 318–373
  admittance (*Y*), 336, 361
  challenges of, 318
  delta-to-wye (Δ-to-Y) transformations, 338–340
  frequency domain of, 327–357
  household distribution circuit, 319, 359–360
  impedance (*Z*) for, 330–331, 333–338, 340–343, 361
  Kirchhoff's laws for, 332–333
  mesh-current method, 345–347
  node-voltage method for, 344–345

Norton equivalent circuit for,
340–342
parallel impedances, 335–338
passive circuit elements, 327–331
phasor diagrams for, 357–359
phasor transforms for, 325–333
phasors, 324–331
responses, 227, 323–324
series impedances, 334–335
sinusoidal sources for, 318–323, 361
source transformations for, 340–343
Thévenin equivalent circuit for,
340–343
transformers, 347–356, 361
voltage to current (*v–i*) relationships, 327–331, 361
Steady-state current component, 324
Steady-state response, 227, 318,
323–324, 361, 511–513, 521,
621–622, 633–638, 650, 678
direct approach to, 635–638, 650
Fourier series approach for,
621–622, 633–638, 650
Fourier transforms for, 678
periodic functions used for,
621–622
periodic voltage and, 633–637
RC circuit periodic voltage
response, 633–637
RLC circuit square-wave voltage
response, 637–638
sinusoidal analysis conditions, 318
sinusoidal sources of, 318, 323–324
square-wave voltage and,
637–638
time constant (*τ*) and, 227
transfer function (*H(s)*) and,
511–513, 521
waveforms of, 635–636
Step function (*Ku(t)*), 447–448, 476
discontinuities of circuits and,
447–448, 476
finite duration representation,
448
unit step function (*u(t)*), 447, 476
Step response, 220, 233–246, 256,
289–296, 300–303, 305, 309–310,
489–490
characteristic equation for, 293, 301
circuit analysis using, 220
comparison of *RC* and *RL*
circuits, 241
critically damped response,
291–293, 300–301
damped radian frequency (*ω_d*),
293, 301
direct approach for, 291–292
general solution for, 241–246, 256

indirect approach for, 290–291
inductor current for, 289–290
inductor voltage versus time,
236–237
integrating-amplifier analysis of, 305
magnetically coupled coils and,
245–246
Laplace transform method for,
489–490
method for, 234, 238, 241–242, 256
Neper frequency (*α*), 293, 301
overdamped response, 291–293,
300–301
parallel, 272, 289–296
resistor-capacitor (*RC*) circuits, 220,
238–246, 256
resistor-inductor (*RL*) circuits, 220,
233–237, 241–246, 256
resistor-inductor-capacitor (*RLC*)
circuits, 289–296, 300–303,
309–310, 489–490
resonant radian frequency (*ω_0*), 293,
301
series-connected, 272, 300–303
symbols for, 274, 289
underdamped response, 291–293,
300–301
Stopband frequency, 538, 564
Straight-line plots, 744–756. *See also*
Amplitude plots; Phase angle
plots
Strain gages, op-amp circuit analysis
for, 151, 171–172
Strength (*K*) of impulse function, 449,
476
Summing-amplifier circuit, 158–159,
172
Supermesh, 109–110
Supernodes, 101–102
Superposition, 92, 129–131, 135,
499–500
circuit analysis using, 92, 129–131,
135
dependent sources and, 130–131
Laplace transform method using,
499–500
s domain applications, 499–500
Surge suppressor analysis, 483, 520
Susceptance (*B*), 336
Switching operations, impulse function
(*Kδ(t)*) for, 514–517
Symmetric two-port circuits, 701–702,
712–713
Symmetry, 104, 198–199, 211, 625–630,
649
capacitors, 198–199, 211
duality as, 104, 198
even-function, 625–626

Fourier coefficient, effects on,
625–630, 649
Fourier series of periodic function
found with, 630
half-wave, 627–628
inductors, 198–199, 211
odd-function, 626–627
quarter-wave, 628–629

T

T-equivalent circuit, 733
Tee (T) interconnection, 76
Terminals, 38, 43–44, 118–126, 152–156,
172, 692–717
circuit behavior and, 118–126,
152–156
current of, 152–156, 172
measurements for circuit construction, 38, 43–44
negative feedback and, 153–154, 172
Norton equivalent circuits and,
120–123
operational amplifier (op amp),
152–156, 172
ports, 692
symbols for, 152
Thévenin equivalent circuits and,
118–126
two-port circuits, 692–717
Terminals, 38, 43–44, 118–126, 152–156,
172, 692–717
Terminated two-port circuits, 703–708
Thévenin equivalent circuits, 92,
118–126, 135, 340–341, 343, 495–496
amplifier circuit analysis using,
125–126
dependent sources and, 122
impedance (*Z*) in, 340–341
frequency-domain circuit simplification, 340–341, 343
Laplace transform method for,
495–496
resistance directly from circuit,
123–126
s domain, 495–496
source transformation for, 121,
340–341, 343
terminal circuit simplification using,
92, 118–123, 135
voltage of, 152–156, 172
Three-phase circuits, 412–443
a-, b-, and c-phase voltage
references, 414
average power measurement in,
430–433, 435
balanced conditions, 412, 416–417,
435

Three-phase circuits (*Continued*)
  basic circuit use and characteristics, 412, 414
  delta (Δ) loads, 426–427
  electric power transmission and distribution, 413, 433–435
  impedance relationships, 422
  instantaneous power in, 427–428
  line current, 417, 418–419, 422–423, 435
  line voltage, 417, 418–419, 435
  neutral terminal for, 415
  phase current, 418–419, 422–423, 435
  phase sequences, 414, 435
  phase voltage, 418–419, 435
  phase windings, 415
  phasor diagrams for, 414, 419
  power calculations in, 425–430, 435
  single-phase equivalent circuit for, 418–421, 435
  unspecified loads, 429–430
  voltage sources, 415
  wye (Y) loads, 425–426
  wye-delta (Y-Δ) circuit analysis, 422–425, 428–429
  wye-wye (Y-Y) circuit analysis, 416–421, 428
Time constant (τ), 223–224, 226–227, 229, 256
  resistor-capacitor (RC) circuits, 229, 256
  resistor-inductor (RL) circuits, 223–224, 226–227, 256
  significance of, 226–227
  steady-state response and, 227
  transient response and, 227
Time domain (t), 446–458, 472–474, 476–477, 545, 559–560, 662–663, 671–672, 674, 679–685, 686
  bandpass filters, 559–560
  convolution in, 674
  final-value theorem for, 472–474, 477
  Fourier transform (f(t)), 662–663, 671–672, 674, 679–685, 686
  frequency domain (s) relationships, 446, 472–474, 477, 545, 559–560
  functional transforms, 447, 452–453, 476
  impulse function (Kδ(t)), 449–451, 476
  initial-value theorem for, 472–474, 477
  Laplace transform (f(t)), 446–447, 453–458, 472–474, 476–477
  low-pass filters, 545

  operational transforms for, 447, 453–458, 476, 674
  Parseval's theorem, 679–685, 686
  sifting property and, 450–451
  step function (Ku(t)), 447–448, 476
  transform pairs, 452–453, 469
  translation in, 674
  unit impulse function (δ(t)), 449, 476
  unit step function (u(t)), 447, 476
Time-invariant circuits, 504–505, 521
Time-invariant instantaneous power, 427–428
Timing signals, 273
Touch screens, 59, 78–80, 183, 209–210
  capacitance of, 183, 209–210
  resistive circuits of, 59, 78–80
Transducers (strain gages), 151, 171–172
Transfer function (H(s)), 500–513, 521, 545, 549, 559, 561, 563, 564–565, 592–593
  bandpass filters, 559, 565
  bandreject filters, 561, 563, 565
  Butterworth filters, 592–593
  circuit analysis and, 501–502, 504–505
  convolution integral and, 505–511, 521
  defined, 500–501
  frequency-selective circuit analysis using, 545, 549, 559, 564–565
  high-pass filters, 549, 565
  Laplace transform method for, 500–513, 521
  low-pass filters, 545, 564
  partial fraction expansion and, 461–470, 502–505
  poles of, 502
  sinusoidal steady-state response and, 511–513, 521
  time-invariant circuits, 504–505, 521
  unit impulse response (h(t)) and, 504–511
  weighting function, 510
  zeros of, 502
Transform pairs, 452–453, 469
Transformers, 347–356, 361, 737–740
  current (i) ratios, 353–355
  dot convention for, 354–355
  equivalent circuits with, 737–740
  frequency domain analysis of, 347–356
  ideal, 351–356, 361, 737–740
  impedance matching, 356
  limiting values of, 351–353
  linear circuits, 347–351, 361

  polarity of voltage and current, 354–355
  reflected impedance (Z_r), 348–349, 361
  self-impedance of, 348
  self-inductance of, 347–348, 351–352
  steady-state analysis of, 347–356, 361
  steady-state analysis, 347–356, 361
  voltage (v) ratios, 353–355
  winding (primary and secondary), 347
Transient current component, 324
Transient response, 227, 677–678
Transition region, 596–597
Transmission parameters, 696, 700
Triangular waveforms, 618, 620, 623–624
Trigonometric identities, 757
Twin-T notch filter, 602
Two-port circuits, 692–717
  cascaded, 708–711
  conversion table for parameters, 699
  hybrid parameters, 696, 697
  immittance, 696
  interconnected, 708–711, 713
  measurements for parameters of, 697, 700
  model assumptions, 692, 712
  parallel, 708–709
  parameters for, 695–703, 712
  reciprocal, 701–702, 712
  relationships among, 698–699, 701
  series-connected, 708–709
  series-parallel, 708–709
  symmetric, 701–702, 712–713
  terminal equations for, 694, 712
  terminated, 703–708
  transmission parameters, 696, 700
  unknown circuit characterization, 693, 711–712
  z parameters, 696–697, 703–706
Two-wattmeter method, 431–432, 435

U

Unbounded response, 250–251, 256
Underdamped response, 282–286, 291–293, 297–299, 300–301, 309
  natural response, 282–286, 297–300, 309
  parallel RLC circuits, 282–286, 291–293, 309
  series RLC circuits, 297–299, 301
  step response, 291–293, 300–301

Unilateral (one-sided) Laplace transform, 446–447
Unit impulse function ($\delta(t)$), 449, 476
Unit impulse response ($h(t)$), 504–511
Unit prefixes, 10–11
Unit step function ($u(t)$), 447, 452, 476, 669
Unknown circuit characterization, 693, 711–712

V

Vampire (standby) power analysis, 375, 399–400
Volt-amp reactive (VAR), unit of, 379, 385, 401
Volt-amps (VA), unit of, 384–385, 401
Voltage (v), 12–17, 34, 40, 42, 46, 50, 58, 70–73, 80, 152–156, 172, 184–186, 199–201, 203–204, 211, 226, 229, 274, 320–322, 332–333, 354–355, 414, 417–419, 435, 631–641, 645–647, 684–685
    a-, b-, and c-phase references, 414
    balanced three-phase circuits, 414, 417–419, 435
    defined, 13
    dot convention for, 354–355
    electric charge and, 12–14
    Fourier series applications, 631–641, 645–647
    frequency domain, 332–333, 354–355
    ideal transformer ratios, 353–355
    induced, 203–204
    inductor relationships, 184–186
    input constraint, 153–154
    Kirchhoff's voltage law (KVL), 40, 42, 50, 332–333
    line, 417, 418–419, 435
    measurement of, 58, 70–73, 80
    mutual inductance and, 199–201, 204, 211
    natural response expressions for, 226, 229, 274
    negative feedback and, 153–154, 172
    Ohm's law for, 33, 50
    op-amp terminals, 152–156, 172
    Parseval's theorem for, 684–685
    phase, 418–419, 435
    phasor notation for, 414
    periodic, 631–637, 639–641, 645–647
    polarity of, 199–201, 204, 211, 354–355
    polarity reference, 14
    power and energy relationship to, 15–17
    resistor power in terms of, 34
    resistor-capacitor (RC) circuits, 229
    resistor-inductor (RL) circuits, 226
    resistor-inductor-capacitor (RLC) circuits, 274, 637–638
    sinusoidal source, 320–322
    square-wave, 633, 637–638
    steady-state analysis and, 320–322, 332–333, 354–355
    unknown found using Kirchhoff's laws, 46
Voltage-divider circuit, 64–66, 79
Voltage division, 68–70, 79, 334
Voltage drop, 199, 516–517
Voltage sources, 28–31, 50, 415
Voltage to current (v–i) relationships, 184–185, 190, 327–330, 361, 484–485, 521
    capacitors, 190, 329–330, 361
    circuit analysis and, 184–185, 190
    inductors, 184–185, 328–329, 361
    Laplace transform method using, 484–485, 521
    phasor domain of, 327–330, 361
    resistors, 327–328, 361
    steady-state analysis and, 327–330, 361
Voltmeter, 70, 72, 80

W

Watt (W), unit of, 384–385, 401
Waveforms, 618, 620, 623–629, 633, 635–636, 650
    even periodic function, 625–626
    half-wave periodic function, 627–628
    odd periodic function, 626–627
    periodic functions for, 618, 620
    periodic voltage, 633, 635–636
    periodic, 618, 620
    quarter-wave periodic function, 628–629
    rectangular, 620
    sinusoidal rectifiers, 618, 620
    square, 620, 633
    square-wave voltage, 633
    steady-state response, 635–636
    symmetry of periodic functions represented as, 625–629
    triangular, 618, 620, 623–624
Wavelength (λ), 7
Weighting function, 510
Wheatstone bridge, 73–75, 80
Winding (primary and secondary), 347, 415
Wye (Y) interconnection, 76
Wye (Y) loads, 425–426
Wye-delta (Y-Δ) circuits, 422–425, 428–429
    analysis of balanced, 422–425
    power calculations for, 428–429
Wye-wye (Y-Y) circuits, 416–421, 428
    analysis of balanced, 416–421
    power calculations for, 428

Z

z parameters, two-port circuits, 696–697, 703–706
Zero frequency, 544
Zeros, 470–472, 476, 502, 743–744, 750–751
    Bode plots and, 743–744, 750–751
    complex, 750–751
    frequency domain ($F(s)$), 470–472, 476
    real, first-order, 743–744
    transfer functions ($H(s)$), 502

NATURAL RESPONSE OF A SERIES *RLC* CIRCUITS

1. Determine the initial capacitor voltage (V_0) and inductor current (I_0) from the circuit.
2. Determine the values of α and ω_0 using the equations in Table 8.4.
3. If $\alpha^2 > \omega_0^2$, the response is overdamped and $i(t) = A_1 e^{s_1 t} + A_2 e^{s_2 t}, \quad t \geq 0$;
If $\alpha^2 < \omega_0^2$ the response is underdamped and $i(t) = B_1 e^{-\alpha t} \cos \omega_d t + B_2 e^{-\alpha t} \sin \omega_d t, \quad t \geq 0$;
If $\alpha^2 = \omega_0^2$, the response is critically damped and $i(t) = D_1 t e^{-\alpha t} + D_2 e^{-\alpha t}, \quad t \geq 0$.
4. If the response is overdamped, calculate s_1 and s_2 using the equations in Table 8.4;
If the response is underdamped, calculate ω_d using the equation in Table 8.4.
5. If the response is overdamped, calculate A_1 and A_2 by simultaneously solving the equations in Table 8.4;
If the response is underdamped, calculate B_1 and B_2 by simultaneously solving the equations in Table 8.4;
If the response is critically damped, calculate D_1 and D_2 by simultaneously solving the equations in Table 8.4.
6. Write the equation for $i(t)$ from Step 3 using the results from Steps 4 and 5; find any desired component voltages.

TABLE 8.4 Equations for analyzing the natural response of series *RLC* circuits

Characteristic equation	$s^2 + \dfrac{R}{L}s + \dfrac{1}{LC} = 0$
Neper, resonant, and damped frequencies	$\alpha = \dfrac{R}{2L} \qquad \omega_0 = \sqrt{\dfrac{1}{LC}} \qquad \omega_d = \sqrt{\omega_0^2 - \alpha^2}$
Roots of the characteristic equation	$s_1 = -\alpha + \sqrt{\alpha^2 - \omega_0^2}, \quad s_2 = -\alpha - \sqrt{\alpha^2 - \omega_0^2}$
$\alpha^2 > \omega_0^2$: overdamped	$i(t) = A_1 e^{s_1 t} + A_2 e^{s_2 t}, t \geq 0$ $i(0^+) = A_1 + A_2 = I_0$ $\dfrac{di(0^+)}{dt} = s_1 A_1 + s_2 A_2 = \dfrac{1}{L}(-RI_0 - V_0)$
$\alpha^2 < \omega_0^2$: underdamped	$i(t) = B_1 e^{-\alpha t} \cos \omega_d t + B_2 e^{-\alpha t} \sin \omega_d t, \quad t \geq 0$ $i(0^+) = B_1 = I_0$ $\dfrac{di(0^+)}{dt} = -\alpha B_1 + \omega_d B_2 = \dfrac{1}{L}(-RI_0 - V_0)$
$\alpha^2 = \omega_0^2$: critically damped	$i(t) = D_1 t e^{-\alpha t} + D_2 e^{-\alpha t}, \quad t \geq 0$ $i(0^+) = D_2 = I_0$ $\dfrac{di(0^+)}{dt} = D_1 - \alpha D_2 = \dfrac{1}{L}(-RI_0 - V_0)$

(Note that the equations in the last three rows assume that the reference direction for the current in every component is in the direction of the reference voltage drop across that component.)

STEP RESPONSE OF A SERIES *RLC* CIRCUITS

1. Determine the initial capacitor voltage (V_0), the initial inductor current (I_0), and the final capacitor voltage (V_f) from the circuit.
2. Determine the values of α and ω_0 using the equations in Table 8.5.
3. If $\alpha^2 > \omega_0^2$, the response is overdamped and $v_C(t) = V_f + A_1' e^{s_1 t} + A_2' e^{s_2 t}, \quad t \geq 0^+$;
If $\alpha^2 < \omega_0^2$, the response is underdamped and $v_C(t) = V_f + B_1' e^{-\alpha t} \cos \omega_d t + B_2' e^{-\alpha t} \sin \omega_d t, \quad t \geq 0^+$;
If $\alpha^2 = \omega_0^2$, the response is critically damped and $v_C(t) = V_f + D_1' t e^{-\alpha t} + D_2' e^{-\alpha t}, \quad t \geq 0^+$.
4. If the response is overdamped, calculate s_1 and s_2 using the equations in Table 8.5;
If the response is underdamped, calculate ω_d using the equation in Table 8.5.
5. If the response is overdamped, calculate A_1' and A_2' by simultaneously solving the equations in Table 8.5;
If the response is underdamped, calculate B_1' and B_2' by simultaneously solving the equations in Table 8.5;
If the response is critically damped, calculate D_1' and D_2' by simultaneously solving the equations in Table 8.5.
6. Write the equation for $v_C(t)$ from Step 3 using the results from Steps 4 and 5; find the inductor voltage and any desired branch currents.

TABLE 8.5 Equations for analyzing the step response of series *RLC* circuits

Characteristic equation	$s^2 + \dfrac{R}{L}s + \dfrac{1}{LC} = \dfrac{V}{LC}$
Neper, resonant, and damped frequencies	$\alpha = \dfrac{R}{2L} \qquad \omega_0 = \sqrt{\dfrac{1}{LC}} \qquad \omega_d = \sqrt{\omega_0^2 - \alpha^2}$
Roots of the characteristic equation	$s_1 = -\alpha + \sqrt{\alpha^2 - \omega_0^2}, \quad s_2 = -\alpha - \sqrt{\alpha^2 - \omega_0^2}$
$\alpha^2 > \omega_0^2$: overdamped	$v_C(t) = V_f + A_1' e^{s_1 t} + A_2' e^{s_2 t}, t \geq 0$ $v_C(0^+) = V_f + A_1' + A_2' = V_0$ $\dfrac{dv_C(0^+)}{dt} = s_1 A_1' + s_2 A_2' = \dfrac{I_0}{C}$
$\alpha^2 < \omega_0^2$: underdamped	$v_C(t) = V_f + B_1' e^{-\alpha t} \cos \omega_d t + B_2' e^{-\alpha t} \sin \omega_d t, t \geq 0$ $v_C(0^+) = V_f + B_1' = V_0$ $\dfrac{dv_C(0^+)}{dt} = -\alpha B_1' + \omega_d B_2' = \dfrac{I_0}{C}$
$\alpha^2 = \omega_0^2$: critically damped	$v_C(t) = V_f + D_1' t e^{-\alpha t} + D_2' e^{-\alpha t}, t \geq 0$ $v_C(0^+) = V_f + D_2' = V_0$ $\dfrac{dv_C(0^+)}{dt} = D_1' - \alpha D_2' = \dfrac{I_0}{C}$

(Note that the equations in the last three rows assume that the reference direction for the current in every component is in the direction of the reference voltage drop across that component.)